Lecture Notes in Computer Science 5702

Commenced Publication in 1973
Founding and Former Series Editors:
Gerhard Goos, Juris Hartmanis, and Jan van Leeuwen

Editorial Board

David Hutchison
 Lancaster University, UK
Takeo Kanade
 Carnegie Mellon University, Pittsburgh, PA, USA
Josef Kittler
 University of Surrey, Guildford, UK
Jon M. Kleinberg
 Cornell University, Ithaca, NY, USA
Alfred Kobsa
 University of California, Irvine, CA, USA
Friedemann Mattern
 ETH Zurich, Switzerland
John C. Mitchell
 Stanford University, CA, USA
Moni Naor
 Weizmann Institute of Science, Rehovot, Israel
Oscar Nierstrasz
 University of Bern, Switzerland
C. Pandu Rangan
 Indian Institute of Technology, Madras, India
Bernhard Steffen
 University of Dortmund, Germany
Madhu Sudan
 Microsoft Research, Cambridge, MA, USA
Demetri Terzopoulos
 University of California, Los Angeles, CA, USA
Doug Tygar
 University of California, Berkeley, CA, USA
Gerhard Weikum
 Max-Planck Institute of Computer Science, Saarbruecken, Germany

Xiaoyi Jiang   Nicolai Petkov (Eds.)

Computer Analysis of Images and Patterns

13th International Conference, CAIP 2009
Münster, Germany, September 2-4, 2009
Proceedings

Volume Editors

Xiaoyi Jiang
University of Münster
Department of Mathematics and Computer Science
Einsteinstrasse 62, 48149, Münster, Germany
E-mail: xjiang@uni-muenster.de

Nicolai Petkov
University of Groningen
Institute of Mathematics and Computing Science
Nijenborgh 9, 9747 AG Groningen, The Netherlands
E-mail: n.petkov@rug.nl

Library of Congress Control Number: 2009932884

CR Subject Classification (1998): I.5, I.4, I.3, I.2.10, I.2.6, H.2.8, I.2.7

LNCS Sublibrary: SL 6 – Image Processing, Computer Vision, Pattern Recognition, and Graphics

ISSN	0302-9743
ISBN-10	3-642-03766-6 Springer Berlin Heidelberg New York
ISBN-13	978-3-642-03766-5 Springer Berlin Heidelberg New York

This work is subject to copyright. All rights are reserved, whether the whole or part of the material is concerned, specifically the rights of translation, reprinting, re-use of illustrations, recitation, broadcasting, reproduction on microfilms or in any other way, and storage in data banks. Duplication of this publication or parts thereof is permitted only under the provisions of the German Copyright Law of September 9, 1965, in its current version, and permission for use must always be obtained from Springer. Violations are liable to prosecution under the German Copyright Law.

springer.com

© Springer-Verlag Berlin Heidelberg 2009
Printed in Germany

Typesetting: Camera-ready by author, data conversion by Scientific Publishing Services, Chennai, India
Printed on acid-free paper   SPIN: 12743087   06/3180   5 4 3 2 1 0

Preface

It was an honor and a pleasure to organize the 13th International Conference on Computer Analysis of Images and Patterns (CAIP 2009) in Münster, Germany.

CAIP has been held biennially since 1985: Berlin (1985), Wismar (1987), Leipzig (1989), Dresden (1991), Budapest (1993), Prague (1995), Kiel (1997), Ljubljana (1999), Warsaw (2001), Groningen (2003), Paris (2005), and Vienna (2007). Initially, this conference series served as a forum for getting together scientists from East and West Europe. Nowadays, CAIP enjoys a high international visibility and attracts participants from all over the world.

For CAIP 2009 we received a record number of 405 submissions. All papers were reviewed by two, and in most cases, three reviewers. Finally, 148 papers were selected for presentation at the conference, resulting in an acceptance rate of 36%. All Program Committee members and additional reviewers listed here deserve a great thanks for their timely and competent reviews. The accepted papers were presented either as oral presentations or posters in a single-track program. In addition, we were very happy to have Aljoscha Smolic and David G. Stork as our invited speakers to present their work in two fascinating areas. With this scientific program we hope to continue the tradition of CAIP in providing a forum for scientific exchange at a high quality level.

A successful conference like CAIP 2009 would not be possible without the support of many institutions and people. First of all, we like to thank all the authors of submitted papers and the invited speakers for their contributions. The Steering Committee members were always there when advice was needed. The Publicity Chair, Dacheng Tao, and many colleagues helped to promote the conference, which resulted in the large number of submissions as the basis for the excellent scientific program. We are grateful to our sponsors for their direct and indirect financial support. Also, the cooperation with Münster City Marketing was very pleasant and helpful. Finally, many thanks go to the members of the Local Organizing Committee.

We like to thank Springer for giving us the opportunity of continuing to publish CAIP proceedings in the LNCS series.

Founded in 793, Münster belongs to the historical cities of Germany. It is most famous as the site of signing the Treaty of Westphalia ending the Thirty Years' War in 1648. Today, it is acknowledged as a city of science and learning (and the capital city of bicycles, Germany's Climate Protection Capital, and more). With its almost 40,000 students, the University of Münster is among the largest universities in Germany. It was our great pleasure to offer the participants the platform in this multi-faceted city for a lively scientific exchange and many other relaxed hours. Finally, to the readers of this proceedings book: enjoy!

September 2009

Xiaoyi Jiang
Nicolai Petkov

CAIP 2009 Organization

General Chairs

Xiaoyi Jiang   Univesity of Münster, Germany
Nicolai Petkov   University of Groningen, The Netherlands

Publicity Chair

Dacheng Tao   Nanyang Technological University, Singapore

Steering Committee

André Gagalowicz (France)
Reinhard Klette (New Zealand)
Walter Kropatsch (Austria)

Nicolai Petkov (The Netherlands)
Gerald Sommer (Germany)

Program Committee

Aksoy, Selim (Turkey)
Antonacopulos, Apostolos (UK)
Bellon, Olga (Brazil)
Borgefors, Gunilla (Sweden)
Breuel, Thomas (Germany)
Brun, Luc (France)
Bunke, Horst (Switzerland)
Burger, Martin (Germany)
Campisi, Patrizio (Italy)
Chan, Yung-Kuan (Taiwan)
Chen, Yung-Fu (Taiwan)
Cheng, Da-Chuan (Taiwan)
Chetverikov, Dimitri (Hungary)
Chung, Kuo-Liang (Taiwan)
Corrochano, Eduardo Bayro (Mexico)
Dawood, Mohammad (Germany)
Del Bimbo, Alberto (Italy)
Dengel, Andreas (Germany)
Denzler, Joachim (Germany)
di Baja, Gabriella Sanniti (Italy)
du Buf, Hans (Portugal)
Ercil, Aytul (Turkey)

Escolano, Francisco (Spain)
Ferrer, Miquel (Spain)
Fink, Gernot (Germany)
Fisher, Robert (UK)
Flusser, Jan (Czech Republic)
Foggia, Pasquale (Italy)
Fred, Ana (Portugal)
Fu, Yun (USA)
Gagalowicz, André (France)
Gao, Xinbo (China)
Ghosh, Anarta (Denmark)
Gimelfarb, Georgy (New Zealand)
Goldgof, Dmitry (USA)
Grigorescu, Cosmin (The Netherlands)
Haan, Gerard de (The Netherlands)
Haar Romeny, Bart ter
   (The Netherlands)
Haindl, Michal (Czech Republic)
Hancock, Edwin (UK)
He, Changzheng (China)
He, Xiaofei (China)
Hlavac, Vaclav (Czech Republic)

Hornegger, Joachim (Germany)
Huang, Yo-Ping (Taiwan)
Huang, Yung-Fa (Taiwan)
Imiya, Atsushi (Japan)
Kampel, Martin (Austria)
Kiryati, Nahum (Israel)
Klette, Reinhard (New Zealand)
Koschan, Andreas (USA)
Kropatsch, Walter (Austria)
Kwok, James (HK, China)
Lados, Joseph (Spain)
Latecki, Longin Jan (USA)
Leonardis, Ales (Slovenia)
Li, Xuelong (UK)
Liao, Hong-Yuan Mark (Taiwan)
Lie, Wen-Nung (Taiwan)
Lin, Guo-Shiang (Taiwan)
Ma, Matthew (USA)
Michaelsen, Eckart (Germany)
Lovell, Brian C. (Australia)
Nielsen, Mads (Denmark)
Omachi, Shinichiro (Japan)
Perner, Petra (Germany)
Pietikainen, Matti (Finland)
Pinz, Axel (Austria)
Pitas, Ioannis (Greece)
Radeva, Petia (Spain)
Real, Pedro (Spain)
Roerdink, Jos B.T.M.
  (The Netherlands)
Roli, Fabio (Italy)
Rosenhahn, Bodo (Germany)

Sablatnig, Robert (Austria)
Sabourin, Robert (Canada)
Sagerer, Gerhard (Germany)
Saito, Hideo (Japan)
Sanfeliu, Arberto (Spain)
Sarkar, Sudeep (USA)
Schreer, Oliver (Germany)
Serratosa, Francesc (Spain)
Shen, Jialie (Singapore)
Shulcloper, José Ruiz (Cuba)
Silva, Luciano (Brazil)
Sommer, Gerald (Germany)
Song, Mingli (China)
Suen, Ching Y. (Canada)
Sugimoto, Akihiro (Japan)
Tao, Dacheng (Singapore)
Tombre, Karl (France)
Tönnies, Klaus (Germany)
Torsello, Andrea (Italy)
Tsai, Chwei-Shyong (Taiwan)
Valveny, Ernst (Spain)
Vento, Mario (Italy)
Viergever, Max A. (The Netherlands)
Wachenfeld, Steffen (USA)
Wang, Patrick (USA)
Westenberg, Michel (The Netherlands)
Whelan, Paul F. (Ireland)
Xiong, Hui (USA)
Ye, Jieping (USA)
You, Jia Jane (HK, China)
Zha, Hongbin (China)
Zhang, David (HK, China)

Additional Reviewers

Albarelli, Andrea
Almsick, Markus van
Arnold, Mirko
Artner, Nicole M.
Azzopardi, George
Ballan, Lamberto
Becciu, Alessandro
Bellon, Olga
Benezeth, Yannick
Bertini, Marco

Biedert, Ralf
Blauensteiner, Philipp
Brun, Anders
Bulò, Samuel Rota
Candamo, Joshua
Caulfield, Darren
Chiang, Jui-Chiu
Choi, Jongmoo
Christophe, Rosenberger
Conte, Donatello

Cotsaces, Costas
Didaci, Luca
Donner, Rene
Dragon, Ralf
El-Abed, Mohamad
Elbrandt, Tobias
Erdem, Aykut
Fischer, Andreas
Fourey, Sébastien
Franco, Fernando

Freni, Biagio
Frinken, Volkmar
Fumera, Giorgio
Fuster, Andrea
Garcia, Edel
Garea, Eduardo
Giacinto, Giorgio
Giotis, Ioannis
Grahl, Miranda
Gritti, Tommaso
Gu, Yuhua
Ham, Mabel Iglesias
Hanheide, Marc
Hauberg, Søren
Haxhimusa, Yll
Hsu, Chih-Yu
Indermühle, Emanuel
Ion, Adrian
Ji, Ming
Jiang, Ruyi
Kortkamp, Marco
Korzhova, Valentina
Kramer, Kurt
Lauze, Francois
Lezoray, Olivier
Li, Na
Liaw, Jiun-Jian
Lin, Huei-Yung
Lisanti, Giuseppe
Liu, Shing-Hong
Liwicki, Marcus
Luengo, Cris
Ma, Wei
Ma, Xiangyin
Manohar, Vasant
Meng, Fang
Morales, Sandino
Morris, John
Oliveira, Manuel
Olsen, Søren
Ortiz, Esther Antunez
Pala, Pietro
Papari, Giuseppe
Pastarmov, Yulian
Pedersen, Kim Steenstrup
Pei, Yuru
Percannella, Gennaro
Pernici, Federico
Peters, Christian
Piras, Luca
Platel, Bram
Pletschacher, Stefan
Puig, Domenec
Queirolo, Chauã
Rabie, Ahmad
Ramella, Giuliana
Rattani, Ajita
Reuschling, Christian
Riesen, Kaspar
Rodolà, Emanuele
Rosenberger, Christophe
Sansone, Carlo
Schauerte, Boris
Schulze, Christian
Segundo, Maurcio Pamplona
Seidenari, Lorenzo
Serra, Giuseppe
Shafait, Faisal
Shan, Caifeng
Shreve, Matt
Sporring, Jon
Stoettinger, Julian
Stricker, Didier
Subramanian, Easwara
Swadzba, Agnes
Tscherepanow, Marko
Vajda, Szilard
Vaudrey, Tobi
Vrubel, Alexandre
Wachsmuth, Sven
Yeh, Chia-Hung
Zambanini, Sebastian
Zeng, Gang
Zhang, Luming

Local Organizing Committee

Hildegard Brunstering
Daniel Abdala
Mohammad Dawood
Michael Fieseler

Lucas Franek
Fabian Gigengack
Sergej Lewin
Kai Rothaus

Daniel Tenbrinck
Pakaket Wattuya

Sponsoring Institutions

University of Münster, Germany
International Association for Pattern Recognition
Olympus Soft Imaging Solutions GmbH
Philips

Table of Contents

Invited Talks

An Overview of 3D Video and Free Viewpoint Video 1
  Aljoscha Smolic

Computer Vision and Computer Graphics Analysis of Paintings and
Drawings: An Introduction to the Literature......................... 9
  David G. Stork

Biometrics

Head Pose Estimation by a Stepwise Nonlinear Regression 25
  Kevin Bailly, Maurice Milgram, and Philippe Phothisane

Model-Based Illumination Correction for Face Images in Uncontrolled
Scenarios 33
  Bas Boom, Luuk Spreeuwers, and Raymond Veldhuis

A New Gabor Phase Difference Pattern for Face and Ear
Recognition 41
  *Yimo Guo, Guoying Zhao, Jie Chen, Matti Pietikäinen, and
  Zhengguang Xu*

Is White Light the Best Illumination for Palmprint Recognition?..... 50
  Zhenhua Guo, David Zhang, and Lei Zhang

Second-Level Partition for Estimating FAR Confidence Intervals in
Biometric Systems 58
  Rongfeng Li, Darun Tang, Wenxin Li, and David Zhang

Smooth Multi-Manifold Embedding for Robust Identity-Independent
Head Pose Estimation 66
  Xiangyang Liu, Hongtao Lu, and Heng Luo

Human Age Estimation by Metric Learning for Regression Problems ... 74
  Yangjing Long

Face Detection Using GPU-Based Convolutional Neural Networks 83
  Fabian Nasse, Christian Thurau, and Gernot A. Fink

Gaussian Weak Classifiers Based on Haar-Like Features with Four
Rectangles for Real-time Face Detection 91
  Sri-Kaushik Pavani, David Delgado Gomez, and Alejandro F. Frangi

Model Based Analysis of Face Images for Facial Feature Extraction 99
  Zahid Riaz, Christoph Mayer, Michael Beetz, and Bernd Radig

Dynamics Analysis of Facial Expressions for Person Identification 107
  Hidenori Tanaka and Hideo Saito

Regression Based Non-frontal Face Synthesis for Improved Identity
Verification 116
  Yongkang Wong, Conrad Sanderson, and Brian C. Lovell

Differential Feature Analysis for Palmprint Authentication 125
  Xiangqian Wu, Kuanquan Wang, Yong Xu, and David Zhang

Combining Facial Appearance and Dynamics for Face Recognition 133
  Ning Ye and Terence Sim

Finger-Knuckle-Print Verification Based on Band-Limited Phase-Only
Correlation... 141
  Lin Zhang, Lei Zhang, and David Zhang

Calibration

Calibration of Radially Symmetric Distortion by Fitting Principal
Component 149
  *Hideitsu Hino, Yumi Usami, Jun Fujiki, Shotaro Akaho, and
  Noboru Murata*

Calibration of Rotating Sensors 157
  Karsten Scheibe, Fay Huang, and Reinhard Klette

Document Analysis

A New Approach for Segmentation and Recognition of Arabic
Handwritten Touching Numeral Pairs 165
  Huda Alamri, Chun Lei He, and Ching Y. Suen

Ridges Based Curled Textline Region Detection from Grayscale
Camera-Captured Document Images 173
  Syed Saqib Bukhari, Faisal Shafait, and Thomas M. Breuel

Kernel PCA for HMM-Based Cursive Handwriting Recognition 181
  Andreas Fischer and Horst Bunke

Improved Handwriting Recognition by Combining Two Forms of
Hidden Markov Models and a Recurrent Neural Network 189
  *Volkmar Frinken, Tim Peter, Andreas Fischer, Horst Bunke,
  Trinh-Minh-Tri Do, and Thierry Artieres*

Embedded Bernoulli Mixture HMMs for Continuous Handwritten Text
Recognition 197
    Adrià Giménez and Alfons Juan

Recognition-Based Segmentation of Nom Characters from Body Text
Regions of Stele Images Using Area Voronoi Diagram 205
    Thai V. Hoang, Salvatore Tabbone, and Ngoc-Yen Pham

A Novel Approach for Word Spotting Using Merge-Split Edit
Distance 213
    Khurram Khurshid, Claudie Faure, and Nicole Vincent

Hierarchical Decomposition of Handwritten Manuscripts Layouts 221
    Vincent Malleron, Véronique Eglin, Hubert Emptoz,
    Stéphanie Dord-Crouslé, and Philippe Régnier

Camera-Based Online Signature Verification with Sequential Marginal
Likelihood Change Detector 229
    Daigo Muramatsu, Kumiko Yasuda, Satoshi Shirato, and
    Takashi Matsumoto

Separation of Overlapping and Touching Lines within Handwritten
Arabic Documents 237
    Nazih Ouwayed and Abdel Belaïd

Combining Contour Based Orientation and Curvature Features for
Writer Recognition 245
    Imran Siddiqi and Nicole Vincent

Features

A Novel Approach to Estimate Fractal Dimension from Closed
Curves 253
    André R. Backes, João B. Florindo, and Odemir M. Bruno

Saliency Based on Decorrelation and Distinctiveness of Local
Responses 261
    Antón Garcia-Diaz, Xosé R. Fdez-Vidal, Xosé M. Pardo, and
    Raquel Dosil

Incorporating Shape Features in an Appearance-Based Object
Detection System 269
    Gurman Gill and Martin Levine

Symmetry Detection for Multi-object Using Local Polar Coordinate 277
    Yuanhao Gong, Qicong Wang, Chenhui Yang, Yahui Gao, and
    Cuihua Li

Detection of Non-convex Objects by Dynamic Programming............ 285
  Andree Große, Kai Rothaus, and Xiaoyi Jiang

Finding Intrinsic and Extrinsic Viewing Parameters from a Single
Realist Painting 293
  Tadeusz Jordan, David G. Stork, Wai L. Khoo, and Zhigang Zhu

A Model for Saliency Detection Using NMFsc Algorithm.............. 301
  Jian Liu and Yuncai Liu

Directional Force Field-Based Maps: Implementation and
Application 309
  JingBo Ni, Melanie Veltman, and Pascal Matsakis

A Spatio-Temporal Isotropic Operator for the Attention-Point
Extraction 318
  Roman M. Palenichka and Marek B. Zaremba

Homological Tree-Based Strategies for Image Analysis 326
  P. Real, H. Molina-Abril, and W. Kropatsch

Affine Moment Invariants of Color Images........................... 334
  Tomáš Suk and Jan Flusser

Graph Representations

Graph-Based k-Means Clustering: A Comparison of the Set Median
versus the Generalized Median Graph 342
  M. Ferrer, E. Valveny, F. Serratosa, I. Bardají, and H. Bunke

Algorithms for the Sample Mean of Graphs 351
  Brijnesh J. Jain and Klaus Obermayer

A Hypergraph-Based Model for Graph Clustering: Application to
Image Indexing 360
  Salim Jouili and Salvatore Tabbone

Hypergraphs, Characteristic Polynomials and the Ihara Zeta
Function 369
  Peng Ren, Tatjana Aleksić, Richard C. Wilson, and
  Edwin R. Hancock

Feature Ranking Algorithms for Improving Classification of Vector
Space Embedded Graphs.. 377
  Kaspar Riesen and Horst Bunke

Graph-Based Object Class Discovery 385
  Shengping Xia and Edwin R. Hancock

Image Processing

The Clifford-Hodge Flow: An Extension of the Beltrami Flow 394
  Thomas Batard and Michel Berthier

Speedup of Color Palette Indexing in Self–Organization of Kohonen
Feature Map 402
  *Kuo-Liang Chung, Jyun-Pin Wang, Ming-Shao Cheng, and
  Yong-Huai Huang*

Probabilistic Satellite Image Fusion 410
  *Farid Flitti, Mohammed Bennamoun, Du Huynh,
  Amine Bermak, and Christophe Collet*

A Riemannian Scalar Measure for Diffusion Tensor Images 419
  Andrea Fuster, Laura Astola, and Luc Florack

Structure-Preserving Smoothing of Biomedical Images 427
  *Debora Gil, Aura Hernàndez-Sabaté, Mireia Burnat,
  Steven Jansen, and Jordi Martínez-Villalta*

Fast Block Clustering Based Optimized Adaptive Mediod Shift 435
  Zulqarnain Gilani and Naveed Iqbal Rao

Color Me Right–Seamless Image Compositing 444
  Dong Guo and Terence Sim

Transform Invariant Video Fingerprinting by NMF 452
  Ozan Gursoy, Bilge Gunsel, and Neslihan Sengor

A Model Based Method for Overall Well Focused Catadioptric Image
Acquisition with Multi-focal Images 460
  Weiming Li, Youfu Li, and Yihong Wu

Colorization Using Segmentation with Random Walk 468
  Xiaoming Liu, Jun Liu, and Zhilin Feng

Edge-Based Image Compression with Homogeneous Diffusion 476
  Markus Mainberger and Joachim Weickert

Color Quantization Based on PCA and Kohonen SOFM 484
  D. Mavridis and N. Papamarkos

On Adapting the Tensor Voting Framework to Robust Color Image
Denoising 492
  *Rodrigo Moreno, Miguel Angel Garcia, Domenec Puig, and
  Carme Julià*

Two-Dimensional Windowing in the Structural Similarity Index for the
Colour Image Quality Assessment 501
  Krzysztof Okarma

Reduced Inverse Distance Weighting Interpolation for Painterly
Rendering.. 509
  Giuseppe Papari and Nicolai Petkov

Nonlinear Diffusion Filters without Parameters for Image
Segmentation.. 517
  Carlos Platero, Javier Sanguino, and Olga Velasco

Color Quantization by Multiresolution Analysis................. 525
  Giuliana Ramella and Gabriella Sanniti di Baja

Total Variation Processing of Images with Poisson Statistics... 533
  Alex Sawatzky, Christoph Brune, Jahn Müller, and Martin Burger

Fast Trilateral Filtering..................................... 541
  Tobi Vaudrey and Reinhard Klette

Image Registration

Joint Affine and Radiometric Registration Using Kernel Operators..... 549
  Boaz Vigdor and Joseph M. Francos

MCMC-Based Algorithm to Adjust Scale Bias in Large Series of
Electron Microscopical Ultrathin Sections 557
  Huaizhong Zhang, E. Patricia Rodriguez, Philip Morrow,
  Sally McClean, and Kurt Saetzler

Image and Video Retrieval

Accelerating Image Retrieval Using Factorial Correspondence Analysis
on GPU 565
  Nguyen-Khang Pham, Annie Morin, and Patrick Gros

Color Based Bags-of-Emotions 573
  Martin Solli and Reiner Lenz

Measuring the Influence of Concept Detection on Video Retrieval 581
  Pablo Toharia, Oscar D. Robles, Alan F. Smeaton, and
  Ángel Rodríguez

Medical Imaging

SEM Image Analysis for Quality Control of Nanoparticles............. 590
  S.K. Alexander, R. Azencott, B.G. Bodmann, A. Bouamrani,
  C. Chiappini, M. Ferrari, X. Liu, and E. Tasciotti

Extraction of Cardiac Motion Using Scale-Space Features Points and
Gauged Reconstruction 598
  Alessandro Becciu, Bart J. Janssen, Hans van Assen, Luc Florack,
  Vivian Roode, and Bart M. ter Haar Romeny

A Non-Local Fuzzy Segmentation Method: Application to Brain
MRI 606
  Benoît Caldairou, François Rousseau, Nicolas Passat, Piotr Habas,
  Colin Studholme, and Christian Heinrich

Development of a High Resolution 3D Infant Stomach Model for
Surgical Planning 614
  Qaiser Chaudry, S. Hussain Raza, Jeonggyu Lee, Yan Xu,
  Mark Wulkan, and May D. Wang

Improved Arterial Inner Wall Detection Using Generalized Median
Computation 622
  Da-Chuan Cheng, Arno Schmidt-Trucksäss, Shing-Hong Liu, and
  Xiaoyi Jiang

Parcellation of the Auditory Cortex into Landmark–Related Regions of
Interest 631
  Karin Engel, Klaus Tönnies, and André Brechmann

Automatic Fontanel Extraction from Newborns' CT Images Using
Variational Level Set 639
  Kamran Kazemi, Sona Ghadimi, Alireza Lyaghat, Alla Tarighati,
  Narjes Golshaeyan, Hamid Abrishami-Moghaddam, Reinhard Grebe,
  Catherine Gondary-Jouet, and Fabrice Wallois

Modeling and Measurement of 3D Deformation of Scoliotic Spine Using
2D X-ray Images 647
  Hao Li, Wee Kheng Leow, Chao-Hui Huang, and Tet Sen Howe

A Comparative Study on Feature Selection for Retinal Vessel
Segmentation Using FABC 655
  Carmen Alina Lupaşcu, Domenico Tegolo, and Emanuele Trucco

Directional Multi-scale Modeling of High-Resolution Computed
Tomography (HRCT) Lung Images for Diffuse Lung Disease
Classification 663
  Kiet T. Vo and Arcot Sowmya

Statistical Deformable Model-Based Reconstruction of a Patient-Specific
Surface Model from Single Standard X-ray Radiograph 672
  Guoyan Zheng

Object and Scene Recognition

Plant Species Identification Using Multi-scale Fractal Dimension
Applied to Images of Adaxial Surface Epidermis 680
  *André R. Backes, Jarbas J. de M. Sá Junior, Rosana M. Kolb, and
  Odemir M. Bruno*

Fast Invariant Contour-Based Classification of Hand Symbols for
HCI 689
  Thomas Bader, René Räpple, and Jürgen Beyerer

Recognition of Simple 3D Geometrical Objects under Partial
Occlusion 697
  Alexandra Barchunova and Gerald Sommer

Shape Classification Using a Flexible Graph Kernel 705
  François-Xavier Dupé and Luc Brun

Bio-inspired Approach for the Recognition of Goal-Directed Hand
Actions 714
  Falk Fleischer, Antonino Casile, and Martin A. Giese

Wide-Baseline Visible Features for Highly Dynamic Scene
Recognition 723
  Aram Kawewong, Sirinart Tangruamsub, and Osamu Hasegawa

Jumping Emerging Substrings in Image Classification 732
  Łukasz Kobyliński and Krzysztof Walczak

Human Action Recognition Using LBP-TOP as Sparse Spatio-Temporal
Feature Descriptor 740
  Riccardo Mattivi and Ling Shao

Contextual-Guided Bag-of-Visual-Words Model for Multi-class Object
Categorization 748
  Mehdi Mirza-Mohammadi, Sergio Escalera, and Petia Radeva

Isometric Deformation Modelling for Object Recognition 757
  *Dirk Smeets, Thomas Fabry, Jeroen Hermans,
  Dirk Vandermeulen, and Paul Suetens*

Image Categorization Based on a Hierarchical Spatial Markov Model . . . 766
  Lihua Wang, Zhiwu Lu, and Horace H.S. Ip

Soft Measure of Visual Token Occurrences for Object Categorization . . . 774
  Yanjie Wang, Xiabi Liu, and Yunde Jia

Indexing Large Visual Vocabulary by Randomized Dimensions Hashing
for High Quantization Accuracy: Improving the Object Retrieval
Quality 783
  Heng Yang, Qing Wang, and Zhoucan He

Pattern Recognition

Design of Clinical Support Systems Using Integrated Genetic Algorithm
and Support Vector Machine 791
   *Yung-Fu Chen, Yung-Fa Huang, Xiaoyi Jiang, Yuan-Nian Hsu, and
Hsuan-Hung Lin*

Decision Trees Using the Minimum Entropy-of-Error Principle 799
   *J.P. Marques de Sá, João Gama, Raquel Sebastião, and
Luís A. Alexandre*

k/K-Nearest Neighborhood Criterion for Improvement of Locally
Linear Embedding 808
   *Armin Eftekhari, Hamid Abrishami-Moghaddam, and
Massoud Babaie-Zadeh*

A Parameter Free Approach for Clustering Analysis 816
   Haiqiao Huang, Pik-yin Mok, Yi-lin Kwok, and Sau-Chuen Au

Fitting Product of HMM to Human Motions 824
   *M. Ángeles Mendoza, Nicolás Pérez de la Blanca, and
Manuel J. Marín-Jiménez*

Reworking Bridging for Use within the Image Domain 832
   Henry Petersen and Josiah Poon

Detection of Ambiguous Patterns Using SVMs: Application to
Handwritten Numeral Recognition 840
   Leticia Seijas and Enrique Segura

Shape Recovery

Accurate 3D Modelling by Fusion of Potentially Reliable Active Range
and Passive Stereo Data 848
   *Yuk Hin Chan, Patrice Delmas, Georgy Gimel'farb, and
Robert Valkenburg*

Rapid Classification of Surface Reflectance from Image Velocities 856
   Katja Doerschner, Dan Kersten, and Paul Schrater

Structure-Preserving Regularisation Constraints for
Shape-from-Shading 865
   Rui Huang and William A.P. Smith

3D Object Reconstruction Using Full Pixel Matching 873
   Yuichi Yaguchi, Kenta Iseki, Nguyen Tien Viet, and Ryuichi Oka

Rapid Inference of Object Rigidity and Reflectance Using Optic Flow... 881
   Di Zang, Katja Doerschner, and Paul R. Schrater

On the Recovery of Depth from a Single Defocused Image............. 889
Shaojie Zhuo and Terence Sim

Segmentation

Modelling Human Segmentation trough Color and Space Analysis 898
Agnés Borràs and Josep Lladós

A Metric and Multiscale Color Segmentation Using the Color
Monogenic Signal 906
Guillaume Demarcq, Laurent Mascarilla, and Pierre Courtellemont

An Interactive Level Set Approach to Semi-automatic Detection of
Features in Food Micrographs 914
Gaetano Impoco and Giuseppe Licitra

Shape Detection from Line Drawings by Hierarchical Matching 922
Rujie Liu, Yuehong Wang, Takayuki Baba, and Daiki Masumoto

A Fast Level Set-Like Algorithm with Topology Preserving
Constraint 930
Martin Maška and Pavel Matula

Significance Tests and Statistical Inequalities for Segmentation by
Region Growing on Graph 939
*Guillaume Née, Stéphanie Jehan-Besson, Luc Brun, and
Marinette Revenu*

Scale Space Hierarchy of Segments 947
Haruhiko Nishiguchi, Atsushi Imiya, and Tomoya Sakai

Point Cloud Segmentation Based on Radial Reflection 955
Mario Richtsfeld and Markus Vincze

Locally Adaptive Speed Functions for Level Sets in Image
Segmentation... 963
Karsten Rink and Klaus Tönnies

Improving User Control with Minimum Involvement in User-Guided
Segmentation by Image Foresting Transform......................... 971
*T.V. Spina, Javier A. Montoya-Zegarra, P.A.V. Miranda, and
A.X. Falcão*

3D Image Segmentation Using the Bounded Irregular Pyramid......... 979
Fuensanta Torres, Rebeca Marfil, and Antonio Bandera

The Gabor-Based Tensor Level Set Method for Multiregional Image
Segmentation... 987
Bin Wang, Xinbo Gao, Dacheng Tao, Xuelong Li, and Jie Li

Embedded Geometric Active Contour with Shape Constraint for Mass
Segmentation 995
  Ying Wang, Xinbo Gao, Xuelong Li, Dacheng Tao, and Bin Wang

An Efficient Parallel Algorithm for Graph-Based Image
Segmentation 1003
  Jan Wassenberg, Wolfgang Middelmann, and Peter Sanders

Stereo and Video Analysis

Coarse-to-Fine Tracking of Articulated Objects Using a Hierarchical
Spring System 1011
  Nicole Artner, Adrian Ion, and Walter Kropatsch

Cooperative Stereo Matching with Color-Based Adaptive Local
Support 1019
  Roland Brockers

Iterative Camera Motion and Depth Estimation in a Video Sequence ... 1028
  Françoise Dibos, Claire Jonchery, and Georges Koepfler

Performance Prediction for Unsupervised Video Indexing 1036
  Ralph Ewerth and Bernd Freisleben

New Lane Model and Distance Transform for Lane Detection and
Tracking 1044
  Ruyi Jiang, Reinhard Klette, Tobi Vaudrey, and Shigang Wang

Real-Time Volumetric Reconstruction and Tracking of Hands in a
Desktop Environment 1053
  Christoph John, Ulrich Schwanecke, and Holger Regenbrecht

Stereo Localization Using Dual PTZ Cameras 1061
  Sanjeev Kumar, Christian Micheloni, and Claudio Piciarelli

Object Tracking in Video Sequences by Unsupervised
Learning 1070
  R.M. Luque, J.M. Ortiz-de-Lazcano-Lobato,
  Ezequiel Lopez-Rubio, and E.J. Palomo

A Third Eye for Performance Evaluation in Stereo Sequence Analysis ... 1078
  Sandino Morales and Reinhard Klette

OIF - An Online Inferential Framework for Multi-object Tracking with
Kalman Filter 1087
  Saira Saleem Pathan, Ayoub Al-Hamadi, and Bernd Michaelis

Real-Time Stereo Vision: Making More Out of Dynamic
Programming 1096
  Jan Salmen, Marc Schlipsing, Johann Edelbrunner,
  Stefan Hegemann, and Stefan Lüke

Optic Flow Using Multi-scale Anchor Points 1104
  Pieter van Dorst, Bart Janssen, Luc Florack, and
  Bart ter Haar Romenij

A Methodology for Evaluating Illumination Artifact Removal for
Corresponding Images 1113
  Tobi Vaudrey, Andreas Wedel, and Reinhard Klette

Nonlinear Motion Detection 1122
  Lennart Wietzke and Gerald Sommer

Texture Analysis

Rotation Invariant Texture Classification Using Binary Filter Response
Pattern (BFRP) 1130
  Zhenhua Guo, Lei Zhang, and David Zhang

Near-Regular Texture Synthesis 1138
  Michal Haindl and Martin Hatka

Texture Editing Using Frequency Swap Strategy 1146
  Michal Haindl and Vojtěch Havlíček

A Quantitative Evaluation of Texture Feature Robustness and
Interpolation Behaviour 1154
  Stefan Thumfart, Wolfgang Heidl, Josef Scharinger, and
  Christian Eitzinger

Applications

Nonlinear Dimension Reduction and Visualization of Labeled Data 1162
  Kerstin Bunte, Barbara Hammer, and Michael Biehl

Performance Evaluation of Airport Lighting Using Mobile Camera
Techniques 1171
  Shyama Prosad Chowdhury, Karen McMenemy, and Jian-Xun Peng

Intelligent Video Surveillance for Detecting Snow and Ice Coverage on
Electrical Insulators of Power Transmission Lines 1179
  Irene Y.H. Gu, Unai Sistiaga, Sonja M. Berlijn, and
  Anders Fahlström

Size from Specular Highlights for Analyzing Droplet Size
Distributions 1188
    Andrei C. Jalba, Michel A. Westenberg, and Mart H.M. Grooten

Capturing Physiology of Emotion along Facial Muscles: A Method of
Distinguishing Feigned from Involuntary Expressions 1196
    Masood Mehmood Khan, Robert D. Ward, and Michael Ingleby

Atmospheric Visibility Monitoring Using Digital Image Analysis
Techniques 1204
    Jiun-Jian Liaw, Ssu-Bin Lian, Yung-Fa Huang, and
    Rung-Ching Chen

Minimized Database of Unit Selection in Visual Speech Synthesis
without Loss of Naturalness 1212
    Kang Liu and Joern Ostermann

Analysis of Speed Sign Classification Algorithms Using Shape Based
Segmentation of Binary Images 1220
    Azam Sheikh Muhammad, Niklas Lavesson, Paul Davidsson, and
    Mikael Nilsson

Using CCD Moiré Pattern Analysis to Implement Pressure-Sensitive
Touch Surfaces 1228
    Tong Tu and Wooi Boon Goh

Enhanced Landmine Detection from Low Resolution IR Image
Sequences 1236
    Tiesheng Wang, Irene Yu-Hua Gu, and Tardi Tjahjadi

Author Index 1245

An Overview of 3D Video and Free Viewpoint Video

Aljoscha Smolic[*]

Disney Research, Zurich
Clausiusstrasse 49
8092 Zurich, Switzerland
smolic@zurich.disneyresearch.com

Abstract. An overview of 3D video and free viewpoint video is given in this paper. Free viewpoint video allows the user to freely navigate within real world visual scenes, as known from virtual worlds in computer graphics. 3D video provides the user with a 3D depth impression of the observed scene, which is also known as stereo video. In that sense as functionalities, 3D video and free viewpoint video are not mutually exclusive but can very well be combined in a single system. Research in this area combines computer graphics, computer vision and visual communications. It spans the whole media processing chain from capture to display and the design of systems has to take all parts into account. The conclusion is that the necessary technology including standard media formats for 3D video and free viewpoint video is available or will be available in the future, and that there is a clear demand from industry and user side for such new types of visual media.

Keywords: 3D video, free viewpoint video, MPEG, 3DTV.

1 Introduction

Convergence of technologies from computer graphics, computer vision, multimedia and related fields enabled the development of new types of visual media, such as 3D video (3DV) and free viewpoint video (FVV) that expand the user's sensation beyond what is offered by traditional 2D video [1]. 3DV, also referred to as stereo, offers a 3D depth impression of the observed scenery, while FVV allows for an interactive selection of viewpoint and direction within a certain operating range, as known from computer graphics. Both do not exclude each other. In contrary, they can be very well combined within a single system, since they are both based on a suitable 3D scene representation (see below). In other words, given a 3D representation of a scene, if a stereo pair corresponding to the human eyes can be rendered, the functionality of 3DV is provided. If a virtual view (i.e., not an available camera view) corresponding to an arbitrary viewpoint and viewing direction can be rendered, the functionality of FVV is provided. The ideal future visual media system will provide full FVV and 3DV at the same time. In order to enable 3DV and FVV applications, the whole processing chain, including acquisition, sender side processing, 3D representation, coding,

---

[*] Work for this paper was performed during the author's prior affiliation with the Fraunhofer Institute for Telecommunications – Heinrich-Hertz-Institut (FhG-HHI), Berlin, Germany.

transmission, rendering and display need to be considered. The 3DV and FVV processing chain is illustrated in Fig. 1. The design has to take all parts into account, since there are strong interrelations between all of them. For instance, an interactive display that requires random access to 3D data will affect the performance of a coding scheme that is based on data prediction.

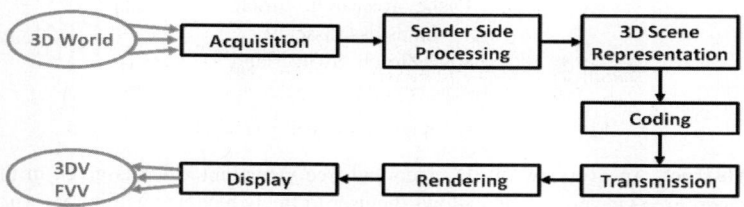

Fig. 1. 3DV and FVV processing chain

2 3D Scene Representation

The choice of a 3D scene representation format is of central importance for the design of any 3DV or FVV system [2]. On the one hand, the 3D scene representation sets the requirements for acquisition and signal processing on sender side, e.g. the number and setting of cameras and the algorithms to extract the necessary data types. On the other hand, the 3D scene representation determines the rendering algorithms (and with that also navigation range, quality, etc.), interactivity, as well as coding and transmission.

In computer graphics literature, methods for 3D scene representation are often classified as a continuum in between two extremes as illustrated in Fig. 2 [3]. These principles can also be applied for 3DV and FVV. The one extreme is represented by classical 3D computer graphics. This approach can also be called geometry-based modeling. In most cases scene geometry is described on the basis of 3D meshes. Real world objects are reproduced using geometric 3D surfaces with an associated texture mapped onto them. More sophisticated attributes can be assigned as well. For instance, appearance properties (opacity, reflectance, specular lights, etc.) can significantly enhance the realism of the models.

The other extreme in 3D scene representations in Fig. 2 is called image-based modeling and does not use any 3D geometry at all. In this case virtual intermediate views are generated from available natural camera views by interpolation. The main advantage is a potentially high quality of virtual view synthesis avoiding any 3D scene reconstruction. However, this benefit has to be paid by dense sampling of the real world with a sufficiently large number of natural camera view images. In general, the synthesis quality increases with the number of available views. Hence, typically a large amount of cameras has to be set up to achieve high-performance rendering, and a tremendous amount of image data needs to be processed therefore. Contrariwise, if the number of used cameras is too low, interpolation and occlusion artifacts will appear in the synthesized images, possibly affecting the quality.

In between the two extremes there exists a number of methods that make more or less use of both approaches and combine the advantages in some way. Some of these representations do not use explicit 3D models but depth or disparity maps. Such maps assign a depth value to each pixel of an image (see Fig 5).

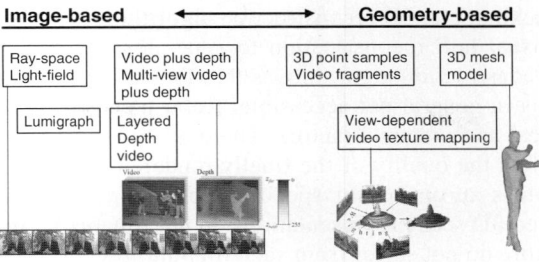

Fig. 2. 3D scene representations for 3DV and FVV

3 Acquisition

In most cases 3DV and FVV approaches rely on specific acquisition systems. Although automatic and interactive 2D-3D conversion (i.e. from 2D video to 3DV or FVV) is an important research area for itself. Most 3DV and FVV acquisition systems use multiple cameras to capture real world scenery [4]. These are sometimes combined with active depth sensors, structured light, etc. in order to capture scene geometry. The camera setting (e.g. dome type as in Fig. 3) and density (i.e. number of cameras) impose practical limitations on navigation and quality of rendered views at a certain virtual position. Therefore, there is a classical trade-off to consider between costs (for equipment, cameras, processors, etc.) and quality (navigation range, quality of virtual views). Fig. 3 illustrates a dome type multi camera acquisition system and captured multi-view video. Such multi-view acquisition is an important and highly actual research area [4].

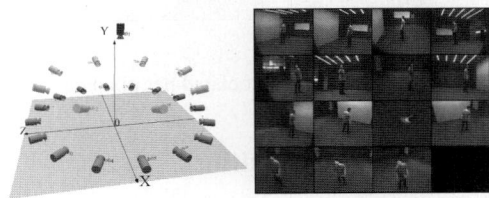

Fig. 3. Multi-camera setup for 3DVO acquisition and captured multi-view video

4 Sender Side Processing

After acquisition, the necessary data as defined by the 3D representation format have to be extracted from the multiple video and other captured data. This sender side processing can include automatic and interactive steps; it may be real-time or offline. Content creation and post processing are included here. Tasks may be divided into low-level computer vision algorithms and higher-level 3D reconstruction algorithms.

Low-level vision may include algorithms like color correction, while balancing, normalization, filtering, rectification, segmentation, camera calibration, feature extraction and tracking, etc. 3D reconstruction algorithms include for instance depth estimation and visual hull reconstruction to generate 3D mesh models. A general problem of 3D reconstruction algorithms is that they are estimations by nature. The true information is in general not accessible. Robustness of the estimation depends on many theoretical and practical factors. There is always a residual error probability which may affect the quality of the finally rendered output views. User-assisted content generation is an option for specific applications to improve performance. Purely image-based 3D scene representations do not rely on 3D reconstruction algorithms, and therefore do not suffer from such limitations.

Fig. 4 illustrates different steps of a 3D reconstruction pipeline. This includes visual hull reconstruction, surface extraction, surface smoothing, and mesh simplification [5]. Fig. 5 illustrates video and associated per pixel depth data.

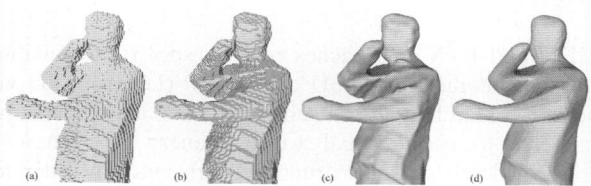

Fig. 4. Different steps of 3D reconstruction

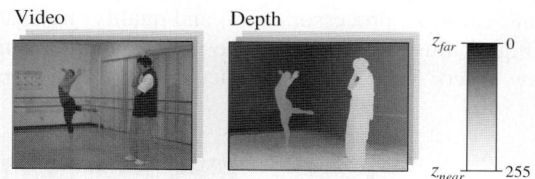

Fig. 5. Video and associated per pixel depth data

5   Coding, Transmission, Decoding

For transmission over limited channels 3DV and FVV data have to be compressed efficiently. This has been widely studied in literature and powerful algorithms are available [6]. International standards for content formats and associated coding technology are necessary to ensure interoperability between different systems. ISO-MPEG and ITU-VCEG are international organizations that released a variety of important standards for digital media including standards for 3DV and FVV. Classical 2-view stereo is already supported by MPEG-2 since the mid 90ies. Current releases of the latest video coding standard H.264/AVC also include a variety of highly efficient modes to support stereo video. This easily extends to multi-view video coding (MVC) with inter-view prediction, a recently released extension of H.264/AVC, which is illustrated in Fig. 6 [7]. It is the currently most efficient way to encode 2 or more videos showing the same scenery from different viewpoints.

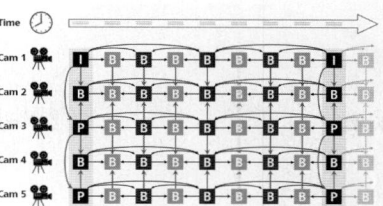

Fig. 6. Multi-view video coding (MVC)

Video plus depth as illustrated in Fig. 5 is already supported by a standard known as MPEG-C Part 3. It is an alternative format for 3DV that requires view synthesis at the receiver (see next section). Video plus depth supports extended functionality compared to classical 2-view stereo such as baseline adaptation to adjust depth impression to different displays and viewing preferences [8]. Currently MPEG prepares a new standard that will provide even more extended functionalities by using multi-view video plus depth or layered depth video [9].

Different model-based or 3D point cloud representations for FVV are supported by various tools of the MPEG-4 standard. Fig. 7 illustrates coding and multiplexing of dynamic 3D geometry, associated video textures and auxiliary data [10].

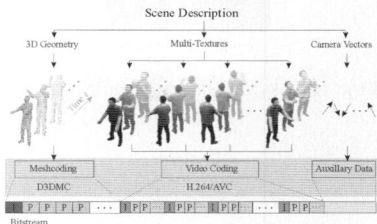

Fig. 7. Coding and multiplexing of dynamic 3D geometry, associated video textures and auxiliary data

6 Rendering

Rendering is the process of generation of the final output views from data in the 3D representation format. Fig. 8 illustrates an interactive 3D scene with a FVV object included. The scene further includes a 360° panorama and a computer graphics object. In this case rendering is done by classical computer graphics methods. The user can navigate freely and watch the dynamic scene from any desired viewpoint and viewing direction.

Fig. 9 illustrates virtual intermediate view synthesis by 3D warping from multiple video plus depth data. Any desired view in between available camera views can be generated this way to support free viewpoint navigation and advanced 3DV functionalities [11]. For instance a multi-view auto-stereoscopic display can be supported efficiently by rendering 9, 16 or more views from a limited number of multi-view video plus depth data (e.g. 2 or 3 video and depth streams).

Fig. 8. Integrated interactive 3D scene with FVV

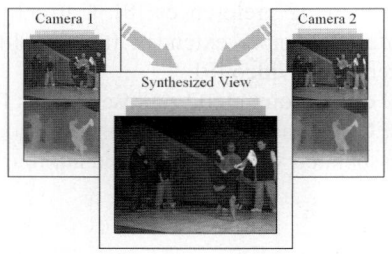

Fig. 9. Intermediate view synthesis from multiple video and depth data

7 Display

Finally the rendered output views are presented to the user on a display. FVV requires interactive input from the user to select the viewpoint. This can be done by classical devices like mouse or joystick. Some systems also track the user (head or gaze) employing cameras and infrared sensors.

In order to provide a depth impression 2 or more views have to be presented to the user appropriately at the same time using a specific 3D display. Such 3D displays ensure that the user perceives a different view with each eye at a time, by filtering the displayed views appropriately. If it is a proper stereo pair, the brain will compute a 3D depth impression of the observed scene.

Currently, various types of 3D displays are available and under development [12]. Most of them use classical 2-view stereo with one view for each eye and some kind of glasses (polarization, shutter, anaglyph) to filter the corresponding view. Then there are so called multi-view auto-stereoscopic displays which do not require glasses. Here, 2 or more views are displayed at the same time and a lenticular sheet or parallax barrier element in front of the light emitters ensures correct view separation for the viewer's eyes.

Fig. 10 illustrates a 2-view auto-stereoscopic display developed by Fraunhofer HHI, which does not require wearing glasses. 2 views are displayed at a time and a lenticular sheet projects them into different directions. A camera system tracks the

Fig. 10. Auto-stereoscopic display made by Fraunhofer HHI

user's gaze direction. A mechanical system orients the display within practical limits according to the user's motion and ensures proper projection of left and right view into direction of the corresponding eye.

8 Summary and Conclusions

This paper provided an overview of 3DV and FVV. It is meant to be supplemental material to the invited talk at the conference. Naturally, different aspects were summarized briefly. For more details the reader is referred to the publications listed below.

3DV and FVV were introduced as extended visual media that provide new functionalities compared to standard 2D video. Both can very well be provided by a single system. New technology spans the whole processing chain from capture to display. The 3D scene representation is determining the whole system. Technology for all the different parts is available, maturating and further emerging.

Growing interest for such applications is noticed from industry and users. 3DV is well established in cinemas. E.g. Hollywood is producing more and more 3D movies. There is a strong push from industry to bring 3DV also to the home, e.g. via Blu-ray or 3DTV. FVV is established as post-production technology. FVV end-user mass market applications are still to be expected for the future.

Acknowledgments. I would like to thank the Interactive Visual Media Group of Microsoft Research for providing the Breakdancers and Ballet data sets, and the Computer Graphics Lab of ETH Zurich for providing the Doo Young multi-view data set.

Background, knowledge and material for this paper were developed during my previous employment at the Fraunhofer Institute for Telecommunications – Heinrich-Hertz-Institut (FhG-HH), Berlin, Germany. Text and illustrations were developed in collaboration with colleagues including Karsten Mueller, Philipp Merkle, Birgit Kaspar, Matthias Kautzner, Sabine Lukaschik, Peter Kauff, Peter Eisert, Thomas Wiegand, Christoph Fehn, Ralf Schaefer.

References

1. Smolic, A., Mueller, K., Merkle, P., Fehn, C., Kauff, P., Eisert, P., Wiegand, T.: 3D Video and Free View-point Video – Technologies, Applications and MPEG Standards. In: ICME 2006, International Conference on Multimedia and Expo, Toronto, Ontario, Canada (July 2006)
2. Smolic, A., Kauff, P.: Interactive 3D Video Representation and Coding Technologies. Proceedings of the IEEE, Special Issue on Advances in Video Coding and Delivery 93(1) (January 2005)
3. Kang, S.B., Szeliski, R., Anandan, P.: The Geometry-Image Representation Tradeoff for Rendering. In: ICIP 2000, IEEE International Conference on Image Processing, Vancouver, Canada (September 2000)
4. Kubota, A., Smolic, A., Magnor, M., Chen, T., Tanimoto, M.: Multi-View Imaging and 3DTV – Special Issue Overview and Introduction. IEEE Signal Processing Magazine, Special Issue on Multi-view Imaging and 3DTV 24(6) (November 2007)
5. Smolic, A., Mueller, K., Merkle, P., Rein, T., Eisert, P., Wiegand, T.: Free Viewpoint Video Extraction, Representation, Coding, and Rendering. In: ICIP 2004, IEEE International Conference on Image Processing, Singapore, October 24-27 (2004)
6. Smolic, A., Mueller, K., Stefanoski, N., Ostermann, J., Gotchev, A., Akar, G.B., Triantafyllidis, G., Koz, A.: Coding Algorithms for 3DTV - A Survey. IEEE Transactions on Circuits and Systems for Video Technology, Special Issue on Multiview Video Coding and 3DTV 17(11) (November 2007)
7. Merkle, P., Smolic, A., Mueller, K., Wiegand, T.: Efficient Prediction Structures for Multiview Video Coding. IEEE Transactions on Circuits and Systems for Video Technology, Special Issue on Multiview Video Coding and 3DTV 17(11) (November 2007)
8. Fehn, C., Kauff, P., Op de Beeck, M., Ernst, F., Ijsselsteijn, W., Pollefeys, M., Vangool, L., Ofek, E., Sexton, I.: An Evolutionary and Optimised Approach on 3D-TV. In: IBC 2002, Int. Broadcast Convention, Amsterdam, Netherlands (September 2002)
9. Smolic, A., Mueller, K., Merkle, P., Vetro, A.: Development of a new MPEG Standard for Advanced 3D Video Applications. In: ISPA 2009, 6th International Symposium on Image and Signal Processing and Analysis, Salzburg, Austria (September 2009)
10. Smolic, A., Mueller, K., Merkle, P., Kautzner, M., Wiegand, T.: 3D Video Objects for Interactive Applications. In: EUSIPCO 2005, Antalya, Turkey, September 4-8 (2005)
11. Mueller, K., Smolic, A., Dix, K., Merkle, P., Kauff, P., Wiegand, T.: View Synthesis for Advanced 3D Video Systems. EURASIP Journal on Image and Video Processing 2008, doi:10.1155/2008/438148
12. Konrad, J., Halle, M.: 3-D Displays and Signal Processing – An Answer to 3-D Ills? IEEE Signal Processing Magazine 24(6) (November 2007)

Computer Vision and Computer Graphics Analysis of Paintings and Drawings: An Introduction to the Literature

David G. Stork

Ricoh Innovations, 2882 Sand Hill Road Suite 115, Menlo Park CA 94025 USA
and Department of Statistics, Stanford University, Stanford CA 94305 USA

Abstract. In the past few years, a number of scholars trained in computer vision, pattern recognition, image processing, computer graphics, and art history have developed rigorous computer methods for addressing an increasing number of problems in the history of art. In some cases, these computer methods are more accurate than even highly trained connoisseurs, art historians and artists. Computer graphics models of artists' studios and subjects allow scholars to explore "what if" scenarios and determine artists' studio praxis. Rigorous computer ray-tracing software sheds light on claims that some artists employed optical tools. Computer methods will not replace tradition art historical methods of connoisseurship but enhance and extend them. As such, for these computer methods to be useful to the art community, they must continue to be refined through application to a variety of significant art historical problems.

Keywords: pattern recognition, computer image analysis, brush stroke analysis, painting analysis, image forensics, compositing, computer graphics reconstructions.

1 Introduction

There is a long history of the use of sophisticated imaging and, in the past several decades *digital* imaging, in the study of art. [1] Shortly after the 19th-century discovery of x-rays such rays were used to reveal underdrawings and *pentimenti*. Later, infra-red photography and reflectography were exploited to similar ends; multispectra, fluoroesence and ultra-violet imaging have become a widespread, and used in revealing pigment composition and more. [2, 3, 4, 5]

In such techniques, the resulting image is generally interpreted by an art scholar. In the past few years, however, we have entered a new era: one where some of the image interpretation relies in great part upon sophisticated algorithms developed from computer vision, the discipline seeking to make computers "see." [6, 7] In some circumstances, computers can analyze certain aspects of perspective, lighting, color, the subtleties of the shapes of brush strokes better than even a trained art scholar, artist, or connoisseur. Rather than replacing connoisseurship, these methods—like other scientific methods such as imaging and material studies [8]—hold promise to enhance and extend it, just as microscopes extend the powers of biologists.

The source of the power of these computer methods arises from the following:

- The computer methods can rely on visual features that are hard to determine by eye, for instance subtle relationships among the structure of a brush stroke at different scales or colors, as in Perugino's *Holy family*, or lighting in de la Tour's *Christ in the carpenter's studio*, or perspective anomalies in van Eyck's *Arnolfini portrait*.
- The computer methods can abstract information from lots of visual evidence, for example, in principle *every* brush stroke executed by van Gogh and his contemporaries—a wealth of information that few scholars even experience, much less fully exploit.
- Computer methods are objective—which need not mean they are "superior" to subjective methods, but rather promise to extend the language of to include terms that are not highly ambiguous. While today an art historian may describe a brush stroke as "bold" or "tentative" or "fluid" someday this scholar may also use technical terms and mathematical measures derived from computer image analysis, terms that other scholars will understand as well.
- Rigorous computer graphics techniques can reveal new three-dimensional views based on two-dimensional artwork, and provide new views into tableaus by dewarping the images reflected in mirrors depicted within a painting.

This brief paper lists some of these new computer techniques and how they have been used in the study of art. The set of topics and reference works here is by no means complete but is meant to show art scholars the power of these method and to encourage art scholars to propose new art historical problems amenable to attach through computer methods. [9, 10, 11, 12, 13, 14] We shall not consider many other areas of computer use in arts, for instance computer art databases and retrieval, nor the task of *imaging* of art—the lighting, spectral filtering, exposure, and so on. Instead, we focus on the application of computer vision, image analysis and computer graphics algorithms to process and understand digital images of scanned art, particularly paintings and drawings.

We begin by describing traditional point- or pixel-based processes such as color adjustment, then consider algorithms based on a number of pixels in a digital image of a painting, and then successively more complex methods of high-level computer vision and graphics, such as dewarping, perspective analysis, lighting analysis, and three-dimensional computer modelling.

2 Point-Based Procedures

Here and below we assume we have a digital image of a painting or drawing, the format required for computer analysis. The conceptually simplest class of computer image methods in the study of art are point- or pixel-based processing, that is, methods to alter the color and brightness of each pixel based solely on the color of that pixel. Such algorithms are better described as image *processing* than as image *analysis*. [15, 16] Multispectral imaging and processing has been used for pigment analysis, color rejuvenation and predicting the effects of curatorial treatment. [17, 18, 19, 20, 21, 22] Pixel-based image processing has been used to adjust the relative weights of different spectral bands to enhance readability, [23] as in the Archimedes palimpsest, [24, 25, 26, 27, 28] and to reveal details and structure in art that otherwise difficult to discern by the unaided eye. [29]

The range of light intensities in the natural world—from a darkened room to bright sunlight—spans as much as a factor of 10^{14} while the dynamic range in oil painting might be a factor of merely 10^2, even in the works of Caravaggio, de la Tour, Joseph Wright of Derby and others who exploited *chiaroscuro*. As such all artists must compress the luminance range in their works. [30] Graham and Field explored the nonlinear compression of the dynamic range in several classes of realist art work, a process that is based on the individual pixel values. [31, 32]

3 Area-Based Procedures

A very large class of image processing algorithms involve *filtering* a source image, where the color or grayscale value of a pixel is a function of the values of pixels in an area or region of the input image. In linear filtering the output value (color or gray level) is a linear combination of the values of the input pixels, while in non-linear filtering allows arbitrary functions of the input pixels. Typically the input image is a photograph of a painting and the output image a digital image processed to reveal some properties that are difficult to discern by unaided eye. Such filtering can remove gradual variations in the color across a painting and leave or even enhance the edges or contours as, for instance, the (nonlinear) Canny edge detector. [33, 34, 35] The Chamfer transform (or distance transform) is useful for quantifying similarity or difference between two shapes, for example when testing the fidelity that artists can achieve using different copying methods. [36, 37]

Another class of non-linear filters are the morphological operators. Such operators are generally used on binary (black and white) images rather than color or grayscale, where the *shape* (rather than the color) is the matter of interest. For example, a *skeletonization* operator yields a single-pixel-wide curve down the center of a black brush stroke, regardless of the varying width of the stroke. [38] Other popular morphological operators implement *erosion, dilation, opening* and *closing*. For instance, Stork, Meador and Noble compared the shapes of different passages of brick work in a painting from the Dutch Golden Age despite variations and irregularities in the width of the painted lines. To this end they preprocessed a high-resolution photograph of the painting using a morphological filter to create an image where the mortar lines were of uniform thickness. They then computed the cross-correlation of this image to search for repeated patterns. [39, 33]

4 Perspective Analysis

The analysis of perspective, scale and geometry has a long and important history in the study of realist art, particularly in art history of the Renaissance. [40] Most of these analytic methods involve simple drawing of perspective lines, finding horizon lines, vanishing points, and so on and can be done without computers. Recently, however, a number of sophisticated computer-based methods for analysis of perspective and geometry have been developed. Criminisi and his colleagues have pioneered rigorous methods for recovering three-dimensional space from single "uncalibrated" images, such as in paintings. These methods have produced three-dimensional virtual spaces of artworks such as Masaccio's *Trinità*. [41, 42, 43, 44] Smith, Stork and Zhang reconstructed the three-dimensional space of the tableau based on multiple views depicted in plane reflections within a single painting. [45] This method also reveals spatial inconsistencies between direct and reflected views and thereby sheds light on the artist's working methods.

While one can use simple commercial software, such as Adobe *Photoshop*, to perform perspective transformation between two images or passages—for instance two arms on the chandelier in van Eyck's *Arnolfini portrait* [46]—but such a technique suffers from a number of drawbacks, the most severe is that the experimenter can arbitrarily choose which portions of one image should match their partners in the other image. Criminisi derived rigorous, principled methods for finding optimal transformations that minimized the shape differences, thus eliminating this drawback. [47, 42, 36] Such analysis of perspective, perspective inconsistencies, and subtleties in shape have shed light on a number of topics, including the question of whether an artist used optical aids. [48, 49, 50, 51, 52, 53, 54, 55, 56, 57, 58]

5 Anamorphic Art

Anamorphic art is distorted art that appears distorted when viewed directly but undistorted when viewed from a special position or in reflection from a curved surface. Such a mirrored cylinder or cone is called an "anamorphoscope." The root word comes from the Greek $\alpha\nu\alpha$, "again," and $\mu o\rho\phi\eta$, "form"—the image in anamorphic art is *formed again*. The earliest surviving deliberate anamorphic image appears to be a sketch of a slant anamorphic eye, drawn by Leonardo around 1485. Perhaps the most celebrated example of such slant anamorphoses is the skull along the bottom in Hans Holbein's *The ambassadors*. [59, 60] There was much experimentation, mathematical analysis and flourishing of anamorphic art in the seventeenth and eighteenth centuries, particularly in France. Today the transformations required by anamorphic art are easily performed by computer. The optics and perspective underlying such art appears in a number of basic texts, [30, 61] but perhaps the most complete and rigorous explanation is given by Hunt, Nickel and Gigault. [62, 63]

6 Dewarping of Curved Art

Many frescos and mosaics on architectural spandrels, barrel vaults and markings on pottery are curved and warped, and art scholars use computer methods to dewarp them to better study the images. [64, 65] Often such digital dewarping requires an estimate of the camera position and the curvature properties of the surface itself. Another class of dewarping centers on dewarping the virtual image appearing in depictions of curved mirrors, such as in Parmigianino's *Self portrait in a convex mirror*, van Eyck's *Arnolfini portrait*, and Campin's *Heinrich von Werl and St. John the Baptist*. Here one models the optics of the spherical or parabolic mirror and adjusts parameters to yield an undistorted image. [66, 41, 67] The mirror properties (curvature and focal length) inferred by such a technique have been used to address claims that artists used such mirrors for optical projectors. [49, 50, 54, 55] On can also use computer graphics methods (cf., Sect. 12, below) to dewarp reflected images, for instance in the analysis of Hans Memling's *van Nieuwenhove Diptych*. [67]

7 Analysis of Lighting and Illumination

Some problems in art history require knowing or estimating the position and direction of lighting in a tableau. Such knowledge can be used for determining the studio conditions when the painting was executed; significant differences in the lighting on different subjects within a tableau may indicate the different studio conditions or presence of different hands, for example. Moreover, this information may indicate whether the artist used optical aids: if the light source was local rather than distant solar illumination, for instance, then it is highly unlikely projections were used. [68]

If the illuminant can be assumed to be small and relatively distant from the tableau, the simple method of cast-shadow analysis is particularly effective in locating the illuminant: one merely draws a line from a point on a cast shadow, through its associated occluder. This line should pass through the position of the illuminant. [69] Several such lines, from a set of occluder-shadow pairs, will intersect at the position of the illuminant.

A more sophisticated method, occluding-contour analysis, derives from forensic analysis of digital photographs, and is based on the pattern of light along an object's outer boundary or occluding contour. [70] Stork and Johnson recently extended the technique to apply to the case of *diffuse* illumination, where the pattern of illumination along an occluding boundary is described as a weighted set of spherical harmonics. If two figures in a painting differ significantly in their sets of coefficients, then they were likely painted under different studio lighting conditions. They studied the lighting in different figures by realist portraitist Garth Herrick and showed that different subjects were executed under different illumination conditions. [71] Stork and Kale modeled the physics flat surfaces, and thereby inferred the position of the illuminant from the floor in Georges de la Tour's *Christ in the carpenter's studio* and Caravaggio's *The calling of*

St. Matthew. [72, 73] This analysis showed that the light source in these works was likely *local*, rather than distant solar illumination, a result that rebuts claims that these works were executed by means of optical projections of very bright tableaus. Bayesian statistical methods can be used to integrate estimates derived from different sources, for example cast shadows and occluding contours, thereby refining and improving overall estimates. [74, 70]

Computer shape-from-shading methods infer the properties of illumination given a known or assumed three-dimensional model of objects and a shaded image of those objects, such as in a highly realistic painting. One can assume a generic three-dimensional model (for instance the face in *Girl with a pearl earring*) and refine both the model and the direction to illumination. [75] Another method for estimating the lighting in a realist tableau is to create a full computer graphics model of the scene and adjust the positions of the virtual illuminants so that the rendered image matches the painting as closely as possible. This has been used to estimate the direction of illumination in Vermeer's *Girl with a pearl earring*, and Georges de la Tour's *Christ in the carpenter's studio* [75, 76], as described in Sect. 12.

8 Analysis of Brush Strokes and Marks

One of the most extensively explored areas of computer analysis of art is the analysis of marks and drawing tools. The basic approach is to use techniques of statistical pattern recognition to learn visual properties of brush strokes that correspond to a particular painter or marking tool. [33, 38, 77, 78, 79, 80, 81, 82, 83, 84, 85, 86, 87, 88, 89, 90, 91, 92, 93, 94, 95] In related techniques, Hedges analyzed the changes in marks in Renaissance copperplate prints as the plates were cleaned; his method yielded an image-based "clock" for estimating the age of such prints. [96, 97] Shahram, Stork and Donoho developed the `De-pict` algorithm, which removed successive layers of brush strokes in a digital image of a painting, such as van Gogh's *Self portrait in a grey felt hat*. When such a sequence of images is displayed in "reverse" (i.e., showing the sequence of brush strokes as likely added to the painting), scholars can see and better understand the aesthetic choices made by the artist. [98]

A somewhat separate class of mark analyses are those for the analysis of dripped painting, particularly in the works of American Abstract Expressionist Jackson Pollock. Here the analyses are generally based on fractals, a mathematical structure that exhibits regularities at different scales or sizes. [99] Taylor and his colleagues first proposed that Pollock's drip paintings exhibited fractal structure, [100] though a number of scholars have questioned the recent claim that simple fractal dimension suffices to distinguish genuine Pollocks from forgeries or other apparently random images. [101, 102, 103, 104, 105, 106] Recent work has returned to the application of traditional image measures for the analysis of Pollock's works, for instance curvature, connected components, and statistical pattern classification methods. [107, 108]

9 Optical Analysis of Art

It is well known that some artists used optical aids during the execution of passages in some of their works, e.g., Canaletto, Thomas Eakins, photo-realists such as Richard Estes and Robert Bechtle, and many others. Computer image analysis has addressed claims that some artists used optical aids when executing some passages in some of their works, for instance that Lorenzo Lotto secretly used a concave mirror projector to execute *Husband and wife*, [109] that a wide range of artists as early as 1430 secretly traced optical images, [110] that Jan Vermeer traced images projected in a camera obscura, [111] and so on. While there are a number of perspective and lighting analyses brought to bear on such claims, [112, 113, 114, 57, 115, 42] as well as textual and material analyses, [116, 117] the first and only analysis of paintings done by sophisticated computer ray tracing programs was by Stork and Robinson. [118, 119] This research analyzed the aberrations and other properties of the setup in purported use of optics for Lorenzo Lotto and ultimately questioned the claim he used optics.

10 Analysis of Craquelure

Craquelure is the pattern of fine cracks on the surface of paintings and several scholars have used computer image analysis to characterize the patterns for art imagery retrieval. [120, 121, 122] There remain opportunities for additional algorithms, for instance to detect and classify changes to craquelure due to injury to paintings.

11 Analysis of Composition

Many artists have characteristic compositional styles and it is natural that computer methods be applied to learning and classifying these styles. A particularly attractive oeuvre is that of the neo-plastic abstractionist Piet Mondrian, where the formal elements are simple (horizontal and vertical lines, rectangles of a small number of colors, etc.) and the two-dimensional composition of central importance. A few scholars have approached this problem for Mondrian [123, 124] but there are additional painters whose works might yield information through these methods, such as the large Abstract Expressionist works of Franz Kline.

12 Computer Graphics

Computer graphics allows scholars to understand realist artists' working methods by exploring "what if" scenarios. Note that creating a three-dimensional model based on a two-dimensional painting is formally "ill-posed," that is, an infinite number of three-dimensional tableaus are consistent with a given two-dimensional projection. [125] As such, the creation of a *tableau virtuel* is part art, part science. Nevertheless, the weal the (2D) image information such as occlusion and physical constraints such as objects toughing or being supported on a given floor, and lighting consistency, strongly constrain the three-dimensional models. It is important that the assumption—for instance that bodies have normal proportions, that faces are approximately left-right symmetric, and so on—not bias or favor one conclusion over another.

Stork and Furuichi built a full three-dimensional model of Georges de la Tour's *Christ in the carpenter's studio* and adjusted the location of the virtual illuminant in the *tableau virtuel* until the digitally rendered image matched the painting as closely as possible. In this way, these authors found the illuminant was likely at the position of the candle, rather than in place of the figures, and thereby rebutted the claim that painting was executed by means of optical projections. [76] Savarese and colleagues built a very simple model of the convex mirror and a planar model of the tableau in the left panel of Hans Memling's *van Nieuwenhove Diptych* to test the consistency between the tableau and the image depicted in the convex mirror. The discrepancies between the mirror and the simple tableau suggested that Memling added the mirror later, as an afterthought. [67] Johnson and colleagues built a computer model of Vermeer's *Girl with a pearl earring* to estimate the direction of illumination. [75] Most recently, Stork and Furuichi created a model of both the tableau in Diego Velàsquez' *Las meninas* as well as the viewer's space to explore the relationship between these two spaces, for instance whether the position of the viewer corresponded to that of the king and queen visible in the plane mirror. [126]

13 Websites

There are a few of websites addressing computer image analysis of art.

- Computer image analysis in the study of art:
  www.diatrope.com/stork/FAQs.html
- Digital painting analysis: digitalpaintinganalysis.org
- IAPR computer vision in cultural heritage applications:
  iapr-tc19.prip.tuwien.ac.at/
- Antonio Criminisi's publications:
  research.microsoft.com/~antcrim/papers.htm
- Christopher W. Tyler's research on perception of art by humans and machines: www.diatrope.com/projects
- Computers in the history of art: www.chart.ac.uk

Acknowledgments. My thanks to all the scholars who provided references for this compilation.

References

1. Pelagotti, A., Mastio, A.D., Rosa, A.D., Piva, A.: Multispectral imaging of paintings. IEEE Signal Processing Magazine 25(4), 27–36 (2008)
2. Martinez, K., Cupitt, J., Saunders, D., Pillay, R.: Ten years of art imaging research. Proceedings of the IEEE 90(1), 28–41 (2002)

3. Barni, M., Pelagotti, A., Piva, A.: Image processing for the analysis and conservation of paintings: Opportunities and challenges. IEEE Signal Processing Magazine 22(5), 141–144 (2005)
4. Berezhnoy, I.E., Postma, E.O., van den Herik, J.: Computer analysis of van Gogh's complementary colours. Pattern Recognition Letters 28(6), 703–709 (2006)
5. Maitre, H., Schmitt, F., Lahanier, C.: 15 years of image processing and the fine arts. In: Proceedings International Conference on Image Processing (ICIP), vol. 1, pp. 557–561 (2001)
6. Stork, D.G., Coddington, J. (eds.): Computer image analysis in the study of art, vol. 6810. SPIE/IS&T, Bellingham (2008)
7. Stork, D.G., Coddington, J., Bentkowska-Kafel, A. (eds.): Computer vision and image analysis of art. SPIE/IS&T, Bellingham (forthcoming, 2010)
8. Kirsh, A., Levenson, R.S.: Seeing through paintings: Physical examination in art historical studies. Yale U. Press, New Haven (2000)
9. Stork, D.G.: Computer vision, image analysis and master art, Part I. IEEE Multimedia 13(3), 16–20 (2006)
10. Stork, D.G., Johnson, M.K.: Computer vision, image analysis and master art, Part II. IEEE Multimedia 14(3), 12–17 (2006)
11. Stork, D.G., Duarte, M.: Computer vision, image analysis and master art, Part III. IEEE Multimedia 14(1), 14–18 (2007)
12. Stork, D.G.: Imaging technology enhances the study of art. Vision Systems Design 12(10), 69–71 (2007)
13. van de Wetering, E.: Thirty years of the Rembrandt Research Project: The tension between science and connoisseurship in authenticating art. IFAR Journal 4(2), 14, 24 (2001)
14. Leung, H., Wong, S.T.S., Ip, H.H.S.: In the name of art. IEEE Signal Processing Magazine 25(4), 49–52 (2008)
15. Huang, X., Mohan, A., Tumblin, J.: Deep shadows in a shallow box. In: Stork, D.G., Coddington, J. (eds.) Computer image analysis in the study of art, vol. 6810, pp. 681003-1–9. IS&T/SPIE, Bellingham (2008)
16. Berezhnoy, I.E., Postma, E.O., van den Herik, H.J.: Computerized visual analysis of paintings. In: Proceedings of the 16th International Conference of the Association for History and Computing, Amsterdam, The Netherlands, pp. 28–32. Royal Netherlands Academy of Arts and Sciences (2005)
17. Berns, R.S.: Rejuvenating Seurat's palette using color and imaging science: A simulation. In: Herbert, R.L. (ed.) Seurat and the making of La Grande Jatte, pp. 214–227. Art Institute of Chicago, Chicago (2004)
18. Zhao, Y., Berns, R.S., Taplin, L.A., Coddington, J.: An investigation of multispectral imaging for the mapping of pigments in paintings. In: Stork, D.G., Coddington, J. (eds.) Computer image analysis in the study of art, vol. 6810, pp. 681007-1–9. IS&T/SPIE, Bellingham (2008)
19. Martinez, K., Goodall, M.: Colour clustering analysis for pigment identification. In: Stork, D.G., Coddington, J. (eds.) Computer image analysis in the study of art, vol. 6810, pp. 681008-1–8. IS&T/SPIE, Bellingham (2008)
20. Pappas, M., Pitas, I.: Digital color restoration of old paintings. IEEE Transactions on Image Processing 9(2), 291–294 (2000)
21. Chahine, H., Cupitt, J., Saunders, D., Martinez, K.: Investigation and modelling of color change in paintings during conservation treatment. In: Imaging the past: Electronic imaging and computer graphics in museums and archaeology. Occasional papers of the British Museum, vol. 114, pp. 23–33 (1996)

22. Mastio, A.D., Piva, A., Barni, M., Cappellini, V., Stefanini, L.: Color transplant for reverse ageing of faded artworks. In: Stork, D.G., Coddington, J. (eds.) Computer image analysis in the study of art, vol. 6810, pp. 681006-1–12. IS&T/SPIE, Bellingham (2008)
23. Verri, G., Comelli, D., Cather, S., Saunders, D., Piqué, F.: Post-capture data analysis as an aid to the interpretation of ultraviolet-induced fluorescence images. In: Stork, D.G., Coddington, J. (eds.) Computer image analysis in the study of art, vol. 6810, pp. 681002-1–12. IS&T/SPIE, Bellingham (2008)
24. Easton Jr., R.L., Knox, K.T., Christens-Barry, W.A.: Multispectral imaging of the Archimedes palimpsest. In: Proceedings of the 32nd Applied Imagery Pattern Recognition Workshop, AIPR 2003, El Segundo, CA, pp. 111–116. IEEE, Los Alamitos (2003)
25. Netz, R., Noel, W.: The Archimedes Codex: How a Medieval prayer book is revealing the true genius of Antiquity's greatest scientist. Da Capo Press, Philadelphia (2007)
26. Knox, K.T.: Enhancement of overwritten text in the Archimedes Palimpset. In: Stork, D.G., Coddington, J. (eds.) Computer image analysis in the study of art, vol. 6810, pp. 681004-1–11. IS&T/SPIE, Bellingham (2008)
27. Walvoord, D., Bright, A., Easton Jr., R.L.: Multispectral processing of combined visible and x-ray fluoroescence imagery in the Archimedes palimpsest. In: Stork, D.G., Coddington, J. (eds.) Computer image analysis in the study of art, vol. 6810, pp. 681004-1–11. IS&T/SPIE, Bellingham (2008)
28. Walvoord, D., Easton Jr., R.L.: Digital transcription of the Archimedes palimpsest. IEEE Signal Processing Magazine 25(4), 100–104 (2008)
29. Minturn, K.: Digitally-enhanced evidence: MoMA's reconfiguration of Namuth's Pollock. Visual Resources 17(1), 127–145 (2001)
30. Falk, D., Brill, D., Stork, D.: Seeing the Light: Optics in nature, photography, color, vision and holography. Wiley, New York (1986)
31. Graham, D.J., Field, D.: Global nonlinear compression of natural luminances in painted art. In: Stork, D.G., Coddington, J. (eds.) Computer image analysis in the study of art, vol. 6810, pp. 68100K-1–11. IS&T/SPIE, Bellingham (2008)
32. Graham, D.J., Field, D.: Variations in intensity statistics for representational and abstract art, and for art from the eastern and western hemispheres. Perception 37(9), 1341–1352 (2008)
33. Stork, D.G., Meador, S., Noble, P.: Painted or printed? Correlation analysis of the brickwork in Jan van den Heyden's View of Oudezijds Voorburgwal and the Oude Kerk in Amsterdam. In: Rogowitz, B.E., Pappas, T.N. (eds.) Electronic Imaging: Human vision and electronic imaging XIV, vol. 7240, pp. 72401O1–10. SPIE/IS&T, Bellingham (2009)
34. Lettner, M., Diem, M., Sablatnig, R., Kammerer, P., Miklas, H.: Registration of multi-spectral manuscript images as prerequisite for computer aided script description. In: Grabner, M., Grabner, H. (eds.) Proceedings of the 12th Computer Vision Winter Workshop, St. Lambrecht, Austria, pp. 51–58 (2007)
35. Ketelsen, T., Simon, O., Reiche, I., Merchel, S., Stork, D.G.: Evidence for mechanical (not optical) copying and enlarging in Jan van Eyck's Portrait of Niccolò Albergati. In: Optical Society of American Annual Meeting, Rochester, NY, OSA (2004)
36. Criminisi, A., Stork, D.G.: Did the great masters use optical projections while painting? Perspective comparison of paintings and photographs of Renaissance chandeliers. In: Kittler, J., Petrou, M., Nixon, M.S. (eds.) Proceedings of the 17th International Conference on Pattern Recognition, vol. IV, pp. 645–648 (2004)

37. Duarte, M., Stork, D.G.: Image contour fidelity analysis of mechanically aided enlargements of Jan van Eyck's Portrait of Cardinal Niccolò Albergati. Leonardo 42 (in press, 2009)
38. Lettner, M., Sablatnig, R.: Estimating the original drawing trace of painted strokes. In: Stork, D.G., Coddington, J. (eds.) Computer image analysis in the study of art, vol. 6810, pp. 68100C–1–10. IS&T/SPIE, Bellingham (2008)
39. Noble, P., Stork, D.G.: Jan van der Heyden's View of Oudezijds Voorburgwal and the Oude Kerk in Amsterdam, examined and restored (submitted, 2010)
40. Damisch, H.: The origin of perspective. MIT Press, Cambridge (1995)
41. Criminisi, A., Kemp, M., Zisserman, A.: Bringing pictorial space to life: Computer techniques for the analysis of paintings. In: Bentkowska-Kafel, A., Cashen, T., Gardner, H. (eds.) Digital art history: A subject in transition, pp. 77–100. Intellect Books, Bristol (2005)
42. Criminisi, A.: Machine vision: The answer to the optical debate? Optical Society of American Annual Meeting, Rochester, NY (abstract) (2004)
43. Kemp, M., Criminisi, A.: Computer vision and painter's vision in Italian and Nederlandish art of the fifteenth century. In: Carpo, M., Lemerle, F. (eds.) Perspective, projections and design technologies of architectural representation, pp. 31–46. Routledge, New York (2008)
44. Kemp, M., Criminisi, A.: Paolo Uccello's 'Rout of San Romano': Order from chaos. NEW Magazine 1(1), 99–104 (2005)
45. Smith, B., Stork, D.G., Zhang, L.: Three-dimensional reconstruction from multiple reflected views within a realist painting: An application to Scott Fraser's Three way vanitas. In: Beraldin, J.A., Cheok, G.S., McCarthy, M., Neuschaefer-Rube, U. (eds.) Electronic imaging: 3D imaging metrology, vol. 7239, pp. 72390U1–10. SPIE/IS&T, Bellingham (2009)
46. Hockney, D., Falco, C.M.: Quantitative analysis of qualitative images. In: Electronic Imaging. SPIE Press, Bellingham (2005)
47. Criminisi, A.: Accurate visual metrology from single and multiple uncalibrated images. ACM Distinguished Dissertation Series. Springer, London (2001)
48. Stork, D.G.: Were optical projections used in early Renaissance painting? A geometric vision analysis of Jan van Eyck's Arnolfini portrait and Robert Campin's Mérode Altarpiece. In: Latecki, L.J., Mount, D.M., Wu, A.Y. (eds.) SPIE Electronic Imaging: Vision Geometry XII, pp. 23–30. SPIE, Bellingham (2004)
49. Stork, D.G.: Optics and the old masters revisited. Optics and Photonics News 15(3), 30–37 (2004)
50. Stork, D.G.: Optics and realism in Renaissance art. Scientific American 291(6), 76–84 (2004)
51. Stork, D.G.: Did Jan van Eyck build the first 'photocopier' in 1432? In: Eschbach, R., Marcu, G.G. (eds.) SPIE Electronic Imaging: Color Imaging IX: Processing, Hardcopy and Applications, pp. 50–56. SPIE, Bellingham (2004)
52. Stork, D.G.: Did Hans Memling employ optical projections when painting Flower still-life? Leonardo 38(2), 57–62 (2005)
53. Stork, D.G.: Asymmetry in 'Lotto carpets' and implications for Hockney's optical projection theory. In: Rogowitz, B.E., Pappas, T.N., Daly, S.J. (eds.) SPIE Electronic Imaging: Human vision and electronic imaging X, vol. 5666, pp. 337–343. SPIE, Bellingham (2005)
54. Stork, D.G.: Optique et réalisme dans l'art de la Renaissance. Revue Pour la Science 327, 74–86 (2005)
55. Stork, D.G.: Spieglein, Spieglein and der Wand. Spektrum der Wissenschaft: Forschung und Technik in der Renaissance Spezial 4/2004(4), 58–61 (2005)

56. Stork, D.G.: Mathematical foundations for quantifying shape, shading and cast shadows in realist master drawings and paintings. In: Ritter, G.X., Schmalz, M.S., Barrera, J., Astola, J.T. (eds.) SPIE Electronic Imaging: Mathematics of data/image pattern recognition, compression and encryption with applications IX, vol. 6314, pp. 63150K–1–6. SPIE, Bellingham (2006)
57. Tyler, C.W.: 'Rosetta stone?' Hockney, Falco and the sources of 'opticality' in Renaissance art. Leonardo 37(5), 397–401 (2004)
58. Kulkarni, A., Stork, D.G.: Optical or mechanical aids to drawing in the early Renaissance? A geometric analysis of the trellis in Robert Campin's Mérode Altarpiece. In: Niel, K.S., Fofi, D. (eds.) SPIE Electronic Imaging: Machine vision applications II, vol. 7251, pp. 72510R1–9. SPIE/IS&T, Bellingham (2009)
59. Leeman, F.: Hidden images: Games of perception, anamorphic art, and illusion. Abrams, New York (1976)
60. Baltrušaitis, J.: Anamorphic art. Abrams, New York (1977)
61. Stork, D.G.: Anamorphic art and photography: Deliberate distortions that can be easily undone. Optics and Photonics News 3(11), 8–12 (1992)
62. Hunt, J.L., Nickel, B.G., Gigault, C.: Anamorphic images. American Journal of Physics 68(3), 232–237 (2000)
63. DeWeerd, A.J., Hill, S.E.: Comment on 'Anamorphic images,' by J. L. Hunt, B. G. Nickel and Christian Gigault. American Journal of Physics 74(1), 83–84 (2006)
64. Farid, H.: Reconstructing ancient Egyptian tombs. In: Fisher, B., Dawson-Howe, K., Sullivan, C.O. (eds.) Virtual and augmented architecture: The international symposium on virtual and augmented reality, pp. 23–34 (2001)
65. Farid, H., Farid, S.: Unfolding Sennedjem's tomb. KMT: A modern journal of ancient Egypt 12(1), 46–59 (2001)
66. Criminisi, A., Kemp, M., Kang, S.B.: Reflections of reality in Jan van Eyck and Robert Campin. Historical methods 3(37), 109–121 (2004)
67. Savarese, S., Spronk, R., Stork, D.G., DelPozo, A.: Reflections on praxis and facture in a devotional portrait diptych: A computer analysis of the mirror in Hans Memling's Virgin and Child and Maarten van Nieuwenhove. In: Stork, D.G., Coddington, J. (eds.) Computer image analysis in the study of art, vol. 6810, pp. 68100G–1–10. SPIE/IS&T, Bellingham (2008)
68. Stork, D.G.: Color and illumination in the Hockney theory: A critical evaluation. In: Proceedings of the 11th Color Imaging Conference (CIC11), Scottsdale, AZ, vol. 11, pp. 11–15. IS&T (2003)
69. Stork, D.G.: Did Georges de la Tour use optical projections while painting Christ in the carpenter's studio? In: Said, A., Apolstolopoulos, J.G. (eds.) SPIE Electronic Imaging: Image and video communications and processing, vol. 5685, pp. 214–219. SPIE, Bellingham (2005)
70. Stork, D.G., Johnson, M.K.: Estimating the location of illuminants in realist master paintings: Computer image analysis addresses a debate in art history of the Baroque. In: Proceedings of the 18th International Conference on Pattern Recognition, Hong Kong, vol. I, pp. 255–258. IEEE Press, Los Alamitos (2006)
71. Stork, D.G., Johnson, M.K.: Lighting analysis of diffusely illuminated tabeaus in realist paintings: An application to detecting 'compositing' in the portraits of Garth Herrick. In: Delp III, E.J., Dittmann, J., Memon, N.D., Wong, P.W. (eds.) Electronic Imaging: Media forensics and security XI, vol. 7254, pp. 72540L1–8. SPIE/IS&T, Bellingham (2009)

72. Stork, D.G.: Locating illumination sources from lighting on planar surfaces in paintings: An application to Georges de la Tour and Caravaggio. In: Optical Society of American Annual Meeting, Rochester, NY. Optical Society of America (2008)
73. Kale, D., Stork, D.G.: Estimating the position of illuminants in paintings under weak model assumptions: An application to the works of two Baroque masters. In: Rogowitz, B.E., Pappas, T.N. (eds.) Electronic Imaging: Human vision and electronic imaging XIV, vol. 7240, pp. 72401M1–12. SPIE/IS&T, Bellingham (2009)
74. Duda, R.O., Hart, P.E., Stork, D.G.: Pattern classification, 2nd edn. John Wiley and Sons, New York (2001)
75. Johnson, M.K., Stork, D.G., Biswas, S., Furuichi, Y.: Inferring illumination direction estimated from disparate sources in paintings: An investigation into Jan Vermeer's Girl with a pearl earring. In: Stork, D.G., Coddington, J. (eds.) Computer image analysis in the study of art, vol. 6810, pp. 68100I–1–12. SPIE/IS&T, Bellingham (2008)
76. Stork, D.G., Furuichi, Y.: Image analysis of paintings by computer graphics synthesis: An investigation of the illumination in Georges de la Tour's Christ in the carpenter's studio. In: Stork, D.G., Coddington, J. (eds.) Computer image analysis in the study of art, vol. 6810, pp. 68100J–1–12. SPIE/IS&T, Bellingham (2008)
77. Berezhnoy, I.E., Postma, E.O., van den Herik, H.J.: AUTHENTIC: Computerized brushstroke analysis. In: International Conference on Multimedia and Expo., Amsterdam, The Netherlands, vol. 6, pp. 1586–1588. IEEE Press, Los Alamitos (2005)
78. Jones-Smith, K., Mathur, H.: Fractal analysis: Revisiting Pollock's drip paintings. Nature 444, E9–E10 (2006)
79. Li, J., Wang, J.Z.: Studying digital imagery of ancient paintings by mixtures of stochastic models. IEEE Transactions on Image Processing 13(3), 340–353 (2004)
80. Lyu, S., Rockmore, D., Farid, H.: A digital technique for art authentication. Proceedings of the National Academy of Sciences 101(49), 17006–17010 (2004)
81. Melzer, T., Kammerer, P., Zolda, E.: Stroke detection of brush strokes in portrait miniatures using a semi-parametric and a model based approach. In: Proceedings of the 14th International Conference on Pattern Recognition, vol. 1, pp. 474–476 (1998)
82. Sablatnig, R., Kammerer, P., Zolda, E.: Hierarchical classification of painted portraits using face- and brush stroke models. In: Jain, A.K., Venkatesh, S., Lovell, B.C. (eds.) Proceedings of the 14th International Conference on Pattern Recognition, vol. I, pp. 172–174. IEEE Press, Los Alamitos (1998)
83. Wang, J.Z., Wiederhold, G., Firschein, O., Wei, S.X.: Content-based image indexing and searching using Daubechies' wavelets. International Journal on Digital Libraries 1(4), 311–328 (1998)
84. Kammerer, P., Lettner, M., Zolda, E., Sablatnig, R.: Identification of drawing tools by classification of textural and boundary features of strokes. Pattern Recognition Letters 28(6), 710–718 (2007)
85. Tastl, I., Sablatnig, R., Kropatsch, W.G.: Model-based classification of painted portraits. In: Pinz, A. (ed.) Pattern Recognition 1996: Proceedings of the 20th ÖAGM Workshop. OCG Schriftenreihe, vol. 90, pp. 237–250. Oldenbourg Wien, München (1996)
86. Sablatnig, R., Kammerer, P., Zolda, E.: Structural analysis of paintings based on brush strokes. In: McCrone, W.C., Weiss, R.J. (eds.) Fakebusters: Scientific detection of fakery in art, Hansen, Stoughton, Massachusetts, pp. 222–244 (1999)

87. Kammerer, P., Zolda, E., Sablatnig, R.: Computer aided analysis of underdrawings in infrared reflectograms. In: Arnold, D., Chalmers, A., Nicolucci, F. (eds.) Proceedings of the 4th International Symposium on Virtual Reality, Archaeology and Intelligent Cultural Heritage, Brighton, United Kingdom, pp. 19–27 (2003)
88. Lettner, M., Kammerer, P., Sablatnig, R.: Texture analysis of painted strokes. In: Burger, W., Scharinger, J. (eds.) Digital Imaging in Media and Education, Proceedings of the 28th Workshop of the Austrian Association for Pattern Recognition (OAGM/AAPR), Hagenberg, Austria, vol. 179, pp. 269–276. Schriftenreihe der OCG (2004)
89. Lettner, M., Sablatnig, R.: Texture based drawing tool classification in infrared reflectograms. In: Hanbury, A., Bischof, H. (eds.) Proceedings of the 10th Computer Vision Winter Workshop, Zell an der Pram, Austria, pp. 63–72 (2005)
90. Lettner, M., Sablatnig, R.: Texture based drawing tool classification. In: Chetverikov, D., Czuni, L., Vinzce, M. (eds.) Joint Hungarian-Austrian Conference on Image Processing and Pattern Recognition, Proceedings of the 29th Workshop of the Austrian Association for Pattern Recognition (OAGM/AAPR), vol. 189, pp. 171–178. Schriftenreihe der OCG (2005)
91. Lettner, M., Sablatnig, R.: Stroke trace estimation in pencil drawings. In: Arnold, D., Chalmers, A., Nicolucci, F. (eds.) Proceedings of the 8th International Symposium on Virtual Reality, Archaeology and Cultural Heritage, Brighton, UK, pp. 43–47 (2007)
92. Vill, M.C., Sablatnig, R.: Drawing tool recognition by stroke ending analysis. In: Stork, D.G., Coddington, J. (eds.) Computer image analysis in the study of art, vol. 6810, pp. 68100B-1–11. IS&T/SPIE, Bellingham (2008)
93. van den Herik, H.J., Postma, E.O.: Discovering the visual signature of painters. In: Kasabov, N. (ed.) Future directions for intelligent systems and information sciences: The future of image technologies, brain computers, WWW and bioinformatics, pp. 129–147. Physica-Verlag, Heidelberg (2000)
94. Vill, M.C., Sablatnig, R.: Stroke ending shape features for stroke classification. In: Perš, J. (ed.) Proceedings of the Computer Vision Winter Workshop 2008, Moravske Toplice, Slovenia, pp. 91–98 (2008)
95. Johnson, C.R., Hendriks, E., Berezhnoy, I.J., Brevdo, E., Hughes, S.M., Daubechies, I., Li, J., Postma, E., Wang, J.Z.: Image processing for artist identification. IEEE Signal Processing magazine 25(4), 37–48 (2008)
96. Hedges, S.B.: A method for dating early books and prints using image analysis. Proceedings of the Royal Society A 462(2076), 3555–3573 (2006)
97. Hedges, S.B.: Image analysis of Renaissance copperplate prints. In: Stork, D.G., Coddington, J. (eds.) Computer image analysis in the study of art, vol. 6810, pp. 681009-1–20. IS&T/SPIE, Bellingham (2008)
98. Shahram, M., Stork, D.G., Donoho, D.: Recovering layers of brushstrokes through statistical analysis of color and shape: An application to van Gogh's Self portrait with grey felt hat. In: Stork, D.G., Coddington, J. (eds.) Computer image analysis in the study of art, vol. 6810, pp. 68100D-1–8. SPIE/IS&T, Bellingham (2008)
99. Mandelbrot, B.: The fractal geometry of nature. W. H. Freeman, New York (1982)
100. Taylor, R.P., Micolich, A.P., Jonas, D.: Fractal analysis of Pollock's drip paintings. Nature 399, 422 (1999)
101. Coddington, J., Elton, J., Rockmore, D., Wang, Y.: Multifractal analysis and authentication of Jackson Pollock's paintings. In: Stork, D.G., Coddington, J. (eds.) Computer image analysis in the study of art, vol. 6810, pp. 68100F-1–12. IS&T/SPIE, Bellingham (2008)

102. Alvarez-Ramirez, J., Ibarra-Valdez, C., Rodríguez, E., Dagdug, L.: 1/f-noise structures in Pollock's drip paintings. Physica A 387(1), 281–295 (2008)
103. Fernandez, D., Wilkins, A.J.: Uncomfortable images in art and nature. Perception 37(7), 1098–1113 (2008)
104. Graham, D.J., Field, D.: Statistical regularities of art images and natural scenes: Spectra, sparseness and nonlinearities. Spatial vision 21(1-2), 149–164 (2007)
105. Mureika, J.R., Dyer, C.C., Cupchik, G.C.: On multifractal structure in non-representational art. Physical Review 72(4), 046101 (2005)
106. Redies, C., Hasenstein, J., Denzler, J.: Fractal-like image statistics in visual art: Similarity to natural scenes. Spatial Vision 21(1-2), 137–148 (2007)
107. Irfan, M., Stork, D.G.: Multiple visual features for the computer authentication of Jackson Pollock's drip paintings: Beyond box-counting and fractals. In: Niel, K.S., Fofi, D. (eds.) SPIE Electronic Imaging: Machine vision applications II, vol. 7251, pp. 72510Q1–11. SPIE/IS&T, Bellingham (2009)
108. Stork, D.G.: Learning-based authentication of Jackson Pollock's paintings. SPIE Professional (2009)
109. Hockney, D., Falco, C.M.: Optical insights into Renaissance art. Optics and Photonics News 11(7), 52–59 (2000)
110. Hockney, D.: Secret knowledge: Rediscovering the lost techniques of the old masters. Viking Studio, New York (2001)
111. Steadman, P.: Vermeer's camera: Uncovering the truth behind the masterpieces. Oxford U. Press, Oxford (2002)
112. Stork, D.G.: Did early Renaissance painters trace optical projections? Evidence pro and con. In: Latecki, L.J., Mount, D.M., Wu, A.Y. (eds.) SPIE Electronic Imaging: Vision geometry XIII, vol. 5675, pp. 25–31. SPIE, Bellingham (2005)
113. Stork, D.G.: Tracing the history of art: Review of Early Science and Medicine: Optics, instruments and painting, 1420-1720: Reflections on the Hockney-Falco theory. Nature 438(7070), 916–917 (2005)
114. Stork, D.G., Duarte, M.: Revisiting computer image analysis and art. IEEE Multimedia 14(3), 108–109 (2007)
115. Tyler, C.W., Stork, D.G.: Did Lorenzo Lotto use optical projections when painting Husband and wife? In: Optical Society of American Annual Meeting, Rochester, NY, Optical Society of America (2004)
116. Dupré, S. (ed.): Early Science and Medicine: A Journal for the Study of Science, Technology and Medicine in the Pre-modern Period: Optics, instruments and painting 1420–1720: Reflections on the Hockney-Falco Thesis, vol. X(2). Brill Academic Publishers, Leiden (2005)
117. Lüthy, C.: Reactions of historians of science and art to the Hockney thesis. In: Summary of the European Science Foundation's conference of 12–15 November, 2003. Optical Society of American Annual Meeting, Rochester, NY (Abstract) (2004)
118. Stork, D.G.: Aberration analysis of the putative projector for Lorenzo Lotto's Husband and wife. In: Optical Society of American Annual Meeting, San Jose, CA (Abstract) (2007)
119. Robinson, M.D., Stork, D.G.: Aberration analysis of the putative projector for Lorenzo Lotto's Husband and wife: Image analysis through computer ray-tracing. In: Stork, D.G., Coddington, J. (eds.) Computer image analysis in the study of art, vol. 6810, pp. 68100H–1–11. SPIE/IS&T, Bellingham (2008)
120. Abas, F.S., Martinez, K.: Craquelure analysis for content-based retrieval. In: Proceedings of the 14th Conference on Digital Signal Processing, Santorini, Greece, pp. 111–114 (2002)

121. Abas, F.S.: Analysis of craquelure patterns for content-based retrieval. Ph.D. thesis, University of Southampton, Southampton, UK (2004)
122. Bucklow, S.: A stylometric analysis of craquelure. Computers and Humanities 31(6), 503–521 (1998)
123. Colagrossi, A., Sciarrone, F., Seccaroni, C.: A method for automating the classification of works of art using neural networks. Leonardo 36(1), 69–96 (2003)
124. Heitz, F., Maitre, H., de Couessin, C.: Application of autoregressive models to fine arts painting analysis. Signal Processing 13(1), 1–14 (1987)
125. Remondino, F., El-Hakim, S.F., Grün, A., Zhang, L.: Turning images into 3-D models. IEEE Signal Processing magazine 25(4), 55–65 (2008)
126. Stork, D.G., Furuichi, Y.: Computer graphics synthesis for interring artist studio practice: An application to Diego Velázquez's Las meninas. In: McDowall, I.E., Dolinsky, M. (eds.) Electronic imaging: The engineering reality of virtual reality, vol. 7238, pp. 7238061–9. SPIE/IS&T, Bellingham (2009)

Head Pose Estimation by a Stepwise Nonlinear Regression

Kevin Bailly[1,2], Maurice Milgram[1,2], and Philippe Phothisane[1,2]

[1] UPMC Univ Paris 06, F-75005, Paris, France
[2] CNRS, UMR 7222, ISIR, Institut des Systèmes Intelligents et de Robotique, F-75005, Paris, France

Abstract. Head pose estimation is a crucial step for numerous face applications such as gaze tracking and face recognition. In this paper, we introduce a new method to learn the mapping between a set of features and the corresponding head pose. It combines a filter based feature selection and a Generalized Regression Neural Network where inputs are sequentially selected through a boosting process. We propose the Fuzzy Functional Criterion, a new filter used to select relevant features. At each step, features are evaluated using weights on examples computed using the error produced by the neural network at the previous step. This boosting strategy helps to focus on hard examples and selects a set of complementary features. Results are compared with three state-of-the-art methods on the Pointing 04 database.

1 Introduction

Head pose estimation is a complex problem and is currently an area of great interest for the face analysis and image processing communities. Numerous methods have been proposed (please refer to [1] for a complete survey on this area). The nonlinear regression is one popular approach. It consists of estimating the mapping from an image or feature data to a head pose direction. Given a set of labelled image, the relation is learnt using a regression tool such as a multi layer perceptron [2]. Thus, a continuous head pose can be estimated for any unseen face image.

Formally, the problem can be stated as follows. We have a set of face images $\mathbf{x_i} \in \mathbb{R}^d$ where d is the image dimension. To each example $\mathbf{x_i}$ an angle $y_i \in \mathbb{R}$ is associated that we want to predict. Data are divided into a training set A, a validation set V and a test set E. A set F of features H_k ($1 \leq k \leq N$) can be computed for each $\mathbf{x_i}$ such that $h_{k,i} = H_k(\mathbf{x_i})$. F can be extremely large (typically more than 10 000 elements in our case). The main objective of our method is to select a subset $FS \subset F$ and to learn the functional relation between the values of this set of features and the head pose.

This paper addresses two main issues: select image features and learn the required mapping (image to head pose). To achieve these goals, two contributions are proposed. (i) The boosted stepwise feature selection is combined with the neural network training (ii) The Fuzzy Functional Criterion (FFC), a new filter used to measure the functional dependency between two variables.

2 Regression Based Head Pose Estimation

2.1 Boosted Input Selection Algorithm for Regression

We introduce a new forward feature selection method for regression named BISAR (Boosted Input Selection Algorithm for Regression). It combines a filter with the boosting paradigm : (i) A new filter, the Fuzzy Functional Criterion (FFC) is designed to select *relevant features*. This criterion measures the functional dependency of the output y on a feature H_k. (ii) A new boosting strategy selects incrementally new *complementary inputs* for the regressor. The prediction performance is used to adjust impact of each training example on the choice of next feature. The detailed algorithm can be found at the end of this section.

2.2 Image Features

First, we describe the four kinds of features we have used. The first three descriptors correspond to the popular Haar like features [3] as depicted in Fig. 1. We choose these features because of their good performance obtained in related areas such as face detection [3] and image alignment [4]. Moreover they are simple and very fast to compute using the integral image [3].

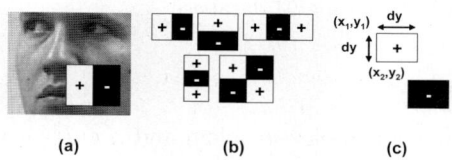

Fig. 1. Image features. (a) a feature example. (b) Representation of the Haar like features used in our experiment. (c) The proposed features are the difference between the sum of the pixels within two non connected rectangular regions.

Another kind of features, we propose here, is the difference between the sum of the pixels within two non connected rectangular regions. Features are parameterized by four values x_1, y_1, dx and dy. x and y correspond to the horizontal and vertical positions of the feature in the image respectively. dx and dy are the width and height of a rectangle. Two extra parameters x_2 and y_2 are needed to describe features of the fourth category. They correspond to the position of the second rectangle in the image (Fig. 1).

2.3 Fuzzy Functional Criterion

The Fuzzy Functional Criterion is like an energy function. The basic idea is: we want to evaluate the relation between two variables u and v (u is a measure and v a target to be predicted). We call P the logical proposition "u_1 and u_2 are close" and Q the logical proposition "v_1 and v_2 are close". If there exists a functional smooth relation between u and v (i.e. $v = F(u)$ where F is a smooth continuous function), then we can say that the logical implication $P \Rightarrow Q$ is true. If the variable v takes value in a discrete subset of the real numbers set, we have to adapt the criterion slightly. Using the truth table of the logical implication, we find it is equivalent to "$\neg P$ or Q" or $\neg(P \text{ and } \neg Q)$. We take the fuzzy logic formulation of P and Q based on classical triangular shaped functions [5] denoted by L defined as:

$$L_a(e) = \begin{cases} 1 - \frac{|e|}{a} & \text{if } |e| < a \\ 0 & \text{otherwise} \end{cases} \quad (1)$$

To quantify the fact that "u_1 and u_2 are close" is true, we just take the value $L_a(|u_1 - u_2|)$, where a is the spread of the triangular function and its value is discussed later. We can do the same for v and write:

$$Z = 1 - L_a(|u_1 - u_2|)(1 - L_b(|v_1 - v_2|)) \quad (2)$$

Equation 2 is a fuzzy evaluation of our implication "$P \Rightarrow Q$". To build our criterion, we have to sum up Z over all (u_1, u_2, v_1, v_2) (the constant "1" can be dropped):

$$FFC = \sum_i \sum_j -L_a(|u_i - u_j|)(1 - L_b(|v_i - v_j|)) \quad (3)$$

The sum is taken over all quadruples. In our case, u_i is the value of some feature H_k for an example $\mathbf{x_i}$ and v_i is the corresponding target value y_i (the head pose). We also introduce weights on examples, w_i with $\sum w_i = 1$. These weights will be useful in the boosting process. So we can reformulate our criterion as:

$$FFC(H_k) = -\sum_i w_i \sum_j L_a(|H_k(\mathbf{x_i}) - H_k(\mathbf{x_j})|)(1 - L_b(|y_i - y_j|)) \quad (4)$$

Modification on these weights does not imply to compute from scratch a new FFC at each step: it is possible to compute an intermediary form of FFC and to modulate it with the weights on examples based on the regression error. Equation 4 can be rewritten as:

$$FFC(H_k) = -\sum_i w_i Q(i) \quad (5)$$

where:

$$Q(i) = \sum_j L_a(|H_k(x_i) - H_k(x_j)|)(1 - L_b(|y_i - y_j|)) \quad (6)$$

Q can be precomputed once. To be selected, a feature H_k must maximize the criterion $FFC(H_k)$. Two parameters (a and b) control the spread of the L functions and are used to normalize the criterion over all features. In many problems, the range of u and v variables are bounded and can be determined. We shall tune a and b as a fixed proportion of these ranges. Experiments have shown that the criterion is not too sensitive to this tuning. Another way to cope with the normalization issue is to replace u values by their rank. In this option, one can see the rank as a simple normalization: if there are M examples, the number $rank(u)/M$ is always in the interval [0 1]. Notice that the difference of ranks divided by the number of examples has the same distribution for all features. It is also important to see that the proximity of values is taken into account in our criterion which is not the case for other criteria such as mutual information.

2.4 Regressor

The second element of our system is the regressor and we have used a Generalized Regression Neural Network (GRNN) which is known for its good properties and which does not need any training step. The design of each GRNN is done by using a very simple algorithm, quite similar to that presented in [6].

GRNN is a two layer network. The first layer (hidden layer) has radial basis neurons, each one representing one prototype. The second layer linearly combines the targets and the outputs of the first layer i.e. the output of the network is the normalized sum of the product of "target × output" over all the hidden neurons. It means that for this network, one does not make a real training step but instead just keeps all prototypes (in the hidden layer) and all targets (in the second layer weights). This network can be viewed as a Parzen window procedure applied to an interpolation task.

2.5 Boosting Strategy

At the beginning, the GRNN starts with only one input cell which corresponds to the best feature given the criterion according to a uniform weighting function. In the second step, a new GRNN is trained with two input cells. The new input corresponds to the second feature which is selected by FFC among all features (except features already selected) using weights provided by the error of the first GRNN. It means that examples that are poorly processed by the first GRNN will receive higher weights than others. The resulting weights, when fed to the criterion, will lead to choose a new feature fitted to these examples. It is clear that this boosting paradigm enhances the regression system.

We test two boosting mechanisms. The first one lies on a memoryless process. Weights at iteration t are only related to the absolute error err_k^{t-1} of the previous iteration. In our case, we adopted this simple relation:

$$w_k^t = \frac{(err_k^{t-1})^2}{\sum_k (err_k^{t-1})^2}; \qquad (7)$$

The second boosting strategy is cumulative and each example weight depends on the regression model error and the previous weights. Weights update is inspired by [7]. In the latter, the threshold on the error is a constant and the accumulation factor depends on the regression error. In our approach the accumulation factor is fixed and the threshold depends on the median error. Our strategy is defined by

$$\tilde{w}_k^t = \begin{cases} w_k^{t-1} & \text{if } err_k^{t-1} < median_k(err_k^{t-1}) \\ max\{\alpha w_k^{t-1}, w_{max}\} & \text{otherwise} \end{cases} \quad (8)$$

$$w_k^t = \frac{\tilde{w}_k^t}{\sum_k (\tilde{w}_k^t)} \quad (9)$$

α corresponds to an accumulation factor and w_{max} is a constant used to avoid overfitting. Typically, α is set to 1.1 and w_{max} to 0.1. The next section presents a comparison of these boosting approaches.

---

1. initialization : $t = 0$
   - Set initial values for $w^0 = (\frac{1}{M}, \ldots, \frac{1}{M})$ and $FS \leftarrow \emptyset$
   - Compute Q used to evaluate the FFC criterion :
   $$Q(i) = \sum_j L_a(|h_{k,i} - h_{k,j}|).(1 - L_b(|y_i - y_j|))$$
2. iteration : $t = 1 \ldots T$
   - Evaluate the FFC criterion for each feature :
   $$FFC(H_k) = -\sum_i w_i Q(i)$$
   $$b_t = \arg\max_{k \in \{1,\ldots,N\}}(FCC(H_k))$$
   - Add the best feature H_{b_t} to the set of selected features. $FS \leftarrow FS \cup H_{b_t}$ and $F \leftarrow F \backslash H_{b_t}$.
   - Set the regressor (GRNN) taking $(h_{b_1,i}, \ldots, h_{b_t,i})$ as input and y_i as output for each example $\mathbf{x_i}$.
   - Compute the new weight vector
   $$w_k^{t+1} = \frac{err_k^t}{\sum_k (err_k^t)}$$
   - Stop if $t \geq T$ or $err_k^{t+1} - err_k^t \geq 0$ on V

Fig. 2. BISAR algorithm

3 Experimental Results

We performed simulation using Pointing 04. It was used to evaluate head pose estimation systems in the Pointing 2004 Workshop and in the International Evaluation on Classification of Events Activities and Relationships (CLEAR 2006). The corpus consists of 15 sets of images. Each set contains 2 series of 93 images of the same person at 93 different poses. There are 15 people, male and female, wearing glasses or not and with various skin color and facial hair. The pose is determined by 2 angles (h,v), ranging from -90° to +90° for the horizontal angle (pan) and from -60° to +60° for the vertical orientation (tilt).

The first series is used for learning (in our case 13 people are in the training set and the remaining 2 people are in the validation set). Tests are carried out on the second series.

We present pan and tilt estimation results obtained on the test set. The error is defined as the mean of absolute deviation from the ground truth.

3.1 Face Localization

The face localization is a simple algorithm based on skin color. First, frontal faces are detected in the training database using the Viola-Jones face detector [3]. Pixels within the detected area are used to build an histogram on H and S channels of HSV color space. Other pixels are randomly picked outside the face area in order to built a background histogram. Face pixels are detected in the test image set with a Bayesian rule based on these skin and non skin histograms. The face bounding box is proportional to the standard deviation of skin pixels along X and Y axis. In this method, only frontal faces of the training set are needed and no manual labelling is required. The main drawback is the accuracy. For example, pixels of the neck can be included in the bounding box. This poor localization can greatly affect head pose results. So, results will be presented for both manual and automatic face localization.

3.2 Boosting Effect

Fig. 3 depicts the evolution of pan error on the manually cropped test database and presents the effect of boosting technique. The red solid curve corresponds to the error without any boosting, i.e. weights on examples are constant and equal. Mean absolute error after 500 iterations is equal to 8.3°. The green dotted line highlights the effect of memoryless boosting (cf. equation 7). After 500 iterations, pan error is 7.4°, which is 10% better. Results obtained with cumulative boosting (cf. equation 8) are represented by the blue dashed curve. This technique is less effective than the memoryless boosting but results are nevertheless better than without boosting, i.e. weights on example remains constant during iterations.

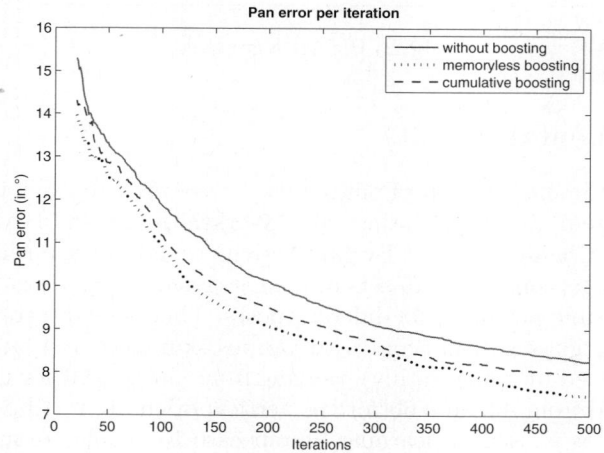

Fig. 3. Influence of the different boosting strategies on the test dataset

Results presented in the rest of this article are obtained with the memoryless boosting strategy.

3.3 Comparison with Other Methods

Performance are compared with three state-of-the-art methods for this database [1]. The first one proposed by Voit et al. [2] uses a multi layer perceptron (MLP) to estimate the pose. First, head is localized using a linear boundary decision classifier to detect skin color. Head area corresponds to the bounding box surrounding the biggest skin color connected component. Inputs of the MLP are the gray levels and gradient magnitudes of resized images (64x64 pixels). Continuous outputs of the network are the estimated head angle values. In Gourier et al. [8], head poses are computed using Auto Associative Memories (one per orientation). Head pose is estimated by selecting the auto-associative network with highest likelihood score. Face was previously located using a skin color based method quite similar to [2]. The last method is proposed by Tu et al.[9]. The appearance variation caused by head pose changes is characterized by tensor model. Given an image patch from the testing set, it is projected into the tensor subspace, and head pose can be estimated from the tensor coefficients obtained from a High Order Singular Value Decomposition (HOSVD).

Table 1. Results on tilt and pan estimation and comparison with other methods

Method	Tilt error	Classification	Pan error	Classification
Manual face localization				
BISAR	**8.5°**	69.2%	7.4°	60.0%
High-Order SVD [9]	8.6°	**75.7%**	**6.2°**	**72.4%**
Automatic face localization				
BISAR	**11.4°**	**59.6%**	11.5°	49.3%
High-order SVD [9]	17.9°	54.8%	12.9°	49.3%
Auto associative memories [8]	12.1°	53.8%	**7.3°**	**61.3%**
Multi layer perceptron [2]	12.8°	53.1%	12.3°	41.8%
Human performance [8]	9.4°	59.0%	11.8°	40.7%

Table 1 summarizes results obtained by the BISAR method and the three related approaches. Score are obtained on the test set using the best regressor on the validation set. To obtain a classification measure, we have mapped the output of our GRNN into the head orientation classes provided in the data. When faces are manually localized, BISAR results are the same on tilt estimation as the HOSVD method and less accurate on the pan estimation. But HOSVD results fall when the face is automatically cropped. This method is very sensitive to the localization accuracy. Auto associative memories outperform BISAR on pan estimation but the latter obtains a better score for tilt prediction. In comparison with the multi layer perceptron, BISAR has better results on both tilt and pan estimation

Results are also compared to human capability reported in [8]. One can notice that BISAR and human performance are quite similar.

4 Conclusion

We have presented a new method to estimate head pose in still images. This method relies on BISAR, a new algorithm used to select a set of image features and to learn the functional mapping from these features to head pose using a generalized regression neural network. The latter is designed incrementally with the help of a boosting strategy and the Fuzzy Functional Criterion (FFC), a new filter used to measure the functional dependency between two variables. This approach has been tested on the Pointing 04 database. Comparison with three related methods has demonstrated state of the art results.

References

1. Murphy-Chutorian, E., Trivedi, M.: Head pose estimation in computer vision: A survey. IEEE Transactions on Pattern Analysis and Machine Intelligence 99 (2008)
2. Voit, M., Nickel, K., Stiefelhagen, R.: Neural network-based head pose estimation and multi-view fusion. In: Stiefelhagen, R., Garofolo, J.S. (eds.) CLEAR 2006. LNCS, vol. 4122, pp. 291–298. Springer, Heidelberg (2007)
3. Viola, P., Jones, M.: Robust real-time object detection. International Journal of Computer Vision (2002)
4. Wu, H., Liu, X., Doretto, G.: Face alignment via boosted ranking model. In: IEEE Conference Computer Vision and Pattern Recognition, pp. 1–8 (2008)
5. Zadeh, L.A.: Soft computing and fuzzy logic. IEEE Softw. 11(6), 48–56 (1994)
6. Wasserman, P.D.: Advanced Methods in Neural Computing. John Wiley & Sons, Inc., New York (1993)
7. Shrestha, D.L., Solomatine, D.P.: Experiments with adaboost.rt, an improved boosting scheme for regression. Neural Computation 18(7), 1678–1710 (2006)
8. Gourier, N., Maisonnasse, J., Hall, D., Crowley, J.L.: Head pose estimation on low resolution images. In: Stiefelhagen, R., Garofolo, J.S. (eds.) CLEAR 2006. LNCS, vol. 4122, pp. 270–280. Springer, Heidelberg (2007)
9. Tu, J., Fu, Y., Hu, Y., Huang, T.S.: Evaluation of head pose estimation for studio data. In: Stiefelhagen, R., Garofolo, J.S. (eds.) CLEAR 2006. LNCS, vol. 4122, pp. 281–290. Springer, Heidelberg (2007)

Model-Based Illumination Correction for Face Images in Uncontrolled Scenarios

Bas Boom, Luuk Spreeuwers, and Raymond Veldhuis

University of Twente, EEMSC, Signals & Systems,
P.O. box 217, 7500 AE, Enschede, The Netherlands
{b.j.boom,l.j.spreeuwers,r.n.j.veldhuis}@utwente.nl

Abstract. Face Recognition under uncontrolled illumination conditions is partly an unsolved problem. Several illumination correction methods have been proposed, but these are usually tested on illumination conditions created in a laboratory. Our focus is more on uncontrolled conditions. We use the Phong model which allows us to model ambient light in shadow areas. By estimating the face surface and illumination conditions, we are able to reconstruct a face image containing frontal illumination. The reconstructed face images give a large improvement in performance of face recognition in uncontrolled conditions.

1 Introduction

One of the major problems with face recognition in uncontrolled scenarios is the illumination variation, which is often larger than the variations between individuals. We want to correct for these illumination variations in a single face image. In literature, several methods have been proposed to make face images invariant to illumination. These methods can be divided into two categories: The first category contains methods that perform preprocessing based on the local regions, like Histogram Equalization [1] or (Simplified) Local Binary Patterns [2,3]. These methods are direct and simple, but fail to model the global illumination conditions. The methods in second category estimate a global physical model of the illumination mechanism and its interaction with the facial surface. One of the earlier methods in this category is the Quotient Image [4], which estimates illumination in a single image allowing the computation of a quotient image. More recent correction methods [5,6] are also able to deal with shadows and reflections using an addition error term. In our experience, these methods work on images with illumination conditions created in a laboratory, but fail in uncontrolled scenarios. In [7], 3D morphable models are used to simulate the illumination conditions in a single images, calculating shadows and reflections properly. The disadvantage of this method is the computational cost for a single image. In [8], a illumination normalization is proposed for uncontrolled conditions which requires a color image together a 3D range image.

We developed a new method for illumination correction in [9] which used only a single grey level image, but this method improved the recognition for face images taken under uncontrolled conditions. During these experiments, we discovered that both our method and [5] have problems modelling shadow areas which still contain some reflection. This often occurs in face images taken under uncontrolled conditions. Furthermore, we observed that the found surface normals were not restricted by the geometrical constrains. In this paper, we tried to solve these issues by improving our previous method.

2 Illumination Correction Method

2.1 Phong Model

To model the shadow areas that contain some reflections, we use the Phong model, which explains these areas using the ambient reflection term. In our previous work and in [5], the Lambertian model with a summed error term was used to model the shadows. This however fails when both the intensities of the light source on the face and the reflections in the shadow areas vary. The Phong model in combination with a shadow expectation is able to model these effects. If we assume a single diffuse light source l, the Phong model is given by the following Equation:

$$b(\mathbf{p}) = c_a(\mathbf{p})i_a + c_d(\mathbf{p})\mathbf{n}(\mathbf{p})^T\mathbf{s}_l i_d + \text{specular reflections} \qquad (1)$$

The image $b(\mathbf{p})$ at location $\mathbf{p}$ can be modelled using three parts namely: the ambient, diffuse and specular reflections. The ambient reflections exist of the albedo $c_a(\mathbf{p})$ and the intensity of the ambient light i_a. The ambient reflections are still visible if there is no diffuse light, for instance in shadow areas which are not entirely dark. The diffuse reflections are similar to the Lambertian model, where the surface normals $\mathbf{n} \in \mathcal{R}^3$ define the direction of the reflection and together with the albedo $c_d(\mathbf{p})$ give the shape $\mathbf{h}(\mathbf{p}) = c_d(\mathbf{p})\mathbf{n}(\mathbf{p})^T$. The diffuse light can be modelled by a normalized vector $\mathbf{s} \in \mathcal{R}^3$, which gives the light direction and the intensity of the diffuse light i_d. The final term contains the specular reflections, which explain the highlights in the image, but because this phenomenon is usually only present in a very small part of the image we will ignore this term.

The shadow can be modelled as a hard binary decision. If a light source can not reach a certain region, it makes a shadow. This holds expect for areas which contain the transition between light and shadow. Using a 3D range map of a face, we can compute the shadow area given a certain light direction using a ray tracer. Computing these shadow areas for multiple images, allows us to calculate an expectation of shadow $e_l(\mathbf{p})$ on the position $\mathbf{p}$ for the light directions $\mathbf{s}_l$. This gives us a user independent shadow model given the light direction.

$$b(\mathbf{p}) = c(\mathbf{p})i_a + c(\mathbf{p})\mathbf{n}(\mathbf{p})^T\mathbf{s}_l i_d e_l(\mathbf{p}) \qquad (2)$$

$$\mathbf{b} = \mathbf{c}i_a + H\mathbf{s}_l i_d \star \mathbf{e}_l \qquad (3)$$

In Equation 2, we simplified the Phong model and we added the expectation term $e_l(\mathbf{p})$ to model shadows. We also use the same albedo term for ambient and diffuse illumination, which is common practice [7]. In Equation 3, we vectorized all the terms, where the $\star$ denotes the Cartesian product. Our goal is to find the face shape and the light conditions given only a single image.

2.2 Search Strategy for Light Conditions and Face Shape

An method to estimate both the face shape and the light conditions is to vary one of the variables and calculate the others. In our case, we chose to vary the light direction allowing us to calculate the other variables. After obtaining the other variable, e.g. light intensity, surface and albedo, we use an evaluation criteria to see which light direction gives the best estimates. The pseudo-code of our correction method is given below:

- For a grid of light directions s_l
  - Estimate the light intensities i_a and i_d
  - Estimate the initial face shape
  - Estimate the surface using geometrical constrains and a 3D surface model
  - Computing the albedo and its variations
  - Evaluation of the found parameters
- Refine the search to find the best light direction.
- Reconstruct a face images under frontal illumination.

We start with a grid where we vary the azimuth and elevation of the light direction with 20 degrees. The grid allows us to locate the global minimum, from there we can refine the search using the downhill simplex search method [10] to find the light direction with an accuracy of ± 2 degrees. Using the found parameters like light conditions and face shape, we can reconstruct a face image under frontal illumination, which can be used in face recognition. In the next sections, we will discuss the different components mentioned in the pseudo-code.

2.3 Estimate the Light Intensities

Given the light direction s_l and the shadow expectation $e_l(\mathbf{p})$, we can estimate the light intensities using the mean face shape $\overline{\mathbf{h}}(\mathbf{p})$ and mean albedo $\overline{c}(\mathbf{p})$. The mean face shape and albedo are determined using a set of face images together with there 3D range maps. This gives us the following linearly solvable equation, allow us to obtain the light intensities $\{i_a, i_d\}$:

$$\{i_a, i_d\} = \arg\min_{\{i_a, i_d\}} \sum_{\mathbf{p}} \|b(\mathbf{p}) - \overline{c}(\mathbf{p})i_a - \overline{\mathbf{h}}^T \mathbf{s}_l i_d e_l(\mathbf{p})\|^2 \qquad (4)$$

Because this is an over-determined system, we can use the mean face shape and mean albedo to estimate the light intensities, which still gives a very accurate estimation. However, this might normalize the difference in intensity of the skin color. If the light intensities are negative, we skip the rest of the computations.

2.4 Estimate the Initial Face Shape

To estimate the initial face shape given the light conditions $\{\mathbf{s}_l, \mathbf{e}_l(\mathbf{p}), i_a, i_d\}$, we use the following two assumptions: Firstly, the Phong model must hold, which gives us the following equations:

$$b(\mathbf{p}) = c(\mathbf{p})i_a(\mathbf{p}) + h_x(\mathbf{p})s_{x,l}i_d e_l(\mathbf{p}) + h_y(\mathbf{p})s_{y,l}i_d e_l(\mathbf{p}) + h_z(\mathbf{p})s_{z,l}i_d e_l(\mathbf{p}) \quad (5)$$

Secondly, the face shape should be similar to the mean face shape. This can be measure by taking the Mahalanobis distance between the face shape $\mathbf{h}(\mathbf{p})$ and the mean face shape $\overline{\mathbf{h}}(\mathbf{p})$. Using Lagrange multipliers, we can minimize the distance with Equation 5 as a constrain. This allows us to find an initial face shape $\hat{\mathbf{h}}(\mathbf{p})$, which we will improve in the next steps using a surface model together with geometrical constrains.

2.5 Estimate Surface Using Geometrical Constrains and a 3D Surface Model

Given an estimate of the face shape $\hat{\mathbf{h}}$, we want to determine the surface $\mathbf{z}$, which is a depth map of the face image. Given a set of 3D range images of faces, we can calculate depth maps $\{\mathbf{z}^t\}_{t=1}^T$ and we can obtain the mean surface $\overline{\mathbf{z}}$ and a covariance matrix $\Sigma_\mathbf{z}$. Using Principal Component Analysis (PCA), we computer the subspace by solving the eigenvalue problem:

$$\Lambda_\mathbf{z} = \Phi^T \Sigma_\mathbf{z} \Phi \qquad \hat{\mathbf{z}} = \overline{\mathbf{z}} + \sum_{k=0}^K \Phi_k u_\mathbf{z}(k) \quad (6)$$

where $\Lambda_\mathbf{z}$ are the eigenvalues and Φ are the eigenvectors of the covariance matrix $\Sigma_\mathbf{z}$, which allows to express the surface in variations $\mathbf{u_z}$ for the mean surface $\overline{\mathbf{z}}$. We also know that $h_{zx}(\mathbf{p}) = \frac{h_z(\mathbf{p})}{h_x(\mathbf{p})} = \nabla_x z(p)$ and $h_{zy}(\mathbf{p}) = \frac{h_z(\mathbf{p})}{h_y(\mathbf{p})} = \nabla_y z(\mathbf{p})$ holds, where ∇_x and ∇_y denote the gradient in x and y direction. This allows us to calculate the variations of the surface $\mathbf{u_z}$ using the following equation:

$$\mathbf{u_z} = \arg\min_{\mathbf{u_z}} ||\nabla_x \overline{\mathbf{z}} + \nabla_x \Phi \mathbf{u_z} - \hat{\mathbf{h}}_{zx}||^2 + ||\nabla_y \overline{\mathbf{z}} + \nabla_y \Phi \mathbf{u_z} - \hat{\mathbf{h}}_{zy}||^2 \quad (7)$$

The surface $\hat{\mathbf{z}}$ can be found using Equation 6 and from this surface we can also find the surface normals $\mathbf{n}(\mathbf{p})$. In this case, the surface normals are restricted by geometrical constrains. Using only the geometrical constrains does not have to be sufficient to determine the face surface, therefore, we use the surface model to ensure the convergence.

2.6 Computing the Albedo and Its Variations

In the previous sections, we obtained the surface normals $\mathbf{n}(\mathbf{p})$ and the illumination conditions $\{\mathbf{s}_l, \mathbf{e}_l(\mathbf{p}), i_a, i_d\}$. This allows us to calculate the albedo $\mathbf{c}$ from Equation 2. In order to find out whether the albedo is correct, we also create a PCA model of the albedo. Given a set of face images together with their 3D range maps, we estimated the albedo, see [9]. Vectorizing the albedo $\{\mathbf{c}^t\}_{t=1}^T$ allows us to calculate a PCA model and find the variations $\mathbf{u_c}$, which is also used for the surface model. Using the variations $\mathbf{u_c}$, we calculated also a projection of albedo $\hat{\mathbf{c}}$ to PCA model. The projection $\hat{\mathbf{c}}$ does not contain all details necessary for the face recognition. For this reason, we use the albedo $\hat{\mathbf{c}}$ from the PCA model in the evaluation criteria, while we use the albedo $\mathbf{c}$ obtained from Equation 2 in the reconstructed image.

2.7 Evaluation of the Found Parameters

Because we calculate the face shape for multiple light directions, we have to determine which light direction results in the best face shape. Furthermore, the downhill simplex algorithms also needs an evaluation criteria to be able to find the light direction more accurately. Using the found light conditions and face shape, we can reconstruct an image $\mathbf{b}_r$ which should be similar to the original image. This can be measured using the sum of the square differences between the pixels values. Minimizing this may cause overfitting of our models at certain light directions. For this reason, we use the maximum a posterior probability estimator given by $P(\mathbf{u_c}, \mathbf{u_z}|\mathbf{b})$, which can be minimized by the following equations, see [7]:

$$E = \frac{1}{\sigma_b} \sum_p \|b(\mathbf{p}) + b_r(\mathbf{p})\|^2 + \sum_{k=1}^{K} \frac{u_\mathbf{z}^2(k)}{\lambda_\mathbf{z}(k)} + \sum_{j=1}^{J} \frac{u_\mathbf{c}^2(j)}{\lambda_\mathbf{c}(j)} \qquad (8)$$

In this case, σ_b controls the relative weight of the prior probability, which is the most important factor to minimize. $\lambda_\mathbf{z}$ and $\lambda_\mathbf{c}$ are the eigenvalues of the surface and albedo. The light directions that minimizes Equation 8, give us the parameters from which we can reconstruct a face image with frontal illumination.

3 Experiments and Results

We correct for illumination by estimating both the illumination conditions and the face surface. In this section, we will show some of the estimate surfaces together with their corrected images. The main purpose of the illumination correction is to improve the performance of the face recognition method. Our goal is therefore to demonstrate that our face recognition method indeed benefits from the improvement in the illumination correction. For this purpose, we use the FRGCv1 database where we have controlled face images in the enrollment and uncontrolled face images as probe images.

3.1 3D Database to Train the Illumination Correction Models

For our method, a database is needed that contains both face images and 3D range maps to compute the surface, shape, shadow and albedo models. In this case, we used the Spring 2003 subset of Face Recognition Grand Challenge (FRGC) database, which contains face images together with their 3D range maps. These face images contain almost frontal illumination and no shadows, making this subset of the database ideal to compute the surface and albedo. The exact method to retrieve the albedo is describe in [9].

3.2 Recognition Experiment on FRGCv1 Database

The FRGCv1 database contains frontal face images taken under both controlled and uncontrolled illumination conditions as is shown in Figure 1. The first image in Figure 1 is taken under controlled conditions, while the other images are taken under uncontrolled conditions. For the first person, we show that our method is able to correct for different unknown illumination conditions. In case of the last image, we observe more highlighted areas directly under the eyes, this is caused by the reflection of the glasses which are not modelled by our method.

In order to test if this illumination correction method improves the performance in face recognition, we performed the following experiment to see if illumination conditions are removed in the images taken under uncontrolled conditions. In this case, we use the images with uncontrolled illumination as probe image and make one user template for every person with the images taken under controlled conditions. To train our face recognition method, we randomly divided both the controlled and uncontrolled set of the FRGCv1 database into two parts, each containing approximately half of the face images. The first halves are used to train the face recognition method. The second half of the controlled set is used to compute the user templates, while the second half of the uncontrolled set is used as probe images. We repeat this experiment 20 times using different images in both halves to become invariant against statistical fluctuations. The Receiver operating characteristic (ROC) in Figure 2 is obtained using the PCA-LDA likelihood ratio [11] for face recognition. The first three lines in Figure 2 are also stated in our previous work [9], where the last line depicts the improvements obtained using the method described in this paper. The ROC curves in Figure 2 shows only the improvements due to illumination correction. In the

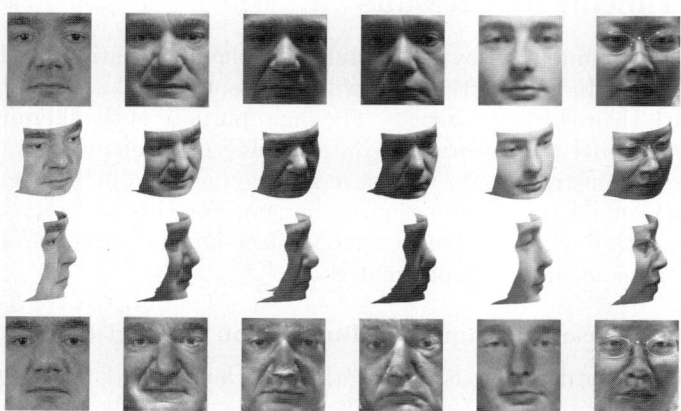

Fig. 1. First row contains the original images from the FRGCv1 database, second and third row show the resulting surface, the fourth row depicts the reconstructed frontal illumination

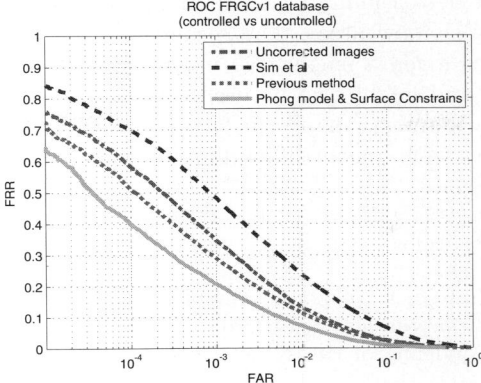

Fig. 2. ROC on the FRGCv1 database with a comparison to our previous work [9] and to work of Sim et al [5]

case of our previous method, the False Reject Rate (FRR) becomes significantly better at a False Accept Rate (FAR) smaller than 1%, while the last line is overall better. Most other illumination correction methods like [6, 8] evaluated their method only on a database create in a laboratory or do not perform a recognition experiment, which makes the comparison with other methods difficult.

4 Discussion

In figure 1, we observe that the phenomenon, where the shadow areas are still not completely dark, often occurs in uncontrolled illumination conditions. Improving our model on this point gave also improvements in the recognition results, which was the main purpose of our illumination correction. We choose to ignore other illumination effects like specular reflections, because we expected a small performance gain in face recognition and a large increase in computation time.

The second improvement is a restriction to the face shape by computing the surface instead of the surface normals, which also slightly improved the face recognition results. Another benefit is that we obtained an estimation of the surface of the face, which might be handy in other applications. In our research, the focus has not been on the quality of the estimated surfaces. Although we expect that this can be an interesting result to improve for instance 3D face acquisition and recognition.

5 Conclusion

We present two major improvements for our illumination correction method for face images, where the purpose of our method is to improve the recognition results of images taken under uncontrolled illumination conditions. The first improvement uses a better illumination model, which allows us to model the ambient light in the shadow areas. The second improvement computes an estimate of the surface given a single face image. This surface gives us a more accurate face shape and might also be useful in other applications. Because of both improvements, the performance in face recognition becomes significantly better for face images with uncontrolled illumination conditions.

References

1. Shan, S., Gao, W., Cao, B., Zhao, D.: Illumination normalization for robust face recognition against varying lighting conditions. In: IEEE International Workshop on Analysis and Modeling of Faces and Gestures, AMFG 2003, October 17, pp. 157–164 (2003)
2. Heusch, G., Rodriguez, Y., Marcel, S.: Local binary patterns as an image preprocessing for face authentication. In: 7th International Conference on Automatic Face and Gesture Recognition, FGR 2006, April 10-12, pp. 6–14 (2006)
3. Tao, Q., Veldhuis, R.N.J.: Illumination normalization based on simplified local binary patterns for a face verification system. In: Biometrics Symposium 2007 at The Biometrics Consortium Conference, Baltimore, Maryland, USA, September 2007, pp. 1–6. IEEE Computational Intelligence Society, Los Alamitos (2007)
4. Shashua, A., Riklin-Raviv, T.: The quotient image: class-based re-rendering and recognition with varying illuminations. IEEE Transactions on Pattern Analysis and Machine Intelligence 23(2), 129–139 (2001)
5. Sim, T., Kanade, T.: Combining models and exemplars for face recognition: An illuminating example. In: Proc. CVPR Workshop on Models versus Exemplars in Computer Vision (December 2001)
6. Basri, R., Jacobs, D.: Lambertian reflectance and linear subspaces. IEEE Transactions on Pattern Analysis and Machine Intelligence 25(2), 218–233 (2003)
7. Blanz, V., Vetter, T.: A morphable model for the synthesis of 3d faces. In: IGGRAPH 1999: Proceedings of the 26th annual Conference on Computer Graphics and Interactive Techniques, pp. 187–194 (1999)
8. Al-Osaimi, F.R., Bennamoun, M., Mian, A.: Illumination normalization for color face images. In: Bebis, G., Boyle, R., Parvin, B., Koracin, D., Remagnino, P., Nefian, A., Meenakshisundaram, G., Pascucci, V., Zara, J., Molineros, J., Theisel, H., Malzbender, T. (eds.) ISVC 2006. LNCS, vol. 4291, pp. 90–101. Springer, Heidelberg (2006)
9. Boom, B., Spreeuwers, L., Veldhuis, R.: Model-based reconstruction for illumination variation in face images. In: 8th International Conference on Automatic Face and Gesture Recognition, FGR 2008 (2008)
10. Nelder, J., Mead, R.: A simplex method for function minimization. The Computer Journal 7, 308–313 (1965)
11. Veldhuis, R., Bazen, A., Booij, W., Hendrikse, A.: Hand-geometry recognition based on contour parameters. In: Proceedings of SPIE Biometric Technology for Human Identification II, Orlando, FL, USA, March 2005, pp. 344–353 (2005)

A New Gabor Phase Difference Pattern for Face and Ear Recognition

Yimo Guo[1,2], Guoying Zhao[1], Jie Chen[1], Matti Pietikäinen[1], and Zhengguang Xu[2]

[1] Machine Vision Group, Department of Electrical and Information Engineering,
University of Oulu, PO Box 4500, 90014, Finland
[2] School of Information Engineering, University of Science and Technology Beijing,
Beijing, 100083, China

Abstract. A new local feature based image representation method is proposed. It is derived from the local Gabor phase difference pattern (LGPDP). This method represents images by exploiting relationships of Gabor phase between pixel and its neighbors. There are two main contributions: 1) a novel phase difference measure is defined; 2) new encoding rules to mirror Gabor phase difference information are designed. Because of them, this method describes Gabor phase difference more precisely than the conventional LGPDP. Moreover, it could discard useless information and redundancy produced near quadrant boundary, which commonly exist in LGPDP. It is shown that the proposed method brings higher discriminative ability to Gabor phase based pattern. Experiments are conducted on the FRGC version 2.0 and USTB Ear Database to evaluate its validity and generalizability. The proposed method is also compared with several state-of-the-art approaches. It is observed that our method achieves the highest recognition rates among them.

1 Introduction

Over the last decades, biometrics has gained increasing attention because of its broad applications ranged from identification to security. As one of its main research topics, face recognition has been developing rapidly. Numerous face recognition methods have been put forward and adopted in real-life advanced technologies. Meanwhile, ear recognition has also raised interest in research and commercial communities, since human ear is one of the representative human identifiers, and ear recognition would not encounter facial expression and aging problems [1]. As we know, the main task of vision-based biometrics is to extract compact descriptions from images that would subsequently be used to confirm the identity [2]. For face recognition, the key problem is also to represent objects effectively and improve the recognition performance, which is the same with that in ear recognition.

In this paper, a new image representation method is presented based on the Local Gabor Phase Difference Pattern (LGPDP) [8]. It captures the Gabor phase difference in a novel way to represent images. According to the Gabor function and new phase difference definition, we design encoding rules for effective feature extraction. The proposed encoding rules have the following characteristics: 1) divide the quadrants precisely and encode them by a 2-bit number; and 2) avoid useless patterns that might be produced near the quadrant boundary, which is an unsolved problem in the LGPDP. From the results of experiments conducted on the FRGC ver 2.0 database for face recognition and USTB ear database for ear recognition, the proposed method is observed to further improve the capability of capturing information from the Gabor phase. The extension of its application from face recognition to ear recognition also achieves impressive results, which demonstrates its ability as a general image representation for biometrics. To our best knowledge, this is the first utilization of Gabor phase difference in ear image representation.

2 Background of Methodology

By now, various image representation methods have been proposed for vision-based biometrics. For face recognition, these methods can be generally divided into two categories: holistic matching methods and local matching methods. The Local binary pattern (LBP) [16], Gabor features and their related methods [3]-[6] have been considered as promising ways to achieve high recognition rates in face recognition. One of the influential local approaches is the histogram of Gabor phase patterns (HGPP) [7]. It uses global and local Gabor phase patterns for representation taking advantage of the fact that the Gabor phase can provide useful information as well as the Gabor magnitude. To avoid the sensitivity of Gabor phase to location variations, the Local Gabor Phase Difference Pattern (LGPDP) is put forward later [8]. Unlike the HGPP that exploits Gabor phase relationships between neighbors, LGPDP encodes discriminative information in an elaborate way to achieve a better result. However, its encoding rules would result in a loose quadrant division. This might produce useless and redundancy patterns near the quadrant boundary, which would bring confounding effect and reduce efficiency.

For ear recognition, research involves 2D ear recognition, 3D ear recognition, earprint recognition and so on. Although many approaches have been proposed, such as the PCA for ear recognition [9], Linear Discriminant Analysis and their kernel based methods [10][11][12][13], 2D ear recognition remains a challenging task in real world applications as most of these methods are based on the statistical learning theory, which inspires the usage of local based approaches later [14].

It has been confirmed that the Gabor phase could provide discriminative information for classification and the Gabor phase difference has sufficient discriminative ability [6][8]. However, the Gabor phase difference should be exploited elaborately to avoid useless information. Therefore, we are motivated to give a new phase difference definition and design new encoding rules in order to extract features from Gabor phase differences effectively.

3 Gabor Phase Based Image Representation Method

3.1 Gabor Function as Image Descriptors

In this section, we present the new local feature based image representation method, which also captures Gabor phase differences between the referencing pixel and its neighboring pixels at each scale and orientation, but using a new phase difference definition and encoding rules.

Gabor wavelets are biologically motivated convolution kernels in the shape of plane waves, restricted by a Gaussian envelope function [15]. The general form of a 2D Gabor wavelet is defined as:

$$\Psi_{\mu,v}(z) = \frac{\|k_{\mu,v}\|^2}{\sigma^2} e^{-\|k_{\mu,v}\|^2 \|z\|^2 / 2\sigma^2} \left[e^{ik_{\mu,v}z} - e^{-\sigma^2/2} \right]. \tag{1}$$

In this equation, $z = (x, y)$ is the variable in a complex spatial domain, $\|\bullet\|$ denotes the norm operator, σ is the standard deviation of the Gaussian envelope determining the number of oscillations. The wave vector $k_{\mu,v}$ is defined as $k_{\mu,v} = k_v e^{i\phi_\mu}$, where $k_v = k_{max} / f^v$ and $\phi_\mu = \pi\mu/8$, k_{max} is the maximum frequency of interest, and f is the spacing factor between kernels in the frequency domain.

3.2 The Novel Local Gabor Phase Difference Pattern

Because of the biological relevance with human vision, Gabor wavelets can enhance visual properties which are useful for image understanding and recognition [4]. Thus, Gabor filter based representations are expected to be robust to unfavorable factors, for example, the illumination. The idea that uses Gabor function as image descriptor is to enhance discriminative information by the convolution between the original image and a set of Gabor kernels with different scales and orientations.

A Gabor wavelet kernel is the product of an elliptical Gaussian envelope and a complex plane wave. The Gabor kernels in Equation (1) are all self-similar since they can be generated by scaling and rotation via the wave vector $k_{\mu,v}$. We choose eight orientations $\mu : \{0,1,...,7\}$ and five scales $v : \{0,1,...4\}$, thus make a total of 40 Gabor kernels. The values of other parameters follow the setting in [4]: $\sigma = 2\pi$, $k_{max} = \pi/2$, $f = \sqrt{2}$. The Gabor-based feature is obtained by the convolution of the original image $I(z)$ and each Gabor filter $\Psi_{\mu,v}(z)$:

$$O_{\mu,v}(z) = I(z) * \Psi_{\mu,v}(z). \tag{2}$$

$O_{\mu,v}(z)$ is the convolution result corresponding to the Gabor kernel at orientation μ and scale v. The magnitude and phase spectrum of the 40 $O_{\mu,v}(z)$ are shown in Fig. 1. The magnitude spectrum of $O_{\mu,v}(z)$ is defined as:

$$M_{O_{\mu,v}}(z) = \sqrt{\text{Re}(O_{\mu,v}(z))^2 + \text{Im}(O_{\mu,v}(z))^2}, \tag{3}$$

where $\text{Re}(\cdot)$ and $\text{Im}(\cdot)$ denote the real and imaginary part of the Gabor transformed image respectively. Usually, $\|O_{\mu,v}(z)\|$ as the magnitude part of $O_{\mu,v}(z)$ is adopted in the feature selection [4][17]. But in our case, we choose the phase part of $O_{\mu,v}(z)$ to utilize the discriminative power of the Gabor phase which was confirmed in [18]. The phase spectrum of $O_{\mu,v}(z)$ is defined as:

$$\phi_{O_{\mu,v}}(z) = \arctan\left[\frac{\text{Im}(O_{\mu,v}(z))}{\text{Re}(O_{\mu,v}(z))}\right]. \tag{4}$$

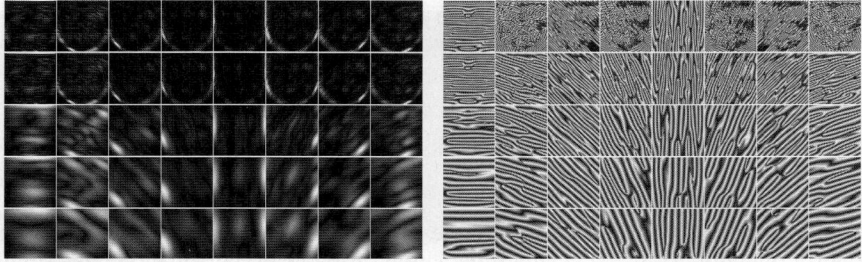

Fig. 1. Illustrative Gabor magnitude spectrum (left) and Gabor phase spectrum (right)

Our method is based on the local Gabor phase difference pattern which captures discriminative information from the Gabor phase for image representation [8]. In the LGPDP, the absolute values of Gabor phase differences, ranged from 0 to 2π, are calculated for each pixel in the image. Then they are reformulated to a 1-bit number: 1 denotes phase differences from 0 to $\pi/2$, and 0 denotes phase differences from $\pi/2$ to 2π. This pattern has two shortages: 1) the division of quadrant is loose, so only 3/4 of the quadrant would be encoded as an 1-bit number; 2) as phase differences are taken by $|\Delta\theta_{\mu,v}(z)|$, Gabor phase differences near 2π are almost useless. To increase the efficiency of the Gabor phase based pattern, in our method, we define the phase difference as: $\min\{|\Delta\theta_{\mu,v}(z)|, 2\pi - |\Delta\theta_{\mu,v}(z)|\}$. Thus, the values of Gabor phase differences are ranged in $[0,\pi]$, and would be reformulated to a 2-bit number by encoding rules:

$$C_p = \begin{cases} 00, 0 \leq \min\{|\Delta\theta_{\mu,\nu}(z)|, 2\pi - |\Delta\theta_{\mu,\nu}(z)|\} < \frac{\pi}{4} \\ 01, \frac{\pi}{4} \leq \min\{|\Delta\theta_{\mu,\nu}(z)|, 2\pi - |\Delta\theta_{\mu,\nu}(z)|\} < \frac{\pi}{2} \\ 10, \frac{\pi}{2} \leq \min\{|\Delta\theta_{\mu,\nu}(z)|, 2\pi - |\Delta\theta_{\mu,\nu}(z)|\} < \frac{3\pi}{4} \\ 11, \frac{3\pi}{4} \leq \min\{|\Delta\theta_{\mu,\nu}(z)|, 2\pi - |\Delta\theta_{\mu,\nu}(z)|\} \leq \pi \end{cases}$$

where C_p denotes the new coding. In this way, the range of the Gabor phase difference is concentrated from $[0,2\pi]$ to $[0,\pi]$. Thus, each $\pi/4$ of the half quadrant can be encoded to be a 2-bit number, which is more precise than the LGPDP which divides the quadrant into two unequal parts. Meanwhile, the new definition of phase difference would discard useless information near the quadrant boundary. In this way, the coding of eight neighbors can be combined to be a 16-bit binary string for each pixel and converted to a decimal number ranged in [0,255]. This process is described in Fig. 2. The eight 2-bit numbers are concatenated into a 16-bit number without weight, so that the histogram would not be strongly dependent on the ordering of neighbors (clockwise in LGPDP). Each of these values represents a mode how the Gabor phase of the reference pixel is different from that of its neighbors and what is the range between them. Fig. 2 gives an example of the pattern. The visualizations of the new pattern and the LGPDP are illustrated in Fig. 3. The μ and v are selected randomly.

The histograms (256 bins) of Gabor phase differences at different scales and orientations are calculated and concatenated to form the image representation. As a single histogram suffers from losing spatial structure information, images are decomposed into sub-regions, from which local features are extracted. To capture both the global and local information, these histograms are concatenated to an extended histogram for each scale and orientation. The discriminative capability of this pattern could be observed from the results of histogram distance comparison ($\mu=90$, $v=5.47$), listed in Table 1 and 2. $S1x$ ($x=1,2$) and $S2y$ ($y=1,2$) are four images for two subjects.

Table 1. The histogram distances of four images for two subjects using the proposed pattern

Subjects	S11	S12	S21	S22
S11	0	2556	3986	5144
S12	--	0	3702	5308
S21	--	--	0	2826
S22	--	--	--	0

Table 2. The histogram distances of four images for two subjects using the LGPDP

Subjects	S11	S12	S21	S22
S11	0	3216	3630	4166
S12	--	0	3300	3788
S21	--	--	0	2932
S22	--	--	--	0

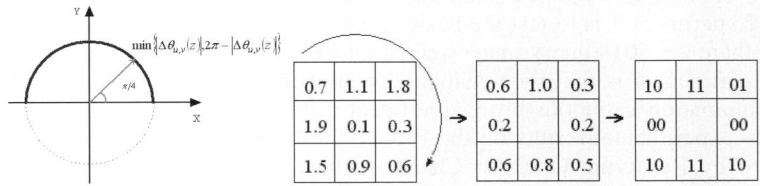

Fig. 2. Quadrant bit coding and an example of the new Gabor phase difference pattern. The first 8-neighborhood records coefficients of π that describe neighborhood Gabor phases. The second one records coefficients of π that describe neighborhood Gabor phase differences using the new definition. The third one records corresponding binary numbers according to the new encoding rules. The binary string is 1011010010111000. The decimal number corresponding to this string is then normalized to the range of [0,255].

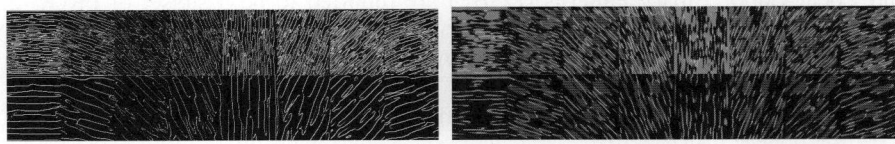

Fig. 3. Illustrative samples of resultant results when convoluting an image with Gabor filters ($v=5.47,8.20$) using the proposed pattern (left) and the LGPDP (right).

4 Experiments

The proposed method is tested on the FRGC ver 2.0 database [19] and the USTB ear database [13] for face recognition and ear recognition, respectively. The classifier is the simplest classification scheme: nearest neighbour classifier in the image space with Chi square statistics as a similarity measure.

4.1 Experiments on the FRGC ver 2.0 Database

To evaluate the performance of the proposed method in face recognition, we conduct experiments on the FRGC version 2.0 database which is one of the most challenging face databases [19]. The images are normalized and cropped to the size of 120×120 using the provided eye coordinates. Some samples are shown in Fig. 4.

Fig. 4. Face images from the FRGC 2.0 database

In FRGC 2.0 database, there are 12776 images taken from 222 subjects in the training set and 16028 images in the target set. We conduct Experiment 1 and Experiment 4 protocols to evaluate the performance of different approaches. In Experiment 1, there are 16028 query images taken under the controlled illumination condition. The goal of Experiment 1 is to test the basic recognition ability of approaches. In Experiment 4, there are 8014 query images taken under the uncontrolled illumination condition. Experiment 4 is the most challenging protocol in FRGC because uncontrolled large illumination variations bring significant difficulties to achieve high recognition rate. The experimental results on the FRGC 2.0 database in Experiment 1 and 4 are evaluated by Receiving Operator Characteristics (ROC), which is face verification rate (FVR) versus false accept rate (FAR). Tables 3 and 4 list the performance of different approaches on face verification rate (FVR) at false accept rate (FAR) of 0.1% in Experiment 1 and 4.

From experimental results listed in Table 3 and 4, the proposed method achieves the best performance, which demonstrates its basic abilities in face recognition. Table 5 exhibits results of the comparison with some well-known approaches. The images are uniformly divided into 64 sub-regions in local based methods. The database used in experiments for Gabor + Fisher Linear Discriminant Analysis (FLDA) and Local Gabor Binary Patterns (LGBP) are reported to be a subset of FRGC 2.0 [20], while the whole database is used for others. It is observed that our pattern has high discriminative ability and could improve face recognition performance.

Table 3. The FVR values in Experiment 1 of the FRGC 2.0 database

Methods	FVR at FAR = 0.1% (in %)		
	ROC 1	ROC 2	ROC 3
BEE Baseline [19]	77.63	75.13	70.88
LBP [16]	86.24	83.84	79.72
Our method	98.38	95.14	93.05

Table 4. The FVR values in Experiment 4 of the FRGC 2.0 database

Methods	FVR at FAR = 0.1% (in %)		
	ROC 1	ROC 2	ROC 3
BEE Baseline [19]	17.13	15.22	13.98
LBP [16]	58.49	54.18	52.17
Our method	82.82	80.74	78.36

Table 5. The ROC 3 on the FRGC 2.0 in Experiment 4

Methods	ROC 3, FVR at FAR = 0.1% (in %)
BEE Baseline [19]	13.98
Gabor + FLDA [20]	48.84
LBP [16]	52.17
LGBP [20]	52.88
LGPDP [8]	69.92
Our method	78.36

4.2 Experiments on the USTB Ear Database

For ear recognition, experiments are conducted on a subset (40 subjects) of the USTB ear database that contains 308 images of 77 subjects [13]. These images are taken under 3 viewing conditions (azimuth $\alpha \in \{-30, 0, 30\}$) and different illumination conditions. The original images, with a resolution of 300×400, are cropped to grayscale images with a resolution of 270×360. Sample images for two subjects are shown in Fig. 5. In this experiment, three images of one subject are taken as the training set and the remaining one serves as the testing set. Considering that complex information is contained in the ear print area, we divide images into sub-regions with 5-pixel overlapping.

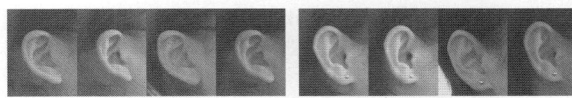

Fig. 5. Ear images from the USTB ear database for two subjects

As in other local feature based methods, recognition performance can be improved by image division. Here we divide images uniformly into nine sub-regions with small overlapping. The spatially enhanced histogram is defined as the combination of features extracted from each sub-region. In this way, the texture of image could be locally encoded by micro-patterns and the ear shape could be recovered by the construction of feature histograms.

To evaluate the performance of the proposed method in ear recognition, it is compared with some widely-used methods: Principal Components Analysis (PCA) for ear recognition, Fisher Discriminant Analysis, rotation invariant descriptor, Local binary pattern and LGPDP. The average recognition rates (in %) using cross validation are listed in Table 6. From experimental results, we can observe that the proposed pattern performs well in ear recognition, which demonstrate its efficiency and generalizability as an image representation for biometrics.

Table 6. Experimental results of ear recognition

Methods	Recognition rate (in %)
PCA [9]	78.68
FDA [10]	85.71
Rotation invariant descriptor [12]	88.32
LBP [16]	89.79
LGPDP [8]	89.53
Our method	92.45

5 Conclusions

In this paper, we propose a new Gabor phase based image representation method, which is based on the local Gabor phase difference pattern (LGPDP). There are two disadvantages of the conventional LGPDP: 1) 3/4 of the quadrant is encoded as an 1-bit number because of its loose quadrant division; and 2) Gabor phase difference patterns near 2π are almost useless because the phase difference is defined as the absolute value of phase distance between neighbors, which might bring confounding effects to image representation. Therefore, we propose a new local feature for effective image representation, which could discard useless information by defining the phase difference measure in a novel way. Moreover, new encoding rules are designed to provide more precise quadrant division than the LGPDP. In virtue of these two contributions, the discriminative ability of the Gabor phase based pattern can be significantly improved. This method is evaluated on both the FRGC version 2.0 database and the USTB ear database. It is also compared with several state-of-the-art approaches and achieves the highest recognition rates among them. The experimental results could demonstrate its capability and generalizability as an image representation.

Acknowledgments. The authors would like to thank the Academy of Finland for their support to this work.

References

1. Iannarelli, A.: Ear Identification. Forensic Identification Series. Paramount Publishing Company, Fremont (1989)
2. Burge, M., Burger, W.: Ear Recognition. In: Jain, A.K., Bolle, R., Pankanti, S. (eds.) Biometrics: Personal Identzjication in Networked Society, pp. 273–286. Kluwer Academic Publishing, Dordrecht (1998)
3. Lyons, M.J., Budynek, J., Plante, A., Akamatsu, S.: Classifying Facial Attributes using a 2-d Gabor Wavelet Representation and Discriminant Analysis. In: IEEE International Conference on Automatic Face and Gesture Recognition, pp. 1357–1362 (2000)
4. Liu, C., Wechsler, H.: Gabor Feature based Classification using the Enhanced Fisher Linear Discriminant Model for Face Recognition. IEEE Transactions on Image Processing 4, 467–476 (2002)
5. Zhang, W., Shan, S., Gao, W., Chen, X., Zhang, H.: Local Gabor Binary Pattern Histogram Sequence (LGBPHS): A Novel Non-Statistical Model for Face Representation and Recognition. In: International Conference on Computer Vision, pp. 786–791 (2005)

6. Zhang, W., Shan, S., Chen, X., Gao, W.: Are Gabor Phases Really Useless for Face Recognition? In: International Conference on Pattern Recognition, pp. 606–609 (2006)
7. Zhang, B., Shan, S., Chen, X., Gao, W.: Histogram of Gabor Phase Pattern (HGPP): A novel object representation approach for face recognition. IEEE Transactions on Image Processing 1, 57–68 (2007)
8. Guo, Y., Xu, Z.: Local Gabor Phase Difference Pattern for Face Recognition. In: International Conference on Pattern Recognition, pp. 1–4 (2008)
9. Chang, K., Bowyer, K., Sarkar, S., Victor, B.: Comparison and Combination of Ear and Face Images in Appearance-Based Biometrics. IEEE Transactions on Pattern Analysis and Machine Intelligence 9, 1160–1165 (2003)
10. Liu, Y., Mu, Z., Yuan, L.: Application of Kernel Function Based Fisher Discriminant Analysis Algorithm in Ear Recognition. Measurements and Control, 304–306 (2006)
11. Shailaja, D., Gupta, P.: A Simple Geometric Approach for Ear Recognition. In: International Conference on Information Technology, pp. 164–167 (2006)
12. Fabate, A., Nappi, M., Riccio, D., Ricciardi, S.: Ear Recognition by means of a Rotation Invariant Descriptor. In: International Conference on Pattern Recognition (2006)
13. Yuan, L., Mu, Z.: Ear Recognition based on 2D Images. In: IEEE International Conference on Biometrics: Theory, Applications, and Systems, pp. 1–5 (2007)
14. Guo, Y., Xu, Z.: Ear Recognition using a New Local Matching Approach. In: IEEE International Conference on Image Processing, pp. 289–292 (2008)
15. Wiskott, L., Fellous, J.-M., Kruger, N., Malsburg, C.v.d.: Face Recognition by Elastic Bunch Graph Matching. In: Intelligent Biometric Techniques in Fingerprint and Face Recognition, ch. 11, pp. 355–396 (1999)
16. Ahonen, T., Hadid, A., Pietikäinen, M.: Face Description with Local Binary Pattern. IEEE Transactions on Pattern Analysis and Machine Intelligence 28, 2037–2041 (2006)
17. Tao, D., Li, X., Wu, X., Maybank, S.J.: General Tensor Discriminant Analysis and Gabor Features for Gait Recognition. IEEE Transactions on Pattern Analysis and Machine Intelligence 29, 1700–1715 (2007)
18. Qing, L., Shan, S., Chen, X., Gao, W.: Face Recognition under Varying Lighting based on the Probabilistic Model of Gabor Phase. In: International Conference on Pattern Recognition, vol. 3, pp. 1139–1142 (2006)
19. Phillips, P.J., Flynn, P.J., Scruggs, T., Bowyer, K.W., Chang, J., Hoffman, K., Marques, J., Min, J., Worek, W.: Overview of the Face Recognition Grand Challenge. In: IEEE Conference on Computer Vision and Pattern Recognition, pp. 947–954 (2005)
20. Lei, Z., Liao, S., He, R., Pietikäinen, M., Li, S.Z.: Gabor Volume based Local Binary Pattern for Face Representation and Recognition. In: IEEE Conference on Automatic Face and Gesture Recognition (2008)

Is White Light the Best Illumination for Palmprint Recognition?

Zhenhua Guo, David Zhang, and Lei Zhang

Biometrics Research Centre, Department of Computing
The Hong Kong Polytechnic University, Hong Kong
{cszguo,csdzhang,cslzhang}@comp.polyu.edu.hk

Abstract. Palmprint as a new biometric has received great research attention in the past decades. It owns many merits, such as robustness, low cost, user friendliness, and high accuracy. Most of the current palmprint recognition systems use an active light to acquire clear palmprint images. Thus, light source is a key component in the system to capture enough of discriminant information for palmprint recognition. To the best of our knowledge, white light is the most widely used light source. However, little work has been done on investigating whether white light is the best illumination for palmprint recognition. In this study, we empirically compared palmprint recognition accuracy using white light and other six different color lights. The experiments on a large database show that white light is not the optimal illumination for palmprint recognition. This finding will be useful to future palmprint recognition system design.

Keywords: Biometrics, Palmprint recognition, $(2D)^2 PCA$, Illumination.

1 Introduction

Automatic authentication using biometric characteristics, as a replacement or complement to traditional personal authentication, is becoming more and more popular in the current e-world. Biometrics is the study of methods for uniquely recognizing humans based on one or more intrinsic physical or behavioral traits [1]. As an important member of the biometric characteristics, palmprint has merits such as robustness, user-friendliness, high accuracy, and cost-effectiveness. Because of these good properties, palmprint recognition has received a lot of research attention and many systems have been proposed.

In the early stage, most of works focus on offline palmprint images [2-3]. With the development of digital image acquisition devices, many online palmprint systems have been proposed [4-12]. There are mainly three kinds of online palmprint image acquisition systems: desktop scanner [4-6], Charge Coupled Device (CCD) camera or Complementary Metal-Oxide-Semiconductor (CMOS) camera with passive illumination [7-8], and CCD or CMOS with active illumination [9-12].

Desktop scanner could provide high quality palmprint images [4-6] with different resolutions. However, it may suffer from the slow speed and requires the full touch with whole hand which may bring sanitary issues during data collection. Using CCD or CMOS with uncontrolled ambient lighting [7-8] does not have the above problems.

However, the image quality may not be very good so that the recognition accuracy may not be high enough. Because CCD or CMOS camera mounted with active light could collect image data quickly with good image quality and does not require the full touch with the device, this kind of system have been attracting much attention [9-12]. Although all of these studies [9-12] used white light source to enhance the palmprint line and texture information, to the best of knowledge, no work has been done to systematically validate whether white light is the optimal light for palmprint recognition. This study focused on this problem through a series of experiments on a large multispectral palmprint database we established [17].

In general, there are mainly two kinds of approaches to pattern recognition and analysis: structural and statistical methods. Because the statistical methods are more computational effective and are straightforward to implement, many algorithms have been proposed, such as Principal Component Analysis (PCA) [5-6, 14], Locality Preserving Projection [7, 13]. In this study, we employ the $(2D)^2 PCA$ method [15-16] to extract palmprint features in order for feature extraction and matching. The $(2D)^2 PCA$ method can alleviate much the small sample size problem in subspace analysis and can better preserve the image local structural information than PCA.

The rest of this paper is organized as follows. Section 2 describes our data collection. Section 3 briefly introduces the $(2D)^2 PCA$ algorithm. Section 4 presents the experimental results and Section 5 concludes the paper.

2 Multispectral Palmprint Data Collection

It is known that Red, Green, and Blue are the three primary colors (refer to Fig. 1), the combination of which could result in many different colors in the visible spectrum. We designed a multispectral palmprint data collection device which includes the three primary color illumination sources (LED light sources). By using this device we can simulate different illumination conditions. For example, when the red and green LEDs are switched on simultaneously, the yellow like light could be generated. Totally our device could collect palmprint images under seven different color illuminations: red, green, blue, cyan, yellow, magenta and white.

The device is mainly composed of a monochrome CCD camera, a lens, an A/D converter, a light controller and multispectral light sources. To fairly study the illumination effect, the lens, A/D converter and CCD camera are selected according to

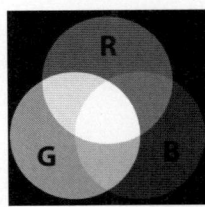

Fig. 1. Additive Color Mixing: adding red to green yields yellow, adding all three primary colors together yields white

previous palmprint scanning [18] to remove the influence of scanner devices, and resolutions. The illuminator is a LED array, which is arranged in a circle to provide a uniform illumination. The peak spectrums of red, green, and blue LEDs are 660nm, 525nm, and 470 nm respectively. The LED array can switch to different light in about 100ms. The light controller is used to switch on or off the different color LEDs.

As shown in Fig. 2, in data collection the user is asked to put his/her palm on the device. The device could collect a multispectral palmprint cube, including seven different palmprint images, in less 2 seconds. Fig. 3 shows examples of the collected images by different illuminations.

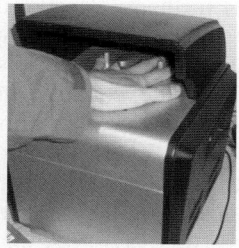

Fig. 2. The prototype device

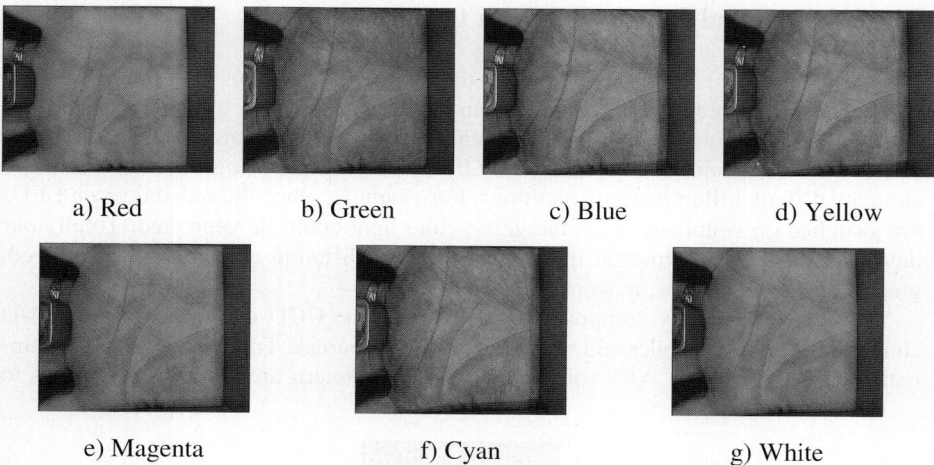

a) Red b) Green c) Blue d) Yellow

e) Magenta f) Cyan g) White

Fig. 3. A sample of collected image of one palm with different illuminations

3 Feature Extraction

In this study, we employ the $(2D)^2$PCA method [15-16] to extract palmprint features. The $(2D)^2$PCA method can much alleviate the small sample size problem in subspace analysis and can well preserve the image local structural information.

Suppose we have M subjects and each subject has S sessions in the training data set, i.e. S multispectral palmprint cube were acquired at different times for each subject. Then, we denote by X_{ms}^b the bth band image for the mth individual in the sth session. X_{ms}^b is an $I_r * I_c$ matrix, where I_r and I_c represent the numbers of rows and columns of the image. The covariance matrices along the row and column directions are computed as:

$$G_1^b = \frac{1}{MS}\sum_{s=1}^{S}\sum_{m=1}^{M}(X_{ms}^b - \overline{X^b})^T(X_{ms}^b - \overline{X^b})$$
$$G_2^b = \frac{1}{MS}\sum_{s=1}^{S}\sum_{m=1}^{M}(X_{ms}^b - \overline{X^b})(X_{ms}^b - \overline{X^b})^T \quad (1)$$

where $\overline{X^b} = \frac{1}{MS}\sum_{s=1}^{S}\sum_{m=1}^{M}X_{ms}^b$.

The project matrix $V_1^b = [v_{11}^b, v_{12}^b, ..., v_{1k_1^b}^b]$ is composed of the orthogonal eigenvectors of G_1^b corresponding to the k_1^b largest eigenvalues, and the projection matrix $V_2^b = [v_{21}^b, v_{22}^b, ..., v_{2k_2^b}^b]$ consists of the orthogonal eigenvectors of G_2^b corresponding to the largest k_2^b eigenvalues. k_1^b and k_2^b can be determined by setting a threshold to the cumulant eigenvalues:

$$\sum_{j_c=1}^{k_1^b}\lambda_{1j_c}^b / \sum_{j_c=1}^{I_c}\lambda_{1j_c}^b \geq C_u, \sum_{j_r=1}^{k_2^b}\lambda_{2j_c}^b / \sum_{j_r=1}^{I_r}\lambda_{2j_c}^b \geq C_u \quad (2)$$

where $\lambda_{11}^b, \lambda_{12}^b, ..., \lambda_{1I_c}^b$ are the first I_c biggest eigenvalues of G_1^b, $\lambda_{21}^b, \lambda_{22}^b, ..., \lambda_{2I_r}^b$ are the first I_r biggest eigenvalues of G_2^b, and C_u is a pre-set threshold.

For each given band b^{th}, the test image T^b is projected to $\widetilde{T^b}$ by V_1^b and V_2^b. The distance of the projection result to the mth individual is defined as:

$$d_{ms}^b = \left\| V_2^{bT}T^bV_1^b - \widetilde{X_{ms}^b} \right\| \quad (3)$$

where $\widetilde{X_{ms}^b} = V_2^{bT}X_{ms}^bV_1^b$ is the projection data from the training set. Then the classification decision of a test band image is made as:

$$c^b = \arg\min_{m} d_{ms}^b, m = 1, 2, ..., M, s = 1, 2, ..., S \quad (4)$$

4 Experiment Results

We collected multispectral palmprint images from 250 subjects using the developed data acquisition device. The subjects were mainly volunteers from our institute. In the database, 195 people are male and the age distribution is from 20 to 60 years old. We collected the multispectral palmprint images on two separate sessions. The average time interval between the two occasions is 9 days. On each session, the subject was asked to provide 6 samples of each of his/her left and right palms. So our database contains 6,000 images for each band from 500 different palms. For each shot, the device collected 7 images from different bands (Red, Green, Blue, Cyan, Yellow, Magenta, and White) in less than two seconds. In palmprint acquisition, the users are asked to keep their palms stable on the device. The resolution of the images is 352*288 (<100 DPI).

After obtaining the multispectral cube, a local coordinate of the palmprint image is established [9] from the blue band, and then a Region of Interest (ROI) is cropped from each band based on the local coordinate. For the convenience of analysis, we normalized these ROIs to a size of 128*128. To remove the global intensity and contrast effect, all images are normalized to have a mean of 128 and standard deviation of 20.

The whole database is partitioned into two parts, training set and test set. The training set is used to estimate the projection matrix and is taken as gallery samples. The test samples are matched with the training samples, and Eq. 4 is used to decide the recognition output. The ratio of the number of correct matches to the number of test samples, i.e. the recognition accuracy, is used as the evaluation criteria. To reduce the dependency of experimental results on training sample selection, we designed the experiments as follows. Firstly, the first three samples in the first session are chosen as training set and the remaining samples are used as test set. Secondly, the first three samples in the second session are chosen as training set, and the remaining samples are used as test set. Finally, the average accuracy is computed.

Fig. 4 shows the accuracies of different spectrum with different cumulant Eigenvalue threshoslds, C_u. Several findings could be found from Fig. 4.

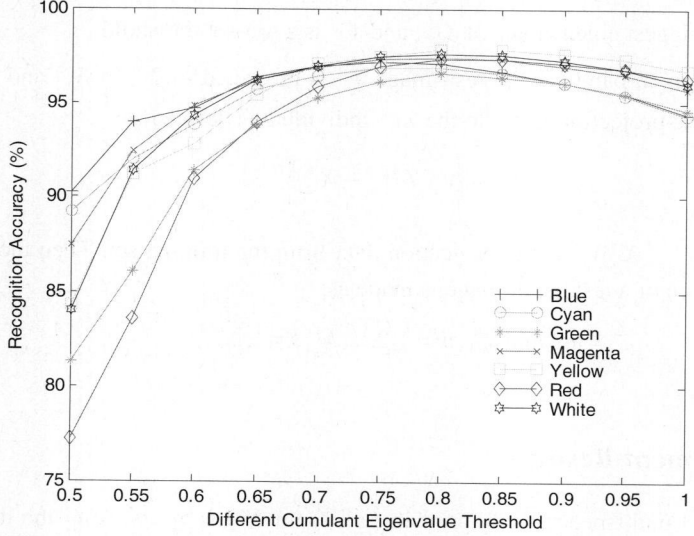

Fig. 4. Recognition Accuracy under different C_u

First, recognition accuracy is dependent with threshold. As threshold increases, the accuracy increases to a peak, then drops a little. The highest accuracy with the threshold for each color is listed in Table 1.

Table 1. The highest accuracy with threshold for each color

Color	Highest accuracy	Corresponding C_u
Blue	97.2000	0.75
Cyan	96.8777	0.75
Green	96.6334	0.80
Magenta	97.4333	0.80
Yellow	**97.8777**	0.80
Red	97.3555	0.80
White	97.6334	0.80

Second, no single spectrum could compete with all the others for all thresholds. This is mainly because different light could enhance different features of palms, while these different features have different intensity distributions which are in favor of different parameters.

Third, among the three primary colors, Red has a little higher accuracy than Blue and Green. This is mainly because Red could not only capture most of the palm line information, but also capture some palm vein structures as shown in Fig. 3. This additional palm vein information helps classify those palms with similar palm lines. It could also explain why those composite colors Magenta, Yellow, White get better accuracy than Cyan.

Finally, White color could not get higher accuracy than Yellow color. This is probably because Blue and Green collect redundant information for palm skin. As shown in Fig. 3, the palmprint images under Blue and Green illumination are more similar to each other than to the image under Red illumination. The redundancy makes White color fail to capture more information than the Yellow color, and sometimes the accuracy drops a little.

5 Conclusion

Palmprint recognition has been attracting lots of research attention in the past decade and many novel data collection devices have been proposed. Because the good image quality and capture speed, CCD or CMOS camera mounted with active lighting source is the most popular device configuration. All these devices use white light as the illumination source but there was no systematic analysis on whether the White light is the optimal light source for palmprint recognition. This paper made a good effort on this problem by establishing a large multispectral palmprint database using our developed device. With the database we empirically evaluated the recognition accuracies of palmprint images under seven different colors. Our experimental results showed that the White color is not the optimal color for palmprint recognition and the Yellow color could achieve higher accuracy than the White color. In the future, other feature extraction methods, such as structural and texture coding methods will be used to further investigate the best illumination conditions of palmprint recognition.

Acknowledgments

The work is partially supported by the CERG fund from the HKSAR Government, Hong Kong RGC General Research Fund (PolyU 5351/08E), the central fund from Hong Kong Polytechnic University, Science Foundation of Shenzhen City (CXQ2008019), the Natural Science Foundation of China (NSFC) under Contract No. 60620160097, No. 60803090, and the National High-Tech Research and Development Plan of China (863) under Contract No. 2006AA01Z193.

References

1. Jain, A., Bolle, R., Pankanti, S.: Biometrics: Personal Identification in Network Society. Kluwer Academic Publishers, Boston (1999)
2. Duta, N., Jain, A.K., Mardia, K.V.: Matching of palmprint. Pattern Recognition Letter 23, 477–485 (2002)
3. Zhang, D., Shu, W.: Two novel characteristics in palmprint verification: datum point invariance and line feature matching. Pattern Recognition 32, 691–702 (1999)
4. Lin, C., Chuang, T., Fan, K.: Palmprint verification using hierarchical decomposition. Pattern Recognition 38, 2639–2652 (2005)
5. Ribaric, S., Fratric, I.: A biometric identification system based on Eigenpalm and Eigenfinger features. IEEE Trans. on Pattern Analysis and Machine Intelligence 27, 1698–1709 (2005)
6. Connie, T., Jin, A.T.B., Ong, M.G.K., Ling, D.N.C.: An automated palmprint recognition system. Image and Vision Computing 23, 501–515 (2005)
7. Wang, J.-G., Yau, W.-Y., Suwandy, A., Sung, E.: Person recognition by fusing palmprint and palm vein images based on "Laplacianpalm" representation. Pattern Recognition 41, 1514–1527 (2008)
8. Han, Y., Sun, Z., Wang, F., Tan, T.: Palmprint recognition under unconstrained scenes. In: Yagi, Y., Kang, S.B., Kweon, I.S., Zha, H. (eds.) ACCV 2007, Part II. LNCS, vol. 4844, pp. 1–11. Springer, Heidelberg (2007)
9. Zhang, D., Kong, W., You, J., Wong, M.: Online Palmprint Identification. IEEE Trans. Pattern Analysis and Machine Intelligence 25, 1041–1050 (2003)
10. Han, C.: A hand-based personal authentication using a coarse-to-fine strategy. Image and Vision Computing 22, 909–918 (2004)
11. Michael, G.K.O., Connie, T., Teoh, A.B.J.: Touch-less palm print biometrics: novel design and implementation. Image and Vision Computing 26, 1551–1560 (2008)
12. Wu, J., Qiu, Z., Sun, D.: A hierarchical identification method based on improved hand geometry and regional content feature for low-resolution hand images. Signal Processing 88, 1447–1460 (2008)
13. Hu, D., Feng, G., Zhou, Z.: Two-dimensional locality preserving projections (2DLPP) with its application to palmprint recognition. Pattern Recognition 40, 339–342 (2007)
14. Kumar, A., Zhang, D.: Personal authentication using multiple palmprint representation. Pattern Recognition 38, 1695–1704 (2005)

15. Zuo, W., Zhang, D., Wang, K.: Bidirectional PCA With Assembled Matrix Distance Metric for Image Recognition. IEEE Transactions on Systems, Man, and Cybernetics - Part B: Cybernetics 36, 863–872 (2006)
16. Zhang, D., Zhou, Z.: (2D)2PCA: 2-Directional 2-Dimensional PCA for Efficient Face Representation and Recognition. Neurocomputing 69, 224–231 (2005)
17. Han, D., Guo, Z., Zhang, D.: Multispectral palmprint recognition using wavelet-based image fusion. In: International conference on signal processing, pp. 2074–2077 (2008)
18. Wong, M., Zhang, D., Kong, W.-K., Lu, G.: Real-time palmprint acquisition system design. IEE Proc.-Vis. Image Signal Process. 152, 527–534 (2005)

Second-Level Partition for Estimating FAR Confidence Intervals in Biometric Systems

Rongfeng Li[1,2], Darun Tang[1,2], Wenxin Li[1], and David Zhang[2]

[1] Key Laboratory of Machine Perception, Peking University, Beijing 100871, China
[2] Department of Computing, The Hong Kong Polytechnic University, Hong Kong
{rongfeng,tangcatcat,lwx}@pku.edu.cn, csdzhang@polyu.edu.hk

Abstract. Most biometric authentication algorithms make use of a similarity score that defines how similar two templates are according to a threshold and the accuracy of the results are expressed in terms of a False Reject Rate (FRR) or False Accept Rate (FAR) that is estimated using the training data set. A confidence interval is assigned to any claim of accuracy with 90% being commonly assumed for biometric-based authentication systems. However, these confidence intervals may not be as accurate as is presumed. In this paper, we report the results of experiments measuring the performance of the widely-used subset bootstrap approach to estimating the confidence interval of FAR. We find that the coverage of the FAR confidence intervals estimated by the subset bootstrap approach is reduced by the dependence between two similarities when they come from two individual pairs shared with a common individual. This is because subset bootstrap requires the independence of different subsets. To deal with this, we present a second-level partition to the similarity score set between different individuals, producing what we call a subset false accept rate (SFAR) bootstrap estimation. The experimental results show that the proposed procedures greatly increase the coverage of the FAR confidence intervals.

Keywords: Biometric, performance evaluation, bootstrap, confidence interval.

1 Introduction

A biometric authentication system verifies the identity of an individual by using biometric features such as palmprints, fingerprints, the face, or iris. Systems do this by matching template samples of an individual's features and accepting or rejecting them. Most biometric authentication algorithms make use of a similarity score that defines how similar two templates are. The score makes use of a threshold but the setting of that threshold still allows systems to erroneously either accept or reject a particular match. The accuracy of a biometric authentication system, expressed in terms of these two error rates, the False Reject Rate (FRR) and False Accept Rate (FAR) [1][2][4][5][7], is estimated using the training data set and a confidence interval is assigned to any claim of accuracy with a confidence interval of 90% being commonly assumed for biometric-based authentication systems. We use the statistics coverage to evaluate how much real-world FRR/FAR can be cover with the estimated FRR/FAR confidence intervals.

Most methods for estimating error rate confidence intervals are either parametric or non-parametric. The parametric methods are based on the probability distribution of the similarity score between two patterns. Various such methods [1][10][11][12][13] assume the score distributions conform to an i.i.d. Gaussian distribution but in practical applications the distribution of the score is often unknown. In contrast, non-parametric methods do not require knowledge of distribution of the similarity score. The most widely used of the non-parametric methods is bootstrap, which has been shown to be robust when handling unknown similarity score distributions [6]. A thorough description of bootstrap methods can be found in [3]. Bolle et al. [6][8] applied bootstrap to estimating FRR and FAR confidence intervals and found it superior to the parametric methods. The bootstrap method still requires the i.i.d. score distribution while the dependence between similarities which come from a same pair of individual will lead to the unconformity of i.i.d.. Subset bootstrap [9] partition the dataset into subset by the individual pairs so that it can handle this kind of dependency between similarities scores of the same individual pairs. Two-level bootstrap [14] improved on subset bootstrap in that it allowed faster convergence. However, there is another kind of dependence between two similarities when they come from two individual pairs which share a common individual. For example we have A,B,C three persons. The similarity between A, B and the similarity between A,C are dependent because (A,B) and (A,C) share the common person A. This kind of dependence can not be avoided by both of the subset bootstrap approach and the two-level bootstrap. This reduces the coverage of FAR confidence interval, for the subset bootstrap required the independence of different subsets as we know in [9].

In this work, we first give brief introduction to the subset bootstrap method We then propose a second-level partition to the individual pairs which separate the dependent individual pairs. This then allows us to estimate the FAR confidence intervals using what we refer to as Subset FAR (SFAR). Finally, we compared the proposed methods against subset bootstrap, using Recognition Algorithm Testing Engine (RATE) [15], which is an online testing system for the biometric algorithms, and found that the proposed procedures greatly increase the coverage of the FAR confidence intervals.

The rest of this paper is organized as follows: Section 2 formulates the problem of establishing a suitable confidence interval of the FAR and provides a formulation for the problem of subset bootstrap estimation [6]. Section 3 presents a second-level partition to the individual pairs and proposes the use of Subset FAR (SFAR) to estimate the FAR confidence intervals. Section 4 provides the experimental results of second-level partition and SFAR approach for FAR confidence interval estimation. Section 5 offers our Conclusion.

2 Subset Bootstrap Estimation

2.1 Preliminaries

Most biometric authentication algorithms define a similarity score s which is used in deciding how similar two templates are. A threshold t is specified to decide the rejection and acceptance which will cause the False Reject Rate (FRR) and False Accept Rate

(FAR). FRR and FAR are expressions of statistical random variables for each threshold t. Real-world applications use the training data set to estimate these random variables and thereby describe the accuracy of the biometric authentication system. As 100% accuracy of the FRR and FAR is impossible, claims of accuracy are set within a $(1-\alpha)100\%$ confidence interval. The $(1-\alpha)100\%$ confidence level FRR confidence interval [$FRR_u(t)$, $FRR_d(t)$] and FAR confidence interval [$FAR_u(t)$, $FAR_d(t)$] are defined as follows:

$$P(FRR_u(t) < FRR(t) < FRR_d(t)) > 1-\alpha \quad (1)$$

$$P(FAR_u(t) < FAR(t) < FAR_d(t)) > 1-\alpha \quad (2)$$

α is the probability that the real-world FRR/FAR are outside the estimated interval.

2.2 Subset Bootstrap Estimation

Subset bootstrap estimation [9] can be formulated as follows.

Suppose we have N individuals and we acquire d sample templates from each individual, giving Nd templates. These generate $Nd(d-1)/2$ self-similar match scores (a comparison of an individual's own templates) and $N(N-1)d^2/2$ mismatch scores between the templates of different individuals.

We denotes the collection of mismatch scores between different individuals by S and partition S into $N(N-1)/2$ subset according to different individual index pairs. That is

$$S = S[1] \cup S[2] \cup ... \cup S[M], \quad M = N(N-1)/2 \quad (3)$$

The Subset Bootstrap estimation of the FAR confidence interval in is described as following steps:

Step1) Divide the mismatch scores set S into M subset $S[i]$, $i=1,2,...,M$ according different individual index pairs so that $M=N(N-1)/2$

Step2) Do B times (k=1 to B):

i) Generate random integer array $r_1, r_2, ..., r_M$ with replacement from $\{1,2,...,M\}$

ii) Generate the bootstrap resample set $S_k = \bigcup_{i=1}^{M} S[r_i]$

iii) Calculate the $FAR_k(t)$ using the equations:

$$FAR_k(t) = \frac{1}{d^2 N(N-1)/2} \sum_{s \in S_k} I\{s < t\} \quad (4)$$

iv) Sort the B bootstrap estimates $FAR_k(t)$ k=1,2,...,B by

$$FAR_1^*(t) \leq FAR_2^*(t) \leq ... \leq FAR_B^*(t)$$

Step 3) Eliminate the bottom $\alpha/2$ and the top $\alpha/2$ of the B bootstrap estimations. The margin of the leftover estimations gives the $(1-\alpha)100\%$ confidence intervals. That is $[FAR_{\lceil (\alpha/2)B \rceil}^*(t), FAR_{\lfloor (1-\alpha/2)B \rfloor}^*(t)]$

3 Second-Level Partition and SFAR Estimation

The subset bootstrap FAR confidence interval estimation partition the mismatch score set to N(N-1)/2 subsets and assuming they are independent. However, if we denote that S[i,j] is the similarity score set of two different individual i,j. for three different individuals i, j, k, we cannot assume S[i,j] and S[j,k] are independent because they have a common individual j.

In this section we will present a second-level partition to S so that we can handle the second-level subsets separately to avoid the dependence.

Suppose m is a integer which can divide N(N-1)/2 exactly , K=N(N-1)/2m.
The second-level partition is as follows:

$$S^{(k)} = S[i_1^{(k)}, j_1^{(k)}] \cup S[i_1^{(k)}, j_1^{(k)}] \cup ... \cup S[i_m^{(k)}, j_m^{(k)}] \qquad (5)$$

$$S = \bigcup_{k=1}^{K} S^{(k)} \qquad (6)$$

Here, k=1,2,…,K.

For each k=1,2,…, K, if the index $i_1^{(k)}, j_1^{(k)}, i_2^{(k)}, j_2^{(k)}, ..., i_m^{(k)}, j_m^{(k)}$ are exactly 2m different integers, we call the partition (6) an independent partition. We indicate Subset False Accept Rate(SFAR) for each subset $S^{(k)}$:

$$SFAR^{(k)}(t) = P(s > t \mid s \in S^{(k)}) \qquad (7)$$

Notice that

$$FAR(t) = P(s > t \mid s \in S) = \sum_{k=1}^{K} P(S > t \mid s \in S^{(k)}) P(s \in S^{(k)} \mid s \in S) \quad \text{Thus,}$$

$$c \qquad (8)$$

Then, we can estimate the FAR confidence interval by the K SFAR confidence intervals.

For each k=1,2,…, K, the subset bootstrap estimation in Section 2 can work independently according to the partition (3). Thus, we have K SFAR bootstrap 90% confidence intervals: $[SFAR^{(k)}_d(t), SFAR^{(k)}_u(t)]$, k=1,2,…,K

Let

$$FAR_d(t) = \frac{1}{K} \sum_{k=1}^{K} SFAR^{(k)}_d(t) \qquad (9)$$

$$FAR_u(t) = \frac{1}{K}\sum_{k=1}^{K} SFAR^{(k)}_u(t) \qquad (10)$$

Then [$FAR_d(t), FAR_u(t)$] is the 90% confidence interval of FAR(t).

The steps of the second-level partition and SFAR estimation can be illustrated in Fig. 1.

When implementing the second-level partition (6), the immediate question is how to partition it independently such that: $i_1^{(k)}, j_1^{(k)}, i_2^{(k)}, j_2^{(k)}, ..., i_m^{(k)}, j_m^{(k)}$ are 2m different integers. One way is to use the following partition:

1. if N is an odd number: $S^{(k)} = \bigcup_{\substack{i+j \equiv k-1 (\bmod N) \\ 1 \leq i \leq j \leq N}} S[i,j]$, m=(N-1)/2

2. if N is an even number: $S^{(k)} = \left(\bigcup_{\substack{i+j \equiv k (\bmod N-1) \\ 1 \leq i < j \leq N}} S[i,j] \right) \cup \left(\bigcup_{\substack{2i \equiv k (\bmod N-1) \\ 1 \leq i \leq N}} S[i,N] \right)$, m=N/2

With this method, we can avoid the dependence between two similarities when they come from two individual pairs which share a common individual. The coverage is higher than 90% when the training data size is above 100, as shown in Section 4.

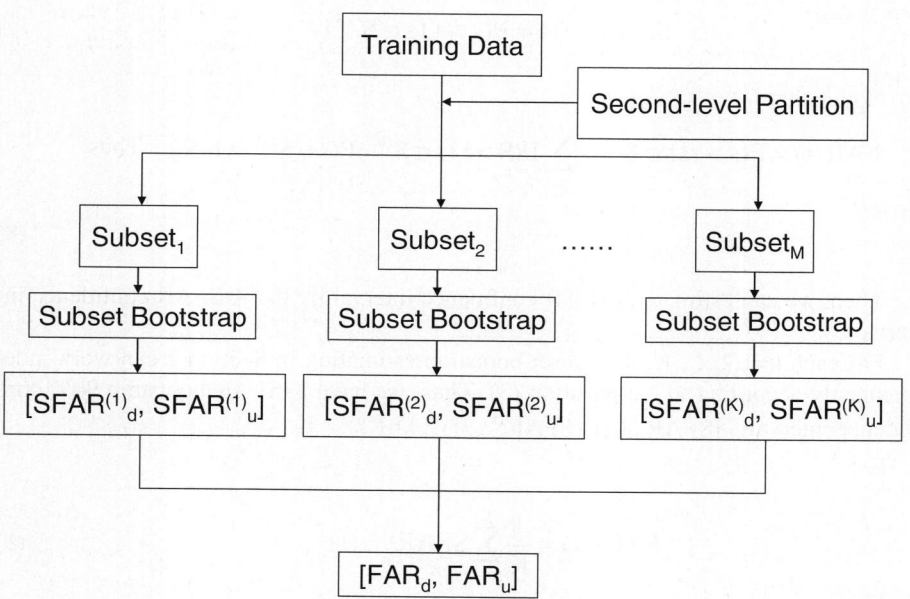

Fig. 1. Steps of the second-level partition and SFAR estimation

4 Experimental Results

Recognition Algorithm Testing Engine (RATE) [15] is an online performance evaluation system for pattern recognition algorithms. It was developed by the AI-lab of Peking University [16] and offers three different palmprint databases DATA1, DATA2, DATA3, allowing the users to submit their biometric algorithm and see the testing results. DATA1 were acquired from the AI-lab [16] and DATA2 and DATA3 were acquired from Biometric Centre of the Hong Kong Polytechnic University [17]. Images of the palmprint in these databases, which are the central parts extracted from the original images, are 128×128 and 200dpi. Table 1 shows the number of individuals and templates in the databases.

Table 1. Number of individuals and templates in the databases

	DATA0	DATA1	DATA2
N	51	213	261
d	15	4	10

N denotes the number of individuals
d denotes the number of templates for each individual

We submit 4 different palmprint recognition algorithms "fft", "surface", "texture" and "wavelet" developed by Li [18][19][20]. We choose the 40% sample size from the dataset as a training set to estimate the confidence interval. It is repeated 100 times. To calculate the coverage, we use the method described in Section 2, choosing 80% sample size as a testing set and including the training set. Tables 2 shows the coverage of the FAR confidence interval using the subset bootstrap method described in Section 2.

Table 2. Coverage of the FAR confidence interval using the subset bootstrap

	DATA1	DATA2	DATA3
fft	69%	18%	7%
surface	48%	11%	6%
texture	24%	51%	1%
wavelet	49%	21%	15%

We can see the coverage decreases while the sample size increases. This is probably because the dependence between two similarities when they come from two individual pairs which share a common individual increasing along with the sample size.

Tables 3 shows the results for FAR confidence interval estimated when using second-level partition and SFAR estimation, as described in Section 3.

The coverage of the FAR confidence interval estimated using SFAR bootstrap is much higher than when using subset bootstrap. The coverage was more than 90% on DATA3, which has a training set larger than 100.

Table 3. Coverage of the FAR confidence interval using second-level partition and SFAR estimation

	DATA1	DATA2	DATA3
fft	21%	60%	99%
surface	100%	77%	92%
texture	92%	99%	100%
wavelet	84%	96%	90%

5 Conclusion

In this paper, we introduced a second-level partition to the mismatch similarity score set and present a SFAR bootstrap confidence interval estimation that allows us to avoid dependence between two similarities when they come from two individual pairs which shared a common individual. Experimental results show that applying the second-level partition greatly improves the coverage of confidence intervals when compared with conventional subset bootstrap estimation.

In fact, our approach takes the FAR as the average of SFARs. But it is doubtful that the average of the SFARs confidence interval will automatically generate the confidence interval of FAR. An issue for future work is to demonstrate the average of the SFAR confidence intervals will generate the FAR confidence interval.

The experiment in this paper is based on palmprint database and with small sample size. Thus another future work is to extend this work to other biometric databases and larger sample sizes.

References

[1] Shen, W., Surette, M., Khanna, R.: Evaluation of Automated Biometrics-Based Identification and Verification Systems. Proceedings of the IEEE 85(9), 1464–1478 (1997)
[2] Golfarelli, M., Maio, D., Maltoni, D.: On the Error-Reject Trade-Off in Biometric Verification Systems. IEEE Transactions on Pattern Analysis and Machine Intelligence 19(7) (July 1997)
[3] Politis, D.N.: Computer-intensive methods in statistical analysis. IEEE Signal Processing Magazine 15(1), 39–55 (1998)
[4] Wayman, J.L.: Error rate equations for the general biometric system. IEEE Robotics & Automation Magazine 6(1), 35–48 (1999)
[5] Phillips, P.J., Martin, A., Wilson, C.L., Przybocki, M.: An Introduction to Evaluating Biometric Systems. Computer 33(2), 56–63 (2000)
[6] Bolle, R.M., Pankanti, S., Ratha, N.K.: Evaluation techniques for biometrics-based authentication systems (FRR). In: 15th International Conference on Pattern Recognition, Proceedings, September 2000, vol. 2(3-7), pp. 831–837 (2000)
[7] Dahel, S.K., Xiao, Q.: Accuracy performance analysis of multimodal biometrics. In: Information Assurance Workshop, IEEE Systems, Man and Cybernetics Society, June 18-20, pp. 170–173 (2003)

8. Bolle, R.M., Ratha, N.K., Pankanti, S.: IBM Thomas J. Watson Research Center. In: Proceedings of the 17th International Conference on An Evaluation of Error Confidence Interval Estimation Methods, August 23-26, vol. 3, pp. 103–106 (2004)
9. Bolle, R.M., Ratha, N.K., Pankanti, S.: Error analysis of pattern recognition systems—the subsets bootstrap. Computer Vision and Image Understanding 93(1), 1–33 (2004)
10. Schmid, N.A., O'Sullivan, J.A.: Performance Prediction Methodology for Biometric Systems Using a Large Deviations Approach. IEEE Transactions on Signal Processing 52(10) (October 20, 2004)
11. Gamassi, M., Lazzaroni, M., Misino, M., Piuri, V., Sana, D., Scotti, F.: Quality Assessment of Biometric Systems: A Comprehensive Perspective Based on Accuracy and Performance Measurement. IEEE Transactions on Instrumentation and Measurement 54(4) (August 2005)
12. Snelick, R., Uludag, U., Mink, A., Indovina, M., Jain, A.: Large-scale evaluation of multimodal biometric authentication using state-of-the-art systems. IEEE Transactions on Pattern Analysis and Machine Intelligence 27(3), 450–455 (2005)
13. Hube, J.P.: Using Biometric Verification to Estimate Identification Performance. In: Biometric Consortium Conference, 2006 Biometrics Symposium: Special Session on Research, pp. 1–6 (2006)
14. Poh, N., Bengio, S.: Estimating the Confidence Interval of Expected Performance Curve in Biometric Authentication using Joint Bootstrap. In: IEEE International Conference on Acoustics, Speech and Signal Processing, April 15-20, vol. 2, pp. II-137 – II-140 (2007)
15. http://rate.pku.edu.cn
16. http://ai.pku.edu.cn
17. http://www.comp.polyu.edu.hk
18. Li, W., Zhang, D., Xu, Z.Q.: Palmprint Recognition by Fourier Transform. Journal of Software 5(13), 879–886 (2002)
19. Li, W., You, J., Zhang, D.: A Texture-based Approach to Palmprint Retrieval for Personal Identification. Accepted by IEEE Trans. on Multimedia (2003)
20. Kong, W.K., Zhang, D., Li, W.: Palmprint Feature Extraction Using 2-D Gabor Filters. Pattern Recognition 36(10), 2339–2347 (2003)

Smooth Multi-Manifold Embedding for Robust Identity-Independent Head Pose Estimation

Xiangyang Liu[1,2], Hongtao Lu[1], and Heng Luo[1]

[1] MOE-Microsoft Laboratory for Intelligent Computing and Intelligent Systems, Department of Computer Science and Engineering, Shanghai Jiao Tong University, Shanghai, 200240, China
[2] College of Science, Hohai University, Nanjing, 210098, China
{liuxy,htlu,hengluo}@sjtu.edu.cn

Abstract. In this paper, we propose a supervised Smooth Multi-Manifold Embedding (SMME) method for robust identity-independent head pose estimation. In order to handle the appearance variations caused by identity, we consider the pose data space as multiple manifolds in which each manifold characterizes the underlying subspace of subjects with similar appearance. We then propose a novel embedding criterion to learn each manifold from the exemplar-centered local structure of subjects. The experiment results on the standard databases demonstrates that the SMME is robust to variations of identities and achieves high pose estimation accuracy.

1 Introduction

Head pose estimation from images or videos is a classical problem in computer vision [1]. Robust identity-independent head pose estimation plays a significant role in many human-centered computing applications such as view-independent face detection systems and multi-view face recognition systems.

After neuroscientists emphasized manifold ways of visual perception [2], many researchers indicated that the variations of head pose can be visualized as data points lying on a low-dimensional manifold in the image space of a high dimensionality [3,4]. However, how to extract effective pose features for the low-dimensional manifold, and synchronously ignore appearance variations like changes in identity, scale, illumination, etc [5], remain to be challenging problems due to the nonlinear and high data dimensionality. The focus of this paper is to seek the optimal low-dimensional manifold describing the intrinsical pose variations and to provide a robust identity-independent pose estimator.

The changes of pose images due to identity changes are usually larger than that caused by different poses of same person. Thus, it is difficult to obtain the identity-independent manifold embedding which preserves the pose differences. In this paper, we present a Smooth Multi-Manifold Embedding (SMME) method, which considers the pose data space as multiple manifolds. Each manifold characterizes the underlying subspace of the local structure of subjects with similar appearance. We propose a novel embedding criterion to learn each manifold from the exemplar-centered local structure of subjects. The embedding method is supervised by both pose and identity information. Each learned manifold with a unique geometric structure is smooth and discriminative. The proposed SMME method aims to provide intra-class compactness and inter-class separability in low-dimensional pose space. For new images of a new subject, we first locate their nearest exemplar, then embed them into the corresponding manifold, and finally decide the pose angle by its k nearest neighbors in the projected subspace.

2 Related Work

The effective manifold learning methods [4,5,6,7,8] for head pose estimation seek a low-dimensional continuous manifold, and new images can then be embedded into these manifolds to estimate the pose. The embedding can be learned by many approaches, such as Locally Embedded Analysis (LEA) [4], and Locality Preserving Projections (LPP) [6]. To incorporate the pose labels that are usually available during training phase, Balasubramanian et al. [7] presented a framework based on pose information to compute a biased neighborhood. Yan et al. [8] proposed a synchronized manifold embedding method. They all demonstrated their effectiveness for head pose estimation. However, many methods proposed to capture the structure of the pose manifold are local. Thus, they fail to handle new samples without the consistent local information. In addition, they use a single manifold to represent the pose space. In this paper, we use multi-manifold to represent the feature space by a novel embedding method.

Several multi-subspace methods have been proposed in the literature [9,10,11]. Kim et al. [9] presented locally linear discriminant analysis for face recognition with a single model image. Vidal et al. [10] proposed an algebraic geometric approach to estimate a mixture of subspaces. Tipping et al. [11] proposed a mixture model of probabilistic principal component analyzers for face recognition. The parameters of the mixture model are determined using an EM algorithm. They have high computing complexity for the iterative solution methods.

The major contribution of this paper is to introduce the Affinity Propagation (AP) [12] method to obtain local structures of subjects with similar appearance which are used to construct multiple manifolds. Another contribution is the novel formation of the discriminative embedding using the exemplars solved in a closed-form instead of a iterative method.

3 Multi-Manifold Embedding for Head Pose Estimation

Assume that the training data are $X = [x_1^1, x_2^1, \cdots, x_P^1, \cdots, x_1^S, x_2^S, \cdots, x_P^S]_{M \times N}$, where $x_p^s \in R^M$, $s = 1, 2, \cdots, S$, $p = 1, 2, \cdots, P$, S is the number of subjects, and P is the number of poses for a subject α^s, and there are $N = S \times P$ samples in total. The pose angle of the sample x_p^s is denoted as β_p. We aim to seeking a discriminative embedding that mapping the original M dimensional image space into an m dimensional feature space with $m \ll M$.

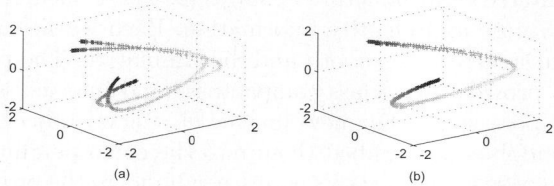

Fig. 1. The 3-dimensional embedding of the pose data by LLE. (a) 2 subjects with dissimilar individual appearance. (b) 2 subjects with similar individual appearance.

3.1 Motivations

The changes of pose images due to identity changes are usually larger than that caused by different poses of same subject. Thus, for head pose estimation, it is crucial to obtain the identity-independent manifold embedding which preserves the pose differences. The SMME method is motivated by two observations: (1) The appearance variations caused by identity lead to translation, rotation and warp changes of the subject's embeddings. Two subjects with similar individual appearance almost lie on a same continuous manifold by Locally Linear Embedding (LLE) [13] shown in Fig. 1-(b). Otherwise, Fig. 1-(a) shows that the embeddings may not be close from two subjects with dissimilar individual appearance. (2) It is difficult to make sure that the pose data lie on a single continuous manifold for the individual variations.

3.2 Smooth Multi-Manifold Embedding

Taking account of the effect caused by the appearance variations from different subjects, we first group subjects in the training data set into clusters (represented by their exemplar), and then seek a discriminative embedding for each cluster supervised by both pose and identity information. Finally, we estimate the pose by the k nearest neighbors in the low-dimensional embedding space.

Clustering Using Affinity Propagation. Frey and Dueck [12] proposed the Affinity Propagation (AP) algorithm which is capable of finding an optimal set of clusters with representative exemplars. Compared with other clustering methods, AP do not preset the number of clusters and has good clustering performance. In our scheme, AP is used to seek the local structures of subjects with similar embeddings in the low-dimensional pose space.

For two head images x_p^s and $x_{p'}^{s'}$, we compute the similarity as follows

$$sim(x_p^s, x_{p'}^{s'}) = -\|x_p^s - x_{p'}^{s'}\|^2. \tag{1}$$

Then, we define the similarity of the two subjects α^i and α^k as

$$s(i,k) = \sum_p sim(x_p^i, x_p^k). \tag{2}$$

The parameter responsibility of AP can be determined experimentally by cross-validation. The output is the clusters $\{X^1, X^2, \cdots, X^K\}$ with the corresponding exemplars $\{x^1, x^2, \cdots, x^K\}$. Later experiments show that each cluster with a exemplar can be used to seek a discriminative embedding.

Embedding Method. For each local structure we seek a low-dimensional embedding to provide intra-class compactness and inter-class separability in the low-dimensional pose subspace. The optimization of the projection is synchronous as follows: (1) Intra-class Compactness: For each pose, the projection minimizes the distances between the embeddings of the exemplar and the other subjects. (2) Inter-class Separability: For each subject, the projection maximizes the distances between the embeddings of the different poses.

To obtain a low-dimensional pose space that is good for pose estimation, it is desirable to minimize the intra-class compactness. We formulate it as the distances between the embeddings of the exemplar and the other subjects for each pose. Namely, we should minimize

$$\sum_p \sum_{i \in X^c} \|y_p^i - y_p^c\|^2, \qquad (3)$$

where y_p^c is the embedding of the head image x_p^c that is the exemplar of the cluster X^c with the pose angle β_p.

At the same time, we promote the inter-class separability of different poses by maximizing the distances between the embedding of the different poses for each subject. Namely, we maximize

$$\sum_s \sum_{i \neq j} \|y_i^s - y_j^s\|^2 T_{ij}, \qquad (4)$$

where T_{ij} is a penalty for poses i and j. We introduce a heavy penalty to penalize the poses i and j when they are close to each other, this is given as $T_{ij} = exp(-\|\beta_i - \beta_j\|^2)/\sum_i exp(-\|\beta_i - \beta_j\|^2)$. To combine (3) and (4) simultaneously, we minimize the following objective

$$J = \frac{\sum_p \sum_{i \in X^c} \|y_p^i - y_p^c\|^2}{\sum_s \sum_{i \neq j} \|y_i^s - y_j^s\|^2 T_{ij}}, \qquad (5)$$

where J is the objective to seek the embedding y_p^s of the head pose x_p^s.

Fig. 2 (a) shows the intrinsical embeddings from a local structure of four subjects (three subjects denoted by circles and an exemplar denoted by star). The optimization for the projection is to minimize the distances between the exemplar and the other subjects with a same pose and maximize the distances between different poses of a subject. Fig. 2 (b) shows the corresponding embeddings which minimized the distances denoted by the dashed lines and maximized the distances denoted by the solid lines. The objective of the embedding is to generate many pose clusters each corresponding to a specific pose angle.

In this paper, we employ a linear projection approach, namely, the embedding is achieved by seeking a projection matrix $W \in R^{M \times m}$ ($m \ll M$) such that

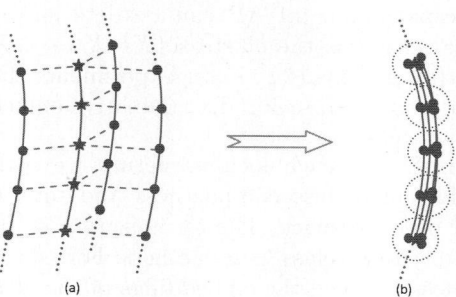

Fig. 2. Illustration of the embedding method. (a) shows the intrinsical embeddings from a local structure of four subjects (three subjects denoted by circles and an exemplar denoted by star). (b) shows the embeddings which minimized the distances denoted by the dashed lines and maximized the distances by the solid lines.

$y_p^s = W^T x_p^s$, where $y_p^s \in R^m$ is the low-dimensional embedding of $x_p^s \in R^M$. Then, W is obtained by the following optimization

$$W^* = \arg\min_W \frac{\sum_p \sum_{i \in X^c} \|W^T x_p^i - W^T x_p^c\|^2}{\sum_s \sum_{i \neq j} \|W^T x_i^s - W^T x_j^s\|^2 T_{ij}}. \quad (6)$$

It is not difficult to see that the objective function can be transformed into

$$W^* = \arg\max_W \frac{Tr(W^T S_2 W)}{Tr(W^T S_1 W)}, \quad (7)$$

where $Tr(\cdot)$ means the trace of a square matrix, and

$$S_1 = \sum_p \sum_{i \in C} (x_p^i - x_p^c)(x_p^i - x_p^c)^T, \quad S_2 = \sum_s \sum_{i \neq j} (x_i^s - x_j^s)(x_i^s - x_j^s)^T T_{ij}. \quad (8)$$

The objective function in (7) can be solved with the generalized eigenvalue decomposition method as $S_2 W_i = \lambda_i S_1 W_i$, where the vector W_i is the eigenvector corresponding to the i-th largest eigenvalue λ_i, and it constitutes the i-th column vector of the projection matrix W.

4 Experiments and Results

The proposed SMME method was validated using the FacePix database [14], which contains 5430 head images spanning $-90°$ to $+90°$ in yaw at $1°$ intervals. We also collected head pose images from the Pointing'04 database [15] for testing. The images were equalized and sub-sampled to 32x32 resolution, and preprocessed by the Laplacian of Gaussian (LoG) filter to capture the edge map that is directly related to pose variations [7].

To evaluate the performance of our system, we use the Mean Absolute Error (MAE) [1] which is computed by averaging the difference between expected pose and estimated pose for all images. To test the generalization ability, we use the leave-one-out strategy [8] (one subject in turn as the testing data and all the remaining subjects for the embedding learning).

4.1 Embedding Space

We use the proposed SMME method on the data sets mentioned above to show the embeddings. Fig. 3-(a) shows two 3-dimensional manifold embeddings from two clusters of 4 subjects with pose variations from $[-75° + 75°]$ at $4°$ intervals. The result has much better smoothness, intra-class compactness and inter-class separability in the low-dimensional embedding space. And the embedding manifold curves have different geometrical structures and different locations which indicates the multi-manifold representation is benefit for pose estimation.

Fig. 3-(b) shows the distance difference between the image and embedding space for similar poses of the same subject and different subjects with the same pose (We fix the subject 1 with pose $30°$, and locate another points by the distance from it). We can see that the distance between images from different subjects with the same pose becomes less than the distance between images from the same subject with similar poses in the low-dimensional embedding space. It indicates that the SMME provides better discriminability for pose estimation.

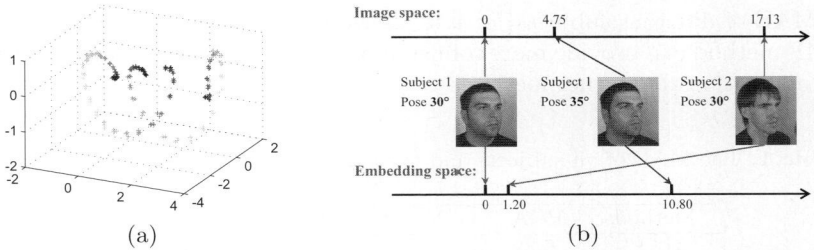

Fig. 3. Illustration of smoothness and discriminability of the embedding space. (a) shows two 3-dimensional manifold embeddings from two cluster. (b) shows the distance difference between the image and embedding space.

4.2 Comparison of SMME with Other Methods

We compare SMME with other pose estimation methods: the global-based PCA method, the local-based manifold learning LPP methods [6] and Marginal Fisher Analysis (MFA) [16] methods. Fig. 4 (a) shows the pose estimation results in different dimensionalities. It shows that the proposed SMME method significantly improves the estimation performance compared to other methods. Fig. 4 (b) shows the MAE with pose variations from $[-90° + 90°]$ at $1°$ intervals. The result shows that the accuracy of the proposed SMME method is still in general

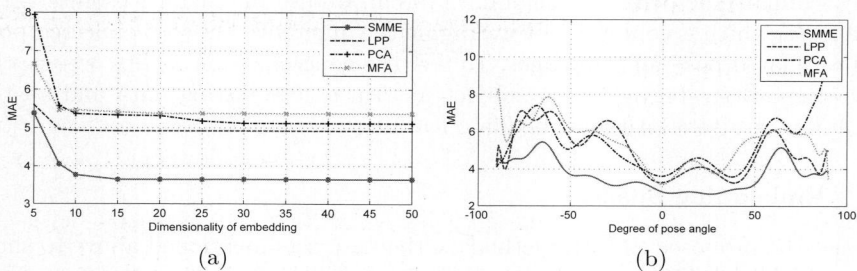

Fig. 4. Comparison of our method against other methods. (a) The MAE in different dimensionality. (b)The MAE under different poses.

better than other methods. We notice that the MAE curve of SMME is much more flat than other methods within a relative wide range of the frontal view $[-50° + 50°]$, this implies that SMME is more robust in $[-50° + 50°]$.

4.3 Robustness against Different Identities

In order to test the robustness of SMME against different identities, we use the samples of one subject in turn as the testing data and use all the remaining subjects for embedding learning to compute the MAE of each subjects. The proposed SMME method achieves the average MAE of 3.64° and the variance (for MAE of different subjects) of 1.13 shown in Table 1, which shows that the SMME method can provide more robust and accurate identity-independent head pose estimation than other methods.

Table 1. The MAE of all subjects and the variance of MAE for different subjects

Methods	PCA	MFA	LPP	SMME
MAE	5.32	5.41	4.96	**3.64**
Variance	4.66	4.79	3.21	**1.13**

5 Conclusions

In this paper, we present the SMME method for robust head pose estimation which provides better intra-class compactness and inter-class separability in low-dimensional pose subspace than traditional methods. For identity-independent head pose estimation, we achieved the MAE of about 3° on the standard databases, and even lower MAE can be achieved on larger data sets. In addition, the method has been demonstrated as more robust to individual variations for new identities than the traditional methods. In future, we plan to evaluate the proposed method in terms of feasibility for more complex real world scenarios, and develop auto-adaptive multi-manifold embedding method.

Acknowledgment

This work is supported by the National Laboratory of Pattern Recognition under grant 09-4-1, the National High Technology Research and Development Program of China (No. 2008AA02Z310) and 973 Program 2009CB320900.

References

1. Murphy-Chutorian, E., Trivedi, M.: Head pose estimation in computer vision: a survey. IEEE Transactions on PAMI, 442–449 (2008)
2. Sebastian, H., Lee, D.: The manifold ways of perception. Science 290(12), 2268–2269 (2000)
3. Tenenbaum, J., Silva, V., Langford, J.: A global geometric framework for nonlinear dimensionality reduction. Science 290(5500), 2319–2323 (2000)
4. Fu, Y., Huang, T.: Graph embedded analysis for head pose estimation. In: Proc. of International Conference on Automatic Face and Gesture Recognition (2006)
5. Wang, X., Huang, X., Gao, J., Yang, R.: Illumination and person-insensitive head pose estimation using distance metric learning. In: Forsyth, D., Torr, P., Zisserman, A. (eds.) ECCV 2008, Part II. LNCS, vol. 5303, pp. 624–637. Springer, Heidelberg (2008)
6. Raytchev, B., Yoda, I., Sakaue, K.: Head pose estimation by nonlinear manifold learning. In: ICPR (2004)
7. Balasubramanian, V., Ye, J., Panchanathan, S.: Biased manifold embedding: a framework for person-independent head pose estimation. In: CVPR (2007)
8. Yan, S., Wang, H., Fu, Y., Yan, J., Tang, X., Huang, T.S.: Synchronized submanifold embedding for person-independent pose estimation and beyond. IEEE Transactions on Image Processing (2008)
9. Kim, T., Kittler, J.: Locally linear discriminant analysis for multimodally distributed classes for face recognition with a single model image. IEEE Transactions on PAMI 27(3), 318–327 (2005)
10. Vidal, R., Ma, Y., Sastry, S.: Generalized principal component analysis (GPCA). IEEE Transactions on PAMI 27(12), 1945–1959 (2005)
11. Tipping, M., Bishop, C.: Mixtures of probabilistic principal component analyzers. Neural computation 11(2), 443–482 (1999)
12. Frey, B., Dueck, D.: Clustering by passing messages between data points. Science 315(514), 972–977 (2007)
13. Roweis, S., Saul, L.: Nonlinear dimensionality reduction by locally linear embedding. Science 290(5500), 2323–2326 (2000)
14. Little, D., Krishna, S., Black, J., Panchanathan, S.: A methodology for evaluating robustness of face recognition algorithms with respect to variations in pose angle and illumination angle. In: ICASSP, vol. 2 (2005)
15. Gourier, N., Hall, D., Crowley, J.: Estimating Face orientation from Robust Detection of Salient Facial Structures. In: VODG, pp. 281–290 (2004)
16. Yan, S., Xu, D., Zhang, B., Zhang, H., Yang, Q., Lin, S.: Graph embedding and extensions: a general framework for dimensionality reduction. PAMI, 40–51 (2007)

Human Age Estimation by Metric Learning for Regression Problems

Yangjing Long

Max Planck Institute for Mathematics in the Sciences
27 Inselstrasse, Leipzig, Germany
long.yangjing@hotmail.com

Abstract. The estimation of human age from face images is an interesting problem in computer vision. We proposed a general distance metric learning scheme for regression problems, which utilizes not only data themselves, but also their corresponding labels to strengthen the credibility of distances. This metric could be learned by solving an optimization problem. Furthermore, the test data could be projected to this metric by a simple linear transformation and it is feasible to be combined with manifold learning algorithms to improve their performance. Experiments are conducted on the public FG-NET database by Gaussian process regression in the learned metric to validate our framework, which shows that the performance is improved over traditional methods.

Keywords: Age Estimation, Metric Learning, Regression.

1 Introduction

The estimation of human age from face images is an interesting problem in computer vision. As an important hint for human communication, facial images comprehend lots of useful information including gender, expression, age, pose, etc. Unfortunately, compared with other cognition problems, age estimation from face images is still very challenging. This is mainly due to the fact that, aging progress is influenced by not only personal gene but also many external factors. Physical condition, living style etc. may accelerate or slower aging process. Besides, since aging process is slow and with long duration, collecting sufficient data for training is a fairly strenuous work.

[10,17] formulated human ages as a quadratic function. Yan et al. [27,28] modeled the age value as the square norm of a matrix where age labels were treated as a non-negative interval instead of a certain fixed value. However, all of them regarded age estimation as a regression problem without special concern about the own characteristics of aging variation. As Deffenbacher [8] stated, the aging factor has its own essential sequential patterns. For example, aging is irreversible, which is expressed as a trend of growing older along the time axis. Such general evolution of aging course is beneficial to age estimation, especially when training data are limited and distributed unbalanced over each age range.

Geng et al. [13,12] firstly made some pioneer research on seeking for the underlying aging patterns by projecting each face in their aging pattern subspace (AGES).

Guo et al. [16] proposed a scheme based on Orthogonal Locality Preserving Projections (OLPP) [5] for aging manifold learning and get the state-of-art results. In [16], SVR (Support Vector Regression) is used to estimate ages on such a manifold and the result is locally adjusted by SVM. However, they only tested their OLPP-based method on a private large database consisting of only Japanese people, and no dimension reduction work was done to exact the so-called aging trend on the public available FG-NET database [1]. A possible reason is that, FG-NET database may not supply enough samples to recover the intrinsic structure of data. The lack of sufficient data is a prominent barrier in age estimation.

We propose a new framework aiming to learn a special metric for regression problems. Age is predicted based on the learned metric rather than the traditional Euclidean distance. We accomplish this idea by formulating an optimization problem, which approximates a special designed distance that scaled by a factor determined according to the labels of data. In this way, the metric measuring the similarity of samples is strengthened. More importantly, since labels are incorporated to depict the underlying sample distribution tendency, which signifies the inclusion of more information, a smaller amount of training data is required. Unlike the nonlinear manifold learning where it is repeated to find its low dimensional embedding, a merit of our framework is that, a full metric over the input space is learned and expressed as a linear transformation, and it is easy to project a novel data into this metric. Moreover, the proposed framework may also be used as a pre-processing step to assist those unsupervised manifold learning algorithms to find a better solution.

2 Metric Learning for Regression

Let $S = (X_i, y_i)$ ($1 \leq i \leq N$) denotes a training set of N observations with inputs $X_i \in R^d$ and their corresponding non negative labels y_i. Our goal is to rearrange these data in high-dimensional space with a distinct trend as what their labels characterize. In other words, we hope to find a linear transformation $T: R^d \rightarrow R^d$, after applying which, the distances between each pair-wise observation may be measured as:

$$\hat{d}(X_i, X_j) = \| T(X_i - X_j) \|^2 \qquad (1)$$

2.1 Problem Formulation

Metrics is a general concept, as a function giving a generalized scalar distance between two argument patterns [11]. Straightforwardly, different distances are also possible to depict the tendency of a data set. Similar to Weinberger et al. [25] and Xing et al. [26], we consider learning a distance metric of the form

$$d_A(X_i, X_j) = \sqrt{(X_i - X_j)^T A(X_i - X_j)} \qquad (2)$$

But unlike their works for classification problems, in regression problems, every two observations are of different classes. Better metrics over their inputs are expected and a new metric learning strategy ought to be established.

Suppose given certain well-defined distance $\hat{d}_{ij}=\hat{d}(X_i,X_j)$ ideally delineating the data trend, our target is to approximate $\hat{d}_{ij}$ by $d_A(X_i,X_j)$ minimizing the energy function

$$\varepsilon(A)=\sum_{i,j}\left(d_A(X_i,X_j)^p-(\hat{d}_{ij})^p\right)^2 \qquad (3)$$

To promise that A is a metric, A is restricted to be symmetric and positive semi-definite. For simplicity, p is assigned to be 2. This metric learning task is formulated as an optimization problem with the form below

$$\min\sum_{i,j}\left((X_i-X_j)^T A(X_i-X_j)-(\hat{d}_{ij})^2\right)^2 \qquad (4)$$

satisfying the matrix A is symmetric and positive semi- definite. And there exists a unique lower triangular L with positive diagonal entries such that $A=LL^T$ [15]. Hence learning the distance metric A is equivalent to finding a linear transform L^T projecting observation data from the original Euclidean metric to a new one by

$$\tilde{X}=L^T X \qquad (5)$$

2.2 Distance with Label Information

In practical application, Euclidean distance is not always capable to guarantee the rational relationship among input data. Although manifold learning algorithms may discover the intrinsic low-dimensional parameterizations of the high dimensional data space, at the outset, it also requires Euclidean distance to apply k-Nearest Neighbors to know the local structure of the original space. On the other hand, manifold learning demands a large amount of samples, which is not available in some circumstances.

For many regression and classification problems, it is in fact a waste of information if only data X_i is utilized but with their associated labels y_i ignored in the training stage. Balasubramanian et al. [2] proposed a biased manifold embedding framework to estimate head poses. In their work, the distance between data is modified by a factor of the dissimilarities fetched from labels. The basic form of this modified distance is

$$d'(i,j)=\frac{\beta\times P(i,j)}{\max_{m,n}P(m,n)-P(i,j)}\times d(i,j) \qquad (6)$$

where $d(i,j)$ is the Euclidean distance between two samples X_i and X_j. $P(i,j)$ is the difference of poses between X_i and X_j.

Through incorporating the label information to adjust Euclidean distance, the modified distances are prone to give rise to the true tendency of data variation i.e. if the distance of two observations is large, then the distance of their labels is also large, vice versa. Hence it is intuitively that the biased distance is a good choice for $\hat{d}_{ij}$ in Eq.(3):

$$\hat{d}(i,j)=\left(\frac{\beta\times|L(i,j)|}{C-L(i,j)}\right)^p\times d(i,j) \qquad (7)$$

Analogously, $L(i,j)$ is the label difference between two data. C is a constant greater than any label value in a train set and p is selected to make data easier to discriminate. $d(i,j)$ is the Euclidean distance between two samples $X_i\psi$and X_j.

2.3 Optimization Strategy

Since the energy function is not convex, it is a non-convex optimization and consequently it is impossible to find a closed form solution. The metric A is with the property to be symmetric and positive semi-definite, so it is natural to compute a numerical solution to Eq.(4) using the Newton's method. Similar to [26], in each iteration, a gradient descent step is employed to update A. The iteration algorithm is summarized as follows:

1. Initialize A and step length α;
2. Enforce A to be symmetric by $A \leftarrow (A+A^T)/2$;
3. The Singular Value Decomposition of $A=L^T \Delta L$, where the diagonal matrix Δ consists of the eigenvalues $\lambda_1,...,\lambda_n$ of A and columns of L contains the corresponding eigenvectors;
4. Ensure A to be positive semi-definite by $A \leftarrow L^T \Delta' L$, where $\Delta'=\mathrm{diag}(\max(\lambda_1,0),...,\max(\lambda_n,0))$;
5. Update $A' \leftarrow A - \alpha \nabla_A \varepsilon(A)$, where $\nabla_A \varepsilon(A)$ is the gradient of the energy function in Eq.(3) w.r.t. A;
6. Compare the energy function $\varepsilon(A)$ with $\varepsilon(A')$ in Eq.(3), if $\varepsilon(A)<\varepsilon(A')$, then augment the step length α with a momentum to accelerate the optimization process; otherwise, shrink α to assure a local minimum is not overpassed.
7. If A has converged or the maximum iteration times are reached, terminate; otherwise go back to Step 2.

3 Gaussian Process Regression

Given a training set $S = (X_i, y_i)$ ($1 \leq i \leq N$) as described in Section II and a sample X^* for query, GPR predicts its output y^* by putting a Gaussian process prior on this function $f(\cdot)$, assuming that all sample points evaluated from the function have a multivariate Gaussian density [20].

Let $X=[X_1,...,X_N]$ and $Y=[y_1,...,y_N]^T$, the Gaussian predictive distribution of y^* is derived of the form

$$p(y^*|X^*,X,Y,\Theta) \sim N(\mu(X^*),V(X^*)) \tag{8}$$

The mean prediction and covariance matrix in Eq.(8) are

$$\mu(X^*)=k(X^*,X)[\mathbf{K}+\sigma^2 I]^{-1} Y \tag{9}$$

$$V(X^*)=k(X^*,X^*)-k(X^*,X)^T[\mathbf{K}+\sigma^2 I]^{-1}k(X^*,X) \tag{10}$$

where $k(\cdot,\cdot)$ is the covariance function, $\mathbf{K}$ is the covariance matrix of $\mathbf{X}$ and σ^2 is the variance of noise.

Another way to perceive and thus rewrite Eq.(9) is to treat the mean prediction as a linear combination of N kernel functions:

$$\mu(X^*) = \sum_{c=1}^{N} \alpha_c k(X^*, x_c) \tag{11}$$

where $\alpha=(\mathbf{K}+\sigma^2\mathbf{I})^{-1}Y$

Gaussian kernel is a good choice for the covariance function

$$k(X_i,X_j)=v^2\exp(-\|X_i-X_j\|^2/2l^2+\sigma^2\sigma_{X_iX_j}) \quad (12)$$

In respect that the proposed learned metric encodes label information implicitly, it is bestowed as the similarity measure and Eq.(12) becomes

$$k(X_i,X_j)=v^2\exp(-(X_i-X_j)^T A(X_i-X_j)/2l^2+\sigma^2\sigma_{X_iX_j}) \quad (13)$$

4   Experimental Results

Age estimation is carried on the public FG-NET Aging Database [1] by the regression strategy on the basis of the proposed metric. The database contains totally 1002 color or gray images from 82 people. Each person has around 10 face images with the ranges from 0 to 69 with labeled ground truth. These images are taken under varying lighting condition, poses and expressions. Each image is labeled by 68 points characterizing its shape features. Similar to [13,16,27,28], input features are selected to be the parameters of AAMs [6].

Firstly we hope to testify that the proposed metric is able to disinter some internal patterns of human's aging progression. We randomly choose 300 images out of all the 1002 images in FG-NET Database as training samples, and the rest as test samples. The parameters in Eq.(7) are chosen as $C=100$, $\beta=1$ and $p=1$. The energy function is converged after 50 iterations or so. Figure 1(a) and 1(b) portrays the positional relationship among training samples in the hyper-space measured by Euclidean distance and the learned metric A. The 2D view is acquired by Multi-Dimensional Scaling (MDS) [7]. Figure 2 plots the relative position of the remaining 702 image samples for test. Contrast to Figure 1, manifold learning algorithms like Isomap, LLE and OLPP fails to predicate the aging trend sometimes. Furthermore, though only 30% of the entire data set is directed for learning the aging trend is effectually set up.

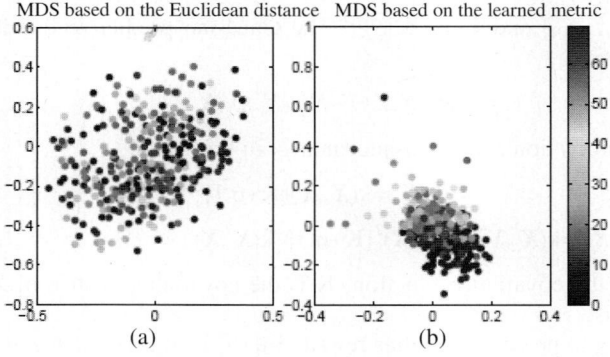

Fig. 1. 2-D view of the clustering effects of the 300 training samples by metric learning. It illustrates the 2 dimensional embedding of the training data sampled from FG-NET Aging Database by MDS. Points of age from 0 to 69 are marked from blue to red. It is seen that, the distance calculated based on our learned metric in Figure (b) preserves local proximity of samples with close labels better than that based on the traditional Euclidean distance in Figure (a).

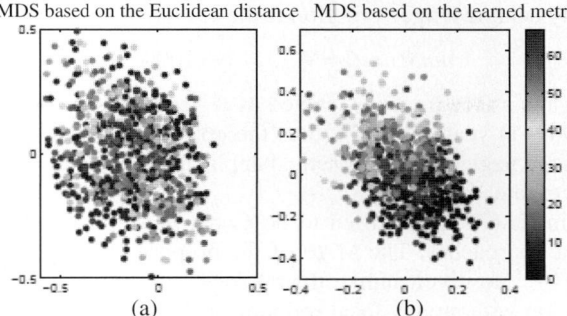

Fig. 2. 2-D view of the clustering effects of the 702 testing samples by metric learning, corresponding to Figure 1. It is obvious that, the actual aging trend is, to some extended, manifested in the hyper-space based on our learned metric.

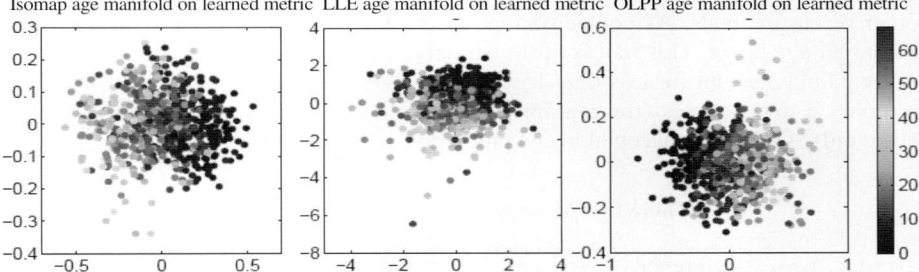

Fig. 3. 2D age manifolds. This figure illustrates the 2 dimensional embedding of FG-NET Aging Database by Isomap, LLE and OLPP algorithms based on our learned metric.

As in Eq.(5), the original parameters from AAMs can be linearly transformed into a hyper-space based on our learned metric, by multiplying L^T satisfying $A=LL^T$. Figure 3 draws the 2D aging manifold inputted with the transformed data. Compared to Figure 1, the linear transform L^T is salutary for other manifold algorithms to find an improved aging trend.

Then, age estimation of our methodology is compared with the performance of some state-of-art approaches. The Leave- One-Person-Out mode [13,16,27,28] is the mechanism for experimentation, i.e. each time we choose one person for testing and all others for training. The same as in [13,16,27, 28], two criteria are adopted for performance evaluation. One is the Mean Absolute Value (MAE), which is defined as

$$MAE = \sum_{i=1}^{N} |\widetilde{age}_i - age_i| / N \qquad (14)$$

where for each X_i, $\widetilde{age}_i$ is its labeled ground truth and age_i is the estimated age. N is the number of testing images.

Another widely acknowledged criterion is the cumulative score at error level l [13]

$$CumScore(l)=N_{error \leq l}/N\times 100\% \qquad (15)$$

In respect that, when a face image is labeled as O years old, the person is customarily thought to be $[O,O+1)$ years old [27], thus the error less than a specified number of years is by and large neglectable in practical application. Eq.(15) is an indicator of the algorithmic correct rate.

The parameters in Eq.(7) are rectified to be $C=80$, $\beta=1$ and $p=0.6$. Table 1 lists the MAE of different approaches. The MAE of the proposed method is almost the same as the best one [16]. However, unlike their LARR, we simply predict ages in a new metric by regression without any local refinement. LARR slides the estimated age up and down by checking different age values to see if it can come up with a better prediction [16]. The parameters defining the search range is determined manually, which is at least not convenient and automatic enough, and may be laborious and not feasible in some real-world applications. Table 2 details Table 1 with separate MAEs over different age range. The MAE of our method in younger people is slightly higher than other recent methods. As compensation, an outstanding improvement is achieved in the larger age range. This trait is fairly attractive considering the fact that, people over 30 years old account for less than 15% of the whole FG-NET database. Even if there are only a few samples (for example, there are only 8 images out of 1002 over 60 years old), a relatively acceptable age prediction can be obtained.

Table 1. MAE comparison of different methods

Reference	Method	MAE
[13]	AGES	6.77
[12]	KAGES	6.18
[27]	RUN1	5.78
[28]	RUN2	5.33
[16]	LARR	5.07
Proposed	Metric learning+GPR	5.08

Table 2. MAEs over various age ranges on FG-NET Database for the proposed method, GPR and RUN. In the first column, the value in the parenthesis stands for the proportion (percentage) for each age group out of the whole database.

Age Range	Proposed	GPR	RUN[27]
0-9(37.0%)	2.99	3.55	2.51
10-19(33.8%)	4.19	4.34	3.76
20-29(14.4%)	5.34	5.09	6.38
30-39(7.9%)	9.28	9.04	12.51
40-49(4.6%)	13.52	14.65	20.09
50-59(1.5%)	17.79	19.77	28.07
60-69(0.8%)	22.68	31.76	42.50
Average	5.08	5.45	5.78

5 Conclusions

In this paper, a new metric learning framework is proposed to resolve regression problems. It is feasible to be applied to many other problems in machine learning or computer vision. No assumptions about the structure or distribution of the samples are made, and a relatively small quantity of training samples is required to learn their underlying variation trend. Experiments shows the effectiveness of the learned metric to restore the intrinsic infrastructure of input sample data and encouraging performance is acquired on a widely used public face aging database.

References

1. FG-NET Aging Database, http://www.fgnet.rsunit.com
2. Balasubramanian, V.N., Ye, J., Panchanathan, S.: Biased manifold embedding: A framework for person-independent head pose estimation. In: IEEE Conf. CVPR, pp. 1–7 (2007)
3. Bar-Hillel, A., Weinshall, D.: Learning distance function by coding similarity. In: Proc. ICML, pp. 65–72 (2007)
4. Belkin, M., Niyogi, P., Sindhwani, V.: Manifold regularization: a geometric framework for learing from labeled and unlabeled examples. Journal of Machine Learning Research 7, 2399–2434 (2006)
5. Cai, D., He, X., Han, J., Zhang, H.J.: Orthogonal laplacianfaces for face recognition. IEEE Trans. Image Processing 15, 3608–3614 (2006)
6. Cootes, T., Edwards, G., Taylar, C.: Active appearance models. IEEE Trans. Pattern Analysis & Machine Intelligence 23(6), 681–685 (2001)
7. Cox, T., Cox, M.: Multidimensional Scaling. Chapman & Hall, Lodon (1994)
8. Deffenbacher, K.A., Vetter, T., Johanson, J., O'Toole, A.J.: Facial aging, attractiveness, and aistinctiveness. Perception 27 (1998)
9. Donoho, D.L., Grimes, C.E.: When does geodesic distance recover the true hidden parametrization of families of articulated images? In: Proc. European Symposium on Artificial Neural Networks (2002)
10. Draganova, A.L.C., Christodoulou, C.: Comparing different classifiers for automatic age estimation. IEEE Trans. Systems, Man, and Cybernetics 34(1), 621–628 (2004)
11. Duda, R.O., Hart, P.E., Stork, D.G.: Pattern Classification, 2nd edn. John Wiley & Sons, Inc., New York (2001)
12. Geng, X., Smith-Miles, K., Zhou, Z.-Z.: Facial age estimation by nonlinear aging pattern subspace. In: Proc. ACM Conf. Multimedia (2008)
13. Geng, X., Zhou, Z.H., Zhang, Y., Li, G., Dai, H.: Learning from facial aging patterns for automatic age estimation. In: Proc. ACM Conf. Multimedia, pp. 307–316 (2006)
14. Goldberger, J., Roweis, S., Hinton, G., Salakhutdinov, R.: Neighbourhood components analysis. In: NIPS (2005)
15. Golub, G.H., Loan, C.F.V.: Matrix Computations. Johns Hopkins Univ. Press (1996)
16. Guo, G., Fu, Y., Dyer, C., Huang, T.S.: Image-based human age estimation by manifold learning and locally adjusted robust regression. IEEE Trans. on Image Processing 17, 1178–1188 (2008)
17. Lanitis, A., Taylor, C.J., Cootes, T.: Toward automatic simulation of aging effects on face images. IEEE Trans. Pattern Analysis and Machine Intelligence 24(4), 442–455 (2002)
18. Neal, R.M.: Monte carlo implementation of gaussian process models for bayesian regression and classification. Technical Report CRG-TR-97-2

19. Nilsson, J., Sha, F., Jordan, M.I.: Regression on manifolds using kernel dimension reduction. In: IEEE Conf. ICML, pp. 265–272 (2007)
20. Raumussen, C.E., Williams, C.K.: Gaussian Processes for Machine Learning. MIT Press, Cambridge (2006)
21. Roweis, S.T., Saul, L.K.: Nonlinear dimensionality reduction by locally linear embedding. Science 290(5500), 2323–2326 (2000)
22. Scholkopf, B., Smola, A.J.: Learning with Kernels: Support Vector Machines, Regularization, Optimization, and Beyond. MIT Press, Cambridge (2002)
23. Sugiyama, M., Hachiya, H., Towell, C., Vijayakumar, S.: Geodesic gaussian kernels for value function approximation. Autonomous Robots 25, 287–304 (2008)
24. Tenebaum, J.B., de Silva, V., Langford, J.C.: A global geometric framework for nonlinear dimensionally reduction. Science 290(5500), 2319–2323 (2000)
25. Weinberger, K., Blitzer, J., Saul, L.: Distance metric learning for large margin nearest neighbor classification. In: Proc. NIPS, pp. 1475–1482 (2006)
26. Xing, E., Ng, A., Jordan, M.I., Russell, S.: Distance metric learning with application to clustering with side-information. In: Proc. NIPS (2002)
27. Yan, S., Wang, H., Huang, T.S., Tang, X.: Ranking with uncertain labels. In: IEEE Conf. Mulitimedia and Expo, pp. 96–99 (2007)
28. Yan, S., Wang, H., Tang, X., Huang, T.S.: Learning autostructured regressor from uncertain nonnegative labels. In: IEEE Conf. ICCV, pp. 1–8 (2007)

Face Detection Using GPU-Based Convolutional Neural Networks

Fabian Nasse[1], Christian Thurau[2], and Gernot A. Fink[1]

[1] TU Dortmund University, Department of Computer Science, Dortmund, Germany
[2] Fraunhofer IAIS, Sankt Augustin, Germany

Abstract. In this paper, we consider the problem of face detection under pose variations. Unlike other contributions, a focus of this work resides within efficient implementation utilizing the computational powers of modern graphics cards. The proposed system consists of a parallelized implementation of convolutional neural networks (CNNs) with a special emphasize on also parallelizing the detection process. Experimental validation in a smart conference room with 4 active ceiling-mounted cameras shows a dramatic speed-gain under real-life conditions.

1 Introduction

The past years yielded increasing interest in transfering costly computations to Graphics Processing Units (GPUs). Due to parallel execution of commands this often results in a massive speedup. However, it also requires a carefull adaption and parallelization of the algorithm to be implemented. As noted in [1], convolutional neural networks (CNNs) [8,2,4,3] offer state of the art recognizers for a variety of problems. However, they can be difficult to implement and can be slower than other classifiers, e.g. traditional multi-layer perceptrons. The focus of this paper resides within the implementation details of parallelizing CNNs for the task of face detection and pose estimation and evaluating its run-time performance. In contrast to [1], where GPU optimized CNNs were adapted for document processing, the considered face detection task requires additional considerations: In detail, the contributions of this paper are (a) extension of the face recognition (and pose estimation) system in [3] by parallelizing important parts of the computational process and implementing it on a graphics card, and (b) further enhancing the system by an optimized detector. Experimental validation takes place in a multi-camera environment and shows accurate detection and high performance under real-life conditions.

The remainder of this paper is organized as follows: In Section 2 we give a brief introduction to convolutional networks and how they are used in this work. Section 3 explains in detail the process of face detection using CVs, and Section 4 shows how this process can be efficiently parallelized. In Section 5 we present experimental results of the proposed optimized face detection approach. Finally we conclude this paper in Section 6.

2 Convolutional Neural Networks

In the following we will briefly introduce *convolutional neural networks (CNNs)* [8,2,4,3]. In a nutshell, a CNN classifies an input pattern by a set of several concatenated operations, i.e. convolutions, subsamplings and full connections. Figure 1 shows a simple example for a CNN as it was used in [8]. For practical reasons the net is organized in successive layers (L_1 to L_4). On the left side we see an input image, in our case a monochromatic 20×20 pixel image. Each subsequent layer consist of several fields of the same size which represent the intermediate results within the net. Each directed edge stands for a particular operation which is applied on a field of a preceding layer and its result is stored into another field of a successive layer. In the case that more than one edge directs to a field, the results of the operations are summed. After each layer a bias is added to every pixel (which may be different for each field) and the result is passed through a sigmoid function, e.g. $s(x) = a \cdot tanh(b \cdot x)$, to finally perform a mapping onto an output variable.

Each convolution uses a different two-dimensional set of filter coefficients. Note that in case of a convolution the size of the successive field shrinks because the border cases are skipped. For subsampling operations a simple method is used which halves the dimensions of an image by summing up the values of disjunct 2x2-subimages and weighting each result value with the same factor. The term "'full connection"' describes a function, in which each output value is the weighted sum over all input values. Note that a full connection can be described as a set of convolutions where each field of the preceding layer is connected with every field of the successive layer and the filters have the same size as the input image. Thus, we do not treat full connections as a separate case here. The last layer forms the output vector. In the given example the output consists of a single value which finally classifies a given input image.

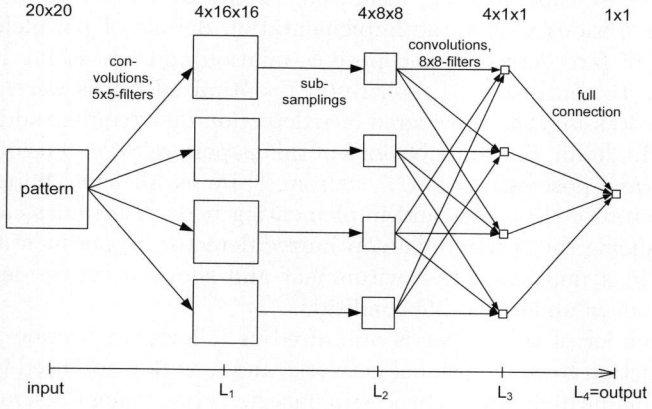

Fig. 1. Structure of a simple convolutional net

An important attribute of CNNs is the availability of an efficient training method. Since CNNs are based on the classical neural networks, each pixel of a field can be represented by a neuron and the all afore mentioned operations can be modeled by connections between neurons. For training, a modified version of a standard backpropagation algorithm can be applied. For further details on CNNs we recommend [2].

3 The Detection Process

For object/face detection we are usually interested in detecting all occurancies of an object in a given image. For a trained CNN, we can use a simple sliding window approach over a variety of scaled input images, see also Figure 2(a). The window is shifted above the image to get one result at each position. To search inside a specified size range the process is repeated with different scaling factors. In the given example the image is downscaled each time with the factor $1/\sqrt{2}$. By choosing this factor we make the assumption that the trained CNN is robust against variation in size at the range between two scaling steps.

One of the key advantages of CNNs are the inherent possibility for parallelizing the computational process when used as a detector. If a neural net is repeatedly applied on overlapping image areas redundancies in calculation occur. This is the case for convolutions as well as subsamplings in all layers. A significant speed gain is reached by avoiding these redundancies. We accomplished this by applying each operation within the net on the whole image at once instead repeatingly on all subimages. In the case of a subsampling operation four different offsets have to be considered depending on whether the 2x2-subimages start with odd or even coordinates in horizontal or vertical direction respectively. For the example given in figure 1 and an input image of size 320x240 this leads to a scheme as shown in figure 2(b). To assemble the four output images to one the coordinates have to be multiplied by two and the spatial offset given in the last row must

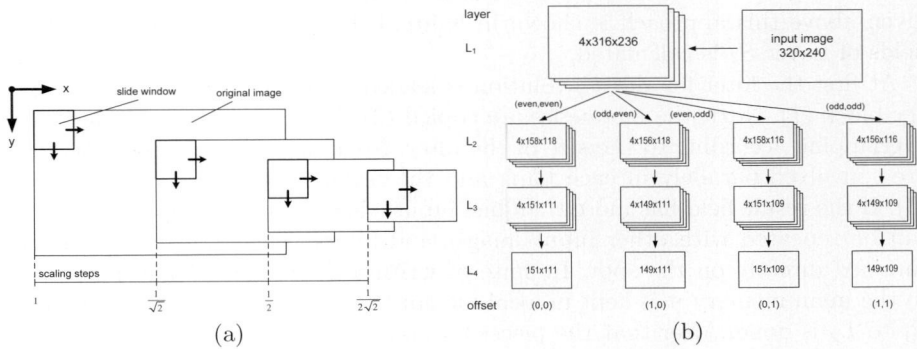

Fig. 2. Figure 2(a) shows a slide-window on an input image with different scaling factors. Figure 2(b) shows the scheme of the location process for the CNN from Figure 1 and an input image of size 320x240 px.

be added. Note that at the expense of precision the calculation of some paths in the tree may be skipped. The assumption by doing so is that the trained CNN is robust against small spatial shifts.

4   Parallelization of the Detection

The shown detection process was implemented and evaluated for the Nvidia GeForce 8800 GT GPU using the CUDA-architecture, which allows using the processing power of the graphics card for the purposes of scientific research. Note that also the following method is dedicated to CUDA-devices, similar approaches can be applied to other multi-core devices. According to the number of cores the architecture consists of multiple threads each with its own set of registers. An important task for the design of the parallel algorithm is to take the memory hierarchy into account. Thus, figure 3 shows an overview. The host system (i.e. the pc) as well as the threads have access to the global main memory of the graphics card. Furthermore the threads are grouped into so called blocks. The threads of a block share a small and fast memory bank.

The different bandwidths of the memory interfaces leads to the following strategy: The data transfer between the host system and the main memory has a low bandwidth. Thus, the communication here has to be minimized. At the beginning the net parameters are loaded once from the host system into the main memory. The same is done for every new input image. At the end of the detection process the results are written back to the host. For the threads the access time to the memory banks is much faster than to the main memory. Therefore most of the calculation should be done with the data stored in the local memory banks. Because for a convolution the pixel values have to be read several times this brings a significant speed gain. Normally an input image is to large to be stored in one memory bank, thus it is divided into equidistant rectangles. Each rectangles is loaded from the main memory into another memory bank and the partial images are treated by the threads of the according block. For the example given above this approach is shown in figure 4. In this cutout one of the four fields of Layer S_2 is calculated.

At first the filter for the convolution is loaded from the main memory in every block (1). Next, the rectangles are copied (2). Note, that the subimages are overlapping according to the size of the filter. Next the filter and the subimage are convolved parallely in each bank and the result is stored in local memory (3). If the result field has more than one input edge (as in L_3) step one to three can be repeated with other input images and filters. Than the results can be summed directly on the spot. In spite of writing the results of Layer S_1 back to the main memory it is kept in local memory and the subsampling from layer L_1 to L_2 is done. After that the pieces are assembled in the main memory (4). How many steps can be accomplished without writing data back to the main memory or sharing data between blocks depends on the net structure and can be optimized for a particular net. Not mentioned so far is the addition of a bias and the use of a sigmoid function after each layer. For every thread the bias

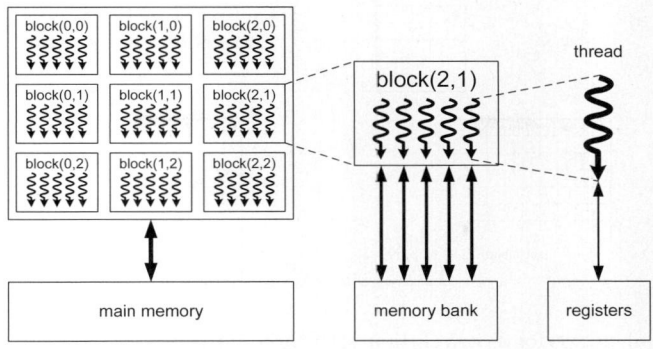

Fig. 3. Memory hierarchy for the CUDA architecture

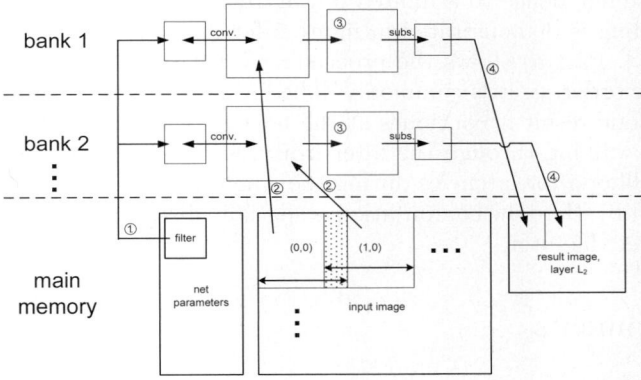

Fig. 4. Parallelizazion by dividing the input image into several rectangles

value is kept in a register (not shown in fig. 3). The sigmoid function is applied before writing the result value from a register back to local memory. Therefore no additional reading operation on the local memory is required. Nvidia GPUs support a complex instruction set with any kind of trigonometric or hyperbolic functions respectively. Nevertheless, if you use a hardware device without or with slow support for a particular function required for the sigmoid function, we recommend the use of a taylor approximation, e.g. $tanh(x/2) \approx (d-6)/(d+6)$ with $d = 6 + x \cdot (6 + x \cdot (3 + x))$. Another important aspect not mentioned yet is how a convolution between a subimage and a filter is handled in detail. Note, that especially for convolutions concurrencies can form the bottleneck of the feasible speed gain. According to fig. 3 each subimage is treated by several threads with reading access to the same memory bank. Thus, each thread computes a subset of result values. Hence to the parallel execution concurrency between reading accesses occur. Each thread needs to read the whole filter and threads who compute neighboring values need access to overlapping image areas.

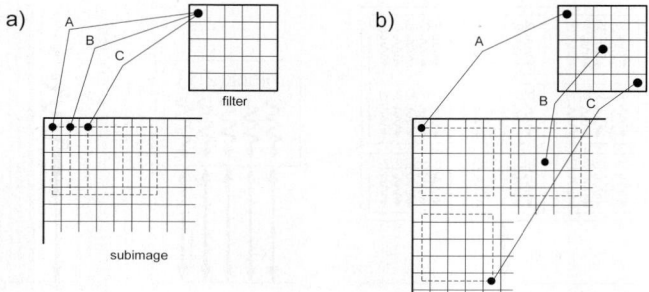

Fig. 5. Reading access for a convolution with three threads A,B and C. a) lexicographic access. b) spatially distributed access.

To reduce latency hence to a limited fan in, reading accesses should be spread. This proceeding is demonstrated in figure 5 for an example with three threads (A, B and C). Part a) shows the proceeding when the computation is done in lexicographic order, as it is commonly done for a single process. When a thread has finished one result it continues at the next position in line of the subimage and restarts walking through the filter from the upper left corner. This causes a strong likelihood for latencies during the whole process. A better approach is shown in part b). Here the computation is spatially distributed over the subimage as well as over the filter.

5 Experiments

In order to test our accelerated detector under real-time-conditions, we implemented a recognition system for faces with variation of pose based on the researches of [3]. According to fig. 6 a) the CNN we used consists of 42.750 parameters and it was trained with 6.000 non-faces and 6.000 faces with annotated poses (6 b). The system was applied in a smart conference room with 4 active ceiling-mounted cameras (6 c,d).

Although the main focus of this paper resides within an efficient implementation, we still want to briefly report on the detection rates here. Given a frontal/side view of a persons face the system is able to detect multiple persons with an average accuracy of 80%-90% percent and occasional occuring false positives (evaluated on a per frame basis for four longer sequences containing multiple persons). For a better detection we added additional training material for this particular environment, resulting in a well functional and usable system. For sake of completeness we also evaluated the proposed system on three standard data-sets [7,5,6]. We get the following average detection rates : 81% [7], 75% [5], and 83% [6] (with an average of 8 false positives). Note that we did not try to maximize detection rates for these data-sets since the applicability of convolutional neural networks for face detection was already sufficiently shown in [3].

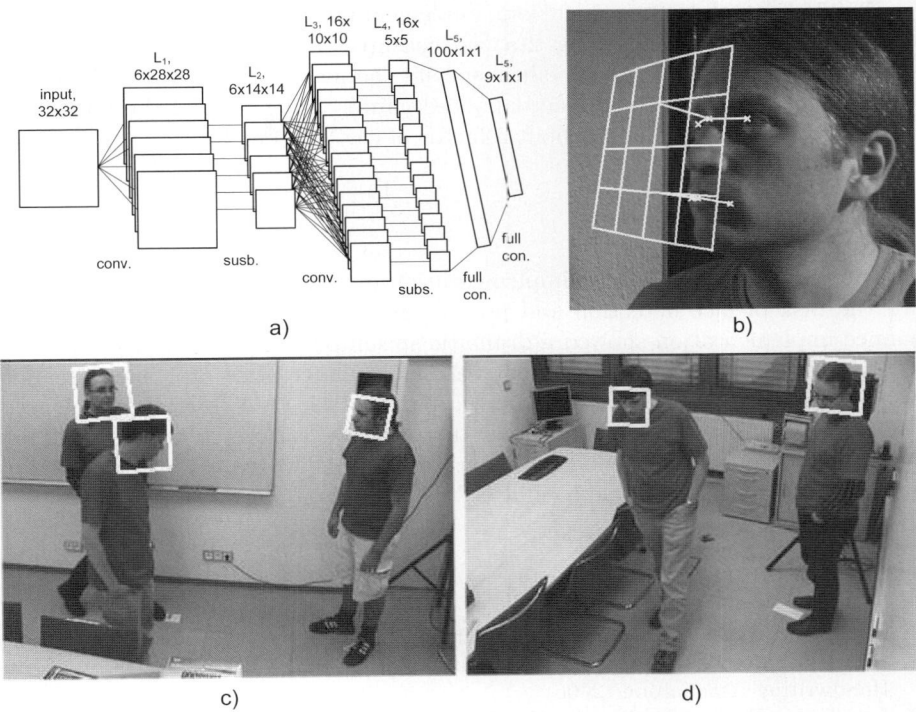

Fig. 6. Face detection system. a) CNN structure. b) annotation example. c) and d) indoor-environment for testing.

		0	1	2	3	4	5	6	7	8
800x600	GPU	318	488	581	619	647	672	695	715	733
	CPU	4123	6165	7047	7554	7761	7858	7878	7885	7888
640x480	GPU	209	312	373	407	434	456	477	497	-
	CPU	2637	3933	4490	4811	4915	4959	4996	5044	-
378x278	GPU	73	109	137	162	184	203	222	-	-
	CPU	851	1276	1470	1541	1557	1575	1594	-	-

Fig. 7. Runtime measurements for CPU and GPU (average milliseconds per frame)

To get an insight of the feasible acceleration by using a graphics card, we compared the parallel method with a corresponding single-CPU implementation (without a specific processor optimization). The GPU (Nvidia GeForce 8800 GT) comes with 14 Multiprocessors each composed of 8 processors with a clock rate about 600 MHz, while the CPU (Intel Pentium 4) comes with a clock rate about 3,4 GHz. Hence, the expected speed gain by full parallelization is about a factor of 19,76. In practice the gain is smaller, because of above mentioned latencies and additional overhead (e.g. loading data into the memory banks).

Fig. 7 shows the runtimes for our implementation. We tested three different image sizes and according to fig. 2(a) we measured the runtime after each scaling step (up to eight). The given values specify the averagely elapsed milliseconds per frame. As we can see the actual speed gain is depending on the image size and number of scaling steps about a factor of ca. 11 up to 13.

6 Conclusions

We presented an parallelized implementation of convolutional neural networks for the task of face detection and pose estimation. The proposed high performance implementation showed a dramatic speedup compared to a conventional CPU based implementation. Given reasonable image sizes and image scaling steps we can expect speed gains about a factor of 11-13. Note that these speed gains are very likely to increase with the next generations of GPUs (since we effectively used an older generation of graphics cards, we expect further speedup using the currently available Nvidia GTX 280 cards).

References

1. Chellapilla, K., Pur, S., Simard, P.: High performance convolutional neural networks for document processing. In: Tenth International Workshop on Frontiers in Handwriting Recognition (2006)
2. LeCun, Y., Bottou, L., Bengio, Y., Haffner, P.: Gradient-based learning applied to document recognition. Proceedings of the IEEE 86(11), 2278–2324 (1998)
3. Osadchy, M., LeCun, Y., Miller, M.: Synergistic face detection and pose estimation with energy-based models. Journal of Machine Learning Research 8(2007), 1197–1215 (2007)
4. Osadchy, R., Miller, M., LeCun, Y.: Synergistic face detection and pose estimation with energy-based models. In: Advances in Neural Information Processing Systems (NIPS 2004), pp. 1197–1215. MIT Press, Cambridge (2005)
5. Rowley, H., Baluja, S., Kanade, T.: Rotation invariant neural network-based face detection. In: 1998 IEEE Computer Society Conference on Computer Vision and Pattern Recognition, Proceedings, pp. 963–963 (1998)
6. Schneiderman, H., Kanade, T.: A statistical method for 3d object detection applied to faces and cars. In: IEEE Conference on Computer Vision and Pattern Recognition, 2000. Proceedings, vol. 1, pp. 746–751 (2000)
7. Sung, K.K., Poggio, T.: Example-based learning for view-based human face detection. IEEE Transactions on Pattern Analysis and Machine Intelligence 20(1), 39–51 (1998)
8. Vaillant, R., Monrocq, C., LeCun, Y.: An original approach for the localisation of objects in images. In: International Conference on Artificial Neural Networks, pp. 26–30 (1993)

Gaussian Weak Classifiers Based on Haar-Like Features with Four Rectangles for Real-time Face Detection

Sri-Kaushik Pavani, David Delgado Gomez, and Alejandro F. Frangi*

Center for Computational Imaging & Simulation Technologies in Biomedicine (CISTIB),
Universitat Pompeu Fabra, Barcelona, Spain
{kaushik.pavani,david.delgado,alejandro.frangi}@upf.edu

Abstract. This paper proposes Gaussian weak classifiers (GWCs) for use in real-time face detection systems. GWCs are based on Haar-like features (HFs) with four rectangles (HF4s), which constitute the majority of the HFs used to train a face detector. To label an image as face or clutter (non-face), GWC uses the responses of the two HF2s in a HF4 to compute a Mahalanobis distance which is later compared to a threshold to make decisions. For a fixed accuracy on the face class, GWCs can classify clutter images with more accuracy than the existing weak classifier types. Our experiments compare the accuracy and speed of the face detectors built with four different weak classifier types: GWCs, Viola & Jones's, Rasolzadeh *et al.*'s and Mita *et al.*'s. On the standard MIT+CMU image database, the GWC-based face detector provided 40% less false positives and required 32% less time for the scanning process when compared to the detector that used Viola & Jones's weak classifiers. When compared to detectors that used Rasolzadeh *et al.*'s and Mita *et al.*'s weak classifiers, the GWC-based detector produced 11% and 9% fewer false positives. Simultaneously, it required 37% and 42% less time for the scanning process.

1 Introduction

Classifiers based on Haar-like features (HFs) [7] have been successful in building face detectors that are both fast and accurate [11]. This is mainly due to the fact that the classifiers based on HFs provide an attractive trade-off between evaluation speed and accuracy. Using a HF-based classifier that takes 60 microprocessor instructions to evaluate, Viola & Jones [11] achieved 1% false negatives and 40% false positives for the face detection problem.

This paper proposes Gaussian weak classifiers (GWCs) as an alternative to the weak classifiers proposed by Viola & Jones [11], Rasolzadeh *et al.* [8] and Mita *et al.* [6]. GWCs are based on Haar-like features with four rectangles (See Fig. 1c), which form the majority of the HFs that are used to train a face detector. GWC, in comparison to existing weak classifiers, are capable of classifying clutter (non-face) images more accurately.

---

* This work was partially funded by grant TEC2006-03617/TCM, from the Spanish Ministry of Innovation & Science, and grants FIT-360000-2006-55 and FIT-360005-2007-9 from the Spanish Ministry of Industry.

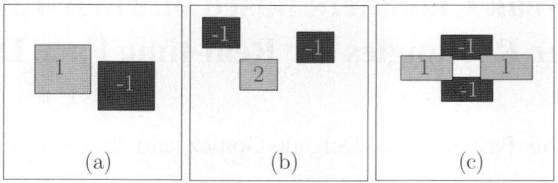

Fig. 1. Examples of Haar-like features with (a) two, (b) three and (c) four rectangles. The numbers within the rectangles indicate the weight assigned to them. Although, in theory, any number of rectangular regions can be used to form a HF, for practical reasons, the number of rectangles used are restricted to two (HF2), three (HF3) or four (HF4). If n is the number of distinct rectangles that can be fit in a template of given size, then the number of HF2s, HF3s, and HF4s that can be constructed is of the order $O(n^2)$, $O(n^3)$, and $O(n^4)$, respectively.

GWCs classify an image as face or clutter in two steps. Firstly, the response of a HF4 is split into two components, each belonging to a HF2. As discussed in Section 4, the motivation behind the split is to take advantage of the compact and Gaussian-like distribution of feature values from the face class images. Secondly, the responses of HF2s are used to compute a Mahalanobis distance which is compared to a threshold to make decisions. In our experiments, we compare the speed and the accuracy of the GWC-based face detector with equivalent detectors that use weak classifiers proposed by Viola & Jones [11], Rasolzadeh et al. [8] and Mita et al. [6]. Our results, presented in Section 5, show that GWC-based face detectors provide the best trade-off between speed and accuracy.

2 Related Work

The seminal paper by Viola & Jones [11] spurred a lot of interest in object detection. Thereafter, several papers were published, mainly focussing on the following three parallel lines of improvements.

1. The geometrical diversity of the HFs was increased to obtain better performance both in terms of accuracy and speed [2][3][5].
2. The AdaBoost [10] procedure used to select weak classifiers in [11] was improved in [3][4][12].
3. The linear weak classifiers used by Viola and Jones were replaced by weak classifiers that provided a better accuracy-speed trade-off. Rasolzadeh et al. [8] demonstrated that more accurate pedestrian detectors can be achieved by increasing the discriminating strength of the individual weak classifiers. Their weak classifiers were obtained through response binning [8], which can be thought of as assigning multiple thresholds to the response of HFs. Viola & Jones, in comparison, use a single threshold on the feature value of the HFs to make decisions. Mita et al. [6] fuse outputs of multiple linear weak classifiers to form more powerful ones. Their weak classifiers produce lower error rates than the Viola and Jones's linear weak classifiers. The GWCs, that are proposed in this paper, fall into the third category of improvements.

3 Weak Classifiers Based on HFs

Haar-like features [7], shown in Fig. 1, consist of two or more rectangular regions enclosed in a template. Such features, when evaluated on an image, produce a feature value as in (1).

$$f_t = \sum_{i=1}^{q} w^{(i)} \cdot \mu^{(i)} \tag{1}$$

where i is an iterator that iterates through all the q rectangles of the HF. The quantity $\mu^{(i)}$ represents the mean intensity of the pixels in image $\mathbf{x}$ enclosed within the ith rectangle. Every rectangle in the HF is assigned a weight that is represented by $w^{(i)}$. The weights are set to *default* integer numbers such that $\sum_{i=1}^{q} w^{(i)} = 0$ is satisfied. For example, the rectangles of a HF2 as in Fig. 1a are assigned default weights 1 and -1. The rectangles of a HF3 as in Fig. 1b are assigned default weights $1, -2$ and 1.

Viola & Jones's weak classifiers ($h_{vj}(\mathbf{x})$) compare the feature value f_t to a threshold θ according to (2).

$$h_{vj}(\mathbf{x}) = \begin{cases} 1, \text{ face}, & f_t \cdot p \leq \theta \cdot p \\ -1, \text{ clutter}, & \text{otherwise} \end{cases} \tag{2}$$

Here, $p \in \{1, -1\}$ is a polarity term, which can be used to invert the inequality relationship between f_t and θ.

Rasolzadeh et al.'s weak classifier [8] ($h_r(\mathbf{x})$) compares f_t to two threshold values (θ_1 and θ_2) as shown in (3).

$$h_r(\mathbf{x}) = \begin{cases} 1, \text{ face}, & \theta_1 \leq f_t \leq \theta_2 \\ -1, \text{ clutter}, & \text{otherwise} \end{cases} \tag{3}$$

Mita et al.'s weak classifier [6] ($h_m(\mathbf{x})$) fuses k Viola and Jones's weak classifiers to make decisions as shown in (4).

$$h_m(\mathbf{x}) = \begin{cases} 1, \text{ face}, & \sum_{i=1}^{k} 2^{k-i} h_{vj}^{(i)}(\mathbf{x}) \geq \theta \\ -1, \text{ clutter}, & \text{otherwise} \end{cases} \tag{4}$$

3.1 Gaussian Weak Classifiers

As stated earlier, we define GWCs using HF4s. A HF4 (Fig. 1c) can be considered to be a combination of two HF2s (Fig. 1a). Therefore, the feature value f_t of a HF4 can be split into two components, $f_1 = \sum_{i=1}^{2} w^{(i)} \cdot \mu^{(i)}$ and $f_2 = \sum_{i=3}^{4} w^{(i)} \cdot \mu^{(i)}$, each belonging to a HF2. Classification by GWC is performed in two steps. Firstly, a Mahalanobis distance d is computed using $\mathbf{f} = [f_1 \; f_2]$ as shown in (5). Secondly, the computed distance is compared to a threshold to make decision on whether the test image belongs to face or clutter as in (6).

Algorithm 1. Computation of $\bar{\mathbf{f}}$, Σ (left), d_t and p (right)

Input: Face class training images: $\mathbf{x}^{(i)}, i \in \{1, \ldots, m\}$	**Input**: Training images (both face and clutter): $\mathbf{x}^{(i)}, i \in \{1, \ldots, m\}$
Input: Weights assigned to images of the face class: $z^{(i)}, i \in \{1, \ldots, m\}$, such that $\sum_{i=1}^{m} z^{(i)} = 1$	**Input**: Training labels: $y^{(i)} \in \{1, -1\}, i \in \{1, \ldots, m\}$
Input: Weak classifier: h	**Input**: Weights for each training image: $z^{(i)} \in \Re, i \in \{1, \ldots, m\}$
begin	**Input**: Weak classifier to be trained: h
Evaluate the HF associated with h on all face class images to get $\mathbf{f}^{(i)} = [f_1^{(i)} \; f_2^{(i)}], i \in \{1, \ldots, m\}$.	**Input**: $\bar{\mathbf{f}}$ and Σ
Compute the weighted mean of feature values: $\bar{\mathbf{f}} = \sum_{i=1}^{m} z^{(i)} \cdot \mathbf{f}^{(i)}$	**begin**
Compute the weighted covariance matrix (Σ): $\Sigma^{(a,b)} = \sum_{i=i}^{m} z^{(i)} (f_a^{(i)} - \bar{f}_a)(f_b^{(i)} - \bar{f}_b)$ where $a, b \in \{1, 2\}$.	Evaluate the HF associated with h on all face class images to get $\mathbf{f}^{(i)} = [f_1^{(i)} \; f_2^{(i)}], i \in \{1, \ldots, m\}$. Compute distance $d^{(i)} \in \mathbf{d}: d^{(i)} = \sqrt[2]{\left(\mathbf{f}^{(i)} - \bar{\mathbf{f}}\right)^T \Sigma^{-1} \left(\mathbf{f}^{(i)} - \bar{\mathbf{f}}\right)}$ Find d_t and p that minimize the training error: $[d_t, p] = \underset{[d_t \in \mathbf{d}, p \in \{-1,1\}]}{\arg\min} \epsilon$ where, $\epsilon = \sum_{i=1}^{m} z^{(i)} |h(\mathbf{x}^{(i)}) - y^{(i)}|$
end	**end**
Output: $\bar{\mathbf{f}}, \Sigma$	**Output**: d_t, p

The quantities $\bar{\mathbf{f}}$ and Σ in (5) are the mean and the covariance matrix obtained from $\mathbf{f}$ computed on a database of face class training images. The computation of these quantities are shown in Algorithm 1 (left). Conceptually, the distance d measures how different a test image is from a mean instance of the images from the face class. In (6), d is compared to a threshold value d_t to decide whether the image belongs to the face or to the clutter class. The quantities d_t and p are determined as shown in Algorithm 1 (right).

$$d = \sqrt[2]{\left(\mathbf{f} - \bar{\mathbf{f}}\right)^T \Sigma^{-1} \left(\mathbf{f} - \bar{\mathbf{f}}\right)} \quad (5)$$

$$h(\mathbf{x}) = \begin{cases} 1, \text{ face}, & d \cdot p \leq d_t \cdot p \\ -1, \text{ clutter}, & \text{otherwise} \end{cases} \quad (6)$$

Comparing (2), (3), (4) to (5) and (6), it can be noted that GWCs are computationally more expensive owing to additional subtraction, matrix multiplication and the square root operation. The question of whether its accuracy is adequate to provide a better accuracy-speed trade-off still remains to be seen.

4 Motivation for Using Gaussian Weak Classifiers

Rasolzadeh *et al.* [8] experimentally observed that the distribution of the feature values of HFs when evaluated on an object and clutter class images resemble a Gaussian distribution. Two-dimensional joint feature spaces spanned by the feature values, f_1 and f_2, from three arbitrarily chosen pairs of HF2s are shown in Fig. 2.

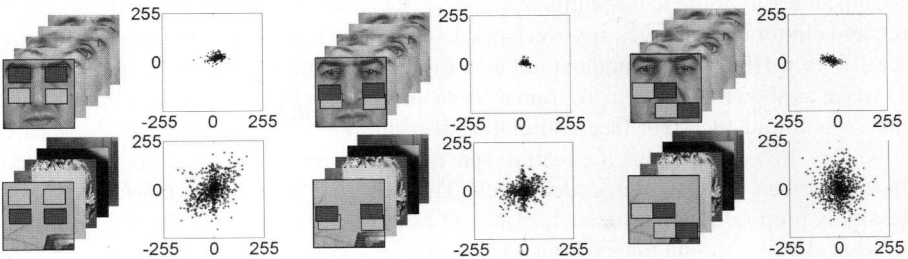

Fig. 2. Joint distribution of feature values, f_1 and f_2, obtained from two HF2s when evaluated on face (top row) and clutter images (bottom row). The HF2s used to generate the plot have been super-imposed on the face and clutter class images. As face class images are correlated with each other, we observe that the distribution of feature values from the face class are more compact than those obtained from the clutter class.

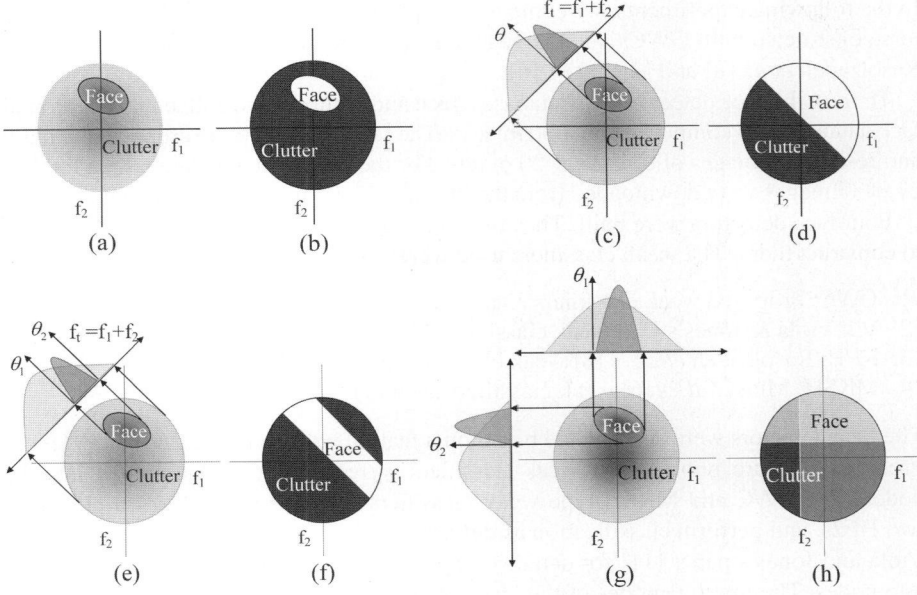

Fig. 3. A geometrical view of the performance of GWC and the weak classifiers used by Viola & Jones [11], Rasolzadeh et al. [8] and Mita et al. [6]. (a) shows a hypothetical joint feature space formed by the feature values f_1 and f_2. (b) shows the partitioned feature space using GWC. (c) shows the projection of 2D space formed by f_1 and f_2 onto a 1D space, which represents the feature value of a HF4 computed in the traditional way. To train a weak classifier used by Viola & Jones, a threshold θ is found to separate face from clutter. (d) shows the corresponding partitioned feature space. (e) shows a geometrical view of the Rasolzadeh et al.'s weak classifiers which use two thresholds to separate face from clutter; the corresponding partitioned feature space is shown in (f). Mita et al.'s weak classifier that fuses two HF2s is illustrated in (g) and (h). The feature values f_1 and f_2 from the two HF2s are used to construct independent weak classifiers whose results are fused to generate the decision space shown in (h).

Consider a hypothetical feature space as in Fig. 3a where the measurements from face and clutter class images are overlapped. Geometrically, the computation of feature value f_t of a HF4, can be understood as a projection of the 2D feature space on to a 1D space as shown in Fig. 3c. To train a Viola and Jones's type weak classifier, a scalar value θ is found such that face and clutter distributions are best separated. As shown in the Fig. 3(d), θ partitions the feature space into two regions; the face region, coded white, and the clutter region, coded black. The decision space of the GWC and weak classifiers proposed by Rasolzadeh et al. and Mita et al. are shown in Figs. 3b, 3f and 3h, respectively. Among the existing weak classifiers, GWC has the potential to extract the maximum discrimination ability from a HF4. As HF4s form the majority of HFs in the feature pool that is used to train the face detector, the classification power of a majority of the weak classifiers can be potentially increased by using GWCs.

5 Results

In the following experiments, we compare the speed and the accuracy of the face detectors constructed with GWCs and the weak classifiers proposed by Viola & Jones [11], Rasolzadeh et al. [8] and Mita et al. [6].

To train the face detector, two databases, face and clutter, were collected. The frontal face database was composed of $5,000$ images. The facial regions were cropped manually and resized to images of size 20×20 pixels. For the clutter image database, a total of $27,000$ images were downloaded from the internet. These images did not contain faces.

Four face detectors were built. They differed only in the type of weak classifier used to construct them. The weak classifiers used were:

1. GWC: Proposed weak classifiers as in (6).
2. VJ: Viola & Jones's [11] weak classifiers as in (2).
3. RPP: Rasolzadeh et al. 's [8] weak classifiers as in (3).
4. MKSH: Mita et al. 's [6] weak classifiers as in (4).

The face detectors were constructed based on a feature pool containing $175,429$ HF4s. From this feature pool, $2,255$ weak-classifiers were selected and arranged into 14 nodes. The GWC and MKSH-type weak classifiers split the response of a HF4 into two HF2s, and perform classification as defined by (6) and (4). Readers are referred to Viola and Jones's paper [11] for details on how the features are selected and arranged into nodes. The first five nodes of the face detectors were assigned one weak classifier each, and the rest were assigned $(nn-5)*50$ weak classifiers. Here, nn stands for node number. Every node of the rejection cascade was trained so that their false rejection rate on a database of face class validation set is at most 0.01.

The accuracy and speed of the face detectors were compared on the MIT+CMU face database [9]. Fig. 4 shows the Receiver Operating Characteristics (ROC) curves obtained by testing the face detectors on this database. Each point on the ROC curve was generated by varying the number of nodes in the face detector. As the false rejection rate of each node was pre-set to a constant value during training, we observe that the face detectors, at each operating point, have similar true positive rates. A rectangular bounding box has been used to group ROC points generated using face detectors working at similar operating points, i.e, with the same number of nodes and weak classifiers.

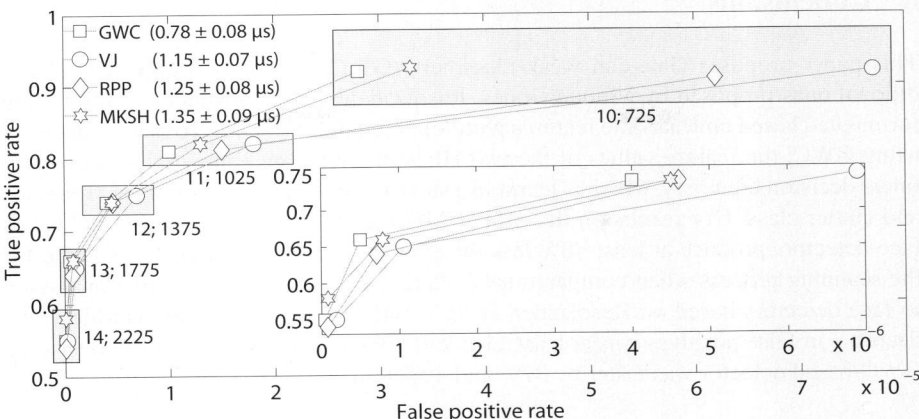

Fig. 4. Comparison of ROC curves of face detectors constructed with GWC, VJ, RPP, and MKSH-type weak classifiers. The detectors were tested at different operating points which were defined by the number of nodes in them. The ROC points generated at equivalent operating points, *i.e*, with same number of nodes and weak classifiers, have been bounded by a rectangular box. The numbers (separated by a semicolon) beside the bounding box indicate the number of nodes and the number of weak classifiers used to build them. The inset shows a zoomed version of the ROC curves for the operating points defined by 12, 13 and 14 nodes. The average time, in microseconds, required to process an image sub-region of the MIT+CMU database (excluding the time required to compute integral images) is listed along with the legend.

During the testing phase, the face detectors perform two tasks: 1) computation of integral image and integral image square (See [11]). In our implementation, the computation of integral images was done using Intel®Integrated Performance Primitives 6.0 [1]. 2) scan through all possible sub-regions of a test image. The time taken to compute the two integral images constituted less than 1% of the total time required to process a 352×288 image at 10 resolutions. The rest of the time, $\sim 99\%$, was spent in scanning the image. The time taken to compute integral images is a common overhead for all four detectors. The scanning time, on the other hand, is dependent on the computational efficiency with which a detector can process clutter images [11], which is dependent on the type of weak classifier used. Assuming that the integral images have been pre-computed, we measured the average time required to label an image sub-region of the MIT+CMU database. The average times are shown along with the legend in Fig. 4.

The GWC-based face detector outperformed the rest significantly both in accuracy and evaluation time. Although GWCs are computationally more expensive to evaluate (See Sec. 3.1), a GWC-based face detector is able to scan through test images faster than those built with the traditional weak classifiers. This is because GWCs, on average, require fewer classifiers to label a clutter image.

6 Conclusions

This paper proposes Gaussian weak classifiers (GWCs) as an alternative to the traditional ones proposed by Viola & Jones, Rasolzadeh et al. and Mita et al. GWCs are formulated based on Haar-like features with four rectangles (HF4s). To make a decision using GWC, the feature values of the two HF2s in a HF4 are compared to a 2D nonlinear decision boundary, which is learnt in a supervised manner using images from face and clutter class. Our results on the MIT+CMU face database show that GWC-based face detectors produce at least 40% less false positives and require 32% less time for the scanning process when compared to Viola & Jones's face detector. In comparison to face detectors based on Rasolzadeh et al.'s and Mita et al.'s weak classifiers, the decrease in false positives was at least 11% and 10% respectively. Simultaneously, the GWC-based detector was faster by 37% and 42% to make decisions.

References

1. Intel® Integrated Performance Primitives 6.0 Home Page (April 9, 2009), http://software.intel.com/en-us/intel-ipp/
2. Jones, M.J., Viola, P.: Fast multi-view face detection. Technical Report MERL-TR2003-96, Mitsubishi Electric Research Laboratories (2003)
3. Li, S.Z., Zhu, L., Zhang, Z., Blake, A., Zhang, H., Shum, H.: Statistical learning of multi-view face detection. In: Heyden, A., Sparr, G., Nielsen, M., Johansen, P. (eds.) ECCV 2002. LNCS, vol. 2353, pp. 67–81. Springer, Heidelberg (2002)
4. Lienhart, R., Kuranov, E., Pisarevsky, V.: Empirical analysis of detection cascades of boosted classifiers for rapid object detection. In: PRS 2003, pp. 297–304 (2003)
5. Lienhart, R., Maydt, J.: An extended set of Haar-like features for rapid object detection. In: ICIP 2002, pp. 900–903 (2002)
6. Mita, T., Kaneko, T., Stenger, B., Hori, O.: Discriminative feature co-occurrence selection for object detection. IEEE TPAMI 30(7), 1257–1269 (2008)
7. Papageorgiou, C.P., Oren, M., Poggio, T.: A general framework for object detection. In: ICCV 1998, pp. 555–562 (1998)
8. Rasolzadeh, B., Petersson, L., Pettersson, N.: Response binning: Improved weak classifiers for boosting. In: IV 2006, pp. 344–349 (2006)
9. Rowley, H.A., Baluja, S., Kanade, T.: Neural Network-Based Face Detection. IEEE TPAMI 20(1), 23–38 (1998)
10. Schapire, R.E.: A brief introduction to boosting. In: IJCAI 1999, pp. 1401–1406 (1999)
11. Viola, P., Jones, M.J.: Rapid object detection using a boosted cascade of simple features. In: CVPR 2001, pp. 511–518 (2001)
12. Viola, P., Jones, M.J.: Fast and robust classification using asymmetric adaboost and a detector cascade. In: NIPS 2002, pp. 1311–1318 (2002)

Model Based Analysis of Face Images for Facial Feature Extraction

Zahid Riaz, Christoph Mayer, Michael Beetz, and Bernd Radig

Technische Universität München,
Boltzmannstr. 3, 85748 Garching, Germany
{riaz,mayerc,beetz,radig}@in.tum.de
http://www9.in.tum.de

Abstract. This paper describes a comprehensive approach to extract a common feature set from the image sequences. We use simple features which are easily extracted from a 3D wireframe model and efficiently used for different applications on a benchmark database. Features verstality is experimented on facial expressions recognition, face reognition and gender classification. We experiment different combinations of the features and find reasonable results with a combined features approach which contain structural, textural and temporal variations. The idea follows in fitting a model to human face images and extracting shape and texture information. We parametrize these extracted information from the image sequences using active appearance model (AAM) approach. We further compute temporal parameters using optical flow to consider local feature variations. Finally we combine these parameters to form a feature vector for all the images in our database. These features are then experimented with binary decision tree (BDT) and Bayesian Network (BN) for classification. We evaluated our results on image sequences of Cohn Kanade Facial Expression Database (CKFED). The proposed system produced very promising recognition rates for our applications with same set of features and classifiers. The system is also realtime capable and automatic.

Keywords: Feature Extraction, Face Image Analysis, Face Recognition, Facial Expressions Recognition, Human Robot Interaction.

1 Introduction

In the recent decade model based image analysis of human faces has become a challenging field due to its capability to deal with the real world scenarios. Further it outperforms the previous techniques which were constrained to user intervention with the system either to manually interact with system or to be frontal to the camera. Currently available model based techniques are trying to deal with some of the future challenges like developing state-of-the-art algorithms, improving efficiency, fully automated system development and verstality under different applications. In this paper we deal with some of these challenges. We focus on feature extraction technique which is fully automatic and verstile enough for different applications like face recognition, facial expressions recognition and gender classification. These capabilities of the system suggest to apply it in interactive secnarios like human machine interaction, security of personalized utilities like tokenless devices, facial analysis for person behavior and person security.

Model-based image interpretation techniques extract information about facial expression, person identitiy and gender classification from images of human faces via facial changes. Models take benefit of the prior knowledge of the object shape and hence try to match themselves with the object in an image for which they are designed. Face models impose knowledge about human faces and reduce high dimensional image data to a small number of expressive model parameters. We integrate the three-dimensional Candide-3 face model [8] that has been specifically designed for observing facial features variations defined by facial action coding system (FACS) [13]. The model parameters together with extracted texture and motion information is utilized to train classifiers that determine person-specific information. A combination of different facial features is used for classifiers to classify six basic facial expressoins i.e. anger, fear, surprise, saddness, laugh and disgust, facial identity and gender classification.

Our feature vector for each image consists of structral, textural and temporal variations of the faces in the image sequence. Shape and textural parameters define active appearance models (AAM) in partial 3D space with shape parameters extracted from 3D landmarks and texture from 2D image. Temporal features are extracted using optical flow. These extracted features are more informative than AAM parameters since we consider local motion patterns in the image sequences in the form temporal parameters.

The remainder of this paper is divided in four main sections. In section 2, related work to our applications is discussed. In section 3 we discuss our approach in detail. In section 4 higher level features extraction from model based image interpretation is described. This includes description from model fitting to face image to feature vector formation. Section 5 discusses about evaluation of our results on the database. Finally we conclude our results with some future directions.

2 Related Work

We initiate with a three step approach that has been suggested by Pantic et al. [1] for facial expression recognition. However, the generality of this approach makes it applicable not only to facial expression estimation but also to apply it for person identification and gender classification at the same time. The first step aims at determining the position and shape of the face in the image by fitting a model. Descriptive features are extracted in the second step. In the third step a classifier is applied to the features to determine high level information from the features. Several face models and fitting approaches have been presented in the recent years. Cootes et al. [5] introduced modeling face shapes with Active Contours. Further enhancements included the idea of expanding shape models with texture information [6]. In contrast, three-dimensional shape models such as the Candide-3 face model considers the real-world face structure rather than the appearance in the image. Blanz et al. propose a face model that considers both, the three-dimensional structur as well as its texture [7]. However, model parameters that describe the current image content need to be determined in order to extract high-level information, a process known as model fitting. In order to fit a model to an image. Van Ginneken et al. learned local objective functions from annotated training images [18]. In this work, image features are obtained by approximating the pixel values in a region around a pixel of interest The learning algorithm use to map images features to objective values is a k-Nearest-Neighbor classifier (kNN) learned from the data. We used similar methodology developed by Wimmer et al. [4] which combines multitude of qualitatively different features [19], determines the most relevant features using machine learning and learns objective functions from annotated images [18]. To extract discriptive features from the image, Michel et al. [14] extracted the location of 22 feature points within the face and determine their motion between an image that shows the neutral state of the face and an image that represents a facial expression. The very similar approach of Cohn et al. [15] uses hierarchical optical flow in order to determine the motion of 30 feature points. A set of training data formed from the extracted features is utilized to learn on a classifier. For facial expressions, some approaches infer the expressions from rules stated by Ekman and Friesen [13]. This approach is applied by Kotsia et al. [16] to design Support Vector Machines (SVM) for classification. Michel et al. [14] train a Support Vector Machine (SVM) that determines the visible facial expression within the video sequences of the Cohn-Kanade Facial Expression Database by comparing the first frame with the neutral expression to the last frame with the peak expression. In order to perform face recognition applications many researchers have applied model based approaches. Edwards et al [2] use weighted distance classifier called Mahalanobis distance measure for AAM parameters. However, they isolate the sources of variation by maximizing the inter class variations using Linear Discriminant Analysis (LDA), a holistic approach which was used for Fisherfaces representation [3]. However they do not discuss face recognition under facial expression. Riaz et al [17] apply similar features for explaining face recognition using bayesian networks. However results are limited to face recognition application only. They used expression invariant technique for face recognition, which is also used in 3D scenarios by Bronstein et al [9] without 3D reconstruction of the faces and using geodesic distance. Park et. al. [10] apply 3D model for face recognition on videos from CMU Face in Action (FIA) database. They reconstruct a 3D model acquiring views from 2D model fitting to the images.

3 Our Approach

In this section we explain in detail the approach adopted in this paper including model fitting, image warping and parameters extraction for shape, texture and temporal information.

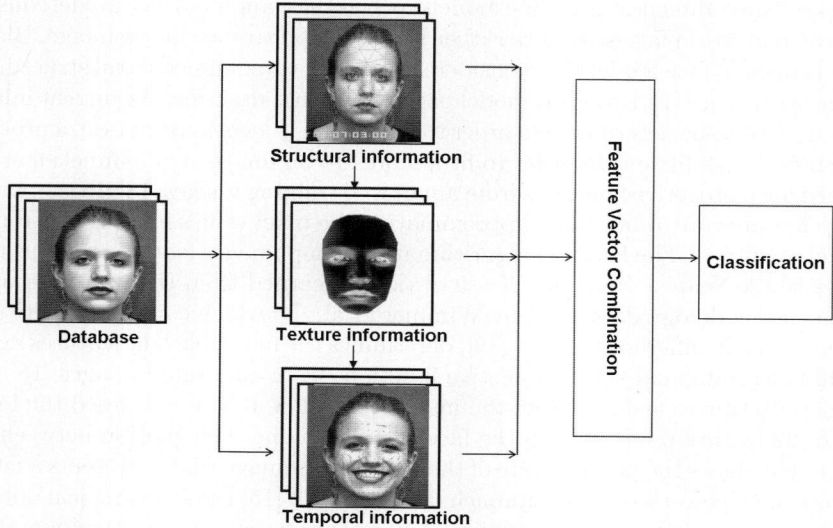

Fig. 1. Our Approach: Sequential flow for feature extraction

We use a wireframe 3D face model known as candide-III [8]. The model is fitted to the face image using objective function approach [4]. After fitting the model to the example face image, we use the projections of the 3D landmarks in 2D for texture mapping. Texture information is mapped from the example image to a reference shape which is the mean shape of all the shapes available in database. However the choice of mean shape is arbitrary. Image texture is extracted using planar subdivisions of the reference and the example shapes. We use delauny triangulations of the distribution of our model points. Texture warping between the trigulations is performed using affine transformation. Principal Component Analysis (PCA) is used to obtain the texture and shape parameters of the example image. This approach is similar to extracting AAM parameters. In addition to AAM parameters, temporal features of the facial changes are also calculated. Local motion of the feature points is observed using optical flow. We use reduced descriptors by trading off between accuracy and run time performance. These features are then used for classification. Our approach achieves real-time performance and provides robustness against facial expressions in real-world scenarios. This computer vision task comprises of various phases shown in Figure 1 for which it exploits model-based techniques that accurately localize facial features, seamlessly track them through image sequences, and finally infer facial features. We specifically adapt state-of-the-art techniques to each of these challenging phases.

4 Determining High-Level Information

In order to initialize, We apply the algorithm of Viola et al. [20] to roughly detect the face position within the image. Then, model parameters are estimated by applying the approach of Wimmer et al. [4] because it is able to robustly determine model parameters in real-time.

To extract descriptive features, the model parameters are exploited. The model configuration represents information about various facial features, such as lips, eye brows or eyes and therefore contributes to the extracted features. These structural features include both, information about the person's face structure that helps to determine person-specific information such as gender or identity. Furthermore, changes in these features indicates shape changes and therefore contributes to the recognition of facial expressions.

The shape x is parametrized by using mean shape x_m and matrix of eigenvectors P_s to obtain the parameter vector b_s [11].

$$x = x_m + P_s b_s \qquad (1)$$

The extracted texture is parametrized using PCA by using mean texture g_m and matrix of eigenvectors P_g to obtain the parameter vector b_g [11]. Figure 2 shows shape model fitting and texture extracted from face image.

$$g = g_m + P_g b_g \qquad (2)$$

Further, temporal features of the facial changes are also calculated that take movement over time into consideration. Local motion of feature points is observed using optical flow. We do not specify the location of these feature points manually but distribute equally in the whole face region. The number of feature points is chosen in a way that the system is still capable of performing in real time and therefore inherits a trade off between accuracy and runtime performance. Figure 3 shows motion patterns for some of the images from database.

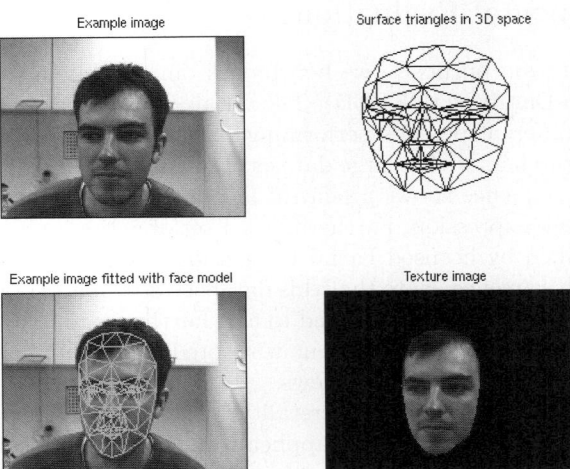

Fig. 2. Texture information is represented by an appearance model. Model parameters of the fitted model are extracted to represent single image information.

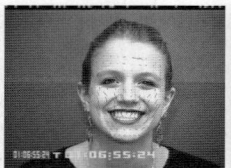 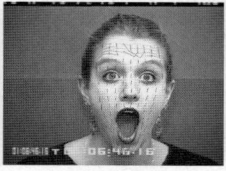  

Fig. 3. Motion patterns within the image are extracted and the temporal features are calculated from them. These features are descriptive for a sequence of images rather than single images.

We combine all extracted features into a single feature vector. Single image information is considered by the structural and textural features whereas image sequence information is considered by the temporal features. The overall feature vector becomes:

$$u = (b_{s1}, ..., b_{sm}, b_{g_1}, ..., b_{g_n}, b_{t1}, ..., b_{tp},) \tag{3}$$

Where b_s, b_g and b_t are shape, textural and temporal parameters respectively.

We extract 85 structural features, 74 textural features and 12 temporal features textural parameters to form a combined feature vector for each image. These features are then used for binary decision tree (BDT) and bayesian network (BN) for different classifications. The face feature vector consists of the shape, texture and temporal variations, which sufficiently defines global and local variations of the face. All the subjects in the database are labeled for classification.

5 Experimental Evaluations

For experimentation purposes, we benchmark our results on Cohn Kanade Facial Expression Database (CKFED). The database contains 488 short image sequences of 97 different persons performing six universal facial expressions [12]. It provides researchers with a large dataset for experimenting and benchmarking purpose. Each sequence shows a neutral face at the beginning and then develops into the peak expression. Furthermore, a set of action units (AUs) has been manually specified by licensed Facial Expressions Coding System (FACS) [13] experts for each sequence. Note that this database does not contain natural facial expressions, but volunteers were asked to act. Furthermore, the image sequences are taken in a laboratory environment with predefined illumination conditions, solid background and frontal face views.

In order to experiment feature verstality we use two different classifiers with same feature set on three different applications: face recognition, facial expressions recognition and gender classification. The results are evaluated using classifiers from weka [21] with 10-fold cross validation. Table 1 shows different recognition rates achieved during experimentations. In all three cases BDT outperforms BN.

Table 1. Comparison of Extracted Features

	BDT	BN
Face Recognition	98.49%	90.66%
Facial Expressions Recognition	85.70%	80.57%
Gender Classification	99.08%	89.70%

6 Conclusions

We introduced a technique to develop a set of feature vectors which consist of three types of facial information. The features set is applied to three different applications: face recognition, facial expressions recognition and gender classification, which produced the reasonable results in all three cases for CKFED. We consider different classifiers for checking the versatility of our extracted features. We use two different classifiers with same specifications which evidence simplicity of our approach however, the results can be further optimized by trying other classifiers. The database consists of frontal views with uniform illuminations. Since the algorithm is working in real time, hence it is suitable to apply it in real time environment keeping in consideration the limitation of database. Further extensions of this work is to enhance the feature sets to include information about pose and lighting variations.

References

1. Pantic, M., Rothkrantz, L.J.M.: Automatic analysis of facial expressions: The state of the art. IEEE Transactions on Pattern Analysis and Machine Intelligence 22(12), 1424–1445 (2000)
2. Edwards, G.J., Cootes, T.F., Taylor, C.J.: Face Recognition Using Active Appearance Models. In: Burkhardt, H., Neumann, B. (eds.) ECCV 1998. LNCS, vol. 1407, pp. 581–695. Springer, Heidelberg (1998)
3. Belheumeur, P.N., Hespanha, J.P., Kreigman, D.J.: Eigenfaces vs Fisherfaces: Recognition using Class Specific Linear Projection. IEEE Transaction on Pattern Analysis and Machine Intelligence 19(7) (July 1997)
4. Wimmer, M., Stulp, F., Tschechne, S., Radig, B.: Learning Robust Objective Functions for Model Fitting in Image Understanding Applications. In: Proceedings of the 17th British Machine Vision Conference, BMVA, Edinburgh, UK, pp. 1159–1168 (2006)
5. Cootes, T.F., Taylor, C.J.: Active shape models – smart snakes. In: Proceedings of the 3rd British Machine Vision Conference, pp. 266–275. Springer, Heidelberg (1992)
6. Cootes, T.F., Edwards, G.J., Taylor, C.J.: Active Appearance Models. In: Burkhardt, H., Neumann, B. (eds.) ECCV 1998. LNCS, vol. 1407, pp. 484–498. Springer, Heidelberg (1998)
7. Blanz, V., Vetter, T.: Face Recognition Based on Fitting a 3D Morphable Model. IEEE Transactions on Pattern Analysis and Machine Intelligence 25(9), 1063–1074 (2003)

8. Ahlberg, J.: An Experiment on 3D Face Model Adaptation using the Active Appearance Algorithm. Image Coding Group, Dept. of Electric Engineering Linköping University
9. Bronstein, A., Bronstein, M., Kimmel, M., Spira, A.: 3D face recognition without facial surface reconstruction. In: Proceedings of European Conference of Computer Vision, Prague, Czech Republic, May 11-14 (2004)
10. Park, U., Jain, A.K.: 3D Model-Based Face Recognition in Video. In: 2nd International Conference on Biometrics, Seoul, Korea (2007)
11. Li, S.Z., Jain, A.K.: Handbook of Face recognition. Springer, Heidelberg (2005)
12. Kanade, T., Cohn, J.F., Tian, Y.: Comprehensive database for facial expression analysis. In: Proceedings of Fourth IEEE International Conference on Automatic Face and Gesture Recognition (FGR 2000), Grenoble, France, pp. 46–53 (2000)
13. Ekman, P., Friesen, W.: The Facial Action Coding System: A Technique for The Measurement of Facial Movement. Consulting Psychologists Press, San Francisco (1978)
14. Michel, P., Kaliouby, R.E.: Real time facial expression recognition in video using support vector machines. In: Fifth International Conference on Multimodal Interfaces, Vancouver, pp. 258–264 (2003)
15. Cohn, J., Zlochower, A., Lien, J.-J.J., Kanade, T.: Feature-point tracking by optical flow discriminates subtle differences in facial expression. In: Proceedings of the 3rd IEEE International Conference on Automatic Face and Gesture Recognition, April 1998, pp. 396–401 (1998)
16. Kotsia, I., Pitaa, I.: Facial expression recognition in image sequences using geometric deformation features and support vector machines. IEEE Transaction on Image Processing 16(1) (2007)
17. Riaz, Z., et al.: A Model Based Approach for Expression Invariant Face Recognition. In: 3rd International Conference on Biometrics, Italy (June 2009)
18. Ginneken, B., Frangi, A., Staal, J., Haar, B., Viergever, R.: Active shape model segmentation with optimal features. IEEE Transactions on Medical Imaging 21(8), 924–933 (2002)
19. Romdhani, S.: Face Image Analysis using a Multiple Feature Fitting Strategy. PhD thesis, University of Basel, Computer Science Department, Basel, CH (January 2005)
20. Viola, P., Jones, M.J.: Robust real-time face detection. International Journal of Computer Vision 57(2), 137–154 (2004)
21. Witten, I.H., Frank, E.: Data Mining: Practical machine learning tools and techniques, 2nd edn. Morgan Kaufmann, San Francisco (2005)

Dynamics Analysis of Facial Expressions for Person Identification

Hidenori Tanaka and Hideo Saito

Graduate School of Science and Technology, Keio University
3-14-1, Hiyoshi, Kouhoku-ku, Yokohama, Kanagawa, 223-8522 Japan
hidenori@hvrl.ics.keio.ac.jp

Abstract. We propose a new method for analyzing the dynamics of facial expressions to identify persons using Active Appearance Models and accurate facial feature point tracking. Several methods have been proposed to identify persons using facial images. In most methods, variations in facial expressions are one trouble factor. However, the dynamics of facial expressions are one measure of personal characteristics. In the proposed method, facial feature points are automatically extracted using Active Appearance Models in the first frame of each video. They are then tracked using the Lucas-Kanade based feature point tracking method. Next, a temporal interval is extracted from the beginning time to the ending time of facial expression changes. Finally, a feature vector is obtained. In the identification phase, an input feature vector is classified by calculating the distance between the input vector and the training vectors using dynamic programming matching. We show the effectiveness of the proposed method using smile videos from the MMI Facial Expression Database.

Keywords: facial expression analysis, AAMs, LK-based feature point tracking, DP matching, person identification.

1 Introduction

Facial expression analysis is utilized in man-machine interfaces such as human-robot interactions. Most previous research in this field has tried to classify facial expressions into fundamental categories based on emotions. However, facial expressions contain not only expressions of emotions but also individual differences over time [1]. In this paper, we focus on the individual differences and propose a new method for analyzing the dynamics of facial expressions to identify persons.

To achieve biometric identification services, we consider that various physical features have to be fused and that facial expression is one of them. Since facial expression might not provide enough discriminating power, this research is considered as a type of soft biometrics. Figure 1 shows an example of using facial expressions for person identification service at a high-class membership club. At the entrance, a robot approaches the members and communicates with them while the identification process is performed.

In most person identification methods using facial images, the variations in facial expressions are one of the factors that lower the discriminating power.

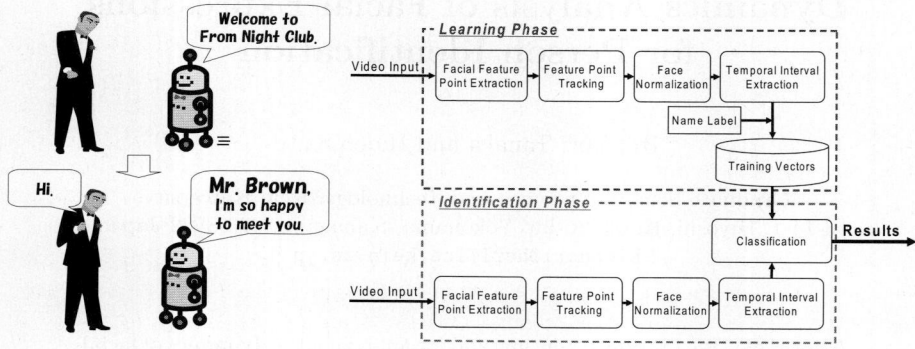

Fig. 1. Person identification service at a high class membership club

Fig. 2. Flow of proposed method

A number of methods [2, 3, 4] have been developed to address this problem. In the case of facial expression videos, however, person identification can be performed by using individual differences behind the facial expressions.

Previous research in the area of person identification using facial expressions is very limited. Ning et al. [5] generated features by summing up the flow fields over time using smile videos. In this method, the dynamics of facial expressions were not well described because features were generated by summing up the flow fields over time. Also, optical flow was calculated for whole facial images and did not accurately show the background regions of the images. This lowers the discriminating power. Further, facial feature points were manually extracted to normalize the facial images. Chen et al. [6] constructed a high-dimensional feature vector that concatenated a sequence of motion flow fields using videos of persons speaking. In this method, the dynamics of facial expressions were described. However, optical flow was calculated for whole facial images and facial feature points were manually extracted.

In contrast, we propose a novel method that analyzes the dynamics of facial expressions to identify persons using Active Appearance Models (AAMs) and accurate facial feature point tracking. The next section provides an overview of the proposed method and details its facial feature point extraction, feature point tracking, facial normalization, temporal interval extraction and identification processes. Section 3 demonstrates the method's effectiveness using smile videos from a published facial expression database. In Section 4, we offer conclusions pertaining to our work.

2 Proposed Method

The proposed method consists of two phases: "the learning phase," and "the identification phase." Figure 2 shows the flow of the method. In the learning phase, facial feature points (eyebrow, eye, nose, mouth, and facial contour parts) are automatically extracted using AAMs in the first frame of the facial expression videos. The feature points are then tracked using the Lucas-Kanade based feature point tracking method (LK-based feature point tracking). Each facial image is normalized using three facial feature points to account for the variations in the object's head pose and those in the distance from face to camera. Temporal intervals are extracted from the difference between the feature points' position in the current frame and that in the previous frame. A feature vector is also generated and stored with the name label. In the identification phase, an input feature vector is generated as in the learning phase and classified by calculating the distance between the input vector and the training vectors using dynamic programming (DP) matching. The next subsections detail each process.

2.1 Facial Feature Point Extraction

To extract facial feature points in the first frame of each video, we use AAMs [7]. AAMs are generative and parametric models of a certain visual phenomenon that shows both shape and grey-level appearance variations. These variations are represented by a linear model.

Initially, grey-level variance independent from shape variance is needed for learning the correlation between shape and grey-level. The training data for AAMs is a set of images and coordinate values of feature points on the images. A shape vector s is composed of coordinate values on feature points. A grey-level vector g is composed of intensity values in a warped image, which is obtained by extracting the face region from an image along its feature points and normalizing its shape into a mean shape $\bar{s}$ of the normalized shapes.

Next, the distribution and correlation between shapes and grey-level is calculated. Principal Component Analysis (PCA) is performed on a set of shape vectors s and grey-level vectors g in training data.

$$s = \bar{s} + P_s c_s, \ g = \bar{g} + P_g c_g \tag{1}$$

where $\bar{s}$ is a mean vector of s, $\bar{g}$ is a mean vector of g, P_s and P_g are orthogonal matrixes where each column vector is a base vector, and c_s and c_g are coefficients of the base vector. Since there may be correlations between the shape and grey-level variations, PCA is performed again. If an input image and the training model are given, we can treat facial feature point extraction as an optimization problem in which we minimize the grey-level difference between an input image and a synthesized image using the parameter vector d^*.

$$d^* = \arg\min_d \mid g_i - g_m \mid^2 \tag{2}$$

where d is a parameter vector controlling both the shape and greylevels of the model, g_i is a warped input image and g_m is a synthesized image. $\mid g_i - g_m \mid$ are iteratively minimized and we get the optimization result d^*. From the vector d^*, the shape vector of an input image is obtained. Figure 3 shows an example of a training image. We put 65 feature points on the image.

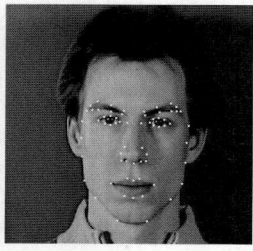

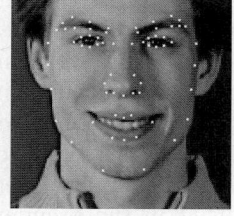

(a) AAM tracking (b) LK-based tracking

Fig. 3. An example of a training image for AAMs

Fig. 4. Comparison of facial feature point tracking results

2.2 Feature Point Tracking

In each video, we need to track the facial feature points that are extracted in the first frame using AAMs. AAM tracking (part of AAM-API) is a training-based feature point tracking method in which it is necessary to make many training images to track the movement of facial feature points accurately. It is also known that AAM tracking based on a large number of datasets has difficulties in tracking accuracy because of local minimums [8]. In case we make training images consisting of a neutral face, the feature point tracking fails (around the mouth) as shown in Fig.4 (a). In our method, we use the feature point tracking method developed by Lien [9], which we call "LK-based feature point tracking," to track the feature points accurately. The goal of feature point tracking is to find the best matching positions between an N x N window R in the t frame and those in the $t + 1$ frame that minimize the cost function E of the weighted sum of squared differences (SSD) as follows:

$$E(\epsilon_x) = \sum_{x \in R} \omega_x \cdot [I_t(x) - I_{t+1}(x - \epsilon_x)]^2 \qquad (3)$$

where $I_t(x)$ denotes the grey value of the pixel position x in the t frame, ϵ_x is the motion vector of x between two consecutive frames, and ω_x is a window function for weighting the squared differences in E, which are defined by LK-based weight. Here, this weight is empirically determined. In our method, the feature point tracking is robustly performed against illumination variations because we use a part of the facial edge (facial feature points). The feature point tracking result using the LK-based feature point tracking method is shown in Fig.4 (b) . From this figure, we can see that the feature points are accurately tracked when the facial expression is changed.

2.3 Facial Normalization

To account for the object's head pose movements and the different distances from the object's face to the camera, each facial image is normalized. In our method, we use three facial feature points (two inner canthi and a philtrum) to align each facial image. These three points are extracted in the first frame of each video using AAMs and then tracked using the LK-based feature point tracking method. These three points are then moved into the aligned positions. Facial images are aligned by 2D affine transformation with respect to these three points. The inner canthi and philtrum of all aligned facial images have the same coordinate values. After alignment, facial images are cropped and resized. In our experiment, the resized image size is 128 x 128 pixels.

2.4 Temporal Interval Extraction

To determine the frames that encompass the duration of facial expression changes, the starting and ending frames are extracted in each video. Frames that do not contain any facial expression changes will be abandoned because these frames will lower the discriminating power. In our method, the starting and ending frames are extracted by differences in the facial feature points between two successive frames as follows:

$$F(x^t) = \mid f(x^t) - f(x^{t-1}) \mid, \ f(x^t) = \sum_{i=1}^{K} \mid x_i^t - x_i^0 \mid \qquad (4)$$

$$\begin{pmatrix} t_s = t & \text{if } F(x^t) > Th_{period} \\ t_e = t & \text{if } t > t_s \text{ and } F(x^t) < Th_{period} \end{pmatrix}$$

where $F(x^t)$ denotes the differences between the coordinate values of the facial feature points in the t frame and those in the $t-1$ frame, K is the total number of facial feature points, x_i^t denotes the coordinate values of the ith feature points in the t frame, t_s and t_e are the starting and ending frames, and Th_{period} is the threshold value. Here, Th_{period} is empirically determined depending on the experimental environment because Th_{period} is mainly affected by the illumination variations in the experimental environment. In our experiment, facial expression changes always start from a neutral face.

After this process, we can obtain a feature vector. It consists of the 2D coordinate sequence of facial feature points and the vector dimensions total 2D coordinate x 65 points x $(t_e\text{-}t_s)$ frames.

2.5 Identification

Identification is performed by classifying an input feature vector. Because temporal intervals of facial expression changes will not be the same between different individuals and will not be constant at all times even for the same person, we need to absorb the variations. There are many matching algorithms to compare the patterns whose temporal intervals are different. In our method, we use conventional dynamic programming (DP) matching to compare an input feature vector with the training feature vectors. In detail, the distance $G(i, j)$ between an input feature vector $A = (a_1, a_2, \ldots, a_i, \ldots, a_{T_1})$ and the training feature vector $B = (b_1, b_2, \ldots, b_j, \ldots, b_{T_2})$ is calculated as follows:

$$G(i,j) = \min \begin{pmatrix} G(i-1,j) + D(i,j) \\ G(i-1,j-1) + 2D(i,j) \\ G(i,j-1) + D(i,j) \end{pmatrix} \quad (5)$$

where $D(i,j)$ denotes the Euclidian distance between a_i and b_j. The calculated distance is normalized by the length of the input vector and the training vector $(T_1 + T_2)$ and the DP distance is obtained. In identification, the input vector is classified by the threshold value.

3 Experiments

To show the effectiveness of our method, we conducted experiments using smile videos from the MMI Facial Expression Database [10]. In our experiments, the resolution of the videos is 720 x 576 pixels and the frame rate is 25 frames/second. Facial expression changes start with a neutral face, move to a smile, and then go back again to the neutral face. We selected 48 smile videos (12 persons, 4 videos/person) from the database to evaluate. All videos were very nearly frontal facial images, and so ideally suit our facial normalization method.

We first evaluate the discriminating power of our method using all facial feature points, comparing them with the previous method. Then, we evaluate the discriminating power of our method for each facial part. For evaluation purposes, we considered that one video was for test data and that the other videos were for training data. To evaluate the discriminating power, we used the equal error rate (EER) and the recognition rate (RR). EER is the probability that the false acceptance rate (FAR) and the false rejection rate (FRR) are equal. In general, the discriminating power is high when EER is lower and RR is higher.

3.1 Discriminating Power of a Whole Face

In this experiment, we first show the discriminating power of our method with all facial feature points (65 points) using smile videos. Figure 5 shows the tracking results of three persons in smile videos. From this figure, we can see individual differences in smile dynamics. From the evaluation results, the EER value was 14.0% and the RR value was 92.5%. This shows that smile dynamics represented by our method have high discriminating power.

To compare the discriminating power of our method with that of the previous method, we applied the optical-flow based method to the same datasets. In this experiment, each facial image was sampled into 64 points evenly spaced over a whole facial image for computing the optical flow field, and optical-flow was calculated against the points. A feature vector was obtained after temporal interval extraction. The identification was performed in the same way as in our method. From the evaluation results, the EER value was 36.8% and the RR value was 62.3%. These results show that the discriminating power of our method is higher than that of the optical-flow based method. Here, the factor that lowers the discriminating power in the optical-flow based method was some

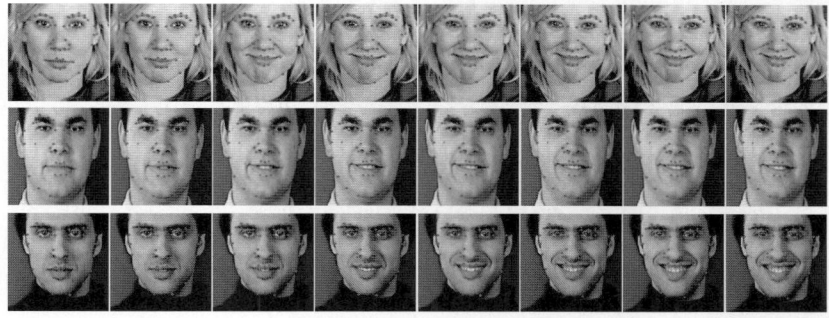

time(frame)

Fig. 5. Tracking results of three persons in smile videos (Every three frames from a neutral face to a smile face)

unexpected flows found in the background regions of the facial image. In contrast, in our method, we extract the facial feature points of the facial image and the background regions of the image do not affect the tracking process.

3.2 Discriminating Power in Each Facial Part

In this experiment, we show the discriminating power of our method in each facial part (eyebrow, eye, nose, mouth, and facial contour parts) using smile videos. We also show the differences in the discriminating power by temporal interval extraction because temporal interval plays an important role in the dynamics of facial expressions. First, we use the same temporal interval as in the previous experiments (temporal interval extraction by all facial feature points). Table 1 shows EER and RR results obtained for each facial part. From this table, we can see that the EER value of the eyebrow part is lowest and the RR value of the eyebrow part is highest. From this result, it can be said that the eyebrow part had higher discriminating power than the other parts in these datasets. On the other hand, we can also see that the EER values of the eye part and those of the mouth part are higher and that the RR values of the eye part and those of the mouth part are lower. From this result, it can be said that the eye and mouth parts have less discriminating power than the other parts, while wrinkling around the corners of the eyes and a rising in the corners of the mouth are characteristic movements in smiles.

Next, we use the temporal interval extracted by each facial feature point. Table 2 shows EER and RR results obtained for each facial part. From this table, we can see that the EER values of all parts in Table 2 are higher than those in Table 1 and the RR values of all parts in Table 2 are mostly lower than those in Table 1. From this result, it can be said that temporal interval extraction by all facial feature points had higher discriminating power than that by each facial

Table 1. EER and RR in each facial part (Temporal interval is extracted by all facial feature points)

facial part	EER[%]	RR[%]
eyebrow	15.2	88.7
eye	24.4	71.7
nose	21.7	70.7
mouth	24.7	64.2
facial contour	17.0	81.1

Table 2. EER and RR in each facial part (Temporal interval is extracted by each facial feature points)

facial part	EER[%]	RR[%]
eyebrow	17.3	77.4
eye	31.0	67.9
nose	22.5	81.1
mouth	27.6	62.3
facial contour	19.0	67.0

feature point. In other words, the combination of facial parts' movements over time had individual differences.

4 Conclusions

In this paper, we proposed a method for analyzing the dynamics of facial expressions to identify persons. We automatically extracted facial feature points and accurately tracked them. We evaluated the discriminating power of our method using 48 smile videos from the MMI Facial Expression Database. The evaluation results showed that the EER value was 14.0% and the RR value was 92.5% and the discriminating power of our method was higher than that of the previous method. We also found that the eyebrow part had higher discriminating power than the other parts of the face and that the eye and mouth parts had less discriminating power than the other parts even though these parts are characteristic parts in smiles. Further, the combination of facial parts' movements over time had individual differences.

In future work, we plan to generate a feature vector while considering the appearance of facial images and evaluate our method using other facial expression videos.

References

1. Cohn, J.F., Schmidt, K., Gross, R., Ekman, P.: Individual differences in facial expression: stability over time, relation to self-reported emotion, and ability to inform person identification. In: ICMI 2002, pp. 491–496 (2002)
2. Tsai, P.H., Jan, T.: Expression-invariant face recognition system using subspace model analysis. In: SMC 2005, vol. 2, pp. 1712–1717 (2005)
3. Ramachandran, M., Zhou, S.K., Jhalani, D., Chellappa, R.: A method for converting a smiling face to a neutral face with applications to face recognition. In: ICASSP 2005, vol. 2, pp. 977–980 (2005)
4. Li, X., Mori, G., Zhang, H.: Expression-invariant face recognition with expression classification. In: CRV 2006, pp. 77–83 (2006)
5. Ning, Y., Sim, T.: Smile, you're on identity camera. In: ICPR 2008, pp. 1–4 (2008)

6. Chen, L.-F., Liao, H.-Y.M., Lin, J.-C.: Person identification using facial motion. In: ICIP 2001, vol. 2, pp. 677–680 (2001)
7. Cootes, T.F., Edwards, G.J., Taylor, C.J.: Active appearance models. IEEE Transactions on PAMI 23(6), 681–685 (2001)
8. Gross, R., Matthews, I., Baker, S.: Generic vs. person specific active appearance models. Image and Vision Computing 23, 1080–1093 (2005)
9. Lien, J.: Automatic recognition of facial expressions using hidden markov models and estimation of expression intensity. PhD thesis, Carnegie Mellon University (1998)
10. Pantic, M., Valstar, M., Rademaker, R., Maat, L.: Web-based database for facial expression analysis. In: ICME 2005, pp. 317–321 (2005)

Regression Based Non-frontal Face Synthesis for Improved Identity Verification

Yongkang Wong[1,2], Conrad Sanderson[1,2], and Brian C. Lovell[1,2]

[1] NICTA, PO Box 6020, St Lucia, QLD 4067, Australia*
[2] The University of Queensland, School of ITEE, QLD 4072, Australia

Abstract. We propose a low-complexity face synthesis technique which transforms a 2D frontal view image into views at specific poses, without recourse to computationally expensive 3D analysis or iterative fitting techniques that may fail to converge. The method first divides a given image into multiple overlapping blocks, followed by synthesising a non-frontal representation through applying a multivariate linear regression model on a low-dimensional representation of each block. To demonstrate one application of the proposed technique, we augment a frontal face verification system by incorporating multi-view reference (gallery) images synthesised from the frontal view. Experiments on the pose subset of the FERET database show considerable reductions in error rates, especially for large deviations from the frontal view.

1 Introduction

Face based identity inference subject to pose variations is a challenging problem, as previously studied and documented in FRVT test reports [1]. In certain applications the only reference (gallery) face images available are in one pose — e.g. frontal passport photos. Under typical surveillance conditions, CCTV cameras are unable to provide good quality frontal face images, largely due to the positioning of the cameras. In such situations an identity inference system based on frontal reference views will tend to have poor accuracy, unless extra processing is used to reduce the pose mismatch between the reference and acquired surveillance images.

The mismatch reduction can be accomplished through transforming the acquired surveillance images to be of the same pose as the reference image, or vice-versa. Recent face transformation methods include techniques based on Active Appearance Models (AAMs) [2,3] and fitting a 2D image into a 3D morphable model [4,5]. The AAM based synthesis approach requires an initialisation stage to label the important facial features (e.g. ~ 60 points). The initialisation can be done manually or automatically, where it may fail to converge. The morphable

---

* NICTA is funded by the Australian Government as represented by the Department of Broadband, Communications and the Digital Economy, as well as the Australian Research Council through the ICT Centre of Excellence program.

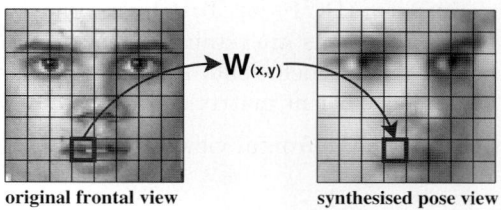

Fig. 1. Conceptual example of block-by-block image synthesis. The transformation matrix $\boldsymbol{W}_{(x,y)}$ is specific to location (x,y).

model based approach estimates the 3D shape and texture from a single image, with the fitting process taking about 4.5 minutes on a 2 GHz Pentium 4 workstation.

In this paper we propose a low-complexity, non-iterative face synthesis technique which transforms a given frontal view image into views at specific poses, without recourse to computationally expensive 3D analysis or fitting techniques that may fail to converge. The method first divides a given image into multiple overlapping blocks, followed by applying a multivariate linear regression model on a low-dimensional representation of each block. A conceptual example of this process is shown in Fig. 1. We demonstrate an application of the technique by augmenting a frontal face verification system with artificial multi-view reference (gallery) images synthesised from the frontal view.

We continue as follows. The details of the face synthesis method are described in Section 2. A preliminary comparison with AAM based image synthesis is given in Section 3. Identity verification experiments on the pose subset of the FERET database are given in Section 4. Conclusions and an outlook are presented in Section 5.

2 Face Synthesis with Multivariate Linear Regression

The proposed face synthesis method is a non-iterative process comprised of five steps: **(1)** block-based image analysis, **(2)** low-dimensional representation of each block, **(3)** transformation with multivariate linear regression, **(4)** block reconstruction, **(5)** block-based image synthesis. The steps are elaborated below, followed by a subsection which explores the effects of several parameters.

1. A given frontal image, $\boldsymbol{X}^{[\text{frontal}]}$, is analysed on an overlapping block-by-block basis, resulting in a set of 2D blocks, $\left\{ \boldsymbol{b}^{[\text{frontal}]}_{(p,q)}, \boldsymbol{b}^{[\text{frontal}]}_{(r,s)}, \cdots \right\}$, where the subscript indicates the position of the block within the image. Based on preliminary experiments (using face images with a size of 64×64 pixels), each block has a size of 8×8 pixels.

2. To ameliorate dimensionality problems described later, each block $\boldsymbol{b}^{[\text{frontal}]}_{(x,y)}$ is represented by a vector of coefficients, $\boldsymbol{v}^{[\text{frontal}]}_{(x,y)}$, resulting from the 2D

Discrete Cosine Transform (DCT) [6]. To achieve dimensionality reduction, only the lower order coefficients are retained (specifically, to reduce the dimensionality from 64 to 16, coefficients are taken from the top-left 4×4 sub-matrix of the 8×8 coefficient matrix).

3. Each vector obtained from the frontal view is then transformed to represent a non-frontal view Θ using:

$$v^{[\Theta]}_{(x,y)} = \left[1 \ \left(v^{[\text{frontal}]}_{(x,y)}\right)^T\right] W^{[\Theta]}_{(x,y)} \qquad (1)$$

where $W^{[\Theta]}_{(x,y)}$ is a transformation matrix specific to view Θ and location (x,y).

Two sets of training vectors, obtained from frontal and non-frontal faces, are required to obtain $W^{[\Theta]}_{(x,y)}$. In each set the vectors are also specific to location (x,y). Let us place the training frontal vectors into an extended matrix $\boldsymbol{A}$ and the training non-frontal vectors into matrix $\boldsymbol{B}$. If both matrices are constrained to have N number of vectors, we can define a linear regression model as follows:

$$\boldsymbol{B}_{(x,y)} = \boldsymbol{A}_{(x,y)} \ W^{[\Theta]}_{(x,y)} \qquad (2)$$

$$\begin{bmatrix} \boldsymbol{b}_1^T \\ \boldsymbol{b}_2^T \\ \vdots \\ \boldsymbol{b}_N^T \end{bmatrix} = \begin{bmatrix} 1 & \boldsymbol{a}_1^T \\ 1 & \boldsymbol{a}_2^T \\ & \vdots \\ 1 & \boldsymbol{a}_N^T \end{bmatrix} \begin{bmatrix} w_{1,1} & \cdots & w_{1,D} \\ w_{2,1} & \cdots & w_{2,D} \\ \vdots & & \vdots \\ w_{D+1,1} & \cdots & w_{D+1,D} \end{bmatrix} \qquad (3)$$

where D is the dimensionality of the vectors and $N > D+1$. Under the sum-of-least-squares regression criterion, $W^{[\Theta]}_{(x,y)}$ can be found using [7]:

$$W^{[\Theta]}_{(x,y)} = \left(\boldsymbol{A}^T_{(x,y)} \boldsymbol{A}_{(x,y)}\right)^{-1} \boldsymbol{A}^T_{(x,y)} \boldsymbol{B}_{(x,y)} \qquad (4)$$

Due to the constraint on N, the higher the dimensionality of the vectors, the more training faces are required. Given that there might be a limited number of such faces, or there might be memory constraints for solving Eqn. (4), it is preferable to keep the dimensionality low.

4. Each synthesised non-frontal vector $v^{[\Theta]}_{(x,y)}$ is converted to a non-frontal block $\boldsymbol{b}^{[\Theta]}_{(x,y)}$ through an inverse 2D DCT. The omitted DCT coefficients are set to zero.

5. A synthesised image $\boldsymbol{X}^{[\Theta]}$ for non-frontal view Θ is constructed from blocks $\left\{\boldsymbol{b}^{[\Theta]}_{(p,q)}, \boldsymbol{b}^{[\Theta]}_{(r,s)}, \cdots\right\}$ through an averaging operation. An auxiliary matrix, $\boldsymbol{C}$, is used for keeping the count of pixels placed at each location. Elements of $\boldsymbol{X}^{[\Theta]}$ and $\boldsymbol{C}$ are first set to zero. A block $\boldsymbol{b}^{[\Theta]}_{(x,y)}$ is placed into $\boldsymbol{X}^{[\Theta]}$ at location (x,y) by adding to the elements already present in $\boldsymbol{X}^{[\Theta]}$. The corresponding elements of $\boldsymbol{C}$ are increased by one. This process is repeated until all the blocks have been placed. Finally, each element of $\boldsymbol{X}^{[\Theta]}$ is divided by the corresponding element in $\boldsymbol{C}$.

2.1 Effects of Vector Dimensionality and Degree of Block Overlap

In this section we show the effects of the degree of dimensionality reduction as well as the amount of block overlap. For evaluation we use frontal and non-frontal faces from subset b of the FERET dataset [8]. The subset has 200 persons in 9 views (frontal, $\pm 60°$, $\pm 40°$, $\pm 25°$ and $\pm 15°$). Each image was size normalised and cropped so that the eyes were at the same positions in all images. The resulting image size was 64×64 pixels. 100 randomly selected persons were used to train the transformation matrices for each pose angle. Frontal images from the remaining 100 persons were then fed to the proposed synthesis technique.

Examples of typical effects of the amount of block overlap are shown in Fig. 2, where a frontal face is transformed to a synthetic $+40°$ view. 25% of the DCT coefficients were retained (i.e. 16 out of 64) for each block. The block overlap varied from 0% to 87.5%. A 50% overlap indicates that each 8×8 pixel block overlapped its neighbours by 4 pixels.

The quality of the synthesised images improves remarkably as the overlap increases. This can be attributed to the considerable increase in the number of transformation matrices (from 64 in the 0% overlap case to 3249 in the 87.5% case), leading to the overall image transformation being much more detailed. Furthermore, mistakes in the synthesis of individual blocks tend to be reduced through the averaging process described in step 5 of the algorithm.

The effect of the degree of dimensionality reduction is shown qualitatively in Fig. 3 and quantitatively in Fig. 4. The optimal amount of retained coefficients appears to be around 25%, which has the effect of removing high frequencies. Using more than 25% of the coefficients results in poorer quality images — this can be attributed to the dimensionality being too high in relation to the available number of training examples.

Fig. 4 shows the relative reduction in pixel based mean square error (MSE) for faces not used during training, which can be used to quantitatively gauge the improvement in image quality. Here a baseline MSE was obtained by comparing the frontal image with the real side image for each person. A secondary MSE was then obtained by comparing real side images with corresponding synthesised side

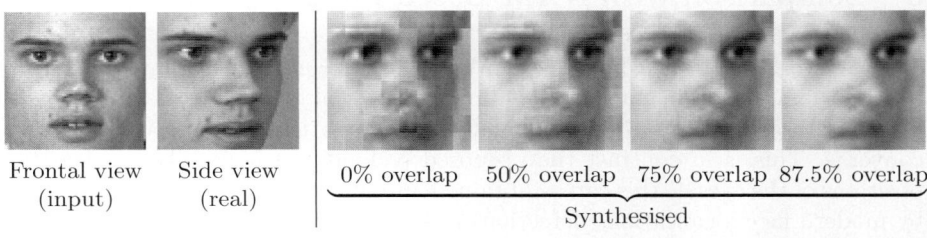

Frontal view (input)   Side view (real)   0% overlap   50% overlap   75% overlap   87.5% overlap
                                          _____Synthesised_____/

Fig. 2. Examples of non-frontal (side view) face synthesis with various degrees of block overlap. The images are synthesised from the frontal input image. The real side view, for the person depicted in the frontal image, is never used by the system. 25% of the DCT coefficients were retained.

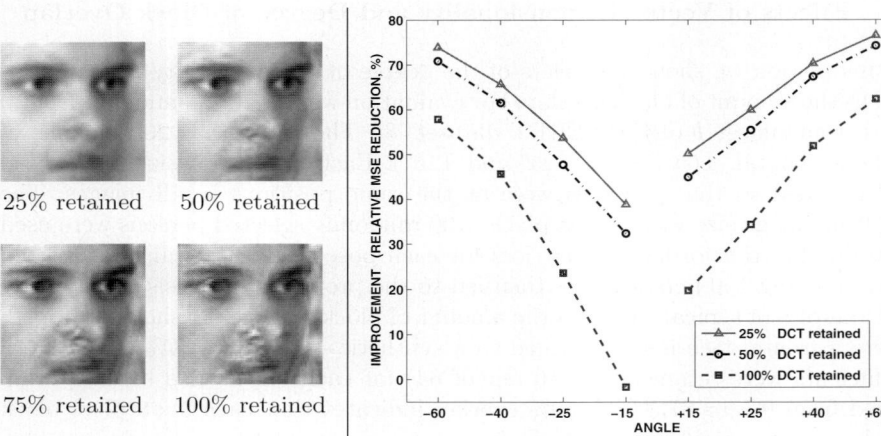

Fig. 3. Examples of face synthesis with various amounts of retained DCT coefficients. There is a maximum of 64 coefficients per block. The block overlap was set to 87.5%.

Fig. 4. Synthesis improvement (quality) for various amounts of retained DCT coefficients. A baseline mean square error (MSE) was obtained by comparing each frontal image with the corresponding real side image. A secondary MSE was then obtained by comparing real side images with corresponding synthesised side images. The improvement is how much smaller the secondary MSE is compared to the baseline (i.e. relative reduction).

images. We define the relative reduction in MSE as how much smaller the secondary MSE is compared to the baseline MSE. The quantitative results presented in Fig. 4 confirm the qualitative results shown in Fig. 3, where using 25% of the DCT coefficients is better than using 100%. It can also be observed that the best improvement occurs for high pose angles ($\pm 60°$).

3 Comparison with AAM Based Image Synthesis

The Active Appearance Model (AAM) based image synthesis approach [2,3] requires a initialisation stage to label important facial features (e.g. ~ 60 points). The initialisation can be done manually or automatically, where it can fail to converge. This is in contrast the proposed technique, which only requires the positions of the eyes — this can be obtained automatically and relatively robustly by modern face localisation (detection) algorithms [9].

Fig. 5 shows a preliminary and qualitative comparison of image synthesis using the proposed regression based technique and the AAM based method described in [3]. We follow the setup described in [3], which is similar to that of Section 2.1. The main difference is that side view images are transformed to a frontal view rather than vice-versa.

| Non-frontal input image | Real frontal view | AAM based synthesis of frontal view | Regression based synthesis of frontal view |

Fig. 5. Comparison of frontal view synthesis from non-frontal input view. Results for the AAM based approach taken from [3].

On first sight, the synthesised image produced by the regression method appears to be less defined than the AAM generated image. However, facial features such as the eyes and nose are actually better approximated when compared directly with the real frontal view image. Specifically, the eyes synthesised by the regression method retain their overall shape (outer edges are pointing downwards) and the nose remains thin. This is not the case for the AAM-based technique, where the eyes lose much of their original shape and the nose is considerably wider.

4 Face Verification with Synthesised Non-frontal Images

This section shows an application of the proposed face synthesis technique. Synthesised faces are used to build a multi-angle model to address the pose mismatch problem described in Section 1. As the baseline we use the PCA/GMM face verification system described in [10], which is easily extendable to multi-angle models while remaining in a probabilistic framework.

The first step is Principal Component Analysis (PCA) based feature extraction. Briefly, a given face image is represented by a matrix containing grey pixel values. The matrix is converted to a face vector, f, by column concatenation. A D-dimensional feature vector, x, is obtained by:

$$x = U^T \left(f - f_\mu \right) \qquad (5)$$

where U contains D eigenvectors (corresponding to the D largest eigenvalues) of the training data covariance matrix and f_μ is the mean of training face vectors [11].

In the verification task we wish to find out whether a given biometric sample belongs to the true claimant or an impostor. A claim for the identity of person C is classified as coming from the that person (i.e. the true claimant) when

$$\frac{p(x|\lambda_C)}{p(x|\lambda_{\text{impostor}})} > t \qquad (6)$$

and as coming from an impostor otherwise. Here t is a decision threshold, λ_C is the model for person C and $\lambda_{\text{impostor}}$ is the approximate impostor model. The distribution of feature vectors is described by a Gaussian Mixture Model (GMM):

$$p(\boldsymbol{x}|\lambda) = \sum_{g=1}^{N_G} w_g \mathcal{N}(\boldsymbol{x}|\boldsymbol{\mu}_g, \boldsymbol{\Sigma}_g) \tag{7}$$

where $\lambda = \{w_g, \boldsymbol{\mu}_g, \boldsymbol{\Sigma}_g\}_{g=1}^{N_G}$ and $\mathcal{N}(\boldsymbol{x}|\boldsymbol{\mu}, \boldsymbol{\Sigma})$ is a D-dimensional Gaussian function with mean $\boldsymbol{\mu}$ and diagonal covariance matrix $\boldsymbol{\Sigma}$. N_G is the number of Gaussians and w_g is the weight for Gaussian g (with constraints $\sum_{g=1}^{N_G} w_g = 1$) [11,12]. Due to the relatively small amount of training data for each person (i.e. one frontal image), a common covariance matrix is used for all Gaussians and all models.

Frontal face models, for each person enrolled in the system, are comprised of one Gaussian. The Gaussian's mean is equal to the PCA-derived feature vector obtained from the frontal face. In a similar manner, the approximate impostor model is comprised of 32 Gaussians, where the Gaussian means are taken to be equal to the PCA-derived feature vectors of 32 randomly selected persons. The weights are all equal.

4.1 Synthesised Multi-angle Models

In order for the system to automatically handle non-frontal views, each frontal face model is extended by concatenating it with models generated from synthesised non-frontal views. Formally, an extended (or multi-angle) model for person C is created using:

$$\lambda_C^{[\text{extended}]} = \lambda_C^{[\text{frontal}]} \sqcup \lambda_C^{[+60°]} \sqcup \lambda_C^{[+40°]} \cdots \sqcup \lambda_C^{[-40°]} \sqcup \lambda_C^{[-60°]} \tag{8}$$

$$= \sqcup_{i \in \Phi} \lambda_C^{[i]} \tag{9}$$

where $\lambda_C^{[\text{frontal}]}$ represents the frontal model, Φ is a set of angles, e.g. $\Phi = \{\pm 15°, \pm 25°, \pm 40°, \pm 60°\}$, and $\sqcup$ is an operator for joining GMM parameter sets, defined as follows. Let us suppose we have two GMM parameter sets, $\lambda^{[a]}$ and $\lambda^{[b]}$, comprised of parameters for $N_G^{[a]}$ and $N_G^{[b]}$ Gaussians, respectively. The $\sqcup$ operator is defined as follows:

$$\lambda^{[\text{joined}]} = \lambda^{[a]} \sqcup \lambda^{[b]} = \left\{\alpha w_g^{[a]}, \boldsymbol{\mu}_g^{[a]}, \boldsymbol{\Sigma}_g^{[a]}\right\}_{g=1}^{N_G^{[a]}} \cup \left\{\beta w_g^{[b]}, \boldsymbol{\mu}_g^{[b]}, \boldsymbol{\Sigma}_g^{[b]}\right\}_{g=1}^{N_G^{[b]}} \tag{10}$$

where $\alpha = N_G^{[a]} / \left(N_G^{[a]} + N_G^{[b]}\right)$ and $\beta = 1 - \alpha$.

4.2 Experiments and Results

The experiments were done[1] using data from two subsets of the FERET dataset. Subset f was used to train the PCA based feature extractor (i.e. $\boldsymbol{U}$ and $\boldsymbol{f}_\mu$) and to obtain the common covariance matrix for the GMMs. Faces for $\lambda_{\text{impostor}}$ were also selected from this subset. Subset b was randomly split into three disjoint groups: group A, group B and an impostor group. Group A had 100 persons and was used to train the transformation matrices for each pose view (i.e. from frontal to dedicated pose). Group B had 80 persons and the remaining 20 persons were

---

[1] The experiments were performed with the aid of the Armadillo C++ linear algebra library, available from http://arma.sourceforge.net

placed in the impostor group. The latter two groups were used in verification tests, which were comprised of 80 true claims and 20×80 = 1600 impostor attacks per view angle.

Two systems were evaluated: (1) frontal models and (2) synthesised multi-angle models. In the latter case each person's model had 9 Gaussians, with each Gaussian representing a particular view (i.e. the original frontal and synthesised ±15°, ±25°, ±40°, ±60° views). For each angle the results are reported in terms of the Equal Error Rate (EER) [12].

The results, shown in Fig. 6, demonstrate considerable error reductions across all pose views. The largest improvement in performance occurs for large deviations from the frontal view (±60°), where the errors are remarkably reduced by an absolute difference of about 15 percentage points, or a relative difference of about 30%.

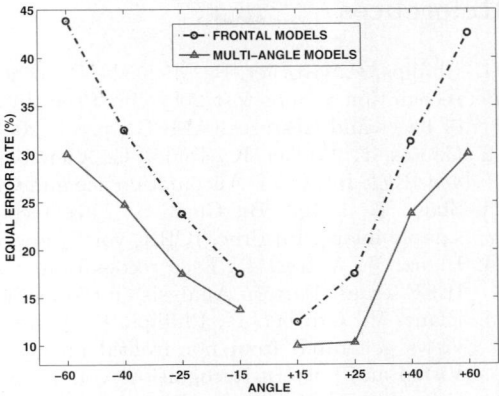

Fig. 6. Verification error rates using frontal and multi-angle models. The latter use synthesised non-frontal faces.

5 Conclusions and Outlook

In this paper we proposed a low-complexity face synthesis technique which transforms a 2D frontal view image into views at specific poses, without recourse to computationally expensive 3D analysis or iterative fitting techniques that may fail to converge (as used by Active Appearance Models [3]). The proposed method first divides a given image into multiple overlapping blocks, followed by synthesising a non-frontal representation through applying a multivariate linear regression model on a low-dimensional representation of each block.

The proposed synthesis method is relatively straightforward, with low computational requirements for both training and image synthesis. Using 100 persons for training, learning the regression matrices took about 3 seconds on a 2 GHz Intel Core 2 processor. Synthesis of 1 test image took less than 0.04 seconds (C++ implementation, gcc 4.1.2, Linux 2.6.26).

To demonstrate one application of the proposed technique, we augmented a frontal face verification system by incorporating multi-view reference (gallery) images synthesised from the frontal view. Experiments on the pose subset of the FERET database indicate considerable reductions in error rates, especially for large deviations from the frontal view.

Improvements in synthesis quality could be obtained through a more precise low-dimensional representation of each block. For example, rather than using the generic 2D DCT, a position dependent local PCA could be employed, which may have the additional advantage of further reducing the dimensionality.

References

1. Phillips, P., Grother, P., Micheals, R., Blackburn, D., Tabassi, E., Bone, M.: Face recognition vendor test 2002. In: Proc. Int. Workshop on Analysis and Modeling of Faces and Gestures (AMFG), p. 44 (2003)
2. Cootes, T., Walker, K., Taylor, C.: View-based active appearance models. In: Proc. 4th IEEE Int. Conf. Automatic Face and Gesture Recognition, pp. 227–232 (2000)
3. Shan, T., Lovell, B., Chen, S.: Face recognition robust to head pose from one sample image. In: Proc. ICPR, vol. 1, pp. 515–518 (2006)
4. Blanz, V., Vetter, T.: Face recognition based on fitting a 3D morphable model. IEEE Trans. Pattern Analysis and Machine Intelligence 25(9), 1063–1074 (2003)
5. Blanz, V., Grother, P., Phillips, P., Vetter, T.: Face recognition based on frontal views generated from non-frontal images. In: Proc. IEEE Int. Conf. Computer Vision and Pattern Recognition, vol. 2, pp. 454–461 (2005)
6. Gonzales, R., Woods, R.: Digital Image Processing, 3rd edn. Prentice-Hall, Englewood Cliffs (2007)
7. Rice, J.: Mathematical Statistics and Data Analysis, 2nd edn. Duxbury Press (1995)
8. Phillips, P., Moon, H., Rizvi, S., Rauss, P.: The FERET evaluation methodology for face-recognition algorithms. IEEE Trans. Pattern Analysis and Machine Intelligence 22(10), 1090–1104 (2000)
9. Viola, P., Jones, M.J.: Robust real-time face detection. International Journal of Computer Vision 57(2), 137–154 (2004)
10. Rodriguez, Y., Cardinaux, F., Bengio, S., Mariethoz, J.: Measuring the performance of face localization systems. Image and Vision Comput. 24, 882–893 (2006)
11. Bishop, C.: Pattern Recognition and Machine Learning. Springer, Heidelberg (2006)
12. Sanderson, C.: Biometric Person Recognition: Face, Speech and Fusion. VDM-Verlag (2008)

Differential Feature Analysis for Palmprint Authentication

Xiangqian Wu[1], Kuanquan Wang[1], Yong Xu[1], and David Zhang[2]

[1] School of Computer Science and Technology, Harbin Institute of Technology
Harbin 150001, China
{xqwu,wangkq,xuyong}@hit.edu.cn
[2] Department of Computing, The Hong Kong Polytechnic University
Kwooloon, Hong Kong
csdzhang@comp.polyu.edu.hk

Abstract. Palmprint authentication is becoming one of the most important biometric techniques because of its high accuracy and ease to use. The features on palm, including the palm lines, ridges and textures, etc., are resulted from the gray scale variance of the palmprint images. This paper characterizes these variance using different order differential operations. To avoid the effect of the illumination variance, only the signs of the pixel values of the differential images are used to encode palmprint to form palmprint differential code (PDC). In matching stage, normalized Hamming distance is employed to measure the similarity between different PDCs. The experimental results demonstrate that the proposed approach outperforms the existing palmprint authentication algorithms in terms of the accuracy, speed and storage requirement and the differential operations may be considered as one of the standard methods for palmprint feature extraction.

1 Introduction

Computer-aided personal recognition is becoming increasingly important in our information society. Biometrics is one of the most important and reliable methods in this field [1,2]. The palmprint is a relatively new biometric feature and has many advantages for personal authentication [3]. Palmprint recognition is becoming a hotspot in biometrics field.

Han et al. [4] used Sobel and morphological operations to extract line-like features from palmprints. Similarly, for verification, Kumar et al. [5] used other directional masks to extract line-like features. Zhang et al. [6,7] used 2-D Gabor filters to extract the phase information (called PalmCode) from low-resolution palmprint images. Wu et al. [8] extract the palm lines and authenticate persons according to the line structure. Jia et al. [9] used a modified finite Radon transform to compute the line direction of palmprint and employed pixel to region matching for verification. Kong and Zhang [10] defined an orientation for each pixel using a bank of Gabor filters and matched palmprint by compute the angular distance (called CompCode). Sun et al. [11] extract orthogonal line ordinal features (called OrdnCode) to represent palmprint. Up to now, the CompCode and OrdnCode are the most effective algorithms for palmprint authentication.

Different algorithms extract different features from palmprint. Actually, all features on palm, such as palm lines, ridge and textures, etc., are resulted from the gray scale

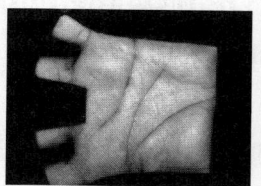 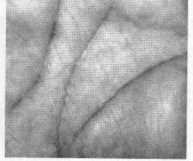

(a) Original Palmprint   (b) Cropped Image

Fig. 1. A palmprint and the preprocessed image

variance of the palmprint image. The derivative of an image naturally and effectively reflect these variance, which motivates us to investigate differential feature of palmprint for personal authentication.

The palmprints used in this paper are from the Polyu Palmprint Database [12]. We use the technique in [6] to preprocess a palmprint and crop the central part of the image to represent the whole palmprint. Figure 1 shows a palmprint and its cropped image.

2 Gaussian Derivative Filters (GDF)

Let $G(x, y, \sigma)$ denote a 2-D Gaussian function with variance σ, which is defined as following:

$$G(x, y, \sigma) = \exp\left(-\frac{x^2 + y^2}{2\sigma^2}\right) \quad (1)$$

The nth-order Gaussian derivative filters (n-GDF) can be obtained by computing the nth-order derivatives of the Gaussian function. For simplification, this paper just considers the derivatives along x and y axis. Therefore, the $1st$ to $3rd$-order Gaussian derivative filters along the x axis are computed as following equations:

$1st$-order GDF:

$$G_{x^{(1)}}(x, y, \sigma) = -\frac{x}{\sigma^2} \exp\left(-\frac{x^2 + y^2}{2\sigma^2}\right) \quad (2)$$

$2nd$-order GDF:

$$G_{x^{(2)}}(x, y, \sigma) = \left(-\frac{1}{\sigma^2} + \frac{x^2}{\sigma^4}\right) \exp\left(-\frac{x^2 + y^2}{2\sigma^2}\right) \quad (3)$$

$3rd$-order GDF:

$$G_{x^{(3)}}(x, y, \sigma) = \left(\frac{3x}{\sigma^4} - \frac{x^3}{\sigma^6}\right) \exp\left(-\frac{x^2 + y^2}{2\sigma^2}\right) \quad (4)$$

The different order Gaussian derivative filters along y axis, $G_{y^{(1)}}(x, y, \sigma)$, $G_{y^{(2)}}(x, y, \sigma)$ and $G_{y^{(3)}}(x, y, \sigma)$, can also be computed by exchanging the positions of variable x and y at the right of the above corresponding equations.

3 Palmprint Differential Code (PDC) Extraction

As mentioned above, all features on palms are resulted from the gray scale variance of palmprint images and the derivative is an effective way to capture these variance. We can compute the derivative of a palmprint by convolving it with the corresponding GDF.

Denote I as a preprocessed palmprint image and denote $G_{x(k)}$ and $G_{y(k)}$ as the kth order Gaussian derivative filters. The kth order derivative of I in x and y directions can be computed as following:

$$I_{x(k)} = I * G_{x(k)} \tag{5}$$

$$I_{y(k)} = I * G_{y(k)} \tag{6}$$

where "$*$" is the convolving operation.

To avoid the effect of the illuminance variance, we only use the signs of pixel values of the filtered images to encode the palmprint:

$$C_{x(k)}(i,j) = \begin{cases} 1, & \text{if } I_{x(k)} > 0; \\ 0, & \text{otherwise}. \end{cases} \tag{7}$$

$$C_{y(k)}(i,j) = \begin{cases} 1, & \text{if } I_{y(k)} > 0; \\ 0, & \text{otherwise}. \end{cases} \tag{8}$$

$C = (C_{x(k)}, C_{y(k)})$ is called the kth order palmprint differential code (k-PDC). Figure 2 shows some examples of different order PDCs. This figure demonstrates some properties of the PDCs. The 1-PDCs contain the most prominent features of palmprint, such as the principal lines, but miss most of the details. The 2-PDCs contain both the remarkable features and the most of the palmprint details. Though the 3-PDCs contain more details of the palmprints, they also contain so much noise which can ruin the palmprint features. From these evidences, we can deduce that the higher order PDCs should contain much more noises and cannot be used for palmprint authentication.

In Figure 2, the last two palmprints are captured from the same palm and the first one is from a different palm. From this figure, we can intuitively find that the PDCs from the same palm are more similar than those from different palm. Therefore, the PDCs can be used to distinguish different palms.

4 Similarity Measurement of PDCs

Denote $C_1 = (C_{x^k}^1, C_{y^k}^1)$ and $C_2 = (C_{x^k}^2, C_{y^k}^2)$ as the k-PDCs of two palmprint images. The normalized Hamming distance between C_1 and C_2 is defined as following:

$$D(C_1, C_2) = \frac{\sum_{i=1}^{M} \sum_{j=1}^{N} \left[C_{x^k}^1(i,j) \otimes C_{x^k}^2(i,j) + C_{y^k}^1(i,j) \otimes C_{y^k}^2(i,j) \right]}{M \times N} \tag{9}$$

where M and N are the dimension of palmprint image and "$\otimes$" is the logical XOR operation.

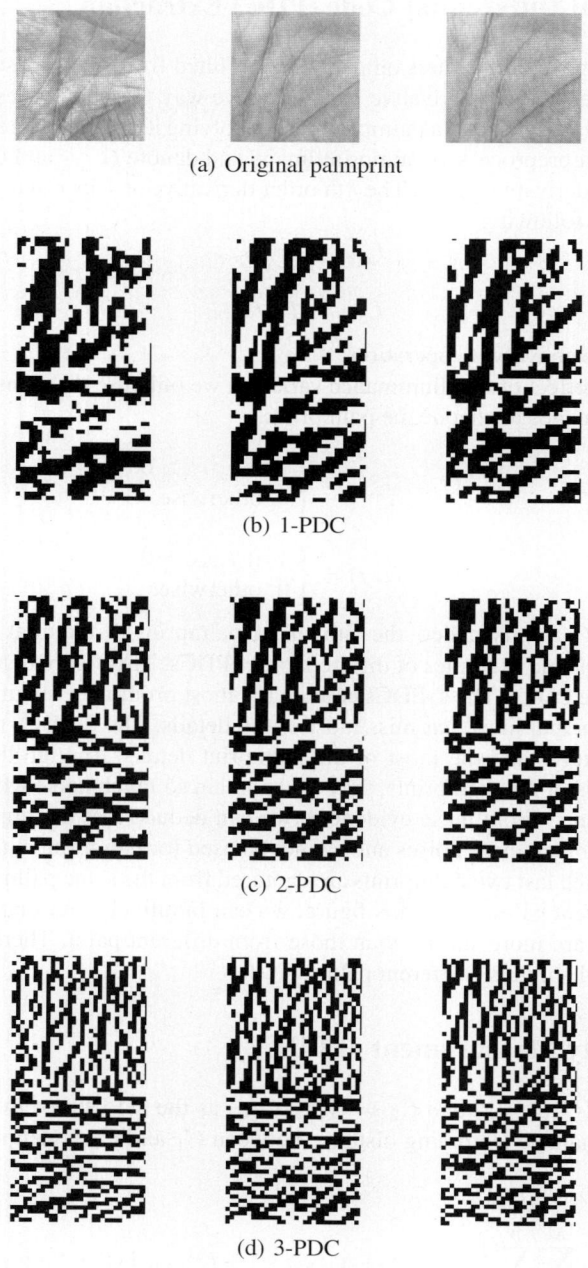

Fig. 2. Some examples of PDC with different order

The similarity between C_1 and C_2 can be measured using a matching score S which computed as following:

$$S(C_1, C_2) = 1 - D(C_1, C_2) \tag{10}$$

Obviously, $0 \leqslant S(C_1, C_2) \leqslant 1$. The larger $S(C_1, C_2)$ is, the more similar C_1 and C_2 are. If C_1 and C_2 are exactly same, which means the perfect matching, $S(C_1, C_2) = 1$. Because of imperfect preprocessing, there may still be a little translation between the palmprints captured from the same palm at different times. To overcome this problem, we vertically and horizontally translate C_1 a few points to get the translated C_1, and then, at each translated position, compute the matching score between the translated C_1 and C_2. Finally, the final matching score is taken to be the maximum matching score of all the translated positions.

Table 1 lists the matching scores between the PDCs shown in Figure 2. From this table, we also can find that the scores between the PDCs from the same palm (> 0.8) are much larger than those from different palms (< 0.6).

Table 1. The matching scores between the PDCs shown in Figure 2

PDC	Column	Left	Middle	Right
1-PDC	Left	1	0.5682	0.5600
	Middle	-	1	0.8452
	Right	-	-	1
2-PDC	Left	1	0.5595	0.5650
	Middle	-	1	0.8169
	Right	-	-	1
3-PDC	Left	1	0.5852	0.5691
	Middle	-	1	0.8218
	Right	-	-	1

5 Experimental Results and Analysis

5.1 Database

We employed the PolyU Palmprint Database [12] for testing and comparison. This database contains 7752 grayscale images captured from 386 different palms by a CCD-based device. About 20 images are captured from each palm. The size of the images is 384×284 pixels. Using the preprocessing technique described in [6], the central 128×128 part of the image was cropped to represent the whole palmprint.

5.2 Matching Test

To investigate the performance of the proposed approach, each sample in the database is matched against the other samples. Therefore, a total of $30,042,876$ ($7752 \times 7751/2$) matchings have been performed, in which 74068 matchings are genuine matchings. Figure 3 shows the genuine and impostor matching score distribution of the different order PDCs. There are two distinct peaks in the distributions of the matching score for

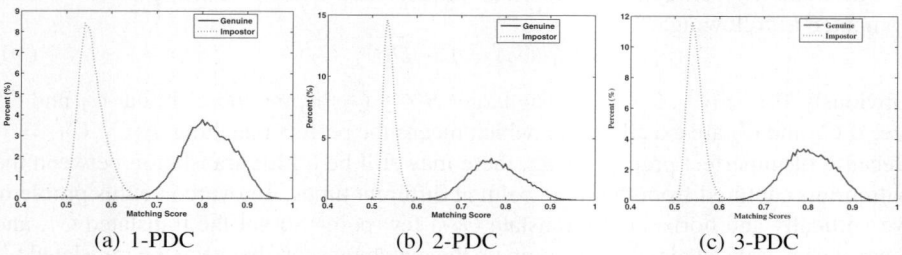

(a) 1-PDC　　　　　　　(b) 2-PDC　　　　　　　(c) 3-PDC

Fig. 3. Matching score distribution

each order PDCs. These two peaks are widely separated and the distribution curve of the genuine matching scores intersects very little with that of impostor matching scores. Therefore, the different order PDCs can effectively discriminate between the palmprints from different palms.

5.3 Accuracy

To evaluate the accuracy of the proposed approach, each sample is matched against the other palmprints and the ROC curves of the different order PDCs are shown in Figure 4. For comparison, the competitive code (CompCode) [10] and ordinal filters (OrdnCode) [11] based method are also implemented and tested on this database. Their ROC curves are also plotted in Figure 4. According to this figure, the 2-PDC's ROC curve is under the curves of the 1-PDC and 3-PDC and the 3-PDC's curve is under that of the 1-PDC. Therefore, among these three order PDCs, the 2-PDC obtains the highest accuracy while the 1-PDC get the lowest accuracy. Also from this figure, the performance of 2-PDC is also better than the CompCode and OrdnCode algorithms. The accuracy of the CompCode algorithm is similar with that of the 3-PDC and the EER of the OrdnCode method is similar with that of the 1-PDC.

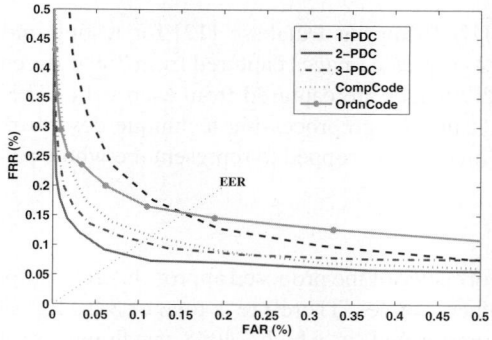

Fig. 4. The ROC curves of different palmprint algorithms

Table 2. Comparisons of different palmprint algorithms

	EER (%)	Feature Size	Extracting Time(ms)	Matching Time(ms)
1-PDC	0.1462	256	9.6	0.09
2-PDC	0.0759	256	10.1	0.09
3-PDC	0.0965	256	10.3	0.09
CompCode	0.1160	384	70	0.18
OrdnCode	0.1676	384	58	0.18

The PDC approach, CompCode algorithm and OrdnCode method are compared in Table 2 in terms of accuracy, feature size and speed. From this table, the proposed PDC approach outperforms other two in all of these aspects.

5.4 Discussion

According to the experimental results, the accuracy of the PDC is higher than the CompCode and OrdnCode, which may be for the following reasons. Both the CompCode and the OrdnCode methods extracted the orientation information of each pixel on the palmprint while the PDC approach captured the grayscale variance tendencies. Obviously, the orientation information is much more sensitive to the rotation of the palm than the grayscale variance tendency. Hence, the PDC approach performs better on the palmprints with some rotation than the CompCode and OrdnCode methods. Although the preprocessing removes most of the rotation between the palmprints of the same palms, there may still remain a little. Therefore, the PDC approach can get a higher accuracy. The proposed approach demonstrates the power of the differential operations for palmprint authentication.

Actually, most of the existing palmprint recognition methods can be looked as the differential operations based methods. Since the Gabor filters can be regarded as the weighted 2nd or 3rd-order Gaussian derivative filters, the Gabor filters based methods, such as CompCode [10], PalmCode [6] and FusionCode [7], etc., can be considered as the differential operations based methods. The orthogonal line ordinal filter used for OrdnCode extraction [11] is a kind of the 1st-differtinal operator. And the Sobel operators based methods [4] are also based on the 1st-differential operations. Therefore, the differential operations may be considered as one of the standard methods for palmprint feature extraction.

6 Conclusions and Future Work

This paper encodes the palmprint image using the different order differential operations. This approach can capture the typical character of the palmprint and can effectively distinguish palmprints from different palms. The 2nd order derivative of palmprint is the most distinguishable. Compared with the existing palmprint methods, the proposed approach can get a higher accuracy with less storage requirement and less response time. The differential operations may be considered as one of the standard methods for palmprint feature extraction.

In the future, we will investigate the PDC with different directions and study the multiscale PDC for palmprint recognition.

Acknowledgment

Portions of the work were tested on the PolyU Palmprint Database. The work is supported by the Natural Science Foundation of China (NSFC) under Contract No. 60873140, 60602038 and 60620160097, the National High-Tech Research and Development Plan of China (863) under Contract No. 2007AA01Z195, the Program for New Century Excellent Talents in University under Contract No. NCET-08-0155 and NCET-08-0156, and the Natural Science Foundation of Hei Longjiang Province of China under Contract No. F2007- 04.

References

1. Zhang, D.: Automated Biometrics–Technologies and Systems. Kluwer Academic Publishers, Dordrecht (2000)
2. Jain, A., Bolle, R., Pankanti, S.: Biometrics: Personal Identification in Networked Society. Kluwer Academic Publishers, Dordrecht (1999)
3. Jain, A., Ross, A., Prabhakar, S.: An introduction to biometric recognition. IEEE Trans. Circuits Syst. Video Technol. 14(1), 4–20 (2004)
4. Han, C., Chen, H., Lin, C., Fan, K.: Personal authentication using palm-print features. Pattern Recognition 36(2), 371–381 (2003)
5. Kumar, A., Wong, D., Shen, H., Jain, A.: Personal verification using palmprint and hand geometry biometric. In: Kittler, J., Nixon, M.S. (eds.) AVBPA 2003. LNCS, vol. 2688, pp. 668–678. Springer, Heidelberg (2003)
6. Zhang, D., Kong, W., You, J., Wong, M.: Online palmprint identification. IEEE Trans. Pattern Anal. Machine Intell. 25(9), 1041–1050 (2003)
7. Kong, A.W.-K., Zhang, D.: Feature-level fusion for effective palmprint authentication. In: Zhang, D., Jain, A.K. (eds.) ICBA 2004. LNCS, vol. 3072, pp. 761–767. Springer, Heidelberg (2004)
8. Wu, X., Zhang, D., Wang, K.: Palm-line extraction and matching for personal authentication. IEEE Trans. Syst., Man, Cybern. A 36(5), 978–987 (2006)
9. Jiaa, W., Huanga, D.S., Zhang, D.: Palmprint verification based on robust line orientation code. Pattern Recognition 41(5), 1504–1513 (2008)
10. Kong, A., Zhang, D.: Competitive coding scheme for palmprint verification. In: IEEE International Conference on Pattern Recognition, pp. 520–523 (2004)
11. Sun, Z., Tan, T., Wang, Y., Li, S.Z.: Ordinal palmprint represention for personal identification. In: Proceedings of the 2005 IEEE Computer Society Conference on Computer Vision and Pattern Recognition, CVPR 2005 (2005)
12. PolyU Palmprint Database, http://www.comp.polyu.edu.hk/~biometrics/

Combining Facial Appearance and Dynamics for Face Recognition

Ning Ye and Terence Sim

School of Computing, National University of Singapore,
COM1, 13 Computing Drive, Singapore 117417
{yening,tsim}@comp.nus.edu.sg
http://www.comp.nus.edu.sg/ face/

Abstract. In this paper, we present a novel hybrid feature for face recognition. This hybrid feature is created by combining the traditional holistic facial appearance feature with a recently proposed facial dynamics feature. We measure and compare the inherent discriminating power of this hybrid feature and the holistic facial appearance feature by the statistical separability between genuine feature distance and impostor feature distance. Our measurement indicates that the hybrid feature is more discriminative than the appearance feature.

Keywords: face, biometrics, appearance, dynamics.

1 Introduction

In face recognition, the holistic facial appearance feature is one of the most popular features used by researchers. A variety of pattern classification approaches have been used on it, including PCA (Principal Component Analysis) [1], LDA (Linear Discriminant Analysis) [2], SVM (Support Vector Machine) [3], evolutionary pursuit [4], nearest feature line [5], ICA (Independent Components Analysis) [6] and probabilistic decision-based neural networks [7], *etc*.

In comparison, for a long time, facial motion (*e.g.* facial expression) has been considered as a hindrance to face recognition. Similar to pose and illumination variation, facial motion is to be removed in traditional methods [8,9]. However, recent advances in studies of facial motion (from both psychology and computer vision communities) have revealed that some types of facial motion are identity-specific [10,11,12,13]. Thus, rather than ignoring facial motion, we can use it to facilitate face recognition.

Motion-based face recognition techniques came to be used only recently. Chen *et al*. [14] used the visual cues observed in speeches to recognize a person. Facial dynamics was extracted by computing a dense motion flow field from each video clip. Zhang *et al*. [15] estimated the elasticity of masseter muscle from a pair of side-view face range images (neutral and mouth-open faces) to characterize a face. They said that this muscular elasticity could be used as a biometric trait. Pamudurthy *et al*. [16] used the facial deformation observed in a subtle smile to recognize a person. A dense displacement vector field was computed from each pair of images (neutral face and subtle smile). This vector field was claimed to be very much unaffected by face makeup. Tulyakov *et al.* [13] extracted the displacement of a set of key points from a pair of face images (neutral and smile) and concatenated the displacement vectors to make a feature vector. They concluded that this feature could be used as a soft biometric trait. Ye and Sim [17] used the dynamics of smile for face recognition. A dense optical flow field was computed from each video clip and summed over time to generate a motion feature vector. They claimed that this feature could be used as a biometric trait.

Aforementioned works have shown that certain types of facial motion are discriminative to some extent. In this paper, we are going to show that by combining appearance and dynamics, we may be able to achieve more. Specifically, we combine the traditional holistic facial appearance feature and the smile dynamics feature proposed by Ye and Sim [17] to generate a novel hybrid feature. Please note that we have NOT built a complete face recognition system which usually has two major components, feature extraction and classification. Instead, in this paper, we focus on feature analysis only. Specifically, we measure and compare the discriminating power of the appearance feature and the hybrid feature. The discriminating power is measured by the separability between the distribution of genuine distance, *i.e.* the distance between any two faces belonging to one subject, and the distribution of impostor distance, *i.e.* the distance between any two faces belonging to two different subjects. To the best of our knowledge, we are the first to combine facial appearance and facial motion/dynamics for face recognition. Our measurement on the discriminating power shows that the hybrid feature is more discriminative than the appearance feature.

2 The Dataset

We merge the smile videos of three databases into one dataset in our research. The three databases are the FEEDTUM video database [18], the MMI face database [19] and the NUS smile video database [17]. The FEEDTUM database contains 18 subjects, with three smile videos per subject. The MMI database contains 17 subjects, with one to 16 smile videos per subject. The NUS database contains 10 subjects, with around 30 smile videos per subject. After eliminating those unusable videos (mainly due to excessive out-of-plane head motion), we end up with a dataset which consists of 45 subjects and 435 videos in total. Each video clip is a frontal-view recording of a subject performing a facial expression from neutral to smile and back to neutral. In pre-processing the data, faces are aligned by the positions of eyes. After alignment, face regions are cropped, converted to gray scale and resized to 81 by 91 pixels.

3 The Features

3.1 Holistic Facial Appearance Feature

Let u^a denote a column vector made by stacking all the pixel values of the first frame of a video clip, which is a static neutral face image (Figure 1(a)). Then

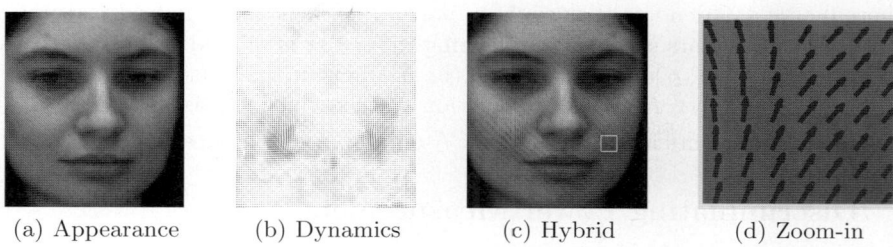

(a) Appearance   (b) Dynamics   (c) Hybrid   (d) Zoom-in

Fig. 1. The three types of features examined in our study: readers may want to zoom in on (b) (c) to see the motion flows clearly

the appearance feature v^a is computed as,

$$v^a = P^a_{k^a}(u^a - \overline{u^a}), \qquad (1)$$

where $P^a_{k^a}$ is the projection matrix which consists of the first k^a principal components after applying PCA on all u^a; $\overline{u^a}$ denotes the mean of u^a.

3.2 Smile Dynamics Feature

Smile dynamics [17] summarize pixel-wise observations of the facial motion from a neutral expression to a smile apex (Figure 1(b)). Mathematically,

$$F(z) = \sum_{t=1}^{N-1} f^t(z), \qquad (2)$$

where z is a pixel; $f^t(z)$ denotes the motion flow observed at z in the t-th frame, which is computed by applying optical flow algorithm [20] on the t-th and $(t+1)$-th frames; the N-th frame contains the smile apex while the first frame contains a neutral face; $F(z)$ is the smile dynamics observed at z. Let u^m denote a column vector made by stacking all pixel-wise $F(z)$ computed from one video clip. Then the smile dynamics feature v^m is computed as,

$$v^m = P^m_{k^m}(u^m - \overline{u^m}), \qquad (3)$$

where $P^m_{k^m}$ is the projection matrix which consists of the first k^m principal components after applying PCA on all u^m; $\overline{u^m}$ denotes the mean of u^m. Please note that this feature is purely motion-based (constructed from motion flows). No appearance information is involved.

3.3 Hybrid Feature

Figure 1(c) illustrates our hybrid feature. Mathematically, we construct the hybrid feature v^h by fusing appearance and smile dynamics as follows,

$$u^h = \begin{pmatrix} (1-w)u^a/\alpha \\ wu^m/\beta \end{pmatrix}, \qquad (4)$$

$$v^h = P^h_{k^h}(u^h - \overline{u^h}), \qquad (5)$$

where $0 \leq w \leq 1$ denotes the weight of smile dynamics in the hybrid feature; α and β are two scalars used for normalizing the scales of u^a and u^m, respectively (in our experiment, α and β are set to the medians of the l^2-norm of all u^a and all u^m, respectively); $P_{k^h}^h$ is the projection matrix which consists of the first k^h principal components after applying PCA on all u^h; $\overline{u^h}$ denotes the mean of u^h.

4 Discriminating Power Measurement

We measure the discriminating power of a feature by the statistical separability between the distribution of genuine distance and the distribution of impostor distance. Given a set of feature vectors with identity labels, the genuine distance set D_G and the impostor distance set D_I are defined as follow,

$$D_G = \{\|v_i - v_j\|_2\}, L(v_i) = L(v_j), i \neq j, \quad (6)$$
$$D_I = \{\|v_i - v_j\|_2\}, L(v_i) \neq L(v_j), i \neq j, \quad (7)$$

where v_i and v_j are two feature vectors; $L(v_i)$ and $L(v_j)$ are the identity labels of v_i and v_j, respectively; $\| \circ \|_2$ denotes the l^2-norm. From our dataset, we manage to compute 5886 genuine distances and 88509 impostor distances, *i.e.* $|D_G| = 5886, |D_I| = 88509$.

The separability of the two distributions underlying those two distance sets indicates the discriminating power of the feature. Bayes' error is the ideal tool for measuring the separability, because it is the theoretical minimum error rate that any classifier can achieve in classifying the two distances. In other words, Bayes' error is determined by the feature. However, computing Bayes' error directly is difficult in practice, because the exact probability density functions are usually unknown, as in our case. In our research, we use the Bhattacharyya coefficient [21], denoted by ρ, to estimate the Bayes' error,

$$\rho = \sum_x \sqrt{p_{D_G}(x) p_{D_I}(x)}, \quad (8)$$

where $p_{D_G}(x)$ and $p_{D_I}(x)$ denote the two discrete probability density functions underlying D_G and D_I, respectively. In our experiment, we approximate $p_{D_G}(x)$ and $p_{D_I}(x)$ using the histograms constructed from D_G and D_I, respectively. Please note that $0 \leq \rho \leq 1$, where $\rho = 0$ implies a complete separation between the two distributions and $\rho = 1$ implies a complete overlap between the two distributions. The smaller the ρ is, the more separable the two distributions are and therefore the more discriminative the feature is. Bhattacharyya coefficient is an upper bound of Bayes' error in two-category classification problems,

$$R = \rho/2 \geq E_{Bayes}. \quad (9)$$

In our experiment, we use R, *i.e.* the upper bound of Bayes' error, as the measurement of the discriminating power of feature v. Note that $0 \leq R \leq 0.5$ where a smaller R indicates stronger discriminating power of v. We use R^a, R^m, R^h to denote the measurement computed from the holistic facial appearance feature (v^a), the smile dynamics feature (v^m) and the hybrid feature (v^h), respectively.

4.1 Appearance Feature *vs.* Smile Dynamics Feature

Figure 2(a) shows R^a and R^m with varying dimensions of the feature vectors (k^a in Eq.(1) and k^m in Eq.(3)). R^a hits its minimum of 0.028 at $k^a = 16$. R^m hits its minimum of 0.127 at $k^m = 13$. And almost at any dimension, R^a is at least three times smaller than R^m. This observation implies that, with respect to our current dataset, the appearance feature can be at least three times more discriminative than the smile dynamics feature.

Figure 2(c) and Figure 2(d) show the distributions of genuine distance and impostor distance computed from the appearance feature vectors and the smile dynamics feature vectors at $k^a = 16$ and $k^m = 13$, respectively. We can see that the overlap in Figure 2(c) is much smaller than the overlap in Figure 2(d). Since the overlap is directly related to the Bayes' error [21], this observation also implies that the appearance feature is more discriminative than the smile dynamics feature.

4.2 Appearance Feature *vs.* Hybrid Feature

Since the appearance feature outperforms the smile dynamics feature considerably, we compare the hybrid feature with the appearance feature only.

Figure 2(b) shows R^h with varying w (the weight of the smile dynamics feature in the combination, see Eq.(4)). We vary w from 0 to 1 with an increment of 0.005 in each step. And since the appearance feature performs best at $k^a = 16$, we fix $k^h = 16$ so that the comparison between the appearance feature and the hybrid feature is fair. In Figure 2(b), we can see that R^h keeps going down as

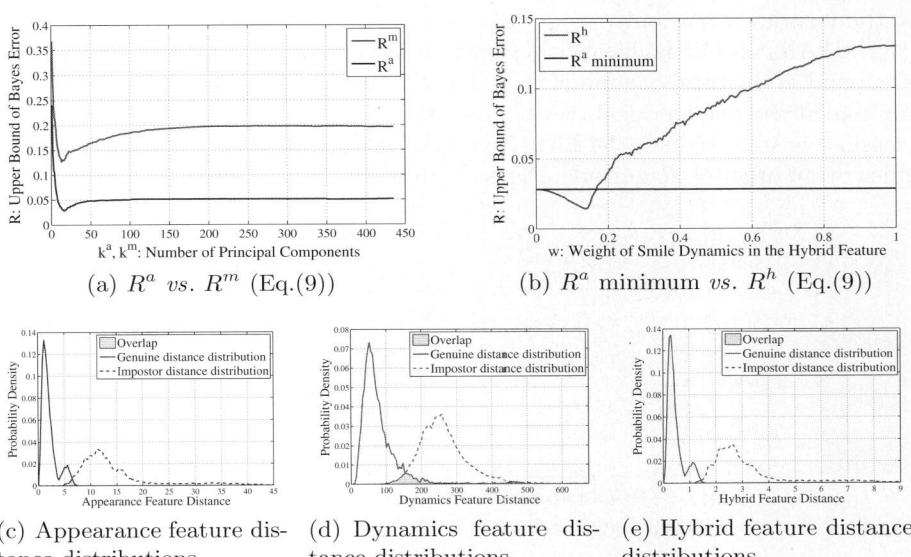

Fig. 2. Discriminating Power Measurement

w grows, until $w = 0.135$. After that, adding more smile dynamics causes R^h to increase. At $w = 0.135$, R^h hits its minimum of 0.014, which is a half of 0.028, the minimum of R^a. This observation implies that with respect to our current dataset, the hybrid feature can be twice more discriminative than the appearance feature.

Figure 2(e) shows the distributions of genuine distance and impostor distance computed from the hybrid feature at $k^h = 16, w = 0.135$. Compared with Figure 2(c), the overlap between the two distributions becomes smaller, which implies stronger discriminating power of the hybrid feature compared with the appearance feature.

5 An Attempt on the Identical Twins Problem

The identical twins problem is *the Holy Grail* in face recognition. In the following experiment, we make an attempt on it. Although our identical twin dataset is too small (only one pair of identical twin brothers) for us to draw any statistically convincing conclusion, the experiment results do encourage us to continue our journey along this way.

We collect around 20 smile video clips from each of a pair of identical twin brothers (Figure 3(a)) and add those videos into the previous dataset. We train two Bayes classifiers on the whole dataset. One of them uses the appearance feature and the other uses the hybrid feature. We test the two classifiers on the same dataset as used for training. For each classifier, two sets of FAR (False Accept Rate) and FRR (False Reject Rate) are computed, one from classifying the data of all the ordinary subjects only and the other from classifying the data of the identical twins only.

The FARs and FRRs have been shown in Table 1. The most interesting results are found in the third column of the table, the two FARs computed from classifying the data of the identical twins. The FAR (twins) of the hybrid-feature-based classifier is smaller than the FAR (twins) of the appearance-feature-based classifier by an order of magnitude. Please note that the FAR (twins) represents the

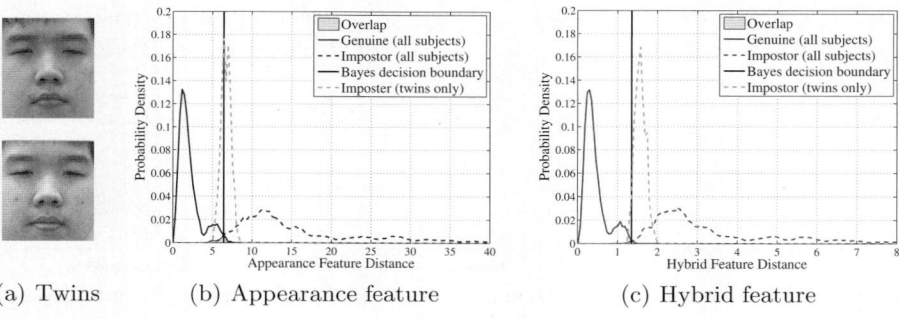

Fig. 3. An attempt on the identical twins problem

Table 1. FRRs and FARs of two Bayes classifiers applied on the identical twins data

	FRR (twins)	FAR (twins)	FRR (ordinary subjects)	FAR (ordinary subjects)
Appearance feature based	0%	25.53%	1.34%	1.51%
Hybrid feature based	0%	2.11%	0.48%	0.48%

chances of taking two identical twins as one person. Visually, by comparing Figure 3(b) and Figure 3(c), we can see that with the hybrid feature, the distribution of impostor distance between the twin brothers shifts towards the right side of the Bayes decision boundary (ideally, the distribution of impostor distance should be all to the right side of the decision boundary so that the FAR is zero).

Readers may ask why we do not train the classifiers on the data of twins only and then test their performance and compute the FAR (twins). The reason is that in the real world, a face recognition system can never know beforehand if the two faces in question are from a pair of identical twins or not. If the system knows that they are identical twins, then it already knows that they are from two different persons. Thus, the system will never choose to use a classifier trained specifically for identical twins. The best we can do is to build one system and try to make it applicable to both ordinary people and identical twins. This is the way we have followed in this trial.

6 Conclusions

In this paper, we propose a novel feature for face recognition by fusing facial appearance and facial dynamics. We measure and compare the discriminating power of this hybrid feature and the appearance feature by the statistical separability between genuine feature distance and impostor feature distance. In terms of Bayes' error, the discriminating power of this hybrid feature can be twice stronger than that of the appearance feature.

In the future, we are going to build a complete face recognition system based on this hybrid face feature. We are also going to study other types of facial motion features, their combination with traditional facial appearance features and their applications in face recognition problems. Moreover, we wish to run our experiment on a larger dataset, especially, if possible, on a larger identical twins dataset.

Acknowledgements

We would like to thank all the volunteers, without whose lovely smiles this research would not be possible. Thanks are extended to Dr. Yu Dan for the many constructive discussions between us. Special thanks are also given to the anonymous reviewers who have devoted their precious time in reading this paper and given us valuable comments. The authors acknowledge the generous support of NUS Research Grant ♯R-252-000-261-422.

References

1. Turk, M., Pentland, A.: Eigenfaces for recognition. Journal of Cognitive Neuroscience 3(1), 71–86 (1991)
2. Belhumeur, P., Hespanha, J., Kriegman, D.: Eigenfaces vs. fisherfaces: recognition using class specific linear projection. IEEE TPAMI 19(7), 711–720 (1997)
3. Phillips, P.J.: Support vector machines applied to face recognition. In: Proc. The Conference on Advances in Neural Information Processing Systems II (1999)
4. Liu, C., Wechsler, H.: Evolutionary pursuit and its application to face recognition. IEEE TPAMI 22(6), 570–582 (2000)
5. Li, S., Lu, J.: Face recognition using the nearest feature line method. IEEE Transactions on Neural Networks 10(2), 439–443 (1999)
6. Bartlett, M.S., Sejnowski, T.J.: Independent components of face images: A representation for face recognition. In: Proc. Joint Symposium on Neural Computation (1997)
7. Lin, S.H., Kung, S.Y., Lin, L.J.: Face recognition/detection by probabilistic decision-based neuralnetwork. IEEE Trans. on Neural Networks 8, 114–132 (1997)
8. Liu, X., Chen, T., Kumar, B.V.K.V.: Face authentication for multiple subjects using eigenflow. Pattern Recognition, Special issue on Biometrics 36, 313–328 (2001)
9. Lee, H.S., Kim, D.: Expression-invariant face recognition by facial expression transformations. Pattern Recognition Letters 29(13), 1797–1805 (2008)
10. Lander, K., Chuang, L., Wickham, L.: Recognizing face identity from natural and morphed smiles. Quarterly Journal of Experimental Psychology 59(5), 801–808 (2006)
11. Pilz, K.S., Thornton, I.M., Bülthoff, H.H.: A search advantage for faces learned in motion. Experimental Brain Research 171(4), 436–447 (2006)
12. Roark, D.A., Barrett, S.E., Spence, M., Abdi, H., O'Toole, A.J.: Memory for moving faces: Psychological and neural perspectives on the role of motion in face recognition. Behavioral and Cognitive Neuroscience Reviews 2(1), 15–46 (2003)
13. Tulyakov, S., Slowe, T., Zhang, Z., Govindaraju, V.: Facial expression biometrics using tracker displacement features. In: Proc. CVPR (2007)
14. Chen, L.F., Liao, H.Y., Lin, J.C.: Person identification using facial motion. In: Proc. ICIP (2001)
15. Zhang, Y., Kundu, S., Goldgof, D., Sarkar, S., Tsap, L.: Elastic face - an anatomy-based biometrics beyond visible cue. In: Proc. ICPR (2004)
16. Pamudurthy, S., Guan, E., Müller, K., Rafailovich, M.: Dynamic approach for face recognition using digital image skin correlation. In: Kanade, T., Jain, A., Ratha, N.K. (eds.) AVBPA 2005. LNCS, vol. 3546, pp. 1010–1018. Springer, Heidelberg (2005)
17. Ye, N., Sim, T.: Smile, you're on identity camera. In: Proc. ICPR (2008)
18. Wallhoff, F.: Facial expressions and emotion database, Technische Universität München (2006), www.mmk.ei.tum.de/~waf/fgnet/feedtum.html
19. Pantic, M., Valstar, M., Rademaker, R., Maat, L.: Web-based database for facial expression analysis. In: Proc. ICME (2005)
20. Lucas, B.D., Kanade, T.: An iterative image registration technique with an application to stereo vision. In: Proc. DARPA Image Understanding Workshop (1981)
21. Duda, R.O., Hart, P.E., Stork, D.G.: Pattern Classification, 2nd edn. Wiley Interscience, Hoboken (2000)

Finger-Knuckle-Print Verification Based on Band-Limited Phase-Only Correlation

Lin Zhang, Lei Zhang, and David Zhang

Biometrics Research Center, Department of Computing,
The Hong Kong Polytechnic Univeristy
{cslinzhang,cslzhang,csdzhang}@comp.polyu.edu.hk

Abstract. This paper investigates a new automated personal authentication technique using finger-knuckle-print (FKP) imaging. First, a specific data acquisition device is developed to capture the FKP images. The local convex direction map of the FKP image is then extracted, based on which a coordinate system is defined to align the images and a region of interest (ROI) is cropped for feature extraction and matching. To match two FKPs, we present a Band-Limited Phase-Only Correlation (BLPOC) based method to register the images and further to evaluate their similarity. An FKP database is established to examine the performance of the proposed method, and the promising experimental results demonstrated its advantage over the existing finger-back surface based biometric systems.

Keywords: Biometrics, finger-knuckle-print, personal authentication.

1 Introduction

Personal authentication is a common concern to both industries and academic research due to its numerous applications. Biometrics, which refers to the unique physiological or behavioral characteristics of human beings, can be used to distinguish between individuals. In the past three decades, researchers have exhaustively investigated the use of a number of biometric characteristics [1].

Recently, it has been noticed that the texture in the outer finger surface has the potential to do personal authentication. Woodward et al. [2-3] set up a 3-D hand database with the Minolta 900/910 sensor and they extracted 3-D features from finger surface to identify a person's identity. However, they did not provide a practical solution to establishing an efficient system using the outer finger surface features. The cost, size and weight of the Minolta 900/910 sensor limit the use of it in a practical biometric system, and the time-consuming 3-D data acquisition and processing limit its use in real-time applications. Later, Kumar and Ravikanth [4-5] proposed another approach to personal authentication by using 2-D finger-back surface imaging. They developed a device to capture hand-back images and then extracted the finger knuckle areas by some preprocessing steps. The subspace analysis methods such as PCA, LDA and ICA were combined to do feature extraction and matching. With Kumar et al.'s design, the acquisition device is doomed to have a large size because nearly the whole hand back area has to be captured, despite the fact that the finger knuckle area only occupies a small portion of the acquired image. Furthermore, subspace analysis methods may be effective for face recognition but they may not be able to effectively extract the distinctive line features from the finger knuckle surface.

This paper presents a novel system for online personal authentication based on finger-knuckle-print (FKP), which refers to the inherent skin pattern of the outer surface around the phalangeal joint of one's finger. A specially designed acquisition device is constructed to collect FKP images. Unlike the systems in [3] and [5] which first capture the image of the whole hand and then extract the finger or finger knuckle surface areas, the proposed system captures the image around the finger knuckle area of a finger directly, which largely simplifies the following data preprocessing steps. Meanwhile, with such a design the size of the imaging system can be greatly reduced, which improves much its applicability. Since the finger knuckle will be slightly bent when being imaged in the proposed system, the inherent finger knuckle print patterns can be clearly captured and hence the unique features of FKP can be better exploited. For matching FKPs, we present an efficient and effective Band-Limited Phase-Only Correlation based method. Compared with the existing finger knuckle surface based biometric systems [2-5], the proposed system performs much better in terms of both recognition accuracy and speed.

The rest of this paper is organized as follows. Section 2 introduces the design and structure of the FKP image acquisition device. Section 3 describes the FKP image preprocessing and ROI extraction methods. Section 4 investigates the BLPOC-based FKP matching. Section 5 reports the experimental results. Finally, conclusions are presented in section 6.

2 The FKP Recognition System

The proposed FKP recognition system is composed of an FKP image acquisition device and a data processing module. The device (referring to Fig. 1-a) is composed of a finger bracket, a ring LED light source, a lens, a CCD camera and a frame grabber. The captured FKP image is inputted to the data processing module, which comprises three basic steps: ROI (region of interest) extraction, feature extraction, and feature matching. Refer to Fig. 1-a, a basal block and a triangular block are used to fix the position of the finger joint. The vertical view of the triangular block is illustrated in Fig. 1-b. Fig. 2-a shows a sample image acquired by the developed device.

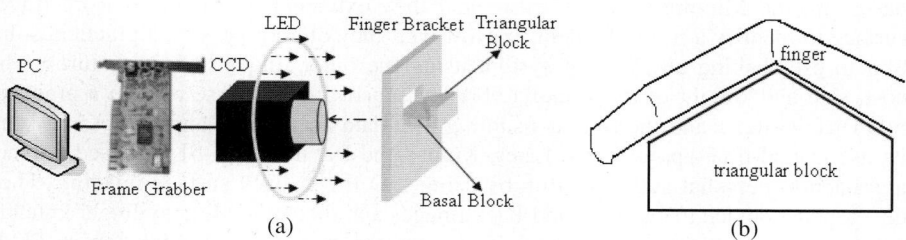

Fig. 1. FKP image acquisition device

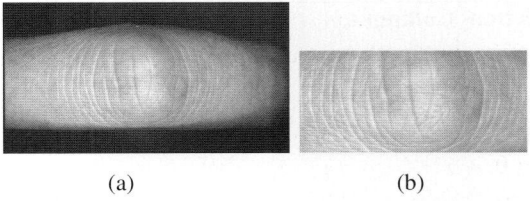

Fig. 2. (a) is a sample FKP image; (b) is the ROI image of (a)

3 ROI Extraction

It is necessary to construct a local coordinate system for each FKP image. With such a coordinate system, an ROI can be cropped from the original image for reliable feature extraction. The detailed steps for setting up such a coordinate system are as follows.

Step 1: determine the X-axis of the coordinate system. The bottom boundary of the finger can be easily extracted by a Canny edge detector. Actually, this bottom boundary is nearly consistent to all FKP images because all the fingers are put flatly on the basal block in data acquisition. By fitting this boundary as a straight line, the X-axis of the local coordinate system is determined.

Step 2: crop a sub-image I_S. The left and right boundaries of I_S are two fixed values evaluated empirically. The top and bottom boundaries are estimated according to the boundary of real fingers and they can be obtained by a Canny edge detector.

Step 3: Canny edge detection. Apply the Canny edge detector to I_S to obtain the edge map I_E.

Step 4: convex direction coding for I_E. We define an ideal model for FKP "curves". In this model, an FKP "curve" is either convex leftward or convex rightward. We code the pixels on convex leftward curves as "1", pixels on convex rightward curves as "-1", and the other pixels not on any curves as "0".

Step 5: determine the Y-axis of the coordinate system. For an FKP image, "curves" on the left part of phalangeal joint are mostly convex leftward and those on the right part are mostly convex rightward. Meanwhile, "curves" in a small area around the phalangeal joint do not have obvious convex directions. Based on this observation, at a horizontal position x (x represents the column) of an FKP image, we define the "convexity magnitude" as:

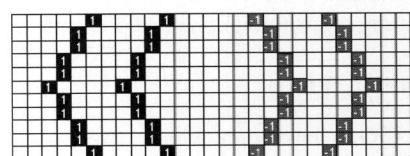

Fig. 3. Illustration for convex direction coding scheme

Fig. 3 illustrates this convex direction coding scheme and the pseudo codes are presented as follows:

```
Convex_Direction_Coding( $I_E$ )
Output: $I_{CD}$ (convex direction code map)
$y_{mid}$ = height of $I_E$ / 2 ;
for each $I_E(i, j)$ :
    if $I_E(i, j) = 0$
        $I_{CD}(i, j) = 0$ ;
    else if $I_E(i+1, j-1) = 1$ and $I_E(i+1, j+1) = 1$
        $I_{CD}(i, j) = 0$ ;
    else if $(I_E(i+1, j-1) = 1$ and $i <= y_{mid})$ or $(I_E(i+1, j+1) = 1$ and $i > y_{mid})$
        $I_{CD}(i, j) = 1$ ;
    else if $(I_E(i+1, j+1) = 1$ and $i <= y_{mid})$ or $(I_E(i+1, j-1) = 1$ and $i > y_{mid})$
        $I_{CD}(i, j) = -1$ ;
    end if
end for
```

$$conMag(x) = abs\left(\sum_W I_{CD}\right) \qquad (1)$$

where W is a window being symmetrical about the axis $X = x$. W is of the size $d \times h$, where h is the height of I_S. The characteristic of the FKP image suggests that $conMag(x)$ will reach a minimum around the center of the phalangeal joint and this position can be used to set the Y-axis of the coordinate system. Let

$$x_0' = \arg\min_x (conMag(x)) \qquad (2)$$

Then $X = x_0'$ is set as the Y-axis.

Step 6: crop the ROI image. Now that we have fixed the X-axis and Y-axis, the local coordinate system can then be determined and the ROI sub-image I_{ROI} can be extracted with a fixed size. Fig. 2-b shows an example of the extracted ROI images.

4 BLPOC-Based FKP Matching

Given two FKP ROIs, a matching algorithm determines their degree of similarity. A BLPOC based FKP matching algorithm is presented in this section.

4.1 Phase-Only Correlation (POC)

Phase-Only Correlation (POC) has been widely used for image registration tasks [6-8], and recently it has been adopted in some biometric systems as a similarity measure [9-11]. POC based method relies on the translation property of the Fourier transform. Let f and g be the two images that differ only by a displacement (x_0, y_0) i.e.

$$g(x, y) = f(x - x_0, y - y_0) \qquad (3)$$

Their corresponding Fourier transforms $G(u,v)$ and $F(u,v)$ will be related by

$$G(u,v) = e^{-j2\pi(ux_0+vy_0)}F(u,v) \quad (4)$$

The cross-phase spectrum $R_{GF}(u,v)$ between $G(u,v)$ and $F(u,v)$ is given by

$$R_{GF}(u,v) = \frac{G(u,v)F^*(u,v)}{|G(u,v)F^*(u,v)|} = e^{-j2\pi(ux_0+vy_0)} \quad (5)$$

where F^* is the complex conjugate of F. By taking inverse Fourier transform of R_{GF} back to the time domain, we will have a Dirac impulse centered on (x_0, y_0).

In practice, we should consider the finite discrete representations. Consider two $M \times N$ images, $f(m,n)$ and $g(m,n)$, where the index ranges are $m=-M_0,\ldots,M_0$ ($M_0 > 0$) and $n=-N_0,\ldots,N_0$ ($N_0 > 0$), and $M = 2M_0+1$ and $N = 2N_0+1$. Denote by $F(u,v)$ and $G(u,v)$ the 2D DFTs of the two images and they are given by

$$F(u,v) = \sum_{m=-M_0}^{M_0}\sum_{n=-N_0}^{N_0} f(m,n)e^{-j2\pi\left(\frac{mu}{M}+\frac{nv}{N}\right)} = A_F(u,v)e^{j\phi_F(u,v)} \quad (6)$$

$$G(u,v) = \sum_{m=-M_0}^{M_0}\sum_{n=-N_0}^{N_0} g(m,n)e^{-j2\pi\left(\frac{mu}{M}-\frac{nv}{N}\right)} = A_G(u,v)e^{j\phi_G(u,v)} \quad (7)$$

where $u=-M_0,\ldots,M_0$, $v=-N_0,\ldots,N_0$, $A_F(u,v)$ and $A_G(u,v)$ are amplitude components, and $\phi_F(u,v)$ and $\phi_G(u,v)$ are phase components. Then, the cross-phase spectrum $R_{GF}(u,v)$ between $G(u,v)$ and $F(u,v)$ is given by

$$R_{GF}(u,v) = \frac{G(u,v)F^*(u,v)}{|G(u,v)F^*(u,v)|} = e^{j\{\phi_G(u,v)-\phi_F(u,v)\}} \quad (8)$$

The POC function $p_{gf}(m,n)$ is the 2D Inverse DFT (IDFT) of $R_{GF}(u,v)$:

$$p_{gf}(m,n) = \frac{1}{MN}\sum_{u=-M_0}^{M_0}\sum_{v=-N_0}^{N_0} R_{GF}(u,v)e^{j2\pi\left(\frac{mu}{M}+\frac{nv}{N}\right)} \quad (9)$$

If the two images f and g are similar, their POC function will give a distinct sharp peak. On the contrary, if they are not similar, the peak value will drop significantly. Thus, the height of the peak value can be used as a similarity measure, and the location of the peak shows the translational displacement between the two images.

4.2 Band-Limited Phase-Only Correlation (BLPOC)

In the POC-based image matching method, all the frequency components are involved. However, high frequency tends to emphasize detail information and can be prone to noise. To eliminate meaningless high frequency components, K. Ito et al. [9] proposed the Band-Limited Phase-Only Correlation (BLPOC).

The BLPOC limits the range of the spectrum of the given FKP image. Assume that the ranges of the inherent frequency band of FKP texture are given by $u=-U_0,\ldots,U_0$ and $v=-V_0,\ldots V_0$, where $0<=U_0<=M_0$, $0<=V_0<=N_0$. Thus, the effective size of spectrum is given by $L_1=2U_0+1$ and $L_2=2V_0+1$. The BLPOC function is defined as

$$p_{gf}^{U_0 V_0}(m,n) = \frac{1}{L_1 L_2} \sum_{u=-U_0}^{U_0} \sum_{v=-V_0}^{V_0} R_{GF}(u,v) e^{j2\pi\left(\frac{mu}{L_1} + \frac{nv}{L_2}\right)} \quad (10)$$

where $m = -U_0, \ldots, U_0$ and $n = -V_0, \ldots, V_0$. When two images are similar, their BLPOC function gives a distinct sharp peak. Also, the translational displacement between the two images can be estimated by the correlation peak position. Experiments indicate that the BLPOC function provides a much higher discrimination capability than the original POC function.

4.3 BLPOC-Based FKP Matching

Given two FKP ROIs $f(m,n)$ and $g(m,n)$, we assume that there is only translational displacement between them. Thus, the BLPOC-based FKP matching process is quite straightforward and is summarized as follows:

FKP_Matching($f(m, n)$, $g(m, n)$)
Output: *matching_score*

register f and g based on BLPOC method;
extract overlapping areas f' and g' from registered f and g, respectively;
if $area(f') / area(f) < threshold$
　　$P = BLPOC(f, g)$;
else
　　$P = BLPOC(f', g')$;
end
matching_score = $\max(P)$;

5　Experimental Results

Database establishment. An FKP database was established. The FKP images were collected from 165 volunteers, including 125 males and 40 females. Among them, 143 subjects are 20~30 years old and the others are 30~50 years old. We collected the images in two separate sessions. In each session, the subject was asked to provide 6 images for each of the left index finger, the left middle finger, the right index finger and the right middle finger. Therefore, 48 images from 4 fingers were collected from each subject. In total, the database contains 7,920 images from 660 different fingers.

Experiment 1. In the first experiment, we took images captured in the first session as the gallery set and images captured in the second session as the probe set. Therefore, there were 660 classes and 3,960 images in the gallery set and the probe set each. Each image in the probe set was matched against the all the images in the gallery set. A match was counted as genuine if the two FKPs were from the same finger; otherwise the match was counted as imposter. As a result, the numbers of genuine matches and imposter matches were 23,760 and 7,828,920, respectively. The EER (Equal Error Rate) wet got was 1.68%, which was quite promising.

Fig. 4a depicts the corresponding FAR (False Acceptance Rate) and FRR (False Rejection Rate) curves. The distance distributions of genuine matching and imposter matching obtained in this experiment are plotted in Fig. 4b. Experimental results indicate that the proposed system has a good capability to verify a person's identity.

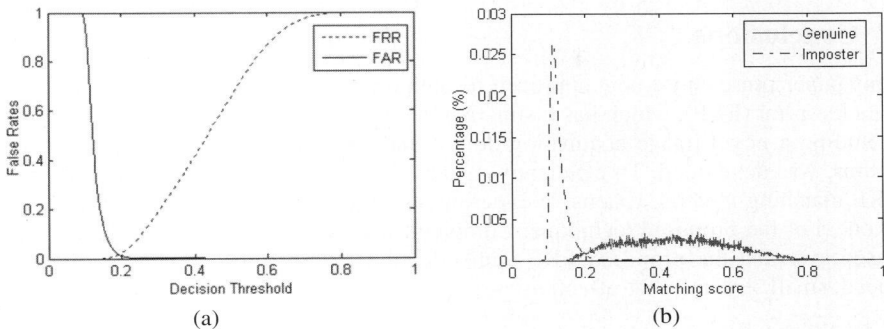

Fig. 4. (a) FAR and FRR curves obtained in experiment 1. (b) Distance distributions of genuine matching and imposter matching obtained in experiment 1.

Experiment 2. The goal of this experiment was to investigate the system's performance when we fuse information from 2 or more fingers of a person. In fact, at such a case the system works as a kind of multi-modal system with a single biometric trait but multiple units. We adopt the SUM fusing rule as follows:

$$s_{sum} = \sum s_i \qquad (11)$$

where s_i is the matching score of the client's i_{th} finger.

Table 1. Experimental results in experiment 2

Method	gallery classes	gallery samples	probe classes	probe samples	Finger types for fusion	EER (%)
Ours	165	990	165	990	l-index, l-middle	0.72
Ours	165	990	165	990	r-index, r-middle	0.31
Ours	165	990	165	990	l-index, r-index	0.40
Ours	165	990	165	990	l-middle, r-middle	0.31
Woodward [3]	132	660	177	531	r-index, r-middle, r-ring	5.50
Kumar [5]	105	420	105	210	index, middle, ring, little	1.39

We tested several different fusions of fingers and the results are presented in Table 1, from which it can be easily observed that by integrating information from more fingers the recognition performance of the system could be largely improved. We also present the results extracted from [3] and [5] in Table 1 for comparison. It is clearly shown that the proposed system performed much better even though we incorporated information from fewer fingers.

Speed. The software was implemented with Visual C#.Net 2005 on a Dell Inspiron 530s PC embedded Intel E6500 process and 2G RAM. The execution time for data preprocessing and ROI extraction was 216ms. The time for one BLPOC-based matching was 4.2ms. Thus, the total execution time for one verification operation was less than 0.3s in our prototype system, which was fast enough for real-time applications.

6 Conclusions

This paper presented a new approach to online personal authentication using finger-knuckle-print (FKP), which has distinctive line features. A cost-effective FKP system, including a novel image acquisition device and the associated data processing algorithms, was developed. To efficiently match the FKPs, we proposed a BLPOC based FKP matching method. Extensive experiments demonstrated the efficiency and effectiveness of the proposed technique. Compared with other existing finger back surface based systems, the proposed FKP authentication has merits of high accuracy, high speed, small size and cost-effective.

Acknowledgments

The work is partially supported by the Edward Sai Kim Hotung Fund (5-ZH52), the HK-PolyU Internal Competitive Research Grant (G-YH54), and the Hong Kong RGC General Research Fund (PolyU 5351/08E).

References

1. Jain, A.K., Flynn, P., Ross, A. (eds.): Handbook of Biometrics. Springer, Heidelberg (2007)
2. Woodard, D.L., Flynn, P.J.: Personal identification utilizing finger surface features. In: Proc. CVPR, vol. 2, pp. 1030–1036 (2005)
3. Woodard, D.L., Flynn, P.J.: Finger surface as a biometric identifier. Computer Vision and Image Understanding 100(3), 357–384 (2005)
4. Ravikanth, C., Kumar, A.: Biometric authentication using finger-back surface. In: Proc. CVPR, pp. 1–6 (2007)
5. Kumar, A., Ravikanth, C.: Personal authentication using finger knuckle surface. IEEE Trans. Information Forensics and Security 4(1), 98–109 (2009)
6. Kuglin, C.D., Hines, D.C.: The phase correlation image alignment method. In: Proc. Int. Conf. on Cybernetics and Society, pp. 163–165 (1975)
7. Chen, Q.S., Defrise, M., Deconinck, F.: Symmetric phase-only matched filtering of Fourier-Mellin transforms for image registration and recognition. IEEE Trans. PAMI 16(12), 1156–1168 (1994)
8. Srinivasa Reddy, B., Chatterji, B.N.: An FFT-based technique for translation, rotation, and scale-invariant image registration. IEEE Trans. IP 5(8), 1266–1271 (1996)
9. Ito, K., Nakajima, H., Kobayashi, K., Aoki, T., Higuchi, T.: A fingerprint matching algorithm using phase-only correlation. IEICE Trans. Fundamentals E87-A(3), 682–691 (2004)
10. Zhang, J.X., Ou, Z.Y., Wei, H.L.: Fingerprint matching using phase-only correlation and Fourier-Mellin transforms. In: Proc. Int. Conf. ISDA, vol. 2, pp. 379–383 (2006)
11. Miyazawa, K., Ito, K., Aoki, T., Kobayashi, K., Nakajima, H.: An effective approach for iris recognition using phase-based image matching. IEEE Trans. PAMI 20(10), 1741–1756 (2008)

Calibration of Radially Symmetric Distortion by Fitting Principal Component

Hideitsu Hino[1], Yumi Usami[1], Jun Fujiki[2], Shotaro Akaho[2], and Noboru Murata[1]

[1] Waseda University, Okubo 3–4–1, Shinjuku-ku, Tokyo, 169–8555, Japan
hideitsu.hino@toki.waseda.jp, yumi.usami@murata.eb.waseda.ac.jp,
noboru.murata@eb.waseda.ac.jp
[2] National Institute of Advanced Industrial Science and Technology,
Tsukuba, Ibaraki 305–8568, Japan
{jun-fujiki,s.akaho}@aist.go.jp

Abstract. To calibrate radially symmetric distortion of omnidirectional cameras such as fish-eye lenses, calibration parameters are usually estimated so that lines, which are supposed to be straight in the 3D real scene, are mapped to straight lines in the calibrated image. In this paper, this problem is treated as a fitting problem of the principal component in uncalibrated images, and an estimation procedure of calibration parameters is proposed based on the principal component analysis. Experimental results for synthetic data and real images are presented to demonstrate the performance of our calibration method.

Keywords: camera calibration, fish-eye lens, principal component analysis.

1 Introduction

There is a growing interest in panoramic imagery, which has many potential applications such as security, 3D reconstruction, medical surgery and single view metrology. For those applications, cameras with fish-eye lenses are often employed because they can realize a quite large field of view. Since images taken with these cameras usually have significant distortions, they have to be transformed into perspective projection plane images in a distortion free manner. There are a lot of sources of distortions, and they can be classified into three categories: radial distortion, decentering distortion, and thin prism distortion [1]. In the case of perfectly centered fish-eye lenses, their geometrical distortion is mainly caused by the shape of lenses, and thus the radial distortion is dominant. Calibration of the radial distortion has been actively studied [2],[3],[4],[5],[6], and most of those studies are based on the fact that straight lines in the real scene should be straight in the perspective projection plane [7].

This paper deals with the problem of estimating the non-linear transformation that maps points in the distorted observation plane to points in the undistorted (calibrated) plane. In this paper, we discuss the radial distortion with a perfectly centered lens model. Let r_d be the distance between a point (x, y) and the center in the distorted observation plane, and let r_u be the distance in the undistorted plane, that is, the distance in the perspective projection plane between the point corresponding to (x, y) and the center. We can formulate the calibration problem as determination of a calibration function $f : r_d \mapsto r_u$. We model this function f by a linear combination of basis functions as $f(r) = \sum_{n=1}^{N} c_n f_n(r)$, and estimate its coefficients $\{c_n\}_{n=1}^{N}$. We will show that this estimation is formulated as a linear fitting problem in the perspective projection plane, and reduced to solving two eigen value problems of matrices.

The rest of this paper is organized as follows. In section 2, the camera model treated in this paper is explained. In section 3, an algebraic method of camera calibration is proposed. The method is validated by experiments with synthetic data and real world image taken by a camera with a fish-eye lens in section 4, and section 5 is devoted to discussion and concluding remarks.

2 Fish-Eye Imaging Model

In this section, we briefly explain the camera model adopted in this paper. Hereafter, we assume that the center of the radially symmetric distortion coincides with the center of the observation image. We note that the calibration of distortion center can be done using the method proposed in [3], for example. Let $\boldsymbol{x} \in \mathbb{R}^2$ be a point in the observation plane, and $\boldsymbol{\phi}(\boldsymbol{x}) \in \mathbb{R}^2$ be a point in the corresponding perspective projection plane. Under the assumption of radially symmetric distortion, the mapping $\boldsymbol{\phi}(\boldsymbol{x})$ is represented by a function $f(r) : \mathbb{R} \to \mathbb{R}$ which only depend on the radius r:

$$\boldsymbol{\phi}(\boldsymbol{x}) = f(r) \frac{\boldsymbol{x}}{||\boldsymbol{x}||} = \frac{f(r)}{r} \boldsymbol{x}, \quad ||\boldsymbol{x}|| = r. \tag{1}$$

Figure 1 depicts the correspondence between the observation plane and the perspective projection plane. In this paper, we consider the case that the calibration function can be approximated by a linear combination of basis functions $\{f_n(r)\}_{n=1}^{N}$ as

$$f(r) = \sum_{n=1}^{N} c_n f_n(r). \tag{2}$$

Consequently, the calibration of the radially symmetric distortion is conducted by determining the coefficients $c_i, i = 1, \ldots, N$ using given data points. Usually, a 3D real scene contains a number of straight lines, and these lines are observed with distortion caused by the fish-eye lens. We suppose that there are S distorted lines in the observation plane, and denote data points on one of those lines as $\{\boldsymbol{x}_{[d]}^s\}_{d=1}^{D_s}, s = 1, \ldots, S$, which means the s-th line includes D_s data points. We also suppose that S is more than N, the number of basis of the calibration function.

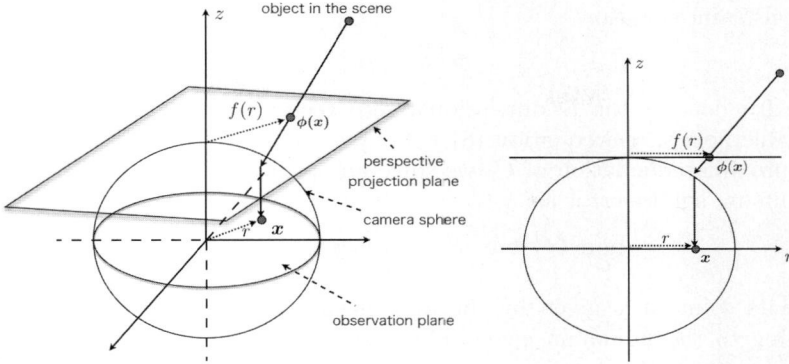

Fig. 1. Camera model. Left: Correspondence between an object in scene, observation plane, and perspective projection plane. Right: Cross section view.

3 Calibration of Radially Symmetric Distortion

In this section, we derive a non-iterative method for estimating the coefficient vector $\boldsymbol{c} = (c_1, \ldots, c_N)$ of the calibration function (2). In the following discussion, all the vectors are regarded as column vectors.

We first consider points $\{\boldsymbol{x}_{[d]}\}_{d=1}^{D}$ on a line. We can rewrite the mapping as

$$\boldsymbol{\phi}(\boldsymbol{x}_{[d]}) = \sum_{n=1}^{N} c_n \, \boldsymbol{\phi}_n(\boldsymbol{x}_{[d]}) = P(\boldsymbol{x}_{[d]})\boldsymbol{c}, \tag{3}$$

where $\boldsymbol{\phi}_n(\boldsymbol{x}) = f_n(\|\boldsymbol{x}\|)\frac{\boldsymbol{x}}{\|\boldsymbol{x}\|}$ and

$$P(\boldsymbol{x}_{[d]}) = \underset{(2\times N)}{\bigl(\boldsymbol{\phi}_1(\boldsymbol{x}_{[d]}) \; \cdots \; \boldsymbol{\phi}_N(\boldsymbol{x}_{[d]})\bigr)}. \tag{4}$$

Because all points $\boldsymbol{\phi}(\boldsymbol{x}_{[d]})$, $d = 1, \ldots, D$ should lie on the same line, we get D constraints using a normal vector $\boldsymbol{a}$ and an intercept b of the line as

$$\begin{aligned}(\boldsymbol{a})^\top \boldsymbol{\phi}(\boldsymbol{x}_{[d]}) + b &= (\boldsymbol{a})^\top P(\boldsymbol{x}_{[d]})\boldsymbol{c} + b \\ &= (\boldsymbol{c} \otimes \boldsymbol{a})^\top \bigl(\text{vec}P(\boldsymbol{x}_{[d]})\bigr) + b = 0, \quad d = 1, \ldots, D,\end{aligned} \tag{5}$$

where $\text{vec}X$ denotes a column vector expansion (column span) of a matrix X. Using homogeneous coordinates

$$\boldsymbol{\Phi}(\boldsymbol{x}) = \bigl(\text{vec}P(\boldsymbol{x})^\top \;\; 1\bigr)^\top \in \mathbb{R}^{2N+1}, \tag{6}$$

and letting $\boldsymbol{C} = \bigl((\boldsymbol{c} \otimes \boldsymbol{a})^\top \;\; b\bigr)^\top \in \mathbb{R}^{2N+1}$, the constraint (5) is written as

$$\boldsymbol{\Phi}^\top(\boldsymbol{x}_{[d]})\boldsymbol{C} = 0. \tag{7}$$

Combining all the points on the line as $\boldsymbol{\Phi} = \left(\boldsymbol{\Phi}(\boldsymbol{x}_{[1]}) \cdots \boldsymbol{\Phi}(\boldsymbol{x}_{[D]})\right)$, we get a line-constraint equation

$$\boldsymbol{\Phi}^\top \boldsymbol{C} = \boldsymbol{0}_D, \qquad (8)$$

where $\boldsymbol{0}_D$ denotes the D dimensional zero vector. In real images, there exist inevitable noises, and equation (8) only approximately holds. Hence, to obtain an approximated solution of $\boldsymbol{C}$, we solve an optimization problem in the sense of minimum square error as

$$\min_{\boldsymbol{C}} \boldsymbol{C}^\top \boldsymbol{\Phi} \boldsymbol{\Phi}^\top \boldsymbol{C} \quad \text{subject to} \quad ||\boldsymbol{C}|| = 1, \qquad (9)$$

where its solution is given by the unit eigen vector of a matrix $\boldsymbol{\Phi}\boldsymbol{\Phi}^\top$ corresponding to the minimum eigen value. Solving optimization problems (9) for all S lines, we get S estimates $\boldsymbol{C}^s = \left((\boldsymbol{c} \otimes \boldsymbol{a}^s)^\top \; b^s\right)^\top, s = 1, \ldots, S$. We note that the unknown coefficient $\boldsymbol{c}$ is common for all S estimates. When we write $A_{(2 \times S)} = \left(\boldsymbol{a}^1 \cdots \boldsymbol{a}^S\right)$, from the property of the Kronecker product, an identity

$$\underset{(2N \times S)}{\boldsymbol{c} \otimes A} = \left(\boldsymbol{c} \otimes \boldsymbol{a}^1 \cdots \boldsymbol{c} \otimes \boldsymbol{a}^S\right) \qquad (10)$$

holds. We further transform this matrix $\boldsymbol{c} \otimes A$ by an operator $\mathcal{H}$ for $(2N \times S)$ matrix defined as follows. Let M be a matrix with even dimensional rows

$$\underset{(2N \times S)}{M} = \begin{pmatrix} \boldsymbol{m}_1 & \boldsymbol{m}_2 & \cdots & \boldsymbol{m}_{2N} \end{pmatrix}^\top,$$

where $\boldsymbol{m}_i, i = 1, \ldots, 2N$ are S dimensional column vectors. Then the operator $\mathcal{H}$ is defined by

$$\underset{(N \times S)}{\mathcal{H}(M)} := \left((\boldsymbol{m}_1 + \boldsymbol{m}_2) \; \cdots \; (\boldsymbol{m}_{2N-1} + \boldsymbol{m}_{2N})\right)^\top = (I_n \otimes \boldsymbol{1}_2^\top)M, \qquad (11)$$

where I_n is an $n \times n$ identity matrix and $\boldsymbol{1}_n$ is an n dimensional column vector with one for all entries. With this operator $\mathcal{H}$, we see that

$$\underset{(N \times S)}{\mathcal{H}(\boldsymbol{c} \otimes A)} = (I_n \otimes \boldsymbol{1}_2^\top)(\boldsymbol{c} \otimes A) = (I_n \boldsymbol{c}) \otimes (\boldsymbol{1}_2^\top A) = \underset{(N \times 1)}{\boldsymbol{c}} \; \underset{(1 \times S)}{\mathcal{H}(A)}. \qquad (12)$$

Because of the estimation error for $\{\boldsymbol{C}^s\}_{s=1}^S$, equation (12) will hold only approximately. We need to find a decomposition of $\mathcal{H}(\boldsymbol{c} \otimes A)$ into $\boldsymbol{c}$ and $\mathcal{H}(A)$ which approximate (12). When we let $\boldsymbol{\Psi} := \mathcal{H}(\boldsymbol{c} \otimes A)$, then the decomposition of $\boldsymbol{\Psi}$ is, up to an arbitrary non-zero multiplicative factor, approximated by the solution of the following optimization problem

$$\max_{\boldsymbol{c}} \boldsymbol{c}^\top \boldsymbol{\Psi} \boldsymbol{\Psi}^\top \boldsymbol{c} \quad \text{subject to} \quad ||\boldsymbol{c}|| = 1, \qquad (13)$$

where its solution is given by the unit eigen vector of a matrix $\boldsymbol{\Psi}\boldsymbol{\Psi}^\top$ corresponding to the largest eigen value. We note that Loan and Pitsianis [8] have proposed a general method for approximating a matrix $A \in \mathbb{R}^{m \times n}$ with $m = m_1 m_2$ and $n = n_1 n_2$ by Kronecker product of matrices $B \in \mathbb{R}^{m_1 \times n_1}$ and $C \in \mathbb{R}^{m_2 \times n_2}$ so that the Frobenius norm $||A - B \otimes C||_F$ is minimized. For our aim, however, the decomposition method $\mathcal{H}(\boldsymbol{c} \otimes A) \simeq \boldsymbol{c}\mathcal{H}(A)$ is simpler and sufficient.

4 Experimental Results

With simulated data and real images taken by a camera with a fish-eye lens, we carried out a number of experiments to assess the performance of our calibration algorithm.

4.1 Simulation Result

The proposed calibration method is based on the fact that straight lines in the 3D real scene must be straight in the perspective projection plane. Therefore, one of the measure of the performance of the calibration method is the linearity of calibrated points. That is, how accurately the points on a line in the 3D real scene are projected to points on a line in the perspective projection plane. This linearity can be measured by the contribution ratio of the first eigen value to the sum of all eigen values of the variance-covariance matrix of the projected points on a line.

We generated points on lines in $[-3, 3] \times [-3, 3]$ region on the perspective projection plane (Fig. 2 (a), plotted with ○). Then, using the inverse of a calibration function $f(r) = 0.7r + 0.7r^3$, the points on the lines are mapped on the observation plane (plotted with △). In this preliminary experiment, we used the same basis functions $\{f_1(r) = r, f_2(r) = r^3\}$ for calibration. The estimated parameter should be nearly equal to $\bm{c} = (c_1, c_2) = (0.7, 0.7)$ up to constant. In a column named as "no noise" of Table 1, we show the average contribution ratio of the first eigen value of the variance-covariance matrix of the calibrated points, and the ratio of the parameters c_1/c_2. The figures in Table 1 are means and one standard deviations of 100 trials.

An example of the calibration is shown in Fig. 2 (b). In the figure, the points on the observation plane and the points on the perspective projection plane are simultaneously plotted on the same plane with different marks. The observed points are mapped to the perspective projection plane (plotted with ●) by the estimated calibration function.

To see the effect of the noise in the observation images, we also conduct experiments with noisy observations. Gaussian noises with zero-mean and σ^2 variance are added to the observed points. The noise level σ is set to $1/300$ for an experiment with small Gaussian noises, and set to $1/100$ for an experiment with large Gaussian noises. Examples of the calibration with noised data are also shown in Fig. 2, (c) is an example of a small noise, and (d) is an example of a large noise experiment. The contribution ratios and the estimated parameter ratios for experiments with noisy data are shown in Table 1. From these results, we see that the contribution ratio is nearly equal to one even with large Gaussian noises, so the linearity of the calibrated points are well recovered by the proposed method. We also see from Fig. 2 (d) that calibrated points show some dispersions when there are large noises in the observed points. In the concluding remark, we will briefly note the idea to make our method robust.

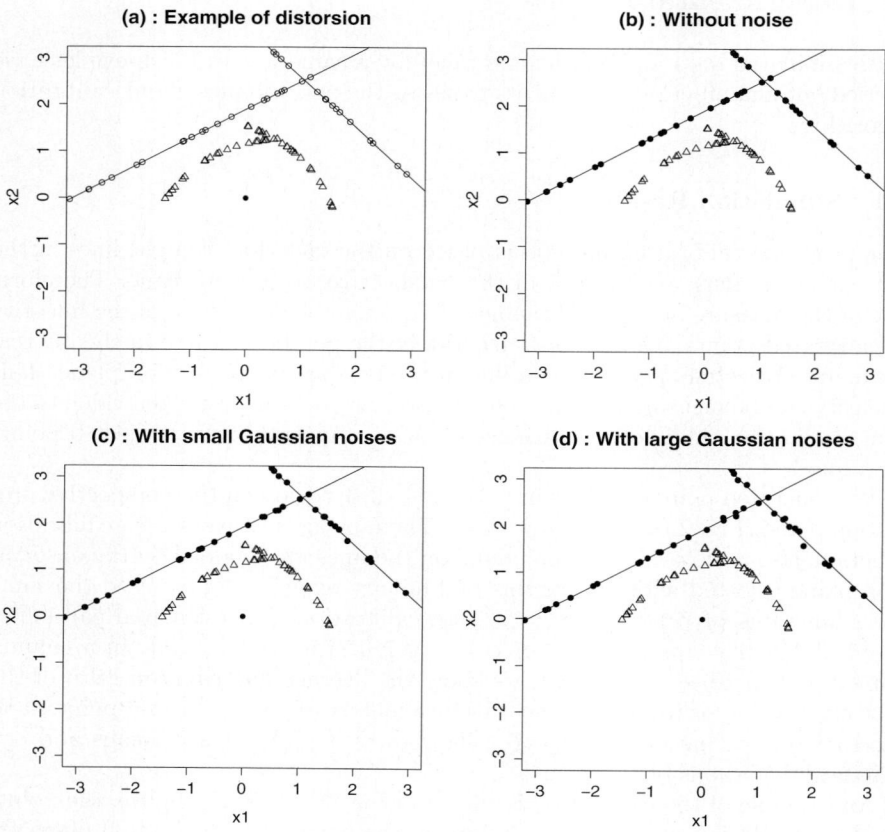

Fig. 2. Samples of calibration results for various noise levels. The observation plane and the perspective projection plane are superimposed on the same plane. (a) : Points on lines in perspective projection plane (○) are mapped to points on observation plane (△). (b) : Observed points (△) are mapped to the perspective projection plane (•) with the estimated calibration function. (c) : Small Gaussian noise is added to the observed points. (d) : Large Gaussian noise is added to the observed points.

Table 1. Contribution ratio of the first eigen value and parameter estimation

	no noise	small Gaussian	large Gaussian
Contribution ratio	1 ± 0	0.99997 ± 0.00002	0.99918 ± 0.00525
Parameter ratio	0.99994 ± 0.00012	1.01390 ± 0.08042	1.07099 ± 0.59017

4.2 Real Images

We demonstrate an experiment with real image taken by a fish-eye lens to confirm the validity of our proposed method. The fish-eye lens used here is FUJINON

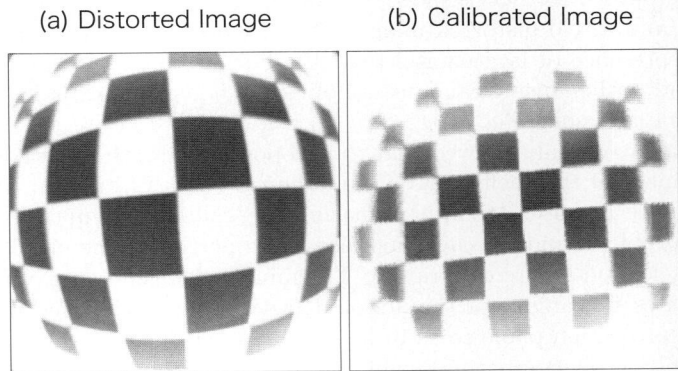

Fig. 3. (a) : An image taken by fish-eye camera. (b) : A calibrated image with our proposed method.

YV2.2x1.4A2 with field angle 185 degrees, mounted on Watec WAT-221S digital camera. A fish-eye image taken with this fish-eye camera for calibration is shown in Fig. 3 (a). The original resolution of the fish-eye image is 752 × 596, though, for the presentation purpose, we cropped the region of interest of the image. From this fish-eye image, 2 image curves and total amount of 46 sample points on these lines are chosen using the KLT feature tracker[9]. The coefficients c of the calibration function are estimated using the proposed calibration method. In this experiment, we adopted a polynomial model as

$$f(r) = c_1 r + c_5 r^5, \qquad (14)$$

which acquires the best estimation performance in our preliminary experiments. The estimated parameter was $c = (c_1, c_2) = (0.13726, 0.99054)$. Then, we apply these estimated parameters to undistort the fish-eye image, and the calibrated image is obtained as shown in Fig. 3 (b).

5 Conclusion

In this paper, we proposed a novel calibration method for radially symmetric distortion lenses based on the principal component fitting in a projected plane. The most closely related work is presented in [3], in which they made use of the coplanarity condition of observed points on a line to calibrate radially symmetric distortion. A distinguished feature of our method is adopting the division model [10]. In the division model, the coplanarity condition used in [3] is difficult to apply because we need to solve quadratic equations for calibration parameters. The proposed method is simple and completely algebraic and contains no iteration, so the solution is free from local optima. The comparative study on the proposed method and other existing calibration methods remains as our future work. The proposed method is shown to work properly by experiments with synthetic data and a real distorted image. The proposed method did not perform well in the presence of large noise, thus an effort to make our method robust to noises and outliers using random sampling methods and statistical tests (see [11] for example) is ongoing.

In this paper, we only showed that the proposed calibration method can find a good estimate of the coefficients of an assumed bases $\{f_n(r)\}_{n=1}^N$ in equation (2). The choice of bases for approximating the calibration function is a very important problem, and we can make use of properties of the eigen values for the model selection before calibration. The minimum eigen value of the matrix $\mathbf{\Phi\Phi}^\top$ indicates the ability of the adopted bases to map observed data to the perspective projection plane so as to lie on a straight line. We can estimate the adequacy of bases by comparing the minimum eigen values of the matrices $\mathbf{\Phi\Phi}^\top$ for different basis sets. We can expect and select the combination of bases from candidates $\{f_n(r)\}_{n=1}^N$ such that the minimum eigen value is the smallest. This model selection scheme must be useful for finding good bases to calibrate real images which are distorted by unknown mechanism, and thorough investigation of the optimality of the selected bases is our future work.

References

1. Weng, J., Cohen, P., Herniou, M.: Camera calibration with distortion models and accuracy evaluation. IEEE Trans. PAMI 14(10), 965–980 (1992)
2. Shah, S., Aggarwal, J.K.: Intrinsic parameter calibration procedure for a (high-distortion) fish-eye lens camera with distortion model and accuracy estimation. PR 29, 1775–1788 (1996)
3. Tardif, J.P., Sturm, P., Roy, S.: Self-calibration of a general radially symmetric distortion model. In: Leonardis, A., Bischof, H., Pinz, A. (eds.) ECCV 2006. LNCS, vol. 3954, pp. 186–199. Springer, Heidelberg (2006)
4. Ying, X.H., Hu, Z.Y., Zha, H.B.: Fisheye lenses calibration using straight-line spherical perspective projection constraint. In: Narayanan, P.J., Nayar, S.K., Shum, H.-Y. (eds.) ACCV 2006. LNCS, vol. 3852, pp. 61–70. Springer, Heidelberg (2006)
5. Kang, S.B.: Radial distortion snakes. In: IAPR Workshop on Machine Vision Applications (MVA 2000), Tokyo, Japan, pp. 603–606 (2000)
6. Thirthala, S., Pollefeys, M.: The radial trifocal tensor: A tool for calibrating the radial distortion of wide-angle cameras. In: CVPR 2005, pp. 321–328 (2005)
7. Devernay, F., Faugeras, O.: Straight lines have to be straight: automatic calibration and removal of distortion from scenes of structured environments. Mach. Vision Appl. 13(1), 14–24 (2001)
8. Loan, C.V., Pitsianis, N.: Approximation with Kronecker products. Technical report, Ithaca, NY, USA (1992)
9. Tomasi, C., Kanade, T.: Detection and Traking of Point Features. Technical report, CMT (1991)
10. Fitzgibbon, A.W.: Simultaneous linear estimation of multiple view geometry and lens distortion. In: CVPR 2001, pp. 125–132 (2001)
11. Sugaya, Y., Kanatani, K.: Outlier removal for motion tracking by subspace separation. IEICE Trans. Information and Systems 86, 1095–1102 (2003)

Calibration of Rotating Sensors

Karsten Scheibe[1], Fay Huang[2], and Reinhard Klette[3]

[1] German Aerospace Center (DLR), Berlin, Germany
[2] CSIE, National Ilan University, Yi-Lan, Taiwan
[3] Department for Computer Science, The University of Auckland, New Zealand

Abstract. This paper reports about a method for calibrating rotating senors, namely, rotating sensor-line cameras and laser range-finders. Both together are used to reconstruct accurately 3D environments, such as, for example, large buildings. One of the important steps in the 3D reconstruction pipeline is the fusion of data. This requires an understanding of spatial relationships among the acquired data. Sensor calibration is the key to accurate 3D models.

1 Sensors

Since the 1990s, theoretical studies by various authors (e.g., [2,4,6]) pointed out that the use of a rotating sensor-line camera, where panoramas are recorded line by line, each line with its own projective center, allows us to control conditions for improved stereo analysis or stereo viewing. Basically, this was the start into a new category of digital panoramas defined by super-high resolution and geometric accuracy. Actually, sensor-line cameras had been designed for digital aerial imaging (using a *push-broom technique* [1]) since the early 1980s.

A *laser range-finder* (LRF) or *laser scanner* determines distances to opaque objects; it is also known as LIDAR (laser imaging detection and ranging). It records distances at accurate horizontal and vertical angular increments, between projection center and visible surfaces, which generates a *range-scan*. A produced range-scan is a (noisy) "cloud" of points in 3D space which represent visible surfaces at discrete positions. Figure 1 shows a partial view of a building (at Tamaki campus, The University of Auckland), a partial range scan, and two frames of its 3D animation. The range scan shows millions of measured 3D points by means of a gray-level depth map. These isolated points need to be mapped into meshed (e.g., triangulated) surfaces, geometrically filtered, and rendered using (e.g.) high resolution color panoramic images captured by a rotating sensor-line camera. Various algorithms have been developed for visualizing clouds of 3D points. In this paper, rather than on visualization, we elaborate on the geometric calibration issue of sensors and specify some details that readers of [2] have been asking for.

The scan geometries of range-finder and rotating sensor-line camera are very similar, and this supports accurate rendering of 3D surfaces, generated from range-scans, using color panoramic images, recorded with a rotating sensor-line camera. The fusion problem of image data and LRF depth data has also been discussed in [5], but for the use of a regular digital camera attached to the LRF's turntable. The same technology is illustrated with Figure 1; the build-in camera only provides low quality color data.

Fig. 1. A building at Tamaki campus, The University of Auckland. Left: color image of a build-in camera. Middle: partial depth map. Right: 3D animated depth map.

In this paper we assume two independently rotating panoramic high-accuracy sensors, capturing data at different times.

Different camera calibration approaches have been proposed for panoramic imaging sensors. A model equipped with fish-eye lens on a rotating rig is able to capture high-resolution full spherical panorama images. Hirota et al. [3] proposed a method to calibrate such spherical panorama images. Smadja et al. [7] dealt with calibrating cameras capturing cylindrical panoramic images, and their method is similar to the one discussed in this paper. However, these two calibration methods assume single-center projection geometry. This case is hardly do achieve in practice; rotation occurs off-axis, resulting into multiple projection centers. This paper discusses calibration for the multi-center case.

Rotating Sensor-Line Cameras. As the camera is rotated 360 degrees around an axis, the trajectory of the camera's projection center defines ideally a circle, called *base circle*, illustrated by a bold dashed line in Figure 2. We assume that the plane of the base circle is perpendicular to the rotation axis, the camera's optical axis remains coplanar to the base circle at all of its positions during the rotation, and the sensor-cell array is configured parallel to the rotation axis. Through such a 360°-rotation, the sensor-cell array of the camera describes (in some abstract sense) a cylindric surface. This *image cylinder* identifies the locations of the rotating tri-linear (i.e., R,G, and B sensor line)

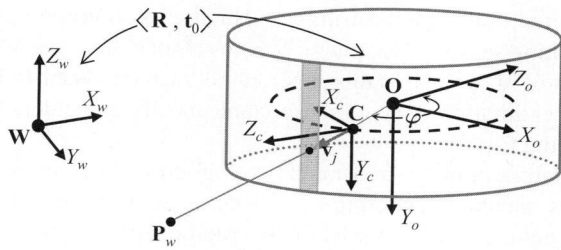

Fig. 2. Local sensor coordinate system (with origin at **O**), camera coordinate system (with origin at **C**), and world coordinate system (with origin at **W**)

sensor. The rotation axis is the axis of the image cylinder, and point **O** on the axis denotes the center of base circle. The base circle has a radius R, which is called the *off-axis distance*. To distinguish different camera positions on the base circle, the subscript i is used, where i also indicates the ith column of the resulting panoramic image. The optical axis of a camera at position $\mathbf{C}_i$ forms a *principle angle* ω with the ray emitting from **O** and passing through $\mathbf{C}_i$. Finally, f denotes the camera's focal length. We assume that f and ω remain constant during a rotation of a sensor-line camera.

Laser Range-Finder. Today's laser scanners may have different scan geometries (i.e., how the rays are progressing and positioned during a single scan of a 3D scene). We refer to the following scan geometry: angular increments between rays are uniformly defined in two dimensions, which are vertically by a rotating deflecting mirror, and horizontally by rotating the whole measuring system (e.g., the vertical scan range of the IMAGER 5003 is 310°, and the horizontal scan range is 360°). This scan geometry is similar to the one known for theodolites, which are traditional instruments for measuring (manually) both horizontal and vertical angles.

2 Coordinate Systems for Sensors

In case of a camera with a single (tri-linear) sensor-line, we use index j to identify different pixel locations. An *image vector* $\mathbf{v}_j$ points from the (current) camera's projection center $\mathbf{C}$ to the image point (sensor element, pixel) of index j. We have

$$\mathbf{v}_j = \begin{pmatrix} 0 \\ j\tau - y_0 \\ f \end{pmatrix}$$

where τ is the height of the pixel (assuming squared pixel), and y_0 denotes the central point of the image (intersection point of the sensor-line with the camera's optical axis).

A local 3D sensor coordinate system (with origin at **O**) is used to describe the orientation and position of the sensor-line camera in relation to a defined world coordinate system (with origin at **W**).[1] The Y_o-axis of the sensor coordinate system coincides with the rotation axis (pointing downward; see Figure 2). Let **R** denote the rotation matrix, and $\mathbf{t}_0$ denotes the translation vector between sensor and world coordinates systems.

Rotation angle φ is defined to be the angle between the Z_o-axis and line segment $\overline{\mathbf{OC}}$. A rotation matrix $\mathbf{R}_{\varphi(i)}$ is used to describe the camera's orientation at $\mathbf{C}_i$ with respect to the local sensor coordinate system. A 3D point with respect to the world coordinate system $\mathbf{P}_w$ can be expressed by its corresponding image vector $\mathbf{v}_j$ as follows:

$$\mathbf{P}_w = \mathbf{t}_0 + \mathbf{R}\mathbf{R}_{\varphi(i)} \left[\lambda \mathbf{R}_\omega \begin{pmatrix} 0 \\ j\tau - y_0 \\ f \end{pmatrix} + R \begin{pmatrix} 0 \\ 0 \\ 1 \end{pmatrix} \right]$$

where matrix $\mathbf{R}_\omega$ specifies the additional rotation of the sensor-line when $\omega \neq 0$.

---

[1] It is possible to assume that the world coordinate system coincides with the range-finder's coordinate system, or, if some special calibration object is used, then it is preferred to assume that the world coordinate system is defined by this calibration object.

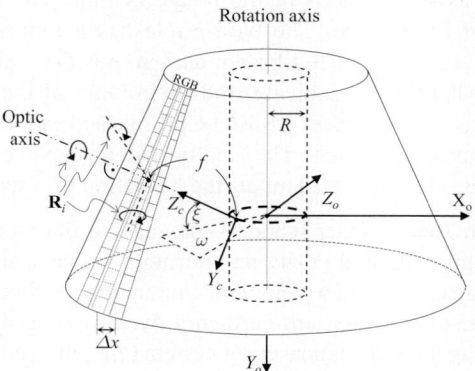

Fig. 3. The sensor coordinate system of the rotating line camera: the optical axis identifies the central point x_0, y_0 and is tilted by $\mathbf{R}_\xi$ and $\mathbf{R}_\omega$; each tri-linear RGB line has a constant distance Δ between a central point (on the green line) and the red or blue line. The tilt of the sensor-line with respect to the optical axis is specified by $\mathbf{R}_i(\alpha, \beta, \delta)$.

Figure 3 illustrates the general case of a rotating RGB sensor-line camera. In applications we also have to model the following deviations:

- At any discrete moment i of time, the sensor-line is tilted (within the local coordinate system) by three angles which define a time-dependent rotation matrix $\mathbf{R}_i(\alpha, \beta, \delta)$; this defines the *inner pose* of a sensor-line about all three axes with respect to the central point x_0, y_0.
- The red and blue lines have an offset $\boldsymbol{\Delta}$ with respect to the central point on the green line.
- The optical axis is rotated by ξ about the X_o-axis.
- The optical axis is rotated by the fixed principle angle ω about the Y_o-axis.
- The sensor-line is rotating with an eccentricity or a desired off-axis distance $R > 0$.

Practically, the inner pose $\mathbf{R}_i(\alpha, \beta, \delta)$, the central point (x_0, y_0), and the off-set Δ are sufficient to describe the positioning of the sensor line in any case. The image vector $\mathbf{v}_j$ is split into two terms as follows:

$$\mathbf{v}_j = \mathbf{v}_{j,\Delta} + \mathbf{v}_f = (\Delta_x - x_0, j\tau + \Delta_y - y_0, 0)^T + (0, 0, f)^T$$

Altogether, the coordinate transform is now the following:

$$\mathbf{P}_w = \mathbf{t}_0 + \mathbf{R}\mathbf{R}_{\varphi(i)} \left[\lambda \mathbf{R}_\xi \mathbf{R}_\omega \left[\mathbf{R}_i \begin{pmatrix} \Delta_x - x_0 \\ j\tau + \Delta_y - y_0 \\ \Delta_z \end{pmatrix} + \begin{pmatrix} 0 \\ 0 \\ f \end{pmatrix} \right] + \begin{pmatrix} 0 \\ 0 \\ R \end{pmatrix} \right] \quad (1)$$

This equation is used for sensor calibration.

3 Sensor Calibration

The common camera or sensor parameter calibration approach is a point-based approach, aiming at a minimization of differences between ideal and actual projections of known 3D points, such as calibration marks on a calibration object, or localized points in the 3D scene. By taking many images of calibration marks, we are able to apply a least-square error (LSE) optimization procedure.

Parameters and Objective Functions. In the sequel, we describe a standard least-square approach, as known from photogrammetry, but adapted to a rotating sensor-line camera. This approach determines unknown extrinsic parameters of the sensor, which are matrices $\mathbf{R}$, $\mathbf{R}_\xi$, $\mathbf{t}_0$, and off-axis distance R and principle angle ω.

It also determines the intrinsic parameters, which are matrix $\mathbf{R}_i(\alpha, \beta, \delta)$, describing the tilt of the sensor, the focal length f, and the sensor's central point x_0, y_0; the latter one also written as vector Δ.

The rotation angle φ of the rotating sensor (sensor-line camera) may be measured using an internal measuring system of the turntable. Modern technology allows that the angle is determined with an accuracy of 1/1000 degree at least.

Note that the frequently needed recalculation of a "focal length" (i.e., of the camera constant) aims at an exact determination of the (typically unknown) virtual projection center of a pinhole-type model, namely the distance between the entrance pupil to a virtual sensor plane which fulfills the linear imaging assumption.

An *observation* is a recorded calibration mark (with physically measured coordinates, identified with a point (X, Y, Z) such as, e.g., the centroid of the mark) at corresponding image coordinates i and j (i.e., pixel (i, j) for the rotating sensor-line camera, when projecting point (X, Y, Z) into the cylindric panorama). Note that two observations are derivable for one calibration mark because of using two collinearity equations (i.e., one observation is given by two collinearity equations and the corresponding residues).

We have a linear system of n equations with m unknown; the sth observation is given by l_s. The sum of all equations can be written in this form:

$$\sum_{s=0}^{n} l_s = a_{11} \cdot x_1 + a_{12} \cdot x_2 + \ldots + a_{sm} \cdot x_m$$

Observations are considered to be the residues of an iterative Taylor approximation of kth order (which defines a Newton method):

$$l = F(u) - \nabla F^k(\hat{u})\Delta u$$

For the determination of extrinsic parameters and the calibration of intrinsic parameters of a sensor, we place various calibration marks "around the sensor" in the scene. Some of them are projected into image data (depending on visibility), and we assume that all projected calibration marks can be uniquely identified in resulting image data (e.g., in the panoramic image).

Assume that we have m unknowns in total (i.e., elements in matrices, vectors, and parameters), and given are n observations, with $n \geq m$.

General Error Criterion. We use Equation (1) to model the geometric mapping of 3D points into the sensor coordinate system. By substituting $\mathbf{A} = \mathbf{R}\mathbf{R}_{\varphi(i)}$, $\mathbf{B} = \mathbf{R}_\xi \mathbf{R}_\omega$, and $\mathbf{C} = \mathbf{B}\mathbf{R}_i$ (with matrix elements $\mathbf{A} = a_{11}, ..., a_{33}$, $\mathbf{B} = b_{11}, ..., b_{33}$ and so forth), where $\mathbf{v}_{j,\Delta}$ is the image vector:

$$\mathbf{v}_{j,\Delta} = \begin{pmatrix} \Delta_x - x_0 \\ j\tau + \Delta_y - y_0 \\ \Delta_z \end{pmatrix}$$

This vector is also written in vectorial components as $\mathbf{v}_{j,\Delta} = (\mathbf{v}_x, \mathbf{v}_y, \mathbf{v}_z)^T$.

After those substitutions, the general mapping equation is now given as follows:

$$\mathbf{P}_w = \mathbf{t}_0 + \mathbf{A}(\lambda \mathbf{B}(\mathbf{R}_i \mathbf{v}_{j,\Delta} + f\mathbf{z}^\circ) + R\mathbf{z}^\circ)$$
$$\mathbf{A}^{-1}(\mathbf{P}_w - \mathbf{t}_0) - R\mathbf{z}^\circ = \lambda \mathbf{C}\mathbf{v}_{j,\Delta} + \mathbf{B}f\mathbf{z}^\circ$$

We rewrite this for all three components of this equation, using $\tilde{\mathbf{P}} = \mathbf{P}_w - \mathbf{t}_0$:

$$a_{11}\tilde{\mathbf{P}}_x + a_{21}\tilde{\mathbf{P}}_y + a_{31}\tilde{\mathbf{P}}_z = \lambda(c_{11}\mathbf{v}_x + c_{12}\mathbf{v}_y + c_{13}\mathbf{v}_z + b_{13}f)$$
$$a_{12}\tilde{\mathbf{P}}_x + a_{22}\tilde{\mathbf{P}}_y + a_{32}\tilde{\mathbf{P}}_z = \lambda(c_{23}\mathbf{v}_x + c_{22}\mathbf{v}_y + c_{23}\mathbf{v}_z + b_{23}f)$$
$$a_{13}\tilde{\mathbf{P}}_x + a_{23}\tilde{\mathbf{P}}_y + a_{33}\tilde{\mathbf{P}}_z - R = \lambda(c_{31}\mathbf{v}_x + c_{32}\mathbf{v}_y + c_{33}\mathbf{v}_z + b_{33}f)$$

The matrix of coefficients $a_{11}, ..., a_{33}$ is finally transposed because of the inversion of the matrix $\mathbf{A}$. (Recall that, for a rotation matrix, $\mathbf{E} = \mathbf{R} \cdot \mathbf{R}^T$ is the unit matrix, and, consequently, $\mathbf{R}^{-1} = \mathbf{R}^T$.)

By dividing these equations we may eliminate the scaling factor λ, and we obtain, from the left-hand sides of those three equations, the following two equations:

$$F_{x/z} := \frac{a_{11}(\tilde{\mathbf{P}}_x) + a_{21}(\tilde{\mathbf{P}}_y) + a_{31}(\tilde{\mathbf{P}}_z)}{a_{13}(\tilde{\mathbf{P}}_x) + a_{23}(\tilde{\mathbf{P}}_y) + a_{33}(\tilde{\mathbf{P}}_z) - R}$$

and

$$F_{y/z} := \frac{a_{12}(\tilde{\mathbf{P}}_x) + a_{22}(\tilde{\mathbf{P}}_y) + a_{32}(\tilde{\mathbf{P}}_z)}{a_{13}(\tilde{\mathbf{P}}_x) + a_{23}(\tilde{\mathbf{P}}_y) + a_{33}(\tilde{\mathbf{P}}_z) - R}$$

For the right-hand sides we obtain that

$$G_{x/z} := \frac{c_{11}\mathbf{v}_x + c_{12}\mathbf{v}_y + c_{13}\mathbf{v}_z + b_{13}f}{c_{31}\mathbf{v}_x + c_{32}\mathbf{v}_y + c_{33}\mathbf{v}_z + b_{33}f}$$

and

$$G_{y/z} := \frac{c_{23}\mathbf{v}_x + c_{22}\mathbf{v}_y + c_{23}\mathbf{v}_z + b_{23}f}{c_{31}\mathbf{v}_x + c_{32}\mathbf{v}_y + c_{33}\mathbf{v}_z + b_{33}f}$$

These are the general collinearities, and we also express them by $F_{x/z} = G_{x/z}$ and $F_{y/z} = G_{y/z}$ in short form.

By linearization of these equations it is now possible to estimate iteratively the unknown parameters

$$\mathbf{u} = (t_{x0}, t_{y0}, t_{z0}, \psi, \phi, \kappa, R)$$

for the left-hand sides $F_{x/z}$ and $F_{y/z}$, and

$$\mathbf{u} = (\xi, \alpha, \beta, \delta, \omega, f, y_0, x_0)$$

for the right-hand sides $G_{x/z}$ and $G_{y/z}$, respectively. The three unknowns ψ, ϕ, κ specify the rotation angles about the x-, y-, and z-axis, respectively. (Note that $\mathbf{R} = \mathbf{R}_x(\psi) \cdot \mathbf{R}_y(\phi) \cdot \mathbf{R}_z(\kappa)$.) The upper index k is the number of the iteration step. The linearization is given as follows:

$$\nabla(G_{x/z} - F_{x/z}) = \left(\frac{\partial G_{x/z}}{\partial u_1} - \frac{\partial F_{x/z}}{\partial u_1}, \frac{\partial G_{x/z}}{\partial u_2} - \frac{\partial F_{x/z}}{\partial u_2}, ..., \frac{\partial G_{x/z}}{\partial u_m} - \frac{\partial F_{x/z}}{\partial u_m} \right)$$

$$F_{x,z}^k - G_{x,z}^k = \nabla(G_{x,z} - F_{x,z})^k \cdot \Delta \mathbf{u}$$
$$l = \mathbf{M} \cdot \Delta \mathbf{u}$$

For $n = m$, the solution is uniquely given by

$$\Delta \mathbf{u} = \mathbf{M}^{-1} \cdot l$$

assuming linear independence between equations.

For $n > m$ observations (i.e., a typical adjustment problem), we apply now the method of least-square error minimization. The error is given as follows:

$$v = \mathbf{M} \cdot \Delta \hat{\mathbf{u}} - l$$

The error function (which needs to be minimized) is defined as follows:

$$\begin{aligned} \min &= v^T v \\ &= (\mathbf{M} \cdot \Delta \hat{\mathbf{u}} - l)^T (\mathbf{M} \cdot \Delta \hat{\mathbf{u}} - l) \\ &= \Delta \hat{\mathbf{u}}^T \mathbf{M}^T \mathbf{M} \cdot \Delta \hat{\mathbf{u}} - 2l^T \mathbf{M} \cdot \Delta \hat{\mathbf{u}} + l^T l \end{aligned}$$

For identifying the minimum, we differentiate and have the resulting function equal to zero:

$$\frac{\partial (v^T v)}{\partial \Delta \hat{\mathbf{u}}} = 2\Delta \hat{\mathbf{u}}^T \mathbf{M}^T \mathbf{M} - 2l^T \mathbf{M} = 0$$

This leads to the following solution:

$$\Delta \hat{\mathbf{u}} = \left(\mathbf{M}^T \mathbf{M} \right)^{-1} \mathbf{M}^T l \qquad (2)$$

The Jacobian matrix M contains all first-order partial derivatives, and l are the residues as defined above. This is solved by means of iterations; the vector $\Delta \mathbf{u}$ contains the corrections of of each unknown. A minimum is found if the unknowns do not change significantly anymore (e.g., $\sum_{s=0}^{m} |\Delta \mathbf{u}_s| < \varepsilon$, with $\varepsilon = 10^{-9}$).

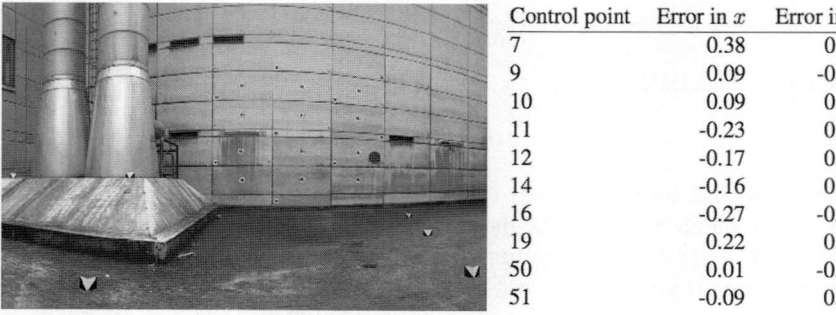

Control point	Error in x	Error in y
7	0.38	0.03
9	0.09	-0.44
10	0.09	0.08
11	-0.23	0.54
12	-0.17	0.28
14	-0.16	0.05
16	-0.27	-0.35
19	0.22	0.30
50	0.01	-0.12
51	-0.09	0.11

Fig. 4. Left: calibration marks are distributed around a courtyard at TFH Berlin. Right: deviations between calculated image coordinates and their actual reference coordinates in this courtyard (listed values in subpixels).

4 Conclusions

Some kind of human intervention is in general required for this calibration approach for identifying the projections of those 3D points in a real scene (e.g., the projected points) used as calibration marks, possibly supported by some SIFT feature detector or moment-based sub-pixel accuracy point locator. If a specially designed calibration object is used, this process can be supported by an automatic calibration mark detection algorithm, where marks are located with sub-pixel accuracy (using, e.g., centroid calculation within a mark's region, or intersection points of approximated straight lines when using a checkerboard).

The described least-square approach was used in many applications of panoramic sensors, and is so far our recommended way for calibrating all the mentioned parameters, possibly also including a tilt of the rotation axis of the sensor. Our experimental results in Figure 4 show that this approach is capable to achieve accuracy with less than one pixel error.

References

1. Gupta, R., Hartley, R.I.: Linear Pushbroom Cameras. PAMI 19(9), 963–975 (1997)
2. Huang, F., Klette, R., Scheibe, K.: Panoramic Imaging: Sensor-Line Cameras and Laser Range-Finders. Wiley, Chichester (2008)
3. Hirota, T., Nagahara, H., Yachida, M.: Calibration of Rotating Line Camera for Spherical Imaging. In: Narayanan, P.J., Nayar, S.K., Shum, H.-Y. (eds.) ACCV 2006. LNCS, vol. 3851, pp. 389–398. Springer, Heidelberg (2006)
4. Ishiguro, H., Yamamoto, M., Tsuji, S.: Omni-directional Stereo. PAMI 14(2), 257–262 (1992)
5. Jiang, W., Lu, J.: Panoramic 3D Reconstruction by Fusing Color Intensity and Laser Range Data. IEEJ Transactions on Electronics Information and Systems 127(4), 568–576 (2007)
6. Li, Y., Shum, H., Tang, C., Szeliski, R.: Stereo Reconstruction from Multiperspective Panoramas. PAMI 26(1), 45–62 (2004)
7. Smadja, L., Benosman, R., Devars, J.: Cylindrical Sensor Calibration using Lines. In: Proc. ICIP 2004, pp. 1851–1854 (2004)

A New Approach for Segmentation and Recognition of Arabic Handwritten Touching Numeral Pairs

Huda Alamri, Chun Lei He, and Ching Y. Suen

CENPARMI (Center for Pattern Recognition and Machine Intelligence),
Computer Science and Software Engineering Department, Concordia University,
Montreal, Quebec, Canada
Tel.: (514)-848-2424-Ext.: 7950
Fax : (514)-8482830
{hu_alam,cl_he,suen}@encs.concordia.ca

Abstract. In this paper, we propose a new approach on segmentation and recognition of off-line unconstrained Arabic handwritten numerals, which failed to be segmented with connected component analysis. In our approach, the touching numerals are automatically segmented when a set of parameters is chosen. Models with different sets of parameters for each numeral pair are designed for recognition. Each image in each model is recognized as an isolated numeral. After normalizing and binarizing the images, gradient features are extracted and recognized using SVMs. Finally, a post-processing is proposed by based on the optimal combinations of the recognition probabilities for each model. Experiments were conducted on the CENPARMI Arabic, Dari, and Urdu touching numeral pair databases [1,12].

Keywords: Numeral pair segmentation, Arabic Digit Recognition, Gradient features.

1 Introduction

Recognition of handwritten numeral strings is a very important branch of the handwritten recognition field. This is due to its heavy involvement in many important applications, such as automatic processing of bank cheques, postal code recognition, reading of tax forms, and the recognition of other specifically designed forms [10]. The varieties in handwriting styles, the low quality of some types of paper, and all other factors have made the recognition of handwritten numeral strings quite challenging. Various approaches and techniques have been developed to solve the problems of segmentation and recognition of numeral strings. In general, we can classify all of the different approaches into one of the following categories: holistic approaches [15], segmentation-then-recognition [11], and segmentation-based recognition [5]. Holistic approaches attempt to recognize the whole numeral string as a one unit. On the other hand, the second two categories segment a numeral string image into a sequence of constituents and each constituent is classified into one of the ten classes {0,1,...,9}. However, the segmentation process is not a trivial task since overlapped or touched digits may frequently exist in the numeral strings. In our work, we focus on solving the problem of segmenting numeral strings that are completely touching and cannot be segmented through basic connected component analysis. Given an input image of a numeral touching pair, we extract the two regions that could represent the digits in the image, each one individually, by searching a 2-dimentional parameter space. The search results in a set of candidate regions for each digit in the image. Then, the extracted regions are scaled and preprocessed to be ready for the recognition stage. For classification, we have developed an SVM classifier for recognition. The classifier has been separately trained on the CENPARMI Arabic, Urdu and Dari Isolated Digit Databases. To test the proposed segmentation method, we have used a subset of the CENPARMI Arabic, Dari, and Urdu Numeral String Databases [1,12]. The used subsets include only samples of complete touching digits. Examples of Arabic touching pairs are shown in Figure 1.

This paper is organized as follows: Related work is reviewed in Section 2. Then, the segmentation approach is presented in Section 3. The isolated digit classification, gradient feature extraction, training and recognition are carried out in Section 4. Post processing is explained in Section 5. Databases are presented in Section 6 followed by experiments and results in Section 7. Finally, conclusions are presented in Section 8.

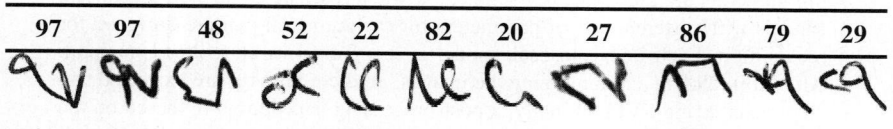

Fig. 1. Samples of Touched Pairs from the CENPARMI Arabic Numeral Dataset and the equivalent Latin Digits

2 Related Work

In the last decade, there has been more research addressing the problem of Arabic handwritten digit recognition [2,8]. More recent works have reported high recognition rates. In 2004, Harifi et al. [6] proposed an asymmetrical segmentation pattern to obtain a feature vector for the recognition of handwritten Persian/Arabic digits. A recognition rate of 97.6 % was reported on a database of around 730 digits written by 73 participants. Sameh et al. (2009) [3] proposed a multiple feature/resolution scheme for Arabic numeral recognition using Hidden Markov Models (HHMs). The multiple features included gradient, structural, and concavity features. The proposed scheme was tested on a database of 21,120 digits written by 44 writers with 48 samples per digit and an average recognition rate of 99% was achieved. For numeral string segmentation problems, many approaches have been proposed [7,9,11].

However, for the problem of touching or overlaped cases, few works has been published. In 2006, Dipankar et al. [5] proposed an approach for the segmentation and recognition of unconstrained off-line Bangla handwritten numerals. A projection profile-based heuristic technique was used to segment the numerals. The method was tested on 500 Bangla numerals. A total of 89% of the touching pairs were segmented correctly with a 10% rejection ratio. Wang et al. (2008), [15] proposed a model-based holistic approach to recognize handwritten numeral touching pairs. The models of numeral pairs were generated as the combinations of two corresponding numerals. Each numeral was modeled as a set of polygonal lines and the corresponding best recognition rate of 93.6% was achieved on 1000 of these test images from the NIST SD19 database.

3 Segmentation of Touching Pairs

In this work, we focus on solving the problem of numeral touching pair segmentation and recognition. Our Segmentation of touching pairs begins by extracting the bounding box to eliminate the unwanted white area around the image, thereby reducing the search space, and placing the touching digits in the center of the image. We can think of each digit in the image as being surrounded by a rectangular box. We used the searching technique presented in [15] to find a rectangular box in a 2-dimensional space. In order to do that, we needed to determine its dimensions, height and width. Therefore, we considered two parameters. The first parameter, let it be α, is the ratio of the first digit width (W_f) to the whole image width (W_i). The second parameter, let it be β, is the horizontal distance between the first digit and the second one. Using a set of values for α and β, we could calculate the dimensions for the rectangular boxes [15]. As shown in Figure 2, an image with (W_i) width, we calculate the first digit width (W_f), and the second digit width (W_s) as follows:

$$W_f = \alpha \cdot W_i \tag{1}$$

$$W_s = W_i - \alpha \cdot W_i - \beta \tag{2}$$

The height of the bounding box around each digit (Hf and Hs) can be defined as the vertical distance between the lower-pixel and the higher-pixel in the corresponding area of the image (see Figure 2). Accordingly, by searching different values of (α_i, β_j), we generate segmented candidate images. For feasible computation, $i = 0,...,5$, and $j = 0,...,5$. Therefore, 25 different models are extracted to individually represent the two digits in the image, where each model represents all the images with the same (α_i, β_j).

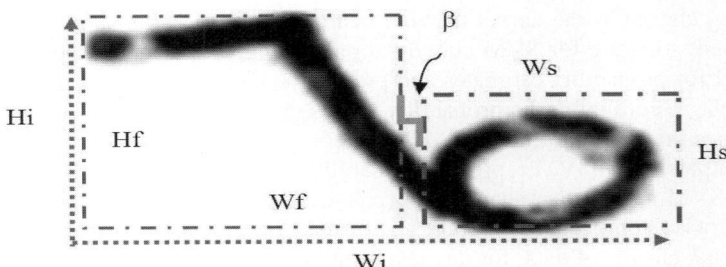

Fig. 2. Example of Two Rectangular Boxes Representing the Digits in a Touching Pair

4 Isolated Digit Recognition

Recognition of each image in each model is the same as recognizing the isolated numeral in Arabic. In image pre-processing, we perform noise removal, grayscale normalization, size normalization, and binarization of the grayscale images. For feature extraction, gradient features [13] are extracted on grayscale images. The Robert's Cross operator uses the diagonal directions to calculate the gradient vector. For example, the gradient magnitude and direction of pixel $g(m,n)$ are calculated as follows:

$$\frac{\partial I}{\partial x} = \Delta u = g(m,n) - g(m+1,n+1) \tag{3}$$

$$\frac{\partial I}{\partial y} = \Delta v = g(m,n+1) - g(m+1,n) \tag{4}$$

$$\theta(m,n) = \arctan\left(\frac{\Delta v}{\Delta u}\right) \tag{5}$$

$$s(m,n) = \sqrt{\Delta u^2 + \Delta v^2} \tag{6}$$

where $\theta(m,n)$ and $s(m,n)$ specify the direction and gradient magnitude of pixel (m,n), respectively.

The direction of the gradient on each pixel is quantized to 32 levels with an interval of $\pi/16$. The normalized character image is divided into 81 (9×9) blocks. After extracting the strengths and directions in each image, the spatial resolution is reduced from 9×9 to 5×5 by down sampling every two horizontal and every two vertical blocks with a 5×5 Gaussian filter. Similarly, the directional resolution is reduced from 32 to 16 levels by down sampling with a weight vector $[1\ 4\ 6\ 4\ 1]^T$, to produce a feature vector of size 400(5×5×16). Moreover, the transformation $y = x^{0.4}$ is applied to make the distribution of the features Gaussian-like. Finally, we scale the feature vectors by a constant factor so that the values of feature components range from 0 to 1.

For the classification stage, Support Vector Machines (SVMs)[14] were chosen as the classifier. SVMs with different kernel functions can transform a non-linear separable problem into a linear separable one by projecting data into the feature space, and then SVMs can find the optimal separating hyperplane. The Radial Basis Function (RBF) was chosen as the kernel in this research. LibSVM [4] is simple, easy-to-use, and efficient software for SVM classification and regression. It was possible to train our model for probability estimates and predict the test samples with probability values as well. According to the probability values, post-processing can be implemented and enhance the recognition results.

The outputs of LibSVM represent probabilities information in its classification results. LibSVM applies one against one (pair-wise) strategy in a multi-class problem. In the pairwise approach, K^2 support vector machines are trained for a k-class problem. Given k classes of data, for any test sample x, the goal is to estimate p_i, which is obtained from all r_{ij}. In this case, r_{ij} is a one-against-one class probability and is obtained from the known training data by solving the following optimization problem:

$$\min_{p} \frac{1}{2} \sum_{i=1}^{k} \sum_{j: j \neq i} (r_{ji} p_i - r_{ij} p_j)^2$$

$$\text{subject to} \sum_{i=1}^{k} p_i = 1, \ p_i \geq 0, \forall i.$$

where $p_i = p(y = i \mid x)$, $i = 1,...,k$, $r_{ij} \approx p(y = i \mid y = i \text{ or } j, x)$.

5 Post Processing Module

In each model, all segmented images are recognized by the designed classifier, as described in the previous section. Since the output of each image has confidence values (probabilities) on all classes, a post processing is applied based on the given probabilities of different models. This module is able to verify and improve the final recognition rates. When the highest probability of the given input in all 25 models is greater than a certain threshold (*Th*), the final result is based on the recognition result with the highest probability. Otherwise, the final recognition result is based on ranking the outputs of all the other models by using different ranking schemes. Finally, the output with the highest rank is chosen. One ranking scheme ranks the results based on the majority votes (*M_V*). It chooses the result with the highest number of votes from 25 models. The second ranking scheme considers the estimated probability from each result of the different models and chooses the result with the highest probability (*H_P*) as the final one. The third voting scheme is a combination of both the probability estimation and the majority vote schemes (*H_P + M_V*).

6 Databases

In order to test the proposed approach, we applied it on the CENPARMI Arabic, Urdu, and Dari Numeral String Databases [1,12]. Each database includes samples of handwritten words, characters, isolated digits, numeral strings, dates and special symbols. We conducted the recognition experiments on the isolated digit and numeral string datasets from these databases. The total number of isolated digits in each database is: 30,983 in Arabic, 29,069 in Dari and 60,329 in Urdu. Each database was divided into training and testing sets. The total number of samples in both sets for each language is presented in Table 1. For segmentation, we only selected the touching pairs from the numeral string databases. We applied the segmentation method based on connected component analysis on the numeral string databases, and the ones that could not be segmented were selected to test our method. A total of 721 touching pairs were found in the databases. There were 400 pairs from the Dari database, 189 pairs from the Urdu database and 132 pairs from the Arabic database. Table 2 shows the basic isolated digits in each language and the matched Latin numerals. As shown, three languages share some numerals.

Table 1. No. of Samples in Various Isolated Numeral Databases

	Arabic	Dari	Urdu
Training Set	24784	23255	47151
Testing Set	6199	5814	13178

Table 2. The Basic Isolated Digits in Each Language and The Equivalent English Digits

	0	1	2	3	4	5	6	7	8	9
Arabic	`	\	C	۳	٤	o	٦	V	ʌ	٩
Dari	`	/	۲	۳	۴	۵	۶	V	ʌ	۹
Urdu	◆	\	۲	۳	۴	۵	۶	۷	ʌ	۹

7 Experiments and Results

First, our SVMs classifier was individually trained and tested on the isolated digit database of each language. There were two parameters in RBF to be optimized: **c** and λ. Where **c>0** was the penalty parameter of the error term, and λ was the parameter in RBF. These two parameters were optimally chosen by the cross-validation. For example, when lg(c) = 1 and lg(λ) = -7, the performance on the Arabic training set achieved the highest recognition rate (98.0471%). Thus, we set c = 2 and λ = 0.0078125, and then we tested it on the testing set. The recognition results on the testing sets in each database were 98.48%, 98.66%, and 98.61% for Arabic, Dari and Urdu databases, respectively (see Table 3).

Table 3. Recognition Rates on Testing Sets of Three Isolated Numeral Databases

	Arabic	Dari	Urdu
Recognition Rate on Test Set	98.48%	98.66%	98.61%

For the segmentation, the first step was choosing a set of values for the parameters α and β. For feasible computation, a set of five values was chosen for each parameter. The chosen sets are [0.3, 0.4, 0.5, 0.6, 0.7] for α and [-6, -3, 0, 3, 6] for β. We searched each image using all the different compilations of the chosen values. As a result, 25 different models are detected and extracted to individually represent each numeral. Afterwards, each model as passed to the isolated digit classifier for the classification. The highest recognition rates were: Arabic 85.50% for the model 12 (α = 0.5, β = 0.0), Urdu 84.57% for model (α = 0.5, β = 0.0) 12 and Dari 79.07% for model 13 (α = 0.5, β = -3). Table 4 illustrates the recognition results for six different models in each language.

For post processing, as described in Section 4, the model that gives the highest recognition rate is chosen as the default final output if its estimated probability is higher than a certain threshold *Th*. Otherwise, the post processing module will be activated and a scoring scheme will be applied to choose the final recognition output. We applied three different scoring schemes. First, we applied *(M_V)* and *(H_P)* scoring schemes, using the outputs from all of the 25 models. Both scoring schemes were able to improve the overall recognition rate, and the highest recognition rates were: 88.55 % in Arabic, 88.03% in Dari and 89% in Urdu datasets. Then, we applied a scoring scheme based on a combination of both of the *(M_V)* and the *(H_P)* using only the outputs for five models; M^5, M^{10}, M^{12}, M^{16}, and M^{21}. The choice of the best model, the different values for *(Th)* and the five models were all validated with the experimental results on the Arabic dataset. The Urdu and Dari datasets were used to test the final choices for the different parameters. From Table 5, we can see that the best recognition rates were: 92.22%, 90.43%, and 86.09% in Arabic, Urdu and Dari, respectively. They are achieved by applying the scoring scheme *(H_P+ M_V)* on the chosen five models and by applying *(Th = 0.95)*.

Table 4. The Recognition Results for Six Different Models for Each Language

	M_5	M_{10}	M_{12}	M_{13}	M_{16}	M_{21}
	α = 0.3	α = 0.4	α = 0.5	α = 0.5	α = 0.6	α = 0.7
	β = -6	β = -3	β = 0.0	β = -3	β = -6	β = -6
ARABIC	60.31%	77.48%	85.50%	84.73%	83.21%	62.21%
URDU	68.09%	80.31%	84.57%	84.04%	72.87%	60.90%
DARI	56.64%	72.81%	78.82%	79.07%	75.70%	62.91%

Table 5. Recognition Rates on Three Different Numeral (Touching Pair) Databases

Language	Th	The Best Result(M^{12})	25 Models H_P	25 Models M_V	Combination of 5 Models H_P+ M_V
ARABIC	0.500	84.732%	86.6412%	87.7863%	88.1679%
	0.8755	84.732%	88.1679%	88.1679%	91.6031%
	0.9000	84.732%	88.5496%	88.5496%	91.6031%
	0.95	84.732%	88.5496%	87.4046%	92.2214%
URDU	0.500	84.5745%	85.6383%	85.1064%	86.7021%
	0.9000	84.5745%	88.0319%	86.4362%	87.2340%
	0.9044	84.5745%	88.2979%	86.4362%	89.2340%
	0.95	84.5745%	89.8936%	86.4362%	90.4255%
DARI	0.500	79.0727%	82.3308%	80.0752%	82.2055%
	0.8795	79.0727%	85.7143%	81.3283%	86.0902%
	0.9000	79.0727%	88.0319%	88.0319%	85.5890%
	0.95	79.0727%	85.4637%	80.9524%	84.7118%

8 Conclusion

In this paper, we have addressed the problem of segmentation and recognition of handwritten numeral touching strings in three Arabic-script databases. For the segmentation process, we have generated 25 models with different segmentation parameters. We have developed an SVM classifier, which has been trained on isolated numerals, to recognize each image in each model. Moreover, we have designed a post processing module to verify and improve the recognition results with the confidence values from different models and to predict the final recognition results. As a result, the average recognition rate was 98% in isolated digits, and the performances on these digits as completely touching numeral pairs were 92.22%, 90.43%, and 86.09% in Arabic, Urdu and Dari, respectively. All of the images failed to be segmented and recognized with connected component analysis. In the future, different combination schemes may be designed and applied.

References

1. Alamri, H., Sadri, J., Nobile, N., Suen, C.: A Novel Comprehensive Database for Arabic Off-Line Handwriting Recognition. In: 11th International Conference on Frontiers in Handwriting Recognition, Montreal, pp. 664–669 (2008)
2. Al-Omari, F.A., Al-Jarrah, O.: Handwritten Indian numerals recognition system using probabilistic neural networks. Advanced Engineering Informatics 18, 9–16 (2004)
3. Awaidah, S.M., Mahmoud, S.A.: A multiple feature/resolution scheme to Arabic (Indian) numerals recognition using hidden Markov models. Signal Processing 89, 1176–1184 (2009)
4. Chang, C.-C., Lin, C.-J.: LIBSVM: a library for support vector machines (2001), http://www.csie.ntu.edu.tw/~cjlin/libsvm
5. Das, D., Yasmin, R.: Segmentation and Recognition of Unconstrained Bangla Handwritten Numeral. Asian Journal of Information Technology 5(2), 155–159 (2006)
6. Harifi, A., Aghagolzade, A.: A New Pattern for Handwritten Persian/Arabic Digit Recognition. International Journal of Information Technology 1(4), 293–296 (2004)
7. Liu, C.L., Sako, H., Fujisawa, H.: Effects of Classifier Structure and Training Regimes on Integrated Segmentation and Recognition of Handwritten Numerals Strings. IEEE Transactions on Pattern Analysis and Machine Intelligence, PAMI 26(11), 1395–1407 (2004)
8. Lorigo, L., Govindaraju, V.: Offline Arabic handwriting recognition: a survey. IEEE Transactions on Pattern Analysis and Machine Intelligence 28, 712–724 (2006)
9. Oliveira, L.S., Sabourin, R., Bortolozzi, F., Suen, C.Y.: Automatic Segmentation of Handwritten Numerical Strings: A Recognition and Verification Strategy. IEEE Transactions on Pattern Analysis and Machine Intelligence 24(11), 1438–1454 (2002)
10. Plamondon, R., Srihari, S.N.: On-line and Off-line Handwriting Recognition: A Comprehensive Survey. IEEE Transaction on Pattern Analysis and Machine Intelligence 22(1), 63–84 (2000)
11. Sadri, J., Suen, C.Y., Bui, T.D.: Automatic Segmentation of Unconstrained Handwritten Numeral Strings. In: International Workshop on Frontiers in Handwriting Recognition (IWFHR), Tokyo, pp. 317–322 (2004)
12. Shah, M.I., Sadri, J., Nobile, N., Suen, C.: A New Multipurpose Comprehensive Database for Handwritten Dari Recognition. In: 11th International Conference on Frontiers in Handwriting Recognition, Montreal, pp. 635–640 (2008)
13. Shi, M., Fujisawa, Y., Wakabayashi, T., Kimura, F.: Handwritten numeral recognition using gradient and curvature of gray scale image. Pattern Recognition 35(10), 2051–2059 (2002)
14. Vapnik, V., Lerner, A.: Pattern recognition using generalized portrait method. Automation and Remote Control 24, 774–780 (1963)
15. Wang, Y., Liu, X., Jia, Y.: A holistic approach to handwritten numeral pair recognition based on generative models of numeral pairs. In: 11th International Conference on Frontiers in Handwriting Recognition, Montreal, pp. 54–58 (2008)

Ridges Based Curled Textline Region Detection from Grayscale Camera-Captured Document Images

Syed Saqib Bukhari[1], Faisal Shafait[2], and Thomas M. Breuel[1,2]

[1] Technical University of Kaiserslautern, Germany
[2] German Research Center for Artificial Intelligence (DFKI), Kaiserslautern, Germany
bukhari@informatik.uni-kl.de
faisal@iupr.dfki.de
tmb@informatik.uni-kl.de
http://www.iupr.com

Abstract. As compared to scanners, cameras offer fast, flexible and non-contact document imaging, but with distortions like uneven shading and warped shape. Therefore, camera-captured document images need preprocessing steps like binarization and textline detection for dewarping so that traditional document image processing steps can be applied on them. Previous approaches of binarization and curled textline detection are sensitive to distortions and loose some crucial image information during each step, which badly affects dewarping and further processing. Here we introduce a novel algorithm for curled textline region detection directly from a grayscale camera-captured document image, in which matched filter bank approach is used for enhancing textline structure and then ridges detection is applied for finding central line of curled textlines. The resulting ridges can be potentially used for binarization, dewarping or designing new techniques for camera-captured document image processing. Our approach is robust against bad shading and high degrees of curl. We have achieved around 91% detection accuracy on the dataset of CBDAR 2007 document image dewarping contest.

Keywords: Curled Textline Finding, Camera-Captured Document Images, Grayscale Document Images, Ridges, Anisotropic Gaussian Smoothing.

1 Introduction

Since decades, scanners are being used for capturing document images. Nowadays, cameras are considered as a potential substitute of scanners due to their high production and low cost. Together with high production and low cost, cameras also offer long-ranged, non-contact and fast capturing as compared to scanners. These features speed up the traditional document image capturing and open doors for new versatile applications, like mobile OCR, digitizing thick books, digitizing fragile historical documents, etc.

Document imaging with hand-held camera is typically done in an uncontrolled environment, which induces several types of distortions in the captured image. Some of major distortions are: motion blur, low resolution, uneven light shading, under-exposure, over-exposure, perspective distortion and non-planar shape. Because of these distortions, traditional document image processing algorithms can not be applied directly to camera-captured document images.

Tremendous research is being devoted to make camera-captured document images suitable for traditional algorithms, by using dewarping techniques. Binarization and curled textline detection are the main steps of dewarping. Researchers use well known thresholding techniques [1] for binarization and have developed many new techniques for curled textline detection. Previous techniques of curled textline detection can be classified into two main categories: a) heuristic search [2,3,4,5,6,7] and b) active contours (snakes) [8,9].

Heuristic search based approaches start from a single connected component and search other components of a textline in a growing neighborhood region. These approaches use complex and rule-based criteria for textline searching. Active contours (snakes) have been used in [8,9] for curled textline segmentation. These approaches start with initializing open-curved snakes over connected components and result in textlines detection by deforming snakes in vertical directions only. In general, both heuristic search and active contours (snakes) approaches rely on adaptive thresholding for the binarization of camera-captured document image before textline detection. Binarization may give poor results, especially under the problems of uneven shading, low resolution, motion blur and under- or over-exposure, as shown in Figure 1. Poor binarization (Figure 1(b)) can negatively affect the textline detection results and later on text recognition results.

In this paper we introduce a novel approach for curled textlines regions detection directly from grayscale camera-captured document images. Therefore, our method does not use binarization before textline detection. The method starts by enhancing the curled textline structure of a grayscale document image using multi-oriented multi-scale anisotropic Gaussian smoothing based on matched filter bank approach as in [10]. Then ridges detection [11,12] technique is applied on

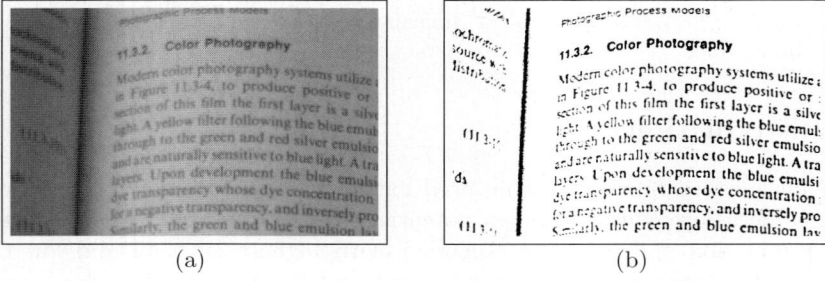

(a)   (b)

Fig. 1. a) Grayscale camera-captured document image portion from CDBAR 2007 dewarping contest dataset. b) Poor binarization (adaptive thresholding) result; which has lost most of its details.

the smoothed image. Resulting ridges represent the complete unbroken central line regions of curled textlines.

We make the following contributions in this paper. The method presented here works directly on grayscale camera-captured document image and is therefore independent of binarization and its errors under the problems of uneven shading, low resolution, motion blur and under- or over-exposure. Together with the independence of binarization, our method is robust against the problems of low resolution, motion blur and under- or over-exposure and detects textlines accurately under these problems. Furthermore, unlike previous approaches of curled textline detection, our method is also robust against high degrees of curl, variable directions of curl, different line spacing and font sizes problems.

2 Curled Textline Detection

As a first step, we apply matched filter bank approach from [10] for curl textlines enhancement of grayscale camera-captured document images. The main idea is to use oriented anisotropic Gaussian smoothing filter (Equation 2) to generate the set of Gaussian smoothing windows from the ranges of horizontal and vertical standard deviations (scales) and orientation.

$$g(x,y;\sigma_x,\sigma_y,\theta) = \frac{1}{2\pi\sigma_x\sigma_y}exp\{-\frac{1}{2}(\frac{(xcos\theta + ysin\theta)^2}{\sigma_x^2} + \frac{(-xsin\theta + ycos\theta)^2}{\sigma_y^2})\}$$

The reason of considering ranges for scales and orientation is because of variable font sizes and high degrees of multi-oriented curls within an image, respectively. To achieve this, we define an automatic way of selecting the ranges for σ_x, σ_y and θ. The same range is selected for both σ_x and σ_y, which is a function of the height of the document image (H), that is aH to bH with $a < b$. The suitable range for θ is from -45 to 45 degrees. The set of filters is defined by selecting all possible combinations of σ_x, σ_y and θ, from their ranges. It covers all typical font sizes and curl conditions in an image. The set of filters is applied on each pixel of grayscale document image and then the maximum resulting value among all the resulting values is selected for the final smoothed image. Multi-oriented multi-scale anisotropic Gaussian smoothing enhances the curled textlines structure well, which is clearly visible in Figure 2(b).

The next step is to detect textlines from the smoothed image. Some researchers [13,14] use adaptive thresholding of the smoothed image to detect the textlines or the portions of textlines and then perform heuristic postprocessing steps to join neighboring portions. Our method breaks this tradition and uses ridges detection for finding textlines form smoothed image. Firstly, adaptive thresholding of smoothed grayscale image gives poor results and misses textlines or portions of textlines, as shown in Figure 2(c). Secondly, we do not want to use any type of heuristic postprocessing steps.

Ridges are primary type of features, which provides important information about the existence of objects together with the symmetrical axes of these objects. Since decades, ridges detection has been used popularly for producing

Fig. 2. a) Grayscale camera-captured document image from CDBAR 2007 dewarping contest dataset. b) Smoothed image : Enhanced curled textlines structure. c) Poor binarization (adaptive thresholding) result of smoothed image. d) Detected ridges from smooth image which are mapped over document image with different colors; represent curled textlines regions. e) Curled textlines detection result : Detected ridges are mapped over the binarized image (given in the dataset) and assign the corresponding ridges label to connected components.

rich description of significant features from smoothed grayscale images [11] and speech-energy representation in time-frequency domain [12]. In this paper we are interested in detecting the central line regions of curl textlines. Ridges detection can do this efficiently by estimating the symmetrical axes from smoothed textlines structure. We have seen in the previous section that multi-oriented multi-scale anisotropic Gaussian smoothing generates well enhanced textlines structure. Therefore, ridges detection over smoothed image can produce unbroken central lines structure of textlines. Here, Horn-Riley [11,12] based ridges detection approach is used. This approach is based on the informations of local direction of gradient and second derivatives as the measure of curvature. From this information, which is calculated by Hessian matrix, ridges are detected by finding the zero-crossing of the appropriate directional derivatives of smoothed image. Detected Ridges over the smoothed image of Figure 2(b) are shown in Figure 2(d). It is clearly visible in the Figure 2(d) that each ridge covers the complete central line region of a textline, which results in textlines detection.

3 Experiments and Performance Evaluation

To demonstrate the performance of presented approach, we evaluate it on the real-world hand-held camera-captured document images dataset used in CB-DAR 2007 for document image dewarping contest [15]. This dataset consists of 102 grayscale and their binarized images, with ground truth of binarized images in color coded format. Previously, researchers have used only the binarized images and corresponding ground truth from this dataset, for the development and evaluation of their algorithms. We use grayscale images from this dataset. But, there is no direct evaluation method for grayscale textline detection results. Therefore, the ridges detected from grayscale image are mapped over its cleaned-up binarized image and the corresponding ridges label are assigned to connected components, as shown in Figure 2(e). Now we can compare this textlines detection result with ground truth. Here we use two different standard textline detection evaluation methods [16,17]. Evaluation method presented in [17] is designed for handwritten textline segmentation without any background noise, therefore we are using 91 manually cleaned-up document images from the CB-DAR 2007 dataset[1].

Descriptions of performance evaluation metrics for textline segmentation based on [18,16] are as follows. Consider we have two segmented images, the ground truth G and hypothesized snake-segmentation H. We can compute a weighted bipartite graph called "pixel-correspondence graph" [19] between G and H for evaluating the quality of the segmentation algorithm. Each node in G or H represents a segmented component. An edge is constructed between two nodes such that the weight of the edge equals the number of foreground pixels in the intersection of the regions covered by the two segments represented by the nodes. The matching between G and H is perfect if there is only one edge incident on each component of G or H, otherwise it is not perfect, i.e. each node in G or H may have multiple edges. The edge incident on a node is significant if the value of (w_i/P) meets some thresholding criteria, where w_i is the edge-weight and P is the number of pixels corresponding to a node (segment).

On the basis of the above description the performance evaluation metrics are:

- **Correct segmentation** (N_c)**:** the total number of G and H components' pairs which have only one significant edge in between.
- **Total oversegmentations** (N_{tos})**:** the total number of significant edges that ground truth lines have, minus the number of ground truth lines.
- **Total undersegmentations** (N_{tus})**:** the total number of significant edges that segmented lines have, minus the number of segmented lines.

---

[1] The cleaned-up data set and ground truth are available by contacting the authors.

Table 1. Performance evaluation results based on [16]

Number of ground truth lines (N_g)	2713
Number of segmented lines (N_s)	2704
Number of correct segmentation (N_c)	2455
Total oversegmentations (N_{tos})	105
Total undersegmentations (N_{tus})	99
Oversegmented components (N_{oc})	99
Undersegmented components (N_{us})	93
Missed components (N_{mc})	6
Correct segmentation accuracy ($100 * N_c/N_g$)	90.50%

- **Oversegmented components** (N_{oc}): the number of ground truth lines having more than one significant edge.
- **Undersegmented components** (N_{uc}): the number of segmented lines having more than one significant edge.
- **Missed components** (N_{mc}): the number of ground truth components that matched the background in the hypothesized segmentation.

Performance evaluation results of our textline segmentation algorithm, based on the above metrics, are given in a Table 1.

According to the methodology [17], the matching score is equal to or above a specified acceptance threshold (i.e. 95%), where matching score is defined as:

$$MatchScore(i,j) = T(G_j \cap R_i \cap I)/T(G_j \cup R_i \cup I) \quad (1)$$

where I is the set of all image pixels, G_j and R_i are the sets of all pixels covering the jth ground truth region and i^{th} result region respectively. Based on the matching scores, detection rate (DR), recognition accuracy (RA) and combine performance metric FM are calculated as follows:

$$DR = w_1 \frac{o2o}{N} + w_2 \frac{g\_o2m}{N} + w_3 \frac{g\_m2o}{N} \quad (2)$$

$$RA = w_4 \frac{o2o}{M} + w_5 \frac{d\_o2m}{M} + w_6 \frac{d\_m2o}{M} \quad (3)$$

where N and M are the total number of ground truth and result elements respectively, $w_1(1), w_2(0.25), w_3(0.25), w_4(1), w_5(0.25), w_6(0.25)$ are predetermined weights, $o2o$ is one to one match, g_o2m is one ground truth to many detected, g_m2o is many ground truth to one detected, d_o2m one detected to many ground truth and d_m2o is many detected to one ground truth. Performance evaluation results based on equations (1), (2) and (3) are given in Table 2.

Table 2. Performance evaluation results based on [17]

Ground truth elements (N)	2749
Detected elements (M)	2717
One to one match ($o2o$)	2503
One ground truth to many detected (g_o2m)	64
Many ground truth to one detected (g_m2o)	127
One detected to many ground truth (d_{o2m})	62
Many detected to one ground truth (d_{m2o})	130
Detection rate (DR)	92.79%
Recognition accuracy (RA)	93.89%
$FM = (2*DR*RA)/(DR+RA)$	93.34%
Correct segmentation accuracy ($100*o2o/N$)	91.05%

4 Discussion

We have proposed a novel approach for curled textlines regions detection from grayscale camera-captured document images. We introduced a combination of multi-scale multi-oriented Gaussian smoothing and ridges detection for finding curled textlines directly from grayscale intensity values of camera-captured document images. Therefore, our approach does not use binarization and is independent of binarization errors. We have achieved around 91% of one-to-one correct segmentation textline detection accuracy on the dataset of CDBAR 2007 document image dewarping contest, which proves the effectiveness of our method. The 9% of errors are mainly because of oversegmentation, that is more than one textline detected for a single ground truth textline. These oversegmentation errors can be easily overcome by grouping ridges in horizontal neighborhood region, as a post-processing step. Our approach is robust against uneven shading, low resolution, motion blur, under- or over-exposure, high degrees of curl, variable directions of curl, different line spacing and font sizes problems and detects textlines under these problems. Therefore, our method can be integrated with vareity of camera devices (from low to high image quality) and can be used under vareity of image capturing environment (from rough and uncontrolled to highly controlled environment). The textline detection results of presented method can be used for image dewarping, developing efficient binarization techniques and introducing new grayscale OCR algorithms.

References

1. Shafait, F., Keysers, D., Breuel, T.M.: Efficient implementation of local adaptive thresholding techniques using integral images. In: Proc. Document Recognition and Retrieval XV, San Jose, CA, USA, vol. 6815, p. 81510 (2008)
2. Zhang, Z., Tan, C.L.: Correcting document image warping based on regression of curved text lines. In: Proc. 7th Int. Conf. on Document Analysis and Recognition, Edinburgh, Scotland, pp. 589–593 (2003)

3. Lu, S.J., Tan, C.L.: The restoration of camera documents through image segmentation. In: Proc. 7th IAPR workshop on Document Analysis Systems, Nelson, New Zealand, pp. 484–495 (2006)
4. Fu, B., Wu, M., Li, R., Li, W., Xu, Z.: A model-based book dewarping method using text line detection. In: Proc. 2nd Int. Workshop on Camera Based Document Analysis and Recognition, Curitiba, Barazil, pp. 63–70 (2007)
5. Gatos, B., Pratikakis, I., Ntirogiannis, K.: Segmentation based recovery of arbitrarily warped document images. In: Proc. 9th Int. Conf. on Document Analysis and Recognition, Curitiba, Barazil, pp. 989–993 (2007)
6. Stamatopoulos, N., Gatos, B., Pratikakis, I., Perantonis, S.J.: A two-step dewarping of camera document images. In: Proc. 8th IAPR Workshop on Document Analysis Systems, Nara, Japan, pp. 209–216 (2008)
7. Ulges, A., Lampert, C.H., Breuel, T.M.: Document image dewarping using robust estimation of curled text lines. In: Proc. 8th Int. Conf. on Document Analysis and Recognition, Seoul, Korea, pp. 1001–1005 (2005)
8. Bukhari, S.S., Shafait, F., Breuel, T.M.: Segmentation of curled textlines using active contours. In: Proc. 8th IAPR Workshop on Document Analysis Systems, Nara, Japan, pp. 270–277 (2008)
9. Bukhari, S.S., Shafait, F., Breuel, T.M.: Coupled snakelet model for curled textline segmentation of camera-captured document images. In: Proc. 10th Int. Conf. on Document Analysis and Recognition, Barcelona, Spain (2009)
10. Bukhari, S.S., Shafait, F., Breuel, T.M.: Script-independent handwritten textlines segmentation using active contours. In: Proc. 10th Int. Conf. on Document Analysis and Recognition, Barcelona, Spain (2009)
11. Horn, B.K.P.: Shape from shading: A method for obtaining the shape of a smooth opaque object from one view. PhD Thesis, MIT (1970)
12. Riley, M.D.: Time-frequency representation for speech signals. PhD Thesis, MIT (1987)
13. Li, Y., Zheng, Y., Doermann, D., Jaeger, S.: Script-independent text line segmentation in freestyle handwritten documents. IEEE Transactions on Pattern Analysis and Machine Intelligence 30(8), 1313–1329 (2008)
14. Du, X., Pan, W., Bui, T.D.: Text line segmentation in handwritten documents using Mumford-Shah model. In: Proc. Int. Conf. on Frontiers in Handwriting Recognition, Montreal, Canada, pp. 1–6 (2008)
15. Shafait, F., Breuel, T.M.: Document image dewarping contest. In: Proc. 2nd Int. Workshop on Camera Based Document Analysis and Recognition, Curitiba, Brazil, pp. 181–188 (2007)
16. Shafait, F., Keysers, D., Breuel, T.M.: Performance evaluation and benchmarking of six page segmentation algorithms. IEEE Transactions on Pattern Analysis and Machine Intelligence 30(6), 941–954 (2008)
17. Gatos, B., Antonacopoulos, A., Stamatopoulos, N.: ICDAR 2007 handwriting segmenentation contest. In: Proc. 9th Intelligence Conf. on Document Analysis and Recognition, Curitiba, Brazil, pp. 1284–1288 (2007)
18. Shafait, F., Keysers, D., Breuel, T.M.: Pixel-accurate representation and evaluation of page segmentation in document images. In: Proc. Int. Conf. on Pattern Recognition, Hong Kong, China, August 2006, pp. 872–875 (2006)
19. Breuel, T.M.: Representations and metrics for off-line handwriting segmentation. In: Proc. 8th Int. Workshop on Frontiers in Handwriting Recognition, Ontario, Canada, pp. 428–433 (2002)

Kernel PCA for HMM-Based Cursive Handwriting Recognition

Andreas Fischer and Horst Bunke

Institute of Computer Science and Applied Mathematics
University of Bern, Neubrückstrasse 10, CH-3012 Bern, Switzerland
{afischer,bunke}@iam.unibe.ch

Abstract. In this paper, we propose Kernel Principal Component Analysis as a feature selection method for offline cursive handwriting recognition based on Hidden Markov Models. In contrast to formerly used feature selection methods, namely standard Principal Component Analysis and Independent Component Analysis, nonlinearity is achieved by making use of a radial basis function kernel. In an experimental study we demonstrate that the proposed nonlinear method has a great potential to improve cursive handwriting recognition systems and is able to significantly outperform linear feature selection methods. We consider two diverse datasets of isolated handwritten words for the experimental evaluation, the first consisting of modern English words, and the second consisting of medieval Middle High German words.

1 Introduction

Offline recognition of unconstrained cursively handwritten text is still a widely unsolved problem in computer science and an active area of research. In contrast to online recognition, where time information about the writing process is available, offline recognition is performed solely on text images. Here, a high recognition accuracy can be achieved for small, specific domains such as address or bankcheck reading. When it comes to unconstrained text recognition, only few systems exist that are able to cope with a high variety of writing styles and an extensive vocabulary. For a survey, see [1]. A widely used type of recognizer suited for this task is Hidden Markov Models (HMM). Examples of HMM-based recognition systems can be found in [2,3].

In this paper, we aim at improving handwriting recognition by feature selection. [1] Feature selection can be used to improve the quality of the features used, thereby often reducing the number of features in order to eliminate noise. Standard methods include removing linear correlations among the features by means of Principal Component Analysis (PCA) or Linear Discriminant Analysis (LDA) and, as a more general approach, finding independent feature sources by means of Independent Component Analysis (ICA) [4]. In [3], PCA and ICA were used to improve HMM-based cursive handwriting recognition.

---

[1] Note that the term 'feature selection' is used in a broad sense in this paper including feature space transformation.

The main restriction of the feature selection methods mentioned is their assumption of linearity. PCA does not take into account nonlinear correlations among the features and ICA performs a linear transform in order to maximize the independency of the features in the new feature space. With the introduction of kernel methods [5], nonlinear transformations have come within the reach of efficient computability. Kernelizable algorithms are executed in high dimensional feature spaces based only on the dot product, using kernel functions that can often be computed efficiently even for nonlinear feature space mappings.

Since the introduction of kernel PCA (KPCA) about a decade ago [6], this powerful nonlinear technique has been successfully applied to several pattern recognition tasks including, for example, face detection [7] and palmprint recognition [8]. In [9], KPCA was used for Chinese character recognition. Here, images of single characters are recognized that are, in contrast, not available for Roman cursive handwriting recognition where whole words and sentences are considered without character segmentation. In the domain of speech recognition, which is closely related to cursive handwriting recognition, a successful application was reported in [10] for an HMM-based recognition system.

In this paper, we apply KPCA feature selection with a radial basis function (RBF) kernel to HMM-based cursive handwriting recognition. On a set of nine geometrical features presented in [2] and two diverse datasets, we demonstrate that KPCA is able to significantly increase the accuracy of handwriting recognition and outperform standard linear feature selection methods, i.e. PCA and ICA. The first datased considered for experimental evaluation consists of isolated, modern English words taken from the publicly available IAM database [11] and the second dataset consists of medieval Middle High German words taken from the Parzival database recently presented in [12].

The remainder of this paper is organized as follows. In Section 2, HMM-based recognition is introduced, Section 3 presents the feature selection methods applied, Section 4 describes the underlying two datasets and discusses the experimental results, and Section 5 draws some conclusions.

2 HMM Recognition

We use an HMM-based system for cursive handwriting recognition as presented in [2]. In the following, the different stages of the recognition process are briefly discussed, i.e. image preprocessing, feature extraction, and HMM application.

2.1 Preprocessing and Feature Extraction

The input for the recognizer are binary images of cursively handwritten words. They have been isolated from textline images after correction of the skew and the slant, vertical scaling with respect to the upper and lower baseline, and horizontal scaling with respect to the mean black-white transitions. This normalization allows the recognizer to cope with different writing styles.

From the normalized word images, a sequence of feature vectors is extracted with a sliding window approach. A window with width of one pixel and height of the image is moved from left to right over the word and at each step, a set of nine geometrical features $\mathbf{x} \in \mathbb{R}^9$ is calculated from the enclosed pixels. Three global features capture the fraction of black pixels, the center of gravity, and the second order moment. The remaining six local features consist of the position of the upper and lower contour, the gradient of the contours, the number of black-white transitions, and the fraction of black pixels between the contours [2].

2.2 Hidden Markov Models

In the HMM-based approach, each letter is modeled with a number of hidden states s_i arranged in a linear topology. The states emit observable feature vectors with output probability distributions $p_{s_i}(\mathbf{x})$ given by a mixture of Gaussians. Starting from the first state s_1, the model either rests in a state or changes to the next state with transition probabilities $P(s_i, s_i)$ and $P(s_i, s_{i+1})$, respectively. After an appropriate initialization, the output probability distributions and the transition probabilities are trained with the Baum-Welch algorithm [13].

For isolated word recognition, all possible words are modeled by an HMM built from the trained letter HMMs and the most probable word is chosen with the Viterbi algorithm [13] for a given word image. In this paper, we follow a closed vocabulary assumption by taking only words into account that exist in the dataset and treat them with equal probability, i.e. no language model is used.

3 Feature Selection

After feature extraction using a sliding window, feature selection is applied to the feature vector sequence. In general, the ground truth of the training samples contains only the transcription, i.e. the sequence of letters, but not the start and end position of the letters within the feature vector sequence. Therefore, class labels are not given for the single feature vectors and feature selection is constrained to unsupervised methods. In this section, PCA and ICA are presented as two standard linear methods and KPCA as the new nonlinear method of interest for handwriting recognition.

3.1 Principal Component Analysis

Principal component analysis (PCA) applies a linear transform $\mathbf{y} = W\mathbf{x}$ to the original feature vectors $\mathbf{x} = (x_1, \ldots, x_n) \in \mathbb{R}^n$ in order to remove linear correlation among the new features $\mathbf{y} = (y_1, \ldots, y_m) \in \mathbb{R}^m$ and to reduce the dimension to $m \leq n$ features that capture most of the variance. For data centered around the mean vector, the mapping is given by an orthogonal transform where each row in matrix W is an eigenvector of the covariance matrix in the original feature space. The eigenvectors, called principal components $p_1, \ldots, p_n$, can be found by eigenvalue decomposition and can be ordered by variance such that $\sigma_1^2 \geq \ldots \geq \sigma_n^2$. By choosing only the first $m \leq n$ principal components, the dimensionality is reduced while maximum variance is captured.

The omitted dimensions with low variance can be regarded as noise. For HMM-based recognition, the output probability distributions of the HMM states are often modeled with diagonal covariance matrices in order to reduce the number of parameters to train. The features obtained with PCA conform to this implicit assumption of uncorrelated features. Reducing the feature space dimension also reduces the computational complexity of HMM-based recognition.

3.2 Independent Component Analysis

In contrast to PCA, Independent component analysis (ICA) aims not only at finding linearly uncorrelated components, but also statistically independent components by optimizing the parameters of a linear transform $\mathbf{y} = W\mathbf{x}$. Statistically independent features are uncorrelated, but the reverse does not hold true in general. For joint normal distribution, however, uncorrelated features are already independent. In this case, the application of ICA is not expected to further improve the feature space when compared to PCA.

In fact, measures of non-Gaussianity, a property expressing how different a given distribution is from a normal distribution, are often used as contrast functions to optimize the parameters of the matrix W. A widely used measure of non-Gaussianity is negentropy given by

$$J(y_i) = H(y_{i,Gauss}) - H(y_i)$$

where $H(y_i) = -\int p(y_i) \log p(y_i) dy_i$ is the entropy with density function $p(y_i)$ and a Gaussian variable $y_{i,Gauss}$ with the same covariance and mean as the single feature y_i. From information theory it is known that maximum negentropy means maximum independency for uncorrelated features. Thus, independent components can be found in the directions of maximum negentropy. An efficient solution for ICA is given by the FastICA algorithm [4] where uncorrelated features with unit variance are found first, thereby reducing the problem to optimizing an orthogonal matrix W, before the independent components are found with respect to an approximated negentropy function.

Independent features are expected to capture distinct properties of the underlying patterns and therefore should be well-suited for pattern recognition. Furthermore, the feature space can also be reduced to $m \leq n$ dimensions in order to eliminate noise and to reduce the computational complexity.

3.3 Kernel PCA

Kernel methods [5] are based on the idea of mapping feature vectors into a higher dimensional feature space $\phi : \mathbb{R}^n \to \mathcal{F}$. Instead of an explicit mapping, only the dot product of two feature vectors is calculated as the kernel function $\kappa(\mathbf{x}_1, \mathbf{x}_2) = \phi(\mathbf{x}_1) \cdot \phi(\mathbf{x}_2)$. Often, this dot product can be calculated efficiently even for nonlinear mappings ϕ. Many kernelizable algorithms do not rely on the feature vectors, but can be rewritten in terms of dot products only, also known as the kernel trick. If valid kernel functions are used other than the standard dot product, the algorithm is effectively performed in the feature space $\mathcal{F}$.

PCA is kernelizable as shown in [6]. For kernel PCA (KPCA), the projection of a mapped feature vector $\phi(\mathbf{x})$ onto a principal component $\mathbf{p}_i$ of the feature space is given by

$$\mathbf{p}_i \cdot \phi(\mathbf{x}) = \sum_{j=1}^{N} \alpha_j^i \kappa(\mathbf{x}_j, \mathbf{x})$$

where $\mathbf{x}_j$ is one out of N training feature vectors and α^i is an eigenvector of the training kernel matrix $K[l,m] = \kappa(\mathbf{x}_l, \mathbf{x}_m), 1 \leq l, m \leq N$. The final transform of $\mathbf{x}$ is then given by $\mathbf{y} = (\mathbf{p}_1 \cdot \phi(\mathbf{x}), \ldots, \mathbf{p}_N \cdot \phi(\mathbf{x}))$. Note that the centering of the data around the mean vector in the feature space can also be done in terms of dot products only and the principal components can still be ordered such that the first $m \leq N$ components capture most of the variance in $\mathcal{F}$.

In this paper, we apply KPCA with the well-known radial basis function (RBF) kernel

$$\kappa(\mathbf{x}_1, \mathbf{x}_2) = \exp(-\gamma \cdot ||\mathbf{x}_1 - \mathbf{x}_2||^2), \gamma > 0$$

While linear correlations are removed in the feature space $\mathcal{F}$, the transform of the original feature vector $\mathbf{x} \in \mathbb{R}^n$ into the new feature vector $\mathbf{y} \in \mathbb{R}^m$ is nonlinear. As a main advantage over PCA and ICA, nonlinear correlations are taken into account. The reduced dimension $m \leq N$ can also be higher than the original dimension n making it possible to effectively increase the feature dimension.

4 Experimental Results

KPCA feature selection for HMM-based recognition is applied to two datasets of isolated, handwritten words and the results are compared with PCA and ICA.

The first dataset contains words taken from the IAM database presented in [11]. This publicly available database contains handwritings of English sentences from several hundred writers. The second dataset contains words taken from the recently presented Parzival database [12] that consists of word images of Medieval handwritings from the 13th century. In order to keep the computational effort within reasonable bounds, a subset including all different writing styles is considered for both databases. Table 1 summarizes the statistics of the datasets and Figures 1a and 1b show some sample word images after preprocessing.

4.1 Experimental Setup

For both datasets, the words are first divided into a training, validation, and a test set. Half of the words, i.e. each second word, is used for training and a

Table 1. Statistics of the datasets

Dataset	Word Instances	Word Classes	Letters
IAM	12,265	2,460	66
Parzival	11,743	3,177	87

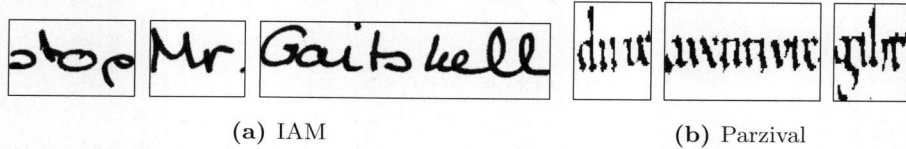

(a) IAM (b) Parzival

Fig. 1. Sample images of the datasets

quarter of the words for validation and testing, respectively. The HMMs are then trained with an optimized number of states for each letter based on the mean width of the letters [2,12]. The optimal number of Gaussian mixtures for the HMM output probability densities, ranging from 1 to 30, is found with respect to the word accuracy on the validation set.

For the feature selection, the optimal dimensionality of the new feature vectors is also found with respect to the validation accuracy, ranging from 1 to 9 for PCA as well as for ICA, and up to 19 for KPCA. The RBF kernel parameter γ is validated over a logarithmic range from 10^{-2} up to over 10^3. While for PCA and ICA the whole training set is used, the training of the KPCA is constrained to the feature vector sequences of each 100th word due to computational reasons, similarly to [10]. Still, several thousand feature vectors are considered in order to find the principal components of the feature space.

4.2 Results

The results of the experimental evaluation on the test set are given in Table 2. For both datasets, the word accuracy is given for the reference system [2] using the nine geometrical features without feature selection as well as for the three feature selection methods PCA, ICA, and KPCA, respectively. The optimal KPCA parameter values are indicated in the three right-most columns, i.e. the dimension of the feature vectors, the γ parameter of the RBF kernel, and the number of Gaussian mixtures (GM) of the HMM recognizer.

The improvements of the word accuracy are all statistically significant (t-test with $\alpha = 0.05$) except for PCA when compared to the Parzival reference system. In particular, ICA achieved statistically better results than PCA and KPCA achieved statistically better results than ICA. Figure 2 illustrates the

Table 2. Experimental results (word accuracy). Improvements marked with a star (*) are statistically significant over the value in the preceding column (t-test, $\alpha = 0.05$).

Dataset	Ref.	PCA	ICA	KPCA	Dim.	γ	GM
IAM	73.91	75.77*	77.56*	79.65*	13	10^2	28
Parzival	88.69	88.76	89.88*	91.38*	7	10^3	22

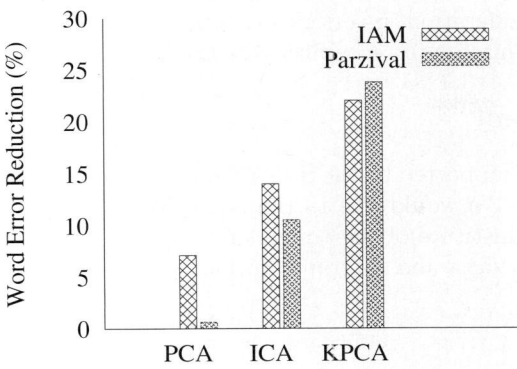

Fig. 2. Experimental results (word error reduction)

word error rate reduction of the feature selection methods. For both datasets, KPCA reduced the reference word error rate by over 20 percent.

Regarding the KPCA parameters, a dimension increase from 9 to 13 dimensions was optimal or the IAM dataset. The γ parameter had a high impact on the word accuracy and was validated over a rather large logarithmic scale.

5 Conclusions

The nonlinear KPCA feature selection method was successfully applied to cursive handwriting recognition and compared with two standard linear methods, namely PCA and ICA. Unsupervised feature selection was performed on feature vector sequences of nine geometrical features extracted by a sliding window. Then, an HMM-based recognizer was applied to two diverse datasets of isolated, handwritten word images, one with modern English words (IAM database) and one with medieval Middle High German words (Parzival database).

In accordance with previous work, the experimental results show that feature selection has a great potential to improve cursive handwriting recognition. The word accuracy of an HMM recognizer could be improved with statistical significance using PCA and ICA. With the independent ICA features, significantly better results were achieved than with the linearly uncorrelated PCA features.

As a main contribution, we show that KPCA is able to further improve the word accuracy significantly. By using KPCA with an RBF kernel, the linear PCA transform in the feature space corresponds to an efficient nonlinear transform of the original feature vectors, thus overcoming the limitation of linearity of the standard feature selection methods. For both datasets under consideration, the original word error rate was reduced by over 20 percent with KPCA features outperforming PCA and ICA significantly. To the knowledge of the authors, this is the first time that KPCA has been applied to cursive handwriting recognition.

Future work includes the extension of the scope of features, kernels, and recognizers under consideration, investigating other kernel methods such as kernel ICA, and the combination of recognizers trained in different feature spaces.

Acknowledgment

This work has been supported by the Swiss National Science Foundation (Project CRSI22_125220/1). We would like to thank M. Stolz and G. Viehauser of the German Language Institute of the University of Bern for providing us with the Parzival manuscript data and its transcription.

References

1. Plamondon, R., Srihari, S.: Online and off-line handwriting recognition: A comprehensive survey. IEEE Trans. Pattern Analysis and Machine Intelligence 22, 63–84 (2000)
2. Marti, U.V., Bunke, H.: Using a statistical language model to improve the performance of an HMM-based cursive handwriting recognition system. Journal of Pattern Recognition and Art. Intelligence 15, 65–90 (2001)
3. Vinciarelli, A., Bengio, S.: Off-line cursive word recognition using continuous density HMMs trained with PCA and ICA features. In: 16th Int. Conf. on Pattern Recognition, vol. 3, pp. 81–84 (2002)
4. Oja, E.: Independent component analysis: algorithms and applications. Neural Networks 13, 411–430 (2000)
5. Müller, K., Mika, S., Rätsch, G., Tsuda, K., Schölkopf, B.: An introduction to kernel-based learning algorithms. Neural Networks 12(2), 181–202 (2001)
6. Schölkopf, B., Smola, A., Müller, K.R.: Kernel principal component analysis. In: Advances in Kernel Method – Support Vector Learning, pp. 327–352. MIT Press, Cambridge (1999)
7. Liu, Y.H., Chen, Y.T., Lu, S.S.: Face detection using kernel PCA and imbalanced SVM. In: Jiao, L., Wang, L., Gao, X.-b., Liu, J., Wu, F. (eds.) ICNC 2006. LNCS, vol. 4221, pp. 351–360. Springer, Heidelberg (2006)
8. Ekinci, M., Aykut, M.: Palmprint recognition by applying wavelet-based kernel PCA. Journal of Computer Science and Technology 23(5), 851–861 (2008)
9. Shi, D., Ong, Y.S., Tan, E.C.: Handwritten chinese character recognition using kernel active handwriting model. In: Proc. IEEE Int. Conf. on Systems, Man and Cybernetics, vol. 1, pp. 251–255. IEEE, Los Alamitos (2003)
10. Takiguchi, T., Ariki, Y.: Robust feature extraction using kernel PCA. In: Proc. IEEE Int. Conf. on Acoustics, Speech, and Signal Processing, pp. 509–512. IEEE, Los Alamitos (2006)
11. Marti, U.V., Bunke, H.: The IAM-database: an English sentence database for off-line handwriting recognition. Int. Journal on Document Analysis and Recognition 5, 39–46 (2002)
12. Wüthrich, M., Liwicki, M., Fischer, A., Indermühle, E., Bunke, H., Viehhauser, G., Stolz, M.: Lanugage model integration for the recognition of handwritten medieval documents. Accepted for publication in Proc. IEEE Int. Conf. on Document Analysis and Recognition (2009)
13. Rabiner, L.: A tutorial on hidden Markov models and selected applications in speech recognition. Proceedings of the IEEE 77(2), 257–285 (1989)

Improved Handwriting Recognition by Combining Two Forms of Hidden Markov Models and a Recurrent Neural Network

Volkmar Frinken[1], Tim Peter[1], Andreas Fischer[1], Horst Bunke[1], Trinh-Minh-Tri Do[2], and Thierry Artieres[2]

[1] Institute of Computer Science and Applied Mathematics, University of Bern, Neubrückstrasse 10, CH-3012 Bern, Switzerland
tim.peter@swissdrg.org, {frinken,afischer,bunke}@iam.unibe.ch
[2] LIP6 - Université Pierre et Marie Curie
104 Avenue du Président Kennedy, 75016 Paris, France
{Trinh-Minh-Tri.Do,thierry.artieres}@lip6.fr

Abstract. Handwritten word recognition has received a substantial amount of attention in the past. Neural Networks as well as discriminatively trained Maximum Margin Hidden Markov Models have emerged as cutting-edge alternatives to the commonly used Hidden Markov Models. In this paper, we analyze the combination of these classifiers with respect to their potential for improving recognition performance. It is shown that a significant improvement can in fact be achieved, although the individual recognizers are highly optimized state-of-the-art systems. Also, it is demonstrated that the diversity of the recognizers has a profound impact on the improvement that can be achieved by the combination.

1 Introduction

The automatic recognition of handwritten text has been a focus of intensive research for several decades [1]. Yet the problem is far from being solved, especially in the field of unconstrained handwritten word recognition. This problem occurs, for example, in automatic postal address or form reading [2], or in the transcription of unconstrained text after text lines have been extracted from a page and segmented into individual words [3].

A common and widely used approach to handwritten word recognition is based on Hidden Markov Models (HMM). However, novel recognizers have been developed recently that clearly outperform HMMs. Among the most promising ones are recognizers based on Maximum Margin Hidden Markov Models (MMHMM) [4] and novel recurrent neural network architectures (NN) [5].

The availability of these novel types of recognizers raises the question whether it is possible to exploit their different nature and to build a multiple classifier system (MCS) to further increase the performance. The potential of MCSs in various pattern recognition tasks has been impressively demonstrated [6,7], yet it is unclear for a specific task what the optimal combination strategy is and what performance increase one can expect.

The research that has been done to combine classifiers for offline word recognition, e.g. [8,9,10], focuses mostly on the combination of similar underlying base recognizers. By contrast, we aim at utilizing base recognizers of different architectures (i.e. HMM, NN and MMHMM). From such a combination, a higher degree of diversity and, consequently, a greater reduction of the error rate can be expected.

The rest of the paper is structured as follows. The three base recognizers, HMM, MMHMM, and NN, are described in Section 2. Various combination methods investigated in this paper are introduced in Section 3. An experimental evaluation is presented in Section 4 and the paper concludes with Section 5.

2 Preprocessing and Base Classifiers

The task at hand is recognizing scanned images of hand-written words. In our case the words come from the IAM database [11]. They have been binarized via a gray scale value threshold, corrected (skew and slant) and normalized (height and width). After these steps, a horizontally sliding window with a width of one pixel is used to extract nine geometric features at each position. The 0th, 1st and 2nd moment of the black pixels' distribution within the window, the top-most and bottom-most black pixel, the inclination of the top and bottom contour, the number of vertical black/white transitions, and the average grey scale value between the top-most and bottom-most black pixel. For details on these steps, we refer to [12].

2.1 Hidden Markov Models

Hidden Markov Modes are a statistical model to analyze sequences and are therefore well suited for handwritten text recognition. Assuming that a statistical process created the observed sequence, one is interested in identifying the internal states of the creating process at each time step. These internal states can then be mapped to a sequence of letters which constitute the recognized word. A detailed description of the HMM based recognizer we use in the work described in this paper can be found in [12].

2.2 Maximum Margin Hidden Markov Models

The traditional Maximum Likelihood Estimation learning strategy for Hidden Markov Models does not focus on minimizing the classification error rate and a number of attempts have been made to develop discriminative learning methods for HMMs. Recently, large margin learning of HMMs has been investigated [13] and shown to significantly outperform these discriminant methods [14]. Yet none of the already proposed large margin training algorithms actually handles the full problem of maximum margin learning for HMM parameters. Methods are limited to simple HMMs only or are dedicated to learn mean parameters only, etc.

We use here a new technique for maximum margin learning of HMM models whose efficiency has been investigated on both on-line handwritten digit recognition and on off-line handwritten word recognition tasks (see [4,15] for details).

2.3 Neural Networks

The neural network based recognizer used in this paper is a recently developed recurrent neural network, termed *bidirectional long short-term memory* (BLSTM) neural network [16]. The output of the network is a sequence of output activations. Each node represents one character and its activation indicates the probability that the corresponding character is present at that position. Given an input word, the probability of that word being the origin on the sequence can be computed using the Connectionist Temporal Classification Token Passing algorithm [5].

2.4 Diversity Analysis

Combining recognizers can only improve the recognition accuracy if a wrong output from one recognizer is corrected with the correct output from a different recognizer. Obviously, this is only possible if the recognizers are not too similar. A diversity measure indicates how different the individual recognizers of an ensemble are. Of course, a high diversity does not automatically guarantee an improvement of the recognition rate. Nevertheless, it still contains useful information about promising combinations. There exists two types of diversity measures, pairwise and non-pairwise [17]. In this paper, we focus on two popular pairwise measures, namely *Correlation* and *Disagreement*.

To formally describe these measures, consider two recognizers R_1 and R_2 and the four possible recognition probabilities: both correct (a), R_1 wrong and R_2 correct (b), R_1 correct and R_2 wrong (c), and both wrong (c). With these values, the diversity measures can be calculated:

$$\text{Correlation}_{R_1,R_2} = \frac{ad - bc}{\sqrt{(a+b)(c+d)(a+c)(b+d)}}$$
$$\text{Dissagreement}_{R_1,R_2} = b + c$$

To extend these diversity measures to ensembles containing more than two recognizers, the mean of all pairwise diversity measures is used.

3 Combination Methods

We investigated eight methods for combining two or three recognizers, namely *Voting, Borda Count, Exponentiated Borda Count, Average Likelihood, Class Reduced Average Likelihood, Maximum Likelihood, Maximum Likelihood Margin,* and *Meta Voting* [7]. All these combination methods have been tested in a simple *unweighted* version and two weighted versions. The *perf-weighted* version uses the performance, i.e. the recognition rate on an independent validation set for each base recognizer as a weight. The *ga-weighted* version optimizes the weights on the validation set using a genetic algorithm [8].

Voting considers the output of each recognizer and returns the most frequent word. In case of a tie, the word is chosen randomly among the recognizers' outputs. In the weighted versions, a recognizer's vote counts as much as its weight. For combining recognizers with *Borda Count*, each recognizer's n-best list is used and the word at position i on the list is given the score $n - i + 1$. For each word in the dictionary, the score is summed up over each recognizer's list and the word with the highest score is returned. The weighted versions apply weights to each list in the final summation. We also investigated a novel method, called *Exponentiated Borda Count*, where the score function changes to $(n - i + 1)^p$, with p being a constant that is optimized on an independent validation set. This method accounts for the fact that the probability of the correct word being at postion i seems to decrease faster than linear as i increases.

While *Borda Count* and *Exponentiated Borda Count* need just a ranked list of classes from each recognizer, other methods require that a recognizer returns a likelihood value for each word in the dictionary. However, since the values returned by the MMHMMs and HMM are fundamentally different from the values returned by the NN, all output values are normalized. The *Maximum Likelihood* and *Average Likelihood* combination methods are common approaches that consider the maximum (average) of the normalized likelihood values among all recognizers for each word in the dictionary and return the word having the largest value. The *Class Reduced Average Likelihood* follows the same procedure, but considers only the words occurring in the 1-best lists of the recognizers. The weighted versions multiply weights to the normalized likelihoods before taking the maximum and the average, respectively.

The *Maximum Likelihood Margin* method uses only the top two words from each recognizer and their normalized likelihood values. The difference of the top word's value to the following word's value, the likelihood margin, decides which word is returned. Obviously, a weighted version can be introduced by multiplying weights to the margins.

Meta Voting, in which the best versions (found on the validation set) of all seven of the previously described combination methods vote on the output was also investigated. These outputs were combined with *unweighted* and *ga-weighted* voting.

4 Experimental Evaluation

We experimentally evaluated the impact of classifier combination on a subset of the IAM Handwriting Database [11][1]. The words consist of the 26 lower case letters of the Roman alphabet. The complete set of words was split up in a test (9,674 words), a validation (5,164 words), and a training set (34,557 words) contributed by different writers each. (Hence, a writer independent recognition task was considered.) On the training set, an HMM, an MMHMM and an NN

---

[1] http://www.iam.unibe.ch/fki/databases/iam-handwriting-database

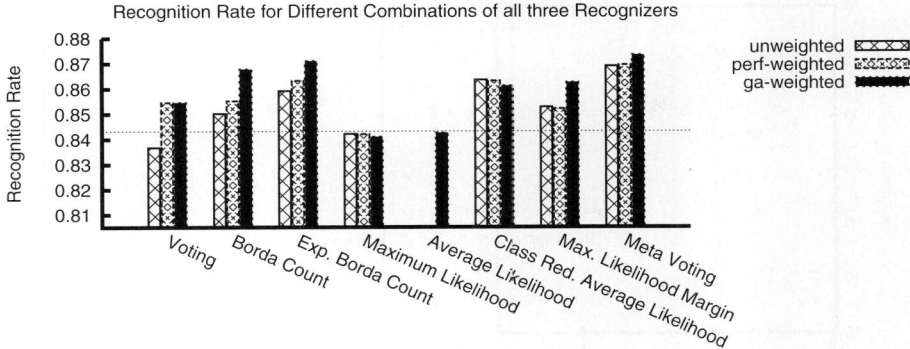

Fig. 1. The recognition rate of all combination methods for combining HMM, MMHMM and NN compared to the best base classifier's performance (horizontal line). Bars not shown are below the range of the diagram.

were trained and validated separately. For testing, a dictionary containing all words that occur in the three sets was used.

The performance of the three individual classifiers is shown in Fig. 2 (three leftmost columns). A 20-best list for each recognizer was used to analyze the different combination methods. We first show the resulting recognition accuracies for the combination of all three recognizers (the combination results involving two recognizers are similar). In Fig. 1, the different recognition rates are compared to the best base recognizer (NN, recognition rate of 0.8432), indicated by the horizontal dotted line. The other two recognizers with a recognition rate of 0.7475 (HMM) and 0.7605 (MMHMM) perform below the range of the diagram. First of all, it becomes obvious that the choice of a proper combination method is of paramount importance, as some combinations lead to a deterioration of the recognition rate while others achieve a substantial increase. Secondly, the versions with

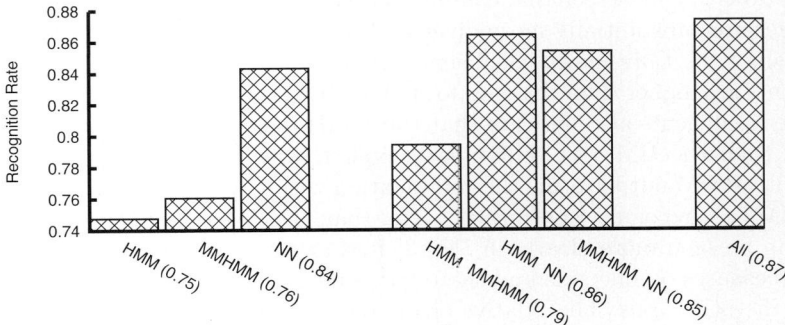

Fig. 2. The recognition rate of the base recognizer as well as the best combination methods. All differences are statistically significant.

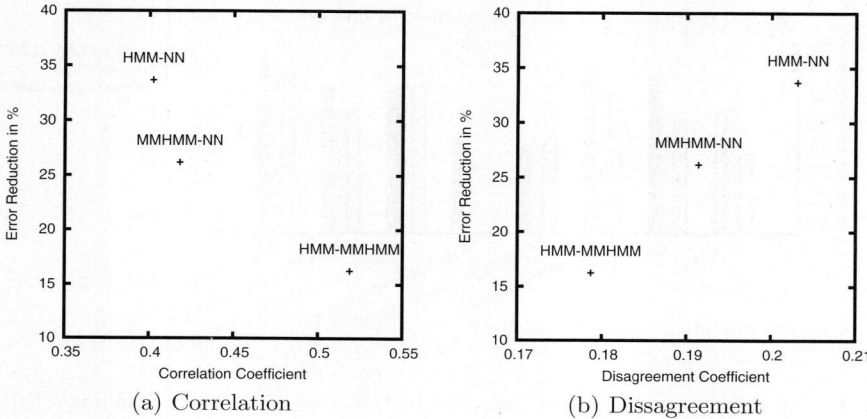

Fig. 3. A scatter plot to compare combinations of two recognizers. The x-axis is the diversity coefficient while the y-axis displays the reduction of error rate (average error rate of the two base classifiers vs. error rate of the best combination).

optimized weights outperform the other versions almost always. All improvements over the NN base recognizer are statistically significant[2]. It turns out that *Meta Voting* with optimized weights is the best combination method for the case of three recognizers.

Next, also combinations of two recognizers are analyzed. In this case, only the method performing best on an independent validation set is considered for each combination. A direct comparison can be seen in Fig. 2. The neural network is the best performing base recognizer and achieves a higher recognition rate than the HMM and MMHMM combination. Combining the NN with the HMM or with the MMHMM increases the recognition rate. The combination of all three recognizers outperforms all other systems significantly.

The following table shows the different diversity measures applied to all pairwise combinations. It can be seen that combinations including the neural network based recognizer are substantially more diverse than the HMM and MMHMM combination, since the *Correlation* coefficients are much lower while the *Disagreement* coefficients are higher than the ones for the MMHMM and HMM combination.

These coefficients also indicate that the MMHMM is slightly more similar to the NN than the HMM is. This might explain why the neural network combined with HMM outperforms the combination with the MMHMM although the MMHMM base recognizer performes better than the HMM base recognizer. This effect can be seen more clearly in Fig. 3. It shows a scatter plot of the two diversity measures on one axis and the relative reduction of the recognizer's error rate on the other axis. The relative error reduction is the decrease of the averaged error rate of the base recognizers as compared to the error rate of the best

---
[2] A standard z-test was conducted for $\alpha = 0.05$.

combination method. These results clearly indicate that combinations having a higher correlation or a smaller disagreement among the base recognizers do not reduce the error rate as much as combinations with diverse recognizers.

Combined Recognizers	Correlation	Disagreement
MMHMM & HMM	0.5187	0.1787
NN & HMM	0.4022	0.2031
NN & MMHMM	0.4181	0.1914

5 Conclusion

We investigated various methods to combine classic Hidden Markov Models with novel Maximum Margin Hidden Markov Models and Neural Networks for handwritten text recognition. Methods to combine two of these recognizers as well as all three recognizers were tested and compared. Among the most promising combination methods are weighted versions of *Exponentiated Borda Count*, *Maximum Likelihood Margin* and *Meta Voting*. It turned out that a significant increase of the recognition rate can be achieved, although already highly optimized state-of-the-art recognizers are used as the base recognizers. In fact, using the combination of all three recognizers resulted in an error reduction of 19.33% as compared to the best base classifier (NN).

Diversity analysis shows that combinations including the neural network based recognizer are substantially more diverse than combinations including the HMM and MMHMM based recognizer. We furthermore demonstrated that the lower a combination's correlation and disagreement coefficient is, the higher is its error rate reduction. The neural network's inherent difference to the other two recognizers, therefore, enables the construction of a combined system with high performance.

Future work will include experiments with recognizer combinations using additional base recognizers. These could be modified or enhanced versions of the recognizers considered in this paper as well as recognizers based on completely different techniques, e.g. dynamic time warping. Also combination methods, such as bagging, boosting and entropy based decision rules are of potential interest.

Acknowledgments

This work has been supported by the Swiss National Center of Competence in Research (NCCR) on Interactive Multimodal Information Management (IM2).

References

1. Vinciarelli, A.: A Survey On Off-Line Cursive Word Recognition. Pattern Recognition 35(7), 1433–1446 (2002)
2. Brakensiek, A., Rigoll, G.: Handwritten Address Recognition Using Hidden Markov Models. In: Dengel, A.R., Junker, M., Weisbecker, A. (eds.) Reading and Learning. LNCS, vol. 2956, pp. 103–122. Springer, Heidelberg (2004)

3. Kim, G., Govindaraju, V., Srihari, S.N.: An architecture for handwritten text recognition systems. Int'l Journal on Document Analysis and Recognition 2(1), 37–44 (1999)
4. Do, T.-M.-T., Artieres, T.: Maximum Margin Training of Gaussian HMMs for Handwriting Recognition. In: 10th Int'l Conference on Document Analysis and Recognition (2009)
5. Graves, A., Fernández, S., Gomez, F., Schmidhuber, J.: Connectionist Temporal Classification: Labelling Unsegmented Sequential Data with Recurrent Neural Networks. In: 23rd Int'l Conf. on Machine Learning, pp. 369–376 (2006)
6. Kittler, J., Roli, F. (eds.): MCS 2000. LNCS, vol. 1857. Springer, Heidelberg (2000)
7. Kuncheva, L.I.: Combining Pattern Classifiers: Methods and Algorithms. Wiley-Interscience, Hoboken (2004)
8. Günter, S., Bunke, H.: Ensembles of Classifiers for Handwritten Word Recognition. Int'l Journal on Document Analysis and Recognition 5(4), 224–232 (2003)
9. Verma, B.K., Gader, P.D., Chen, W.-T.: Fusion of multiple handwritten word recognition techniques. Pattern Recognition Letters 22(9), 991–998 (2001)
10. Wang, W., Brakensiek, A., Rigoll, G.: Combination of Multiple Classifiers for Handwritten Word Recognition. In: Int'l Workshop on Frontiers in Handwriting Recognition, pp. 117–122 (2002)
11. Marti, U.-V., Bunke, H.: The IAM-Database: An English Sentence Database for Offline Handwriting Recognition. Int'l Journal on Document Analysis and Recognition 5, 39–46 (2002)
12. Marti, U.-V., Bunke, H.: Using a Statistical Language Model to Improve the Performance of an HMM-Based Cursive Handwriting Recognition System. Int'l Journal of Pattern Recognition and Artificial Intelligence 15, 65–90 (2001)
13. Yu, D., Deng, L.: Large-Margin Discriminative Training of Hidden Markov Models for Speech Recognition. In: First IEEE Int'l Conference on Semantic Computing, pp. 429–438 (2007)
14. Sha, F., Saul, L.K.: Large Margin Hidden Markov Models for Automatic Speech Recognition. In: NIPS. MIT Press, Cambridge (2007)
15. Do, T.-M.-T., Artieres, T.: Large Margin Trainng for Hidden Markov Models with Partially Observed States. In: 26th Int'l Conference on Machine Learning (2009)
16. Graves, A., Liwicki, M., Fernández, S., Bertolami, R., Bunke, H., Schmidhuber, J.: A Novel Connectionist System for Unconstrained Handwriting Recognition. IEEE Transaction on Pattern Analysis and Machine Intelligence (accepted for publication)
17. Kuncheva, L.I., Whitaker, C.J.: Measures of Diversity in Classifier Ensembles and Their Relationship with the Ensemble Accuracy. Machine Learning 51(2), 181–207 (2003)

Embedded Bernoulli Mixture HMMs for Continuous Handwritten Text Recognition*

Adrià Giménez and Alfons Juan

DSIC/ITI, Univ. Politècnica de València,
E-46022 València, Spain
{agimenez,ajuan}@dsic.upv.es

Abstract. Hidden Markov Models (HMMs) are now widely used in off-line handwritten text recognition. As in speech recognition, they are usually built from shared, embedded HMMs at symbol level, in which state-conditional probability density functions are modelled with Gaussian mixtures. In contrast to speech recognition, however, it is unclear which kind of real-valued features should be used and, indeed, very different features sets are in use today. In this paper, we propose to by-pass feature extraction and directly fed columns of raw, binary image pixels into embedded Bernoulli mixture HMMs, that is, embedded HMMs in which the emission probabilities are modelled with Bernoulli mixtures. The idea is to ensure that no discriminative information is filtered out during feature extraction, which in some sense is integrated into the recognition model. Good empirical results are reported on the well-known IAM database.

Keywords: HMMs, Bernoulli Mixtures, Handwritten Text Recognition.

1 Introduction

Handwritten Text Recognition (HTR) is now usually approached by using technology imported from speech recognition; that is, HMM-based text image modelling and n-gram language modelling [1,2,3]. In contrast to speech recognition, however, there is no a de-facto standard regarding the kind of features that should be computed when transforming the input (image) signal into a sequence of feature vectors [1,2].

In [4,5], an isolated handwritten word recogniser is proposed in which binary image pixels are directly fed into word-conditional *embedded Bernoulli mixture HMMs,* that is, embedded HMMs in which the emission probabilities are modelled with Bernoulli mixtures. As in [6], the basic idea is to ensure that no discriminative information is filtered out during feature extraction, which in some sense is integrated into the recognition model. In this paper, this idea is extended to general, continuous handwritten text recognition; that is, whole sentences instead of isolated words. Good empirical results are reported on the well-known IAM database.

The paper is organised as follows. In Sections 2, 3 and 4, we review plain Bernoulli mixtures, Bernoulli mixture HMMs and embedded Bernoulli mixture HMMs. While

---

* Work supported by the EC (FEDER) and the Spanish MEC under the MIPRCV "Consolider Ingenio 2010" research programme (CSD2007-00018), the iTransDoc research project (TIN2006-15694-CO2-01), and the FPU grant AP2005-1840.

their estimation by maximum likelihood is described in Section 5. In Section 6, empirical results are reported. Concluding remarks are discussed in Section 7.

2 Bernoulli Mixture

Let o be a D-dimensional feature vector. A finite mixture is a probability (density) function of the form:

$$p_\Theta(o) = \sum_{k=1}^{K} \pi_k\, p_{\Theta'}(o \mid k) , \qquad (1)$$

where K is the number of mixture components, π_k is the kth component coefficient, and $p_{\Theta'}(o \mid k)$ is the kth component-conditional probability (density) function. The mixture is controlled by a parameter vector Θ comprising the mixture coefficients and a parameter vector for the components, Θ'. It can be seen as a generative model that first selects the kth component with probability π_k and then generates o in accordance with $p_{\Theta'}(o \mid k)$.

A Bernoulli mixture model is a particular case of (1) in which each component k has a D-dimensional Bernoulli probability function governed by its own vector of parameters or *prototype* $p_k = (p_{k1}, \ldots, p_{kD})^t \in [0,1]^D$,

$$p_{\Theta'}(o \mid k) = \prod_{d=1}^{D} p_{kd}^{o_d} (1 - p_{kd})^{1-o_d} , \qquad (2)$$

where p_{kd} is the probability for bit d to be 1. Note that this equation is just the product of independent, unidimensional Bernoulli probability functions. Therefore, for a fixed k, it can not capture any kind of dependencies or correlations between individual bits.

Consider the example given in Fig. 1. Three binary images (**a, b** and **c**) are shown as being generated from a Bernoulli prototype depicted as a grey image (black=1, white=0, grey=0.5). The prototype has been obtained by averaging images **a** and **c,** and it is the best approximate solution to assign a high, equal probability to these images. However, as individual pixel probabilities are not conditioned to other pixel values, there are $2^6 = 64$ different binary images (including **a, b** and **c**) into which the whole probability mass is uniformly distributed. It is then not possible, using a single Bernoulli prototype, to assign a probability of 0.5 to **a** and **c,** and null probability to any other image such as **b.** Nevertheless, this limitation can be easily overcome by using a Bernoulli mixture and allowing a different prototype to each different image shape. That is, in our example, a two-component mixture of equal coefficients, and prototypes **a** and **b,** does the job.

3 Bernoulli Mixture HMM

Let $O = (o_1, \ldots, o_T)$ be a sequence of feature vectors. An HMM is a probability (density) function of the form:

$$p_\Theta(O) = \sum_{I, q_1, \ldots, q_T, F} a_{Iq_1} \left[\prod_{t=1}^{T-1} a_{q_t q_{t+1}} \right] a_{q_T F} \prod_{t=1}^{T} b_{q_t}(o_t) , \qquad (3)$$

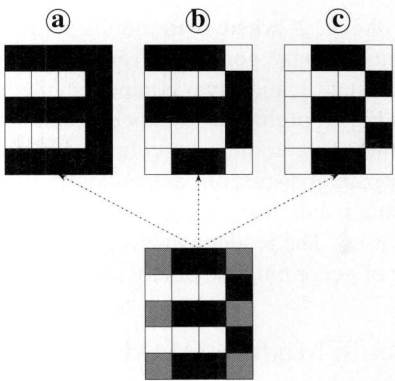

Fig. 1. Three binary images (*a, b* and *c*) are shown as being generated from a Bernoulli prototype depicted as a grey image (black=1, white=0, grey=0.5)

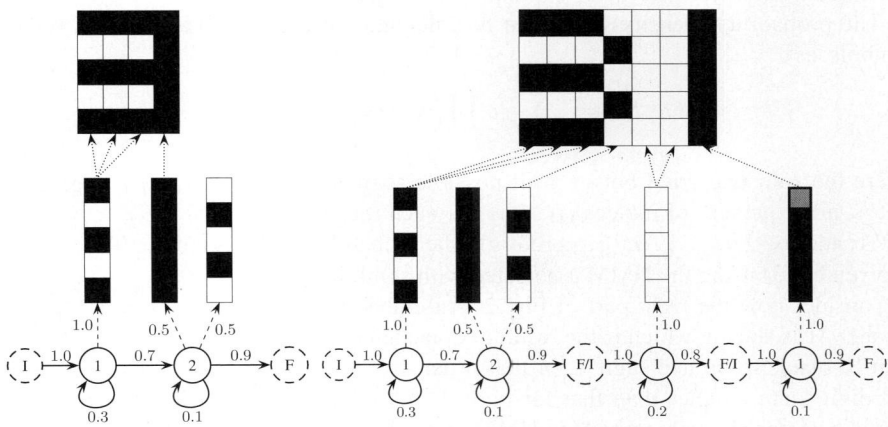

Fig. 2. Example of embedded Bernoulli mixture HMMs for the numbers 3 (left) and 31 (right), and binary images generated from them. Note that a shared Bernoulli mixture HMM for digit 3 is used.

where I and F are the special states for *start* and *stop* respectively, a_{ij} is the state-transition probability between states i and j, b_j is the observation probability (density) function in state j, and q_t denotes the state at time t.

A Bernoulli mixture HMM is an HMM in which the probability of observing o_t, when $q_t = j$, is given by a Bernoulli mixture probability function for the state j:

$$b_j(\boldsymbol{o}_t) = \sum_{k=1}^{K} \pi_{jk} \prod_{d=1}^{D} p_{jkd}^{o_{td}} (1 - p_{jkd})^{1-o_{td}}, \qquad (4)$$

where $\boldsymbol{\pi}_j$ are the priors of the jth state mixture components, and $\boldsymbol{p}_{jk}$ is the kth component prototype in state j.

Consider the left part of Fig. 2, where a Bernoulli mixture HMM for number 3 is shown, together with a binary image generated from it. It is a two-state model with a single prototype attached to state 1, and a two-component mixture assigned to state 2. In contrast to the example in Fig. 1, prototypes do not account for whole digit realisations, but only for single columns. This column-by-column emission of feature vectors attempts to better model horizontal distortions at symbol level and, indeed, it is the usual approach in both speech and handwriting recognition when continuous-density (Gaussian mixture) HMMs are used. The reader can easily check that, by direct application of Eq. (3), the probability of generating the binary image is 0.02835.

4 Embedded Bernoulli Mixture HMM

Let C be the number of different characters (symbols) from which words are formed, and assume that each character c is modelled with a different HMM of parameter vector $\boldsymbol{\lambda}_c$. Let $\boldsymbol{\Lambda} = \{\boldsymbol{\lambda}_1, \ldots, \boldsymbol{\lambda}_C\}$, and let $O = (\boldsymbol{o}_1, \ldots, \boldsymbol{o}_T)$ be a sequence of feature vectors generated from a word formed by a sequence of symbols $S = (s_1, \ldots, s_L)$, with $L \leq T$. The probability (density) of O can be calculated, using embedded HMMs for its symbols, as:

$$p_{\boldsymbol{\Lambda}}(O \mid S) = \sum_{i_1, \ldots, i_{L+1}} \prod_{l=1}^{L} p_{\boldsymbol{\lambda}_{s_l}}(\boldsymbol{o}_{i_l}, \ldots, \boldsymbol{o}_{i_{l+1}-1}), \quad (5)$$

where the sum is carried out over all possible segmentations of O into L segments, that is, all sequences of indices $i_1, \ldots, i_{L+1}$ such that $1 = i_1 < \cdots < i_L < i_{L+1} = T + 1$; and $p_{\boldsymbol{\lambda}_{s_l}}(\boldsymbol{o}_{i_l}, \ldots, \boldsymbol{o}_{i_{l+1}-1})$ refers to the probability (density) of the lth segment, as given by (3) using the HMM associated with symbol s_l.

Consider now the right part of Fig. 2. An embedded Bernoulli mixture HMM for number 31 is shown, which is the result of concatenating Bernoulli mixture HMMs for digit 3, blank space and digit 1, in that order. Note that the HMMs for blank space and digit 1 are simpler than that for digit 3. Also note that the HMM for digit 3 is shared between the two embedded HMMs shown in the Figure. The binary image of the number 31 shown above can only be generated from the segmentation represented as arrows connecting prototypes to image columns. This is due to the fact that all but the rightmost prototype are $0 - 1$ column prototypes that can only emit themselves as binary columns. It is straightforward to check that, according to (5), the probability of generating this image of number 31 is 0.020412.

As with conventional HMMs, the exact probability (density) of an observation can be efficiently computing by dynamic programming. For each time t, symbol s_l and state j from the HMM for symbol s_l, define $\alpha_{lt}(j)$ as:

$$\alpha_{lt}(j) = p_{\boldsymbol{\Lambda}}(O_1^t, q_t = (l, j) \mid S), \quad (6)$$

that is, the probability (density) of generating O up to its tth element and ending at state j from the HMM for symbol s_l. This definition includes (5) as the particular case in which $t = T$, $l = L$ and $j = F_{s_L}$; that is,

$$p_{\boldsymbol{\Lambda}}(O \mid S) = \alpha_{LT(F_{s_L})}. \quad (7)$$

To compute $\alpha_{LT(F_{s_L})}$, we must first take into account that, for each position l in S except for the first, the initial state of the HMM for s_l is joined with final state of its preceding HMM, i.e.

$$\alpha_{lt}(I_{s_l}) = \alpha_{l-1t}(F_{s_{l-1}}) \qquad \begin{aligned} 1 < l \leq L \\ 1 \leq t \leq T \end{aligned}. \tag{8}$$

Having (8) in mind, we can proceed at symbol level as with conventional HMMs. In the case of final states, we have:

$$\alpha_{lt}(F_{s_l}) = \sum_{j=1}^{M_{s_l}} \alpha_{lt}(j) a_{s_l j F_{s_l}} \qquad \begin{aligned} 1 \leq l \leq L \\ 1 \leq t \leq T \end{aligned}, \tag{9}$$

while, for regular states, $1 \leq j \leq M_{s_l}$, we have:

$$\alpha_{lt}(j) = \left[\sum_{i \in \{I_{s_l}, 1, \ldots, M_{s_l}\}} \alpha_{lt-1}(i) a_{s_l i j} \right] b_{s_l j}(\boldsymbol{o}_t), \tag{10}$$

with $1 \leq l \leq L$ and $1 < t \leq T$. The base case is for $t = 1$:

$$\alpha_{l1}(i) = \begin{cases} a_{s_1 I_{s_1} i} b_{s_1 i}(\boldsymbol{o}_1) & l = 1, 1 \leq i \leq M_{s_1} \\ 0 & \text{otherwise} \end{cases}. \tag{11}$$

5 Maximum Likelihood Estimation

Maximum likelihood estimation of the parameters governing an embedded Bernoulli mixture HMM does not differ significantly from the conventional Gaussian case, and it can be carried out using the well-known EM (Baum-Welch) re-estimation formulae [3,7]. Let $(O_1, S_1), \ldots, (O_N, S_N)$, be a collection of N training samples in which the nth observation has length T_n, $O_n = (\boldsymbol{o}_{n1}, \ldots, \boldsymbol{o}_{nT_n})$, and was generated from a sequence of L_n symbols ($L_n \leq T_n$), $S_n = (s_{n1}, \ldots, s_{nL_n})$. At iteration r, the E step requires the computation, for each training sample n, of their corresponding forward and backward probabilities (see (6) and [3,7]), as well as the expected value for its tth feature vector to be generated from kth component of the state j in the HMM for symbol s_l,

$$z_{nltk}^{(r)}(j) = \frac{\pi_{s_{nl}jk}^{(r)} \prod_{d=1}^{D} p_{s_{nl}jkd}^{(r)}{}^{o_{ntd}} \left(1 - p_{s_{nl}jkd}^{(r)}\right)^{1-o_{ntd}}}{b_{s_{nl}j}^{(r)}(\boldsymbol{o}_{nt})},$$

for each t, k, j and l.

In the M step, the Bernoulli prototype corresponding to the kth component of the state j in the HMM for character c has to be updated as:

$$\boldsymbol{p}_{cjk}^{(r+1)} = \frac{1}{\gamma_{ck}(j)} \sum_n \frac{\sum_{l:s_{nl}=c} \sum_{t=1}^{T_n} \xi_{nltk}^{(r)}(j) \boldsymbol{o}_{nt}}{P(O_n \mid S_n, \lambda_1^C)}, \tag{12}$$

where $\gamma_{ck}(j)$ is a normalisation factor,

$$\gamma_{ck}(j) = \sum_n \frac{\sum_{l:s_{nl}=c} \sum_{t=1}^{T_n} \xi_{nltk}^{(r)}(j)}{P(O_n \mid S_n, \lambda_1^C)}, \qquad (13)$$

and $\xi_{nltk}^{(r)}(j)$ the probability for the tth feature vector of the nth sample, to be generated from the kth component of the state j in the HMM for symbol s_l,

$$\xi_{nltk}^{(r)}(j) = \alpha_{nlt}^{(r)}(j) z_{nltk}^{(r)}(j) \beta_{nlt}^{(r)}(j). \qquad (14)$$

Similarly, the kth component coefficient of the state j in the HMM for character c has to be updated as:

$$\pi_{cjk}^{(r+1)} = \frac{1}{\gamma_c(j)} \sum_n \frac{\sum_{l:s_{nl}=c} \sum_{t=1}^{T_n} \xi_{nltk}^{(r)}(j)}{P(O_n \mid S_n, \lambda_1^C)}, \qquad (15)$$

where $\gamma_c(j)$ is a normalisation factor,

$$\gamma_c(j) = \sum_n \frac{\sum_{l:s_{nl}=c} \sum_{t=1}^{T_n} \alpha_{nlt}^{(r)}(j) \beta_{nlt}^{(r)}(j)}{P(O_n \mid S_n, \lambda_1^C)}. \qquad (16)$$

To avoid null probabilities in Bernoulli prototypes, they can be smoothed by linear interpolation with a flat (uniform) prototype, **0.5**, and using $\delta = 10^{-6}$,

$$\tilde{\boldsymbol{p}} = (1-\delta)\,\boldsymbol{p} + \delta\,\boldsymbol{0.5}. \qquad (17)$$

6 Experiments

Experiments have been carried out using the IAM database [8]. This corpus contains forms of unconstrained handwritten English text. All texts were extracted from the LOB corpus. A total of 657 writers contributed. Different datasets were obtained by using segmentation techniques and, in particular, we have used the handwritten text lines dataset. More precisely, we have used the partition described in [1]. This partition is split into three sets. The training set consists of 6161 samples written by 283 writers; to the validation set 56 writers have contributed with 920 samples, while the test set contains 2781 samples by 161 writers.

All input gray level images were preprocessed before transforming them into sequences of feature vectors. Preprocessing consisted of three steps: gray level normalisation, deslanting, and size normalisation of ascenders and descenders [2]. As in [4,5], feature extraction has been carried out by rescaling the image to height 20 while respecting the original aspect ratio, and applying an Otsu binarisation to the resulting image. Therefore, the observation sequence is in fact a binary image of height 20.

The language model is derived from three different English text corpora (LOB corpus, Brown corpus and Wellington corpus) in a similar way as described in [1]. The underlying vocabulary consists about 34000 words, that is, words with at least 4 occurrences. Note, that the vocabulary is not closed over the test set. Thus, the most optimistic result that we can obtain is a Word Error Rate (WER) of 4.51%.

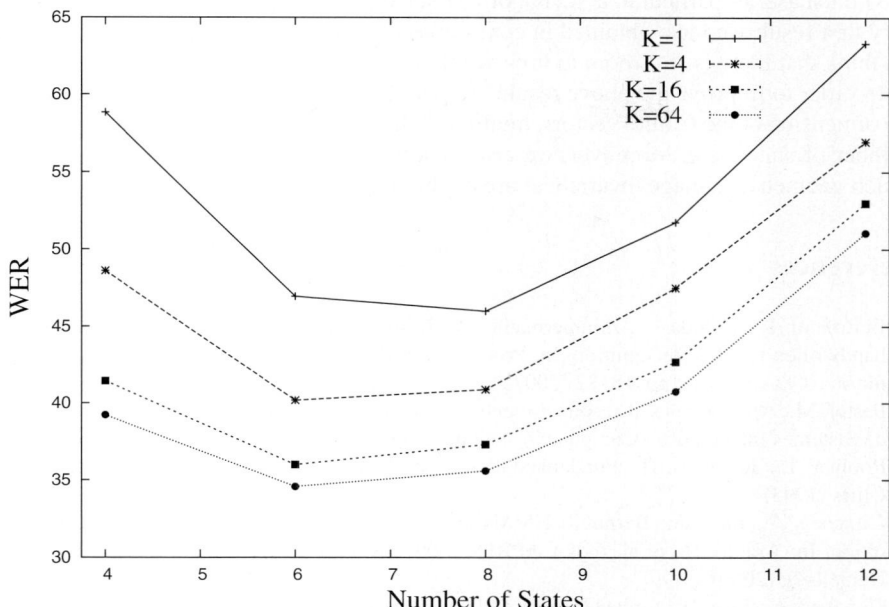

Fig. 3. *WER* as a function of the number of states, for varying number of components (K)

The validation set has been used in order to adjust an appropriate number of states per HMM, $Q \in \{4,6,8,10,12\}$, and the number of mixture components per state, $K \in \{1,4,16,64\}$. For $K = 1$, the recogniser was initialised by first segmenting the training set using a "neutral" model, and then using the resulting segments to perform a Viterbi initialisation. For $K > 1$, it was initialised by splitting the mixture components of the trained model with $K/4$ mixture components per state. The results obtained are shown in Fig. 3. A Grammar Scale Factor of 4 and a Word Insertion Penalty of -2 was used in all cases.

From the results in Fig. 3, it becomes clear that appropriate values for Q and K are 6 and 64 respectively. Using these values, our continuous handwritten text recogniser attains a comparatively good 34.6% of WER on the validation set. Also using these values of Q and K, the text recogniser was applied to the test set and a WER of 42.1% was obtained. This result is not as good as the 35.5% WER reported in [1], which is obtained with a similar system based on Gaussian HMMs, geometrical features and flexible number of states. It is also worth noting that our 42.1% of WER is the very first result we have obtained in continuous handwritten text recognition, and thus we think that there is still room to significantly improve this WER.

7 Concluding Remarks and Future Work

Embedded Bernoulli mixture HMMs have been proposed for Continuous Handwritten Text Recognition and comparatively good results have been obtained on the well-known

IAM database. In particular, a 42.1% of test-set WER has been achieved, which is the very first result we have obtained in continuous handwritten text recognition, and thus we think that there is still room to improve it.

In order to improve the above result, we plan to study a number of ideas including: the dimension of the feature vectors, multiple-column feature vectors, flexible (variable) number of states, etc. Moreover, we also plan to use new Bernoulli-based models in which geometrical image invariances are explicitly modelled.

References

1. Bertolami, R., Uchida, S., Zimmermann, M., Bunke, H.: Non-uniform slant correction for handwritten text line recognition. In: Proc. 9th Int. Conf. on Document Analysis and Recognition (ICDAR 2007), pp. 18–22 (2007)
2. Pastor, M.: Aportaciones al reconocimiento automático de texto manuscrito. PhD thesis, Dep. de Sistemes Informàtics i Computació, València, Spain (October 2007)
3. Rabiner, L., Juang, B.-H.: Fundamentals of speech recognition. Prentice-Hall, Englewood Cliffs (1993)
4. Giménez, A., Juan, A.: Bernoulli HMMs at Subword Level for Handwritten Word Recognition. In: Araujo, H., et al. (eds.) IbPRIA 2009. LNCS, vol. 5524, pp. 497–504. Springer, Heidelberg (2009)
5. Giménez, A., Juan, A.: Embedded Bernoulli Mixture HMMs for Handwritten Word Recognition. In: Proc. of the 10th Int. Conf. on Document Analysis and Recognition (ICDAR 2009), Barcelona, Spain (July 2009) (accepted)
6. Saon, G., Belaïd, A.: High Performance Unconstrained Word Recognition System Combining HMMs and Markov Random Fields. Int. J. of Pattern Recognition and Artificial Intelligence 11(5), 771–788 (1997)
7. Young, S., et al.: The HTK Book. Cambridge University Engineering Department, Cambridge (1995)
8. Marti, U.V., Bunke, H.: The IAM-database: an English sentence database for off-line handwriting recognition. Int. J. on Document Analysis and Recognition 5(1), 39–46 (2002)

Recognition-Based Segmentation of Nom Characters from Body Text Regions of Stele Images Using Area Voronoi Diagram

Thai V. Hoang[1,2,*], Salvatore Tabbone[2], and Ngoc-Yen Pham[1]

[1] MICA Research Center, Hanoi University of Technology, Hanoi, Vietnam
[2] Université Nancy 2, LORIA, UMR 7503, 54506 Vandoeuvre-lès-Nancy, France
{vanthai.hoang,tabbone}@loria.fr,ngoc-yen.pham@mica.edu.vn

Abstract. Segmentation of Nom characters from body text regions of stele images is a challenging problem due to the confusing spatial distribution of the connected components composing these characters. In this paper, for each vertical text line, area Voronoi diagram is employed to represent the neighborhood of the connected components and Voronoi edges are used as nonlinear segmentation hypotheses. Characters are then segmented by selecting appropriate adjacent Voronoi regions. For this purpose, we utilize the information about the horizontal overlap of connected components and the recognition distances of candidate characters provided by an OCR engine. Experimental results show that the proposed method is highly accurate and robust to various types of stele.

Keywords: Recognition-based segmentation, area Voronoi diagram, stele images, Nom characters, horizontal overlap, segmentation graph.

1 Introduction

Stone steles in Vietnam usually contain Nom characters, a derivative of Chinese which was used before the 20th century describing important historical events. Today, the exploitation of these steles is necessary to better understand the history and form a solid base for future development. Automatic processing of stele images composes of three sub-problems: extraction of body text regions, segmentation of Nom characters, and finally representation of these Nom characters in a database allowing search for information. The first one has been tackled using the information about the thickness of connected components [1]. Fig. 1(a)-1(b) show an example of a stele image and its extracted body text region respectively.

As shown in Fig. 1(b), Nom characters are engraved on stone steles in vertical text line from right to left and each character may be composed of several connected components. In each text line, the gaps between characters is indistinguishable from the gaps between connected components belonging to one

---

[*] This work is supported by CNRS and EFEO under the framework of an International Scientific Cooperation Program (PICS: 2007-2009).

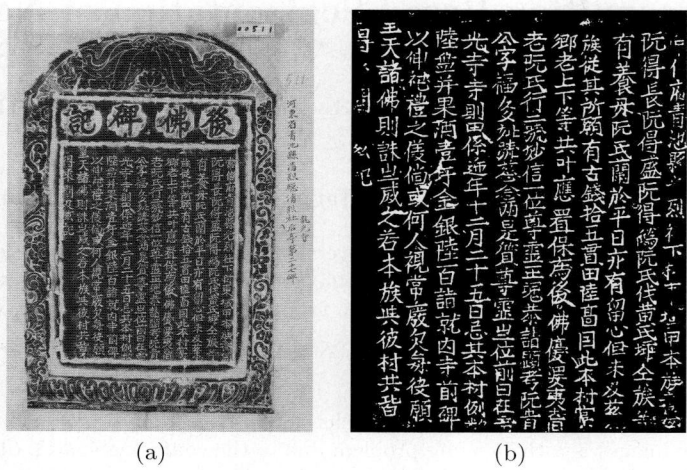

Fig. 1. A stele image (a) and its extracted body text region (b)

character. This makes segmentation of Nom characters a challenging problem to be dealt with in this paper.

There exists many methods in literature for character segmentation. Lu [2], Lu and Shridhar [3] reviewed methods for character segmentation in machine printed documents and handwritten works. Casey and Lecolinet [4] classified existing methods into three strategies: *dissection, recognition-based,* and *holistic.* The first strategy decomposes the image into a sequence of sub-images using general features like character height, width and white space between characters. In the second strategy, the system searches the image for components that match classes in its alphabet. The third strategy seeks to recognize the words as a whole avoiding the need to segment into characters. This strategy is inappropriate for Nom characters as each Nom character has its own meaning.

Tseng and Chen [5] proposed a method for Chinese character segmentation by first generating bounding boxes for character strokes then using knowledge-based merging operations to merge these bounding boxes into candidate boxes and finally applying dynamic programming algorithm to determine optimal segmentation paths. However, the assumption of similarity on character sizes makes this method unsuitable for Nom characters written vertically. Viterbi algorithm and background thinning method were used in [6,7] to locate nonlinear segmentation hypotheses separating handwritten Chinese characters. These methods are also inappropriate for Nom characters as there may exist horizontal gaps inside a character separating its connected components and connected components from neighboring characters on the same line may horizontally overlap.

Area Voronoi diagram has been used by some researchers for document image analysis. For example, Kise et al. [8] and Lu et al. [9] used area Voronoi diagram for page segmentation and word grouping in document images respectively using the distance and area ratio between neighboring connected components.

However, these methods work only for alphanumeric documents in which each character is represented as a connected components.

In this paper, we propose an efficient method combining *dissection* and *recognition-based* strategies. For each vertical text line extracted from the body text region using vertical projection profile, area Voronoi diagram is employed to represent the neighborhood of connected components and Voronoi edges are used as nonlinear segmentation hypotheses. Adjacent Voronoi regions are first grouped using the information about the horizontal overlap of connected components. The remaining Voronoi edges are used as vertices in a segmentation graph in which the arcs' weights are the recognition distances of the corresponding candidate characters using an OCR engine. The vertices in the shortest path detected from the segmentation graph represent the optimal segmentation paths.

The remainder of this paper is organized as follows. Section 2 briefly gives a basic definition of area Voronoi diagram. Section 3 presents a method to group adjacent Voronoi regions using horizontal overlap of connected components. Section 4 describes the details of the algorithm determining optimal segmentation paths using recognition distances of candidate characters. Experimental results are given in Section 5, and finally conclusions are drawn in Section 6.

2 Area Voronoi Diagram

Let $G = \{g_1, \ldots, g_n\}$ be a set of non-overlapping connected components in the two dimensional plane, and let $d(p, g_i)$ be the Euclidean distance between a point p and a connected component g_i defined by $d(p, g_i) = \min_{q \in g_i} d(p, q)$, then Voronoi region $V(g_i)$ and area Voronoi diagram $V(G)$ are defined by:

$$V(g_i) = \{p \mid d(p, g_i) \leq d(p, g_j), \forall j \neq i\}$$
$$V(G) = \{V(g_1), \ldots, V(g_n)\}$$

The Voronoi region of each image component corresponds to a portion of the two dimensional plane. It consists of the points from which the distance to the corresponding component is less than or equal to the distance to any other image components. The boundaries of Voronoi regions, which are always curves, are called *Voronoi edges*.

To construct area Voronoi diagram, we utilize the approach represented in [10] that first labels the image components and then applies morphological operations to expand their boundaries until two expanding labels are met. Fig. 2(a)-2(d) show the area Voronoi diagram for a column text line extracted from the body text region in Fig. 1(b). Original text line I is given in Fig. 2(a). Fig. 2(b) demonstrates the Euclidean distance map in which the gray value of each pixel is proportional to the distance from that pixel to the nearest connected component. The area Voronoi diagram V of I is shown in Fig. 2(c). Voronoi regions with their corresponding connected components are given in Fig. 2(d).

As shown in Fig. 2(d), each connected component is represented by one Voronoi region and Voronoi edges can be used as nonlinear segmentation hypotheses. The process of character segmentation is then considered as the process

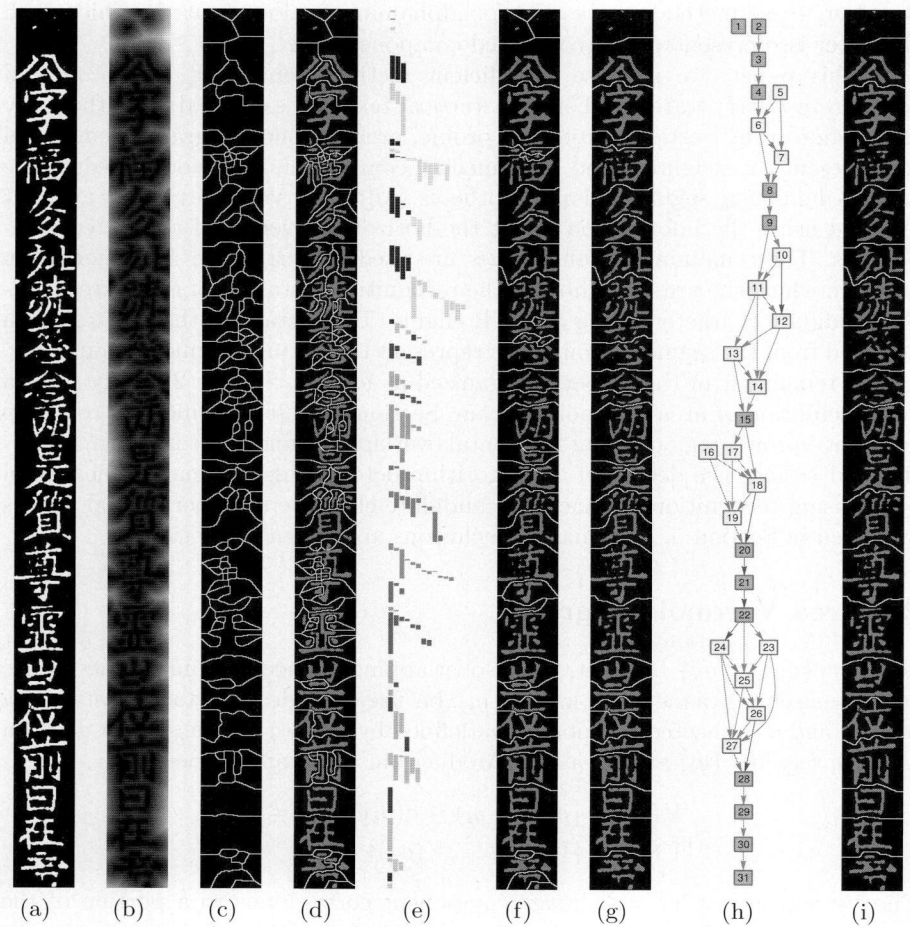

Fig. 2. Steps in segmenting Nom characters from one vertical text line

of grouping adjacent Voronoi regions representing one character. In this paper, we propose to group adjacent Voronoi regions in two steps:

- *Step 1*: Voronoi regions are first grouped using the criteria based on the degree of horizontal overlap of their corresponding connected components. We argue that connected components from one vertical text line overlapped horizontally to a certain degree should belong to one character. This algorithm is described in Section 3.
- *Step 2*: As each Nom character may be composed of several connected components and there may exist horizontal gaps between these connected components, the above algorithm does not guarantee a grouped Voronoi region for each character. We further group Voronoi regions using the recognition distances of candidate characters in Section 4.

3 Grouping of Voronoi Regions Using Horizontal Overlapping Profile of Connected Components

In order to calculate the degree of horizontal overlap of two connected components, we define T_i and B_i as the top and bottom coordinates of the bounding box of the connected component g_i ($B_i > T_i$). The degree of horizontal overlap VO_{ij} of two connected components g_i and g_j is calculated as:

$$VO_{ij} = \frac{max\{min(B_i - T_j, B_j - T_i), 0\}}{min(B_i - T_i, B_j - T_j)} \qquad (1)$$

The numerator of (1) is interpreted as the length of the overlapping segment and VO_{ij} is the proportion of the shorter connected component being overlapped. Thus two connected components g_i and g_j which have $VO_{ij} \geq VO_{thr}$ are considered as belonging to one character and their corresponding Voronoi regions are grouped. Fig. 2(e) provides the horizontal projections of the bounding boxes of connected components in Fig. 2(a) with each vertical line corresponds to one connected component. The grouped Voronoi regions are shown in Fig. 2(f) and adjacent lines correspond to each grouped Voronoi region are labeled using the same color in Fig. 2(e).

By observation of Fig. 2(d) we realize that if each Nom character is represented by one group of Voronoi regions, these grouped region should span from the left border of the text line to its right border. From this viewpoint, those Voronoi regions that do not span from left to right in Fig. 2(f) need to be further grouped to one of its adjacent regions. We propose to use the distance d_{ij} between neighboring connected components g_i and g_j as the criterion of grouping:

$$d_{ij} = \min_{p_i \in g_i, p_j \in g_j} d(p_i, p_j)$$

where $d(p_i, p_j)$ is the Euclidean distance between p_i and p_j. Thus, for each Voronoi region i to be further grouped, we select a region j from a set of its adjacent regions D by $j = \text{argmin}_{k \in D} d_{ik}$ and then group region i with region j. The resulting grouped Voronoi regions are provided in Fig. 2(g).

4 Recognition-Based Determination of Optimal Segmentation Paths

The validity of segmentation hypotheses in Fig. 2(g) is verified by feeding candidate characters into an OCR engine and using their recognition distances to determine the optimal segmentation paths. A segmentation graph is constructed by using the segmentation hypotheses as its vertices and recognition distance of the candidate character corresponding to vertex i and vertex j as the weight of the arc connecting i and j. The segmentation graph in Fig. 2(h) has 31 vertices corresponding to 29 segmentation hypotheses in Fig. 2(g) plus the top and bottom lines. Assuming that a lower value of recognition distance corresponds to higher confidence of the OCR engine in the candidate character, the optimal

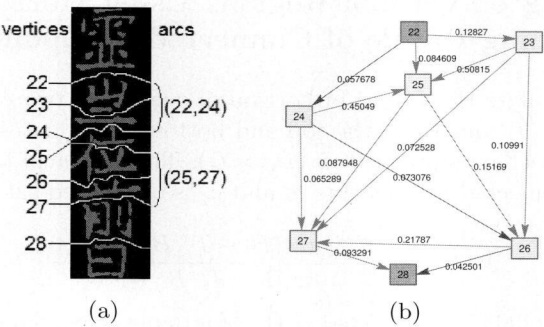

Fig. 3. (a) A line segment of Fig. 2(g), (b) Its corresponding segmentation graph

segmentation paths are thus determined by finding the shortest path in the segmentation graph. For the graph in Fig. 2(h), we need to find the shortest path from vertex 1 to 31.

In updating the weights of the graph, instead of feeding all candidate characters into the OCR engine, we only feed candidate characters that have the character-like feature $H \leq H_{thr}W$ where H and W are the height and width of the candidate character. By doing this, candidate character covering at least two Nom characters are mostly discarded. The arcs shown in Fig. 2(h) correspond to all candidate characters that have the character-like feature.

As there exists no OCR engine for Nom characters, we employ a Chinese OCR engine [11] admitting that not all Nom characters are recognizable by the engine. However, as the proportion of Nom characters that are not Chinese characters are small, the shortest path algorithm can overcome the case in which one non-Chinese character lies between two Chinese characters. An example of a line segment of Nom characters along with its segmentation graph are given in Fig. 3(b) and 3(a) respectively. The optimal segmentation paths in this case contains vertices $\{22, 24, 26, 28\}$.

For the full segmentation graph in Fig. 2(h), directly applying a shortest path algorithm to the may result in error due to the inappropriateness of the OCR engine, we propose here a three-steps algorithm to find the shortest path of the segmentation graph:

- Find cut vertices (sky blue vertices in Fig. 2(h)): a cut vertex is a vertex the removal of which disconnect the remaining graph, the shortest path of the graph must contain cut vertices.
- Find arcs having corresponding candidate characters of high confidence (read arcs in Fig. 2(h)): candidate characters that have recognition distances less than RD_{thr} are said to be of high confidence and the arcs corresponding to these characters should be included in the shortest path.
- Find the shortest paths of the remaining subgraphs.

The final optimal segmentation paths are shown in Fig. 2(i).

5 Experimental Results

In order to determine the appropriate values of VO_{thr} and H_{thr}, a learning set of 16 Nom text lines containing 280 characters have been used for the evaluation of the proposed algorithm's accuracy at different threshold values. According to Fig. 4, the values of VO_{thr} and H_{thr} are selected as 0.4 and 1.25 respectively corresponding to the maxima of the curves. The value of RD_{thr}, which depends on the OCR engine, is selected experimentally as 0.06.

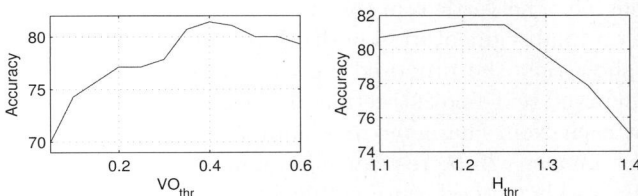

Fig. 4. Algorithm's accuracy at different values of VO_{thr} and H_{thr}

Out of 205 stele images provided by EFEO (The French School of Asian Studies) for evaluation, only 40 images are eligible for the proposed algorithm. The remaining images are regarded as insufficient because of their too noisy body text regions or their poor resolution. We have randomly selected 20 stele images containing 4998 Nom characters for experiment. The ground truth for these characters are defined by hand. Table 1 summarizes the experimental results. The accuracy which is defined as the percentage of characters that are segmented correctly has the value 89.14%. There are two sources of error: one is missing and the other is incorrect grouping. The error of missing concerns with characters that are classified as background noise. In incorrect grouping, each segmented character is not composed of all the connected components from one Nom character, its connected components may come from background noise or neighboring Nom characters.

Table 1. Performance of Nom character segmentation

	EFEO database
Accuracy(%)	89.14
Missing(%)	0.84
Incorrect grouping(%)	10.02

A considerable amount of error comes from the steles which contain characters of various sizes and curved text lines in their body text regions. The layout of these steles thus cannot be aligned into vertical straight lines. This results in error in the extraction of text lines from body text regions using vertical projection profile and consequently the segmented characters will be inaccurate.

6 Conclusions

In this paper, area Voronoi diagram has demonstrated to be effective in representing the neighborhood of connected components in digital images. Voronoi edges then function as nonlinear segmentation hypotheses that need to be validated. Adjacent Voronoi regions have been first grouped using the information about the vertical overlap of connected components. The remaining Voronoi edges are used as vertices in a segmentation graph in which the weight of each arc is the recognition distance of the corresponding candidate character provided by an OCR engine. The vertices in the shortest path of the segmentation graph represent the optimal segmentation paths. Experimental results on a number of stele images show that the proposed method is highly accurate. Further work will employ a curved text line extraction algorithm [12] to increase the accuracy and represent each Nom character in a database for later retrieval. Moreover, poor-resolution images will be re-scanned and noise in body text regions will be removed to make this method more applicable.

References

1. Hoang, T.V., Tabbone, S., Pham, N.Y.: Extraction of Nom text regions from stele images using area Voronoi diagram. In: Proceedings of ICDAR 2009 (to appear, 2009)
2. Lu, Y.: Machine printed character segmentation - An overview. Pattern Recognition 28(1), 67–80 (1995)
3. Lu, Y., Shridhar, M.: Character segmentation in handwritten words - An overview. Pattern Recognition 29(1), 77–96 (1996)
4. Casey, R.G., Lecolinet, E.: A survey of methods and strategies in character segmentation. IEEE Trans. Pattern Anal. Mach. Intell. 18(7), 690–706 (1996)
5. Tseng, L.Y., Chen, R.C.: Segmenting handwritten Chinese characters based on heuristic merging of stroke bounding boxes and dynamic programming. Pattern Recognition Letters 19(10), 963–973 (1998)
6. Tseng, Y.H., Lee, H.J.: Recognition-based handwritten Chinese character segmentation using a probabilistic Viterbi algorithm. Pattern Recognition Letters 20(8), 791–806 (1999)
7. Zhao, S., Chi, Z., Shi, P., Yan, H.: Two-stage segmentation of unconstrained handwritten Chinese characters. Pattern Recognition 36(1), 145–156 (2003)
8. Kise, K., Sato, A., Iwata, M.: Segmentation of page images using the area Voronoi diagram. Comput. Vis. Image Underst. 70(3), 370–382 (1998)
9. Lu, Y., Wang, Z., Tan, C.-L.: Word grouping in document images based on voronoi tessellation. In: Marinai, S., Dengel, A.R. (eds.) DAS 2004. LNCS, vol. 3163, pp. 147–157. Springer, Heidelberg (2004)
10. Lu, Y., Xiao, C., Tan, C.L.: Constructing area Voronoi diagram based on direct calculation of the Freeman code of expanded contours. IJPRAI 21(5), 947–960 (2007)
11. Léonard, J.: COCR2: A small experimental Chinese OCR, http://users.belgacom.net/chardic/cocr2.html
12. Likforman-Sulem, L., Zahour, A., Taconet, B.: Text line segmentation of historical documents: a survey. IJDAR 9(2-4), 123–138 (2007)

A Novel Approach for Word Spotting Using Merge-Split Edit Distance

Khurram Khurshid[1], Claudie Faure[2], and Nicole Vincent[1]

[1] Laboratoire CRIP5 – SIP, Université Paris Descartes, 45 rue des Saints-Pères, 75006, Paris, France
{khurram.khurshid,nicole.vincent}@mi.parisdescartes.fr
[2] UMR CNRS 5141 - GET ENST, 46 rue Barrault
75634 Paris Cedex 13, France
cfaure@enst.fr

Abstract. Edit distance matching has been used in literature for word spotting with characters taken as primitives. The recognition rate however, is limited by the segmentation inconsistencies of characters (broken or merged) caused by noisy images or distorted characters. In this paper, we have proposed a Merge-split edit distance which overcomes these segmentation problems by incorporating a multi-purpose merge cost function. The system is based on the extraction of words and characters in the text and then attributing each character with a set of features. Characters are matched by comparing their extracted feature sets using Dynamic Time Warping (DTW) while the words are matched by comparing the strings of characters using the proposed Merge-Split Edit distance algorithm. Evaluation of the method on 19th century historical document images exhibits extremely promising results.

Keywords: Word Spotting, Edit Distance, Dynamic Time warping.

1 Introduction

Word spotting on Latin alphabets has received considerable attention over the last few years. A wide variety of techniques have been proposed in literature but the field still remains inviting and challenging specially if the document base comprises low quality images of historical documents. There are plenty of issues and problems related to ancient printed documents which are discussed in detail in [4] and [5]. These include physical issues such as quality of the documents, the marks and strains of liquids, inks and dust etc; and semantic issues such as foreground entity labeling. In this paper though, we will concentrate only on the segmentation related problems. The main focus of this paper will be the improvements in word spotting to make it clear of a good character segmentation requirement, so that if we do not get a good segmentation, we can still obtain better word retrieval rates. A brief account of some state of the art methods in word spotting that inspired our own work is given here:

Adamek et al. [2] introduced an approach based on the matching of word contours for holistic word recognition in historical handwritten manuscripts. The closed word contours are extracted and their multiscale contour-based descriptors are matched using a Dynamic Time Warping (DTW) based elastic contour matching technique. In [12], Rath et al. argued that word spotting using profile feature matching gives better results. Words are extracted from the text and are represented by a set of four profile features. Feature matching is performed using the scale invariant DTW algorithm. In [9], we showed that, for historical printed documents, instead of using word features, character features give a better and distinct representation of a word. Two words are compared by matching their character features using DTW [8]. The method can be improved by adding Edit distance matching for the words. It means that the features of two characters are matched using DTW while track of these character matching costs between the two strings is kept by Edit distance. Addition of the Edit distance stage significantly reduced the number of false positives while improving the recognition rate as later shown in results. However, if the character segmentation is not good there will be a significant drop in retrieval rate as the classic edit distance does not cater for merged and broken characters.

Different variations of Edit distance have been proposed in literature for different matching applications [3,7,11,13]. Kaygin et al. introduced a variation of Edit distance for shape recognition in which polygon vertices are taken as primitives and are matched using the modified Edit distance. The operations of inserting and deleting a vertex represent the cost of splitting and combining the line segments respectively [7]. A minimal edit distance method for word recognition has been proposed by [13]. Complex substitution costs have been defined for the Edit distance function and string-to-string matching has been done by explicitly segmenting the characters of the words. Another recent variant of Edit distance has been proposed in [11] where apart from the classic substitution costs, two new operations namely combination and split are supported.

This paper presents an effective way for word spotting in historical printed documents using a combination of our merge-split Edit distance variant and an elastic DTW algorithm. Main aim of this work is to be able to match words without the need of having perfect character segmentation *a priori*. Paper is divided into multiple sections beginning with the overview of the proposed system in the next section. This is followed by description of method and its different stages and the results achieved.

2 Proposed Method

The model is based on the extraction of a multi-dimensional feature set for the character images. As opposed to [12] where features are extracted at word level, we define features at character level, thus giving more precision in word spotting [8].

Figure 1 illustrates the different stages of our system. Document image is binarized using NICK algorithm [10]. Text in the document image is separated from graphics and words in the text area are extracted by applying a horizontal RLSA [14] and finding the connected components in that RLSA image. Characters in a word image are found by performing its connected component analysis followed by a 3-step post-processing stage. For each character, we define 6 feature sequences which represents a character's overall form. Query word is searched within candidate words by a combination of our Merge-split Edit distance at word level and DTW at character level which allows us to spot even the words with improper character segmentation.

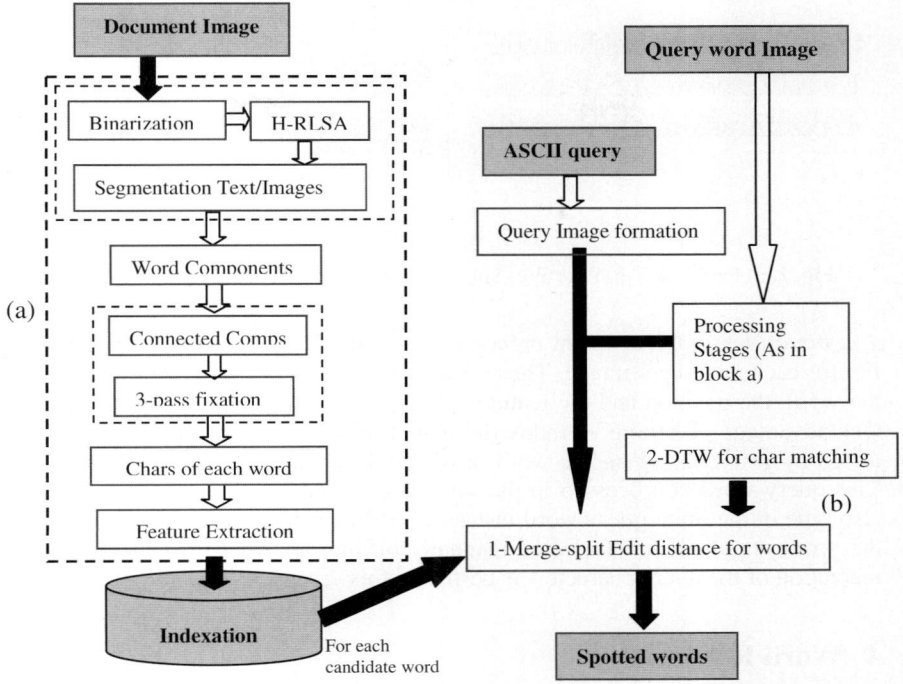

Fig. 1. a) Document image Indexing involving character segmentation and feature extraction. b) Word retrieval using a combination of DTW and Merge-split edit distance.

3 Document Indexing

In this section, we only give an overview of the indexing process involving character segmentation and feature extraction.

The connected components (CCs) of the binary image do not always correspond to characters. A character may be broken into multiple components or multiple characters may form a single component. For that, we have a 3-pass processing stage for the CCs in a word component to get the characters. In the first pass, components on the top of each other are assigned to the same character (fig 2a). In the 2nd pass, overlapping components are merged into a single character (fig 2a). In the 3rd pass, the punctuation marks (like ',' '.') are detected using size and location criteria and are removed from the word (fig 2b). After the three passes, there still remain some improperly extracted characters which are of main interest in this paper (fig 2c,3).

Each character is represented by a sequence of six feature vectors. The length of each of the six vectors associated with a character is equal to the width of the character bounding box. The six features we use are *Vertical projection profile* (on gray level image), *Upper character profile position*, *Lower character profile position*, *Vertical histogram*, *Number of ink/non-ink transitions* and *Middle row transition state*

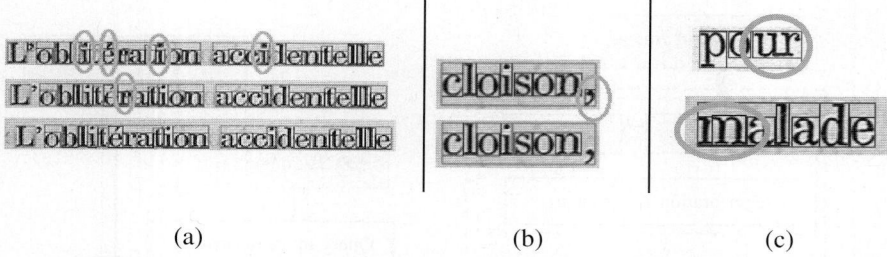

Fig. 2. a) Pass 1 & 2 b) Pass 3 c) Some examples of the remaining unfixed characters

(on binary image). Document processing/indexing is done offline, creating an index file for each document image. The coordinates of each word, number of characters in the word, the position and the features of each character are stored in the index files. One advantage of having an index file is that the query word can now be crisply selected by just clicking on the word in the GUI of our document processing system. This query word is processed in the same way to extract character features. We can also type in the input query word instead of a click-selection. In that case, features of the prototype characters (selected manually offline) are used for matching. The segmentation of the query characters is perfect in this case.

4 Word Retrieval

For word spotting, we have proposed a two step retrieval system. In the first step, a *length-ratio filter* finds all eligible word candidates for the query word. For two words to be considered eligible for matching, we have set bounds on the ratio of their lengths. If this ratio does not lie within a specific interval, we do not consider this word as a candidate. Through this step, we are on average able to eliminate more than 65% of the words. In the second step, the query word and candidate word are matched using a multi-stage method in which the characters of the two words are matched using the Merge-Split Edit distance algorithm while features of the two characters are matched using elastic DTW method coupled with Euclidean distance. The need for an algorithm catering for the merge and split of characters arises because we may not have 100% accurate segmentation of characters all the time (fig 3).

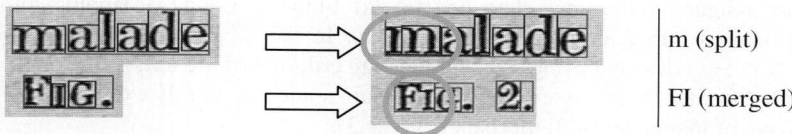

Fig. 3. Query words and some instances of them which were not spotted without Merge-split

4.1 Word Matching – Merge Split Edit Distance

To address character segmentation issues, we have introduced two new merge-based character matching operations **Merge1** and **Merge2** that enable to model a merge and split capability respectively in Edit distance, thus overcoming the limitation of having a perfect character segmentation by catering for the broken and merged characters during word matching.

Consider 2 words A and B; A, the query word, having s characters while B, the test word, having t characters. We treat both words as two series of characters, $A = (a_1 \ldots a_s)$ and $B = (b_1 \ldots b_t)$. To determine the distance/cost between these two character series, we find the Edit matrix W which shows the cost of aligning the two subsequences. Apart from the three classic Edit operations, we have introduced two new operations $a_i \rightarrow (b_j+b_{j+1})$ and $(a_i+a_{i+1}) \rightarrow b_j$ which represent merge1 and merge2 respectively. Merge1 function allows one character of the query word to be matched against two characters of the current test word, while merge2 function allows one character of the test word to be matched against a combination of two query word characters thus modeling a split b_j. Combination of two characters is done by concatenating the 6 feature sequences of both of them. The entries of matrix W initialized by +infinity are calculated as:

$$W(0,0) = 0$$
$$W(0,j) = W(i, j-1) + \gamma(\Lambda \rightarrow b_j)$$
$$W(i,0) = W(i-1, j) + \gamma(a_i \rightarrow \Lambda)$$

$$W(i,j) = \min \begin{cases} W(i-1, j-1) + \gamma(a_i \rightarrow b_j) \\ W(i-1, j-1) + \gamma(a_i \rightarrow (b_j + b_{j+1})) & (for\ j < t) \\ W(i-1, j-1) + \gamma((a_i + a_{i+1}) \rightarrow b_j) & (for\ i < s) \\ W(i-1, j) + \gamma(a_i \rightarrow \Lambda) \\ W(i, j-1) + \gamma(\Lambda \rightarrow b_j) \end{cases} \quad (1)$$

Here, Λ is an empty character with all feature vector values set to 0. $\gamma(a_i \rightarrow b_j)$ is the cost of changing a_i to b_j. $\gamma(a_i \rightarrow \Lambda)$ is the cost of deleting a_i and $\gamma(\Lambda \rightarrow b_j)$ is the cost of inserting b_j.

$\gamma(a_i \rightarrow (b_j+b_{j+1}))$ shows the cost of changing character a_i of query word to two character components b_j+b_{j+1} of the test word. It means that if the character b_j was broken into two components b_j and b_{j+1}, we would be able to match this with a_i using Merge1 function. The feature sequences of b_j and b_{j+1} are concatenated and are matched with the feature vectors of a_i to get W(i,j). Once W(i,j) is calculated, we copy the same value of W(i,j) to the cell W(i,j+1) signifying that we had used the merge function.

Similarly, $\gamma((a_i+a_{i+1}) \rightarrow b_j)$ shows the cost of changing two characters of query word to one character of test word. It means that if b_j was infact a component having two characters merged into one, we would be able to detect and match that with a_i+a_{i+1} using our Merge2 function. Here, instead of splitting the feature vectors of b_j (which is more difficult as we do not know exactly where to split), we merge the query word characters, thus emulating the split function. W(i,j) is calculated the same way and it is copied to the cell W(i+1,j) signifying the use of the split function.

4.2 Character Matching Cost - DTW

All the character matching costs are calculated by matching the feature sequences of two character components using DTW. The advantage of using DTW here is that it is able to account for the nonlinear stretch and compression of characters. Hence two same characters differing in dimension will be matched correctly.

Two characters X and Y of widths m and n respectively are represented by vector sequences $X = (x_1 \ldots x_m)$ and $Y = (y_1 \ldots y_n)$ where x_i and y_j are vectors of length 6 (= No. of features). To determine the DTW distance between these two sequences, we find a matrix D of m x n order which shows the cost of aligning the two subsequences. The entries of the matrix D are found as:

$$D(i, j) = \min \begin{Bmatrix} D(i, j-1) \\ D(i-1, j) \\ D(i-1, j-1) \end{Bmatrix} + d(x_i, y_j) \qquad (2)$$

Here for d (x_i, y_j), we have used the Euclidean distance in the feature space:

$$d(x_i, y_j) = \sqrt{\sum_{k=1}^{p}(x_{i,k} - y_{j,k})^2} \qquad (3)$$

where p represents the number of features which in our case is six.

The entry D(m, n) of the matrix D contains the cost of matching the two characters. To normalize this value, we divide the final cost D(m, n) by the average width of the two characters.

$$\text{Final char-cost} = D(m,n) / [(m+n)/2] \qquad (4)$$

4.3 Final Word Distance

Once all the values of W are calculated, the warping path is determined by backtracking along the minimum cost path starting from (s, t) while taking into the account the number of merge/split functions used in the way. The normalization factor K is found by subtracting the number of merge/split functions from the total number of steps in the warping path. The final matching cost of the two words is:

$$\text{Final word-cost} = W(s,t) / K \qquad (5)$$

Two words are ranked similar if this final matching cost is less than an empirically determined threshold.

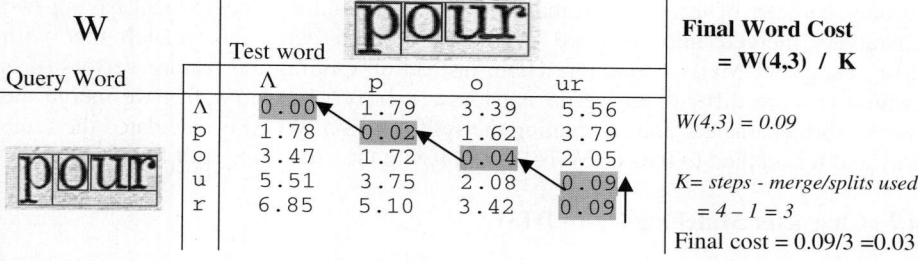

Fig. 4. Calculating Edit Matrix W for two similar words of lengths 4 and 3, and their final cost

Figure 4 shows an example of matching two words of lengths 4 and 3. The query word is well segmented while the last two characters in the test word are merged into one. While finding W, we see that one merge operation is used for matching 'ur' with 'u', thus decrementing the value of K by one.

5 Experimental Results

The comparison of character-feature based matching technique with other methods (method in [12] and correlation based matching method) has already been done in [9]. Here we compare our merge-split Edit distance based matching method with a similar method but using classic Edit distance in place (we will refer to it as [Ed]) and also with [9] to show the improvement in results. We also give the results of a professional OCR software Abbyy FineReader [1] on the same data set. Experiments were carried out on the document images provided by BIUM [6]. For experiments, we chose 48 pages from 12 different books (4 images from each book), having a total of more than 17,000 words in all. For testing, 60 different query lexiques having 435 instances in total were selected based on their varied lengths, styles and also context of the book. Table 1 summarizes the results of word spotting.

Table 1. Result Summary of Word matching algorithms

	Method in [9]	Classic Edit [Ed]	Our method	ABBYY [1]
#query word instances	435	435	435	435
#words detected perfectly	401	406	427	422
#words missed	34	29	8	13
#False positives	51	16	4	0
Precision %	88.71%	96.20%	99.01%	100%
Recall %	92.18%	93.34%	98.16%	97.01%
F-Measure	90.41%	94.75%	98.58%	98.48%

We can see from the above table that the presented merge-split Edit distance based method achieves a much higher recognition rate while maintaining a better precision than [9] and [Ed]. This is due to the fact that we are able to spot even those words where the segmentation of characters is not good. The layout of the pages used was not difficult to manage for a professional OCR. So on the same data-set, [1] produced a precision of 100% but we still obtained a better recall rate than it.

We also analyzed the effect of word length on the Edit distance threshold, learning that for smaller words (with lesser number of characters), a lower threshold gives better precision and recall rates, while for longer words, a higher threshold value proves to be more effective. This is because for longer words, it is more unlikely to find similar words; so we can allow a bit more relaxation in threshold value. Another thing we learned is that the time taken to search a query increases linearly with the number of documents searched.

Considering optimum threshold values (found empirically), our system achieves a precision of 99% while obtaining an overall recognition rate of 98.16%.

6 Conclusion

We have proposed a new approach for word spotting based on the matching of character features by employing a combination of DTW and Merge-Split Edit distance. The main objective of this approach is to cater for the improper segmented characters during the matching process. Results obtained using this method are very encouraging. The number of query words missed is very less as compared to [9] and [Ed] which shows the prospects of taking this method even further by improving different stages and adding more features to achieve even higher percentages.

References

1. ABBYY FineReader professional v6.0
2. Adamek, T., O'Connor, N.E., Smeaton, A.F.: Word matching using single closed contours for indexing handwritten historical documents, IJDAR (2007)
3. Ambauen, R., Fischer, S., Bunke, H.: Graph Edit Distance with Node Splitting and Merging and its Application to Diatom Identification. In: Hancock, E.R., Vento, M. (eds.) GbRPR 2003. LNCS, vol. 2726, pp. 259–264. Springer, Heidelberg (2003)
4. Antonacopoulos, A., Karatzas, D., Krawczyk, H., Wiszniewski, B.: The Lifecycle of a Digital Historical Document: Structure and Content. In: ACM Symposium on DE (2004)
5. Baird, H.S.: Difficult and urgent open problems in document image analysis for libraries. In: 1st International workshop on Document Image Analysis for Libraries (2004)
6. Digital Library of Bibliotheque Interuniversitaire de Medecine, Paris, http://www.bium.univparis5.fr/histmed/medica.htm
7. Kaygin, S., Bulut, M.M.: Shape recognition using attributed string matching with polygon vertices as the primitives. Pattern Recognition Letters (2002)
8. Keogh, E., Pazzani, M.: Derivative Dynamic Time Warping. In: First SIAM International Conference on Data Mining, Chicago, IL (2001)
9. Khurshid, K., Faure, C., Vincent, N.: Feature based word spotting in ancient printed documents. In: Proceedings of PRIS (2008)
10. Khurshid, K., Siddiqi, I., Faure, C., Vincent, N.: Comparison of Niblack inspired binarization techniques for ancient document images. In: 16th International conference DDR (2009)
11. Manolis, C., Brey, G.: Edit Distance with Single-Symbol Combinations and Splits. In: Proceedings of the Prague Stringology Conference (2008)
12. Rath, T.M., Manmatha, R.: Word Spotting for historical documents. IJDAR (2007)
13. Waard, W.P.: An optimised minimal edit distance for hand-written word recognition. Pattern Recognition Letters (1995)
14. Wong, K.Y., Casey, R.G., Wahl, F.M.: Document analysis system. IBM J. Res. Development (1982)

Hierarchical Decomposition of Handwritten Manuscripts Layouts

Vincent Malleron[1,2], Véronique Eglin[1], Hubert Emptoz[1],
Stéphanie Dord-Crouslé[2], and Philippe Régnier[2]

[1] Université de Lyon, CNRS,
INSA-Lyon, LIRIS, UMR5205,
F-69621, France
vincent.malleron@liris.cnrs.fr
[2] Université de Lyon, CNRS,
LIRE, UMR 5611
F-69007, France
Stephanie.Dordcrousle@ens-lsh.fr

Abstract. In this paper we propose a new approach to improve electronic editions of literary corpus, providing an efficient estimation of manuscripts pages structure. In any handwriting documents analysis process, structure recognition is an important issue. The presence of variable inter-line spaces, of inconstant base-line skews, overlappings and occlusions in unconstrained ancient 19th handwritten documents complicates the structure recognition task. Text line and fragment extraction is based on the connexity labelling of the adjacency graph at different resolution levels, for borders, lines and fragments extraction.

Keywords: text lines and fragments extraction, graph, handwriting.

1 Introduction

Our work takes place in an human science project which aims at the realization of an electronic edition of the "dossiers de Bouvard et Pécuchet" corpus. This corpus is composed by french 19th century manuscripts gathered by Gustave Flaubert who intends to prepare the redaction of a second volume to his novel "Bouvard et Pécuchet". Corpus contents are diversified in term of sense as well as in term of shape (different writers, styles and layouts). Besides, the corpus is mainly composed by text fragments (Newspapers extracts, various notes, etc.) put together by Flaubert. To produce the electronic edition we must consider the particular framework of the corpus : structure informations must be known in order to reproduce as well as possible its primary state. The main goal of this work is to retrieve as many structural information as possible in order to provide a good estimation of handwritten document pages structure. The document structure is mainly composed by pages, fragments, lines, words and characters. Our proposition consists in representing the overall page layout with and oriented adjacency graph that contains all information relative to the page content. This paper is organized as follows : section 2 details previous works on text line extraction and structure recognition, section 3 presents our approach for text lines and fragments extraction, section 4 provides results and perspectives and section 5 gives concluding remarks.

2 Related Works

Handwritten text line segmentation and structure recognition are still challenging tasks in ancient handwritten document image analysis. Most of works based on text line segmentation can be roughly categorized as bottom-up or top-down approaches. In the top-down methodology, a document page is first segmented into zones, and a zone is then segmented into lines, and so on. Projection based methods is one of the most successful top-down algorithm for printed documents and it can be applied on handwritings only if gaps between two neighboring handwritten lines are sufficient [1]. Connected component based methods is a popular bottom-up method : connected components are grouped into lines, and lines into blocks. In [2] and [3] the algorithms for text lines extraction are based on connected components grouping coupled with Hough Transform are exposed. Likforman-Sulem et al. [5] give an overview of all text lines segmentation methods. In [6], Yi Li et al. propose a curvilinear text lines extraction algorithm based on level-set method. This algorithm uses no prior knowledge, and achieves high accuracy text line detection.

Most of works on document structure recognition are performed on machine-printed texts. In [7] T.M.Breuel presents algorithms for machine-printed documents layout analysis. Theses algorithms are noise resistant and adapted to different layouts and languages. Kise et.al [8] propose an algorithm for machine printed pages layout analysis based on Voronoi diagrams and successfully achieve segmentation of document components in non-manhattan documents. Lemaitre et. al in [4] propose an application of Voronoi tessellation for handwritten documents recognition. It is not trivial to extend machine printed documents algorithms to handwritten documents, especially when handwritten text lines are curvilinear and when neighboring handwritten text lines may be close or touch each other. In [9], L.O'Gorman's Docstrum allows to retrieve layout structure of handwritten pages with regular layouts. S.Nicolas et al. in [10] and [11] propose an Hidden Markov Model based algorithm for manuscript page segmentation : 6 classes are extracted on a page representing background, textual components, erasures, diacritics, interline and interword spaces.

3 Our Graph Based Approach

3.1 Connected Components Distance Measure

We introduce at first our connected components distance measure : let's consider two handwritten shapes A and B that can represent words, word fragments or characters. The distance between A and B is given by the smallest edge to edge

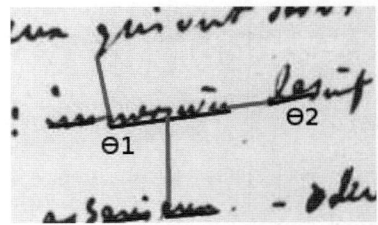

Fig. 1. Connected components distance

distance (d_{edges}). This measure is more representative of page structure than a simple Euclidean distance from center to center and provide an estimation of interwords and interline spaces (figure 1).

To be consistent with orientation variation in handwritten documents we weight our measure with and orientation coefficient. This coefficient, based on the fact that orientation remain mainly constant in a line or a fragment is computed using orientation of the two connected components : $\theta 1$ and $\theta 2$.

$$\Delta\Theta = \alpha * (1 + |\frac{|\theta 1| - |\theta 2|}{|\theta 1| + |\theta 2|}|) \qquad (1)$$

Orientation is estimated using Hough transform at low resolution. We compute an orientation map of extracted hough lines and each connected component gets its orientation from the associate line in orientation map. Our distance can be summarize with the following statement :

$$D(A, B) = \Delta\Theta * min(d_{edges}(A, B)) \qquad (2)$$

Figure 1 shows baselines in blue and minimal edge to edge distances between connected components in red. As $D(A, B)$ represents the distance between two connected components and not between their contour edges, the three distance properties can be simply demonstrated. For the graph construction and labelling we use this distance to find nearest neighbours of each connected component in different orientations ranges to build the adjacency graph and perform lines and fragments extraction.

3.2 Graph Construction

For each connected component we search for the nearest neighbour in four direction of space : top, down, left and right. Exact directions are provided by the orientation estimation described in 3.2 : nearest neighbour research is performed around hough direction, orthogonal hough direction, inverse hough direction and inverse orthogonal hough direction. Once the four neighbours are computed for each connected component we build a weighted directed graph $G = (V, A)$. $V = \{v_1, v_2, ..., v_n\}$ with v_i representing a connected component of our page. Outdegree of G is 4 : each vertex is the tail of up to four arcs ($e_i = (v_i, v_j)$),

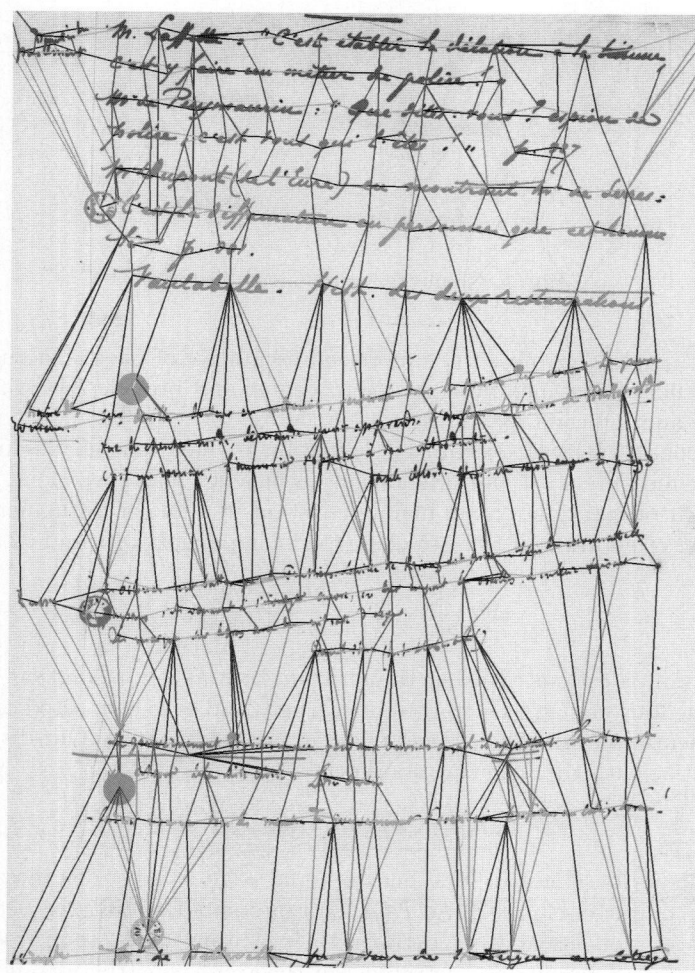

Fig. 2. Reprojected graph, 1f188r

representing the link between the connected component and its neighbours : $A = \{e_1, e_2, ..., e_n\}$ is the arc set of our graph. Arc weights are provided by the real distance between connected components. This graph can be reprojected on a manuscript image as shown on figure2. Right arcs (direct hough direction) are colored in cyan, left arcs (inverse hough direction) are colored in black, top arcs (orthogonal hough direction) are colored in red or blue depending of the weight and down arcs (inverse orthogonal hough direction) are colored in blue. When two arcs are superposed only tops and rights arcs are represented. We can also observe that several arcs can converge to a single connected component. This single connected component is generally a long connexity (a long connected word, a straight line or underline).

3.3 Graph Labelling for a Multi Level Layout Structure Analysis

Arcs weight analysis. In order to set thresholds for borders, text lines and fragments extraction we compute an histogram of arcs weights only in top and down directions. Figure 3 represents the histogram of cumulated distances. Interline spaces clearly appears between 30 and 120 whereas interfragment spaces are above 120. Threshold values for text lines, borders and fragments extraction are computed from this histogram by considering the highest local gradient of the histogram.

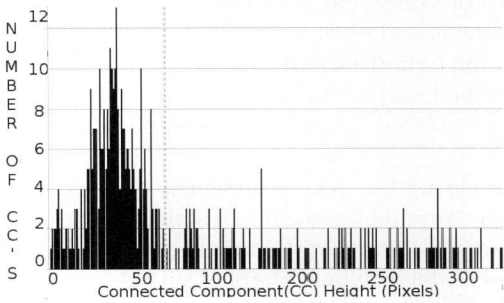

Fig. 3. Arcs weights histogram

Text borders extraction. The purpose of this step is to label each vertex of G with his corresponding label in the five classes described in figure 4. Adjacency function is a simple function, computed on an arc which returns a value corresponding to the direction of the arc (DirectHough = 1, InverseHough = 2, OrthHough = 3, InvOrthHough = 4, NoArc = 0). To extract text borders, adjacency function also used the maximum range value computed above. In practice, the graph labelling is based on a simple evaluation of the neighbourood of a vertex. If the outdegree of the vertex is equal to 4, it represents an inner-text connected component. If the outdegree is less than 4, the label is computed given the result

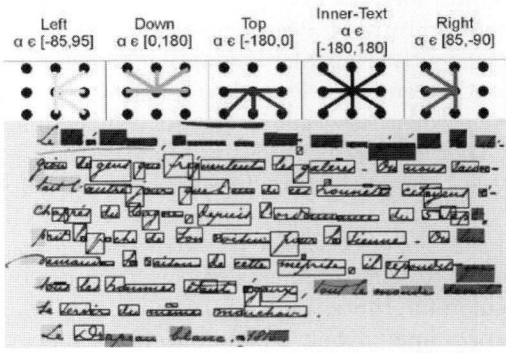

Fig. 4. Borders extraction results

of adjacency function on each arc. We use the following color scale on figure 4 to show the results of border extraction : yellow for left components, red for rights, blue for tops, green for downs and black for inner-text ones.

Text lines extraction. The text line is an important structural element : the knowledge of text-lines positions and shapes is an essential step that allows to perform the alignment between handwritten images fragments and ASCII transcriptions. It also gives us a partial knowledge of page structure and therefore a good a priori for fragments extraction. The result of border extraction is used as an initialization. To be consistent with latine script direction a line starts with a left border component and ends with a right border component. Line extraction is performed using a shortest path research between left and right borders on the previously described graph. A postprocessing step is performed to include a missed component positioned between two components of the same line.

Fragments extraction. The text fragment is the highest structural element extracted in the page. Textual fragment extraction has been developed to fit with the particular framework of our corpus, mainly composed by text fragments. Text fragments extraction is done by line grouping with simple rules based on interline space and orientation variation coupled with fragment contours computation by a best path research between top, down, left and right vertices. The path starts and ends on the same vertice and describe the fragment contour. Cost function for best path computation is based on distance values and transition costs between tops, downs, lefts and rights vertices. Threshold values are given by the distance histogram described above. Figure 2 shows interline space variation : large interline spaces arcs are colored in blue and green whereas small interline spaces arcs are colored in yellow or red. Figure 5 presents the results of fragments extraction.

4 Results and Perspectives

The proposed algorithm for text lines extraction has been tested on a sample set of the "dossiers de Bouvard et Pécuchet" Corpus. A few number of ground-truthed images are currently available. Evaluation has been performed on pages of different shapes and layout in order to be representative of the corpus. A line is considered as wrong when less than 75 per cent of the line is included in the line bounding box. Overlapping between lines makes the two overlapped lines count as wrong. Table 1 shows the text lines extraction results for sample pages of the corpus.

Figure 5 shows results of text fragments extraction on a sample page of our corpus. Five of the six textual fragments of this page are correctly extracted. Our algorithm based on interlinespace and baseline orientation comparison performs well on page of simple layout. Some limitations appears on page of more complex layouts : errors can occurs when two fragments are adjacent with a small orientation variation or when topological configuration of space cannot be

Table 1. Line extraction results

Page	Wrong Lines	Correct Lines
1 f 179 r	1	15
1 f 007 v	4	27
228 f 020 r	2	21
4 f 234 r	0	24
1 f 188 r(fig 5)	1	24
Simple Layout Pages	14	198
Complex Layout Pages	39	216

describe with our distance. Those limitations can be seen on figure 5. Our graph based approach remains insensible to classic connected components approach limitations such as connected components overlapping or inclusions.

Graph based representation is an intuitive multiscale representation which allows us to describe and extract the layout of complex handwritten pages. It also allows to identify special configurations of space, such as underlines or included connected components. Theses configurations could be used to improve the results of lines and fragments extraction.

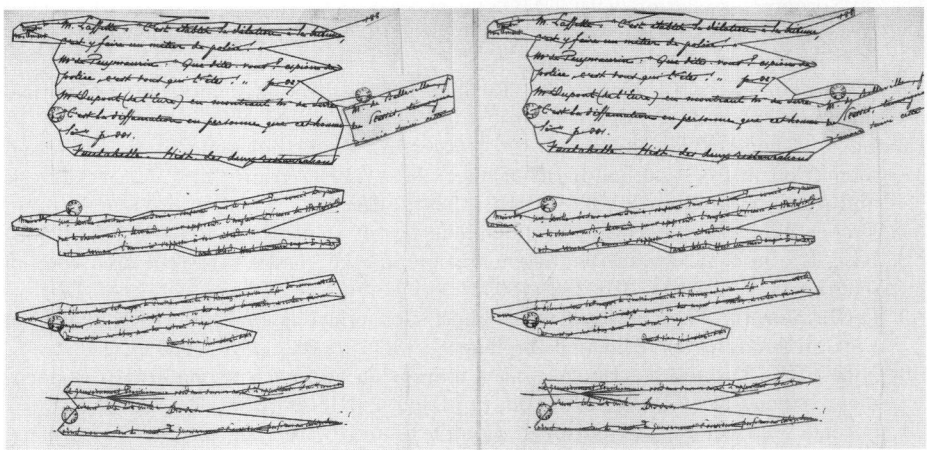

Fig. 5. Fragment extraction, 1f188r : Ground Truth (Left), Simulation (Right)

5 Concluding Remarks

In this paper, we proposed a dedicated text lines and fragments segmentation approach for author's draft handwritings. Knowing that fragments extraction on an humanist corpus is usually a costly and time-consuming hand-made task, it is necessary to provide useful tools for the pre-extraction of fragments in drafts documents. Experiments of our approach show that our proposition is really consistent regarding the complexity of many page layouts in the corpus. Our edge to edge distance allow us to face some classic limitation of connected components based approach like the included connected component problem. Our methodology should be compared to conventional text lines segmentation methods, such as [6] or [5]. Due to the difference in segmentation goals, the comparison required some adaptations.

These studies had the support of the Rhone-Alpes region in the context of cluster project. We also want to acknowledge the consortium of the "Bouvard et Pécuchet" ANR Project.

References

1. Yu, B., Jain, A.K.: A robust and fast skew detection algorithm for generic documents. PR 29(10), 1599–1629 (1996)
2. Louloudis, G., Gatos, B., Pratikakis, I., Halatsis, C.: Text line detection in handwritten documents. PR 41(12), 3758–3772 (2008)
3. Likforman-Sulem, L., Hanimyan, A., Faure, C.: A hough based algorithm for extracting text lines in handwritten documents. In: ICDAR 1995, Washington, DC, USA, p. 774. IEEE Computer Society, Los Alamitos (1995)
4. Lemaitre, A., Coüasnon, B., Leplumey, I.: Using a neighbourhood graph based on voronoï tessellation with DMOS, a generic method for structured document recognition. In: Liu, W., Lladós, J. (eds.) GREC 2005. LNCS, vol. 3926, pp. 267–278. Springer, Heidelberg (2006)
5. Likforman Sulem, L., Zahour, A., Taconet, B.: Text line segmentation of historical documents: a survey. IJDAR 9(2-4), 123–138 (2007)
6. Li, Y., Zheng, Y.F., Doermann, D., Jaeger, S.: Script-independent text line segmentation in freestyle handwritten documents. IEEE Trans. on Pattern Analysis and Machine Intelligence 30(8), 1313–1329 (2008)
7. Breuel, T.M.: High performance document layout analysis. In: SDIUT 2003 (2003)
8. Kise, K., Sato, A., Iwata, M.: Segmentation of page images using the area voronoi diagram. CVIU 70(3), 370–382 (1998)
9. O'Gorman, L.: The document spectrum for page layout analysis. IEEE Trans. on Pattern Analysis and Machine Intelligence 15(11), 1162–1173 (1993)
10. Nicolas, S., Paquet, T., Heutte, L.: A markovian approach for handwritten document segmentation. In: ICPR 2006, pp. III: 292–III: 295(2006)
11. Nicolas, S., Paquet, T., Heutte, L.: Complex handwritten page segmentation using contextual models. In: DIAL 2006, pp. 46–59 (2006)
12. Etemad, K., Doermann, D., Chellappa, R.: Page segmentation using decision integration and wavelet packets. In: ICPR 1994, pp. B:345–B:349 (1994)

Camera-Based Online Signature Verification with Sequential Marginal Likelihood Change Detector

Daigo Muramatsu[1], Kumiko Yasuda[2], Satoshi Shirato[2], and Takashi Matsumoto[2]

[1] Department of Electrical and Mechanical Engineering, Seikei University
3-3-1 Kichijouji-kitamachi, Musashino-shi, Tokyo, 180-8633, Japan
muramatsu@st.seikei.ac.jp
[2] Department of Electrical Engineering and Bioscience, Waseda University
3-4-1, Okubo Shinjuku-ku, Tokyo, 169-8555, Japan
{yasuda06,shirato08,takashi}@matsumoto.elec.waseda.ac.jp

Abstract. Several online signature verification systems that use cameras have been proposed. These systems obtain online signature data from video images by tracking the pen tip. Such systems are very useful because special devices such as pen-operated digital tablets are not necessary. One drawback, however, is that if the captured images are blurred, pen tip tracking may fail, which causes performance degradation. To solve this problem, here we propose a scheme to detect such images and re-estimate the pen tip position associated with the blurred images. Our pen tracking algorithm is implemented by using the sequential Monte Carlo method, and a sequential marginal likelihood is used for blurred image detection. Preliminary experiments were performed using private data consisting of 390 genuine signatures and 1560 forged signatures. The experimental results show that the proposed algorithm improved performance in terms of verification accuracy.

1 Introduction

Online signature verification is a biometric person authentication technology that uses data obtained while a signature is being written, and it is a promising candidate for several reasons. First, handwritten signatures are widely accepted as means of authentication in many countries for various purposes. Second, because online signature verification can incorporate dynamic information about a handwritten signature, it can achieve higher accuracy than verification using static signatures [1]. Finally, a person can modify his or her signature if it is stolen. This is a notable feature because physiological biometrics such as fingerprints or irises cannot be modified or renewed[1].

Several data acquisition devices are used for online signature verification, for example, pen-operated digital tablets, Tablet PCs, PDAs, data acquisition pens,

---

[1] In order to solve this problem, several template protection methods have been proposed [2].

and cameras. Among them, pen-operated digital tablets are the most common device for data acquisition in online signature verification. However, because tablets are not ubiquitous, they must be specially provided for online signature verification. On the other hand, web cameras have become relatively widespread these days. Therefore, online signature verification systems using web cameras for data acquisition[3,4] are very promising.

However, they also have some drawbacks. A camera-based online signature verification system obtains online signature data as trajectories of the pen tip position by tracking the pen tip in video images (time-series images). If one or more of the images are blurred, the system may fail to obtain the correct pen tip position. If the detected pen tip position is not correct, this can cause performance degradation. To reduce this degradation, we propose a method involving detecting such blurred images and re-estimating the pen tip positions in the blurred images. We used a sequential marginal likelihood for blurred image detection. We assumed that the estimated pen tip position at time t associated with the blurred image is questionable. Thus, the pen tip position is re-estimated using estimated pen tip positions around that time. By these processes, the online signature data are corrected, and the corrected data are used for verification.

Preliminary experiments were performed to evaluate the proposed algorithm. Experimental results showed that the equal error rate (EER) was improved from 3.8% to 3.2%.

2 Overview of the System

Figure 1 shows an overview of the system. There are two phases in the algorithm:

(i) Enrollment phase
The user signs for enrollment. The raw online signature data is obtained from the images captured by a camera by tracking the pen tip. Then, the system checks the obtained online signature data set, and parts of the raw signature data are re-estimated if necessary. After preprocessing and feature extraction, a set of extracted time-series features is enrolled as a reference signature.
(ii) Verification phase
The user submits his/her identity and signs for verification. The raw online signature data is obtained from camera images, and if necessary, parts of the raw signature data are re-estimated, similarly to the enrollment phase. After preprocessing and feature extraction, time-series features are compared with reference signatures and several dissimilarity scores are computed. These scores are combined, and a decision is made by using the combined score.

The enrollment and verification phases involve all or some of the following stages: (a) pen tip tracking, (b) data evaluation, (c) re-estimation, (d) preprocessing, (e) feature extraction, (f) dissimilarity calculation, and (g) decision making. We describe each stage below.

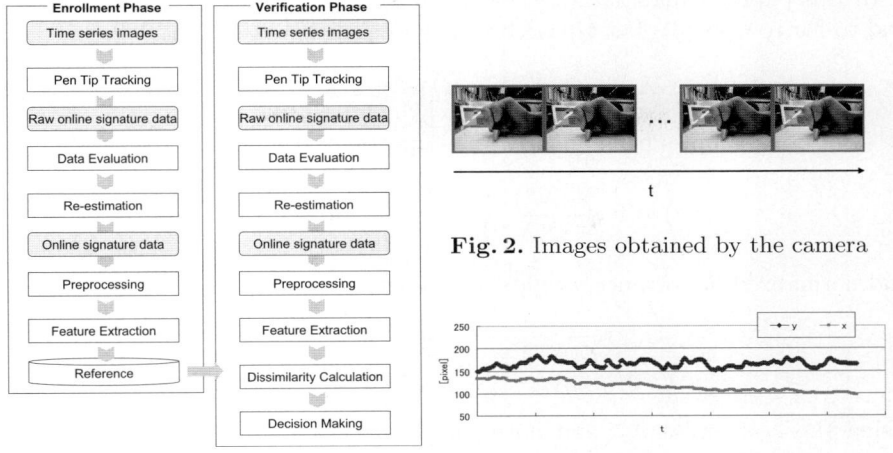

Fig. 1. Overview of the system

Fig. 2. Images obtained by the camera

Fig. 3. Trajectories of raw signature data

2.1 Pen Tip Tracking

In this system, a camera is placed to the side of the hand, and time-series images, such as those shown in Fig.2, are captured. A rectangular area (width=w, height=h) centered around the pen tip is considered as a template for pen tip tracking. An initial point of the template is given manually in this paper.

Let Y_t be a grayscale image at time t and $Y_{1:t} = \{Y_1, Y_2, ..., Y_t\}$ be the image datasets acquired up to time t. Let X_t be the center coordinates of a rectangle (width=w, height=h) in image Y_t.

When, image datasets are given, we estimate the pen tip position $\hat{X}_t = (x_t, y_t)$ as

$$\hat{X}_t = \int X_t P(X_t|Y_{1:t}) dX_t, \qquad (1)$$

where $P(X_t|Y_{1:t})$ is the posterior distribution of the pen tip position at time t, which follows from the Bayes formula

$$P(X_t|Y_{1:t}) = \frac{P(Y_t|X_t)P(X_t|Y_{1:t-1})}{P(Y_t|Y_{1:t-1})}. \qquad (2)$$

Here, the likelihood function $P(Y_t|X_t)$ is defined in this paper by

$$P(Y_t|X_t) = \left(\frac{\lambda_t}{2\pi}\right)^{\frac{hw}{2}} \exp\left[-\frac{\lambda_t}{2} \times score_{SSD}\right], \qquad (3)$$

where $score_{SSD}$ in (3) is the sum of square difference (SSD) between the template and a rectangular area (whose center coordinates are X_t) in image Y_t, and λ_t is a hyperparameter to be learned online.

In this paper, sequential Monte Carlo (SMC) [5] is used to approximate (1) and to learn λ_t in (3). Let $Q(X_t)$ be a proposal distribution, and let

$$X_t^n \sim Q\left(X_t|Y_{1:t-1}\right), n=1,2,...,N \qquad (4)$$

be a set of N samples from $Q(X_t)$. The importance weights are defined by

$$\omega(X_t^n) = \frac{P(Y_t|X_t^n)P(X_t^n|Y_{1:t-1})}{Q(X_t^n|Y_{1:t-1})}, n=1,2,...,N. \qquad (5)$$

and normalized importance weights are calculated by

$$\tilde{\omega}(X_t^n) = \frac{\omega(X_t^n)}{\sum_{n=1}^N \omega(X_t^n)}. \qquad (6)$$

Using the N samples X_t^n and normalized importance weights, equation(1) is approximated by

$$\hat{X}_t = \sum_{n=1}^N \tilde{\omega}(X_t^n)X_t^n. \qquad (7)$$

Figure 3 shows trajectories of obtained raw signature data. Details of the pen tracking algorithm are described in [4].

2.2 Data Evaluation

As stated in section 1, a system can fail to detect the pen tip correctly if the captured image is blurred. Figure 4 and 5 show samples of a non-blurred image and a blurred image. In Fig. 5, we can observe that the pen tip is not detected correctly, whereas it is correctly detected in Fig.4. The incorrect pen tip position causes performance degradation. To avoid this type of degradation, first, we need to know which image is blurred. To detect such an image automatically, we used a sequential marginal likelihood defined by

$$P(Y_t|Y_{1:t-1}) = \int P(Y_t|X_t)P(X_t|Y_{1:t-1})dX_t. \qquad (8)$$

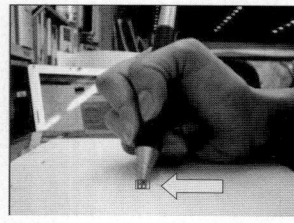

Fig. 4. Non-blurred image

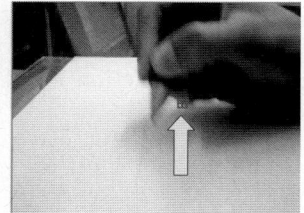

Fig. 5. Blurred image

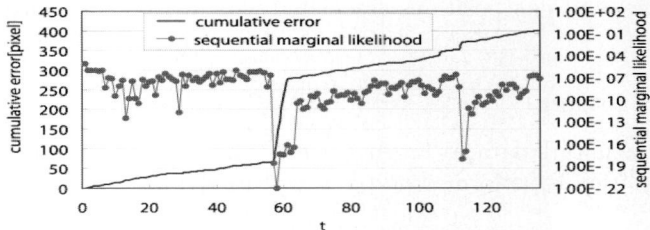

Fig. 6. Cumulative error and $P(Y_t|Y_{1:t-1})$

This likelihood has been successfully used for change detection problems[6,7]. In order to explain why (8) is used in our attempt to perform blurred image detection, let us consider a general likelihood function $P(Y_t|z)$, where z denotes any quantity associated with the model structure underlying the data Y_t. Recall that the given data $Y_t := Y_t^*$ gives information about z. More precisely, the likelihood function $P(Y_t = Y_t^*|z)$ quantifies the appropriateness of z for the given data Y_t^*. Another interpretation is that $P(Y_t = Y_t^*|z)$ gives the degree of appropriateness of the data with respect to z. In (8), $z = Y_{1:t-1}$, so that it quantifies the degree of appropriateness of the new data $Y_t := Y_t^*$ with respect to the preceding sequential data $Y_{1:t-1}$. Figure 6 shows the cumulative error of pen tip tracking together with the changes of sequential marginal likelihood $P(Y_t|Y_{1:t-1})$. The system fails to detect the pen tip position at $t = 57$ because the image Y_t is blurred. Thus, the cumulative error increases abruptly at $t = 57$. We can also observe a discernible dip in $P(Y_t|Y_{1:t-1})$ at $t = 57$. Therefore, $P(Y_t|Y_{1:t-1})$ can be useful for this blurred image detection. We assume that

$$\text{image } Y_t = \begin{cases} \text{blurred} & \text{if } P(Y_t|Y_{1:t-1}) < TRD_{lk} \\ \text{non-blurred} & \text{otherwise} \end{cases}. \quad (9)$$

Here, TRD_{lk} is a threshold value. The issue of how to set the threshold is a difficult problem[6]. Therefore, in this paper, we set the threshold empirically.

In the case of a blurred image, we assumed that the estimated pen tip position $\hat{X}_t$ associated with the blurred image is questionable. Then, this estimated position is re-estimated in the next stage.

2.3 Re-estimation

The re-estimation stage is performed after the pen tip positions at all times are estimated.

Let Y_{t_b} be a blurred image, $Y_{t_b-\Delta t_p}$ be the last non-blurred image prior to t_b, and $Y_{t_b+\Delta t_f}$ be the closest non-blurred image after t_b. Let the estimated pen tip positions at each time be $\hat{X}_{t_b}$, $\hat{X}_{t_b-\Delta t_p}$, and $\hat{X}_{t_b+\Delta t_f}$, respectively. The pen tip position at time t_b is re-estimated and updated by

$$\hat{X}_{t_b} \leftarrow \frac{\Delta t_p \hat{X}_{t_b+\Delta t_f} + \Delta t_f \hat{X}_{t_b-\Delta t_p}}{\Delta t_p + \Delta t_f}. \quad (10)$$

Only the estimated pen positions associated with the blurred image Y_t are re-estimated.

2.4 Preprocessing

After online signature data $\hat{X}_t = (x_t, y_t), t = 1, 2, ..., T$ are obtained, the following transformation is performed on the signature data:

$$\overline{x}_t = \frac{x_t - x_g}{x_{max} - x_{min}}, \overline{y}_t = \frac{y_t - y_g}{y_{max} - y_{min}}, \quad (11)$$

where

$$x_g = \frac{1}{T}\sum_{t=1}^{T} x_t, \quad y_g = \frac{1}{T}\sum_{t=1}^{T} y_t.$$

$$x_{min} = \min_{t} x_t, \; x_{max} = \max_{t} x_t, \; y_{min} = \min_{t} y_t, \; y_{max} = \max_{t} y_t.$$

2.5 Feature Extraction

The following three additional features are calculated:

$$L_t = \sqrt{\overline{x}_t^2 + \overline{y}_t^2}, \; \theta_t = \tan^{-1}\frac{\overline{y}_t}{\overline{x}_t}, \; V\theta_t = \begin{cases} 0 & \text{if } t = 1 \\ \tan^{-1}\frac{\overline{y}_t - \overline{y}_{t-1}}{\overline{x}_t - \overline{x}_{t-1}} & \text{otherwise} \end{cases} \quad (12)$$

2.6 Distance Calculation

Five types of time-series feature data $(\overline{x}_t, \overline{y}_t, L_t, \theta_t, V\theta_t)$ are considered in this paper, and five distances $D_n, n = 1, 2, ..., 5$ are calculated independently by using dynamic time warping[8]. See [4] for details.

Then, dissimilarity scores $ds_n, n = 1, 2, ..., 5$ are calculated by

$$ds_n = \frac{D_n}{T_{ref}, Z_n}. \quad (13)$$

Here, T_{ref} is the data length of a reference signature, and Z_n are the normalization constants. In this paper, Z_n is calculated by using reference signatures.

2.7 Decision Making

A combined dissimilarity score $Score_{vf}$ is calculated by

$$Score_{vf} = \sum_{n=1}^{5} ds_n, \quad (14)$$

and a final decision is made based on the following rule:

$$\text{input signature is} \begin{cases} Accepted & \text{if } Score_{vf} \leq THR_{vf} \\ Rejected & \text{otherwise} \end{cases} \quad (15)$$

3 Experiment

3.1 Data Collection

Online signature data was collected from thirteen students. All of the students were right-handed. Each student wrote ten genuine signatures in first, second, and third sessions. In total, thirty genuine signatures were collected from each student. As for forgery data, four different students imitated genuine signatures to produce forgery data. The forgers could see videos of genuine signatures captured by the camera. Thus, they could see dynamic information of the genuine signatures that they had to imitate. Each forger produced 30 forgeries for each genuine user, and a total of 120 forgeries for each genuine user were collected.

The first 5 genuine signatures collected in the first session were used for enrollment, and the remaining 25 genuine signatures and 120 forgeries were used for evaluation.

3.2 Experimental Settings and Results

The cameras had a resolution of 320×240 pixels. The number of samples N was 4000, and the width and height of the rectangular area were $w = 10$ and $h = 10$. The experimental results are shown in Fig. 7. For comparison purposes, a system that did not have the re-estimation stage was evaluated as a baseline system. From Fig.7, we can observe that the performance of the proposed system was improved, as indicated by the equal error rate, which dropped to 3.2%.

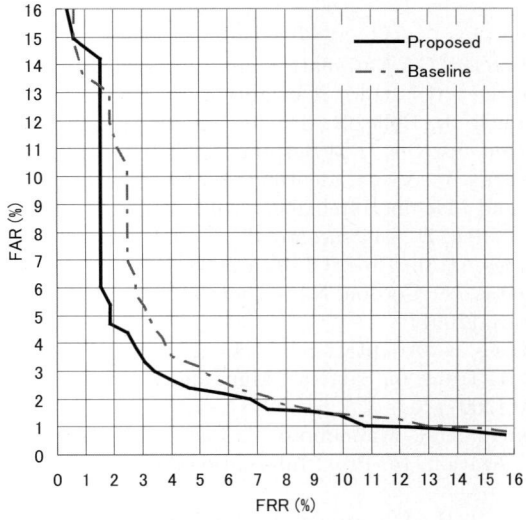

Fig. 7. Error trade-off curve

4 Conclusion and Future Work

In camera-based online signature verification, blurred images cause failure of pen tip detection, and this failure results in performance degradation. In this paper, we propose a camera-based online signature verification algorithm that can detect blurred images and re-estimate the pen tip position associated with the blurred images. Data evaluation and re-estimation stages are included in the proposed algorithm. By including these stages, the equal error rate was improved from 3.8% to 3.2%, indicating that the proposed algorithm is promising. Error rates of biometric systems are dependant on the database, and especially on its number of users. Thus, we need to evaluate the proposed system against a bigger database. In this proposed algorithm, linear interpolation is used for re-estimation, and dynamical information such as velocity and acceleration is not considered. Therefore, improvement of the re-estimation method will be the topic of a future project. Several dissimilarity scores are combined by a summation rule in this paper. More useful fusion strategies described in [9], will also be the subject of future work.

Acknowledgements

This work was supported by the New-Generation Information Security Technologies R&D project (Ministry of Economy, Trade and Industry of Japan).

References

1. Plamondon, R., Lorette, G.: Automatic signature verification and writer identification - the state of the art. Pattern Recognition 22(2), 107–131 (1989)
2. Ratha, N.K., Connell, J., Bolle, R.: Enhancing security and privacy of biometric-based authentication systems. IBM Systems Journal 40(3), 614–634 (2001)
3. Munich, M.E., Perona, P.: Visual identification by signature tracking. IEEE Trans. Pattern Analysis and Machine Intelligence 25(2), 200–217 (2003)
4. Yasuda, K., Muramatsu, D., Matsumoto, T.: Visual-based online signature verification by pen tip tracking. In: Proc. CIMCA 2008, pp. 175–180 (2008)
5. Doucet, A., de Freitas, N., Gordon, N.: Sequential Monte Carlo Methods in Practice. Springer, Heidelberg (2001)
6. Matsumoto, T., Yosui, K.: Adaptation and change detection with a sequential Monte Carlo scheme. IEEE Trans. on Systems, Man, and Cybernetics – part B: Cybernetics 37(3), 592–606 (2007)
7. Matsui, A., Clippingdale, S., Matsumoto, T.: Bayesian sequential face detection with automatic re-initialization. In: Proc. International Conference on Pattern Recognition (2008)
8. Rabiner, L., Juang, B.-H.: Fundamentals of speech recognition. Prentice-Hall, Englewood Cliffs (1993)
9. Ross, A.A., Nandakumar, K., Jain, A.K.: Handbook of Multibiometrics. Springer Science+Business Media, LLC, Heidelberg (2006)

Separation of Overlapping and Touching Lines within Handwritten Arabic Documents

Nazih Ouwayed and Abdel Belaïd

LORIA, University Nancy 2, Vandoeuvre-Lès-Nancy, France
{nazih.ouwayed,abelaid}@loria.fr
http://www.loria.fr/equipes/read/

Abstract. In this paper, we propose an approach for the separation of overlapping and touching lines within handwritten Arabic documents. Our approach is based on the morphology analysis of the terminal letters of Arabic words. Starting from 4 categories of possible endings, we use the angular variance to follow the connection and separate the endings. The proposed separation scheme has been evaluated on 100 documents contains 640 overlapping and touching occurrences reaching an accuracy of about 96.88%.

Keywords: Handwriting line segmentation, Arabic documents, Overlapping and Touching lines, Calligraph morphology.

1 Introduction

The text line and word extraction from a handwritten document, is seen as a labored task. The difficulty rises from the characteristics of the handwritten documents especially when they are ancient. These documents present irregular spacing between the text lines. The lines can overlap or touch when their ascenders and descenders regions belong to each other (see figure 1). Furthermore, the lines can be skewed which constitutes new orientations.

In the literature, several methods have been proposed dealing with skewed lines [5,6,7,8,9]. Few methods have been proposed for the separation of connected words in the adjacent lines. From them, an Independent Component Analysis (ICA [4]) segmentation algorithm is proposed by Chen et al. in [1]. The ICA converts the original connected words into a blind source matrix and calculates the weighted value matrix before the values are re-evaluated using a fast model.

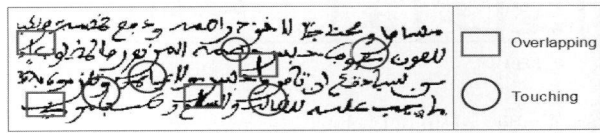

Fig. 1. Extract of a document representing some overlapping and touching lines

The readjusted weighted value matrix is applied to the blind source matrix in order to separate the word components. Louloudis et al. propose in [2] a skeleton-based algorithm. In [3], Takru and Leedham propose a technique that employs structural knowledge on handwriting styles where overlap is frequently observed. All above approaches are applicable to Latin scripts which are not easy adaptable to Arabic because of specific morphology.

In this paper, we propose a novel method considering the morphology of the terminal letters in the PAWs. The rest of the paper is organized as follows: overlapping and touching types in the Arabic document are listed in section 2. The Arabic script morphology is discussed in section 3. In section 4, our separation approach is detailed. Experiments results are showed in section 5 and last section concludes the paper.

2 Overlapping/Touching Types

The Arabic alphabet is composed of 28 letters. Among of them, 21 letters have an ascender (alif ا, Ta ط, DHa ظ, kaf ك, lam ل, heh ه), right descender (ra ر, zïn ز, sïn س, shïn ش, Sad ص, Dad ض, qaf ق, mïm م, nün ن, waw و, yeh ي) or left descender (jïm ج, ḥa ح, kha خ, 'ain ع, ghain غ) causing the connecting lines (left or right indicates the descender starting, see figure 2.a). The connection in Arabic documents can happen in two cases: when the interlines spacing is small (see figure 2.b) or when we use a calligraph with big jambs (descenders) as Diwani (see figure 2.c).

Statistically, we have found 4 overlapping/touching types in handwritten Arabic documents. In the first type (see table 1.a), a right descender with a loop overlaps/touches a vertical ascender. In the second type (see table 1.b), a left descender with a loop overlaps/touches a vertical ascender. In the third type (see table 1.c), a right descender overlaps/touches the curving top of lower letter. In

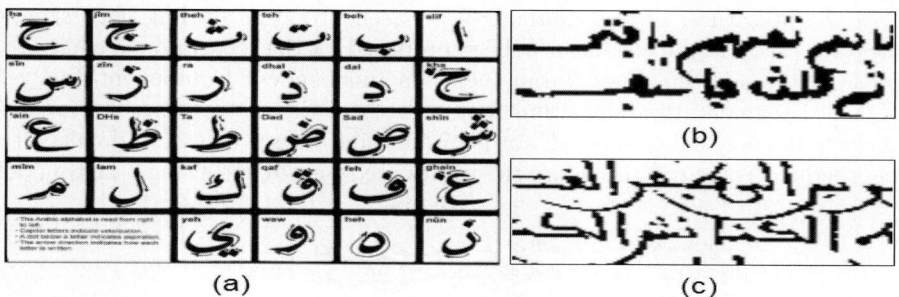

Fig. 2. (a) The Arabic alphabet chart (arrow indicates the writing direction, source: [14]), (b) Connection is due to the small interlines spacing, (c) Overlapping/touching due to the script type (Diwani)

Table 1. The 4 overlapping/touching types in the handwritten Arabic documents

Types		Terminal Letters	Overlapping/ Touching Zones	Samples
a	Top:	ر, ز, س, ش, ص, ض, ن, ق, و, ي		
	Bottom:	ا, ط, ظ, ك, ل		
b	Top:	ج, ح, خ, ع, غ		
	Bottom:	ا, ط, ظ, ك, ل		
c	Top:	ر, ز, م, و		
	Bottom:	ص, ض, ه		
d	Top:	ج, ح, خ, ع, غ		
	Bottom:	ه		

the fourth type (see table 1.d), a left descender overlaps/touches the curving top of lower letter. In each type, the top letters stretches out to overlap/touch the bottom letters.

3 Arabic Morphology Analysis

In all cases of connection, we notice the presence of a descender connecting a lower terminal letter (see table 1, column overlapping/touching zones). These descenders are clustered in two categories : (a,c) when the descender comes from right and (b,d) when the descender comes from left. To face this connection problem, the analysis will be focused on the connection zones (see figure 3).

The zones are determined considering a rectangle around the intersection point S_p of the two connected components which size is fixed manually (see section 4.2). The starting ligature point B_p is the highest point in the zone close to the baseline. The descender direction is determined according to B_p relative to S_p (see section 4.2).

According to these characteristics, our idea consists to follow the skeleton pixels within the zone using the starting point B_p and the right descender direction. The follow-up will then cross the intersection point S_p and continues in the right direction that we have to determine.

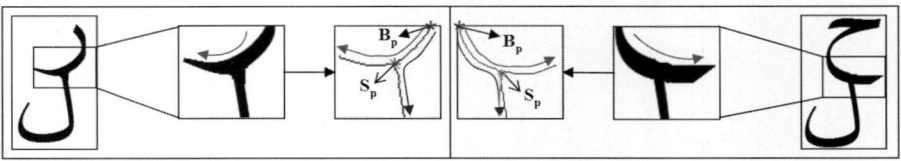

Fig. 3. Overlapping/touching connected components zones and descenders direction (rectangles represent the connection zones, right direction indicated by red arrow and the false by blue)

3.1 Right Follow-Up Direction

The determination of the right direction follow-up requires the study of curves in the skeleton image (i.e. each zone in the figure 3 has two curves).

In the literature, there are two main categories of curve detection methods [11]. In the first category, the Generalized Hough Transform (GHT) is used [12]. This technique is not efficient in our case because the present curves have few points and GHT needs much points for correct detection. In the second category, the chains (some of connected skeleton pixels) of points, or segments yielded by the polygonal approximation of such chains are used. The basic idea is to compute an estimation of the curvature for these chains [13]. This technique is insufficient because it does not study the continuity and the morphology of the curve.

The proposed method is based on the skeleton pixels follow-up and the angular variance. The follow-up starts from B_p and continues to the intersection point S_p. At this point S_p, the follow-up continues in multiple directions (see figure 4.a). The follow-up continues in each direction to extract the possible curves (C_{1+2}, C_{1+3} and C_{1+4}). The next step is to find the curve that represents the descender terminal letter. By experience, we found that the Arabic terminal letters have a minimum angular variance. This is explained by the fact that the terminal Arabic letters have the same orientation angle along the descender curve.

3.1.1 Angular Variance

The angular variance represents the dispersion of the orientation angles along the curve. It is estimated using the statistical variance formula:

$$Var(\Theta) = \sum_{i=1}^{n}(\theta_i - \mu)^2 \qquad (1)$$

where Θ is the angles variation vector of the curve and μ is the average of Θ.

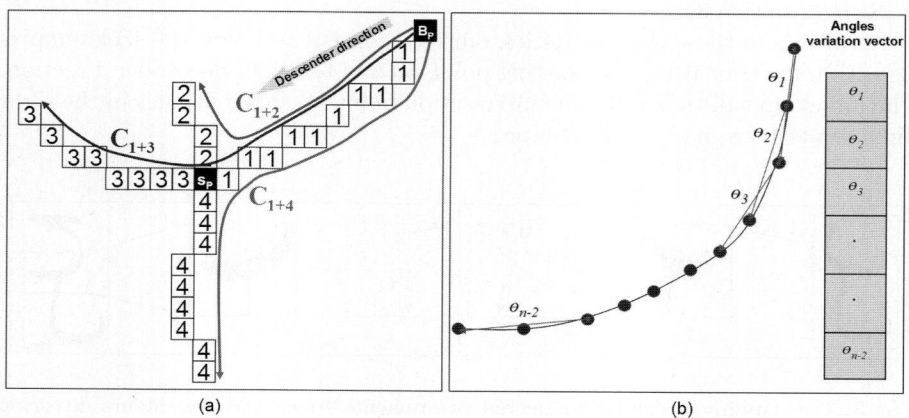

Fig. 4. (a) Example of Arabic overlapping connected components ("ra ر" overlaps "alif ا"), (b) Angles variation vector estimation algorithm

The angles variation vector Θ of the curve is estimated using an iterative algorithm that calculates the angle θ_i between two successive pixels p_i and p_{i+2} using the formula below (see figure 4.b):

$$\theta_i = \left| \text{Arctan}\left(\frac{\text{dy}_{i,i+2}}{\text{dx}_{i,i+2}}\right) \right| \qquad (2)$$

Because of the symmetric branches, the angle value must be always positive. For example, in figure 4, the angular variances are: $Var(C_{1+2}) = 703.19$, $Var(C_{1+3}) = 299$, $Var(C_{1+4}) = 572.37$. In this example, the minimum angular variance $Var(C_{1+3})$ is given by the right follow-up direction.

4 Proposed Method

The method involves four steps as follows:

4.1 Step 1: Overlapping and Touching Connected Components Detection

The present paper is a continuation of our work published in [9]. The lines are extracted in [9] from the handwritten Arabic documents. Some adjacent lines can be connected in one or more connected components. This components belonging to two adjacent lines are considered as connected (see figure 5.a).

4.2 Step 2: Curve Detection

To detect the curves, the skeleton is first extracted using a thinning algorithm descried in [10]. Then, the intersection points of each connected component are detected (see figure 5.b). An intersection point is a pixel that has at least three neighbor pixels. As in Arabic script, the overlapping or touching may occur at just one intersection point S_p near the minima axis (valley between two connected lines in the projection histogram of the document, see figure 5.c). For this, S_p is the nearest point of the minima axis (see figure 5.d). Once the S_p is located, we look for the connected components zone. The center of this zone is S_p and its width (resp. height) is equal to $w_{ccx}/4$ (resp. $h_{ccx}/4$) where w_{ccx} (resp. h_{ccx}) is the width (resp. height) of the overlapping or touching connected component (4 is a determined experimentally). Since this zone is extracted from the initial document, it is cleaned by removing the small connected components. To do it, the connected components are labeled and we keep only the connected component containing S_p (see figure 5.e). After the pre-processing, the skeleton of the zone is extracted and an skeleton follow-up algorithm is applied using the descender follow-up starting point B_p (the pixel that has the minimum y_i in the zone) and the direction follow-up (right to left if $x(B_p) > x(S_p)$ and left to right if $x(B_p) < x(S_p)$, see figure 5.f).

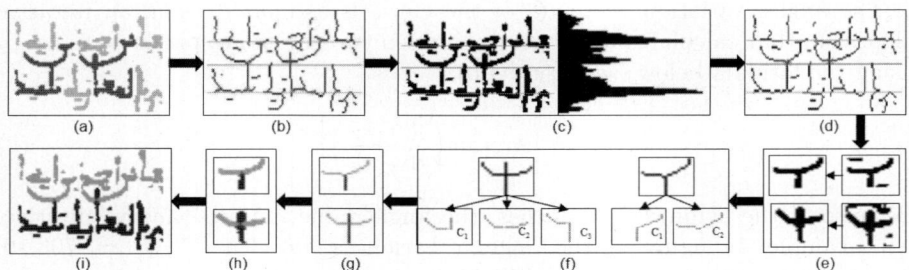

Fig. 5. Separation of overlapping and touching connected components approach steps

4.3 Step 3: Curve Angular Variance Estimation

The angular variance of each curve is estimated using the algorithm detailed in the section 3.1.1. In the figure 5.f, the first touching components have $V(C_1) = 538.2099$ and $V(C_2) = 754.2284$. The C_1 having the minimum angular variance is the descender. The second overlapping components have $V(C_1) = 1160.8$, $V(C_2) = 438.4$ and $V(C_3) = 1208$. The C_2 having the minimum angular minimum variance is the descender.

4.4 Step 4: Pixels Assignment

In the figure 5.g, there are two curves and three different pixels types (intersection point "by red", first connected component "by green" and second connected component "by blue"). This step consists in assigning each black pixel in the initial image (see figure 5.e) to its appropriate curve. To do it, the image is scanned pixel by pixel and the 48-connected pixels of each image pixel p_i are regarded in the skeleton image. The closest branches pixel value is assigned to the initial image pixel (see figure 5.h). Finally, the pixels assigning is done at the initial document in order to obtain the final result (see figure 5.i).

5 Experimental Results and Discussion

The approach was applied to 100 handwritten Arabic documents belonging to the Tunisian National Library, National Library of Medicine in the USA, National Library and Archives of Egypt that contain 640 overlapping and touching connected components. The tests were prepared after a manual indexing step of the overlapping and touching connected components of each document. Then, these components are clustered in 4 types (see section 2): 253 occurrences of type (a), 194 occurrences of (b), 117 occurrences of (c), 76 occurrences of (d) have been detected. The Table 2 describes the results for each type. The weighted mean of these results is equal to 96.88%. The 3.12% rate error is due to the intersection point detection algorithm because in some cases the overlapping/touching do not hold near the minima, and to the angular variance criterion because in some

Table 2. Results of the separation of overlapping and touching connected components approach

Overlapping/ touching types	Occurrences	Connections missed	Separations failed	Correctly separations rate %
a	253	2	3	98.02%
b	194	4	2	96.90%
c	117	3	1	95.73%
d	76	1	2	94.75%

Fig. 6. Samples of our results

cases the minimum angular variance can occur for a false direction. Figure 6 illustrates the effectiveness of the algorithm on a sample of 12 representative connected components chosen from 640 occurrences.

6 Conclusion and Future Trends

An original method of separation overlapping and touching connected components in the adjacent text lines from the handwritten Arabic documents has been proposed in this paper. The proposed method is based on the Arabic calligraph where overlapping and touching is most frequently observed. The approach is armed by statistical informations about Arabic writing structures. Experiments showed the efficiency and the performance of our approach. The future step of this work is related to the segmentation of the lines into single words.

References

1. Chen, Y., Leedham, G.: Independent Component Analysis Segmentation Algorithm. In: 8th International Conference on Document Analysis and Recognition, pp. 680–684 (2005)
2. Louloudis, G., Gatos, B., Halatsis, C.: Line And Word Segmentation of Handwritten Documents. In: 11th International Conference on Frontiers in Handwriting Recognition, Canada, pp. 599–603 (2008)

3. Takru, K., Leedham, G.: Separation of touching and overlapping words in adjacent lines of handwritten text. In: International Workshop on Frontiers in Handwriting Recognition, pp. 496–501 (2002)
4. Hyvarinen, A.: Survey on Independent Component Analysis. Helsinki University of Technology, Finland (1999)
5. Lüthy, F., Varga, T., Bunke, H.: Using hidden Markov models as a tool for handwritten text line segmentation. In: 9th Int. Conf. on Document Analysis and Recognition, pp. 8–12 (2007)
6. Zahour, A., Likforman-Sulem, L., Boussellaa, W., Taconet, B.: Text Line Segmentation of Historical Arabic Documents. In: Proceedings of the Ninth International Conference on Document Analysis and Recognition, Brazil, pp. 138–142 (2007)
7. Bukhari, S.S., Shafait, F., Breuel, T.M.: Segmentation of Curled Text Lines using Active Contours. In: Proceedings of Eight IAPR Workshop on Document Analysis Systems, pp. 270–277 (2008)
8. Shi, Z., Govindaraju, V.: Line Separation for Complex Document Images Using Fuzzy Run length. In: Proc. of the Int. Workshop on Document Image Analysis for Libraries, Palo, Alto, CA (2004)
9. Ouwayed, N., Belaïd, A.: Multi-oriented Text Line Extraction from Handwritten Arabic Documents. In: The Eighth IAPR International Workshop on Document Analysis Systems (DAS 2008), Japan, pp. 339–346 (2008)
10. Lam, L., Lee, S.-W., Suen, C.Y.: Thinning Methodologies-A Comprehensive Survey. IEEE Transactions on Pattern Analysis and Machine Intelligence 14(9), 869–885 (1992)
11. Dori, D., Liu, W.: Stepwise recovery of arc segmentation in complex line environments. International Journal on Document Analysis and Recognition 1(1), 62–71 (1998)
12. Ballard, D.H.: Generalizing the Hough Transform to detect arbitrary shapes. Pattern Recognition 13(2), 111–122 (1981)
13. Rosin, P.L., West, G.A.: Segmentation of Edges into Lines and Arcs. Image and Vision Computing 7(2), 109–114 (1989)
14. Stanford University, USA, http://www.stanford.edu/

Combining Contour Based Orientation and Curvature Features for Writer Recognition

Imran Siddiqi and Nicole Vincent

Laboratoire CRIP5 –SIP, Paris Descartes University,
45 Rue des Saint Pères, 75006 France
{imran.siddiqi,nicole.vincent}@mi.parisdescartes.fr

Abstract. This paper presents an effective method for writer recognition in handwritten documents. We have introduced a set of features that are extracted from two different representations of the contours of handwritten images. These features mainly capture the orientation and curvature information at different levels of observation, first from the chain code sequence of the contours and then from a set of polygons approximating these contours. Two writings are then compared by computing the distances between their respective features. The system trained and tested on a data set of 650 writers exhibited promising results on writer identification and verification.

Keywords: Writer Recognition, Freeman Chain Code, Polygonization.

1 Introduction

Among the expressive behaviors of human, handwriting carries the richest information to gain insight into the physical, mental, and emotional states of the writer. Each written movement or stroke reveals a specific personality trait, the neuro-muscular movement tendencies being correlated with specific observable personality features [2]. This explains the stability in the writing style of an individual and the variability between the writings of different writers, making it possible to identify the author for which one has already seen a written text. This automatic writer recognition serves as a valuable solution for the document examiners, paleographers and forensic experts. In the context of handwriting recognition, identifying the author of a given sample allows adaptation of the system to the type of writer [10].

Writer recognition comprises the tasks of writer identification and verification. Writer Identification involves finding the author of a query document given a reference base with documents of known writers. Writer verification on the other hand determines whether two samples have been written by the same person or not. The techniques proposed for writer recognition are traditionally classified into text-dependent [15,17] and text-independent methods, which can make use of global [5,11] or local [4,6] features. Combining the global and local features is also known to improve the writer recognition performance [7,13,15]. Lately, the methods that compare a set of patterns (a writer specific or a universal code book) to a questioned writing have shown promising results as well [4,13]. These methods however rely on handwriting segmentation and defining an optimal segmentation remains a problem.

In this paper[1], we present a system for offline writer recognition using a set of simple contour based features extracted by changing the scale of observation as well as the level of detail in the writing. We first compute a set of features from the chain code sequence representing the contours, at a global as well as at a local level. We then extract another set of features from the line segments estimating the contours of handwritten text. Finally we perform a comparative evaluation of the different types of features and explore their various combinations as presented in the following.

2 Feature Extraction

For feature extraction, we have chosen to work on the contours of the text images as they eliminate the writing instrument dependency while preserving the writer-dependent variations between character shapes. We start with the gray scale input image, binarize it using the Otsu's global thresholding algorithm and extract the contours of the handwritten text. We then represent the contours in two ways: by a sequence of Freeman chain codes and by a set of polygons obtained by applying a polygonization algorithm to the contours. We then proceed to the extraction of features that capture the orientation and curvature information of writing; the two most important visual aspects that enable humans instinctively discriminate between two writings. These features have been discussed in the sections to follow where we first discuss the chain code based features and then we present how we have modeled some loss of details keeping only a general and simple view of the writing.

2.1 Chain Code Based Features

Chain codes have shown effective performance for shape registration [1] and object recognition [3] and since the handwritten characters issued by a particular writer have a specific shape, chain code based features are likely to perform well on tasks like writer recognition. We therefore compute a set of features from the chain code sequence of the text contours first at a global level and then from the stroke fragments within small observation windows.

2.1.1 Global Features
At the global level, in order to capture the orientation information, we start with the well-known histogram of all the chain codes/slopes (slope density function $f1$) of the contours where the bins of the histogram represent the percentage contributions of the eight principal directions. In addition, we also find the histograms of the first (and second) order differential chain codes that are computed by subtracting each element of the chain code from the previous one and taking the result modulo *connectivity*. These histograms ($f2$ & $f3$) represent the distribution of the angles between successive text pixels ($f2$) and the variations of these angles ($f3$) as the stroke progresses.

---

[1] This study has been carried out in the framework of the project ANR GRAPHEM: ANR-07-MDCO-006-04.

The distributions of chain codes and their differentials give a crude idea about the writing shapes but they might not be very effective in capturing the fine details in writing; we thus propose to count not only the occurrences of the individual chain code directions but also the chain code pairs, in a histogram *f4*, illustrated for two writings in figure 1. The bin *(i,j)* of the (8x8) histogram represents the percentage contribution of the pair *i,j* in the chain code sequence of the contours. Employing the same principle, we also compute the (8x8x8) histogram of chain code triplets *f5*. It is important to precise that all the 64 possible pairs and 512 possible triplets cannot exist while tracing the contours and we can have a total of 44 pairs and 236 triplets.

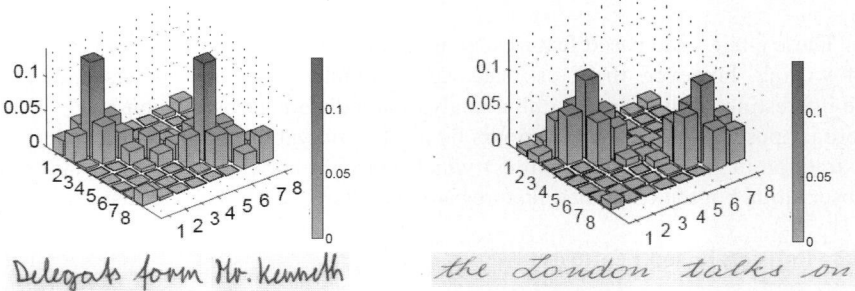

Fig. 1. Two writings and their respective distributions of chain code pairs

Finally, for the purpose of comparison, we use an estimate of curvature computed from the histograms of contour chain code, presented in [3] for object recognition. A correlation measure between the distribution of directions on both sides (forward and backward) of a contour pixel p_c is used to approximate the curvature at p_c and it is the distribution of these estimates which is used to characterize an author (*f6*).

2.1.2 Local Features

The features *f1 – f6*, although computed locally, capture the global aspects of writing thus the relative stroke information is lost. We therefore chose to carry out an analysis of small stroke fragments as well. Employing an adaptive window positioning algorithm [13], we divide the writing into a large number of small square windows, the window size being fixed empirically to 13x13. Within each window, we find the percentage contribution of each of the eight directions (chain codes), the percentages being quantized into ten percentiles. These contributions are counted in a histogram (*f7*): the bin *(i,j)* is incremented by one if the direction *i* is represented in the *jth* percentile. Figure 2 shows the windows positioned over an image and the contribution of one of the windows to the histogram where the three direction codes present in the window lead to three contributions to the histogram. The process is carried out for all the windows and the distribution is finally normalized. This distribution (*f7*) thus, can be considered as a window-based local variant of *f1*. Using the same idea, we also compute *f2* and *f3* locally, represented by the distributions *f8* and *f9* respectively. These distributions have been discussed in detail in [14].

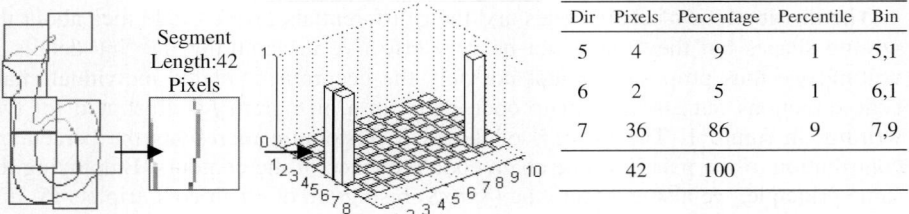

Fig. 2. Windows positioned over the chain-coded image and the contribution of one of the windows to the distribution $f7$

These chain code based features compute the orientation and curvature information of writing, however, these estimates are computed from raw pixels and it would be interesting to carry out a similar analysis at a different observation level. We therefore propose to estimate the contours by a set of polygons and then proceed to feature extraction (a set of global features) which not only corresponds to a distant scale of observation but the computed features are also more robust to noise distortions.

2.2 Polygon Based Features

These features are aimed at keeping only the significant characteristics of writing discarding the minute details. Employing the sequential polygonization algorithm presented in [16], we carry out an estimation of the contours by a set of line segments. The algorithm requires a user defined parameter T that controls the accuracy of approximation. Larger values of T create longer segments at the cost of character shape degradation and vice versa. Figure 3 shows the polygon estimation of the contours of a handwritten word for different values of T. For our system, we have used a value of T equal to 2, chosen empirically on a validation set. We then extract a set of features from these line segments.

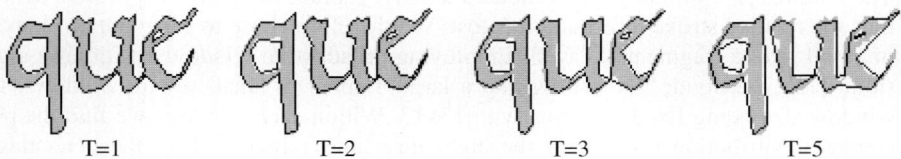

Fig. 3. Polygonization at different values of T

We first compute the slope of each of the line segments and employ their distribution ($f10$) for characterizing the writer. Each line is identified as belonging to one of the bins (classes) illustrated in figure 4. These bins are chosen in such a way that the lines having nearly the same orientations as the principal directions (vertical, horizontal etc) fall in their respective classes. For example, all the segments in the range $-12°$ to $12°$ are classified as (nearly) horizontal and so on.

Not only the number of slopes in a particular direction is important but their corresponding lengths as well, so in order to complement the distribution *f10*, we also compute a length-weighted distribution of slopes (*f11*), where for each segment at slope *i*, the bin *i* in *f11* is incremented by the length of the segment. The distribution is finally normalized by the total length of segments in the image.

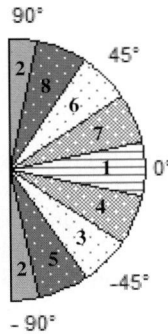

Bin	Class
1	Horizontal (H)
2	Vertical (V)
3	Left Diagonal (LD)
4	LD Inclined towards Horizontal (LDH)
5	LD Inclined towards vertical (LDV)
6	Right Diagonal (RD)
7	RD Inclined towards Horizontal (RDH)
8	RD Inclined towards Vertical (RDV)

Fig. 4. Division of slopes (-90° to 90°) into bins and the corresponding segment classes

We next estimate the curvature by computing the angle between two connected segments and use the distribution of these angles as our next feature (*f12*). The angle bins are divided in a similar fashion as for the slopes. Similarly, in order to take into account the lengths of the segments forming a particular angle, a length weighted version of *f12*, *f13* is also computed.

Finally, irrespective of the orientation, it would be interesting to analyze the distribution of the lengths of segments in a writing. Generally, smooth strokes will lead to longer and fewer segments while shaky stokes will result in many small segments, thus the straight segment lengths could be useful in distinguishing the writings of different authors. We therefore use the distribution of these lengths (*f14*) as a writer specific feature, the number and partitioning of bins being determined by analyzing the distribution of lengths in writings of the validation set.

We thus extract a set of fourteen (normalized) distributions to represent a document image. The distributions for which the number of bins is not discussed explicitly have been partitioned empirically on the validation set. Table 1 summarizes the proposed features with the dimensionalities of each.

Table 1. Dimensionalities of the proposed features

Feature	*f1*	*f2*	*f3*	*f4*	*f5*	*f6*	*f7*	*f8*	*f9*	*f10*	*f11*	*f12*	*f13*	*f14*
Dim	8	7	8	44	236	11	80	70	80	8	8	8	8	10

3 Writer Recognition

The dissimilarity between two writing samples is defined by computing a distance between their respective features. We tested a number of distance measures including:

Euclidean, χ^2, Bhattacharyya and Hamming distance, χ^2 distance reading the best results in our evaluations. Writer Identification is performed by computing the distance between the query image Q and all the images in the data set, the writer of Q being identified as the writer of the document that reports the minimum distance. For writer verification, the Receiver Operating Characteristic (ROC) curves are computed by varying the acceptance threshold, verification performance being quantified by the Equal Error Rate (EER). The identification and verification results have been presented in the following section.

4 Experimental Results

For the experimental study of our system, we have chosen the IAM [8] data set which contains samples of unconstrained handwritten text from 650 different writers. This data set contains a variable number of handwritten images per writer with 350 writers having contributed only one page. In order to fit in all the 650 writers in our experiments, we keep only the first two images for the writers having more than two pages and split the image roughly in half for writers who contributed a single page thus ensuring two images per writer, one used in training while the other in testing.

We carried out an extensive series of experiments evaluating the performance of individual features and their various combinations. We will report only a sub-set of results presented for the three types of features and their combinations in figure 5. It can be observed that the combined performance of each type of features is more or less the same with the combination of polygon based features performing slightly better (82.3% against 81.5% & 81.1%). Combining two types of features boosts the identification rates to around 84-85% which rises to 89% when using all the features. Similarly for writer verification we achieve an equal error rate of as low as 2.5%.

Table 2. Comparison of writer identification methods

		Writers	Samples/ writer	Performance
Marti et al. (2001)	[9]	20	5	90.7%
Bensefia et al. (2004)	[4]	150	2	86%
Schlapbach and Bunke (2006)	[12]	100	5/4	98.46%
Bulacu and Schomaker (2007)	[7]	650	2	89%
Our method		150	2	94%
		650	2	89% / 86.5%

We finally present a comparative overview of the results of recent studies on writer identification task on the IAM data set (Table 2). Although identification rates of as high as 98% have been reported, they are based on a smaller number of writers. Bulacu and Schomaker [7] currently hold the best results reading 89% on 650 writers and we have achieved the same identification rate with the proposed features. It is however, important to precise that we distinguish the training and test sets while in [7] the authors have used a leave-one-out approach on the entire data set. Thus, in order to present a true comparison we also carried out a similar experimentation and achieved an identification rate of 86.5%. We hope to improve the performance of the system by optimizing the selection of the proposed features.

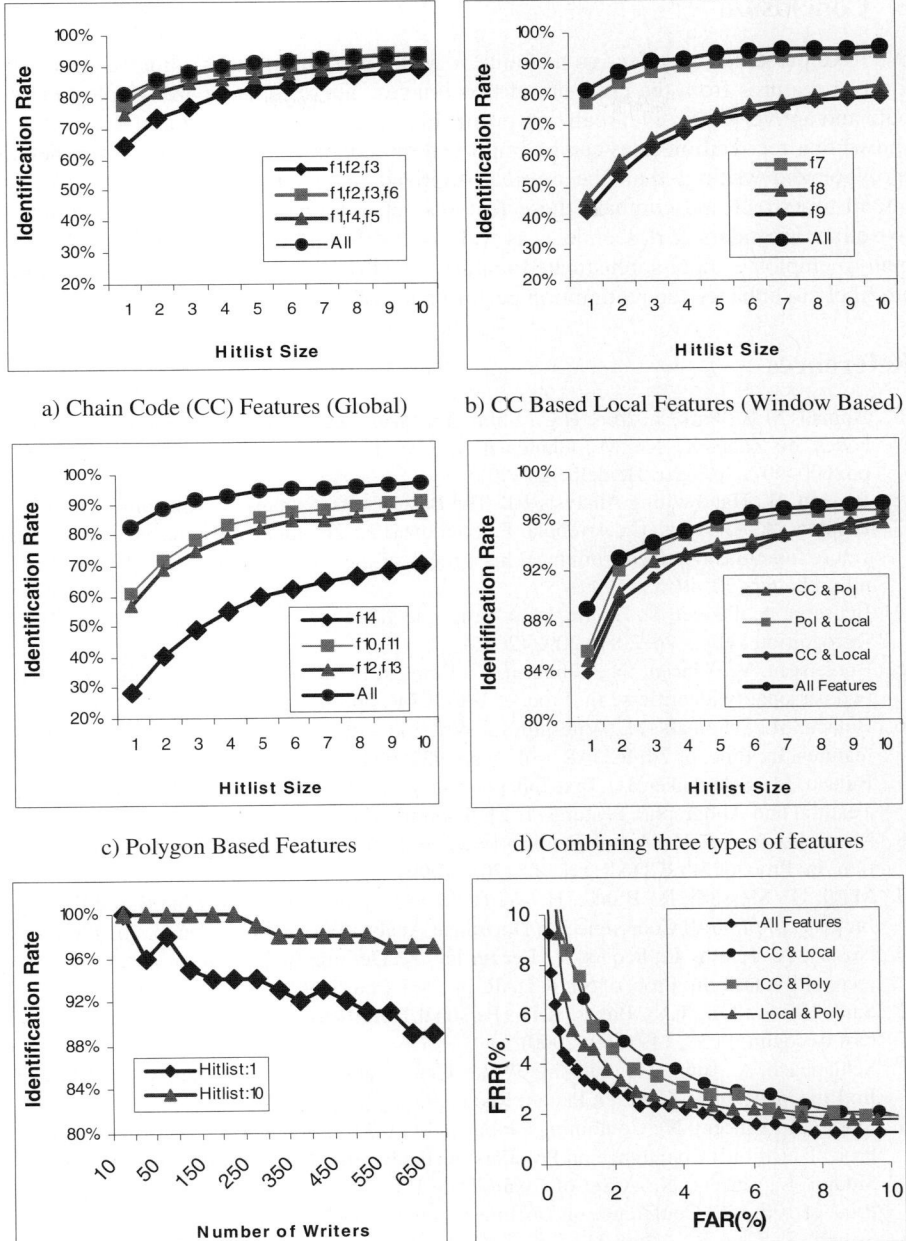

Fig. 5. Writer Identification and Verification Results

5 Conclusion

We have presented an effective method for writer recognition that relies on extracting a set of features from the contours of text images. These features are simple to compute and are very effective, realizing promising results on writer recognition. We have shown that recognition rates can be improved by modeling a vision with fewer details (polygonized writing) than the original digitized image. In our future research, we intend to extract and compare these features separately for different parts of words (baseline, ascenders & descenders) as well as at different image resolutions. We also plan to employ a more sophisticated feature selection mechanism and these additions are likely to enhance the recognition performance of the system.

References

1. Ahmad, M.B., Park, J., Min, H.C., Shim, Y.: Shape registration based on modified chain codes. In: Zhou, X., Xu, M., Jähnichen, S., Cao, J. (eds.) APPT 2003. LNCS, vol. 2834, pp. 600–607. Springer, Heidelberg (2003)
2. Baggett, B.: Handwriting Analysis 101-The Basic Traits. Empressé Publishing (2004)
3. Bandera, A., Urdiales, C., Arrebola, F., Sandoval, F.: 2D object recognition based on curvature functions obtained from local histograms of the contour chain code. Pattern Recognition Letters 20, 49–55 (1999)
4. Bensefia, A., Paquet, T., Heutte, L.: A writer identification and verification system. Pattern Recognition Letters 26, 2080–2092 (2005)
5. Bouletreau, V., Vincent, N., Sabourin, R., Emptoz, H.: Handwriting and signature: one or two personality identifiers? In: Proc. of 14th ICPR, pp. 1758–1760 (1998)
6. Bulacu, M., Schomaker, L., Vuurpijl, L.: Writer identification using edge-based directional features. In: Proc. of 7th ICDAR, vol. 1, pp. 937–994 (2003)
7. Bulacu, M., Schomaker, L.: Text-Independent Writer Identification and Verification Using Textural and Allographic Features. IEEE Transactions on PAMI, 701–717 (2007)
8. Marti, U., Bunke, H.: A full English sentence database for off-line handwriting recognition. In: Proc. of 5th ICDAR, pp. 705–708 (1999)
9. Marti, U., Messerli, R., Bunke, H.: Writer Identification Using Text Line Based Features. In: Proc. of 6th Int'l Conference on Document Analysis and Recognition, p. 101 (2001)
10. Nosary, A., Heutte, L., Paquet, T., Lecourtier, Y.: Defining writer's invariants to adapt the recognition task. In: Proc. of 5th ICDAR, pp. 765–768 (1999)
11. Said, H.E.S., Tan, T.N., Baker, K.D.: Personal Identification Based on Handwriting. Pattern Recognition 33, 149–160 (2000)
12. Schlapbach, A., Bunke, H.: Off-line Writer Identification Using Gaussian Mixture Models. In: Proc. of 18th Int. Conf. on Pattern Recognition, pp. 992–995 (2006)
13. Siddiqi, I., Vincent, N.: Combining Global and Local Features for Writer Identification. In: Proc of 11th Int'l Conference on Frontiers in Handwriting Recognition (2008)
14. Siddiqi, I., Vincent, N.: A Set of Chain Code Based Features for Writer Recognition. In: Proc. of 10th Int'l Conference on Document Analysis and Recognition (2009)
15. Srihari, S., Cha, S., Arora, H., Lee, S.: Individuality of handwriting. J. of Forensic Sciences 47(4), 1.17 (2002)
16. Wall, K., Danielsson, P.: A fast sequential method for polygonal approximation of digitized curves. Computer Vision, Graphics and Image Processing 28, 220–227 (1984)
17. Zois, E.N., Anastassopoulos, V.: Morphological waveform coding for writer identification. Pattern Recognition 33(3), 385–398 (2000)

A Novel Approach to Estimate Fractal Dimension from Closed Curves

André R. Backes[1], João B. Florindo[2], and Odemir M. Bruno[2]

[1] Instituto de Ciências Matemáticas e de Computação (ICMC)
Universidade de São Paulo (USP)
Avenida do Trabalhador São-carlense, 400
13560-970 São Carlos SP Brazil
backes@icmc.usp.br
[2] Instituto de Física de São Carlos (IFSC)
Universidade de São Paulo (USP)
Avenida do Trabalhador São-carlense, 400
13560-970 São Carlos SP Brazil
jbflorindo@ursa.ifsc.usp.br, bruno@ifsc.usp.br

Abstract. An important point in pattern recognition and image analysis is the study of properties of the shapes used to represent an object in an image. Particularly, an interesting measure of a shape is its level of complexity, a value that can be obtained from its fractal dimension. Many methods were developed for estimating the fractal dimensions of shapes but none of these are efficient for every situation. This work proposes a novel approach to estimate the fractal dimension from shape contour by using Curvature Scale Space (CSS). Efficiency of the technique in comparison to the well-known method of Bouligand-Minkowski. Results show that the use of CSS yields fractal dimension values robust to several shape transformations (such as rotation, scale and presence of noise), so providing interesting results for a process of classification of shapes based on this measure.

Keywords: shape analysis, curvature, complexity, fractal dimension.

1 Introduction

Shape plays an important role in pattern recognition and image analysis. It is the characteristic that provides the most relevant information about an object, thus allowing its characterization and identification. Literature presents many approaches to extract information from the geometric aspect of the shape [1,2] and one of these approaches is the complexity.

Complexity is a term that lacks of formal definition, although it is widely used in literature for many applications. In shape analysis, it is straight related to shape irregularity and space occupation level. One interesting approach to estimate shape complexity is using the Fractal dimension, a property from fractal objects, which describes, by using a fractionary value, how complex and self-similar an object is [3,4,5].

Another property straight connected to shape aspect is the curvature. Given a shape contour, curvature outputs a curve associated to concave and convex regions of the contour, i.e., maximum and minimum local points of the curve correspond to the direction changes in the shape contour [6]. Curvature is a property associated to the human perception of shapes, and it presents good tolerance to rotation and scale variations.

In this paper, we propose a novel approach to estimate the fractal dimension based on the curvature. Using Curvature Scale Space (CSS) [4,7,8] it is possible to analyze how shape changes along the scale, so providing an estimation of its complexity that is tolerance to shape transformations.

So, the paper starts with a brief review of complexity and fractal dimension (Section 2). The theory about Curvature Scale Space is presented in Section 3. In Section 4, we present our approach to estimate fractal dimension from the curvature of a shape contour. The proposed approach is compared with Bouligand-Minkowski fractal dimension method [5] and results are shown in Section 5. Finally, in Section 6, the conclusions and future works for the method are discussed.

2 Shape and Complexity

Since Mandelbrodt [9], fractal geometry has presented significant advances in the study and representation of shapes, which cannot be perfectly modeled by the Euclidean geometry, such as objects commonly found in the Nature.

A fractal can be defined as an object constructed by applying a simple construction rule iteratively over an original simple object. At each iteration, the original object is transformed (for example, copied and scaled) to become part of a more complex object [9,5]. This process generates fractal objects, which are self-similar at an infinite level. However, objects found in the Nature do not present self-similarity at an infinite level. However, they can be approximated by a fractal entity without a significant loss of information.

The fractal dimension is a very useful property from fractal objects. Unlike the Euclidean concept of integer dimension, fractal dimension is a real number expressing the level of self-similarity of a fractal object. The fractal dimension can also measure the spatial occupation and complexity of a shape, even when this shape is not a true fractal object, as the objects studied in this work [10]. In fact, the literature [9,5] shows that the larger the value of the fractal dimension is, the most complex is the shape.

3 Curvature Scale Space

Considering a curve as a parametric vector $C(t) = (x(u), y(u))$, curvature is defined in terms of derivative as

$$k(t) = \frac{x^{(1)}(t)y^{(2)}(t) - x^{(2)}(t)y^{(1)}(t)}{(x^{(1)}(t)^2 + y^{(1)}(t)^2)^{3/2}},$$

where $^{(1)}$ and $^{(2)}$ denotes the first and second derivatives, respectively.

Curvature is a powerful tool in shape analysis that allows the study of the behavior of a curve by its changes in orientation. However, in many situations it is interesting to analyze the curvature properties along the scales. These situations can be handled by applying a multi-scale transform over the original signal. A frequently used multi-scale transform is the space-scale transform [4,7,8], which employs the convolution of a Gaussian function $g(t,\sigma)$ over the signal to reduce the effects of noise and high frequency information before curvature measurement:

$$X(t,\sigma) = x(t) * g(t,\sigma),$$
$$Y(t,\sigma) = y(t) * g(t,\sigma),$$

where

$$g(t,\sigma) = \frac{1}{\sigma\sqrt{2\pi}} \exp^{\frac{-t^2}{2\sigma^2}},$$

is the Gaussian function with standard deviation σ. By using convolution properties, it is possible to compute derivatives components from Gaussian function, instead of the original signal:

$$X^{(1)}(t,\sigma) = x^{(1)}(t) * g(t,\sigma) = (x(t) * g(t,\sigma))^{(1)} = x(t) * g^{(1)}(t,\sigma),$$
$$X^{(2)}(t,\sigma) = x^{(2)}(t) * g(t,\sigma) = (x(t) * g(t,\sigma))^{(2)} = x(t) * g^{(2)}(t,\sigma).$$

In this way, it is possible to rewrite curvature equation considering the scale evolution as

$$k(t,\sigma) = \frac{X^{(1)}(t,\sigma)Y^{(2)}(t,\sigma) - X^{(2)}(t,\sigma)Y^{(1)}(t,\sigma)}{(X^{(1)}(t,\sigma)^2 + Y^{(1)}(t,\sigma)^2)^{3/2}},$$

By increasing linearly the value of σ, a set of curves is achieved. These curves compose an image named Curvature Scale Space (CSS), which is widely used to represent curves and shapes in different applications.

4 Proposed Approach

Through multi-scale transform, it is possible to analyze a curve along scale. Each scale is represented by a different smoothing level computed using the Gaussian function, which reduces the amount of high frequencies information and noise. As the smoothing level increases, the curve information is reduced and more similar to a circle the curve becomes (Figure 1). This process reflects in the complexity of the curve, so that, curvature scale space can be used to estimate the fractal dimension of a curve.

Considering the curvature of a given curve at scale σ, we define the amount of curvature as the sum of the module of all curvature coefficients as

$$S(\sigma) = \sum_{t} |k(t,\sigma)|$$

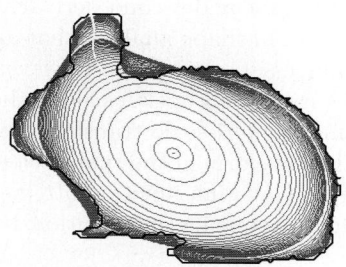

Fig. 1. Smoothing effect performed over a curve using different σ values

where $S(\sigma)$ represents the resulting orientation of the curve at scale σ.

As in CSS method, σ is linearly increased, so resulting in a set of $S(\sigma)$ values. By accumulating these values, it is possible to create a relation of dependence among scales, where posterior scales are influenced by previous scales. Thus, the accumulated amount of curvature is defined as

$$A(\sigma) = \sum_{i=0}^{\sigma} S(i).$$

It is possible to quantify the degree of space filling of a curve by considering its order of growth. By analyzing the behavior of $\sigma \times A(\sigma)$ curve in the logarithmic scale, we note that it obeys a power law. Literature [5] shows that this power law can be used to estimate the fractal dimension of the original curve:

$$D = \lim_{\sigma \to 0} \frac{\log A(\sigma)}{\log \sigma}.$$

From the log-log plot of $\sigma \times A(\sigma)$, it is possible to estimate a straight line with slope α, where $D = \alpha$ is the estimated fractal dimension for the closed curve $C(t)$.

5 Experiments

The proposed method was experimented with shapes under different types of transformations. Figure 2 shows the original set of shape images considered. This was performed, so that, properties such as rotation, scale and noise tolerance could be evaluated. We also compared results obtained by our proposed method with the Bouligand-Minkowski method, which is considered by literature one of the most accurate methods to estimate the fractal dimension from shapes [5].

5.1 Rotation and Scale Tolerance

Variations in size and orientation are a very common issue in image analysis applications. A robust method must be tolerant to these transformations in

A Novel Approach to Estimate Fractal Dimension from Closed Curves

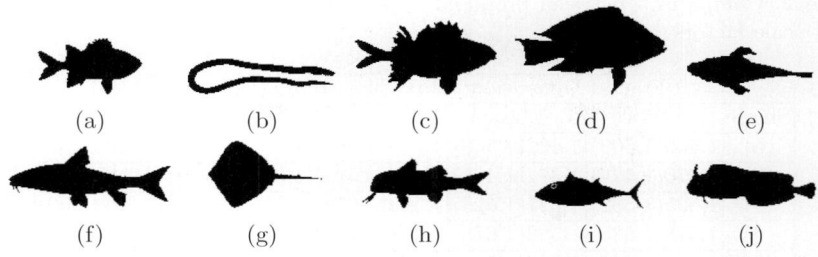

Fig. 2. Set of fish images considered during the experiments

order to characterize and identify an object. Thus, given an original image, different rotated and scaled versions of the image were generated and the fractal dimension computed using the proposed method and the Bouligand-Minkowski method. For this experiment, 5 rotation angles (7°, 35°, 132°, 201° and 298°) and 5 scales factor (1.25×, 1.50×, 1.75×, 2.00× and 2.25×) were used.

Table 1 presents the fractal dimension values computed for a given image under rotation transform. We note that, independently of the angle employed in the rotation transform, the estimated fractal dimension remains constant. This is most due to the fact that curvature is not sensitive to orientation changes in the shape. Curvature value is a measure of curve orientation at a specific point, i.e., the changes in the direction inside the curve. Bouligand-Minkowski method is also tolerant to rotation. This method uses the Euclidean distance to compute the influence area of the shape, a distance measure that is not affected by transformations in orientation.

In Table 2, we have the fractal dimension values as the scale factor increases. As we can see, the proposed approach is also tolerant to scales changes. The proposed approach uses the sum of the module of the curvature coefficients as the basis to estimate the shape complexity. Figure 3 shows that this sum does

Table 1. Comparison between proposed approach and Bouligand-Minkowski for different orientations of a shape

Shape	Fractal Dimension: Curvature (Bouligand-Minkowski)				
	7°	35°	132°	201°	298°
(a)	1.448(1.448)	1.452(1.448)	1.449(1.448)	1.445(1.447)	1.459(1.448)
(b)	1.372(1.374)	1.375(1.375)	1.366(1.374)	1.395(1.374)	1.376(1.374)
(c)	1.526(1.525)	1.529(1.525)	1.535(1.525)	1.530(1.525)	1.519(1.525)
(d)	1.577(1.579)	1.581(1.579)	1.573(1.578)	1.580(1.578)	1.582(1.579)
(e)	1.390(1.389)	1.395(1.389)	1.378(1.389)	1.382(1.388)	1.382(1.389)
(f)	1.445(1.446)	1.465(1.446)	1.445(1.447)	1.471(1.447)	1.447(1.447)
(g)	1.429(1.432)	1.422(1.434)	1.433(1.433)	1.432(1.433)	1.433(1.433)
(h)	1.404(1.403)	1.397(1.404)	1.399(1.403)	1.408(1.403)	1.392(1.403)
(i)	1.337(1.338)	1.354(1.339)	1.337(1.339)	1.342(1.338)	1.337(1.339)
(j)	1.474(1.471)	1.467(1.471)	1.463(1.471)	1.500(1.470)	1.453(1.471)

Table 2. Comparison between proposed approach and Bouligand-Minkowski for different scale factors

Shape	Fractal Dimension: Curvature (Bouligand-Minkowski)				
	1.25×	1.50×	1.75×	2.00×	2.25×
(a)	1.459(1.509)	1.452(1.558)	1.455(1.596)	1.458(1.626)	1.459(1.652)
(b)	1.370(1.424)	1.374(1.465)	1.370(1.494)	1.374(1.514)	1.372(1.531)
(c)	1.525(1.582)	1.527(1.625)	1.522(1.659)	1.519(1.686)	1.522(1.709)
(d)	1.578(1.635)	1.582(1.678)	1.578(1.710)	1.580(1.737)	1.579(1.758)
(e)	1.390(1.457)	1.380(1.510)	1.377(1.552)	1.378(1.587)	1.385(1.616)
(f)	1.448(1.507)	1.440(1.553)	1.434(1.591)	1.436(1.623)	1.442(1.649)
(g)	1.406(1.504)	1.442(1.556)	1.444(1.598)	1.423(1.634)	1.413(1.663)
(h)	1.400(1.466)	1.398(1.516)	1.390(1.556)	1.388(1.590)	1.394(1.619)
(i)	1.339(1.407)	1.336(1.461)	1.332(1.504)	1.335(1.540)	1.337(1.571)
(j)	1.465(1.536)	1.466(1.586)	1.455(1.625)	1.449(1.656)	1.458(1.682)

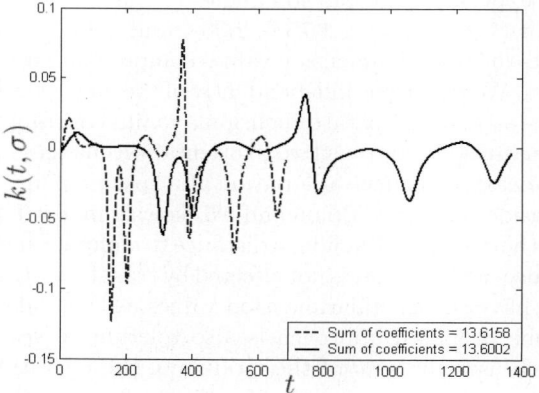

Fig. 3. Curvature and the sum of its coefficients for a contour and its scaled version (2.00×)

not change as the size of the object changes. As the size of the shape increases, also increases the number of points in its contour. However, the relation among minimum and maximum local points and the direction changes in the contour is preserved, so as the estimated fractal dimension. The same is not valid for the Bouligand-Minkowski method, where the influence area depends on a dilation radius which depends on the shape size.

5.2 Noise Tolerance

At first, different noise levels were added to a given a contour shape. The noise consisted of a random signal with the same length of the contour and whose values range from $[-d, +d]$, where d is an integer number which defines the noise intensity (Figure 4).

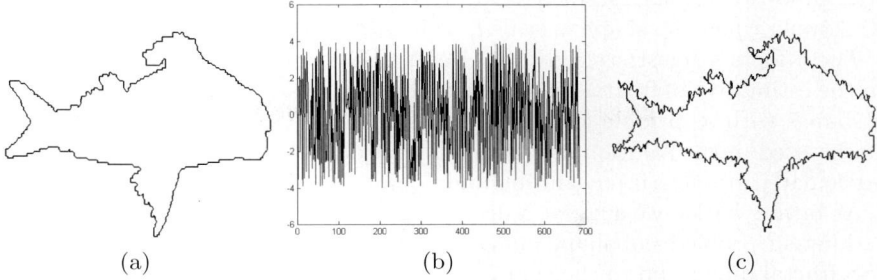

Fig. 4. Process of noisy shapes generation: (a) Original contour; (b) Random signal ranging from $[-d, +d]$, for $d = 4$; (c) Resulting noisy contour

Table 3. Comparison between proposed approach and Bouligand-Minkowski for different noise levels

Shape	Fractal Dimension: Curvature (Bouligand-Minkowski)				
	$d=0$	$d=1$	$d=2$	$d=3$	$d=4$
(a)	1.458(1.447)	1.458(1.449)	1.458(1.449)	1.462(1.453)	1.456(1.456)
(b)	1.374(1.372)	1.372(1.378)	1.377(1.378)	1.367(1.383)	1.374(1.394)
(c)	1.519(1.525)	1.520(1.526)	1.517(1.528)	1.516(1.531)	1.528(1.532)
(d)	1.580(1.578)	1.579(1.577)	1.580(1.579)	1.584(1.582)	1.576(1.583)
(e)	1.378(1.388)	1.379(1.390)	1.377(1.392)	1.381(1.394)	1.382(1.400)
(f)	1.436(1.446)	1.432(1.447)	1.442(1.449)	1.426(1.453)	1.450(1.459)
(g)	1.423(1.433)	1.437(1.434)	1.434(1.436)	1.456(1.434)	1.360(1.438)
(h)	1.388(1.403)	1.388(1.404)	1.390(1.408)	1.392(1.413)	1.392(1.416)
(i)	1.336(1.338)	1.342(1.341)	1.338(1.345)	1.352(1.350)	1.337(1.351)
(j)	1.450(1.471)	1.444(1.472)	1.455(1.472)	1.453(1.476)	1.472(1.478)

Table 3 shows the estimated fractal dimension as the noise levels increases. Four different noise levels, $d = \{1, 2, 3, 4\}$, were considered for this experiment and $d = 0$ represents the original shape (without noise). Results show that proposed method presents a good noise tolerance in comparison to the Bouligand-Minkowski method. Most of its tolerance is due to the use of a Gaussian filter during the calculus of the CSS, which acts reducing the importance of this type of information as the smoothing level increases. On the other hand, high noise levels may change the geometric patterns in the shape and the smoothing process is not capable of correcting such changes. This explains the small changes in the fractal dimension value.

6 Conclusion

From the results, we note that the estimative of fractal dimension by CSS shows a good precision when in comparison to the well-known method of

Bouligand-Minkowski. Besides, CSS presents some advantages over Bouligand-Minkowski when the shape is scaled, as it commonly occurs in real situations.

These facts suggest strongly the use of CSS as an interesting new approach for the estimative of fractal dimension and its application to the characterization of shapes with a variable level of complexity and aspect. Such characterization can be used in the solution of a large amount of problems in Computer Science, particularly in pattern recognition and computer vision.

As future works, we suggest a deeper study of CSS fractal dimension applied to different problems of shape analysis and characterization. One can apply the CSS fractal dimension to the characterization of the shape used, for instance, in a classification process. Several data sets can be classified by this technique and the results can be analyzed.

Acknowledgments

A.R.B. acknowledges support from FAPESP (2006/54367-9). J.B.F. acknowledges support from CNPq (870336/1997-5) and FAPESP (2006 / 53959-0). O.M.B. acknowledges support from CNPq (306628 / 2007-4).

References

1. Loncaric, S.: A survey of shape analysis techniques. Pattern Recognition 31(9), 983–1001 (1998)
2. da, R., Torres, S., Falcão, A.X., da, L., Costa, F.: A graph-based approach for multiscale shape analysis. Pattern Recognition 37, 1163–1174 (2003)
3. Backes, A.R., Bruno, O.M.: A new approach to estimate fractal dimension of texture images. In: Elmoataz, A., Lezoray, O., Nouboud, F., Mammass, D. (eds.) ICISP 2008. LNCS, vol. 5099, pp. 136–143. Springer, Heidelberg (2008)
4. da, L., Costa, F., Cesar Jr., R.M.: Shape Analysis and Classification: Theory and Practice. CRC Press, Boca Raton (2000)
5. Tricot, C.: Curves and Fractal Dimension. Springer, Heidelberg (1995)
6. Wu, W.Y., Wang, M.J.: On detecting the dominant points by the curvature-based polygonal approximation. CVGIP: Graphical Models Image Process 55, 79–88 (1993)
7. Witkin, A.P.: Scale space filtering: a new approach to multi-scale descriptions. In: ICASSP - IEEE International Conference on Acoustics, Speech, and Signal Processing, GRETSI, pp. 79–95 (2003)
8. Mokhtarian, F., Abbasi, S.: Matching shapes with self-intersections: application to leaf classification. IEEE Transactions on Image Processing 13(5) (2004)
9. Mandelbrot, B.B.: The Fractal geometry of nature. Freeman, New York (1968)
10. Carlin, M.: Measuring the complexity of non-fractal shapes by a fractal method. PRL: Pattern Recognition Letters 21(11), 1013–1017 (2000)

Saliency Based on Decorrelation and Distinctiveness of Local Responses

Antón Garcia-Diaz, Xosé R. Fdez-Vidal, Xosé M. Pardo, and Raquel Dosil

Universidade de Santiago de Compostela, Grupo de Visión Artificial,
Departamento de Electrónica e Computación, Campus Sur s/n,
15782 Santiago de Compostela, Spain
{anton.garcia,xose.vidal,xose.pardo,raquel.dosil}@usc.es

Abstract. In this paper we validate a new model of bottom-up saliency based in the decorrelation and the distinctiveness of local responses. The model is simple and light, and is based on biologically plausible mechanisms. Decorrelation is achieved by applying principal components analysis over a set of multiscale low level features. Distinctiveness is measured using the Hotelling's T^2 statistic. The presented approach provides a suitable framework for the incorporation of top-down processes like contextual priors, but also learning and recognition. We show its capability of reproducing human fixations on an open access image dataset and we compare it with other recently proposed models of the state of the art.

Keywords: saliency, bottom-up, attention, eye-fixations.

1 Introduction

It is well known, from the analysis of visual search problems, that vision processes have to face a huge computational complexity [1] . The Human Visual System (HVS) tackles this challenge through the selection of information with several mechanisms, starting from foveation. In the basis of this selection is the visual attention, including its data-driven component leading to the so called bottom-up saliency. In the last decades, the interest in the understanding of saliency mechanisms and the appraisal of its relative importance in relation to the top-down (knowledge-based) relevance, has constantly raised. Besides, attention models can facilitate a solution of technical problems, ranging from robotics navigation [2] to image compression or object recognition [3].

Recently, several approaches to bottom-up saliency have been proposed based on similarity and local information measures. In these models local distinctiveness is obtained either from self-information [4][5], mutual information [6][7], or from dissimilarity [8], using different decomposition and competition schemes.

In this paper, we propose a new model of bottom-up saliency, simple and with low computational complexity. To achieve this, we take into account the decorrelation of neural responses when considering the behavior of a population of neurons subject to stimuli of a natural image [9]. This is believed to be closely related to the important role of non classical receptive fields (NCRF) in the functioning of HVS. Therefore, we start from a multiscale decomposition on two feature dimensions: local orientation energy and color. We obtain the decorrelated responses applying PCA to the multiscale features. Then, we measure the statistical distance of each feature to the center of the distribution as the Hotelling's T^2 distance. Finally, we apply normalization and Gaussian smoothing to gain robustness. The resulting maps are firstly summed, delivering local energy and color conspicuities, and then they are normalized and averaged, producing the final saliency map.

In order to achieve a psychophysical validation, most models of bottom-up saliency assess their performance in predicting human fixations, and compare their results with those provided by other previous models. The most frequent comparison method consists in the use of the receiver-operator-curve (ROC) and the corresponding value of the area under the curve (AUC), as a measure of predictive power [5][6][8]. The use of Kullback-Leibler (K-L) divergence to compare priority and saliency maps is also found on related literature [10]. From the use of both methods, we obtain results that match or improve those achieved with models of the state of the art. Moreover, ruling out the use of center-surround differences we definitely improve the results respect to those obtained with a previous proposal [11].

The paper is developed as follows. Section 2 is devoted to overview the visual attention model. In Section 3 we present and discuss the experimental work carried out, and the achieved results. Finally, Section 4 summarizes the paper.

2 Model

Our model takes as input a color image codified using the Lab color model. Unlike other implementations of saliency [6][12] this election is based on a widely used psychophysical standard. We decompose the luminance image by means of a Gabor-like bank of filters, in agreement with the standard model of V1. Since orientation selectivity is very weakly associated with color selectivity, the opponent color components a and b simply undergo a multiscale decomposition. Hence, we employ two feature dimensions: color and local energy. By decorrelating the multiscale responses, extracting from them a local measure of variability, and further performing a local averaging, we obtain a unified and efficient measure of saliency.

2.1 Local Energy and Color Maps

Local energy is extracted applying a bank of log Gabor filters [13] to the luminance component. In the frequency domain, the log Gabor function takes the expression:

$$logGabor(\rho, \alpha; \rho_i, \alpha_i) = e^{-\frac{(\log(\rho/\rho_i))^2}{2(\log(\sigma_{\rho i}/\rho_i))^2}} e^{-\frac{(\alpha-\alpha_i)^2}{2(\sigma_\alpha)^2}}. \tag{1}$$

where (ρ, α) are polar frequency coordinates and (ρ_i, α_i) is the central frequency of the filter. Log Gabor filters, unlike Gabor, have no DC or negative frequency components, therefore avoiding artifacts. Their long tail towards the high frequencies improves its localization. In the spatial domain, they are complex valued functions (with no analytical expression), whose components are a pair of filters in phase quadrature, f and h. Thus for each scale s and orientation o, we obtain a complex response. Its modulus is a measure of the local energy of the input associated to the corresponding frequency band [14] [15] .

$$e_{so}(x,y) = \sqrt{(L * f_{so})^2 + (L * h_{so})^2}. \qquad (2)$$

Regarding the color dimension, we obtain a multiscale representation both for a and b, from the responses to a bank of log Gaussian filters.

$$logGauss(\rho) = e^{-\frac{(\log(\rho))^2}{2(\log(2^n \sigma))^2}}. \qquad (3)$$

Thus, for each scale and color opponent component we get a real valued response. The parameters used here were: 8 scales spaced by one octave, 4 orientations (for local energy), minimum wavelength of 2 pixels, angular standard deviation of $\sigma_\alpha = 37.5°$, and a frequency bandwidth of 2 octaves.

2.2 Measurement of Distinctiveness

Observations from neurobiology show decorrelation of neural responses, as well as an increased population sparseness in comparison to what can be expected from a standard Gabor-like representation [16]. Accordingly we decorrelate the multiscale information of each sub-feature (orientations and color components) through a PCA on the corresponding set of scales. On the other hand, variability and richness of structural content have been proven as driving attention [17]. Therefore, we have chosen a measure of distance between local and global structure to represent distinctiveness. Once scales are decorrelated, we extract the statistical distance at each point as the Hotelling's T^2 statistic:

$$\mathbf{X} = \begin{pmatrix} x_{11} & \cdots & x_{1N} \\ \vdots & \vdots & \vdots \\ x_{S1} & \cdots & x_{SN} \end{pmatrix} \rightarrow (PCA) \rightarrow \mathbf{T}^2 = \left(T_1^2, \cdots, T_N^2\right). \qquad (4)$$

That is, being S the number of scales (original coordinates) and N the number of pixels (samples), we compute the statistical distance of each pixel (sample) in the decorrelated coordinates. Given the covariance matrix ($\mathbf{W}$), $\mathbf{T}^2$ is defined as:

$$T_i^2 = (\mathbf{x_i} - \bar{\mathbf{x}})'\mathbf{W}^{-1}(\mathbf{x_i} - \bar{\mathbf{x}}). \qquad (5)$$

This is the key point of our approach to the integration process. This multivariate measure of the distance from a feature vector associated to a point in the image, to the average feature vector of the global scene, is in fact, a measure of the local feature contrast [18].

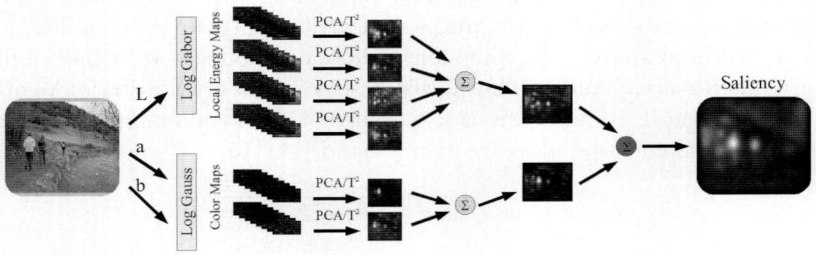

Fig. 1. Bottom-up Saliency Model

Final Map. The final saliency map is obtained normalizing to [0,1], smoothing, and summing the extracted maps, first within each feature dimension and next with the resulting local energy conspicuity and color conspicuity maps. In this way we obtain a unique measure of saliency for each point of the image.

Computational Complexity. The whole process involves two kinds of operations. Firstly, filtering for decomposition and smoothing, has been realized in the frequency domain, as the product of the transfer functions of the input and the filters, using the Fast Fourier Transform (FFT) and its inverse (IFFT). This implies a computational complexity of $O(N \log(N) + N)$, being N the number of pixels of the image. The other operation is PCA with a complexity of $O(S^3 + S^2 N)$, being S the number of scales (dimensionality) and N the number of pixels (samples) [19]. We are interested in the dependency on the number of pixels, being $O(N)$, since the number of scales remains constant. Therefore, the overall complexity is given by $O(N \log(N))$.

3 Experimental Results and Discussion

In this work we demonstrate the efficiency of the model predicting eye fixations in natural images. In Section 3.1 we show the performance of the model in terms of AUC values from ROC analysis, on a public image dataset. We compare these results with those obtained by other models representative of the state of the art. Moreover, we discuss the details of this procedure, as well as the difficulties and limitations that it poses. This last issue motivates Section 3.2, where we use a metric based on the K-L divergence.

3.1 ROC Values

In this experiment we employ an image dataset published by Bruce & Tsotsos. It is made up of 120 images, and the corresponding fixation data for 20 different subjects. A detailed description of the eye-tracking experiment can be found in [4].

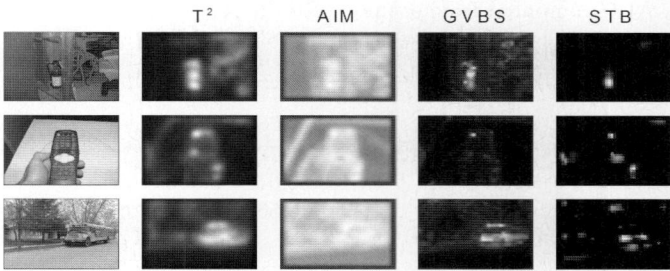

Fig. 2. Results of the model (T^2) with three of the images used. Results for other models, have been obtained with the corresponding open access version.

Table 1. AUC average values from computed like in [6]. ([*]published by the authors)

Model	AUC	std
T^2-Based	0.791	0.080
Gao et al. [6][*]	0.769	---
GBVS [8]	0.688	0.119

The details of AUC computation from ROC analysis face a difficulty from the beginning. Bruce & Tsotsos construct one unique ROC curve for the complete set of images, with the corresponding AUC. The uncertainty is provided, based on the proposal of Cortes & Mohri [20]. They give a value for their model and the model of Itti et al. [12]. On the other hand Gao et al. compare these results with those obtained by their model, but with another procedure. They construct one ROC curve and extract the corresponding AUC for each image. They take the average of the AUC as the overall value, but they don't provide any estimation of uncertainty. The same averaging procedure is employed by Harel et al. but on a different set of images [8].

When computing AUC with the average procedure we find a problem: standard deviation is larger than the differences between models, although these differences can be also large. This is reflected in table 1. The model of Gao et al. should have a similar value of standard deviation, since partial graphical values are similar to the models of Bruce & Tsotsos and Itti et al. [6].

Instead, we can proceed like Bruce & Tsotsos, who obtain a much tighter value for a 95% uncertainty, while the AUC is very similar (slightly lower). Thus, this would make it possible to rank the models by their overall behavior on the whole dataset. Results are shown in table 2. Our model has equivalent results to Bruce & Tsotsos, improving the performance of all of the other models.

However, this approach hides a problem. We are analyzing all the scenes as a unique sample, instead of considering each scene separately. Hence, the approach of Bruce & Tsotsos means to loose the inter-scene variance, performing a global assessment. Then, a question arises: does the kind of scene affect the ranking of models?. That is, could there be scene-biased models? If the kind of scene is

Table 2. AUC computed from a unique ROC curve (*values provided by [5])

Model	AUC	std
T^2-Based	0.776	0.008
AIM [9]*	0.781	0.008
Itti et al. [12]*	0.728	0.008
GBVS [8]	0.675	0.008
STB [21]	0.569	0.011

important, probably this dataset is not representative enough of natural images. In fact, urban and man-made scenes are clearly predominant. For instance, there is no landscape, and there is only one image with an animal (but in an indoor environment).

This fact could help to explain the results reported by Harel et al. [8], that show a higher (and excellent) performance compared to models of Bruce & Tsotsos and Itti et al., using a different image dataset. We must notice here that these other images were gray-level (without color information), and were mainly images of plants.

3.2 K-L Divergence Values

In this Section we employ the K-L divergence to compare priority maps from fixations with saliency maps, similarly to [10]. As priority maps we use the density maps computed by Bruce & Tsotsos to reflect the foveated region with each fixation.

The priority map can be interpreted as a measure of the probability of each point to attract gaze, and the saliency map can be viewed, in turn, as a prediction of that probability. Hence, it makes sense to compare both distributions through the K-L divergence. It is worth noting that, instead of gray levels probabilities [5], we compare distributions of probabilities in the space.

Table 3. K-L comparison. Other models have been executed using their default values.

Model	K-L	std
T^2-Based	1.3	0.3
AIM [9]	1.7	0.3
GBVS [8]	2.1	0.6
STB [21]	13.0	1.6

With this aim, we obtain probability maps simply dividing a given map by the sum of its gray-level values. We denote by $h_i = h(x,y)$ and $m_i = m(x,y)$ the priority map from fixations and the saliency map from each model (taken as probability distributions) respectively. Then, being N the number of pixels, we compute the K-L divergences:

$$D_{KL}(h,m) = \sum_{i=1}^{N} h_i \cdot \log \frac{h_i}{m_i} \quad (6)$$

It is possible to use other procedures to construct the priority maps, to take into account other parameters like the duration of fixations, and not merely their positions [22]. Therefore, this metric should be viewed as an interesting complement to ROC analysis.

Nevertheless, the results shown in table 2, lead to a similar interpretation to that derived from analysis of ROC curves. Standard deviations are similar in relative order of magnitude. Our model exhibits an slightly better performance than the model of Bruce & Tsotsos [5], and again clearly better than the model proposed by Harel et al [8]. The implementation proposed by Walther [21] of the model of Itti & Koch [23] exhibits again the worst result.

4 Conclusions

In this work we have shown a simple and light model of saliency, that resorts to the decorrelation of the responses to a Gabor-like bank of filters. This mechanism is biologically plausible and could have an important role in the influence of NCRF when V1 cells are subjected to natural stimuli [9][16].

We have validated the model, comparing its performance with others, in the prediction of human fixations. Using ROC analysis we have obtained a result equivalent to that achieved by Bruce & Tsotsos [5], after the optimization of their decomposition process. With the same model in a previous work they obtained an AUC value of 0.7288 [4], clearly lower. On the other hand, using a different metric based on K-L divergence, that takes into account the area of foveation of each fixation, the model performs slightly better than the approach of Bruce & Tsotsos. Other models [6][8][21] deliver worse results in both comparisons.

Similarly to Bruce & Tsotsos [5], we avoid any parameterization of the process, beyond the initial decomposition of the image. However, this decomposition remains ordered and suitable for the incorporation, from the beginning, of top-down influences. Finally, our model of saliency presents a lower computational complexity than models that are benchmarks for psychophysical plausibility.

Acknowledgments. This work has been granted by the Ministry of Education and Science of Spain (project AVISTA TIN2006-08447), and the Government of Galicia (project PGIDIT07PXIB206028PR).

References

1. Tsotsos, J.K.: Computational foundations for attentive Processes. In: Itti, L., Rees, G., Tsotsos, J.K. (eds.) Neurobiology of Attention, pp. 3–7. Elsevier, Amsterdam (2005)
2. Frintrop, S., Jensfelt, P.: Attentional Landmarks and Active Gaze Control for Visual SLAM. IEEE Transactions on Robotics, Special Issue on Visual SLAM 24(5) (2008)
3. Harel, J., Koch, C.: On the Optimality of Spatial Attention for Object Detection, Attention in Cognitive Systems. In: WAPCV (2008)

4. Bruce, N., Tsotsos, J.K.: Saliency Based on Information Maximization. In: NIPS, vol. 18, pp. 155–162 (2006)
5. Bruce, N., Tsotsos, J.K.: Saliency, attention, and visual search: An information theoretic approach. Journal of Vision 9(3), 1–24 (2009)
6. Gao, D., Mahadevan, V., Vasconcelos, N.: On the plausibility of the discriminant center-surround hypothesis for visual saliency. Journal of Vision 8(7), 13, 1–18 (2008)
7. Gao, D., Mahadevan, V., Vasconcelos, N.: The discriminant center-surround hypothesis for bottom-up saliency. In: NIPS (2007)
8. Harel, J., Koch, C., Perona, P.: Graph-Based Visual Saliency. In: NIPS, vol. 19, pp. 545–552 (2007)
9. Olshausen, B.A., Field, D.J.: How Close Are We to Understanding V1? Neural Computation 17, 1665–1699 (2005)
10. Le Meur, O., Le Callet, P., Barba, D., Thoreau, D.: A coherent computational approach to model bottom-up visual attention. IEEE Transactions on Pattern Analysis and Machine Intelligence 28, 802–817 (2006)
11. Garcia-Diaz, A., Fdez-Vidal, X.R., Pardo, X.M., Dosil, R.: Local energy variability as a generic measure of bottom-up salience. In: Yin, P.-Y. (ed.) Pattern Recognition Techniques, Technology and Applications, In-Teh, Vienna, pp. 1–24 (2008)
12. Itti, L., Koch, C., Niebur, E.: A model of saliency-based visual attention for rapid scene analysis. IEEE Transactions on Pattern Analysis and Machine Intelligence 20(11), 1254–1259 (1998)
13. Field, D.J.: Relations Between the Statistics of Natural Images and the Response Properties of Cortical Cells. Journal of the Optical Society of America A 4(12), 2379–2394 (1987)
14. Kovesi, P.: Invariant Measures of Image Features from Phase Information. Ph.D. Thesis, The University or Western Australia (1996)
15. Morrone, M.C., Burr, D.C.: Feature Detection in Human Vision: A Phase-Dependent Energy Model. Proceedings of the Royal Society of London B 235, 221–245 (1988)
16. Vinje, W.E., Gallant, J.L.: Sparse coding and decorrelation in primary visual cortex during natural vision. Science 287, 1273–1276 (2000)
17. Zetzsche, C.: Natural Scene Statistics and Salient Visual Features. In: Itti, L., Rees, G., Tsotsos, J.K. (eds.) Neurobiology of Attention, pp. 226–232. Elsevier, Amsterdam (2005)
18. Nothdurft, H.C.: Salience of Feature Contrast. In: Itti, L., Rees, G., Tsotsos, J.K. (eds.) Neurobiology of Attention, pp. 233–239. Elsevier, Amsterdam (2005)
19. Sharma, A., Paliwal, K.K.: Fast principal component analysis using fixed-point algorithm. Pattern Recognition Letters 28, 1151–1155 (2007)
20. Cortes, C., Mohri, M.: Confidence intervals for the area under the ROC curve. In: NIPS, vol. 17, p. 305 (2005)
21. Walther, D., Koch, C.: Modeling attention to salient proto-objects. Neural Networks 19, 1395–1407 (2006)
22. Ouerhani, N.: Visual Attention: Form Bio-Inspired Modelling to Real-Time Implementation, PhD thesis, University of Neuchatel (2004)
23. Itti, L., Koch, C.: A saliency-based search mechanism for overt and covert shifts of visual attention. Vision Research 40, 1489–1506 (2000)

Incorporating Shape Features in an Appearance-Based Object Detection System

Gurman Gill and Martin Levine

Center for Intelligent Machines
Dept. of Electrical and Computer Engineering
McGill University, Montreal, Canada

Abstract. Most object detection techniques discussed in the literature are based solely on texture-based features that capture the global or local appearance of an object. While results indicate their ability to effectively represent an object class, these features can be detected repeatably only in the object interior, and so cannot effectively exploit the powerful recognition cue of contour. Since generic object classes can be characterized by shape and appearance, this paper has formulated a method to combine these attributes to enhance the object model. We present an approach for incorporating the recently introduced shape-based features called k-Adjacent-Segments (kAS) in our appearance-based framework based on *dense* SIFT features. Class-specific kAS features are detected in an arbitrary image to form a shape map that is then employed in two novel ways to augment the appearance-based technique. This is shown to improve the detection performance for all classes in the challenging 3D dataset by 3-18% and the PASCAL VOC 2006 by 5%.

1 Introduction

Appearance-based techniques for object detection have been very popular in the literature because many objects are well represented by their texture. Texture is typically captured using local features obtained by extracting image patches either randomly [1], from a dense grid of image locations [2] or by employing a salient interest point detector [3][4]. Certain objects, however, are better characterized by their *shape* rather than their appearance. Recently, efforts have been made to develop shape-based features that exploit contour cues and represent the underlying object using these contours [5][6][7][8]. Shotton et al. [5] and Opelt et al. [6] independently proposed to construct contour fragments tailored to a specific class. The idea is to extract class-discriminative boundary fragments and use the object's boundary to detect object regions in an image. [7] introduced a class of shape-based salient regions that captures local shape convexities in the form of circular arcs of edgels. Ferrari et al. presented a family of local contour features that are obtained by connecting k adjacent segments (kAS) detected on a *contour segment network* (*CSN*) of the image [8].

Recently, the complementary information provided by shape-based and appearance-based features has been combined into a common framework [9][10][11]. This permits recognition of generic object classes that could be characterized by shape, appearance or both. In [9], a shape context operator applied to points derived from the Canny edge detector is used to describe the global shape of an object while Difference-of-Gaussian [12] features described using PCA-SIFT represent the local features. Opelt et al. [10] extend their boundary fragment model [6] by adding the appearance patches and learning the best features from a pool of patches and boundaries. Similarly, Shotton et al. [11] combine their contour fragments [5] with texton features obtained by clustering the response of training images to a bank of filters.

This paper presents an approach to incorporate shape-based features (called k-Adjacent-Segments [8]) within our appearance-based framework based on *densely sampled* SIFT features [2]. The aim is to exploit shape cues to capture variations in generic object classes and strengthen the object model.

The main idea is to determine those kAS features that are *discriminative* with respect to a given object class. These features can then be detected in an arbitrary image to produce a *shape map* that indicates the locations containing class-specific shape information. This paper describes two novel ways of exploiting the shape map to enhance the representational power of the appearance-based object model. The first overlays the shape map onto the image to augment the edges on the object and uses the resulting edge-augmented image to compute SIFT descriptors. The second assigns a *shape* score to each detected window which measures how closely the shape of the detected object matches that of the actual object.

The paper begins with a brief description of our appearance-based object model [2] in section 2. Section 3 describes the method for generating the shape map and the two methods for employing it in conjunction with the appearance-based approach. In section 4, the shape-based augmentation is evaluated on challenging image databases 3D dataset [4] and two classes from PASCAL VOC 2006 [13]. Lastly, section 5 presents the conclusions.

2 Appearance-Based Object Model

In this section, we briefly describe our appearance-based object model [2] that is created using training images from both the object and background classes. In this model, the object is divided into different spatial regions, where each one is associated with an object part. The appearance of each object part is represented by a dense set of local features that are described using a SIFT descriptor [12].

The distribution of these features across the training images is then described in a lower dimensional space using *supervised* Locally Linear Embedding [14]. Each object part (spatial region) is essentially represented by a spatial cluster in the embedding space. Additionally, spatial relationships between object parts are established and used during the detection stage to localize instances of the object class in a novel image.

The unique aspect of this model is that all object parts *and* the background class are created simultaneously in the *same* underlying lower dimensional space. Spatial clusters represent the former while the latter is represented by a single background cluster in the embedding space.

Our approach obtains viewpoint invariance by dividing the view-sphere into a discrete number of view segments. Several spatial clusters represent the object in each view segment. All spatial clusters can be represented in just a single embedding space or, alternatively, multiple view models can be constructed by representing spatial clusters of different views in multiple embedding spaces.

The object model can be applied to an image containing an arbitrary amount of background clutter to detect instances of the object class present in any view. The detection scheme is based on hypothesizing object parts in a test image and then finding those that are spatially consistent to estimate the location of the object. Experiments on the challenging 3D dataset [4] and PASCAL VOC 2006 [13] produced encouraging results in [2]. They also indicated that the performance based on multiple embedding spaces is always superior to that based on a single embedding space. In this paper, we report results based on multiple embedding spaces only.

3   Incorporating Shape-Based Features

In this section, we discuss incorporating a family of local contour features called k-Adjacent-Segments (kAS) [8] into our appearance-based framework [2]. kAS consists of k approximately straight contour segments that mainly cover object boundary without including nearby edge clutter. Each kAS has a well-defined location and scale and is described by a low dimensional translation and scale invariant descriptor. The descriptor encodes the geometric properties of the segments and of their spatial arrangement.

The aim of this paper is to augment the appearance-based local SIFT features with these local contour features. We accomplish this primarily by finding *class-specific* kAS: those kAS features that occur more frequently on a particular object than elsewhere. We next present the method for finding *class-specific* kAS and then describe two novel ways in which these features are incorporated in the appearance-based framework.

The methodology adopted to find *class-specific* kAS follows the method of Vidal et al. for finding *informative* image patches called fragments [1]. A kAS feature, described by feature vector f, is said to be detected in an image I if the distance $D(f, I)$ of the feature vector f to the kAS feature vectors in the image I is below a certain threshold θ. In contrast to using mutual information (as in [1]), this paper uses the likelihood ratio $\mathbb{L}(\theta)$ [3] to measure the *discriminability* of the kAS feature vector f, given threshold θ. Essentially, $\mathbb{L}(\theta)$ represents the ratio of the frequency of detecting feature f in object training images to the frequency of detecting it in background training images.

The maximum likelihood value $M(f) = max(\mathbb{L}(\theta))$ is the information score associated with the feature f. The larger this value, the more discriminative is the feature. The detection threshold $\theta_f = arg\{max_\theta(\mathbb{L}(\theta))\}$ for feature f is computed so as to maximize the discriminability of the feature with respect to the class. This is the maximum distance D the feature can have within an image for it to be detected.

According to the procedure outlined above, all kAS features in each object training image are assigned an information score. We regard all features whose information score is above a certain threshold θ_I as discriminative features.

Discriminative features are usually similar across the object training images. These are therefore clustered using a clique partitioning algorithm [8] to yield *representative* discriminative kAS (RDkAS) for the object class. Note that RDkAS are computed for

(a) Shape map showing the segments constituting the detected RD3AS features

(b) The test image is edge-augmented with the segments in the shape map

Fig. 1. Example of detecting RDkAS on a test image of a car viewed from the back side. Some features in the background are also detected. (Best viewed with magnification)

each view-segment separately and then pooled to form a set of discriminative features representing the object across all views.

Based on the detection threshold θ_f, each RDkAS is tested for detection in a given image. Typically, a small fraction of the RDkAS is detected. Recall that each RDkAS is a collection of k segments (by definition of kAS). Figure 1(a) shows the shape map produced by the segments constituting the detected RDkAS. The shape map shows the segments displayed according to their *strength* (all segments are associated with the average strength of the edgels forming the segment [8]). We use the shape map in *two* novel ways as discussed in the next section.

3.1 Employing Shape Map

The first method for employing the shape map simply overlays the detected segments onto the original image so as to emphasize the edge information. Note that in this method all segments are overlaid in black, regardless of their strength. This is done to highlight edge information independent of contrast in the image. The SIFT descriptors are then computed over the edge-augmented image (figure 1(b)) instead of the original image. The resulting SIFT descriptor is more distinctive than pure appearance-based features and also robust to background variations. Edge augmentation is applied to all training images and the object model is then constructed in exactly the same way as described in section 2 (see [2] for details).

The detection algorithm is applied onto the edge-augmented test images with the output being a set of object windows, each with a detection score. We refer to this detection score as the "Appearance with kAS Score" (AkS) to distinguish it from the Appearance Score (AS) obtained using the original formulation in section 2.

The second method for employing the shape map assigns a *shape score* to the object windows generated by the detection algorithm. The scoring scheme is based on the intuition that most of the discriminative shape features found in the test image must belong to the true positive. On the other hand, a false positive should not contain *many* discriminative shape features. Consequently, a detected window can be assigned a shape score (SS) defined as the ratio of number of segments within the detected window w and the total number of segments in the test image I, each weighted by the strength of the respective segments (see first term in equation 1).

This intuition is based on the assumption that the object is present in the image. When this is not true, even a false-positive can get a high score. To account for this, we first make an observation regarding the segments detected in a *non*-object image: These segments do not often lie in a localized region and the output windows usually contains a small number of them. Consequently, the *number* of segments in a given detection window w would be less in non-object images than object images. Thus the number of segments inside a false-positive would be less than the number of segments within a true-positive. Therefore, based on a threshold on this number, the shape score could be suitably modified as explained next.

During the training stage, the *average* number of segments $\hat{s}$ detected on the object is computed. Let n_w denote the number of segments in the window w detected in a test image. The shape score for this window is modified based on the intuition that n_w must be comparable to or greater than $\hat{s}$ for an object to exist. The final form of the shape score is given by:

$$SS(w) = \frac{\sum_{t \in w} p(t)}{\sum_{t \in I} p(t)} \cdot min\left(1, \frac{n_w}{\hat{s}}\right) \qquad (1)$$

where $p(.)$ denotes the strength of the segment. In practice, $\hat{s}$ is computed by constructing an histogram of the number of segments detected across all training object instances and finding its mode.

It is worth noting that the shape score is computed over the window locations provided by the appearance-based algorithm (which can be implemented with or without edge-augmentation). The cumulative detection score for each window location can be written as a linear combination of the *appearance* and *shape* scores: $DS = w_1 * AS + w_2 * SS$, where w_1 and w_2 are weights. As a matter of notation, if the edge-augmentation step was used, the scores are represented with a subscript k as AkS and SS_k. We obtain DS using $(w_1, w_2) = (1, 0); (1, 1)$. In other words, we will compare results obtained using the following detection scores: $(AS, AS + SS, AkS, AkS + SS_k)$. The detection score DS permits us to examine the effect of employing the two ways of incorporating shape information, either individually or cumulatively. This is explored next.

4 Experiments and Results

For these experiments, all of the parameters involved in the appearance-based technique were kept exactly the same as in [2]. The distribution of training and test images was also unchanged. The aim was only to evaluate whether any performance enhancement prevailed due to the incorporation of kAS features. The detection and classification performance were, respectively, evaluated using Average Precision (AP) from the Recall-Precision Curve (RPC) and area-under-ROC-curve (AUC) [13].

Throughout the experiments presented in this section, only 3AS features were detected. Other kAS features were not used due to the following reasons. 2AS features are not very discriminative while 4AS features turn out to be very object-specific and thus

Table 1. Comparison of the detection performance (AP) obtained using the different detection scores in the multiple embedding spaces

	Bicycle	Car	Cell phone	Iron	Mouse	Shoe	Stapler	Toaster
AS	74.1	72.6	18.9	36.2	7.4	28.9	28.6	72.6
$AS + SS$	78.6	79.9	**22.5**	43.1	13.5	39.5	**40.4**	**79.0**
AkS	79.8	72.3	10.7	**43.3**	4.7	37.4	25	71.0
$AkS + SS_k$	**85.7**	**84.0**	19.3	40.7	**16.3**	**46.8**	35.5	74.4

Table 2. Comparison of the classification performance (AUC) obtained using the different detection scores in the multiple embedding spaces

	Bicycle	Car	Cell phone	Iron	Mouse	Shoe	Stapler	Toaster
[4]	82.9	73.7	77.5	79.4	83.5	68.0	75.4	73.6
AS	97.4	95.8	73.9	**90.2**	77.0	86.3	74.8	95.3
$AS + SS$	98.6	**98.1**	**74.4**	87.0	**79.6**	81.1	**88.5**	97.0
AkS	98.3	97.2	71.6	87.5	61.6	76.7	87.3	94.8
$AkS + SS_k$	**98.8**	**98.1**	69.9	87.1	60.3	**89.2**	81.9	**97.4**

difficult to detect in novel images (such an observation was also made in [8]). Moreover, since the segments of the discriminative kAS features form the shape map and they are mostly shared between 3AS and 4AS, there is not much difference between detecting 3AS and 4AS.

In this paper, we have chosen $\theta_I = 2$ as the threshold to determine whether a feature is discriminative. As mentioned in section 3, all discriminative features are clustered to yield *representative* discriminative kAS (RDkAS). The number of RDkAS differs with view and the object class. For example, some classes such as car have 60 RDkAS in a given view-segment while others like iron have 20. Since each RDkAS is tested for detection in an image, it is likely that a larger number of RDkAS features will detect more segments *not* belonging to the object. At the same time, it has been observed that the number of segments found *on* the object does not typically change. This is because segments can be shared across different kAS features and a segment is considered detected even if one of the kAS features is detected. Taking this into consideration and to maintain uniformity across all object classes, only the top 10 RDkAS in a given view-segment of the object class were selected. Next, we investigate the role played by edge-augmentation and shape score on the 3D dataset [4].

Effect of edge-augmentation (comparing AS with AkS): From table 1, it can be observed that the edge-augmentation step improves the detection performance for three classes (bicycle, iron and shoe) and reduces it for four (cell phone, stapler, mouse and

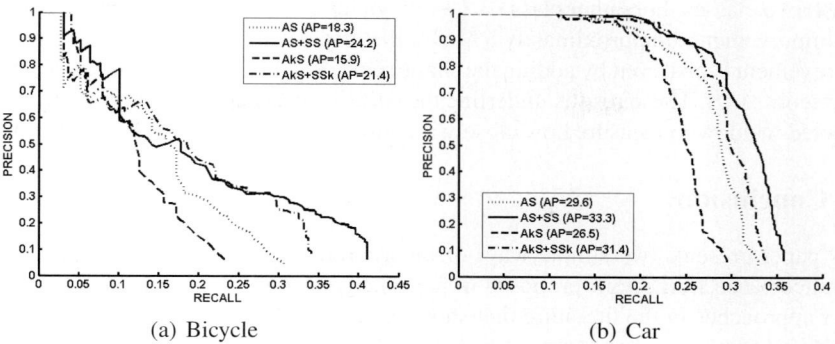

Fig. 2. Comparing RPC curves and Average Precision (AP) obtained by different detection scores for two classes from the VOC 2006 dataset [13]. For reference, AP for car class in [13] is 16 (MIT_fergus) and 21.7 (MIT_torralba).

toaster). The primary reason for the decrease in the detection performance in the latter cases can be attributed to the simplistic edge structure of these objects. Discriminative features of one object class are easily found in others and thus edges are augmented in all of these classes. As a result, the edge-augmentation step makes the local SIFT features even less distinctive than when solely appearance is used. In contrast, augmenting edges in the classes having complex edge structure produces more distinctive SIFT features, which leads to improvement in detection performance.

Effect of shape score (comparing AS with $AS + SS$ and AkS with $AkS + SS_k$): Table 1 shows that adding shape score (SS) to the original appearance score (AS) improves the detection performance of *all* of the object classes. Similar improvement in the detection performance is observed by adding the shape score (SS_k) to the edge-augmented appearance score (AkS) except for the iron class, which shows a slight reduction in performance. On the average, there is an improvement of 7% by $AS + SS$ over AS and 8% by $AkS + SS_k$ over AkS.

Effect of edge-augmentation and shape score (comparing AS with $AkS + SS_k$): Table 1 shows that the detection performance obtained with $AkS + SS_k$ is better than AS for *all* object classes. The improvement is conspicuous for the bicycle (11.6%), car (11.4%) and shoe (17.9%) classes.

It is clear from the above analysis that the shape score definitely improves the detection performance whereas edge-augmentation only improves it for classes having distinctive edge structure. The classification performance (table 2) is also improved for all classes except for iron. Specifically, $AS + SS$ and $AkS + SS_k$ each achieve the best performance in three classes and the same performance for the car class. The method presented in this paper outperforms [4] (which report only the classification results) on 6 classes by a margin varying between $6 - 24\%$ (based on $AkS + SS_k$).

Lastly, we compare the results obtained on the VOC 2006 dataset [13]. As in [2], the appearance-based model learnt from the 3D dataset is applied to the VOC 2006 dataset. Figure 2 shows the RPC curves corresponding to each combination of the detection scores for the two classes. For either class, $AS + SS$ yields the best performance with an average improvement of approximately 5% over the original appearance score AS. Similar improvement is observed by adding the shape score SS_k to the edge-augmented appearance score AkS. These results underline the success in assigning a shape score to each detected window to measure how closely it conforms to the shape of the actual object.

5 Conclusion

This paper presents two simple ways of incorporating shape information into the appearance based framework proposed in [2]. The results are in agreement with various other approaches in the literature that show improvement in detection performance by combining shape- and appearance-based features. The efficacy of assigning a shape score to a detected window based on its conformity to the shape of the actual object is demonstrated for various object classes. Further, it has been shown that augmenting the edges of certain object classes that have distinctive edge structure also improves detection performance.

References

1. Vidal-Naquet, M., Ullman, S.: Object Recognition with Informative Features and Linear Classification. In: ICCV, pp. 281–288 (2003)
2. Gill, G., Levine, M.: Multi-view Object Detection using Spatial Consistency in a Low Dimensional Space. Accepted in DAGM (September 2009)
3. Dorkó, G., Schmid, C.: Selection of Scale-Invariant Parts for Object Class Recognition. In: ICCV, pp. 634–640 (2003)
4. Savarese, S., Fei-Fei, L.: 3D Generic Object Categorization, Localization and Pose Estimation. In: ICCV, October 2007, pp. 1–8 (2007)
5. Shotton, J., Blake, A., Cipolla, R.: Contour-Based Learning for Object Detection. In: ICCV, vol. 1, pp. 503–510 (2005)
6. Opelt, A., Pinz, A., Zisserman, A.: A Boundary-Fragment-Model for Object Detection. In: Leonardis, A., Bischof, H., Pinz, A. (eds.) ECCV 2006. LNCS, vol. 3952, pp. 575–588. Springer, Heidelberg (2006)
7. Jurie, F., Schmid, C.: Scale-Invariant Shape Features for Recognition of Object Categories. In: CVPR, vol. II, pp. 90–96 (2004)
8. Ferrari, V., Fevrier, L., Jurie, F., Schmid, C.: Groups of Adjacent Contour Segments for Object Detection. IEEE Transactions PAMI 30(1), 36–51 (2008)
9. Zhang, W., Yu, B., Zelinsky, G., Samaras, D.: Object Class Recognition using Multiple Layer Boosting with Multiple Features. In: CVPR, pp. II:323–II:330 (2005)
10. Opelt, A., Zisserman, A., Pinz, A.: Fusing Shape and Appearance Information for Object Category Detection. In: BMVC, vol. 1, pp. 117–126 (2006)
11. Shotton, J., Blake, A., Cipolla, R.: Efficiently Combining Contour and Texture Cues for Object Recognition. In: BMVC (2008)
12. Lowe, D.G.: Object Recognition from Local Scale-Invariant Features. In: ICCV, vol. 2, pp. 1150–1157 (1999)
13. Everingham, M., Zisserman, A., Williams, C.K.I., Van Gool, L.: The PASCAL Visual Object Classes Challenge (VOC 2006) Results (2006),
    http://www.pascal-network.org/challenges/VOC/voc2006/results.pdf
14. de Ridder, D., Duin, R.: Locally Linear Embedding for Classification. Technical Report PH-2002-01, Pattern Recognition Group, Delft Univ. of Tech., Delft (2002)

Symmetry Detection for Multi-object Using Local Polar Coordinate[*]

Yuanhao Gong[1], Qicong Wang[1], Chenhui Yang[1], Yahui Gao[2], and Cuihua Li[1]

[1] Department of Computer Science, Xiamen University, Fujian, China
[2] School of Life Sciences, Xiamen University, Fujian, China
chyang@xmu.edu.cn

Abstract. In this paper, a novel method is presented, which detects symmetry axes for multi-object. It uses vertical line detection method in local polar coordinate. The approach costs little computation and can get efficient results, which means that it can be used in database applications.

1 Introduction

Many images, including synthetic figures and natural pictures, have symmetries Fig.1. It is an important feature for a lot of practical applications, including object alignment, recognition and segmentation. The image might be affected by illumination variation or changes in other conditions, which makes symmetry axes hard to detect.

(a) papercut   (b) butterfly   (c) steps   (d) toy cars

Fig. 1. Some images with kinds of symmetries

Actually, there are many kinds of symmetries and they can be classified into different classes. According to the theory of wallpaper groups [1], there are exactly seventeen different plane symmetry groups, which can be classified into translation symmetry, reflection symmetry, rotation symmetry and glide reflection symmetry. From another perspective, they can also be divided into perfect symmetry and almost symmetry. The third kind of classification is local symmetry and global symmetry. The general shapes of objects in images maybe

---

[*] Supported by National Natural Science Foundation (No.40627001) and the 985 Innovation Project on Information Technique of Xiamen University (2004-2008).

have global symmetries, but not perfect symmetries when their details show up. From the view of components in images, there are symmetries for sigle object Fig.1(a,b,c) and multi-object Fig.1(d).

In the paper, we propose a new method to detect symmetry axes for multi-object in one image.

1.1 Previous Work

Early papers focused on detecting exact symmetry[2], which limited their applications in more complex objects' analysis. Marola adopted a maximization of a specially defined coefficient of symmetry to find all the axes of symmetry of symmetric and almost symmetric planar images with nonuniform gray-level (intensity images)[3]. Zabrodsky et al proposed a method for approximate symmetry detection and defined SD (Symmetry Distance) [4]. Reisfeld et al. described a method that does not require prior segmentation or knowledge. An objection to this method was that it was dependent on the contrast of the feature in addition to its geometric shape[5]. The symmetry of a one-dimensional function can be measured in the frequency domain as the fraction of its energy that resides in the symmetric Fourier basis functions. The approach was extended into two dimensions [6]. C.Sun et al used an orientation histogram for symmetry detection [7,10]. Peter Kovesi started to use Log Gabor functions and local phase information for symmetry detection [8]. Atsushi Imiya et al proposed a randomized method to speed up the detection [9]. Qiong Yang et al proposed a method by combining PCA with the even-odd decomposition principle [11]. Prasad et al addressed the bilateral symmetry detection problem in images. In their work, edge-gradient information was used to achieve robustness to variation in illumination [12]. Jingrui et al proposed an optimization-based approach for automatic peak number detection, based on which they design a new feature to depict image symmetry property [13]. Jun Wu et al used phase information of original images instead of gradient information [14]. Thrun et al proposed an approach which identifies all of probable symmetries and used them to extend the partial 3-D shape model into the occluded space[15]. Keller et al presented an algorithm that detected rotational and reflectional symmetries of two-dimensional objects using the angular correlation (AC) [16]. Loy et al present a method for grouping feature points on the basis of their underlying symmetry and their characterizing the symmetries present in an image[17]. Niloy et al presented a new algorithm that processed geometric models and efficiently discovers and extracts a compact representation of their Euclidean symmetries. Based on matching simple local shape signatures in pairs, the method used these matches to accumulate evidence for symmetries in an appropriate transformation space[18]. Joshua et al described a planar reflective symmetry transform (PRST) that captured a continuous measure of the reflectional symmetry of a shape with respect to all possible planes and used Monte Carlo framework to improve computation[19,20,21]. Niloy et al formulated symmetrizing deformations as an optimization process that couples the spatial domain with a transformation configuration space, where symmetries were expressed more naturally and compactly as parameterized point pair mappings[22].

Zhitao et al gave a comparison between gradient based methods and phase based methods after analyzing typical image symmetry detectors[25]. Park et al gave an evaluation methodology of methods used in [17] and Digital Papercutting (SIGGRAPH 2005) and another evaluation methodology of methods used in [17] and Detecting Rotational Symmetries (ICCV05)[26] for rotation symmetry.

2 Our Method

Previous works mainly focus on a single object in images. Even the approach used in [14] can detect symmetry axes for mulit-objects, it treats the image only as a set of pixels, not considering its geometric information. Therefore, it costs too much computation that it can't be used in any database application. In this paper, local geometric information is used to find symmetry axes of multi-objects in an image. Our method costs less compuation, which makes sure that it can be used in many database applications.

2.1 Object Localization

Mathematical morphology is used to locate objects in our work. Mathematical morphology is well-studied for image processing. The basic idea in binary morphology is to probe an image with a simple, pre-defined shape, drawing conclusions on how this shape fits or misses the shapes in the image. This simple "probe" is called structuring element. Let E^r denote a structuring element. In our work, we let E^r be an open disk of radius r, centered at the origin. The following steps are used to locate objects. ■ Edge detection with Canny to turn the input image $I(x, y)$ into binary image $f(x, y)$. ■ Closing operation with $f(x, y)$ by E^5 . ■ Fill holes in objects. ■ Opening operation with $f(x, y)$ by E^5. ■ Closing operation with $f(x, y)$ by E^5. ■ Label connected components for multi-objects Fig.2(b). The change of the parameter 5 here wouldn't affect the final result because of basic properties of morphology operations. More about the method's robustness with parameters' variance can be found in the 'robust' subsection.

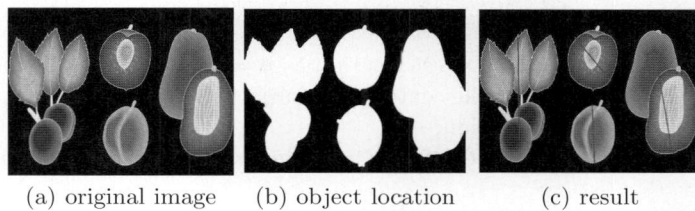

(a) original image   (b) object location   (c) result

Fig. 2. An example

2.2 Local Polar Coordinate

Our work is based on the Theorem 1. Since it is well-accepted, the proof process is omitted.

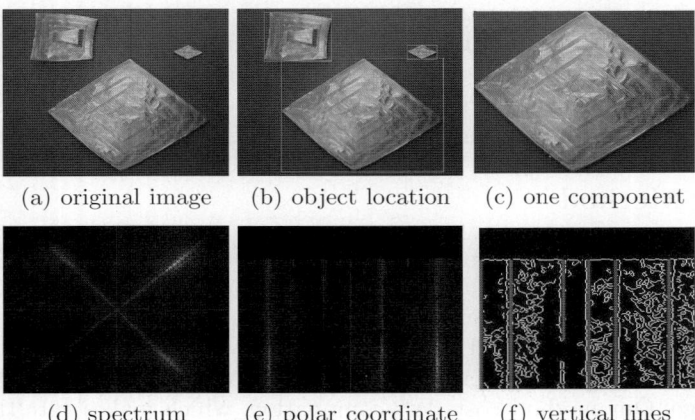

(a) original image (b) object location (c) one component
(d) spectrum (e) polar coordinate (f) vertical lines

Fig. 3. Pipeline of our method

Theorem. Let $y = k(x-x_c)+y_c$ be a symmetry axis through the object's center (x_c, y_c) in the original image, then $y = kx$ will be a symmetry axis through the origin in the corresponding magnitude spectrum of Fourier Transformation. ∎

Theorem 1 is the footstone of our approach. Let $I(x, y)$ be the input image. Let $f(x, y)$ be its labeled binary image and there are n components in $f(x, y)$. For each pixel of a labeled connected component, its value is j ($j = 1, \cdots n$). The connected component is denoted by $f_j(x, y)$. The object's center is calculated using $x_c = \dfrac{\sum x \times f_j(x,y)}{\sum f_j(x,y)}$ and $y_c = \dfrac{\sum y \times f_j(x,y)}{\sum f_j(x,y)}$. It's also easy to find two integers m, n such that a rectangle $R(x, y)$ with left-top coordinate $(x_c - \frac{m}{2}, y_c + \frac{n}{2})$ and right-bottom coordinate $(x_c + \frac{m}{2}, y_c - \frac{n}{2})$ only contains the jth object. That is:

$$R(x,y) = \begin{cases} I(x,y) & \text{if } f(x,y) = j \\ 0 & \text{otherwise} \end{cases} \quad (1)$$

We can asume the symmetry axes equation as $y = k(x - x_c) + y_c$ in the original image, because it must cross the center of the object. Our main task is to estimate the parameter k in the equation.

We calculate the gradient G of $R(x, y)$ to get robustness from illumination's variation.

$$G(x,y) = \frac{\partial R(x,y)}{\partial x} + i\frac{\partial R(x,y)}{\partial y} \quad (2)$$

Then we get the magnitude $M(x, y)$ of Fourier Transformation of $G(x, y)$

$$M(x,y) = |FFT(G(x,y))| \quad (3)$$

From theorem 1, we know that $y = kx$ is a symmetry axes in $M(x, y)$. A reasonable idea is to use a line detection method to find lines in $M(x, y)$. But the method can't guarantee that all of these lines cross the original. Georgios et al suggest using the presentation of $M(x,y)$ in the polar coordinate with original at (x_c, y_c) to overcome the problem [23]. Let $H(\gamma, \theta)$ be the presentation of $M(x,y)$ in the polar coordinate with original at (x_c, y_c). They use the singular value decomposition (SVD) method to find a proper radius in the polar coordinate and then treat it as a peak detection problem. But it's obvious that Georgios et al's method is sensitive to noise. In our study, we pay special attention on θ. For a fixed θ', $H(\gamma, \theta)$ is relatively stable. So line detection techniques can be used for parameters estimation.

For more accurate k, some pixels around the origin should be removed by a threshold δ_1. If $\gamma < \delta_1$, $H(\gamma, \theta) = 0$. Then we use Standard Hough Transform to detect lines. It's easy to see that only vertical lines cross the origin. So we set another threshold δ_2 to make sure all of the lines are vertical. For a line ended with two points (x_a, y_a) and (x_b, y_b), if $|x_a - x_b| < \delta_2$, the line has to be removed from the result. For each line ended with point (x_a, y_a) and point (x_b, y_b) in the result, we construct a set E_0 containing $(x_a + x_b)/2$.

Each vertical line in the result, which is in correspond to a symmetry axis and also a number in E_0, might be missed by the line detection method. From above, it is known that if $k \in E_0$ then $k + pi \ (mod \ 2\pi)$ should also have been in E_0. This property of E_0 can be used to find back the missed lines and get more accurate k. The following algorithm is used for this purpose.

Algorithm
(1)Initialize E with $\emptyset$ and give a threshold δ_3.
(2)$\forall k \in E_0$,
  if $\exists k' \ s.t. |k' - (k+\pi)|(mod \ 2\pi) < \delta_3$
    then $\frac{k+k'\pm\pi}{2}|(mod \ 2\pi)$ should be added into E.
  otherwise, k and $k + \pi (mod \ 2\pi)$ should be added into E.
Then, the symmetry axes can be estimated as $y = y_c + k(x - x_c)$, $k \in E$.

2.3 Robust

Our method is not sensitive to parameters' variation. Let E be the result and E' be the result with some parameter changed or the image noised. Then a measurement is defined as:

$$m = \frac{1}{s} \sum_{i=1}^{s} \frac{|e_i - e'_i|}{e_i + e'_i} \quad (4)$$

where $e_i \in E$ is the ith largest in E and $e'_i \in E'$ is the ith largest in E'. s is the number of elements in E.

There are three parameters in our method, $\delta_1, \delta_2, \delta_3$. For δ_1, it is a threshold to remove inaccuracy data. It's obvious that δ_1 doesn't affect the result if it's not small enough. In our experiment, δ_1 varies from 0.1 to 0.6, shown in Fig.4(a). For δ_2, it is used to make sure only vertical lines detected. It doesn't affect the result if it's not large enough. In our experiment, δ_2 varies from 3 to 33, shown

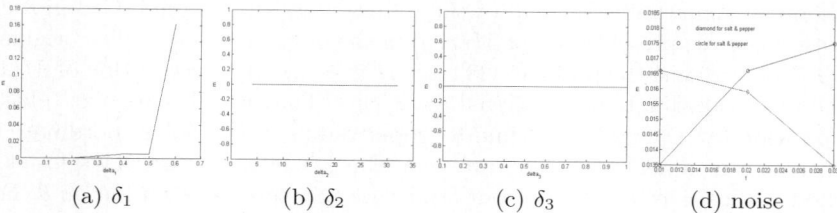

Fig. 4. Results affected by parameters and two kinds of noises

in Fig.4(b). For δ_3, it is used to merge the two lines which are in close to each other. In our experiment, δ_3 varies from 0.1 to 1, shown in Fig.4(c). The effection of two kinds of noises, Gaussian noise and proper salt noise, is also presented in Fig.4(d), where abscissa t stands for noise intensity.

2.4 Fast

The algorithm is implemented in a Matlab 2008a's script on a PC with an intel CPU E2200. The computation time is shown in Tab.1. When being used in database applications, the method should be implemented in C++ and the time costs will be significantly reduced.

Table 1. Time costs

image size (number of objects)	640*638(3)	800*800(5)	498*433(2)
time	0.316690	0.729246	0.210623

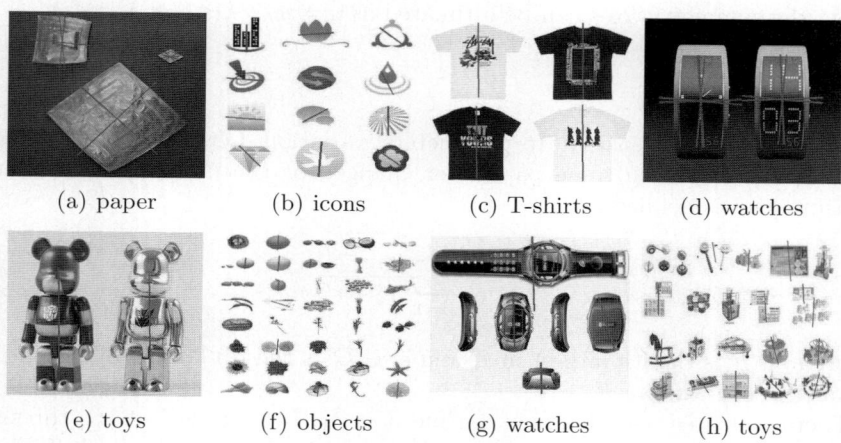

Fig. 5. Experiment results

2.5 Efficient

Experiment results can be found in Fig.5. Actually, there are two kinds of evidence for symmetry detection in our method. One is shape (or called contour) information and the other is texture information. For objects with less texture Fig.5(a, b), their symmetry axes are decided by shape information. For complex objects Fig.5(f, g, h), texture information is the main factor. Our method achieves implicit balance between the two kinds of information because of the gradient information we adopted Eq.2.

3 Conclusion and Future Work

In this paper, we propose a new symmtry detection method for multi-object using local coordinate and a vertical line detection method. Our method has been proved to be robust, fast and efficient.

In the future work, the method will be tested on our image database. Furthermore, symmetry detection for multi-object with overlap is hard for segmentation and also symmtry detection. More study is needed in the area because our method proposed in this paper couldn't solve the problem.

References

1. Grunbaum, B., Shephard, G.C.: Tilings and Patterns. W.H.Freeman and Company, New York (1987)
2. Atallah, M.J.: On Symmetry Detection. IEEE Transactions on Computers c-34(7) (July 1985)
3. Marola, G.: On the Detection of the Axes of Symmetry of Symmetric and Almost Symmetric Planar Images. IEEE Trans on PAMI 11(1), 104–108 (1989)
4. Zabrodsky, H., Peleg, S., Avnir, D.: Symmetry as a Continuous Feature. IEEE Transactions on Pattern Analysis and Machine Intelligence 17(12) (December 1995)
5. Reisfeld, D., Wolfson, H., Yeshurun, Y.: Context-free attentional operators: The generalized symmetry transform. IJCV 14(2), 119–130 (1995)
6. Gofman, Y., Kiryati, N.: Detecting Symmetry in Grey Level Images: The Global Optimization Approach. In: Proceeding of ICPR (1996)
7. Sun, C., Sherrah, J.: 3D Symmetry Detection Using The Extended Gaussian Image. IEEE Transactions on Pattern Analysis and Machine Intelligence 19(2) (February 1997)
8. Kovesi, P.: Symmetry and Asymmetry From Local Phase. In: Sattar, A. (ed.) Canadian AI 1997. LNCS, vol. 1342, pp. 2–4. Springer, Heidelberg (1997)
9. Imiya, A., Ueno, T., Fermin, I.: Symmetry Detection by Random Sampling and Voting Process. In: 10th International Conference on Image Analysis and Processing, Italy, Venice, pp. 400–405 (1999)
10. Sun, C., Si, D.: Fast Reflectional Symmetry Detection Using Orientation Histograms. Real-Time Imaging 5, 63–74 (1999)
11. Yang, Q., Ding, X.: Symmetrical PCA in Face Recognition. In: IEEE ICIP (2002)
12. Shiv Naga Prasad, V., Yegnanarayana, B.: Finding Axes of Symmetry From Potential Fields. IEEE Transaction on Image Processing 13(12) (December 2004)

13. He, J., Li, M., Zhang, H.-J., Tong, H., Zhang, C.: Automatic peak number detection in image symmetry analysis. In: Aizawa, K., Nakamura, Y., Satoh, S. (eds.) PCM 2004. LNCS, vol. 3333, pp. 111–118. Springer, Heidelberg (2004)
14. Wu, J., Yang, Z.: Detecting Image Symmetry Basedon Phase Information. In: Proceedings of the Fourth International Conference on Machine Learning and Cybernetics, Guangzhou, August 2005, pp. 18–21 (2005)
15. Thrun, S., Wegbreit, B.: Shape From Symmetry. In: Proceedings of the Tenth IEEE International Congerence on Computer Vision (ICCV) (2005)
16. Keller, Y., Shkolnisky, Y.: A Signal Processing Approach to Symmetry Detection. IEEE Transactions on Image Processing 15(8) (August 2006)
17. Loy, G., Eklundh, J.-O.: Detecting symmetry and symmetric constellations of features. In: Leonardis, A., Bischof, H., Pinz, A. (eds.) ECCV 2006. LNCS, vol. 3952, pp. 508–521. Springer, Heidelberg (2006)
18. Mitra, N.J., Guibas, L.J., Pauly, M.: Partial and Approximate Symmetry Detection for 3D Geometry. In: Proceedings of ACM SIGGRAPH 2006 (2006)
19. Podolak, J., et al.: A Planar-Reflective Symmetry Transform for 3D Shapes. In: Proceedings of ACM SIGGRAPH 2006 (2006)
20. Podolak, J., Golovinskiy, A., Rusinkiewicz, S.: Symmetry-Enhanced Remeshing of Surfaces. In: Eurographics Symposium on Geometry Processing (2007)
21. Golovinskiy, A., Podolak, J., Funkhouser, T.: Symmetry-Aware Mesh Processing, Princeton University TR-782-07 (April 2007)
22. Mitra, N.J., Guibas, L.J., Pauly, M.: Symmetrization. In: ACM SIGGRAPH 2007 (2007)
23. Tzimiropoulos, G., Stathaki, T.: Symmetry Detection Using Frequency Domain Motion Estimation Techniques. In: Acoustics, Speech and Signal Processing 2008 (2008)
24. Gong, Y., et al.: A Novel Symmetry Detection Method for Images and Its application for Motion Deblurring. In: MMIT 2008 (2008)
25. Xiao, Z., Wu, J.: Analysis on Image Symmetry Detection Algorithms. In: Fourth International Conference on Fuzzy Systems and Knowledge Discovery, FSKD 2007 (2007)
26. Park, M., et al.: Performance Evaluation of State-of-the-Art Discrete Symmetry Detection Algorithms. In: CVPR 2008 (2008)

Detection of Non-convex Objects by Dynamic Programming

Andree Große, Kai Rothaus, and Xiaoyi Jiang

Department of Mathematics and Computer Science, University of Münster,
Einsteinstraße 62, D-48149 Münster, Germany
Tel.: +49-251-83-33759
{andree.grosse,kai.rothaus,xjiang}@uni-muenster.de

Abstract. In this work we present the RACK algorithm for the detection of optimal non-convex contours in an image. It represents a combination of an user-driven image transformation and dynamic programming. The goal is to detect a closed contour in a scene based on the image's edge strength. For this, we introduce a graph construction technique based on a "rack" and derive the image as a directed acyclic graph (DAG). In this graph, the shortest path with respect to an adequate cost function can be calculated efficiently via dynamic programming. Results demonstrate that this approach works well for a certain range of images and has big potential for most other images.

Keywords: Contour Detection, Shortest Path, Dynamic Programming, Rack.

1 Introduction

Finding contours of non-convex objects is an important and challenging task in the field of image analysis. We address this task with a new approach that will be called the RACK algorithm and is an extension of the well-known dynamic programming (DP) algorithm for detecting closed contours in color images. The idea is to specify the general shape of the desired object by using a "rack" (see Fig. 5(c)). This rack is used to model the image as a directed acyclic graph that is likely to contain a path matching the object contour. With the image's edge strength as graph weights, an optimal contour can be defined as a the shortest path with respect to some cost function. We construct a cost function that is independent of the path length but can be globally minimized via dynamic programming.

The RACK algorithm forms a combination of different well-known approaches in image analysis. With a user-driven image transformation, the contour detection process in a 2D-image is reduced to an 1D-optimization problem that can be solved via DP. This transformation is profoundly anisotropic and a direct application would result in a loss of precision. To avoid this, the image is modeled as a DAG representing this transformation.

The algorithm is useful for detecting objects that are distinguishable from the background by some variation in color or brightness along the object border. This

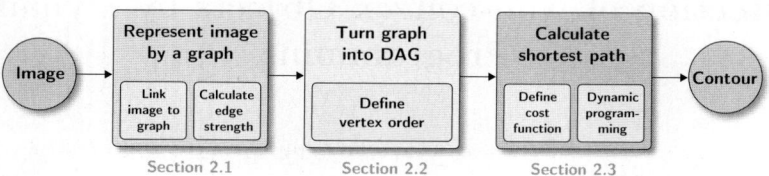

Fig. 1. Outline of the RACK algorithm

may include objects that consist of multiple segments with completely different texture. The RACK algorithm can be applied in situations where some user interaction is acceptable, e.g. image processing tools, or when a priori knowledge about the image (like the position or general shape of the object) is available.

The method of dynamic programming was introduced by Bellman [1]. An overview on edge-based image segmentation in general can be found in [2]. Other methods that utilize DP for contour detection include Intelligent Scissors [3] or boundary templates. While the RACK algorithm calculates globally optimal contours for a user-defined search direction, Intelligent Scissors form an interactive contour tracing tool that uses 2-D dynamic programming to compute piece-wise optimal contour segments. Corridor Scissors [4] are similar to this approach, but use a search corridor instead of interactive contour tracing.

The remainder of this work is organized as follows. In Sec. 2, we will describe our algorithm according to the three steps shown in Fig. 1: The image is first modeled as an undirected graph and then turned into a DAG. As a third step, the shortest path under certain constraints is calculated in this DAG, which represents an optimal contour. After presenting further extension to the general RACK algorithm, some results are shown in Sec. 3. Finally, we conclude with a discussion in Sec. 4.

2 The Rack Algorithm

As a motivation for the RACK algorithm, we may regard the popular method for detecting star-shaped contours by transforming the image into the polar coordinate space. Let z_0 denote the origin of the polar coordinate system. A point (x, y) in Cartesian coordinates can be expressed as a complex number and has the polar coordinates (r, θ) with $z_0 + re^{i\theta} = x + iy$. An optimal contour with strictly increasing θ-values relates to a star-shaped contour in Cartesian space.

This approach has two disadvantages: It can only detect star-shaped contours and the transformation process results in an uneven resolution of different image regions. To avoid the latter problem, we first model the image and its edge strengths as a graph (Sec. 2.1). We then define the RACK transformation (similar to polar coordinates) which can handle non-convex shapes. The graph is turned into a DAG that represents this transformation (Sec. 2.2). An optimal contour is defined as the shortest path in this DAG with respect to some cost function (Sec. 2.3). Finally, we discuss the time complexity (Sec. 2.4) and present some extensions (Sec. 2.5).

2.1 Graph Representation of an Image

Figure 2 illustrates how an image of size $M \times N$ is modeled by a *grid graph* of size $(M+1) \times (N+1)$. A contour in the image can be expressed by a path in this graph. As each vertex represents a boundary point between adjacent pixels, a closed path will divide the image into two regions, the *object* and the *background*.

A weight w is assigned to each graph edge which indicates the image's edge strength at the corresponding position in the image, e.g. the value of the Sobel operator. For simplicity we use the distance of the color values in the CIELab color space for all examples in this work. We assume that the weight is normalized, i.e. we have a function $w : E \rightarrow [0, 1]$. Figure 2(c) illustrates the weight for each graph edge as grayscale highlights, where black indicates areas of high edge strength. This grid graph therefore is a weighted undirected graph $G = (V, E, w)$.

We assume that a weight of 0 means highest image edge strength and a value of 1 indicates completely homogeneous areas. Hence, finding a contour with high edge strengths is equivalent to calculating a path in the graph with low weights.

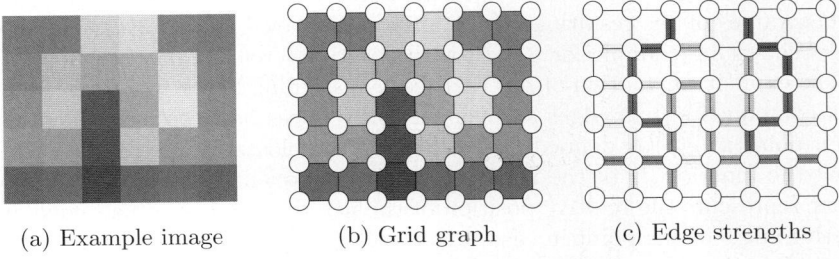

(a) Example image   (b) Grid graph   (c) Edge strengths

Fig. 2. Graph representation of an image

2.2 Construction of a DAG (The Rack Transformation)

In order to be able to search a path with low weights in a grid graph via DP, the graph has to be turned into a DAG. This is done by postulating a vertex order ρ (1) similar to the angle θ of polar coordinates. This vertex order is calculated based on a user defined intuitive structure (2): the *rack*. Finally, the graph is modified to contain only paths with strictly increasing ρ values (3).

(1) Vertex order ρ. Similar to polar coordinates we construct a function $\rho : \mathbb{R}^2 \longrightarrow [0, 1)$ and regard paths with strictly increasing ρ values. This function implicitly defines the shape of the contours that can be detected by the RACK algorithm. In order to receive a broad range of contours, the following characteristics should be given: (i) ρ is smooth and has a small number of local extrema, and (ii) the ρ values for all vertices in the grid graph are evenly spread in $[0, 1)$. In the next paragraph, we explain how to realize these properties using a *rack*.

(2) The Rack. To construct a ρ function that is able to represent non-convex contours, we define another graph that will be called a *rack*. This structure

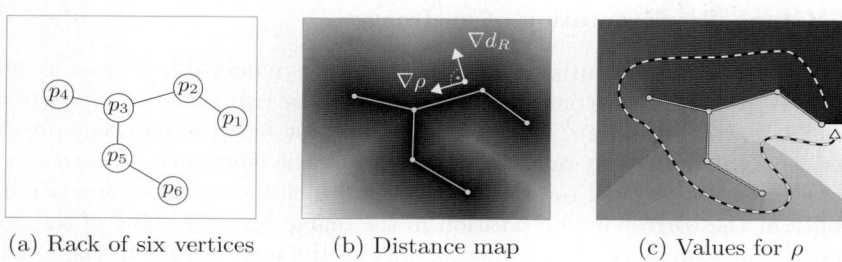

(a) Rack of six vertices (b) Distance map (c) Values for ρ

Fig. 3. The rack transformation

is a user defined planar graph in continuous space without cycles. Figure 3(a) shows such a rack with six vertices. Similar to polar coordinates where contours are searched that traverse around the origin, a rack is used to find contours that enclose this rack. The ρ function corresponding to a rack R will be defined via the *distance function* d_R (see Fig. 3(b)). ρ is constructed in such a way that at each point x where d_R is differentiable, ρ is also differentiable and $\langle \nabla \rho(x), \nabla d_R(x) \rangle = 0$. The values of the resulting ρ function are displayed in Fig. 3(c). The dashed arrow shows an example for a path with strictly increasing ρ values.

The explicit calculation of ρ is realized by assigning linear functions f_k to the rack components (half edges and connection vertices between half edges). For a given point p, $\rho(p)$ is defined by calculating the closest point on the rack and using the function f_k of the according rack component. Hence, the value of f_k either represents the relative position of the nearest point on a half edge or the relative angle of the point in case that the closest rack component is a vertex.

(3) Turn graph into DAG. Instead of actually transforming the image, the ρ function will be used as a *vertex order* to create a DAG. Figure 4(a) illustrates the values of ρ for a rack with three points.

Each vertex v in a grid graph lies at a certain position (x_v, y_v) in image coordinates. Hence, we can define $\rho(v) := \rho(x_v, y_v)$ as illustrated in Fig. 4(b). The edge set E of the graph is then reduced to graph edges (a, b) with $\rho(a) < \rho(b)$, i.e. ρ induces a *(complete) strict partial order* "$<$" on the vertices. The resulting graph is shown in 4(c), where graph edges crossing the rack and between the "end" and "start" of the ρ function have also been removed. This new edge set will be denoted with $\tilde{E}$.

2.3 Calculate the Shortest Path

Together with the weight function w from Sec. 2.1, we now have a directed acyclic weighted graph $\mathcal{G} = (V, \tilde{E}, w)$ that will be referred to as an *image graph*. Figure 5(a) shows such an image graph based on a simple rack for the example from Fig. 2. We define an optimal contour as the shortest "closed" path in $\mathcal{G}$ with respect to some cost function. This cost function should fulfill the important property to be independent of the path length. The *weighted cost* for a path $(v_1, \ldots, v_n)$ will therefore be defined as

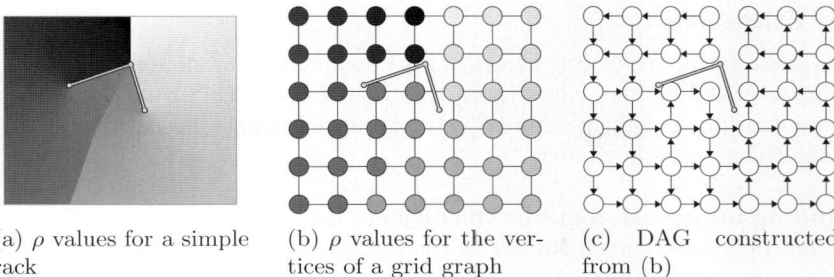

(a) ρ values for a simple rack (b) ρ values for the vertices of a grid graph (c) DAG constructed from (b)

Fig. 4. DAG from vertex order ρ

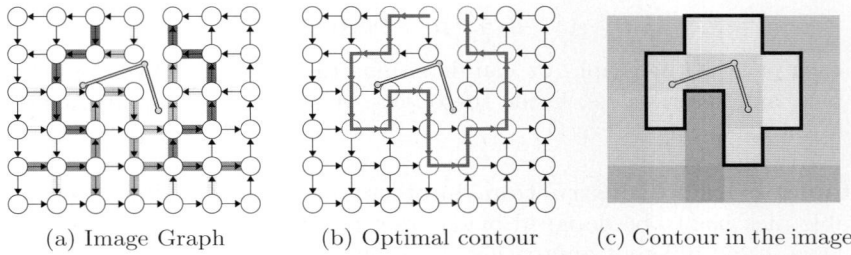

(a) Image Graph (b) Optimal contour (c) Contour in the image

Fig. 5. Optimal contour in the DAG

$$J(v_1,\ldots,v_n) := \sum_{i=1}^{n-1}(\rho(v_{i+1}) - \rho(v_i))w(v_i, v_{i+1}). \qquad (1)$$

The *optimal contour* $\mathbf{p}^*$ for $\mathcal{G}$ is defined as

$$\mathbf{p}^* := \arg\min\left\{J(v_1,\ldots,v_n) \mid \forall i : (v_i, v_{i+1}) \in \tilde{E}, v_1 \text{ and } v_n \text{ are adjacent}\right\}. \qquad (2)$$

The imprecise term "adjacent" in Eq. (2) is used for simplicity. In our implementation, we construct additional vertices that allow us to express all closed paths enclosing the rack as paths $(v_1,\ldots,v_n)$ with $\rho(v_1) = 0$ and $\rho(v_n) = 1$, where v_1 and v_n represent the same points in the image.

2.4 Computational Complexity

For a rack with K components, the calculation of ρ for all vertices is in $O(K|V|)$. To calculate the the global minimum for Eq. (2) via dynamic programming, the vertices in V have to be sorted by ρ, which can be done in $O(|V|\log|V|)$ time. The simple shortest path with respect to Eq. (1) can then be calculated in $O(|V|)$. In order to receive a closed path as in Eq. (2), the DP algorithm needs to be performed multiple times. It appears that the method proposed by Farin and de With [5] can be applied to our graph. Their algorithm has a worst-case complexity of $O(|V|\log|V|)$ and a practical computation time of $O(|V|)$ for most cases. Hence, the whole RACK algorithm can perform in $O(|V|\log|V|)$.

2.5 Extensions

In its basic form, the RACK algorithm is capable of detecting objects that can be characterized by high edge strength at their border. For situations where the user's intention is slightly different, we introduced additional parameter to give the algorithm more flexibility.

Including information about the objects interior. In many cases an object does not only have some contrast to the background, but also posses a more or less homogeneous texture. Let $h(\mathbf{p})$ be an measure for the homogeneity of the area inside a path $\mathbf{p}$. This could for example be received by texture analysis. An modified cost function that includes a measure of the object's homogeneity is

$$J_\psi(\mathbf{p}) := (1 - \psi)J(\mathbf{p}) + \psi h(\mathbf{p}), \ \psi \in [0, 1]. \tag{3}$$

For each path in the graph, we can determine the region that will lie between the path and the rack, i.e. inside the object. Hence, J_ψ can be minimized via DP in the same way as Eq. (1).

Automatic Rack Construction. In order to segment the desired object, a suitable rack has to be designed by the user or from some a-priori knowledge. We have tested different approaches to automatically construct a rack for the most prominent structure in the image or to improve an initial rack. A detailed discussion of this would go beyond the scope of this paper, but some results can be found at `http://cvpr.uni-muenster.de/research/rack/`.

3 Results

Figure 6 shows the results for different test images. The images 6(a)-(f) are taken from the *Berkeley segmentation dataset* [6]. Each image shows the rack that has been used and the detected contour. The images (a), (f), (g) are well suited for the RACK approach. The object is distinguishable by an edge along its border, although it might consist of different textures as in (g). The contour found in example (b) contains local errors due to the fact that the object consists of very different textures and at the same time the background is quite scattered. The scene (c) represents a problematic type of image for the RACK algorithm: most parts of the object border have lower edge strength than the textures of the background and the object's interior. Image (d) presents a rack which requires many vertices to specify the desired object. In the case (e), the contour also has local errors due to high edge strengths inside the object. This situation would profit from more sophisticated edge measures based on texture analysis.

All in all it can be observed that the algorithm produces good results on images where the border of the desired object mostly consists of segments with high edge strength. On objects where several parts of the border have low contrast to the background, the general shape can be found in most cases but, the contour contains local errors. Most problematic are images with textures of high edge strength throughout the whole image area – in this case, normally no useful results can be obtained.

Fig. 6. Segmentation results for the RACK algorithm

4 Discussion and Conclusions

We have developed an algorithm to combine dynamic programming with a user-driven image transformation. The RACK algorithm can detect globally optimal closed contours in $O(|V|\log|V|)$ time. It combines an acceptable amount of user interaction with generally good segmentation results. This is useful for images where some user input is indispensable to specify the desired object.

An advantage of the RACK algorithm compared to boundary-template based tools is that the object of interest can be sketched very roughly. While an approximation of the object border might be very tedious on complex contours, placing a rack somewhere more or less centered inside the object requires far less precision. As the border of most real objects has limited complexity while still being slightly non-starshaped, a simple rack generally is both necessary and sufficient. Compared to tools like Intelligent Scissors, which use the sum of local edge costs for the optimal path calculation, the weighted sum of the RACK algorithm in Eq. (1) does less likely create "shortcuts" in the contour.

On the other hand, the RACK algorithm is not suited for objects with holes, and complex structures like tree branches would require a very detailed rack. Another drawback is that the user has no direct influence on the contour detection process, i.e. there are no mechanism to adjust single parts of the contour that do not match the user's expectation.

There is some room for improvement, mainly in the area of the edge measure. Currently we simply used the Sobel operator on the different CIELab canals to calculate the image's edge strength. On complex images, the performance of the algorithm could probably be greatly improved by using preprocessing techniques like denoising or texture analysis.

Due to the graph representation of the image, the contour might sometimes traverse almost perpendicular to the rack. To receive a smoother contour, the direction of search can be further restricted: Instead of considering all graph edges (a, b) with $\rho(a) < \rho(b)$, we can enforce the additional constraint

$$\left\langle \frac{b-a}{\|b-a\|_2}, \frac{\nabla \rho(a)}{\|\nabla \rho(a)\|_2} \right\rangle \geq \omega, \quad \omega \in [0, 1].$$

Another interesting idea would be to let the user construct the rack as a continuous stroke instead of a graph to receive a smoother ρ function.

References

1. Bellman, R.: Dynamic Programming. Princeton University Press, Princeton (1957)
2. Jain, R.C., Kasturi, R., Schunck, B.G.: Machine Vision. McGraw-Hill, Inc., New York (1995)
3. Mortensen, E.N., Barrett, W.A.: Intelligent scissors for image composition. In: SIGGRAPH 1995: Proceedings of the 22nd annual conference on Computer graphics and interactive techniques, pp. 191–198. ACM Press, New York (1995)
4. Farin, D., Pfeffer, M., de With, P.H.N., Effelsberg, W.: Corridor scissors: A semi-automatic segmentation tool employing minimum-cost circular paths. In: ICIP (2004)
5. Farin, D., de With, P.H.N.: Shortest circular paths on planar graphs. In: 27th Symposium on Information Theory in the Benelux (2006)
6. Martin, D.R., Fowlkes, C., Tal, D., Malik, J.: A database of human segmented natural images and its application to evaluating segmentation algorithms and measuring ecological statistics. Technical Report UCB/CSD-01-1133, EECS Department, University of California, Berkeley (January 2001)
7. Appleton, B., Sun, C.: Circular shortest paths by branch and bound. Pattern Recognition 36(11), 2513–2520 (2003)
8. Canny, J.: A computational approach to edge detection. IEEE Trans. Pattern Anal. Mach. Intell. 8(6), 679–698 (1986)
9. Sun, C., Pallottino, S.: Circular shortest path in images. Pattern Recognition 36(11), 709–719 (2003)

Finding Intrinsic and Extrinsic Viewing Parameters from a Single Realist Painting

Tadeusz Jordan[1], David G. Stork[2,3], Wai L. Khoo[1], and Zhigang Zhu[1]

[1] CUNY City College, Department of Computer Science,
Convent Avenue and 138th Street, New York NY 10031
tedjj123@gmail.com, wlkhoo@gmail.com, zhu@cs.ccny.cuny.edu
[2] Ricoh Innovations, 2882 Sand Hill Road, Suite 115, Menlo Park CA 94025-7054
[3] Stanford University, Department of Statistics, Stanford CA 94305
artanalyst@gmail.com

Abstract. In this paper we studied the geometry of a three-dimensional tableau from a single realist painting – Scott Fraser's *Three way vanitas* (2006). The tableau contains a carefully chosen complex arrangement of objects including a moth, egg, cup, and strand of string, glass of water, bone, and hand mirror. Each of the three plane mirrors presents a different view of the tableau from a virtual camera behind each mirror and symmetric to the artist's viewing point. Our new contribution was to incorporate single-view geometric information extracted from the direct image of the wooden mirror frames in order to obtain the camera models of both the real camera and the three virtual cameras. Both the intrinsic and extrinsic parameters are estimated for the direct image and the images in three plane mirrors depicted within the painting.

Keywords: camera calibration, perspective geometry, art analysis.

1 Introduction

The problem of reconstructing a three-dimensional scene from multiple views is well explored, and a number of general methods, such as those based on correlation, relaxation, dynamic programming, have been developed and fully characterized [5, 6]. Three-dimensional reconstruction and metrology can be based on single views as well [3, 1]. Criminisi and his colleagues [2] have recently applied such techniques to the analysis of paintings, for instance reconstructing the virtual spaces in Masacio's *Holy Trinity* (c. 1425), Piero della Francesca's *Flagellation of Christ* (c. 1453), Hendrick V. Steenwick's *St. Jerome in his study* (1630), Jan Vermeer's *A lady at the Virginals with a gentleman* (1662-1665), and others. These methods reveal both the high geometric accuracies in some passages, and the geometric inconsistencies in others, properties that are nearly impossible to determine by eye. Such analyses shed new light on these works and the artists' working methods, for instance revealing whether an artist likely used geometrical aids during the execution of their work.

Recently Smith, Stork and Zhang reconstructed the three-dimensional space depicted in a highly realistic modern painting, Scott Fraser's *Three way vanitas* (Fig. 1) using traditional multiple-view reconstruction methods applied to the direct view and a view visible in a depicted mirror [11]. Even though using reflected images by mirrors is a very popular approach for stereo vision in computer vision [7, 9, 10, 12], it was the first time to analyze a painting with such a setup. However, there were some limitations in that previous work as well as unexplored opportunities. For instance, the images of the frames of the mirrors provide geometric constraints about the centers of projection of the images depicted within each mirror, and the earlier scholarship did not incorporate that information when reconstructing the three-dimensional space.

Section 2 describes the painting, previous scholarship and an overview of our new approach to camera parameter estimation. Section 3 introduces some notations and constraints used in the paper. Sections 4 to 7 describe the details of estimating both intrinsic and extrinsic parameters of the main and virtual cameras: finding image center, estimating focal length, representing mirror planes and locating virtual cameras. Section 8 gives a brief summary and discussion.

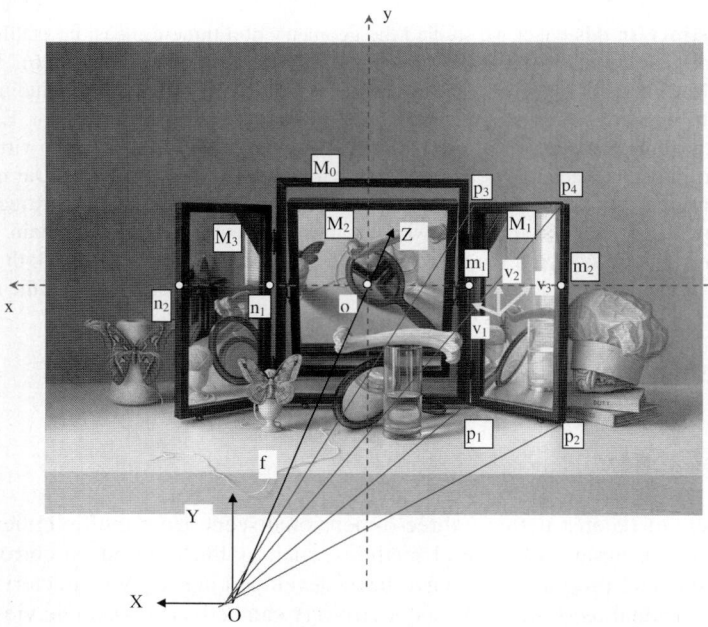

Fig. 1. The work, the notations and the constraints

2 The Work and Problem Addressed

Fig. 1 shows the work we consider, Scott Fraser's *Three way vanitas* (2006). This painting was commissioned as part of The Object Project, in which fifteen artists were commissioned to create works, each containing five specified objects: hand mirror, bone, moth, ball of string and drinking glass [4]. Our method for estimating the camera models for the virtual cameras is based on the single-image information of the mirror frames in the primary image of the painting; the image seen from the artist's viewing point (the main camera). We construct geometric invariants such as horizontal/vertical lines and vanishing points from the locations of the vertices of its frame, to estimate the focal length, the center of projection of the main "camera", and the location and orientation of each mirror. Then, we use the pose information of each mirror to compute the location and orientation of the virtual camera by finding its coordinate system that is symmetric to the artist's viewing point in the plane of the mirror. Using the knowledge of the mirror frames, both the intrinsic and extrinsic parameters between a pair of stereo cameras can be found. This is difficult if basic fundamental matrix method is used, as in [11], which used only limited number of points on the table in both the real image and the mirror images.

3 Notations and Constraints

We label the mirrors, reading right to left, M_i, their associated reflected images I_i, and corresponding centers of projection C_i, for i = 1, 2, 3. The middle frame is labeled as M_0. Let's use the mirror M_1 as an example. Each of the four vertices of the frame rectangle, P_i has a corresponding image point $p_i = (a_i, b_i, f)$, i = 1, 2, 3, 4. These image points can also be viewed as vectors from the optical center O to those points. Since $P_1P_2 \parallel P_3P_4$, the direction of the parallel lines can be estimated as

$$v_1 = (p_3 \times p_4) \times (p_1 \times p_2) . \qquad (1)$$

Since $P_1P_3 \parallel P_2P_4$, the direction of the parallel lines can be estimated as

$$v_2 = (p_4 \times p_2) \times (p_3 \times p_1) . \qquad (2)$$

The normal of the rectangular mirror surface is

$$v_3 = v_1 \times v_2 . \qquad (3)$$

Here are a few notes:

1. Point (a_i, b_i) should be measured in the xoy coordinate system that is aligned with the main camera coordinate system O-XYZ. Therefore the image center o should be estimated first for using the above equations.
2. The focal length f is unknown and should be estimated first to use Eqs. (1) – (3).
3. If the projections of a pair of parallel lines are not parallel in the image, such as p_1p_2 and p_3p_4, the direction of the pair can be calculated through their vanishing point in the image plane, (v_{1x}, v_{1y}, f), therefore

$$v_1 \cong (v_{1x}, v_{1y}, f) . \qquad (4)$$

Otherwise, the vanishing point is in infinite, and the third dimension of the direction vector such as v_2 should be zero. This is the advantage to use Eqs. (1) to (3) to find directions of parallel lines instead of using vanishing point estimation.

In the following, we will describe methods in estimating the following parameters of the real camera and the three virtual cameras created by the three mirrors: 1). the image center $o(c_x,c_y)$; 2). the focal length f; 3). the plane equation of each mirror; and

4). the pose of the three virtual cameras related to the real camera (the artist's eye). The four cameras share the same focal length and the same image center, but the images of the three mirrored cameras are reversed in the x direction. The intrinsic geometric constraints (*assumptions*) we use are the following:

1. All three mirrors are rectangular.
2. The two flanking plane mirrors have the same inherent shape and are arranged vertically, each rotated by an unknown angle, and that the central plane mirror is viewed frontally, tipped forward by an unknown angle.
3. The back edges of the two flanking mirrors are at the same distance.
4. The aspect ratio of the image is 1:1.

In addition to the above constraints, by analyzing the images of the frames and mirrors, we have also observed that the left and right flanking mirrors (M_3 and M_1) and the central frame (M_0) are vertical, and the middle inset mirror (M_1) is only tilted in the y direction.

4 Finding Image Center

To find the image center, we will need to have both a set of 3D horizontal lines and a set of vertical lines whose projections are not parallel in the image. By finding the vanishing point of the two pairs of horizontal edges of the left and right flanking mirrors, (vl_x, vl_y) and (vr_x, vr_y), the y coordinate of the image center can be determined as $c_y = (vl_y + vr_y)/2$. By finding the vanishing point of the two vertical edges of the inset mirror, (vv_x, vv_y), the y coordinate of the image center can be determined as $c_x = vv_x$. In Fig. 2, using the digital image coordinate system (x_i, y_i) as shown in Fig. 2, we obtain the results in Table 1:

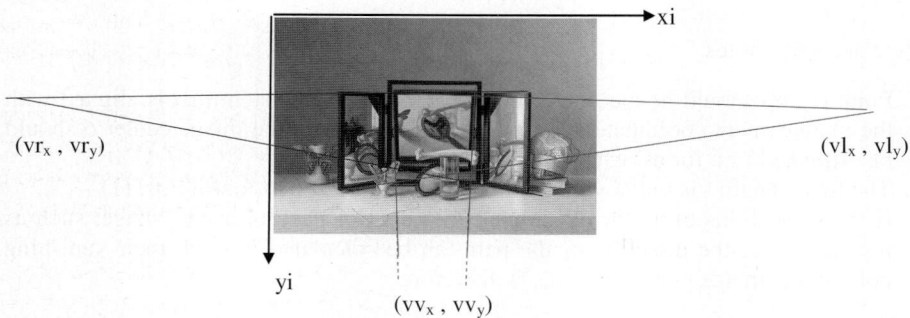

Fig. 2. Estimating image center using vanishing points

Table 1. Vanishing points portrayed in Fig. 2

Using left mirror	Using right mirror	Using middle mirror
(3219.8, 572.1)	(-1360.0, 519.7)	(868.6, 6524.3)

5 Estimating Focal Length

The focal length estimation is the most critical step. Once we find f, the rest of steps will be rather straightforward. We can find the focal length f by using the fact that the left and right flanking mirrors have exactly the same width, i.e. $W_1 = W_2$ in Fig. 3.

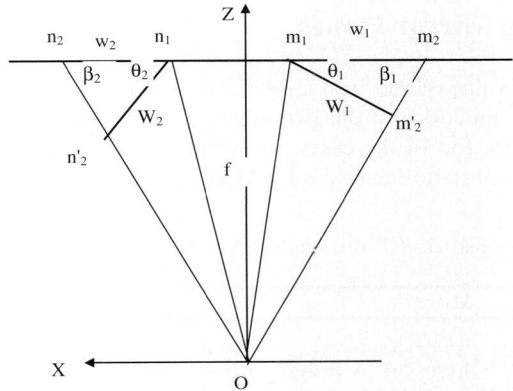

Fig. 3. Focal length estimation using the equal-width constraint

We scaled the mirrors into pixel representation so the distance of the two back edges (m_1 and n_1) are f (refer to Fig. 1). The directions of $n_1n'_2$ and $m_1m'_2$ can be represented by the vanishing points, $v_m = (v_{mx}, v_{my}, f) = (2228.6, 7.2, f)$, generated from the two sets of horizontal edges. The dot product of normalized v_m and the X-axis (1, 0, 0) should be $v_m / |v_m| \cdot (1, 0, 0) = \cos\theta_1$, assuming that v_m is in the horizontal plane as the X-axis (from assumption No 2). We can derive a similar relation for v_n. Therefore, the angles θ_1 and θ_2 can be represented as functions of f:

$$\theta_1 = \tan^{-1}(f / |v_{mx}|), \quad \theta_2 = \tan^{-1}(f / |v_{nx}|) \quad . \tag{5}$$

From the image, location m_1, m_2 and n_1, n_2 can be measured (in the x direction, refer to Fig. 1). Therefore, the two distances w_1 and w_2 can be calculated as $w_1 = |m_2 - m_1|$, $w_2 = |n_2 - n_1|$. The two angles β_1 and β_2 can be also represented as functions of f:

$$\beta_1 = \tan^{-1}(f / |m_2|), \quad \beta_2 = \tan^{-1}(f / |n_2|) \quad . \tag{6}$$

Using the sine law with triangles $n_1n_2n'_2$ and $m_1m_2m'_2$, and using the fact that $W_1 = W_2$, we can derive that:

$$f = \sqrt{\frac{a_n^2 v_{nx}^2 - a_m^2 v_{mx}^2}{a_m^2 - a_n^2}} \tag{7}$$

where $a_m = \dfrac{w_1}{|m_2| + |v_{mx}|}, a_n = \dfrac{w_2}{|n_2| + |v_{nx}|}$.

Using the measurements from the image, we obtain f = 2050.7 (pixels) Note that this method does not work if the two mirrors are symmetric. We calculated the two angles using Eq. (5): $\theta_1 = \tan^{-1}(f/|v_{mx}|) = 42.62°$, $\theta_2 = \tan^{-1}(f/|v_{nx}|) = 39.36°$. The two mirrors have different flanking angles, so they are not symmetric, therefore we can obtain the focal length estimation.

6 Representing Mirror Planes

Once we obtain the value of the focal length f, we can use the general equations (1) to (3) to calculate the directions of the two edges of each mirror M_i, and its normal n_i. These equations work for all the cases no matter if we can find vanishing points or not. The results of the rotation matrices for M_i (i=0, 1, 2, 3) are listed in Table 2.

Table 2. The rotation matrices (2nd row) and angles (3rd row, in degrees) of M_i (i = 0, 1, 2, 3)

M_0	M_1	M_2	M_3
0.9999 -0.0054 0.0129 0.0053 0.9999 0.0112 -0.0130 -0.0114 0.9999	0.7363 0.0157 -0.66765 0.0201 0.9988 0.0451 0.6764 -0.0468 0.7350	1.0000 -0.0028 0.0054 0.0009 0.9455 0.3257 -0.0060 -0.3257 0.9454	0.7743 -0.0046 0.6328 0.0028 1.0000 -0.0038 -0.6328 -0.0012 0.7743
(0.66, -0.76, 0.31)	(3.59, 43.16, 1.26)	(19.38, -0.32, 0.16)	(0.29, -39.84, 0.35)

From the rotation matrices it can be seen that the three angles of the middle frame M_0 are almost zero degrees, the middle inset mirror M_2 is mainly tipped forward, the right mirror M_1 and the left mirror M_2 mainly have flanking angles consistent with the results obtained by using Eq. (5). The results highly agree with our assumptions of the intrinsic geometry constraints in Section 2.

The plane equation of the mirror can be represented as

$$\mathbf{n}_i \bullet \mathbf{P}_c + d_i = 0 \ . \tag{8}$$

where $P_c = (X_c, Y_c, Z_c)^t$ is represented in the camera coordinate system O-XYZ. In order to find d_i, and further find the virtual camera parameters mirrored by each mirror, we will build a world coordinate system on each mirror. For example, for mirror M_1, after we find the three vectors and then normalized them into column unit vectors, still represented in v_1, v_2 and v_3, we can define a world coordinate system using the middle of the mirror plane as its origin, and the three vectors as its three coordinate axes. The transformation between the camera coordinate system and the world coordinate system of the *i*th mirror M_i (i = 0, 1, 2, 3) can be represented as

$$P_c = \mathbf{R}_{1i} P_w + \mathbf{T}_{1i} \ . \tag{9}$$

where $P_c = (X_c, Y_c, Z_c)^t$, $P_w = (X_w, Y_w, Z_w)^t$, $R_{1i} = (r_{pq})3\times3$, which is ($v_1$, v_2, v_3) for M_1, and T_{1i} to be determined. The projection of P_c into the image of the main camera is $(x, y) = (f X_c/Z_c, f Y_c/Z_c)$.

To find the translational vectors for all the three mirrors, we use the dimension of the middle frame M_0 as a reference. To calculate the translations for mirrors M_1 and M_3, we use their heights h= 566.464 pixels, which can be measured in image with respect to the height H of the middle frame, since they are at about the same distance. Similarly, to calculate the translation for the middle inset mirror, we use its width w = 461.5911 pixels, which can be measured against the width W.

Table 3. The translations and the distances of M_0 to M_3

Mirror	T_{1i} (i = 0, 1, 2, 3)	d_i (i = 0, 1, 2, 3)
M_0	(-51.3, -176.1, 2048.5)	2045.6
M_1	(-430.1, -215.7, 1929.7)	1802.8
M_2	(-50.9, -182.6, 2051.1)	1879.5
M_3	(340.5, -214.6, 1968.0)	1801.1

7 Locating Virtual Cameras

After we obtain the plane equation for each mirror (Eq. (8), the mirrored coordinate system, i.e., the virtual camera C_i, can be easily obtained, as in [9]. Here we use a coordinate transformation method to find the relation between each virtual camera C_i and the real camera C, by finding the rotation matrix R_i and translational vector T_i (i = 1, 2, 3). In our implementation, we use Eq. (9) to represent the origin and the three axes of the camera coordinate system in the world coordinate system $X_w Y_w Z_w$ of each mirror M_i. Since $X_w O Y_w$ is the mirror plane, the mirrored origin and axes can be simply obtained by changing the signs of their Z_w components. Then we do a similar procedure as in Eq. (9) to find the transformation (characterized by R_{2i} and T_{2i}, i = 1, 2, 3) between the world coordinate system M_i and the virtual camera C_i:

$$P_w = \mathbf{R}_{2i} P_i + \mathbf{T}_{2i}. \tag{10}$$

Combining Eqs (9) and (10), we can find the transformation between the real camera and the virtual camera:

$$P_c = \mathbf{R}_i P_i + \mathbf{T}_i. \tag{11}$$

$$\mathbf{R}_i = \mathbf{R}_{1i} \mathbf{R}_{2i}, \mathbf{T}i = \mathbf{R}_{1i} \mathbf{T}_{2i} + \mathbf{T}_{1i}. \tag{12}$$

Table 4 shows the estimated results. Using the full 6 degree-of-freedom (DOF) relation between each virtual camera and the main camera, stereo reconstruction can be performed, and the accuracy of the painting can be analyzed.

Table 4. The transformations between virtual camera C_i (i = 1, 2, 3) and the main camera

M_i	M_1			M_2			M_3		
R_i	-0.0846	0.0610	0.9945	-0.9999	-0.0035	-0.0103	-0.1992	-0.0049	-0.9800
	-0.0610	0.9959	-0.0663	0.0035	0.7878	-0.6159	0.0049	1.0000	-0.0059
	-0.9945	-0.0663	-0.0806	0.0103	-0.6159	-0.7877	0.9800	-0.0059	-0.1991
T_i	(-2299.7, 153.4, 2498.7)			(20.4, 1224.4, 3553.8)			(2200.2, 13.3, 2692.3)		

8 Conclusions and Discussions

Our method for estimating the camera models for the virtual cameras was based on single-view analysis. The relative 3D structures of the rectangles mirrors and frames are estimated by using their perspective analysis with a few assumptions. The camera calibration for the cameras is then successfully performed. Its results can then be used for 3D reconstruction and painting analysis, which is described in an accompanying paper [8]. The 3D estimates of both the frames and regular objects are consistent among the single-view analysis, and results from multiple stereo triangulations. However, there are some stereo and perspective inconsistencies across four views. These could either be the accuracies of the perspective distortions, orientations and sizes of the mirrors, or of the locations of objects inside the mirrors, which cannot be easily observed by eye. As such, our work extends the new discipline of computer vision applied to the study of fine art.

Acknowledgements

We thank Scott Fraser for information about his working methods and permission to reproduce his painting. The work is also partially supported by NSF under Grant No. CNS-0551598.

References

[1] Criminisi, A.: Accurate visual metrology from single and multiple uncalibrated images. ACM Distinguished Dissertation Series. Springer, London (2001)
[2] Criminisi, A., Kemp, M., Zisserman, A.: Bringing pictorial space to life: computer techniques for the analysis of paintings. In: Bentkowska-Kafel, A., Cashen, T., Gardner, H. (eds.) Digital Art History: A Subject in Transition, pp. 77–100 (2005)
[3] Criminisi, A., Reid, I., Zisserman, A.: Single view metrology. IJCV 40(2), 123–148 (1999)
[4] Doherty, M.S.: Object Project: Five objects, fifteen artists. Evansville Museum of Arts, History and Science. Evansville, IN (2007)
[5] Faugeras, O.: Three-Dimensional Computer Vision: A Geometric Viewpoint. MIT Press, Cambridge (1993)
[6] Hartley, R., Zisserman, A.: Multiple View Geometry in Computer Vision, 2nd edn. Cambridge University Press, Cambridge (2003)
[7] Huang, P.-H., Lai, S.-H.: Contour-based structure from reflection. In: CVPR 2006, vol. 1, pp. 379–386 (2006)
[8] Khoo, W., Jordan, T., Stork, D., Zhu, Z.: Reconstruction of a three-dimensional tableau from a single realist painting. In: 15th Int. Conf. Virtual Systems & Multimedia, September 9-12 (2009)
[9] Kumar, R.K., Ilie, A., Frahm, J.-M., Pollefeys, M.: Simple calibration of non-overlapping cameras with a mirror. In: CVPR 2008 (2008)
[10] Nene, S.A., Nayar, S.K.: Stereo with mirrors. In: ICCV 1998, pp. 1087–1094 (1998)
[11] Smith, B., Stork, D.G., Zhang, L.: Three-dimensional reconstruction from multiple reflected views within a realist painting: an application to Scott Fraser's Three way vanitas. SPIE/IS&T, Bellingham (2009)
[12] Zhu, Z., Lin, X., Shi, D., Xu, G.: A Single camera stereo system for obstacle detection. In: 4th Int. Conf. Information Systems Analysis and Synthesis, July 12-16, vol. 3, pp. 230–237 (1998)

A Model for Saliency Detection Using NMFsc Algorithm

Jian Liu and Yuncai Liu

Institute of Image Processing and Pattern Recognition
Shanghai Jiaotong University
No.800 Dong Chuan Road, Shanghai, P.R. China
{Sword_lj,Whomliu}@sjtu.edu.cn

Abstract. Saliency mechanism has been considered crucial in the human visual system and helpful to object detection and recognition. This paper addresses an information theoretic model for visual saliency detection. It consists of two steps: first, using the Non-negative Matrix Factorization with sparseness constraints (NMFsc) algorithm to learn the basis functions from a set of randomly sampled natural image patches; and then, applying information theoretic principle to generate the saliency map by the Salient Information (SI) which is calculated from the coefficients represented by basis functions. We compare our model with the previous methods on natural images. Experimental results show that our model performs better than existing approaches.

Keywords: Saliency Detection, NMFsc algorithm, Salient Information.

1 Introduction

Most of the traditional object detectors require training sets to detect specific object categories [1]. However, because of the training complexity and innumerable categories of visual patterns, there is a significant bottleneck to expand these models to generalized tasks. Taking notice of the biological vision, human can rapidly orientate towards salient objects in a cluttered visual scene, which is due to the existence of a saliency mechanism. Furthermore, with this mechanism, human vision has several advantages (e.g. robustness, flexibility and etc.) over computer vision. From this point, a general purpose saliency detector is needed to combine with traditional model.

During the past decade, several computational models have been invented to simulate human visual attention. Inspired by the feature-integration theory [2], Itti [3] proposed one of the earliest bottom-up selective attention model by utilizing color, intensity and orientation of images. Oliva [4] applied the global distributions of low-level features to detect the potential location of target objects. Harel [5] presented a graph-based visual saliency model (GBVS), which employed the graph theory to concentrate mass on activation maps and to form them from raw features. Bruce [6] introduced the idea of using Shannon's self-information to measure the perceptual saliency. More recently, Hou [7] proposed a Spectral Residual (SR) approach to calculate the saliency map based on Fourier Transform.

In this paper, we build on the work of Bruce [6] in extending the information theoretic principle for estimating the saliency maps. Specifically, we apply the NMFsc algorithm [8] to learn the basis functions from images patches instead of ICA, and define the saliency by the *Salient Information* (SI), which is calculated from the coefficients represented by basis functions, as the residual of self-information and entropy.

The organization of the paper is as follows: in Section 2, we introduce the NMFsc algorithm and use it to learn the basis functions from image patches. Then, estimate the saliency map from the residual of self-information and entropy that are obtained from the basis coefficients. Section 3 presents the experimental results of comparison between our model and three previous methods, and the conclusions and discussions are given in Section 4.

2 The Model

2.1 Image Representation via NMFsc

Sparse coding strategy [9] suggests that a good objective for an efficient coding of natural scenes should maximize the sparseness of the representation. It is shown that the NMFsc algorithm is a successful method in producing a representation of multidimensional data as sparse coding. Moreover, many evidences have showed that NMFsc has several advantages for simulating the behavior of the simple-cells than Independent Component Analysis (ICA)[10].

Only considering of linearity, the basic idea of sparse coding by NMFsc is relatively simple: a vectorized image patch, $\mathcal{V}(x)$, can be represented in terms of a linear superposition of (not necessarily orthogonal) basis functions $a_i(x)$. The linear generative form is:

$$\mathcal{V}(x) = \sum_{i=1}^{n} a_i(x) s_i \qquad (1)$$

where the coefficients, s_i, are stochastic and dynamic variables that change from the input images. Sparseness is a property independent of scale, and implies that s_i have probability densities with a high peak at zero and heavy tails.

Arranging all the input $\mathcal{V}(x)$ into the columns of matrix $\mathcal{V}$; the basis functions $a_i(x)$ into the matrix $\mathcal{A}$; and the corresponding coefficients s_i into the matrix S, so the linear representative form of the model is given:

$$\mathcal{V} = \mathcal{A}S \qquad (2)$$

$$\mathcal{W} = \mathcal{A}^{-1} \qquad (3)$$

$$S = \mathcal{W}\mathcal{V} \qquad (4)$$

where all the entries of both $\mathcal{A}$ and S to be non-negative.

With a purpose of making the features sparse, the optional sparseness measure is given based on the relationship between the L_1 norm and the L_2 norm, and it is defined as:

$$sparseness(\boldsymbol{x}) = \frac{\sqrt{n} - (\sum |x_i|)/\sqrt{\sum x_i^2}}{\sqrt{n} - 1} \quad (5)$$

where n is the dimensionality of $\boldsymbol{x}$. Particularly, we only set the sparseness of S to 0.85, so the $\boldsymbol{x}$ in Eq. 5 denotes each row of S.

2.2 Learning the Basis Functions

Since there exists a 'color opponent-component' system in human brain, which facilitate that red/green, green/red, blue/yellow and yellow/blue are color pairs inhibited by each other [11], we decompose a RGB image into three parallel channels as red/green ($\mathcal{RG}$), blue/yellow ($\mathcal{BY}$) and intensity ($\mathcal{I}$):

$$O_1 : \mathcal{RG} = \mathcal{R} - \mathcal{G} \quad (6)$$

$$O_2 : \mathcal{BY} = \mathcal{B} - \mathcal{Y} \quad (7)$$

$$O_3 : \mathcal{I} = (r + b + g)/3 \quad (8)$$

where $\mathcal{R} = r-(b+g)/2, \mathcal{G} = g-(r+b)/2, \mathcal{B} = b-(r+g)/2, \mathcal{Y} = r+g-|r-g|/2-b$.

Before learning the basis functions, we take measures to the original nature images. At first, we randomly sample $n \times n$ image patches M times from a set of nature images (we set $n = 8$ and $M = 100000$ in this paper), and then separate them into three channels. After that, for each channel, we convert each image patch into one column. Thus, the image patches are represented by three matrixes with the size of $N \times M (N = n^2)$. Finally, we train the three matrixes respectively, the results of three $N \times N$ ($N = 64$) basis functions are shown in Fig. 1.

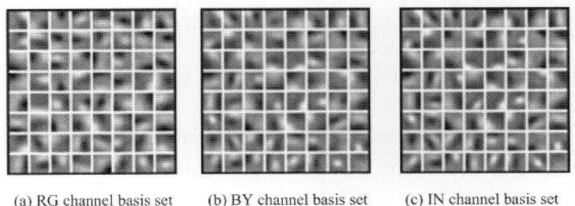

(a) RG channel basis set   (b) BY channel basis set   (c) IN channel basis set

Fig. 1. Basis functions trained from natural images patches

2.3 Estimating Saliency Map

Guided by the insights of previous work [6,12], we propose a measure of maximizing response information to estimate the saliency.

Suppose a $N \times M$ sample matrix $\mathcal{V} = [v_1, v_2, v_3 \ldots v_m]$, where v is a vectorized image patch. We project $\mathcal{V}$ into basis space by the inverse matrix of the learned basis set $(\mathcal{W})$ with a size of $N \times N$ (Eq. 4). Then we obtain the coefficients matrix S, which describes the contribution from each basis function or the responses of simple-cells in the point of view on biological vision.

Finally, we compute the average response probability function from each row in S as:

$$p_i = \frac{\sum_{j=1}^{m} |s_{ij}|}{\sum_{j=1}^{m} \sum_{i=1}^{n} |s_{ij}|} \qquad (9)$$

According to Shannon's information theory [13], we can measure the information of possible events whose probabilities of occurrence are $p_1, p_2, p_3 \ldots p_n$ from the quantities of the form

$$H(p) = -\sum_{i}^{n} p_i \log(p_i) \qquad (10)$$

where H is recognized as *entropy*, and the *self-information* is defined as:

$$I(p_i) = -\log(p_i) \qquad (11)$$

which is inversely proportional to the information of observing event.

For each p in (Eq. 9), we calculate the self-information $I(p_i)$ and the entropy $H(p)$, considering the *Salient Information* (*SI*) as:

$$SI(i) = -\log(p_i) - H(p) \qquad (12)$$

where $SI(i)$ denotes the uniqueness of the ith basis vector relative to all of the basis functions.

It is clearly shown that a feature is salient only when it can offer more unique information. With this general intuition of saliency, we neglect the non-uniqueness basis described as below the zero $(SI(i) < 0)$, and set them to zero $(SI(i) = 0)$. After that, we quantify the *Salient Response*: $SR = [r_1, r_2, r_3 \ldots r_m]$ as follows:

$$r_j = \max_{i=1 \rightarrow n} (SI(i) \cdot s_{ij}) \qquad (13)$$

then, we transform SR into image space with a Gaussian filter and finally generate the *Saliency Map* (*sM*).

Given a color image, there will be three saliency maps $(s\mathcal{M}_{BY}, s\mathcal{M}_{RG}, s\mathcal{M}_I)$ that are needed to combine into one conspicuous map $\overline{s\mathcal{M}}$ as:

$$\overline{s\mathcal{M}} = (s\mathcal{M}_{BY} + s\mathcal{M}_{RG} + s\mathcal{M}_I)/3 \qquad (14)$$

where the maximum of $\overline{s\mathcal{M}}$ defines the most salient object in the image.

3 Experimental Results and Analysis

In our experiment, the basis functions are trained from the nature image database collected by [14] and all the other the test images are taken from previous work [7].

We sequentially sample all over the possible image patches with a 8 × 8 window and compare our method with Bruce's approach (Self-info) [6], SR algorithm [7] and Itti's model (using the SaliencyToolBox (STB) [15]) at MATLAB2007a environment on Windows platform.

3.1 Evaluating the Result

In order to appropriately evaluate our model, we not only use the *Hit Rate* (HR) and *False Alarm Rate* (FAR) approach introduced in [7], but also test the *Receiver Operating Characteristic* (ROC) curve [16] performance on a human eye fixations database called DOVES, which is collected by van der Linde [17] and is available at the website[1].

HR/FAR Performance: In this part of the experiment, 10 subjects are provided to specify salient objects in 100 images by drawing contours. For each image, a binary label is given to indicate the pixels whether or not belongs to the salient objects, and a binary image $\mathcal{B}_i(x)$ is obtained as:

$$\mathcal{B}_i(x) = \begin{cases} 1 & pixels \in salient\ objects \\ 0 & otherwise \end{cases}$$

Considering all the subjects, the average $\overline{\mathcal{B}_i(x)}$ can be calculated as:

$$\overline{\mathcal{B}_i(x)} = \frac{1}{m}\sum_{i=1}^{m}\mathcal{B}_i(x) \quad (m=10) \tag{15}$$

Given the generated saliency map $\overline{s\mathcal{M}_i}$, the HR and FAR are defined by:

$$HR = \frac{1}{n}\sum_{i=1}^{n}E\{\overline{\mathcal{B}_i(x)} \cdot \overline{s\mathcal{M}_i}\} \quad (n=100) \tag{16}$$

$$FAR = \frac{1}{n}\sum_{i=1}^{n}E\{(1-\overline{\mathcal{B}_i(x)}) \cdot \overline{s\mathcal{M}_i}\} \quad (n=100) \tag{17}$$

where $E\{\mathcal{I}(x)\}$ is the average intensity of the saliency map.

In order to make the comparison, we use the *ratio* (r) of the HR and FAR instead of adjusting the intensity of the saliency map which can also evaluate the performance of salient objects detection in images.

$$r = \frac{HR}{FAR} \tag{18}$$

The results of the *ratio* and the *Average Time Cost* (ATC) are shown in Table 1.

From the results, we note that our method performs the overall the best among the three methods, achieving the highest *ratio* of (HR) and (FAR).

---
[1] http://live.ece.utexas.edu/research/doves

Table 1. HR/FAR performance of the three methods

	Our method	Self-info	SR	STB
r	2.997	2.913	2.895	1.785
ATC(s)	0.3373	6.4753	0.0618	1.3427

ROC Analysis: In this part of the experiment, we use DOVES as a benchmark for comparison. The data set contains a collection of eye movements from 29 human observers as they viewed 101 natural calibrated images of size 1024×768. We down-sample each image to an appropriate scale (256×192, $\frac{1}{4}$ of the original size). Fig. 2 shows an actual image with the resulting saliency maps, and mean ROC area scores of three models are indicated in Table 2.

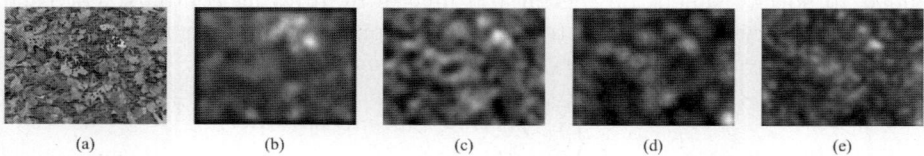

(a) (b) (c) (d) (e)

Fig. 2. (a) An image from the data set with the fixations indicated using red solid circles (b) The saliency map of our method (c) The saliency map generated by Bruce's approach (d) The saliency map of SR (e) The saliency map generated by STB

Table 2. ROC scores of the three methods

	Our method	Self-info	SR	STB
ROC	0.6545	0.6142	0.5691	0.5903

According to the definition of ROC curve, it provides a measure summarizing performance across all possible thresholds. For perfect prediction, the ROC area under the curve will be 1.0; for opportunistic performance, the value will be 0.5, and when the system is predicting worse than chance, the area will be less than 0.5. Our implementation of the mean ROC area is 0.6545, which is higher than the performance of the other three methods. Thus, it is shown that our method can predict human fixations more reliably.

In addition, we show the performance metric of mean ROC and 'inter-subject ROC' [5]. To compute an inter-subject ROC score, we use the fixations from 5 subjects on a single image, and calculate the mean (across subjects) ROC metric between a single subject's fixations and a heat map generated from the remaining fixations of subjects, which is called a 'leave-one-out' procedure. Fig. 3 demonstrates the effective performance of our method contrasted against the other three approaches, for our curve is more close to the upper bound.

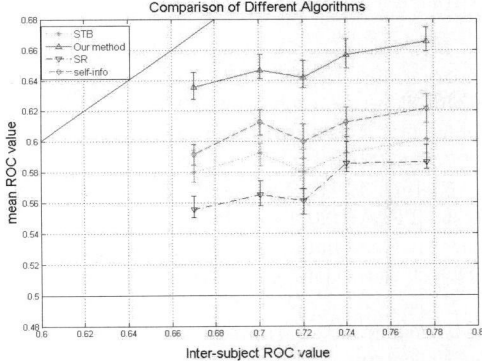

Fig. 3. Comparison of predictive power among three methods. Our curve is more close to the upper bound.

4 Conclusions and Discussions

We presented a model to estimate saliency map with the purpose of object detection. The method uses NMFsc algorithm to represent an image as a linear combination of sparse basis based on sparse coding strategy, and to generate the saliency map from the informative responses of basis functions, which is defined as *Salient Information* (SI). We experimentally compared our method with the previous methods. The experimental results show that our model performs more accurately than the other methods to find the salient objects in images (see Fig. 4).

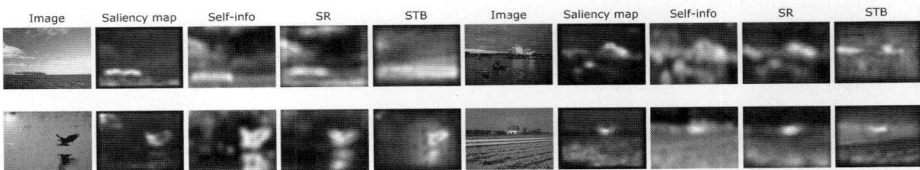

Fig. 4. The results of our model in comparison with Self-info approach, SR method and Itti's method(STB)

In future work, we plan to incorporate motion information into our method, which is also an important feature to attract human attention. Furthermore, we are extending our model to multidimensional space for simulating the nonlinear properties of complex-cells in V1. In addition, it is necessary to combine with the traditional model for developing a more robust object detector in general purpose application.

References

1. Fergus, R., Perona, P., Zisserman, A.: Object Class Recognition by Unsupervised Scale-invariant Learning. In: IEEE Computer Society Conference on Computer Vision and Pattern Recognition (2003)
2. Treisman, A., Gelade, G.: A Feature-Integration Theory of Attention. Cognitive Psychology 12(1), 97–136 (1980)
3. Itti, L., Koch, C., Niebur, E.: A Model of Saliency-Based Visual Attention for Rapid Scene Analysis. TPAMI 20(11), 1254–1259 (1998)
4. Oliva, A., Torralba, A., Castelhano, M., Henderson, J.: Top-down control of visual attention in object detection. In: IEEE International Conference on Image Processing (2003)
5. Harel, J., Koch, C., Perona, P.: Graph-Based Visual Saliency. In: Advances in Neural Information Processing Systems (2007)
6. Bruce, N., Tsotsos, J.: Saliency Based on Information Maximization. In: Advances in Neural Information Processing Systems, vol. 18 (2006)
7. Hou, X., Zhang, L.: Saliency Detection: A Spectral Residual Approach. In: IEEE Computer Society Conference on Computer Vision and Pattern Recognition (2007)
8. Hoyer, P.: Non-negative matrix factorization with sparseness constraints. Journal of Machine Learning Research 14, 1457–1469 (2004)
9. Olshausen, B., Field, D.: Natural image statistics and efficient coding. Network: Computation in Neural Systems 14, 333–339 (1996)
10. Hoyer, P.: Non-negative sparse coding. In: Proc. IEEE Workshop on Neural Networks for Signal Processing, vol. 12, pp. 557–565 (2002)
11. Engel, S., Zhang, X., Wandell, B.: Colour Tuning in Human Visual Cortex Measured With Functional Magnetic Resonance Imaging. Nature 388(6637), 68–71 (1997)
12. Hou, X., Zhang, L.: Dynamic Visual Attention: Searching for coding length increments. In: Advances in Neural Information Processing Systems (2008)
13. Shannon, C.: A mathematical theory of communication. Bell System Technical Journal 27, 93–154 (1948)
14. Frey, H.P., König, P., Einhäuser, W.: The Role of First-and Second-Order Stimulus Features for Human Overt Attention. Perception and Psychophysics 69, 153–161 (2007)
15. Walther, D., Koch, C.: Modeling attention to salient proto-objects. Neural Networks 19, 1395–1407 (2006)
16. Tatler, B., Baddeley, R., Gilchrist, I.: Visual correlates of fixation selection: effects of scale and time. Vision Research 14, 643–659 (2005)
17. van der Linde, I., Rajashekar, U., Bovik, A., Cormack, L.: DOVES: A Database of Visual Eye Movements. Spatial Vision (2008) (in press)

Directional Force Field-Based Maps: Implementation and Application

JingBo Ni, Melanie Veltman, and Pascal Matsakis

Computing and Information Science, University of Guelph, ON, N1G 2W1, Canada
{jni,mveltman,pmatsaki}@uoguelph.ca

Abstract. A directional relationship (e.g., right, above) to a reference object can be modeled by a directional map – an image where the value of each point represents how well the relationship holds between the point and the object. As we showed in previous work, such a map can be derived from a force field created by the object (which is seen as a physical entity). This force field-based model, defined by equations in the continuous domain, shows unique characteristics. However, the approximation algorithms that were proposed in the case of 2-D raster data lack efficiency and accuracy. We introduce here new algorithms that correct this flaw, and we illustrate the potential of the force field-based approach through an application to scene matching.

Keywords: spatial relationships, force fields, directional maps, scene matching.

1 Introduction

Research on the modeling of spatial relationships raises two questions: (a) How to identify the relationships between two given objects [1,2]? (b) How to identify, in a scene, the object that best satisfies a given relationship to a reference object [3]? The second question defines an object localization task. One theory supported by cognitive experiments is that people accomplish this task by parsing space around the reference object into *good* regions (where the object being sought is more likely to be), *acceptable* and *unacceptable* regions (where the object being sought cannot be) [4,5]. These regions form a so-called *spatial template* [5,6], which assigns each point in space a value between 0 (unacceptable region) and 1 (good region).

When focusing on *directional* (also called *projective* [7] or *cardinal* [8]) relationships (e.g., front, south, above), spatial templates can be referred to as *directional maps* [9] (or as *fuzzy landscapes* [3]). A directional map is an image where the value of each point reflects the degree to which the point satisfies some directional relationship to a reference object.

The map as defined in [3] takes the object's shape into account and depends essentially on angular deviation (two characteristics supported by cognitive studies). We call it the *standard map*, $S^{\delta R}$, where δ represents the directional relationship and R the reference object. Two algorithms have been designed for fast calculation of $S^{\delta R}$: the first one is based on a propagation technique [3] and the second one on partitioning the image into parallel lines [10].

Matsakis et al. [9] proposed another model (Section 2), which relies on the idea of considering the reference object as a physical entity that creates a force field. All directional maps induced by the object can then be derived from the force field. Compared with standard maps, force field-based maps better cope with outliers, elongated objects and concavities [9]. However, the algorithm for force field computation described in [9] is slow. Moreover, although the directional maps induced by the object can be derived from the force field in negligible time, their calculation lacks accuracy. Indeed, the maps depend on a supremum that can only be estimated, and which, in [9], is often greatly overestimated. We introduce here new algorithms that correct these flaws (Section 3) and we illustrate the potential of the force field-based approach through an application to scene matching (Section 4). Note that directional maps can be used for many tasks, including spatial reasoning, object localization and identification, structural and model-based pattern recognition [11,12,13,14].

2 Force Field-Based Maps

The notations used in the rest of the paper are as follows. $\mathbb{Z}_+$ is the set of positive integers. $\mathbb{P}$ is the Euclidean plane. For any points p and q of $\mathbb{P}$, pq is the vector from p to q with norm $|pq|$. $\vec{\delta}$ is the unit vector pointing at direction $\delta \in [0, 2\pi)$. The radian measure in $[0,\pi]$ of the angle between two nonzero vectors $\vec{u}$ and $\vec{v}$ is denoted by $\angle(\vec{u}, \vec{v})$. An *object* R is a subset of $\mathbb{P}$, bounded, closed, with area $|R| \neq 0$. We have $R(p)=1$ if $p \in R$ and $R(p)=0$ if $p \notin R$.

We assume that point q exerts on point p a force of magnitude $1/|pq|^r$ in the direction of pq, where r is a given real number. The force $\Phi_r^R(q)$ that q exerts on R is the integration of the forces that q exerts on all the points of R:

$$\Phi_r^R(q) = \int_{p \in \mathbb{P}} \frac{R(p)}{|pq|^r} \frac{pq}{|pq|} dp = \int_{p \in R} \frac{pq}{|pq|^{r+1}} dp. \qquad (1)$$

Φ_r^R is called the *force field* created by R [9]. Note that the algorithm for force field computation described in [9] is rather slow. A much more efficient algorithm is introduced in Section 3.1. The *force field-based map* $\Phi_r^{\delta R}$ in direction δ can be defined by, e.g., $\Phi_r^{\delta R}(q) = \mu(\angle(\Phi_r^R(q), \vec{\delta}))$, with $\mu(x) = \max\{0, 1-2x/\pi\}$, or:

$$\Phi_r^{\delta R}(q) = \max\{0, (\Phi_r^R(q) \cdot \vec{\delta}) / (\sup_{p \in \mathbb{P}} \Phi_r^R(p) \cdot \vec{\delta})\}. \qquad (2)$$

We focus here on (2), which explicitly takes account of distance information (when $r \neq 0$). Unfortunately, $\sup_{p \in \mathbb{P}} \Phi_r^R(p) \cdot \vec{\delta}$ cannot be easily determined (unless $r=0$). In [9], Matsakis et al. replaced it with an upper bound that was determined analytically. However, this upper bound is often much higher than the supremum itself. As a result, the calculated value for $\Phi_r^{\delta R}(q)$ is often unreasonably low. In Section 3.2, we show that $\sup_{p \in \mathbb{P}} \Phi_r^R(p) \cdot \vec{\delta}$ can be better estimated using a heuristic search algorithm.

3 Implementation

Here, we see an object R in a digital image $\mathbb{G}$ as a surface covered by pixels (of size 1×1), not as a discrete cloud of points like in [9]. The force that one pixel exerts on another is defined based upon the following considerations. First, we draw a horizontal axis X and a vertical axis Y. For any two distinct points p and q of $\mathbb{P}$, let $\theta \in [0,2\pi)$ be the angle between pq and X, p_X and q_X be the projections of p and q on X, and p_Y and q_Y be their projections on Y. Thus, the magnitude of the force that q exerts on p can also be computed as: $1/|pq|^r = |\cos\theta|^r/|p_X q_X|^r$ if $\theta \in \Theta_h = [0,\pi/4] \cup [3\pi/4, 5\pi/4] \cup [7\pi/4, 2\pi)$; and $1/|pq|^r = |\sin\theta|^r/|p_Y q_Y|^r$ if $\theta \in \Theta_v = (\pi/4, 3\pi/4) \cup (5\pi/4, 7\pi/4)$. See Fig. 1(a). Now, consider p and q two distinct pixels in $\mathbb{G}$ centered at points (x_p, y_p) and (x_q, y_q). We define that the force pixel q exerts on pixel p is in the direction of $(x_q - x_p, y_q - y_p)$, and has the magnitude: $F_r(p,q) = |\cos\theta|^r f_r(|x_q - x_p|)$ if $\theta \in \Theta_h$; and $F_r(p,q) = |\sin\theta|^r f_r(|y_q - y_p|)$ if $\theta \in \Theta_v$, where θ is the angle between $(x_q - x_p, y_q - y_p)$ and X. Now, let us only consider the case that $\theta \in \Theta_h$. The value of $1/|p_X q_X|^r$ mentioned above in fact is the magnitude of the force between the projections of the points p and q on X, p_X and q_X, which are points too. When p and q are pixels, their projections on X, I and J, are unit line segments (instead of points), as shown in Fig. 1(b). It is therefore natural to define $f_r(s)$, $s \in \mathbb{Z}_+$, as the sum of the forces that the points of J exert on the points of I, i.e.,

$$f_r(s) = \int_0^1 \int_0^1 1/(y-x+s)^r \, dxdy. \quad (3)$$

Function f_r is well defined on $\mathbb{Z}_+$ when $r<2$, and the double integral can be solved analytically. When $p=q$, we set $F_r(p,q)=0$ indicating that the forces that a pixel exerts on itself are balanced out. Having defined the force between two pixels, the force that one pixel q exerts on a raster object R (i.e., a set of pixels) can be computed as:

$$\Phi_r^R(q) = \sum_{p \in \mathbb{G}} R(p) F_r(p,q)(x_q - x_p, y_q - y_p)/|(x_q - x_p, y_q - y_p)|. \quad (4)$$

3.1 An Algorithm for Approximating Φ_r^R

For any image $\mathbb{G}$ of size $N = m \times n$, Equation (4) calculates $\Phi_r^R(q)$ in $\mathcal{O}(N)$, i.e., it calculates the entire force field Φ_r^R in $\mathcal{O}(N^2)$. Here, we propose an algorithm for fast approximating Φ_r^R. Let q be the origin of $\mathbb{P}$, i.e., $q=(0,0)$. By rewriting (1) using the polar coordinates (θ, ℓ) of p, we have:

$$\Phi_r^R(q) = \Phi_r^R(0,0) = \int_0^{2\pi} \int_0^{+\infty} R(\theta, \ell)(-\cos\theta, -\sin\theta)/\ell^{r-1} d\ell d\theta. \quad (5)$$

By letting $\ell = s/|\cos\theta|$ for $\theta \in \Theta_h$ and $\ell = s/|\sin\theta|$ for $\theta \in \Theta_v$, (5) becomes:

$$\Phi_r^R(q) = \int_{\theta \in \Theta_h} |\cos\theta|^{r-2}(-\cos\theta, -\sin\theta)[\int_0^{+\infty} R(\theta, s/|\cos\theta|)/s^{r-1} ds] d\theta \\ + \int_{\theta \in \Theta_v} |\sin\theta|^{r-2}(-\cos\theta, -\sin\theta)[\int_0^{+\infty} R(\theta, s/|\sin\theta|)/s^{r-1} ds] d\theta. \quad (6)$$

In (6), point $p = (\theta, s/|\cos\theta|)$ (or $(\theta, s/|\sin\theta|)$) lies on the line that starts from q and points at direction θ, and $1/s^{r-1}$ is the magnitude of the force between the projects of p and q on axis X (or Y). According to the discussion made above, we can transform (6) to

$$\Phi_r^R(q) = \sum_{\theta \in \Theta_h} |\cos\theta|^{r-2} (-\cos\theta, -\sin\theta)[\sum_{s \in \mathbb{Z}_+} R(p_{\theta,s}) f_{r-1}(s)]\Delta\theta \atop + \sum_{\theta \in \Theta_v} |\sin\theta|^{r-2} (-\cos\theta, -\sin\theta)[\sum_{s \in \mathbb{Z}_+} R(p_{\theta,s}) f_{r-1}(s)]\Delta\theta \quad , \quad (7)$$

for handling raster data. In (7), θ belongs to a set $\{2\pi k/K\}_{k \in 0..K-1}$ of K reference directions, and therefore $\Delta\theta$ is $2\pi/K$. $p_{\theta,0}=q$, $p_{\theta,1}$, $p_{\theta,2}$, etc., are the pixels successively encountered on the rasterization of the line $\Lambda_\theta(\omega)$, which starts from an edge pixel ω of $\mathbb{G}$ and points at direction θ. See Fig. 1(c). Since R is crisp, which means each $R(p_{\theta,s})$ is either 1 or 0, expression $\sum_{s \in \mathbb{Z}_+} R(p_{\theta,s}) f_{r-1}(s)$ in (7) can then be written as:

$$\sum_{s \in \mathbb{Z}_+} R(p_{\theta,s}) f_{r-1}(s) = \sum_{i=1}^{M} \sum_{s=a_i}^{b_i} f_{r-1}(s) = \sum_{i=1}^{M} \int_{a_i}^{b_i+1} \int_0^1 1/(y-x)^{r-1} dxdy \ . \quad (8)$$

M here is the number of segments of R on the line $\Lambda_\theta(\omega)$ encountered after q. When R is convex, $M \leq 1$ for any θ and q. In (8), the rightmost double integral can again be analytically solved. Equations (7,8) then calculate $\Phi_r^R(q)$ in $\mathcal{O}(KM)$, and calculate Φ_r^R in $\mathcal{O}(KMN)+\mathcal{O}(KN)=\mathcal{O}(KMN)$, where $\mathcal{O}(KN)$ time is required to rasterize the lines $\Lambda_\theta(\omega)$ (for all θ and ω) in $\mathbb{G}$, and to determine the segments of R on those lines. In practice, the value of M is usually small, however, in the worst case, the value of M can reach $\sqrt{N}$, which raises the complexity up to $\mathcal{O}(KN\sqrt{N})$. When R is fuzzy, the manipulation of R can always be reduced to that of its level-cuts, which are crisp.

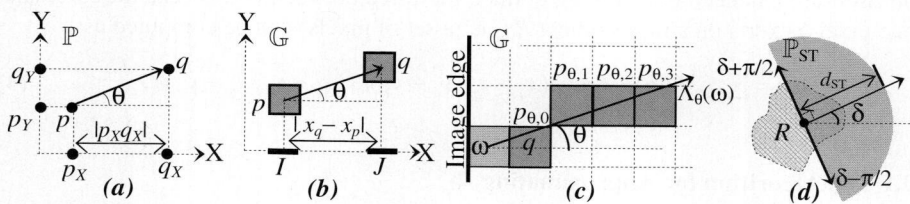

Fig. 1. (a) The force between two points; (b) The force between two pixels; (c) The rasterization of the line $\Lambda\theta(\omega)$; (d) The searching territory $\mathbb{P}_{ST}$.

3.2 Estimation of $\sup_{p \in \mathbb{P}} \Phi_r^R(p) \cdot \vec{\delta}$

When $r=0$, we have $\sup_{p \in \mathbb{P}} \Phi_0^R(p) \cdot \vec{\delta} = |R|$ [9]. For $r \neq 0$, we develop an algorithm which searches in a pre-determined territory for the point $p_{\max}$, such that $\Phi_r^R(p_{\max}) \cdot \vec{\delta}$ forms a good approximation of $\sup_{p \in \mathbb{P}} \Phi_r^R(p) \cdot \vec{\delta}$. Equation (2) can then be replaced with:

$$\Phi_r^{\delta R}(q) = \max\{0, \min\{1, (\Phi_r^R(q) \cdot \vec{\delta})/(\Phi_r^R(p_{\max}) \cdot \vec{\delta})\}\} \ . \quad (9)$$

For easy illustration, we express $\mathbb{P}$ in terms of polar coordinates (θ, ℓ) and set the origin $(0,0)$ at the centroid of R. The searching territory (Fig. 1(d)) is defined as: $\mathbb{P}_{ST} = [\delta-\pi/2, \delta+\pi/2] \times [0, d_{ST}] \subset \mathbb{P}$. $d_{ST} = \alpha + \beta/\sqrt{r}$, where $\alpha = 2 \int_\theta \int_\ell R(\theta, \ell) |\ell\cos(\theta-\delta)|/|R| d\ell d\theta$ and $\beta = 2 \int_\theta \int_\ell R(\theta, \ell) |\ell\sin(\theta-\delta)|/|R| d\ell d\theta$, is the pre-determined searching distance. The searching algorithm is given as follows:

```
δ'=δ; d=d_ST;         /*The initial searching direction and distance are δ and d_ST.*/
s_p=5;                /*The searching speed is set to 5*/
Δδ=0.001;             /* the minimum angle difference is set to 0.001*/
FOR each iteration    /*The number of iterations is set to 5*/
   δ_a=δ'−π/2, δ_b=δ'+π/2;
   WHILE δ_b−δ_a ≥ Δδ      /*Search on the curve defined by [δ_a,δ_b]×{d} */
      p_a=(δ_a,d) and p_b=(δ_b,d);
      IF Φ_r^R(p_a)·δ̄ < Φ_r^R(p_b)·δ̄ : δ_a=δ_a+(δ_b−δ_a)/s_p; ELSE: δ_b=δ_b−(δ_b−δ_a)/s_p;
   δ'=δ_b, d_a=0 and d_b=d_ST;     /*δ' is the adjusted searching direction*/
   WHILE d_b − d_a ≥ 1      /*Search on the line defined by {δ'}×[d_a,d_b] */
      p_a=(δ',d_a) and p_b=(δ',d_b);
      IF Φ_r^R(p_a)·δ̄ < Φ_r^R(p_b)·δ̄ : d_a=d_a+(d_b−d_a)/s_p; ELSE: d_b=d_b−(d_b−d_a)/s_p;
   d =d_b;                  /*d is the adjusted searching distance*/
   IF sup < Φ_r^R(p_b)·δ̄ : sup = Φ_r^R(p_b)·δ̄ ; /* sup is initially set to 0 */
   ELSE: RETURN p_max= p_b;   /* Nothing to update means we found p_max = p_b */
RETURN p_max= p_b;           /*After all iterations, we let p_max = p_b */
```

3.3 Experiments

Let Φ_r^R be the exact force field calculated according to (4), and $^K\Phi_r^R$ be the force field computed using (7,8). The difference between Φ_r^R and $^K\Phi_r^R$ is measured by the difference ratio (DR), which takes on values in [0,1], and is 0 iff $\Phi_r^R = {}^K\Phi_r^R$:

$$DR = (\sum_q |\Phi_r^R(q) - {}^K\Phi_r^R(q)|) / \sum_q (|\Phi_r^R(q)| + |{}^K\Phi_r^R(q)|). \quad (10)$$

The force field $^K\Phi_r^R$ approximates Φ_r^R. The accuracy of the approximation increases with K (Fig. 2(b)) and is quite high (DR is less than 0.5%) even when K is relatively small (K=90). The accuracy also depends on the image size N and on r (Figs. 2(b,c)).

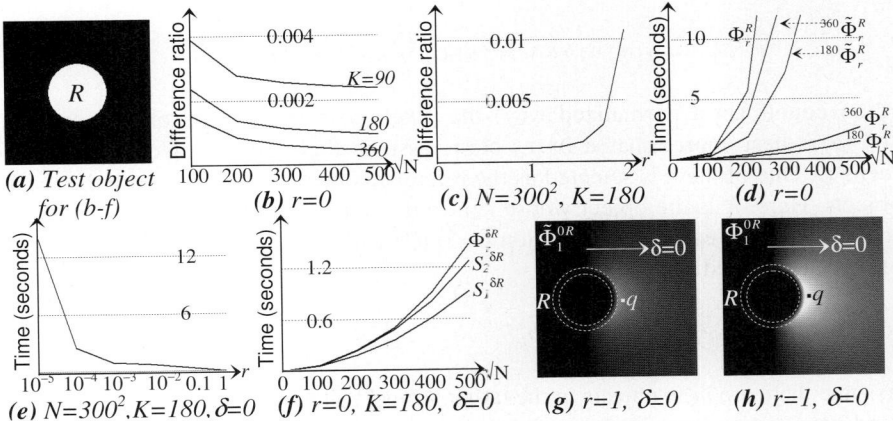

Fig. 2. Experiments. $S_1^{\delta R}$ and $S_2^{\delta R}$ in (f) are the standard maps generated using the first [3] and the second algorithms [10] (Section 1). All algorithms were implemented in C.

Let $^K\tilde{\Phi}_r^R$ be the force field computed using the algorithm proposed in [9]. Fig. 2(d) shows that the processing times of $^K\tilde{\Phi}_r^R$ and $^K\Phi_r^R$ both increase with K and N, but at different rates. Computing $^K\Phi_r^R$ is far more efficient than computing $^K\tilde{\Phi}_r^R$ and Φ_r^R. Once the force field is calculated, directional maps $\Phi_r^{\delta R}$ in various δ can be computed using (9). The Computation is generally fast, unless r takes some arbitrarily small positive value (like 10^{-5}) (Fig. 2(e)). Fig. 2(f) shows that the computation of $\Phi_r^{\delta R}$ ($^K\Phi_r^R$ plus $\Phi_r^{\delta R}$) is comparably efficient to that of the standard map $S^{\delta R}$.

Let $\tilde{\Phi}_r^{\delta R}$ be the force field-based map generated by the second transformation proposed in [9]. Consider the case that R is a concentric shell, $r=1$, and $\delta=0$. The values of $\tilde{\Phi}_1^{0R}$ are all fairly low ($\tilde{\Phi}_1^{0R}$ appears dark in Fig.2 (g)) due to the overestimation problem mentioned in Section 2. This raises some serious issues. For example, according to intuition, pixel q (in Fig.2 (g)) is perfectly close to and to the right of R. However, $\tilde{\Phi}_1^{0R}$ somewhat denies this perception since $\tilde{\Phi}_1^{0R}(q) = 0.5$. In the map generated using (9), Φ_1^{0R} (Fig.2 (h)), such issue does not exist and $\Phi_1^{0R}(q) = 1$.

4 Application

Here, we illustrate the potential of the force field-based approach through a scene matching task. Consider a target scene depicting a number of objects. As an example, the scene in Fig. 3(a) contains 21 disconnected objects, Figs. 3(b,c) show two (hand-drawn) query scenes, and the task is to determine if there exists a match between query and target. Note that for our purposes, a 'match' exists when there are objects in the target scene whose relative positions correspond to those found between the objects in the query scene. Furthermore, we want matching to be invariant to scaling and rotation. What follows is a description of how this task can be performed.

Consider a reference object R and a number of located objects L_i with $i=1..n$. Object R's view histogram in direction δ, h_R^δ, is a function from $\{t_k = k/P\}_{k=0..P}$ to $[0,1]$:

$$h_R^\delta(t_k) = \sum_{\{q \mid |\Phi_r^{\delta R}(q)-t_k| = \min_j |\Phi_r^{\delta R}(q)-t_j|\}} [(\sum_{i=1}^n L_i(q))/\sum_{i=1}^n |L_i|]. \tag{11}$$

$h_R^\delta(t_k)$ counts (in a normalized way) the pixels q in the located objects such that $\Phi_r^{\delta R}(q)$ is best approximated by t_k. Now, assume there are n objects in the query scene Q. One of them is selected as the reference object R, the others are the located objects. Here, R is the object whose centroid is closest to the centroid of the entire scene. Then, for each view direction $\delta_i = 2\pi i/D$ with $i=0..D-1$, four view histograms of R are computed in Q:

$$Q_R^{\delta_i} = (h_R^{\delta_i}, h_R^{\delta_i + \pi/2}, h_R^{\delta_i + \pi}, h_R^{\delta_i + 3\pi/2}). \tag{12}$$

Assume there are $m \geq n$ objects in the target scene T. The matching between T and Q is conducted in an exhaustive way:

FOR each object O in T:
  Let O be the reference object;
  List all the possible ways of drawing $n-1$ objects from the other $m-1$ objects in T;
  FOR each drawing:
    Let the $n-1$ objects be the located objects, which, together with the reference object O, form a sub-scene T' of T;
    Compute O's view histograms in the sub-scene: $T'_O = (h_O^0, h_O^{\pi/2}, h_O^{\pi}, h_O^{3\pi/2})$;
    FOR each $Q_R^{\delta_i}$:
      Compute the similarity degree between T'_O and $Q_R^{\delta_i}$, $sim(T'_O, Q_R^{\delta_i})$.
      Let $\delta_{max}=\delta_k$ such that $sim(T'_O, Q_R^{\delta_k}) = \max_i \{ sim(T'_O, Q_R^{\delta_i}) \}$, and **Let** $sim(T'_O, Q_R^{\delta_{max}})$ be the degree of matching between Q and T'.

The similarity degree between T'_O and $Q_R^{\delta_i}$, $sim(T'_O, Q_R^{\delta_i})$, is computed as:

$$sim(T'_O, Q_R^{\delta_i}) = \min_{j=0..3} \{ d(h_O^{j\pi/2}, h_R^{\delta_i + j\pi/2}) \}, \quad (13)$$

where $d(h_1,h_2)=\max\{0, 1-\sum_k |h_1(t_k)-h_2(t_k)|\}$ measures the similarity between the view histograms h_1 and h_2. Note that $d(h_1,h_2)=1$ *iff* $h_1=h_2$. Finally, we present the sub-scenes T' that best match Q (the sub-scenes with highest degrees $sim(T'_O, Q_R^{\delta_{max}})$).

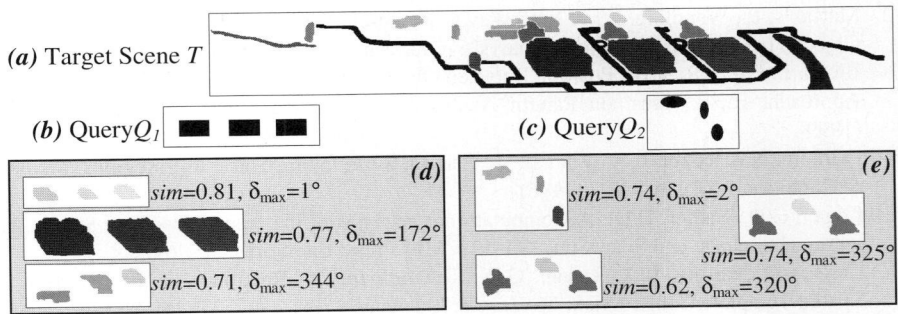

Fig. 3. A scene matching task. *(a)* A hand-segmented laser radar range image of the powerplant at China Lake, CA. The image was used by Matsakis, Keller *et al.* in [15]. *(d)* The subscenes of T that best match Q_1, and *(e)* those that best match Q_2, where *sim* is the similarity degree between a query and a sub-scene (after counterclockwise rotation by δ_{max}).

As shown by Fig. 3, the proposed algorithm generates reasonable results. In this experiment, $P=100$, $D=360$, and each force field-based map was computed using the algorithm defined by (7,8,9) with $r=0$ and $K=90$. Under this configuration, for each of the query scenes in Figs. 3(b,c), the matching algorithm finishes within one minute. Of course, smaller values of P, D and K can be chosen to compromise precision for speed. Readers may ask whether standard maps can be applied to scene matching tasks like the one presented here. The answer is negative. Since standard maps rely merely on angular deviation, they lack representation power for dimension and distance information, which is critical to these tasks.

5 Conclusions

In [9], Matsakis *et al.* developed a new quantitative model of the directional relationships to a reference object. The model relies on the idea that all directional maps induced by the object can be derived from a force field. However, the proposed algorithms lack efficiency and accuracy. In this paper, we have introduced algorithms that correct this flaw and we have demonstrated the potential of the force field-based approach through an application to scene matching. In future work, we will further explore the idea of using directional maps as a tool for pattern recognition and scene understanding.

Acknowledgements

The authors want to express their gratitude for support from the Natural Science and Engineering Research Council of Canada (NSERC), grant 262117.

References

1. Krishnapuram, R., Keller, J.M., Ma, Y.: Quantitative Analysis of Properties and Spatial Relations of Fuzzy Image Regions. IEEE Trans. on Fuzzy Systems 1(3), 222–233 (1993)
2. Matsakis, P., Wendling, L.: A New Way to Represent the Relative Position between Areal Objects. IEEE Trans. on Pattern Analysis and Machine Intelligence 21(7), 634–643 (1999)
3. Bloch, I.: Fuzzy Relative Position Between Objects in Image Processing: A Morphological Approach. IEEE Trans. on Pattern Analysis and Machine Intelligence 21(7), 657–664 (1999)
4. Franklin, N., Henkel, L.A., Zangas, T.: Parsing Surrounding Space into Regions. Memory & Cognition 23(4), 397–407 (1995)
5. Logan, G.D., Sadler, D.D.: A Computational Analysis of the Apprehension of Spatial Relations. Language and Space, pp. 493–529. MIT Press, Cambridge (1996)
6. Carlson-Radvansky, L.A., Logan, G.D.: The Influence of Reference Frame Selection on Spatial Template Construction. Memory & Language 37(3), 411–437 (1997)
7. Gapp, K.-P.: Angle, Distance, Shape, and Their Relationship to Projective Relations. In: Proc. 17th Conf. Cognitive Science Soc., pp. 112–117 (1995)
8. Frank, A.U.: Qualitative Spatial Reasoning: Cardinal Directions as an Example. Int. J. of Geographical Information Systems 10(3), 269–290 (1996)
9. Matsakis, P., Ni, J., Veltman, M.: Directional Relationships to a Reference Object: A Quantitative Approach based on Force Fields. Submitted to ICIP 2009 (2009)
10. Matsakis, P., Ni, J., Wang, X.: Object Localization based on Directional Information: Case of 2D Raster Data. In: Proc. 18th Int. Conf. on Pattern Recognition, vol. 2, pp. 142–146 (2006)
11. Bloch, I., Saffiotti, A.: On the Representation of Fuzzy Spatial Relations in Robot Maps. In: Bouchon-Meunier, B., Foulloy, L., Yager, R.R. (eds.) Intelligent Systems for Information Processing, pp. 47–57. Elsevier, NL (2003)
12. Colliot, O., Camara, O., Bloch, I.: Integration of Fuzzy Spatial Relations in Deformable Models-Application to Brain MRI Segmentation. Pattern Recognition 39, 1401–1414 (2006)

13. Krishnapuram, R., Medasani, S., Jung, S.-H., Choi, Y.-S., Balasubramaniam, R.: Content-based Image Retrieval Based on a Fuzzy Approach. IEEE Trans. on Knowledge and Data Engineering 16(10), 1185–1199 (2004)
14. Smith, G.B., Bridges, S.M.: Fuzzy Spatial Data Mining. In: Proc. NAFIPS, pp. 184–189 (2002)
15. Matsakis, P., Keller, J., Wendling, L., Marjamaa, J., Sjahputera, O.: Linguistic Description of Relative Positions in Images. TSMC Part B 31(4), 573–588 (2001)

A Spatio-Temporal Isotropic Operator for the Attention-Point Extraction

Roman M. Palenichka and Marek B. Zaremba

Université du Québec en Outaouais, Gatineau, Québec, Canada
{palenich,zaremba}@uqo.ca

Abstract. It is proposed to extract multi-location image features at maxima points of a spatio-temporal attention operator, which indicates locations with high intensity contrast, region homogeneity, shape saliency and temporal change. The scale-adaptive estimation of local change (motion) and its aggregation with the region shape saliency contribute to robust detection of moving objects. Experiments on the accuracy of interest-point detection have proved the operator consistency and its high potential for object detection in image sequences.

Keywords: feature extraction, visual attention, motion detection, local scale.

1 Introduction

The object recognition from images and image sequences (video) is usually preceded by a feature extraction procedure since the decision is based on object-relevant image descriptors called object features. There are two major ways of image feature extraction for object recognition: global approach with a single (and long) vector of global features [1, 2] and multi-location approach with a set (or a relationship graph) of local feature vectors [3]. Since the image may contain many different objects of interest (moving and stationary) with different spatial extents, many local image features have to be extracted for object recognition. The second approach is preferred in practice because of its high descriptive power, stability to distortions, and resistance to occlusions.

The use of a visual attention operator is a time-efficient solution of object recognition problems by feature extraction in multiple image locations containing potential objects of interest [4-9]. The greater part of the attention operators is based on the image spatial differentiation and integration [6-8], while some of them involve measurements of local symmetry [5]. An important characteristic of many operators is the local scale concept and its selection method [8-11].

More recently, a temporal domain extension of the visual attention approach has been made to rapidly detect moving object and temporal changes. An example of the integration of temporal attention with spatial saliency information is the attention pyramid method [12]. The temporal differentiation in the form of a novelty map is used to detect motion, which is linearly aggregated with the spatial saliency map. However, no local scale concept is involved. A new operator for the detection of space-time interest points has been proposed [13]. It is based on the idea of the Harris and Forstner interest-point operators [5] and detects local structures in space-time domain where the image has significant local variations both in space and time. However, the direct integration of temporal differentiation with spatial derivative functions in the 3x3 operator matrix provides little contribution to effective motion detection. The local scale selection is complicated by the involvement of other scale variables that results in a computationally extensive method. Another existing spatio-temporal operator is the spatio-velocity contrast sensitivity function (CSF) as an attention model of the human visual system [14]. The advantageous feature of this model is the involvement of relative (object-to-retina) movement of stimuli in the attention modeling. Its utility in the multi-location feature extraction is limited by the absence of a time-efficient implementation scheme.

In order to properly design our spatio-temporal operator, formal requirements for attention-point conditions of image locations are formulated in the form of attention tokens (Section 2). The proposed morphological (region-based) scale definition with its fast estimation algorithm contributes to reliable detection of interest points. In our approach, temporal change detection is scale-adaptive and is performed in locations with high local contrast and region homogeneity. In this way, a particular operator of the multi-scale spatio-temporal attention – Spatio-Temporal Isotropic Attention (STIA) operator – has been designed to rapidly extract object-relevant features from a sequence of images (Section 3). This method is an extension to temporal axis of another spatial attention operator, which is based on the multi-scale matched filtering [15]. Experimental results of the interest-point detection (Section 4) confirm the operator consistency and its high potential for image feature extraction.

2 Design Requirements for Spatio-Temporal Visual Attention

In order to achieve an effective feature extraction, the proposed STIA operator has to be designed in accordance with some basic requirements for its reliability and effectiveness. Similarly to other attention operators, the object features are estimated at the STIA local maxima called attention points [9-12]. Moreover, the operator's values are proportional to the relevance or saliency of image fragments centered at the maxima points. The local scale determines the fragment size. We adopted the scale morphological definition [10] where the scale estimation consists in selecting the greatest structuring element S_r centered at (i,j) and inscribed into the fragment's central homogeneous region. The scale value is the diameter r of the structuring element in pixels. The attention points are extracted as consecutive local maxima of the attention operator in a viewing area A and observation period T as follows:

$$(x, y, t)_p = \arg \max_{(i,j) \in A, k \in T} \left\{ \Phi[f(i,j,k), \rho(i,j,k)], (i,j) \notin \Gamma_{p-1} \right\}, \qquad (1)$$

where $f(i,j,k)$ is the image intensity in the point (i,j) of the kth frame, $\Phi[f(i,j,k),\rho(i,j,k)]$ is the STIA function at the local scale $\rho(i,j,k)$, and $(x,y,t)_p$ are the coordinates of the pth local maximum. The region Γ_{p-1} in Eq. (1) is the masking region that excludes areas around the previously determined maxima from further extraction. By the feature-extraction problem statement, this operator has to perform a data-driven multi-scale attention since the object-relevant features can be extracted only after the operator computation. The data-driven definition of the visual attention as well as the required stability to distortions and invariance to image transformations imply the following basic conditions in the operator's design.

1) *Sensitivity to local regional contrast and homogeneity* is necessary for a reliable feature extraction in variable conditions of image acquisition. This requirement is equivalent to the high *signal-to-noise ratio* condition for the stability of signal detection.
2) *Multi-scale and isotropic image analysis* is requested since objects of interest (homogeneous regions) may have different sizes and orientations.
3) *Selectivity to high radial symmetry* of a region shape allows object-relevant details to be distinguished from other less relevant locations with the same high contrast and homogeneity. It determines the *local uniqueness* for the attention points. The operator's maxima have to be located on the median lines of homogeneous regions including *corners* as opposed to simple edges.
4) *Scale-adaptive detection of temporal changes* is necessary for motion integration with the multi-scale spatial saliency of the corresponding location. Temporal changes indicate image locations of high relevance, which usually correspond to moving objects of interest.
5) *Fast implementation* capability implies the computational complexity of the multi-scale operator has to be independent from the scale size.

3 Multi-scale Spatio-Temporal Isotropic Operator

The spatio-temporal isotropic operator proposed in this article is a generic attention operator, which assumes local maxima in image fragments of a variable size at the center of homogeneous regions with a high contrast and temporal change presence. It fulfills all the design conditions disclosed in Section 2.

The local scale ρ has to be estimated before the attention function $\Phi[f(i,j,k), \rho(i,j,k)]$ and is computed by maximizing the local isotropic contrast with a homogeneity constraint over a scale range R:

$$\rho(i,j,k) = \arg\max_{1 \leq r \leq R} \{c(i,j,k,r) - \alpha \cdot d(i,j,k,r)\}, \tag{2}$$

where $c(i,j,k,r)$ is an estimate of the local isotropic contrast at (i,j,k), $d(i,j,k,r)$ is an estimate of the region non-homogeneity, r is the diameter of the disk structuring element S_r in the uniform scale system [10], and α is the homogeneity weight coefficient. The quantification of the attention requirements (Section 2) provides attention tokens as functions of the local scale, image coordinates and time, which are aggregated in the STIA operator with weight coefficients:

$$\Phi[f(i,j,k), \rho(i,j,k)] = c(i,j,k,\rho) - \alpha \cdot d(i,j,k,\rho) + \beta \cdot e(i,j,k,\rho), \tag{3}$$

where the variables $c(i,j,k,\rho)$ and $d(i,j,k,\rho)$ and α have the same meaning as in Eq. (2), ρ is the local scale value, $e(i,j,k,\rho)$ is the temporal change estimate between kth and (k-1)th frames, and β is the temporal change coefficient. An estimate for the optimal values of α and β in the maximum-likelihood sense can be computed assuming appropriate distributions of the three terms in Eq. (3) as random independent variables [9].

A Spatio-Temporal Isotropic Operator for the Attention-Point Extraction

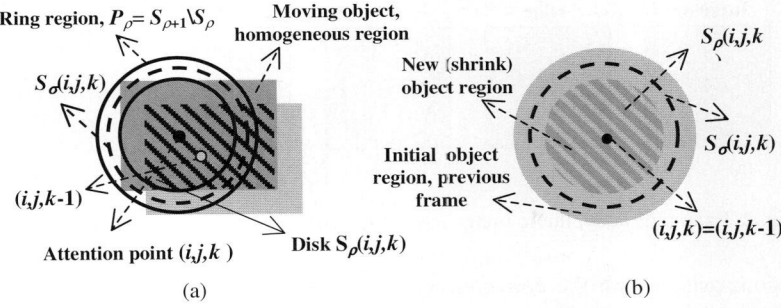

Fig. 1. Estimation of the local isotropic contrast and temporal change: (a) case of a rigid motion; (b) case of a region shrink deformation

The first term in Eq. (2) is defined in such a way that it includes two tokens: local contrast and shape symmetry. Two isotropic estimation regions are involved in the estimation of the local isotropic contrast based on the structuring element S_ρ: *disc region* $Q_\rho = S_\rho$ and *ring region* $P_\rho = S_{\rho+1} \setminus S_\rho$ (Fig. 1). The local isotropic contrast $c(i,j,k,\rho)$ at the ρth scale in (i,j,k) is estimated as the mean square deviation in the ring region $P_\rho(i,j)$ with respect to the mean intensity of the disk region $Q_\rho(i,j)$:

$$c^2(i,j,k,\rho) = \frac{1}{|P_\rho|} \sum_{m,n \in P_\rho(i,j)} (f(i,j,k,\rho) - f(m,n,k))^2, \qquad (4)$$

where $f(i,j,k,\rho)$ stands for the mean value of $f(i,j,k)$ in $Q_\rho(i,j)$ and $|P_\rho|$ is the total number of grid points in $P_\rho(i,j)$. The intensity deviation in the ring P_ρ will be proportional to the amount of contact edge points of the greatest inscribed structuring element S_ρ since the region $P_\rho(i,j)$ will include some background points near the region border. It measures the level of radial symmetry around (i,j) (Fig. 2). The number of background pixels in $P_\rho(i,j)$ is equal to zero for the interior region while a round object such as the disk in Fig. 2e gives the maximum of the radial symmetry measure in Eq. (4). As an estimate of the region non-homogeneity, $d(i,j,k,\rho)$, the mean intensity deviation within the disk region $Q_\rho(i,j)$ have been used in Eq. (3). Another option for estimating homogeneity is to use the sum of the intensity first derivatives as it is implemented in the existing operators of the Harris and Forstner type [5, 13]. The third term in Eq. (3) is the temporal change token, which corresponds to the fourth requirement for our spatio-temporal operator design (Section 2). A temporal differentiation is performed at the first step: $h(i,j,k) = |f(i,j,k) - f(i,j,k-1)|$. At the second step, a scale-adaptive linear filtering (e.g., a simple averaging) is applied to the difference image $h(i,j,k)$ in order to scale-adaptively estimate the temporal change $e(i,j,k,\rho)$. The diameter σ of the structuring element $S_\sigma(i,j)$ for the averaging at the point (i,j) is larger than the local scale ρ at (i,j) in order to capture different types of temporal changes (Fig. 1). The diameter difference, $\sigma - \rho$, is proportional to the potential change amount or motion velocity. The integration of the temporal differentiation pixels over the disc region $S_\sigma(i,j)$ assumes local maxima at the centers of moving homogeneous regions (Fig. 1).

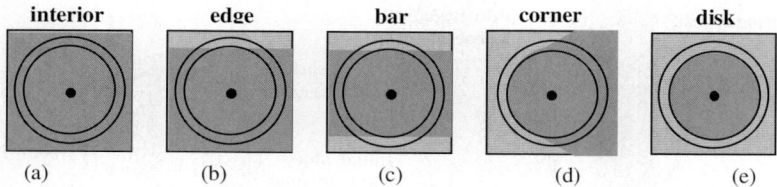

Fig. 2. Examples of synthetic image fragments with various levels of radial symmetry

In comparison with the conventional differentiation approach, the scale-adaptive estimation of the temporal change and its integration with the spatial saliency in Eq. (3) has two advantages. First, it provides the location of a moving-object region at its geometrical center whereas the direct differentiation usually gives the region edges in the motion direction. Second, its response to weak temporal changes will be strong due to the scale-adaptive change integration within $S_\sigma(i,j)$.

The general approach to fast recursive computation of the function $\Phi[f(i,j,k)$, $\rho(i,j,k)]$ is to represent it in the form of combinations of linear filters having the impulse responses as *primitive kernel functions* [16]. The kernel of the averaging filter used in the recursive implementation of the STIA operator is a zero-order primitive kernel function since it is a constant. The STIA complexity was reduced to $O(R)$ operations per pixel by combining recursive implementations of the averaging filters, where R is the total number of scales and $O(.)$ is the standard big "O" function.

4 Experimental Results

4.1 Accuracy and Invariance of the Attention-Point Extraction

The experiments with the STIA operators have been related to biometrical identification, gesture recognition and video surveillance. The goal of the experiments was to investigate the accuracy (invariance) of attention points coordinates when images were subjected to similarity transformations (translation, scaling and rotation) and to intensity linear transformations simultaneously with random distortions. The test images were transformed by one of the transformations at a time and the STIA operator was applied to find the attention points. To estimate the extraction error, the attention point coordinates were back-transformed to the initial image plane and the mean of the displacements was calculated with respect to the initial coordinates.

As an example, a test image of hand (Fig. 3) has been submitted with different poses (transformation parameters are known) of the palm on the image plane (rotations, scaling and translations are included). The results of the rotational invariance testing for this example are given in Fig. 4a. Some results of the accuracy testing are disclosed in Fig. 4b. The level of random distortions is characterized by the contrast-to-noise ratio where the noise level is its mean square deviation. The final error estimate is averaged over single-point errors from 16 first local maxima of the STIA operator. In these examples, the STIA operator was implemented for a large scale range using eighty consecutive and uniform scales ($R=80$) since the hand palm region

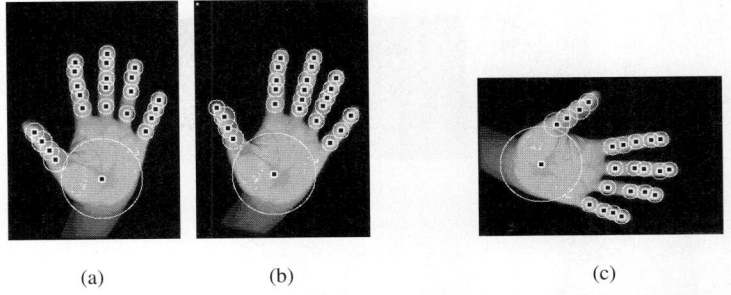

Fig. 3. Examples of rotated images and extracted attention points

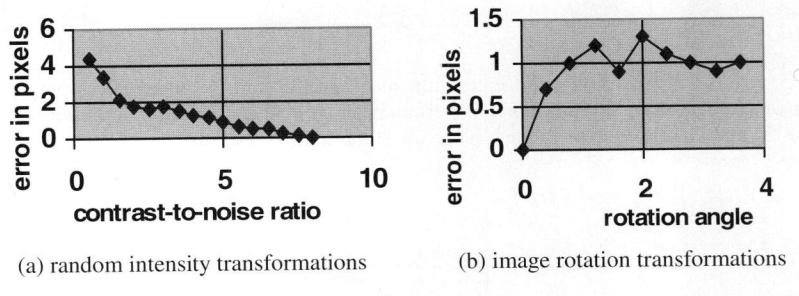

Fig. 4. Experimental evaluation of the STIA operator accuracy in the attention-point extraction

has the local size significantly greater than that of the fingers. The following weight coefficients have been involved in the STIA: $\alpha=0.9$ and $\beta=2.9$.

4.2 Experiments on Moving Object Detection

The goal of the experiments was the accuracy evaluation of the attention points extraction for moving object detection using the STIA approach. For this purpose, first global STIA maxima satisfying the saliency tests and the local maxima condition were chosen as the attention points for the evaluation of the correct detection rate and false alarm rate. The motion will be correctly detected (i.e., true positive result), if the current attention point is located at the moving region center and not at its border. The false alarm case in the motion detection using the STIA operator is the attention-point positions on stationary image fragments (with high intensity saliency), ghost (inexistent) objects and regional border locations. The experiments were conducted on moving hand sequences (Fig. 5). The following weight coefficients have been involved in the STIA: $\alpha=0.7$ and $\beta=3.5$.

The performance of this method was numerically evaluated in the form of the receiver operating characteristic (ROC) curve, i.e., the plot of the true positive rate versus the false alarm rate (Fig. 6). The shape of the ROC curve in Fig. 6 and the large area under the curve confirm the favorable relationship between the two rates: true positive vs. false positive.

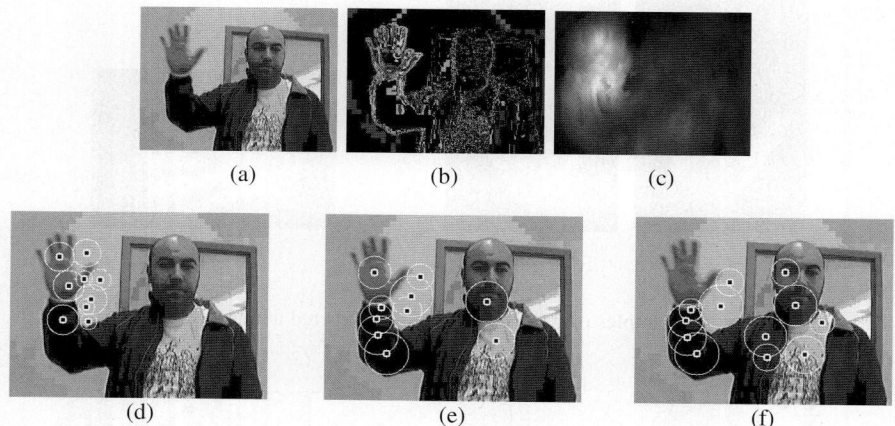

Fig. 5. Results of attention-point extraction in a hand gesture sequence: (a) previous frame from image sequence; (b) temporal differentiation; (c) STIA operator; attention points with lower (d) and larger (e) scale ranges; (f) result of the MIMF operator

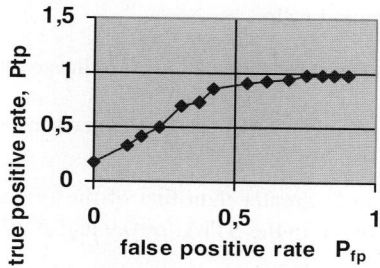

Fig. 6. ROC of the attention-point extraction using the STIA operator

Theoretical and experimental comparison of the STIA operator with the similar existing attention operators [6, 11, 15] shows the STIA advantage in more effective estimation of the local contrast using Eq. 4. It is capable to detect multi-level homogeneous regions with intensity and shape saliency whereas the MIMF operator [15] can affectively extract only two-level homogeneous regions and edges (Fig. 5f).

5 Conclusions

A novel spatio-temporal attention operator for features extraction from image sequences is proposed. It scale-adaptively integrates the local change estimate with multi-scale saliency tokens such as the intensity contrast, region homogeneity, and radial symmetry. The operator has the following advantages in the context of multi-scale and multi-location feature extraction: invariance to translations, scaling and rotations; invariance to intensity linear transformations; reliable extraction of motion features; fast recursive implementation for multi-scale image analysis.

References

1. Duda, R.O., Hart, P.E., Stork, D.G.: Pattern Classification. John Wiley & Sons, New York (2001)
2. Evgeniou, T., et al.: Image representations and feature selection for multimedia database search. IEEE Trans. KDE 15(4), 911–920 (2003)
3. Schmid, C., Mohr, R.: Local gray-value invariants for image retrieval. IEEE Trans. PAMI 19(5), 530–535 (1997)
4. Reisfeld, D., Wolfson, H., Yeshurun, Y.: Context-free attentional operators: the generalized symmetry transform. Int. J. of Comp. Vis. 14, 119–130 (1995)
5. Harris, C., Stephens, M.: A combined corner and edge detector. In: Proc. 4th ALVEY Vision Conference, pp. 147–151 (1988)
6. Lindeberg, T.: Detecting salient blob-like image structures and their scale with a scale-space primal sketch: A method for focus of attention. Int. J. of Comp. Vision 11, 283–318 (1993)
7. Kadir, T., Brady, M.: Saliency, scale and image description. Int. J. of Comp. Vision 45(2), 85–103 (2001)
8. Itti, L., Koch, C., Niebur, E.: A model of saliency-based visual attention for rapid scene analysis. IEEE Trans. PAMI 20(11), 1254–1259 (1998)
9. Tagare, H.D., Toyama, K., Wang, J.G.: A maximum-likelihood strategy for directing attention during visual search. IEEE Trans. PAMI 23(5), 490–500 (2001)
10. Maragos, P.: Pattern spectrum and multi-scale shape representation. IEEE Trans. PAMI 11(7), 701–717 (1989)
11. Lowe, D.G.: Distinctive image features from scale-invariant key-points. Int. J. of Comp. Vision 60(2), 91–110 (2004)
12. Conception, V., Wechsler, H.: Detection and localization of objects in time-varying imagery using attention, representation and memory pyramids. Pattern Recognition 29(9), 1543–1557 (1996)
13. Laptev, I.: On space-time interest points. Int. J. of Comp. Vision 64(2/3), 107–123 (2005)
14. Laird, J., Rosen, M., Pelz, J., Montag, E., Daly, S.: Spatio-velocity CSF as a function of retinal velocity using unstabilized stimuli. In: SPIE Proc., vol. 6057, pp. 1–12 (2006)
15. Palenichka, R.M., Zaremba, M.B.: Multi-scale isotropic matched filtering for individual tree detection in LiDAR images. IEEE Trans. GRS 45(12), 3944–3957 (2007)
16. Di Gesù, V., Palenichka, R.M.: Fast recursive computation of local axial moments. Signal Processing 81, 265–273 (2001)

Homological Tree-Based Strategies for Image Analysis[*]

P. Real[1], H. Molina-Abril[1,2], and W. Kropatsch[2]

[1] Departamento de Matematica Aplicada I, Universidad de Sevilla
{real,habril}@us.es
http://ma1.eii.us.es/

[2] Vienna University of Technology, Faculty of Informatics, PRIP Group
{habril,krw}@prip.tuwien.ac.at
http://www.prip.tuwien.ac.at/

Abstract. Homological characteristics of digital objects can be obtained in a straightforward manner computing an algebraic map ϕ over a finite cell complex K (with coefficients in the finite field $\mathbf{F}_2 = \{0, 1\}$) which represents the digital object [9]. Computable homological information includes the Euler characteristic, homology generators and representative cycles, higher (co)homology operations, etc. This algebraic map ϕ is described in combinatorial terms using a mixed three-level forest. Different strategies changing only two parameters of this algorithm for computing ϕ are presented. Each one of those strategies gives rise to different maps, although all of them provides the same homological information for K. For example, tree-based structures useful in image analysis like topological skeletons and pyramids can be obtained as subgraphs of this forest.

Keywords: Cell complex, chain homotopy, digital volume, homology, gradient vector field, image pyramid, tree, skeleton.

1 Introduction

A finite cell complex K is a graded set formed of cells, with an operator ∂ describing the boundary of each cell in terms of linear combination of its faces. The finite linear combination (with coefficients in $\mathbb{F}_2 = \{0, 1\}$) of cells form a graded vector space called chain complex associated to K and denoted by $C_*(K; \mathbb{F}_2)$. In [9] the solution to the homology computation problem (calculating n-dimensional holes) of K is described in the following terms: to find a concrete linear map $\phi : C_*(K; \mathbb{F}_2) \to C_{*+1}(K; \mathbb{F}_2)$, increasing the dimension by one and satisfying that $\phi\phi = 0$, $\phi\partial\phi = \phi$ and $\partial\phi\partial = \partial$. In [10], a map ϕ of this kind is called homology gradient vector field (hgvf). This datum ϕ is, in fact, a chain homotopy operator on K (a purely homological algebra notion) and it is immediate

---

[*] This work has been partially supported by PAICYT research project FQM-296, "Andalusian research project" PO6-TIC-02268, Spanish MEC project MTM2006-03722 and the Austrian Science Fund under grant P20134-N13.

to establish a strong algebraic link between the cell complex associate to K and its homology groups $(H_0(K), H_1(K), H_2(K))$.

This approach is followed here as a solution to the homology computation problem. We will codify the deformation process ϕ to a minimal homological expression in terms of graphs. In [9] the input of the algorithm for computing homology information is a filtered finite chain complex. This input filter and the choice of a homological face for each cell-step are the two parameters of this algorithm. Different useful strategies in digital imagery (segmentation, analysis, topological skeleton, multiresolution analysis,...) can be performed running this algorithm. The final result depends on suitable choices of these two factors. Concerning the output, we will express the resulting homology gvfs of the algorithm as a three-level forest data structure which geometrically represent the underlying acyclic submodule $\text{Im } \phi = \{x \in C(K; \mathbb{F}_2) : x = \phi(y), \text{for some } y\}$. This connectivity encoding method is susceptible to be generalized to any subdivision representation scheme, higher dimension and coefficient ring. Different modalities of this process are presented here showing its versatility.

2 Obtaining Homological Information

Some notions about algebraic topology must be introduced. A q–*chain* a of a three-dimensional cell complex K is a formal sum of cells of $K^{(q)}$ ($q = 0, 1, 2, 3$). Let us consider the ground ring as the finite field $\mathbf{F}_2 = \{0, 1\}$. The q–chains form a group with respect to the component–wise addition; this group is the qth *chain complex* of K, denoted by $C_q(K)$. There is a chain group for every integer $q \geq 0$, but for a complex in $\mathbf{R}^3$, only the ones for $0 \leq q \leq 3$ may be non-trivial. The boundary map $\partial_q : C_q(K) \to C_{q-1}(K)$ applied to a q–cell σ gives us the collection of all its $(q-1)$–faces which is a $(q-1)$–chain. By linearity, the boundary operator ∂_q can be extended to q–chains, and satisfies $\partial_{q-1}\partial_q = 0$. From now on, a cell complex will be denoted by (K, ∂). A chain $a \in C_q(K)$ is called a q–*cycle* if $\partial_q(a) = 0$. If $a = \partial_{q+1}(a')$ for some $a' \in C_{q+1}(K)$ then a is called a q–*boundary*. Define the qth *homology group* to be the quotient group of q–cycles and q–boundaries, denoted by $H_q(K)$.

Let (K, d) be a finite cell complex. A linear map of chains $\phi : C_*(K) \to C_{*+1}(K)$ is a *combinatorial gradient vector field* (or, shortly, combinatorial gvf) on K if the following conditions hold: (1) For any cell $a \in K_q$, $\phi(a)$ is a $q+1$-cell b; (2) $\phi^2 = 0$. Removing the first condition, ϕ will be called an *algebraic gradient vector field*. An algebraic gvf satisfying the conditions $\phi d \phi = \phi$ and $d\phi d = d$ will be called a *homology gvf* [9]. If ϕ is a combinatorial gvf which is only non-null for a unique cell $a \in K_q$ and satisfying the extra-condition $\phi \partial \phi = \phi$, then it is called a (combinatorial) *integral operator* [6]. An algebraic gvf ϕ is called *strongly nilpotent* if it satisfies the following property: Given any $u \in K^{(q)}$, if $\phi(u) = \sum_{i=1}^{r} v_i$ then $\phi(v_i) = 0$ for all $i = 1, \ldots, r$. We say that a linear map $f : C_*(K) \to C_*(K)$ is *strongly null* over an algebraic gradient vector field ϕ if given any $u \in K^{(q)}$, if $\phi(u) = \sum_{i=1}^{r} v_i$ then $f(v_i) = 0$ for all $i = 1, \ldots, r$.

Using homological algebra arguments, it is possible to deduce that a homology gvf ϕ determines a strong algebraic relationship connecting $C(K)$ and its homology vector space $H(K)$. Let us define a *chain contraction* (f, g, ϕ) : $(C, \partial) => (C', \partial')$ between two chain complexes as a triple of linear maps such that $f : C_* \to C'_*$, $g : C'_* \to C_*$ and $\phi : C_* \to C_{*+1}$ and they satisfy the following conditions: (a) $id_C - gf = \partial\phi + \phi\partial$; (b) $fg = id_{C'}$; (c) $f\phi = 0$; (d) $\phi g = 0$; (e) $\phi\phi = 0$. Given a chain contraction (f, g, ϕ), elementary homological algebra results are: (a) Ker ϕ = Im$g \bigoplus$ Imϕ (direct sum); (b) Imϕ is acyclic (i.e, it has null homology); (c) Ker ∂ = Im$g \bigoplus$ Im∂ (direct sum).

Proposition 1 (see [9]). *Let (K, ∂) be a finite cell complex. A homology gvf $\phi : C_*(K) \to C_{*+1}(K)$ over K gives rise to a chain contraction (π, ι, ϕ) from $C(K)$ onto its chain subcomplex isomorphic to the homology of K, where $\pi = id_{C(K)} + \phi\partial + \partial\phi$ is a projection and ι is the inclusion map. Reciprocally, if (f, g, ϕ) is a chain contraction from $C(K)$ to its homology $H(K)$, then ϕ is a homology gvf.*

Given a cell complex (K, ∂), an ordered set of cells $\mathcal{K} = \langle c_1, \ldots, c_m \rangle$ is called a *filter* for K if $\{c_1, \ldots, c_m\} = \bigcup_{q \in \mathbb{Z}} K^{(q)}$, and for each $j = 1, \ldots, m$, all the faces of c_j are contained in the subset $\{c_1, \ldots, c_{j-1}\}$. A straightforward filter for K can be consider all the 0–cells, then all the 1–cells, and so on. Another important example is to consider a spanning tree strategy as a filter for a one-dimensional cell complex or a topological graph $G = (V, E)$. It is a classical result that the edges of $E \setminus E'$ determine the homologically different 1-cycles for G. A generalization of a spanning tree technique for higher dimensional cell complexes [11] can be applied for determining filters with interesting homological properties. The algorithm proposed in [10,9] for calculating in an incremental way a $\mathbb{F}_2$-homology gvf is the following one:

Algorithm 1 (see [10,9]). *Let (K, ∂) be a finite cell complex with the filter $\mathcal{K}_m = \langle c_0, \ldots, c_m \rangle$. For each $i = 0, \ldots, m$, we represent the cell subcomplex of K by the filter $\mathcal{K}_i := \langle c_0, \ldots, c_i \rangle$, with the boundary map ∂_i. Let $\mathcal{H}$ be the homology chain complex (that is, a chain complex with the zero boundary map) associated to (K, ∂).*

$\phi_0(c_0) := 0$.
For $i := 1$ to m do
    define $\bar{c}_i := c_i + \phi_{i-1}\partial_i(c_i)$
    $\forall e_{s_j} \in \mathcal{K}_{i-1}$ such that $\partial_i(c_i) = \sum_{j=1}^r \lambda_j e_{s_j}$
    define $\bar{e}_{s_j} := (id_{C(\mathcal{K}_{i-1})} - \phi_{i-1}\partial_{i-1} - \partial_{i-1}\phi_{i-1})(e_{s_j})$ $\forall j = 1, \ldots, r$
    $\phi_i(c_i) := 0$,
    If $\partial_i \bar{c}_i = 0$ then
        For $j := 0$ to $i - 1$ do
            $\phi_i(c_j) := \phi_{i-1}(c_j)$.
    Otherwise choose an element $\bar{e}_{s_k} \neq 0$ and define $\tilde{\phi}(\bar{e}_{s_k}) := \bar{c}_i$
    and zero otherwise.
        For $j := 0$ to $i - 1$ do
            $\phi_i(c_j) := (\phi_{i-1} + \tilde{\phi}(id_{C(\mathcal{K}_i)} + \phi_{i-1}\partial_i + \partial_{i-1}\phi_{i-1}))(c_j)$,
OUTPUT: a homology gradient vector field ϕ_m for K.

The key idea of this algorithm is the same as in [4]: in the ith step, the element c_i of the filter is added and a homology class is created or destroyed. Let us note that the homology of (K, ∂) is given by the set $\mathcal{H} := \text{Im}\,(1_{\mathcal{K}_m} + \phi_m \partial_m + \partial_m \phi_m)$.

Algorithm 1 essentially depends on two parameters (framed in the algorithm): the filter and the choice of a homological face for each cell step. Concerning the filter parameter, different orders of the cells as an input of the algorithm give rise to different homology gradient vector fields, although they all provide the same homological results. The homological face choice parameter involves the selected cell $e_{s_k}^-$ used to define $\tilde{\phi}$, in order to "kill" the cell $\bar{c}_i$.

Focussing our interest in the output of the Algorithm 1 and based on experimental results, it can be conjectured that the underlying structure of the homology gvf ϕ_m is a suitable generalization to cell complexes of the notion of the spanning forest for a topological graph [11]. The following result is the key for determining the graph-based nature of a homology gradient vector field. This can be easily proved using induction on i.

Proposition 2. *In Algorithm 1, the homology gradient vector field ϕ_m is strongly nilpotent and the map $\pi_m = 1 + \partial \phi_m + \phi_m \partial$ is strongly null over ϕ_m. Moreover, if $\phi_m(c) = \sum_{j=1}^{t} \bar{c}_j$ ($\bar{c}_j$ being a cell of K), then $\bar{c}_j = \phi_m(\bar{e}_j)$, for some $\bar{e}_j \in C(K)$ and $\forall j = 1, \ldots, t$.*

In fact, these properties establishing the combinatorial nature of ϕ_m and the fact that $\text{Im}\phi_m$ is a acyclic vector subspace guarantee that ϕ in each level (levels 0,1 and 2) can be represented as a kind of spanning forest $(T(\phi_m)_0, T_1(\phi_m), T_2(\phi_m))$ for the cell complex K called homological forest of the homology gvf ϕ_m. This structure is a mixed graph in each level (that is, some edges are undirected and others are arcs). We use here the notation $e \in h(v)$, whenever the element e appears as a summand of the linear map h applied to v.

1. Let us form the forest $T_0(\phi_m) = (V_0, E_0)$ (called the vertex homological forest), in which $V_0 = V_0^0 \cup V_0^1$ (red and blue vertices of $T_0(\phi_m)$), with $V_0^0 = K_0$, $V_0^1 = \{e \in K_1 \,/\, e \in \phi(v),$ for some $v \in K_0\}$ and E_0 is composed of all the unordered pairs $\{v, e\}$, where v is a vertex of K appearing as a summand in the boundary of e (that is, $v \in \partial e$).
2. Let us form the forest $T_1(\phi_m) = (V_1, E_1)$ (called the edge homological forest), in which $V_1 = V_1^0 \cup V_1^1$ (red and blue vertices of $T_1(\phi_m)$), with $V_1^0 = (K_1 \backslash V_0^1)$, $V_1^1 = \{e \in K_2 \,/\, e \in \phi(v),$ for some $v \in V_1^0\}$ and E_1 is composed of all the unordered pairs $\{v, e\}$, where v is an edge of $V_1^0 \subset K_1$ appearing as a summand in the boundary of the 2-cell e.
3. Let us form the forest $T_2(\phi_m) = (V_2, E_2)$ (called the face homological forest), in which $V_2 = V_2^0 \cup V_2^1$ (red and blue vertices of $T_2(\phi_m)$), with $V_2^0 = (K_2 \backslash V_1^1)$, $V_2^1 = \{K_3\}$ and E_2 is composed of all the unordered pairs $\{v, e\}$, where v is a 2-cell belonging to $K_2 \setminus V_1^1$ appearing as a summand in the boundary of the 3-cell e.

The edges of the set E_i ($i = 0, 1, 2$) connect a red vertex with a blue one. Moreover, we establish an arc in $T_i(\phi_m)$ ($i = 0, 1, 2$) from the red vertex v to the blue vertex e, if $e \in \phi_m(v)$.

Proposition 3. *The filter and the homological face choice at each cell step in Algorithm 1, can be determined in order to obtain as an output a homological forest of the homology gvf ϕ_m.*

3 Homological Strategies

The versatility of Algorithm 1 is shown throughout this section. Taking into account different variations of the two parameters, useful results for segmentation, analysis, recognition, compression, etc. are presented here.

3.1 Topological Skeleton

The aim of topological thinning is to reduce the image content to its essentials. Thinning an image consists in eliminating border voxels until only a skeleton of the original image remains. We consider a border point of one direction if its adjacent point in this direction is white.

Considering a filter over the cells in a way that each step of Algorithm 1 takes a border cell c_i and one of its faces $\bar{e}_{s_k}$ satisfying some conditions: we are performing a topological thinning of the initial object. The main idea is to apply the chain homotopy operator ϕ corresponding to cell collapsing [6] on each direction of the 3D volume, reducing the initial cell complex. Applying this operator to border points satisfying some requirement, we obtain a medial axis skeleton of the 3D initial volume. This requirements can be seen as a selection of those border cells $\bar{e}_{s_k}$ which belong to a higher dimension cell c_i, being $\bar{e}_{s_k}$ a free face of c_i (non-shared with other cells). In that case we will assign $\tilde{\phi}(\bar{e}_{s_k}) = c_i$.

In fact cell collapse operations are seen here from a purely algebraic point of view. Each one of them is a chain homotopy equivalence algebraically connecting the object, before and after the collapsing process. The complete thinning will be considered as a composition of these chain homotopies. In order to preserve the shape of the object, we need additional criteria for deleting points which prevent excessive shrinking.

A special kind of skeleton called Reeb graph can be also obtained using Algorithm 1. Reeb graphs are skeleton graphs that provide a way to understand the intrinsic topological structure of a shape. If we proceed filtering the volume by subdividing it into 2 dimensional slices, the minimum homological representation of each connected component in the slice will be the result of applying the

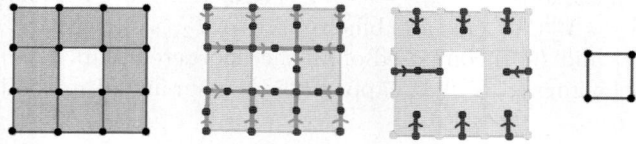

Fig. 1. Topological skeleton strategy showing the T_0 and T_1 forests and the skeleton

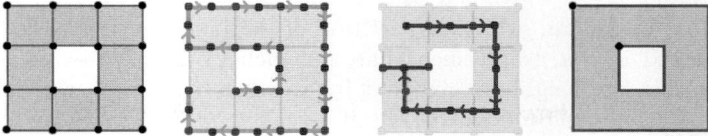

Fig. 2. Segmentation-based strategy showing the T_0 and T_1 forests and the final result

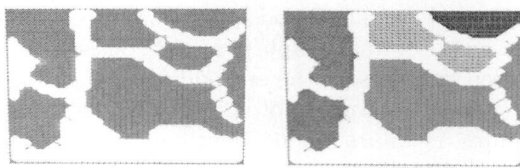

Fig. 3. Segmentation based on connected components of a trabecular bone volume (courtesy of the Institute of Biomechanics, Zurich, Swiss)

algorithm to this 2D images. After processing all the cells belonging to each slice, cells joining different slices will be taken into account. These cells will join each graph with its correspondences (same connected component) in the slices above and below. The result is the Reeb graph with loops codifying the topological structure of the volume.

3.2 Segmentation Strategy

The goal of segmentation is to partition the image into regions with homogeneous properties. A segmentation of an initial volume V can be achieved using the homology gvf algorithm. The filter restrictions used for this aim, is completely different to the one used in Section 1. In this case inner cells of the complex will be included in the algorithm before the ones in the border. Selecting a shared-face $\bar{e}_{s_k}$ of an inner cell c_i and considering $\tilde{\phi}(\bar{e}_{s_k}) = c_i$, the resultant complex after this iteration will contain a cell which is the merge of both cells sharing u_s.

Repeating this process until no more inner cells exists on the complex, the final result will be a big cell which is the result of merging all the cells belonging to the same region. This segmentation can be done in terms of different criteria, like connected components, color segmentation, etc. An example of this algorithm segmenting a 3–dimensional image is shown in Fig. 3.

3.3 Topological Pyramids

Image Pyramids are hierarchical structures widely used in Image Analysis [1]. They are made of multiple copies of the same cellular structure, in which the number of cell is reduced from copy to copy (levels of the pyramid). Irregular graph pyramids are defined as a stack of successively reduced graphs [8] where each level represents an arbitrary partition of the pixel set into cells. An image is transformed into a graph such that, for each pixel a vertex is associated, and pixels that are neighbors are joint by an edge. The graph which represent the pixels is denoted by $G = (V, E)$. It is called *primal graph* and it divides the plane into faces. Let denote each face by a new vertex and connect the faces that are neighbors (sharing the same edge) by a new edge. These vertices and edges added, compose a new graph $\overline{G}$, which is the *dual graph* of G. The edges of $\overline{G}$ represent the borders of the cells of G, including so called pseudo edges needed to represent neighborhood relations to a cell completely enclosed by another cell. A level of a dual graph pyramid consists of a pair $(G_k, \overline{G_k})$ of plane graphs, in order to represent the embedding of the graph in the plane. A cell on level $k + 1$ (parent) is a union of neighboring cells on level k (children). Every parent computes its values independently of other cells on the same level. The union of cells is controlled by so called contraction kernels and the only operations used are removal and contraction [8]. The contraction operation is defined informally as the removal of the i-cell and the merging of two $(i - 1)$-cells, effectively removing one of these $(i - 1)$-cells. E.g. when contracting an edge, the two bounding vertices of this edge are merged into a single vertex, removing one of the vertices. The connecting edge is also removed in the process. Because a contraction operation merges two $(i - 1)$-cells, only cells of dimension $i \geq 1$ may be contracted. Intuitively and in a general way for an n-dimensional space, the removal of an i-cell consists in removing this cell and in merging its two incidents $(i + 1)$-cells: so removal can be defined for $0\ldots(n - 1)$-cells. According to [2],[3] the removal operation is the dual counterpart to a contraction.

As mentioned in [5], the two operations used to construct an irregular graph pyramid are integral operators satisfying the chain-homotopy property. Following this lemma, an irregular graph pyramid can be directly built using Algorithm 1. We can combine one of the existing algorithms for constructing an image pyramid [7], but updating at each step the resulting gvf following Algorithm 1. For example, given two cells c_i and c_j sharing a face u_s, and defining $\tilde{\phi}(u_s) = c_i$, the cell u_s will be removed and c_i and c_j will be merged, performing in that way a removal operation. Similar situation occurs with cell contractions. The advantage of using this algorithm for building a pyramid is the complete topological control during the whole process, and the possibility of directly compute topological invariants at each level of the pyramid.

4 Conclusions and Future Work

Roughly speaking, in order to progress in homological knowledge for a finite cell complex, we determine here two operators on $C_*(K; \mathbb{F}_2)$: a boundary operator, $\partial : C_*(K; \mathbb{F}_2) \to C_{*-1}(K; \mathbb{F}_2)$, decreasing the dimension by one and measuring in algebraic terms the boundary of any set of cells; and a "coboundary" operator $\phi : C_*(K; \mathbb{F}_2) \to C_{*+1}(K; \mathbb{F}_2)$, increasing the dimension by one and measuring in some sort the degree of acyclicity (a space is acyclic is it has the same homology than a point) of any set of cells. In particular, we have seen that the map $1_K + \partial\phi + \phi\partial$ applied to every cell provides us representative cycles for all the homology generators of K. In fact, having this information at hand, it is possible to compute finer topological invariants for K (relations between the homology generators, cohomology algebra, (co)homology operations, ...) as well as partially or full homologically equivalent graph features (topological skeletons, Reeb graphs, contraction kernels, ...). In combinatorial terms, we specify for the acyclicity operator ϕ a homological forest. As future work, we plan to explore the possibilities of this forest for 3D and 4D geo-topological (saving essential geometric and topological information) representation as well as its robustness with regards to small changes (deformation, noise,...) in the object.

References

1. Burt, P., Hong, T.-H., Rosenfeld, A.: Segmentation and estimation of image region properties through cooperative hierarchical computation. IEEE Transactions on Systems, Man and Cybernetics, 802–809 (December 1981)
2. Damiand, G., Lienhardt, P.: Removal and contraction for n-dimensional generalized maps. In: Nyström, I., Sanniti di Baja, G., Svensson, S. (eds.) DGCI 2003. LNCS, vol. 2886, pp. 408–419. Springer, Heidelberg (2003)
3. Brun, L., Kropatsch, W.G.: Introduction to combinatorial pyramids. In: Bertrand, G., Imiya, A., Klette, R. (eds.) Digital and Image Geometry. LNCS, vol. 2243, pp. 108–128. Springer, Heidelberg (2002)
4. Delfinado, C.J.A., Edelsbrunner, H.: An Incremental Algorithm for Betti Numbers of Simplicial Complexes on the 3-Sphere. Comput. Aided Geom. Design 12, 771–784 (1995)
5. Gonzalez-Diaz, R., Ion, A., Iglesias-Ham, M., Kropatsch, W.G.: Irregular Graph Pyramids and Representative Cocycles of Cohomology Generators. In: Torsello, A., Escolano, F., Brun, L. (eds.) GbRPR 2009. LNCS, vol. 5534, pp. 263–272. Springer, Heidelberg (2009)
6. Gonzalez-Diaz, R., Jimenez, M.J., Medrano, B., Molina-Abril, H., Real, P.: Integral Operators for Computing Homology Generators at Any Dimension. In: Ruiz-Shulcloper, J., Kropatsch, W.G. (eds.) CIARP 2008. LNCS, vol. 5197, pp. 356–363. Springer, Heidelberg (2008)
7. Haxhimusa, Y.: The structurally Optimal Dual Graph Pyramid and its application in image partitioning. Dissertations in Artificial Intelligence, vol. 308 (2007)
8. Kropatsch, W.G.: Equivalent contraction kernels and the domain of dual irregular pyramids. Tech. report PRIP-TR-42, Vienna University of Technology
9. Molina-Abril, H., Real, P.: Cell AT-models for digital volumes. In: Torsello, A., Escolano, F., Brun, L. (eds.) GbRPR 2009. LNCS, vol. 5534, pp. 314–323. Springer, Heidelberg (2009)
10. Molina-Abril, H., Real, P.: Advanced Homological information on 3D Digital volumes. In: SSPR 2008. LNCS, vol. 5342, pp. 361–371. Springer, Heidelberg (2008)
11. Suuriniemi, S., Tarhasaari, T., Kettunen, L.: Generalization of the spanning-tree technique. IEEE Transactions on Magnetics 38(2), 525–528 (2002)

Affine Moment Invariants of Color Images*

Tomáš Suk and Jan Flusser

Institute of Information Theory and Automation,
Academy of Sciences of the Czech Republic,
Pod vodárenskou věží 4, 182 08 Prague 8, Czech Republic
{flusser,suk}@utia.cas.cz
http://www.utia.cas.cz

Abstract. A new type of affine moment invariants for color images is proposed in this paper. The traditional affine moment invariants can be computed on each color channel separately, yet when the channels are transformed together, by the same affine transform, additional invariants can be computed. They have low order and therefore high robustness to noise. The new invariants are compared with another set of invariants for color images using second powers of the image function. The basic properties of the new features are tested on real images in a numerical experiment.

1 Introduction

The pattern recognition of objects on images distorted by affine transform has been studied for many years. Affine moment invariants showed suitability for this purpose, however they are typically computed from binary or gray-level images. In case of color images, additional features can be computed using color.

Certain satellites are capable of capturing a large number (let us say n) of spectral bands (often infrared, less commonly ultraviolet or visible), the image from one spectral band is called *channel*. Ordinary color photographs are much more readily accessible, and since they contain 3 visible spectral bands they can be considered multi-channel images with $n = 3$. Typically, the color is used for better segmentation and invariants are computed from binary silhouettes of the segmented objects, in certain cases the first principal component with zeroed background is used.

In pattern recognition the following rule applies and must be satisfied: the number i of independent invariants equals the number m of independent measurements of a certain object (the number of moments in our case) minus the number t of independent constraints(see e.g. [1]). In most cases it equals the number of parameters of the transformation, in case of the affine transform, $t = 6$, thus this rule can be rewritten as $i = m - t$.

If we decide to compute the invariants directly from channels, we can use each channel separately and obtain triple (n-fold) features. Nevertheless, if we use m

---

* This work has been supported by the grant No. 102/08/1593 of the Grant Agency of the Czech Republic.

moments of each n channels and t parameters of the transform, then we obtain $i = n(m - t)$ independent invariants of separate channels.

In case of multi-channel images, where there is no geometric transform between channels, we can suppose a two-dimensional affine transform with identical parameters in each channel. In this case, we could obtain $nm - t$ independent invariants, and thus $t(n - 1)$ additional invariants, t for each additional channel.

Mindru et al published a series of conference contributions, e.g. [2] and [3] including a survey in the journal paper [4] with the combined invariants to the affine transform and illumination changes, where moments computed from a certain power of the channels were utilized. Their generalized color moment of order $s = p + q$ and degree $d = \alpha + \beta + \gamma$ of a certain object Ω is then defined

$$M_{pq}^{\alpha\beta\gamma} = \int\int_{\Omega} x^p y^q (R(x,y))^\alpha (G(x,y))^\beta (B(x,y))^\gamma dx dy , \qquad (1)$$

where R, G, and B are three color channels. The authors use these moments for the construction of combined invariants to the affine transform of coordinates and contrast changes. In their most complex version, they suppose a general affine transform of RGB values.

In the case of an infinite set of moments of all orders, only the moments where $d = 1$ are independent, e.g. if we know $M_{pq}^{\alpha\beta\gamma}$ for $s = 0$ and $\alpha, \beta, \gamma = 0, 1, \ldots, 255$, we could theoretically reconstruct the complete 3D histogram. This redundancy decreases as the maximum order of moments decreases, for low-order moments this method may yield meaningful results. However, even for low orders, using higher powers of brightness in individual channels is more sensitive to nonlinearity of the contrast changes and may lead to misclassification.

The problematic issue in certain applications is not illumination changes, but exclusively or almost exclusively the geometric distortion. In such cases using traditional moments with $d = 1$ is more suitable and construction of a commensurate feature set is the theme of this contribution.

2 Moment Invariants of Multi-channel Images

Affine transformation is a general linear transform of space coordinates of an image. It can be expressed as

$$\begin{aligned} u &= a_0 + a_1 x + a_2 y \\ v &= b_0 + b_1 x + b_2 y. \end{aligned} \qquad (2)$$

An exact model of photographing a planar scene by a pin-hole camera whose optical axis is not perpendicular to the scene is a projective transform. Since the projective transform is not linear, its Jacobian is a function of spatial coordinates and projective moment invariants from a finite number of moments cannot exist. The perspective effect is negligible for small objects and large camera-to-scene distances and thus the affine transform can be used as good approximation of the projective transform.

The geometric moment m_{pq} of the order $s = p + q$ of an integrable image function $f(x, y)$ is defined as

$$m_{pq}^{(f)} = \int_{-\infty}^{\infty} \int_{-\infty}^{\infty} x^p y^q f(x, y) dx dy \ . \tag{3}$$

If the coordinates are translated so their origin appears the centroid of the image $x_c^{(f)} = m_{10}^{(f)}/m_{00}^{(f)}$, $y_c^{(f)} = m_{01}^{(f)}/m_{00}^{(f)}$, they are called *central moments*

$$\mu_{pq}^{(f)} = \int_{-\infty}^{\infty} \int_{-\infty}^{\infty} (x - x_c)^p (y - y_c)^q f(x, y) dx dy \ . \tag{4}$$

The first few affine moment invariants, the derivation of which can be found in e.g. [5], [6], or [7], are as follows:

$I_1 = (\mu_{20}\mu_{02} - \mu_{11}^2)/\mu_{00}^4$
$I_2 = (-\mu_{30}^2\mu_{03}^2 + 6\mu_{30}\mu_{21}\mu_{12}\mu_{03} - 4\mu_{30}\mu_{12}^3 - 4\mu_{21}^3\mu_{03} + 3\mu_{21}^2\mu_{12}^2)/\mu_{00}^{10}$
$I_3 = (\mu_{20}\mu_{21}\mu_{03} - \mu_{20}\mu_{12}^2 - \mu_{11}\mu_{30}\mu_{03} + \mu_{11}\mu_{21}\mu_{12} + \mu_{02}\mu_{30}\mu_{12} - \mu_{02}\mu_{21}^2)/\mu_{00}^7$
$I_4 = (-\mu_{20}^3\mu_{03}^2 + 6\mu_{20}^2\mu_{11}\mu_{12}\mu_{03} - 3\mu_{20}^2\mu_{02}\mu_{12}^2 - 6\mu_{20}\mu_{11}^2\mu_{21}\mu_{03} - 6\mu_{20}\mu_{11}^2\mu_{12}^2$
$\quad + 12\mu_{20}\mu_{11}\mu_{02}\mu_{21}\mu_{12} - 3\mu_{20}\mu_{02}^2\mu_{21}^2 + 2\mu_{11}^3\mu_{30}\mu_{03} + 6\mu_{11}^2\mu_{21}\mu_{12}$
$\quad - 6\mu_{11}^2\mu_{02}\mu_{30}\mu_{12} - 6\mu_{11}^2\mu_{02}\mu_{21}^2 + 6\mu_{11}\mu_{02}^2\mu_{30}\mu_{21} - \mu_{02}^3\mu_{30}^2)/\mu_{00}^{11}.$

The index (f) can be omitted, if the invariants are computed from one channel only. The theory of algebraic invariants (e.g. [8] among many others) offers using simultaneous invariants, i.e. invariants from moments of more than one order. They preserve their invariance even if we compute the moments of different orders on different objects, the only constraint is the common affine transform of the objects. We can choose two channels (let us label them a and b), take an arbitrary simultaneous invariant, e.g. I_3, and substitute second-order moments computed on one channel and third-order moments computed on the other channel

$I_{C23}^{(a,b)} = (\mu_{20}^{(a)}\mu_{21}^{(b)}\mu_{03}^{(b)} - \mu_{20}^{(a)}(\mu_{12}^{(b)})^2 - \mu_{11}^{(a)}\mu_{30}^{(b)}\mu_{03}^{(b)} + \mu_{11}^{(a)}\mu_{21}^{(b)}\mu_{12}^{(b)} + \mu_{02}^{(a)}\mu_{30}^{(b)}\mu_{12}^{(b)} -$
$\quad - \mu_{02}^{(a)}(\mu_{21}^{(b)})^2)/\mu_{00}^7,$

where $\mu_{00} = \mu_{00}^{(a)} + \mu_{00}^{(b)}$. The letter C in the index represents a common invariant of more channels (or color invariant) and the numbers stand for orders of the moments. We can also utilize algebraic invariants of two or more binary forms with the same order. An example of such an invariant is of the second order

$$I_{C2}^{(a,b)} = (\mu_{20}^{(a)}\mu_{02}^{(b)} + \mu_{20}^{(b)}\mu_{02}^{(a)} - 2\mu_{11}^{(a)}\mu_{11}^{(b)})/\mu_{00}^4.$$

If we use moments of one channel only (i.e. $a = b$), we obtain essentially I_1. Another such invariant is of the third order

$$I_{C3}^{(a,b)} = (\mu_{30}^{(a)}\mu_{03}^{(b)} - 3\mu_{21}^{(a)}\mu_{12}^{(b)} + 3\mu_{21}^{(b)}\mu_{12}^{(a)} - \mu_{30}^{(b)}\mu_{03}^{(a)})/\mu_{00}^5.$$

If we use moments of one channel only, it becomes zero. The third-order invariant of degree two from one channel does not exist, while that from two channels does exist. An example of three-channel fourth-order symmetric invariant is as follows

$$\begin{aligned}I_{C4}^{(a,b,c)} = (&\mu_{40}^{(a)}\mu_{22}^{(b)}\mu_{04}^{(c)} + \mu_{40}^{(a)}\mu_{22}^{(c)}\mu_{04}^{(b)} + \mu_{40}^{(b)}\mu_{22}^{(a)}\mu_{04}^{(c)} + \mu_{40}^{(b)}\mu_{22}^{(c)}\mu_{04}^{(a)} + \\ +&\mu_{40}^{(c)}\mu_{22}^{(a)}\mu_{04}^{(b)} + \mu_{40}^{(c)}\mu_{22}^{(b)}\mu_{04}^{(a)} - 2\mu_{40}^{(a)}\mu_{13}^{(b)}\mu_{13}^{(c)} - 2\mu_{40}^{(b)}\mu_{13}^{(a)}\mu_{13}^{(c)} - 2\mu_{40}^{(c)}\mu_{13}^{(a)}\mu_{13}^{(b)} - \\ -&2\mu_{31}^{(a)}\mu_{31}^{(b)}\mu_{04}^{(c)} - 2\mu_{31}^{(a)}\mu_{31}^{(c)}\mu_{04}^{(b)} - 2\mu_{31}^{(b)}\mu_{31}^{(c)}\mu_{04}^{(a)} + 2\mu_{31}^{(a)}\mu_{22}^{(b)}\mu_{13}^{(c)} + 2\mu_{31}^{(a)}\mu_{22}^{(c)}\mu_{13}^{(b)} + \\ +&2\mu_{31}^{(b)}\mu_{22}^{(a)}\mu_{13}^{(c)} + 2\mu_{31}^{(b)}\mu_{22}^{(c)}\mu_{13}^{(a)} + 2\mu_{31}^{(c)}\mu_{22}^{(a)}\mu_{13}^{(b)} + 2\mu_{31}^{(c)}\mu_{22}^{(b)}\mu_{13}^{(a)} - \\ -&6\mu_{22}^{(a)}\mu_{22}^{(b)}\mu_{22}^{(c)})/\mu_{00}^{9}.\end{aligned}$$

In this case $\mu_{00} = \mu_{00}^{(a)} + \mu_{00}^{(b)} + \mu_{00}^{(c)}$. The term "symmetric" here means that the formula is equivalent for each channel, if we permute the channels, we obtain the same formula.

We need some common μ_{00} and centroid of all channels, the simplest unification of that from individual channels is

$$\begin{aligned}m_{00} &= m_{00}^{(a)} + m_{00}^{(b)} + \ldots, \\ x_c &= (m_{10}^{(a)} + m_{10}^{(b)} + \ldots)/m_{00}, \qquad y_c = (m_{01}^{(a)} + m_{01}^{(b)} + \ldots)/m_{00}.\end{aligned} \quad (5)$$

The central moments are centered with respect to the common centroid

$$\mu_{pq}^{(a)} = \int_{-\infty}^{\infty}\int_{-\infty}^{\infty} (x-x_c)^p (y-y_c)^q a(x,y)dxdy \qquad p,q = 0,1,2,\ldots. \quad (6)$$

Now, channel centroids $x_c^{(k)} = m_{10}^{(k)}/m_{00}^{(k)}$, $y_c^{(k)} = m_{01}^{(k)}/m_{00}^{(k)}$, k=a,b,... can differ from the common centroid x_c, y_c. First-order moments need not be zero and we can use them for the construction of additional invariants, e.g.

$$I_{C1}^{(a,b)} = (\mu_{10}^{(a)}\mu_{01}^{(b)} - \mu_{10}^{(b)}\mu_{01}^{(a)})/\mu_{00}^{3}.$$

The value is zero, if computed from two channels only, m_{00}, x_c and y_c (5) must include a certain third channel for a non-zero result. Another example of a simultaneous invariant of first and second orders

$$I_{C12}^{(a,b)} = (\mu_{20}^{(a)}(\mu_{01}^{(b)})^2 + \mu_{02}^{(a)}(\mu_{10}^{(b)})^2 - 2\mu_{11}^{(a)}\mu_{10}^{(b)}\mu_{01}^{(b)})/\mu_{00}^{5}.$$

Even zero-order two-channel affine invariant does exist:

$$I_{C0}^{(a,b)} = \mu_{00}^{(a)}/\mu_{00}^{(b)}.$$

The rule concerning the number of invariants goes through the wringer, indices 0,0 really satisfy five constraints at the same time - all except scaling. If we compute affine moment invariants from a color photograph, we can use 12 additional invariants computed from more than one channel. Generally, if we have a multi-channel image with more than three channels, we can use 6 additional invariants per channel.

3 Numerical Experiment

The goal of this experiment is to show properties of the new moment invariants in pattern recognition. We have photographed a series of cards used in a game called mastercards (also pexeso), where the objective is to find the same pairs of cards turned face-down. Cards from each of the ten pairs are shown on Fig. 1.

Fig. 1. The mastercards. First row from left: Girl, Old scratch, Tyre-ride, Room-bell and Fireplace, second row: Winter cottage, Spring cottage, Summer cottage, Bell and Star.

Each card was captured eight times, rotated by approximately 45° between consecutive snaps. An example of the rotation of a pair of cards is shown on Fig. 2. Any small deviation from the perpendicular direction during capturing carried a small projective distortion that can be approximated by the affine transform. The cards were snapped on a dark background and segmented by region growing while small objects (less than 10000 pixels) were removed. The first snap of each card was used as a representative of its class and the following seven snaps were recognized by the minimum-distance classifier, so the theoretical maximum number of errors is 140.

Our feature set includes I_1, I_2, I_3, I_4 and $I_{C12}^{(a,a)}$ from each channel, $I_{C0}^{(R,G)}$, $I_{C0}^{(B,G)}$, $I_{C1}^{(R,B)}$, $I_{C12}^{(R,B)}$, $I_{C2}^{(R,G)}$ and $I_{C2}^{(B,G)}$, i.e. 21 invariants. The complete set should include additional 3 invariants, e.g. $I_{C23}^{(R,B)}$, $I_{C3}^{(R,G)}$ and $I_{C3}^{(B,G)}$, we omitted them because of the comparison with the same number of invariants from the other set. The moments are always centered to the common center of all 3 channels. The invariants are normalized to magnitude by the following procedure. The moments are first normalized to scaling

$$\tilde{\mu}_{pq} = \frac{\mu_{pq}}{\mu_{00}^{\frac{p+q}{2}+1}}, \qquad (7)$$

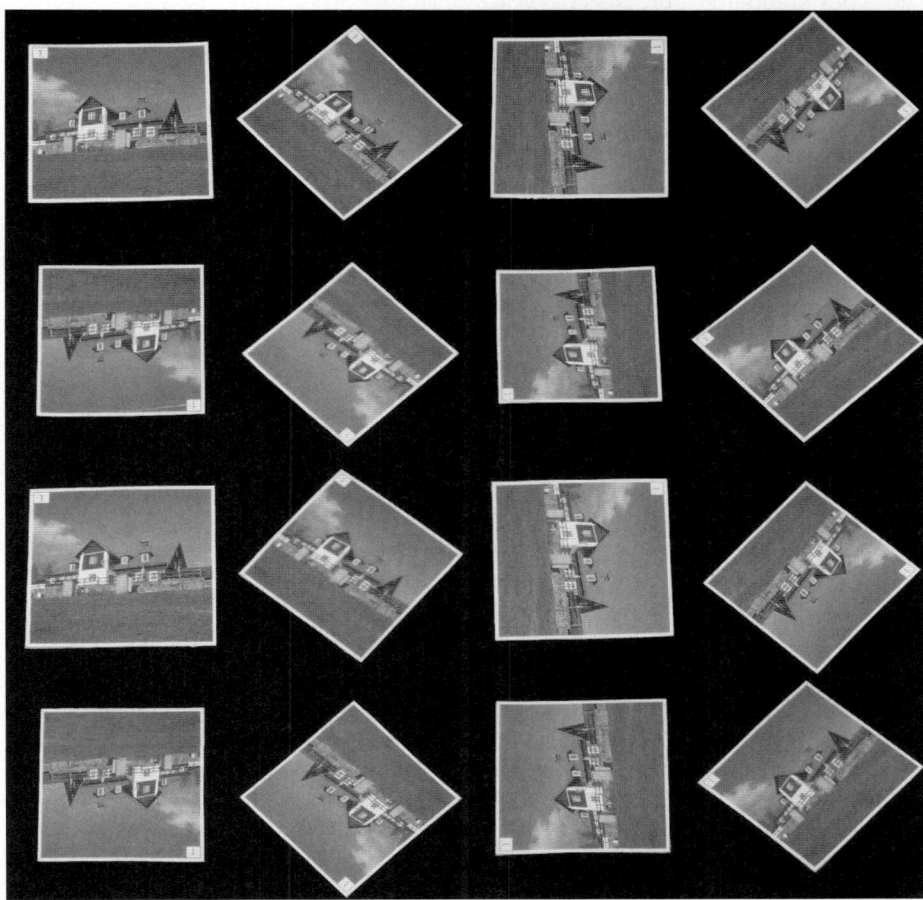

Fig. 2. The card "Summer cottage" with all rotations. The third and fourth rows contain the other card from the pair.

then to the magnitude

$$\hat{\mu}_{pq} = \pi^{\frac{p+q}{2}} \left(\frac{p+q}{2} + 1 \right) \tilde{\mu}_{pq} , \qquad (8)$$

and then the invariants are computed and normalized to the degree

$$\hat{I} = \text{sign}(I)|I|^{\frac{1}{r}} , \qquad (9)$$

where r is the degree of the invariant, i.e. the number of moments in one term. The minimum distance classifier was used. The cards were classified correctly, without an error. The classification of a card as the other card from the pair was not considered an error. Nevertheless, we can see the feature space of the zeroth-order invariants on Fig. 3 and some clusters are divided into two subclusters, i.e. the cards from that pair were distinct, while other pairs create compact clusters. Together 102 (73%) cards were assigned to the correct card from the pair, while 38 (27%) cards were assigned to the other card from the pair, but this datum depends not only on the quality of the features, but also on the actual differences of both cards.

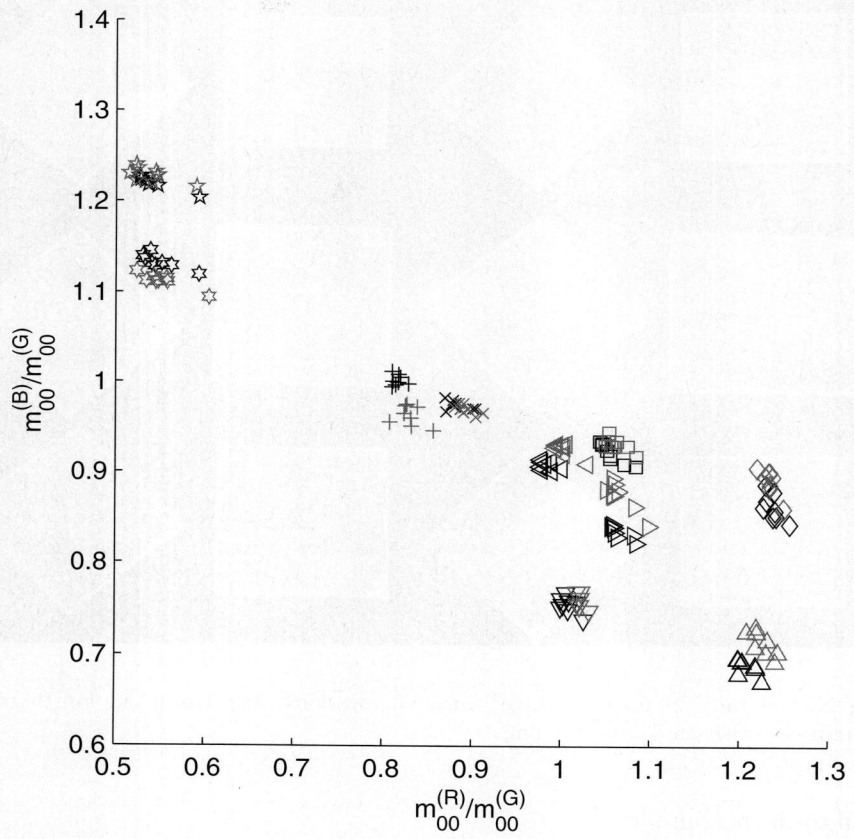

Fig. 3. The mastercards. Legend: ▽– Girl, △– Old scratch, □– Tyre-ride, ◇– Roombell and ▷– Fireplace, ×– Winter cottage, ◁– Spring cottage, +– Summer cottage, ✮– Bell and ✩– Star. A card from each pair is expressed by the black symbol while the other card is expressed by the magenta (gray) symbol.

Next, we carried out the experiment with one-channel invariants I_1, I_2, I_3 and I_4 only, i.e. 12 features together, centered to the centroid of each channel separately. The cards were classified with 3 errors (2.1%), which means the new invariants bring better discriminability and consequently improve the recognition.

For comparison, the same experiment was repeated with the invariants from [2], labeled "GPD invariants" in [4]. The number of invariants was the same as in the first experiment, i.e. 21, they were normalized by the standard deviation through the whole set and provided classification of the objects with 6 errors (4.3%). This can still be considered a good result, but obviously, new invariants can improve the recognition, when the illumination changes are small.

4 Conclusion

A new type of affine moment invariants for multi-channel images was proposed primarily for color images. They yield better results than the current invariants combining affine invariance with invariance to illumination changes. Nevertheless, the new invariants can be normalized to simple contrast changes as well.

The new invariants have a low order and therefore high robustness to noise, and include even zeroth-order invariants, which does not exist in the case of one-channel invariants.

References

1. Van Gool, E.P.L., Moons, T., Oosterlinck, A.: Vision and lie's approach to invariance. Invited paper for Image and Vision Computing 13, 259–277 (1995)
2. Mindru, F., Moons, T., Van Gool, L.: Color-based moment invariants for viewpoint and illumination independent recognition of planar color patterns. In: International Conference on Advances in Pattern Recognition, ICAPR 1998, pp. 113–122 (1998)
3. Mindru, F., Moons, T., van Gool, L.: Recognizing color patterns irrespective of viewpoint and illumination. In: Proc. IEEE Conf. Computer Vision Pattern Recognition, CVPR 1999, vol. 1, pp. 368–373 (1999)
4. Mindru, F., Tuytelaars, T., Van Gool, L., Moons, T.: Moment invariants for recognition under changing viewpoint and illumination. Computer Vision and Image Understanding 94, 3–27 (2004)
5. Flusser, J., Suk, T.: Pattern recognition by affine moment invariants. Pattern Recognition 26, 167–174 (1993)
6. Suk, T., Flusser, J.: Graph method for generating affine moment invariants. In: ICPR 2004, 17th International Conference on Pattern Recognition, pp. 192–195. IEEE Computer Society, Los Alamitos (2004)
7. Suk, T., Flusser, J.: Tables of affine moment invariants generated by the graph method. Research Report 2156, Institute of Information Theory and Automation (2005)
8. Hilbert, D.: Theory of Algebraic Invariants. Cambridge University Press, Cambridge (1993)

Graph-Based k-Means Clustering: A Comparison of the Set Median versus the Generalized Median Graph

M. Ferrer[1], E. Valveny[2], F. Serratosa[3], I. Bardají[1], and H. Bunke[4]

[1] Institut de Robòtica i Informàtica Industrial, CSIC-UPC
C. Llorens Artigas 4-6, 08028 Barcelona, Spain
{mferrer,ibardaji}@iri.upc.edu
[2] Centre de Visió per Computador, Universitat Autònoma de Barcelona
Edifici O Campus UAB, 08193 Bellaterra, Spain
ernest@cvc.uab.cat
[3] Departament d'Informàtica i Matemàtiques, Universitat Rovira i Virgili
Av. Països Catalans 26, 43007 Tarragona, Spain
francesc.serratosa@urv.cat
[4] Institute of Computer Science and Applied Mathematics, University of Bern
Neubrückstrasse 10, CH-3012 Bern, Switzerland
bunke@iam.unibe.ch

Abstract. In this paper we propose the application of the generalized median graph in a graph-based k-means clustering algorithm. In the graph-based k-means algorithm, the centers of the clusters have been traditionally represented using the set median graph. We propose an approximate method for the generalized median graph computation that allows to use it to represent the centers of the clusters. Experiments on three databases show that using the generalized median graph as the clusters representative yields better results than the set median graph.

1 Introduction

Clustering with graphs is a well studied topic in the literature, and various approaches have been proposed up to now. The classical paradigm in those approaches is to treat the entire clustering problem as a graph, that is, each element to be clustered is represented as a node and the distance between two elements is modeled by a certain weight on the edge linking the nodes [1]. Some other recent approaches propose to perform clustering directly on graph-based data. For instance in [2], the graph edit distance and the weighted mean of a pair of graphs were used to cluster graph-based data under an extension of self-organizing maps (SOMs). In [3], the authors investigated the clustering of attributed graphs by means of Function-Described Graphs (FDGs) to obtain representatives of clusters. Trees have also been used for clustering purposes. For instance, in [4] the clustering of shock trees using the tree edit distance was introduced. Finally, the extension of the k-means clustering algorithm to graph based representations was introduced in [5].

In this later approach the set median graph [6] has been used to represent the center of each cluster. Nevertheless, the concept of the generalized median graph [6] seems to be more adequate to represent the data of each cluster. Given a set of graphs, the generalized median graph [6] is defined as the graph that has the minimum sum of distances to all graphs in the set. It can be seen as the representative of the set. Thus it has a large number of potential applications in many classical algorithms for learning, clustering and classification, usually executed in the vector domain. However, its computation is exponential both in the number of input graphs and their size [7]. A number of algorithms for the generalized median graph computation have been reported in the past [6,8,9], but in general they suffer from either a large complexity or are restricted to special types of graphs.

In this paper we propose, for the first time, the use of the generalized median graph as the representative of a cluster in a graph-based version of the k-means algorithm. To deal with the high time and space complexity of the median graph computation, a new approximate method based on graph embedding in vector spaces is also proposed. First, we map each graph into a vector space using an approach similar to [10]. The median of the set of vectors obtained with this mapping can be easily computed in the vector space. Then, using the two closest points in the vector space and the weighted mean of a pair of graphs [11] we obtain an approximation of the median graph as the final result.

The experiments reported in this paper focus on running the k-means algorithm using the set median and the generalized median as the cluster representatives and comparing the two approaches to each other. To this end, three different databases (two of them containing real-world data) have been used. The results are evaluated through two standard clustering performance measures (the Rand index and the Dunn index). The results show that the generalized median graph yields better results than the set median graph when it is taken as the representative of a cluster. Furthermore, our procedure potentially allows us to transfer any machine learning algorithm that uses a median from the vector to the graph domain.

The rest of this paper is organized as follows. In the next section we introduce the basic concepts used in the paper. Then in Section 3 the proposed method for the median computation is described. Section 4 reports a number of experiments and present results achieved with our method. Finally, in Section 5 we draw some conclusions.

2 Background

2.1 The Graph-Based k-Means Clustering Algorithm

The k-means clustering algorithm is one of the most simple and straightforward methods for clustering data [12]. The usual way is to represent the data items as a collection of n numeric values usually arranged into a vector form in the space $\mathbb{R}^n$. Then, the *Euclidean distance* in this space and the *centroid* of a set of vectors are used to compute the mean of the data in the cluster.

A graph-based version of the classic k-means clustering algorithm has been presented in [5]. The main differences consist in the distance and the centroid computation. In the former, the graph edit distance [13] is used instead of the Euclidean distance. In the latter, in order to obtain a representative of each cluster, the set median graph (see definition below) is used instead of the centroid.

2.2 Median Graph

The median graph has been proposed as a useful tool to compute a representative of a set of graphs [6]. Let U be the set of graphs that can be constructed using a given set of labels L. Given $S = \{g_1, g_2, ..., g_n\} \subseteq U$, we can distiguish between the **set median graph** $\hat{g}$, and the **generalized median graph** $\bar{g}$ of S:

$$\hat{g} = arg \min_{g \in S} \sum_{g_i \in S} d(g, g_i), \quad \bar{g} = arg \min_{g \in U} \sum_{g_i \in S} d(g, g_i)$$

where d denotes a distance or a dissimilarity measure between graphs, in our case the graph edit distance [13,14].

The set median graph $\hat{g}$ is a graph g belonging to the training set S that suitably represents it. However, if we extend the search space to the whole set U, it is natural to think that a better representative (the generalized median graph) can be obtained.

The computation of the generalized median graph is a higly complex task, as any graph in U is a potential candidate. This makes its computation exponential in both the number and size of graphs [7]. The existing exact algorithms can only be applied to small sets of graphs with a very small number of nodes. Approximate algorithms are therefore needed [6,9]. Thus, graph embedding techniques have been recently used to solve graph matching problems more efficiently.

2.3 Graph Embedding

Graph embedding [15] aims to convert graphs into another structure, for example, real vectors, and then operate in the associated space to make easier some typical graph-based tasks, such as matching and clustering. A first group of embedding techniques are based on spectral graph theory. For instance, a relatively early approach based on the adjacency matrix of a graph is proposed in [16]. Another similar approach has been presented in [17], where the authors use the coefficients of some symmetric polynomials constructed from the spectral features of the Laplacian matrix, to convert the graphs into a vectorial form. Finally, in a recent approach [18], the idea is to embed the nodes of a graph into a metric space and view the graph edge set as geodesics between pairs of points on a Riemannian manifold. In this work we will use another class of graph embedding procedures based on the selection of some prototypes and graph edit distance computation. This approach, which we explain in more detail in the next section, was first presented in [10], and it is based on the work proposed in [19]. The basic intuition is that the description of the regularities in observations of classes and objects is the basis to perform pattern classification. Thus, based on the selection of a number of prototypes, each object is embedded into a vector space by taking its distance to all these prototypes.

3 Median Graph via Embedding

In this section we propose a novel approach for the approximate computation of the median graph based on graph embedding in a vector space. A similar approach has been presented in [20]. Nevertheless in the present procedure, only two graphs (instead of three as in [20]) are used to recover the median graph, which simplifies this task. This new procedure consists of three steps.

In a first step, graphs are embedded into a vector space using a variation of the novel approach proposed in [10]. In that work, a set T of prototypes is used to embed each graph in a vector space. In our case, the set of prototypes is exactly the same set $S = \{g_1, g_2, ..., g_n\}$ of training graphs that are used to compute the median graph. We therefore compute the graph edit distance between every pair of graphs in the set S. Since computing the graph edit distance is a NP-complete problem, in this work we have used the suboptimal methods presented in [21,22]. The resulting distances are arranged in a distance matrix. Each row (column) of the matrix can be seen as an n-dimensional vector. Since each row (column) of the distance matrix is assigned to one graph, such an n-dimensional vector is the vectorial representation of the corresponding graph.

Once all the graphs have been embedded in the vector space, the median vector is computed. To this end we use the concept of *Euclidean Median*. Given a set X, the *Euclidean Median* is a point $y \in \mathbb{R}^n$ that minimizes the sum of the Euclidean distances to all the points in the set. The Euclidean median has been chosen as the representative in the vector domain for two reasons. The first reason is that the median of a set of objects is one of the most promising ways to obtain the representative of such a set. The second is that, since the median graph is defined in a way very close to the median vector, we expect the median vector to represent accurately the vectorial representation of the median graph, and then, from the median vector to obtain a good approximation of the median graph. In this work we have used the most common approximate algorithm for the computation of the Euclidean median, that is, the Weiszfeld's algorithm [23].

Finally, in order to obtain the median graph, the last step is to transform the Euclidean median into a graph. Such a graph will be considered as an approximation of the median graph of the set S. To this end we will use a procedure based on the weighted mean of a pair of graphs [11].

The *weighed mean* of two graphs g and g' is a graph g'' such that $d(g, g'') = a$, $d(g'', g') = b$ and $d(g, g') = a + b$ for any two constants a and b with $0 \leq a, b \leq d(g, g')$. That is, g'' is a graph in between the graphs g and g' along the edit path between them. Figure 1 illustrates this idea.

To transform the median vector obtained in step 2 into a graph, we propose a strategy that uses two points in the vector space. The idea is the following (see Figure 2). Once the median vector v_m is computed, we choose its two closest points (v_1 and v_2 in Figure 2). Then, we compute the median vector of these two

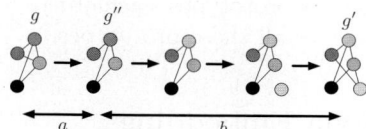

Fig. 1. Example of the weighted mean of a pair of graphs

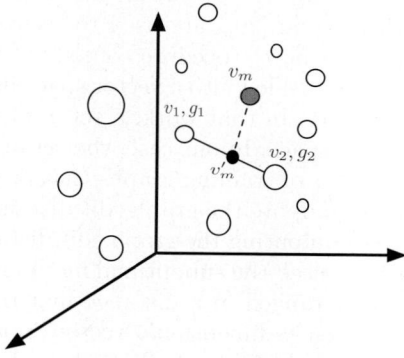

Fig. 2. Illustration of the two-point based procedure

points obtaining v'_m. This point v'_m is used to obtain the approximate median graph. To this end, we first compute the distance of v_1 and v_2 to v'_m, and then, with these distances we apply the weighted mean of a pair of graphs, between g_1 and g_2 (which correspond to v_1 and v_2 respectively), to obtain g'_m, the approximate median graph.

4 Application to Graph-Based k-Means Clustering

In this section we propose to use the approximate method for the median graph computation to obtain the representatives of the clusters in a graph-based k-means algorithm.

4.1 Experimental Setup

To perform the clustering experiments, we used the Molecule, the Webpage and the GREC datasets from [24]. For each dataset, the experiments consisted in computing the centers of the clusters using the set median (SM) and the generalized median (GM) with the method introduced in Section 3. The number k of clusters were set according to the number of classes in the dataset. Table 1 summarizes some basic parameters of each dataset. In order to evaluate the obtained results we performed 10 repetitions of each experiment. The

Table 1. Number of classes and number of elements per class for each database

Dataset	#Classes	Elements/Class
Molecules	2	100
Webpages	6	30
GREC	32	20

clustering performance was evaluated using two standard clustering performance measures, namely the *Rand index* and the *Dunn index*.

The *Rand index* R [25] measures how closely the clusters created by the clustering algorithm match the ground truth. It produces measures with values in the interval $[0, 1]$, with 1 meaning a perfect match between the result of the clustering algorithm and the ground truth.

The *Dunn index* D [26] is a measure of the compactness and separation of the clusters. It is not an accuracy measure like the *Rand Index*. It is rather based on the assumption that in a "perfect" clustering, items in the same cluster should be similar (i.e. should have a small distance) and items in different clusters should be dissimilar (i.e. should have a large distance). Higher values of the *Dunn Index* indicate a better clustering. Unlike the *Rand Index*, the *Dunn Index* is not bounded in the interval $[0, 1]$ but in the interval $[0, \infty)$.

4.2 Results

The results for this experiment are summarized in Tables 2 and 3. In each table the minimum, mean and maximum values for the *Rand Index* (Table 2) and the *Dunn Index* (Table 3) for each dataset are shown. In both tables, the best results are marked in **bold** face.

Results based on the *Rand Index* show that in almost all cases the *GM* method obtains better results than the set median graph. More concretely, seven out of the nine best results in Table 2 correspond to the *GM* method. Since the *Rand Index* is a measure of how similar the clusters are to the ground truth, these overall results demonstrate the idea that the median graph is a good representative of a given set, better than the set median graph.

Results based on the *Dunn Index* are shown in Table 3. Differently from the *Rand Index*, which is bound in between 0 and 1, the *Dunn Index* is not bounded. Thus, for the *Rand Index* it is relatively easy to interpret the value, because 0

Table 2. Minimum, average and maximum values of the Rand index for different datasets

	Minimum		Average		Maximum	
	SM	GM	SM	GM	SM	GM
Molecule	0.5072	**0.5545**	0.5620	**0.5952**	0.6205	**0.6860**
Webpages	0.6841	**0.8332**	0.8083	**0.8773**	0.8558	**0.9133**
GREC	**0.9410**	0.9340	0.9506	**0.9513**	**0.9602**	0.9566

Table 3. Minimum, average and maximum values of the Dunn index for different datasets

	Minimum		Average		Maximum		
	SM	GM	SM	GM	SM	GM	GT
Molecule	0.0113	**0.0272**	**0.034**	0.0288	**0.0909**	0.0431	0.0182
Webpages	**0.2039**	0.1028	**0.2448**	0.2027	**0.6046**	0.5784	0.1835
GREC	0.0411	**0.0423**	0.0503	**0.0507**	**0.0651**	0.0569	0.0619

means a completely uncorrelated result with respect to the groundtruth and 1 means a perfect match between the result and the groundtruth independently of the dataset used. However, the same reasoning is not possible for the *Dunn Index*. That is, we cannot say how good a result x for the *Dunn Index* is unless the *Dunn Index* for the groundtruth is given. For this reason, we have also computed the *Dunn Index* for the groundtruth (GT).

The results for each method are shown in Table 3. In this case the majority of the best results correspond to the set median. At first glance, these results could be interpreted in the sense that the set median reflects better the ideal cluster. Actually, however, they show that the set median graph obtains a better separation of the data into compact clusters. Yet, the results of the *Dunn Index* for the groundtruth show very low values. That means that the original datasets have low separability and compactness. In this sense, the *GM* method has more similar results to the *GT* than the set median. That means that it is able to better capture the original information of the clusters.

5 Conclusions

In this paper we have presented, for the first time, the use of the generalized median graph to obtain the centers of the clusters in a graph-based k-means algorithm using real-world data. To deal with the high computational requirements of the median graph computation, a new approximate method based on graph embedding in vector spaces has also been presented.

We performed a series of clustering experiments using three different databases. To evaluate the results, two standard clustering performance measures, namely the *Rand Index* and the *Dunn Index* have been used. Results in terms of the *Rand Index* show that with the median graph we obtain clusters closer to the groundtruth than using the set median graph. In addition, results given by the *Dunn Index* show that, although the set median graph obtains higher scores, the median graph obtains again results closer to the groundtruth.

With these results, we have shown that the median graph can be a better representative of a set of graphs. Furthermore, this new approximate procedure potentially allows the use of the median graph in other applications such as classification using real data.

Acknowledgements

This work has been supported by the Spanish research programmes Consolider Ingenio 2010 CSD2007-00018, TIN2006-15694-C02-02 and TIN2008-04998.

References

1. Jain, A.K., Murty, M.N., Flynn, P.J.: Data clustering: A review. ACM Comput. Surv. 31(3), 264–323 (1999)
2. Günter, S., Bunke, H.: Self-organizing map for clustering in the graph domain. Pattern Recognition Letters 23(4), 405–417 (2002)
3. Serratosa, F., Alquézar, R., Sanfeliu, A.: Synthesis of function-described graphs and clustering of attributed graphs. International Journal of Pattern Recognition and Artificial Intelligence 16(6), 621–656 (2002)
4. Luo, B., Robles-Kelly, A., Torsello, A., Wilson, R.C., Hancock, E.: Clustering shock trees. In: Proc. 3rd IAPR Workshop Graph-Based Representations in Pattern Recognition, pp. 217–228 (2001)
5. Schenker, A., Bunke, H., Last, M., Kandel, A.: Graph-Theoretic Techniques for Web Content Mining. World Scientific Publishing, USA (2005)
6. Jiang, X., Münger, A., Bunke, H.: On median graphs: Properties, algorithms, and applications. IEEE Trans. Pattern Anal. Mach. Intell. 23(10), 1144–1151 (2001)
7. Bunke, H., Münger, A., Jiang, X.: Combinatorial search versus genetic algorithms: A case study based on the generalized median graph problem. Pattern Recognition Letters 20(11-13), 1271–1277 (1999)
8. Hlaoui, A., Wang, S.: Median graph computation for graph clustering. Soft Comput. 10(1), 47–53 (2006)
9. Ferrer, M., Serratosa, F., Sanfeliu, A.: Synthesis of median spectral graph. In: Marques, J.S., Pérez de la Blanca, N., Pina, P. (eds.) IbPRIA 2005. LNCS, vol. 3523, pp. 139–146. Springer, Heidelberg (2005)
10. Riesen, K., Neuhaus, M., Bunke, H.: Graph embedding in vector spaces by means of prototype selection. In: Escolano, F., Vento, M. (eds.) GbRPR 2007. LNCS, vol. 4538, pp. 383–393. Springer, Heidelberg (2007)
11. Bunke, H., Günter, S.: Weighted mean of a pair of graphs. Computing 67(3), 209–224 (2001)
12. Mitchell, T.M.: Machine Learning. McGraw-Hill, New York (1997)
13. Bunke, H., Allerman, G.: Inexact graph matching for structural pattern recognition. Pattern Recognition Letters 1(4), 245–253 (1983)
14. Sanfeliu, A., Fu, K.: A distance measure between attributed relational graphs for pattern recognition. IEEE Transactions on Systems, Man and Cybernetics 13(3), 353–362 (1983)
15. Indyk, P.: Algorithmic applications of low-distortion geometric embeddings. In: IEEE Symposium on Foundations of Computer Science, pp. 10–33 (2001)
16. Luo, B., Wilson, R.C., Hancock, E.R.: Spectral embedding of graphs. Pattern Recognition 36(10), 2213–2230 (2003)
17. Wilson, R.C., Hancock, E.R., Luo, B.: Pattern vectors from algebraic graph theory. IEEE Trans. Pattern Anal. Mach. Intell. 27(7), 1112–1124 (2005)
18. Robles-Kelly, A., Hancock, E.R.: A Riemannian approach to graph embedding. Pattern Recognition 40(3), 1042–1056 (2007)

19. Pekalska, E., Duin, R.P.W., Paclík, P.: Prototype selection for dissimilarity-based classifiers. Pattern Recognition 39(2), 189–208 (2006)
20. Ferrer, M., Valveny, E., Serratosa, F., Riesen, K., Bunke, H.: An approximate algorithm for median graph computation using graph embedding. In: Proceedings of 19th ICPR, pp. 287–297 (2008)
21. Neuhaus, M., Riesen, K., Bunke, H.: Fast suboptimal algorithms for the computation of graph edit distance. In: Yeung, D.-Y., Kwok, J.T., Fred, A., Roli, F., de Ridder, D. (eds.) SSPR 2006 and SPR 2006. LNCS, vol. 4109, pp. 163–172. Springer, Heidelberg (2006)
22. Riesen, K., Neuhaus, M., Bunke, H.: Bipartite graph matching for computing the edit distance of graphs. In: Escolano, F., Vento, M. (eds.) GbRPR 2007. LNCS, vol. 4538, pp. 1–12. Springer, Heidelberg (2007)
23. Weiszfeld, E.: Sur le point pour lequel la somme des distances de n points donnés est minimum. Tohoku Math. Journal (43), 355–386 (1937)
24. Riesen, K., Bunke, H.: IAM graph database repository for graph based pattern recognition and machine learning. In: da Vitoria Lobo, N., Kasparis, T., Roli, F., Kwok, J.T., Georgiopoulos, M., Anagnostopoulos, G.C., Loog, M. (eds.) S+SSPR 2008. LNCS, vol. 5342, pp. 287–297. Springer, Heidelberg (2008)
25. Rand, W.M.: Objective criteria for the evaluation of clustering methods. Journal of the American Statistival Association 66, 846–850 (1971)
26. Dunn, J.: Well separated clusters and optimal fuzzy partitions. Journal of Cibernetics 4, 95–104 (1974)

Algorithms for the Sample Mean of Graphs

Brijnesh J. Jain and Klaus Obermayer

Berlin Institute of Technology, Germany
{jbj,oby}@cs.tu-berlin.de

Abstract. Measures of central tendency for graphs are important for protoype construction, frequent substructure mining, and multiple alignment of protein structures. This contribution proposes subgradient-based methods for determining a sample mean of graphs. We assess the performance of the proposed algorithms in a comparative empirical study.

1 Introduction

Measures of central tendency (MCT) for structures like the sample mean of graphs [5] or the generalized median graph [6] find their applications in central clustering of structures in computer vision [3], multiple alignment of protein structures [4], and frequent substructure mining [9]. A key problem is the exponential complexity of determining a MCT for graphs. Thus to apply a MCT in a practical setting, approximate algorithms are usually preferred. Almost all algorithms reported in the literature have been devised for the generalized median graph, where the underlying graph edit distance can be discontinuous [1].

In this contribution, we focus on the problem of sample mean of graphs. A sample mean is a graph that minimizes the sum of squared distances (SSD) to the sample graphs. The underlying distance measure is a pointwise minimizer of a set of geometric distance function. This formulation of a sample mean amounts in a MCT that summarizes a sample of graphs by recording the relative frequencies of common vertices and edges within their structural context. Since the SSD is locally Lipschitz [5] it admits the concept of generalized gradient from nonsmooth analysis and provides access to efficient techniques from nonsmooth optimization [8]. Using theoretical results from [5], we propose different subgradient algorithms for approximating a sample mean of structures. The proposed algorithms are variants of batch subgradient, incremental subgradient, and guide tree methods. We perform a comparative empirical study to assess the performance of the proposed algorithms.

2 The Sample Mean of Graphs

This section introduces the sample mean of graphs and provides some results proved in [5]. To approach the structural version of the sample mean in a principled way, we first introduce the concept of $\mathcal{T}$-space.

Let $\mathbb{E}$ be a d-dimensional Euclidean vector space. An (*attributed*) graph is a triple $X = (V, E, \alpha)$ consisting of a finite nonempty set V of *vertices*, a set $E \subseteq V \times V$ of *edges*, and an *attribute function* $\alpha : V \times V \to \mathbb{E}$, such that $\alpha(i, j) \neq \mathbf{0}$ for each edge and $\alpha(i, j) = \mathbf{0}$ for each non-edge. Attributes $\alpha(i, i)$ of vertices i may take any value from $\mathbb{E}$. For simplifying the mathematical treatment, we assume that all graphs are of order n, where n is chosen to be sufficiently large. Graphs of order less than n can be extended to order n by including isolated vertices with attribute zero.

A graph X is completely specified by its *matrix representation* $\boldsymbol{X} = (\boldsymbol{x}_{ij})$ with elements $\boldsymbol{x}_{ij} = \alpha(i, j)$ for all $1 \leq i, j \leq n$. By concatenating the columns of $\boldsymbol{X}$, we obtain a *vector representation* $\boldsymbol{x}$ of X. Let $\mathcal{X} = \mathbb{E}^{n \times n}$ be the Euclidean space of all $(n \times n)$-matrices and let $\mathcal{T}$ denote a subset of the set $\mathcal{P}^n$ of all $(n \times n)$-permutation matrices. Two matrices $\boldsymbol{X} \in \mathcal{X}$ and $\boldsymbol{X'} \in \mathcal{X}$ are said to be equivalent, if there is a permutation matrix $P \in \mathcal{T}$ such that $\boldsymbol{P}^\mathsf{T} \boldsymbol{X} \boldsymbol{P} = \boldsymbol{X'}$. The quotient set $\mathcal{X}_\mathcal{T} = \mathcal{X}/\mathcal{T} = \{[\boldsymbol{X}] : \boldsymbol{X} \in \mathcal{X}\}$ is the $\mathcal{T}$-*Space* over the *representation space* $\mathcal{X}$. A $\mathcal{T}$-space is a relaxation of the set $\mathcal{G}_\mathcal{T} = \mathcal{G}/\mathcal{T}$ of all *abstract graphs* $[\boldsymbol{X}]$, where $\boldsymbol{X}$ is a matrix representation of graph X. In the remainder, we identify $\mathcal{X}$ with $\mathbb{E}^N$ ($N = n^2$) and consider vector- rather than matrix representations of abstract graphs. By abuse of notation, we sometimes identify X with $[\boldsymbol{x}]$ and write $\boldsymbol{x} \in X$ instead of $\boldsymbol{x} \in [\boldsymbol{x}]$.

Next, we equip a $\mathcal{T}$-space with a metric related to the Euclidean metric. Suppose that d is an Euclidean metric on $\mathcal{X}$ induced by some inner product. Then the distance function $D(X, Y) = \min\{d(\boldsymbol{x}, \boldsymbol{y}) : \boldsymbol{x} \in X, \boldsymbol{y} \in Y\}$ is a metric with the same geometric properties as d. A pair $(\boldsymbol{x}, \boldsymbol{y}) \in X \times Y$ of vector representations is called *optimal alignment* if $D(X, Y) = d(\boldsymbol{x}, \boldsymbol{y})$.

Using the metric D on $\mathcal{X}_\mathcal{T}$, we can now define the concept of sample mean. Suppose that $\mathcal{D}_\mathcal{T} = (X_1, \ldots, X_k)$ is a sample of k abstract graphs from $\mathcal{G}_\mathcal{T} \subseteq \mathcal{X}_\mathcal{T}$. A *sample mean* of $\mathcal{D}_\mathcal{T}$ is any solution of the optimization problem

$$(P) \quad \begin{array}{l} \min \quad F(X) = \frac{1}{2} \sum_{i=1}^k D(X, X_i)^2 \\ \text{s.t.} \quad X \in \mathcal{X}_\mathcal{T} \end{array}.$$

The cost function F is the *sum of squared distances* (SSD) to the sample graphs. Here, the problem is to find a solution from an uncountable infinite set $\mathcal{X}_\mathcal{T}$. A simpler problem is to restrict the set $\mathcal{X}_\mathcal{T}$ of feasible solutions to the finite sample $\mathcal{D}_\mathcal{T} \subseteq \mathcal{X}_\mathcal{T}$. A *set mean graph* of $\mathcal{D}_\mathcal{T}$ is defined by $X = \arg\min\{F(X) : X \in \mathcal{D}_\mathcal{T}\}$.

We summarize the most important results from [5] for deriving subgradient-based algorithms for solving problem (P).

Theorem 1. *Let $\mathcal{D}_\mathcal{T} = (X_1, \ldots, X_k) \subseteq \mathcal{G}_\mathcal{T}$ be a sample of k abstract graphs.*

1. *Problem (P) has a solution. The solutions are abstract graphs from $\mathcal{G}_\mathcal{T}$.*
2. *The SSD function F is locally Lipschitz.*
3. *A vector representation $\boldsymbol{m}$ of a sample mean $M \in \mathcal{X}_\mathcal{T}$ of $\mathcal{D}_\mathcal{T}$ is of the form*

$$\boldsymbol{m} = \frac{1}{k} \sum_{i=1}^k \boldsymbol{x}_i,$$

where $d(\boldsymbol{x}_i, \boldsymbol{m}) = D(X_i, M)$ for all $i \in \{1, \ldots, k\}$. We call $(\boldsymbol{x}_1, \ldots, \boldsymbol{x}_k) \in X_1 \times \cdots \times X_k$ an optimal multiple alignment of $\mathcal{D}_\mathcal{T}$.

4. Let $(\boldsymbol{x}_1, \ldots, \boldsymbol{x}_k)$ be an optimal multiple alignment of $\mathcal{D}_\mathcal{T}$. Then

$$\sum_{i=1}^{k} \sum_{j=i+1}^{k} \langle \boldsymbol{x}_i, \boldsymbol{x}_j \rangle \geq \sum_{i=1}^{k} \sum_{j=i+1}^{k} \langle \boldsymbol{x}'_i, \boldsymbol{x}'_j \rangle$$

for all vector representations $\boldsymbol{x}'_1 \in X_1, \ldots, \boldsymbol{x}'_k \in X_k$.

The first statement ensures that problem (P) can be solved and has feasible solutions. Since the SSD satisfies the locally Lipschitz condition, we can apply generalized gradient techniques from nonsmooth optimization for minimizing the SSD [8]. The third statement shows that a vector representation of a structural sample mean is the standard sample mean of certain vector representations of the sample graphs. In addition, we see that problem (P) is a discrete rather than a continuous optimization problem, where a solution can be chosen from the finite set $X_1 \times \cdots \times X_k = \{(\boldsymbol{x}_1, \ldots, \boldsymbol{x}_k) : \boldsymbol{x}_i \in X_i\}$. The latter property combined with the fourth statement can be exploited for constructing search algorithms or meta-heuristics like genetic algorithms. The fourth statement asks for maximizing the sum of pairwise similarities (SPS). The standard sample mean of a vector representation maximizing the SPS is a vector representation of a structural sample mean. Apart from this, the fourth property provides a geometric characterization stating that an optimal multiple alignment has minimal volume within the subspace spanned by the vector representations. In the case that D is derived from the maximum common subgraph problem, the fourth property says that an optimal multiple alignment maximizes the sum of common edges of the sample graphs. This in turn indicates that computation of the sample mean has potential applications in frequent substructure mining.

3 Algorithms

This section proposes different algorithms for approximating a sample mean of a sample $\mathcal{D}_\mathcal{T} = (X_1, \ldots, X_k)$ of k graphs.

Generic Subgradient Method. Suppose that we want to minimize a locally Lipschitz function f on $\mathcal{X}$. Then f admits a generalized gradient at each point. The generalized gradient coincides with the gradient at differentiable points and is a convex set of points, called subgradients, at non-differentiable points. The basic idea of subgradient methods is to generalize the methods for smooth problems by replacing the gradient by an arbitrary subgradient. Algorithm 1 outlines the basic procedure of a generic subgradient method.

At differentiable points, direction finding generates a descent direction $\boldsymbol{d}$ by exploiting the fact that the direction opposite to the gradient of f is locally the steepest descent direction. At non-differentiable points, direction finding amounts in generating an arbitrary subgradient. The problem is that a subgradient at a non-differentiable point is not necessarily a direction of descent.

Algorithm 1. (Generic Subgradient Method)

---
01   set $t := 0$ and choose starting point $\boldsymbol{x}^t \in \mathcal{X}$
02   **repeat**
03   DIRECTION FINDING:
04   determine $\boldsymbol{d} \in \mathcal{X}$ and $\eta > 0$ such that $f(\boldsymbol{x}^t + \eta \boldsymbol{d}) < f(f(\boldsymbol{x}^t))$
05   LINE SEARCH:
06   find step size $\eta_* > 0$ such that $\eta_* \approx \arg\min_{\eta > 0} f(\boldsymbol{x}^t + \eta \boldsymbol{d})$
07   UPDATING:
08   set $\boldsymbol{x}^{t+1} := \boldsymbol{x}^t + \eta_* \boldsymbol{d}$
09   set $t := t + 1$
10   **until** some termination criterion is satisfied

But according to Rademacher's Theorem, the set of non-differentiable points is a set of Lebesgue measure zero. Line search determines a step size $\eta_* > 0$ with which the current solution $\boldsymbol{x}^t$ is moved along direction $\boldsymbol{d}$ in the updating step. Subgradient methods use predetermined step sizes $\eta_{t,i}$, instead of some efficient univariate smooth optimization method or polynomial interpolation as in gradient descent methods. One reason for this is that a subgradient determined in the direction finding step is not necessarily a direction of descent. Thus, the viability of subgradient methods depend critically on the sequence of step sizes. Updating moves the current solution $\boldsymbol{x}^t$ to the next solution $\boldsymbol{x}^t + \eta_* \boldsymbol{d}$. Since the subgradient method is not a descent method, it is common to keep track of the best point found so far, i.e., the one with smallest function value. For more details on subgradient methods and more advanced techniques to minimize locally Lipschitz functions, we refer to [8].

Batch Subgradient Methods

BSG – Batch Subgradient Method. Minimizing the SSD can be achieved using a batch subgradient algorithm. Successive estimates $\boldsymbol{m}^t \in M^t$ of the vector representations of a sample mean are computed using the following formula

$$\boldsymbol{m}^{t+1} = \boldsymbol{m}^t - \eta^t \left(\boldsymbol{m}^t - \boldsymbol{d}\right) = \left(1 - \eta^t\right) \boldsymbol{m}^t - \eta^t \sum_{i=1}^{k} \boldsymbol{x}_i$$

where η^t is the step size and $(\boldsymbol{x}_i, \boldsymbol{m}^t)$ are optimal alignments for all $i \in \{1, \ldots, k\}$. The direction graph represented by the vector $\boldsymbol{d}$ is a subgradient of F at M^t.

BAM – Batch Arithmetic Mean. As a variant of the BSG algorithm, batch arithmetic mean emulates the standard formulation of the sample mean

$$\boldsymbol{m} = \frac{1}{k} \sum_{i=1}^{k} \boldsymbol{x}_i,$$

where $\boldsymbol{x}_1, \ldots, \boldsymbol{x}_k$ are optimally aligned with some randomly chosen vector representation $\boldsymbol{x}_i$. BAM terminates after one iteration through the sample. This procedure is justified by Theorem 1.3.

BEM – Batch Expectation-Maximization. The batch expectation-maximization method repeatedly applies the BAM algorithm until two consecutive solutions do not differ by more than a prespecified threshold. The E-step aligns the vector representations of the sample graphs against the vector representation of the current sample mean as in the BAM method. The M-step readjusts the vector representation of the current sample mean given the alignment of the sample graphs in the E-Step.

Incremental Subgradient Methods

ISG – Incremental Subgradient Method. The elementary incremental subgradient method randomly chooses a sample graph X_i from $\mathcal{D}_\mathcal{T}$ at each iteration and updates the estimates $\boldsymbol{m}^t \in M^t$ of the vector representations of a sample mean according to the formula

$$\boldsymbol{m}^{t+1} = \boldsymbol{m}^t - \eta^t \left(\boldsymbol{m}^t - \boldsymbol{x}_i \right),$$

where η^t is the step size and $(\boldsymbol{x}_i, \boldsymbol{m}^t)$ is an optimal alignment.

IAM – Incremental Arithmetic Mean. As a special case of the ISG algorithm, the incremental arithmetic mean method emulates the incremental calculation of the standard sample mean. First the order of the sample graphs from $\mathcal{D}_\mathcal{T}$ is randomly permuted. Then a sample mean is estimates according to the formula

$$\boldsymbol{m}^1 = \boldsymbol{x}_1$$
$$\boldsymbol{m}^k = \frac{k-1}{k} \boldsymbol{m}^{k-1} + \frac{1}{k} \boldsymbol{x}_k \qquad \text{for } k > 1$$

where $(\boldsymbol{x}_i, \boldsymbol{m}^{i-1})$ is an optimal alignment for all $i > 1$. As BAM, this procedure is justified by Theorem 1.3 and requires only one iteration through the sample.

IMJ – Incremental Median Joining. While IAM randomly chooses the next sample graph for determining a sample mean, incremental median joining orders the sample graphs with increasing distance to the set mean graph. To determine the set median graph and the order with which the graphs are chosen, this procedure requires all pairwise distances between the sample graphs.

Guide Tree Methods. Guide tree methods perform agglomerative clustering on the sample graphs to construct a dendogram, called *guide tree*. The leaves of the guide tree represent the individual graphs from $\mathcal{D}_\mathcal{T}$. The inner nodes represent a weighted mean of its child nodes such that the root node represents a sample mean of $\mathcal{D}_\mathcal{T}$. Starting with the leaves, a guide tree determines the order of how two child nodes in the tree are merged to a weighted mean represented by their parent node. As IMJ, guide trees require all pairwise distances between the sample graphs. Algorithm 2 outlines a generic guide tree method.

We augment each node N^α of the guide tree by a weight α. Leaves have weight 1. The weight of an inner node is the sum of the weights of its child nodes. The root node has weight k. If we regard the nodes as representations of clusters, then the weights represent the cardinality of the respective clusters.

Algorithm 2. (Generic Guide Tree Method)

01	set $\mathcal{N} = \{N_1^1, \ldots, N_k^1\}$, where $N_i^1 = X_i$
01	calculate pairwise distance matrix $\boldsymbol{D} = (d_{ij})$ of $\mathcal{N}$
01	**repeat**
06	Find pair $(r, s) = \arg\min_{i,j} d_{ij}$ with lowest value in $\boldsymbol{D}$
07	remove N_r^α and N_s^β from $\mathcal{N}$
07	compute weighted mean $N_{rs}^\gamma = \left(\alpha N_r^\alpha \oplus \beta N_s^\beta\right)/\gamma$, where $\gamma = \alpha + \beta$
07	recalculate distances of nodes from $\mathcal{N}$ to new node N_{rs}^γ
07	insert new node N_{rs}^γ into $\mathcal{N}$
10	**until** $|\mathcal{N}| = 1$

Guide tree methods differ in the way they recalculate the distance matrix. Here, we consider two approaches, the neighbor joining and weighted centroid method.

GNJ – Neighbor Joining. The neighbor joining method is frequently used in bioinformatics, for the construction of phylogenetic trees and for multiple alignment of protein structures. We therefore refer to [10] for a detailed description of how the distances are recalculated.

GWC – Weighted Centroid Method. The weighted centroid method recalculates the distances to the new node using the distance function D.

4 Experiments

To assess the performance and to investigate the behavior of the sample mean algorithms described in Section 3, we conducted an empirical comparison on random graphs, letter graphs, and chemical graphs. For computing approximate subgradients we applied the graduated assignment (`GA`) algorithm. For datasets consisting of small graphs, we also applied a depth first search (`DF`) algorithm that guarantees to return an exact subgradient.

Data

Random Graphs. The first data set consists of randomly generated graphs. We sampled k graphs by distorting a given initial graph according to the following scheme: First, we randomly generated an initial graph M_0 with 6 vertices and edge density 0.5. Next, we assigned a feature vector to each vertex and edge of M_0 drawn from a uniform distribution over $[0,1]^d$ ($d = 3$). Given M_0, we randomly generated k distorted graphs as follows: Each vertex and edge was deleted with 20% probability. A new vertex was inserted with 10% probability and randomly connected to other vertices with 50% probability. Uniform noise from $[0,1]^d$ with standard deviation $\sigma \in [0,1]$ was imposed to all feature vectors. Finally, the vertices of the distorted graphs were randomly permuted. We generated 500 samples each consisting of $k = 10$ graphs. For each sample the noise level $\sigma \in [0,1]$ was randomly prespecified.

Table 1. Average SSD μ and standard deviations σ

		BSG	BAM	BEM	ISG	IAM	IMJ	GNJ	GWC
Random Graphs						set mean = 42.97 ($\pm$7.5)			
DF	μ	30.0	31.5	31.4	29.6	29.3	29.1	29.7	29.1
	σ	$\pm$5.5	$\pm$6.0	$\pm$5.9	$\pm$5.3	$\pm$5.2	$\pm$5.1	$\pm$5.3	$\pm$5.1
GA	μ	33.1	34.4	34.5	34.5	33.8	32.2	33.2	31.9
	σ	$\pm$6.7	$\pm$7.3	$\pm$7.0	$\pm$6.6	$\pm$6.0	$\pm$6.8	$\pm$6.8	$\pm$5.9
Letter Graphs (A, medium)						set mean = 60.5 ($\pm$16.6)			
DF	μ	45.1	46.0	45.2	42.3	42.4	42.2	43.2	42.5
	σ	$\pm$11.5	$\pm$11.7	$\pm$11.5	$\pm$10.1	$\pm$10.3	$\pm$10.1	$\pm$10.3	$\pm$10.1
GA	μ	44.5	46.7	44.6	43.9	44.2	44.0	46.0	44.1
	σ	$\pm$11.6	$\pm$12.8	$\pm$11.6	$\pm$11.1	1$\pm$1.3	$\pm$11.3	$\pm$12.4	$\pm$10.7
Chemical Graphs						set mean = 338.0 ($\pm$115.0)			
GA	μ	286.0	289.3	288.6	262.2	269.2	267.6	282.7	276.6
	σ	$\pm$112.9	$\pm$114.0	$\pm$114.3	$\pm$113.6	$\pm$113.7	$\pm$115.2	$\pm$116.2	$\pm$113.6

Letter Graphs. The letter graphs were taken from the IAM Graph Database Repository.[1] The graphs represent distorted letter drawings from the Roman alphabet that consist of straight lines only. Lines of a letter are represented by edges and ending points of lines by vertices. Each vertex is labeled with a two-dimensional vector giving the position of its end point relative to a reference coordinate system. Edges are labeled with weight 1. We considered the 150 letter graphs representing the capital letter A at a medium distortion level. We generated 100 samples each consisting or $k = 10$ letter graphs drawn from a uniform distribution over the dataset of 150 graph letters representing letter A at a medium distortion level.

Chemical Graphs. The chemical compound database was taken from the URL[2]. The dataset contains 340 chemical compounds, 66 atom types, and 4 types of bonds. On average a chemical compound consists of 27 vertices and 28 edges. Atoms are represented by vertices and bonds between atoms by edges. As attributes for atom types and type of bonds, we used a 1-to-k binary encoding, where $k = 66$ for encoding atom types and $k = 4$ for encoding types of bonds. We generated 100 samples each consisting of $k = 10$ chemical graphs drawn from a uniform distribution over the dataset of 340 chemical graphs.

Results. Table 1 shows the average SSD obtained by the different sample mean algorithms. Average SSD of the set mean graphs serve as reference values. Table 2 presents the average number of graph matching problems solved by each algorithm in order to approximate a sample mean. The approximated sample means

---

[1] URL = http://www.iam.unibe.ch/fki/databases/iam-graph-database
[2] http://www.xifengyan.net/software/gSpan.htm

Table 2. Average time μ and standard deviations σ. Time is measured in number of graph matching problems solved to obtain an approximate solution.

		BSG	BAM	BEM	ISG	IAM	IMJ	GNJ	GWC
		Random Graphs							
DF	μ	48.7	9.0	36.6	68.4	9.0	54.0	54.0	90.0
	σ	±28.5	±0.0	±6.5	±18.9	±0.0	±0.0	±0.0	±0.0
GA	μ	61.3	9.0	40.9	71.8	9.0	54.0	54.0	90.0
	σ	±24.7	±0.0	±6.0	±19.2	±0.0	±0.0	±0.0	±0.0
		Letter Graphs (A, medium)							
DF	μ	34.7	9.0	30.6	112.5	9.0	54.0	54.0	90.0
	σ	±20.8	±0.0	±10.4	±42.7	±0.0	±0.0	±0.0	±0.0
GA	μ	36.8	9.0	31.1	161.7	9.0	54.0	54.0	90.0
	σ	±23.2	±0.0	±9.5	±65.3	±0.0	±0.0	±0.0	±0.0
		Chemical Graphs							
GA	μ	44.3	9.0	40.9	64.5	9.0	54.0	54.0	90.0
	σ	±15.6	±0.0	±3.2	±21.7	±0.0	±0.0	±0.0	±0.0

have lower average SSD than the corresponding set mean graphs indicating that all subgradient methods yield reasonable SSD solutions. By Theorem 1.4, the subgradient-based algorithms yield multiple alignment of chemical graphs with larger sum of common edges than a multiple alignment against the set mean graph. This result indicates that the sample mean is a potential candidate for frequent substructure mining. In almost all cases sample mean algorithms employing depth first (DF) return better approximations than the same algorithms using an approximate graph matching procedure. On average, incremental methods perform best and batch methods worse. The reason is that batch methods align graphs independently against the current estimate of a sample mean. BSG and ISG suffer from slow convergence. Merging graphs in the order of their similarities turned out to be a computationally inefficient heuristic without gaining improved solutions. Incremental arithmetic mean (IAM) best trades solution quality against computation time.

5 Conclusion

We presented different versions of subgradient-based methods for approximating a sample mean of graphs. Incremental subgradient methods performed best with respect to solution quality while batch method performed worse. It turned out that incremental arithmetic mean is an efficient method that best trades solution quality against computation time and therefore is a good candidate for potential applications such as central clustering or frequent subgraph mining.

References

1. Ferrer, M.: Theory and algorithms on the median graph. Application to graph-based classification and clustering, PhD Thesis, Univ. Autònoma de Barcelona (2007)
2. Gold, S., Rangarajan, A.: Graduated Assignment Algorithm for Graph Matching. IEEE Trans. Pattern Analysis and Machine Intelligence 18, 377–388 (1996)
3. Günter, S., Bunke, H.: Self-organizing map for clustering in the graph domain. Pattern Recognition Letters 23, 401–417 (2002)
4. Jain, B., Stehr, H., Lappe, M.: Multiple Alignment of Contact Maps. In: IJCNN 2009 Conference Proceedings (2009)
5. Jain, B., Obermayer, K.: On the sample mean of graphs. In: IJCNN 2008 Conference Proceedings, pp. 993–1000 (2008)
6. Jiang, X., Munger, X.A., Bunke, H.: On Median Graphs: Properties, Algorithms, and Applications. IEEE Trans. PAMI 23(10), 1144–1151 (2001)
7. Luo, B., et al.: Clustering shock trees. In: GbR 2001, pp. 217–228 (2001)
8. Mäkelä, M., Neittaanmäki, P.: Nonsmooth Optimization: Analysis and Algorithms with Applications to Optimal Control. World Scientific, Singapore (1992)
9. Mukherjee, L., et al.: Generalized median graphs and applications. Journal of Combinatorial Optimization 17, 21–44 (2009)
10. Saitou, N., Nei, M.: The neighbor-joining method: a new method for reconstructing phylogenetic trees. Mol. Biol. Evol. 4(4), 406–425 (1987)

A Hypergraph-Based Model for Graph Clustering: Application to Image Indexing

Salim Jouili and Salvatore Tabbone

LORIA UMR 7503 - University of Nancy 2
BP 239, 54506 Vandoeuvre-lès-Nancy Cedex, France
{salim.jouili,tabbone}@loria.fr

Abstract. In this paper, we introduce a prototype-based clustering algorithm dealing with graphs. We propose a hypergraph-based model for graph data sets by allowing clusters overlapping. More precisely, in this representation one graph can be assigned to more than one cluster. Using the concept of the graph median and a given threshold, the proposed algorithm detects automatically the number of classes in the graph database. We consider clusters as hyperedges in our hypergraph model and we define a retrieval technique indexing the database with hyperedge centroids. This model is interesting to travel the data set and efficient to cluster and retrieve graphs.

1 Introduction

Graphs give a universal and flexible framework to describe the structure and the relationship between objects. They are useful in many different application domains like pattern recognition, computer vision and image analysis. For example in the context of content-based image retrieval, the user formulate a visual query. The user's target is seldom represented by a whole image which should not be processed like a one unit, because it is generally composed by a set of visual regions carrying out some semantics. Then, the graphs, by their natures, propose an adjusted solution for this task. Moreover, to reduce the number of graphs to be computed for matching or indexing tasks it is generally required to cluster objects. By this way, clustering similar images becomes equivalent to look for those graph representations that are similar to each other in a database. In this context, it is natural to apply clustering techniques to graphs. Clustering large set of graphs is still widely unexplored and is one of the most challenging problems in structural pattern recognition. In the recent years, some investigations on graph clustering and the organization of graph databases have been revitalized in [6,10,13,22]. Graph clustering problems rely in the organization of large structural databases, in discovering shape categories and view structure of objects, or in the construction of nearest neighbor classifiers. In this perspective, we propose a hypergraph model to cluster a set of graphs. A hypergraph [3] H=(ϑ, ξ) consists of a set of vertices ϑ and a set of hyperedges ξ; each hyperedge is a subset of vertices. We can note that the difference between an edge in a graph and a hyperedge in a hypergraph is that the former is always a subset of one or two vertices, and in the latter, the subset of vertices can be of arbitrary cardinality. In our model, we represent each graph by a vertex and each cluster by a hyperedge. The *degree* of a vertex is the number of hyperedges it belongs to, and the degree of a hyperedge is the number of vertices it contains. We denote the maximum degree of a vertex v by $\Delta_\vartheta(v)$ and the maximum degree of a hyperedge h by $\Delta_\xi(h)$. Recently, the hypergraph has been used, in the pattern recognition domain, for object representation [15], similarity measures [5], and object clustering [1]. In this paper we establish a hypergraph-based model for a graph database, we process as follows: firstly, a clustering technique based on the prototype selection is proposed to cluster the graph set into k independent clusters (k is detected automatically using a given threshold). Secondly, these clusters will be overlapped to define the final hypergraph structure. The idea of clusters overlapping is in the same vein as the works in [4,24] but the representation is different here. In fact from a set of experiments, we have remarked that the hypergraph structure provides a framework to retrieve and to browse graphs. This also leads to high clustering rate and improves the retrieval performance.

2 The Proposed Hypergraph Model

Since, we have focus our work in that one graph can belongs to several clusters, we consider that the proposed hypergraph is connected (*1-edge-connected*). Therefore, each graph G_i in the proposed structure is assigned to $\Delta_\vartheta(G_i)$ clusters and each cluster C_j contains $\Delta_\xi(C_j)$ graphs. However, a key problem in structuring a set of graphs into a hypergraph is the determination of the number of clusters (hyperedges) and the determination of related graphs (similar graphs) that can be grouped as hyperedges. In this perspective, we consider that the number of hyperedges is equal to the size of a representative set, defined on a selection of the most representative graphs in the whole set. We denote each selected graph as a hyperedge centroid. The selection of these graphs is similar to the problem of Prototype Selection [2,17,23]. K. Riesen and al. [17] enumerate some techniques to select prototypes from a training set. These techniques require a specification of the number of prototypes and there are no premises for determining automatically this number. Therefore, if we are in a unsupervised context where no information about the number of representative graphs is available, this number will be determined empirically. In this perspective, Spath [23] proposes an algorithm using leaders and distance based threshold where the number of selected prototype is inversely proportional to the selected threshold. However, the Leader algorithm [23] is sensitive to the selection of the initial prototype which is selected randomly among the input data. To overcome this problem, we introduce a representative graphs (hyperedge centroids) selection based on a peeling-off strategy. This method can be viewed as an improvement of the Leader algorithm and the K-Centers. After the selection of the hyperedge centroids, we define the hypergraph structure by assigning each graph to the corresponding hyperedges. Then the browsing and the retrieval of the graphs will be transposed into the hypergraph structure.

Hyperedge centroids selection. As stated above, the hyperedge centroids selection is similar to the Prototype Selection problem. Therefore, we aim to select a set of graphs which capture the most significant aspects of a set of graphs. We introduce an improvement for the Leader algorithm [23]. The proposed algorithm proceeds as follows:

1. Select the median graph [11] G_m from the unassigned graphs in the whole set of graphs S. Then the furthest graph G_{p_k} (which has not been previously assigned) to G_m, becomes the centroids of the cluster C_k. In the first iteration, the graph G_{p_k} is the initial selected prototype.
2. Distances of every unassigned graph $g_i \in S \setminus \{G_{p_k}\}$ are compared with that of the last selected prototype G_{p_k}. If the distances $d(g_i, G_{p_k})$ and $d(g_i, g_j \in C_k)$ are less than a predefined threshold T, the graph g_i is assigned to the cluster C_k with the centroid G_{p_k}, and g_i is tagged as assigned.
3. Recompute the median graph G_{m_k} of C_k, if $G_{m_k} \neq G_{p_k}$, replace G_{p_k} by G_{m_k}. If any replacements is done, go to the next step, otherwise all g_j are tagged as unassigned, $\forall g_j \in C_k$, then return to step 2.
4. While S contains an unassigned graphs return to step 1, otherwise stop.

Given a threshold T, the algorithm clusters the set of graphs with an intra-class inertia (I_i) less or equal to T. This property is performed on the step 2. In addition, this algorithm ensures the selection of the prototypes which are given by the centers of the resulted clusters. Futhermore, it guarantees a certain separability between classes of one partition. By using an edit distance d, we can formulate the between-class inertia (I_b) of a partition C composed of two classes C_1, C_2 by the Ward [25] criterion:

$$I_b(C_1, C_2) = \frac{\eta_1 \times \eta_2}{\eta_1 + \eta_2} d^2_{g_{c1}, g_{c2}} \qquad (1)$$

where g_{c_i} is the centroid of the class C_i and η_i is the number of members of C_i. The analysis of this formula shows that there is a strong dependence between the interclass inertia and the centroid. However, we know that the distance between two centroids is higher than the threshold T and $I_i \leq T$. Moreover, by fixing the initial selected prototype as the furthest graph to the median graph of the graph set, multiple runs of the algorithm produce identical results. We denote this clustering algorithm by D-hypergraph (disconnected hypergraph).

The hypergraph-based representation. Let S be the whole set of graphs and P be the set of selected prototypes P ($P \subset S$). Classical clustering techniques find for each graph $g \in S \setminus P$ its nearest neighbor $p_i \in P$ and add the graph to the cluster C_i corresponding to the prototype p_i. In fact, if a graph g presents a similar distances to two prototypes p_i and p_j, g is added to the cluster with the nearest prototype even though the difference between the two distances is very minor. Moreover, the provided clusters are disjoint and can be exploited for a retrieval task as used in [18,19,20,21], but it will be difficult to find an algorithm for browsing the whole set of graphs through disjoint clusters.

On the contrary, we propose a hypergraph-based model which allows the overlapping of clusters. In fact, henceforth the clusters will be viewed as hyperedges of hypergraph and the graphs as the vertices. Firstly, for each selected prototype p_i a hyperedge h_i is defined with a centroid p_i. Secondly, every hyperedge is defined as follows : each graph $g \in S \setminus P$ is added to the hyperedges with the nearest prototypes to g (their distances to g is less than the threshold T used in the previous algorithm). We denote this procedure by C-hypergraph (connected hypergraph).

Figure 1(a) illustrates our motivation. In the leftmost part of the figure $d_i = d(p_i, g_1)$ and we suppose that d_1 and d_2 are less or equal than T, so the graph g_1 shares some informations with p_1 and p_2 (informations are illustrated in colors). With the hypergraph model we will able to assign g_1 to the both hyperedges h_1 and h_2. The rightmost part of the figure 1(a) describes how two hyperedges (clusters) can overlap with one graph in common. Here, $\Delta_\vartheta(g_1)=2$ and $\Delta_\xi(h_1)=\Delta_\xi(h_2)=2$.

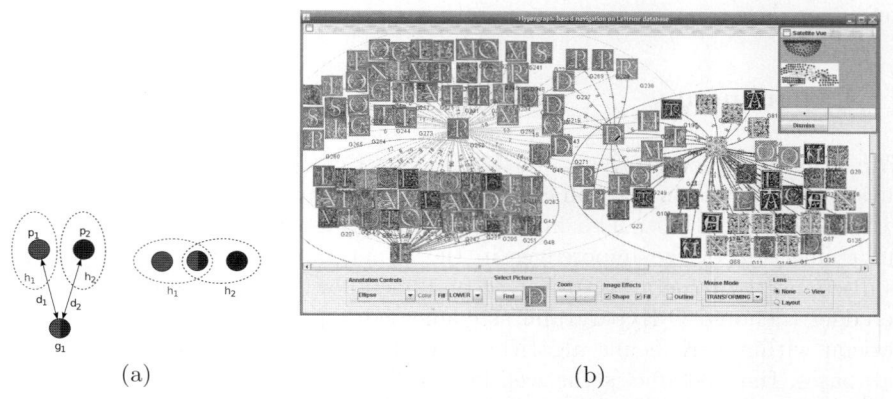

Fig. 1. Illustration of the hypergraph-based representation

Once all the hyperedges are defined from the graphs, we recompute, for each hyperedge, the generalized median graph which will be the new hyperedge centroids. The aim of this step is to update the hyperedge centroid after the hypergraph construction step and to maintain it including as much information as possible of the graphs in the corresponding hyperedge. We have chosen to use the generalized median graph to define the centroid of a cluster (unlike the Minimum Common Supergraph [6]) because it is less expensive in a viewpoint of computational time.

Interrogation and Navigation of hypergraph-based model of a set of graphs. Classically, interrogation of a set of graphs consists in searching the most similar graphs to a given query. This retrieval task ranks the graphs in an increasing distance order from the query. As remarked by a few works in the literature [18,19,20,21], this method do not exploit sophistically the distances, and the authors propose a clustering-based retrieval technique to improve the retrieval results. Here, we introduce a procedure which involves the hypergraph-based model presented previously. The main idea is to find the most similar hyperedge centroid to a given graph query. Then, we look for the most similar graphs within the hyperedge which it centroid is the most similar to the query. We can describe the retrieval procedure into the hypergraph model as follows:

1. For a query graph g_q, compute the set of distances between g_q and each hyperedge centroid.
2. Get the nearest hyperedge centroid p_i to g_q.
3. Retrieve the most similar graphs g_j to g_q, where $g_j \in h_i$ and h_i is the hyperedge with the centroid p_i.

This hypergraph-based model can be exploited to travel through the hypergraph. Once the previous retrieval procedure is performed, the user can browse the set of graphs, through a graphical interface (see figure 1(b), where the clusters (hyperedges) are represented by overlapped ellipses), by performing a walk among the hyperedges.

3 Experiments

The clustering evaluation. In this first part of the experiments, our contribution is evaluated in a graph clustering context. Here, our contribution is used within two algorithms. The first one is the prototype-based clustering without connection of the hyperedges in the hypergraph (denoted D-Hypergraph as disconnected hypergraph). The second one allows the overlapping of the hyperedges (denoted C-Hypergraph as connected hypergraph). We drawn a comparison within a K-means algorithm. To this aim we have used three image databases, the first one is the well-known COIL database [14] which contains different views of 3D objects. The images in COIL are converted into graphs by feature points extraction using the Harris interest points [9] and Delaunay triangulation. The second is the well-known GREC [16,7] database which consists of graphs representing symbols from architectural and electronic drawings. Here the ending points (ie corners, intersections and circles) are represented by nodes which are connected by undirected edges and labeled as lines or arcs. Finally, we have performed the clustering evaluation on an ornamental letters data set which contains lettrine (graphical object) extracted from digitized ancient documents [1]. Since one lettrine contains a lot of information (i.e. texture, decorated background, letters), the graphs are extracted from a region-based segmentation [8] of the lettrine. The nodes of the graph are represented by the regions and the edges describe their adjacency relationships. The graph distance measure used on the clustering is the graph matching measure based on the node signatures [12]. The clustering results are evaluated by the *Rand index*, the *Dunn index* and

---

[1] Provided by the CESR - University of Tours on the context of the ANR Navidomass project http://l3iexp.univ-lr.fr/navidomass/

the *Davies-Bouldin index*. The *Rand index* measures how closely the clusters created by the clustering algorithm match the ground truth. The *Dunn index* is a measure of the compactness and separation of the clusters and unlike the *Rand index*, the *Dunn index* is not normalized. The *Davies-Bouldin index* is a function of intra-cluster compactness and inter-cluster separation. We note that a good clustering provides a smaller *Davies-Bouldin index* and a higher *Rand* and *Dunn indexes*. In this experiment the threshold T, used by our method, is defined as the mean of distances between graphs in the same database. The number of classes k used by the K-means is defined in accordance with the ground truth.

Table 1 shows the results of the three cluster validity indexes. From these results, it is clear that our disconnected hypergraph produces clusters more compact and well separated. We note that when the C-Hypergraph is performed the *Dunn index* take the value 0, because some graphs share clusters and the minimum between-class distance becomes 0. Moreover, in a viewpoint of similarity to the ground truth, our model provides better results for the GREC and the Lettrine database, and we can remark also that the *Rand* index of the C-Hypergraph for the three databases are higher than the *Rand* index of the D-Hypergraph. Therefore, the connected hypergraph fits better the ground truth and encourages us to exploit the hypergraph-based structure for the graph retrieval problem.

Table 1. Clustering evaluation and comparison with K-means (Nc: the number of detected clusters)

	K-means	D-Hypergraph	C-Hypergraph
COIL Database	k=100	T=18.66, Nc=276	T=18.66
Rand Index	0.75	0.74	**0.75**
Dunn Index	0.03	**0.04**	0.00
DB Index	0.98	**0.88**	1.25
GREC Database	k=22	T=6.20, Nc=21	T=6.20
Rand Index	0.86	0.88	**0.91**
Dunn Index	0.01	**0.04**	0.00
DB Index	0.83	**0.76**	0.94
Lettrine Database	k=4	T=53.20, Nc=4	T=53.20
Rand Index	0.64	0.68	**0.69**
Dunn Index	0.10	**0.13**	0.00
DB Index	0.81	**0.61**	0.92

Evaluation of the retrieval with the hypergraph-based model. In this part of the experiments, we investigate the retrieval in the hypergraphs by performing the algorithm detailed previously on the Ornamental letters database. We provide a comparison with a classical retrieval task in which the graph query is compared to all the graphs in the database and then the most similar (the nearest ones) are retrieved. In the proposed approach, the hyperedges centroids are the entries of the database. That is to say, firstly the query graph is compared only to the hyperedge centroids. Then, the retrieval is performed among

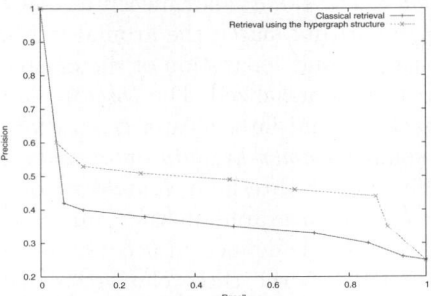

Fig. 2. Precision-Recall curves: comparison between classical retrieval and hypergraph-based retrieval

the graphs which belong to the hyperedge with the nearest centroid to the query. We used the receiver-operating curve (ROC) to measure retrieval performances. The ROC curves are formed by Precision rate against Recall rate, and drawn in the figure 2. By analyzing the two curves, we can remark that the results are better when the retrieval is performed only in one hyperedge. Furthermore, the hypergraph-based model is less time-consuming than the classic technique since it does not compare the query with all graphs in the set but only with graphs in an appropriate clusters.

4 Conclusion

In this paper we have investigated how the hypergraph structure can be used for the purpose of graph database representation. We have proposed a prototype-based method to cluster graphs and to select automatically the prototypes which collect as much information as possible from the graph set without a predefined number of clusters. The major task of this work is to allow the multi-assignment of one graph, i.e. one graph can be assigned to more than one cluster. We have also shown that our hypergraph-based model improve the retrieval and can be used to navigate into a graph database.

Acknowledgments. This research is partially supported by the French National Research Agency project NAVIDOMASS referenced under ANR-06-MCDA-012 and Lorraine region. The authors would like to thank Ines Mili for useful programming contribution.

References

1. Agarwal, S., Lim, J., Zelnik-Manor, L., Perona, P., Kriegman, D.J., Belongie, S.: Beyond pairwise clustering. In: IEEE CVPR, pp. 838–845 (2005)
2. Babu, T.R., Murty, M.N.: Comparaison of genetic algorithm based prototype selection schemes. Pattern Recognition 34, 523–525 (2001)

3. Berge, C.: Graphes et Hypergraphes. Paris Dunod (1970)
4. Bezdek, J.C., Pal, M.R., Keller, J., Krisnapuram, R.: Fuzzy Models and Algorithms for Pattern Recognition and Image Processing. Kluwer Academic Publishers, Norwell (1999)
5. Bunke, H., Dickinson, P.J., Kraetzl, M., Neuhaus, M., Stettler, M.: Matching of hypergraphs - algorithms, applications, and experiments. In: Bunke, H., Kandel, A., Last, M. (eds.) Applied Pattern Recognition. Studies in Computational Intelligence, vol. 91, pp. 131–154. Springer, Heidelberg (2008)
6. Bunke, H., Foggia, P., Guidobaldi, C., Vento, M.: Graph clustering using the weighted minimum common supergraph. In: Hancock, E.R., Vento, M. (eds.) GbRPR 2003. LNCS, vol. 2726, pp. 235–246. Springer, Heidelberg (2003)
7. Dosch, P., Valveny, E.: Report on the second symbol recognition contest. In: Liu, W., Lladós, J. (eds.) GREC 2005. LNCS, vol. 3926, pp. 381–397. Springer, Heidelberg (2006)
8. Felzenszwalb, P.F., Huttenlocher, D.P.: Efficient graph-based image segmentation. International Journal of Computer Vision 59(2), 167–181 (2004)
9. Harris, C., Stephens, M.: A combined corner and edge detection. In: Proceedings of The Fourth Alvey Vision Conference, pp. 147–151 (1988)
10. Hlaoui, A., Wang, S.: A graph clustering algorithm with applications to content-based image retrieval. In: ICMLC 2003, vol. 3, pp. 1855–1861 (2003)
11. Jiang, X., Munger, A., Bunke, H.: On median graphs:properties, algorithms, and applications. IEEE TPAMI 23(10), 1144–1151 (2001)
12. Jouili, S., Tabbone, S.: Graph matching using node signatures. In: IAPR Workshop on GbRPR. LNCS, vol. 5534, pp. 154–163. Springer, Heidelberg (2009)
13. Luo, B., Wilson, R.C., Hancock, E.R.: Spectral feature vectors for graph clustering. In: Caelli, T.M., Amin, A., Duin, R.P.W., Kamel, M.S., de Ridder, D. (eds.) SPR 2002 and SSPR 2002. LNCS, vol. 2396, pp. 83–93. Springer, Heidelberg (2002)
14. Nene, S., Nayar, S., Murase, H.: Columbia object image library (coil-100). Technical report, Columbia Univ. (1996)
15. Ren, P., Wilson, R.C., Hancock, E.R.: Spectral embedding of feature hypergraphs. In: da Vitoria Lobo, N., Kasparis, T., Roli, F., Kwok, J.T., Georgiopoulos, M., Anagnostopoulos, G.C., Loog, M. (eds.) S+SSPR 2008. LNCS, vol. 5342, pp. 308–317. Springer, Heidelberg (2008)
16. Riesen, K., Bunke, H.: IAM graph database repository for graph based pattern recognition and machine learning. In: da Vitoria Lobo, N., Kasparis, T., Roli, F., Kwok, J.T., Georgiopoulos, M., Anagnostopoulos, G.C., Loog, M. (eds.) S+SSPR 2008. LNCS, vol. 5342, pp. 287–297. Springer, Heidelberg (2008)
17. Riesen, K., Neuhaus, M., Bunke, H.: Graph embedding in vector spaces by means of prototype selection. In: Escolano, F., Vento, M. (eds.) GbRPR 2007. LNCS, vol. 4538, pp. 383–393. Springer, Heidelberg (2007)
18. Robles-Kelly, A., Hancock, E.R.: Graph edit distance from spectral seriation. IEEE TPAMI 27(3), 365–378 (2005)
19. Sebastian, T.B., Klein, P.N., Kimia, B.B.: Shock-based indexing into large shape databases. In: Heyden, A., Sparr, G., Nielsen, M., Johansen, P. (eds.) ECCV 2002. LNCS, vol. 2352, pp. 731–746. Springer, Heidelberg (2002)
20. Sengupta, K., Boyer, K.: Organizing large structural modelbases. IEEE TPAMI 17(4), 321–332 (1995)
21. Shapiro, L.G., Haralick, R.M.: Organization of relational models for scene analysis. IEEE TPAMI PAMI-4(6), 595–602 (1982)

22. Shokoufandeh, A., Dickinson, S.J.: A unified framework for indexing and matching hierarchical shape structures. In: Arcelli, C., Cordella, L.P., Sanniti di Baja, G. (eds.) IWVF 2001. LNCS, vol. 2059, pp. 67–84. Springer, Heidelberg (2001)
23. Spath, H.: Cluster analysis algorithms for data reduction and classification of objects. Ellis Horwood Limited, West Sussex (1980)
24. Torsello, A., Bulò, S.R., Pelillo, M.: Beyond partitions: Allowing overlapping groups in pairwise clustering. In: ICPR, pp. 1–4. IEEE, Los Alamitos (2008)
25. Ward, J.H.: Hierarchical grouping to optimize an objective function. Journal of the American Statistical Association 58(301), 236–244 (1963)

Hypergraphs, Characteristic Polynomials and the Ihara Zeta Function

Peng Ren[1], Tatjana Aleksić[2], Richard C. Wilson[1], and Edwin R. Hancock[1]

[1] Department of Computer Science, The University of York, York, YO10 5DD, UK
{pengren,wilson,erh}@cs.york.ac.uk
[2] University of Kragujevac, Faculty of Science, 34000 Kragujevac, Serbia
taleksic@kg.ac.rs

Abstract. In this paper we make a characteristic polynomial analysis on hypergraphs for the purpose of clustering. Our starting point is the Ihara zeta function [8] which captures the cycle structure for hypergraphs. The Ihara zeta function for a hypergraph can be expressed in a determinant form as the reciprocal of the characteristic polynomial of the adjacency matrix for a transformed graph representation. Our hypergraph characterization is based on the coefficients of the characteristic polynomial, and can be used to construct feature vectors for hypergraphs. In the experimental evaluation, we demonstrate the effectiveness of the proposed characterization for clustering hypergraphs.

1 Introduction

There has recently been an increasing interest in hypergraph-based methods for representing and processing visual information extracted from images. The main reason for this is that hypergraph representations allow nodes to be multiply connected by edges, and can hence capture multiple relationships between features. To the best of our knowledge, the first attempt for representing visual objects using hypergraphs dates back to Wong et al.'s [9] framework for 3D object recognition. Here a 3D object model based on a hypergraph representation is constructed, and this encodes the geometric and shape information of polyhedrons as vertices and hyperedges. Object synthesis and recognition tasks are performed by merging and partitioning the hyperedge and vertex set. The method is realized by using set operations and the hypergraphs are not characterized in a mathematically consistent way. Later Bretto et al. [2] have introduced a hypergraph model for image representation, where they successfully solved the problems of image segmentation, noise reduction and edge detection. However, their method also relies on a crude form of set manipulation. Recently Bunke et al. [3] have developed a hypergraph matching algorithm for object recognition, where consistency checks are conducted on hyperedges. The computational paradigm underlying their method is based on tree search operations. Hypergraphs have also been represented using matrices. For instance, Zass et al. [10] have presented a matrix representation for regular hypergraphs and used them for correspondence matching. However, the method has not been investigated for irregular hypergraphs.

One common feature of existing hypergraph-based methods is that they exploit domain specific and goal directed representations, and do not lend themselves to generalization. The reason for this lies in the difficulty in formulating a hypergraph in a mathematically uniform way for computation. There has yet to be a widely accepted and uniform way for representing and characterizing hypergraphs, and this remains an open problem with exploiting hypergraphs for machine learning. Moreover, to be easily manipulated, hypergraphs must be represented in an mathematically consistent way, using structures such as matrices or vectors.

Since Chung's [4] definition of the Laplacian matrix for k-uniform hypergraghs, there have been several attempts to develop matrix representations of hypergraphs [5][7][11]. These hypergraph representations have found widespread applications in categorical data analysis. Recently, we have shown that matrix representation are also suitable for image processing [6], and have proposed an improved hypergraph Laplacian based on the developments of Zhou et al.'s method [11]. However, this method is based on a relatively impoverished spectral characterization and overlooks much of the detail of hypergraph-structure.

In this paper, we make a first attempt to characterize hypergraphs using characteristic polynomials. Specifically, we use the Ihara coefficients, which are the polynomial coefficients of the reciprocal Ihara zeta function for a hypergraph. The Ihara zeta function for a hypergraph has been detailed by Storm [8]. Based on this work, we establish feature vectors using the Ihara coefficients. We apply our feature vectors to clustering hypergraphs extracted from images of different object views and demonstrate their effectiveness in hypergraph characterization.

2 Hypergraph Laplacian Spectrum

A hypergraph is normally defined as a pair $H(V, E_H)$ where V is a set of elements, called nodes or vertices, and E is a set of non-empty subsets of V called hyperedges. The incidence matrix $\boldsymbol{H}$ of $H(V, E_H)$ is a $|E_H(H)| \times |V(H)|$ matrix with the (i, j)th entry 1 if the vertex $v_j \in V(H)$ is contained in the hyperedge $e_i \in E_H$ and 0 otherwise. For $H(V, E_H)$, one of the alternative definitions of the adjacency matrix and the corresponding Laplacian matrix is $\boldsymbol{A}_H = \boldsymbol{H}\boldsymbol{H}^T - \boldsymbol{D}_H$ and $\boldsymbol{L}_H = \boldsymbol{D}_H - \boldsymbol{A}_H = 2\boldsymbol{D}_H - \boldsymbol{H}\boldsymbol{H}^T$ respectively, where $\boldsymbol{D}_H$ is the diagonal vertex degree matrix whose diagonal element $d(v_i)$ is the summation of the ith row of $\boldsymbol{H}$ [6]. The eigenvalues of $\boldsymbol{L}_H$ is referred to as the hypergraph Laplacian spectrum and are straightforward characteristics from $H(V, E_H)$. We will use this spectral characterization for experimental comparison in Section 5.

3 Ihara Zeta Function for a Hypergraph

The definition of the Ihara zeta function for a hypergraph $H(V, E_H)$ is as follows [8]:

$$\zeta_H(u) = \prod_{p \in P_H} \left(1 - u^{|p|}\right)^{-1}. \tag{1}$$

Here P_H is the set of the equivalence classes of prime cycles in the hypergraph $H(V, E_H)$. A prime cycle in a hypergraph is a closed path with no backtracking, that is, no hyperedge is traversed twice in the prime cycle.

The Ihara zeta function for a hypergraph in the form of (1) is generally an infinite product. However, one of its elegant features is that it can be collapsed down into a rational function, which renders it of practical utility. To recast the hypergraph Ihara zeta function as a rational function, the graph representation of hypergraph is needed. There are several ways in which a hypergraph can be transformed into a graph representation. One of the most useful representations is the clique expansion, to which we turn our attention in more detail in the following section. Agarwal et al. [1] has reviewed the alternative graph representations of hypergraphs and detailed their relationships with each other with a particular emphasis on machine learning. To obtain the rational expression for the hypergraph Ihara zeta function, we make use of the bipartite graph representation of hypergraphs. To establish the associated bipartite graph, we use a dual representation in which each hyperedge is represented by a new vertex. The new vertex is incident to every original vertex encompassed by the corresponding hyperedge. The new vertex set and together with the original vertex set constitute the associated bipartite graph; the new vertices corresponding to hyperedges in one partition and the original vertices to the second partition. To provide an example, the bipartite graph associated with the hypergraph in Fig. 1 is shown in Fig. 2.

The Ihara zeta function for the hypergraph $H(V, E_H)$ can be equivalently expressed in a rational form as follow [8]:

$$\zeta_H(u) = (1-u)^{\chi(BG)} \det\left(\boldsymbol{I}_{|V(H)|+|E_H(H)|} - \sqrt{u}\boldsymbol{A}_{BG} + u\boldsymbol{Q}_{BG}\right)^{-1} \quad (2)$$

where $\chi(BG) = |V|$ is the Euler Number which equals the difference between the cardinalities of vertex set and edge set of the associated bipartite graph, $\boldsymbol{A}_{BG}$ its adjacency matrix and $\boldsymbol{Q}_{BG} = \boldsymbol{D}_{BG} - \boldsymbol{I}_{|V(H)|+|E_H(H)|}$.

4 Perron-Frobenius Operator for Hypergraphs

From (1) it is clear that the Ihara zeta function for a hypergraph can be rewritten in the form of the reciprocal of a polynomial. Although the Ihara zeta function can be evaluated efficiently using (2), the task of enumerating the coefficients of the polynomial appearing in the denominator of the Ihara zeta function (2) is difficult, except by resorting to software for symbolic calculation. To efficiently compute these coefficients, we adopt the determinant expression of the Ihara zeta function for a hypergraph $H(V, E_H)$ [8]:

$$\zeta_H(u) = \det(\boldsymbol{I}_H - u\boldsymbol{T}_H)^{-1}. \quad (3)$$

Here $\boldsymbol{T}_H$ is a square matrix which is referred to as the Perron-Frobenius operator of the hypergraph $H(V, E_H)$. It is the adjacency matrix of the oriented line graph associated with $H(V, E_H)$.

The establishment of the oriented line graph associated with $H(V, E_H)$ commences by constructing an $|e_i|$-clique by connecting each pair of vertices in the e_i through an edge for each hyperedge $e_i \in E_H$. The resulting graph is denoted by $GH(V, E_G)$. For the example hypergraph in Fig. 1, the associated $GH(V, E_G)$ is shown in Fig. 3. In this example, the oriented edges derived from the same hyperedge are colored the same while from different hyperedges are colored differently.

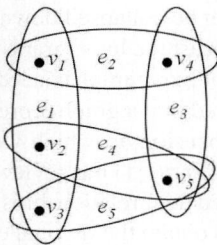

Fig. 1. Hypergraph **Fig. 2.** Bipartite Graph

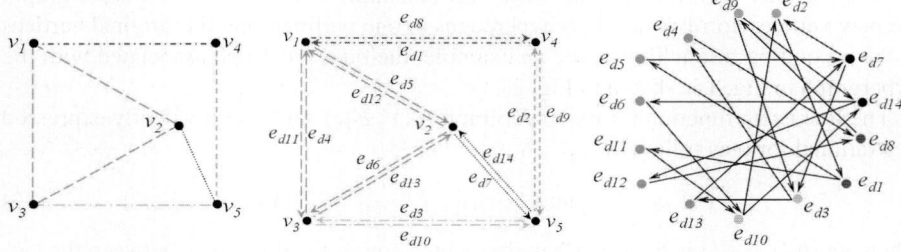

Fig. 3. Clique **Fig. 4.** Di-clique **Fig. 5.** Oriented Line Graph

For the graph $GH(V, E_G)$, the associated symmetric digraph $DGH(V, E_d)$ can be obtained by replacing each edge of $GH(V, E_G)$ by an arc (oriented edge) pair in which the two arcs are inverse to each other. For $GH(V, E_G)$ in Fig. 3, the associated $DGH(V, E_d)$ is shown in Fig. 4. Finally, we can establish the oriented line graph of the hypergraph based on the symmetric digraph. The vertex set and edge set of the the oriented line graph are defined as follows:

$$\begin{cases} V_{ol} = E_d(DGH) \\ E_{ol} = \{(e_d(u,v), e_d(v,w)) \in E_d \times E_d \, ; \, u \cup w \not\subset E_H \}. \end{cases} \quad (4)$$

The oriented line graph of the hypergraph in Fig. 1 is shown in Fig. 5. Here what we should note is that the oriented edges in the same clique of DGH can not establish an oriented edge in the oriented line graph. For instance, in Fig. 4 the terminus of the arc e_{d5} points to the origin of the arc e_{d6}. However, there is no oriented edge between vertices e_{d5} and e_{d6} in Fig. 5 because they are derived from the same hyperedge e_1 in Fig. 1. Therefore, this constraint prevents connections between any nodes with the same color in Fig. 5.

To establish feature vectors from the hypergraph Ihara zeta function for the purposes of characterizing hypergraphs in machine learning, it is natural to consider taking function samples as the vector elements. Although the function values at most of the sampling points will perform well in distinguishing hypergraphs, there is the possibility of sampling at poles giving rise to meaningless infinities. Hence, the feature vectors consisting of function samples are potentially unstable representations of hypergraphs, since the distribution of poles is unknown beforehand.

On the other hand, from (3) it is clear that the reciprocal of the hypergraph Ihara zeta function is the characteristic polynomial of the Perron-Frobenius operator $\boldsymbol{T}_H$ and it can be deployed as:

$$\zeta_H^{-1}(u) = c_0 + c_1 u + \cdots + c_{M-1} u^{M-1} + c_M u^M \quad (5)$$

where M is the dimensionality of the square matrix $\boldsymbol{T}_H$. Each coefficient can be computed from the elementary symmetric polynomials of the eigenvalue set $\{\tilde{\lambda}_1, \tilde{\lambda}_2 \ldots \tilde{\lambda}_n\}$ of $\boldsymbol{T}_H$ as follows:

$$c_r = (-1)^r \sum_{k_1 < k_2 < \ldots < k_r} \tilde{\lambda}_{k_1} \tilde{\lambda}_{k_2} \ldots \tilde{\lambda}_{k_r}. \quad (6)$$

The characteristic polynomial coefficients in (5) do not give rise to infinities. Furthermore, these coefficients highly relate to the hypergraph-structure since the Ihara zeta function records the information about prime cycles in the hypergraphs. We refer to the set of characteristic polynomial coefficients as Ihara coefficients. We use the Ihara coefficients as the elements of the feature vector for a hypergraph and then apply them to clustering hypergraphs.

5 Experimental Evaluation

We apply the proposed feature vectors to two hypergraph datasets extracted from images of different object views. The first set of hypergraphs are extracted from house images in the CMU, MOVI and Chalet sequences (samples are shown in Fig. 6(a)) and the second set are extracted from images of eight objects in COIL dataset (samples are shown in Fig. 6(b)). To establish hypergraphs on the objects, we first extract feature points using the Harris detector as the vertices of hypergraphs. Let $\boldsymbol{c}(v_i)$ denote the spatial coordinate of the feature point v_i in an image, $I(v_i)$ denote the intensity of v_i. For each image, we construct the hypergraph using the method introduced in [6], where an element of incidence matrix is denoted as:

$$H(i,j) = \begin{cases} 1 & \text{if } \|\boldsymbol{c}(v_i) - \boldsymbol{c}(v_j)\| \leq Th_{j1} \text{ and if } |I(v_i) - I(v_j)| \leq Th_{j2} \\ 0 & \text{otherwise.} \end{cases} \quad (7)$$

Here Th_{j1} is a fixed value which represents the distance threshold for neighborhood and is set to 1/4 the size of the image, and Th_{j2} is the similarity threshold, which is determined by the standard deviation of the intensities of the feature points in the neighborhood of v_j.

We compute the Ihara coefficients as introduced in Section 4, generating the feature vector in the form of $\boldsymbol{v}_I = [c_3,\ c_4,\ \ln(|c_{M-3}|),\ \ln(|c_{M-2}|),\ \ln(|c_{M-1}|),\ \ln(|c_M|)]^T$. The last four components of the feature vector are manipulated in a logarithmic way to avoid problems of dynamic range. Fig. 7 shows the PCA projection of the hypergraphs from the Chalet images based on Ihara coefficients. Each point in the pattern space is marked with a view number which corresponds to the camera angle. The coefficients

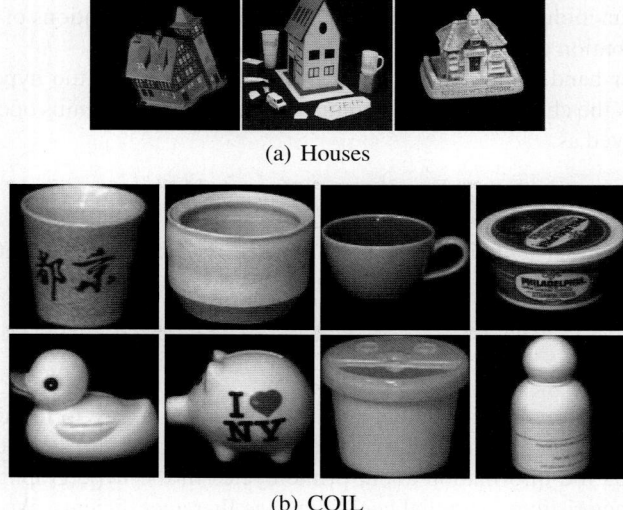

(a) Houses

(b) COIL

Fig. 6. Dataset

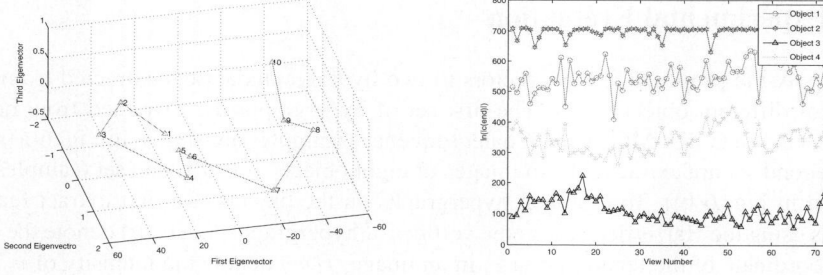

Fig. 7. Within-class Trajectory   **Fig. 8.** Ihara Coefficient Plot

Table 1. Rand Indices

Feature Vector	Number of Object Classes			
	5	6	7	8
Spectra	0.8574	0.8564	0.8454	0.8449
Ihara Coefficients	0.9355	0.8859	0.8716	0.8812

produce a clear trajectory and the neighboring images in the sequence are generally close together in the eigenspace.

Fig. 8 illustrates the behavior of the coefficient $\ln(|c_M|)$ of the first four objects in the COIL dataset. The four dotted lines represent the coefficient $\ln(|c_M|)$ of the four

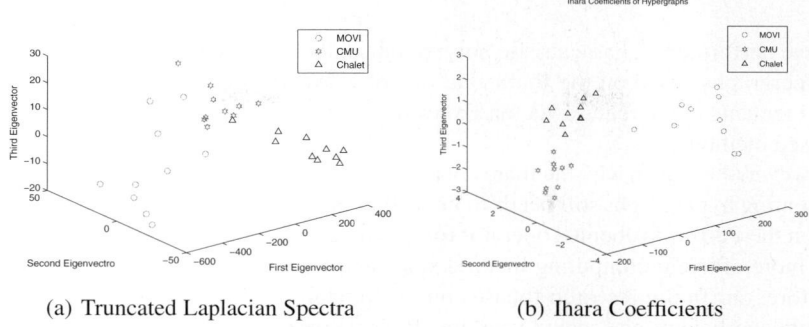

(a) Truncated Laplacian Spectra (b) Ihara Coefficients

Fig. 9. Clusters for Three Classes of Houses

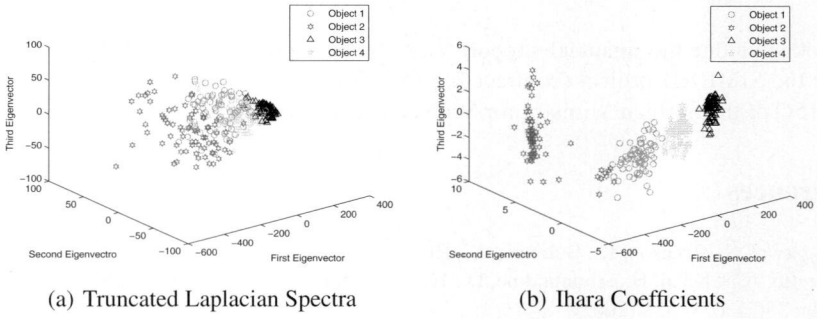

(a) Truncated Laplacian Spectra (b) Ihara Coefficients

Fig. 10. Clusters for Four Objects in COIL Dataset

objects separately. The coefficients of the different objects are well separated, thus indicating that the objects are well clustered. We then embed the feature vectors into a three-dimensional space using PCA for visualization. Figs. 9(a) and 9(b) indicate the results of the three classes of houses in Fig. 6(a) by using the feature vector consisting of the first through to sixth nonzero Laplacian eigenvalues and the proposed feature vector consisting of the Ihara coefficients respectively. Figs. 10(a) and 10(b) indicate the results for the first four objects in COIL dataset by using the feature vector consisting of the first through to sixth nonzero Laplacian eigenvalues and the proposed feature vector consisting of the Ihara coefficients respectively. From Fig. 9 and 10 we can see the proposed method is superior to the truncated Laplacian spectra in clustering hypergraphs.

To take the quantitative evaluation of the feature vectors one step further, we concentrate our attention on the COIL dataset, and evaluate the clustering performance obtained with different numbers of object classes. After performing PCA on the feature vectors, we locate the clusters using the K-means method and calculate the Rand index for the resulting clusters. The Rand indices for the truncated Laplacian spectra and for the Ihara coefficients are listed in Table 1. It is clear that the Ihara coefficients outperform the truncated Laplacian spectra for all numbers of object classes studied.

6 Conclusion and Future Work

We have performed a characteristic polynomial analysis on hypergraphs and characterize hypergraphs based on the Ihara zeta function. We have used the Ihara coefficients as the elements in the feature vector. Experimental results show the effectiveness of the proposed method.

However, the reason why the Ihara coefficients are superior to the Laplacian spectra in representing hypergraphs still needs to be further investigated. Moreover, the manipulations on the Perron-Frobenius operator for a hypergraph are computationally expensive and a more efficient computing method is needed for obtaining the Ihara coefficients. Therefore, our future research focuses on theoretically explaining the effectiveness of the Ihara coefficients and seeking a more efficient computation method.

Acknowledgments

We acknowledge the financial support from the FET programme within the EU FP7, under the SIMBAD project (contract 213250). Tatjana Aleksić is supported by Grant 144015G of the Serbian Ministry for Science and The British Scholarship Trust.

References

1. Agarwal, S., Branson, K., Belongie, S.: Higher-order learning with graphs. In: ICML (2006)
2. Bretto, A., Cherifi, H., Aboutajdine, D.: Hypergraph imaging: an overview. Pattern Recognition 35(3), 651–658 (2002)
3. Bunke, H., Dickinson, P., Neuhaus, M., Stettler, M.: Matching of hypergraphs — algorithms, applications, and experiments. Studies in Computational Intelligence 91, 131–154 (2008)
4. Chung, F.: The laplacian of a hypergraph. AMS DIMACS Series in Discrete Mathematics and Theoretical Computer Science 10, 21–36 (1993)
5. Li, W., Sole, P.: Spectra of regular graphs and hypergraphs and orthogonal polynomials. European Journal of Combinatorics 17, 461–477 (1996)
6. Ren, P., Wilson, R.C., Hancock, E.R.: Spectral embedding of feature hypergraphs. In: da Vitoria Lobo, N., Kasparis, T., Roli, F., Kwok, J.T., Georgiopoulos, M., Anagnostopoulos, G.C., Loog, M. (eds.) S+SSPR 2008. LNCS, vol. 5342, pp. 308–317. Springer, Heidelberg (2008)
7. Rodriguez, J.A.: On the laplacian eigenvalues and metric parameters of hypergraphs. Linear and Multilinear Algebra 51, 285–297 (2003)
8. Storm, C.K.: The zeta function of a hypergraph. Electronic Journal of Combinatorics 13 (2006)
9. Wong, A.K.C., Lu, S.W., Rioux, M.: Recognition and shape synthesis of 3d objects based on attributed hypergraphs. IEEE Transactions on Pattern Analysis and Machine Intelligence 11(3), 279–290 (1989)
10. Zass, R., Shashua, A.: Probabilistic graph and hypergraph matching. In: CVPR (2008)
11. Zhou, D., Huang, J., Scholkopf, B.: Learning with hypergraphs: Clustering, classification, and embedding. In: NIPS (2007)

Feature Ranking Algorithms for Improving Classification of Vector Space Embedded Graphs

Kaspar Riesen and Horst Bunke

Institute of Computer Science and Applied Mathematics, University of Bern, Neubrückstrasse 10, CH-3012 Bern, Switzerland
{riesen,bunke}@iam.unibe.ch

Abstract. Graphs provide us with a powerful and flexible representation formalism for pattern recognition. Yet, the vast majority of pattern recognition algorithms rely on vectorial data descriptions and cannot directly be applied to graphs. In order to overcome this severe limitation, an embedding of the underlying graphs in a vector space $\mathbb{R}^n$ is employed. The basic idea is to regard the dissimilarities of a graph g to a number of prototype graphs as numerical features of g. In previous works, the prototypes are selected beforehand with selection strategies based on some heuristics. In the present paper we take a more fundamental approach and regard the problem of prototype selection as a feature selection problem, for which many methods are available. With several experimental results we show the feasibility of graph embedding based on prototypes obtained from feature selection algorithms.

1 Introduction

A crucial question in pattern recognition is how one describes the objects under consideration adequately. In statistical pattern recognition an object **x** is formally represented as a vector of n measurements, or features, i.e. $\mathbf{x} = (x_1, \ldots, x_n) \in \mathbb{R}^n$. Due to the mathematical wealth of operations available in a vector space, a huge amount of pattern recognition algorithms for objects given in terms of feature vectors have been developed in recent years. Yet, the use of feature vectors implicates two limitations. First, as vectors always represent a predefined set of features, all vectors in a particular application have to preserve the same length. Furthermore, there is no direct possibility to describe binary relationships among different parts of an object.

Both constraints can be overcome by graph based representations [1]. As a matter of fact, graphs are not only able to describe properties of an object but also binary relationships among different parts. Furthermore, graphs are not constrained to a fixed size, i.e. the number of nodes and edges is not limited a priori and can be adapted to the size or the complexity of each individual object under consideration.

One of the major drawbacks of graphs is that there is little mathematical structure in the graph domain. That is, most of the basic mathematical operations available for vectors do not exist for graphs (e.g. computing the sum or the product of two graphs). Nevertheless, in the last decades a number of graph matching methods have been proposed, which can be employed to measure the dissimilarity, or distance, of graphs (see [1] for a survey). In this paper the edit distance of graphs is used as a basic dissimilarity measure [2]. Yet, being able to measure dissimilarities between graphs is still not sufficient for most standard pattern recognition algorithms, and edit distance based classification is basically limited to nearest-neighbor approaches.

A promising direction to overcome the lack of algorithmic tools for graph based pattern recognition is graph embedding. A prominent class of graph embedding is based on spectral methods (e.g. [3,4]). The basic idea is to represent graphs by the eigendecomposition of their adjacency or their Laplacian matrix. The resulting representation exhibits interesting properties. However, spectral methods are not fully able to cope with larger amounts of structural noise, such as missing or spurious nodes or edges. Furthermore, most spectral approaches are only applicable to unlabeled graphs or graphs with constrained label alphabets.

The present paper considers a new class of graph embedding procedures inspired by the idea of mapping feature vectors into dissimilarity spaces [5]. This idea was recently generalized to the domain of graphs [6]. The key idea of this approach is to use the distances of an input graph g to n prototype graphs $\mathcal{P} = \{p_1, \ldots, p_n\}$ as a vectorial description of g. Due to the general applicability of graph edit distance (which is used to derive the pairwise dissimilarities), the proposed embedding approach is able to cope with structural errors as well as with various kinds of graphs with unconstrained label alphabets.

Apparently, the definition of the prototype set $\mathcal{P}$ is a critical issue since the graphs in $\mathcal{P}$ affect the resulting vectors in the embedding space. Thus, a good selection of prototypes is crucial to succeed with the algorithm to be applied. Commonly, the prototypes are selected from a training set $\mathcal{T}$ before the embedding is carried out. In previous works, this prototype selection uses some heuristics based on graph edit distances between the members of $\mathcal{T}$ [6]. In the present paper, however, a new approach is proposed where all available elements from the training set are used as prototypes, i.e. $\mathcal{P} = \mathcal{T}$, and well known feature subset selection algorithms [7,8] are applied to the vector space embedded graphs. In other words, rather than selecting the prototypes beforehand, the embedding is carried out first and then the problem of prototype selection is reduced to a feature subset selection problem. This process is much more principled than the previous approach [6] and allows us to completely avoid the difficult problem of heuristic prototype selection.

2 Dissimilarity Space Embeddings Applied on Graphs

Assume we have a labeled set of sample graphs, $\mathcal{T} = \{p_1, \ldots, p_N\}$, and a graph dissimilarity measure $d(g, g')$. We compute the dissimilarity of a given input graph g to each graph $p_i \in \mathcal{T}$. Note that g can be an element of $\mathcal{T}$ or any other graph. This leads to N dissimilarities, $d_1 = d(g, p_1), \ldots, d_N = d(g, p_N)$, which can be arranged in an N-dimensional vector $(d_1, \ldots, d_N)$. Note that in theory $\mathcal{T}$

can be any kind of graph set. The samples in $\mathcal{T}$ can even be synthesized graphs. However, for the sake of convenience, $\mathcal{T}$ is often defined as a set of existing training graphs.

Definition 1 (Graph Embedding). *Let $\mathcal{G}$ be a finite or infinite set of graphs and $\mathcal{T} = \{p_1, \ldots, p_N\} \subset \mathcal{G}$ be a set of prototype graphs. Then, the mapping $\varphi_N^{\mathcal{T}} : \mathcal{G} \to \mathbb{R}^n$ is defined as the function $\varphi_N^{\mathcal{T}}(g) \mapsto (d(g, p_1), \ldots, d(g, p_N))$, where $d(g, p_i)$ is any graph dissimilarity measure between graph g and the i-th prototype graph.*

Obviously, the range of function $\varphi_N^{\mathcal{T}}$ is a vector space where each dimension corresponds to a prototype p_i and the coordinate values of the embedded graph g are the distances from g to p_i.

The embedding procedure proposed in this paper makes use of graph edit distance [2] as a basic dissimilarity model. The key idea of graph edit distance is to define edit operations on the graphs to be matched. A standard set of edit operations is given by *insertions*, *deletions*, and *substitutions* of both nodes and edges. A sequence of edit operations $e_1, \ldots, e_k$ that transform a graph g into another graph g' is commonly referred to as an *edit path* between g and g'. Obviously, for every pair of graphs (g, g'), there exist a number of different edit paths transforming g into g'. Let $\Upsilon(g, g')$ denote the set of all such edit paths. To find the most suitable edit path out of $\Upsilon(g, g')$, one introduces a cost for each edit operation, measuring the strength of the corresponding operation. The idea of such cost functions is to define whether or not an edit operation represents a strong modification of the graph. The edit distance of graphs is then defined as the cost of the minimum cost edit path.

Since the exact computation of graph edit distance is exponential in the number of nodes for general graphs, the complexity of the graph embedding procedure introduced in Def. 1 is exponential as well. However, there exist a number of efficient approximation algorithms for graph edit distance computation (e.g. [9] with cubic time complexity). Consequently, the embedding of one particular graph is established by means of N distance computations with polynomial time.

3 Feature Subset Selection

3.1 General Approach

The graph embedding framework presented so far uses all available training graphs from $\mathcal{T}$. This approach has two severe shortcomings. First, the dimensionality of the resulting vector space is equal to the size N of the training set $\mathcal{T}$. Consequently, if the training set is large, the mapping results in (possibly too) high dimensional feature vectors. Secondly, the presence of similar prototypes as well as outlier graphs in the training set $\mathcal{T}$ is most likely. Therefore, redundant and noisy or irrelevant information will be captured in the feature vectors. The inclusion of irrelevant and redundant information may harm the performance of pattern recognition algorithms [10].

To overcome this problem, in previous works on the novel graph embedding framework, prototype selection methods are employed [6]. That is, a prototype set $\mathcal{P}$ with $n \leq N$ prototype graphs is defined as a subset of the training set of graphs $\mathcal{T}$, i.e. $\mathcal{P} \subseteq \mathcal{T}$. In [6] different prototype selection algorithms are discussed. These prototype selection strategies use some heuristics based on the underlying dissimilarities in the original graph domain. It turns out that none of them is globally best, i.e. the quality of a prototype selector depends on the underlying data set.

In this paper we propose a new approach where we use all available elements from the training set of prototypes, i.e. $\mathcal{P} = \mathcal{T}$, and subsequently apply feature subset selection methods to the resulting feature vectors. Feature subset selection aims at selecting a suitable subset of features such that the performance of a classification or clustering algorithm is improved [8,11]. By means of *forward selection* search strategies, the search starts with an empty set and iteratively adds useful features to this set. Conversely, *backward elimination* refers to the process of iteratively removing useless features starting with the full set of features. Also *floating search* methods are available, where alternately useful features are added and useless features are removed. For a review on searching strategies for feature subset selection we refer to [11].

In order to evaluate the quality of subsets of features, two general strategies, viz. *filters* and *wrappers*, exist. The former approach evaluates the individual features by giving them a score according to general characteristics of the training set [10]. In the latter approach the classification accuracy of the underlying classification algorithm is used as a criterion for feature subset evaluation [8]. In the present paper a combination of both approaches is carried out. The combination consists of two stages, the first involving a filter and the second a wrapper step.

In the filter step feature-ranking techniques are applied. Commonly, a fixed number of top ranked features, or alternatively, only features whose score exceeds a predefined threshold on the criterion, are selected for further analysis. In the present paper the feature ranking is used to define nested subsets of features $\mathcal{F}_1 \subset \mathcal{F}_2 \subset \ldots \subset \mathcal{F}$, where $|\mathcal{F}_1| = 1$, $\mathcal{F}$ denotes the full set of features $\{x_1, \ldots, x_N\}$, and $|\mathcal{F}_{i+1}| = |\mathcal{F}_i| + 1$. An optimal subset of features can eventually be found by varying a single parameter, viz. the number of features. For this second step, the classifier's accuracy serves us as a quality criterion, i.e. a wrapper strategy is applied. The classifier used in the wrapper step (and also in our final classification system) is the support vector machine (SVM). Of course, any other statistical classifier could be used for this purpose as well. However, we feel that the SVM is particularly suitable because of its theoretical advantages and its superior performance that has been empirically confirmed in many practical classification problems.

3.2 Feature Ranking Algorithms

For feature ranking, two different methods are employed, viz. mRMR [12] and SVM-RFE [13]. The former uses a forward selection strategy, while the latter implements a backward feature elimination strategy. Both procedures are described in detail below.

mRMR Ranking Method. This feature ranking algorithm is based on mutual information, which is a widely used measure to analyze the dependency of variables. The mutual information of an individual feature $x \in \mathcal{F}$ and the ground truth y is defined in terms of their probability density functions $p(x)$, $p(y)$, and $p(x,y)$:

$$I(x,y) = \int \int p(x,y) \log \frac{p(x,y)}{p(x)p(y)} dx dy$$

For discrete feature variables, the integral operation reduces to summation. In terms of mutual information, the purpose of feature selection is to find a feature subset $\mathcal{F}_i \subset \mathcal{F}$ whose features jointly have the largest relevance to the target class y [12]. In the present paper, the mean value of all mutual information values between individual feature $x \in \mathcal{F}_i$ and class y is used as a quality criterion for a certain feature subset $\mathcal{F}_i$. Formally,

$$A = \frac{1}{|\mathcal{F}_i|} \sum_{x \in \mathcal{F}_i} I(x,y) \ .$$

A feature subset $\mathcal{F}_i$ that maximizes criterion A is referred to as *maximal-relevance* feature subset, or MR for short. Obviously, pairwise features selected by this criterion can have a large degree of redundancy, i.e. the dependency among the features $x \in \mathcal{F}_i$ can be large. In order to select mutually independent features, we introduce the following criterion B for a certain feature subset $\mathcal{F}_i$.

$$B = \frac{1}{|\mathcal{F}_i|^2} \sum_{x,x' \in \mathcal{F}_i} I(x,x')$$

A feature subset $\mathcal{F}_i$ that minimizes criterion B is referred to as *minimal-redundancy* feature subset, or mR for short.

In order to define our final feature ranking, an incremental greedy search method is employed where both criteria A and B are optimized simultaneously. Consider the case where a feature set $\mathcal{F}_{n-1}$ with $n-1$ features is already defined. The task is now to select the n-th feature from the remaining features $\mathcal{F} \setminus \mathcal{F}_{n-1}$. That is, the respective incremental algorithm selects the feature that optimizes the following condition [12]:

$$\max_{x' \in \mathcal{F} \setminus \mathcal{F}_{n-1}} \left[\frac{I(x',y)}{\frac{1}{n-1} \sum_{x \in \mathcal{F}_{n-1}} I(x',x)} \right] .$$

We refer to this feature ranking algorithm as *mRMR* [12].

SVM-RFE Ranking Method. In [13] a feature ranking method based on SVM classifiers is introduced. The basic idea of this approach is to use the weights $\mathbf{w}$ of a classifier's hyperplane $g(\mathbf{x}) = \langle \mathbf{w}, \mathbf{x} \rangle + b$, where $\mathbf{w} \in \mathbb{R}^n$ and $b \in \mathbb{R}$, as feature ranking criterion. The rationale is that the inputs that are weighted by the largest value have the most influence on the classification decision. Consequently, if the classifier performs well, those inputs with the largest weights correspond to the most informative features [13]. In the case of SVMs, the term to be optimized is $J = \min_{\mathbf{w} \in \mathbb{R}^n} \frac{1}{2}||\mathbf{w}||^2$. We therefore use $(w_i)^2$ as feature ranking criterion. Particularly, the authors in [13] propose to use a recursive feature elimination procedure (RFE) to define the feature ranking. RFE is an instance of backward feature elimination that iteratively proceeds as follows.

1. Train the SVM, i.e. optimize the weights w_i with respect to J.
2. Compute the ranking criterion $(w_i)^2$ for all features.
3. Remove the feature with smallest ranking criterion.

Obviously, top ranked features correspond to features eliminated last. We refer to this method as *SVM-RFE*.

4 Experimental Evaluation

The experimental evaluation of the proposed embedding procedure is based on graph based pattern classification. We consider four different graph data sets from the IAM graph database repository for graph based pattern recognition and machine learning[1]. Each of our graph sets is divided into three disjoint subsets, viz. a training, a validation, and a test set. In our experiments we use graphs representing distorted letter line drawings out of 15 classes (Letter), molecular compounds with activity against HIV or not (AIDS), fingerprint images out of four classes (Fingerprint), and webpages that originate from 20 different categories (Webpage). For a thorough description of these data sets we refer to [14].

The classifier in the embedding space is an SVM with RBF-kernel. That is, besides the weighting parameter C, which controls whether the maximization of the margin or the minimization of the error is more important, the meta parameter $\gamma > 0$ in the kernel function $k(\mathbf{x}, \mathbf{x}') = exp\left(-\gamma ||\mathbf{x} - \mathbf{x}'||^2\right)$ has to be optimized. Both parameters are tuned on the independent validation set.

One of the few classifiers directly applicable to arbitrary graphs is the k-nearest-neighbor classifier (k-NN) in conjunction with graph edit distance. This classifier in the graph domain will serve us as our first reference system. The second reference method is, similarly to our novel procedure, an SVM with RBF-kernel applied to dissimilarity space embedded graphs. However, in contrast to our novel approach, the prototypes are defined beforehand by a heuristic selector [6]. The prototype selector applied is based on k-medoids clustering. After application of the clustering procedure, the cluster centers are selected as prototypes. The final number of prototypes (i.e. the dimensionality) is also evaluated by means of the target classifier in a wrapper approach. We refer to this method as *k-Med*.

---

[1] Note that all data sets used in the present work are publicly available under http://www.iam.unibe.ch/fki/databases/iam-graph-database

Table 1. Experimental Results

Data Set	Ref. Systems		Proposed Method	
	k-NN	k-Med	mRMR	SVM-REF
Letter	90.00	92.40	92.27	92.00
AIDS	97.27	97.27	97.40	98.00
Fingerprint	77.60	77.70	82.40	82.20
Webpage	80.26	81.92	80.26	81.79

In Table 1 the classification results of all reference systems and the proposed approach using the feature subset selection algorithms are given. Comparing the results of our novel approach with the results achieved by the first reference system, it clearly turns out that the novel procedure is much more powerful than the traditional k-NN in the graph domain. In seven out of eight cases the embedding in conjunction with feature subset selection algorithms outperforms the first reference system's classification results and in one case equal accuracies are achieved. Note that at least one of the improvements per data set achieved by mRMR or SVM-RFE is statistically significant (Z-test, $\alpha = 0.05$).

Regarding the results achieved by the second reference system (k-Med), we observe that our approach using feature subset selection algorithms rather than heuristic prototype selection achieves better results on half of the data sets (AIDS and Fingerprint). Note that two out of four improvements are statistically significant, but none of the deteriorations. Moreover, we note that k-Med outperforms the first reference system only in two out of four cases with statistical significance (Letter and Webpage). Our novel procedure, however, achieves improvements with statistical significance on all data sets. Hence one can conclude that the novel approach using feature subset selection strengthens the former framework of graph embedding based on prototype selection.

5 Conclusions

Graphs are a versatile alternative to feature vectors, and are known to be a powerful and flexible representation formalism. However, graph based object representation suffers from a rather limited repository of algorithmic tools in the graph domain. A solution to this problem is graph embedding in real vector spaces. In the present paper a novel approach to graph embedding, using prototypes and dissimilarities, is proposed. The basic idea of the embedding method is to map a graph g to a vector space by arranging the edit distances of g to a set of prototypes as a vector. Previous work on graph embedding depends on the selection of suitable prototypes. With the method proposed in this paper we avoid the critical task of prototype selection by taking all available graphs from the training set as prototypes and eventually apply powerful feature subset selection strategies, viz. mRMR and SVM-RFE. From the experimental evaluation one can draw the following conclusions. Classifiers using the embedded graphs outperform classification systems using the original graph edit distances. Moreover, the more fundamental approach of feature subset selection rather than heuristic prototype selection is clearly beneficial. That is, by means of our novel approach the powerful embedding framework is further improved.

Acknowledgements

This work has been supported by the Swiss National Science Foundation (Project 200021-113198/1).

References

1. Conte, D., Foggia, P., Sansone, C., Vento, M.: Thirty years of graph matching in pattern recognition. Int. Journal of Pattern Recognition and Artificial Intelligence 18(3), 265–298 (2004)
2. Bunke, H., Allermann, G.: Inexact graph matching for structural pattern recognition. Pattern Recognition Letters 1, 245–253 (1983)
3. Luo, B., Wilson, R., Hancock, E.: Spectral embedding of graphs. Pattern Recognition 36(10), 2213–2223 (2003)
4. Caelli, T., Kosinov, S.: Inexact graph matching using eigen-subspace projection clustering. Int. Journal of Pattern Recognition and Artificial Intelligence 18(3), 329–355 (2004)
5. Pekalska, E., Duin, R.: The Dissimilarity Representation for Pattern Recognition: Foundations and Applications. World Scientific, Singapore (2005)
6. Riesen, K., Bunke, H.: Graph classification based on vector space embedding. Int. Journal of Pattern Recognition and Artificial Intelligence (2008) (accepted for publication)
7. Langley, P.: Selection of relevant features in machine learning. In: AAAI Fall Symposium on Relevance, pp. 140–144 (1994)
8. Kohavi, R., John, G.: Wrappers for feature subset selection. Artificial Intelligence 97(1-2), 273–324 (1997)
9. Riesen, K., Bunke, H.: Approximate graph edit distance computation by means of bipartite graph matching. Image and Vision Computing 27(4), 950–959 (2009)
10. Wang, Y., Tetko, I., Hall, M., Frank, E., Facius, A., Mayer, K., Mewes, H.W.: Gene selection from microarray data for cancer classification – a machine learning approach. Computational Biology and Chemistry 29, 37–46 (2005)
11. Pudil, P., Novovicova, J., Kittler, J.: Floating search methods in feature-selection. Pattern Recognition Letters 15(11), 1119–1125 (1994)
12. Peng, H., Long, F., Ding, C.: Feature selection based on mutual information: Criteria of max-dependency, max-relevance, and min-redundancy. IEEE Trans. on Pattern Analysis and Machine Intelligence 27(8), 1226–1238 (2005)
13. Guyon, I., Weston, J., Barnhill, S., Vapnik, V.: Gene selection for cancer classification using support vector machines. Machine Learning 46(1-3), 389–422 (2002)
14. Riesen, K., Bunke, H.: IAM graph database repository for graph based pattern recognition and machine learning. In: da Vitoria Lobo, N., Kasparis, T., Roli, F., Kwok, J.T., Georgiopoulos, M., Anagnostopoulos, G.C., Loog, M. (eds.) S+SSPR 2008. LNCS, vol. 5342, pp. 287–297. Springer, Heidelberg (2008)

Graph-Based Object Class Discovery

Shengping Xia[1] and Edwin R. Hancock[2]

[1] ATR Lab, School of Electronic Science and Engineering,
National University of Defense Technology, Changsha, Hunan, P.R. China 410073
[2] Department of Computer Science, University of York, York YO10 5DD, UK

Abstract. We are interested in the problem of discovering the set of object classes present in a database of images using a weakly supervised graph-based framework. Rather than making use of the "Bag-of-Features (BoF)" approach widely used in current work on object recognition, we represent each image by a graph using a group of selected local invariant features. Using local feature matching and iterative Procrustes alignment, we perform graph matching and compute a similarity measure. Borrowing the idea of query expansion, we develop a similarity propagation based graph clustering (SPGC) method. Using this method class specific clusters of the graphs can be obtained. Such a cluster can be generally represented by using a higher level graph model whose vertices are the clustered graphs, and the edge weights are determined by the pairwise similarity measure. Experiments are performed on a dataset, in which the number of images increases from 1 to 50K and the number of objects increases from 1 to over 500. Some objects have been discovered with total recall and a precision 1 in a single cluster.

1 Introduction

In the statistical text analysis community, latent topic models such as probabilistic Latent Semantic Analysis (pLSA) [1] and Latent Dirichlet Allocation (LDA) [2] have had significant impact as methods for "semantic" clustering. Given a collection of documents, with each document represented by a "Bag-of-Words (BoW)" vector, the models are able to learn common topics such as "biology" or "astronomy".

Given the success of these models, several papers in computer vision [3][4][5][6] have applied them to the visual domain, replacing text words with visual features [7][8]. This approach is usually referred to as the "Bag-of-Features (BoF)" method. Rather than discovering topics, the BoF method aims to discover visual categories, such as cars or bikes in the image database. However, in the visual domain there are strong geometric relations within images, which do not exist in the text domain. There have been several attempts to learn visual categories in an unsupervised manner by jointly modeling the appearance of local patches and their spatial arrangement [9][10][11][12]. Examples include the rotation, translation and scale invariant pLSA (RTSI-pLSA) model proposed by Li et al [9] and the geometric LDA (gLDA) model proposed by Philbin et al [10]. These methods can be regarded as extensions of the basic BoF based method. However, there are three basic problems which may compromise their modeling or recognition performance: a) local invariant features in the vision domain do not operate at the same semantic level as words in text domain; b) local features are not object specific; and c) visual vocabulary needs to be incrementally adapted as more data becomes available.

Although the BoF method has demonstrated impressive levels of performance and provides arguably the most successful paradigm for object discovery and recognition, because of the shortcomings listed above, in this paper we offer an alternative to the BoF model. We regard a group of local features together with their spatial arrangement as a visual entity. If such a visual entity is of a certain semantic meaning, e.g. corresponding to a car, then it is placed at the word-level in text domain. Since each visual entity is represented by structured data, a more versatile and expressive representational tool is provided by attributed graphs [13]. Hence we simply term such a visual entity as a graph, which takes the place of a BoF based vector. We thus demonstrate how to implement object discovery and recognition without using the BoF model.

2 Ingredients of a Scalable Search Engine

Image representation. For each image in the dataset local invariant features are detected. A variety of feature detectors have been developed [14][15][16][17], and these include SIFT [14] and SURF (Speeded Up Robust Features) [15]. Here we use method previously described in [18] to extract a manageable number of salient SIFT features. Using this method, the SIFT features of an image $\mathcal{I}$ are ranked in order according to their decreasing matching frequency. We select the $\mathcal{T}$ top ranked SIFT features, denoted as $\mathcal{V} = \{V^t, t = 1, 2, ..., \mathcal{T}\}$, where $V^t = ((\vec{X}^t)^T, (\vec{D}^t)^T, (\vec{U}^t)^T)^T$. Here, $\vec{X}^t$ is the location, $\vec{D}^t$ is the direction vector and $\vec{U}^t$ is the set of descriptors of a SIFT feature. In our experiments, $\mathcal{T}$ is set to 40. If there are less than this number of feature points present then all available SIFT features in an image are selected.

We regard the above selected local features together with their spatial arrangement as a semantic visual entity, which is placed at the word-level in text domain. This kind of structured data can be represented by using attributed graphs G [13] (hereafter simply graphs). We can obtain a set of graphs $\mathbb{G} = \{G_l, l = 1, 2, ..., N\}$ from a set of images.

Pairwise graph matching for spatial verification and similarity measure. As shown in [8][11][12], the recognition or retrieval results can be significantly improved using the geometry of spatial feature arrangement to verify consistency. In our approach, on the other hand, each image is represented by a graph. As a result the spatial verification problem becomes one of pairwise graph matching (PGM). We perform PGM with the aim of finding a maximum common subgraph (MCS) between two graphs G_l and G_q, and the result is denoted as $MCS(G_l, G_q)$. There are a plethora of available methods for finding matching features consistent with a given set of geometric constraints, and the problem has been proven to be NP-hard. RANSAC provides one popular set of methods, however their implementation is slow [19]. In [20], pairwise graph matching is achieved by combining SIFT feature matching and iterative Procrustes alignment [21]. The method can not only be used to align the feature points, but can also be used to discard those features that do not satisfy the spatial arrangement constraints. Given $MCS(G_l, G_q)$ obtained by PGM, they define a similarity measure between the graphs G_l and G_q as follows:

$$R(G_l, G_q) = \|MCS(G_l, G_q)\| \times (\exp(-e(X_l, X_q)))^{\kappa}. \quad (1)$$

Here a) $\|MCS(G_l, G_q)\|$ is the cardinality of the MCS of G_l and G_q, b) κ is the number of roughly mismatched feature pairs by SIFT matching, which is used to amplify the influence of the geometric dissimilarity between X_l and X_q, and c) X_l and X_q are respectively the position coordinates in graphs G_l and G_q corresponding to the vertexes of $MCS(G_l, G_q)$.

This similarity measure is significantly different from the BoF similarity measure which is based on the L1 or L2 distance between vectors [8][11][12][22], and captures both the similarity of local appearance and global spatial consistency.

Clustering of feature descriptors. For BoF based methods, the vector quantization of feature descriptors has been used by Sivic and Zisserman [8]. Here small vocabularies were generated using the k-means clustering method. It was subsequently shown in [12][22][23] that for large scale cases a more discriminative vocabulary is necessary. In [22], hierarchical k-means (HKM) and in [12] a KD-forest approximation were explored as possible refinements of the method. The aim of using these clustering methods is to obtain the visual vocabulary for construction of a BoF vector. Each image is then represented by a high dimensional tf-idf (Term Frequency-Inverse Document Frequency) [12][22][23] weighted BoF vector.

Although we represent images by graphs, we still require means of clustering the local feature descriptors. We use a SOM neural net based tree clustering method (termed a RSOM tree) proposed in [24] for learning a large corpus of SIFT descriptors. The RSOM tree can be incrementally trained since it utilizes a self-organizing divide-and-conquer architecture. It is also important to stress that though the leaf nodes in an RSOM tree are a quantization of the descriptors, we do not regard such a quantization as a visual vocabulary. We simply use the RSOM tree to efficiently retrieve candidate matching graphs, in the manner detailed in the following paragraphs.

Search engine. Given a graph set $\mathbb{G} = \{G_q, q = 1, 2, ..., N\}$, for each graph $G_l \in \mathbb{G}$, and the remaining graphs in the set ($\forall G_q \in \mathbb{G}$), we obtain the pairwise graph similarity measures $R(G_l, G_q)$ defined in Equation (1). Using the similarity measures we rank the graphs G_q in decreasing order. The K top-ranked graphs are defined as the generalized K-nearest neighbor graphs (KNNG) of graph G_i, denoted as $\mathbb{K}\{G_l\}$.

With increasing size of the graph dataset, it becomes time consuming to obtain $\mathbb{K}\{G_l\}$ if a sequential search strategy is adopted. Fortunately, for a large graph set, most of the similarity measures are low. For a single graph G_l, if we can efficiently find a subset $\mathbb{G}'$ with significant similarity values as a pre-filtering stage, then we only need to perform pairwise graph matching on this subset. To this end, we use the above mentioned RSOM clustering tree.

To obtain $\mathbb{K}\{G_l\}$ for each sample graph using a trained RSOM tree we proceed as follows. Given a graph G_l, we find the winner of the leaf nodes for each descriptor of this graph and define the union of all graphs for the winners as follows:

$$UG\{G_l\} = \{ G_q \mid U_q^j \in G_q, U_q^j \in WL\{U_l^t\}, U_l^t \in G_l \}. \quad (2)$$

where $WL\{U_l^t\}$ is the winner of the leaf nodes for descriptor U_l^t. The frequency of graph G_q, denoted as H_q, represents the number of roughly matched descriptors between two graphs. Since we aim to obtain $\mathbb{K}\{G_l\}$, we need not process all graphs in the set $UG\{G_l\}$. We rank the graphs in $UG\{G_l\}$ according to decreasing frequency H_q. From the ranked list, we select the first K graphs, denoted by $\mathbb{K}'\{G_l\}$ as follows:

$$\mathbb{K}'\{G_l\} = \{ G_q \mid G_q \in UG\{G_l\}, H_q > H_{q+1}, q = 1, 2, ..., K.\}. \tag{3}$$

For each graph G_q in $\mathbb{K}'\{G_l\}$, we obtain the similarity measure according to Equation (1) and then $\mathbb{K}\{G_l\}$ can be obtained. Using this method, we can efficiently obtain $\mathbb{K}\{G_l\}$. As a result, it is not necessary to use a search engine constructed from BoF vectors, which is the usual practice in the text domain. Hence, the method can be easily adapted to incremental learning settings.

3 Object Discovery and Model Learning

This section commences by presenting a new graph clustering method developed by borrowing the widely used idea of query expansion from text query. We then explain how the method can be used to discover object classes and learn object class models.

3.1 Similarity Propagation Based Graph Clustering

In the text retrieval literature, a standard method for improving performance is query expansion, where a number of the highly ranked documents from the original query are reissued as a new query. This allows the retrieval system to use relevant terms not present in the original query. In [11][12], query expansion was imported into the visual processing domain using spatial constraints to improve the accuracy of each returned image. Our query, on the other hand, expansion method is based on the RSOM tree and the set $\mathbb{K}\{G_l\}$ for each graph, obtained in the training stage. The method is based on a pairwise similarity propagation algorithm for graph clustering (SPGC). Stated simply, the method is as follows. A group of graphs are referred to as siblings of a given graph G_l provided they satisfy the following condition:

$$S\{G_l\} = \{G_q \in \mathbb{K}\{G_l\} \mid R(G_l, G_q) \geq R_\tau\} \triangleq S_{R_\tau}\{G_l\}. \tag{4}$$

We use the definition to recursively obtain the family tree for the graph G_l, and this is formally defined as follows.

Family Tree of a Graph (FTOG). For any given similarity threshold R_τ, an FTOG of G_l with k generations and denoted as $M\{G_l, k\}$, is defined as follows:

$$M\{G_l, k\} = M\{G_l, k-1\} \bigcup_{G_q \in L\{G_l, k-1\}} S_{R_\tau}\{G_q\}. \tag{5}$$

where, if $k = 1$, $M\{G_l, 1\} = M\{G_l, 0\} \bigcup S\{G_l\}$ and $M\{G_l, 0\} = \{G_l\}$; and the process stops when $M\{G_l, k\} = M\{G_l, k+1\}$. An FTOG, whose graphs satisfy the restriction defined in Equation (4), can be regarded as a cluster of graphs. However, it must be stressed that this is not a clustering method based on a central prototype. Instead, graphs are clustered using the similarity propagation strategy.

3.2 Weakly Supervised Object Discovery and Model Learning

The clustering process is controlled by the threshold R_τ. By varying the parameter, we can represent images using canonical graphs constructed from a number of selected local invariant features so that most of the graphs belonging to an identical object form a single FTOG.

In this case, the FTOG is class specific. From the perspective of fault tolerance, the precision does not necessarily need to be 100%.

According to our experiments, if $R_\tau \geq 10.0$ is large then the corresponding FTOG will have high precision (close to 1). With this setting, we can obtain a number of disjoint FTOG's from a large graph dataset (in each FTOG, there are at least 2 graphs.). We use these FTOG's as cluster seeds. For each FTOG, in a weak supervision stage, we manually assign a groundtruth object label (or name) to a cluster. We can also manually adjust the threshold to obtain a better cluster containing more graphs belonging to the same object class. We denote the similarity threshold for the corresponding FTOG's as $R_{\tau_1}(M_l\{G_q\})$. In this way each cluster corresponds to an object class discovered from the data, and the corresponding FTOG is the class model. However, a single object may be split between multiple clusters.

If we regard each graph in an FTOG as a vertex in a higher level graph, for a pair of vertexes an edge, weighted by the similarity measure, is defined iff their similarity measure is subject to the given similarity constraint, an FTOG can be further regarded as a class specific graph (CSG) model.

Given that c FTOGs have been detected for a single object-class in a dataset, i.e. $\exists G_{l_i}, i = 1, 2, ..., c$, subject to $M\{G_{l_i}, g\} \cap M\{G_{l_j}, g\} = \emptyset, i \neq j, i, j \in \{1, 2, ..., c\}$, then we uniquely label the corresponding FTOG's as $L_1, L_2, ..., L_c$. We denote the set of clusters for a single discovered object model as follows:

$$C_{R_\tau} = \{ M_{R_\tau}\{G_l, \infty\}\} \triangleq \{ M_l \mid l \in \{L_1, L_2, ..., L_c\} \}. \tag{6}$$

A set of class specific FTOGs of an object can also be regarded as class specific graph models and still termed CSG model. Ideally a single object has one corresponding FOTG, that is $c = 1$. However, in an incremental learning setting, each object will tend to have more than one FTOG. With an increasing size of the dataset it is likely that two disjoint FTOG's will become merged when intermediate graphs are encountered. In an incremental learning setting, a new graph G_l is added to its discovered model according to the following rules:

1. If $\exists$ more than one graph G_q, s.t. $R(G_l, G_q) \geq R_{\tau_0}$, G_l is processed as a redundant duplicate graph of G_q, in our settings, $R_{\tau_0} = 18$.
2. If $\exists G_{q_0}$, s.t. $R(G_l, G_{q_0}) \geq R_{\tau_1}(M_l\{G_{q_0}\})$, G_l is incremented as an irreducible graph of $M_l\{G_{q_0}\}$; If there is another graph $G_{q_1} \in M_l\{G_{q_1}\}$, $M_l\{G_{q_0}\}$ and $M_l\{G_{q_1}\}$ come from different classes, G_l then is marked as an ambiguous graph. If $M_l\{G_{q_0}\}$ and $M_l\{G_{q_1}\}$ belong to the same class, then we merge these two FTOG's.
3. If $\max\{R(G_l, G_q)\} < R_{\tau_1}(M_l\{G_q\})$, create a new FTOG $M_l\{G_l\}$.

Once a CSG model is trained, for a test graph G_l, we can obtain $K_\tau\{G_l\}$ and use a k-nearest neighbor weighted voting recognition strategy, using the similarity measure $R(G_l, G_q)$ as a weight.

4 Experimental Results

We have collected 53536 training images, some examples of which are shown in Figure 1, as training data. The data spans more than 500 objects including human faces and natural scenes. For each of these images, we extract ranked SIFT features using the method presented in [18]. Of these at most 40 highly ranked SIFT features are selected to construct a graph. We have collected over 2,140,000 SIFT features and 53536 graphs for the training set. We have trained an RSOM clustering tree with 25334 leaf nodes for the SIFT descriptors using the incremental RSOM training method. In this stage, we have also obtained $K\{G_l\}$ for each of the graphs. We use the 68 objects (detailed in Fig. 1) to test our object class discovery method.

The object discovery results for the 68 object problem are shown in Fig. 2, it is clear that for most of the objects sampled under controlled imaging conditions, ideal performance has been achieved. For 35 objects in COIL 100, 35 models are individually discovered with total recall and precision of unit in one FTOG. For 13 objects, 6 models are discovered. Each group of objects in Fig. 3 (A)(B)(C)(D) are actually identical in shape but color. Since it only uses gray scale information in SIFT, our method fails in this case. We hence regard these objects in the four groups as being correctly discovered according to shape.

Unfortunately, in most practical situations, the images of an object are likely to be obtained with large variations of imaging conditions and are more likely to be clustered into several FTOGs. As a result, each object gives rise to multiple clusters. For objects 51 to 58 there are no more than 30 images with large variations in viewing conditions,

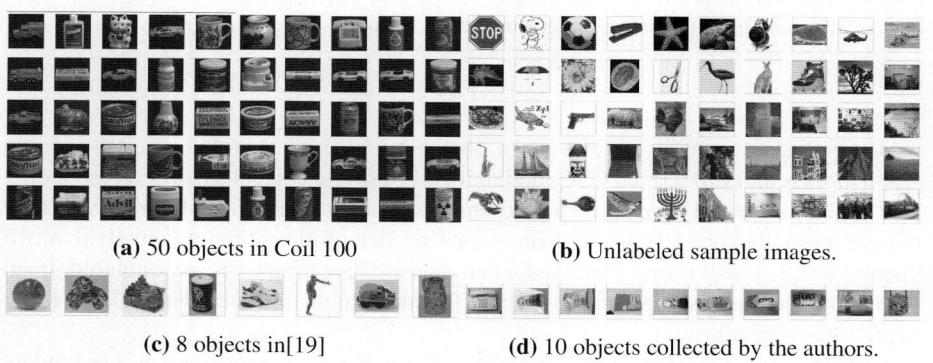

(a) 50 objects in Coil 100   (b) Unlabeled sample images.

(c) 8 objects in[19]   (d) 10 objects collected by the authors.

Fig. 1. Image data sets.a: 3600 images of 50 objects in COIL 100, labeled as A1~A50; b: 29875 unlabeled images from many other standard datasets, e.g. Caltech101 [3] and Google images, covering over 450 objects and used as negative samples; c: 161 images of 8 objects used in [19], labeled as C1 to C8; d: 20000 images of 10 objects collected by us, labeled as D1 to D10. For each of the objects in D1 to D9, we collect 1500 images which traverse a large variation of imaging conditions, and similarly 6500 images for D10. For simple description, the 4 dada sets are denoted as A to D. The objects in Figure 1a,Figure 1c and Figure 1d are numbered from left to right and then from top to bottom as shown in the corresponding figures, e.g. A1 to A50 in Figure 1a. As a whole, the 68 objects are also identified as Object 1 to Object 68.

ID Par	1~50 Except 3, 39	3	39	51	52	53	54	55	56	57	58	59	60	61	62	63	64	65	66	67	68
N_i	72	72	72	29	20	16	16	16	16	28	20	1500	1500	1500	1500	1500	1500	1500	1500	1500	6500
N_d	72	64	66	22	16	15	15	16	16	27	20	1491	1483	1500	1500	1500	1475	1487	1500	1467	6488
p	1.0	1.0	1.0	1.0	1.0	1.0	1.0	1.0	1.0	1.0	1.0	1.0	1.0	1.0	1.0	1.0	1.0	1.0	1.0	1.0	1.0
r	1.0	.875	.917	.759	.80	.938	.938	1.0	1.0	.964	1.0	.994	.989	1.0	1.0	1.0	.983	.991	1.0	.978	.998
N_c	1	2	2	3	1	3	3	3	3	4	2	6	8	4	5	3	9	3	3	4	1

Fig. 2. Results of object discovery. In the above table, ID is the Object ID; N_i is the number of the initial images of an object; N_d is the number of images discovered by using our method; N_d^+ is the number of correctly discovered images; p is the precision defined as N_d^+/N_d; r is recall defined as N_d^+/N_i. N_c is the number of discovered clusters of each object.

Fig. 3. 6 groups of objects are overlapping-clustered into 6 clusters

and the images are not representative enough to perform ideal recognition. However, for objects 59 to 68, the images clustered together are sufficient to form an effective object model which can be used for recognition. For object 68, since there are thousands of images, the different views form a single cluster.

It is worth pointing out that all of these experiments are implemented in an incremental learning setting. We commenced by using 3600 images of 50 objects in the COIL 100 database as the first batch of images. From which, the models of these 50 objects are discovered. With the increase of additional training images, the discovered models of the 50 objects have not been changed. We than trained a second batch of samples containing instances of Object 51 to Object 68, and their corresponding models are discovered. The number of images is than increased to over 50K by including over 450 other objects. From the third stage database, we re-discovered the same models of the above 68 objects. Compare to the up-to-date BoF based methods, the size of the RSOM clustering tree is dynamic, the scale of the image corpus is also dynamic. The discovered model keeps stable and can be refined in the incremental settings.

5 Conclusion

In this paper, we have presented a scalable framework for learning object classes (object discovery). The method is graph-based and makes use of the RSOM tree clustering of local feature descriptors and graph clustering using similarity propagation (SPGC). As such it is therefore distinct from current state-of-the-art "Bag-of-Feature" based methods. Using the object models learned using our technique we can potentially simultaneously effect object detection, recognition and annotation. We will explore these problems in future work.

Acknowledgements

We acknowledge financial support from the FET programme within the EU FP7, under the SIMBAD project (contract 213250), and from the ATR Lab Foundation, under the project 91408001020603.

References

1. Hofmann, T.: Unsupervised learning by probabilistic latent semantic analysis. Machine Learning 43, 17–196 (2001)
2. Blei, D., Ng, A., Jordan, M.: Latent dirichlet allocation. In: NIPS (2002)
3. Fei-Fei, L., Perona, P.: A bayesian hierarchical model for learning natural scene categories. In: CVPR (2005)
4. Quelhas, P., Monay, F., Odobez, J., Gatica, D., Tuytelaars, T., Van Gool, L.: Modeling scenes with local descriptors and latent aspects. In: ICCV, pp. 883–890 (2005)
5. Sivic, J., Russell, B.C., Efros, A.A., Zisserman, A., Freeman, W.: Discovering object categories in image collections. In: ICCV (2005)
6. Russell, B.C., Efros, A.A., Sivic, J., Freeman, W.T., Zisserman, A.: Using multiple segmentations to discover objects and their extent in image collections. In: CVPR (2006)
7. Csurka, G., Bray, C., Dance, C., Fan, L.: Visual categorization with bags of keypoints. In: Workshop on Statistical Learning in Computer Vision, ECCV, pp. 1–22 (2004)
8. Sivic, J., Zisserman, A.: Video google: A text retrieval approach to object matching in videos. In: ICCV, pp. 1470–1477 (2003)
9. Li, Y., Wang, W.-Q., Gao, W.: A robust approach for object recognition. In: Zhuang, Y.-t., Yang, S.-Q., Rui, Y., He, Q. (eds.) PCM 2006. LNCS, vol. 4261, pp. 262–269. Springer, Heidelberg (2006)
10. Philbin, J., Sivic, J., Zisserman, A.: Geometric lda: A generative model for particular object discovery. In: BMVC (2008)
11. Chum, O., Philbin, J., Sivic, J., Isard, M., Zisserman, A.: Total recall: Automatic query expansion with a generative feature model for object retrieval. In: ICCV (2007)
12. Philbin, J., Chum, O., Isard, M., Sivic, J., Zissermans, A.: Object retrieval with large vocabularies and fast spatial matching. In: CVPR (2007)
13. Chung, F.: Spectral graph theory. American Mathematical Society, Providence (1997)
14. Lowe, D.: Distinctive image features from scale-invariant key points. IJCV 60(2), 91–110 (2004)
15. Bay, H., Tuytelaars, T., Van Gool, L.: SURF: Speeded up robust features. In: Leonardis, A., Bischof, H., Pinz, A. (eds.) ECCV 2006, Part I. LNCS, vol. 3951, pp. 404–417. Springer, Heidelberg (2006)
16. Kadir, T., Brady, M., Zisserman, A.: An invariant method for selecting salient regions in images. In: Proc. Eighth ECCV, vol. 1(1), pp. 345–457 (2004)
17. Mikolajczyk, K., Schmid, C.: A performance evaluation of local descriptors. PAMI 27(10), 1615–1630 (2005)
18. Xia, S.P., Ren, P., Hancock, E.R.: Ranking the local invariant features for the robust visual saliencies. In: ICPR 2008 (2008)
19. Rothganger, F., Lazebnik, S., Schmid, C., Ponce, J.: 3d object modeling and recognition using local affine-invariant image descriptors and multi-view spatial constraints. IJCV 66(3), 231–259 (2006)

20. Xia, S., Hancock, E.R.: 3D object recognition using hyper-graphs and ranked local invariant features. In: da Vitoria Lobo, N., Kasparis, T., Roli, F., Kwok, J.T., Georgiopoulos, M., Anagnostopoulos, G.C., Loog, M. (eds.) S+SSPR 2008. LNCS, vol. 5342, pp. 117–126. Springer, Heidelberg (2008)
21. Schonemann, P.: A generalized solution of the orthogonal procrustes problem. Psychometrika 31(3), 1–10 (1966)
22. Jegou, H., Harzallah, H., Schmid, C.: A contextual dissimilarity measure for accurate and efficient image search. In: CVPR (2007)
23. Nister, D., Stewenius, H.: Scalable recognition with a vocabulary tree. In: CVPR (2006)
24. Xia, S.P., Liu, J.J., Yuan, Z.T., Yu, H., Zhang, L.F., Yu, W.X.: Cluster-computer based incremental and distributed rsom data-clustering. ACTA Electronica sinica 35(3), 385–391 (2007)

The Clifford-Hodge Flow: An Extension of the Beltrami Flow[*]

Thomas Batard and Michel Berthier

Laboratory Mathematics, Image and Applications
University of La Rochelle
17042 La Rochelle Cedex, France
{tbatar01,mberthie}@univ-lr.fr

Abstract. In this paper, we make use of the theory of Clifford algebras for anisotropic smoothing of vector-valued data. It provides a common framework to smooth functions, tangent vector fields and mappings taking values in $\mathfrak{so}(m)$, the Lie algebra of SO(m), defined on surfaces and more generally on Riemannian manifolds. Smoothing process arises from a convolution with a kernel associated to a second order differential operator: the Hodge Laplacian. It generalizes the Beltrami flow in the sense that the Laplace-Beltrami operator is the restriction to functions of minus the Hodge operator. We obtain a common framework for anisotropic smoothing of images, vector fields and oriented orthonormal frame fields defined on the charts.

Keywords: Anisotropic diffusion, Vector-valued data, Clifford algebras, Differential Geometry, Heat equations.

1 Introduction

Most multivalued image smoothing process are based on PDE's of the form

$$\frac{\partial I^i}{\partial t} = \sum_{j,k=1}^{2} f_{jk} \frac{\partial^2 I^i}{\partial j\, \partial k} + \text{first-order part}$$

where $I\colon (x,y) \longmapsto (I^1(0,x), \cdots, I^n(0,x))$ is a n-channels image, and f_{jk} are real functions. We refer to [4] for an overview on related works. From a theoretical point of view, the set of right terms, for $i = 1 \cdots n$, may be viewed as a second-order differential operator acting on sections of a vector bundle over a Riemannian manifold, namely a **generalized Laplacian** H [6]. As a consequence, it ensures existence and unicity of a kernel $K_t(x,y,H)$ generating a solution of the corresponding heat equation. For arbitrary Riemannian manifold X and vector bundle E, it is usually impossible to find a closed form for this kernel. However, for many problems, the use of an approximate solution

---

[*] This work was partially supported by the ONR Grant N00014-09-1-0493, and by the "Communauté d'Agglomération de La Rochelle".

is sufficient. On the vector bundle $C^\infty(X)$ of smooth functions on X, there is a canonical generalized Laplacian: the Laplace-Beltrami operator. Considering each component I_k of a nD-image as a function over a well-chosen Riemannian manifold, we obtain the Beltrami flow of Sochen et al. [2]. The aim of this paper is to extend this flow to tangent vector fields on X and mappings from X to $\mathfrak{so}(m)$. The former is devoted to anisotropic smoothing of vector fields on the charts of X, i.e. some domains of $\mathbb{R}^m$, the latter to anisotropic smoothing of mappings from the charts to $SO(m)$, such as oriented orthonormal frame fields [1],[3],[5]. A natural extension is to consider the Hodge Laplacian Δ operating on sections of the Clifford bundle $Cl(X)$ of X. Indeed, functions, tangent vector fields and mappings from X to $\mathfrak{so}(m)$ and $SO(m)$ may be viewed as sections of $Cl(X)$ and the Hodge Laplacian, restricted to functions, corresponds to minus the Laplace-Beltrami operator. In this paper, we use existing estimates formula of heat kernels to compute an approximation, for small t, of the solution of the **heat equation**:

$$\frac{\partial s}{\partial t} + \Delta s = 0 \qquad (1)$$

where $s(x) = s(0, x)$ is a given section of $Cl(X)$.

In Section 2., we introduce the notion of heat kernel of generalized Laplacian, and give some of its properties. Section 3. is devoted to the Hodge Laplacian on Clifford bundles. In Section 4., we study the case $m = 2$ and illustrate it by anisotropic smoothings of a color image and a vector field related to this image.

2 Heat Kernels of Generalized Laplacians

We refer to [7] for an introduction to differential geometry. For a smooth vector bundle E over a manifold X, the symbol $\Gamma(E)$ denotes the space of smooth sections of E. For $x \in X$, E_x denotes the fiber over x.

Definition 1. Generalized Laplacian
Let E be a vector bundle over a Riemannian manifold (X, g). A **generalized Laplacian** on E is a second-order differential operator H, that may be written

$$H = -\sum_{ij} g^{ij}(x) \partial_i \partial_j + \text{first-order part}$$

in any local coordinates system, where $(g^{ij}(x))$ is the inverse of the matrix $g(x) = (g_{ij}(x))$.

In this paper, we make use of the following result:
To any generalized Laplacian H, one may associate an operator $e^{-tH} : \Gamma(E) \longrightarrow \Gamma(E)$, for $t > 0$, which is a smoothing operator with the property that if $I \in \Gamma(E)$, then $I(t, x) = e^{-tH} I(x)$ satisfies the heat equation $\partial I / \partial t + HI = 0$.

We shall define e^{-tH} as an integral operator of the form

$$(e^{-tH} I)(x) = \int_X K_t(x, y, H) I(y) dy$$

where $K_t(x, y, H) \colon E_y \longrightarrow E_x$ is a linear map depending smoothly on x, y and t. This kernel K is called the **heat kernel for H** [8].

In the following theorem, we give some results on approximations of the heat kernel and solutions of the heat equation. See [6] for details.

Theorem 1
Let $n = dim(X)$. Let ϵ chosen smaller than the injectivy radius of X. Let $\Psi \colon \mathbb{R}_+ \longrightarrow [0, 1]$ be a smooth function such that $\Psi(s) = 1$ if $s < \epsilon^2/4$ and $\Psi(s) = 0$ if $s > \epsilon^2$.

Let $\tau(x, y) \colon E_y \longrightarrow E_x$ be the parallel transport along the unique geodesic curve joining y and x, and $d(x, y)$ its length.

There exist functions J and sections Φ_i such that the kernel $K_t^N(x, y, H)$ defined by

$$\left(\frac{1}{4\pi t}\right)^{\frac{n}{2}} e^{-d(x,y)^2/4t} \, \Psi(d(x,y)^2) \sum_{i=0}^{N} t^i \Phi_i(x, y, H) J(x, y)^{-\frac{1}{2}}$$

satisfies

1. For every $N > n/2$, the kernel $K_t^N(x, y, H)$ is asymptotic to $K_t(x, y, H)$:

$$\left\|\partial_t^k [K_t(x, y, H) - K_t^N(x, y, H)]\right\|_l = O(t^{N-n/2-l/2-k})$$

where $\|\ \|_l$ is a norm on C^l sections.

2. Let us denote by k_t^N the operator defined by

$$(k_t^N I)(x) = \int_X K_t^N(x, y, H) I(y) dy$$

Then for every N, $\lim_{t \to 0} \|k_t^N I - I\|_l = 0$.

In what follows, we need the following properties:

(P1). $J(x, y) = 1 + O(\|\mathbf{y}\|^2)$ where $\mathbf{y}$ are the normal coordinates of y around x.

(P2). $\Phi_0(x, y) = \tau(x, y)$.

For the applications we propose in this paper, the base manifold is of dimension 2. Therefore, the kernel K_t^N is asymptotic to the heat kernel for $N \geq 1$. However, for the sake of simplicity, we restrict to the computation of the leading term $K_t^0(x, y)$, as it is done for the short time Beltrami kernel [2]. Moreover, we approximate $J(x, y)$ by 1. As a consequence, the smoothing will be performed by the operator k_t^0 defined by

$$(k_t^0 u)(x) = \int_X K_t^0(x, y) u(y) dy$$

and the discrete convolution will be made with small masks because of (P1).

3 The Hodge Laplacian on Clifford Bundles

The Hodge Laplacian Δ is a generalized Laplacian acting on differential forms of a Riemannian manifold. It is defined by $\Delta = d\delta + \delta d$, where d is the exterior derivative operator and δ its formal adjoint [9]. In particular, when applied to 0-forms, i.e. functions, Δ corresponds to the Laplace-Beltrami operator.

The Hodge Laplacian can be applied to tangent vector fields on the manifold, i.e. sections of the tangent bundle, by considering the Clifford bundle of the manifold. Let us first recall briefly some properties of Clifford algebras. We refer to [8] for details.

Definition 2. Clifford algebra
Let V be a vector space of finite dimension m over a field $\mathbb{K}$, and equipped with a quadratic form Q. Let $(e_1, \cdots, e_m)$ be a Q-orthonormal basis of V. The Clifford algebra of (V, Q), denoted by $Cl(V, Q)$, is an algebra over $\mathbb{K}$ of dimension 2^m of basis

$$(1, e_1, \cdots, e_m, e_1 e_2, \cdots, e_{m-1} e_m, \cdots, e_1 \cdots e_m) \tag{2}$$

with the relations $e_i^2 = -Q(e_i)$ and $e_i e_j = -e_j e_i$.

In particular, V and $\mathbb{K}$ are embedded into $Cl(V, Q)$. In (2), $e_1, \cdots, e_m$ denote the images of elements of the orthonormal basis mentionned above, whereas 1 denotes the image of the unit element of $\mathbb{K}$.

As a consequence of Definition 2, we can identify the exterior algebra $\bigwedge V^*$ of the dual space of V and $Cl(V,Q)$ by mapping $e^{i_1} \wedge e^{i_2} \wedge \cdots \wedge e^{i_k}$ to $e_{i_1} e_{i_2} \cdots e_{i_k}$, for $i_1 < \cdots < i_k$.

Definition 3. Clifford bundle of a Riemannian manifold
Given a Riemannian manifold (X, g), the tangent space $T(p)$ over $p \in X$ is an euclidean vector space $(T(p), g(p))$. Then, it generates a Clifford algebra $Cl(T(p), g(p))$ over $\mathbb{R}$. The set $\{Cl(T(p), g(p)), p \in X\}$ is called the **Clifford bundle** of (X, g), and will be noted $Cl(X)$.

From the embedding of $T(p)$ into $Cl(T(p), g(p))$, we see that any tangent vector field on a Riemannian manifold may be viewed as a section of its Clifford bundle. Moreover, from the embedding of $\mathbb{R}$ into $Cl(T(p), g(p))$, functions may also be viewed as sections of the Clifford bundle. More precisely, let $(e_1, \cdots, e_m)$ be an orthonormal frame field on (X, g). Then, any section s of $Cl(X)$ takes the form

$$s(p) = s_1(p)1(p) + s_2(p)e_1(p) + s_{m+1}(p)e_m(p) + \cdots + s_{2^m}(p)e_1 \cdots e_m(p)$$

We deduce the identification between differential forms on a Riemannian manifold and sections of its Clifford bundle. It is given by

$$\text{k-form}: \sum_{\substack{i_1 < \cdots < i_k \\ 1 \leq i_k \leq m}} \omega_{i_1 \cdots i_k} e^{i_1} \wedge \cdots \wedge e^{i_k} \longleftrightarrow \sum_{\substack{i_1 < \cdots < i_k \\ 1 \leq i_k \leq m}} \omega_{i_1 \cdots i_k} e_{i_1} \cdots e_{i_k}$$

$$\text{0-form}: \qquad\qquad f \longleftrightarrow f1$$

From this identification, the Hodge Laplacian may be applied to sections of the Clifford bundle, and in particular on vector fields and functions. Then, for $I_0 \in \Gamma(Cl(X))$, we define the **Clifford-Hodge flow** of I_0 as the solution of the heat equation

$$\frac{\partial I}{\partial t} + \Delta I = 0 \qquad (3)$$

of initial condition $I_{|t=0} = I_0$.

4 Application: The Case $m = 2$

First, let us introduce the group Spin(2) and its Lie algebra $\mathfrak{spin}(2)$. Under the identification of $\mathbb{R}^2$ and its embedding into $Cl(\mathbb{R}^2, \|\,\|_2)$, the rotation of angle θ is the map

$$r(\theta): v = v_1 e_1 + v_2 e_2 \longmapsto (cos(\theta) + sin(\theta) e_1 e_2)(v_1 e_1 + v_2 e_2)$$

The set

$$\left\{ cos(\theta) + sin(\theta) e_1 e_2, \theta \in [0, 2\pi[\right\}$$

is called the **spin group**, and is denoted by Spin(2).

We have Spin(2) $\simeq S^1 \simeq$ SO(2), from which we deduce that $\mathfrak{spin}(2) \simeq so(2)$. The Lie algebra of Spin(2) is the one-dimensional subspace generated by $e_1 e_2$, and the exponential map exp of Spin(2) is the ordinary exponential function. Therefore, we have $exp(\theta\, e_1 e_2) = cos(\theta) + sin(\theta) e_1 e_2$.

Given a function, vector field, or SO(2)-valued mapping defined on $\Omega \subset \mathbb{R}^2$, where (φ, Ω) is a chart of a Riemannian manifold (X, g) of dimension 2, we aim at smoothing it in a anisotropic way with respect to the geometry of X. From the previous Sections, we just need to consider it as a section of $Cl(X)$, as follows.

By definition, a function on a manifold is given by its values on charts. Therefore, given a function f on Ω, it defines the local section $f\, 1$ of $Cl(X)$.

Let $v = (v_1, v_2)$ be a vector field on Ω. Using the chart, v may be viewed as a local tangent vector field on X of coordinates (v_1, v_2) in the frame field $(\partial/\partial x, \partial/\partial y)$ given by a coordinates system. Let $(\tilde{v}_1, \tilde{v}_2)$ be its coordinates in the orthonormal frame field (e_1, e_2), then v is the section $\tilde{v}_1 e_1 + \tilde{v}_2 e_2$ of $Cl(X)$.

Let R be a SO(2)-valued mapping on Ω, e.g. a direction field or an oriented orthonormal frame field. From the identification of SO(2) and Spin(2) seen above, R may be viewed as a Spin(2)-valued mapping on X. The Hodge Laplacian does not preserve the Spin(2) structure, then we can not smooth a Spin(2)-valued mapping by applying the Clifford-Hodge flow on it. However, it preserves the $\mathfrak{spin}(2)$ structure. As a consequence, a Spin(2)-valued mapping may be smoothed through the smoothing of a corresponding $\mathfrak{spin}(2)$-valued mapping and the use of the exponential map. This is the Clifford algebra framework counterpart of the problem of orientations diffusion [10].

Let us give some precisions about the parallel transport map on $Cl(X)$, needed to smooth sections of $Cl(X)$. Indeed, as mentionned in Section 2., we compute an approximating solution of (3) by convolving I_0 with the leading term of the heat kernel $K_t(x, y, \Delta)$. It is determined by geodesic distances on X and the parallel transport map on $Cl(X)$.

Proposition 1. Parallel transport on $Cl(X)$
Let $\gamma : J \subset \mathbb{R} \longrightarrow X$ be C^1 curve where $\gamma(0) = p$, and $Y_0 = Y_0^1 \, 1(p) + Y_0^2 \, e_1(p) + Y_0^3 \, e_2(p) + Y_0^4 \, e_1 e_2(p) \in Cl(T(p), g(p))$. Then, the parallel transport of Y_0 along γ is $Y(t) = Y_1(t) \, 1(\gamma(t)) + Y_2(t) \, e_1(\gamma(t)) + Y_3(t) \, e_2(\gamma(t)) + Y_4(t) \, e_1 e_2(\gamma(t))$ where

$$Y_1(t) = Y_0^1 \qquad Y_4(t) = Y_0^4$$

and Y_2, Y_3 satisfy

$$Y_2(t) + i\, Y_3(t) = exp\left(i \int_0^t \frac{\partial \gamma_1}{\partial s}(s) \Gamma_{11}^2(s) + \frac{\partial \gamma_2}{\partial s}(s) \Gamma_{21}^2(s) \, ds\right)(Y_0^2 + i\, Y_0^3)$$

Note that $\partial \gamma_i / \partial s$ and the Christoffel symbols Γ_{ij}^k are given with respect to the orthonormal frame field (e_1, e_2).

Let $I : \Omega \subset \mathbb{R}^2 \longrightarrow \mathbb{R}^n$ be a nD image of components $(I_1, \cdots, I_n)$. I defines a surface S embedded in $\mathbb{R}^{n+2}$, parametrized by

$$\varphi : (x, y) \longmapsto (x, y, I_1(x, y), I_2(x, y), \cdots, I_n(x, y))$$

Any definite positive scalar product h on $\mathbb{R}^{n+2}$ induces a metric g on S and makes the couple (S, g) be a Riemannian manifold of dimension 2 of chart (φ, Ω).

For color images (n=3), a suitable choice for h is

$$h = \begin{pmatrix} 1 & 0 \\ 0 & 1 \end{pmatrix} \oplus \begin{pmatrix} \lambda & 0 & 0 \\ 0 & \lambda & 0 \\ 0 & 0 & \lambda \end{pmatrix}$$

where λ is a positive function. In the Beltrami framework of Sochen et al. [2], each component I_k is considered as a function over (S, g). Then, the so-called Beltrami flow is obtained by solving the PDE's $\partial I_k / \partial t = \Delta_g I_k$, where Δ_g is the Laplace-Beltrami operator. Then, considering the functions I_k as sections of $Cl(S)$ of the form $\tilde{I}_k = I_k 1$, the Clifford-Hodge flow (3) of $\tilde{I}_k$ is equivalent to the Beltrami flow of I_k.

Fig. 1 is an illustration of the Clifford-Hodge flow on functions, that provides an anisotropic smoothing of images. Fig. 1(a) is a natural color image taken from the Berkeley image segmentation database [11]. Fig. 1(b) is the result of the Clifford-Hodge flow after 5 iterations, for $\lambda = 0.01$ and $t = 0.3$.

Fig. 2 is an illustration of the Clifford-Hodge flow on tangent vector fields, that provides an anisotropic smoothing of vector fields on the charts. Fig. 2(a) is the field of unit vectors indicating the directions of lowest variations around the hat of Fig. 1(b). Fig. 2(b) is the result of the Clifford-Hodge after 99 iterations, for $\lambda = 0.01$ and $t = 0.3$. We see that it tends to preserve the vector field on high edges of the image, conversely to vanish it on low edges.

(a) Original image (b) Anisotropic Clifford-Hodge flow

Fig. 1. Clifford-Hodge flow on functions

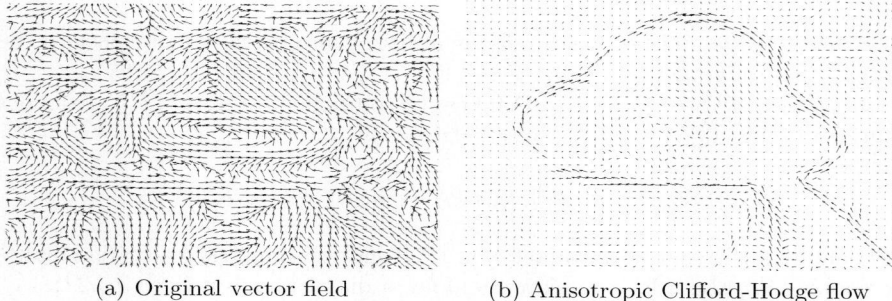

(a) Original vector field (b) Anisotropic Clifford-Hodge flow

Fig. 2. Clifford-Hodge flow on tangent vector fields

5 Conclusion

We introduced the Hodge Laplacian for anisotropic smoothing of vector-valued data. From this point of view, the anisotropy is related to the metric of a Riemannian manifold. Smoothing process is provided by the solution of the corresponding heat equation. It results from a convolution with a kernel generalizing the gaussian kernel to non-flat spaces by computation of geodesic distances and transport parallel map. Strictly speaking, the Hodge Laplacian acts on differential forms of a Riemannian manifold. Identifying the bundle of differential forms and the Clifford bundle, the Hodge Laplacian may be applied to tangent vector fields and functions. It allows anisotropic smoothing of vector fields and nD images defined on a domain of $\mathbb{R}^2$. Moreover, from the identification of Spin(2) and SO(2), and the embedding of $\mathfrak{spin}(2)$ into $Cl(\mathbb{R}^2, \| \ \|_2)$, we have shown that Clifford bundles also provide a framework to smooth SO(2)-valued mappings such as direction fields and oriented orthonormal frame fields. Finally, we obtain a general framework for anisotropic smoothing of data of different natur on $\mathbb{R}^2$. From Section 3. and $\mathfrak{so}(m) \simeq \mathfrak{spin}(m) \subset Cl(\mathbb{R}^m, \| \ \|_2)$, the generalization for $m \geq 3$ is straightforward. In particular, smoothing of volumetric data may be envisaged.

References

1. Tang, B., Sapiro, G., Casselles, V.: Diffusion of General Data on Non-Flat Manifolds via Harmonic Maps Theory: The Direction Diffusion Case. Int. J. Comp. Vis. 36(2), 149–161 (2000)
2. Spira, A., Kimmel, R., Sochen, N.: A Short-time Beltrami Kernel for Smoothing Images and Manifolds. IEEE Trans. Image Processing 16, 1628–1636 (2007)
3. Gur, Y., Sochen, N.: Regularizing Flows over Lie Groups. J. Math. Imag. Vis. 33, 195–208 (2009)
4. Tschumperlé, D., Deriche, R.: Vector-valued Image Regularization with PDE's: A Common Framework for Different Applications. IEEE Trans. Pattern Analysis and Machine Intelligence 27, 506–517 (2005)
5. Tschumperlé, D., Deriche, R.: Orthonormal vector sets regularization with PDE's and applications. Int. J. Comp. Vis. 50(3), 237–252 (2002)
6. Berline, N., Getzler, E., Vergne, M.: Heat Kernels and Dirac Operators. Springer, Heidelberg (2004)
7. Spivak, M.: A Comprehensive Introduction to Differential Geometry. Publish or Perish, Inc., Houston (1979)
8. Lawson, H.B., Michelson, M.-L.: Spin Geometry. Princeton University Press, Princeton (1989)
9. De Rham, G.: Variétés différentiables: Formes, Courants, Formes harmoniques. Hermann, Paris (1973)
10. Perona, P.: Orientation Diffusions. IEEE Trans. Image Processing 7, 457–467 (1998)
11. Martin, D., Fowlkes, C., Tal, D., Malik, J.: A database of human segmented natural images and its application to evaluating segmentation algorithms and measuring ecological statistics. In: Proc. 8th Int'l Conf. Comp. Vis., vol. 2, pp. 416–423 (2001)

Speedup of Color Palette Indexing in Self–Organization of Kohonen Feature Map

Kuo-Liang Chung*, Jyun-Pin Wang, Ming-Shao Cheng, and Yong-Huai Huang

Department of Computer Science and Information Engineering
National Taiwan University of Science and Technology
No. 43, Section 4, Keelung Road, Taipei, Taiwan 10672, R.O.C.

Abstract. Based on the self–organization of Kohonen feature map (SOFM), recently, Pei et al. presented an efficient color palette indexing method to construct a color table for compression. Taking their palette indexing method as a representative, this paper presents two new strategies, the pruning–based search strategy and the lookup table (LUT)–based update strategy, to speed up the learning process in the SOFM. Based on four typical testing images, experimental results demonstrate that our proposed two strategies have 35% execution–time improvement ratio in average. In fact, our proposed two strategies could be used to speed up the other SOFM–based learning processes in different applications.

Keywords: Color palette indexing, lateral update interaction, learning process, lookup table, SOFM, speedup, winning neuron.

1 Introduction

In order to achieve good compression performance by using image compression standards, such as JPEG–LS [3], JPEG–2000 [4], and PNG [5], how to design a good color palette table is an important issue. In 1981, Kohonen presented a pioneer work, self–organization of Kohonen feature map (SOFM) [1], which is a powerful unsupervised neuron learning model. The SOFM has been studied extensively in the applications such as color quantization, color palette indexing design, clustering, data mining, pattern recognition, and so on.

In this paper, we take Pei et al.'s color palette indexing method [2] as the representative to demonstrate two computational bottlenecks existed in the SOFM. In order to alleviate the two bottlenecks, for each training vector, we first present a pruning–based search strategy to speed up the process for finding the winning neuron in each iteration; further, a lookup table (LUT) strategy is presented to speed up the lateral update interaction between the winning neuron and its neighboring neurons. Based on four typical testing images, experimental results demonstrate that our proposed two strategies have 35% execution–time improvement ratio in average while preserving the same result as in the SOFM. Precisely

---

* Corresponding author. Email: k.l.chung@mail.ntust.edu.tw. Supported by the National Science Council of R. O. C. under the contract NSC97–2221–E–011–128.

speaking, the first strategy has 19% execution–time improvement ratio; the second strategy has 16% execution–time improvement ratio.

2 The Palette Indexing Method by Pei et al. and Two Computational Bottlenecks

In this section, we first introduce the palette indexing method by Pei et al. [2], then we point out two computational bottlenecks in their SOFM–based learning model.

2.1 SOFM–Based Learning Model

As shown in Fig. 1, the 1–D SOFM used in Pei et al.'s method has two layers, namely the input layer and the output layer. The output layer has N neurons $u_1, u_2, ..., $ and u_N; the initial triple weight of the i–th neuron u_i in the neural network, $1 \leq i \leq N$, is set to

$$\mu_i = [r_i(0), g_i(0), b_i(0)]^T \\ = [(i-1)*256/N, (i-1)*256/N, (i-1)*256/N]^T \quad (1)$$

where $r_i(0)$, $g_i(0)$, and $b_i(0)$ indicate the red value, green value, and blue value at the 0–th iteration, i.e. at the initial iteration in the i–th neuron; '256' indicates the maximal allowable number of indices in the palette. From the input layer, we feed each pixel $x = [r_x, g_x, b_x]^T$ in the input color image as the training vector to feed into the SOFM for training a good color palette table.

In the training process of the SOFM, M training vectors are required in each sweep. In order to maximize the randomness of the input training vectors, avoid biased training, and avoid some clusters being overtrained during the training process, Pei et al. presented an effective butterfly–jumping sequence to generate the input training vectors for each training sweep. Based on the butterfly–jumping sequence, all pixels in the $W \times H$ input image is separated into $\frac{W \times H}{M}$ training sets $S_1, S_2, ...,$ and $S_{\frac{W \times H}{M}}$. Each set contains M training vector to be used in a training sweep.

For the m–th training sweep, $1 \leq m \leq \frac{W \times H}{M}$, each training vector x in S_m is fed into the SOFM to search the best matched neuron u_c, $1 \leq c \leq N$, i.e.

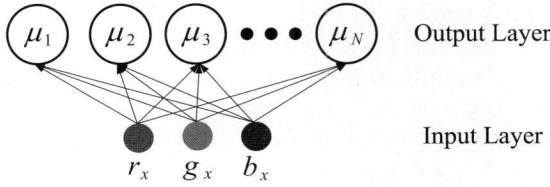

Fig. 1. The used 1–D SOFM

the winning neuron, in the output layer based on the following square Euclidean distance function:

$$c = \arg \min_{1 \leq i \leq N} \|x - \mu_i\|^2 = \arg \min_{1 \leq i \leq N} [(r_x - r_i)^2 + (g_x - g_i)^2 + (b_x - b_i)^2]. \quad (2)$$

Based on the input vector x and the winning neuron u_c with weight $\mu_c = [r_c, g_c, b_c]^T$, the output layer therefore updates the weight of the winning neuron and simultaneously performs the lateral update interaction between the winning neuron u_c and its neighboring neurons by the following learning function:

$$\mu_i(m, n+1) = \begin{cases} \mu_i(m,n) + \alpha(m) \cdot g(i,c,\sigma(m)) \cdot [x - \mu_i(m,n)], & \text{if } |i-c| \leq \lfloor \sigma(m) \rfloor \\ \mu_i(m,n), & \text{otherwise} \end{cases} \quad (3)$$

where m and n denotes the sweep number and the iteration number, respectively; $g(i, c, \sigma(m))$ is the neighboring function providing the lateral interaction between u_i and u_c; $\sigma(m)$ and $\alpha(m)$ denote the width of neighboring function and scalar gain function used in the m–th sweep, respectively. $g(i, c, \sigma(m))$, $\sigma(m)$, and $\alpha(m)$ are defined by

$$g(i, c, \sigma(m)) = \exp(-|i-c|^2/\sigma^2(m)), \quad (4)$$
$$\sigma(m) = \sigma(0) \times k_1^m, \text{ and} \quad (5)$$
$$\alpha(m) = \alpha(0) \times k_2^m, \quad (6)$$

respectively, where k_1 and k_2 are set to values in $[0.8, 0.99]$. By Eq. (5) and (6), $\sigma(m)$ and $\alpha(m)$ are decreasing functions in terms of sweep number m. $\sigma(0)$ is set to a value in $[1, 10]$ and $\alpha(0)$ is set to a value less than one initially. In our implementation, we set $\sigma(0) = 10$, $\alpha(0) = 0.1$, $k_1 = 0.8$, and $k_2 = 0.8$ initially.

After updating the weights of neurons in the output layer, the next training vector in S_m will be fed into the SOFM to search the best matched neuron by Eq. (2) and update the weights of relevant neurons by Eq. (3). The above training process is performed iteratively until $\alpha(m)$ becomes small enough. The stopping rule of the learning process is set to that if the condition $\alpha(m) \leq 4.05648 \times 10^{-5}$ is held in the m-th sweep. From the stopping rule, we find that the number of sweeps N_s can be set to 35 since $\alpha(35)$ satisfies the stopping rule.

2.2 Two Computational Bottlenecks

From the description in last subsection, we now discuss two concerned computational bottlenecks. The first computational bottleneck occurs in the process for searching the winning neuron. In each sweep, we have M training vectors and for each training vector, Eq. (2) is performed to calculate N square Euclidean distances to select the minimal one, and then determine the winning neuron. In Eq. (2), it needs three subtractions, two additions, and three square operations for calculating each square Euclidean distance and from the N calculated Euclidean distances, $N - 1$ comparisons are required to find the winning neuron with minimal Euclidean distance. Overall, each sweep needs to calculate Eq. (2) M times. Let T_{add}, T_{sub}, T_{sq}, and T_{cmp} denote the time required to perform one addition, one subtraction, one square operation, and one comparison, respectively. From the state of the art in VLSI technology, one addition, one subtraction, and one square operation can be performed using almost the same time. Therefore, we can use T_{add} to instead of each one of T_{sub} and T_{sq} since $T_{add} = T_{sub} = T_{sq}$. Further, one comparison is composed of one subtraction and one sign test. Thus, we assume $T_{cmp} = 2T_{add}$ for convenience. By setting the number of sweeps performed in the SOFM to be 35, i.e. $N_s = 35$, we have the following proposition.

Proposition 1. *For each training vector, finding the winning neuron takes T_W = $(10N - 2) \times T_{add}$ $(= N \times (3T_{sub} + 2T_{add} + 3T_{sq}) + (N-1) \times T_{cmp})$ time. For each sweep, it takes $M \times T_W$ time to find M winning neurons. For an input color image, it takes $T_W^I = N_s \times M \times T_W$ time to find all winning neurons. For the case $N_s = 35$, we have $T_W^I = (350MN - 70M) \times T_{add}$.*

The second computational bottleneck occurs in the process for updating the weights of winning neuron and its neighbors. For each determined winning neuron, $(2\lfloor \sigma(m) \rfloor + 1)$ neurons must be updated by Eq. (3) where $\lfloor \cdot \rfloor$ denotes the floor operation. From Eq. (3), it needs three additions and four multiplications to sum up the four terms $\mu_i(m,n)$, $[x - \mu_i(m,n)]$, $g(i, c, \sigma(m))$, and $\alpha(m)$; the term $[x - \mu_i(m,n)]$ needs three subtractions; the neighboring function $g(i, c, \sigma(m))$ defined in Eq. (4) needs one subtraction, one multiplication, two square operations, one division, and one exponentiation operation.

Further, each sweep also needs two multiplications to calculate $\sigma(m)$ and $\alpha(m)$ by Eq. (5) and Eq. (6), respectively. Let T_{mul}, T_{exp}, and T_{div} denote the time required to perform one multiplication, one exponentiation operation, and one division, respectively. In our experiments, we have $T_{mul} = T_{add}$, $T_{exp} = 160T_{add}$, and $T_{div} = 10T_{add}$. We thus have the following proposition.

Proposition 2. *For each winning neuron in the m–th sweep, it takes T_U^m = $(2\lfloor \sigma(m) \rfloor + 1) \times (14T_{add} + T_{div} + T_{exp})$ $(= (2\lfloor \sigma(m) \rfloor + 1) \times (3T_{add} + 5T_{mul} + 4T_{sub} + 2T_{sq} + T_{div} + T_{exp}))$ time to update the weights of $(2\lfloor \sigma(m) \rfloor + 1)$ neurons. For the m–th sweep, it takes $M \times T_U^m + 2T_{add}$ $(= M \times T_U^m + 2T_{mul})$ time for updating M winning neurons. For an input color image, it takes $T_U^I = M \sum_{m=1}^{N_s} T_U^m + N_s \times 2T_{add}$ time to update the weights of all concerned neurons. For the case $N_s = 35$, we have $\sum_1^{35} \lfloor \sigma(m) \rfloor = 33$ and it leads to $T_U^I = (18584M + 70) \times T_{add}$ $(= (1414M + 70) \times T_{add} + 101M \times (T_{div} + T_{exp}))$.*

3 The Proposed Faster Learning Process

The above two propositions lead to two motivations of our research. From Proposition 1, Subsection 3.1 presents an efficient pruning–based search strategy to speed up the process for finding the winning neuron in each iteration. From Proposition 2, Subsection 3.2 presents a lookup table (LUT) strategy to speed up the lateral update interaction between the winning neuron and its neighboring neurons.

3.1 Pruning–Based Search Strategy for Finding Winning Neurons

By Proposition 1, in order to reduce the time complexity, we build up a mean pyramid structure for the i–th neuron, $1 \leq i \leq N$. The base of pyramid saves the three entries of μ_i, i.e. r_i, g_i, and b_i; the apex of pyramid saves the mean of μ_i, i.e. $\overline{\mu_i} = \frac{1}{3}(r_i + g_i + b_i)$. Totally, N mean pyramids are constructed for the N neurons. For the input training vector $x = [r_x, g_x, b_x]^T$, we also construct a mean pyramid and $\overline{x} = \frac{1}{3}(r_x + g_x + b_x)$.

Let $D_b^2(x, \mu_i)$ $(= (r_x - r_i)^2 + (g_x - g_i)^2 + (b_x - b_i)^2)$ denotes the squared Euclidean distance between x and μ_i; let $D_t^2(\overline{x}, \overline{\mu_i})$ $(= (\overline{x} - \overline{\mu_i})^2 = \frac{1}{9}(r_x - r_i + g_x - g_i + b_x - b_i)^2)$ denotes the squared Euclidean distance between the $\overline{x}$ and $\overline{\mu_i}$, respectively. We have the following theorem.

Theorem 1. $D_b^2(x, \mu_i) \geq 3D_t^2(\overline{x}, \overline{\mu_i})$.

Proof. Let $d_r = r_x - r_i$, $d_g = g_x - g_i$, and $b_x = x_b - b_i$. It yields to $3(d_r^2 + d_g^2 + d_b^2) - (d_r + d_g + d_b)^2 = (d_r - d_g)^2 + (d_r - d_b)^2 + (d_g - d_b)^2 \geq 0$. It further yields to $d_r^2 + d_g^2 + d_b^2 \geq 3 \times \frac{1}{9}(d_r + d_g + d_b)^2$. We thus have $D_b^2(x, \mu_i) \geq 3D_t^2(\overline{x}, \overline{\mu_i})$. We complete the proof. Following the similar proving technique, it yields to $D_b^2(X, Y) \geq pD_t^2(\overline{X}, \overline{Y})$ where $X = [x_1, x_2, ..., x_p]^T$, $Y = [y_1, y_2, ..., y_p]^T$, $\overline{X} = \frac{1}{p}\sum_{i=1}^{p} x_i$, and $\overline{Y} = \frac{1}{p}\sum_{i=1}^{p} y_i$.

By Theorem 1, for the input training vector x, our proposed pruning–based search strategy for finding the winning neuron is shown below.

Step 1: Construct the mean pyramid for the input training vector $x(= [r_x, g_x, b_x]^T)$.
Step 2: Take the mean of x, i.e. $\overline{x}$, as the key, we find the closed $\overline{\mu_c}$ ($c = \arg\min_{1 \leq i \leq N} D_t^2(\overline{x}, \overline{\mu_i})$). Then we calculate the squared Euclidean distance between x and μ_c as the value of $D_{\min}^2$ $(= D_b^2(x, \mu_c))$.
Step 3: Initially, we set $j = 1$.
  3.1: If $j > N$, then stop Step 3. If $D_t^2(\overline{x}, \overline{\mu_j}) \geq \frac{1}{3}D_{\min}^2$, then we reject the neuron u_j, set $j = j + 1$, and go to step 3.1; otherwise, go to step 3.2.
  3.2: Calculate $D_b^2(x, \mu_j)$. If $D_b^2(x, \mu_j) \geq D_{\min}^2$, then we reject the neuron u_j, set $j = j + 1$, and go to step 3.1; otherwise, set u_j to be the winning neuron, set $D_{\min}^2 = D_b^2(x, \mu_j)$, set $j = j + 1$, and go to step 3.1.

The speedup efficiency of our proposed algorithm is dependent of the number of discarded neurons before computing the actual squared Euclidean distances between bases of the mean pyramids and the given input training vectors x. Moreover, in each iteration, only the neurons involved in the learning function defined by Eq. (3) are needed to update their mean pyramids.

Return to Subsection 2.1, Eq. (1) shows that initially for each neuron u_i, the values of r_i, g_i, and b_i in μ_i are the same, and thus we can construct the mean pyramid of μ_i without any operations. In the m–th sweep, for each input training vector x, Step 1 takes $3T_{add}$ $(= 2T_{add} + T_{mul})$ time to construct mean pyramid. In Step 2, the term $c = \arg\min_{1 \leq i \leq N} D_t^2(\overline{x}, \overline{\mu_i})$ takes $N \times 2T_{add}$ $(= N \times (T_{sub} + T_{sq}))$

time to calculate $D_t^2(\overline{x}, \overline{\mu_i})$'s and $(2N-2) \times T_{add}$ $(=(N-1) \times T_{cmp})$ time to find the minimal one from $D_t^2(\overline{x}, \overline{\mu_i})$'s. Further, $8T_{add}$ $(= 3T_{sub} + 2T_{add} + 3T_{sq})$ time is needed for calculating the distance $D_b^2(x, \mu_c)$. In Step 3.1, the term $D_t^2(\overline{x}, \overline{\mu_j}) \geq \frac{1}{3}D_{\min}^2$ takes $2T_{add}$ $(= T_{cmp})$ time to compare $D_t^2(\overline{x}, \overline{\mu_i})$ and $\frac{1}{3}D_{\min}^2$. Step 3.2 takes $8T_{add}$ $(=3T_{sub} + 2T_{add} + 3T_{sq})$ time to calculate $D_b^2(x, \mu_j)$ and takes $2T_{add}$ $(= T_{cmp})$ time for checking the condition $D_b^2(x, \mu_j) \geq D_{\min}^2$. For $1 \leq j \leq N$, we totally perform Step 3.1 N times and Step 3.2 $R \times N$ times for $0 \leq R \leq 1$. After finding the winning neuron, the learning function in Eq. (3) is used to update the weights of $2\lfloor \sigma(m) \rfloor + 1$ neurons, and thus $(2\lfloor \sigma(m) \rfloor + 1) \times 3T_{add}$ $(= (2\lfloor \sigma(m) \rfloor + 1) \times (2T_{add} + T_{mul}))$ time is needed for updating $2\lfloor \sigma(m) \rfloor + 1$ mean pyramids. In our experiments, we find that $R = 0.15$ in average. Thus, we have the following proposition.

Proposition 3. *For each iteration in the m-the sweep, the proposed pruning–based search strategy takes $T_{W'}^m = (7.5N + 6\_\sigma(m)\rfloor + 12) \times T_{add}$ time for finding the winning neuron. For the m-the sweep, it takes $M \times T_{W'}^m$ time to find M winner neurons. For an input color image, it takes $T_{W'}^I = M \sum_{m=1}^{N_s} T_{W'}^m$ time for finding all winner neurons. For the case $N_s = 35$, we have $\sum_1^{35} \lfloor \sigma(m) \rfloor = 33$ and it leads to $T_{W'}^I = (262.5MN + 618M) \times T_{add}$.*

The computational advantage of our proposed pruning–based search strategy can be verified from Proposition 1 and Proposition 3. Suppose that the SOFM has 256 neurons and each sweep contains 256 training vectors, i.e. $N = 256$ and $M = 256$, and then we have $T_W^I = 22919680 T_{add}$ and $T_{W'}^I = 17361408 T_{add}$. Thus, the theoretical execution–time improvement ratio of our proposed pruning–based search strategy over the traditional search strategy is 0.24 $(= \frac{T_W^I - T_{W'}^I}{T_W^I})$. From Proposition 2, we can obtain $T_U^I = 4757574 T_{add}$ and it leads to that the time required for searching all winning neurons over the total time required in the SOFM–based training process is 83% $(= \frac{T_W^I}{T_W^I + T_U^I})$. Consequentially, we have the result.

Theorem 2. *The theoretical execution–time improvement ratio of our proposed pruning–based search strategy over the SOFM–based training process is 0.20 $(= 0.24 \times 83\%)$.*

3.2 LUT–Based Lateral Update Interaction

From Eqs. (3)–(6), it is observed that $g(i_1, c_1, \sigma(m)) \cdot \alpha(m)$ and $\sigma(m)$ relative to the sweep number m and independent of the input image. Further, $g(i_1, c_1, \sigma(m))$ is equal to $g(i_2, c_2, \sigma(m))$ if the condition $|i_1 - c_1| = |i_2 - c_2|$ is held. Thus, two LUTs, L_1 and L_2, where L_1 is an 1–D LUT relative to m and L_2 is a 2–D LUT relative to m and $|i - c|$, can be constructed in advance to reduce the computational effort required in the lateral update interaction. In L_1, each entry $L_1(i_1)$, $1 \leq i \leq N_s$ $(= 35)$, is set to $\lfloor \sigma(i_1) \rfloor$. In L_2, each entry $L_2(i_1, i_2)$ can be calculated by

$$L_2(i_1, i_2) = \begin{cases} \alpha(i_1) \cdot \exp(-i_2^2/\sigma^2(i_1)), & \text{if } i_2 \leq \lfloor \sigma(i_1) \rfloor \\ 0, & \text{otherwise} \end{cases} \quad (7)$$

where $1 \leq i_1 \leq N_s$, $0 \leq i_2 \leq \sigma(1)$, and $\sigma(1) = 8$ $(= \sigma(0) \times k = 10 \times 0.8)$.

For each training vector in the m–th sweep, the weight of the winning neuron and its neighbors in current iteration are updated by

$$\mu_i(m, n+1) = \begin{cases} \mu_i(m,n) + L_2(m, |i-c|) \cdot [x_i(n) - \mu_i(m,n)], & \text{if } |i-c| \leq L_1(m) \\ \mu_i(m,n), & \text{otherwise} \end{cases} \quad (8)$$

In Eq. (8), it takes $10T_{add} + T_{abs}$ $(= 3T_{add} + 3T_{mul} + 4T_{sub} + T_{abs})$ time to update the weights of each concerned neuron where T_{abs} denote the time required to perform one absolute operation. Further, the absolute operation is composed of one sign testing process for the input values and one sign flipping process if the result of sign testing process is negative. We thus set $T_{abs} = 2T_{add}$ for convenience and $10T_{add} + T_{abs}$ can be replaced by $12T_{add}$. For each iteration in the m–th sweep, $2\lfloor \sigma(m) \rfloor + 1$ neurons are needed to be updated by Eq. (8). Thus, we have the following proposition.

Proposition 4. *Based on the proposed LUT–based update strategy, each iteration in the m–th sweep takes $T_{U'}^m = (24\lfloor \sigma(m) \rfloor + 12) \times T_{add}$ time to update the weights of concerned neurons. For the m–th sweep, it takes $M \times T_{U'}^m$ time for updating M winner neurons. For an input color image, it takes $M \sum_1^{N_s} T_{U'}^m$ time to update the weights of all concerned neurons. For the case $N_s = 35$, we have $\sum_1^{35} \lfloor \sigma(m) \rfloor = 33$ and it leads to $T_{U'}^I = 1212M \times T_{add}$.*

The computation advantage of our proposed LUT–based update strategy can be verified from Proposition 2 and Proposition 4. Suppose $N = 256$ and $M = 256$, we have $T_U^I = 4757574 T_{add}$ and $T_{U'}^I = 310272 T_{add}$. Thus, the theoretical execution–time improvement ratio of our proposed LUT–based update strategy over the traditional update strategy is 0.93 $(= \frac{T_U^I - T_{U'}^I}{T_U^I})$. Due to $T_W^I = 22919680 T_{add}$ and $T_U^I = 4757574 T_{add}$, the time required in the update process of all concerned neurons over the total time required in the SOFM training process is 17% $(= \frac{T_U^I}{T_W^I + T_U^I})$. Consequentially, we have the result.

Theorem 3. *The theoretical execution–time improvement ratio of the LUT–based update strategy over the SOFM training process is 0.16 $(= 0.93 \times 17\%)$.*

4 Experimental Results

In this section, some experimental results are demonstrated to show the computational advantage of our proposed pruning–based search strategy and the LUT–based update strategy. All the concerned experiments are performed on the Intel Core2 Duo Processor E7400 with 2.8 GHz and 2 GB RAM. The operating system is MS–Windows XP and the program developing environment is Borland C++ Builder 6.0. Surprisingly, the theoretical execution–time improvement ratios of our proposed two strategies mentioned in Theorem 2 and Theorem 3 are close to the practical ones.

Four popular testing images, namely Baboon, Lena, Airplane, and Pepper, are used to evaluate the performance of the four concerned methods. Our experimental results demonstrate that under the same peak signal-to-noise ratio (PSNR), our proposed pruning–based search strategy has 18.7% average execution–time improvement ratio and it is close to the theoretical execution–time improvement ratio 20%; our proposed LUT–based update strategy has 15.6% average execution–time improvement ratio and it is close to the theoretical execution–time improvement ratio 16%. Overall, our proposed two strategies has 34.3% execution–time improvement ratio in average while preserving the same result as in the traditional method.

5 Conclusions

In the SOFM, the iterative learning process is the kernel process, but it is very time–consuming. In order to alleviate this time–consuming problem, taking Pei et al.'s SOFM–based palette color indexing method as a representative, we have presented the proposed novel two new strategies, the pruning–based search strategy and the LUT–based update strategy, to speed up the learning process significantly while keeping the same image quality. Based on four testing images, experimental results indicate that our proposed two strategies have 35% execution–time improvement ratio in average. It is a future research issue to apply our results to speed up other SOFM–based applications.

References

1. Kohonen, T.: Construction of similarity diagrams for phonemes by a self-organizing algorithm. Technical Report TKK-F-A463, Helsinki University of Technology, Espoo, Finland (1981)
2. Pei, S.C., Chuang, Y.T., Chuang, W.H.: Effective palette indexing for image compression using self-organization of Kohonen feature map. IEEE Transactions on Image Processing 15(9), 52–61 (2006)
3. Information Technology–Lossless and Near-Lossless Compression of Continuous-Tone Still Images, ISO/IEC Standard 14 495-1 (1999)
4. Information Technology–JPEG 2000 Image Coding System. ISO/IEC Standard 15 444-1 (2000)
5. Roleof, G.: PNG: The Definitive Guide, 2nd edn. Greeg Roelof, San Jose (2003), http://www.libpng.org/pub/png/book/

Probabilistic Satellite Image Fusion

Farid Flitti[1], Mohammed Bennamoun[1], Du Huynh[1], Amine Bermak[2], and Christophe Collet[3]

[1] The University of western Australia*, Auatralia
[2] The Hong Kong University of Science and Technology, Hong Kong
[3] Strasbourg University, France

Abstract. Remote sensing satellite images play an important role in many applications such as environment and agriculture lands monitoring. In such images the scene is usually observed with different modalities, e.g. wavelengths. Image Fusion is an important analysis tool that summarizes the available information in a unique composite image. This paper proposes a new transform domain image fusion (IF) algorithm based on a hierarchical vector hidden Markov model (HHMM) and the mixture of probabilistic principal component analysers. Results on real Landsat images, quantified subjectively and using objective measures, are very satisfactory.

1 Introduction

Remote sensing earth monitoring has gained an increasing interest in the last decades. A lot of applications in agriculture, forest resources management and water studies benefit nowadays from such tools. Usually satellite imagers observe earth in various spectral domains to provide a richer description of the scene. However, the benefit from available diverse information requires the ability to provide an efficient summary of the observed scene that avoids the exhaustive examination of all individual spectral bands. Image Fusion (IF) techniques aim to summarise the whole collected information and a built a unique composite image [1]. The book of Blum and Liu [1] provides a good survey of this research area. The authors identify two main categories of IF techniques depending on whether the combination of the input images is done in the original input space (i.e., gray level) or in a Multiscale Analysis (MA) domain. MA based techniques combine the MAs of the inputs to obtain a composite multiscale representation, of which the reconstruction gives the desired fused image. Generally, non-MA based techniques seek for a mapping (linear or not) from an ND space to a 1D space, N being the number of input images [2,3,4,5,6].

A collection of local mappings, based on either principal component analysis (PCA) or self organized maps (SOM), was used in [7] for the fusion of multispectral Landsat images and compared to global PCA and SOM. The author concluded that using local projections gives a better contrast in the fusion result. We propose

---

* This work is sponsored by the Australian Research Council (ARC) DP0771294.

to use the mixture of probabilistic PCA, (MPPCA), on the Laplacian Pyramids (LP) of the input images, with a unique Hidden Markov Tree (HMT) Model as regularisation of the mixture, to provide a fusion model that accounts for all different kind of correlations between the MA coefficients. In fact, MA coefficients exhibits three kinds of correlations, namely inter-scale, intra-scale and inter input images [8,9]. This model was used successfully for multi-component image segmentation [1] and is extended in this work to provide an intuitive way to infer the fusion result composite MA representation. In [10] a forest of HMT modeling the input image MAs separately was used. This model neither encloses a projection model nor considers the existing correlation between the different input images. However, a projection model is suitable for IF, and correlation between observations of the input images exists always. In [8] we proposed an IF algorithm for astronomical images with different data driven model and no included mapping. Besides, this work aimed the detection and fusion of the astronomical object, therefore only significant MA coefficients, corresponding to real objects, were used for the fusion. This is completely different from the context and the methods of this work.

The paper is organized as follows. In section 2, the LP analysis and its statistical properties are recalled. The regularised MPPCA model is detailed in section 3. The proposed IF algorithm is presented in section 4. Section 5 is reserved to experimental results and conclusions.

2 The Laplacian Pyramid

A basic iteration of the Laplacian Pyramid LP [11] involves two steps: reduction and expansion. Given the approximation c_j of the original signal at resolution j, the reduction step computes the approximation c_{j+1} by lowpass filtering and downsampling c_j. The expansion step predicts c_j by upsampling and filtering c_{j+1} and then computes the details coefficients w_{j+1} as the difference between the c_j signal and its prediction[1]. Despite its simplicity, this approach is efficient for image modeling. The reconstruction proceeds with iterations of one step. Given the details and approximation at scale $j + 1$, the approximation c_j, at scale j, is computed by upsampling and filtering c_{j+1} and adding the result to w_{j+1}.

For an input band $\mathcal{F}_i$ of size T^2, one operates separately on the rows and then on the columns as for 1D signal analysis. Thus, for each scale j, we obtain a lattice W_j of size 4^{R-j}, where $R = ln(T)/ln(2)$. The set of scales $\{W_j\}_{0 \leq j \leq R}$ composes the pyramid $\mathcal{W}_i$ of details coefficients corresponding to $\mathcal{F}_i$. Combining all pyramids of the N input bands leads to a unique Vectorial Pyramid (VP) $\mathcal{W}$, where detail coefficients, $w_j^1(k), \cdots, w_j^N(k)$, at the space location k and the scale j form a unique multidimensional vector $\mathcal{W}_{jk}$. The objective of image fusion is to combine this VP in a single composite pyramid for which the reconstruction gives a unique image more suitable for human perception and giving an efficient summary of the whole information existing in all input images.

To simplify notations, we adopt the following convention. As the VP pyramid is naturally partitioned into *scales*, VP= $S = S_0 \cup S_1 \ldots \cup S_R$, from the highest

---

[1] Thi procedure is initialised by choosing c_0 equal to the original signal.

to the lowest resolution, the coefficient vector $\mathcal{W}_{jk}$ at the scale j and the spatial location k is denoted $\mathcal{W}_s$, where $s \in S_j$. Therefore, we note:

$$\mathcal{W}_s = \mathcal{W}_{jk} \equiv (w_s^1, \cdots, w_s^N) = (w_j^1(k), \cdots, w_j^N(k)) \quad (1)$$

The detail coefficients of a single LP exhibit some well known properties [9,12], namely the non gaussianity, persistence and local correlation. In addition, the VP exhibits an inter-band correlation between the details coefficients belonging to the same vector in the VP., *i.e.*, coefficients from the LPs of different bands and situated at the same scale, and the same spacial location. In the next section the regularised MPPCA model is detailed and linked to the properties of VP.

3 The Regularised MPPCA Model

The PPCA model. The PPCA is a statistical formulation of the well known PCA [13]. It links each $N \times 1$ observed vector $\mathcal{W}_s$ to a $q \times 1$ latent vector $\mathbf{t}_s$, $q < N$, as follows:

$$\mathcal{W}_s = A\mathbf{t}_s + \mu + \epsilon \quad (2)$$

where A is a $N \times q$ matrix, μ the observed data mean and ϵ is an isotropic Gaussian noise, *i.e.*, $\mathcal{N}(0, \sigma^2 I)$, I being the $N \times N$ identity matrix. Thus, the probability distribution of $\mathcal{W}_s$ given $\mathbf{t}_s$ is Gaussian $\mathcal{N}(\mathcal{W}_s; A\mathbf{t}_s + \mu, \sigma^2 I)$.

Using a Gaussian *prior*, $\mathcal{N}(\mathbf{t}_s; 0, I)$, for $\mathbf{t}_s$, the marginal distribution of $\mathcal{W}_s$ is the Gaussian $\mathcal{N}(\mathcal{W}_s; \mu, C)$, where $C = \sigma^2 I + AA^t$ is an $N \times N$ matrix [14]. Bayes rule gives the *a posteriori* probability of $\mathbf{t}_s$ [14]:

$$p(\mathbf{t}_s|\mathcal{W}_s) = \mathcal{N}(\mathbf{t}_s; M^{-1}A^t(\mathcal{W}_s - \mu), M^{-1}) \quad (3)$$

where $M = \sigma^2 I - A^t A$ is a $q \times q$ matrix.

The maximization of the data log-likelihood for the scale S^j, $\mathcal{L} = \sum_{s \in S^j} \ln\{p(\mathcal{W}_s)\}$ gives the following parameter estimators [14]:

$$\hat{\mu} = \frac{\sum_{s \in S^j} \mathcal{W}_s}{|S^j|} \; ; \; \hat{\sigma}^2 = \frac{1}{D-q} \sum_{j=q+1}^{D} \lambda_j \; ; \; \hat{A} = U_q (\Lambda_q - \sigma^2 I)^{\frac{1}{2}} R. \quad (4)$$

where λ_j are the eigenvalues of the data covariance matrix, $\Sigma = \frac{1}{|S^j|} \sum_{s \in S^j} (\mathcal{W}_s - \mu)(\mathcal{W}_s - \mu)^t$, given in descending order ($\lambda_1 \geq \cdots \geq \lambda_q$), Λ_q is a diagonal matrix of the q largest eigenvalues, U_q the matrix of the corresponding eigenvectors, and R is an arbitrary orthogonal rotation matrix.

The regularised MPPCA. The mixture of Probabilistic Principal Component Analyzers (MPPCA) [14] was introduced to model complex data structures as a mixture of local PPCAs. For a K component MPPCA, the distribution of the observations is $P(\mathcal{W}_s) = \sum_{i=1}^{K} \Pi_i P(\mathcal{W}_s|X_s = \gamma_i)$, where the local PPCA corresponding to the cluster γ_i is characterized by the mean μ_i, the variance σ_i^2, the projection matrix A_i and the prior Π_i. The label variable X_s indicates to which cluster the observation vector $\mathcal{W}_s$ belongs. Note that in this formulation the *prior* Π_i is not informative about the location of s.

The EM algorithm is used to iteratively estimate the mixture parameters [14]:

$$\hat{\Pi}_i = \frac{\sum_{s \in S^j} \Upsilon_{si}}{|S^j|}; \quad \hat{\mu}_i = \frac{\sum_{s \in S^j} \Upsilon_{si} \mathcal{W}_s}{\sum_{s \in S^j} \Upsilon_{si}}, \quad (5)$$

where $\Upsilon_{si} = P(X_s = \gamma_i | \mathcal{W}_s) = \frac{P(\mathcal{W}_s | X_s = \gamma_i) \Pi_i}{P(\mathcal{W}_s)}$; $\hat{A}_i$ and $\hat{\sigma}_i^2$ are given, in the same way of Eq. 4, by eigen decomposition of the *a posteriori* responsibility-weighted covariance matrix:

$$\Sigma_i = \frac{\sum_{s \in S^j} \Upsilon_{si} (\mathcal{W}_s - \hat{\mu}_i)(\mathcal{W}_n - \hat{\mu}_i)^t}{\sum_{s \in S^j} \Upsilon_{si}}. \quad (6)$$

The use of the MPPCA model in each scale of the VP deals with the non gaussianity and the inter-band correlation properties. To model persistence and local correlation, we change the non informative *prior* by a Hidden Markov Tree (HMT). The observation probability becomes:

$$P(\mathcal{W}_s) = \sum_{i=1}^{K} P(X_s = \gamma_i) P(\mathcal{W}_s | X_s = \gamma_i), \quad (7)$$

where $X = \{X_s; s \in S\}$ is a HMT and each class γ_i is spanned by a local PPCA. The likelihood of $\mathcal{W}_s$, $s \in S^j$, wrt the local PPCA corresponding to the class γ_i, $P(\mathcal{W}_s | X_s = \gamma_i)$, is given by $\mathcal{N}(\mathcal{W}_s; \mu_i, C_i)$, where the matrix C_i is obtained in similar manner to Eq. 4 by eigen-decomposition of the weighted covariance matrix:

$$\Sigma_i = \frac{\sum_{s \in S^j} P(X_s = \gamma_i | \mathcal{W})(\mathcal{W}_s - \hat{\mu}_i)(\mathcal{W}_s - \hat{\mu}_i)^t}{\sum_{s \in S^j} P(X_s = \gamma_i | \mathcal{W})}; \quad (8)$$

with $\hat{\mu}_i = \frac{\sum_{s \in S^j} P(X_s = \gamma_i | \mathcal{W}) \mathcal{W}_s}{\sum_{s \in S^j} P(X_s = \gamma_i | \mathcal{W})}$.

The HMT a priori. The HMT is probabilistic model that define a set of statistical independencies in order to facilitate the computation of the joint distribution of the involved variables, *i.e.*, the observation vectors $\mathcal{W}_s$ and their labels X_s, $s \in S$, also called hidden states (HS). Each $\mathcal{W}_s$ has a unique HS X_s. Each X_s, apart from X_r, has a unique direct *parent* X_{s^-}, on the path to the root. Every node X_s, apart from the *leaves* (*i.e.*, the terminal ones), has four.

The hidden state X_s takes two possible values : $\gamma_2 = 1$ if $\mathcal{W}_s$ is significant(*i.e.*, with large magnitude) and $\gamma_1 = 0$ otherwise. Given the HMT model, the vector $\mathcal{W}_s$ is independent from all the remaining variables conditionally to its hidden state X_s, *i.e.*, $P(\mathcal{W}_s | X) = P(\mathcal{W}_s | X_s)$. Thus, the likelihood of each vector is expressed as a mixture of two densities as follows:

$$P(\mathcal{W}_s) = \sum_{i=1}^{2} P(\mathcal{W}_s | X_s = \gamma_i) P(X_s = \gamma_i). \quad (9)$$

This actually captures the non gaussianity property of the detail coefficient distribution. In practice, the observations are introduced on J scales in the quadtree, for the other scale the likelihood is set to be equal to 1 : $\forall\ j > J$, $\forall\ \mathcal{W}_s \in S_n$, $\forall\ i$: $f_{i,j}(\mathcal{W}_s) = 1$.

To deal with the inter-scale dependencies, each hidden state is assumed to be independent from all its ascendants given its *parent*, therefore the joint probability of the hidden states becomes [15,9]:$P(X) = P(X_r) \prod_{s \neq r} P(X_s|X_{s-})$, where X_r is the root node. It is important to note that since each four neighbor hidden states share the same parent at the next higher scale, the HMT allows also the capture of local inter-scale correlation.

From the assumptions above, it can be easily established that the joint distribution $P(X, \mathcal{W})$ can be expressed as [15,9]:

$$P(X,\mathcal{W}) = P(X_r) \prod_{s \in S, s \neq r} P(X_s|X_{s-}) \prod_{s \in S} P(\mathcal{W}_s|X_s), \qquad (10)$$

where $P(X_r)$ stands for the *a priori* and $P(X_s|X_{s-})$ stands for the *parent* to *child* transition probability.

This formulation of $P(X, \mathcal{W})$ is interesting as it allows to compute exactly and efficiently $P(X_s = \gamma_i|\mathcal{W})$, $s \in S$, the Marginal *a posteriori* probability of the hidden state X_s to be equal to i given all the coefficients in the VP $\mathcal{W}$ by alternating two passes on the HMT [16]

4 The Proposed Image Fusion Algorithm

The proposed approach is illustrated in (Fig.1) for the case of two input bands. Each detail vector $\mathcal{W}_s$, $s \in S^j$, $j = 0, \cdots, J$, is modeled using a regularised MMPCA (Eq. 9), where each local PPCA defines a local mapping (Eq. 2) on a lower dimensionality space. Therefore, a natural way for fusing the VP is then to infer the value of the hidden vector $\hat{\mathbf{t}}_s$ associated with the local PPCAs, for each $\mathcal{W}_s$, $s \in S^j$, $j = 0, \cdots, J$. The best estimate $\hat{\mathbf{t}}_s^i$ of $\mathbf{t}_s$ given the ith PPCA and the detail vector $\mathcal{W}_s$ is the mean of the a posteriori distribution of $\mathbf{t}_s$ given this local PPCA (Eq. 3). In addition, due to the fact that projection axis could be oriented in the N-dimensional space which may cause an undesired artefact in the output, we constrain the the sign of all elements of the projection vectors to be positive [7]. Accordingly, the global estimate given the regularised MMPCA is computed as follows:

$$\hat{\mathbf{t}}_s = \sum_{i=1}^{2} \|M_i^{-1} A_i^t\|(\mathcal{W}_s - \mu_i) P(X_s = \gamma_i|\mathcal{W}) \qquad (11)$$

where $M_i = \sigma_i^2 I - A_i^t A_i$ and $\|.\|$ means that the elements of the matrix are turned to positive. We keep only the first element in $\hat{\mathbf{t}}_s$ corresponding to the PPCA axis with the largest eigenvalue. This is a reasonable operation since the representation in the LP domain is more parsimonious, especially in the case of two input images. This was confirmed by our experiments.

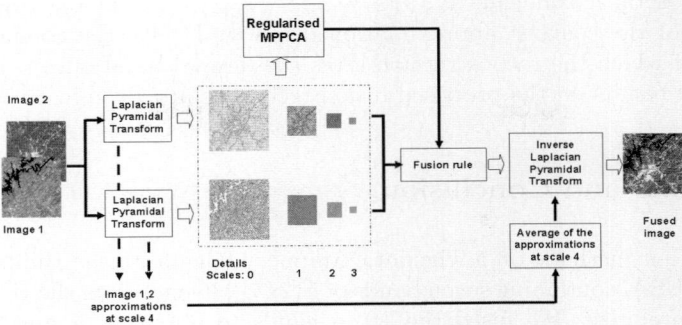

Fig. 1. A flowchart of the image fusion algorithm. The Laplacian Pyramid analyzes the $N = 2$ input images separately (on the left). This leads to a multiscale pyramid of detail coefficients for each band, up to a fixed scale J, $J = 4$ in this example. Then, all pyramids are converted to a unique Vectorial Pyramid (VP) which is modeled using the Regularised MPPCA Model. The *a posteriori* probability of the local PPCA's is used to fuse the VP, and the resulting pyramid, with the average of coarsest approximations are used to reconstruct the fused output image.

In addition, to be more robust against additive noise, we consider that each observed details coefficient is the sum of a noise free (*i.e.*, true) coefficient and a gaussian noise. Thus, we operate a soft shrinking of the observed coefficients [17] before the fusion using Eq. 11. Each shrinked coefficient is given by [17]:$\tilde{\mathcal{W}}_s = \frac{\sigma_s}{\sigma_s + \sigma_n}\mathcal{W}_s$, where σ_s is the standard deviation of the signal (noise free detail coefficient) and σ_n is the standard deviation of the additive noise which is assumed to be gaussian. This estimator of the noise free detail coefficient $\tilde{\mathcal{W}}_s$, $s \in S^j$, $j = 0, \cdots, J$, is called *oracle* since the variances of the noise and the signal are usually unknown as the observation is the addition of the signal and the noise. In the context of the regularised MPPCA model (see the PPCA model in Eq. 2) we propose empirical estimations of these standard deviations. We choose σ_s as the variance of the first axis of the corresponding PPCA, which is equal to the square root of the largest eigenvalue $\sqrt{\lambda_1^i}$. The noise variance σ_n is given by the sum of the variances of the remaining axis and the variance σ_i of the additive noise in the local PPCA model used in the projection. Therefore, the noise free coefficient given the ith local PPCA is:

$$\tilde{\mathcal{W}}_s = \frac{\sqrt{\lambda_1^i}}{\sigma_i + \sum_{j=1}^{q}\sqrt{\lambda_j^i}}\mathcal{W}_s. \qquad (12)$$

The new fusion formula using the noise free detail coefficients is then:

$$\tilde{\mathbf{t}}_s = \sum_{i=1}^{2}\frac{\sqrt{\lambda_1^i}}{\sigma_i + \sum_{j=1}^{q}\sqrt{\lambda_j^i}}\|M_i^{-1}A_i^t\|(\mathcal{W}_s - \mu_i)P(X_s = \gamma_i|\mathcal{W}). \qquad (13)$$

The fused detail coefficients $\tilde{t}_s$, $s \in S^j$, $j = 0, \cdots, J$, with the results of the averaging of the coarsest approximation (see Fig. 1) give the composite fused pyramid of which the reconstruction gives the desired fused output image. Experimental results on the proposed fusion technique are given in next section.

5 Results and Conclusion

We tested our methode on a Thematic Mapper image from the Huntsville area, Alabama, USA, containing seven bands of 512x512 images from the U.S. Landsat series of satellites[2]. We fused the seven bands to obtain a unique fused gray level image. We compared the fusion performances of our method, the simple averaging, and the local and global PCA and SOM in [7]. Visually the results of our method and the average look much better than the results in [7], with a slight advantage to our method, Fig. 2. The superiority of our fusion to the result in [7], even though both use collection of linear projections (especially for the case of local PCAs), could be explained be explained by the fact that we perform our fusion in the MA domain where the representation is more compact while this is done in the gray level (GL) domain in [7]. Therefore the first eigenvalue of the PCA is more predominant in the MA domain than the GL domain, and the projection on the first axis gives less error with our technique. The other aspect of this comparison is the unexpected well performance of the averaging. This could be explained by the high redundancy of the information enclosed in the seven bands. However, the superiority of our fusion is demonstrated in the presence of noise. The last two images in Fig. 2 show the better performance of our method in presence of noise. To avoid that the noise fits perfectly to the Regularised PPCA model which is the core of our method, we simulate the noise as the addition of Gaussain and uniform noises with the same energy. The corresponding Signal to Noise Ratio (SNR) is $5db$. Finally we used the average cross-entropy (CE) between the fused images and the inputs [1] as an objective measure to compare the results. The less the average, the better the results. For the noise free test, we obtain a CE equal to 1.23 with our method while the CE is 1.27 for the result of the averaging. We compute the CE of both techniques in the case of noised inputs. The CE for our method is 1.66 with our algorithm and 1.94 with the averaging. Thus, the objective measure confirms the performance of our proposed algorithm.

In conclusion, an image fusion algorithm, based on the use of a Hidden Markov Tree regularisation of the mixture of PPCA modeling the detail coefficients of the Laplacian Pyramid transform, was presented. This model accounts of all kind of correlation of the input images and allows a Bayesian inference of the fusion image. Results on Landsat images, evaluated using both subjective and objective measures, showed the effectiveness of the proposed algorithm and its robustness.

---

[2] The authors would like to thank Professor P. Scheunders for providing the the Landsat image with seven bands.

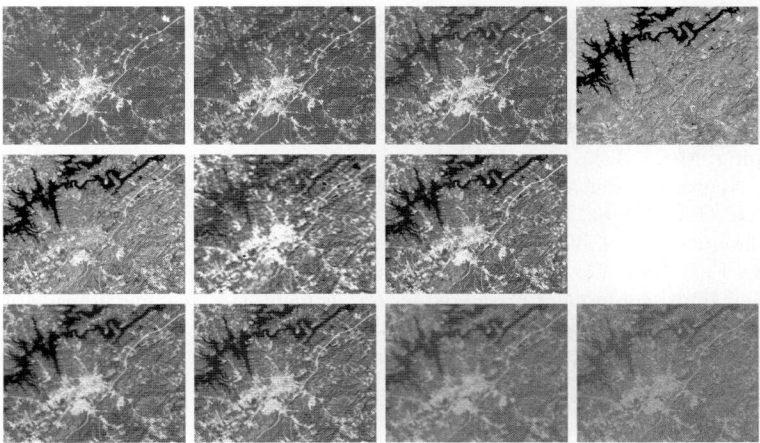

Fig. 2. Landsat remote sensing images. From the top left to bottom right, the seven originals original bands, the fusion result using the proposed method and the fusion result using the mean. The two last images are the results of the fusion after adding noise (SNR 5*db*) to all seven bands using our method and the averaging, respectively.

References

1. Blum, R., Liu, Z. (eds.): Multi-Sensor Image Fusion and Its Applications. CRC Press, Taylor & Francis Group (2006)
2. Rockinger, O., Fechner, T.: Pixel-level image fusion: the case of image sequences. In: Proc. SPIE, vol. 3374, pp. 378–388 (1998)
3. Therrien, C.W., Krebs, W.K.: An adaptive technique for the enhanced fusion of low-light visible with uncooled thermal infrared imagery. In: IEEE Int. Conf. on Image Processing, Santa Barbara, CA, pp. 405–408 (1997)
4. Sharma, R.K., Pavel, M., Leen, T.K.: Bayesian sensor image fusion using local linear generative models. Opt. Eng. 40(7), 1364–1376 (2001)
5. Broussard, R.P., Rogers, S.K., Oxley, M.E., Tarr, G.L.: Physiologically motivated image fusion for object detection using a pulse coupled neural network. IEEE Trans. Neural Network 10(3), 554–563 (1999)
6. Choi, M.: A new intensity-hue-saturation fusion approach to image fusion with a tradeoff parameter. IEEE Trans. on Geoscience and Remote Sensing 44, 1672–1682 (2006)
7. Scheunders, P.: Multispectral image fusion using local mapping techniques. In: Proc. ICPR 2000, Int. Conf. on Pattern Recognition, Barcelona, Spain, September 3-7, pp. 311–314 (2000)
8. Flitti, F., Collet, C., Slezak, E.: Wavelet domain astronomical multiband image fusion and restoration using Markov quadtree and copulas. In: Proc. EUSIPCO, Antalya, Turkey (2005)
9. Crouse, M., Nowak, R., Baraniuk, R.: Wavelet-based statistical signal processing using hidden markov models. IEEE Trans. Image Processing 46 (1998)
10. Yang, J., Blum, R.S.: A Statistical Signal Processing Approach to Image Fusion using Hidden Markov Models. In: Multi-Sensor Image Fusion and Its Applications, pp. 256–287. CRC Press, Taylor & Francis Group (2006)

11. Burt, P.J., Adelson, E.H.: The laplacian pyramid as a compact image code. IEEE Trans. Communications 31, 532–540 (1983)
12. Mallat, S.: A wavelet tour of signal processing. Academic Press, London (1998)
13. Tipping, M.E., Bishop, C.: Probabilistic principal component analysis. Journal of the Royal Statistic Society Series B, 61 (Part 3), 611–622
14. Tipping, M.E., Bishop, C.: Mixtures of Probabilistic Principal Component Analysers. Neural Computation, 443–482 (1999)
15. Laferté, J.M., Pérez, P., Heitz, F.: Discrete markov image modeling and inference on the quad-tree 9, 390–404 (2000)
16. Flitti, F., Collet, C.: Markovian regularization of latent-variable-models mixture for new multi-component image reduction/segmentation scheme. Journal of Signal, Image and Video Processing 1, 191–201 (2007)
17. Donoho, D.L., Johnstone, I.: Ideal spatial adaptation by wavelet shrinkage. Biometrika 81, 425–455 (1994)

A Riemannian Scalar Measure for Diffusion Tensor Images

Andrea Fuster[1], Laura Astola[2], and Luc Florack[2]

[1] Department of Biomedical Engineering, Eindhoven University of Technology,
The Netherlands
a.fuster@tue.nl
[2] Department of Mathematics and Computer Science,
Eindhoven University of Technology, The Netherlands
l.j.astola@tue.nl, l.m.j.florack@tue.nl

Abstract. We study a well-known scalar quantity in differential geometry, the Ricci scalar, in the context of Diffusion Tensor Imaging (DTI). We explore the relation between the Ricci scalar and the two most popular scalar measures in DTI: Mean Diffusivity and Fractional Anisotropy. We discuss results of computing the Ricci scalar on synthetic as well as real DTI data.

1 Introduction

Diffusion Tensor Imaging (DTI) is a magnetic resonance imaging technique that measures diffusion of water molecules in tissue in a non-invasive way [1,2,3]. It is mostly used for the study of brain white matter. DTI is used in clinical research e.g. for localization of brain tumors in relation to white matter tracts, assisting in surgical planning and in the assessment of white matter maturation in children. Information about the white matter architecture and integrity can be extracted from the so-called diffusion tensor.

Different scalar measures (scalar quantities constructed from the diffusion tensor) have been proposed in the literature, aiming to reveal information about the underlying diffusion process and tissue structure. Scalar measures are also studied to relate the state of white matter to different pathologies. The most common ones are the mean diffusivity (MD) and fractional anisotropy (FA) [4].

Mean diffusivity can be interpreted as average diffusion per imaged voxel. In general, MD is smaller in areas with organized tissue. This allows mean diffusivity images to show main white matter tracts but no detailed structure. Fractional anisotropy measures the anisotropy of the diffusion. Roughly speaking, there is more diffusion along elongated structures than across them. High FA values typically indicate the presence of such an elongated structure while low values relate to isotropic diffusion and scarcity or absence of structures. However, low FA values are also found in voxels with complicated fiber architecture such as crossings [5]. Both MD and FA give relevant information about white matter but have limitations as well. It is therefore also worth to study other scalar DTI measures which could improve the clinical utility of DTI.

In this paper we consider a well-known scalar quantity in differential geometry, the Ricci scalar, in the context of DTI. The Ricci scalar has been used in 2D image processing for curvature analysis [6]. The goal of this research is to evaluate whether the Ricci scalar can provide additional information on white matter structures w.r.t. the usual scalar measures. We found promising preliminary results on simulated and phantom data showing high negative values of the Ricci scalar at voxels with crossing structures.

2 Theory

In the DTI model diffusion is represented by the diffusion tensor D. This is a symmetric, positive definite second-order tensor in dimension three constructed from the DTI measurements. On the other hand, a metric tensor g (see, for example, [7]) is a second-order symmetric positive definite tensor field on a manifold which defines the inner product of two tangent vectors v, w as

$$\langle v, w \rangle = g_{ij} v^i w^j \tag{1}$$

The relation between a Riemannian metric and its inverse is[1]

$$g_{ik} g^{kj} = \delta_i^j \tag{2}$$

The diffusion tensor can naturally be associated to (the inverse of) a Riemannian metric tensor in dimension three [8]:

$$D^{ij} = g^{ij}, \quad i, j = 1, 2, 3 \tag{3}$$

In this way large diffusion in a certain direction corresponds to a short distance in the metric space. A number of authors have studied DTI in the Riemannian framework [9,10,11,12].

Different scalar quantities can be constructed from contractions of the metric and its curvature tensors. The simplest of those is the so-called Ricci scalar, intrinsically related to the geometry of the metric space. The Ricci scalar is given by (see, for example, [13]):

$$R = g^{ij} R_{ij} \tag{4}$$

where R_{ij} is the Ricci curvature tensor (see Appendix). Unlike the metric, the Ricci tensor is not positive definite, allowing for both positive and negative values of the Ricci scalar. This is a major difference with respect to the usual DTI scalar measures, which are always positive. In fact, the Ricci scalar in dimension two is twice the Gaussian curvature. For example, the Ricci scalar on a unit sphere is a positive quantity: $R = 2$. On the other hand, on a saddle surface the Ricci scalar is negative everywhere (except at the origin). In dimension three the Ricci scalar does not completely characterize the curvature but represents instead the average of the characterizing curvatures.

---

[1] In this paper we use Einstein's summation convention, $a_i a^i = \sum_i a_i a^i$.

2.1 Ricci Scalar and Mean Diffusivity

The Ricci scalar is related to the mean diffusivity (MD):

$$\text{MD} = \frac{1}{3}(\lambda^1 + \lambda^2 + \lambda^3) \tag{5}$$

where the λ's are the eigenvalues of the diffusion tensor. Indeed, we can write:

$$R = g^{ij} R_{ij} \stackrel{(3)}{=} D^{ij} R_{ij} = \tilde{D}^{ij} \tilde{R}_{ij} = \lambda^1 \tilde{R}_{11} + \lambda^2 \tilde{R}_{22} + \lambda^3 \tilde{R}_{33} \tag{6}$$

The last two equalities follow from the fact that R is a scalar quantity and thus independent of the used coordinate system; we choose the coordinate system in which the diffusion tensor is diagonal, and denote tensors in this coordinate system with a tilde. Comparing (6) to (5) it is clear that we can think of the Ricci scalar as a kind of curvature-weighted mean diffusivity.

2.2 Ricci Scalar and Fractional Anisotropy

The Ricci scalar is intrinsically different from the fractional anisotropy in the following sense. A zero FA value indicates (perfect) isotropic diffusion (see Fig. 1 top left) while a non-zero value indicates anisotropy to some degree, i.e., the presence of structures obstructing the diffusion process. On the other hand, a zero Ricci scalar can indicate both isotropy or anisotropy (see Fig. 1 top left and right). The same is true for a non-zero Ricci scalar (see Fig. 1 bottom left and right).

More precisely, a homogeneous isotropic diffusion corresponds to the case where

$$\lambda^1 = \lambda^2 = \lambda^3 = \lambda \tag{7}$$

In this case the fractional anisotropy

$$\text{FA} = \frac{\sqrt{3}}{\sqrt{2}} \frac{\sqrt{(\lambda^1 - \text{MD})^2 + (\lambda^2 - \text{MD})^2 + (\lambda^3 - \text{MD})^2}}{\sqrt{(\lambda^1)^2 + (\lambda^2)^2 + (\lambda^3)^2}} \tag{8}$$

is just zero. The Riemannian metric and inverse metric associated to the isotropic diffusion tensor are Euclidean:

$$g_{ij} = \frac{1}{\lambda} \begin{pmatrix} 1 & 0 & 0 \\ 0 & 1 & 0 \\ 0 & 0 & 1 \end{pmatrix}, \quad g^{ij} = \lambda \begin{pmatrix} 1 & 0 & 0 \\ 0 & 1 & 0 \\ 0 & 0 & 1 \end{pmatrix} \tag{9}$$

The Riemann tensor of a Euclidean metric is zero. It is clear from equation (10) that the Ricci tensor will also be zero, and so will be the Ricci scalar. In this way we have shown that homogeneous isotropic diffusion implies a zero Ricci scalar. However, this is not the only case where the Ricci scalar is zero. Another situation in which this happens is when the diffusion tensor field is homogeneous and anisotropic. Note that FA in this case would be different from zero.

We conclude that the Ricci scalar and fractional anisotropy certainly have a different character. The zeros (non-zeros) of the Ricci scalar cannot be always related to isotropic (anisotropic) diffusion as in the case of FA.

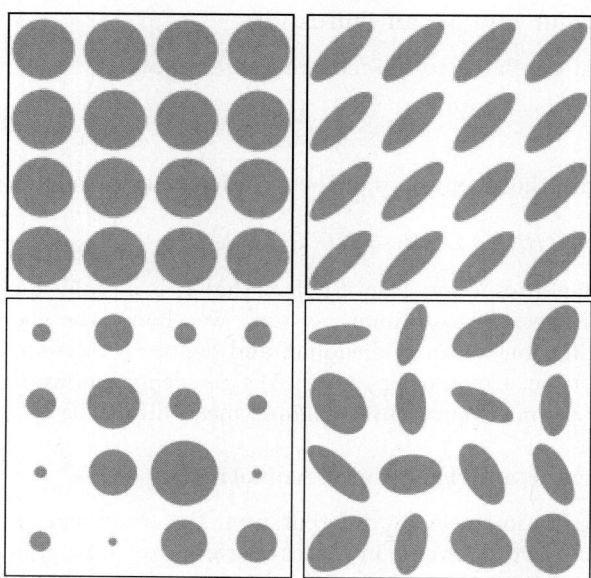

Fig. 1. Top left: Homogeneous isotropic tensor field: $R = \mathrm{FA} = 0$. Top right: Homogeneous anisotropic case: $R = 0$, $\mathrm{FA} \neq 0$. Bottom left: Inhomogeneous isotropic case: $R \neq 0$, $\mathrm{FA} = 0$. Bottom right: Inhomogeneous anisotropic case: $R \neq 0$, $\mathrm{FA} \neq 0$.

3 Experiments

In order to explore the geometric significance of the Ricci scalar, we have experimented with simulated, phantom and real data.

Simulated Data

To get a quick insight in what the Ricci scalar can detect in a tensor field, we refer to Fig. 2, where we have simulated a crossing of orthogonally oriented sets of tensors, modeling homogeneous diffusion tensors corresponding to two fiber bundles. The voxels where the Ricci scalar is non-zero are colored according to its sign. In the crossing region of this tensor field, the Ricci scalar tends to be large and negative. Since the Ricci scalar involves second order derivatives (see Appendix), the minimum size of the region to be considered depends on the scale of the Gaussian differential operator [14] [15] [16].

Phantom Data

We computed Ricci scalars on a real phantom data consisting of cylinder containing a water solution, three sets of crossing synthetic fiber bundles and three supporting pillars on the boundary. In Fig. 3 we see that in the region where the fiber bundles cross Ricci scalars have relatively large negative values, despite

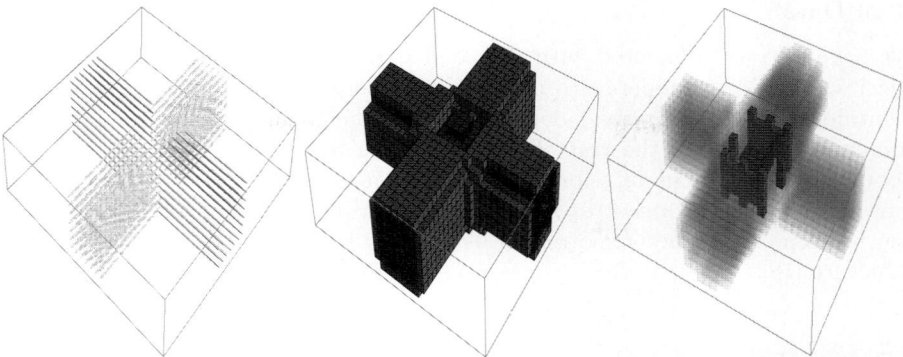

Fig. 2. Left: A simulated crossing of tensors. Middle: Ricci scalars computed on the tensor field, blue for negative- and red for positive values. Right: Voxels where the absolute value of Ricci scalar is the highest.

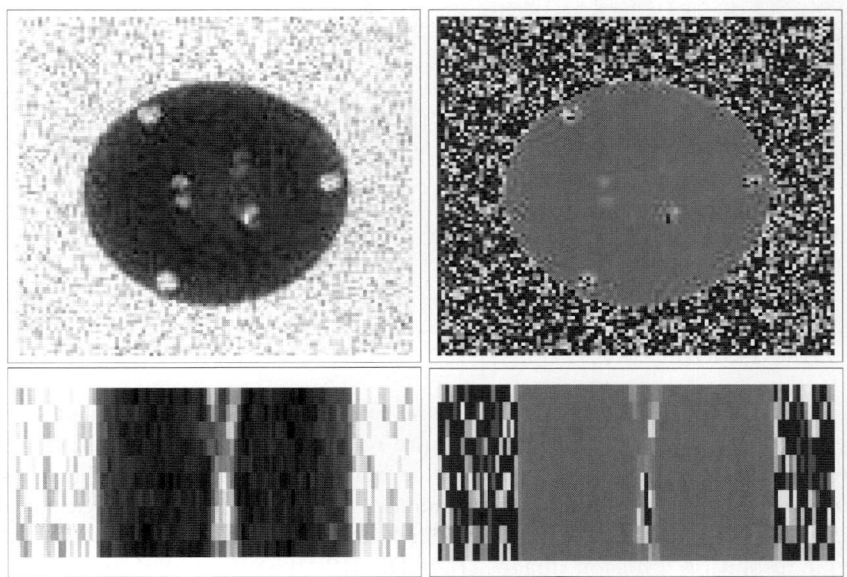

Fig. 3. Top left: Mean diffusivity on a trans-axial slice of the cylinder. Top right: A temperature map of Ricci scalars on the same slice. Blue indicates negative values. Bottom left: Mean diffusivity on an axial slice containing crossing fiber bundles. Bottom right: Ricci scalars on the same slice. Large negative values found in the crossing.

of the noisy nature of the DTI-data. This might be explained by the fact that crossings can be related to saddle shaped structures [17], for which the Ricci scalar has a negative value.

Real Data

We have also experimented with real DTI data of a rat brain. We plotted the Ricci scalars in a temperature map, to emphasize the differences in sign. We identified positive (negative) outliers of the Ricci scalar data with maximum (minimum) values of the rest of the data. The Ricci scalar gives information about the variations in diffusion tensor orientations unlike FA, which will identify tensors with similar anisotropy even though their orientation may differ. This can be seen e.g. in the boxed region in Fig. 4, which is known to have complex structure [18].

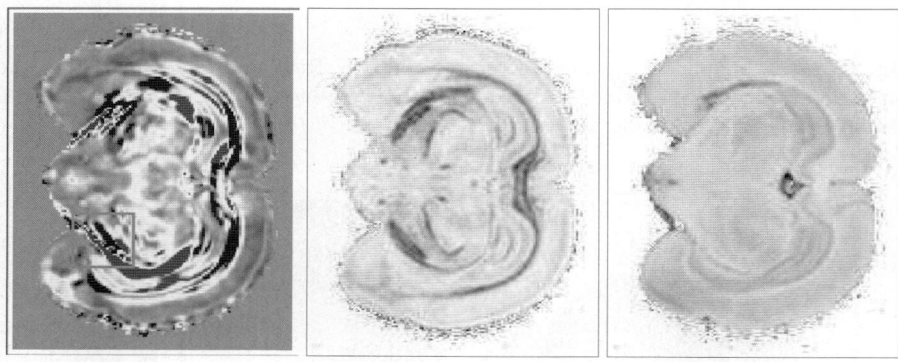

Fig. 4. Left: Ricci scalars on a slice of the rat brain DTI image. Middle: Fractional anisotropy. Right: Mean diffusivity.

4 Discussion

The Ricci scalar on real DTI images show rough structures in a similar way to FA and MD. The more complex curvature related information cannot be fully appreciated with a 2D visualization, since both the Ricci scalar and DTI data are intrinsically 3D. Work in progress includes the integration of the Ricci scalar into a 3D DTI visualization toolkit, which can also render the so-called high angular resolution diffusion images (HARDI) [19]. The simulated and phantom data images show large negative values of the Ricci scalar at fiber crossings. The theoretical explanation for this fact should be refined. Therefore, more work is needed to investigate whether the Ricci scalar can be used in the detection of fiber crossings in DTI data. If this indeed is the case, besides in fiber tracking it would be helpful for voxel classification [20], where the regions with single orientation could be identified as the regular DTI data (second order tensors), and higher order models (e.g. fourth order tensors) could be used in regions where inhomogeneous fiber population is anticipated. It could also be useful in the so-called splitting tracking method in HARDI framework [21], by indicating the potential bifurcation points of fiber bundles with large negative values.

Acknowledgements

L. Astola and L. Florack gratefully acknowledge The Netherlands Organization for Scientific Research (NWO) for financial support. We also want to thank Ellen Brunenberg at Eindhoven University of Technology for providing the rat HARDI data and Pim Pullens at Maastricht University for providing the phantom HARDI data.

References

1. Basser, P., Mattiello, J., Turner, R., Le Bihan, D.: Diffusion tensor echo-planar imaging of human brain. In: Proceedings of the SMRM, p. 584 (1993)
2. Basser, P.J., Mattiello, J., Lebihan, D.: Estimation of the effective self-diffusion tensor from the NMR spin echo. Journal of Magnetic Resonance Series B 103, 247–254 (1994)
3. Basser, P.J., Mattiello, J., Lebihan, D.: MR diffusion tensor spectroscopy and imaging. Biophysical Journal 66, 259–267 (1994)
4. Basser, P., Pierpaoli, C.: Microstructural and physiological features of tissues elucidated by quantitative-diffusion-tensor MRI. Journal of Magnetic Resonance. Series B 111(3), 209–219 (1996)
5. Özarslan, E., Vemuri, B.C., Mareci, T.H.: Generalized scalar measures for diffusion MRI using trace, variance, and entropy. Magn. Reson. Med. 53(4), 866–876 (2005)
6. Saucan, E., Appleboim, E., Wolanski, G., Zeevi, Y.: Combinatorial ricci curvature for image processing. Presented at MICCAI 2008 Workshop Manifolds in Medical Imaging: Metrics, Learning and Beyond (2008)
7. Nakahara, M.: Geometry, Topology and Physics, 2nd edn. Graduate Student Series in Physics. Taylor & Francis, Abington (2003)
8. de Lara, M.C.: Geometric and symmetry properties of a nondegenerate diffusion process. The Annals of Probability 23(4), 1557–1604 (1995)
9. O'Donnell, L., Haker, S., Westin, C.-F.: New approaches to estimation of white matter connectivity in diffusion tensor MRI: Elliptic pDEs and geodesics in a tensor-warped space. In: Dohi, T., Kikinis, R. (eds.) MICCAI 2002. LNCS, vol. 2488, pp. 459–466. Springer, Heidelberg (2002)
10. Lenglet, C., Deriche, R., Faugeras, O.: Inferring white matter geometry from diffusion tensor MRI: Application to connectivity mapping. In: Pajdla, T., Matas, J.G. (eds.) ECCV 2004. LNCS, vol. 3024, pp. 127–140. Springer, Heidelberg (2004)
11. Prados, E., Lenglet, C., Pons, J.P., Wotawa, N., Deriche, R., Faugeras, O., Soatto, S.: Control theory and fast marching techniques for brain connectivity mapping. In: Proceedings of the IEEE Conference on Computer Vision and Pattern Recognition, vol. 1, pp. 1076–1083. IEEE, New York (2006)
12. Astola, L., Florack, L.M.J., ter Haar Romeny, B.M.: Measures for pathway analysis in brain white matter using diffusion tensor images. In: Karssemeijer, N., Lelieveldt, B. (eds.) IPMI 2007. LNCS, vol. 4584, pp. 642–649. Springer, Heidelberg (2007)
13. Carroll, S.: Spacetime and Geometry: An Introduction to General Relativity. Addison-Wesley, Reading (2003)
14. Florack, L., Astola, L.: A multi-resolution framework for diffusion tensor images. In: CVPR Workshop on Tensors in Image Processing and Computer Vision, Anchorage, Alaska, The United States. CVPR, vol. 20, Springer, Heidelberg (2008)

15. ter Haar Romeny, B.M.: Front-End Vision and Multi-Scale Image Analysis: Multi-Scale Computer Vision Theory and Applications, written in Mathematica, vol. 27. Kluwer Academic Publisher, Dordrecht (2003)
16. Florack, L.M.J.: Image Structure, vol. 10. Kluwer Academic Publisher, Dordrecht (1997)
17. Faas, F., Van Vliet, L.: A crossing detector based on the structure tensor. In: Blanc-Talon, J., Philips, W., Popescu, D., Scheunders, P. (eds.) ACIVS 2007. LNCS, vol. 4678, pp. 212–220. Springer, Heidelberg (2007)
18. Brunenberg, E., Prckovska, V., Platel, B., Strijkers, G., ter Haar Romeny, B.M.: Untangling a fiber bundle knot: Preliminary results on STN connectivity using DTI and HARDI on rat brains. In: Proceedings of the 17th Meeting of the International Society for Magnetic Resonance in Medicine (ISMRM), Honolulu, Hawaii (2009)
19. Tuch, D., Reese, T., Wiegell, M., Makris, N., Belliveau, J., Wedeen, V.: High angular resolution diffusion imaging reveals intravoxel white matter fiber heterogeneity. Magnetic Resonance in Medicine 48(4), 577–582 (2002)
20. Alexander, D., Barker, G., Arridge, S.: Detection and modeling of non-gaussian apparent diffusion coefficient profiles in human brain data. Magnetic Resonance in Medicine 48, 331–340 (2002)
21. Deriche, R., Descoteaux, M.: Splitting tracking through crossing fibers: Multidirectional Q-ball tracking. In: Proceedings of the 4th International Symposium on Biomedical Imaging: From Nano to Macro (ISBI 2007), Arlington, Virginia, USA (2007)

Appendix

The Ricci tensor is a symmetric second-order tensor given by:

$$R_{ij} = R^k{}_{ikj} = \frac{\partial \Gamma^k_{ij}}{\partial x^k} - \frac{\partial \Gamma^k_{ik}}{\partial x^j} + \Gamma^k_{kl}\Gamma^l_{ij} - \Gamma^k_{jl}\Gamma^l_{ik} \qquad (10)$$

where $R^i{}_{jkl}$ is the Riemann tensor and the Γ's are the Christoffel symbols:

$$\Gamma^i_{jk} = \frac{1}{2}g^{il}(\partial_k g_{lj} + \partial_j g_{lk} - \partial_l g_{jk}) \qquad (11)$$

Structure-Preserving Smoothing of Biomedical Images

Debora Gil[1], Aura Hernàndez-Sabaté[1], Mireia Burnat[2], Steven Jansen[3], and Jordi Martínez-Villalta[2]

[1] Computer Vision Center, Computer Science Department, UAB, Bellaterra, Spain
[2] CREAF, Campus UAB, Bellaterra, Spain
[3] Institute of Systematic Botany and Ecology, Ulm University, Germany
{debora,aura}@cvc.uab.es

Abstract. Smoothing of biomedical images should preserve gray-level transitions between adjacent tissues, while restoring contours consistent with anatomical structures. Anisotropic diffusion operators are based on image appearance discontinuities (either local or contextual) and might fail at weak inter-tissue transitions. Meanwhile, the output of block-wise and morphological operations is prone to present a block structure due to the shape and size of the considered pixel neighborhood.

In this contribution, we use differential geometry concepts to define a diffusion operator that restricts to image consistent level-sets. In this manner, the final state is a non-uniform intensity image presenting homogeneous inter-tissue transitions along anatomical structures, while smoothing intra-structure texture. Experiments on different types of medical images (magnetic resonance, computerized tomography) illustrate its benefit on a further process (such as segmentation) of images.

Keywords: non-linear smoothing, differential geometry, anatomical structures segmentation, cardiac magnetic resonance, computerized tomography.

1 Introduction

By the sensitivity of medical imaging scanners to tissue physical and chemical properties, the appearance of anatomical structures should be uniform. However, the presence of radiological noise (among other artifacts) disturbs structures homogeneity and suggests an image smoothing before further segmentation of anatomical structures. Medical imaging smoothing should homogenize the intensity inside anatomical structures, while preserving intensity changes at their boundaries without altering their shape. Existing smoothing methods for preserving image features (edges and corners) might be grouped into block-wise and differential operators.

Block-wise operators (like median [1], morphological [1], mean shift [2] or Kuwahara inspired [3]) replace the pixel intensity by a function (usually statistical [2,3]) of neighboring values. Since they can be related to image level-sets evolution (rather than image intensity evolution) they naturally preserve contrast changes. The counterpart is that evolution of image contours alters their shape. Contours in filtered images deform according to the shape of the structure element defining the pixel neighborhood. In many cases [3,1], even the smoothed image might present a block-wise appearance congruent with the shape of such structure element.

Non-linear anisotropic filtering methods [4,5,6] use the formulation of heat diffusion to evolve image intensity. Operators are designed to slow down diffusion across structures and features of interest, which are determined by measures of image appearance discontinuity. Common trends are either the norm of image derivatives (1st order for edges [4] and 2nd order for ridges [6]) or global contextual discontinuities [5]. In order to ensure stability of the diffusion process [7], heat diffuses on the whole image plane, which implies convergence to a uniform intensity image [4,7]. This fact forces adding close-to-data constraints or relying on a given number of iterations (termination problem) to ensure preservation of the image most relevant features [4,8].

In this paper we introduce a differential operator, the Structure-Preserving Diffusion, SPD, which restricts diffusion to a smooth approximation of image contours. Differential geometry arguments [9] ensure stability of the diffusion process. A main contribution is that SPD homogenizes gray-level along regular image contours without altering their shape. In this manner, SPD converges (i.e. the iterative scheme stabilizes) towards a non-uniform image presenting a uniform gray-level inside anatomical structures, while preserving inter tissue transitions.

2 Structure-Preserving Diffusion

Solutions to the heat diffusion equation with initial condition a given image, $I_0(x, y)$, provide a time (scale) dependant family, $I(x, y, t)$, of smoothed versions of $I_0(x, y)$. Heat diffusion is given in divergence form as:

$$I_t(x, y, t) = \mathrm{div}(J\nabla I), \quad I(x, y, 0) = I_0(x, y) \tag{1}$$

where $\nabla I = (I_x, I_y)$ is the image gradient, div is the divergence operator and J is a 2-dimensional symmetric (semi) positive defined tensor that locally describes the way gray level re distributes.

Any symmetric matrix, considered as linear map, diagonalizes [10] in an orthonormal basis:

$$J = Q\Lambda Q^t = \begin{pmatrix} \xi_1 & -\xi_2 \\ \xi_2 & \xi_1 \end{pmatrix} \begin{pmatrix} \lambda_1 & 0 \\ 0 & \lambda_2 \end{pmatrix} \begin{pmatrix} \xi_1 & \xi_1 \\ -\xi_2 & \xi_2 \end{pmatrix}$$

for $\lambda_1 > \lambda_2 >= 0$, J eigenvalues and $\xi = (\xi_1, \xi_2)$ and its perpendicular $\xi^\perp = (-\xi_2, \xi_1)$ J eigenvectors. Symmetric semi-positive defined tensors define a metric in Euclidean space. The unitary vectors associated to the metric are an ellipse with axis of length λ_1, λ_2 oriented along $\xi, \xi^\perp$. The shape of such ellipse describes the preferred diffusion of heat. In this sense, we can talk about isotropic diffusion (equal eigenvalues) and anisotropic diffusion (distinct and strictly positive eigenvalues). By general theory of partial differential equations [7], equation (1) has a unique solution provided that λ_1, λ_2 do not vanish. However, in such case, $I(x,y,t)$ converges to a constant image [4], so that the diffusion time (iterations in numeric implementations) is a critical issue for restoring an image preserving meaningful structures (termination problem [5]).

In [9], the authors showed that, for null eigenvalues, existence and uniqueness of solutions to (1) is guaranteed as long as the eigenvector of positive eigenvalue defines a differentiable curve. In this case, J represents the projection matrix onto the positive eigenvector and diffusion restricts to its integral curves. It follows that $I(x,y,t)$ converges towards a collection of curves of uniform gray level [9], so that the iterative scheme stabilizes at a non-uniform intensity image. Levelcurves of the steady state approximate the original image contours, provided that the positive eigenvector represents their tangent space.

The second moment matrix [11] or structure tensor [12] provides a good description of local image structures. The structure tensor matrix describes the gradient distribution in a local neighborhood of each pixel by averaging the projection matrices onto the image gradient:

$$ST(\rho,\sigma) = g_\rho * \left[\begin{pmatrix} I_x(\sigma) \\ I_y(\sigma) \end{pmatrix} (I_x(\sigma), I_y(\sigma)) \right] = \begin{pmatrix} g_\rho * I_x^2(\sigma) & g_\rho * I_x(\sigma)I_y(\sigma) \\ g_\rho * I_x(\sigma)I_y(\sigma) & g_\rho * I_y^2(\sigma) \end{pmatrix}$$

Image derivatives are computed using gaussian kernels, g_σ, of variance σ (differentiation scale):

$$I_x(\sigma) = g(\sigma)_x * I \text{ and } I_y(\sigma) = g(\sigma)_y * I$$

The projection matrix onto $(I_x(\sigma), I_y(\sigma))$ is averaged using a gaussian of variance ρ (integration scale). Since $ST(\rho,\sigma)$ is the solution to the heat equation with initial condition the projection matrix, its eigenvectors are differentiable (smooth) vector fields that represent image level sets normal (principal eigenvector, ξ) and tangent (secondary eigenvector, $\xi^\perp$) spaces. In the absence of corners (like anatomical contours in bottom right image in fig.1), the vector $\xi^\perp$ is oriented along image consistent contours (in the sense of regular differentiable curves [13]). At textured or noisy regions, $\xi^\perp$ is randomly distributed (upper right image in fig.1).

Our Structure-Preserving Diffusion is given by:

$$I_t = \text{div}(Q\Lambda Q^t \nabla I), \quad I(x,y,0) = I_0(x,y) \tag{2}$$

with:

$$Q = (\xi^\perp, \xi) \text{ and } \Lambda = \begin{pmatrix} 1 & 0 \\ 0 & 0 \end{pmatrix}$$

for ξ the principal eigenvector of $ST(\rho,\sigma)$. By $\xi^\perp$ distribution (fig.1), SPD smoothes image gray values along regular structures (bottom right image in fig.1) and performs like a gaussian filter at textured and noisy regions (upper

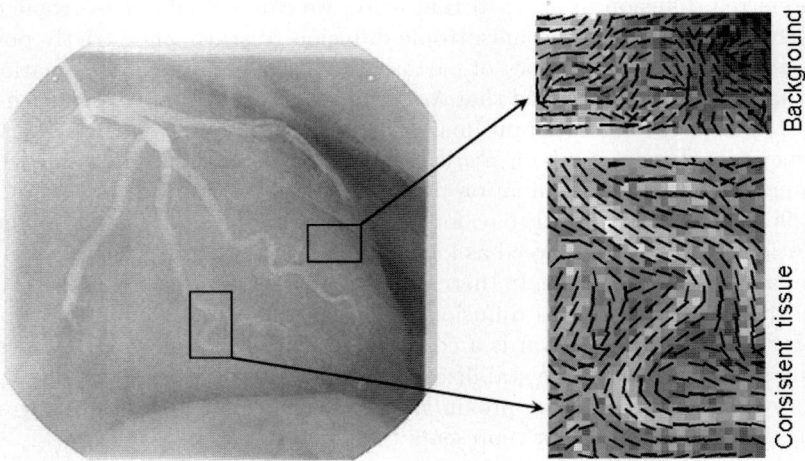

Fig. 1. Vector field representing level curves of an angiography for a vessel (bottom-right image) and a background structure-less area (upper-right image)

right image in fig.1). Its geometric nature makes our restricted diffusion evolution equation converge to a non trivial image that preserves the original image main features as curves of uniform gray level [9]. In this manner, SPD output achieves a uniform response to local image descriptors suitable for a further detection and segmentation of image (anatomical) regions.

3 Results

The goal of our experiments is to show the improvement in quality of SPD images (compared to other filtering approaches) for a further identification of anatomical structures. For the limited length of this communication we have compared SPD to 2 representative techniques: anisotropic and median filtering. SPD has been applied until stabilization of the iterative process, anisotropic diffusion was stopped after 20 iterations and median filter used a 5×5 window. Smoothing techniques have been applied to cardiac Magnetic Resonance (MR) images in short axis (SA) and long axis (LA) views. Image regions segmented using a k-means unsupervised clustering have been compared to manual segmentations in terms of region overlap between two segmentations, X and Y. In particular we have used the Jaccard measure [14] given by $JC := |X \cap Y|/|X \cup Y|$, for $|\cdot|$ the number of pixels.

Fig.2 shows gray-level images and region segmentation with the corresponding manual contours for LA (top rows) and SA (bottom rows) views for, from left to right, non-processed, SPD, Anisotropic Filtering (AF), and Median Filtering (MF). We have segmented three regions: blood (shown in white), myocardial walls (shown in gray) and background (shown in black). Spurious pixels wrongly classified in original views are removed in all filtered images. However, the geometry

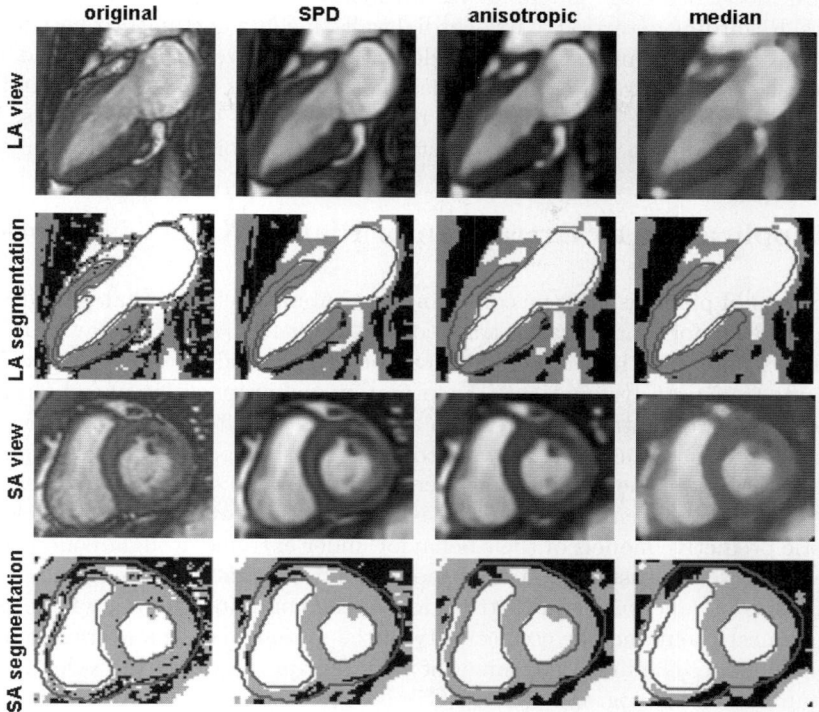

Fig. 2. Performance of smoothing approaches on cardiac magnetic resonance images

Table 1. Jaccard Measure Ranges

	Original		SPD		Anisotropic		Median	
	Blood	Wall	Blood	Wall	Blood	Wall	Blood	Wall
LA	0.7 ± 0.2	0.2 ± 0.3	0.8 ± 0.1	0.5 ± 0.2	0.8 ± 0.1	0.4 ± 0.3	0.8 ± 0.1	0.3 ± 0.3
SA	0.7 ± 0.2	0.5 ± 0.3	0.7 ± 0.2	0.7 ± 0.1	0.7 ± 0.2	0.5 ± 0.3	0.7 ± 0.2	0.5 ± 0.3

of anatomical structures varies across the smoothing type. For LA views the inferior wall of the Left Ventricle (LV) is identified as background in AF images and merged with the adjacent tissue in MF images. Also we observe that the Right Ventricle (RV) blood pool (close to LV inferior wall) merges with LV blood in AF and MF images. SPD images are the only ones preserving the original anatomical structures. Concerning SA views, trabeculae (upper part of LV) and RV wall (left-side of images) are missing in MF images and distorted in AF images. Like in LA views, SPD restores the proper geometry.

Table 1 reports JC statistical ranges for blood and myocardial wall in LA and SA views. For original images blood detection rate is similar in both views, while myocardial wall detection significantly decreases in LA views. For smoothed images, blood detection is similar for all methods and close to original image ranges. Regarding detection of myocardial walls, median and anisotropic smoothing behave similarly and only improve detections (an increase of JC around 0.12) in the case of LA views. SPD images classification rate is the highest one in both views, increasing the average JC 0.3 in LA views and 0.15 in SA views. It is worth to mention the significant reduction in JC variability for SPD images.

4 Application to Extraction of Plant's Xylem Network

The xylem of plants is a tissue consisting of a tubular network that provides the main pathway for long distance transport of water from roots to leaves [15]. Its properties determine how much water can be transported, as well as, the vulnerability to transport dysfunctions (the formation and propagation of emboli) associated to important stress factors, such as droughts and frost. Recent studies [16] link the topology of the xylem network to its overall transport properties including its vulnerability to the propagation of embolism through the system. Thus, modelling the xylem system for representative plant species would help in developing realistic predictive models of their behavior under extreme drought conditions, a key element to forecast vegetation responses under climate change [17].

The size ($\sim$ microns) and distribution of the conduits, the size of the pores that connect them and the connectivity of the system (i.e., the average number of neighbors per conduit) determine the hydraulic properties of the xylem. X-ray computed micro-tomography (micro-CT) is one of the few imaging techniques allowing high resolution imaging of the 3D xylem network [18]. Left images in fig.3 show a tomographic section of a wooden segment from *Fraxinus americana* and

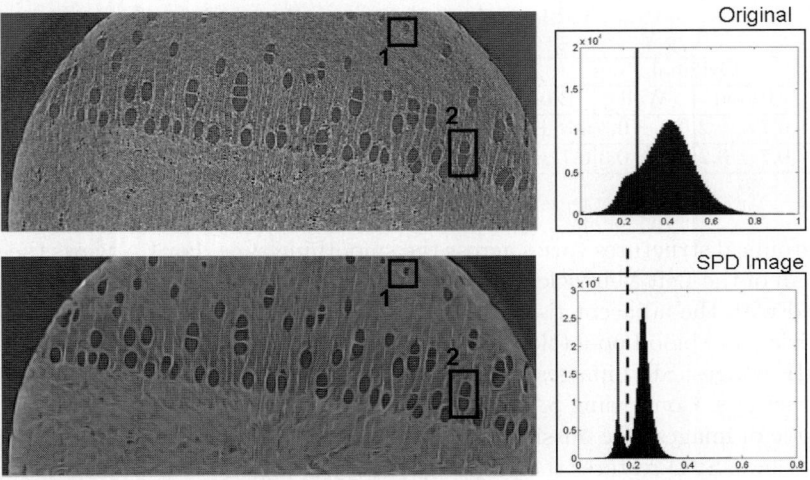

Fig. 3. Benefits of SPD on micro-CT slices of wooden segment: non-processed and SPD images (left) and their intensity histograms (right)

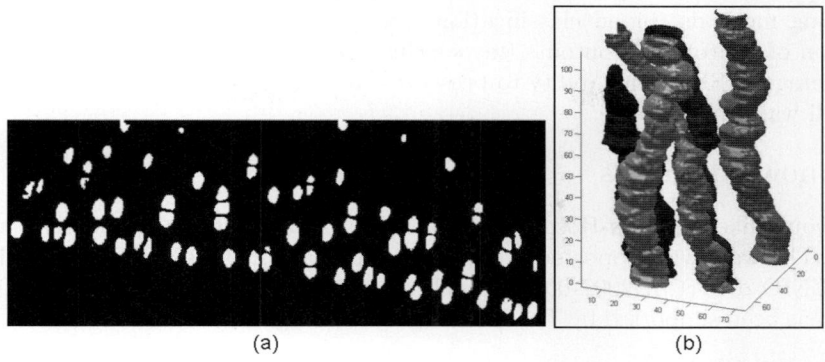

Fig. 4. 3D-Reconstruction of xylem: binary CT-slice, (a) and labelled tubes, (b)

its processed SPD output. Xylem conduits correspond to darker elliptic structures and might appear either isolated or in small (connected) groups separated by a (lighter) thin cell wall. Right plots in fig.3 show gray-level intensity histograms for non-processed (up) and SPD (bottom) images. In the SPD processed image, even the smallest conduits (like the one in square 1) are clearly outlined from the background and there is no loss (i.e. conduit merging) in their connectivity (see the two neighbors in square 2). Furthermore, SPD homogenization of structure intensity produces a bi-modal distribution in histograms separating xylem tubes from background. We have used this property to obtain a detailed 3D reconstruction of the xylem system by simple image processing operators.

Otsu's thresholding method applied to each CT-slice histogram gives the gray-value (vertical dashed line in bottom-right histogram of fig.3) that best separates the two distributions. Morphological operations on binary images are used to remove small structures and close tube holes. Fig.4(a) shows the final binary image representing xylem tubes from SPD image in fig.3. A labelling of the binary 3D block provides the xylem network (as shown in fig.4(b)) and allows the computation of the network connectivity by morphological opening with a structure element of size the maximum separation between connected tubes. These results provide one of the first direct measurements of the connectivity of the xylem network (consistent with previous manual measurements attempts) in any plant species.

5 Discussion and Conclusions

We have presented a diffusion scheme that converges to non-uniform images which present homogenous inter-tissue transitions at consistent anatomical structures and are smooth everywhere else. In order to illustrate independence with respect anatomical shape and medical image modality, we have applied to MR LA and SA views of the heart and to micro-CT of woody segments.

Experiments on MR show that SPD smoothes textured regions similarly to existing methods (blood classification rate in Table 1), while it enhances detection of anatomical contours (myocardial walls statistics in Table 1). A main advantage of SPD is its ability to preserve thin structures (like RV walls in fig.2 or cell walls in fig.4(a)).

Acknowledgments

We would like to thank Hospital de Sant Pau for providing magnetic resonance data. This work was supported by the Spanish projects PI071188, CONSOLIDER-INGENIO 2010 (CSD2007-00018) and Explora (CGL2007-28784-E). The first author has been supported by The Ramon y Cajal Program.

References

1. Pratt, W.: Digital Image Processing. Wiley, Chichester (1998)
2. Comaniciu, D., Meer, P.: Mean shift analysis and applications. In: Int.Conf. Comp. Vis., ICCV (1999)
3. Papari, G., Petkov, N., Campisi, P.: Artistic edge and corner enhancing smoothing. IEEE Trans. Imag. Proc. 16(10), 2449–2462 (2007)
4. Weickert, J.: A Review of Nonlinear Diffusion Filtering. In: ter Haar Romeny, B.M., Florack, L.M.J., Viergever, M.A. (eds.) Scale-Space 1997. LNCS, vol. 1252, pp. 3–28. Springer, Heidelberg (1997)
5. Chen, K.: Adaptive smoothing via contextual and local discontinuities. IEEE Trans. Pat. Ana. Mach. Intel. 27(10), 1552–1567 (2005)
6. Manniesing, R., Viergever, M., Niessen, W.: Vessel enhancing diffusion a scale space representation of vessel structures. Med. Imag. Ana. 10, 815–825 (2005)
7. Evans, L.: Partial Differential Equations. Berkeley Math. Lect. Notes (1993)
8. Saha, P.K., Udupa, J.: Sacle-based diffusive image filtering preserving boundary sharpness and fine structures. IEEE Trans. Med. Imag. 20, 1140–1155 (2001)
9. Gil, D.: Geometric Differential Operators for Shape Modelling. PhD thesis, Universitat Autonoma de Barcelona (2004)
10. Lang, S.: Linear Algebra. Addison-Wesley, Reading (1971)
11. Mikolajczyk, K., Tuytelaars, T., Schmid, C., et al.: A Comparison of Affine Region Detectors. Int. J. Comp. Vis. 65, 43–72 (2005)
12. Jähne, B.: Spatio-temporal image processing:Theory and Scientific Applications. Springer, Heidelberg (1993)
13. Spivak, M.: A Comprehensive Introduction to Differential Geometry, vol. 1. Publish or Perish, Inc., Houston (1999)
14. Cárdenes, R., Bach, M., Chi, Y., Marras, I., de Luis, R., Anderson, M., Cashman, P., Bultelle, M.: Multimodal evaluation for medical image segmentation. In: Kropatsch, W.G., Kampel, M., Hanbury, A. (eds.) CAIP 2007. LNCS, vol. 4673, pp. 229–236. Springer, Heidelberg (2007)
15. Tyree, M., Zimmermann, M.: Xylem structure and the ascent of sap. Springer, Heidelberg (2002)
16. Loepfe, L., Martinez-Vilalta, J., Piñola, J., et al.: The relevance of xylem network structure for plant hydraulic efficiency and safety. J. Th. Biol. 247, 788–803 (2007)
17. Ciais, P., Reichstein, M., Viovy, N., et al.: Europe-wide reduction in primary productivity caused by the heat and drought in 2003. Nature 437, 529–533 (2005)
18. Trtik, P., Dual, J., Keunecke, D., et al.: 3d imaging of microstructure of spruce wood. J. Struct. Biol. 159, 45–56 (2007)

Fast Block Clustering Based Optimized Adaptive Mediod Shift

Zulqarnain Gilani and Naveed Iqbal Rao

Department of Electrical Engineering, College of Telecommunication Engineering,
National University of Sciences and Technology, Pakistan
{zulqarnain.gilani,naveedi}@mcs.edu.pk

Abstract. We present an optimal approach to unsupervised color image clustering, suited for high resolution images based on mode seeking by mediod shifts. It is shown that automatic detection of total number of clusters depends upon overall image statistics as well as the bandwidth of the underlying probability density function. An optimized adaptive mode seeking algorithm based on reverse parallel tree traversal is proposed. This work has contribution in three aspects. 1) Adaptive bandwidth for kernel function is proposed based on the overall image statistics; 2) A novel reverse parallel tree traversing approach for mode seeking is presented which drastically reduces number of computational steps as compared to traditional tree traversing. 3) For high resolution images block clustering based optimized Adaptive Mediod Shift (AMS) is proposed where mode seeking is done in blocks and then the local modes are merged globally. The proposed method has made it possible to perform clustering on variety of high resolution images. Experimental results have shown our algorithm time efficient and robust.

Keywords: Color Image Clustering, Reverse Parallel Tree Traversing, Adaptive Mediod Shift Algorithm.

1 Introduction

All computer vision tasks/applications depend upon identification of true number of clusters and assignment of given data points to their respective clusters [1]. Techniques proposed for partition of data into clusters can be mainly categorized into supervised [2] and unsupervised approaches [3, 4]. Considering solutions to real life problems, more research emphasis has been given to later approaches. In unsupervised approaches, Mean Shift [5,6,7] is a popular non-parametric clustering algorithm based on the idea of associating each data point to a mode of the underlying probability density function but the limitation is the requirement of prerequisite definition of mean. To overcome the shortcomings in Mean Shift, an algorithm for mode seeking based on mediod shift [8] was proposed in which instead of locating the mean, data points themselves were searched. The computations performed during an earlier clustering were invariant to incidence or exitance of samples. Moreover, the need for heuristic terminating conditions in mean shift is eliminated. The method faced two serious drawbacks; first, the bandwidth of the underlying density function was globally fixed and had to be set manually for every data set; second, high computational complexity depending upon the number of data points. Furthermore, it employed the conventional sequential method of tree traversing for seeking of modes. Due to these drawbacks, this algorithm can not be used for high resolution images.

A variation to this approach, "Adaptive Mediod Shift (AMS)" was introduced in [9,10] by converting the fixed global bandwidth of underlying density function to adaptive local bandwidth using k-fixed point sample estimator as k-nearest neighbors. A value of $k=10$ is fixed based on empirical results. Fixed k parameter is good for statistically compact natural images (Figure 1) but increases Root Mean Square (RMS) error in dispersed images. Another common drawback of both the methods is a huge drain on active memory resources for high resolution images.

The contribution of our work is in three folds. First, we give a mathematical insight for the correctness of k=10 for most of the natural images and propose an adaptive data driven k parameter based on the image statistics which reduces the RMS error in the resultant clustered image. Second, a new concept of reverse tree traversal has been introduced, in which the complexity is a function of the height of each independent tree. We prove that in best case as well as worst case scenario the computational complexity of our proposed method is far less than that of AMS. Third, we propose a fast block clustering based optimized AMS where mode seeking is done in blocks and then the local modes are merged globally. This method improves the efficiency of AMS and does away with the memory drain problem. Experiments have proved our algorithm to be robust and time efficient.

2 Overview of Adaptive Mediod Shift Algorithm

Mode seeking by mediod shifts was originally proposed in [8] wherein the bandwidth of the density function is globally fixed. AMS was later proposed in [9,10] which introduces a variable bandwidth for each data point using the concept of k- nearest neighbors [11,12]. The authors have used a value of $k=10$ after performing tests on several images. Thus, the variable bandwidth h_i is defined as,

$$h_i = \left\| x_i - x_{i,k} \right\|_1, \; i=1,2,3.....,n \qquad (1)$$

where x_i is a data point, $x_{i,k}$ are the k-nearest data points and n is the total number of data points. x_i is defined as:

$$\{x_i\} \in \Re^d, \; i = 1, \cdots, n \qquad (2)$$

The bandwidth of the underlying density function is directly proportional to k and as the value of k increases the number of modes (clusters) found decreases and vice-versa. The method calculates a weighted medoid y_{k+1} for every sample data point until the mode is obtained (as proposed originally by [8]),

$$y_{k+1} = \arg\min_{y \in \{x\}} \sum_i \left\| x_i - y \right\|^2 K\left(\left\| \frac{x_i - y_k}{h_i} \right\|^2 \right) \qquad (3)$$

where $K[.]$ is a parzen function. Tree traversal is used to find the unique root (or mode) for all points found in (4). This process is both sequential and iterative. Each data point is visited and a corresponding unique mode is searched using sequential tree traversal. The heuristic is terminated when $y_{k+1}= y_k$. The computational complexity of this algorithm ranges from $\Theta(dN^2 +N^{2.38})$ to $\Theta(dN^2 +N^3)$ depending upon the implementation [13].

3 Adaptive k Parameter

A fixed k parameter selects the nearest k-neighbors of a data point and uses them to calculate the bandwidth of the underlying density function as is clear from (3). However, this does not take into account the spatial diversity of each data set and hence, while a fixed value might be optimal for some natural images, in others, it creates a huge bandwidth increasing the RMS error in the resulting clusters. RMS error for a cluster [14] is defined as,

$$RMS_i = \sum_{j=1}^{p} \frac{1}{p-1} \left\| X_{m_i} - X_j \right\|_2 , i=1,2,...,M \quad (4)$$

where M is the total number of clusters found using AMS, X is the data set from (2), m_i is the mode of the ith cluster and p is the number of data points in a cluster. If $p=1$ RMS error is considered to be zero. RMS error for the complete image/ data set is Mean (RMS_i).

We introduce an adaptive k parameter to cater for the dispersion of data. First the Euclidian distance d between each data point and its next nearest neighbor is calculated as:

$$d_i = floor \left\| x_i - x_i^{\max} \right\|_2 , i=1,2,...,n \quad (5)$$

where n is the total number of data points, $x_i^{\max}$ is data point having maximum correlation with the current data point and is given by,

$$x_i^{\max} = \mathrm{argmax}(E[x_i X]) \quad (6)$$

and E[XY] is the correlation [14] of X and Y given by,

$$E[XY] = \sum_i \sum_n x_i y_n p_{X,Y}(x_i, y_n) \quad (7)$$

Minimum value of $d_i =1$ if the next nearest neighbor of a data point falls with its first neighborhood. By using function *floor* in (6) all distance values are rounded off to the next higher integer value. Next the k parameter is calculated by taking inverse of the distance found in (6) and generalizing it over the first neighborhood (kernel of 3 x3) by multiplying with a factor of 9. Most of the natural images are statistically compact as shown in Fig 1(a). Hence, for such images $d=1$ and the inverse generalized over first neighborhood comes out to be 9. This is the reason why a value of $k=10$ works optimally for most of the natural images. For statistically disperse images our method will result in a smaller value of k yielding optimally 'alike' clusters and reducing the RMS error of the resultant clusters.

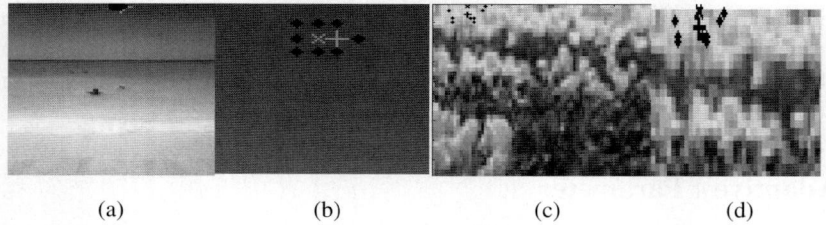

Fig. 1. (a) Image-1, a statistically compact image with small spatial color variance (b) Zoomed in version of Image-1 with current pixel shown as 'x'. (c) Image-2, a statistically well dispersed image with more spatial color variance. (d) Zoomed in version of Image-1 with current pixel shown as 'x'.

Figure 1 (a) and (c) show two statistically different images. Figure 1 (a) shows (Image-1) shows a compact image with most of the neighbors falling within first neighborhood (3 x 3 kernel). This is evident from its 'zoomed in' version in Figure 1(c) where the current pixel is shown in 'x', the nearest neighbor in '+' and the next 8 neighbors in '.'. Image-2 in Figure 1(c) is disperse which is apparent from its 'zoomed in' version in Figure 1(d) where none of the nearest neighbors of the current pixel 'x' fall within a kernel of 3x3. The statistical properties of both images are proven by Figure 1(e) which depicts the distance to next neighbor for both images.

The k parameter can be made adaptive for each data point or the complete data set. In case of adaptive k for each data point we find that the distance d_i of each data point to its nearest neighbor can be different and hence affects the bandwidth of its underlying function. Therefore, each data point can have its own adaptive bandwidth. The inverse of its distance to the nearest neighbor can be generalized over first neighborhood (kernel of 3x3) to obtain the adaptive k-parameter, defined as:

$$k_i = round(\frac{1}{d_i} \times 9), \quad i=1,2,3,...,n \tag{8}$$

where n is the total number of data points and $d_i > 0$ and given by (6). The adaptive bandwidth can be found by replacing k in (1) with k_i and thus we have:

$$h_i = \left\| x_i - x_{i,k_i} \right\|_1 \tag{9}$$

For an adaptive k parameter for the complete data set we take help from the 'law of large numbers' [14]. It can be assumed that the mean of all distances d_i for a very large data set can represent the distance to the next nearest neighbor of the complete image. In this case the k-parameter is given by:-

$$k = round\left(\frac{1}{Mean(d_i)} \times 9\right), \quad i=1,2,3,...,n \tag{10}$$

$Mean(d_i) > 0$. The adaptive bandwidth is then given by (1).

4 Reverse Parallel Tree Traversing

The complete data set/ image can be considered as a forest of free trees (clusters) each with a unique root 'r' (i.e. the Mode). Mode seeking has been performed through tree traversal using *depth first* search technique in [8,9,10] going from each child to its parent node. A unique feature of the tree formed by minimizing (4) for all data points is that each parent node points directly to its child nodes and vice-versa [15]. Using this feature we perform reverse parallel tree traversing in two steps.

4.1 Step-1

Let 'I' be a set of locations where the argument below is minimized based on (4).

$$I = \left\{ i : i \in X \ and \ \forall_j i = \arg\min_{y \in \{x\}} \left[\sum_j \left(\|x_j - y\|^2 K\left(\left\|\frac{x_j - y_k}{h_j}\right\|^2 \right) \right) \right], j=1,2,3,...,n \right\} \quad (11)$$

where X is the data set and the argument is minimized for all data points. Then 'M' is a set of modes given by:-

$$M = \{m : m \subset I \ and \ m \in I_i = i\} \quad (12)$$

where i is the index location and I is given by (12).

4.2 Step-2

For each mode found in Step-1 carry out reverse Tree Traversal going from each parent node to all of its child nodes, to find all data points affiliated with the mode. Let $Y_{k,i} = M_i$, i=1, 2, 3, ..., N_m, where N_m is the total number of modes found in step-1. Then,

$$Y_{k+1,i} = \{y_k : y_{k,i} \in I\} \quad (13)$$

where k= 1,2,3,...,h_i and h_i is the height of the ith tree. $Y_{k+1,i}$ is a vector of all locations where $Y_{k,i}$ matches the set I. The traversing for a mode stops when $Y_{k+1,i} = Y_k$.

4.3 Computational Complexity

Let h_i denote the height of a tree associated with mode i where i=1 ,2 3, ...,N_m. From (14) it is clear that the computational complexity of finding all data points associated with a mode i is a function of the height of the tree created by that mode. Hence the complexity of finding association of all data points will simply be the sum of the heights of all free trees representing the set I (12). Let,

$$\psi = \sum_{i=1}^{N_M} h_i \quad (14)$$

Then the complexity of reverse tree traversing is only,

$$complexity = \Theta(1+\Psi) \quad (15)$$

where the complexity of step-1 of finding all modes as is clear from (13) is only 1. If $N_m \gg 1$ then the complexity can be simplified as $\Theta(\psi)$. We now compare the complexity of both the methods of tree traversing and prove that the complexity of reverse parallel tree traversing is always less than that of forward sequential traversing.

Best Case Scenario. There are no clusters in the data and all data points are the modes themselves as shown in figure 2(c). In this case the complexity of sequential forward traversing will be $\Theta(N)$, where N is the total number of data points. Since there are no trees in the data $h_i=0$ and ψ in (16) is also zero. Hence $\Theta(1+\psi) \ll \Theta(N)$.

Worst Case Scenario. There is only one central mode and all data samples point to it through their neighbor as shown in figure 2(d). In the case of sequential traversing each data point will iteratively point to its neighbor and the complexity will be $\Theta = \sum_{n=1}^{N} n$, where N is the total number of data points. However, in case of reverse traversing the tree will consist of one root and the height will be $h=N$. Therefore (16) yields the complexity as $\Theta(1+N)$ and if $N \gg 1$ the comparison would result in $\Theta(N) \ll \Theta = \sum_{n=1}^{N} n$.

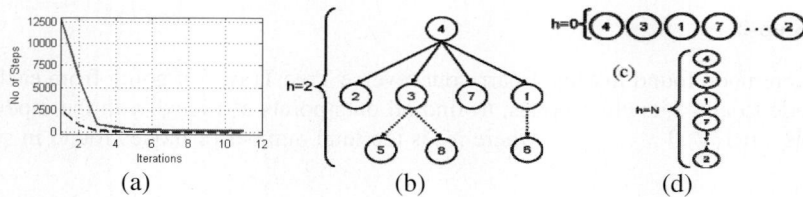

Fig. 2. (a) Steps Comparison between forward sequential tree traversing (solid line) and reverse parallel traversing for the image-1(Fig 1a). (b) Reverse Tree Traversing for an arbitrary mode with height h=2. (c)Best case scenario for mode seeking. All data points are the modes themselves. (d)Worst case scenario for mode seeking. There is only one central mode and all data samples point to it through their neighbor.

Figure 2(a) shows the comparison between sequential forward and parallel reverse traversing for steps needed to converge (4) for all iterations for image-1 (Figure 1a). The sequential forward traversing technique converges in 22323 steps whereas it takes only 3758 steps for parallel reverse traversing to find all modes in 11 iterations. Figure 2(b) depicts the reverse traversing method where the search begins from the root /mode. Height h of the tree is 2. Mode is found in the first step and each height is traversed in a single step thus finding all data samples pointing to the mode in 3 steps only. Figures 2(c) and (d) depict the best and the worst case scenarios for mode seeking discussed in section 4.3.

5 Fast Block Clustering Based AMS

The technique is based on using AMS in blocks instead of processing the complete data. The image/data is divided into blocks of size $m \times n$.

$$B_k^d \subseteq \mathfrak{R}^d \qquad , k=1,2,\ldots,b \qquad (16)$$

where b is the total number of blocks, d is the number of dimensions. AMS is run on each block separately. A local data set x is made as in (2) and mapped to the global data set X. Mapping $x_i \to X$,

$$X = \{x_i : x_i \in B_k \text{ and } X \in \mathfrak{R}\}, i=1,2,\ldots, N_B \qquad (17)$$

where N_B is the total number of data points in a block, x_i is the local data point and X is the global data set. Mode seeking is performed for each block and the local modes are then mapped to global modes M. In order to avoid over segmentation, a global mode merging is performed on all modes found during block clustering.

The computational complexity of the original algorithm proposed in [8] was a function of total number of data points in the image. However, the complexity of our proposed method is a function of only the total number of data points in a block and is given by, $b \times \Theta(N_B^{2.38})$ where b is the total number of blocks. Furthermore, the number of data points in a block is always much less than the total number of data points. Thus $N_B \ll N$.

6 Results

Experiments have been carried out on 500 images from Flicker, Corel and Labelme data sets. Both, statistically compact as well as disperse images of different varieties have been chosen. The images have been chosen such that a comparison analysis between fix k parameter and adaptive k can be carried out in detail. Natural images as well as daily life images have been used for experimentation.

Fig 3(a) shows 10 natural statistically compact as well as disperse images. Time taken to cluster these images using conventional AMS is shown in (c) whereas time taken by our proposed method is shown in (d). The difference in time is in the order or 5000-9000 seconds. Time taken to cluster 500 experimental images is shown in (b). Most of the images yielded clustering results within 4-5 seconds. (e) depicts the mean of all next neighborhood distances for all the images. A value of 1 indicates that the images were statistically compact. Thus a value of $k=9$ is suited for such images. RMS error in the resulting clusters for all images is shown in (f). RMS error increases with fix $k=10$ (solid line) whereas for variable k most of the images have low RMS error. Finally, a comparison of computational steps taken in case of forward sequential and reverse parallel tree traversing is depicted in (g). In case of traditional tree traversing (4) converges in 15000 to 30000 steps while less than 5000 steps are required for this purpose in our proposed method. (h) Displays a compact natural image while (i) is the result of performing block based AMS clustering on the image.

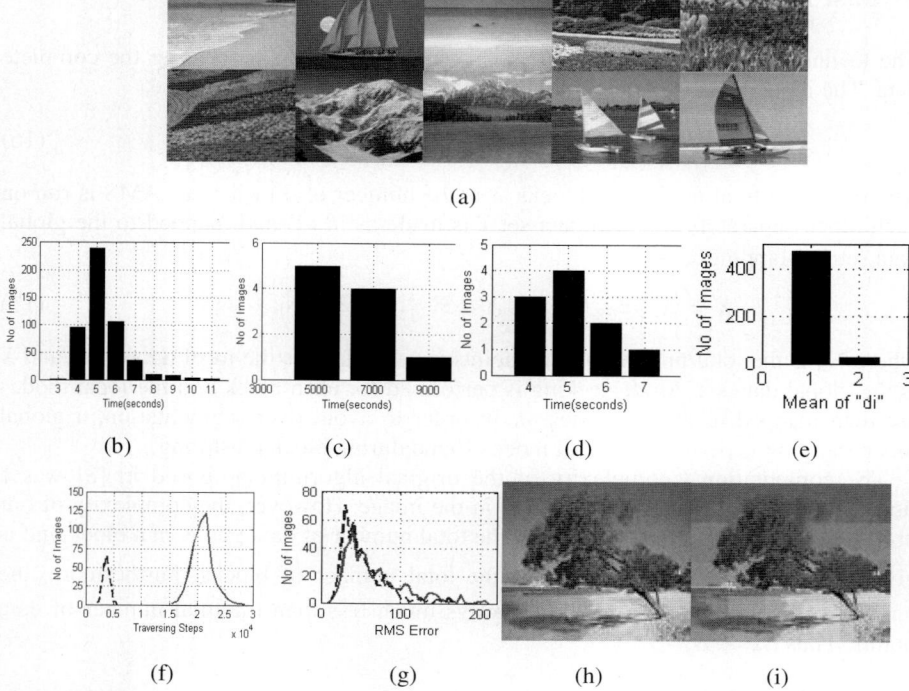

Fig. 3. (a) 10 natural compact and disperse images (b) Time taken to cluster 500 images using our approach (c) Time taken to cluster 10 images using conventional AMS and (d) our proposed method. (e) Mean of all next neighborhood distances (f) RMS Error with fix and variable k (g) computational steps in forward and reverse parallel tree traversing.(h) A natural image (i) AMS Color clustering result.

References

1. Xu, R., Wunsch II, D.: Survey of Clustering Algorithms. IEEE Transactions on Neural Networks 16(3) (May 2005)
2. Cherkassky, V., Mulier, F.: Learning From Data: Concepts, Theory,and Methods. Wiley, New York (1998)
3. Bishop, C.: Neural Networks for Pattern Recognition. Oxford Univ. Press, New York (1995)
4. Duda, R., Hart, P., Stork, D.: Pattern Classification, 2nd edn. Wiley, New York (2001)
5. Fukunaga, K., Hostler, L.D.: The estimation of the gradient of a density function, with applications in pattern recognition. IEEE Transactions on IT IT-21(1), 32–40 (1975)
6. Cheng, Y.: Mean shift, mode seeking, and clustering. PAMI 17(8), 790–799 (1995)
7. Comaniciu, D., Peter, M.: Mean shift: A robust approach toward feature space analysis. PAMI 24(5), 603–619 (2002)
8. Sheikh, Y.A., Khan, E.A., Kanade, T.: Mode seeking by medoidshift. IEEE Trans., ICCV, 1–8 (2007)

9. Asghar, A., Rao, N.I.: Color Image Segmentation Using Multilevel Clustering Approach DICTA, Canberra, pp. 519–524 (December 2008)
10. Asghar, A., Rao, N.I.: Semantics Sensitive Segmentation and Annotation of Natural Images. In: 2008 IEEE International Conference on SITIS, Indonesia, pp. 387–394 (2008)
11. Breiman, L., Meisel, W., Purcell, E.: Variable kernel estimates of multivariate densities. Technometrics 19, 135–144 (1977)
12. Georgescu, B., Shimshoni, I., Peter, M.: Mean shift based clustering in high dimensions: A texture classification example. IEEE Trans., ICCV (2003)
13. Vedaldi, A., Soatto, S.: Quick shift and kernel methods for mode seeking. In: Forsyth, D., Torr, P., Zisserman, A. (eds.) ECCV 2008, Part IV. LNCS, vol. 5305, pp. 705–718. Springer, Heidelberg (2008)
14. Garcia, A.L.: Probability and Random Processes for Electrical Engineers, ch. 5.2, 2nd edn. Pearson Education, London
15. Thomas, H.C.: Introduction to Algorithms, ch. 6, 2nd edn. MIT Press, Cambridge (2001)

Color Me Right–Seamless Image Compositing

Dong Guo and Terence Sim

School of Computing
National University of Singapore
Singapore, 117417

Abstract. This paper introduces an approach of creating an image composite by seamlessly blending a region of interest from an image onto another while faithfully preserving the color of regions specified by user markup. With different regions marked for color-preserving, our approach provides users the flexibility in creating different composites. The experiment results demonstrate the effectiveness of the proposed approach in creating seamless image composite with color preserved.

Keywords: Seamless image compositing, Color preserving, Gradient-based image editing.

1 Introduction and Background

Digital image compositing is a process of blending a region of interest (ROI) from a source image onto a target image. An example of image compositing is shown in Fig. 1[1]. An ROI containing a window frame is selected by a user, as shown inside the yellow line in Fig. 1(a). The ROI is then blended onto the target image (Fig. 1(b)) to produce the final composite (Fig. 1(c)).

Poisson image editing (PIE) [1] proposed by Pérez et al. is a method to seamlessly blend ROI onto the target image. In PIE, the gradient of ROI is pasted onto that of the target image. Then, the result image is reconstructed from the gradient domain by solving a Poisson equation. The main advantage of PIE is the seamless boundary around the pasted ROI in result images. To achieve this goal, the gradient inside ROI is to be kept unchanged to make the result visually similar as the source image. On the other hand, a hard constraint along the boundary is enforced to make the boundary of pasted ROI agree with the target image. PIE keeps the relative values in ROI; however, the absolute values (the color) of ROI may shift in the process of blending. In some cases, the color shift of ROI is desirable (some examples demonstrated in [1,2]) because this makes the tone of pasted ROI similar to the target image. However, the amount of the color change depends on the difference of boundaries of ROI in the source and that in target image, which is uncontrollable by users. Despite the case of tone matching, color shift is usually undesirable. As shown in Fig. 1(e),

---

[1] All the images in this paper are colored and high-resolution. For a better understanding, the reader may be interested in viewing these images on the monitor.

Fig. 1. An example of image compositing. (a) The source image. ROI is the region inside yellow line. (b) The target image. (c) The final composite of our proposed approach by pasting the window from (a) onto (b). (d) User markup (the red scribbles) used in our approach. (e) Composite of Poisson image editing. Note the color shift in the ROI. (f) Composite of Poisson image editing with color constraint by user markup in (d). Note the halo effects around the scribbles. See the text for details.

PIE produced a large color shift in the result. The window and the flower in ROI was expected to be the same as in the source image; however, the whole ROI became reddish.

An improvement of PIE, proposed by Jia *et al.* [2], solved a problem of PIE producing unnatural blurring artifacts when the boundary of ROI does not meet the target image very well. However, in this paper, we do not address this problem. Instead, the aim of this paper is to find a way of producing seamless image composite while preserving the color.

Image matting, such as [3,4], and image segmentation techniques, such as [5,6], are commonly used to extract objects from an image. The extracted objects can be pasted onto a target image to produce a composite. These techniques are suitable for foreground and background being clearly separated. However, in many cases, the foreground interacts with the surroundings, without obvious boundaries, *e.g.* the shadow region below the window frame shown in Fig 1. These techniques are not suitable.

Another approach, interactive digital photomontage [7], aims to assemble images of the same or similar scenes together. Regions from different images are picked out by users and the seams among different regions are minimized with Graph-cut. The final composite keeps the color from each source images. However, if the source images are different from each other, it is difficult to find invisible seams to stitch the images together.

In this paper, our approach take the advantage of gradient-based editing to produce a seamless composite. We use weighted least squares in reconstructing the final composite from the gradient domain. With the boundary of ROI and the user markup being the hard constraints, the colors are faithfully preserved in specified regions. The idea of employing weighted least squares to enforce the similarity within a region can be also found in some previous works, such as colorization [8], tone adjustment [9] and edge-preserving image decomposition [10].

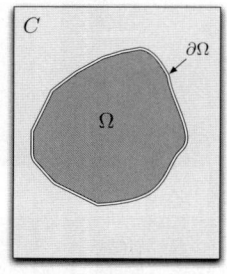

Fig. 2. Illustration of notations

Contribution. In this paper, we present an approach of creating an image composite by seamlessly blending an ROI onto a target image with the color preserved in specified regions. In addition, with different markups, users have the flexibility in creating different composites by choosing different regions for color-preserving. The solution of our approach can be formulated as a sparse linear system which can be efficiently solved.

2 Color-Preserving Image Compositing

The goal of image compositing is to blend the region of interest (ROI) from the source onto the target image. In the final composite, only the region pasted from the source image is unknown, while the rest is directly from the target. As illustrated in Fig. 2, we use Ω denote the unknown region in the final composite C, with $\partial \Omega$ being the boundary. The value of $(C - \Omega)$ is from the target image t. We use r denote the pixel value of Ω, which are the unknowns to be solved; s denotes the corresponding values of the same image.

2.1 Poisson Image Editing

In Poisson image editing [1], the difference between gradient of r and that of s is minimized, while the value r on the boundary $\partial \Omega$ is enforced to be the same with t. Thus, r is the solution of the following minimization problem,

$$\min_r \int |\nabla r - \nabla s|^2 \quad \text{with} \quad r|_{\partial \Omega} = t|_{\partial \Omega}, \tag{1}$$

where ∇ is the gradient operator. The solution to Eq. (1) is given by the following Poisson equation with Dirichlet boundary conditions,

$$\triangle r = \triangle s \quad \text{with} \quad r|_{\partial \Omega} = t|_{\partial \Omega}, \tag{2}$$

here $\triangle$ is the Laplacian operator.

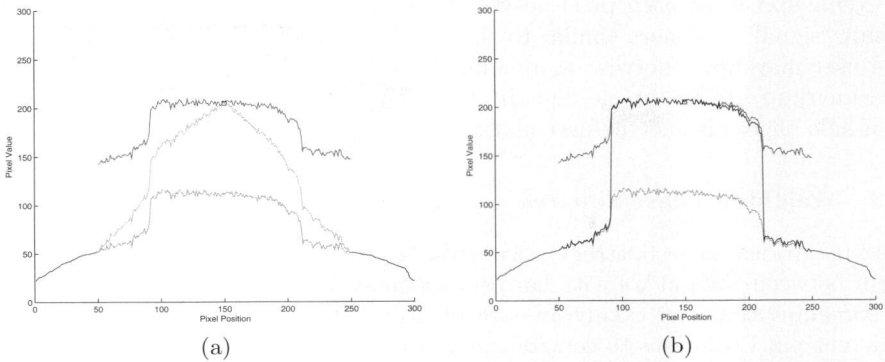

Fig. 3. An illustration of Poisson image editing (PIE) and the proposed approach in 1D case. (a) Red: the source "signal". Blue: the target "signal". Green: result of PIE. Note the color shift from r down to s. Cyan: the result of PIE fixing the middle value of r to s and the boundary. This creates a halo effect. (b) Red, green, blue: same as in (a). Black: the result of our proposed approach. Note that the middle part of r is well preserved, while r is seamlessly connected to t.

We use a 1D example illustrate this idea, as shown in Fig. 3(a). The horizontal axis denotes the pixel position, while the vertical axis denotes the pixel value. The region of blue lines is $(C - \Omega)$, and the region of the green line is Ω. The red line is the source "signal", s. The middle part of s, which has larger value larger than the surroundings, indicates the object to be pasted. Usually, the ROI selected is slightly larger than the object inside. The result of PIE is shown as the green line, which fits two ends of the blue lines and keeps the gradient of s. It appears that PIE successfully blended s onto t without visible seams. However, the absolute value of reconstructed "signal" r shifts quite a bit from the source s. In image domain, this results in a color shift of the ROI in the process of blending, as shown in Fig. 1(e).

2.2 User Markup Constraints

To correct the color shift, an intuitive way is adding additional constraints: user markup indicating the foreground that should be the same as in the source image, s. If there are holes in the foreground, extra user markup should be involved indicating the holes (as background) as well. These constraints are added as hard constraints, *i.e.*

$$r|_{M_1} = s|_{M_1} \text{ and } r|_{M_2} = t|_{M_2}, \quad (3)$$

where M_1 and M_2 denote markup of foreground and background, respectively.

In the 1D example, a point in the middle of Ω is used as user markup of foreground. The result of Eq. (2) with additional constraints of Eq. (3) is shown as the cyan line in Fig. 3(a). With the user markup as hard constraints, the error of the objective function usually becomes much larger. In PIE, this error is evenly spread on each pixel position, resulting in a gradual change in r. The result "signal" r is more similar to the red line s when the position is nearer to the user markup, otherwise more similar to the green line (the result without user markup). In the image domain, the gradual change will undesirably result in a halo effect around the user markup.

2.3 Weighted Least Squares

The main idea to reconstruct r from the gradient domain is to minimize the error between ∇r and ∇s with hard constraints of $\partial \Omega$ and user markup. Instead of spreading the error evenly on each pixel in PIE, we add different weights to different pixel positions to control the desired similarity between ∇r and ∇s.

Ideally, the color of the regions from the foreground should be preserved during compositing; and color preserving should be restricted to these regions only. For a "region", we mean an area where the colors are similar. Within a region, small gradients in the source image shouldd be more likely to be kept small in the result. In other words, the larger the gradient of a pixel in s is, the more this pixel should share the error from the overall minimization error.

In 2D image domain, r is given by the solving the following minimization problem,

$$\min_r \sum_{p \in \Omega} \left(\alpha(p)(r_x(p) - s_x(p))^2 + \beta(p)(r_y(p) - s_y(p))^2 \right) \quad (4)$$

with $r|_{\partial \Omega \cup M_2} = t|_{\partial \Omega \cup M_2}$ and $r|_{M_1} = s|_{M_1}$,

where $\{.\}_x$ and $\{.\}_y$ denote the partial derivative of $\{.\}$ along x and y coordinates respectively, while $\alpha(.)$ and $\beta(.)$ are two weights controlling the similarity between the gradient of r and s. The hard constraints are the same as discussed before. The weights, α, β, are defined according to above analysis, *i.e.*

$$\alpha(p) = \frac{1}{|s_x(p)|^\gamma + \epsilon}, \quad \beta(p) = \frac{1}{|s_y(p)|^\gamma + \epsilon} \quad (5)$$

where ϵ is a small constant preventing division by zero, and the exponent γ is the sensitivity of enforcing the similarity. A similar way of defining the two weights can be found in [9] and [10].

The solution of Eq. (4) is given by the following linear system,

for each $p \in \Omega$,

$$\sum_{q \in \mathcal{N}(p)} \frac{r(p) - r(q)}{|s(p) - s(q)|^\gamma + \epsilon} = \sum_{q \in \mathcal{N}(p)} \frac{s(p) - s(q)}{|s(p) - s(q)|^\gamma + \epsilon}, \quad (6)$$

with hard constraints,

$$\begin{cases} r(p) = t(p) & p \in \partial \Omega \cup M_2 \\ r(p) = s(p) & p \in M_1 \end{cases}, \quad (7)$$

(a)   (b)

Fig. 4. A new image composite result created with a different user markup. (a) User markup. One more stroke on the side of the window frame, compared to Fig. 1(d). (b) Image composite with user markup of (a). Note that the color of the inside of the window frame has been changed to blue from yellow(Fig. 1(c)).

where $\mathcal{N}(p)$ denotes four neighbors of p. By substituting the hard constraints (7) into (6), the linear system can be solved.

In the 1D example previously discussed, we still use only the middle point as user markup. As shown in Fig. 3(b), the resultant curve r (the black line) is obtained via Eq. (6) by setting $\gamma = 4.0$. The middle part of r is kept almost the same to s, while two ends of r are seamlessly connected to t. As expected, the value of the region marked by user is preserved and no halo effects exist. A 2D example has been shown in Fig. 1(c). The final composite was seamless and natural, while the color was well preserved.

3 Experiments and Results

Fig. 5 shows an example of image compositing by blending a bear from the sea to a swimming pool. In the result of Poisson image editing, the pasted bear in the final composite appeared pale. By adding additional constraints of user markup on the body of the bear, the result of PIE produced a halo effect around the markup. In contrast, the final composite of our proposed approach preserved the look and feel of the bear in the source image. Also, the boundary of the pasted bear was seamless.

Another example is shown in Fig. 6. The motorcyclist consists of a lot of different colors. Thus, a bit more user markup are required to cover each component. As shown in Fig. 5a, nine strokes were marked on different parts, including one indicating the background where a hole exists between the motorcyclist's arms and the motorcycle. In the result of PIE, the motorcyclist became reddish. PIE with markup still produced obvious halo effect. The result of our proposed approach faithfully preserved the color of the motorcyclist.

With our proposed approach, a user can create different image composites by choosing different regions for color-preserving. We show this with the wall example in Fig. 1. The source and target image are the same as before. The new

Fig. 5. Image compositing of a bear. (a) The source image with selected ROI (yellow boundary)and user markup (red strokes). (b) The target image. (c) Result of Poisson image editing. Note the color shift of the bear from original brown to pale. (d) Result of PIE with constraints by user markup. Note the halo effects around the head of the bear. (e) The result of our proposed approach without color shift or halo effect. The images are from [1].

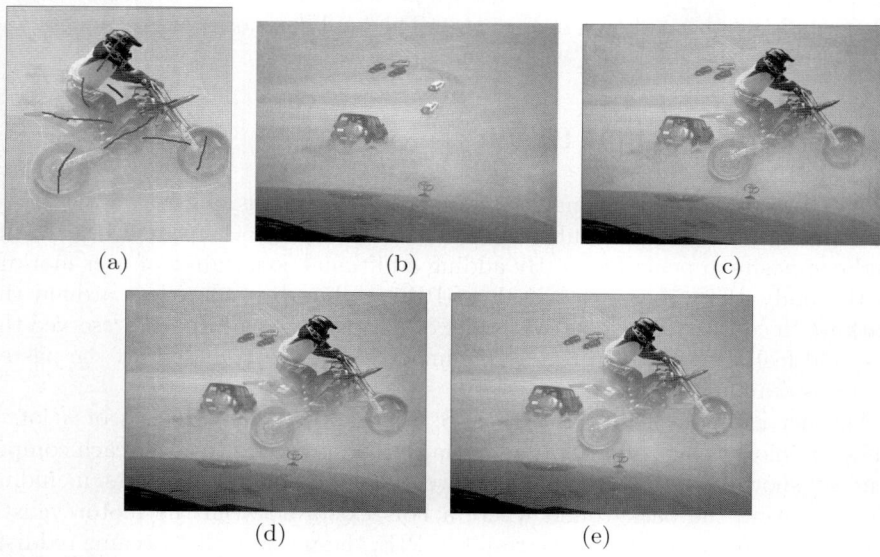

Fig. 6. Image compositing of a motorcyclist. (a) The source image with ROI and user markup of the foreground (red) and background (blue). (b) The target image. (c) Result of PIE. Note the color shift on the helmet. (d) Result of PIE with constraints by user markup. Note the halo effect around the foreground. (e) Result of our proposed approach.

user markup (Fig. 4(a)) contained one more stroke on the side of the window frame. With the new markup, the side of the window frame was kept the same as the source image, with the color being blue, instead of the previous "yellow". Still, there is not any artifact produced in the final composite.

4 Conclusion and Future Work

In this paper, we have presented an approach of creating seamless image composite with the color of specified region faithfully preserved. The main idea is simple yet powerful. By adding different weights to the minimization term, smaller gradient is more likely to be kept small. Thus the color within a region is kept similar in the final composite. By adding user markup, the color in specified regions is preserved in the final composite. With different markups, users have the flexibility in creating different image composites by choosing different regions for color-preserving. Experiment results have demonstrated the effectiveness of our approach.

Since the user should add background markup to the holes of the object in ROI, the marking process could be time-consuming if the object consists of many holes. One future work could be automatically detecting the holes and then keep these hole regions similar to the background in the final composite. One possible solution may be adding global color constraints to keep similar color being similar in the final composite.

The authors acknowledge the generous support of NUS (Research Grant #R-252-000-383-112).

References

1. Pérez, P., Gangnet, M., Blake, A.: Poisson image editing. ACM Trans. Graphics 22(3), 313–318 (2003)
2. Jia, J., Sun, J., Tang, C.K., Shum, H.Y.: Drag-and-drop pasting. ACM Trans. Graphics (2006)
3. Chuang, Y.Y., Curless, B., Salesin, D.H., Szeliski, R.: A bayesian approach to digital matting. In: Proc. CVPR (2001)
4. Sun, J., Jia, J., Tang, C.K., Shum, H.Y.: Poisson matting. ACM Trans. Graphics (2004)
5. Li, Y., Sun, J., Tang, C.K., Shum, H.Y.: Lazy snapping. ACM Trans. Graphics 23(3), 303–308 (2004)
6. Mortensen, E.N., Barrett, W.A.: Intelligent scissors for image composition. In: Proc. ACM SIGGRAPH (1995)
7. Agarwala, A., Dontcheva, M., Agrawala, M., Drucker, S., Colburn, A., Curless, B., Michael Cohen, D.S.: Interactive digital photomontage. ACM Trans. Graphics (2004)
8. Levin, A., Lischinski, D., Weiss, Y.: Colorization using optimization. ACM Trans. Graphics 23(3), 689–694 (2004)
9. Lischinski, D., Farbman, Z., Uyttendaele, M., Szeliski, R.: Interactive local adjustment of tonal values. ACM Trans. Graphics 25(3), 646–653 (2006)
10. Farbman, Z., Fattal, R., Lischinski, D., Szeliski, R.: Edge-preserving decompositions for multi-scale tone and detail manipulation. ACM Trans. Graphics 27(3), 1–10 (2008)

Transform Invariant Video Fingerprinting by NMF

Ozan Gursoy, Bilge Gunsel, and Neslihan Sengor

Istanbul Technical University, Electrical-Electronics Eng. Faculty,
Dep. of Electronics and Communications Engineering
34469 Maslak Istanbul, Turkey
{ozangursoy,gunselb,sengorn}@itu.edu.tr

Abstract. Video fingerprinting is introduced as an effective tool for identification and recognition of video content even after putative modifications. In this paper, we present a video fingerprinting scheme based on non-negative matrix factorization (NMF). NMF is shown to be capable of generating discriminative, parts-based representations while reducing the dimensionality of the data. NMF's representation capacity can be fortified by incorporating geometric transformational duplicates of the base vectors into the factorization. Factorized base vectors are used as content based, representative features that uniquely describe the video content. Obtaining such base vectors by transformational NMF (T-NMF) is furthermore versatile in recognizing the attacked contents as copies of the original instead of considering them as a new content. Thus a novel approach for fingerprinting of video content based on T-NMF is introduced in this work and experimental results obtained on TRECVID data set are presented to demonstrate the robustness to geometric attacks and the improvement in the representation.

Keywords: Video fingerprinting, non-negative matrix factorization, transformation invariance.

1 Introduction

Increasing amount of video data, which is encouraged by the improvements in sharing, storage and accessing capabilities, emerged interest on video search and retrieval techniques. Besides the techniques based on cryptographic hashes and digital watermarking, video fingerprinting is proposed as superior to its predecessors in extracting and transforming characteristic features of a video. It is possible to identify a video content easily and uniquely by a small amount of data which is called "video fingerprint". Unlike to other techniques, video fingerprinting has not to be applied at production level and is not affected by natural distortions (compression, coding) or malicious attacks (logo addition, geometric distortions) unless they do alter the visual content [1].

Most of the video fingerprinting systems in literature are proposed in recent years [2]. While in [1], key-frame based local fingerprints are calculated from the orientation histograms of the local points, in [3] transformation domain features are extracted as video fingerprints where distorted content is included in the transformation. Although first method is robust to Divx compression and has a moderate performance on scaling and motion blur, it is not robust to spatial blurring. Key-frame repeatability is also another issue that has to be considered in [1]. Second approach stores the transformation domain representations obtained by 2D oriented PCA (2D-OPCA) as video fingerprints. Robustness of this method to geometric transformations basically relies on finding the affine covariant regions before applying 2D-OPCA.

Our video fingerprinting scheme generates fingerprints using a variation of non-negative matrix factorization (NMF). NMF was first proposed by Paatero and Tapper [4] but improved and became popular by the work of Lee and Seung [5]. In our previous works, [6,7,8] we have shown that besides its rank reduction and content representation capabilities, NMF is highly suited for the video content representation by non-negativity constraint which results in parts-based representations [5].

In this paper, we propose a new video fingerprinting scheme that benefits from the transformation-invariant NMF which is proposed by Eggert et al. [9]. The matrices for geometric transformations such as shift, flip, rotate and scale are calculated and applied by transformational NMF (T-NMF) so that resultant base vectors are not only robust to global illumination changes but also will be capable of representing geometrically distorted samples. The proposed method, in a novel way, enables to add robustness against geometric attacks to the global descriptors of a video frame.

This paper is organized as follows: Section 2 describes the conventional and transformational NMF together with the formulations of the transformation matrices, adaptation of base vectors that are generated by T-NMF for robust video fingerprinting is explained in section 3, experimental results of the proposed video fingerprinting scheme is detailed in section 4 and conclusion and future work is summarized in section 5.

In all equations, italic, double indices represent a matrix element, an uppercase letter with one index represents a column vector, a bold uppercase letter is equivalent to the whole matrix and a lowercase italic letter with an index determines a vector element.

2 Transformation-Invariant NMF

2.1 Conventional NMF

NMF aims to decompose the data matrix $\mathbf{V} \in R^{p \times q}$ into two matrices, $\mathbf{W} \in R^{p \times r}$ and $\mathbf{H} \in R^{r \times q}$. $\mathbf{W}$ contains the base/mixing vectors for linear approximation of $\mathbf{V}$, and $\mathbf{H}$ is the coefficient matrix used in adding up the base vectors of $\mathbf{W}$ to reconstruct the matrix $\mathbf{V}$. The non-negativity constraints of NMF allow only additive combinations of the bases. The additive property ensures a powerful content representation capability of NMF, which is due to the base vectors property of representing local components of the original data.

The conventional NMF iteratively updates $\mathbf{W}$ and $\mathbf{H}$ matrices in order to satisfy the minimization of a cost function. A simple and efficient cost function is the squared reconstruction error and it is given by Eq.(1).

$$F = \frac{1}{2}\|\mathbf{V} - \mathbf{WH}\| = \frac{1}{2}\sum_i (V_i - R_i)^2, \quad R_i = \sum_j W_j H_{ji} \;. \tag{1}$$

Minimization of the cost function with respect to the **H** and **W** matrices separately is modified by Lee and Seung [5] and their approach brings forth the multiplicative update rules given in Eq.(2) and Eq.(3), which are easy to implement and work well in practice.

$$H_{ji} \leftarrow H_{ji} \odot \frac{V_i^T W_j}{R_i^T W_j} .\qquad(2)$$

$$W_j \leftarrow W_j \odot \frac{\sum_i H_{ji} V_i}{\sum_i H_{ji} R_i} .\qquad(3)$$

2.2 Transformational NMF

NMF's representation capacity can be fortified by incorporating geometric transformational duplicates of the base vectors into the factorization. Eggert et al. [9] introduced the contribution of transformation matrices to the conventional NMF's cost function in order to gain robustness to translational changes in the input data. Thus the cost function and update rules of **H** and **W** matrices, which are given in Eq.(4), Eq.(5) and Eq.(6), respectively, are modified with the introduction of transformation matrix $\mathbf{T^m}$ where m refers to the considered transformation. A penalizing term is added to the cost function in Eq.(4) to prevent trivial solutions and achieve more localized, parts-based and sparse representations [9].

$$F = \frac{1}{2}\sum_i (V_i - R_i)^2 + \sum_{i,j,m} H_{ji}^m, \quad R_i = \sum_j \sum_m H_{ji}^m \mathbf{T^m} \hat{W}_j, \quad \hat{W}_j = \frac{W_j}{\|W_j\|} .\qquad(4)$$

$$H_{ji}^m \leftarrow H_{ji}^m \odot \frac{V_i^T \mathbf{T^m} \hat{W}_j}{R_i^T \mathbf{T^m} \hat{W}_j} .\qquad(5)$$

$$W_j \leftarrow W_j \odot \frac{\sum_i \sum_m \left[H_{ji}^m V_i^T \mathbf{T^m} + \left[H_{ji}^m R_i^T \mathbf{T^m} \hat{W}_j \right] \nabla_{W_j}(\hat{W}_j) \right]}{\sum_i \sum_m \left[H_{ji}^m R_i^T \mathbf{T^m} + \left[H_{ji}^m V_i^T \mathbf{T^m} \hat{W}_j \right] \nabla_{W_j}(\hat{W}_j) \right]} .\qquad(6)$$

In the next section the proposed video fingerprinting scheme by focusing on the influence of the $\mathbf{T^m}$ matrices will be explained.

3 Extraction of Robust Video Fingerprints

We propose a key-frame based video fingerprinting scheme that can be integrated by any of the key-frame selection scheme proposed in literature [1,10,11]. In the proposed method, the video fingerprints are obtained as follows: For each selected key video frame v^i, a set of M geometrically transformed versions, $\{v^{mi}\}_{m=1}^{M}$, together with the original frame form the input matrix **V** so it contains geometrically translated versions of a sample frame. T-NMF is then applied to **V** matrix using the equations given in the previous section. Factorized base matrix, $\mathbf{W}^i$, where upper index specifies the selected key frame, is used as content based, representative feature that belongs to v^i and uniquely describes its content.

By the influence of the $\mathbf{T}^m$ matrices, geometrically distorted, similar contents in the input data are not reflected to the base vectors so they do not contain redundant information. Instead, the basic content, which is included in most of the transformational versions of the input frame, is revealed in the base vectors, thus it is possible to build transformed video inputs only with necessary information.

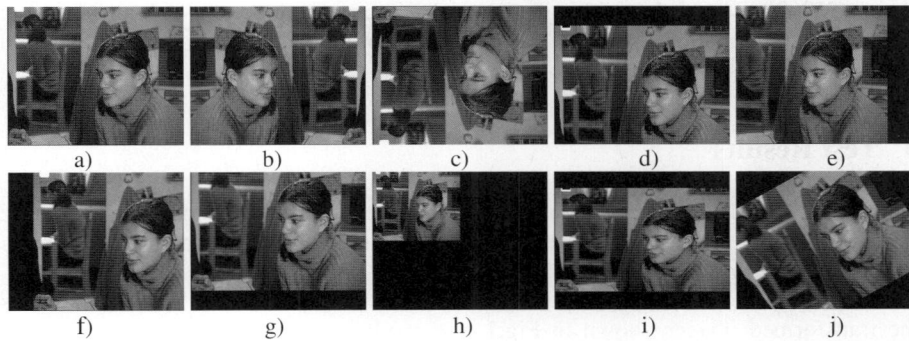

Fig. 1. Geometrically transformed input images: a)untransformed, b)flip horizontal, c)flip vertical, d)shift down, e)shift left, f)shift right, g)shift up, h)pic-in-pic (scale 2-D), i)ratio (scale 1-D), j)rotate

The M different $\mathbf{T}^m$ matrices generate Mxr transformed versions of the base vectors that are added up by $\mathbf{H}^m$ coefficient matrices to reconstruct the input samples. In terms of the robustness of our video fingerprinting scheme to geometric distortions, we will focus on the structure and functionality of the $\mathbf{T}^m$ matrices. The geometric transformations that are used to generate the video fingerprint of a video frame based on both NMF and T-NMF are shown in Fig.1. Each image in Fig.1 forms one column of the input matrix **V** for the factorizations.

Fig.2 gives structures of some of the $\mathbf{T}^m$ matrices that are used in the factorization. An important property of the $\mathbf{T}^m$ matrices is they are orthogonal matrices so, the geometrically inverse of the transformation, which is also achieved by a $\mathbf{T}^m$, can be simply obtained with the transpose of the associated $\mathbf{T}^m$ matrix. This property explains the effect of the $\mathbf{T}^m$ matrices during the update process as in these equations V_i and R_i vectors are back-transformed geometrically.

The proposed video fingerprinting scheme not only takes the heritage of the conventional NMF as a good content representation and robustness to global luminance distortions, but also combines robustness to geometric attacks by fairly translating the geometric distortions to the extracted representation by the help of the $\mathbf{T}^m$ matrices.

$$\begin{bmatrix} E & Z & \cdots & Z \\ Z & E & & \vdots \\ \vdots & & \ddots & Z \\ Z & \cdots & Z & E \end{bmatrix} \quad \begin{bmatrix} Z & \cdots & Z & I \\ \vdots & & I & Z \\ Z & \cdot\cdot\cdot & & \vdots \\ I & Z & \cdots & Z \end{bmatrix} \quad \begin{bmatrix} I^{\rightarrow k} & Z & \cdots & Z \\ Z & I^{\rightarrow k} & & \vdots \\ \vdots & & \ddots & Z \\ Z & \cdots & Z & I^{\rightarrow k} \end{bmatrix} \quad \begin{bmatrix} I^{\leftarrow k} & Z & \cdots & Z \\ Z & I^{\leftarrow k} & & \vdots \\ \vdots & & \ddots & Z \\ Z & \cdots & Z & I^{\leftarrow k} \end{bmatrix}$$

a)   b)   c)   d)

Fig. 2. Structures of some of the $\mathbf{T}^m$ matrices used in the factorization: a)flip horizontal b)flip vertical c)shift k columns left d)shift k columns right. **E**, **Z** and **I** correspond to exchange, full zero and identity matrices, respectively. The $\mathbf{I}^{\rightarrow k}$ shows an identity matrix that is shifted k columns in the direction of the arrow.

The video fingerprinting database is constructed by storing the representative base matrix, $\mathbf{W}^j$, for every selected key frame of the reference video clips, alongside the m different transformation matrices, $\mathbf{T}^m$, which do not change from frame to frame and are used in reconstruction-based matching process.

4 Test Results

NMF and T-NMF representation capabilities under geometric and non-geometric distortions are compared by means of both reconstruction errors and reconstructed image qualities on video frames that are taken from TRECVID 2007-08 database[12].

After NMF and T-NMF is applied separately to the input matrix **V** which includes the transformed images shown in Fig.1, the resultant base vectors are presented in Fig.3. For this experiment 10 different geometric transformations are considered and rank r is set to 2 for both of the factorizations. The difference of the visual quality between NMF (Fig.3.a and Fig.3.b) and the T-NMF (Fig.3.c and Fig.3.d) base vector images can be observed explicitly.

a)   b)   c)   d)

Fig. 3. Base vectors of a-b) conventional NMF and c-d) transformational NMF that are generated by the factorization of 10 geometrically transformed input images. Rank is set to 2 for both of the factorizations.

To evaluate performance of the introduced video fingerprinting algorithm first we have examined the reconstruction capability of the T-NMF. Thus a sample video frame v^i is first projected and h^{ij} is calculated using the $\mathbf{W}^j$, and then back-projected/reconstructed as r^{ij} by the multiplication of $\mathbf{W}^j$ and h^{ij}. Projection step can be carried out either by calculating the pseudo-inverse or by the iterative updates with fixed $\mathbf{W}^j$. We preferred the second approach as the calculation of pseudo-inverse may cause negative values in the projection vector which is not suited to the characteristic of the data. The NMF and T-NMF reconstruction vectors, r^{ij}, and the reconstruction error, $recErr(i,j)$, for a sample vector v^i and a base matrix $\mathbf{W}^j$ are calculated by Eq.(7) and Eq.(8), respectively, where k varies from 1 to the rank and l varies from 1 to the total number of pixels in a sample video frame.

$$r^{ij} = \sum_k \mathbf{W}_k^j \cdot h_k^{ij}, \qquad r^{ij} = \sum_k \sum_m \mathbf{T}^m \mathbf{W}_k^j \cdot h_k^{mij} . \tag{7}$$

$$recErr(i,j) = \sum_l \left\| v_l^i - r_l^{ij} \right\|^2 . \tag{8}$$

The reconstructed images of the untransformed, flip vertical, shift left, ratio and rotate transformations obtained from Fig.1 are demonstrated in Fig.4. The images on each row are formed by T-NMF and NMF base vectors, respectively.

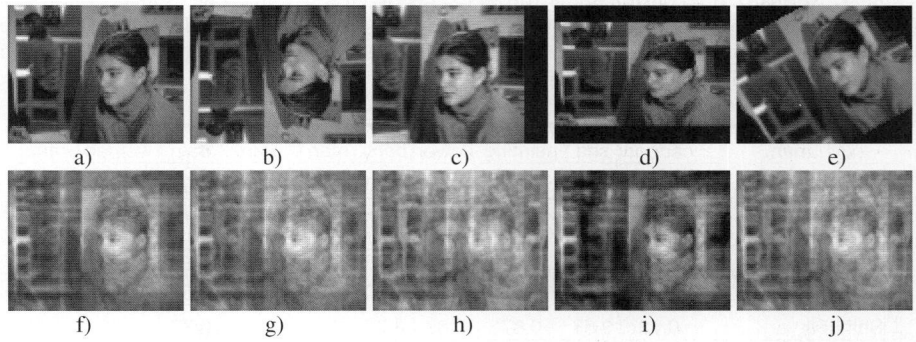

Fig. 4. Reconstructed images: a-f) untransformed, b-g) flip vertical, c-h) shift left, d-i) ratio, e-j) rotate. First row is generated by T-NMF base vectors and the second row is generated by conventional NMF. The input frames are chosen from Fig.1. a), c), e), i) and j), respectively.

Video frames from 10 different video scenes are chosen to demonstrate the discriminative property of the extracted video fingerprints. Fig.5 illustrates selected frames. Note that some of the frames have semantically very close visual content (Fig.5.f and Fig.5.g).

We have ranked the matching scores in terms of the normalized reconstruction error. Table 1 presents the results. For this experiment key frame "man2" (Fig.5.g) is selected as query example and name of the transformations applied to the input video frame "man2" are listed at the first column of Table 1. All the other columns correspond to different $\mathbf{W}^j$ matrices that are used to calculate the reconstruction error scores for the input image as described in Eq.(8). While upper part of Table 1 contains the results that are obtained by T-NMF base matrices, reconstruction errors of NMF base matrices are listed at the lower part.

Fig. 5. Original key frames extracted from different video scenes: a)cartoon, b)sea, c)painting, d)flower, e)baby, f)man1 g)man2 h)train i)woman, j)canal. Note that although the frames at f) and g) are taken from different scenes, they have semantically close visual contents.

Table 1. Normalized reconstruction error(*recErr*) between test frame Fig.5.g)man2, its transformed frames (flip vertical, gamma correlation (G) with a factor of 1.75, shift left 15%, rotate left 30 degrees, blurring (B) and noise (N)), and 10 different key frame. Base vectors obtained by NMF and T-NMF are used for projection.

	man2	cartoon	sea	painting	flower	baby	man1	man2	train	woman	canal
T N M F	Untransformed	0,97	0,80	1,18	1,15	0,93	0,77	**0,00**	0,97	1,06	0,96
	G+B+N	1,19	0,85	1,61	1,46	1,06	0,97	**0,16**	1,14	1,30	1,18
	FlipV	1,01	0,89	1,21	1,26	1,16	1,05	**0,00**	1,23	1,23	1,19
	FlipV+G+B+N	1,14	0,91	1,53	1,51	1,35	1,51	**0,18**	1,52	1,51	1,44
	ShiftL50	0,78	0,66	0,87	0,87	0,77	0,66	**0,04**	0,77	0,89	0,79
	ShiftL50+G+B+N	0,95	0,67	1,08	1,08	0,88	0,90	**0,18**	0,88	1,16	0,97
	RotateL30	1,00	0,74	1,03	1,02	0,90	0,74	**0,04**	1,00	1,05	0,92
	RotateL30+G+B+N	1,20	0,75	1,26	1,24	1,01	1,00	**0,21**	1,23	1,35	1,06
N M F	Untransformed	1,21	0,95	1,38	1,28	1,09	0,84	**0,41**	1,05	1,21	1,16
	G+B+N	1,66	1,01	1,82	1,59	1,23	1,09	**0,48**	1,21	1,47	1,37
	FlipV	1,10	1,06	1,30	1,34	1,25	1,17	**0,46**	1,43	1,37	1,32
	FlipV+G+B+N	1,26	1,26	1,64	1,70	1,53	1,76	**0,86**	1,86	1,72	1,71
	ShiftL50	0,94	0,79	0,92	0,95	0,83	0,75	**0,74**	0,84	0,96	0,90
	ShiftL50+G+B+N	1,21	0,95	1,14	1,20	0,91	1,08	**1,08**	0,98	1,26	1,12
	RotateL30	1,03	1,02	1,06	1,09	0,97	0,97	**0,65**	1,06	1,15	1,00
	RotateL30+G+B+N	1,32	1,18	1,29	1,37	1,13	1,30	**0,92**	1,33	1,50	1,15

The bold faced column in Table 1 contains the lowest reconstruction error scores as it is expected to get best scores when base matrix, $\mathbf{W}^j$, of the video frame "man2" is used to calculate the reconstruction error for the selected query images. Although good reconstruction error scores can be obtained for non-geometrical (G+B+N) transformations or untransformed query images with NMF base matrices, reconstruction errors for geometric transformations are not discriminative. On the other hand reconstruction error scores of T-NMF base matrices are not affected by geometric transformations and are always the lowest.

5 Conclusions

In this paper, a novel robust video fingerprinting scheme based on transformational NMF is proposed. The proposed method takes advantage of the indexed geometric transformation matrices while generating content representative base vectors.

Preliminary results show that **T** matrices increase the reconstruction performance in terms of the Euclidean cost and influence the base vectors to resemble the untransformed input image more than conventional NMF bases. The experimental results also reveal that the proposed method is robust against non-geometric distortions as well as geometric transformations.

Future work will be focusing on storage, search and matching strategies of the extracted fingerprints which will lead us to extent the test dataset.

References

1. Massoudi, A., Lefebvre, F., Demarty, C.-H., Oisel, L., Chupeau, B.: A Video Fingerprint Based on Visual Digest and Local Fingerprints. In: Proc. of IEEE Int. Conf. on Image Processing, pp. 2297–2300 (2006)
2. Law-To, J., Chen, L., Joly, A., Laptev, I., Buisson, O., Gouet-Brunet, V., Boujemaa, N., Steintiford, F.: Video Copy Detection: A Comparative Study. In: Proc. of the 6th ACM Int. Conf. on Image and Video Retrieval, Netherlands, pp. 371–378 (2007)
3. Lee, S., Yoo, C.D.: Robust Video Fingerprinting Based on 2D-OPCA of Affine Covariant Regions. In: Proc. of IEEE Int. Conf. on Image Processing, USA, pp. 2156–2159 (2008)
4. Paatero, P., Tapper, U.: Positive Matrix Factorization: A Non-negative Factor Model with Optimal Utilization of Error Estimates of Data Values. Environmetrics 5, 11–126 (1994)
5. Lee, D., Seung, H.: Learning the Parts of Objects by Non-negative Matrix Factorization. Nature 401, 788–791 (1999)
6. Bucak, S.S., Gunsel, B.: Video Content Representation by Incremental Non-negative Matrix Factorization. In: Proc. of IEEE Int. Conf. on Image Processing, USA, pp. 113–116 (2007)
7. Bucak, S.S., Gunsel, B.: Incremental Clustering via Nonnegative Matrix Factorization. In: Proc. of 19th Int. Conf. on Pattern Recognition, ICPR, USA (2008)
8. Bucak, S.S., Gunsel, B.: Incremental Subspace Learning via Non-negative Matrix Factorization. Pattern Recognition 42, 788–798 (2009)
9. Eggert, J., Wersing, H., Körner, E.: Transformation-invariant Representation and NMF. In: Proc. IEEE Int. Joint Conf. on Neural Networks (2004)
10. Gunsel, B., Ferman, A., Tekalp, A.M.: Temporal Video Segmentation Using Unsupervised Clustering and Semantic Object Tracking. Electronic Imaging 64, 592–604 (1998)
11. Wolf, W.: Key Frame Selection by Motion Analysis. In: Proc. IEEE Int. Conf. on Acoustics, Speech and Signal Processing, vol. 2, pp. 1228–1231 (1996)
12. TREC Video Retrieval Evaluation,
    http://www-nlpir.nist.gov/projects/trecvid/

A Model Based Method for Overall Well Focused Catadioptric Image Acquisition with Multi-focal Images

Weiming Li[1], Youfu Li[1,*], and Yihong Wu[2]

[1] Department of Manufacturing Engineering and Engineering Management,
City University of Hong Kong, 83 Tat Chee Avenue, Kowloon, Hong Kong
[2] National Laboratory of Pattern Recognition, Institute of Automation,
Chinese Academy of Sciences, P.O. Box 2728, Beijing, 100080, China
weimingli2008@hotmail.com, meyfli@cityu.edu.hk,
yhwu@nlpr.ia.ac.cn

Abstract. Based on an analysis on the spatial distribution property of virtual features, we propose that the shapes of the best focused image regions in multi-focal catadioptric images can be modeled by a series of neighboring concentric annuluses. Based on this model, an over-all well focused image can be obtained by combining the best focused regions from a set of multi-focal images in a fast and reliable manner. A robust algorithm for estimating the model parameters is presented. Experiments with real catadioptric images under a variety of scenes verify the validity of the model and the robust performance of the algorithm.

Keywords: Catadioptric system, multi-focal images, well-focused image.

1 Introduction

Being able to capture a wide field of view, catadioptric systems consisting of curved mirrors and conventional lens based cameras have been increasingly popular in many computer vision applications. A frequently observed problem in a catadioptric image is that while some image regions are well focused, some other regions might not be focused as well, which leads to difficulties in the subsequent image processing procedures. While most existing works on catadioptric systems have focused on mirror design[1-3], calibration[4,5], or applications [6], little attention has been paid towards understanding the effects of curved mirrors on the formation of a well-focused image. In this work, we explore the related properties of geometric optics and propose a model based method that combines a set of multi-focal images to obtain an over-all well focused image within which all objects are clearly focused.

Among the several related works, an earlier one was done by Baker and Nayar [1]. The authors studied the defocus blur phenomenon and illustrate the *field curvature effects* [7] by a numerical simulation on a set of sample points. Ishiguro [8] discussed the case of compact catadioptric systems. The author stated that if the depth of field (DOF) of the camera could cover the region where the virtual features of all object

---
* Corresponding author.

points reside, the image would appear clearly focused for all objects. Recently, Swaminathan [9] explicitly derived the positions of virtual features and found that the infinite range of scene depth is limited to a finite extent of virtual features which is named caustic volume. Therefore, according to the previous researches, as long as the camera DOF is wide enough to contain the entire caustic volume, all objects can be clearly captured by just one single catadioptric image.

However, there are still many cases where the camera DOF is not wide. One example is compact catadioptric systems[8], where the cameras are typically mounted at a close distance to the mirror. According to geometrical optics [7], a close object distance leads to a small DOF. Another example is the systems where the cameras work with large apertures [10] to allow efficient photography, which also leads to a small DOF. In these situations, the object points whose virtual features are beyond the DOF would still appear out-of-focus. Since one single shot cannot capture a clear image for all objects, a reasonable option is to combine a set of multi-focal images.

Methods using multi-focal images to obtain an over-all well focused image have been extensively studied for lens based cameras in micro or close-up photography to extend the DOF[11]. Such approaches typically segment the best focused image patches from multi-focal images and merge them into an output image. As local focus measurement has to be evaluated through the entire set of images and optimized with spatial consistency constraints, these methods can be time consuming and error prone.

Different from these methods, we explore the geometric optics in catadioptric systems [9] and propose a model based method, which models the shapes of the best focused image regions in multi-focal images by a series of neighboring concentric annuluses. Based on this model, segmenting the best focused image regions is simplified to estimating the model parameters, which is much more robust to noise. The model parameters are also found to be independent of the scene structure. Therefore once the model is estimated, no computation is needed for different scenes, which endows this method with a fast performance.

An algorithm for estimating the model parameters is presented. This algorithm does not need the mirror shape nor the system parameters to be known as prior, thus is not prone to calibration errors. The focal distance settings for the set of multi-focal images are also not necessary to be known, which makes this algorithm highly automatic. Experiments with real catadioptric images under a variety of scenes verify the validity of the proposed model and the robust performance of the algorithm.

2 The Proposed Model

2.1 Shape of Well Focused Image Region in One Catadioptric Image

We start by discussing the model for one catadioptric image. Considering the imaging process in a catadioptric system, 3D object points are first reflected by the curved mirror and form virtual features. Then the camera captures these virtual features to form an output image. Therefore the well focused image regions correspond to the object points whose virtual features are within the camera DOF.

Recently, Swaminathan [9] explicitly derives the locations of virtual features and finds that virtual features of all object points are located within a finite space which is named caustic volume. Fig. 1 plots an illustrative sketch for the imaging process. As is shown, caustic volume is the space between the mirror surface and a virtual surface which is the caustic volume boundary (CVB). In this work, we further find that the spatial distribution of virtual features within the caustic volume is non-uniform and the majority of virtual features can be regarded to locate on the CVB. Based on the work [9], we explicitly compute the locations of virtual features for a variety of quadric mirror based catadioptric systems, whose eccentricities range from 0.8 to 1.2, heights from the mirror apex to the camera lens range from 5.9mm to 114.7mm, and the diameters of the mirror range from 4.6mm to 136.0mm. This set of parameters covers a typical production list of catadioptric sensor producers such as ACCOWLE Co. Ltd [12]. According to the computation, when an object point is farther than 1 meter to the system, it will have its virtual feature locate within a narrow neighborhood of the CVB whose width is 5% of the distance from CVB to the mirror surface. In most applications, as the points of interests usually lie at a certain distance, the great majority of virtual features can be considered to be located on the CVB.

Following this conclusion, the virtual features that can be clearly focused should be located on the CVB surface and between the front end and rear end of the DOF, which can be modeled as two parallel planes perpendicular with the optical axis. As the system and the CVB is rotational symmetric, the projection of this region onto the CCD sensor is an annulus. Therefore, this work uses this annulus to model the shape of well focused region in a catadioptric image.

Since the CVB and DOF are determined only by the mirror shape and the system setup, the shape of well focused region is independent of the 3D scene structures, which is essentially different from the situation in a conventional dioptric system.

2.2 Shapes of Best Focused Image Regions in Multi-focal Images

Denote the CVB as $V \subset \mathbb{R}^3$. Let $I = \{I_i\}$ be a set of N multi-focal catadioptric images, which is taken at the same view point for the same scene, yet with different focal distance settings $\{f(I_i)\}$, where $f(I_1) < f(I_2) < ... < f(I_N)$. Note that when a camera with image-space telecentric feature [13] is used, the scene content in each image is the same. Denote the DOF of I_i as $d(I_i)$, which is as shown in Fig. 1. To make sure all scene objects are clearly recorded in at least one of the images, I should satisfy the following conditions: (1) $d(I_p) \bigcap d(I_q) \neq \emptyset$ for $\forall I_p, I_q \in I$, $q = p+1$, and (2) $V \subset \bigcup_{i=1}^{N} d(I_i)$. Condition (1) can be met by making the focal distance increment at sufficiently small steps. Condition (2) can be met by letting $d(I_1)$ lie before V and $d(I_N)$ lie behind V, which can be easily examined by visually checking whether both I_1 and I_N are over-all out-of-focus images.

The images of interests in I are a subset of M partly focused images $I' = \{I'_k\} \subset I$ that satisfy $d(I'_k) \bigcap V \neq \emptyset$. Also assume $f(I'_1) < f(I'_2) < ... < f(I'_M)$. As V is a limited space, M can be expected to be a small number. Denote the circular

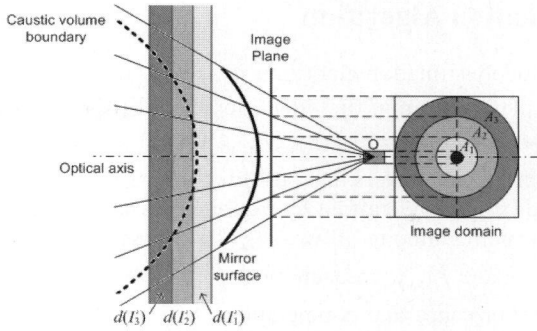

Fig. 1. A sketch of the imaging process of a typical catadioptric system. Since almost all virtual features locate on the caustic volume boundary, the shapes of the best focused image regions in multi-focal catadioptric images can be modeled by a set of neighboring concentric annuluses.

image region corresponding to the mirror surface as $s(I')$. Note that since $I' \subset I$, $s(I) = s(I')$. It can be seen that any part in the image domain of $s(I')$ would be clearly recorded in at least one of the images in I'.

Extending the model in section 2.1, the best focused regions in I' can be modeled by M neighboring concentric annuluses. From the central to the peripheral, let $s(I')$ be divided into M concentric annular areas $A = \{A_k\}$, where $\bigcup_{k=1}^{M} A_k = s(I')$ and $A_p \cap A_q = \emptyset$, for $\forall A_p \neq A_q, A_p, A_q \in A$. Determined by the order of $f(I'_k)$, A_k is the best focused in image I'_k as shown in Fig. 1. Each annulus A_k can be described by the two concentric circles enclosing it, whose radius are r_{k-1} and r_k. Therefore, the model $A = \{A_k\}$ can be parameterized as a set of M+1 radius $\{r_0, r_1, r_2, ..., r_M\}$, where $r_0 < r_1 < r_2 <, ..., < r_M$. Here, r_0 is radius of the circular area in the image central part where the scene is occluded by the camera, which is plotted as the black circle in Fig. 1. r_M is radius of the circular view boundary of the mirror surface in the image. As r_0 and r_M are determined by the system setup, there are M-1 free parameters to be estimated, which is denoted as $R = \{r_1, r_2, ..., r_{M-1}\}$. The final out put image $\hat{I}$ is simply obtained by combining the best focused area in I'_k so that $\hat{I}(A_k) = I'_k(A_k)$.

Note that as R is only determined by the system setup and focal distance settings of I'_k, the same R is applicable for images of arbitrary scenes in spite of their different 3D structures. This is verified by experiments in section 4.

3 Model Estimation Algorithm

The input of the model estimation algorithm is a set of multi-focal images I as introduced in section 2. The algorithm first identifies the set of partly focused images I'. Then the model parameters $R = \{r_1, r_2, ..., r_{M-1}\}$ are estimated using an optimization approach that search the parameters that best fits the data observation.

A partly focused image is identified by examining whether it contains a number of best focused image points among all images. To evaluate the degree of focus for a point $p \in s(I)$ in image I_i, a measurement $F(p; I_i)$ is needed. Considering the balance between performance and computational efficiency, we adopt the high order statistics (HOS) [14] as $F(p; I_i)$. Denote $P = \{p_j\} \subset s(I)$ as a set of uniformly distributed sample points. Then I_w contains p_j as a best-focused point (BFP) when $F(p_j; I_w) = \max_{i=1}^{N} F(p_j; I_i)$. Therefore, partly focused images are identified as the images that have a significantly larger number of BFP. In this work, one half of the largest number of BFP contained in an image is used as the threshold.

Given the set of M partly focused images $I' = \{I'_i\}$, the model parameters to be estimated are: $R = \{r_1, r_2, ..., r_{M-1}\}$. Here we define a model estimation error function $E(R)$ between the model estimation R and the image observation I' as:

$$E(R) = \sum_{k=1}^{M} \sum_{p_j \in P \cap A_k} \delta(\arg(\max_{i=1}^{M} F(p_j; I'_i)) \neq k) \quad (1)$$

where $\delta(s) = 1$ when s is true and $\delta(s) = 0$ when s is false. $P = \{p_j\} \subset s(I')$ is a set of uniformly distributed sample points in $s(I')$. In this function, $\arg(\max_{i=1}^{M} F(p_j; I'_i))$ gives index of the image where point p_j in area A_k is considered best focused by the focus measurement in I'. On the other hand, according to the model in section 2.2, any point p_j in area A_k must be best focused in image I'_k. Therefore, $E(R)$ actually counts the number of sample points whose image observations are not consistent with the model predictions. Then the optimal model estimation $\hat{R}$ is given by minimizing $E(R)$ so that: $E(\hat{R}) = \min_{R}(E(R))$.

The algorithm obtains $\hat{R}$ with the following procedures:

(1) For each p_j, find $B(p_j) = \arg(\max_{i=1}^{M} F(p_j; I'_i))$.
(2) Denote the distance from a pixel p to the image center as $rad(p)$. For all the pixels in P that satisfy $B(p_j) = k$, the average of $rad(p_j)$ is supposed to be a value $a(k)$ between r_{k-1} and r_k. As $r_0 < r_1 < r_2 <, ..., < r_M$, some erroneous estimations of $B(p_j)$ can be rejected by using the following iteration:

(a) Estimate $a(k)$ for each A_k.
(b) Find the points that simultaneously satisfy the following two conditions: (i) $a(k) < rad(p_j) < a(k+1)$; (ii) $B(p_j) < k$ or $B(p_j) > k+1$.
(c) If no points are found in (b), finish the loop. If else, go to (d).
(d) Eliminate the points found in (b) and go back to (a).

(3) For the model $R = \{r_1, r_2, ..., r_{M-1}\}$ to be estimated, $r_k \in (a(k), a(k+1))$. Since the solution space for R is not large, we traverse the entire solution space to find a global optimal estimation $\hat{R}$ that minimize $E(R)$.

As ordering constraints is used in step (2) and an optimal estimation is searched, this method still works well with the presence of noise.

After the model parameters are estimated, the left work is to combine the best focused regions into a single image $\hat{I}$ so that $\hat{I}(A_k) = I'_k(A_k)$. Note that, as $d(I'_p) \bigcap d(I'_q) \neq \emptyset$ for $\forall I'_p, I'_q \in I'$, the image pixels in I'_k within a neighborhood of A_k can still be supposed to be well focused.

4 Experiments

The proposed method is tested with a real catadioptric system, where a Canon PowerShot S50 digital camera is mounted toward a hyperbolic mirror surface with the camera optical axis coincident with the mirror axis. The circular image region of the mirror surface has a radius of 850 pixels. We observe that when the F number is 3.2 and the shutter speed is 1/400 second, it is difficult to capture one overall well focused image with any focal distance setting, which indicates the camera DOF is not wide enough to contain the CVB. With the above setting, 10 multi-focal images $I = \{I_i\}$ are captured, whose focal distances are manually set from 10cm to 28cm with an interval of about 2cm. Other number of images and focal distance settings can also be used as long as the obtained images satisfy the two conditions in section 2.2. Guaranteed by the image-space telecentric feature [13], the scene contents in the images are exactly identical despite of the different focal distances.

With the method in section 3, I_4, I_5 and I_6 are found to contain a significantly higher number of BFP and are therefore selected as the set of partly focused images $I' = \{I'_1, I'_2, I'_3\} = \{I_4, I_5, I_6\}$, as shown in Figure 2. As three partly focused images are involved, two model parameters $\{r_1, r_2\}$ need to be estimated. With the proposed method, a set of 5076 uniformly distributed grid sample points are used and r_1=418(pixels), r_2=619(pixels). Different numbers of noise points from 800 to 4800 are added, for which the image where each noise point is best focused is randomly assigned. Though the number of noise points increases to as large as 94% of the sample points, the maximum deviation of the estimated model parameter is 3.97% for r_1 and 1.46% for r_2, which illustrates the robustness of the model estimation algorithm.

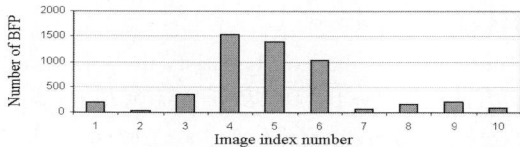

Fig. 2. Number of the best focused points (BFP) in each input image

By combining the best focused regions, an overall well-focused image is obtained as the output image. Harris corner extraction algorithm extracts at least 24.1% more corner points from this image than those from any single partly focused image.

The model estimation algorithm is tested in six different scenes with the same system setting and image capturing procedure. For each scene, three partly focused images are extracted. In Figure 3, one of the partly focused images for each scene is presented in the first row, under each of which a subfigure displays the sample points. According to results from the focus degree measurement function (given in section 3), a sample point is plotted blue if best focused in I'_1, green if in I'_2, and red if in I'_3. The model $\{r_1, r_2\}$ is superimposed on each figure as the two circles. For 85.8 % of the sample points, the model gives the same results as those given by the focus measurement function. It can be intuitively observed that $\{r_1, r_2\}$ are quite consistent with the raw focus measurement. The inconsistent cases are due to the erroneous estimations from the focus measurement function, which may result from local texture less regions, sensor internal reflections, and low SNR in dark regions. In comparison, the proposed model based method can well deal with these local noises.

Notice that though the scene structure changes significantly, the shapes of the best focused regions and the model parameters remain the same. The maximum model parameter deviation is 2.25% for r_1 and 1.59% for r_2 respectively. Therefore, the model parameters can be deemed to be independent from the scene 3D structures.

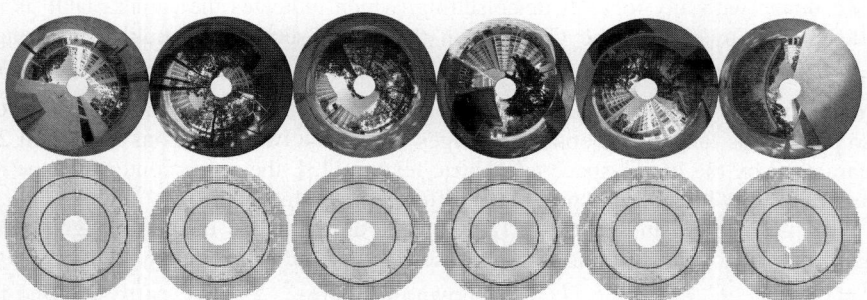

Fig. 3. Model parameters estimated under six different scenes

This feature guarantees that, once estimated, the model parameters are still applicable for other scenes as long as the system setups and imaging parameters do not change. Many modern cameras are able to store the imaging parameters on-chip, which much facilitate the employment of this method. Therefore, an overall well-focused image can be simply obtained by cropping and putting together the annular regions without additional computational load, making the procedure operate very fast.

5 Concluding Remarks and Discussions

This work presents a model based method that obtains an over-all well focused catadioptric image by combining multi-focal images in a fast and reliable manner. The resulting images would benefit the applications where high image resolutions are desirable over a large field of view such as open environment image acquisition, large area 3D structure measurement, and artistic landscape photography. Experiments with real catadioptric images verify the model and the algorithm. In practice, many off-the-shelf cameras are equipped with the focus-bracketing function, with which the proposed method can be easily employed by many existing catadioptric systems to improve the image quality in terms of focus. Note that focus-bracketing requires time for multi-exposure, which might hinder the approach from dynamic environments. However, this problem would be alleviated as newer cameras are ever increasing their frame rate such as Casio EX-F1, which can capture 60fps at 6MP.

Acknowledgement

This work was supported by grants from the Research Grants Council of Hong Kong [Project No. CityU117106 and CityU117507].

References

1. Baker, S., Nayar, S.K.: A theory of Single-viewpoint Catadioptric Image Formation. Int. J. Comput. Vision 35(2), 1–22 (1999)
2. Swaminathan, R., Grossberg, M.D., Nayar, S.K.: Non-single Viewpoint Catadioptric Cameras: Geometry and Analysis. Int. J. Comput. Vision 66(3), 211–229 (2006)
3. Geyer, C., Daniilidis, K.: Catadioptric Projective Geometry. Int. J. Comput. Vision 45(3), 223–243 (2001)
4. Ying, X., Hu, Z.: Catadioptric Camera Calibration Using Geometric Invariants. IEEE Pattern Anal. Mach. Intell. 26(10), 1260–1271 (2004)
5. Geyer, C., Daniilidis, K.: Paracatadioptric Camera Calibration. IEEE Pattern Anal. Mach. Intell. 24(5), 687–695 (2002)
6. Yagi, Y.: Omnidirectional Sensing and Its Applications. IEICE Trans. Inf. Syst. 82(3), 568–579 (1999)
7. Hecht, E.: Optics. Addison-Wesley, Reading (1987)
8. Ishiguro, H.: Development of Low Cost Compact Omni-directional Vision Sensors and Their Applications. In: Benosman, R., Kang, S.B. (eds.) Panoramic Vision: Sensors, Theory, and Applications, pp. 2–38. Springer, Heidelberg (2001)
9. Swaminathan, R.: Focus in Catadioptric Imaging Systems. In: 11th IEEE International Conference on Computer Vision, pp. 1–7. IEEE Press, New York (2007)
10. Hasinoff, S.W., Kutulakos, K.N.: Light-Efficient Photography. In: Forsyth, D., Torr, P., Zisserman, A. (eds.) ECCV 2008, Part IV. LNCS, vol. 5305, pp. 45–59. Springer, Heidelberg (2008)
11. Agarwala, A., et al.: Interactive Digital Photomontage. In: SIGGRAPH, pp. 294–302. ACM, New York (2004)
12. ACCOWLE Co. Ltd.: Production list, http://www.accowle.com/english/products.html
13. Telecentric lens, http://en.wikipedia.org/wiki/Telecentric_lens
14. Kim, C.: Segmenting a Low-Depth-of-Field Image Using Morphological Filters and Region Merging. IEEE Trans. Image Processing 14(10), 1503–1511 (2005)

Colorization Using Segmentation with Random Walk

Xiaoming Liu[1], Jun Liu[1], and Zhilin Feng[2]

[1] College of Computer Science and Technology, Wuhan University of Science and Technology, Wuhan 430081, China
[2] Zhijiang College, Zhejiang University of Technology, Hangzhou 310024, China
lxmspace@gmail.com

Abstract. Traditional monochrome image colorization techniques require considerable user interaction and a lot of time. The segment-based colorization works fast but at the expense of detail loss because of the large segmentation; while the optimization based method looks much more continuous but takes longer time. This paper proposed a novel approach: Segmentation colorization based on random walks, which is a fast segmentation technique and can naturally handle multi-label segmentation problems. It can maintain smoothness almost everywhere except for the sharp discontinuity at the boundaries in the images. Firstly, with the few seeds of pixels set manually scribbled by the user, a global energy is set up according to the spatial information and statistical grayscale information. Then, with random walks, the global optimal segmentation is obtained fast and efficiently. Finally, a banded graph cut based refine procedure is applied to deal with ambiguous regions of the previous segmentation. Several results are shown to demonstrate the effectiveness of the proposed method.

Keywords: colorization, random walks, graph cut, image segmentation.

1 Introduction

Colorization is the technique for adding color to monochrome images. It is an active and challenging area of research problem. In addition to the intentions of coloring old pictures and movies, colorization has also been applied into several other applications such as color changing (editing). This technique may also be used for efficiently encoding and decoding by separating the grayscale and color information.

However, colorization is traditionally very time-consuming. For example, in order to colorize an image, an artist typically begins by segmenting the image into regions, and then assign a color to each region. Unfortunately, fuzzy or complex region boundaries are very difficult to cope with for automatic segmentation methods. Thus, the artist often have to delineate complicated boundaries between different regions manually.

Due to the shortcoming of automatic segmentation methods, several researchers have proposed user guided interactive methods. Levin et al.[1] proposed a simple yet effective user-guided colorization method. In this method the user is required to scribble the desired colors in the interiors of the various regions. These constraints are formulated as a least-squares optimization problem that automatically propagates the scribbled colors to produce a completely colorized image. Other algorithms based on color scribbles have subsequently been proposed [2-4]. While these approaches hasve produced some impressive colorizations from a small amount of user input, they may still require a large amount of carefully placed scribbles for complex images. Qu et al. [5] proposed a colorization technique that propagates color over regions exhibiting pattern-continuity as well as intensity-continuity. The method works effectively on colorizing black-and-white manga which contains intensive amount of strokes, hatching, halftoning and screening. In [6], Takahiko et al. utilize seeded region growing segmentation alike method to colorize with weighted propagation of seed pixels. Irony et al. [7] presented a novel method to colorize grayscale images by transferring color from a segmented example image. Oscar et al. proposed a colorization method based on Bayesian segmentation [8].

Our work is based on the observations: 1) Most part of a grayscale image can be faithfully colorized with little user scribbles. 2) Most existing colorization method needs special deal to avoid the color mixing problem near object border. The reason lies in the fact that the important geometry information (edge) is often ignored during the colorization in previous methods. In this paper, we introduce a new colorization method. Our method utilizes edge information explicitly during colorization. Specially, similar to the method of Levin et al. [1], our approach starts with a few user scribbles. Then, the grayscale image is segmented into regions using random walk method [9, 10] with the constraints of user scribbles. The random walk is a multi-label, interactive segmentation method [9, 10], achieves segmentation by solving a PDE equation of discrete Direchlet problem. Since random walk is a probabilistic based segmentation method, and the final segmentation of pixels with low probability is not faithful, we utilize banded graph cut to refine these regions and produce the final colorization.

2 Colorization with Random Walks

The whole process of our method begins by a user scribbling on regions of interest. The proposed method consists of two steps: a quick color propagation step and a boundary refine step. The first step, color propagation works at a coarse scale, which utilizes the random walks to segment the image into regions. The second step, boundary refinement, utilizes banded graph cut to segment problematic regions.

Our method inherits the advantages of region-based and boundary-based methods in two steps. The first step is intuitive and quick for faithful segmentation and colorization, while the second step is efficient for accurate boundary control.

Random walks [9, 10] is used for segmentation in our method, so we give a review of it here. Suppose that the image is represented by a graph $G = (V, E)$, with vertices (nodes) $v \in V$ and edges $e \in E \subseteq V \times V$. An edge, e, spanning two vertices, v_i and v_j, is denoted by e_{ij}. A weighted graph assigns a value to each edge called weight. The weight of an edge, e_{ij}, is denoted by $w(e_{ij})$ or w_{ij}. The degree of a vertex is $d_i = \sum w(e_{ij})$ for all edges e_{ij} incident on v_i. In order to interpret w_{ij} as the bias affecting a random walker's choice, $w_{ij} > 0$ is required. The graph is assumed to be connected and undirected (i.e., $w_{ij} = w_{ji}$). The random walk problem is to assign probabilities to each node the probabilities which reaches each seeded nodes. The random walk probabilities problem has the same solution as the combinatorial Dirichlet problem [9]. Below, we give a short review of the combinatorial Dirichlet problem.

Given a weighted graph, a set of marked (labeled) nodes, V_M, and a set of unmarked nodes, V_U, such that $V_M \cup V_U = V$ and $V_M \cap V_U = \varnothing$, we would like to label each node $v_i \in V_U$ with a label from the set $C = \{c^1, c^2, \ldots, c^k\}$ having cardinality $k = |C|$. Assume that each node $v_i \in V_M$ has already been assigned a label, $x_i \in C$. The random walks approach to this problem given in [9, 10] is to assign to each node, $v_i \in V_U$, the probability, x_i^s, that a random walk starting from that node first reaches a marked node, $v_j \in V_M$, assigned to label c^s. The segmentation is then completed by assigning each free node to the label for which it has the highest probability, i.e., $x_i = \max_s x_i^s$.

It is known [9, 10] that the minimization of

$$E_{spatial} = x^{sT} L x^s \qquad (1)$$

for an $n \times 1$, real-valued vector x^s, defined over the set of nodes yields the probability, x_i^s, that a random walks starting from node v_i, first reaches a node $v_j \in V_M$ with label c^s (set to $x_j^s = 1$), as opposed to first reaching a node, $v_j \in V_M$, with label $c^{q \neq s}$ (set to $x_j^q = 0$), where L represents the combinatorial Laplacian matrix [9].

In this paper, we will concentrate on how to augment the minimization problem with user input. We refer readers to [9] for a detailed formulation of problem and how to solve it. In the original random walk algorithm, only pixel spatial information is utilized for the segment. If the user input is not just a few pixels for each label, the important statistical information of seeds should be used. Later, in [11], label prior information of pixel intensity is integrated in random walks method to free user interaction. The minimizing problem becomes

$$E_{Total}^s = E_{spatial}^s + \gamma E_{aspatial}^s \qquad (2)$$

where γ is a free parameter and $E_{aspatial}^s$ is the aspatial energy function.

In Equation (2), $E_{spatial}^s$ encodes the spatial information of a node, as in the original random walks method. $E_{aspatial}^s$ encodes the aspatial information of a node, represented by the density estimation. In [11], simple Gaussian kernel is used to produce the densities corresponding to each of the k labels. In our method, to use the statistical information of each class label c^s, we cluster the intensity in seeds of each class c^s by the K-means method firstly. The mean intensity of the clusters of each labels are denoted as $\{K_n^s\}$. The K-means method is initialized to have 4 clusters in our experiments. Then, for each node v_i, we compute the minimum distance from its intensity $I(i)$ to each label c^s as $d_i^s = \min_n \| I(i) - K_n^s \|$. Therefore, the probability, λ_i^s, that node v_i is generated from the distribution corresponding to label c^s is generated through

$$\lambda_i^s = \frac{1}{Z^s} e^{(-\frac{d_i^s}{\sigma})} \tag{3}$$

where σ is a free parameter and Z^s is a normalizing constant for label c^s equal to

$$Z^s = \sum_{q=1}^{k} e(-\frac{d_i^q}{\sigma}) \tag{4}$$

As in [11], the aspatial function is defined as

$$E_{aspatial}^s(x^s) = \sum_{q=1, q \neq s}^{k} x^{qT} \Lambda^q x^q + (x^s - 1)^T \Lambda^s (x^s - 1) \tag{5}$$

To energy in Equation (2) can be minimized as in the original random walks algorithm.

3 Deambiguous Refining

Although the random walk segment preserves the object boundary as accurately as possible, for it is a probabilistic method, there still exist some errors, especially around ambiguous and low contrast edge boundaries. Therefore, we utilize banded graph cut to refine the ambiguous boundary. Firstly, the ambiguous regions are located by their probabilities. A pixel is considered as ambiguous if the difference between its two biggest probabilities is less than a threshold δ. The threshold δ depends on the number of labels (typically $\frac{1}{10k}$ for k different labels). The refine process is dealted as a binary labeling problem. The denoted ambiguous regions are first merged into big regions: if two ambiguous regions are in neighborhood and their largest two probabilities labels are the same, they are merged. For each ambiguous region R_t, let its largest two probabilities belong to label u and v, the banded graph cut problem is constructed on pixels belongs to u and v. The labeling problem is to assign a unique label x_i for each node $i \in R_t$, i.e. $x_i \in \{u(=1), v(=0)\}$. The solution $X = \{x_i\}$ can be obtained by minimizing a Gibbs energy E(X):

$$E(X) = \sum_{i \in R_t} E_1(x_i) + \lambda \sum_{(i,j) \in \varepsilon} E_2(x_i, x_j) \tag{6}$$

where $E_1(x_i)$ is the likelihood energy, encoding the cost when the label of node i is x_i, and $E_2(x_i, x_j)$ is the prior energy, denoting the cost when the labels of adjacent nodes i and j are x_i and x_j respectively.

After the random walks procedure, a straightforward choice for likelihood energy E_1 is to use the output of random walks. $E_1(x_i)$ is defined as follows:

$$\begin{cases} E_1(x_i = 1) = 0 & E_1(x_i = 0) = \infty \quad \forall i \in U \\ E_1(x_i = 1) = \infty & E_1(x_i = 0) = 0 \quad \forall i \in V \\ E_1(x_i = 1) = \dfrac{p_i^v}{p_i^u + p_i^v} & E_1(x_i = 0) = \dfrac{p_i^u}{p_i^u + p_i^v} \quad \forall i \in R \end{cases} \quad (7)$$

Here, U is the set of pixels belonging to label u, and V is the set of pixels belonging to label v, and p_i^u and p_i^v is the calculated probabilities of node i belonging to label u and v respectively during the random walk process. The first two equations guarantee that the nodes in U and V will always have the label consistent with confident random walks output. The third equation encourages the nodes in ambiguous region to have the label with larger probabilities.

We use E_2 to represent the energy due to the gradient along the object boundary. We define E_2 as a function of the intensity gradient between two nodes i and j:

$$E_2(x_i, x_j) = |x_i - x_j| \cdot g(I_{ij}) = |x_i - x_j| \cdot \frac{1}{\|G(i) - G(j)\|^2 + 1} \quad (8)$$

where $g(x) = \dfrac{1}{x+1}$ and I_{ij} is the intensity difference between two pixels i and j. Note that $|x_i - x_j|$ allows us to capture the gradient information only along the segmentation boundary. In other words, E_2 is a penalty term when adjacent nodes are assigned with different labels. The more similar the intensities of the two nodes are, the larger E_2 is, and thus the less likely the edge is on the object boundary.

To minimize the energy E(X) in Equation (6), we use the maxflow algorithm in [12]. This algorithm is specially designed for vision problems. Although it is time consuming to run graph cut on a whole image, we run graph cut only on ambiguous region and with banded graph cut, running time is significantly reduced.

4 Experiment Results

We now present examples of our image colorization technique. The proposed algorithm has been implemented in Matlab. For timing we used an Intel Pentium-M 1.7GHz CPU and 1G RAM running under Windows XP. Fig.1 shows examples of images colorization using our proposed algorithm. The algorithm run time for the examples in Fig.1, measured once the images were loaded into memory, is less than 30 us per pixel.

Figs.2 and 3 compare our method with the one recently proposed by Levin et al. [1]. The method minimizes the difference between a pixel's color and the weighted average color of its neighboring pixels. The weights are provided by the luminance channel. The minimization is an optimization problem, subject to constraints supplied by the user as chrominance scribbles. First, in Fig.2, we observe that we can achieve a

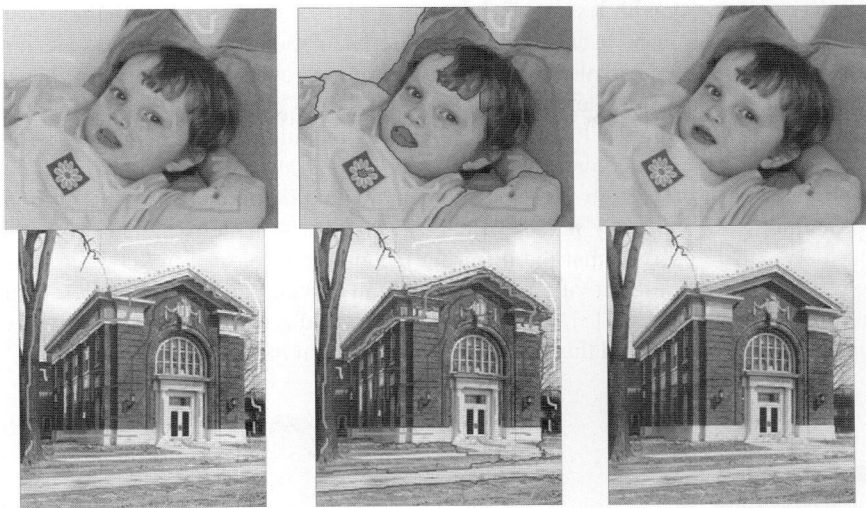

Fig. 1. Image colorization examples. Given a grayscale image, the user marks chrominance scribbles (left), our method segment the image with user provided scribbles (middle), and colorized images (right).

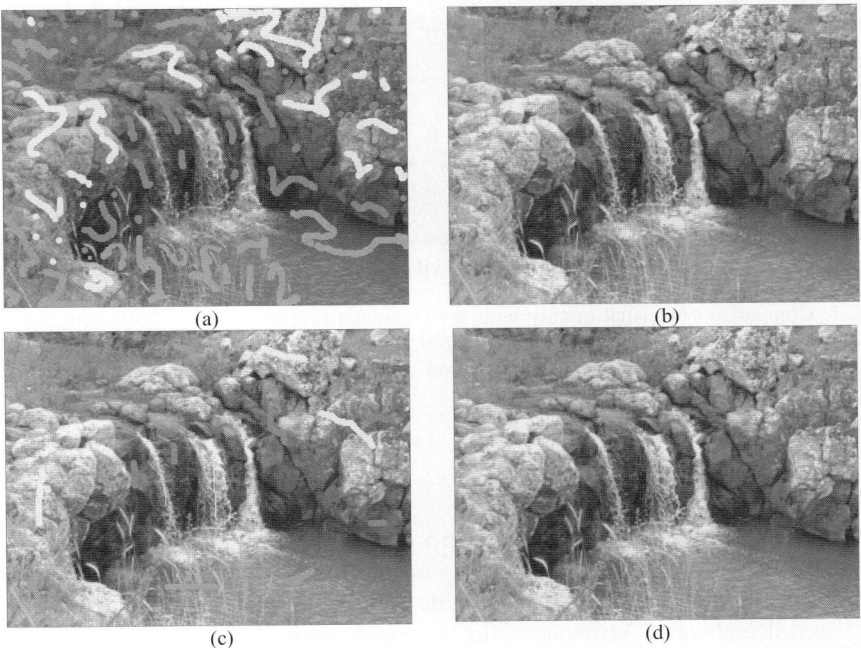

Fig. 2. Comparison of user effort with the technique proposed in [1]. (a) Given grayscale image with user scribbles in [1]; (b) the colorization result in [1]; (c) user scribbles of our method; (d) colorization result of our method. We observe the similar quality at a significantly reduced user effort.

similar visual quality with less user input than in [1]. In our method, with the usage of statistical information of user scribbles, our segmentation based method can utilize spatial and aspatial information simultaneously and so even pixels with large spatial distance but similar grayscale value can be colored with user scribbles. On the whole, the method proposed in [1] performs very well on many images, yet it performs poorly sometimes when colorizing pixels relatively far away from the provided color constrains, as can be seen in Fig.3. As our method explicitly deal with object border information, even far away pixels will receive color from the scribbles in our approach. We also observed that the inspiring technique developed in [1] often fails at strong edges, since these provide zero or very limited weight/influence in their formulation, also pointed out by [3]. While in our method, due to the explicitly process of edge, our method can avoid this problem, as can be seen in Fig.3.

Fig. 3. Comparison of visual quality with the technique proposed in [1]. (a) Given grayscale image; (b) the user marks chrominance scribbles; (c) colorization result of our method; (d) Levin et al. approach with fast implementation of multigrid solver; (e) Levin et al. approach with exact Matlab least squares solver.

5 Conclusion

Despite considerable progress in image processing since 1970, colorization remains a manually intensive and time consuming process. In this paper, we have introduced a new colorization method based on random walks segmentation method. With graph cut based deambiguous refinement, the proposed approach needs less manual effort than previous techniques and can colorize a grayscale image within seconds.

References

1. Anat, L., Dani, L., Yair, W.: Colorization using optimization. ACM Transaction on Graph 23(3), 689–694 (2004)
2. Sapiro, G.: Inpainting the colors. In: IEEE International Conference on Image Processing, ICIP 2005 vol. 2, pp. 698–701 (2005)
3. Yatziv, L., Sapiro, G.: Fast image and video colorization using chrominance blending. IEEE Transactions on Image Processing 15(5), 1120–1129 (2006)
4. Suganuma, K., Sugita, J., Takahashi, T.: Colorization using harmonic templates. In: International Conference on Computer Graphics and Interactive Techniques. ACM, New York (2008)
5. Qu, Y., Wong, T.T., Heng, P.A.: Manga colorization. ACM Transactions on Graphics (TOG) 25(3), 1214–1220 (2006)
6. Horiuchi, T., Hirano, S.: Colorization algorithm for grayscale image by propagating seed pixels. In: 2003 International Conference on Image Processing. IEEE, Los Alamitos (2003)
7. Irony, R., Cohen-Or, D., Lischinski, D.: Colorization by example. In: Proc. of Eurographics Symposium on Rendering 2005 (2005)
8. Dalmau-Cedeno, O., Rivera, M., Mayorga, P.P.: Computing the alpha-Channel with Probabilistic Segmentation for Image Colorization (2007)
9. Grady, L., Funka-Lea, G.: Multi-label image segmentation for medical applications based on graph-theoretic electrical potentials. In: ECCV. Springer, Heidelberg (2004)
10. Grady, L.: Random Walks for Image Segmentation. IEEE Transactions on Pattern Analysis and Machine Intelligence, 1768–1783 (2006)
11. Grady, L.: Multilabel random walker image segmentation using prior models. In: CVPR (2005)
12. Boykov, Y.Y., Jolly, M.P.: Interactive graph cuts for optimal boundary and region segmentation of objects in N-D images. In: Eighth IEEE International Conference on Computer Vision, ICCV 2001, Proceedings (2001)

Edge-Based Image Compression with Homogeneous Diffusion

Markus Mainberger and Joachim Weickert

Mathematical Image Analysis Group,
Faculty of Mathematics and Computer Science, Campus E1.1
Saarland University, 66041 Saarbrücken, Germany
{mainberger,weickert}@mia.uni-saarland.de

Abstract. It is well-known that edges contain semantically important image information. In this paper we present a lossy compression method for cartoon-like images that exploits information at image edges. These edges are extracted with the Marr–Hildreth operator followed by hysteresis thresholding. Their locations are stored in a lossless way using JBIG. Moreover, we encode the grey or colour values at both sides of each edge by applying quantisation, subsampling and PAQ coding. In the decoding step, information outside these encoded data is recovered by solving the Laplace equation, i.e. we inpaint with the steady state of a homogeneous diffusion process. Our experiments show that the suggested method outperforms the widely-used JPEG standard and can even beat the advanced JPEG2000 standard for cartoon-like images.

Keywords: image compression, partial differential equations (PDEs), Laplace equation, contour coding.

1 Introduction

Edges are not only semantically important for humans, they also play a central role in image processing and computer vision. Edge detection can be regarded as an intermediate step from a pixel-based to a semantic image representation. Since it is more compact to describe an image by a few contours than by many pixels, an edge representation is also of potential interest for image compression.

One of the classical edge detectors is based on the Marr-Hildreth operator [1], which extracts edges as zero-crossings of the Laplacian of a Gaussian-smoothed version of the image. Numerous theoretical and experimental papers have been written that investigate if reconstructions from these zero-crossings are possible [2,3,4,5,6,7,8,9,10,11,12]. Unfortunately, it turned out that specifying only the zero-crossing locations is insufficient for typical real-world images. Thus, it has been suggested that one should supplement additional information such as the image gradient, grey values adjacent to the edge, subsampled image data, or scale information. However, none of the before mentioned publications has led to an image compression method that yields results which are competitive to modern compression standards such as JPEG [13] or JPEG2000 [14]: Either the required data cannot be encoded in a sufficiently compact way, or the results turn out to be of inferior quality.

The goal of the present paper is to address this problem. We provide a proof of concept that information on the edge location together with the adjacent grey or colour values is sufficient to reconstruct cartoon-like images in high quality when the unspecified locations are filled in by the steady state of a homogeneous diffusion process. Moreover, by using state-of-the-art techniques for encoding the edge locations and the adjacent grey/colour values, our method may even outperform the quality of leading compression standards such as JPEG2000.

Our semantic image compression approach can be regarded as a specific implementation of a recent result on optimal point selection for compression with homogeneous diffusion: In [15] it is proven that one should preferently store pixels with large modulus of the Laplacian. For a piecewise constant image this comes down to the pixels left and right of an edge contour. It should be noted that contours can be encoded more efficiently than the same number of individual pixels, and they avoid visually unpleasant singularities of the fundamental solution of the Laplace equation. Moreover, by using homogeneous diffusion, our method is simpler and potentially faster than recent compression methods based on nonlinear anisotropic diffusion processes [16].

The structure of our paper is as follows: In Section 2, we present the encoding algorithm of our lossy compression scheme. Section 3 explains how to decode and reconstruct the image information by means of interpolation with homogeneous diffusion. Experiments and a comparison to JPEG and JPEG2000 are presented in Sect. 4. Finally, the paper is concluded with a summary in Sect. 5.

2 Encoding

Step 1: Detecting Edges. Our encoding of an image starts with an edge detection. We use the Marr-Hildreth edge detector [1] combined with a hysteresis thresholding as suggested by Canny [17].

To this end, we extract the zero-crossings of the Laplacian of a Gaussian-smoothed image. To remove zero-crossings that have no obvious perceptual significance, we apply hysteresis thresholding: Pixels obtained by the Marr-Hildreth edge detector with a gradient magnitude that is larger than a lower threshold are considered as edge candidates. All edge candidates with a gradient magnitude that is larger than a higher threshold become seed points for relevant edges and are automatically edge pixels of the final edge image. In order to keep edge pixels still connected as much as possible, we recursively add all edge candidates that are adjacent to final edge pixels.

For cartoon-like images this algorithm gives well-localised contours that are often closed. However, we are not necessarily bound to the suggested edge detector: For images that contain blurry edges, closed contours may be less important, and the original Canny edge detector [17] can give better results. Choosing alternative edge detectors can also be advantageous regarding noisy images. In our experiments we focus on the edge detector described above.

Fig. 1. Zoom (120 × 160) into the image *comic* **(a)** Original image. **(b)** Edge image. **(c)** Colours next to the edges and image boundary. **(d)** Interpolated version using homogeneous diffusion.

Step 2: Encoding the Contour Location. The result of the edge detection step can be regarded as a bi-level edge image as is illustrated in Fig. 1(b). This bi-level image encodes indirectly the location that is used for interpolation later on. We store this image by using the *JBIG (Joint Bi-level Image Experts Group)* standard [18]. It has been developed as a specialised routine for lossless as well as for lossy compression of bi-level images, particularly with regard to textual images for fax transmission.

In this paper we are interested in lossless compression with JBIG. We use the JBIG-KIT [19], which is a free C implementation of the JBIG encoder and decoder. The JBIG standard relies on a context-based arithmetic coding. Two prediction steps are applied beforehand to except pixels from arithmetic coding that can be encoded more efficiently by other methods. In our case only the so-called typical prediction applies since we do not use the progressive mode of JBIG.

Step 3: Encoding the Contour Pixel Values. Next, we consider the pixel values we want to store. Since edges usually split areas of different brightness or colour, we do not use the pixel values that lie directly on the edge, but store the values from both sides of the edge instead (see Fig. 1(c) and [7]). Additionally, all pixel values from the border of the image domain are stored such that these values are also available as Dirichlet boundary for the diffusion-based filling-in later on.

Compression methods are often based on the fact that most differences between consecutive values within a signal are small. In our case we know that the pixel values along a contour usually change only gradually. Thus, it is reasonable to store the pixel values by the order of their occurrence along the edge.

The extracted pixel values can be uniformly quantised to 2^q different values, where $q \in \{1, \ldots, 8\}$. The parameter q can be chosen by the user depending on compression requirements. For RGB colour images, all channels are quantised equally.

A second compression parameter is the sampling distance d: For $d > 1$ a subsampling on the pixel values is performed, i.e. only every d-th value along an edge is stored. As already stated, the pixel values change only marginally along an edge. In contrast, the pixel values of distinct edges might differ by a significant amount. Thus, it is indispensable to subsample each edge separately to obtain feasible reconstruction results.

The quantised and subsampled pixel value signal is then compressed by a *PAQ* compression method [20]. PAQ describes a family of lossless data compression archivers, which are licensed under GPL. At the expense of run time and memory requirements, the compression rates of PAQ outperform those of other compression programmes such as the well-established data compression algorithm *bzip2*. PAQ uses a context mixing algorithm, which is related to Prediction by Partial Matching (PPM) [21]. The compression algorithm is divided into a predictor and an arithmetic coder. For the prediction, PAQ is provided with a large number of models conditioned on different contexts often tuned to special file formats. In this paper we use *PowerPAQ* which is a command line and GUI front end using PAQ8o6 for 32 and 64 bit Linux [22].

Step 4: Storing the Encoded Data. Now that we have encoded the contour location and pixel values, we want to consider the final image format. Obviously, we need to store the quantisation parameter q and the sampling distance d as header data. Furthermore, we store the size of the JBIG data part in order to be able to split the JBIG data from the PAQ data when decoding. The number of channels (1 or 3) has to be stored explicitly for decoding the PAQ part, whereas the image size is automatically encoded in the JBIG data part.

Our entire coded image format is then given by the following representation.

Header					File data	
Size of JBIG data	Number of channels	Quantisation parameter q	Sampling distance d		JBIG data	PAQ data
4 bytes	1 byte	1 byte	3 bytes			

3 Decoding

Step 1: Decoding the Contour Location and Pixel Values. As a first step of the decoding phase, we split our encoded file into the JBIG data and PAQ data part. Both parts are then decoded by using the JBIG and the PAQ method, respectively.

We reconstruct the quantised colours obtained by the PAQ part. The pixel values between the sampled points are computed by using linear interpolation along each edge. Higher order interpolations do not give any advantage since the pixel values hardly vary along an edge.

The decoded pixel values have to be placed at their corresponding positions in the final image: The JBIG data provides a bi-level edge image of the original image size. Given the image size and the edge pixel positions, the pixel values are arranged around the edges in the same order in which they have been encoded.

Step 2: Reconstructing Missing Data. So far, we have decoded the edge locations and the pixel values surrounding the edges. The idea is now to apply interpolation to reconstruct grey/colour values of pixels that lie between edges. We keep the pixel values at the positions obtained in Step 1 and use homogeneous diffusion for interpolation. This is the simplest and computationally most favourable inpainting approach based on partial differential equations (PDEs) [23]. Missing data is reconstructed by computing the steady state of the diffusion equation [24]

$$\partial_t u = \Delta u \, , \tag{1}$$

where the given pixel values are considered to form Dirichlet boundaries. Thus, the reconstructed data satisfies the Laplace equation $\Delta u = 0$ (see also [7]). Such a PDE can be discretised in a straightforward way by finite differences [25].

Interestingly, diffusion-based inpainting from image edges resembles a classical finding in biological vision: Already in 1935 Werner made the hypothesis that a contour-based filling-in process is responsible for the human perception of surface brightness and colour [26].

4 Experiments

Let us now investigate the capabilities of the suggested compression method. Figure 2 shows different test images and their compressed versions using JPEG, JPEG2000, and the suggested PDE-based approach. The compression rates lie between 0.20 and 0.35 bits per pixel (bpp). Since the original colour images use 24 bits per pixel, this comes down to compression ratios of 120:1 to roughly 70:1.

The compression ratio for the PDE-based approach can be influenced by the sampling distance d and the quantisation parameter q but also by the underlying edge image. However, for cartoon-like images, there is most often only one reasonable edge set. All parameters were chosen to give visually pleasant results. Experiments have shown that it suffices to set the quantisation parameter q to 4 or 5, i.e. 16 or 32 different pixel values per channel. The sampling distance should be chosen depending on how detailed the input image is. For simple images such as *comic* we could use $d = 30$ without getting obvious visible degradations in the reconstruction. However, common image manipulation and display programmes are not able to create JPEG2000 images with such high compression rates. Thus, for the sake of comparability all given examples use $d = 5$.

In Figure 2 we observe that JPEG as well as JPEG2000 coding suffers from severe ringing artifacts in regions of edges (see *comic*, and zoomed region thereof). This is a result of their quantisation in the frequency domain and the following back transformation. In the JPEG images block artifacts appear because the discrete cosine transform is computed within blocks of 8×8 pixels. Thus, JPEG cannot even properly describe the smooth gradient in the background of *svalbard* and *comic*. In contrast, our method stores edges explicitly and interpolates in regions between edges. As a result, edges are well-preserved and smooth gradients can still be represented. Visually, this gives the most appealing results.

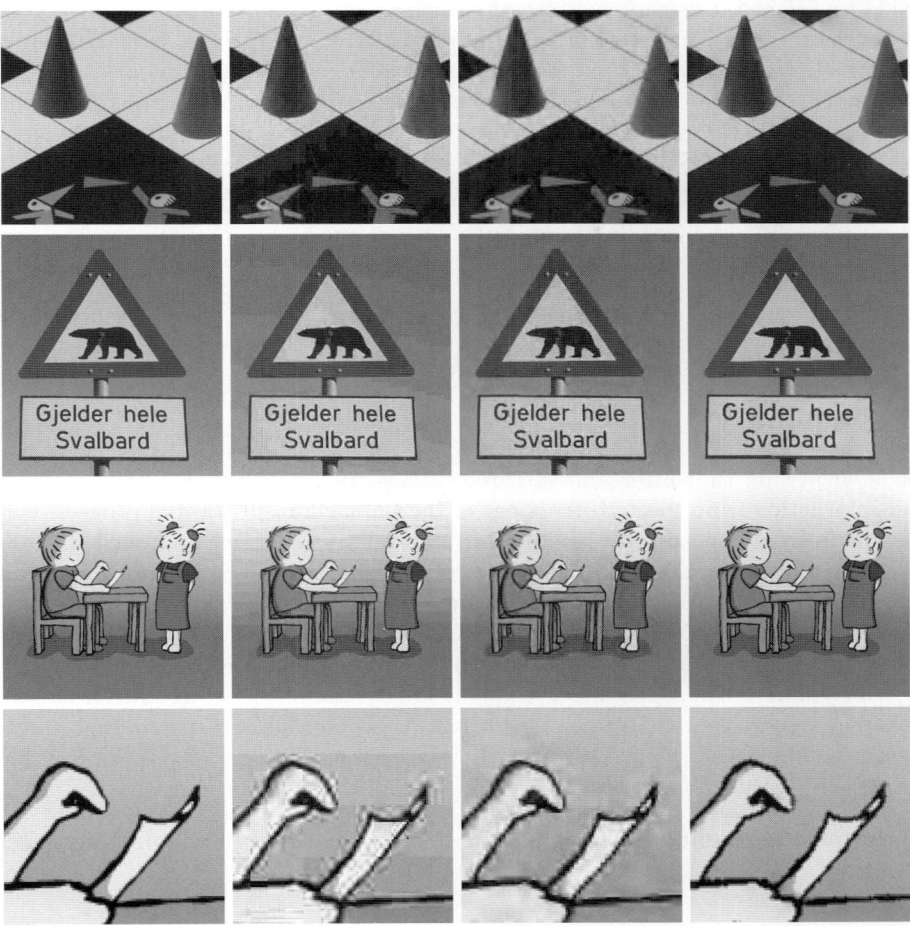

Fig. 2. Comparison of compression methods for different test images. *Rows from top to bottom: coppit* (256 × 256), *svalbard* (380 × 431), *comic* (512 × 512), *comic* (detail) (100 × 100). *Columns from left to right:* original image, JPEG, JPEG2000, and PDE-based compression using contours (sampling distance d from top to bottom: 5, 5, 5; quantisation parameter q: 4, 5, 5).

Table 1. Comparison of the PSNR for different images (see Fig. 2) and different compression methods

image	*coppit*	*svalbard*	*comic*
compression rate	0.34 bpp	0.23 bpp	0.21 bpp
JPEG	26.14	26.91	25.54
JPEG2000	27.56	30.06	27.44
our method	**30.16**	**30.21**	**30.31**

For a quantitative comparison, we use the *peak-signal-to-noise ratio* (PSNR), a common error measure for the qualitative analysis of images. It is defined as follows: Let m be the maximal possible pixel value, which is 255 in our case. Furthermore, let N be the number of image pixels and $(f_i)_{i=1..N}$ and $(g_i)_{i=1..N}$ the pixel values of the original image and its reconstructed/decompressed version, respectively. The PSNR is then defined via the *mean squared error* (MSE):

$$\text{PSNR} := 10 \cdot \log_{10}\left(\frac{m^2}{\text{MSE}}\right) \text{ [dB]} \quad \text{with} \quad \text{MSE} := \frac{1}{N}\sum_{i=1}^{N}(f_i - g_i)^2 \ . \quad (2)$$

Table 1 shows the results for the images presented in Fig. 2 and confirms our visual impression.

5 Summary

It is surprising that after almost three decades of intensive research on image reconstruction from the zero-crossings of the Laplacian, substantial progress can be made by a conceptually simple, but carefully engineered approach. By extracting the information on both sides of the zero-crossings, encoding it in an efficient way, and decoding with homogeneous diffusion inpainting, we have presented a codec that can even beat JPEG2000 for cartoon-like images. This was out of reach for all previous methods on edge-based image reconstruction.

Since none of the steps in our codec is highly complex, we are currently performing research on real-time algorithms for sequential and distributed architectures. Moreover, we are investigating extensions that are also optimised for edges with very smooth slope as well as for highly textured images.

Acknowledgements. We thank Anna Mainberger for providing the image *comic*.

References

1. Marr, D., Hildreth, E.: Theory of edge detection. Proceedings of the Royal Society of London, Series B 207, 187–217 (1980)
2. Logan Jr., B.F.: Information in the zero crossings of bandpass signals. Bell System Technical Journal 56, 487–510 (1977)
3. Yuille, A.L., Poggio, T.A.: Scaling theorems for zero crossings. IEEE Transactions on Pattern Analysis and Machine Intelligence 8(1), 15–25 (1986)
4. Curtis, S.R., Oppenheim, A.V., Lim, J.S.: Reconstruction of two-dimensional signals from threshold crossings. In: Proc. IEEE International Conference on Acoustics, Speech and Signal Processing, Tampa, FL, March 1985, vol. 10, pp. 1057–1060 (1985)
5. Zeevi, Y., Rotem, D.: Image reconstruction from zero-crossings. IEEE Transactions on Acoustics, Speech, and Signal Processing 34, 1269–1277 (1986)
6. Chen, S.: Image reconstruction from zero-crossings. In: Proc. Eighth International Joint Conference on Artificial Intelligence, Milan, Italy, August 1987, vol. 2, pp. 742–744 (1987)

7. Carlsson, S.: Sketch based coding of grey level images. Signal Processing 15, 57–83 (1988)
8. Hummel, R., Moniot, R.: Reconstructions from zero-crossings in scale space. IEEE Transactions on Acoustics, Speech, and Signal Processing 37, 2111–2130 (1989)
9. Grattoni, P., Guiducci, A.: Contour coding for image description. Pattern Recognition Letters 11(2), 95–105 (1990)
10. Mallat, S., Zhong, S.: Characterisation of signals from multiscale edges. IEEE Transactions on Pattern Analysis and Machine Intelligence 14, 720–732 (1992)
11. Dron, L.: The multiscale veto model: A two-stage analog network for edge detection and image reconstruction. International Journal of Computer Vision 11(1), 45–61 (1993)
12. Elder, J.H.: Are edges incomplete? International Journal of Computer Vision 34(2/3), 97–122 (1999)
13. Pennebaker, W.B., Mitchell, J.L.: JPEG: Still Image Data Compression Standard. Springer, New York (1992)
14. Taubman, D.S., Marcellin, M.W. (eds.): JPEG 2000: Image Compression Fundamentals, Standards and Practice. Kluwer, Boston (2002)
15. Belhachmi, Z., Bucur, D., Burgeth, B., Weickert, J.: How to choose interpolation data in images. SIAM Journal on Applied Mathematics (in press, 2009)
16. Galić, I., Weickert, J., Welk, M., Bruhn, A., Belyaev, A., Seidel, H.P.: Image compression with anisotropic diffusion. Journal of Mathematical Imaging and Vision 31(2-3), 255–269 (2008)
17. Canny, J.: A computational approach to edge detection. IEEE Transactions on Pattern Analysis and Machine Intelligence 8, 679–698 (1986)
18. Joint Bi-level Image Experts Group: Information technology – progressive lossy/lossless coding of bi-level images. ISO/IEC JTC1 11544, ITU-T Rec. T.82, Final Committee Draft 11544 (1993)
19. Kuhn, M.: Effiziente Kompression von bi-level Bilddaten durch kontextsensitive arithmetische Codierung. Studienarbeit, Institut für Mathemathische Maschinen und Datenverarbeitung der Friedrich-Alexander-Universität Erlangen-Nürnberg, Germany (July 1995)
20. Mahoney, M.: Adaptive weighing of context models for lossless data compression. Technical Report CS-2005-16, Florida Institute of Technology, Melbourne, Florida (December 2005)
21. Moffat, A.: Implementing the ppm data compression scheme. IEEE Transactions on Communications 38(11), 1917–1921 (1990)
22. Varnavsky, E.: PowerPAQ., http://powerpaq.sourceforge.net/ (last visited March 21, 2009)
23. Masnou, S., Morel, J.M.: Level lines based disocclusion. In: Proc. 1998 IEEE International Conference on Image Processing, Chicago, IL, October 1998, vol. 3, pp. 259–263 (1998)
24. Iijima, T.: Basic theory on normalization of pattern (in case of typical one-dimensional pattern). Bulletin of the Electrotechnical Laboratory 26, 368–388 (1962) (in Japanese)
25. Morton, K.W., Mayers, L.M.: Numerical Solution of Partial Differential Equations. Cambridge University Press, Cambridge (1994)
26. Werner, H.: Studies on contour. The American Journal of Psychology 47(1), 40–64 (1935)

Color Quantization Based on PCA and Kohonen SOFM

D. Mavridis and N. Papamarkos

Image Processing and Multimedia Laboratory
Department of Electrical & Computer Engineering
Democritus University of Thrace

Abstract. A method for initializing optimally Kohonen's Self-Organizing Feature Maps (SOFM) of a fixed zero neighborhood radius for use in color quantization is presented. Standard SOFM is applied to the projection of the input image pixels onto the plane spanned by the two largest principal components and to pixels of the original image defined by the smallest principal component via a thresholding procedure. The neuron values which emerge initialize the final SOFM of a fixed zero neighborhood radius that performs the color quantization of the original image. Experimental results show that the proposed method is able to produce smaller quantization errors than standard SOFM and other existing color quantization methods.

Keywords: Color quantization, neural networks, self-organizing feature maps.

1 Introduction

Color quantization is an image processing technique whose purpose is to reduce the colors of digital images to a small limited number in such a way that a measure of difference between the original image and the color quantized one is minimized.

The color quantization techniques can be classified into major categories. First, it is the class of splitting algorithms where the color space is divided into disjoined regions by consecutively splitting up the color space and then choosing a color to represent the region in the color palette e.g. median-cut [10], variance-based algorithm [19], Octree [1]. They are considered to be fast algorithms, however better quality results are usually obtained by methods of the second major class which are based on cluster analysis of the color space. The methods of SOFM [12], [13], Growing Neural Gas (GNG) [8], Adaptive Color Reduction (ACR) [16], Self-Growing and Self-Organized neural network (SGONG) [2], Fuzzy ART [4], Fuzzy C-Means (FCM) [3], [15] belong to this category. In these methods the cluster representatives, chosen so that a within-cluster distance criterion (average, nearest neighbor, centroid) is minimized and a between-cluster linkage distance (single, complete, average, centroid) is maximized, are continuously adapted to the color content of the original image achieving this way a statistically analogue color distribution and good quality results.

The optimal number of image dominant colors can be estimated initially [2], [16]. The presented method in this paper quantizes an image to a priori fixed number of colors. It is based on a combination of PCA and Kohonen's SOFM.

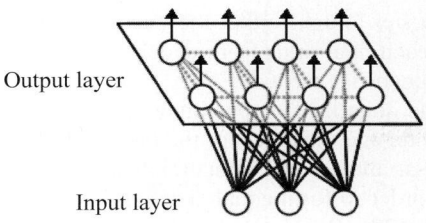

Fig. 1. A two dimensional SOFM neural network with an input layer of three neurons for use in color quantization

2 Self-organizing Feature Map

Kohonen's SOFM [12], [13] (Fig. 1) is a well known unsupervised neural network classifier. It consists of two separate layers of neurons, the input and output layer. The input layer operates as a buffer for the input vectors which are parallelly transmitted to all neurons of the output layer. The number of neurons and the connections between them are user defined and fixed, shaping a linear, planar or multidimensional lattice, independent of the input vectors dimension.

One of the main characteristics of SOFM is that it is topology preserving, meaning that close input signals are mapped to neurons which are close in the lattice structure and conversely, close neurons in the lattice structure come from close input signals in the input space. That makes SOFM a desirable classifier for use in color quantization.

The main steps for using SOFM to perform color quantization to N colors are:

Step 1: Define the output layer structure of the N neurons forming it. Initialize the neuron weights $w_i = [W_{i1}, W_{i2}, W_{i3}]^T$, $0 \leq i < N$ randomly selecting N input vectors (N pixel values $x_k = [r_k, g_k, b_k]^T$ from the input image). Reset variable t which increases on every new input vector $x_k = [r_k, g_k, b_k]^T$ from the input image. Reset epochs variable ep which increases on every m consecutive input vectors. Initialize the neighborhood radius r and define a neighborhood function $\text{nei}(w_i, ep)$, which includes the closest neighboring neurons to w_i, in radius r, at the output layer structure. Initialize learning rate $\varepsilon_1(ep)$ for the winner neuron and learning rate $\varepsilon_2(ep)$ for it's neighboring neurons.

Step 2: Get a new input vector $x_k=[r_k, g_k, b_k]^T$ and find the neuron(s) w_{cj} for which the square of the Euclidean distance in RGB color space is minimal as follows:

$$d(x_k, w_{cj})^2 = \min_i((r_k - W_{i1})^2 + (g_k - W_{i2})^2 + (b_k - W_{i3})^2), 0 \leq i < N. \quad (1)$$

Get the sole solution or pick randomly one of the neurons w_{cj} as winner neuron w_c.

Step 3: The winner neuron w_c and it's neighboring neurons w_b defined from $\text{nei}(w_c, ep)$ are biased towards the input vector $x_k=[r_k, g_k, b_k]^T$ as follows:

$$w_c(t+1) = w_c(t) + \varepsilon_1(ep)(x_k - w_c(t)). \quad (2)$$

$$w_b(t+1) = w_b(t) + \varepsilon_2(ep)(x_k - w_b(t)). \quad (3)$$

Step 4: If m consecutive vectors are inputted, increase ep variable by 1. In this case decrease also the values of learning rates $\varepsilon_1(ep)$ and $\varepsilon_2(ep)$ and the radius of the neighborhood domain.

Step 5: Repeat from step 2 until either the maximum difference of the correspondent weight values of each neuron w_i from epoch to epoch is below a threshold, or a maximum number of epochs is reached.

Step 6: Construct the quantized output image using Euclidean distance metric on the generated color palette and the pixels of the input image.

There exist several variants of SOFM neural network. A criterion of the frequency a neuron is chosen, in order to form equal-sized classes and to avoid underutilization is proposed in [6], [5], [20]. Besides color, local spatial features are considered in ACR [16]. Hybrid methods are also proposed, as in SGONG [2] where the neural network uses the GNG [8] mechanism of growing the neural lattice and the SOFM learning adaptation algorithm.

3 SOFM Initialization Based on Principal Component Analysis

When the SOFM was introduced for the first time, it was shown that the initial values of it's models in the learning process can be selected as random vectors, however the quality and speed of the convergence of the neurons are sensitive to initialization values. In general, it is considered to be a good initialization if the initial weights of the neurons are made to match the input image color distribution more closely. Kohonen in [12] mentions accordingly that the training process could be faster if the point density of the initial models approaches that of the input data and [14] recommends choosing the initial models as a regular array of vectors on the hyperplane spanned by the two largest principal components of input data.

On this basis a two-stage image segmentation is proposed in [11] where the input image is coarsely quantized capturing the dominant colors which are fed afterwards to an SOFM performing the final segmentation. In [17], it is proposed to initialize SOFM distributing the neurons uniformly along the luminance axis of RGB cube. An extension of this method is presented in [21] where a color group of neurons distributed uniformly throughout the RGB cube is added to the previous gray group.

A new initialization scheme (Fig. 2) of SOFM of a fixed zero neighborhood radius for color quantization to N colors is proposed. Let k_1 be the number of neurons to initialize so that their weights approach the input image color distribution, ($k_1 \leq N$), then as proposed in [14] for performing such an initialization, an SOFM with an output layer of k_1 neurons is applied to the projection of the input image pixels onto the plane spanned by the two largest principal components. The weights of the k_1 neurons, after applying the first SOFM, are the intended initialization of k_1 neurons of the final fixed zero neighborhood radius SOFM.

However, the rest k_2 neurons of the final SOFM ($k_1 + k_2 = N$) are initialized based on a contrary in principle method. The smallest principal component of the input image is used, considering this way the color information of pixels with less contribution to the input image color content. Namely, an SOFM with an output layer of k_2 neurons is applied to pixels of the input image defined by the smallest principal component. This selection of the input image pixels is carried out by means of thresholding the projection of the input image pixels onto the smallest principal component. The weights of the k_2 neurons, after applying the SOFM, are the intended initialization

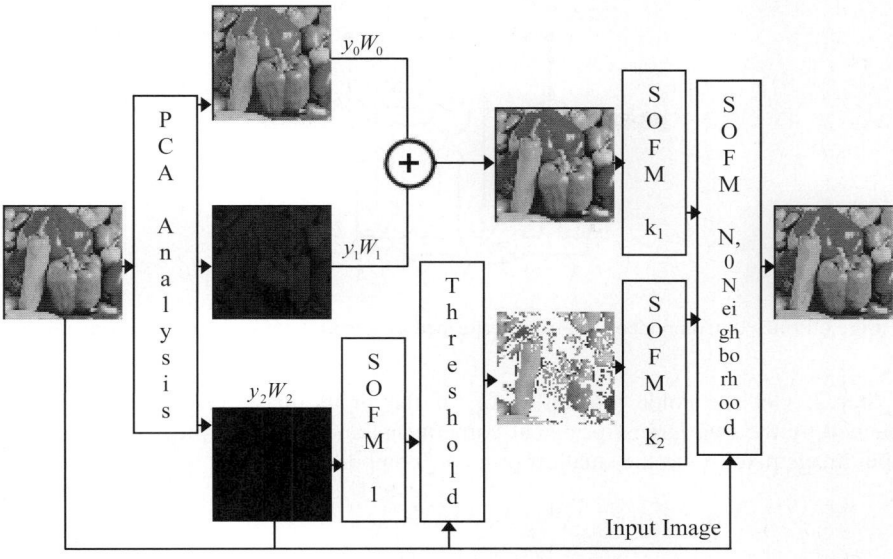

Fig. 2. Depiction of the proposed SOFM initialization method, 1, k_1, k_2 and N ($N=k_1+k_2$), are the number of neurons at the output layer of the respective SOFM neural networks

of the rest k_2 neurons of the final SOFM. Experimental results show that, depending on the values of k_1 and k_2, this initialization scheme can lead to more equal-sized classes compared to standard SOFM and better quality convergence than standard SOFM and other existing color quantization methods.

The main steps for using the proposed method (Fig. 2) to perform color quantization to N colors are:

Step 1: A Principal Component Analysis (PCA) method is performed. A single-layer feedforward neural network is used (Fig. 3), which is trained using the Generalized Hebbian Algorithm (GHA) [18], [9], [7]. This is an unsupervised learning algorithm based on the Hebbian learning rule which states that if two neurons on either side of a synapse are activated simultaneously, the strength of that synapse is selectively increased, otherwise it is weakened or eliminated. Let t be a variable which increases on every new input vector (a new pixel value $x_k = [r_k, g_k, b_k]^T$ from the input image 3-feature vectors X), w the matrix of PCA coefficients, $y = [y_0, y_1, y_2]^T$ the neural network's output computed by the relation $y = wX$, $W_i = [W_{i0}, W_{i1}, W_{i2}]^T$, $0 \le i \le 2$ the neuron weights vectors, n a small enough learning rate parameter. An advantage of using a PCA neural network is that it can be directly applied to large scale problems without explicitly computing the data covariance matrix.

Each principal component can be extracted from the respective neuron's weight vector as follows:

$$W_i(t+1) = W_i(t) + ny_i x_k - ny_i \sum_{j=0}^{i} y_j W_j(t), \quad y_i = W_i^T(t)x_k, \quad x_k \in X, 0 \le i \le 2. \quad (4)$$

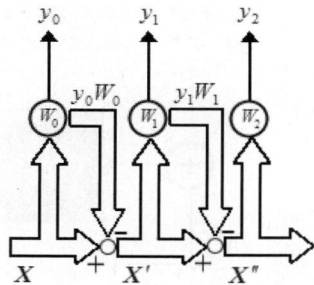

Fig. 3. Principal Component Analysis performed by a neural network implementing GHA

Step 2: Get the color projection $X_{0,1}$ of the input image pixels onto the plane spanned by the two largest principal components and the color projection X_2 of the input image pixels onto the smallest principal component:

$$X_{0,1} = y_0 W_0 + y_1 W_1, \; y_i = W_i^T(t) x_k, \; 0 \le i \le 1 . \qquad (5)$$

$$X_2 = y_2 W_2, \; y_2 = W_2^T(t) x_k . \qquad (6)$$

Step 3: Perform thresholding to the color projection X_2 of the input image pixels onto the smallest principal component, using an SOFM with a single neuron at the output layer.

Step 4: Choose the input image pixels p(x,y) based on the following criterion: the correspondent pixel $p_2(x,y)$ on color projection X_2 has one of it's R, G, B values greater than the respective R, G, B value of the threshold found on step 3.

Step 5: Set parameter value k_1 of the number of neurons initialized using the method involving the two largest principal components and parameter value k_2 of the number of neurons initialized using the smallest principal component.

Step 6: Perform color quantization on the color projection $X_{0,1}$, using an SOFM with k_1 neurons at the output layer.

Step 7: Perform color quantization only on pixels of the input image defined on step 4, using an SOFM with k_2 neurons at the output layer.

Step 8: Initialize the final's SOFM neuron weights using the k_1 and k_2 neuron weights from step 7 and step 8. Perform color quantization on the input image to N colors. Fixing the final SOFM's neighborhood radius to zero ensures that the classes will not influence each other, taking full advantage of the initialization process.

4 Experimental Results

In order to test the performance of the proposed initialization and compare it to other color quantization methods, the following nine images from Ohio State University Signal Analysis and Machine Perception Laboratory webpage (http://sampl.eng.ohio-state.edu) were used: "frymire", "girl", "peppers", "lena", "mandrill", "home3", "monarch", "sails" and "tulips". As a measure of quality of the quantization, Mean Squared Error (MSE) and Peak Signal-to-Noise Ratio (PSNR) (definition e.g. in [2]), commonly used for measuring statistical color quantization errors, were employed.

The methods compared to the proposed SOFM initialization, denoted as PCA-INIT SOFM, are: SOFM [12], [13] as described in section 2, ACR [16], SGONG [2], GNG [8] and FCM [3]. To define an epoch, $m=20$ different sets of 3000 samples are collected from the input image with a maximum of 1000 epochs. The applied settings for the algorithms are as in [2] for comparison reasons and set as follows:

SOFM: A one dimensional grid topology is used for the output layer, $\varepsilon_1(ep) = \varepsilon_2(ep)$ and both learning rates are linearly decreased from 0.02 to 0.0005 reaching their minimum value after 150 epochs from the beginning of the training procedure. The radius of the neighbourhood domain of the winner neuron is initially equal to 67% of the total number of output neurons and it linearly decreases to zero within 150 epochs. The training procedure has 1000 epochs. In each epoch $m=20$ different sets of 3000 samples are collected from the input image.

ACR: The SOFM settings for ACR are the same with SOFM in the previous paragraph. For the splitting and merging conditions: minimum number of pixels in each class is set to greater than 100 and minimum number of samples in each node of the tree is set to greater than 20. No other conditions for splitting (e.g variance based) or merging of the classes are set.

SGONG, GNG and FCM parameters are set as in [2].

PCA-INIT SOFM: After performing PCA, the parameters used for SOFM are set as described in previous SOFM paragraph. An exception is made to the maximum number of epochs which is limited to 300 and the neighborhood radius of the final's SOFM which is fixed to 0.

Table 1. Comparative quantization error results after quantizing the initial images to 16 colors. Parameters k_1 and k_2 for obtaining smaller quantization errors using PCA-INIT SOFM were set as follows: "frymire", "girl", "home3" $k_1=13$ and $k_2=3$, "monarch" $k_1=12$ and $k_2=4$, "mandrill", "sail", "tulips", $k_1=10$ and $k_2=6$, "lena", "peppers" $k_1=8$ and $k_2=8$.

		PCA-INIT SOFM	SOFM	ACR	SGONG	GNG	FCM
frymire	MSE	1335.522	1427.828	1396.897	1449.214	1427.742	1331.241
	PSNR	49.841	49.172	49.391	49.024	49.173	49.873
girl	MSE	63.466	73.567	81.858	69.148	107.079	86.997
	PSNR	79.816	78.339	77.271	78.959	74.586	76.663
lena	MSE	208.903	215.806	216.066	209.772	215.683	211.998
	PSNR	65.598	65.273	65.260	65.557	65.279	65.452
mandrill	MSE	627.452	633.958	629.804	628.259	651.224	633.788
	PSNR	56.905	56.801	56.867	56.892	56.533	56.804
monarch	MSE	254.858	259.082	261.738	258.467	269.133	255.983
	PSNR	64.837	64.673	64.570	64.697	64.292	64.793
peppers	MSE	370.211	379.565	374.679	377.123	392.961	372.182
	PSNR	62.618	62.369	62.498	62.433	62.022	62.565
sail	MSE	279.602	282.726	281.238	280.551	300.569	281.690
	PSNR	64.203	64.092	64.145	64.169	63.480	64.128
tulips	MSE	444.969	559.485	474.775	445.848	467.762	460.935
	PSNR	60.170	57.880	59.522	60.151	59.671	59.818
home3	MSE	170.138	175.094	175.509	179.223	178.552	176.510
	PSNR	67.524	67.237	67.213	67.004	67.042	67.157

In Table 1, quantization error results are presented for color quantization to 16 colors at the first execution of the algorithm. PCA-INIT parameter values of k_1 and k_2 are set, so that best quantization results are obtained. In Table 2, mean values, best and worst quantization error results are shown for five executions of PCA-INIT algorithm. Parameter values of k_1 and k_2 are set as in Table 1.

The method can be used for color quantization to a different number of colors and achieve analogue quality error results. It can also lead to more equal-sized classes compared to standard SOFM as described in section 2.

The proposed method was implemented in software that can be downloaded from http://ipml.ee.duth.gr/~papamark.

Table 2. Quantization error results on quantizing the initial images to 16 colors after executing PCA-INIT SOFM algorithm five times

PCA-INIT SOFM			Mean Value	Worst Value	Best Value
frymire	$k_1=13$	MSE	1381.027	1428.846	1335.522
	$k_2=3$	PSNR	49.508	49.165	49.841
girl	$k_1=13$	MSE	63.479	63.502	63.446
	$k_2=3$	PSNR	79.814	79.811	79.816
lena	$k_1=8$	MSE	211.834	216.214	208.901
	$k_2=8$	PSNR	65.460	65.254	65.598
mandrill	$k_1=10$	MSE	627.475	627.510	627.439
	$k_2=6$	PSNR	56.904	56.904	56.905
monarch	$k_1=12$	MSE	257.164	260.609	254.858
	$k_2=4$	PSNR	64.748	64.614	64.837
peppers	$k_1=8$	MSE	370.267	370.350	370.211
	$k_2=8$	PSNR	62.617	62.615	62.618
sail	$k_1=10$	MSE	279.719	280.188	279.580
	$k_2=6$	PSNR	64.199	64.182	64.204
tulips	$k_1=10$	MSE	453.308	460.280	444.969
	$k_2=6$	PSNR	59.985	59.832	60.170
home3	$k_1=13$	MSE	172.206	175.339	170.129
	$k_2=3$	PSNR	67.405	67.223	67.525

5 Conclusion

A method is proposed for color quantization which is based on PCA initialization of a Kohonen's SOFM of a fixed zero neighborhood radius. A number of the SOFM's neurons is initialized based on the color distribution of the input image, by means of applying standard SOFM on the projection of the input image pixels onto the plane spanned by the two largest principal components. The remaining neurons are initialized considering pixels with minor contribution to the color information of the input image, by means of using the smallest principal component and thresholding. This novel approach, for initializing the neuron weights of an SOFM of a fixed zero neighborhood radius to perform color quantization, is able to lead to smaller quantization errors than standard SOFM and other color quantization methods.

References

1. Ashdown, I.: Octree color quantization. In: Radiocity - A Programmer's Perspective. Wiley, New York (1994)
2. Atsalakis, A., Papamarkos, N.: Color reduction and estimation of the number of dominant colors by using a self-growing and self-organized neural GAs. Engineering Applications of Artificial Intelligence, 19(7), 769–786 (2006)
3. Bezdek, J.C.: Pattern Recognition with Fuzzy Objective Function Algorithms. Plenum Press, New York (1994)
4. Carpenter, G., Grossberg, S., Rosen, D.B.: Fuzzy ART: Fast stable learning and categorization of analog patterns by an adaptive resonance system. Neural Networks 4, 759–771 (1991)
5. Chang, C.-H., Xu, P.F., Xiao, R., Srikanthan, T.: New adaptive color quantization method based on self-organizing maps. IEEE Transactions on Neural Networks 16(1), 237–249 (2005)
6. Dekker, A.H.: Kohonen neural networks for optimal color quantization. Network: Computation in Neural Systems 5, 351–367 (1994)
7. Diamantaras, K.I., Kung, S.Y.: Principal Component Neural Networks. Wiley, New York (1996)
8. Fritzke, B.: Growing cell structures - a self-organizing network for unsupervised and supervised learning. Neural Networks 7(9), 1441–1460 (1994)
9. Haykin, S.: Neural Networks: A Comprehensive Foundation. MacMillan, New York (1994)
10. Heckbert, P.: Color image quantization for frame buffer display. Computer & Graphics 16, 297–307 (1982)
11. Huang, H.Y., Chen, Y.S., Hsu, W.H.: Color image segmentation using a self-organized map algorithm. Journal of Electronic Imaging 11(2), 136–148 (2002)
12. Kohonen, T.: The self-organizing map. Proceedings of IEEE 78(9), 1464–1480 (1990)
13. Kohonen, T.: Self-Organizing Maps, 2nd edn. Springer, Berlin (1997)
14. Kohonen, T.: Self-organized maps of sensory events. Royal Society of London Transactions Series A 361(1807), 1177–1186 (2003)
15. Lim, Y., Lee, S.: On the color image segmentation algorithm based on the thresholding and the fuzzy C-means techniques. Pattern Recognition 23(9), 935–952 (1990)
16. Papamarkos, N., Atsalakis, A., Strouthopoulos, C.: Adaptive color reduction. IEEE Trans. on Systems, Man and Cybernetics Part B: Cybernetics 32(1), 44–56 (2002)
17. Pei, S.-C., Lo, Y.-S.: Color image compression and limited display using self-organization Kohonen map. IEEE Transactions on Circuits Systems for Video Technoogy 8(2), 191–205 (1998)
18. Sanger, T.D.: An Optimality Principle for Unsupervised Learning. In: Advances in Neural Information Processing Systems. Morgan Kaufmann, San Francisco (1989)
19. Wan, S.J., Prusinkiewicz, P., Wong, S.K.M.: Variance based color image quantization for frame buffer display. Color Research and Application 15(1), 52–58 (1990)
20. Wang, C.-H., Lee, C.-N., Hsieh, C.-H.: Sample-size adaptive self-organization map for color images quantization. Pattern Recognition Letters 28, 1616–1629 (2007)
21. Xiao, R., Chang, C.-H., Srikanthan, T.: On the Initialization and Training Methods for Kohonen Self-Organizing Feature Maps in Color Image Quantization. In: Proc. of the First IEEE Internat. Workshop on Electronic Design, Test and Applications, pp. 321–325 (2002)

On Adapting the Tensor Voting Framework to Robust Color Image Denoising

Rodrigo Moreno[1], Miguel Angel Garcia[2], Domenec Puig[1], and Carme Julià[1,⋆]

[1] Intelligent Robotics and Computer Vision Group
Department of Computer Science and Mathematics, Rovira i Virgili University
Av. Països Catalans 26, 43007 Tarragona, Spain
{rodrigo.moreno,domenec.puig,carme.julia}@urv.cat
[2] Department of Informatics Engineering, Autonomous University of Madrid
Cra. Colmenar Viejo Km 15, 28049 Madrid, Spain
miguelangel.garcia@uam.es

Abstract. This paper presents an adaptation of the tensor voting framework for color image denoising, while preserving edges. Tensors are used in order to encode the CIELAB color channels, the uniformity and the edginess of image pixels. A specific voting process is proposed in order to propagate color from a pixel to its neighbors by considering the distance between pixels, the perceptual color difference (by using an optimized version of CIEDE2000), a uniformity measurement and the likelihood of the pixels being impulse noise. The original colors are corrected with those encoded by the tensors obtained after the voting process. Peak to noise ratios and visual inspection show that the proposed methodology has a better performance than state-of-the-art techniques.

1 Introduction

Color image denoising is an important task in computer vision and image processing, as images acquired through color image sensors are usually contaminated by noise. The main goal of color image denoising is to eliminate noise from color images while preserving their features, such as meaningful edges or texture details, as much as possible.

Two main approaches have been followed in color image denoising. The first approach, spatial domain filtering, filters the input image by using the color information of every pixel and its neighbors. The main problem of these filters is their tendency to blur the images. Classical filters, such as mean, median or Gaussian filters, non-local means [1], anisotropic diffusion [2], and conditional random fields [3], among many other methods, follow this approach. The second approach, transform-domain filtering, transforms the input image to a different

---

⋆ This research has been partially supported by the Spanish Ministry of Science and Technology under project DPI2007-66556-C03-03, by the Commissioner for Universities and Research of the Department of Innovation, Universities and Companies of the Catalonian Government and by the European Social Fund.

space, typically to the wavelet domain, filters the transformed image and applies the inverse transformation to the result. Despite its good edge preservation properties, the major criticism to this approach is the introduction of undesirable artifacts. Methods based on wavelets shrinkage [4] and Gaussian scale mixtures [5], among many others, follow the second approach.

This paper proposes a different solution for color image denoising in the spatial domain by adapting the tensor voting framework (TVF) [6] to handle color information. Such an adaptation is necessary as the voting fields proposed in [6] are only valid for problems that can be modeled in terms of the surface reconstruction problem, which is not the case of color image denoising.

The paper is organized as follows. Section 2 details the adaptation of the TVF to image denoising. Section 3 shows a comparative analysis of the proposed method against the state-of-the-art. Finally, Section 4 discusses the obtained results and makes some final remarks.

2 The TVF for Color Image Denoising

The input of the proposed method is the set of pixels of a color image. Thus, positional and color information is available for every input pixel. Positional information is used to determine the neighborhood of every pixel, while color information is used to define the tensors in the encoding step. The next subsections describe the details of the proposed color image denoising method.

2.1 Encoding of Color Information

Before applying the proposed method, color is converted to the CIELAB space. Every CIELAB channel (L, a and b) is then normalized in the range $[0, \pi/2]$. In the first step of the method, the information of every pixel is encoded through three second order 2D tensors, one for each normalized CIELAB color channel. These tensors are represented by 2×2 symmetric positive semidefinite matrices that can be graphically represented by 2D ellipses. There are two extreme cases for the proposed tensors: *stick* tensors, which are *stick*-shaped ellipses with a single eigenvalue, λ_1, different from zero, and *ball* tensors, which are circumference-shaped ellipses whose λ_1 and λ_2 eigenvalues are equal to each other.

Three perceptual measures are encoded in the tensors associated with every input pixel, namely: the most likely normalized noiseless color at the pixel (in the specific channel), a metric of local uniformity (how edgeless its neighborhood is), and an estimation of edginess (how likely finding edges or texture at the pixel's location is). Figure 1 shows the graphical interpretation of a tensor for channel L. The most likely normalized noiseless color is encoded by the angle α between the x axis, which represents the lowest possible color value in the corresponding channel, and the eigenvector corresponding to the largest eigenvalue. For example, in channel L, a tensor with $\alpha = 0$ encodes black, whereas a tensor with $\alpha = \frac{\pi}{2}$ encodes white. In addition, local uniformity and edginess are encoded by means of the normalized $\hat{s_1} = (\lambda_1 - \lambda_2)/\lambda_1$ and $\hat{s_2} = \lambda_2/\lambda_1$ saliencies

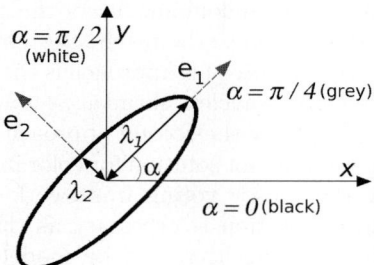

Fig. 1. Encoding process for channel L. Color, uniformity and edginess are encoded by means of α and the normalized $\hat{s}_1 = (\lambda_1 - \lambda_2)/\lambda_1$ and $\hat{s}_2 = \lambda_2/\lambda_1$ saliencies respectively.

respectively. Thus, a pixel located at a completely uniform region is represented by means of three *stick* tensors, one for each color channel. In contrast, a pixel located at an ideal edge is represented by means of three *ball* tensors.

Before applying the voting process, it is necessary to initialize the tensors associated with every pixel. The most likely noiseless colors can be initialized with the colors of the input pixels encoded by means of the angle α between the x axis and the principal eigenvector, as described before. However, since metrics of uniformity and edginess are usually unavailable at the beginning of the voting process, normalized saliency $\hat{s}_1$ is initialized to one and normalized saliency $\hat{s}_2$ is initialized to zero. These initializations allow the method to estimate more appropriate values for the normalized saliencies in the next stages, as described in Subsect. 2.3. Hence, the initial color information of a pixel is encoded through three *stick* tensors oriented along the directions that represent that color in the normalized CIELAB channels: $\mathsf{T}_c(p) = \boldsymbol{t}_c(p)\,\boldsymbol{t}_c(p)^T$, where $\mathsf{T}_c(p)$ is the tensor of the c-th color channel (L, a and b) at pixel p, $\boldsymbol{t}_c(p) = [\cos{(\mathrm{C}_c(p))} \quad \sin{(\mathrm{C}_c(p))}]^T$, and $\mathrm{C}_c(p)$ is the normalized value of the c-th color channel at p.

2.2 Voting Process

In this step, the tensors associated with every pixel are propagated to their neighbors through a convolution-like process. This step is independently applied to the tensors of every channel (L, a and b). Instead of using the voting fields proposed in [6], the proposed voting process uses specially designed tensorial functions referred to as *propagation functions*, which take into account not only the information encoded in the tensors but also the local relations between neighbors. Two propagation functions are proposed for image denoising: a *stick* and a *ball* propagation function. The *stick* propagation function is used to propagate the most likely noiseless color of a pixel, while the *ball* propagation function is used to increase edginess where necessary. The application of these functions leads to *stick* and *ball* votes respectively. *Stick* votes are used to eliminate noise and increase the edginess where the color of the voter and the voted pixels are different. *Ball* votes are used to increase the relevance of the most important edges. The proposed voting process is carried out at every pixel by adding all the *stick* and *ball* votes propagated towards it from its neighbors, by applying the propagation functions. Thus, the total vote received at a pixel p for each color channel c, $\mathsf{TV}_c(p)$, is given by: $\mathsf{TV}_c(p) = \sum_{q \in neigh(p)} \mathsf{S}_c(p,q) + \mathsf{B}_c(p,q)$, where $\mathsf{S}_c(p,q)$ and $\mathsf{B}_c(p,q)$ are the propagation functions that allow a pixel q to cast *stick* and *ball* votes to a neighboring pixel p for channel c respectively.

After applying the voting process at every pixel p, eigenvectors and eigenvalues of $\mathsf{TV}_L(p)$, $\mathsf{TV}_a(p)$ and $\mathsf{TV}_b(p)$ are calculated in order to analyze its local perceptual information. The TVF defines a standard way to interpret the voting results: uniformity increases with the normalized $\hat{s}_1$ saliency and the likelihood of a point belonging to an edge increases as the normalized $\hat{s}_2$ saliency becomes greater than the normalized $\hat{s}_1$ saliency. Additionally, the most likely normalized color at a pixel for each color channel is given by the angle between the first eigenvector of the corresponding tensor and the x axis. These three angles are then used to correct the color of every pixel with the most likely one, reducing in such a way the noise of the image.

2.3 Propagation Functions for Image Denoising

This paper proposes propagation functions that require three measurements for every pair of pixels p and q: the perceptual color difference, ΔE_{pq}; the joint uniformity measurement, $U_c(p,q)$, used to determine if both pixels belong to the same region; and the likelihood of a pixel being impulse noise, $\eta_c(p)$. ΔE_{pq} is calculated through CIEDE2000 [7], while $U_c(p,q) = \hat{s_1}_c(p)\ \hat{s_1}_c(q)$, and $\eta_c(p) = \hat{s_2}_c(p) - \mu_{\hat{s_2}_c}(p)$ if p is located at a local maximum and zero otherwise, where $\mu_{\hat{s_2}_c}(p)$ represents the mean of $\hat{s_2}_c$ over the neighborhood of p.

A *stick* vote is a *stick*-shaped tensor, $\mathsf{ST}_c(q)$, with a strength modulated by three scalar factors. The proposed *stick* propagation function is given by:

$$\mathsf{S}_c(p,q) = GS(p,q)\ \overline{\eta_c}(q)\ SV'_c(p,q)\ \mathsf{ST}_c(q), \qquad (1)$$

with $\mathsf{ST}_c(q)$, $GS(p,q)$, $\overline{\eta_c}(q)$ and $SV'_c(p,q)$ being defined as follows. First, the tensor $\mathsf{ST}_c(q)$ encodes the most likely noiseless color at q. Thus, $\mathsf{ST}_c(q)$ is defined as the tensorized eigenvector corresponding to the largest eigenvalue of the voter pixel, that is, $\mathsf{ST}_c(q) = \boldsymbol{e}_{1c}(q)\ \boldsymbol{e}_{1c}(q)^T$, being $\boldsymbol{e}_{1c}(q)$ the eigenvector with the largest eigenvalue of the tensor associated with channel c at q. Second, the three scalar factors in (1), each ranging between zero and one, are defined as follows. The first factor, $GS(p,q)$, models the influence of the distance between p and q in the vote strength. Thus, $GS(p,q) = G_{\sigma_s}(||p-q||)$, where $G_{\sigma_s}(\cdot)$ is a decaying Gaussian function with zero mean and a user-defined standard deviation σ_s. The second factor, $\overline{\eta_c}(q)$ defined as $\overline{\eta_c}(q) = 1 - \eta_c(q)$, is introduced in order to prevent a pixel q previously classified as impulse noise from propagating its information. The third factor, SV'_c, takes into account the influence of the perceptual color difference, the uniformity and the noisiness of the voted pixel. This factor is given by:

$$SV'_c(p,q) = \overline{\eta_c}(p)\ SV_c(p,q) + \eta_c(p), \qquad (2)$$

where: $SV_c(p,q) = [G_{\sigma_d}(\Delta E_{pq}) + U_c(p,q)]/2$. $SV_c(p,q)$ allows a pixel q to cast a stronger *stick* vote to p either if both pixels belong to the same uniform region, or if the perceptual color difference between them is small. That behavior is achieved by means of the factors $U_c(p,q)$ and the decaying Gaussian function on ΔE_{pq} with a user-defined standard deviation σ_d. A normalizing factor of two is used in order to make $SV_c(p,q)$ to vary from zero to one. The term $\eta_c(p)$ in (2) makes noisy voted pixels, p, to adopt the color of their voting neighbors, q, disregarding local uniformity measurements and perceptual color differences between p and q. The term $\overline{\eta_c}(p)$ in (2) makes SV'_c to vary from zero to one. As expected, the effect of $\eta_c(p)$ and $\overline{\eta_c}(p)$ on the strength of the *stick* vote received at a noiseless pixel p is null.

In turn, a *ball* vote is a *ball*-shaped tensor, $\mathsf{BT}(\mathsf{q})$, with a strength controlled by the scalar factors $GS(p,q)$, $\overline{\eta_c}(q)$ and $BV_c(p,q)$, each varying between zero and one. The *ball* propagation function is given by:

$$\mathsf{B}_c(p,q) = GS(p,q)\, \overline{\eta_c}(q)\, BV_c(p,q)\, \mathsf{BT}(\mathsf{q}), \tag{3}$$

with $\mathsf{BT}(\mathsf{q})$, $GS(p,q)$, $\overline{\eta_c}(q)$ and $BV_c(p,q)$ being defined as follows. First, the tensor represented by the identity matrix is the only possible tensor for $\mathsf{BT}(\mathsf{q})$, since it is the only tensor that complies with the two main design restrictions: a *ball* vote must be equivalent to casting *stick* votes for all possible colors using the hypothesis that all of them are equally likely and, the normalized $\hat{s}_1$ saliency must be zero when only *ball* votes are received at a pixel. Second, $GS(p,q)$ and $\overline{\eta_c}(q)$ are the same as the factors introduced in (1) for the *stick* propagation function. They are included for similar reasons to those given in the definition of the *stick* propagation function. Finally, the scalar factor $BV_c(p,q)$ is given by:

$$BV_c(p,q) = \frac{\overline{G_{\sigma_d}}(\Delta E_{pq}) + \overline{U_c}(p,q) + \overline{G_{\sigma_d}}(\Delta E^c_{pq})}{3}, \tag{4}$$

where $\overline{G_{\sigma_d}}(\cdot) = 1 - G_{\sigma_d}(\cdot)$ and $\overline{U_c}(p,q) = 1 - U_c(p,q)$. $BV_c(p,q)$ models the fact that a pixel q must reinforce the edginess at the voted pixel p either if there is a big perceptual color difference between p and q, or if p and q are not in a uniform region. This behavior is modeled by means of $\overline{G_{\sigma_d}}(\Delta E_{pq})$ and $\overline{U_c}(p,q)$. The additional term $\overline{G_{\sigma_d}}(\Delta E^c_{pq})$ is introduced in order to increase the edginess of pixels in which the only noisy channel is c, where ΔE^c_{pq} denotes the perceptual color difference only measured in the specific color channel c. The normalizing factor of three in (4) allows the *ball* propagation function to cast *ball* votes with a strength between zero and one.

The proposed propagation functions require to apply the voting process twice. The first application is used to obtain an initial estimation of the normalized $\hat{s}_1$ and $\hat{s}_2$ saliencies, as they are necessary to calculate $U_c(p,q)$ and $\eta_c(p)$. For this first application, only perceptual color differences and spatial distances are taken into account. At the second application, the tensors at every pixel are first initialized with the tensors obtained after the first application, and then, (1) and (3) are applied in their full definition, since all necessary data are available.

2.4 Adjustment of the CIEDE2000 Formula

The CIEDE2000 formula [7], which estimates the perceptual color difference between two pixels p and q, ΔE_{pq}, has three parameters, k_L, k_C and k_H, to weight the differences in CIELAB luminance, chroma and hue respectively. They can be adjusted to make the CIEDE2000 formula more suitable for every specific application by taking into account factors such as noise or background luminance, since those factors were not explicitly taken into account in the definition of the formula. These parameters must be greater than or equal to one. Based on [8], the following equations for these parameters are proposed:

$$k_L = F_{B_L} F_{\eta_L}, \quad k_C = F_{B_C} F_{\eta_C}, \quad k_H = F_{B_h} F_{\eta_h}, \qquad (5)$$

where F_{Bm} are factors that take into account the influence of the background color on the calculation of color differences for the color component m (L, C and h) and F_{η_m} are factors that take into account the influence of noise on the calculation of color differences in component m. On the one hand, big color differences in chromatic channels become less perceptually visible as background luminance decreases. Thus, the influence of the background on the CIEDE2000 formula can be modeled by $F_{B_L} = 1$ and $F_{B_C} = F_{B_h} = 1 + 3\,(1 - Y_B)$, where Y_B is the local mean of the background luminance. On the other hand, big color differences become less perceptually visible as noise increases. The influence of noise on CIEDE2000 can be modeled by means of $F_{\eta_m} = MAD(I)_m - MAD(G)_m - J_m$, where I is the image, G is a Gaussian blurred version of I, $MAD(\cdot)_m$ is the median absolute difference (MAD) calculated on component m and J_m are parameters to control the degree of preservation of texture. F_{η_m} is set to 1 in noiseless regions. We have obtained good results with $J_C = J_h = 0$ and $J_L \in [5, 10]$.

3 Experimental Results

One hundred outdoor images from the Berkeley segmentation data set [9] have been contaminated with various amounts of noise according to the methodology proposed in [3] for simulating realistic noise. This methodology aims at accurately reproducing the real noise generated by cameras, taking into account not only the noise generated by CCD sensors, which is mainly Gaussian, but also the necessary processes for converting raw data into images, such as demosaicing and Gamma correction.

The proposed technique has been compared to the methods proposed by [1] (based on non-local means), [2] (based on partial differential equations), and [5] (based on wavelets). These methods will be referred to as NLM, PDE and GSM respectively. The default parameters of NLM and PDE have been used. GSM has been applied with $\sigma = 20$, since its best overall performance was attained with this parameter. GSM has been applied to the three RGB channels independently, since this algorithm was designed for grey level images. The algorithm proposed in this paper, referred to as the tensor voting denoiser (TVD), has been run with standard deviations $\sigma_s = 1.3$ for the Gaussian G_{σ_s}, $\sigma_d = 1.0$ for the Gaussian

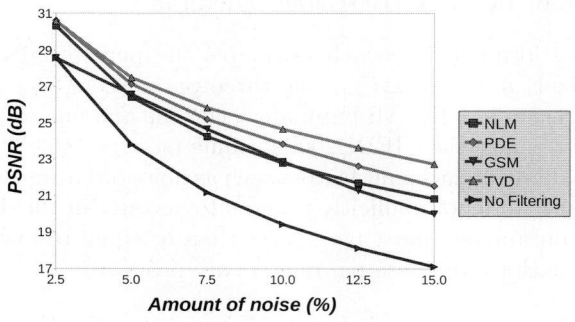

Fig. 2. PSNR vs. amount of noise for tested methods

Fig. 3. Denoising results. The first row shows the original images. The second row shows the noisy images (noise=10%). Rows three to six show the denoised images after applying NLM, PDE, GSM, and TVD respectively. High-resolution images are available at http://deim.urv.cat/~rivi/denoising.html

G_{σ_d}, and the parameters $J_L = 7.0$, $J_C = 0$, and $J_h = 0$. All the algorithms have successively been applied ten times to every input image for different amounts of noise. The output image with the highest PSNR was selected for every amount of noise. The selected images have been used to compare the performance of the algorithms. No pre or post-processing stages have been applied to the images in order to evaluate the ability of the algorithms to remove noise without any help. Figure 2 shows the plot of PSNR vs. amount of noise for all the methods.

NLM, PDE and TVD have almost the same performance for a noise of 2.5%. TVD has the best performance for larger amounts of noise followed by PDE. NLM and GSM have similar performances for amounts of noise greater than 2.5%. Figure 3 shows some denoised images. It can be seen that NLM generates undesirable quantization artifacts and colored spots. PDE generates cross-shaped artifacts. GSM has the worst performance among all the algorithms as the filtered images have a noisy appearance. TVD produces fewer artifacts than the other algorithms, yielding more appealing results.

4 Concluding Remarks

A new method to denoise color images while preserving edges has been proposed based on an adaptation of the TVF originally proposed in [6] for surface reconstruction. New specific encoding and voting processes, and propagation functions have been proposed for image denoising. The CIEDE2000 has also been adjusted for this application. The results show that the proposed adaptation makes the TVF a powerful tool for image denoising. Synthetic realistic noise has been added to natural images and PNSR and visual inspection have been used to determine the performance of the tested algorithms. The proposed method has been compared against some of the state-of-the art color image denoising algorithms, producing better results than them. Future work will include the utilization of the TVF for segmentation of noisy color images.

References

1. Kervrann, C., Boulanger, J.: Local adaptivity to variable smoothness for exemplar-based image regularization and representation. Int. J. Comput. Vis. 79, 45–69 (2008)
2. Tschumperlé, D.: Fast anisotropic smoothing of multi-valued images using curvature-preserving PDE's. Int. J. Comput. Vis. 68(1), 65–82 (2006)
3. Liu, C., Szeliski, R., Kang, S.B., Zitnick, C., Freeman, W.: Automatic estimation and removal of noise from a single image. IEEE Trans. Pattern Anal. Machine Intell. 30(2), 299–314 (2008)
4. Donoho, D.L., Johnstone, I.M.: Ideal spatial adaptation by wavelet shrinkage. Biometrika 81(3), 425–455 (1994)
5. Portilla, J., et al.: Image denoising using scale mixtures of Gaussians in the wavelet domain. IEEE Trans. Image Processing 12(11), 1338–1351 (2003)
6. Medioni, G., Lee, M.S., Tang, C.K.: A Computational Framework for Feature Extraction and Segmentation. Elsevier Science, Amsterdam (2000)

7. Luo, M.R., Cui, G., Rigg, B.: The development of the CIE 2000 colour-difference formula: CIEDE 2000. Color Res. and Appl. 26(5), 340–350 (2001)
8. Chou, C.H., Liu, K.C.: A fidelity metric for assessing visual quality of color images. In: Proc. Int. Conf. Comput. Commun. and Netw., pp. 1154–1159 (2007)
9. Martin, D., Fowlkes, C., Tal, D., Malik, J.: A database of human segmented natural images and its application to evaluating segmentation algorithms and measuring ecological statistics. In: Proc. Int. Conf. Comput. Vis., pp. II:416–II:423 (2001)

Two-Dimensional Windowing in the Structural Similarity Index for the Colour Image Quality Assessment

Krzysztof Okarma

West Pomeranian University of Technology, Szczecin
Faculty of Electrical Engineering,
Chair of Signal Processing and Multimedia Engineering,
26. Kwietnia 10, 71-126 Szczecin, Poland
okarma@zut.edu.pl

Abstract. This paper presents the analysis of the usage of the Structural Similarity (SSIM) index for the quality assessment of the colour images with variable size of the sliding window. The experiments have been performed using the LIVE Image Quality Assessment Database in order to compare the linear correlation of achieved results with the Differential Mean Opinion Score (DMOS) values. The calculations have been done using the value (brightness) channel from the HSV (HSB) colour space as well as commonly used YUV/YIQ luminance channel and the average of the RGB channels. The analysis of the image resolution's influence on the correlation between the SSIM and DMOS values for varying size of the sliding window is also presented as well as some results obtained using the nonlinear mapping based on the logistic function.

Keywords: colour image quality assessment, Structural Similarity.

1 Introduction

The development of some new colour image and video processing algorithms, related to nonlinear filtration, reconstruction, lossy compression etc., requires the reliable quality assessment of the resulting images. Such assessment should not be based only on the analysis of the greyscale images, especially because of the relevant differences between some greyscale image processing algorithms and their colour equivalents. A good example is the median filtration as there are many different approaches for the multichannel median filtering, starting from the Vector Median Filter (VMF) to much more sophisticated methods proposed in recent years.

Nevertheless, many researchers still use not only some metrics developed for the greyscale images but also often some classical ones, which are poorly correlated with the subjective evaluation of images by human observers. The examples of such traditional metrics may be the Mean Square Error (MSE), the Peak Signal to Noise Ratio (PSNR) and many similar ones [2]. Regardless of the fact that for some applications the image quality assessment based only on the luminance channel is enough, some new colour image quality assessment methods are also needed, which should be well correlated with the human perception and relatively easy to calculate. A popular colour image quality metric is the Normalised Colour Difference (NCD) defined in the CIE Lab or CIE Luv colour spaces but its correlation with the Human Visual System (HVS) is at least doubtful.

A promising direction of the research seems to be the no-reference (blind) image quality assessment. Nevertheless, currently available blind methods are rather specialised for the detection of one or two types of distortions e.g. block effects on JPEG compression [4] or image blurring [3] so they cannot be used as the universal techniques, similarly as some reduced-reference metrics [1], which use only a partial information from the original image.

The requirement of the universality of the image quality metrics with their independence on the image content and preferred dynamic range from 0 to 1 (or from −1 to 1 as in the SSIM) has caused the noticeable progress of some modern full-reference image quality metrics in recent years. Apart from some computationally demanding methods based on the Singular Value Decomposition [8] or transforms, the most popular approach seems to be the Structural Similarity index [10] as the extension of the previously proposed Universal Image Quality Index [9]. Nevertheless, there are also some other full-reference metrics, usually sensitive only to some chosen types of distortions e.g. JPEG compression and Gaussian noise or changes of the luminance and the impulse noise.

It is worth noticing that the correlation of results obtained using such a measure with the results of subjective evaluations by human observers should be preferably linear. The application of the nonlinear mapping based on the logistic function, as suggested by the Video Quality Experts Group (VQEG), can increase the correlation coefficient by several percent but the proper choice of the function's coefficients require the usage of some additional optimisation procedures as shown e.g. in the paper [6]. One of the typical methods of image quality measures' verification is the usage of the Differential Mean Opinion Score (DMOS) based on the analysis of the questionnaires filled by human observers for the images with various types of contaminations. Probably the best source of such results is the LIVE database [7] containing the DMOS values for nearly 1000 images with five types of distortions: JPEG2000 and JPEG compression, white Gaussian noise, Gaussian blur and JPEG2000 compressed images transmitted over simulated fast fading Rayleigh channel with bit errors typical for the wireless transmission. All the images are 24-bit colour ones, but many researchers treat them as greyscale ones analysing the luminance channel only or converting them before the analysis.

In this paper the analysis of the correlations between the DMOS values mentioned above and the values of the SSIM index calculated for the HSV (HSB) value (brightness), YUV/YIQ luminance and the average of the RGB channels is presented for the varying size of the sliding window used inside the calculation procedure for the SSIM index. The three channels specified above has been chosen on the basis of some previously obtained results for the constant 11×11

pixels Gaussian window as suggested by the authors of the paper [10] with the use of various colour spaces [5].

2 Application of the Structural Similarity

The Structural Similarity Index considered in the paper originates from the Universal Image Quality Index [9] defined for $M \times N$ pixels window as:

$$Q = \frac{4\sigma_{xy}\bar{x}\bar{y}}{(\sigma_x^2 + \sigma_y^2) \cdot [(\bar{x})^2 + (\bar{y})^2]} = \frac{\sigma_{xy}}{\sigma_x\sigma_y} \cdot \frac{2\bar{x}\bar{y}}{(\bar{x})^2 + (\bar{y})^2} \cdot \frac{2\sigma_x\sigma_y}{\sigma_x^2 + \sigma_y^2}, \quad (1)$$

where x and y denote the original and distorted image respectively, $\bar{x}$ and $\bar{y}$ stand for the average values, σ_x^2, σ_y^2 and σ_{xy} are the respective variances and the covariance for the fragments of both compared images inside the current window.

As the definition (1) is actual for the local quality index, the overall quality index is calculated as the average value from the quality map obtained by using a sliding 8×8 pixels rectangular window for the whole image. The resolution of the quality map is reduced almost unnoticeable in comparison to the original image (by $K-1$ pixels in the horizontal and vertical direction assuming the $K \times K$ pixels sliding window). An interesting feature of that metric is its sensitivity to three common types of distortions: the loss of correlation, luminance distortions and the loss of contrast.

Considering the possibility of division by zero, especially for dark large regions of the same colour, the protecting constants should be used, what leads to the extension of the method into the Structural Similarity with the additional change of the sliding window's shape from rectangular into Gaussian one. The resulting formula for the local quality index is as follows:

$$SSIM = \frac{(2 \cdot \bar{x} \cdot \bar{y} + C_1) \cdot (2 \cdot \sigma_{xy} + C_2)}{(\sigma_x^2 + \sigma_y^2 + C_1) \cdot [(\bar{x})^2 + (\bar{y})^2 + C_2]}, \quad (2)$$

where C_1 and C_2 are chosen in the way that they do not introduce significant changes of the results but prevent the division by zero. The default values suggested by the authors of the paper [10] are $C_1 = (0.01 \times 255)^2$ and $C_2 = (0.03 \times 255)^2$ with the size of the sliding window equal to 11×11 pixels.

3 Calculations and Results

The calculations of the Structural Similarity index have been performed for all the images from the LIVE database using the Gaussian window size from 7×7 to 19×19 pixels for the value (brightness) from the HSV (HSB) colour space as well as for the luminance from the YUV and YIQ and the average of the RGB channels. The application of some other channels, especially CIE Lab luminance has been omitted because of the poor correlation with the DMOS values obtained on some earlier experiments presented in the paper [5].

Table 1. The absolute values of the linear correlation coefficients between the DMOS and SSIM for various size of the sliding window for the YUV/YIQ luminance

Window size	JPEG2000	JPEG	White noise	Gaussian blur	Fast fading Rayleigh	All
7 × 7	0.8833	0.8442	0.9623	0.7822	0.8861	0.7085
9 × 9	0.8931	0.8505	0.9633	0.8230	0.8962	0.7273
11 × 11	0.8974	**0.8503**	0.9644	0.8486	0.9008	0.7364
13 × 13	**0.8988**	0.8477	0.9652	0.8650	**0.9025**	0.7403
15 × 15	0.8986	0.8442	0.9658	0.8758	**0.9025**	**0.7412**
17 × 17	0.8975	0.8406	0.9663	0.8829	0.9016	0.7405
19 × 19	0.8959	0.8370	**0.9666**	**0.8874**	0.9002	0.7387

Table 2. The absolute values of the linear correlation coefficients between the DMOS and SSIM for various size of the sliding window - the smallest images (class A)

Window size	JPEG2000	JPEG	White noise	Gaussian blur	Fast fading Rayleigh	All
Value (Brightness) - HSV (HSB)						
7 × 7	**0.9845**	**0.9065**	0.9900	**0.9893**	**0.9895**	0.7714
9 × 9	0.9802	0.8902	**0.9909**	0.9865	0.9869	**0.7770**
11 × 11	0.9752	0.8767	0.9907	0.9831	0.9818	0.7755
13 × 13	0.9702	0.8661	0.9900	0.9803	0.9760	0.7708
15 × 15	0.9651	0.8577	0.9891	0.9780	0.9705	0.7645
17 × 17	0.9601	0.8511	0.9880	0.9762	0.9655	0.7574
19 × 19	0.9551	0.8458	0.9867	0.9749	0.9612	0.7499
Luminance - YUV/YIQ						
7 × 7	**0.9815**	**0.8982**	0.9842	**0.9883**	**0.9932**	0.7710
9 × 9	0.9753	0.8790	**0.9846**	0.9852	0.9866	**0.7784**
11 × 11	0.9683	0.8627	0.9844	0.9817	0.9777	0.7775
13 × 13	0.9613	0.8493	0.9836	0.9787	0.9686	0.7728
15 × 15	0.9545	0.8385	0.9825	0.9764	0.9605	0.7664
17 × 17	0.9477	0.8296	0.9811	0.9746	0.9534	0.7591
19 × 19	0.9412	0.8222	0.9795	0.9733	0.9476	0.7515
Average - RGB channels						
7 × 7	**0.9828**	**0.8988**	**0.9855**	**0.9879**	**0.9927**	0.7959
9 × 9	0.9772	0.8804	0.9850	0.9850	0.9872	**0.8028**
11 × 11	0.9708	0.8646	0.9838	0.9816	0.9795	0.8017
13 × 13	0.9643	0.8518	0.9822	0.9786	0.9716	0.7970
15 × 15	0.9578	0.8414	0.9802	0.9763	0.9643	0.7906
17 × 17	0.9515	0.8329	0.9780	0.9745	0.9579	0.7834
19 × 19	0.9452	0.8258	0.9756	0.9732	0.9525	0.7758

All the results obtained for the original images present in the database have been eliminated and then the absolute values of the Pearson's correlation coefficients between the DMOS and the SSIM index have been computed (using Matlab's *corrcoef* function) for each type of distortion as well as for the all

Table 3. The absolute values of the linear correlation coefficients between the DMOS and SSIM for various size of the sliding window - the biggest images (class D)

Window size	JPEG2000	JPEG	White noise	Gaussian blur	Fast fading Rayleigh	All
Value (Brightness) - HSV (HSB)						
7×7	0.8772	0.8538	0.9489	0.7405	0.8400	0.6540
9×9	0.8886	0.8658	0.9535	0.7856	0.8566	0.6772
11×11	0.8946	**0.8693**	0.9567	0.8138	0.8657	0.6899
13×13	0.8977	0.8692	0.9593	0.8320	0.8706	0.6970
15×15	0.8991	0.8674	0.9614	0.8439	0.8729	0.7006
17×17	**0.8994**	0.8649	0.9631	0.8519	**0.8737**	0.7021
19×19	0.8990	0.8621	**0.9647**	**0.8572**	0.8736	**0.7022**
Luminance - YUV/YIQ						
7×7	0.8745	0.8450	0.9620	0.7312	0.8412	0.6732
9×9	0.8853	0.8562	0.9633	0.7773	0.8574	0.6954
11×11	0.8909	**0.8589**	0.9646	0.8066	0.8662	0.7073
13×13	0.8935	0.8580	0.9657	0.8257	0.8710	0.7136
15×15	**0.8946**	0.8555	0.9667	0.8386	0.8734	0.7167
17×17	**0.8946**	0.8524	0.9674	0.8474	**0.8742**	**0.7177**
19×19	0.8941	0.8490	**0.9680**	**0.8536**	0.8741	0.7175
Average - RGB channels						
7×7	0.8744	0.8477	0.9676	0.7266	0.8394	0.6965
9×9	0.8857	0.8595	0.9684	0.7732	0.8563	0.7186
11×11	0.8915	**0.8627**	0.9692	0.8029	0.8657	0.7306
13×13	0.8945	0.8622	0.9699	0.8224	0.8709	0.7370
15×15	0.8958	0.8601	0.9703	0.8356	0.8734	0.7402
17×17	**0.8962**	0.8573	0.9706	0.8447	**0.8745**	**0.7414**
19×19	0.8959	0.8542	**0.9708**	**0.8511**	0.8745	**0.7414**

distorted images from the whole database. Obtained results are presented in the Table 1. It is worth noticing that for four from five types of distortions the best results have been obtained for the value (brightness) channel but the overall correlation coefficient is the lowest one. This phenomenon and relatively low overall values of the correlation coefficients are caused by their linearity. Adding the nonlinear mapping e.g. according to the logistic function would increase also the overall correlations to over 90% (as shown further) but the main goal of the paper is to find the possibly highest linear correlation of the SSIM index with varying sliding window size with the DMOS values.

The usage of the 11×11 pixels Gaussian window, as suggested in the paper [10] leads to the best results only for one type of distortions (JPEG compressed images) regardless of the channel used during the computations. The highest overall index has been obtained for 15×15 pixels window in each case so the additional analysis should be performed depending on the resolution of processed images. For that purpose the correlation coefficients have been computed separately for four classes of images present in the LIVE database with the following resolutions:

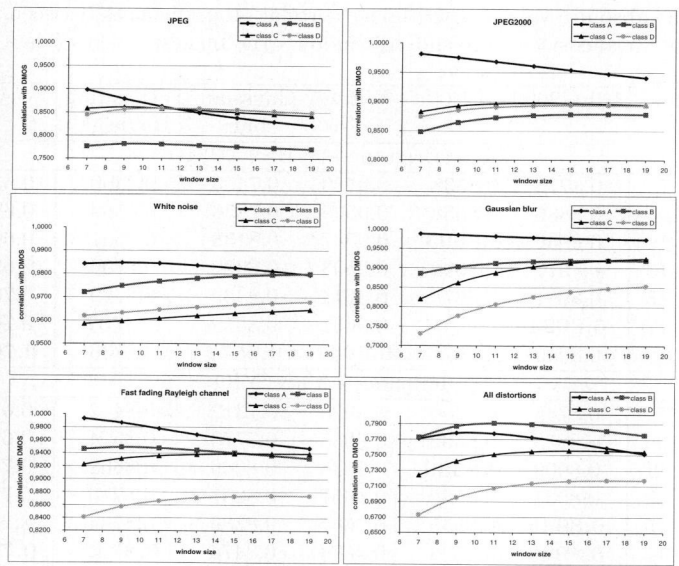

Fig. 1. Correlations between the SSIM and DMOS values for various distortions calculated for the luminance (Y)

- class A: 634 × 438 pixels and 618 × 453 pixels,
- class B: 610 × 488 pixels, 627 × 482 pixels and 632 × 505 pixels,
- class C: 480 × 720 pixels,
- class D: 768 × 512 pixels.

The results obtained for the highly differing classes A and D are presented in Tables 2 and 3. The additional illustration of the results achieved for all four classes using the luminance (Y) channel is presented in Fig. 1.

Analysing the results obtained for the different resolutions it can be easily observed that for the higher resolution images the usage of the wider windows leads to better results. It may lead to the conclusion related to the choice of the default size of the sliding Gaussian window as well suited for some typical 512 × 512 pixels test images but not necessarily for the quality assessment of the high resolution images e.g. from the digital cameras or even the HDTV.

Relatively low values of the linear correlation coefficients are caused not only by the linearity but also by removing all original images from the database. Many authors, even using the logistic function in order to introduce the nonlinearity between their metrics and DMOS values, optimise the coefficients of the mapping function for the whole set of values present in the LIVE database. Such approach cause the unreasonable increase of the correlation values, while a good image quality metric should be well correlated with the human observations primarily for distorted images, as the results obtained for the originals are predictable. As the illustration of the influence of the nonlinear mapping based on the logistic

Table 4. The absolute values of the correlation coefficients between the DMOS and nonlinear mapped SSIM for various size of the sliding window

Window size	JPEG2000	JPEG	White noise	Gaussian blur	Fast fading Rayleigh	All
\multicolumn{7}{c}{Value (Brightness) - HSV (HSB)}
7×7	0.9169	0.9141	0.9504	0.8194	0.9196	0.8173
9×9	0.9303	0.9276	0.9543	0.8564	0.9330	0.8346
11×11	0.9383	0.9336	0.9569	0.8790	0.9412	0.8445
13×13	0.9433	0.9362	0.9588	0.8934	0.9466	0.8504
15×15	0.9464	0.9372	0.9606	0.9033	0.9502	0.8541
17×17	0.9483	**0.9374**	0.9620	0.9102	0.9528	0.8565
19×19	**0.9495**	0.9372	**0.9631**	**0.9150**	**0.9545**	**0.8581**
\multicolumn{7}{c}{Luminance - YUV/YIQ}
7×7	0.9160	0.9102	0.9624	0.8125	0.9244	0.8393
9×9	0.9288	0.9236	0.9638	0.8501	0.9372	0.8547
11×11	0.9364	0.9294	0.9651	0.8734	0.9449	0.8632
13×13	0.9410	0.9318	0.9662	0.8885	0.9499	0.8681
15×15	0.9438	**0.9325**	0.9672	0.8990	0.9532	0.8711
17×17	0.9455	0.9324	**0.9680**	0.9064	0.9555	0.8728
19×19	**0.9464**	0.9318	0.9651	**0.9117**	**0.9572**	**0.8739**
\multicolumn{7}{c}{Average - RGB channels}
7×7	0.9148	0.9100	0.9684	0.8084	0.9208	0.8524
9×9	0.9279	0.9236	0.9693	0.8466	0.9343	0.8678
11×11	0.9357	0.9296	0.9701	0.8702	0.9425	0.8445
13×13	0.9405	0.9321	0.9707	0.8856	0.9478	0.8814
15×15	0.9434	**0.9329**	0.9711	0.8962	0.9513	0.8844
17×17	0.9452	**0.9329**	**0.9717**	0.9038	0.9538	0.8862
19×19	**0.9462**	0.9325	0.9701	**0.9092**	**0.9555**	**0.8873**

function on the correlation between the SSIM and DMOS the optimisation of that function has been performed independently for each window size and type of distortions. The mapping function has been defined as:

$$SSIM_{\mathrm{nonlinear}} = \beta_1 \cdot \mathrm{logistic}(\beta_2, SSIM - \beta_3) + \beta_4 \cdot SSIM + \beta_5 , \quad (3)$$

where

$$\mathrm{logistic}(\tau, x) = \frac{1}{2} - \frac{1}{1 + \exp(x\tau)} . \quad (4)$$

The fitting (optimisation of the five parameters β) has been performed using MATLAB's *fminunc* and *fminsearch* functions and the resulting correlation coefficients obtained after the nonlinear mapping are presented in Table 4.

Analysing the results it can be easily noticed that the best correlation is obtained not for the typical 11×11 pixels windows but for much wider ones. Nevertheless those results are harder for the comparisons because of the different values of the parameters β used for the nonlinear SSIM mapping for each window size and each type of distortion as the result of optimisation.

4 Summary and Future Work

Some experimental results presented in the paper can be treated as a partial verification of the usefulness of the Structural Similarity index and its correlation with the subjective evaluation of the digital image quality. Further analysis would require much more results of the subjective evaluations, also related to some other types of contaminations. In the future work a method for choosing the optimum size and type of the sliding window for the SSIM can be defined, depending on the resolution and the colour model of the image, extending the experiments into chrominance components, also using the nonlinear mapping.

References

1. Carnec, M., Le Callet, P., Barba, P.: An Image Quality Assessment Method Based on Perception of Structural Information. In: Proc. Int. Conf. Image Processing, Barcelona, Spain, vol. 2, pp. 185–188 (2003)
2. Eskicioglu, A.: Quality Measurement for Monochrome Compressed Images in the Past 25 Years. In: Proc. IEEE Int. Conf. Acoust. Speech Signal Process., Istanbul, Turkey, pp. 1907–1910 (2000)
3. Marziliano, P., Dufaux, F., Winkler, S., Ebrahimi, T.: A No-Reference Perceptual Blur Metric. In: Proc. IEEE Int. Conf. Image Processing, Rochester, USA, pp. 57–60 (2002)
4. Meesters, L., Martens, J.-B.: A Single-Ended Blockiness Measure for JPEG-Coded Images. Signal Processing 82(3), 369–387 (2002)
5. Okarma, K.: Colour Image Quality Assessment Using Structural Similarity Index and Singular Value Decomposition. In: Bolc, L., Kulikowski, J.L., Wojciechowski, K. (eds.) ICCVG 2008. LNCS, vol. 5337, pp. 55–65. Springer, Heidelberg (2009)
6. Sendashonga, M., Labeau, F.: Low Complexity Image Quality Assessment Using Frequency Domain Transforms. In: Proc. IEEE Int. Conf. Image Processing, pp. 385–388 (2006)
7. Sheikh, H.R., Wang, Z., Cormack, L., Bovik, A.C.: LIVE Image Quality Assessment Database Release 2, http://live.ece.utexas.edu/research/quality
8. Shnayderman, A., Gusev, A., Eskicioglu, A.: An SVD-Based Gray-Scale Image Quality Measure for Local and Global Assessment. IEEE Trans. Image Processing 15(2), 422–429 (2006)
9. Wang, Z., Bovik, A.: A Universal Image Quality Index. IEEE Signal Processing Letters 9(3), 81–84 (2002)
10. Wang, Z., Bovik, A., Sheikh, H., Simoncelli, E.: Image Quality Assessment: From Error Measurement to Structural Similarity. IEEE Trans. Image Processing 13(4), 600–612 (2004)

Reduced Inverse Distance Weighting Interpolation for Painterly Rendering

Giuseppe Papari and Nicolai Petkov

Institute of Mathematics and Computing Science, University of Groningen

Abstract. The interpolation problem of irregularly distributed data in a multidimensional domain is considered. A modification of the inverse distance weighting interpolation formula is proposed, making computation time independent of the number of data points. Only the first K neighbors of a given point are considered, instead of the entire dataset. Additional factors are introduced, preventing discontinuities on points where the set of local neighbors changes. Theoretical analysis provides conditions which guarantee continuity. The proposed approach is efficient and free from magic numbers. Unlike many existing algorithms based on the k-nearest neighbors, the number of neighbors is derived from theoretical principles. The method has been applied to the problem of vector field generation in the context of artistic imaging. Experimental results show its ability to produce brush strokes oriented along object contours and to effectively render meaningful texture details.

1 Introduction

Interpolating data consists in finding a smooth function whose values are known only on some data points. This problem arises in several fields of science, such as geophysics [1], oceanography [2], meteorology [3], super-resolution [4], or video coding [5]. A large amount of different techniques have been proposed in the literature for this task such as polynomial spline [6], wavelets [4], or variational approaches [3]. In this paper we are interested to the interpolation of irregularly scattered data over a multidimensional domain. The best known techniques for this task are thin plate spline (TSP) [7], multiquadric surfaces (MQS) [8], inverse distance weighting (IDW) [9], and natural neighbor interpolation (NNI) [10].

In TSP and MQS, the interpolant is expressed as a linear combination of radial functions centered on the data sites. While they give very good results, especially for geophysical applications, determining the coefficients of the linear combination is computationally expensive. Moreover, these functions have a scale parameter, which is usually optimized with cross-validation [11], thus requiring further computation. In IDW, the interpolating function is expressed as a weighted average of the data values, where the weights are inverse functions of the distances from the data sites. While IDW is simpler than TSP and MQS, its computation time still increases as the number of data points grows. In NNI, similar weights are computed in terms of the so called *natural neighbor coordinates*, which are interesting from the theoretical point of view [12], but efficient implementations are hardly provided, especially for high dimensional data.

In this paper we propose a simple modification of IDW, whose computation time does not depend on the number of data points. The idea is to limit the average to the first K neighbors and to introduce additional factors in order to prevent discontinuities when the set of local neighbors changes. We also present a simple theoretical analysis of the method, which tells the minimum number of neighbors to consider in order to guarantee continuity everywhere. As an application, we use the proposed interpolation formula for vector field generation in the context of automatic painterly rendering.

2 Proposed Interpolation Formula

Given a set $S = \{\mathbf{r}_1, ..., \mathbf{r}_N\}$ of *data sites* and a set of $F = \{f_1, ..., f_N\}$ of *data values*, the interpolation and extrapolation problem we are interested to consists in finding a continuous function $\tilde{F} : \mathbb{R}^n \mapsto \mathbb{R}$ such that:

$$\tilde{F}(\mathbf{r}_i) = f_i, \qquad \forall \mathbf{r}_i \in S. \tag{1}$$

We will first review the IDW formula (Section 2.1), and then we will present our interpolation technique called *reduced IDW*, which reduces the computational complexity of IDW (Sections 2.2 and 2.3).

2.1 IDW Interpolation Formula

Given a set S of N points in $\mathbb{R}^n$, let $\mathbf{q}_k(\mathbf{r})$ be the k-th nearest neighbor of $\mathbf{r}$ among points of S and let $d_k(\mathbf{r})$ be the Euclidean distance between $\mathbf{r}$ and $\mathbf{q}_k(\mathbf{r})$, as illustrated in Fig. 1. We also construct a piece-wise constant function $f_k(\mathbf{r})$ which takes the value f_i every time $\mathbf{q}_k(\mathbf{r}) = \mathbf{r}_i$. With this notation, the IDW interpolation formula is written as follows:

$$\tilde{f}_K(\mathbf{r}) \triangleq \frac{\sum_{k=1}^{K} f_k(\mathbf{r})[d_k(\mathbf{r})]^{-p}}{\sum_{k=1}^{K}[d_k(\mathbf{r})]^{-p}} \tag{2}$$

where $p > 0$ is an input parameter. The interpolant $\tilde{f}_K$ has cusps on the data sites $\mathbf{r}_k$ for $p \leq 1$ while it is smooth for $p > 1$. Usually the value $p = 2$ is chosen.

It is easy to show that the interpolating function $\tilde{f}_K(\mathbf{r})$ defined in (2) takes the values f_i for $\mathbf{r} = \mathbf{r}_i$, as required in (1). However, it is continuous everywhere only if the number K of neighbors is equal to the total number N of points in S. In fact, for $K < N$, the set of neighbors $q_k(\mathbf{r})$ which are involved in the computation of $\tilde{f}_K(\mathbf{r})$ would change as $\mathbf{r}$ moves in $\mathbb{R}^n$, thus giving rise to discontinuities in the interpolating function. This has the obvious consequence to make the method computationally demanding as N increases.

2.2 Reduced IDW Interpolation Formula

We now propose a simple modification of (2) which makes it continuous everywhere while keeping K independent of N. Let

$$\Gamma_k(\mathbf{r}) \triangleq \{\mathbf{q}_1(\mathbf{r}), ... \mathbf{q}_k(\mathbf{r})\} \tag{3}$$

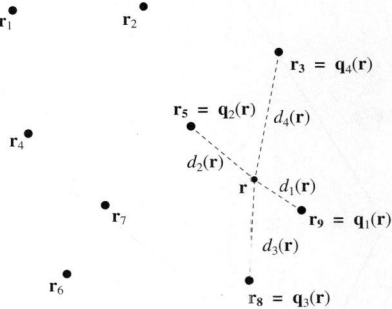

Fig. 1. Illustration of the functions $\mathbf{q}_k(\mathbf{r})$ and $d_k(\mathbf{r})$ defined above

be the set of the first k neighbors of $\mathbf{r}$, which is a piece-wise constant function of $\mathbf{r}$, and let γ_k be the set of the discontinuity points of $\Gamma_k(\mathbf{r})$. We observe that the rank order of the neighbors $q_j(\mathbf{r})$, $(i = 1, .., k)$, does not affect the definition of $\Gamma_k(\mathbf{r})$. Thus, referring to Fig. 2c, we have, e.g., $\Gamma_2(\boldsymbol{\xi}) = \{A, B\} = \{B, A\} = \Gamma_2(\boldsymbol{\eta})$; therefore, the boundary between the regions with $\{\mathbf{q}_1(\mathbf{r}) = A, \mathbf{q}_2(\mathbf{r}) = B\}$ and $\{\mathbf{q}_1(\mathbf{r}) = B, \mathbf{q}_2(\mathbf{r}) = A\}$ does not belong to γ_2. It is easy to show that the interpolating function $\tilde{f}_k(\mathbf{r})$ defined above is discontinuous on γ_k and γ_k only.

Now, let $\delta_k(\mathbf{r})$ be the Euclidean distance between $\mathbf{r}$ and its first neighbor among the points of γ_k, $\delta_k(\mathbf{r}) \triangleq \min_\rho \|\mathbf{r} - \boldsymbol{\rho}\|$ as shown in Fig. 2b. With this notation, we propose the following interpolation formula:

$$\tilde{F}_K(\mathbf{r}) \triangleq \frac{\sum_{k=1}^{K} \tilde{f}_k(\mathbf{r})\delta_k(\mathbf{r})}{\sum_{k=1}^{K} \delta_k(\mathbf{r})} \quad (4)$$

It is easy to see that for points $\mathbf{r} \in \gamma_k$, on which $\tilde{f}_k(\mathbf{r})$ is discontinuous, the function $\delta_k(\mathbf{r})$ vanishes, thus each term in the sum at the numerator of (4) is continuous everywhere. On the other hand, it can be easily shown that if K is sufficiently large, the sum at the denominator is never equal to zero (See Section 2.3 for details). Consequently, the interpolating function $\tilde{F}_K(\mathbf{r})$ defined in (4) is continuous everywhere in $\mathbb{R}^n$. Moreover, since the condition $\tilde{f}_k(\mathbf{r}_i) = f_i, \forall \mathbf{r}_i \in S$ holds for all k, we also have $\tilde{F}_K(\mathbf{r}_i) = f_i$, thus proving that (4) meets the requirement (1) to be an interpolation formula.

2.3 Theoretical Analysis and Automatic Choice of K

In this subsection, we perform a brief theoretical analysis of the behavior of the interpolating function $\tilde{F}_K(\mathbf{r})$ defined in (4), and we show how many neighbors must be taken into account in order to guarantee continuity everywhere.

Let S be a point set in $\mathbb{R}^n$ and let us consider its Voronoi tessellation [13]. Then, let $B_\epsilon(\mathbf{r})$ be a sphere centered in $\mathbf{r}$ with infinitesimal radius ϵ, and let $\mathcal{V}(\mathbf{r})$ be the set of centroids of the Voronoi cells covered by $B_\epsilon(\mathbf{r})$, as shown in Fig. 3 for points $\boldsymbol{\xi}$, $\boldsymbol{\eta}$, and $\boldsymbol{\zeta}$. We indicate by $m(\mathbf{r})$ the cardinality of $\mathcal{V}(\mathbf{r})$.

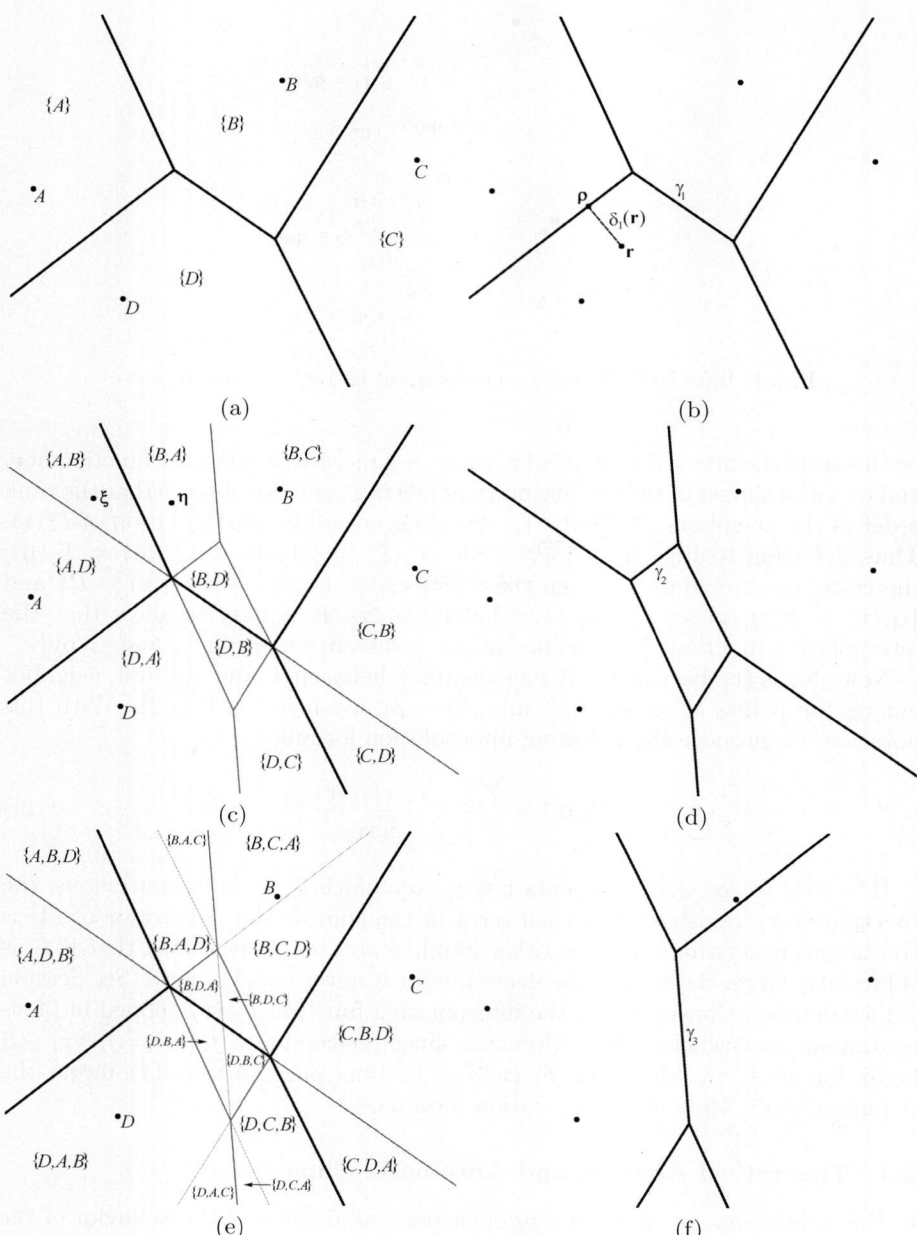

Fig. 2. Illustration of the function $\Gamma_k(\mathbf{r})$ (first column) and the set γ_k (second column), $k = 1, 2, 3$, for a simple set of four points. (a,c,e): values of the function $\Gamma_1(\mathbf{r})$, $\Gamma_2(\mathbf{r})$, $\Gamma_3(\mathbf{r})$. (b,d,f): discontinuity sets γ_1, γ_2, γ_3.

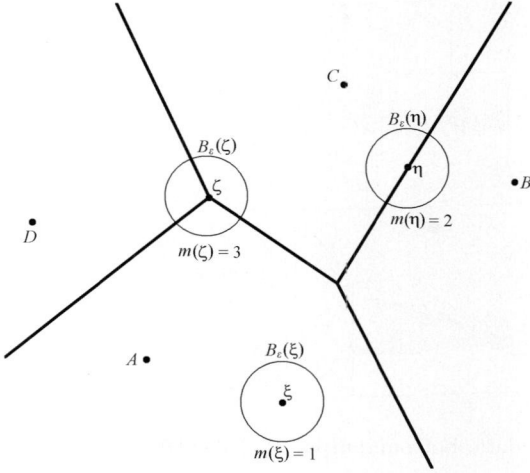

Fig. 3. Illustration of the function $m(\mathbf{r})$ defined above for a simple point set $S = \{A, B, C, D\}$

We will prove that the function Γ_k defined in (3) is continuous in $\mathbf{r}$ for $k = m(\mathbf{r})$. In fact, since ϵ is infinitesimal, it is easy to prove that the first k neighbors of a point $\boldsymbol{\rho} \in B_\epsilon(\mathbf{r})$ are just the centroids of the Voronoi cells covered by $B_\epsilon(\mathbf{r})$. This implies that $\Gamma_{k=m(\mathbf{r})}(\boldsymbol{\rho})$ is constant for $\boldsymbol{\rho} \in B_\epsilon(\mathbf{r})$, thus proving the required continuity. The same does not hold for $k < m(\mathbf{r})$, since $\Gamma_k(\boldsymbol{\rho})$ would be a subset of $\mathcal{V}(\mathbf{r})$, which in general depends on $\boldsymbol{\rho}$.

Since $\Gamma_k(\mathbf{r})$ and the corresponding term $\tilde{f}_k(\mathbf{r})$ of the sum in (4) are discontinuous at the same points, the proposed interpolating function is continuous everywhere iff K is greater of equal to the maximum value taken by $m(\mathbf{r})$:

$$K \geq \max_{\mathbf{r} \in \mathbb{R}^n} m(\mathbf{r}) \qquad (5)$$

It is easy to show that such a maximum is equal to the maximum number of edges of the Voronoi tessellation which intersect at the same cell vertex. In $\mathbb{R}^2$, for randomly scattered data sites, this is usually equal to $K = 3$.

3 Results, Discussion and Conclusions

In order to test the proposed interpolation formula, we used random data sites $S = \{\mathbf{r}_1, ..., \mathbf{r}_N\} \subset \mathbb{R}^2$ and data values $F = \{f_1, ..., f_N\}$. We have interpolated them both with IDW and reduced IDW, for the same value of K ($K = 4$), which has been computed as in (5). The resulting functions are plotted in Fig. 6. As we see, reduced IDW produces a continuous function, whereas IDW gives rise to discontinuities on some edges of the Voronoi tessellation of the input point set.

As an application, we use our interpolation formula for unsupervised painterly rendering. The idea, illustrated in Fig. 5, is first to compute a vector field $\mathbf{v}(\mathbf{r})$,

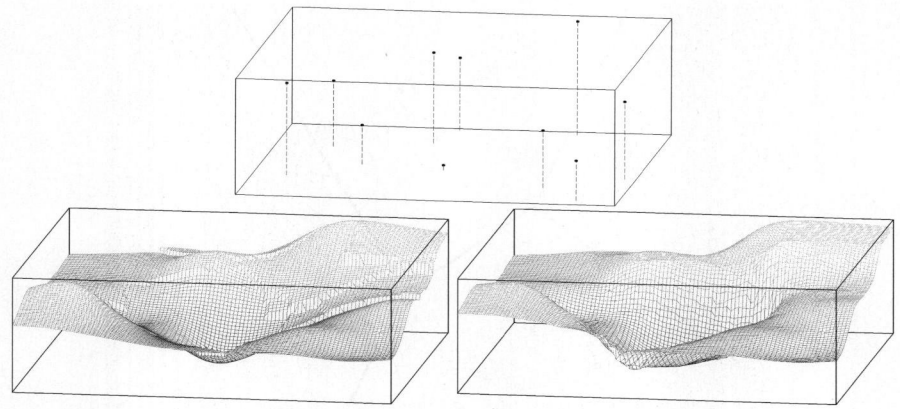

Fig. 4. Top: input data. bottom: output of (left) IDW and (right) reduced IDW.

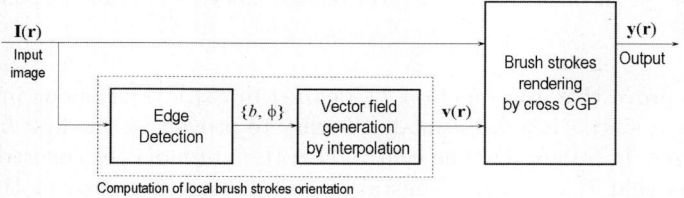

Fig. 5. Proposed approach for automatic painterly rendering

Fig. 6. From left to right: input image, edge map, and output of the proposed operator

which determins the local brush stroke orientation, and then to render elongated curved brush strokes oriented along $\mathbf{v}(\mathbf{r})$. Brush strokes are rendered by means of the operator desribed in [16, 17], which is based on the theory of Glass patterns. With respect to other algorithms for brush strokes simulation, the one proposed in [16, 17] has the advantage to be fast and free from and magic numbers.

In generating $\mathbf{v}(\mathbf{r})$, the first step is edge detection. For simplicity, we use the Canny edge detector [14], where the Gaussian gradient is replaced with the color gradient proposed in [15]. Let $b \triangleq \{\mathbf{r}_1, ..., \mathbf{r}_m\}$ be the set of the detected edge pixels and let $\phi_1, ..., \phi_m$ be the local orientation of each edge point. Let also be

Fig. 7. From left to right: input image, edge map, and output of the proposed operator

$\mathbf{v}_i \triangleq [a\cos\phi_i, a\sin\phi_i]^T$, where the input parameter a controls the brush strokes length. Then the vector field $\mathbf{v}(\mathbf{r})$ is obtained by interpolating the vectors $\mathbf{v}_i$, plcaed on points $\mathbf{r}_i \in b$ with our interpolation formula, first for the x and then for the y component. Examples of how the method works is illustrated in Fig. 6 for a synthetic image and in Fig 7 for natural images.

To summarize, we have proposed a simple modification of IDW interpolation which makes its computational complexity independent on the number N of data sites, while avoiding undesired discontinuities. We observe that the importance of avoiding undesired discontinuities goes beyond the specific computer graphics application described above. An attempt to reduce the complexity of IDW was proposed in [9], by limiting the sum (2) to those neighbors that fall inside a circlecentered on $\mathbf{r}$. However, this would result in several magic numbers and "ad hoc" functions with no theoretical justification. In contrast, the proposed method has no input parameters from the user. In particular, unlike many existing algorithms based on the k-nearest neighbors, the number of neighbors is derived from theoretical principles. Though the IDW formula is quite old, we are not aware of other modification of it propsed in the last fourty years. We have applied our interpolation formula to the problem of vector field generation in the context of artistic imaging. Our results show that the method is able both to produce brush strokes oriented along object contours, even in presence of sharp corners (Fig. 6), and to effectively render meaningful texture details, such as the hair of the lady shown in Fig. 7.

References

1. Hansen, T.M., Mosegaard, K., Pedersen-Tatalovic, R., Uldall, A., Jacobsen, N.L.: Attribute-guided well-log interpolation applied to low-frequency impedance estimation. Geophysics 73(6), R83–R95 (2008)
2. Akmaev, R.A.: A prototype upper-atmospheric data assimilation scheme based on optimal interpolation: 1. theory. Journal of Atmospheric and Solar-Terrestrial Physics 61(6), 491–504 (1999)
3. Panteleev, G.G., Maximenko, N.A., Reiss, C., Yamagata, T.: Variational interpolation of circulation with nonlinear, advective smoothing. Journal of Atmospheric and Oceanic Technology 19(9), 1442–1450 (2002)
4. Lee, W.L., Yang, C.C., Wu, H.T., Chen, M.J.: Wavelet-based interpolation scheme for resolution enhancement of medical images. Journal of Signal Processing Systems 55(1-3), 251–265 (2009)
5. Wedi, T.: Adaptive interpolation filters and high-resolution displacements for video coding. IEEE Trans. Circuits and Systems for Video Technology 16(4), 484–491 (2006)
6. Unser, M., Aldroubi, A., Eden, M.: B-spline signal processing. ii. efficiency design and applications. IEEE Transactions on Signal Processing 41(2), 834–848 (1993)
7. Duchon, J.: Splines minimizing rotation-invariant semi-norms in sobolev spaces. Constructive Theory of Functions of Several Variables 571, 85–100 (1977)
8. Hardy, R.L.: Multiquadric equations of topography and other irregular surfaces. J. Geophys. Res. 76(8), 1905–1915 (1971)
9. Shepard, D.: A two-dimensional interpolation function for irregularly-spaced data. In: Proceedings of the 1968 23rd ACM national conference, pp. 517–524. ACM, New York (1968)
10. Sibson, R.: A brief description of natural neighbour interpolation. Interpreting Multivariate Data, 21–36 (1981)
11. Erdogan, S.: A comparision of interpolation methods for producing digital elevation models at the field scale. Earth Surface Processes and Landforms 34(3), 366–376 (2009)
12. Boissonnat, J.D., Cazals, F.: Natural neighbor coordinates of points on a surface. Computational Geometry: Theory and Applications 19(2-3), 155–173 (2001)
13. Okabe, A., Boots, B., Sugihara, K.: Spatial tessellations: concepts and applications of Voronoi diagrams. John Wiley and Sons, Ltd., Chichester (1992)
14. Canny, J.F.: A computational approach to edge detection. IEEE T-PAMI 8(6), 679–698 (1986)
15. Cumani, A.: Edge detection in multispectral images. CVGIP 53(1), 40–51 (1991)
16. Papari, G., Petkov, N.: Continuous glass patterns for painterly rendering. IEEE Transactions on Image Processing 18(3), 652–664 (2009)
17. Papari, G., Petkov, N.: Glass partterns and artistic imaging. In: Wada, T., Huang, F., Lin, S. (eds.) PSIVT 2009. LNCS, vol. 5414, pp. 1034–1045. Springer, Heidelberg (2009)

Nonlinear Diffusion Filters without Parameters for Image Segmentation

Carlos Platero, Javier Sanguino, and Olga Velasco

Applied Bioengineering Group, Polytechnic University of Madrid, Spain

Abstract. Nonlinear diffusion filtering seeks to improve images qualitatively by removing noise while preserving details and even enhancing edges. However, well known implementations are sensitive to parameters which are necessarily tuned to sharpen a narrow range of edge slopes. In this work, we have selected a nonlinear diffusion filter without control parameters. It has been guided searching the optimum balance between time performance and resulting quality suitable for automatic segmentation tasks. Using a semi-implicit numerical scheme, we have determined the relationship between the slope range to sharpen and the diffusion time. It has also been selected the diffusivity with optimum performances. Several diffusion filters have been applied to noisy computed tomography images and evaluated for their suitability to the medical image segmentation. Experimental results show that our proposal of filter performs quite well in relation to others.

Keywords: nonlinear diffusion filter, segmentation, 3D medical image.

1 Introduction

For improving the segmentation task, a pre-processing filter has to be applied to the original image in order to remove the noise from homogeneous area while keeping clear and sharp the edges. In the field of image processing, several filtering methods are available for this purpose[1]. Convolutions and rank filters (median filter, mean filter, etc.) reduce the image noise, but they did not preserve the details and tended to blur the edges. Nonlinear filters smooth the noise while maintaining clear edges. We have selected a nonlinear diffusion filter without control parameters, searching the optimum balance between time performance and resulting quality suitable for automatic segmentation tasks.

The paper is organized as follows: in section 2, we explain our theoretical framework, which analyzes a diffusivity without tuning parameters and its properties. In section 3, we show the numerical algorithm for nonlinear diffusion and determine the time needed for area smoothing and edge enhancement purposes. Finally, in section 4, some diffusion filters are compared on computed tomography images with a rich variety of features and edge types but also with a significant noise level. Experiments demonstrate that our proposal of filter performs quite well compared to others.

2 Nonlinear Diffusion without Control Parameters

Starting with an initial image $u_0 : \Omega \to \mathbb{R}$ defined over a domain $\Omega \subset \mathbb{R}^m$, another image $u(\boldsymbol{x})$ is obtained as the solution of a nonlinear diffusion equation with initial and Neumann boundary conditions:

$$u_t = \operatorname{div}\left(g(\|\nabla u\|)\nabla u\right), \quad \boldsymbol{x} \in \Omega, \quad t > 0, \tag{1}$$

with $u(\boldsymbol{x},0) = u_0(\boldsymbol{x})$ when $\boldsymbol{x} \in \Omega$ as initial condition and $u_{\boldsymbol{n}} = 0$ when $\boldsymbol{x} \in \partial\Omega$ as boundary condition, with $g(\|\nabla u\|)$ further representing *diffusivity*. We have chosen a diffusivity which balances between sharpen edges over a wide range of selected slopes and reduce noise conservatively with dissipation along feature boundaries. Specifically, the range of sharpned edge slopes is widened as backward diffusion normal to level sets is balanced with forward diffusion tangent to level set. Our family of diffusivity, as in the TV flow case, is free from parameters, but allows backward diffusion along the gradient direction and it is therefore edge enhancing [2][3]:

$$g(\|\nabla u\|) = \frac{1}{\|\nabla u\|^p}, p > 1. \tag{2}$$

The diffusion properties of these filters can be showed when set out in a new orthonormal basis in which one of the axes is determined by the gradient vector $\eta = \nabla u/\|\nabla u\|$ where $\|\nabla u\| \neq 0$, which together with ξ and ζ form the curve/surface at a level perpendicular to η[4]:

$$u_t = g(\|\nabla u\|)(u_{\xi\xi} + u_{\zeta\zeta}) + [\,g(\|\nabla u\|) + g'(\|\nabla u\|)\cdot\|\nabla u\|\,]\,u_{\eta\eta} \tag{3}$$

where $u_{\eta\eta}$ represents the second derivative of u in the direction of η. Thus, tangential diffusion is always forward since $g(\|\nabla u\|) > 0$ and normal diffusion is always backward since

$$[g(\|\nabla u\|) + g'(\|\nabla u\|)\cdot\|\nabla u\|\,] = \frac{1-p}{\|\nabla u\|} < 0. \tag{4}$$

Continuum level analysis of smooth images shows that diffusivity should be chosen so edge slopes with $\|u_\eta\| \leq \alpha_{th}$ must be blurred and edges slopes with $\|u_\eta\| > \alpha_{th}$ must be heightened at a locally maximal rate which leads to sharpening and avoiding staircasing, where α_{th} is the threshold slope on which enhancement is achieved. Open questions are: 1) what is the optimum value of p? and 2) given a particular value of α_{th}, what is the diffusion time required for the selected edge enhancement task? In the continuum domain, this approach gives rise to an ill-posed problem[5,6]. However, in the discrete scheme, under certain data conditions, we can obtain the convergent solutions as referred to in [5]. For more detail see [7].

As we stated above, for enhancement process the $u_{\eta\eta}$ coefficient left to be positive, so this implies that differential operator in (3), loses the necessary conditions for a well-posed problem. An introduction about this topic can be found in a classical Weickert's paper [8] and the references therein cited. However with the Perona-Malik filter [9] in (1) we also get an ill-posed problem, but discretization has a stabilizing effect over this equation [7]. We follow this pattern for the proposed diffusivity (2). Therefore, we apply the Method of Lines to transform the original equation (1) into a semi-discrete problem and a well-posed system of ordinary differential equations using the scale-space framework proposed by Weickert [7]. This also implies some kind of regularization in order to avoid unbounded diffusivity (2), when the gradient tends to 0, and hence to keep the system of ordinary differential equations continuously differentiable as function in u.

With this aim, we use an approximation on finite differences based on the average distance between pixels, which subsequently gives rise to an autonomous system of ordinary differential equations:

$$\dot{u}_i(t) = h^{p-2}\left[\frac{u_{i+1}(t) - u_i(t)}{|u_{i+1}(t) - u_i(t)|^p} - \frac{u_i(t) - u_{i-1}(t)}{|u_i(t) - u_{i-1}(t)|^p}\right] \quad (5)$$

with $h = \Delta x, i = 2, \ldots, n-1$. On carrying out the operation we get an autonomous matrix ordinary differential expression of the type $\frac{d\boldsymbol{U}}{dt}(t) = \boldsymbol{f}(\boldsymbol{U}(t)) = A(\boldsymbol{U}(t))\boldsymbol{U}(t)$. The generalisation to highest dimension is straightforward [10].

3 Numerical Methods

Numerical methods have a decisive effect on the outcome of nonlinear diffusion regarding both quality and computation speed. With spatial discretization, the differential equation has been ported to the pixel grid. An explicit Euler method can be computed by an iterative scheme. For stability it has been assumed that diffusivity is limited, however, edge enhancing flow causes unbounded diffusivity when the gradient tends to 0. A popular solution to this problem is the regularization of the diffusivity by a small positive constant ε, taking $g_\varepsilon(s) = \frac{1}{(s+\varepsilon)^p} \leq \frac{1}{\varepsilon^p}$ with $s \geq 0$, hence $g_\varepsilon \rightarrow g$ when $\varepsilon \rightarrow 0$. The regularization limits the diffusivity and hence the explicit scheme becomes very slow. The stability condition on the time step size can be lifted with a semi-implicit scheme. With this aim, spatial discretization is accomplished with finite difference and the temporal discretization is accomplished with semi-implicit time stepping. All pixels are assumed to have unit aspect ratios and width h and the kth time level is $t = k\tau$,

$$\frac{u_i^{k+1} - u_i^k}{\tau} = h^{p-2}\left[\frac{u_{i+1}^{k+1} - u_i^{k+1}}{|u_{i+1}^k - u_i^k|^p} - \frac{u_i^{k+1} - u_{i-1}^{k+1}}{|u_i^k - u_{i-1}^k|^p}\right]. \quad (6)$$

Using matrix-vector notation, it results an inversion matrix that has to be solved at each iteration:

$$(I - \tau A(U^k))U^{k+1} = U^k \quad (7)$$

where I is the identity matrix and $A(U^k)$ is a matrix of the same size, with the following entries:

$$a_{ij} = \begin{cases} g_{i\sim j}\, h^{-2} & \text{if } j \in N(i) \\ -\sum_{l \in N(i)} g_{i\sim l}\, h^{-2} & \text{if } j = i \\ 0 & \text{otherwise.} \end{cases} \qquad (8)$$

Here, $g_{i\sim j}$ denotes the diffusivity between the pixel i and j, $N(i)$ are the neighbors of pixel i. The matrix $(I - \tau A)$ is a sparse, positive definite and diagonally dominant matrix. Such a tridiagonal matrix can be solved efficiently with the Thomas algorithm. For its implementation in the scope of nonlinear diffusion see [10].

The gray evolution of a pixel depends on the whole of the pixels. For simplification, it would be interesting to observe the interaction with only three pixels within the established dynamic. Applying the semi-implicit Euler method on three pixels, the matrix is inverted giving the expression

$$\begin{bmatrix} u_1^{k+1} \\ u_2^{k+1} \\ u_3^{k+1} \end{bmatrix} = \frac{1}{d} \mathbf{B} \begin{bmatrix} u_1^k \\ u_2^k \\ u_3^k \end{bmatrix} \qquad (9)$$

with

$$\mathbf{B} = \begin{bmatrix} \alpha^p \beta^p + 2r\alpha^p + r\beta^p + r^2 & r(\beta^p + r) & r^2 \\ r(\beta^p + r) & \alpha^p \beta^p + r\alpha^p + r\beta^p + r^2 & r(\alpha^p + r) \\ r^2 & r(\alpha^p + r) & \alpha^p \beta^p + r\alpha^p + 2r\beta^p + r^2 \end{bmatrix} \qquad (10)$$

where $\alpha = |u_2^k - u_1^k| \neq 0$, $\beta = |u_3^k - u_2^k| \neq 0$, $r = \tau h^{p-2}$ and $d = \alpha^p \beta^p + 2r\alpha^p + 2r\beta^p + 3r^2$. For initial arbitrary value of the 3-pixels, and a finite time, the matrix coefficients are all equal to $1/3$. It confirms the stability properties for $r > 0$. However, the issue lies in how to determine the nonlinear diffusion time so that it produces diffusion between the low gradient module pixels without transferring diffusion to the pixels that have a high value of the gradient module. Without loss of overall applicability, in (9) it is imposed $\alpha \gg \beta$, so as to spread forward diffusion between pixels 2 and 3 while keeping the value of the pixel 1. This evolution means that the matrix (9) tends to be $\begin{bmatrix} 1 & 0 & 0 \\ 0 & 1/2 & 1/2 \\ 0 & 1/2 & 1/2 \end{bmatrix}$, which forces $r < 2\alpha^p$. This inequality has been used to determinate the balance between forward and backward diffusion. It has been observed experimentally that this conclusion can be extended to n-pixels [11], obtaining as optimum value $p = 3$. For $h = 1$, we obtained the following expression for the time step:

$$\tau = \frac{\alpha_{th}^p}{5 \cdot n_{iter}} \qquad (11)$$

where n_{iter} denotes the number of iterations (at least four iterations) and α_{th} is the absolute value of the difference between pixels, this being the slope threshold on which enhancement is achieved. Extension to a higher dimension is carried out by applying AOS (Additive Operator Splitting) [10]. Moreover, the numerical method allows parallel and distributed computing.

4 Computational Results

In order to show the performance of our proposal, it has been compared to Gauss ($p = 0$), TV ($p = 1$) and BFB ($p = 2$)[3] filters. In all cases, the time step was

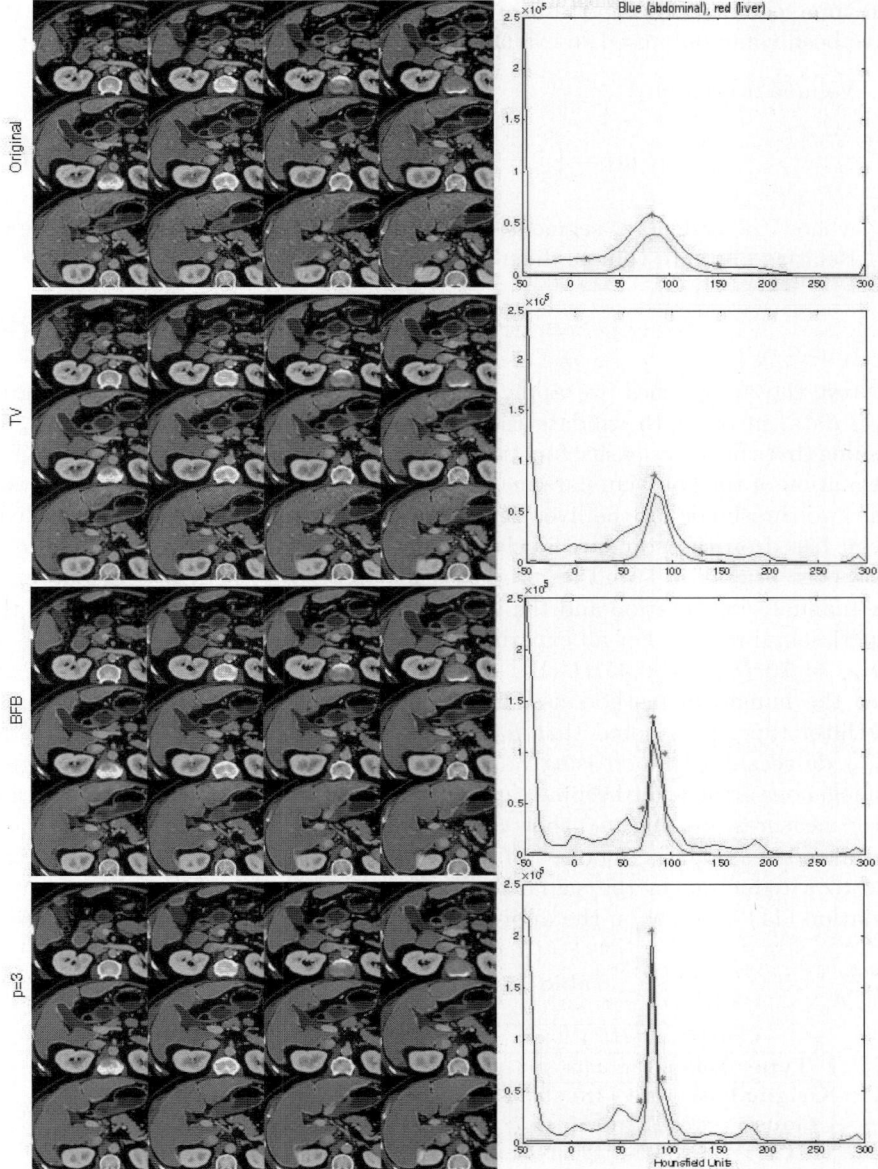

Fig. 1. Some slices over abdominal CT and their histograms a)Original, b)TV, c)BFB, d)$p = 3$

calculated by (11) and a semi-implicit scheme was implemented following (7). The filters have been applied over CT images for liver segmentation task. The images were taken from [12]. It is a training data set that includes both images and binary masks of the segmentations of the structures of interest, produced by human experts. There are 20 CT images. The proposed procedure is to apply the same segmentation algorithm over filtering images. The following two metrics have been commonly used to evaluate the quality of segmentation:

1. Volume overlap m_1:

$$m_1 = \left(1 - \frac{Vol_{seg} \cap Vol_{ref}}{Vol_{seg} \cup Vol_{ref}}\right) \times 100\% \tag{12}$$

where Vol_{seg} denotes segmented volume. Vol_{ref} denotes reference volume.

2. Relative absolute volume difference m_2

$$m_2 = \frac{|Vol_{seg} - Vol_{ref}|}{Vol_{ref}} \times 100\% \tag{13}$$

To test the approaches, we apply a simple threshold technique. It uses gray level data, in order to validate the proposed image processing technique. We assume that the liver density function follows a normal, $N(\mu_{liver}, \sigma_{liver}^2)$[13]. The estimation of the gaussian parameters are obtained through histogram analysis. The two thresholds for the liver have been determined by two offsets from liver mean based on the standard deviation of the gray level of the liver. Figure 1 shows the original CT and its processed images. It also depicts the contour of the manual segmentation and the image histograms. The table summarizes the experimental results. For all experiments were used the following values: $n_{iter} = 5, \alpha_{th} = 70HU, \varepsilon = 0.35HU$ (HU $\equiv$ Hounsfield Units). Firstly, it is observed that the numerical method used is a conservative procedure. Independent of the filter type, it is noted that $\hat{\mu}_{liver}$ remains almost constant. Secondly, the $\hat{\sigma}_{liver}$ decreases with increasing value of p. This means removing the noise from homogenous areas while keeping clear and sharp edges. Finally, the segmentation error measures, m_1 and m_2, show a downward trend with increasing value of p. We have seen that the increase of p inhibits the staircase effect but also gives rise to a reduction in the signal dynamic range. Furthermore, the validity of equation (11) is based on the approximation between the model and numerical

Table 1. Experimental results

Type	$\hat{\mu}_{liver}$ (HU)			$\hat{\sigma}_{liver}$ (HU)			m_1 (%)			m_2 (%)		
	min	mean	max	min	mean	max	min	mean	max	min	mean	max
Original	83	123	181	9.3	25.3	56.0	34.9	42.1	55.3	38.0	57.8	88.2
Gauss	83	125	181	11.7	22.3	46.6	27.4	42.3	78.1	30.3	52.6	70.6
TV	82	126	180	9.2	19.1	39.6	23.1	35.3	42.9	24.6	43.3	65.8
BFB	82	126	180	4.6	14.5	32	16.5	29.6	42.9	16.8	34.3	60.1
$p = 3$	82	126	187	4.6	9.1	21	9.3	22.4	34.3	9.6	23.7	36.5

scheme and on which basis we conclude that a compromise value should be $p = 3$. Experiments demonstrate that our proposal of filter performs quite well in relation to others.

5 Conclusions

A new nonlinear diffusion filter has been developed, which sharpens edges over a wide range of slopes and reduce noise conservatively with dissipation along homogeneous regions. It can be implemented efficiently and absolutely stable with a semi-implicit scheme and ε-regularization. Based on the discrete evolution of three pixels, we have determined the diffusion time required for the edge enhancement from a predetermined threshold. So, edge slopes below the threshold must be blurred and those above the threshold should be sharpened and trying to avoid staircase effect. Using the given time step, some diffusion filters are compared to computed tomography images for segmentation tasks. Experiments demonstrate that our proposal of filter performs quite well compared to others.

References

1. Buades, A., Coll, B., Morel, J.M.: A review of image denoising algorithms with a new one. Multiscale Modeling & Simulation 4, 490–530 (2005)
2. Tsurkov, V.I.: An analytical model of edge protection under noise suppression by anisotropic diffusion. Journal of Computer and Systems Sciences International 39(3), 437–440 (2000)
3. Keeling, S.L., Stollberger, R.: Nonlinear anisotropic diffusion filters for wide range edge sharpening. Inverse Problems 18, 175–190 (2002)
4. Teboul, S., Blanc-Feraud, L., Aubert, G., Barlaud, M.: Variational approach for edge-preserving regularization using couple pdes. IEEE Trasanctions on Image Processing 7, 387–397 (1998)
5. Catte, F., Lions, P.L., Morel, J.M., Coll, T.: Image selective smoothing and edge detection by nonlinear diffusion. SIAM Journal on Applied Mathematics 29(1), 182–193 (1992)
6. Kichenassamy, S.: The perona-malik paradox. SIAM Journal on Applied Mathematics 57(5), 1328–1342 (1997)
7. Weickert, J., Benhamouda, B.: A semidiscrete nonlinear scale-space theory and its relation to the perona-malik paradox. Advances in Computer Vision, 1–10 (1997)
8. Weickert, J.: A Review of Nonlinear Diffusion Filtering. In: ter Haar Romeny, B.M., Florack, L.M.J., Viergever, M.A. (eds.) Scale-Space 1997. LNCS, vol. 1252, pp. 3–28. Springer, Heidelberg (1997)
9. Perona, P., Malik, J.: Scale space and edge detection using anisotropic dffusion. IEEE Transaction on Pattern Analysis and Mach. Intell. (12), 629–639 (1990)
10. Weickert, J., ter Haar Romeny, B., Viergever, M.A.: Efficient and reliable schemes for nonlinear diffusion filtering. IEEE Trasanctions on Image Processing 7(3), 398–410 (1998)

11. Platero, C., Sanguino, J., Tobar, M.C., Poncela, J.M., Asensio, G.: Analytical Approximations for Nonlinear Diffusion Time in Multiscale Edge Enhancement. In: International Conference on Computer Vision Theory and Applications (VISAPP 2009), Lisboa, Portugal, February 5-8 (2009)
12. 3D Segmentation in the Clinic: A Grand Challenge I - Liver Segmentation, http://sliver07.isi.uu.nl/
13. Freiman, M., Eliassaf, O., Taieb, Y., Joskowicz, L., Azraq, Y., Sosna, J.: An iterative Bayesian approach for nearly automatic liver segmentation: algorithm and validation. International Journal of Computer Assisted Radiology and Surgery 3(5), 439–446 (2008)

Color Quantization by Multiresolution Analysis

Giuliana Ramella and Gabriella Sanniti di Baja

Institute of Cybernetics "E. Caianiello", CNR, Pozzuoli (Naples), Italy
{g.ramella,g.sannitidibaja}@cib.na.cnr.it

Abstract. A color quantization method is presented, which is based on the analysis of the histogram at different resolutions computed on a Gaussian pyramid of the input image. Criteria based on persistence and dominance of peaks and pits of the histograms are introduced to detect the modes in the histogram of the input image and to define the reduced colormap. Important features of the method are, besides its limited computational cost, the possibility to obtain quantized images with a variable number of colors, depending on the user's need, and that the number of colors in the resulting image does not need to be a priori fixed.

Keywords: histogram analysis, color quantization, pyramids.

1 Introduction

Color image quantization is a process that enables an efficient compression of color images by reducing the number of distinct colors present in the images. This process, also known as color reduction, is aimed at producing a new version of the input image that, though employing a smaller number of colors, is still visually similar to the original image. Color quantization has a key role both for image display [1], especially in the presence of devices that can display only a limited number of colors due to memory limitations, and for image compression [2], especially when multimedia data of considerably large size have to be recorded or transmitted. Another important application of color quantization regards color based indexing and retrieval from image databases [3].

A number of color quantization methods are available in the literature [1,4-13]. The standard approach is based on the interpretation of color quantization as a clustering problem in the 3D space, where the three axes are the three color channels and the points represent the various colors found in the image. Points are grouped into clusters, by using any clustering technique, and the representative color for each cluster is generally obtained as the average of the points in the cluster [4-6]. Since the problem of finding the optimum clustering is generally NP-hard [4,14], many algorithms have been suggested to build a colormap with an a priori fixed number of colors. One of the first methods to build a reduced colormap is the population or popularity method. This method, suggested by Boyle and Lippman in 1978, has been implemented a few years later and is known as the median cut algorithm [1]. Another well-known method is the octree quantization algorithm [7]. Other approaches are based on histogram analysis [8-10], fuzzy logic [11], neural network [12] and multiresolution analysis [10,13].

Methods such as [1,5,7,9] produce quantized images with low computational load independently of the images they are applied to, but require that the number of colors in the final image is a priori fixed by the user. Methods, producing quantized images with higher quality, build the set of representative colors according to the spatial color distribution in the input image, where the chosen quantization regions are of different size and are selected according to a given criterion. These methods are image-dependent [4,6,8,10,11-13].

Due to the high number of colors generally present in the input image, which makes quantization a complex task, most color quantization methods include a pre-processing step to reduce the data that will be processed by the quantization algorithms, e.g., by reducing the range of each coordinate in the color space [5,10].

In this paper, we suggest a color quantization algorithm that can be framed in the category of image-dependent methods. The algorithm is based on color distribution and on the use of multiresolution image representation. The histograms at different resolutions are examined and the peaks and pits persisting at all resolutions or dominating in the histogram of the input image are used to determine the colormap. Thus, the number of colors in the colormap does not need to be fixed by the user and is automatically detected, based on the number of persistent or dominating peaks and pits. The number of colors depends on the distribution of colors in the input image as well as on the number of resolutions taken into account to select significant peaks and pits. The method allows to obtain different colormaps by using an increasing number of considered resolutions. Clearly, the number of colors diminishes, but color distortion increases, when the number of considered resolutions increases. Finally, we remark that our method does not require any pre-quantization to reduce the data.

Some notions are given in Section 2; the method is described in Section 3 and experimental results are discussed in Section 4. Concluding remarks are given in Section 5.

2 Notions

Different color spaces can be used when working with color images. In this paper, we work with RGB images, even if our method can equally be applied in other color spaces. We interpret colors as three-dimensional vectors, with each vector element having an 8-bit dynamic range.

The human visual system, though able to distinguish a very large number of colors, generally groups colors with similar tonality for image understanding, since a few colors are often enough to this aim [15]. Analogously, color quantization reduces the set of colors in a digital image to a smaller set of representative colors.

The histogram of an image is useful to analyze the distribution of the pixels values within the image, so as to identify values that could be grouped together and could be replaced in the image by a unique representative value. For color images the histogram is a multidimensional structure and its analysis is rather complex [16]. A few attempts to work with multidimensional histograms can be found in the literature, see e.g., [9-10]. In turn, a color image can be interpreted as decomposed into three gray-level images, so that the analysis of the three one-dimensional histograms associated to the three color channels can be done instead of analyzing the complex histogram of colors. We use this kind of independent analysis of the three histograms, as it is mostly done in the literature, though we are aware that some important information contained in the dependence among the channels may be lost.

For one-dimensional histograms, the list of its *modes*, i.e., the intervals of values occurring the most frequently, can be efficiently used to perform quantization. If the peaks in the histogram are detected together with the pits separating them, each set of values from a pit to the successive pit can be seen as corresponding to a mode and all the values can be quantized in a single representative value, e.g., the value of the peak placed in between the delimiting pits. In general, peaks and pits may consist of more than one single value (i.e., they can be shaped as plateaux). Moreover, a generally large number of peaks and pits, not all equally meaningful, exist in the histogram. Thus, some criteria should be adopted to detect peaks and pits in the presence of plateaux and to remove those peaks or pits that are regarded as scarcely meaningful from a perceptual point of view.

Multiresolution representation of an image is useful in different application contexts, as it provides differently condensed representations of the information contents of the image. Pyramids are among the structures most often used for multiresolution representation. They are generally built by using a uniform subdivision rule that summarizes fixed sized regions in the image. A Gaussian pyramid can be built by creating a series of images, which are weighted down using a Gaussian average and scaled down. By doing the same process multiple times, a stack of successively smaller images is created. Also discrete methods exist in the literature to build pyramids. For example, in [17] a subdivision of the image into non-overlapping 2×2 blocks is performed to obtain a smaller image having as many pixels as many are the blocks into which the image is divided. The value of each pixel in the smaller size image is computed by using a multiplicative mask of weights centered on one of the pixels in the corresponding block. The process is repeated to obtain all desired pyramid levels. In this paper, we use the Gaussian pyramid based on a 5×5 support.

3 The Color Quantization Algorithm

Our color quantization method can be sketched as follows. Given a color image with resolution M×N, the pyramid is built. Actually, the number of pyramid levels to be built is fixed by the user, depending on the color distortion that is regarded as acceptable in the quantization. A large number of pyramid levels will produce a colormap with a smaller number of colors, but the resulting image will be less visually similar to the input. Let L denote the number of levels selected for the pyramid. For each color channel and for each pyramid level, the corresponding histogram is computed. Minima and maxima are determined on the histograms. The histograms at different resolution levels for each channel F (where F is R, G or B) are compared to identify the minima and maxima persistent along all levels. The persistent minima and maxima in the histograms and the minima and maxima dominating a significant support in the histogram of the input image are retained, while all other minima and maxima are disregarded. The polygonal line connecting the retained successive minima and maxima is processed to simplify the structure of the histogram. The vertices of the so obtained polygonal line are used to determine the modes of the histogram of the channel F. Then, a representative value for each mode is computed.

We will focus in the following on the process done on the histograms and will describe the process for one channel only.

3.1 Simplifying the Histogram

The histogram generally includes a large number of peaks and pits. The following process is aimed at simplifying the histogram by detecting only the peaks and pits regarded as significant for the quantization.

Let H_i, i=1,...,L be the histograms computed on the L levels of the pyramid for the channel F. On each H_i, peaks and pits are detected, respectively as relative local maxima and relative local minima, i.e., p is a peak (pit) if *height(p-1)≤ height(p)* and *height(p+1) ≤ height(p)* (*height(p)≤ height(p-1)* and *height(p) ≤ height(p+1)*), where *p-1* and *p+1* are the values immediately before and immediately after *p* in the histogram. On H_1, i.e., the histogram of the M×N image, two values v_1 and v_n are identified, respectively coinciding with the first pit to the left coming immediately before the first peak with reasonable height, and with the pit immediately following the last peak with reasonable height. Here, we regard as reasonable the height of a peak if it is at least 1% of the maximal peak height. Only the portion of H_i, i=1,...,L, between v_1 and v_n is taken into account.

Let $v_1, v_2..., v_n$ be the list of peaks and pits (also called vertices) of H_1. Each vertex v_i is associated with three parameters [18], namely the area of the region of the histogram dominated by v_i (actually, the area of the triangle with vertices v_{i-1}, v_i and v_{i+1}), the cosine of the angle formed by the straight lines joining v_{i-1} with v_i and v_i with v_{i+1}, and the distance of v_i from the straight line joining v_{i-1} and v_{i+1}. Let A be the average area of all triangles.

We compare H_1 with the histograms H_j, j=2,...,L and remove from the list of vertices of H_1 any v_i for which the area of the associated triangle is smaller than A and which is not present in all histograms. Once all vertices have been examined, the values of the three parameters for all surviving vertices of H_1 and the average area A are updated. Then, we also remove from the list of vertices of H_1 all successive vertices such that their associated triangles all have area smaller than A and such that the value of at least one of the three parameters is the same for all successive vertices. The last step of histogram simplification is devoted to maintaining in the list of vertices only those that are still peaks or pits.

3.2 Finding the Histogram Modes

Each pair of successive pits identifies a mode of the histogram of F, independently of whether a peak is included between them. The successive step of the process is aimed at computing the representative value for each mode. Three cases are possible:

Case 1 – One single peak is in between the two pits delimiting the mode. Then, the representative value of the mode is the value of the peak.

Case 2 – No peak is included in between the two pits delimiting the mode. Then, if the two pits have different height, the representative value of the mode is the value of the pit with smaller height. Otherwise, if the two pits have the same height, the value of the leftmost pit between the two pits delimiting the mode is used for the representative value of the mode.

Case 3 – A number of peaks exists in between the two pits delimiting the mode. Then, the value of the peak with maximal height is assigned to the representative value of the mode. If a series of successive peaks exists where all peaks have the same height, only the leftmost peak of the series is taken into account in the maximal peak height computation.

The image used as running example (24 bits image "yacht" with 150053 original colors), the modes found on the R, G, and B histograms, and the image resulting after quantization for L=3 are shown in Fig. 1.

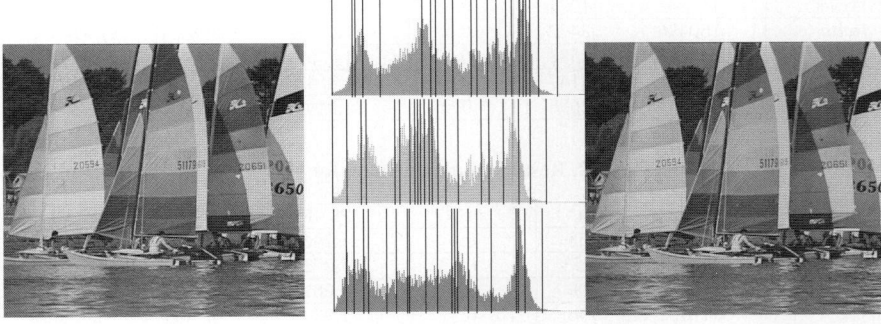

Fig. 1. An input image with 150053 colors, left, the modes for its R, G and B histograms, middle, and the image resulting after color compression for L=3 with 2696 representative colors, right

4 Experimental Results

We have applied our quantization algorithm to a large number of images with different size and color distribution, taken from available databases for color images [19-20]. A small dataset taken from [20] and including seven 512×480 color images (24 bits) is given in Fig. 2. This dataset is used to show the performance of the method in terms of quantitative measures, namely the number of representative colors RC (which accounts for the degree of color compression), the compression ratio CR, computed as the ratio between the size of the output stream and the input stream expressed in bit per pixel (bpp) [21], the Peak Signal to Noise Ratio PSNR [21], the Structural SIMilarity SSIM [22-23], which quantitatively accounts for the similarity between the input image and its quantized version, and the colorloss CL, based on the Euclidean color distance [9,10,24].

Fig. 2. A set of 512×480 color images

The quantization algorithm has been tested for different values of the number L of considered levels of the Gaussian pyramid. In Table 1 and Table 2 the results obtained on the above dataset for L=3 and L=5 are shown. The number of colors in the original images, OC, is also given to better appreciate the obtained color reduction.

Table 1. Results of quantization for L=3

	OC	RC	CR	PSNR	SSIM	CL
cablecar	130416	1472	0,62	30,4991	0,9064	10,3697
cornfield	134514	3901	0,70	31,6933	0,9480	9,1101
flower	111841	1274	0,62	31,9328	0,9093	9,0898
fruits	160476	3422	0,68	31,6469	0,8982	9,2494
pens	121057	1583	0,63	31,1158	0,8966	10,2792
soccer	139156	3149	0,68	31,8542	0,9447	8,9239
yacht	150053	2696	0,66	31,9205	0,9233	8,9934

Table 2. Results of quantization for L=5

	OC	RC	CR	PSNR	SSIM	CL
cablecar	130416	258	0,47	21,2468	0,7860	25,6635
cornfield	134514	1753	0,63	27,2728	0,9025	16,4103
flower	111841	506	0,54	29,0611	0,8575	12,8794
fruits	160476	936	0,57	21,9970	0,7863	28,3888
pens	121057	320	0,49	24,4638	0,7672	20,2832
soccer	139156	594	0,54	24,2289	0,8471	19,0933
yacht	150053	637	0,54	26,6588	0,8347	17,0297

We note that the average compression ratio is equal to 0.66 for quantization obtained with L=3 and 0.54 for L=5, i.e., color information occupies in the average 66% and 54% of its original size after compression, respectively. Of course, the average SSIM and PSNR slightly diminish when passing from L=3 to L=5, though they still indicate a good similarity between the input image and its quantized version. Finally, the average colorloss increases when using L=5, as expected due to the higher compression.

For the running example yacht, the quantized images obtained for L=2, 4, 5 are shown in Fig. 3. The quantization for L=3 has already been given in Fig. 1 right.

L=2, RC=3995   L=4, RC=1020   L=5, RC=637

Fig. 3. Quantized images resulting for different values of L. In all cases, the number RC of representative colors is considerably smaller than the number OC of original colors (OC=150053)

5 Concluding Remarks

We have introduced a lossy method to reduce the number of colors in an image, while maintaining the visual aspect of the quantized image satisfactory enough. The algorithm is based on the analysis of the histograms at different resolution levels of the Gaussian pyramid built from the input image. Only peaks and pits persistent at all resolutions or dominating in the histogram of the input image are considered as significant to identify the modes in the histogram of the input image. Each mode of the histogram is associated with a single representative value.

The use of multiresolution representation has the advantage that different quantizations can be obtained for the same image, depending on the number L of resolution levels that are taken into account for the histogram analysis. The larger is L, the smaller is the number of representative colors. Obviously, the quality of the obtained quantized image decreases when L increases; the value L=3 can be used as a default value. The method is particularly useful for progressive transmission, where an image characterized by strong color reduction, i.e., quantized with a large value of L, can be initially transmitted and better versions can be provided if demanded by the receiver.

The algorithm has a limited computational cost, does not require pre-quantization and does not require to fix a priori the number of representative values. It has been applied to a large set of images, producing satisfactory results both in terms of compression ratio and space saving, and in terms of image quality (evaluated by PSNR, SSIM and colorloss).

Though illustrated with reference to color images in the RGB color space and Gaussian pyramids, the algorithm can be used also in different color spaces and by employing different multiresolution structures. Future work will deal with multidimensional histogram analysis.

References

[1] Heckbert, P.S.: Color Image Quantization for Frame Buffer Display. Proc. ACM SIGGRAPH 1982 16(3), 297–307 (1982)
[2] Plataniotis, K.N., Venetsanopoulos, A.N.: Color Image Processing and Applications. Springer, Heidelberg (2000)
[3] Rui, Y., Huang, T.S.: Image Retrieval: Current Techniques, Promising Directions, and Open Issues. Journal of Visual Communication and Image Representation 10, 39–62 (1999)
[4] Braquelaire, J.-P., Brun, L.: Comparison and Optimization of Methods of Color Image Quantization. IEEE Trans. IP 6(7), 1048–1052 (1997)
[5] Bing, Z., Junyi, S., Qinke, P.: An adjustable algorithm for color quantization. Pattern Recognition Letters 25, 1787–1797 (2004)
[6] Chen, T.W., Chen, Y.-L., Chien, S.-Y.: Fast Image Segmentation Based on K-Means Clustering with Histograms in HSV Color Space. In: Proc. IEEE 10th Workshop on Multimedia Signal Processing, pp. 322–325 (2008)
[7] Gervautz, M., Purgtathofer, W.: Simple Method for Color Quantization: Octree Quantization. Academic, San Diego (1990)
[8] Delon, J., Desolneux, A., Lisani, J.L., Petro, A.B.: A Nonparametric Approach for Histogram Segmentation. IEEE Trans. IP 16(1), 253–261 (2007)

[9] Hsieh, I.S., Fan, K.C.: An adaptive clustering algorithm for color quantization. Pattern Recognition Letters 21, 337–346 (2000)
[10] Kim, N., Kehtarnavaz, N.: DWT-based scene-adaptive color quantization. Real-Time Imaging 11, 443–453 (2005)
[11] Ozdemir, D., Akarun, L.: A fuzzy algorithm for color quantization of images. Pattern Recognition 35, 1785–1791 (2002)
[12] Atsalakis, A., Papamarkos, N.: Color reduction and estimation of the number of dominant colors by using a self-growing and self-organized neural gas. Engineering Applications of Artificial Intelligence 19, 769–786 (2006)
[13] Payne, A.M., Bhaskar, H., Mihaylova, L.: Multi-resolution learning vector quantisation based automatic colour clustering. In: Proc. 11th Int. Conf. on Information Fusion, pp. 1–6 (2008)
[14] Brucker, P.: On the Complexity of Clustering Problems. In: Hem, R., Korte, B., Oettli, W. (eds.) Optimizations and Operations Research, pp. 45–54. Springer, Heidelberg (1977)
[15] Rosch, E.: Principles of categorization. In: Rosch, E., Lloyd, B.B. (eds.) Cognition and categorization, pp. 27–48. Erlbaum, Hillsdale (1978)
[16] Chauveau, J., Rousseau, D., Chapeau-Blondeau, F.: Pair correlation integral for fractal characterization of three-dimensional histograms from color images. In: Elmoataz, A., Lezoray, O., Nouboud, F., Mammass, D. (eds.) ICISP 2008. LNCS, vol. 5099, pp. 200–208. Springer, Heidelberg (2008)
[17] Borgefors, G., Ramella, G., Sanniti di Baja, G., Svensson, S.: On the Multi-scale Representation of 2D and 3D Shapes. Graphical Models and Image Proc. 61, 44–62 (1999)
[18] Arcelli, C., Ramella, G.: Finding contour-based abstractions of planar patterns. Pattern Recognition 26(10), 1563–1577 (1993)
[19] http://www.eecs.berkeley.edu/Research/Projects/CS/vision/bsds/
[20] http://www.hlevkin.com/TestImages/
[21] Salomon, D.: Data Compression: The Complete Reference. Springer, London (2007)
[22] De Simone, F., Ticca, D., Dufaux, F., Ansorge, M., Ebrahimi, T.: A comparative study of color image compression standards using perceptually driven quality metrics. In: Proc. SPIE Conf. on Optics and Photonics, Applications of Digital Image Processing XXXI (2008)
[23] Wang, Z., Lu, L., Bovik, A.C.: Video quality assessment based on structural distortion measurement. Signal Processing: Image Communication 19(2), 121–132 (2004)
[24] Chan, H.C.: Perceived image similarity and quantization resolution. Displays 29, 451–457 (2008)

Total Variation Processing of Images with Poisson Statistics

Alex Sawatzky, Christoph Brune, Jahn Müller, and Martin Burger

Westfälische Wilhelms-Universität Münster,
Institut für Numerische und Angewandte Mathematik,
Einsteinstr. 62, D-48149 Münster, Germany
{alex.sawatzky,christoph.brune,jahn.mueller,martin.burger}@wwu.de
http://imaging.uni-muenster.de

Abstract. This paper deals with denoising of density images with bad Poisson statistics (low count rates), where the reconstruction of the major structures seems the only reasonable task. Obtaining the structures with sharp edges can also be a prerequisite for further processing, e.g. segmentation of objects.

A variety of approaches exists in the case of Gaussian noise, but only a few in the Poisson case. We propose some total variation (TV) based regularization techniques adapted to the case of Poisson data, which we derive from approximations of logarithmic a-posteriori probabilities. In order to guarantee sharp edges we avoid the smoothing of the total variation and use a dual approach for the numerical solution. We illustrate and test the feasibility of our approaches for data in positron emission tomography, namely reconstructions of cardiac structures with ^{18}F-FDG and H_2 ^{15}O tracers, respectively.

Keywords: Denoising, Poisson noise, Total variation, Regularization techniques, Positron emission tomography, Segmentation.

1 Introduction

In this paper we shall discuss some approaches to denoising density images with Poisson statistics, with particular focus on cartoon reconstruction. The latter seems particularly reasonable for low count rates, where the effective SNR is too low to compute further details in the image. Moreover, appropriate cartoons are important for subsequent tasks such as segmentation of objects in the images or further quantitative analysis. For this sake we shall employ variational methods based on penalization by total variation (or related penalization functionals of ℓ^1 or L^1-type), which has become a standard approach for such tasks in the frequently investigated case of additive Gaussian noise.

Variational methods in the case of Gaussian noise [1], [2], can be written as minimizing an energy functional of the form

$$\frac{1}{2}\int_\Omega (u-f)^2 \, d\mu \;+\; \alpha\, R(u) \quad \rightarrow \quad \min_u, \qquad (1)$$

in order to obtain a denoised version u of a given image f, where Ω is the image domain and α is a positive regularization parameter. The first, so-called data fidelity term, penalizes the deviation from the noisy image f and can be derived from the log-likelihood for the noise model (cf. [1], [2]). R is an energy functional that inserts a priori information about the favoured type of smoothness of solutions. The minimization (1) results in suppression of noise in u if R is smoothing, while u is fitted to f. The choice of the regularization term R is important for structure of solutions. Often functionals $R(u) = \int_\Omega |\nabla u|^p$ for $p > 1$ are used, where ∇ denotes the gradient and $|.|$ the Euclidean norm. The simplest choice $p = 2$ results in a scheme equivalent to a linear filter, which can be implemented very efficiently via a fast Fourier transform. However, such regularization approaches always lead to blurring of images, in particular they cannot yield results with sharp edges.

In order to preserve edges and obtain appropriate structures, we use an approach based on total variation (TV) as regularization functional. TV regularization was derived as a denoising technique in [3] and generalized to various other imaging tasks subsequently. The exact definition of TV [4] is

$$R(u) = |u|_{BV} := \sup_{g \in C_0^\infty(\Omega, \mathbb{R}^d),\, \|g\|_\infty \leq 1} \int_\Omega u \operatorname{div} g, \qquad (2)$$

which is formally (true if u is sufficiently regular) $|u|_{BV} = \int_\Omega |\nabla u|$. The space of integrable functions with bounded (total) variation is denoted by $BV(\Omega)$ (cf. [4], [5]). The variational problem (1) with TV as regularization functional is the Rudin-Osher-Fatemi (ROF) model. The motivation for using TV is the effective suppression of noise and the realization of almost homogeneous regions with sharp edges. These features are particularly attractive for a posterior segmentation and quantitative evaluations on structures.

Images with Poisson statistics arise in various applications, e.g. in positron emission tomography (PET), in optical microscopy or in CCD cameras. In most of these cases, the raw data (positron or photon counts) are related to the images via some integral operator, which first needs to be (approximately) inverted in order to obtain an image. Recently, variational methods derived from Bayesian models have been combined with the reconstruction process [6],

$$\int_\Sigma (Ku - g \log Ku)\, d\mu + \alpha |u|_{BV} \to \min_{u \in BV(\Omega)}, \qquad u \geq 0, \qquad (3)$$

where g are the Poisson distributed raw data, Σ is the data domain and K is a linear operator that transforms the spatial distribution of desired object into sampled signals on the detectors. If $K = Id$, (3) becomes a denoising model, where g is the known noisy image. In the absence of regularization ($\alpha = 0$) in (3) the EM algorithm [7], [8], has become a standard reconstruction scheme in problems with incomplete data corrupted by Poisson noise, which is however difficult to be generalized to the regularized case. Robust iterative methods for minimizing (3) have been derived by the authors recently (cf. [9], [10], [11]), but in any case a minimization of (3) requires significant computational effort.

In order to obtain similar results faster, we investigate a natural alternative scheme, namely the postprocessing of reconstructions with EM methods based on a variational denoising model. This posterior denoising step needs to take into account that the reconstructed image still behaves like an image with Poisson noise and thus particular schemes need to be constructed.

In [12], a TV based variational model to denoise an image corrupted by Poisson noise is proposed,

$$\int_\Omega (u - f \log u) \, d\mu + \alpha \, |u|_{BV} \quad \to \quad \min_{u \in BV(\Omega)} , \quad u \geq 0 . \quad (4)$$

A particular complication of (4) compared to (1) is the strong nonlinearity in the data fidelity term and resulting issues in the computation of minimizers. Due to TV the variational problem (4) is non differentiable and the authors in [12] use an approximation of TV by differentiable functionals $\int_\Omega \sqrt{|\nabla u|^2 + \varepsilon}$ for any $\varepsilon > 0$. This approach leads to blurring of edges and due to an additional parameter dependence on ε such algorithms are even less robust. Here, we propose a robust algorithm for (4) without approximation of TV, i.e. we use (2) respectively a dual version. This allows to realize cartoon images with sharp edges. Moreover, we investigate a quadratic approximation of the fidelity term, which yields very similar results as (4) with a more straight-forward numerical solution.

The challenges of this work are that the Poisson based data-fidelity term in (4) and (3) is highly nonlinear and that the functionals to be minimized are non-differentiable. We propose robust minimization schemes using dual approaches for an appropriate treatment of the total variation. To illustrate the behavior of the proposed methods, we evaluate cardiac $H_2\,^{15}O$ and ^{18}F-FDG PET measurements with low SNR.

2 Methods

2.1 EM and EM-TV Reconstruction

In this section we briefly discuss reconstruction methods based on Expectation-Maximization (EM, cf. [8]) and regularized EM methods. We consider the variational problem (3). A standard reconstruction scheme in the absence of regularization ($\alpha = 0$) is the EM method introduced by Shepp and Vardi [7],

$$u_{k+1} = u_k \frac{K^*}{K^*1} \left(\frac{g}{K u_k} \right), \quad (5)$$

an approach which is reasonably easy to implement, K^* is the adjoint operator of K [2]. However, suitable reconstructions can only be obtained for good statistics, and hence either additional postprocessing by variational methods or additional regularization in the functional (cf. [13], [14]) is needed. For the latter we proposed in [9] and [10] a semi-implicit iteration scheme minimizing (3). This scheme can be realized as a nested two step iteration

$$\begin{cases} u_{k+\frac{1}{2}} = u_k \dfrac{K^*}{K^*1}\left(\dfrac{g}{Ku_k}\right) & \text{(EM step)} \\ u_{k+1} = u_{k+\frac{1}{2}} - \tilde{\alpha}\, u_k\, p_{k+1} & \text{(TV step)} \end{cases} \quad (6)$$

with $p_{k+1} \in \partial |u_{k+1}|_{BV}$ and $\tilde{\alpha} := \frac{\alpha}{K^*1}$, where ∂ denotes the subdifferential [2] and generalizes the notion of derivative. The first step in (6) is a single step of the EM algorithm (5). The more involved second step for TV correction in (6) can be realized by solving

$$u_{k+1} = \underset{u \in BV(\Omega)}{\arg\min} \left\{ \frac{1}{2}\int_\Omega \frac{(u - u_{k+\frac{1}{2}})^2}{u_k} + \tilde{\alpha}\,|u|_{BV} \right\}. \quad (7)$$

Inspecting the first order optimality condition confirms the equivalence of this minimization with the TV correction step in (6). Problem (7) is just a modified version of the ROF model, with weight $\frac{1}{u_k}$ in the fidelity term. This analogy creates the opportunity to carry over efficient numerical schemes known for the ROF model and actually to realize cartoon reconstructions with sharp edges. For a detailed analytical examination of EM-TV we refer to [11]. Since the coupled model needs several iteration steps and thus several solutions of (7), it becomes computationally rather involved. Therefore, we study a simple postprocessing strategy based on first computing a reconstruction of visually bad quality via a simple EM algorithm and postprocessing with total variation, which we expect to recover the major structures in the image at least for a certain range of statistics.

2.2 Denoising Images with Poisson Statistics

The straight-forward approach to denoising is based on maximizing the logarithmic a-posteriori probability, i.e. solving (4) (cf. [12]), whose optimality condition is given by

$$u\left(1 - \frac{f}{u} + \alpha p\right) = 0, \qquad p \in \partial |u|_{BV}. \quad (8)$$

Since the reconstruction model (3) coincides in the case of K being the identity operator with (4), we can use the iteration scheme from the previous section, which simply results in (note that $u_{k+\frac{1}{2}} = f$ in this case)

$$u_{k+1} = f - \tilde{\alpha}\, u_k\, p_{k+1}. \quad (9)$$

As noticed above, we can realize this iteration step by solving the modified version of the ROF model (7). Note that (9) is a semi-implicit iteration scheme with respect to the optimality condition (8) and thus actually computes a denoised image in the Poisson case.

The iteration scheme (9) solves the denoising problem (4) by a sequence of modified ROF variational models. In this way one obtains an MAP estimate, but again at the price of high computational effort. Together with the reconstruction via the EM method, the effort is comparable to the incorporated EM-TV scheme.

Hence we introduce a further approximation of the denoising problem, which is based on a second order Taylor approximation of the data fidelity term in (4),

$$u = \arg\min_{u \in BV(\Omega)} \left\{ \frac{1}{2} \int_\Omega \frac{(u-f)^2}{f} + \tilde{\alpha} \, |u|_{BV} \right\} . \tag{10}$$

In this case we can compute the postprocessing by solving a single modified ROF model.

2.3 Computational Approach

Finally we briefly discuss the numerical solution of the minimization problem

$$u = \arg\min_{u \in BV(\Omega)} \left\{ \frac{1}{2} \int_\Omega \frac{(u-v)^2}{w} + \tilde{\alpha} \, |u|_{BV} \right\} , \tag{11}$$

which is the most general form of all schemes above with appropriate setting of v and the weight w. Most computational schemes for the ROF model can be adapted to this weighted modification, here we use a dual approach that does not need any smoothing of the total variation. Our approach is analogous to the one in [15], using a characterization of subgradients of total variations as divergences of vector fields with supremum norm less or equal one. We thus compute the primal variable from the optimality condition with $\tilde{g}$ to be determined as a minimizer of a dual problem

$$u = v - \tilde{\alpha} \, w \, \operatorname{div} \tilde{g} \, , \qquad \tilde{g} = \arg\min_{g, \, \|g\|_\infty \leq 1} \int_\Omega (\tilde{\alpha} \, w \, \operatorname{div} g - v)^2 \, . \tag{12}$$

This problem can be solved with projected gradient-type algorithms, we use

$$g^{n+1} = \frac{g^n + \tau \, \nabla(\tilde{\alpha} \, w \, \operatorname{div} g^n - v)}{1 + \tau \, |\nabla(\tilde{\alpha} \, w \, \operatorname{div} g^n - v)|} \, , \qquad 0 < \tau < \frac{1}{4 \, \tilde{\alpha} \, w} \, , \tag{13}$$

with the damping parameter τ to ensure convergence of the algorithm.

3 Results

We illustrate our techniques at a simple synthetic object, see Fig. 1, and by evaluation of cardiac $H_2{}^{15}O$ and ^{18}F-FDG measurements obtained with positron emission tomography (PET) [16], [17]. In this modality, a specific radioactive tracer, binding to the molecules to be studied, is injected into blood circulation. $H_2{}^{15}O$ is used for the quantification of myocardial blood flow [18]. This quantification needs a segmentation of myocardial tissue, left and right ventricle [18], [19], which is extremely difficult to realize due to very low SNR of $H_2{}^{15}O$ data.

In order to obtain the tracer intensity in the left ventricle we take a fixed 2D layer in a suitable time frame, see Fig. 2. To illustrate the SNR issue we present reconstructions with the EM algorithm (A). As expected, the results

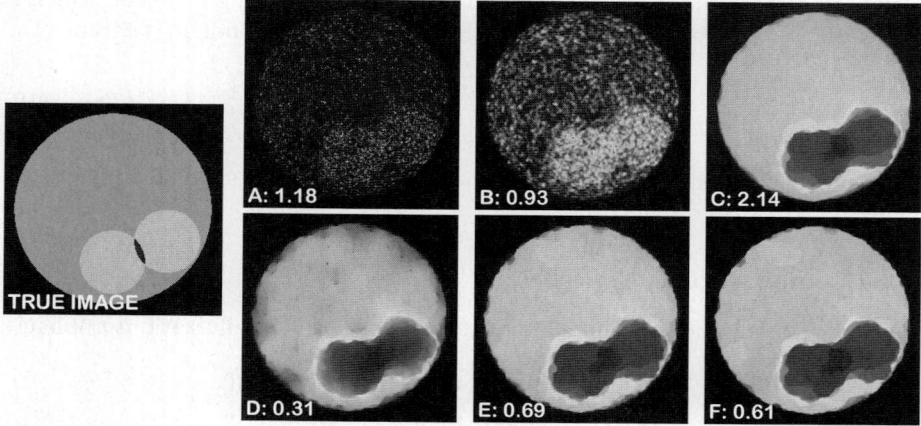

Fig. 1. Synthetic object: results of different reconstruction methods and Kullback-Leibler distances to true image. **A:** EM reconstruction, 18 its. **B:** EM, 18 its, with Gaussian smoothing any 5th step. **C:** A with standard ROF smoothing, (1) with (2). **D:** A with weighted ROF smoothing (10). **E:** A with iterative weighted ROF smoothing (9), 30 its. **F:** Nested EM-TV algorithm (6), 30 its.

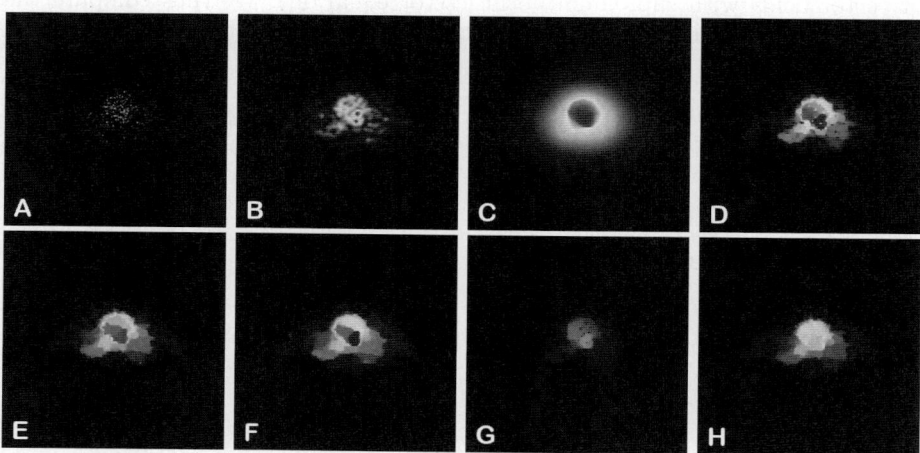

Fig. 2. Cardiac $H_2\,^{15}O$ PET measurements: tracer intensity results of different reconstruction methods in the left ventricle. **A-F:** as in Fig. 1, but all with 20 its. **G, H:** D, E scaled to maximum intensity of F.

suffer from unsatisfactory quality and are impossible to interpret. We hence take EM reconstrutions with Gaussian smoothing (B) as a reference. The next results (C - E) show different approaches of TV smoothing with the (weighted) ROF model. The result C demonstrates the approach with the standard ROF model, the result D is generated with the weighted ROF model (10) and E with

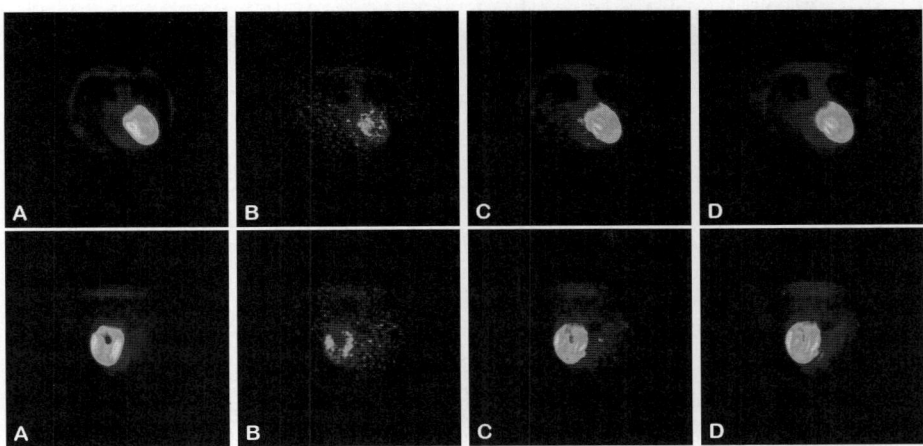

Fig. 3. Cardiac ^{18}F-FDG 3D PET measurements: reconstruction results from two different viewing angles (upper and bottom row). **A:** EM recontruction, 20 its, with Gaussian smoothing any 10th step after 20 minutes data acquisition. **B:** As A but after 5 seconds data acquisition. **C:** B with weighted ROF smoothing (10) and $\tilde{\alpha} = 0.3$. **D:** As C but with $\tilde{\alpha} = 0.5$.

the MAP estimate (4). The approach with the nested EM-TV algorithm (6) is presented in F. The reconstructions G and H are the same results as in D and E appropriate, but the images are scaled to the maximum intensity of F, such that a comparison is possible also for quantitative values. One observes that the results without or with standard smoothing are unsatisfactory, while all total variation techniques yield reasonable reconstructions of the structures. Quantitative values are usually more realistic for the nested EM-TV methods.

In Figure 3, we provide 3D postprocessing results generated with total variation using cardiac ^{18}F-FDG measurements. This tracer is an important radiopharmaceutical and is used for measuring glucose metabolism, e.g in brain, in heart or in cancer. For the illustration of a 3D data set, we take the projections of two fixed viewing angles. The EM reconstruction after a data acquisition of 20 minutes is shown in A as a ground truth for very high count rates. To simulate low count rates, we take the measurements after the first 5 seconds only. The corresponding EM reconstruction is illustrated in B. The results C and D show a postprocessing of B with the weighted ROF model (10) for two different regularization parameters $\tilde{\alpha} = 0.3$ and $\tilde{\alpha} = 0.5$. One observes that the major structures are well reconstructed by this approach also for low count rates.

Acknowledgements. This work has been supported by the German Research Foundation DFG via *SFB 656 Molecular Cardiovascular Imaging* and the project *Regularization with Singular Energies*, as well as by the Federal Ministry of Education and Research BMBF via the project *INVERS*. C.B. acknowledges further support by the Deutsche Telekom Foundation, J.M. by the European

Institute for Molecular Imaging (WWU Münster and SIEMENS Medical). The authors thank Klaus Schäfers (EIMI, WWU Münster) for providing PET data.

References

1. Chan, T.F., Shen, J.: Image Processing and Analysis: Variational, PDE, Wavelet, and Stochastic Methods. SIAM, Soc. for Industrial and Applied Math., Philadelphia (2005)
2. Scherzer, O., Grasmair, M., Grossauer, H., Haltmeier, M., Lenzen, F.: Variational Methods in Imaging. Springer, Heidelberg (2009)
3. Rudin, L.I., Osher, S., Fatemi, E.: Nonlinear total variation based noise removal algorithms. Physica D 60, 259–268 (1992)
4. Acar, R., Vogel, C.R.: Analysis of bounded variation penalty methods for ill-posed problems. Inverse Problems 10, 1217–1229 (1994)
5. Evans, L.C., Gariepy, R.F.: Measure Theory and Fine Properties of Functions. Studies in Advanced Mathematics. CRC Press, Boca Raton (1992)
6. Bertero, M., Lantéri, H., Zanni, L.: Iterative image reconstruction: a point of view. In: Mathematical Methods in Biomedical Imaging and Intensity-Modulated Radiation Therapy (IMRT). CRM series, vol. 7, pp. 37–63 (2008)
7. Shepp, L.A., Vardi, Y.: Maximum likelihood reconstruction for emission tomography. IEEE Transactions on Medical Imaging 1(2), 113–122 (1982)
8. Dempster, A.P., Laird, N.M., Rubin, D.B.: Maximum likelihood from incomplete data via the EM algorithm. J. Roy. Stat. Soc. B 39(1), 1–38 (1977)
9. Sawatzky, A., Brune, C., Wübbeling, F., Kösters, T., Schäfers, K., Burger, M.: Accurate EM-TV algorithm in PET with low SNR. In: IEEE Nuclear Science Symposium Conference Record (2008)
10. Brune, C., Sawatzky, A., Burger, M.: Bregman-EM-TV methods with application to optical nanoscopy. In: Tai, X.-C., et al. (eds.) SSVM 2009. LNCS, vol. 5567, pp. 235–246. Springer, Heidelberg (2009)
11. Brune, C., Sawatzky, A., Wübbeling, F., Kösters, T., Burger, M.: EM-TV methods for inverse problems with Poisson noise (in preparation, 2009)
12. Le, T., Chartrand, R., Asaki, T.J.: A variational approach to reconstructing images corrupted by Poisson noise. J. Math. Imaging Vision 27(3), 257–263 (2007)
13. Natterer, F., Wübbeling, F.: Mathematical Methods in Image Reconstruction. SIAM Monographs on Mathematical Modeling and Computation (2001)
14. Resmerita, E., Engl, H.W., Iusem, A.N.: The EM algorithm for ill-posed integral equations: a convergence analysis. Inverse Problems 23, 2575–2588 (2007)
15. Chambolle, A.: An algorithm for total variation minimization and applications. Journal of Mathematical Imaging and Vision 20, 89–97 (2004)
16. Vardi, Y., Shepp, L.A., Kaufman, L.: A statistical model for positron emission tomography. J. of the American Statistical Association 80(389), 8–20 (1985)
17. Wernick, M.N., Aarsvold, J.N.: Emission Tomography: The Fundamentals of PET and SPECT. Elsevier Academic Press, Amsterdam (2004)
18. Schäfers, K.P., Spinks, T.J., Camici, P.G., Bloomfield, P.M., Rhodes, C.G., Law, M.P., Baker, C.S.R., Rimoldi, O.: Absolute quantification of myocardial blood flow with $H_2\,^{15}O$ and 3-Dimensional PET: An experimental validation. Journal of Nuclear Medicine 43, 1031–1040 (2001)
19. Benning, M., Kösters, T., Wübbeling, F., Schäfers, K.P., Burger, M.: A nonlinear variational method for improved quantification of myocardial blood flow using dynamic $H_2\,^{15}O$ PET. In: IEEE NSS Conference Record (2008)

Fast Trilateral Filtering

Tobi Vaudrey and Reinhard Klette

The *.enpeda..* Project, The University of Auckland, Auckland, New Zealand

Abstract. This paper compares the original implementation of the trilateral filter with two proposed speed improvements. One is using simple look-up-tables, and leads to exactly the same results as the original filter. The other technique is using a novel way of truncating the look-up-table (LUT) to a user specified required accuracy. Here, results differ from those of the original filter, but to a very minor extent. The paper shows that measured speed improvements of this second technique are in the order of several magnitudes, compared to the original or LUT trilateral filter.

1 Introduction

Many smoothing filters have been introduced, varying from the simple mean and median filtering to more complex filters such as anisotropic filtering [2]. These filters aim at smoothing the image to remove some form of noise.

The trilateral filter [1] was introduced as a means to reduce impulse noise in images. The principles of the filter were based on the bilateral filter [7], which is an edge-preserving Gaussian filter. The trilateral filter was extended to be a gradient-preserving filter, including the local image gradient (signal plane) into the filtering process. Figure 1 demonstrates this process using a geometric sketch. This filter has the added benefit that it requires only two user-set parameters (the starting bilateral filter size and a constant that is predefined from [1]), and the rest are self-tuning to the image.

The original paper [1] demonstrated that this filter could be used for 2D images, to reduce contrast of images and make them clearer to a user. It went on to highlight that the

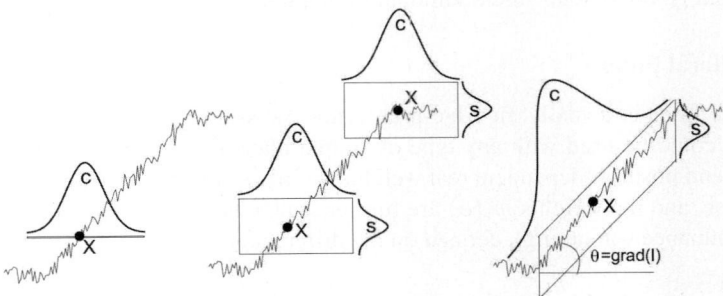

Fig. 1. Illustration of the filtering process using (from left to right) unilateral (Gaussian), bilateral, or trilateral filtering (figure from [1])

filter could also be used to denoise 3D images quite accurately. Recent applications of trilateral filtering have shown that it is also very applicable to biomedical imaging [10]. It decreases noise while still preserving fine details. The trilateral filter has also been used to create residual images (illumination invariant images), to increase the quality of optical flow and stereo matching [9]. Using only one pass produces sufficient results.

Unfortunately, the filter is very slow and requires large local search regions when the image has a low gradient. This issue only gets worse with increasing image sizes, due to the fact that the running time increases super quadratically with image size. This makes large 2D or 3D images very slow to compute. Also, for smaller images, the use of this filter for real-time applications (such as driver assistance systems and security cameras) is limited. In [6] a method was presented for a 15-25 times speed improvement. This paper presents a novel method that provides a speed improvement of several orders of magnitude (100-1000 times faster; from hours down to seconds).

The proposed approach increases the speed of the filter dramatically while still maintaining high accuracy. The filter does not use parallel processing, but can still be parallalised to further increase speed. The approach requires one additional user parameter, required accuracy. Furthermore, we have implemented the trilateral algorithm for standard 2D images, which has been made publicly available [8].

We first introduce the original trilateral filter, followed by a simple speed up technique that does not generate data loss (look up tables). We then present our novel approach, using kernel truncation based on required data accuracy. This is followed by results demonstrating the speed improvements, and the differences in results to the original filter.

2 Definition of Trilateral Filter

An image is defined by $f(\mathbf{x}) \in \mathbb{R}^n$ (n = dimensionality), where $\mathbf{x} \in \Omega$ is the pixel position in image domain Ω. Generally speaking, an n-D (n-dimensional) pixel-discrete image has an image domain defined as, $\emptyset \subset \Omega \subseteq X_n \subset \mathbb{N}^n$ (X_n is our maximum discrete index set of the image domain in dimension n). A smoothing operator will reduce an image to a smoothed version of itself, specifically $S(f) = s$, where s is in the same image domain as f. To introduce the trilateral filter, we must first define the bilateral case; we will then go on to define the traditional trilateral filter using this notation.

2.1 Bilateral Filter

A bilateral filter is actually an edge-preserving Gaussian filter. Of course, the same technique could be used with any type of simple filter (e.g., median or mean). Offset vectors $\mathbf{a}$ and position-dependent real weights $w_1(\mathbf{a})$ (spatial smoothing) define a local convolution, and the weights $w_1(\mathbf{a})$ are further scaled by a second weight function w_2 (range/luminance smoothing), defined on the differences $f(\mathbf{x} + \mathbf{a}) - f(\mathbf{x})$:

$$s(\mathbf{x}) = \frac{1}{k(\mathbf{x})} \int_\Omega f(\mathbf{x} + \mathbf{a}) \cdot w_1(\mathbf{a}) \cdot w_2\Big(f(\mathbf{x} + \mathbf{a}) - f(\mathbf{x})\Big) \, d\mathbf{a} \quad (1)$$

$$k(\mathbf{x}) = \int_\Omega w_1(\mathbf{a}) \cdot w_2\Big(f(\mathbf{x} + \mathbf{a}) - f(\mathbf{x})\Big) \, d\mathbf{a}$$

Function $k(\mathbf{x})$ is used for normalization. The weights w_1 and w_2 are defined by Gaussian functions with standard deviations σ_1 (range) and σ_2 (spatial), respectively (another filter can be substituted, but will provide different results). The smoothed function s equals $S_{BL}(f)$. The bilateral filter requires a specification of parameters σ_1, σ_2, and the size of the used filter kernel $2\mathbf{m}+1$ in f (**m** is the *half kernel size* and is n-dimensional). Of course, the size of the kernel can be selected using σ_1 and σ_2.

2.2 Trilateral Filter

The trilateral filter is a "gradient-preserving" filter. It aims at applying a bilateral filter on the current plane of the image signal. The trilateral case only requires the specification of one parameter σ_1. At first, a bilateral filter is applied on the derivatives of f (i.e., the gradients):

$$g_f(\mathbf{x}) = \frac{1}{k_\nabla(\mathbf{x})} \int_\Omega \nabla f(\mathbf{x}+\mathbf{a}) \cdot w_1(\mathbf{a}) \cdot w_2\left(\|\nabla f(\mathbf{x}+\mathbf{a}) - \nabla f(\mathbf{x})\|\right) \, d\mathbf{a} \quad (2)$$

$$k_\nabla(\mathbf{x}) = \int_\Omega w_1(\mathbf{a}) \cdot w_2\left(\|\nabla f(\mathbf{x}+\mathbf{a}) - \nabla f(\mathbf{x})\|\right) \, d\mathbf{a}$$

To approximate $\nabla f(\mathbf{x})$, forward differences are used, and more advanced techniques (e.g., Sobel gradients, 5-point stencil) are left for future studies. For the subsequent second bilateral filter, [1] suggested the use of the smoothed gradient $g_f(\mathbf{x})$ [instead of $\nabla f(\mathbf{x})$] for estimating an approximating plane

$$p_f(\mathbf{x}, \mathbf{a}) = f(\mathbf{x}) + g_f(\mathbf{x}) \cdot \mathbf{a} \quad (3)$$

Let $f_\triangle(\mathbf{x}, \mathbf{a}) = f(\mathbf{x}+\mathbf{a}) - p_f(\mathbf{x}, \mathbf{a})$. Furthermore, a neighbourhood function

$$N(\mathbf{x}, \mathbf{a}) = \begin{cases} 1 & \text{if } |g_f(\mathbf{x}+\mathbf{a}) - g_f(\mathbf{x})| < c \\ 0 & \text{otherwise} \end{cases} \quad (4)$$

is used for the second weighting. Parameter c specifies the adaptive region and is discussed further below. Finally,

$$s(\mathbf{x}) = f(\mathbf{x}) + \frac{1}{k_\triangle(\mathbf{x})} \int_\Omega f_\triangle(\mathbf{x}, \mathbf{a}) \cdot w_1(\mathbf{a}) \cdot w_2(f_\triangle(\mathbf{x}, \mathbf{a})) \cdot N(\mathbf{x}, \mathbf{a}) \, d\mathbf{a} \quad (5)$$

$$k_\triangle(\mathbf{x}) = \int_\Omega w_1(\mathbf{a}) \cdot w_2(f_\triangle(\mathbf{x}, \mathbf{a})) \cdot N(\mathbf{x}, \mathbf{a}) \, d\mathbf{a}$$

The smoothed function s equals $S_{TL}(f)$.

Again, w_1 and w_2 are assumed to be Gaussian functions, with standard deviations σ_1 and σ_2, respectively. The method requires specification of parameter σ_1 only, which is at first used to be the diameter of circular neighbourhoods at $\mathbf{x}$ in f; let $\overline{g}_f(\mathbf{x})$ be the mean gradient of f in such a neighbourhood. The parameter for w_2 is defined as follows:

$$\sigma_2 = \beta \cdot \left| \max_{\mathbf{x} \in \Omega} \overline{g}_f(\mathbf{x}) - \min_{\mathbf{x} \in \Omega} \overline{g}_f(\mathbf{x}) \right| \quad (6)$$

($\beta = 0.15$ was recommended in [1]). Finally, $c = \sigma_2$.

3 Numerical Speed Improvements

In the previous section, we defined the trilateral filter in a continuous domain. But as we are all aware, the numerical approximation needs to be implemented in real-life. And obviously, this filter takes a lot of processing to work. This section aims at showing how numerical implementations are improved dramatically.

In practice w_1 (the spatial weight) from Equation (2) can be pre-calculated using a look-up-table (LUT), as it only depends on σ_1, $\mathbf{m}$ (kernel size), and $\mathbf{a}$ (offset from central pixel). As this function is Gaussian, the LUT is computed as follows:

$$W_1(\mathbf{i}) = \exp\left(\frac{-\|\mathbf{i}\|^2}{2\sigma_1^2}\right) \tag{7}$$

where $\mathbf{0} \leq \mathbf{i} \leq \mathbf{m}$ (usually $\mathbf{i} \in \mathbb{N}^n$, but can also approximate vectors in $\mathbb{Q}^n$ using interpolation). W_1 is used by simple referencing using $W_1(|\mathbf{a}|)$,[1] which approximates a quarter of the Gaussian kernel (as the Gaussian function is symmetric). Unfortunately, the intensity weight w_2 can not use a look up table, as it depends on the local properties of the image.

Similar principles can be applied to Equation (3). In this equation, w_1 depends on a local adaptive neighbourhood A, depending on the magnitude of the gradients. However, the function is only dependent on the distance from the central pixel $\mathbf{a}$, and since the maximum of A is known, the LUT can be computed as in Equation (7), but where $\mathbf{0} \leq \mathbf{i} \leq \max(A)$. Again, w_2 depends on local information, no LUT can be used. This approach is called the *LUT-trilateral filter*.

From here, to improve speed, there need to be numerical approximations. The presented approach is a smart truncation of the kernel to a defined accuracy $\varepsilon \in \mathbb{Q}^+$, and $0 < \varepsilon < 1$. We know that the function is Gaussian so we shall use this for our truncation. If we want to ignore any values below ε, then only values above this should be used:

$$\varepsilon \leq \exp\left(\frac{-\|\mathbf{i}\|^2}{2\sigma_1^2}\right) \quad \text{which leads to} \quad \|\mathbf{i}\| \leq \sigma_1\sqrt{-2\ln(\varepsilon)} = T$$

where T is the threshold. (Note that $\ln(\varepsilon)$ is strictly negative, so T is strictly positive.) In practice, this means that we can compute a look up table as defined in Equation (7), where $0 \leq \|\mathbf{i}\| \leq T$. This approach could be applied to a bilateral filter, but does not really benefit it. However, when dealing with the trilateral filter, this reduces the number of equations dramatically (as the largest kernel size is equal to the size of the smallest dimension $\min_n(X_n)$ in the image). This truncation will increase the error, but only by, at most, $\varepsilon\left(\min_n(X_n)\right)^2$. We call this method the *fast-trilateral filter*. Note that this filter has only two parameters (which are both logical); σ_1 (the initial kernel diameter) and ε (the required accuracy).

Note that this does not exploit any parallel processing, but is open to massive parallel processing potential, as every pixel is independent within the iteration of trilateral filtering. This is especially noticeable for GPU programming, where the truncated LUT can be saved to texture memory [5].

---

[1] $|\mathbf{a}|$ is here short for $(|a_1|, \ldots, |a_n|)$, and $\|\mathbf{a}\|$ is the L_2-norm.

4 Experimental Results of Filter

We have implemented the trilateral algorithm for standard 2D images, which has been made publicly available [8]. The experiments of this section were performed on a Intel Core 2 Due 3 GHz processor, with 4GB memory, on a Windows Vista platform. Parallel processing was not exploited (e.g., OpenMP or GPU). Of course, further speed improvements can be gained by doing so.

4.1 Dataset

We illustrate our arguments with the 2005 and 2006 Middlebury stereo datasets [4], provided by [3]. We selected a sample set to use for our experiments: *Art, Books, Dolls, Reindeer* and *Baby1*. For each image from this dataset, we use the full resolution image (approx. 1350×1110). We then scale down the image by 50% in both directions, and repeat this 5 times (i.e., 50%, 25%, 13%, 6%, and 3% of original image size), see right part of Figure 2 for example of images used. This allows us to demonstrate running times for differing image sizes.

4.2 Comparison of Running Time

Figure 2 shows the running times of the algorithms on the *Art* images. The results compare two σ_1 values of 3 and 9, and the fast trilateral filter uses $\varepsilon = 10^{-12}$. There is obviously a massive improvement when using trilateral-LUT compared to the original (especially with larger images). With smaller images, the improvement is under 1 magnitude, but increases quickly up to around 1 magnitude improvement (see 1390×1110 results). There is no reason to use the original method instead of the LUT, as there is no accuracy loss with the LUT (the memory usage is negligible compared to calculating image pyramids).

The fast-trilateral filter shows a massive improvement over the other methods (except 43×34, which is not a practical image size). The improvement only gets better as

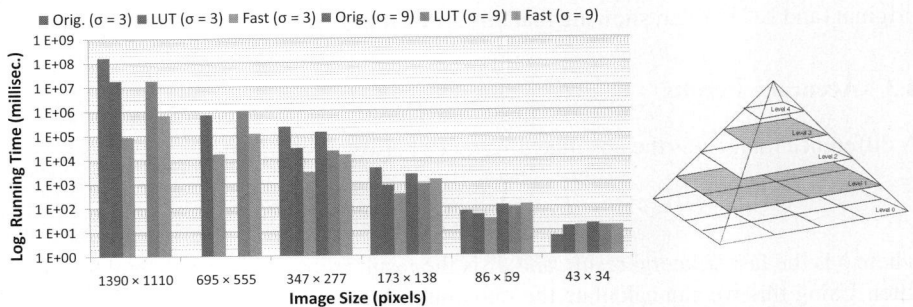

Fig. 2. Running times for *Art* image (left) for different scales of an image (right); displayed in $\log_{10}$ scale. Note: original was not run for $\sigma = 9$ on largest image, nor on second largest image (due to time). For the fast trilateral results, $\varepsilon = 10^{-12}$.

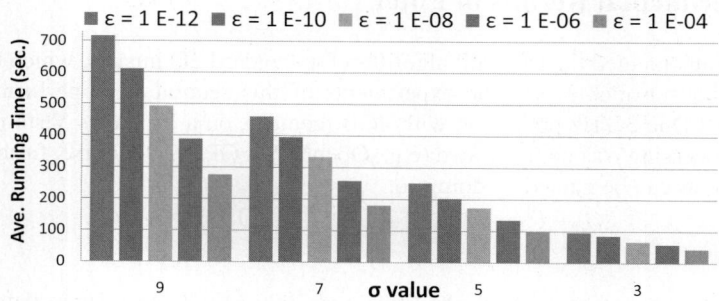

Fig. 3. Average running times for fast-trilateral filter on the dataset of images at maximum resolution. Shows the results for varying kernel sizes σ_1 and accuracy ε.

the size of the image increases; for the largest image size the difference is 46 hours (original) and 5 hours (LUT), compared to 86 seconds for the fast-trilateral filter. That is a dramatic decrease (several orders of magnitude) in computation time.

From these results, we can infer that the improvements will only get better when extending the filter to 3-dimensions (e.g., filtering noisy 3D-meshes), as the number of pixels (or voxels) increases further.

When using the fast-trilateral filter, the user selects the required accuracy. The less accuracy wanted, the faster the filter runs. The comparison in Figure 2 is for the highest accuracy ($\varepsilon = 10^{-12}$), which highlights the improvement over the other filters. To show the effect of reducing the accuracy, compared to running time, we ran the filter across the dataset (at maximum resolution, i.e., approx. 1350×1110) and averaged the running times. Figure 3 shows the results of the fast-trilateral filter for this test using varying kernel sizes (σ_1). This graph shows that the improvements with decreasing accuracy are linear (within each σ_1). A point to note is that when using $\sigma_1 = 9$, the difference in running time goes from 700 seconds ($\varepsilon = 10^{-12}$), down to 290 ($\varepsilon = 10^{-4}$). The next section demonstrates that the results from the fast-trilateral filter are very close to the original (and LUT) filter, showing that this speed improvement is for almost no penalty.

4.3 Accuracy Results

A difference image d is the absolute difference between two images,

$$D(s, s^*) = d \quad \text{with} \quad d(\mathbf{x}) = |s(\mathbf{x}) - s^*(\mathbf{x})| \tag{8}$$

where s is the fast-trilateral result, and s^* is the result from the LUT-trilateral (original) filter. Using this we can calculate the maximum difference $\max_{\mathbf{x} \in \Omega}(d)$, in the image. An example of difference images can be seen in Figure 4. This figure illustrates that there are some subtle differences between the LUT-trilateral filter and the fast-trilateral filter, but they are actually minor. When using $\varepsilon = 10^{-12}$, the differences are too negligible to even count. As for using $\varepsilon = 10^{-4}$, the differences are still very small (maximum error is still less than half an intensity value).

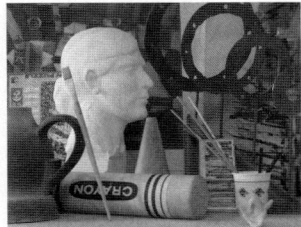

Fig. 4. Left: *Art* image that has been smoothed. Centre and Right: difference image between LUT and fast-trilateral filter, using $\varepsilon = 10^{-4}$ (centre, $\max(d) = 0.36$) and $\varepsilon = 10^{-12}$ (right, $\max(d) = 3.1 \times 10^{-5}$), with $\sigma_1 = 9$. Difference images scaled for visibility, white $\leftrightarrow$ black into $0 \leftrightarrow \max(d)$.

To assess the quality of an image, there needs to be an error metric. A common metric is the *Root Mean Squared* (RMS) *Error*, defined by

$$E_{RMS}(d) = \sqrt{\frac{1}{|\Omega|} \sum_{\mathbf{x} \in \Omega} \left(d(\mathbf{x})^2 \right)} \qquad (9)$$

where $|\Omega|$ is the cardinality of the image domain. The standard RMS error gives an approximate average error for the entire signal, taking every pixel's error independently.

The second metric we use is the *normalised cross correlation* (NCC) percentage

$$C(s, s^*) = \frac{1}{|\Omega|} \sum_{\mathbf{x} \in \Omega} \frac{\left(s(\mathbf{x}) - \mu(s) \right) \left(s^*(\mathbf{x}) - \mu(s^*) \right)}{\sigma(s)\, \sigma(s^*)} \times 100\% \qquad (10)$$

where $\mu(h)$ and $\sigma(h)$ are the mean and standard deviation of image h, respectively. An NCC of 100% means that the images are (almost) identical, and an NCC of 0 means that the images have very large differences.

We calculated the NCC, $\max(d)$ and RMS for the entire dataset, the summary of results can be seen in the table below ($*$ is the don't-care character):

	Average	Minimum at (σ_1, ε)	Maximum at (σ_1, ε)
NCC (%)	100	99.994 at $(9, 10^{-4})$	100 at $(*, *)$
RMS (px)	8.7×10^{-5}	$< 10^{-12}$ at $(3, 10^{-12})$	0.0016 at $(9, 10^{-4})$
$\max(d)$ (px)	0.36	$< 10^{-12}$ at $(3, 10^{-12})$	5.3 at $(9, *)$

From this table it is very apparent that the fast-trilateral filter retains the smoothing properties of the LUT (original) version. The difference is only apparent when using high kernel values σ_1 and also low accuracy values (high ε). Even then, the errors are negligible. In fact, the maximum difference of any individual pixel was only 5.3, with an average maximum of 0.36 pixels.

5 Conclusions and Future Research

In this paper we have covered the original implementation of the trilateral filter. We have suggested two speed improvements. One is using simple look-up-tables, and the other is using a novel way of truncating the look-up-table to a user specified required accuracy.

The speed improvements were shown to be drastic (in the order of several magnitudes) compared to the original or LUT trilateral filter. We identified that the fast-trilateral filter provides very accurate (almost identical) results, compared to the original (and LUT) trilateral filter. This massive speed gain for a very small difference in results is a huge benefit, and thus makes the trilateral filter more usable.

Future work will be to improve speed using parallel architecture (e.g., GPU, Cell Processors, or OpenMP). Also, further applications of the trilateral filter have not been recognised as yet.

Acknowledgment. The authors would like to thank Prasun Choudhury (Adobe Systems, Inc., USA) and Jack Tumblin (EECS, Northwestern University, USA) for their original implementation of the trilateral filter.

References

1. Choudhury, P., Tumblin, J.: The trilateral filter for high contrast images and meshes. In: Proc. Eurographics Symp. Rendering, pp. 1–11 (2003)
2. Perona, P., Malik, J.: Scale-space and edge detection using anisotropic diffusion. IEEE Trans. Pattern Analysis Machine Intelligence 12(7), 629–639 (1990)
3. Hirschmüller, H., Scharstein, D.: Evaluation of stereo matching costs on images with radiometric differences. IEEE Trans. Pattern Analysis Machine Intelligence (to appear)
4. Middlebury data set: stereo data, http://vision.middlebury.edu/stereo/data/
5. Pharr, M., Fernando, R.: Gpu gems 2: programming techniques for high-performance graphics and general-purpose computation. Addison-Wesley, Reading (2005)
6. Shen, J., Fanga, S., Zhaoc, H., Jinc, X., Sun, H.: Fast approximation of trilateral filter for tone mapping using a signal processing approach. Signal Processing 89(5), 901–907 (2008)
7. Tomasi, C., Manduchi, R.: Bilateral filtering for gray and color images. In: IEEE Int. Conf. Computer Vision, pp. 839–846 (1998)
8. Vaudrey, T.: Tobi Vaudrey's homepage, http://www.cs.auckland.ac.nz/~tobi/
9. Vaudrey, T., Klette, R.: Residual images remove illumination artifacts for correspondence algorithms? In: Proc. Pattern Recognition - DAGM 2009. LNCS, vol. 5748. Springer, Heidelberg (2009)
10. Wong, W.C.K., Chung, A.C.S., Yu, S.C.H.: Trilateral filtering for biomedical images. In: IEEE Int. Symp. on Biomedical Imaging: From Nano to Macro (ISBI), pp. 820–823 (2004)

Joint Affine and Radiometric Registration Using Kernel Operators

Boaz Vigdor and Joseph M. Francos

Electrical and Computer Engineering Department, Ben-Gurion University, Israel

Abstract. A new global method for image registration in the presence of affine and radiometric deformations is introduced. The method proposed utilizes kernel operators in order to find corresponding regions without using local features. Application of polynomial type kernel functions results in a low complexity algorithm, allowing estimation of the radiometric deformation regardless of the affine geometric transformation. Preliminary experimentation shows high registration accuracy for the joint task, given real images with varying illuminations.

Keywords: Image Registration, Global Invariants.

1 Introduction

Registration of images is an important task in a vast number of applications such as object recognition, navigation, etc. In many of these applications, the deformations that an object undergoes can be roughly divided into radiometric and geometric. The geometric deformations occur as a result of the object movement in space, as it is projected on the imaging device's coordinate system. The radiometric deformations are a result of several factors. Among them are the changes in illumination sources and the reflectance properties of the object.

Image registration can be divided into two main categories: local and global. The former, such as SIFT [1] and MSER [2] identify small regions by various local features and perform registration by solving the correspondence between matching regions. These methods can tackle changes in global uniform illumination gain, but are sensitive to non-global effects resulting from changes in the illumination source direction. These methods are also sensitive to noisy measurements and require previous knowledge of the object structure.

Global methods, on the other hand, use the entire observation in order to extract various types of invariants to geometric deformations, such as MSA [3], CW [4], or extract the geometric transformation parameters [5]. As of today, there are relatively few studies which address joint registration of geometrically affine and radiometric deformations via moments or invariants. In general, the global methods are less sensitive to noise, but require object segmentation and are sensitive to partial occlusions.

In this paper, a new approach for joint affine and radiometric registration by kernel operators is introduced. In the next section, we elaborate on the registration problem, followed by the kernel operators definition and discussion, experimentation and conclusions.

2 Problem Definition

In the current work we shall assume the following:
1. The observation does not include additional surface not seen in the template, nor it is missing any due to occlusion. In practice, the model is still valid if the occluded area is small in comparison with the image area.
2. The object size is small compared with the distance to the imaging device.
3. The radiometric deformation can be approximated as a linear combination of basis images. This approximation is theoretically valid only for Lambertian surface [6], but in practice, it can sometimes approximate glossy surface quite well.

Under these assumptions, the geometric deformation of a rigid object can be approximated well by an affine transformation. The radiometric deformation can be modeled as a linear combination of template images taken in different controlled illumination conditions. Let $f(\mathbf{x})$ be the observation and $f_i(\mathbf{x}), i = 1, 2, ..., I$ the templates, all bounded and having bounded support functions. Then the radiometric deformation can be modeled as

$$f(\mathbf{x}) = \sum_{i=1}^{I} \alpha_i f_i(\mathbf{x}). \tag{1}$$

Under the previous assumption the geometric deformation can be modeled as an inner composition $\mathbf{x} \to \phi(\mathbf{x}, \theta), \phi \in \Phi$. In this paper we address the case where Φ is the affine group, i.e. $\phi(\mathbf{x}, (A, \mathbf{b})) = A\mathbf{x} + \mathbf{b}$. Combining the illumination model with the geometric one results in

$$f(\mathbf{x}) = \sum_{i=1}^{I} \alpha_i f_i(A\mathbf{x} + \mathbf{b}). \tag{2}$$

The task of registration is defined as estimation of the geometric deformation ϕ and the illumination coefficients α_i. Integrating (2)

$$\int_{R^N} f(\mathbf{x}) d\mathbf{x} = |A^{-1}| \sum_{i=1}^{I} \alpha_i \int_{R^N} f_i(\mathbf{y}) d\mathbf{y} \tag{3}$$

leads to a linear constraint for the illumination coefficients α_i's given the determinant $|A^{-1}|$. We are interested in finding enough independent constraints in the form (3) so that given the Jacobian, the illumination coefficients α_i's can be estimated by standard linear analysis. In order to do so, we need to find new sets of integrable functions g_j, g_{ij} in the form (2)

$$g_j(\mathbf{x}) = \sum_{i=1}^{I} \alpha_i g_{ij}(A\mathbf{x} + \mathbf{b}), j = 1, 2, ..., J (J > I). \tag{4}$$

such that integration would produce large enough a number of independent linear equations in order to estimate the illumination coefficients α_i. We shall utilize kernel operator to synthesize these functions.

3 Kernel Operators

A necessary condition for the desired operators T is to commute with addition and scalar multiplication, i.e. T are linear operators. We exclude the differential operators because of their high sensitivity to noise. The family of kernel operators is a broad family of linear operators which are robust to noise, due to integration over the function domain. Kernel operators are operators in the form of

$$g(\mathbf{y}) = Tf(\mathbf{x}) = \int_{\mathbb{R}^N} k(\mathbf{x}, \mathbf{y}) f(\mathbf{x}) d\mathbf{x}. \tag{5}$$

These operators are clearly linear. We seek kernel functions $k(\mathbf{x}, \mathbf{y})$ that satisfies the following conditions:
1. Map the transformation $\phi(\cdot, \theta)$ into the transformation $\psi(\cdot, \theta)$, i.e.

$$f(\mathbf{x}) = f_i(\phi(\mathbf{x}, \theta)) \Rightarrow g(\mathbf{y}) = (Tf)(\mathbf{y}) = (Tf_i)(\psi(\mathbf{y}, \theta)) = g_i(\psi(\mathbf{y}, \theta)) \tag{6}$$

This condition can be described in the following commutative diagram:

$$\begin{array}{ccc} f & \stackrel{\phi}{\rightarrow} & f_i \\ \downarrow T & & \downarrow T \\ g & \stackrel{\psi}{\rightarrow} & g_i \end{array}$$

2. All the new functions are be integrable, i.e. $\int g(y)dy, \int g_i(y)dy < \infty$.

Theorem 1. *Let $k(\boldsymbol{x}, \boldsymbol{y})$ be a bounded function differentiable almost everywhere. Assume that $f(\boldsymbol{x})$ and $f_i(\boldsymbol{x}), i = 1, 2, ..., N$ are measurable, bounded, and having a bounded support. Let $\phi(\cdot, \theta)$ and $\psi(\cdot, \theta)$ be two differentiable transformation groups. Let $g(\mathbf{y}) = Tf(\mathbf{y}) = \int_{R^M} k(\mathbf{x}, \mathbf{y}) f(\mathbf{x}) d\mathbf{x}$. Then*

$$\begin{array}{c} f(\boldsymbol{x}) = f_i(\phi(\boldsymbol{x}, \theta)) \Rightarrow g(y) = g_i(\psi(\boldsymbol{y}, \theta)), \quad \forall \boldsymbol{x}, \boldsymbol{y}, \theta \\ \Updownarrow \\ k(\boldsymbol{x}, \boldsymbol{y}) = J_\phi(\boldsymbol{x}) k\left(\phi(\boldsymbol{x}, \theta), \psi(\boldsymbol{y}, \theta)\right), \quad \forall x, y, \theta \end{array} \tag{7}$$

where $J_\phi(\boldsymbol{x})$ is the Jacobian of ϕ and the equalities are almost everywhere.

Proof. see [7]

Examples:
1. Let $\phi(x, s) = x + s$, $\psi(y, s) = y + s$. Then any kernel of the form $k(x, y) = \xi(x - y)$ maps ϕ into ψ (Shift Invariant operators), since $k(\phi(x), \psi(y)) = \xi((x + s) - (y + s)) = \xi(x - y) = k(x, y)$.
2. Let $\phi(\mathbf{x}, A) = A\mathbf{x}$, $\psi(\mathbf{y}, A) = A^{-T}\mathbf{y}, \mathbf{x}, \mathbf{y} \in \mathbb{R}^n, A, B \in GL_n(\mathbb{R})$. Any kernel of the form $k(\mathbf{x}, \mathbf{y}) = \xi(\mathbf{x}^T \mathbf{y})$ maps ϕ into ψ since $k(\phi(x), \psi(y)) = \xi\left((Ax)^T (A^{-T}y)\right) = \xi(x^T y) = k(x, y)$ (a special case is the Multi-dimensional Fourier Transform).

The necessary and sufficient condition in Theorem 1 can be used to synthesize appropriate kernel functions for many scenarios. In the current work we shall address the affine transformations and use Kernel operators of the type $k(\mathbf{x}, \mathbf{y}) = \xi(\mathbf{x}^T \mathbf{y})$.

3.1 Creating Partial Support Functions Using Kernel Operators

In the following sections we shall assume that the objects are segmented from the background and that we can use support functions

$$S_f(x) = \begin{cases} 1 & f(x) > 0 \\ 0 & f(x) = 0 \end{cases} \tag{8}$$

of the template and the observation. Initially, both the templates and the observation are translated in order to have their center-of-mass at the origin of the coodrinate system. This reduces the affine transformation into a linear one. The limited registration problem at hand is

$$f(\mathbf{x}) = \sum_{i=1}^{I} \alpha_i f_i(A\mathbf{x}). \tag{9}$$

All the template functions $f_i(x)$ have the same support since they represent the same object under different lighting conditions. According to (2), all these functions are geometrically deformed by the same affine transformation. For simplicity we shall denote the support of these functions as S_{f_1}. Therefore, from the definition of the support function it is clear that

$$S_f(x) = S_{f_1}(Ax). \tag{10}$$

In the method proposed we shall use the supports in order to construct new Partial Support (PS) function that are deformed by the same affine transformation. We shall construct new functions $PS_f^j, PS_{f_1}^j\ j = 1, 2, ..., J$ such that

$$PS_f^j(x) = PS_{f_1}^j(Ax). \tag{11}$$

Each Partial Support pair $PS_f^j, PS_{f_1}^j$ can be used to construct new functions for the registration problem at hand. Multiplying by (2) with PS_f^j results in

$$PS_f^j(\mathbf{x})f(\mathbf{x}) = PS_f^j(\mathbf{x}) \sum_{i=1}^{I} \alpha_i f_i(A\mathbf{x}) = \sum_{i=1}^{I} \alpha_i f_i(A\mathbf{x})PS_{f_i}^j(A\mathbf{x}). \tag{12}$$

Integration yields as many linear equations of the required illumination as the number of PS functions.

The Partial Support functions are created using two types of operations:
1. Kernel operators.
2. Left composition with a measurable bounded function $\omega : \omega(0) = 0$. The left composition operation is suitable to attain additional equations for registration of geometrically deformed objects. For further elaboration see [5].

First, we apply kernel operators of the type $k(\mathbf{x}, \mathbf{y}) = \xi(\mathbf{x}^T \mathbf{y})$ on the support functions S_f, S_{f_1} **twice**

$$PS_f = T(T(S_f)) = T^2(S_f), PS_{f_1} = T^2(S_{f_1}). \tag{13}$$

As seen in Example 2, the kernel $k(\mathbf{x},\mathbf{y}) = \xi(\mathbf{x}^T\mathbf{y})$ maps the linear transformation $A\mathbf{x}$ into $A^{-T}\mathbf{x}$. Since $(A^{-T})^{-T} = A$ the functions obtained are deformed by the original linear transformation, so that $PS_f(\mathbf{x}) = PS_{f_1}(A\mathbf{x})$. Second, In order to achieve many equations, left composition with ω is performed so that

$$\omega \circ PS_f^j(\mathbf{x}) = \omega \circ PS_{f_1}^j(A\mathbf{x}). \tag{14}$$

Each composition function ω creates new functions that are deformed by the original linear transformation. In the current work we used ω having the range set of $\{0,1\}$. The functions obtained can be interpreted as partial supports of the images. In general, other compositions can be used in the same manner. Figure 1 shows a template, a linearly transformed observation and chosen Partial Supports for different kernel operators.

Using a variety of kernel functions and composition functions provides a board family of Partial Support functions, and as a result, many constraints in the form of (3), enabling the estimation of the illumination coefficients α_i via Least-Squares estimation.

3.2 Efficient Computation of Kernel Operator

For two-dimensional images with resolution of $m*n$ pixels, the computational complexity of the kernel operator is $O(n^2m^2)$ if the kernel operator is performed element-wise and the kernel function has to be re-calculated element-wise. In order to reduce this high complexity we can use polynomial kernels $k_N(\mathbf{x},\mathbf{y}) = (\mathbf{x}^T\mathbf{y})^N$. The Operator can be decomposed into the form

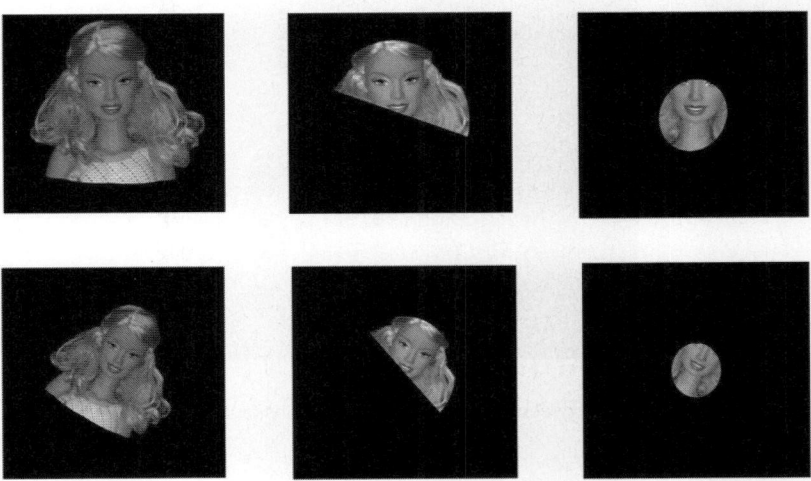

Fig. 1. Template (*top left*) and observation (*bottom left*). Template and observation multiplied by a Partial Support function using a linear kernel (*top middle* and *bottom middle*) and using a quadratic kernel (*top right* and *bottom right*).

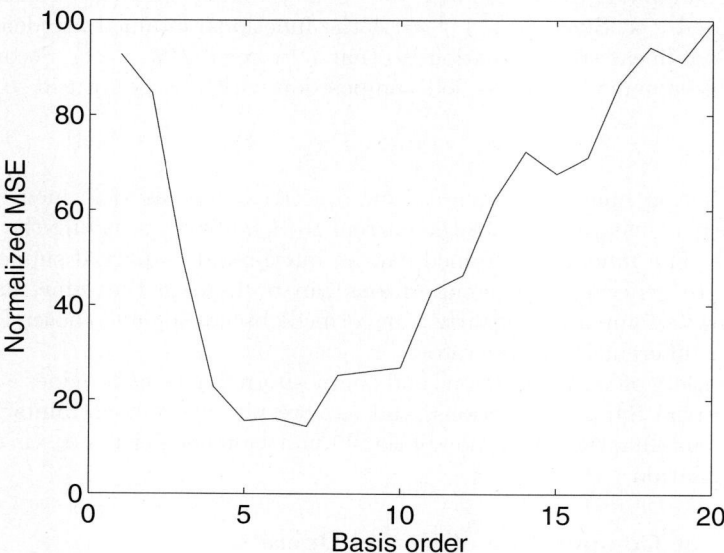

Fig. 2. Normalized Mean Square Error of illumination deformation for increasing basis order

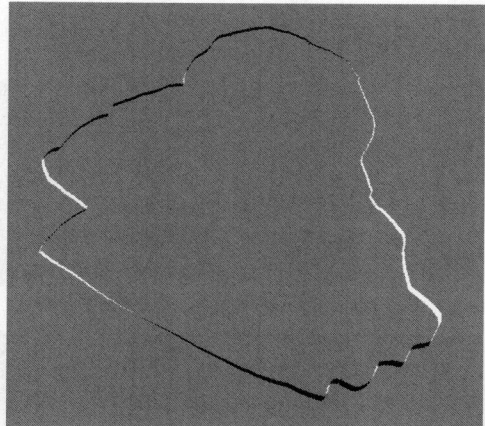

Fig. 3. Difference image between the supports of the observation and estimated Affine registered image

$$Tf(y_1, y_2) = \int_{\mathbb{R}^2} f(x_1, x_2)(x_1 y_1 + x_2 y_2)^N dx_1 dx_2 =$$
$$= \sum_{n=0}^{N} \frac{N!}{n!(N-n)!} y_1{}^n y_2{}^{N-n} \int_{\mathbb{R}^2} x_1{}^n x_2{}^{N-n} f(x_1, x_2) dx_1 dx_2 \quad (15)$$

where the coefficients are moments of $f(x_1, x_2)$ and can be calculated once and offline. Using the polynomial kernels reduces the computational complexity into $O(nm)$.

4 Experimentation

In a designated dark room, we constructed an array of 14 light sources in order to variously illuminate target objects. In the current study, we chose a doll head as the target object. In the learning stage, 149 grey-scale images under various random illumination and constant pose were taken. These images are available in the website [8]. PCA was performed to create 20 basis images. These functions were used as the templates $f_i, i = 1, ..., 20$.

In the test stage, an image, denoted as f_R was chosen. In order to generate an observation f we synthesized random Linear transformations of the form

$$A = \begin{pmatrix} \alpha_1 & 0 \\ 0 & \alpha_2 \end{pmatrix} \begin{pmatrix} \cos\phi & \sin\phi \\ -\sin\phi & \cos\phi \end{pmatrix}$$

where α_1, α_2 and ϕ were uniformly distributed in the intervals $[0.8, 1.2]$ and $[0, 2\pi]$, respectively. 100 random observations were synthesized in this manner. We used polynomial kernels of orders 1,2 and 3 with 24,12 and 22 composition functions, respectively, to a total of 58 linear equations. The illumination coefficients $\hat{\alpha}_i$ were estimated via Least-Squares.

An estimated template was reconstructed as $\hat{f}_R = \sum_{i=1}^{20} \hat{\alpha}_i f_i$. We compared the estimated template $\hat{f}_R$ to the observation prior to the affine transformation f_R in order to assess the illumination estimation. The average Normalized MSE

$$NMSE = \frac{\int \left(f_R - \hat{f}_R\right)^2}{S_f}$$

for increasing basis order is shown in Figure 2. Estimation using 4 to 10 basis images yields the highest accuracy for the illumination registration.

The estimation of the geometric parameters was done similarly to the illumination coefficients using first order moments and assuming 10 basis images. Given the real transformation and the estimated one $\hat{A}$, the matrix $I - A^{-1}\hat{A}$ was calculated for each observation. The bias and variance of the estimated parameters are summarized in the following matrices

$$Bias = \begin{pmatrix} 0 & -.018 & -0.0009 \\ 0.017 & -0.014 \end{pmatrix}, Variance = 10^{-3} \begin{pmatrix} 0.087 & 0.62 \\ 0.029 & 0.03 \end{pmatrix}$$

A sample difference image between an the supports of the observation and the registered estimation image is shown in Figure (3).

5 Conclusions

The utilization of kernel operator family for registration was introduced. The joint registration for affine geometric and radiometric using kernel operators was elaborated and demonstrated by experimentation. Though the registration results for the geometric synthetic case are promising, the relative high bias of the estimated geometric parameters is troubling and requires further study. Nonetheless, the kernel operators show a promising potential in solving the difficult task of joint affine and illumination registration.

References

1. Lowe, D.G.: Distinctive Image Features from Scale-Invariant Keypoints. Int' Journal of Computer Vision 60(2), 91–110 (2004)
2. Matas, J., Chum, O., Urba, M., Pajdla, T.: Robust wide baseline stereo from maximally stable extremal regions. In: British Machine Vision Conference, pp. 384–396 (2002)
3. Rahtu, E., Salo, M., Heikkila, J.: Affine invariant pattern recognition using multiscale autoconvolution. IEEE Transactions On Pattern Analysis and Machine Intelligence 27(6), 908–918 (2005)
4. Yang, Z., Cohen, F.S.: Cross-weighted moments and ane invariants for image registration and matching. IEEE Transactions on Pattern Analysis and Machine Intelligence 21(8), 804–814 (1999)
5. Francos, J.M., Hagege, R., Friedlander, R.B.: Estimation of Multi-Dimensional Homeomorphisms for Object Recognition in Noisy Environments. In: Thirty Seventh Asilomar Conference on Signals, Systems, and Computers (2003)
6. Basri, R., Jacobs, D.W.: Lambertian Refectance and Linear Subspaces. IEEE Transactions on Pattern Analysis and Machine Intelligence 25(2), 218–233 (2003)
7. Vigdor, B., Francos, M.J.: Utilizing Kernel Operators for Image Registration (to appear)
8. http://www.ee.bgu.ac.il/~vigdorb

MCMC-Based Algorithm to Adjust Scale Bias in Large Series of Electron Microscopical Ultrathin Sections

Huaizhong Zhang[1], E. Patricia Rodriguez[2], Philip Morrow[1], Sally McClean[1], and Kurt Saetzler[2]

[1] School of Computing and Information Engineering
[2] School of Biomedical Sciences
University of Ulster, Cromore Road, Coleraine, BT52 1SA, UK
{zhang-h1,pj.morrow,si.mcclean,k.saetzler}@ulster.ac.uk

Abstract. When using a non-rigid registration scheme, it is possible that bias is introduced during the registration process of consecutive sections. This bias can accumulate when large series of sections are to be registered and can cause substantial distortions of the scale space of individual sections thus leading to significant measurement bias. This paper presents an automated scheme based on Markov Chain Monte Carlo (MCMC) techniques to estimate and eliminate registration bias. For this purpose, a hierarchical model is used based on the assumption that (a) each section has the same, independent probability to be deformed by the sectioning and therefore the subsequent registration process and (b) the varying bias introduced by the registration process has to be balanced such that the average section area is preserved forcing the average scale parameters to have a mean value of 1.0.

Keywords: image registration, scale bias, MCMC, hierarchical model, stochastic simulation.

1 Introduction

Quantifying morphological parameters is of increasing importance when it comes to studying the relationship between shape and function of biological structures. This is in particular true when it comes to the analysis of small structures such as synapses and key drivers of synaptic transmission (e.g. vesicles, active zones) from 3D reconstructions [1,2]. Because of the resolution required to unequivocally identify these tiny structures, 3D reconstructions from series of electron microscopical ultrathin sections are still the method of choice. This requires physical sectioning on the limit of what is mechanically possible making this technique more amenable to distortions than normal semi-thin or even thick histological sections. Furthermore, there is no registration method that will consistently succeed in aligning arbitrary image data with arbitrary deformations [2] thus requiring user-input to the semi-automated registration process using intrinsic or imposed fiducial markers [2,3]. Usually, global rigid (translation, rotation) transformations can be employed for the preliminary registration process [4]. However, isotropic and anisotropic scale changes can occur between successive sections, which is particularly true when the volume of interest used for complete reconstructions as well as the number of ultrathin sections that have to be registered is increasing. Moreover, the chosen registration procedure can cause an accumulation of errors whilst processing the sections sequentially [5]. Therefore, the potential influence of the registration bias is amplified putting us in a situation were non-rigid registration schemes need to be employed to restore the relative orientation between consecutive sections. At the same time, however, we have to consider that by using such registration schemes we are altering the scale space of each section and small errors might accumulate across large numbers of sections. These alterations of scale space can then have a substantial influence on the results obtained from subsequent quantitative and morphometric analysis.

We will show that by employing only a simple linear but non-rigid registration scheme, already great care has to be taken such that the altered scale space in each individual section does indeed not accumulate across a large series of sections used for complete 3-D reconstructions. Depending on the underlying structure, the effects can be very dramatic and the scale space of individual sections can be altered substantially in particular where data is collected manually (see Fig. 1). To avoid this source of bias, we propose a statistical equalization scheme that re-distributes small registration errors that cause the described dramatic multiplicative effect on scale space. We implement this scheme by using a hierarchical linear model that estimates the source of bias using MCMC techniques. We will use several complete 3D data sets to show the scale space effects of linear non-rigid transforms and our proposed equalization scheme on quantitative measurements on large serial reconstructions from electron microscopical ultra-thin sections.

This paper is organized as follows. In Section 2, a hierarchical model based on MCMC techniques for addressing the accumulated bias is proposed. Experimental results are given in Section 3. We conclude the paper with a brief discussion and ideas for future work in Section 4.

2 Proposed Approach and the Hierarchical Model

2.1 The MCMC Technique

The primary goal of Monte Carlo techniques is to solve the following problems:

(i) To generate a number of independent samples $\{x^{(r)}\}_{r=1}^{R}$ from the desired distribution $p(x)$.
(ii) To estimate expectations of functions under this distribution or calculate some integrals, such that

$$\Phi = <\phi(x)> = \int d^N x\, p(x)\phi(x) \tag{1}$$

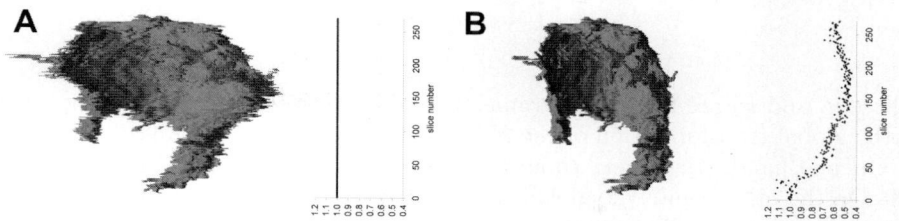

Fig. 1. (A) Shows the original data set. The rigid registration does not compensate for distortions between consecutive sections as can be seen as jitter along the surface of the reconstructed neuron. (B) Shows the data set after non-rigid linear transformations have been applied. The surface is now much smoother, but the scale space of each section has been altered and partly deviates substantially from the ideal value 1.0 representing to original section size (see respective graphs in panels A and B).

where Φ is the expectation value for a random function $\phi(x)$. Solving the sampling problem leads to a straightforward estimation of the expectation value by simply averaging the function values at the sampling points. Based on the Monte Carlo methodology, the MCMC technique aims to simulate the posterior distribution given a prior so that various sampling algorithms can be developed depending on the scheme used for prior analysis. Since consecutive samples are correlated, techniques have been developed to ensure that the Markov Chain will effectively generate independent samples from the given distribution in acceptable time. There are various ways this strategy can be implemented, such as the Gibbs sampler [10] and the Metropolis method [6]. In our case, the unknown mean scale bias can be generated by using MCMC methods to estimate the expectation value by eq. (1).

There are several software implementations for MCMC algorithms. BUGS (Bayesian inference Using Gibbs Sampling) is a freely available software package developed by Spiegelhalter and co-workers [6,9]. We use this tool to generate the prior distribution from some input data that is used to estimate the parameter set needed for compensating the accumulated scale bias.

2.2 Preliminary Analysis and Prior Information

As discussed in Section 1, the equalization scheme aims to adjust the anisotropic registration bias. The following analysis will present a random model that accounts for the additive bias derived from some manually obtained ground truth data sets (e.g. see Fig. 2A).

Since a section transform is affected by the transformations of previous sections, the current section v obtains the error gain field from previous operations. We denote $G_k = T_1 S_{s,1} T_2 S_{s,2}...T_k T_{s,k}$, where $T_i, i = 1...k$ is the translation and rotation between consecutive sections and $S_{s,i}, i = 1...k$ is the section scale transform for anisotropic changes between consecutive sections defined as follows:

$$S_{s,i} = \begin{pmatrix} u_i cos\alpha_i & sin\alpha_i & 0 \\ -sin\alpha_i & w_i cos\alpha_i & 0 \\ 0 & 0 & 1 \end{pmatrix}$$

where u_i and w_i are the scale parameters and α_i is the angle of $S_{s,i}$. Hence, G_k is the global transformation of the kth section.

On one hand, u_i and w_i should not deviate strongly from the ideal value 1.0. On the other hand, the global transformation G_k should not deviate significantly from a rigid linear transform which only accounts for translations and rotations. In addition to these general assumptions, we postulate that the sectioning process itself leads to distortions which are mainly volume conserving. Thus, an object with a circular shaped cross section will be transformed into an elliptically shaped cross section of about the same area. Therefore, we define the eccentricity of the local linear transform to be $e_i = u_i * w_i$ and the corresponding global eccentricity E_k for the global transform G_k by looking at its Eigenvalues in the plane (denoted as $s_{min,k}$ and $s_{max,k}$), leading to $E_i = s_{min,k} * s_{max,k}$.

Because the transformation of the current section impacts on the next one, we have to consider the multiplicative effect of subsequent matrix transformations for serial sections. The corresponding error can be expressed as $X_k = X_{v_1} X_{v_2} ... X_{v_k}$, where X_{v_i} is a random variable accounting for the bias being responsible in the transformation $T_k S_{s,k}$, and X_k is a random variable accounting for the total bias of section v from the previous transforms. Applying a logarithmic operator to the multiplicative formula, we obtain an additive bias model:

$$x_k = x_{v_1} + x_{v_2} + ... + x_{v_k}, x_{v_i} = ln(X_{v_i}) \qquad (2)$$

where x_k is a random variable accounting for the additive bias. Eq. (2) still remains a valid additive model even if we apply a simple transformation to the input data (such as multiplying each value by 10). This means that we successfully converted our multiplicate problem into an additive model.

The following assumptions were made to solve the problem at hand:

(a) **Normal hypothesis for bias variable:** The accumulated bias variable should follow the normal distribution.
(b) **Volume conservation:** Due to the physical sectioning process used to obtain the serial sections it seems reasonable to assume a-priori that the volume of each section is preserved (= area × average section thickness).
(c) **Additive tendency of bias:** In general, the bias will be increasing from the first section to the last one.

2.3 The Hierarchical Model and Its Simulation Strategy

As discussed above, we need to conserve the volume of the transformed section which usually is to keep the value $s_{min,i} * s_{max,i}$ around 1 by altering the local scale parameters, u_i and w_i, slightly at the equalization stage. Here, $s_{min,i}$ and $s_{max,i}$ are the scale parameters of the global transformation G_i. In practice, the values of $s_{min,i}$ and $s_{max,i}$ are adjusted manually (see Fig. 2A) so that this process can correct for the slight bias introduced in each section in a large series of transformations and leads to a reliable 3-D reconstruction that can be used for morphological parameter extraction. Here, we focus on the adjustment of the scale parameters $s_{min,i}$ and $s_{max,i}$ only using above approach as follows:

As input data we use the transformation parameters obtained from manually aligning consecutive sections allowing linear non-rigid transformations only (scale changes and shearing, see 2B for an example). This data has to be subdivided into a discrete number of levels used in our hierarchical model. The number l of levels is estimated and mainly depends on the size of the input data (roughly 30 ~ 50 sections per level). Each measured parameter y_i is assumed to be drawn from one of these levels; $G_i, i = 1\ldots l$ is the ith observation group, where group G_i has a normal distribution with mean λ_i and precision τ (we assume the same variance for different levels and its value is given according to the input data). The unknown percentage of observations P_k falling into level k satisfies the equation $\sum_{k=1}^{l} P_k = 1$ for all observations $P = (P_1, \ldots, P_l)$. The principal model is thus:

$$y_i \sim Normal(\lambda_{T_i}, \tau_{T_i})$$
$$G_i \sim Categorical(P) \qquad (3)$$

The mean value λ_k of all observations y_i that fall into level k will be different reflecting the variability of the input data. We assume that the mean of the subsequent level can be calculated as $\lambda_{k+1} = \lambda_k + \rho * \pi_k, k = 1\ldots l$, where π_k is an informative prior which counteracts the accumulated bias by incorporating the bias information obtained using eq. (2); ρ is a constant which balances the gap between different levels. The other priors in the model are estimated as follows: P follows a uniform distribution in $(0,1)$; λ_i is initialized by the mean of the input data; the precision τ is computed by using the variance σ of the input data using the formula $\tau = 1/\sigma^2$.

The main steps of the simulation procedure are as follows:

(1) Initialize the values for all parameters.
(2) Construct a full conditional distribution for each parameter.
(3) Choose a reasonable length for the 'burn-in' and the total run length.
(4) Compute a summary statistics for the true values of each parameter.
(5) Check the goodness-of-fit of the model.

For the latter, we use a standardized residual $r_i = \frac{y_i - u_k}{\sigma_k}, i = 1\ldots N, k = 1\ldots l$ with mean 0 and variance 1 given a normal data. u_k is the estimated mean of level k of all observations y_i belonging to this level and σ_k is the standard deviation. The mean fourth moment of the standardized residual is as follows:

$$s_{4,y} = \frac{1}{N} \sum_i r_i^4 \qquad (4)$$

If the error distribution is truly normal then s_4 should be close to 3 [6]. Thus, checking this statistic will provide fairly good evidence for the goodness of fit for the simulation model.

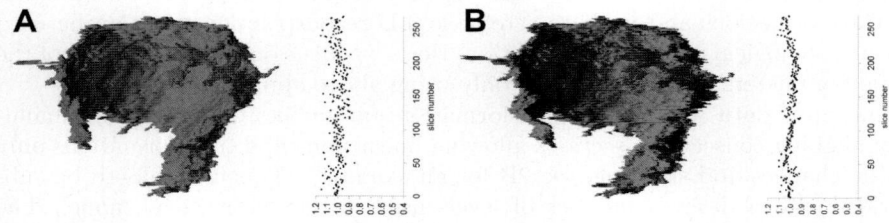

Fig. 2. (A) Shows the manuall corrected data set. (B) Shows the result obtained when applying our MCMC method to estimate and correct for the scale bias in the series of sections.

Table 1. Summary of results and error estimates for parameter s_{min} using manually obtained ground truth data

s_{min}	DS1	DS2	DS3	DS4	DS5	DS6	DS7	DS8
P_1	0.30693	0.4551	0.35113	0.22738	0.25676	0.46508	0.34167	0.24729
P_2	0.35043	0.18716	0.35463	0.39451	0.45	0.23351	0.55294	0.35356
P_3	0.34264	0.35774	0.29424	0.3781	0.29323	0.30141	0.10539	0.39915
λ_1	0.96726	0.97773	0.9556	0.99236	1.01131	0.87664	0.95299	0.96304
λ_2	0.90899	1.02897	0.96434	1.02428	0.96074	0.95533	0.92617	0.83552
λ_3	1.00221	1.10582	0.94686	1.05618	1.05178	0.89238	0.8457	0.92054
S_{ave}	1.1868	1.5113	2.4890	2.4271	3.8669	1.9203	1.6880	1.6120
ϵ	0.0372	0.0301	0.0277	0.0670	0.0512	0.0177	0.0572	0.0620

For assessing the actual simulation results, we compare our simulation results for parameter y_{est} (see e.g. Fig. 2B) with manually adjusted ground truth data y_{gt} (see Fig. 2A) by using the absolute error of parameter y is $\epsilon_y = \left|\frac{y_{\text{gt}} - y_{\text{est}}}{y_{\text{gt}}}\right|$ (see also Tab. 1).

3 Implementation and Results

We tested our approach in 9 data sets of large series of electromicroscopical ultra-thin sections, which were previously used for 3-D reconstructions and subsequent quantitative morphological analysis (see [7,8]). The largest series is used throughout this publication to illustrate our approach (DS9, 270 serial sections). The analysis refers to the results of all datasets. After logarithmic transformation as described previously, the input data is subdivided into 5 levels according to the tendency of the scale mean.

Based on the analysis of the input data, we construct eq. (5) that describes the mean of parameter $s_{\min}$ at 5 different levels ($\lambda_1, \lambda_2, \lambda_3, \lambda_4, \lambda_5$); our hierarchical model will integrate the general model from eq. (3) with the model for the mean value in eq. (5).

$$\lambda_2 = \lambda_1 - 3.5 * \sigma, \lambda_3 = \lambda_2 - 2.5 * \sigma, \lambda_4 = \lambda_3 - \sigma, \lambda_5 = \lambda_4 + \sigma, \theta = 0.05 \quad (5)$$

For parameter, $s_{\max}$, this looks as follows:

$$\lambda_2 = \lambda_1 - 8.5 * \sigma, \lambda_3 = \lambda_2 - 5.5 * \sigma, \lambda_4 = \lambda_3 + \sigma, \lambda_5 = \lambda_4 + 5 * \sigma, \theta = 0.02 \quad (6)$$

The preceeding parameter models (5) and (6) have been used in formula (3) for initializing the simulation. We then executed the simulation using WinBUGS starting with the input data including some unknown registration bias. In WinBUGS, the shape and rate parameters of the Γ-distribution are set to 1000 and 2 in accordance with the average of σ. Using the error value ϵ as described at the end of Section 2.3, we see that the changes between the manually obtained ground truth data and the automatically computed ones (see Fig. 2) are very small ($\epsilon_{s_{min}} = 0.0626; \epsilon_{s_{max}} = 0.0635; \epsilon_E = 0.0749$).

Numerical summaries are not included for the large data set in below table, but are shown graphically in Fig. 2. For this we use the volume conservation parameter E as defined above. The 3D reconstructions used in Fig. 2 show that after applying our method, the shape of the input data can be restored by also improving the surface jitter (see Fig. 1A and Fig. 2B). For the remaining 8 data sets, we will summarize the results obtained for Tab. 1. These data sets were smaller in size and therefore the hierarchical models only comprised 3 levels instead of the 5 levels used for DS9. Due to space limitations, we only show experimental results for the simulated scale Eigenvalue $s_{\min}$. S_{ave} stands for the average value of statistics $s_{4,s_{smin}}$ for all 3 levels.

4 Discussions and Conclusion

We have shown that when non-rigid transformations are introduced into the registration process of large series of histological sections, the scale space of each individual section can be substantially altered depending on the underlying image structure of the data set used. The proposed volume conservation and scale balancing scheme therefore offers a way to treat this problem in a mathematical way. We proposed a two step method, where in step one, the registration process is performed by minimizing the jitter between consecutive sections either manually or with any other image based or landmark based registration method (e.g. ITK [12]). In a second, subsequent step, the scales are analyzed and the data subdivided into several levels depending on the size of the data set. After this manual initialization step, a MCMC based simulation is performed and the scale bias is estimated and corrected for. We have demonstrated the feasibility of our method using 9 large data sets that were generated from electron microscopical ultra-thin sections.

The current method can be improved in several aspects. Firstly, the number of levels used for the subdivision and the initialization of the simulation parameters needs to be completely estimated in an automated fashion using the input data only. Secondly, other registration parameters such as rotation and translation should also be incorporated into the simulation process. This, however, requires that a suitable quantifiable minimization criterion is to be computed either based on image or landmark information. This then converts our registration problem to a multi-objective optimization setting, where competing objectives (minimization of registration error and balancing of scale parameters according to our volume conservation and scale balancing criteria) have to be considered. This implies that our proposed MCMC method can potentially be used to solve the registration process in an algorithm that offers an alternative way to other multi-objective minimization methods such as ant colonies or genetic algorithms [13].

References

1. Fiala, J.C.: Reconstruct: A Free Editor for Serial Section Microscopy. J. Microscopy 218(1), 52–61 (2005)
2. Zitova, B., Flusser, J.: Image registration Methods: A Survey. J. Imag. and Visi. Compu. 21, 977–1000 (2003)
3. Kremer, J.R., Mastronarde, D.N., McIntosh, J.R.: Computer Visualization of Three-Dimensional Image Data using IMOD. J. Stru. Biol. 116(1), 71–76 (1996)
4. Ourselin, S., Roche, A., Subsol, G., Pennec, X., Sattonnet, C.: Reconstructing a 3D Structure Serial Histological Sections. J. Imag. Visi. Compu. 19, 25–31 (2000)
5. Pitiot, A., Bardinet, E., Thompson, P.M., Malandain, G.: Piecewise Affine registration of Biological Images for Volume Reconstruction. J. Medi. Imag. Anal. 10(3), 465–483 (2006)
6. Gilks, W., Richardson, S., Spiegelhalter, D.: Markov Chain Monte Carlo in Practice. Chapman and Hall, London (1997)
7. Sätzler, K., Söhl, L.F., Bollmann, J.H., Gerard, J., Borst, G., Frotscher, M., Sakmann, B., Lübke, J.H.: Three-dimensional reconstruction of a calyx of Held and its postsynaptic principal neuron in the medial nucleus of the trapezoid body. J. Neurosci. 22(24), 10567–10579 (2002)
8. Rollenhagen, A., Sätzler, K., Rodriguez, E.P., Jonas, P., Frotscher, M., Lübke, J.H.R.: Structural Determinants of Transmission at Large Hippocampal Mossy Fiber Synapses. J. Neurosci. 27(39), 10434–10444 (2007)
9. WinBugs Software (2007), http://www.mrc-bsu.cam.ac.uk/bugs/
10. Geman, S., Geman, D.: Stochastic Relaxation, Gibbs Distribution and the Bayesian Restoration of Images. IEEE Tran. Patt. Anal. Mach. Inte. 6(6), 721–741 (1986)
11. Besag, J.: Bayesian computation and stochastic systems. J. Stat. Sci. 10(1), 3–41 (1995)
12. Image registration Toolkit (ITK) (2009), http://itk.org
13. Deb, K.: Multi-objective Optimization Using Evolutionary Algorithms. Wiley Blackwell, Chichester (2008)

Accelerating Image Retrieval Using Factorial Correspondence Analysis on GPU

Nguyen-Khang Pham[1,2], Annie Morin[1], and Patrick Gros[3]

[1] IRISA, Université de Rennes I,
Campus de Beaulieu, 35042 RENNES Cedex, France
{Pham.Nguyen_Khang,Annie.Morin}@irisa.fr
http://www.irisa.fr
[2] Cantho University
1, Ly Tu Trong street, Cantho, Vietnam
pnkhang@cit.ctu.edu.vn
[3] INRIA
Campus de Beaulieu, 35042 RENNES Cedex, France
Patrick.Gros@inria.fr
http://www.inria.fr

Abstract. We are interested in the intensive use of Factorial Correspondence Analysis (FCA) for large-scale content-based image retrieval. Factorial Correspondence Analysis, is a useful method for analyzing textual data, and we adapt it to images using the SIFT local descriptors. FCA is used to reduce dimensions and to limit the number of images to be considered during the search. Graphics Processing Units (GPU) are fast emerging as inexpensive parallel processors due to their high computation power and low price. The G80 family of Nvidia GPUs provides the CUDA programming model that treats the GPU as a SIMD processor array. We present two very fast algorithms on GPU for image retrieval using FCA: the first one is a parallel incremental algorithm for FCA and the second one is an extension of the filtering algorithm in our previous work for filtering step.

Our implementation is able to scale up the FCA computation a factor of 30 compared to the CPU version. For retrieval tasks, the parallel version on GPU performs 10 times faster than the one on CPU. Retrieving images in a database of 1 million images is done in about 8 milliseconds.

Keywords: Factorial correspondence analysis, Image retrieval, SIFT, GPU, CUDA, CUBLAS.

1 Introduction

The goal of Content-Based Image Retrieval (CBIR) systems is to operate on collections of images and, in response to visual queries, to retrieve relevant images. This task is not easy because of two gaps: the *sensory gap* and the *semantic gap* [1]. Recently, image analysis was improved when using local descriptors. Initially, voting-based methods have been used for image retrieval. Images are described as a *set of local descriptors* at interest points. Voting is performed on individual results of descriptor matching [2, 3, 4]. Later, text-based methods such as *tf–idf* (term frequency – inverse document frequency) weighting [5], PLSA (Probabilistic Latent Semantic Analysis) [6], LDA (Latent Dirichlet Allocation) [7] have been adapted to images [8, 9, 10]. In textual data analysis, these methods use bag-of-words models. The input of such methods is a two-way table, often called contingency table crossing documents and words. When adapting these previous methods to images, the documents are the images and we need to define the *"visual words"* [8, 9, 10, 11]. These methods consist of two steps: an *analysis step* and a strictly speaking *search step*. In the *analysis step*, the methods like PLSA, and LDA perform a dimensionality reduction and become very costly in time and/or in memory when dealing with huge image databases. For the *search step*, many efforts have been made for finding efficient similarity search algorithms (with sub-linear complexity) [12, 13]. Based on the cost model in [14] the authors prove that under certain assumptions, above certain dimensionality (e.g. 16), no index structure can perform efficiently a k-nearest neighbors search. Therefore, accelerating sequential searches methods have been proposed instead [15, 16]. For sparse data, inverted file-based methods are alternatives [5, 17]. Another interesting approach is the parallel processing of nearest-neighbor queries in high-dimensional space.

In November 2006, NVIDIA introduced CUDA^{TM}, a general purpose parallel computing architecture, with a new parallel programming model and instruction set architecture, that leverages the parallel compute engine in NVIDIA GPU to solve many complex computational problems in a more efficient way than on a CPU. CUDA comes with a software environment that allows developers to use C as a high-level programming language. In CUDA, the GPU is a device that can execute multiple concurrent threads. The CUDA software stack consists of a hardware driver, an API, its runtime and higher-level mathematical libraries of common usage, e.g., an implementation of Basic Linear Algebra Subprograms (CUBLAS). The CUBLAS library allows access to the computational resources of NVIDIA GPUs.

In this paper, we focus on the parallelization of both steps in image retrieval using FCA. The organization of the paper is the following: section 2 is devoted to a short presentation of FCA. In section 3, we describe a parallel incremental version of the FCA algorithm. Section 4 present a parallel filtering algorithm and in section 5, we use the previous results to large-scale image retrieval. We show some experimental results before concluding.

2 Factorial Correspondence Analysis

FCA is an exploratory data analytic technique designed to analyze simple two-way and multi-way tables containing some measure of correspondence between the rows and columns. It was developed by Benzécri [18] in textual data analysis. FCA on a table crossing documents and words allows us to answer the following questions: Are there proximities between some words ? Are there proximities between some documents ? Are there some relationships between words and documents ? FCA such as factorial methods is based on a singular value decomposition of a matrix and permits the display of words and documents in a low-dimensional space. This reduced subspace is such that the inertia of projected points (documents or words) is maximum.

Let $F = \{f_{ij}\}_{M,N}$ be a contingency table with dimensions $M \times N$ ($N < M$). We normalize F and get $X = \{x_{ij}\}_{M,N}$ by:

$$x_{ij} = \frac{f_{ij}}{s}, \quad s = \sum_{i=1}^{M}\sum_{j=1}^{N} f_{ij} \tag{1}$$

and let's note:

$$p_i = \sum_{j=1}^{N} x_{ij}, \quad \forall i = 1..M, \quad q_j = \sum_{i=1}^{M} x_{ij}, \quad \forall j = 1..N$$
$$P = \mathbf{diag}(p_1 \ldots p_M), \quad Q = \mathbf{diag}(q_1 \ldots q_N) \tag{2}$$

where $\mathbf{diag}(.)$ denotes the diagonal matrix.

To determine the best reduced subspace where the data are projected, we compute the eigenvalues λ, and the eigenvectors μ, of the matrix V of $N \times N$:

$$V = X^T P^{-1} X Q^{-1} \tag{3}$$

where X^T is the transposed matrix of X.

We only keep the first K ($K < N$) largest eigenvalues and the associated eigenvectors[1]. These K eigenvectors define an *orthonormal basis* of the reduced subspace (also called factor space). The number of dimensions of the problem is reduced from N to K. Images are projected in the new reduced space:

$$Z = P^{-1} X A \tag{4}$$

where $P^{-1}X$ represents the row profiles and $A = Q^{-1}\mu$ is the *transition matrix* associated to FCA.

A new document (i.e. the query) $r = [r_1 \; r_2 \; \cdots \; r_N]$ will be projected in the factor space through the transformation formula 4:

$$Z_r = \hat{r} A \text{ where } \hat{r}_i = \frac{r_i}{\sum_{j=1}^{N} r_j} \quad \forall i = 1..N \tag{5}$$

3 Parallel Incremental FCA Algorithm

As mentioned in section 2, the FCA problem seeks to find the eigenvectors and eigenvalues of an particular matrix V (formula 3). In the case of large scale databases, the matrix X is too large to load entirely into memory. In [19], we

---

[1] Like other dimension reduction methods, K is chosen empirically (e.g. by the way of cross-validation).

have proposed an incremental algorithm for FCA which can deal with huge databases on a PC. The algorithm is briefly described as following.

Let's first rewrite formula 3:

$$V = V_0 Q^{-1} \quad \text{where} \tag{6}$$
$$V_0 = X^T P^{-1} X \tag{7}$$

Matrix X is divided into $\mathbf{B}$ blocks by rows (i.e. $X_{[1]}, \ldots, X_{[\mathbf{B}]}$).

Then we compute $P_{[1]}, P_{[2]}, \ldots, P_{[\mathbf{B}]}$ and $Q_{[1]}, Q_{[2]}, \ldots, Q_{[\mathbf{B}]}$ in the same way for Q and P by replacing X with $X_{[i]}$ for all $i \in [1; \mathbf{B}]$ (cf. formula 2). It is clear that:

$$P = \begin{pmatrix} P_{[1]} & & 0 \\ & \ddots & \\ 0 & & P_{[\mathbf{B}]} \end{pmatrix} \quad \text{and} \quad Q = \sum_{i=1}^{\mathbf{B}} Q_{[i]}. \tag{8}$$

If we note:

$$V_{[i]} = X_{[i]}^T P_{[i]}^{-1} X_{[i]} \tag{9}$$

then

$$V_0 = \sum_{i=1}^{\mathbf{B}} V_{[i]}. \tag{10}$$

Two formulas 8 and 10 are key parts for the incremental algorithm. Once V is constructed, the *eigen* problem is performed on a small matrix. Since only some first eigenvectors are used for projection, this problem can be resolved efficiently by some advanced algorithms like LAPACK [20]. The projection step in which images are mapped into the factor space (cf. formula 4) can be performed in the same way of constructing V i.e. the new image representation Z is computed by blocks.

The incremental FCA algorithm described above is able to deal with very large datasets on a PC. However it only runs on one single machine. We have extended it to build a parallel version using a GPU (graphics processing unit) to gain high performance at low cost. The parallel incremental implementation of FCA algorithm using the CUBLAS library (described in algorithm 1) performs matrix computations on the GPU massively parallel computing architecture. Note that in CUDA/CUBLAS, the GPU can execute multiple concurrent threads. Therefore, parallel computations are done in the implicit way.

First, we split a large matrix X into small blocks of rows $X_{[i]}$. For each incremental step, a data block $X_{[i]}$ is loaded to the CPU memory; a data transfer task copies $X_{[i]}$ from CPU to GPU memory; and then formulas 8 and 10 are computed in a parallel manner on GPU. When the incremental step (lines 3–8) finishes, V is also computed on GPU after formula 6 (line 9) before being copied back to CPU memory (line 10). Next, first K eigenvectors and eigenvalue of V are computed in CPU (line 11) and the eigenvectors, μ, are copied to GPU for computing the transition matrix A (line 12). The projection step is performed by

Algorithm 1. Parallel incremental FCA algorithm

1. $Q = 0$
2. $V_0 = 0$
3. **for** $i = 1$ *to* **B do**
4.   load block $X_{[i]}$ into CPU memory
5.   copy $X_{[i]}$ to GPU memory
6.   compute $P_{[i]}$, $Q_{[i]}$ from $X_{[i]}$
7. $Q = Q + Q_{[i]}$
8. $V_0 = V_0 + X_{[i]}^T P_{[i]}^{-1} X_{[i]}$
9. $V = V_0 Q^{-1}$
10. copy V back to CPU memory
11. compute K eigenvalues λ and eigenvectors μ of V on CPU
12. copy μ to GPU memory and compute the transition matrix $A = Q^{-1}\mu$
13. **for** $i = 1$ *to* B **do**
14.   load block $X_{[i]}$ into CPU memory
15.   compute $P_{[i]}$ from $X_{[i]}$
16. $Z_{[i]} = P_{[i]}^{-1} X_{[i]} A$
17.   copy $Z_{[i]}$ back to CPU memory and write it to output

loading every block $X_{[i]}$ on CPU memory; copying it to GPU memory; computing $P_{[i]}$ and the projection of $X_{[i]}$ (i.e. $Z_{[i]}$), by the transition formula 4; copying $Z_{[i]}$ back to CPU memory an writing it to output (lines 13–17). The accuracy of the new algorithm is exactly the same as the original one.

4 Parallel Filtering Algorithm

In previous works [11, 19], we have also proposed an inverted file-based method for accelerating image search using the *the quality of representation* issued from FCA which is defined by:

$$\cos_j^2(i) = \frac{Z_{ij}^2}{\sum_{k=1}^{K} Z_{ik}^2} \qquad (11)$$

where Z_{ij} is the coordinate of the images i on axis j.

The method consists of a filtering step and a refine step. In the filtering step, we first choose the axes on which the projection of query is well represented (the threshold is chosen equal to the *average quality of representation* , i.e. $1/K$) and take corresponding inverted files[2]. These inverted files are then merged to compute the frequency of images in the merged list. Finally, some images with high frequency (e.g. 500 images) are kept for the refining step. The search performance is improved by a factor of 10 in comparison to a sequential scan

---
[2] An inverted file associated to the positive (negative) part of an axis contains all images well represented on this axis and lying on the positive (negative) part.

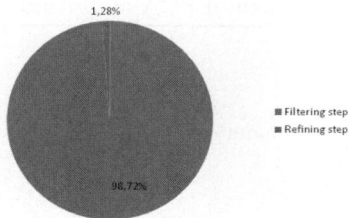

Fig. 1. Computation time for the filtering and refining steps in a 1M database

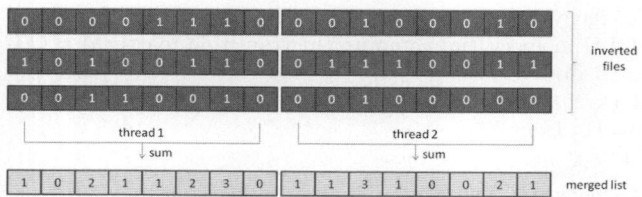

Fig. 2. Parallel computation of frequency of images on GPU: every thread computes for 8 images

without loosing quality. However, in a large database, most time is used for filtering images (cf. Figure 1). This motivates us to parallel this step. An inverted file is represented by a *vector of bits*, F_j^+ (F_j^-)[3]. If an image i belongs to an inverted file F_j^+, then the bit i of F_j^+ is set to 1. Such a representation allows compressing inverted files fitted the GPU memory and computing the merged list in a parallel way. Each thread of GPU computes independently the frequency of 8 images (cf. Figure 2).

5 Numerical Results

We have experimented our methods on the Nistér Stewénius dataset [17], namely N-S dataset. This dataset consists of 2 550 scenes, each of which being taken from 4 different view-points. Hence the dataset contains 10 200 images. We used the **extract_feature** program of Mikolajczyk [21] to extract and compute local descriptors by the Hessian-affine detector and the SIFT descriptor. The number of *visual words* is fixed to 5 000. To evaluate the scalability of our approach, we merged the N-S dataset with one million images downloaded from FlickR. All programs are write in C++, using LAPACK [20] for finding eigenvalues/eigenvectors. The parallel FCA algorithm and filtering algorithm are implemented on a NVIDIA graphic card GTX 280. The *precision* at first 4 returned images, P@4 (including the image used for the query) is used to evaluate different retrieval methods. As there are only 4 relevant images for a given

---

[3] F_j^+ (res. F_j^-) is the inverted file for the positive (res. negative) part of the axis j.

Table 1. Time comparison of FCA computation. (a): times for matrix construction and projection; (b): eigenvalues/eigenvectors computation time; (c) = (a) + (b).

Database size	GPU (s)			CPU (s)			Gains	
	(a)	(b)	(c)	(a)	(b)	(c)	(a)	(c)
100K	4.2	64	68.2	346.0	64	410.0	82.5	6.0
200K	8.0	64	72.0	658.2	64	722.2	82.2	10.0
500K	19.3	64	83.3	1602.3	64	1666.3	83.0	20.0
1M	38.2	64	102.2	3163.4	64	3227.4	82.8	31.6

Table 2. Comparison of search algorithms

Database size	GPU		CPU		Full search	
	time(ms)	P@4	time(ms)	P@4	time(ms)	P@4
100K	1.47	0.761	7.56	0.761	94.32	0.757
200K	2.09	0.749	13.53	0.749	179.76	0.745
500K	1.17	0.731	33.07	0.731	435.00	0.729
1M	7.53	0.719	79.99	0.719	860.884	0.717

query, P@4 is equal to the *recall* at first 4 returned images and also equal to $\frac{1}{4}$ of the N-S score (i.e. number of relevant images in first 4 returned images) [17]. Table 1 presents the computation times obtained by GPU and CPU implementations of the incremental FCA algorithm on different database sizes. The parallel algorithm performs about 80 times faster on the construction of matrix V and yields 30 times faster than the nonparallel one for all computation. It is due to the computation of eigenvalues/eigenvectors on CPU. Note that the computation time of eigenvalues/eigenvectors is constant for all database. For search algorithms, results are shown on table 2. the parallel filtering algorithm performs 10 times faster than the nonparallel filtering algorithm and about 114 times faster than a sequential scan on 1M images database. The numerical test results showed the effectiveness of the new algorithms to deal with very large databases on GPU.

6 Conclusion and Future Works

We have presented two parallel algorithms for images retrieval using FCA on GPU. The first one is an parallel incremental FCA algorithm for huge databases. Our algorithm avoids loading the whole dataset in main memory: only subsets of the data are considered at any one time and update the solution in growing training set. The second one is a parallel filtering algorithm used in the search step. The parallelization is based on GPU to gain high performance at low cost. Numerical results have shown that the parallel FCA algorithm performed 30 times faster the non parallel one. The parallel algorithm for filtering step runs about 10 times faster the CPU version and about 114 times faster a sequential scan without loosing quality. Since the parallel search algorithm performs very fast (e.g. 7.53ms), it is possible to apply our method for applications which require retrieval results in a short time like video copy detection.

References

[1] Smeulders, A.W.M., Worring, M., Santini, S., Gupta, A., Jain, R.: Content-based image retrieval at the end of the early years. IEEE Transactions on Pattern Analysis and Machine Intelligence 22(12), 1349–1380 (2000)
[2] Mikolajczyk, K., Schmid, C.: Indexing based on scale invariant interest points. In: Proc. of ICCV 2001, vol. 1, pp. 525–531 (2001)
[3] Schaffalitzky, F., Zisserman, A.: Automated location matching in movies. Computer Vision and Image Understanding 92, 236–264 (2003)
[4] Lowe, D.G.: Distinctive image features from scale-invariant keypoints. International Journal of Computer Vision 60(2), 91–110 (2004)
[5] Salton, G., Buckley, C.: Term-weighting approaches in automatic text retrieval. Information Processing & Management 24(5), 513–523 (1988)
[6] Hofmann, T.: Probabilistic latent semantic analysis. In: Proc. of the 15th Conference on Uncertainty in Artificial Intelligence (UAI 1999), pp. 289–296 (1999)
[7] Blei, D.M., Ng, A.Y., Jordan, M.I.: Latent dirichlet allocation. Journal of Machine Learning Research 3, 993–1022 (2003)
[8] Sivic, J., Zisserman, A.: Video google: A text retrieval approach to object matching in videos. In: Proc. of ICCV 2003, vol. 2, pp. 1470–1477 (2003)
[9] Bosch, A., Zisserman, A., Muñoz, X.: Scene classification via pLSA. In: Leonardis, A., Bischof, H., Pinz, A. (eds.) ECCV 2006. LNCS, vol. 3954, pp. 517–530. Springer, Heidelberg (2006)
[10] Lienhart, R., Slaney, M.: Plsa on large scale image databases. In: Proc. of International Conference on Acoustics, Speech and Signal Processing, pp. 1217–1220 (2007)
[11] Pham, N.K., Morin, A.: Une nouvelle approche pour la recherche d'images par le contenu. In: Proc. of EGC 2008, vol. RNTI-E-11, pp. 475–486 (2008)
[12] Robinson, J.: The k-d-b-tree: a search structure for large multidimensional dynamic indexes. In: Proc. of the Conference on Management of data, pp. 10–18 (1981)
[13] Guttman, A.: R-trees: A dynamic index structure for spatial searching. In: Proc. of the conference on Management of data, pp. 47–57 (1984)
[14] Berchtold, S., Böhm, C., Keim, D.A., Kriegel, H.P.: A cost model for nearest neighbor search in high-dimensional data space. In: Proc. of ACM symposium on Principles of database systems, pp. 78–86. ACM, New York (1997)
[15] Weber, R., Schek, H.J., Blott, S.: A quantitative analysis and performance study for similarity-search methods in high-dimensional spaces. In: Proc. of VLDB 1998, pp. 194–205 (1998)
[16] Böhm, C., Braunmüller, B., Kriegel, H.P., Schubert, M.: Efficient similarity search in digital libraries. In: Proc. of the Advances in Digital Libraries, pp. 193–199 (2000)
[17] Nistér, D., Stewénius, H.: Scalable recognition with a vocabulary tree. In: Proc. of CVPR 2006, June 2006, vol. 2, pp. 2161–2168 (2006)
[18] Benzecri, J.P.: L'analyse des correspondances. Dunod, Paris (1973)
[19] Pham, N.K., Morin, A., Gros, P., Le, Q.T.: Utilisation de l'analyse factorielle des correspondances pour la recherche d'images à grande échelle. In: Proc. of EGC 2009, vol. RNTI-E-15, pp. 283–294 (2009)
[20] Anderson, E., Bai, Z., Bischof, C., Blackford, S., Demmel, J., Dongarra, J., Du Croz, J., Greenbaum, A., Hammarling, S., McKenney, A., Sorensen, D.: LAPACK Users' Guide. Society for Industrial and Applied Mathematics (1999)
[21] Mikolajczyk, K., Schmid, C.: Scale and affine invariant interest point detectors. International Journal of Computer Vision 60(1), 63–86 (2004)

Color Based Bags-of-Emotions

Martin Solli and Reiner Lenz

ITN, Linköping University
SE-60174 Norrköping, Sweden
Martin.Solli@itn.liu.se, Reiner.Lenz@itn.liu.se

Abstract. In this paper we describe how to include high level semantic information, such as aesthetics and emotions, into Content Based Image Retrieval. We present a color-based emotion-related image descriptor that can be used for describing the emotional content of images. The color emotion metric used is derived from psychophysical experiments and based on three variables: *activity*, *weight* and *heat*. It was originally designed for single-colors, but recent research has shown that the same emotion estimates can be applied in the retrieval of multi-colored images. Here we describe a new approach, based on the assumption that perceived color emotions in images are mainly affected by homogenous regions, defined by the emotion metric, and transitions between regions. RGB coordinates are converted to emotion coordinates, and for each emotion channel, statistical measurements of gradient magnitudes within a stack of low-pass filtered images are used for finding interest points corresponding to homogeneous regions and transitions between regions. Emotion characteristics are derived for patches surrounding each interest point, and saved in a *bag-of-emotions*, that, for instance, can be used for retrieving images based on emotional content.

1 Introduction

Distributions of colors, described in various color spaces, have frequently been used to characterize image content. Similarity between images is then defined based on the distributions of these color descriptors in the images. A limitation of most color spaces, and their usage in Content Based Image Retrieval (CBIR), is that they seldom define similarity in a semantic way. In recent years, the interest for image retrieval methods based on high-level semantic concepts, such as aesthetical measurements or emotions, has increased (see for instance Datta et al. [1]). Here we propose a novel approach using color emotions in CBIR. Our method is based on the assumption that perceived color emotions in images are mainly affected by homogenous emotion regions and transitions between emotion regions, measuring the spatial transitions of emotion values. Our design is motivated by the observation that the human visual system has a reduced color sensitivity in those areas of the visual field that contain high frequency content (see for instance Fairchild [2]). Therefore, we try to avoid "cluttered" image regions characterized by high frequency variations. The method is a continuation of the studies in [3] and [4], where global emotion histograms measure the amount of emotional content in an image. With the *bags-of-emotions* approach, described here, we create a method that also includes the relationship between neighboring emotion values. The proposed algorithm is inspired by the *bag of keypoints* method by Csurka et al. [5]. The method is developed and evaluated with a test database containing 5000 images, both photos and graphics.

2 Related Work

Research on color emotions for single colors and two-color combinations is an established research area. In a series of papers, Ou et al. [6] use psychophysical experiments to derive color emotion models for single colors and two-color combinations. Observers were asked to assess single colors on ten color emotion scales. Using factor analysis they reduce the number of color emotions scales to only three categories, or color emotion factors: *activity*, *weight* and *heat*. They conclude that the three factors agree well with studies done by others, for instance Kobayashi [7] and Sato et al. [8]. In this study we will use those emotion factors when investigating color emotions for multi-colored images.

There are few papers addressing the problem of including color emotions in image retrieval. The methods presented are often focusing on semantic image retrieval in a more general way. Two similar approaches focusing on color content are described by Wang and Yu [9], and Corridoni et al. [10]. Both are using clustering in the color space for segmenting images into regions with homogeneous colors. Regions are then converted to semantic terms, and used for indexing images. Wang et al. [11] present an annotating and retrieval method using a three-dimensional emotional space (with some similarities to the emotion space used in this paper). From histogram features, emotional factors are predicted using a Support Vector Machine. The method was developed and evaluated for paintings. Another approach is presented by Cho and Lee [12], where features are extracted from average colors and wavelet coefficients.

Related to emotions are the concepts of harmony (see an example by Cohen-Or et al. [13]) and aesthetics. In [14][15] Datta et al. study aesthetics in images from an online photo sharing website. These images were peer-rated in the categories *aesthetics* and *originality*. Image features corresponding to visual or aesthetical attributes (like Exposure, Colorfulness, etc.) are extracted, and compared to observer ratings using Support Vector Machines and classification trees. In [16] Datta et al. introduce the phrase "aesthetic gap", and report on their effort to build a real-world dataset for testing and comparison of algorithms.

3 Color Emotions

Color emotions can be described as emotional feelings evoked by single colors or color combinations, typically expressed with semantic words, such as "warm", "soft", "active", etc. As mentioned in the previous section, Ou et al. [6] investigated the relationship between color emotion and color preference. Color emotion models for single-colors and two color-combinations are derived from psychophysical experiments, resulting in three color emotion factors: *activity*, *weight* and *heat*:

$$activity = -2.1 + 0.06 \times \left[(L^* - 50)^2 + (a^* - 3)^2 + \left(\frac{b^* - 17}{1.4}\right)^2\right]^{\frac{1}{2}} \quad (1)$$

$$weight = -1.8 + 0.04(100 - L^*) + 0.45\cos(h - 100°) \quad (2)$$

$$heat = -0.5 + 0.02(C^*)^{1.07}\cos(h - 50°) \quad (3)$$

L^*, a^* and b^* are CIELAB coordinates, h is the CIELAB hue angle and C^* is CIELAB chroma [2]. The presented color emotion model was not developed for multi-colored images of potentially very complex structure. It is therefore easy to construct images were it will fail. We will, however demonstrate that these techniques provide statistical characterizations useful for CBIR.

4 Bags-of-Emotions

A *bag-of-emotions* corresponds to a histogram of the number of occurrences of particular emotion patterns in an image. The main steps for creating a *bag-of-emotions* are:

1. Convert the RGB image to an emotion image with 3 channels: *activity*, *weight* and *heat*
2. For each emotion channel, create a stack of low-pass filtered images (a scale space representation)
3. Derive the gradient magnitude for each image in the stack, and use statistical measurements of gradient magnitudes to detect interest points corresponding to homogeneous emotion areas and transitions between emotion areas
4. Derive emotion characteristics for patches surrounding each interest point
5. Construct a *bag-of-emotions* containing emotion characteristics

Notice the difference to other popular methods for extracting interest points, some of them mentioned in Mikolajczyk et al. [17]. There, keypoints related to corners, etc. are extracted, while we try extract homogeneous regions and transitions between regions.

Since the proposed method should be applicable to any image database, including public search engines and thumbnail databases, we are forced to make some assumptions and simplifications. We assume images are saved in the sRGB color space, and we use the standard illumination D50 when transforming sRGB values to CIELAB values. The image size is restricted to a maximum of 128 pixels (height or width), corresponding to the size of a typical thumbnail image. Images of larger size are scaled with bilinear interpolation. Using Eqs. (1-3), each pixel in the RGB image, $imrgb_n(x, y)$, is converted to a three dimensional color emotion representation, $ime_n(x, y)$. Channel $n = \{1, 2, 3\}$ in $ime_n(x, y)$ corresponds to the emotion *activity*, *weight* or *heat*. For ordinary RGB images with 8bit pixel values, the emotion values are located within the following intervals: *activity*: $[-2.09, 4.79]$, *weight*: $[-2.24, 2.64]$, and *heat*: $[-1.86, 2.43]$. These interval boundaries are used for normalizing individual emotion channels.

4.1 Scale Space Representation

A stack of low-pass filtered images is created for each emotion channel. We describe how simple statistical measurements in each stack are used for finding interest points corresponding to homogeneous emotion areas and transitions between emotion areas. For simplicity, the terminology used is adopted from scale space theory. However, notice that the scale space representation is only used initially. Interest points are detected in another representation, derived from the scale space. The scale space representation used for each channel n, in the emotion image $ime_n(x,y)$, is composed of a set of derived signals $L_n(x,y,t)$, defined by the convolution of $ime_n(x,y)$ with the Gaussian kernel $g(x,y,t)$, such that $L_n(x,y,t) = (g(t) * ime_n)(x,y)$, where t indicates the scale level. Increasing t will add more and more smoothing to $ime_n(x,y)$. In the proposed method the scale space is composed of $d = 4$ images, the original image ($t = 0$) together with three levels of scaling. Scaling is performed with **matlabPyrTools**[1], a Matlab toolbox for multi-scale image processing. The default Gaussian kernel is used as the filter kernel. Notice that the scale space representation is thus a Gaussian pyramid with different image size in each scale level t. Before continuing, the image in each level t is scaled to the same size as the original image, resulting in a stack of low-pass filtered images. For each level t in the stack, we combine the local derivative in x- and y-direction, $L_{nx}(x,y,t)$ and $L_{ny}(x,y,t)$, to obtain the gradient magnitude $M_n(x,y,t) = \sqrt{L_{nx}^2(x,y,t) + L_{ny}^2(x,y,t)}$. In each position (x,y), the mean, $\hat{M}_n(x,y)$, and variance, $V_n(x,y)$, along dimension t is computed

$$\hat{M}_n(x,y) = \frac{1}{d} \sum_{t=1}^{d} M_n(x,y,t) \qquad (4)$$

$$V_n(x,y) = \frac{1}{d-1} \sum_{t=1}^{d} (M_n(x,y,t) - \hat{M}_n(x,y))^2 \qquad (5)$$

For the proposed method, $\hat{M}_n(x,y)$ and $V_n(x,y)$ need to have values within approximately the same interval. Theoretically, with $d = 4$, and the above definition of the gradient magnitude, the possible value intervals are $[0, \sqrt{2}]$ and $[0, 2/3]$ for $\hat{M}_n(x,y)$ and $V_n(x,y)$ respectively. A straight-forward approach is to perform a normalization based on the maximum in each interval. However, real-world values for $V_n(x,y)$ are usually much closer to 0 than 2/3. Consequently, we use a normalization method based on statistics from the test database. The mean $V_n(x,y)$ and $\hat{M}_n(x,y)$ are derived for 5000 images, and the ratio between mean values are used for scaling $V_n(x,y)$.

4.2 Homogeneous Emotion Regions

In a homogeneous emotion region both $\hat{M}_n(x,y)$ and $V_n(x,y)$ should be small. Consequently, we can define $H_n(x,y) = \hat{M}_n(x,y) + V_n(x,y)$ and characterize interest points as local minima in $H_n(x,y)$. Instead of using the common

---
[1] http://www.cns.nyu.edu/~eero/software.php

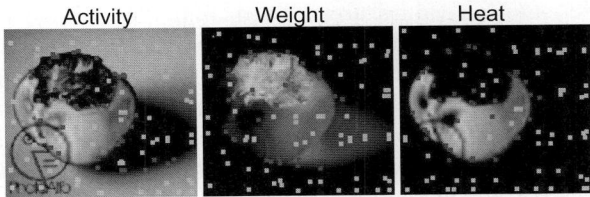

Fig. 1. Detected interest points in channels *activity*, *weigth* and *heat*. Green dots correspond to homogeneous regions, and red dots correspond to emotion transitions. The original RGB image can be seen in Fig. 2. (Color images in the online version)

approach of representing each homogeneous region with a single point, several interest points can be included in the same region, functioning as a simple measure of the region size. A block processing is utilized, where the minimum in each distinct (no overlapping) 15 × 15 block of $H_n(x,y)$ is used. Depending on image size, blocks located at image borders may contain less than 15 × 15 pixels. If the minimum is below a pre-defined threshold value, the location of the minimum is saved as an interest point. The proposed approach ensures that large homogeneous regions will be described by several interest points, and at the same time limits the maximum amount of possible interest points. Since the maximum image size is set to 128 pixels (as motivated in Sec. 4), it is feasible to use a fixed block size. Found interest points for an example image are shown in Fig. 1.

Interest points are obtained for emotion channels $n = \{1, 2, 3\}$ individually. For all interest points, a 7 × 7 window, centered at the point, is extracted from the equivalent position in the RGB image, $imrgb_n(x,y)$. The obtained RGB values, from all emotion channels, are gathered in a temporary image, and used as input to the emotion histogram method presented in [3]. The result is a 64 bins color emotion histogram, hh, describing the emotional content obtained from homogeneous regions in the image.

4.3 Transitions between Emotions

We detect transitions between emotion regions, by looking for points with a strong $\hat{M}_n(x,y)$. If we also favor low values for $V_n(x,y)$, "cluttered" image regions are avoided. Hence, we introduce $T_n(x,y) = \hat{M}_n(x,y) - V_n(x,y)$, and detect interest points using the block processing described in the previous section, now searching for maxima in $T_n(x,y)$ greater than a pre-defined threshold value. Found interest points for an example image are shown in Fig. 1.

Working with each emotion channel separately, a 7 × 7 spatial area surrounding each interest point is extracted from $ime_n(x,y)$, and the orientation of the transition is classified as one of four possible orientations: Horizontal (0°), vertical(90°), or one of the two diagonals (45° or 135°). Based on the orientation, and the direction of the gradient, the extracted area is divided into two equally sized regions. For each extracted area, a pair of emotion values is derived: From the region with the highest mean value, the average value e_h of the three pixels with the highest emotion values are derived, and the average value e_l of the three pixels with the lowest emotion values are derived from the opposite region. For each emotion channel, the distribution of emotion pairs is represented with a 2-dimensional, 4×4 bins, histogram, ht_n, where each dimension represents a quantization of distributions obtained from e_h and e_l respectively. Bin intervals are decided empirically based on statistical measurements of the test database.

We now have a 64 bins histogram, hh, describing the distribution of emotion values for homogeneous image regions, together with three 16 bins histograms, ht_{1-3}, describing emotion transitions for emotion channels $n = \{1, 2, 3\}$. A weighted combination of histograms defines our *bag-of-emotions*. In the next section we give an example of how to use those histograms in image retrieval.

5 Retrieval with Bags-of-Emotions

To illustrate that the proposed method results in useful image descriptors, images are retrieved based on the calculated distance between *bags-of-emotions*. Using the L_2-norm as distance metric between different emotion histograms, or *bags-of-emotions*, we obtain the results shown in Fig. 2. A few more search results based on *bags-of-emotions* are shown in Fig. 3. In the interpretation of these figures one has to take into account that the similarity is based on color emotions. Some colors, for instance shades of red and yellow, are closer in the emotion space than in frequently used color spaces. Another difference can be studied in

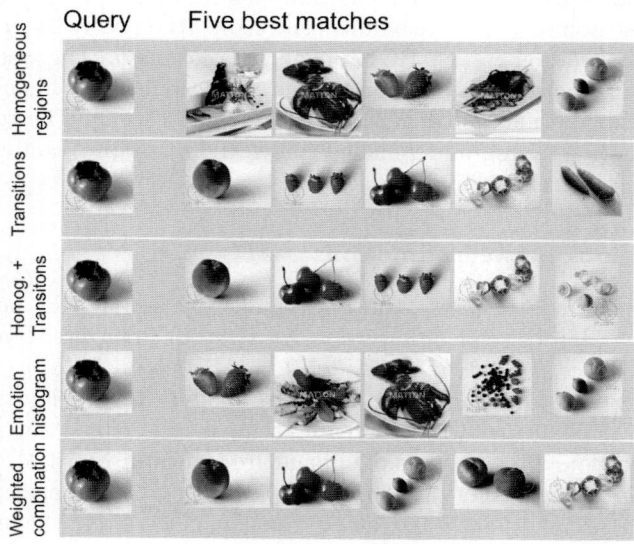

Fig. 2. Retrieval results based on different types of histograms. From row 1 to 5: 1) Homogeneous regions 2) Transitions between regions 3) A combination of homogeneous regions and transitions (*bags-of-emotions*) 4) Emotion histograms from [3] 5) A weighted combination of all histograms used above. (Color images in online version)

Fig. 3. Retrieval based on *bags-of-emotions*. (Color images in the online version)

the bottom right corner of Fig. 3, illustrating that colors rather far from each other in an ordinary color space (like RGB), can be located in the same region in the emotion space (in this example images contain mainly *active* colors). All experiments are conducted with the test database containing 5000 images.

These search results are only illustrations. An objective measure of performance (for instance by measuring Precision and Recall, plotting ROC curves, etc.) is difficult to design since emotion related properties are hard to define numerically. However, the connection between multi-colored images and the three emotion factors used has been evaluated in psychophysical experiments in [4]. The findings show that people do perceive color emotions for multi-colored images in similar ways, and that *activity*, *weight* and *heat* can be used in CBIR. A visual evaluation (made by the authors) of search results indicates that for most images, the *bags-of-emotions* descriptor performs better than the histogram method presented in [3]. Both methods are implemented in our publicly available demo search engine[2], where visitors can compare and evaluate retrieval results.

6 Conclusions

A novel attempt using a color emotion metric in CBIR is presented. The color emotion metric, derived from psychophysical experiments, uses three scales: *activity*, *weight* and *heat*. The presented algorithm is based on the assumption that perceived color emotions in images are mainly affected by homogenous emotion regions and transitions between emotion regions. It was shown that statistical measurements in a stack of low-pass filtered emotion images can be used for finding interest points corresponding to homogeneous emotion regions and transitions between regions. Emotion characteristics for patches surrounding each interest point are saved in a *bag-of-emotions*, that can be used for retrieving images based on emotional content. Experiments with a small database containing 5000 images show promising results, indicating that the method can be used in much larger databases.

This presentation is a first step towards a broader use of emotion related properties in image retrieval. Future research will extend the use of *bags-of-emotions* to

---

[2] http://media-vibrance.itn.liu.se/imemo/

semantic image classification and automatic labeling with color emotion words. In a retrieval task, one can also think of including the ability to design queries based on both single color emotion words and a combination of words (a semantic query). *Bags-of-emotions* can also be used for clustering in large image databases. A specific topic for future research is the use of statistical measurements of images to dynamically adjust the weighting of different types of emotion histograms (hh, ht_{1-3} and the histogram from [3]). Also the use of different distance measures should be investigated further. Examples are positive-definite quadratic form based distances, different kinds of adaptive distance measures, and combinations of those.

References

1. Datta, R., Joshi, D., Li, J., Wang, J.: Image retrieval: Ideas, influences, and trends of the new age. ACM Comput. Surv. 40, 2 (2008)
2. Fairchild, M.: Color Appearance Models. Wiley-IS&T, Chichester (2005)
3. Solli, M., Lenz, R.: Color emotions for image classification and retrieval. In: Proceedings CGIV 2008, pp. 367–371 (2008)
4. Solli, M.: Topics in content based image retrieval - fonts and color emotions. Licentiate thesis No. 1397, Linköping University (2009),
   http://urn.kb.se/resolve?urn=urn:nbn:se:liu:diva-16941
5. Csurka, G., Dance, C.R., Fan, L., Willamowski, J., Bray, C.: Visual categorization with bags of keypoints. In: Proceedings ECCV 2004, pp. 1–22 (2004)
6. Ou, L.C., Luo, M., Woodcock, A., Wright, A.: A study of colour emotion and colour preference. Part I-III. Color Res. Appl. 29, 3, 4, 5 (2004)
7. Kobayashi, S.: Aim and method of the color image scale. Col. Res. Appl. 6, 93–107 (1981)
8. Sato, T., Kajiwara, K., Hoshino, H., Nakamura, T.: Quantitative evaluation and categorising of human emotion induced by colour. Col. Sci. & Tech. 3, 53–59 (2000)
9. Wang, W.N., Yu, Y.L.: Image emotional semantic query based on color semantic description. In: Proceedings of ICMLC 2005, pp. 4571–4576 (2005)
10. Corridoni, J., Del Bimbo, A., Pala, P.: Image retrieval by color semantics. Multimedia Syst. 7, 175–183 (1999)
11. Wang, W.N., Yu, Y.L., Jiang, S.M.: Image retrieval by emotional semantics: A study of emotional space and feature extraction. 2006 IEEE Int. Conf. on Systems, Man and Cybernetics 4, 3534–3539 (2007)
12. Cho, S.B., Lee, J.Y.: A human-oriented image retrieval system using interactive genetic algorithm. IEEE Trans. Syst. Man Cybern. 32, 452–458 (2002)
13. Cohen-Or, D., Sorkine, O., Gal, R., Leyvand, T., Xu, Y.-Q.: Color harmonization. In: ACM SIGGRAPH 2006, vol. 25, pp. 624–630 (2006)
14. Datta, R., Joshi, D., Li, J., Wang, J.: Studying aesthetics in photographic images using a computational approach. In: Leonardis, A., Bischof, H., Pinz, A. (eds.) ECCV 2006. LNCS, vol. 3953, pp. 288–301. Springer, Heidelberg (2006)
15. Datta, R., Li, J., Wang, J.: Learning the consensus on visual quality for next-generation image management. In: 15th ACM Int. Conf. on Multimedia, pp. 533–536 (2007)
16. Datta, R., Li, J., Wang, J.Z.: Algorithmic inferencing of aesthetics and emotion in natural images: An exposition. In: Proceedings of ICIP 2008, pp. 105–108 (2008)
17. Mikolajczyk, K., Tuytelaars, T., Schmid, C., Zisserman, A., Matas, J., Schaffalitzky, F., Kadir, T., Van Gool, L.: A comparison of affine region detectors. IJCV 65, 43–72 (2006)

Measuring the Influence of Concept Detection on Video Retrieval

Pablo Toharia[1], Oscar D. Robles[1], Alan F. Smeaton[2], and Ángel Rodríguez[3]

[1] Dpto. de Arquitectura y Tecnología de Computadores, Ciencias de la Computación e Inteligencia Artificial,
U. Rey Juan Carlos, C/ Tulipán, s/n, 28933 Móstoles, Madrid, Spain
{pablo.toharia,oscardavid.robles}@urjc.es
[2] CLARITY: Center for Sensor Web Technologies, Dublin City University,
Glasnevin, Dublin 9, Ireland
alan.smeaton@dcu.ie
[3] Dpto. de Tecnología Fotónica, U. Politécnica de Madrid,
Campus de Montegancedo s/n, 28660 Boadilla del Monte, Madrid, Spain
arodri@fi.upm.es

Abstract. There is an increasing emphasis on including semantic concept detection as part of video retrieval. This represents a modality for retrieval quite different from metadata-based and keyframe similarity-based approaches. One of the premises on which the success of this is based, is that good quality detection is available in order to guarantee retrieval quality. But how good does the feature detection actually need to be? Is it possible to achieve good retrieval quality, even with poor quality concept detection and if so then what is the "tipping point" below which detection accuracy proves not to be beneficial? In this paper we explore this question using a collection of rushes video where we artificially vary the quality of detection of semantic features and we study the impact on the resulting retrieval. Our results show that the impact of improving or degrading performance of concept detectors is not directly reflected as retrieval performance and this raises interesting questions about how accurate concept detection really needs to be.

1 Introduction and Background

The automatic detection of semantic concepts from video is opening up a completely new modality for supporting content-based operations like search, summarisation, and directed browsing. This approach to managing content compliments using video metadata and using keyframe similarity and is being enabled by improvements in the accuracy, and the number of, such detectors or classifiers by many research groups. This can be seen in the recent development in activities such as TRECVid [1] where it is now realised that retrieval systems based on low-level features like colour and texture do not succeed in describing high-level concepts as a human would do.

Various authors are now making efforts on optimizing automatic detection of semantic concepts for use in applications such as retrieval. However, it is not clear what is the real impact of improving the accuracy of the detection process, i.e. whether a significant improvement in the performance of detection will yield better quality retrieval. There have been some previous studies of the efficiency of using concepts in retrieval [2,3,4]. Recently, Snoek et al. [5] analyzed whether increasing the number of concept detectors as well as their combination would improve the performance of retrieval and found that it does.

Wang and Hua study how to improve the performance of combining video concept detectors when dealing with a large number of them by following a Bottom-Up Incremental Fusion (BUIF) approach [6], but they do not deal with the issue of assessing detectors' real influence in retrieval. Thus it appears there is no work studying the relationship between the quality of detectors and retrieval performance. The work here explores the relationship between concept detection performance and content-based retrieval and to examine whether improving detection will yield an improvement at the retrieval stage, or whether this is worth the effort.

2 Materials and Methods

We now detail how we set up an experimental environment for video retrieval using semantic concepts. Controlled noise in concept detection is introduced so as to improve or worsen it, allowing performance of retrieval to be measured. Section 3 presents experiments together with the analysis and conclusions reached.

2.1 Concept Detection

The first step is to set up a system to extract concepts from shots. In our work we used the TRECVid [7] 2006 rushes collection of 27 hours which gave rise to approximately 2,900 shots. The concepts selected to work with are defined from within LSCOM-Lite, a reduced version of the 449 Large Scale Concept Ontology for Multimedia [8] annotated concepts that form LSCOM.

The concept detection process is broken into several steps. First, a preprocessing stage extracts keyframes that represent the video content. These are then filtered in order to discard shots such as calibration charts, black frames and so on. We then extract low-level features for these keyframes which are then used as the input to the 39 classifiers. More details of the keyframe extraction and filtering stages can be found in [9]. Finally, Support Vector Machines (SVM) provided by Dublin City University from our high level feature detection submission in TRECVid 2006 are used, using low-level primitive features like colour and texture, extracted by the AceToolbox [10]. The concept classifiers each provide a certainty value $C_i \in [-1, 1]$ that each of the shots' keyframes in the original video contains each of the concepts and we use these as baseline examples of the accuracy of a real implementation of concept detection.

2.2 Interactive Concept-Based Retrieval Engine

An interactive video retrieval system is used in order to test the relationship between the quality of detected concepts and retrieval performance. This allows

Fig. 1. Retrieval example using weighted concepts

a user to select which of the available concepts should be used in retrieval, as well as fixing W_i weights for each of the concepts. These are positive if the concept is relevant for the query, and if its absence is relevant to the query it will be negative, else it will be 0. The retrieval engine will assign a value $score_i$ for each shot so a sorted list of results can be presented to the user. Assuming there are N concepts the following is how we obtain a score for each shot:

$$shot_i = \{C_{i1}, C_{i2}, \ldots, C_{iN}\}, \quad C_{ij} \in [-1, 1] \tag{1}$$

$$score_i = \frac{\sum_{i=1}^{N} W_i \cdot C_i}{N}, \quad W_{ij} \in [-1, 1] \tag{2}$$

As was previously stated, other approaches to combining concept features in retrieval are possible, such as proposed by Wang and Hua [6] or by Snoek and Worring [5], but in our present work we were not interested in addressing the detector fusion method. Figure 1 shows a retrieval result based on 8 concepts selected by the user. On the left side, 8 sliding bars allow a user to adjust weights for each concept and a visualization of the top-ranked shots is also shown.

2.3 Degradation and Improvement of Concept Detection

Performing an artificial degradation or improvement of concept detection quality can be achieved by introducing noise into the concept detector output, so the certainty value is increased or decreased as needed. However, rather than depend on the accuracy of automatic detection only, the existence of a ground truth allows us to faithfully simulate improvement and degradation of concept

Table 1. Concepts used in experiments

Concept	Description
Building	Shots of an exterior of a building
Car	Shots of a car
Crowd	Shots depicting a crowd
Outdoor	Shots of Outdoor locations
Person	Shots depicting a person. The face may be partially visible
Road	Shots depicting a road
Sky	Shots depicting sky
Vegetation	Shots depicting natural or artificial greenery, vegetation woods, etc.

detection. To obtain this, a manual process of double annotation of each of the concepts over the whole collection was performed.

To vary detection quality, a percentage P of shots from the collection are randomly selected and their certainty degree is modified for each detector. To improve performance, a value A is added to the certainty value of shots from the ground truth in which the concept is known to be present. If a shot does not contain the concept, the value A will be subtracted from the certainty value. In case of degrading the detectors' performance, the process is reversed.

In measuring the impact of concept detection on retrieval, we use an offline retrieval proces. We use the keyframes of the shots selected to initiate a low-level retrieval process using the low-level image characteristics used as input to concept recognition, to perform keyframe similarity. This generates a content-based ranking of shots for each topic. A concept-based retrieval ranking is also generated using the weights selected by the users and degrading/upgrading the performance of the detectors accordingly. The results of both retrieval rankings are normalized and combined in a 50:50 ratio to give the final retrieval output. While this may seem like diluting the impact of concept retrieval, and concept detection accuracy, it reflects the true way in which video retrieval is carried out in practice. Retrieval performance is evaluated using Mean Average Precision (MAP) over the set of topics and thus by varying the parameters A and P, a change in retrieval MAP should be obtained for each concept.

For our experiments we concentrated on a subset of concepts from LSCOM-Lite, shown in Table 1, chosen because they occur throughout the whole video dataset whereas others occur much less frequently. Our experiments are carried out in two parts, an online part working with non-expert users who perform iterative retrieval, and an automated offline part using results from the user retrieval and performing more exhaustive tests varying concept detection quality. This is shown in Figure 2.

2.4 Experimental Methodology

Our experimental methodology is as follows. In the first stage a user searches for shots given a topic using the interactive system and the concept-based retrieval engine described earlier. Topics have been constructed in such a way that they

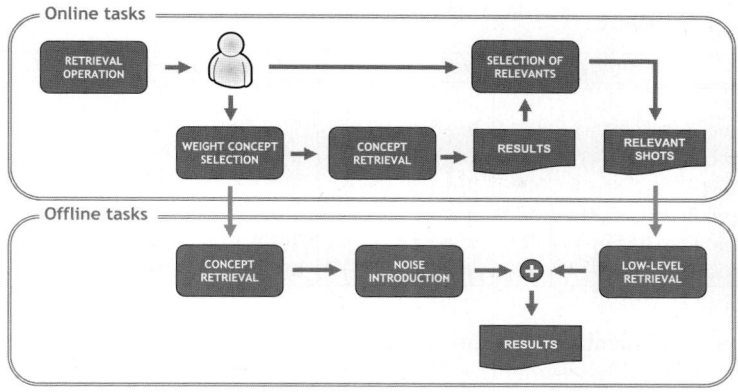

Fig. 2. Experimental framework

require iterations of the retrieval system to refine and adjust topic weights until they are optimal, and that they use concepts both in a positive or negative way. Topics are shown in Table 2, along with their number of relevant shots.

Topics will use available concepts in positive or negative ways, depending on the subject matter. Topic 6 can be associated with negative weighting of the concept "Outdoor", since the aim is that action takes place inside a building. Table 3 shows the ways that the set of 9 users use topics in positive or negative ways. For example for Topic 2, 4 users used the concept "Car" in a positive way and 2 used "Sky" in a negative way. Some aspects of some topics may not be addressable in query formulation with the available concepts and while this may seem a limiting factor, it is also representative of a real world search where there will never be enough appropriate concepts for the variety of user search topics.

For our experiments, 9 users without any professional experience of searching were recruited. Each user was given an introduction to the system, and the 7 topics were presented in a rotating order to avoid bias. Each user adjusted concept weights for each of the topics and a retrieval operation was performed with the user marking relevant shots or adjusting concept weights and performing a new search. Once the sets of relevant shots had been identified, we can calculate retrieval rankings based on combined weighted concept-based and content-based

Table 2. Search topics

Topic	Description: "Find shots containing ..."	# rel. shots
1	...open-sea views	33
2	...2 or more people with plants in an urban area	243
3	...desert-like landscapes	55
4	...village settlements on the coast	73
5	...2 or more people interacting in a natural environment	91
6	...a person talking to an audience inside a building	39
7	...people sailing	42

Table 3. Use of concepts in Topics

	Building	Car	Crowd	Outdoor	Person	Road	Sky	Vegetation
Topic 1:	6(-)	6(-)	3(-)	9(+)	2(-)	5(-)	7(+)	5(-)
Topic 2:	9(+)	4(+)	9(+)	2(+)	7(+)	6(+)	2(-)	9(+)
Topic 3:	7(-)	5(-)	3(-)	9(+)	3(+)	4(-)	9(+)	9(-)
Topic 4:	8(+)	3(+)	2(+)	9(+)	3(+)	5(-)	8(+)	5(+)
Topic 5:	5(-)	3(-)	8(+)	9(+)	8(+)	3(-)	4(+)	8(+)
Topic 6:	5(+)	4(-)	9(+)	9(-)	7(+)	6(-)	6(-)	3(-)
Topic 7:	6(-)	5(-)	5(+)	9(+)	8(+)	7(-)	9(+)	5(+)

techniques and calculate MAP retrieval performance by measuring against an exhaustive manual assessment of shot relevance, our ground truth for retrieval. We can examine the effect of detector quality on retrieval performance by introducing noise into the output of the concept detectors as described in section 2.3. Each variation on the parameters that results in degraded or improved detectors gives a new list of ranked shots which can be evaluated against the ground truth, and MAP calculated. Combining the different options available, we have a total of 9 users, each running 7 queries with improvements and degradations on 8 concepts, to be evaluated.

3 Results and Discussion

3.1 Performance of Retrieval

Table 4 shows the average MAP percentage variations when we degrade or improve the quality of the underlying concept detection above or below the level of concept detection performance obtained from the automatic DCU concept detection. Thus we use the real performance figures as a baseline and vary detection quality above and below this. The MAP performance using unmodified concept detection performance is 0.0254.

What these results tell us, for example, is that when we degrade concept detection performance for all concepts by reducing the certainty value for detection by 0.5 (on a scale of -1 to 1) for 50% of the shots, we get a net drop in MAP performance for retrieval of only 5.69% (bolded entry in Table 4).

Table 5 collects the average Coefficient of Variation values considering the results achieved by all users and among all topics. Coefficient of Variation values

Table 4. MAP variation average for retrieval introducing controlled noise into detector performance

(a) Degradation

P/A	-0.1	-0.3	-0.5
10%	-0.10%	-0.74%	-1.23%
30%	-0.67%	-2.61%	-4.25%
50%	-0.92%	-3.42%	**-5.69%**

(b) Improvement

P/A	+0.1	+0.3	+0.5
10%	0.10%	0.19%	0.53%
30%	0.45%	1.20%	2.16%
50%	0.81%	2.12%	3.98%

Table 5. Avg. Coefficients of Variation considering responses by users, all topics

(a) Per user

Degradation				Improvement			
P/A	-0.1	-0.3	-0.5	P/A	+0.1	+0.3	+0.5
10%	5.277	0.638	0.723	10%	2.454	8.337	5.078
30%	1.156	0.327	0.350	30%	1.960	1.764	1.285
50%	1.299	0.453	0.414	50%	1.081	1.105	0.929

(b) Per topic

Degradation				Improvement			
P/A	-0.1	-0.3	-0.5	P/A	+0.1	+0.3	+0.5
10%	8.361	1.794	1.835	10%	6.052	12.117	7.205
30%	1.881	1.182	1.231	30%	2.511	2.523	2.043
50%	1.620	1.195	1.214	50%	1.372	1.276	0.986

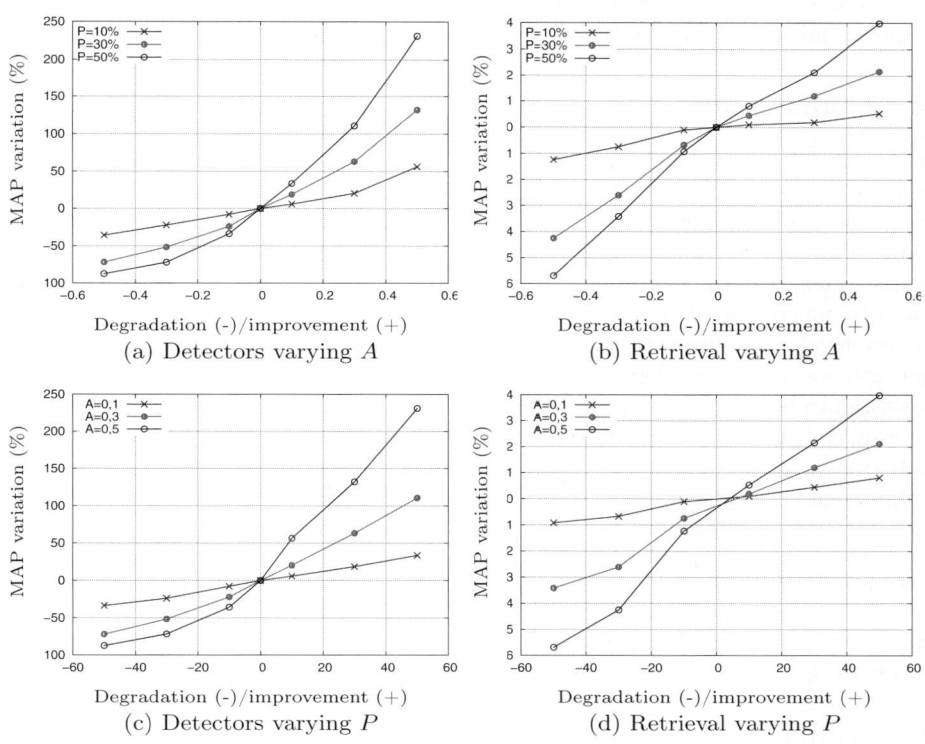

Fig. 3. MAP variations for detection and retrieval, varying A and P

are more stable across users rather than across topics, but the worst cases appear for the lower values of P and A variables because the average variations are very low (Table 4). This can be due to the user interaction with the retrieval engine and to the random controlled noise introduced.

3.2 Detector Performance versus the Retrieval Task

Figure 3 shows MAP variations when fixing one of the parameters, either A or P, for both detection performance and for concept retrieval performance. The x-axis depicts A or P values for improvement (represented as positive values of the scale) or degradation (negative values). The y-axis shows the variation of the MAP in percentages. The curves show similar trends for both A and P transformations for all the tests shown. However, Figures 3(a) and 3(b) (A as parameter) show different range values for positive (improvement) and negative (degradation) intervals, being the variation most noticeable in the improvement transformation. On the other hand, both the tendency and the interval of Figures 3(c) and 3(d) (P as parameter) are very similar. Overall, however, we can say that the impact of detection accuracy is far less pronounced than we would expect, indicating that even poor detection accuracy provides useful retrieval.

4 Conclusions

We have implemented a methodology to analyze the impact of concept detection accuracy on video retrieval on a collection of rushes video. We found that even poor quality detection can yield good retrieval and that as the quality of detection improves the quality of retrieval does not rise accordingly. While this may appear as just an interesting exercise and the results do depend on the set of concepts used, it does represent the state of the art in using concepts in retrieval, as shown in TRECVid, where it is shown that exploiting the dependencies among concepts is non-existent. For future work we plan to further investigate how detection performance is impacted when semantic dependencies among concepts (e.g. "Outdoor/Building" and "Person/Crowd") and this will integrate concept ontologies into our work. Other work will be to extend the number of concepts to see if similar results are obtained for concepts which do not occur as frequently in the video as the ones used here.

Acknowledgments. This work has been partially funded by the the Spanish Ministry of Education and Science (grant TIN2007-67188). BBC video was provided by TRECVid [1]. AS was funded by Science Foundation Ireland under grant 07/CE/I1147.

References

1. Smeaton, A.F., Over, P., Kraaij, W.: High-Level Feature Detection from Video in TRECVid: a 5-Year Retrospective of Achievements. In: Divakaran, A. (ed.) Multimedia Content Analysis, Theory and Applic., pp. 151–174. Springer, Berlin (2009)

2. Hauptmann, A.G., Yan, R., Lin, W.H., Christel, M.G., Wactlar, H.D.: Can high-level concepts fill the semantic gap in video retrieval? a case study with broadcast news. IEEE Transactions on Multimedia 9(5), 958–966 (2007)
3. Christel, M.G., Hauptmann, A.G.: The use and utility of high-level semantic features in video retrieval. In: Leow, W.-K., Lew, M., Chua, T.-S., Ma, W.-Y., Chaisorn, L., Bakker, E.M. (eds.) CIVR 2005. LNCS, vol. 3568, pp. 134–144. Springer, Heidelberg (2005)
4. Wei, X.Y., Ngo, C.W.: Fusing semantics, observability, reliability and diversity of concept detectors for video search. In: Proc. MM 2008, pp. 81–90. ACM, New York (2008)
5. Snoek, C.G.M., Worring, M.: Are concept detector lexicons effective for video search? In: ICME, pp. 1966–1969. IEEE, Los Alamitos (2007)
6. Wang, M., Hua, X.S.: Study on the combination of video concept detectors. In: Proc. MM 2008, pp. 647–650. ACM, New York (2008)
7. Smeaton, A.F., Over, P., Kraaij, W.: Evaluation campaigns and trecvid. In: Proc. MIR 2006, pp. 321–330. ACM Press, New York (2006)
8. Naphade, M.R., Kennedy, L., Kender, J.R., Chang, S.F., Smith, J.R., Over, P., Hauptmann, A.: A light scale concept ontology for multimedia understanding for TRECVID 2005. Tech. Rep. RC23612, IBM T.J. Watson Research Center (2005)
9. Toharia, P., Robles, O.D., Pastor, L., Rodríguez, A.: Combining activity and temporal coherence with low-level information for summarization of video rushes. In: Proc. TVS 2008, pp. 70–74. ACM, New York (2008)
10. O'Connor, N., Cooke, E., le Borgne, H., Blighe, M., Adamek, T.: The acetoolbox: Low-level audiovisual feature extraction for retrieval and classification. In: Proc. EWIMT 2005 (2005)

SEM Image Analysis for Quality Control of Nanoparticles

S.K. Alexander[1], R. Azencott[1], B.G. Bodmann[1], A. Bouamrani[2], C. Chiappini[2], M. Ferrari[2], X. Liu[2], and E. Tasciotti[2]

[1] Department of Mathematics, University of Houston, Houston TX 77004
simon@math.uh.edu, razencot@math.uh.edu
[2] Biomedical Engineering/NanoMedicine, University of Texas, Houston, TX, 77003

Abstract. In nano-medicine, mesoporous silicon particles provide efficient vehicles for the dissemination and delivery of key proteins at the micron scale. We propose a new quality-control method for the nanopore structure of these particles, based on image analysis software developed to automatically inspect scanning electronic microscopy (SEM) images of nanoparticles in a fully automated fashion. Our algorithm first identifies the precise position and shape of each nanopore, then generates a graphic display of these nanopores and of their boundaries. This is essentially a texture segmentation task, and a key quality-control requirement is fast computing speed. Our software then computes key shape characteristics of individual nanopores, such as area, outer diameter, eccentricity, etc., and then generates means, standard deviations, and histograms of each pore-shape feature. Thus, the image analysis algorithms automatically produce a vector from each image which contains relevant nanoparticle quality control characteristics, either for comparison to pre-established acceptability thresholds, or for the analysis of homogeneity and the detection of outliers among families of nanoparticles.

1 SEM Image Data and Quality Control Targets

Quality control in the production of nanostructures poses a challenge for image processing, because it requires interpreting high-resolution images which may be plagued by substantial amounts of noise of various characteristics, and because the material is by design heterogeneous and thus requires flexible analysis algorithms. We present new algorithms for SEM images analysis focused on quality control of porous silicon (pSi) and associated microparticles (PSMs).

Visual inspection of nanoparticles is performed on 24bit SEM images typically of size 1024×768. Each nanoparticle occupies an image surface of approximately 500×500 pixels, has dimensions of the order of 3×3 microns and gathers between 500 and 1000 nanopores having similar shapes. The main goal of our algorithmic image analysis is first to identify the precise positions and shapes of each nanopore and its boundary in a fully automated fashion, and generate a graphic display of these nanopores and boundaries. This is essentially a texture segmentation task, and a key quality control requirement is fast computing

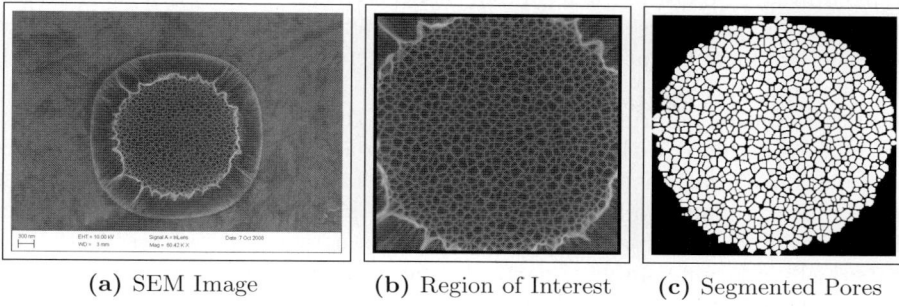

(a) SEM Image (b) Region of Interest (c) Segmented Pores

Fig. 1. Example nanoparticle AN24-GP-12

speed. Then a second algorithm automatically analyzes the shapes of the detected nanopores in order to compute key shape characteristics of nanopores, such as area, outer diameter, eccentricity, boundary thickness, etc.); we thus generate for each nanoparticle a database of roughly 500 to 1000 vectors of individual pore features. At the level of each nanoparticle, we then launch automatic extraction of statistical characteristics of this database, to compute the means, standard deviations, and histograms of each type of shape feature. This defines a vector of nanoparticle characteristics, which thus provides very natural quality control features, either for comparison to pre-established acceptability thresholds, or to analyze homogeneity and outliers among families of nanoparticles.

2 Porous Silicon Microparticles and Nanomedicine Applications

Since the initial proof of its biocompatibility [1] porous silicon has been actively researched as a biomaterial[2,3,4,5,6]. Porous silicon microparticles (PSMs) have demonstrated their efficacy as delivery vectors for therapeutics. PSMs obtained by sonication or ball milling of pSi layers successfully acted as loading and release agents for different drug molecules, encompassing a wide spectrum of solubility and acid/base characteristics[7]. Proteins were also successfully loaded and released from PSM[8]. Oral delivery of pSi has been proven safe[9] and paracellular drug delivery by means of PSMs has been demonstrated in vitro [10]. However, the size and shape polydispersion of PSMs obtained by sonication or ball milling forbids their use as vascular delivery systems. Our group has successfully developed a strategy based on mathematical models [11,12,13], to produce monodisperse porous silicon microparticles of tailored pore size, shape, and porosity (porous silicon elements, PSEs)[14]. We have proven short term safety of PSEs upon injection [15], and demonstrated their suitability as primary vectors in a multi-stage delivery system [16]. The porous silicon elements are produced in a silicon fabrication environment using a top-down approach resulting in selective porosification of bulk silicon wafers. The fabrication process involves multiple steps: thin film deposition, photolithography, selective dry etch, electrochemical etch, etc. The process is subject to batch-to-batch variations that may influence the final product. Variations in lithographic steps may lead to a different PSE size or shape, and hence to variation of pore size and porosity. But to guarantee the PSEs efficacy as primary delivery vectors, their size, shape, and pore sizes must be reproduced within stringent limits to avoid modifying their flow and margination characteristics , altering the payload biodistribution and release profile due to different diffusion characteristics [17] and pSi degradation kinetics [16]. Currently, quality assessment for PSEs is a two step process. Initially a statistically relevant sample of particles from a single production lot is analyzed by expert interactive measurements on SEM images to assess size and shape uniformity. Secondly ten or more production lots are joined to obtain the minimum 10mg sample size necessary for nitrogen absorption/desorption analysis of pore size and porosity. This latter step risks rejection of good quality lots (representing significant time and resources spent) due to necessarily mixing with other lots. The alternative software based algorithmic image analysis we propose here for quality control of PSEs is much faster and generates robust quantitative evaluations of pore sizes and shapes.

3 Image Analysis

3.1 Algorithmic Outline

We first process a high resolution SEM image (e.g. Figure 1a) to compute a graphic display of pore locations, which may, when required by the user, be spatially restricted to circular bands within the nanoparticle. Below are our main algorithmic steps :

- **Step 1: *ROI extraction*** We isolate the nanoparticle of interest, by centering on a region of interest (ROI) labeled I_0. This step is still interactive, but will be easily automatized within a future quality control software. The SEM image as seen in Figure 1a is then cropped, resulting in Figure 1b, and masked to a subimage I_0.
- **Step 2: *Histogram equalization*** We compute the intensity histograms H_1, H_2, ..., on small overlapping image patches of identical sizes R_1, R_2, ..., covering our ROI. Local intensities are then distorted nonlinearly in order to equalize and requantize all local histograms H_1, H_2, ... This step generates an image with more uniform distributions of intensities depicted in Figure 2b.
- **Step 3: *Pore segmentation*** A number of morphological and statistical operations are performed to isolate the individual nanopores within the ROI. This key segmentation algorithm is detailed below.
- **Step 4: *Extraction of pore features*** Following segmentation, each individual pore and its boundary are determined as associated subregions of the ROI image. Classical shape analysis techniques are applied to each one of these subregions in order to compute, for each pore, its area, perimeter,

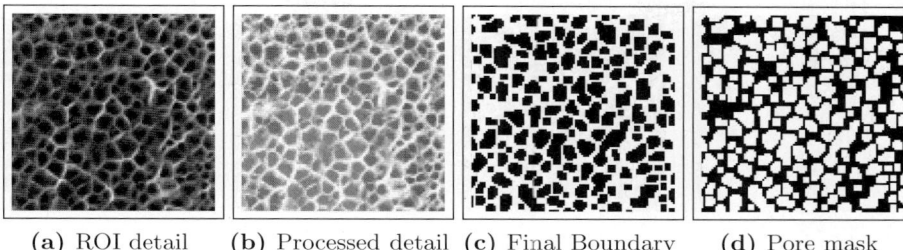

(a) ROI detail   (b) Processed detail   (c) Final Boundary   (d) Pore mask

Fig. 2. Pore shape localization: A detail from the middle of an ROI is shown in (a). The ROI is locally equalized and re-quantized based (b), the pore boundaries are found (c) and finally a mask geometrically identifying all individual pores is created (d).

outer diameter, and boundary thickness, and to evaluate the degree of elongation of the pore, modeled by the eccentricity of an elliptic shape fitted to the pore.

- **Step 5: *Nanoparticle quality control features*** Within the approximately circular nanoparticle, we analyze shape features statistics for all pores, computing the mean, standard deviation, and histogram of each pore shape feature extracted at step 4. This process can be carried out globally for each particle, or repeated for only those pores situated within concentric rings of user selected radius. Thus on each circular ring, the characteristics may be compared giving us the ability to perform both inter- and intra-pore comparisons. Furthermore, the cumulative statistics from one set of particles may be compared with those of another set, as appropriate.

3.2 Segmenting the Nanopores

Our *pore segmentation* algorithm is deliberately localized to better to variations in pore depths, pore shapes, and boundary wall thicknesses. This adaptivity lead us below to implement spatially variable morphological operators, as in [18]. We begin with an image subpatch of the raw ROI I_0, as in Figure 2a, which is then broken into overlapping square regions R_j (with dimensions roughly equal to 30×30 pixels in the cases shown here, determined by a rough estimate of pore size set to be a box of 3 times the typical pore diameter, which will vary with resolution and physical pore size). These regions are histogram equalized and requantized, then merged into a single image I_1 (e.g. Figure 2b).

On each region R_j of I_1 the "skeleton" of R_j is computed relative to the median (local to R_j), computed by a morphological algorithm [19]. Note that due to localized equalization processing, the merged skeleton of I_1 does not suffer from intensity fluctuations present in the original image. We then automatically bridge each skeleton gap between the neighboring ends of any two skeleton branches that are not 4-connected. This extended skeleton is then refined by removing spurs and isolated pixels. The final skeleton is then used to mask I_1

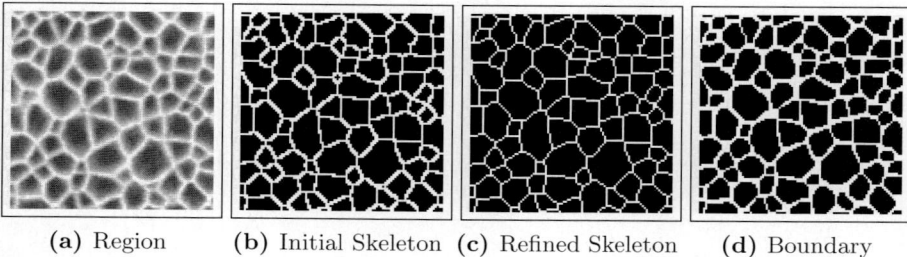

(a) Region (b) Initial Skeleton (c) Refined Skeleton (d) Boundary

Fig. 3. Pore Segmentation steps: (a) locally processed region has (b) initial skeleton which leads to (c) refined skeleton used to find *local intensity threshold* yielding (d) boundary. Finally, the pores are defined as connected components separated by this boundary, as in Figure 2d.

and we compute the mean and standard deviation μ_{skel}, σ_{skel} of pixel intensities over the final skeleton $skel_j$. These values are used to compute a well adapted *local intensity threshold* thr_j on R_j separating intensity values within pores from those on pore boundaries. Using $skel_j$ and thr_j to compute a first version of pore boundaries in R_j, we generate a first map of all pore boundaries (e.g. Figure 2c). This pore boundaries map PB is then cleaned of spurious pixels, and each connected component of $I_1 - PB$ is identified with an individual pore; this defines the final pore map (e.g. Figure 2d). Figure 3 shows details of the skeletonization process and boundary definition on a particularly clear case. The region shown is much larger than R_j boundaries, demonstrating that appropriate partitioning has been performed and does not introduce boundary artifacts.

4 Experimental Results

After a nanoparticle has been processed with the pore segmentation algorithm, the resulting pore map is automatically analyzed as described above in Step 4 to compute a vector of shape features for each pore. An automated statistical analysis of this family of 500 to 1000 vectors of pore shape features computes means and standard deviations of these feature vectors to generate and display a vector of mean quality-control characteristics for the nanoparticle just analyzed.

Our goal was to prove feasibility of software based automated quality control to efficiently monitor the production of nanoparticles, as well as to enable rigorous quantified descriptions of the nanoparticles pore structure. First, we want to quantify the pore-shape homogeneity within a single nanoparticle. Second, we want to compare nanoparticles from the same population to see how consistent the production is. Finally, we would like to quantify the difference between nanoparticles coming from different populations.

To evaluate pore shape variation from the center of the nanoparticle to its natural roughly circular boundary, we mask off several *rings* within the nanoparticle. To compare these sub-populations of pores within the nanoparticle, we need

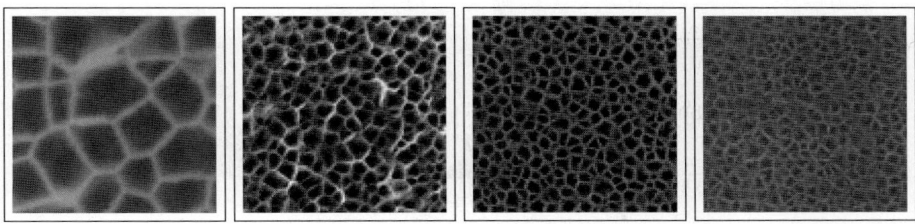

Fig. 4. Several example particles shown with 1:1 pixel dimensions, demonstrating the variability of inputs

Table 1. Estimated values of several features (area, perimeter length, etc.) For pores in the region of interest for several sets of particles. Values are given along with standard error of the mean. The first grouping shows individual sets of similar particles, the three sets in the final grouping are larger collections of particles roughly grouped by size ("Large Pores" LP, and "Extra Large Pores" XLP).

Dataset Name	Long Axis [nm]	Area [nm^2]	Perimeter [nm]	Short Axis [nm]
AG	44 ± 0.18	1100 ± 7.57	100 ± 0.47	34 ± 0.10
AN	75 ± 0.59	3400 ± 47.01	200 ± 1.68	54 ± 0.38
AO	73 ± 0.43	3300 ± 38.71	200 ± 1.32	53 ± 0.32
E18/19	46 ± 0.23	1300 ± 11.07	110 ± 0.63	36 ± 0.14
F2	55 ± 0.80	1800 ± 47.48	140 ± 2.38	40 ± 0.40
F4	55 ± 0.58	1700 ± 26.66	130 ± 1.56	38 ± 0.25
F5	46 ± 0.39	1300 ± 17.79	110 ± 1.11	35 ± 0.23
2 μm	50 ± 0.30	1500 ± 14.53	120 ± 0.83	37 ± 0.19
LP1	51 ± 0.32	1500 ± 15.23	130 ± 0.95	36 ± 0.15
LP2	46 ± 0.13	1300 ± 6.12	110 ± 0.36	35 ± 0.08
XLP1	74 ± 0.32	3400 ± 27.48	200 ± 0.95	54 ± 0.23

to compute robust estimates for the average values of pore shape features over different rings. We first eliminate (and also identify) outliers by quantile analysis on the histograms of pore features values, and thus generate robust estimates of the mean features. Outliers, which are essentially the most oddly shaped pores, can then also displayed for visual inspection by the user.

We have applied our prototype quality control algorithms to SEM imaged samples of various particles production processes. Table 1 collects results from several sets of particles. A typical histogram of three of these features is show in Figure, 5 for one of the data sets.

These results on the statistical distributions of pore characteristics confirms the homogeneity of pore shapes between particles of the same type (i.e generated by the same process). We have studied 8 such groups of between 2 and 4 particles, and within each group, relative variations between particles of mean

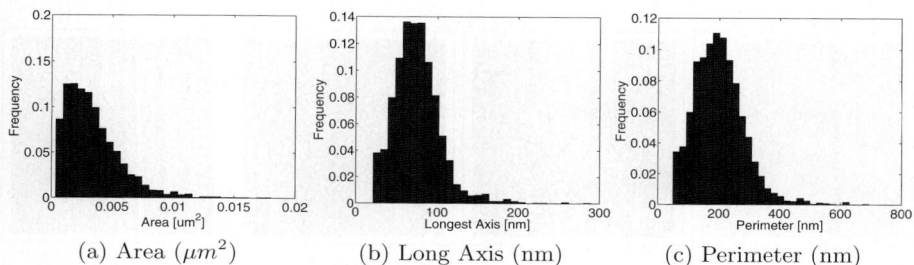

Fig. 5. Relative histograms for three features of the AN set of particles

pore characteristics is small: 5–10% for area, 3–10 % for perimeter, 3–7 % for long axis. The relative accuracy of our estimates of mean pore characteristics for each particle, evaluated by the ratio of standard error over mean, is 0.7–2.5 % for area, 0.5–2 % for perimeter, 0.4–1.4 % for long axis.Thus the accuracy of our estimates is much smaller than the variability between particles of the same type which is a favorable feature for quality control applications.

For difficult SEM images (exhibiting defects such as high oscillations of image intensities, poor adjustment of intensities dynamics, saturation zones, extremely blurred zones) our algorithm still does a very good detection job on 90% of the homogeneous porous area of the particle, and hence the computation of mean pore-shape features remains quite reliable.

For each SEM image of a nanoparticle, roughly 15 seconds of CPU (AMD 4200) are required to generate the tables of mean pore-shape features, the graphic display of pore maps, and the localization of pore outliers. These outputs represent the typical outputs we expect from our future quality control software. Speed of computation could easily be significantly accelerated, since our computing algorithms are highly parallelizable by multi-node treatment

5 Conclusions and Future Research

With these algorithms to analyze individual nanoparticles, we intend to validate our study on a broader collection of SEM images of nanoparticles, to evaluate how processing characteristics during the generation of nanoparticles may or may not affect the pore shapes and pore boundaries. Our research has focused on fast and accurate segmentation/identification of nanopores, combined with extraction of mean pore shapes features and pore homogeneity evaluation. This has immediate applications to specify and implement a future fast and reliable quality control software to monitor quality in nanoparticle production.

References

1. Canham, L.: Bioactive silicon structure fabrication through nanoetching techniques. Adv. Mater. 7, 1033–1037 (1995)
2. Cheng, M., et al.: Nanotechnologies for biomolecular detection and medical diagnostics. Curr. Opin. Chem. Biol. 10, 11–19 (2006)

3. Bayliss, S., et al.: Nature of the silicon-animal cell interface. J. Porus Mater 7, 191–195 (2000)
4. Li, Y., et al.: Polymer replicas of photonic porous silicon for sensing and drug delivery applications. Science 299, 2045–2047 (2003)
5. Buckberry, L., Bayliss, S.: Porous silicon as a biomaterial. Mater World 7, 213–215 (1999)
6. Sun, W., Puzas, J., Sheu, T.J., Fauchet, P.: Porous silicon as a cell interface for bone tissue engineering. Phys. Status Solidi A 204, 1429–1433 (2007)
7. Salonena, J., et al.: Mesoporous silicon microparticles for oral drug delivery: Loading and release of five model drugs. J. Controlled Release 108, 362–374 (2005)
8. Prestidge, C., et al.: Peptide and protein loading into porous silicon wafers. Phys. Status Solidi C 205, 311–315 (2008)
9. Canham, L.: Nanoscale semiconducting silicon as a nutritional food additive. Nanotechnol. 18, 185704 (2007)
10. Foraker, A., et al.: Microfabricated porous silicon particles enhance paracellular delivery of insulin across intestinal caco-2 cell monolayers. Pharm. Res. 20, 110–116 (2003)
11. Decuzzi, P., Lee, S., Bhushan, B., Ferrari, M.: A theoretical model for the margination of particles within blood vessels. Ann. Biomed. Eng. 33, 179–190 (2005)
12. Decuzzi, P., Ferrari, M.: Design maps for nanoparticles targeting the diseased microvasculature. Biomaterials 29, 377–384 (2008)
13. Gentile, F., Ferrari, M., Decuzzi, P.: The transport of nanoparticles in blood vessels: The effect of vessel permeability and blood rheology. Ann. Biomed. Eng. 36, 254–261 (2008)
14. Serda, R., et al.: Porous silicon particles for imaging and therapy of cancer. In: Kumar, C.S. (ed.) Nanomaterials for the Life Sciences. Nanostructured Oxides of Nanomaterials for the Life Sciences, vol. 2, p. 359. Wiley-VCH, Weinheim (2009)
15. Martin, F., et al.: Acute toxicity of intravenously administered microfabricated silicon dioxide drug delivery particles in mice: Preliminary findings. Drugs R D 6, 71 (2005)
16. Tasciotti, E., et al.: Mesoporous silicon particles as a multistage delivery system for imaging and therapeutic applications. Nature 3, 151–157 (2008)
17. Gentile, F., et al.: The effect of shape on the margination dynamics of non-neutrally buoyant particles in two-dimensional shear flows. J. Biomech. (2008)
18. Bouaynaya, N., Schonfeld, D.: Theoretical foundations of spatially-variant mathematical morphology part II: Gray-level images. IEEE Trans. Pattern Analysis Machine Intelligence 30, 837–850 (2008)
19. Soille, P.: Morphological Image Analysis: Principles and Applications, 2nd edn. Springer, Heidelberg (2007)

Extraction of Cardiac Motion Using Scale-Space Features Points and Gauged Reconstruction

Alessandro Becciu, Bart J. Janssen, Hans van Assen, Luc Florack,
Vivian Roode, and Bart M. ter Haar Romeny

Eindhoven University of Technology, Eindhoven 5600 MB, The Netherlands
A.Becciu@tue.nl

Abstract. Motion estimation is an important topic in medical image analysis. The investigation and quantification of, e.g., the cardiac movement is important for assessment of cardiac abnormalities and to get an indication of response to therapy. In this paper we present a new aperture problem-free method to track cardiac motion from 2-dimensional MR tagged images and corresponding sine-phase images. Tracking is achieved by following the movement of scale-space critical points such as maxima, minima and saddles. Reconstruction of dense velocity field is carried out by minimizing an energy functional with regularization term influenced by covariant derivatives gauged by a prior assumption.

MR tags deform along with the tissue, a combination of MR tagged images and sine-phase images was employed to produce a regular grid from which the scale-space critical points were retrieved. Experiments were carried out on real image data, and on artificial phantom data from which the ground truth is known. A comparison between our new method and a similar technique based on homogeneous diffusion regularization and standard derivatives shows increase in performance. Qualitative and quantitative evaluation emphasize the reliability of dense motion field allowing further analysis of deformation and torsion of the cardiac wall.

1 Introduction

In modern society cardiac diseases have emerged as the major cause of death in developed countries [1]. Characterizing the heart's behavior, such as acquiring information from extraction and quantification of cardiac motion, can help in formulating early diagnosis and/or suggesting therapy treatments. Among the available techniques, optic flow of tagged MR acquisitions is a non-invasive method that can be employed to retrieve cardiac movement. Optic flow provides information about the displacement field between two consecutive frames, that is, it measures the apparent motion of moving patterns in image sequences. In several optic flow methods it is assumed that brightness does not change along the displacement field and the motion is estimated by solving the so-called Optic Flow Constraint Equation (OFCE):

$$L_x u + L_y v + L_t = 0 \qquad (1)$$

where $L(x, y, t) : \mathbb{R}^3 \to \mathbb{R}$ is an image sequence, L_x, L_y, L_t are the spatiotemporal derivatives, $u(x, y, t), v(x, y, t) : \mathbb{R}^3 \to \mathbb{R}$ are unknown velocity vectors and x, y and t are the spatial and temporal coordinates respectively. Equation (1) is ill-posed since its solution is not unique, due to the unknown velocities u and v. This has been referred to as the "aperture problem". In order to overcome the problem, Horn and Schunck [2] introduced a gradient constraint in the global smoothness term, finding the solution by minimizing an energy functional. Lately results were impressively improved by Bruhn et al. [3], who combined the robustness of local methods with the full density of global techniques using a multigrid approach. Motion estimation also has been performed by means of feature tracking. Thyrion [4] investigated a technique, where the brightness is preserved and the features are driven to the most likely positions by forces. C. C. Cheng and H. T. Li [5] explored optic flow methods where features are extracted taking into account scatter of brightness, edge acquisition and features orientation. A multi-scale approach to equation (1) has been first proposed by Florack et al. [6] and extension to the technique and an application to cardiac MR images has been investigated by Van Assen et al. and Florack and Van Assen [7,8]. In this paper we estimate 2-dimensional cardiac wall motion by employing an optic flow method based on features points such as maxima, minima and saddles. The features have been calculated in the robust scale-space framework, which is inspired by findings from human visual system. Moreover, our technique does not suffer from the aperture problem and is also not dependent on the constant brightness assumption, since we assume that critical points retrieved at tag crossings, such as from the grid pattern described in section 2, still remain critical points after a displacement, even in presence of fading. Therefore, the algorithm can be robustly applied on image sequences, such as the tagged MR images, where the intensity constancy is not preserved. The reconstruction of the velocity field has been carried out by variational methods and the regularization component is described in terms of covariant derivatives biased by a gauge field. This operation adds vector field information from previous frames and allows a better velocity field reconstruction with respect to the one provided by similar techniques which employ standard derivatives. Tests have been carried out on phantom image sequences with a known ground truth and real images from a volunteer. The outcomes emphasize the reliability of the vector field. In section 2 the image data-set and the preprocessing approach used in the experiments is presented. In section 3 the multi-scale framework and the topological number, introduced as a convenient technique for extracting multi-scale features, are explored. In section 4 and 5, we present the calculation of a sparse velocity vector field and the dense flow's reconstruction technique. Finally, in section 6 and 7 the evaluation, the results and the future directions are discussed.

2   Image Data-Set and Preprocessing Approach

Tagging is a method for noninvasive assessment of myocardial motion. An artificial brightness pattern, represented as dark stripes, is superimposed on images by spatially modulating magnetization with the aim to improve the visualization

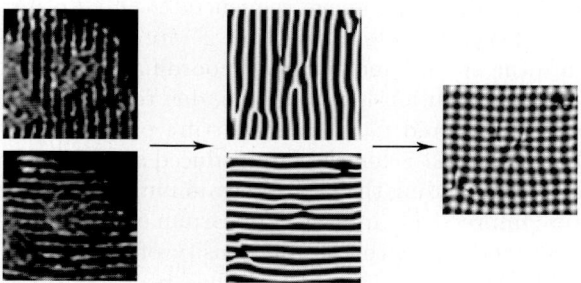

Fig. 1. Column 1: Short axis view of a volunteer's left ventricle. Column 2. The tagged MR images have been filtered in the Fourier domain. Successively inverse Fourier transform and sine function have been applied. Column 3. Image obtained by combination of sine-phase images. The image provides a new pattern from which the feature points have been retrieved.

of intramyocardial motion [9] (Figure 1 column 1). In 1999 Osman et al. [10] introduced the so-called harmonic phase (HARP) technique which overcomes the fading problem by taking into account the spatial information from inverse Fourier transform of filtered images. The experiments have been carried out by employing a similar technique based on Gabor filters [11]. After acquisition of two tagged image series with mutually perpendicular tag lines(Figure 1 column 1), the first harmonic peak has been retained using a band-pass filter in the Fourier domain and the inverse Fourier transform has been applied to the resulting image spectrum. The filtered images present a saw tooth pattern, whose phase varies from 0 to 2π. In the experiments a sine function has been applied to the phase images to avoid spatial discontinuities due to the saw tooth pattern (Figure 1 column 2). A combination of sine phase frames generate a grid from which the critical feature points (maxima, minima, saddles) have been extracted (Figure 1 column 3).

3   Extraction of Scale-Space Critical Points

In the real world, objects are processed by the visual system at different scale levels. Given a static 2-dimensional image $f(x,y) \in \mathbb{L}_2(\mathbb{R}^2)$, its scale space representation $L(x,y;s) \in \mathbb{L}_2(\mathbb{R}^2 \times \mathbb{R}^+)$ is generated by the spatial convolution with a Gaussian kernel $\phi(x,y;s) = \frac{1}{4\pi s}\exp(-\frac{x^2+y^2}{4s})$ such that

$$L(x,y;s) = (f * \phi)(x,y;s) \qquad (2)$$

where x and y are the spatial coordinates, and $s \in \mathbb{R}^+$ denotes the scale. Equation (2) generates a family of a blurred version of the image, where the degree of blurring varies with respect to the scale [12,13,14].

Extraction and classification of critical points is carried out at different scales by computing the so-called topological number [15]. In two-dimensional images the topological number is referred to as the *winding number* and denotes the integrated change of angle of the gradient when traversing a closed curve in a plane. The winding number is always an integer multiple of 2π and its value provides information of the detected critical point. The winding number is zero for regular points, it is $+2\pi$ for extrema, and -2π for saddle points.

4 Sparse Feature Point Velocity Estimation

MR tags have the property to move along with the moving tissue, critical points are located on and between the tag's crossing and therefore also move along with tissue. At a critical point's position the image gradient vanishes. Tag fading, which is a typical artifact in MR images, leaves this property intact, hence critical points satisfy equation (3) over time

$$\nabla L(x(t), y(t), t) = 0 \tag{3}$$

where ∇ represents the spatial gradient and $L(x(t), y(t), t)$ denotes intensity at position x, y and time t. If we differentiate equation (3) with respect to time t and apply the chain rule for implicit functions, we obtain

$$\frac{d}{dt}[\nabla L(x(t), y(t), t)] = \begin{bmatrix} L_{xx}u_i + L_{xy}v_i + L_{xt} \\ L_{yx}u_i + L_{yy}v_i + L_{yt} \end{bmatrix} = 0 \tag{4}$$

where $\frac{d}{dt}$ is the total time derivative, and where we have dropped space-time arguments on the r.h.s. in favor of readability. Equation (4) can also be written as:

$$\begin{bmatrix} u_i \\ v_i \end{bmatrix} = -H^{-1}\frac{\partial \nabla L}{\partial t} \tag{5}$$

where H represents the Hessian matrix of $L(x(t), y(t), t)$. Equation (5) provides the velocity field at critical point positions. The scalars u_i, v_i represent the horizontal and vertical components of a sparse velocity vector at position x_i and y_i, with $i = 1...N$ where N denotes the amount of critical points.

5 Reconstruction of Dense Velocity Field

We aim to reconstruct a dense motion field that provides the most accurate approximation of the true velocity field making use of sparse velocities calculated by equation (5). In literature, examples of velocity field reconstruction as well as image reconstruction techniques based on features can be found in [16,17,18,19]. Given the horizontal and vertical components of the true dense velocity field u_{tf} and v_{tf}, we extract a set of velocity features at scale s_i, such that $u_i = (\phi_i, u_{tf})$ and $v_i = (\phi_i, v_{tf})$, where $(.,.)_{\mathbb{L}_2}$ denotes the $\mathbb{L}_2$ inner product, ϕ_i is the Gaussian kernel $\phi_i(x_i, y_i; s_i) = \frac{1}{4\pi s}\exp(-\frac{(x-x_i)^2+(y-y_i)^2}{4s})$. We look for the functions U and V such that $(\phi_i, U)_{\mathbb{L}_2} \approx (\phi_i, u_{tf})_{\mathbb{L}_2}$ and $(\phi_i, V)_{\mathbb{L}_2} \approx (\phi_i, v_{tf})_{\mathbb{L}_2}$, and minimize the energy functional

$$E(U,V) = \sum_{i=1}^{N} \alpha_i((\phi_i, U)_{\mathbb{L}_2} - u_i)^2 + \alpha_i((\phi_i, V)_{\mathbb{L}_2} - v_i)^2 + \frac{\lambda}{2} \int_{\mathbb{R}^2} \|\nabla U\|^2 + \|\nabla V\|^2 dxdy \tag{6}$$

where $\alpha_i \in \mathbb{R}^+$ is a weighting factor for each feature and the parameter $\lambda \in \mathbb{R}^+$ controls the quality of the approximation. As λ increases, the smoothness degree in vector field increases, reducing the influence of the sparse velocity vectors. The minimization of equation (6) is carried out by solving the corresponding Euler-Lagrange equations. In equation (6) we have chosen a weighting factor $\alpha_i \in \mathbb{R}^+$, such that, $\alpha_i(c_i, \beta) = 1 - \exp(\frac{-\beta}{(c_i-1)^2})$. The weighting factor α ranges from 0 to 1 and depends on an arbitrary parameter $\beta \in \mathbb{R}^+$ and on c_i, the condition number of the Hessian matrix in equation (5). The condition number of a matrix M is defined as $c(M) = \|M^{-1}\| \cdot \|M\|$ [20]. Its value varies from 1 to infinity and gives an indication of sensitivity of a matrix to numerical operations. In case of high condition number, the solution of equation (5) is less accurate and the retrieved velocities are weighted by a small α.

A refinement of equation (6) can be performed by replacing the gradient of the regularization term with a covariant derivative D^{A_h} biased by a gauge field $h \in \mathbb{H}^2(\mathbb{R}^2)$, hence

$$E(U,V) = \sum_{i=1}^{N} \alpha_i((\phi_i, U)_{\mathbb{L}_2} - u_i)^2 + \alpha_i((\phi_i, V)_{\mathbb{L}_2} - v_i)^2 + \frac{\lambda}{2} \int_{\mathbb{R}^2} \|D^{A_h} U\|^2 + \|D^{A_h} V\|^2 dxdy \tag{7}$$

where A_h represents a covector field selected due to gauge field h, such that $D^{A_h} h = (\nabla + A_h) h = 0$. In equation (6) the regularization term selects U and V from all possible approximations to the solution such that they are as smooth as possible (gradient is minimized). In the regularization term of equation (7) gauge field h is used to tune the covariant derivatives, therefore deviations from the gauge field are penalized. This means, that in case the gauge field presents already a vector field that is similar to what we are expecting, the regularization term will take into account this information, and therefore may provide a better reconstruction of U and V. A detailed description of the method for image reconstruction is given in [21].

In the evaluation, we first compute motion field using equation (6) and we use the vector field at frame j as a gauge field h. The gauge field is then applied to equation (7), whose information influences the reconstruction of vector field at frame $j + 1$. In the same way, vector field at frame $j + 1$ computed by equation (7) is used as gauge field h and provides information to construct velocity field at frame $j + 2$. The process is performed for all frames in the sequence.

6 Evaluation

We compare the performance of our optic flow algorithm with reconstruction technique based on covariant derivatives gauged as described in the section 5 with a similar method based on conventional derivatives. The accuracy of the retrieved vector fields has been assessed by analyzing a contracting and expanding artificial phantom of 19 frames and resolution of 99×99 pixels. The artificial phantom presents pattern similar to figure 1 column 3 and deforms according to the analytic function $\{\frac{(x-l)(m-2n\cdot t)}{(l+(m-n\cdot t)t)}, \frac{(y-l)(m-2n\cdot t)}{(l+(m-n\cdot t)t)}\}$, which provides also the vector field's ground truth. The variables x, y, t represent the spatial and temporal coordinates, whereas l, m, n are constant parameters set to 50, 5, and 0.25 respectively. Retrieved vector field and the true vector field of frame 6 are displayed in figure 2, row 1, column 1 and 2 respectively. In the tests we have employed feature points such as maxima, minima and saddles at 4 spatial scales $\sigma = \{1, 1.3, 1.6, 2.\}$ and time scale 1. In order to reduce the influence of velocity outliers during the reconstruction process, sparse velocity vectors extracted using equation (5) have been weighted by employing the weighting function $\alpha(c, \beta)$ dependent on condition number c and parameter β, which we set to 50. Figure 2, row 1 column 3 and 4 illustrates the effects of the weighting function on frame 5 of the real sequence. Moreover, the smoothing parameter λ has been optimized for equation (6) and (7), namely best performance has been achieved with $\lambda = 10^{-2}$ and $\lambda = 10^{-0.5}$ respectively. In order to avoid outlier vectors at the boundaries, the two reconstruction methods have been assessed from frame 5 to frame 9, and 10 pixels distant from the boundaries. Test evaluation has been conducted by comparing extracted flow field with the correspondent ground truth, where accuracy in the results has been described in terms of the so-called *Angular Error*[22].

Fig. 2. Vector Fields. Row 1 Column 1 and 2. Plots depict vector field of frame 6 extracted from artificial phantom and ground truth of frame 6. Vector fields on row 1, column 3 and 4, represent the motion field of frame 5 retrieved from real data before and after the weighting procedure. Weighing factor α penalizes outliers present in row 1 column 1. Row 2 from left to right vector fields of frames 5, 6, 7 and 8, extracted from sequence of sine HARP images of real data. The direction of the velocity vectors is color-encoded, that is, regions in the motion field with the same color show vectors that are pointing in the same direction.

Table 1. Performance of the vector field reconstruction methods based on conventional derivatives and covariant derivatives. The methods have been tested on artificial contracting phantom using maxima, minima and saddles as feature points at spatial scales $\sigma = \{1, 1.3, 1.6, 2.\}$ and time scale 1. Accuracy of the method has been described in terms of average angular error (AAE) and standard deviation (Std) both expressed in degrees. Best performance has been achieved by the employment of covariant derivatives after 1 iteration with AAE = $\{1.88°, 1.53°, 1.33°\}$ for maxima, minima and saddles respectively.

Feature	Maxima		Minima		Saddles	
	AAE	Std	AAE	Std	AAE	Std
Conventional Derivatives	2.35°	1.72°	3.15°	1.47°	1.54°	2.06°
Covariant Derivatives	1.90°	2.11°	1.55°	1.00°	1.34°	1.12°
Covariant Derivatives 1 Iterations	1.88°	1.07°	1.53°	1.00°	1.33°	1.12°

Outcomes, illustrated in table 1, emphasize an increase in performance for our optic flow algorithm with reconstruction technique based on covariant derivatives. Moreover, once we reconstruct the vector field for all image sequence using equation (7), we can employ this new motion field as gauge fields and apply equation (7) again. Outcomes of this process have shown further improvements in the accuracy for our tests. This procedure can be carried out iteratively.

We have also applied our optic flow method on a real sequence of 11 tagged MR images with resolution of 86 × 86 pixels, which depicts the left ventricle of a volunteer in phase of contraction. Filtered vector fields on sine HARP frame 5, 6, 7, and 8 are displayed in figure 2, row 2. Plots are color encoded, where the color gives information of the vector direction.

7 Conclusion

We analyze cardiac motion by employing a new optic flow feature based method with regularization term described by covariant derivatives influenced by a gauge field. We have tested the technique on an artificial contracting and expanding phantom from which we know the ground truth, using maxima, minima and saddles as feature points. Outcomes of comparison with a similar approach, based on conventional derivatives, emphasize high improvements in the accuracy reconstruction provided by our new method. We have also shown that further improvements in the accuracy are achieved, in case the method is repeated one second time with gauge field based on vector field calculated using covariant derivatives. We have applied moreover the technique to a real tagged MR image sequence displaying a heart in phase of contraction. Qualitative results highlight the reliability of the extracted vector field. Finally, in test evaluation we calculate velocity fields at fixed scales, where the most suitable scale is chosen according to the performance with respect to the ground truth. However, deformations of the cardiac walls differs in different regions, therefore feature belonging to two different regions may present best performance at different scales. In future experiments, we will select scales according to the performance of each singular feature, which may provide a better reconstruction of the vector field.

References

1. American Heart Association Statistics Committee and Stroke Statistics Subcommittee: Heart disease and stroke statistics 2009 update. Circulation 119, 480–486 (2009)
2. Horn, B.K.P., Shunck, B.G.: Determining optical flow. AI 17, 185–203 (1981)
3. Bruhn, A., Weickert, J., Kohlberger, T., Schnoerr, C.: A multigrid platform for real-time motion computation with discontinuity-preserving variational methods. IJCV 70(3), 257–277 (2006)
4. Thirion, J.P.: Image matching as a diffusion process: an analogy with Maxwell's demons. Medical Image Analysis 2(3), 243–260 (1998)
5. Cheng, C.C., Li, H.T.: Feature-based optical flow computation. IJIT 12(7), 82–90 (2006)
6. Florack, L.M.J., Niessen, W., Nielsen, M.: The intrinsic structure of optic flow incorporating measurements of duality. IJCV 27(3), 263–286 (1998)
7. van Assen, H.C., Florack, L.M.J., Suinesiaputra, A., Westenberg, J.J.M., ter Haar Romeny, B.M.: Purely evidence based multi-scale cardiac tracking using optic flow. In: MICCAI 2007 workshop on CBM II, pp. 84–93 (2007)
8. Florack, L.M.J., van Assen, H.C.: Dense multiscale motion extraction from cardiac cine MR tagging using HARP technology. In: ICCV workshop on MMBIA (2007)
9. Zerhouni, E.A., Parish, D.M., Rogers, W.J., Yang, A., Sapiro, E.P.: Human heart: Tagging with MR imaging a method for noninvasive assessment of myocardial motion. Radiology 169(1), 59–63 (1988)
10. Osman, N.F., McVeigh, W.S., Prince, J.L.: Cardiac motion tracking using cine harmonic phase (harp) magnetic resonance imaging. Magnetic Resonance in Medicine 42(6), 1048–1060 (1999)
11. Gabor, D.: Theory of communication. J. IEE 93(26), 429–457 (1946)
12. Koenderink, J.J.: The structure of images. Biol. Cybern. 50, 363–370 (1984)
13. ter Haar Romeny, B.M.: Front-End Vision and Multi- Scale Image Analysis: Multiscale Computer Vision Theory and Applications, written in Mathematica. Computational Imaging and Vision. Kluwer Academic Publishers, Dordrecht (2003)
14. Florack, L.M.J.: Image Structure. Computational Imaging and Vision. Kluwer Academic Publishers, Dordrecht (1997)
15. Staal, J., Kalitzin, S., ter Haar Romeny, B.M., Viergever, M.: Detection of critical structures in scale space. In: Nielsen, M., Johansen, P., Fogh Olsen, O., Weickert, J. (eds.) Scale-Space 1999. LNCS, vol. 1682, pp. 105–116. Springer, Heidelberg (1999)
16. Florack, L.M.J., Janssen, B.J., Kanters, F.M.W., Duits, R.: Towards a new paradigm for motion extraction. In: Campilho, A., Kamel, M.S. (eds.) ICIAR 2006. LNCS, vol. 4141, pp. 743–754. Springer, Heidelberg (2006)
17. Janssen, B.J., Florack, L.M.J., Duits, R., ter Haar Romeny, B.M.: Optic flow from multi-scale dynamic anchor point attributes. In: Campilho, A., Kamel, M.S. (eds.) ICIAR 2006. LNCS, vol. 4141, pp. 767–779. Springer, Heidelberg (2006)
18. Janssen, B.J., Kanters, F.M.W., Duits, R., Florack, L.M.J., ter Haar Romeny, B.M.: A linear image reconstruction framework based on sobolev type inner products. IJCV 70(3), 231–240 (2006)
19. Lillholm, M., Nielsen, M., Griffin, L.D.: Feature-based image analysis. IJCV 52(2/3), 73–95 (2003)
20. Numerical Methods for the STEM Undergraduate. Course Notes
21. Janssen, B.J., Duits, R., Florack, L.M.J.: Coarse-to-fine image reconstruction based on weighted differential features and background gauge fields. LNCS, vol. 5567, pp. 377–388. Springer, Heidelberg (2009)
22. Barron, J.L., Fleet, D.J., Beauchemin, S.: Performance of optical flow techniques. IJCV 12(1), 43–77 (1994)

A Non-Local Fuzzy Segmentation Method: Application to Brain MRI

Benoît Caldairou[1], François Rousseau[1], Nicolas Passat[1], Piotr Habas[2], Colin Studholme[2], and Christian Heinrich[1]

[1] LSIIT, UMR 7005 CNRS-Université de Strasbourg, Illkirch, 67412 France
[2] Biomedical Image Computing Group, University of California San Francisco, San Francisco, CA 94143, USA

Abstract. The Fuzzy C-Means algorithm is a widely used and flexible approach for brain tissue segmentation from 3D MRI. Despite its recent enrichment by addition of a spatial dependency to its formulation, it remains quite sensitive to noise. In order to improve its reliability in noisy contexts, we propose a way to select the most suitable example regions for regularisation. This approach inspired by the Non-Local Mean strategy used in image restoration is based on the computation of weights modelling the grey-level similarity between the neighbourhoods being compared. Experiments were performed on MRI data and results illustrate the usefulness of the approach in the context of brain tissue classification.

Keywords: fuzzy clustering, regularisation, non-local processing, image segmentation, MRI.

1 Introduction

Segmentation methods of brain MRI can be categorised into 3 groups: classification methods, region-based methods and boundary-based methods. A very popular one is the K-mean algorithm which has been extended to fuzzy segmentation by Pham *et al.* in [10]. This so called Fuzzy C-Means (FCM) clustering algorithm is a powerful tool for MRI analysis since it authorises voxels to belong to several clusters with varying degrees of membership. Due to its flexibility, this segmentation framework has been intensively extended, for instance by including topological properties [2], DTI handling [1] or prior knowledge.

A main drawback of the standard FCM algorithm remains its sensitivity to noise in medical images. Many pixel-based regularisation term have been proposed such as Tikhonov regularisation [12], Markov Random Field (MRF), *a priori* image model or variational approaches. Inspired by works developed on MRF basics, Pham has proposed [9] a spatial model to improve the robustness to noise of FCM. However, the use of such pixel based regularisation terms assumes a specific image model: for instance, variational approaches can be based on the hypothesis that images are made of smooth regions separated by sharp edges.

Recently, a non-local framework has been proposed to handle more efficiently repetitive structures and textures, for denoising purpose [4] or inverse problems [3,8,11]. In this work, we propose to introduce this non-local framework into the regularisation term of the FCM algorithm.

The sequel of this article is organised as follows. In Section 2, we present the segmentation problem and provide a short overview about FCM and regularisation. Section 3 details the new non-local approach for image segmentation. In Section 4, results obtained on the Brainweb database [5] are presented. Finally, Section 5 discusses these results and brings up to further work.

2 Background

2.1 Fuzzy C-Means (FCM)

The basics of this algorithm have been presented in [10] and make the segmentation equivalent to the minimisation of an energy function:

$$J_{FCM} = \sum_{j \in \Omega} \sum_{k=1}^{C} u_{jk}^q \|\mathbf{y}_j - \mathbf{v}_k\|_2^2. \quad (1)$$

This formulation is used to perform a C-classes segmentation. The parameter u_{jk} represents the membership of the kth class into the jth voxel of the image, the parameter q controls the "fuzziness" of the segmentation (if q gets close to 1, the segmentation becomes more crisp and close to a binary result), $\mathbf{v}_k$ represents the centroid of the kth component and $\mathbf{y}_j$ represents the grey-level of the jth voxel of the image. $\|\mathbf{y}_j - \mathbf{v}_k\|_2$ represents the Euclidean distance between the voxel's grey level and the considered centroid. The proportions are constrained so that: $\sum_{k=1}^{C} u_{jk} = 1$.

Although this method has a fast convergence and provides reliable results in a convenient environment (low level of noise), the performance of this approach strongly decreases for noisy images. In such cases, anatomically hazardous structures may appear, for instance grey matter voxels among white matter volumes. However, FCM has shown to be easily extended and several approaches have been proposed in order to improve the robustness of FCM by introducing the idea of regularisation into the segmentation framework.

2.2 Regularisation

Regularisation is a classic method in inverse problem to determine the most accurate solution among many possible ones [12]. It introduces constraints to eliminate irrelevant solutions. In particular, Pham *et al.* in [9] added a regularisation term in Equation (1) to penalise unlikely configurations that can be met in the image. They called this method: Robust Fuzzy C-Means Algorithm (RFCM). The expression of the obtained energy function is then:

$$J_{RFCM} = \sum_{j \in \Omega} \sum_{k=1}^{C} u_{jk}^q \|\mathbf{y}_j - \mathbf{v}_k\|_2^2 + \frac{\beta}{2} \sum_{j \in \Omega} \sum_{k=1}^{C} u_{jk}^q \sum_{l \in N_j} \sum_{m \in M_k} u_{lm}^q \quad (2)$$

where N_j is the set of the neighbours of voxel j and $M_k = \{1,\ldots,C\}\backslash\{k\}$. The new penalty term is minimised when the membership value for a particular class is large and the membership values for the other classes at neighbouring pixels are small (and *vice versa*).

The parameter β controls the trade-off between the data-term and the smoothing term. Note that if $\beta = 0$, we retrieve the classic FCM algorithm without any regularisation term. If $\beta > 0$, the dependency on the neighbours causes u_{jk} to be large when the neighbouring membership values of the other classes are small. The result is a smoothing effect that causes neighbouring membership values of a class to be negatively correlated with the membership values of the other classes. In [9], Pham *et al.* have proposed to estimate β using cross-validation to obtain near-optimal performances, and they worked with a neighbourhood (N_j) composed of the points 6-adjacent to the current point j. In the work proposed hereafter, we focus on the use of a larger weighted neighbourhood relying on a non-local framework.

3 Non-Local Regularisation

3.1 Non-Local Approach

The Non-Local (NL) Regularisation is a strategy that has been proposed first as a denoising tool [4] and named as NL Mean denoising. Basically, it tries to take advantage of the redundancy of any natural image, broadly speaking a small neighbourhood around a voxel may match neighbourhoods around other voxels of the same image.

The non-local framework proposed by Buades *et al.* [4] relies on a weighted graph w that links together voxels over the image domain. The computation of this graph w is based on the similarity between neighbourhoods of voxels (see illustration in Fig. 1).

In the sequel, we will call such a neighbourhood a *patch* and denote the patch around voxel j as P_j. The similarity of two voxels is defined as the similarity

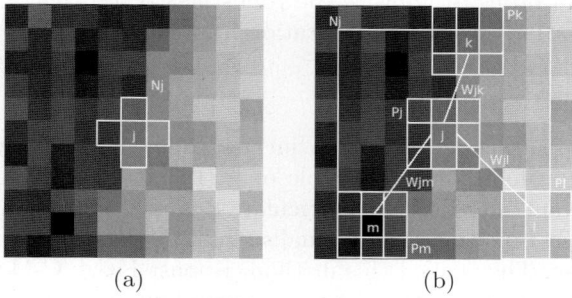

(a)   (b)

Fig. 1. Comparison of the RFCM [9] (a) and the NL approach (b). In this example, the area around the voxel j is more similar to the one of voxel k than the one of voxels m and l. Therefore, the weight w_{jk} will be higher than the weights w_{jm} and w_{jl}.

of the grey-levels contained into P_i and P_j. This similarity can be computed as a Gaussian weighted Euclidean distance, but it has been shown that a simple Euclidean distance is reliable enough [6]. The weight for the voxels i and j is defined as follows:

$$w_{ij} = \frac{1}{Z_i} e^{-\frac{\|\mathbf{y}(P_i)-\mathbf{y}(P_j)\|_2^2}{h^2}} \qquad (3)$$

where Z_i is a normalisation constant and h is a smoothing parameter. The distance between patches is defined as follows:

$$\|\mathbf{y}(P_i) - \mathbf{y}(P_j)\|_2^2 = \sum_{p=1}^{|P_i|} (y^{(p)}(P_i) - y^{(p)}(P_j))^2 \qquad (4)$$

where $\mathbf{y}(P_i)$ is the vector containing the grey-levels of the neighbourhood and $y^{(p)}(P_i)$ is the pth component of this vector.

Note that it is possible to set the parameter h automatically [6] by setting: $h^2 = 2\alpha\sigma^2|P_i|$. The parameter σ^2, namely the standard deviation of the noise, can be computed directly from the image. If the noise in the image is Gaussian, we can set the parameter α to 1 [6]. Otherwise, it can be adjusted to get a more accurate result.

The NL Regularisation approach has already been successfully applied to different kinds of image processing problems. Mignotte [8] used this procedure to constrain a deconvolution process, Bougleux et al. [3] integrated it into the resolution of inverse problems and Rousseau [11] applied it for super-resolution reconstruction techniques.

3.2 Non-Local Fuzzy C-Means Algorithm (NL-FCM)

The key point of the NL approach is the capacity to handle a large neighbourhood without prior knowledge. We show in this work that such methodology can be easily introduced into the FCM framework. We investigate larger neighbourhoods to provide more information for the regularisation. Moreover, the underlying assumption is that voxels who have similar patches in the research area belong to the same tissue as shown in Fig. 1(b). We propose to define a NL version of FCM as follows:

$$J_{NL-FCM} = \sum_{j\in\Omega}\sum_{k=1}^{C} u_{jk}^q \|\mathbf{y}_j - \mathbf{v}_k\|^2 + \frac{\beta}{2}\sum_{j\in\Omega}\sum_{k=1}^{C} u_{jk}^q \sum_{l\in N_j} w_{jl} \sum_{m\in M_k} u_{lm}^q . \qquad (5)$$

Compared to Equation (2), a weight parameter is introduced in order to automatically balance the influence of voxels in the neighbourhood N_j. Note also that contrary to Pham et al. [9] where N_j was a six-neighbourhood system, we investigate in this work larger neighbourhood systems such as the ones used in non-local denoising approach of Buades et al. [4].

The regularisation term of the energy function defined in Equation (5) takes into account the image content in an adaptive and flexible manner to smooth the current segmentation map. In other words, if two voxels neighbourhoods are similar, there might be a chance that they belong to the same tissue and so, the weight w_{jl} increases. Conversely, if two voxels in the original image are quite different, it is normal to decrease the influence of the regularisation term since there is a lower probability that this voxel might have a good influence on the classification of the current one.

The proposed method (and the other ones considered for validation: FCM and RFCM) were optimised through Pham's way [9]: we used the same analytical expressions for the calculation of the centroids and of the membership functions.

4 Results

4.1 Influence of the Non-Local Parameters (α, N_j)

Experiments have been carried out on simulated brain MRI images provided by the Brainweb database [5]. Notice that we perform a 3-class segmentation (Cerebro Spinal Fluid (CSF), Grey Matter (GM), White Matter (WM)) on a T1-weighted image corrupted by a 9 % Rician noise (characteristic from MRI images) [7].

The Brainweb ground truth is used to assess the influence of parameters (α,N_j) of the proposed non-local method. In order to quantify the quality of the segmentation results, we use the following overlap measure:

$$KI = \frac{2.TP}{2.TP + FP + FN} \quad (6)$$

where TP is the amount of true positives, FP is the amount of false positives and FN, the amount of false negatives.

In this work, N_j is considered as a cubic neighbourhood. Results for different sizes of N_j (from $3 \times 3 \times 3$ up to $13 \times 13 \times 13$ voxels) are stated in Fig. 3(a). These experiments emphasise that considering extended neighbourhoods is a way to improve the segmentation results. Nevertheless, increasing N_j above a $5 \times 5 \times 5$ size does not refine the segmentation results and slows the computation down. Therefore, we chose to run the validations in subsection 4.2 with a $5 \times 5 \times 5$ neighbourhood.

We have also investigated the influence of the smoothing parameter α (defined in Section 3.1) on the segmentation results. Fig. 3(b) shows that, in agreement with Buades *et al.* [4], values of α around 1 provide the best results. Moreover, the graph shows that the algorithm is not sensitive to this parameter if its value is set slightly above 1 (α is set to 1.1 for the validations in subsection 4.2).

4.2 Evaluation of the Contribution of the Non-Local Framework

To evaluate the contribution of the non-local framework to the efficiency of the segmentation process, we have also compared the following versions of FCM:

1. classic FCM [10];
2. RFCM [9];

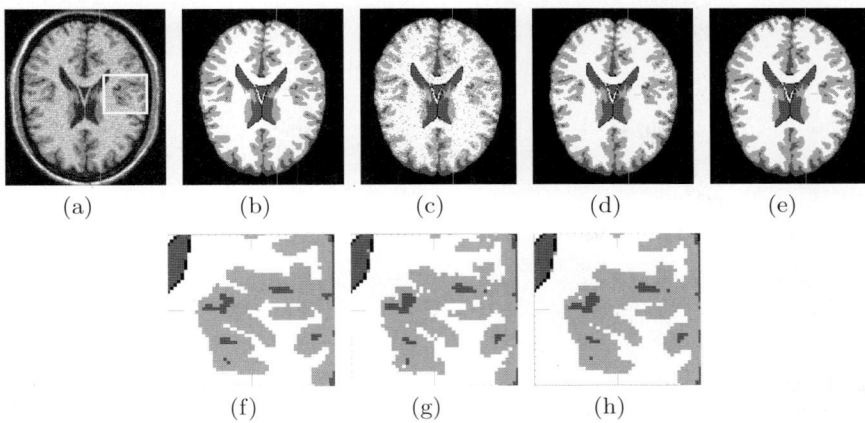

Fig. 2. Results of segmentations using a T1-weighted image with a 9 % Rician noise. (a) Original image with zoom area, (b) Brainweb's ground truth, (c) simple FCM segmentation, (d) RFCM segmentation, (e) NL-FCM segmentation, (f) zoom on Brainweb's ground truth, (g) zoom on RFCM segmentation, (h) zoom on NLFCM segmentation.

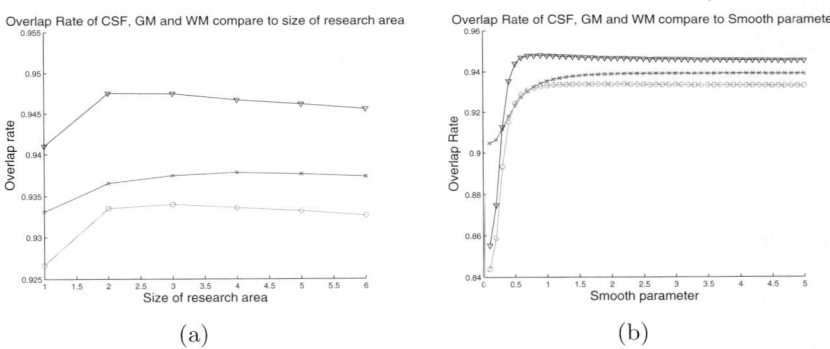

Fig. 3. Influence of the different parameters. Application on a Brainweb T1-weighted image with 9 % Rician noise. (a) Overlap rate to the Size of Search Area N_j and (b) overlap rate to Smooth parameter α. Legend: GM ($\circ$), WM ($\triangledown$), CSF ($\times$).

Table 1. Application of different segmentations on a Brainweb T1-weighted image with a 9 % Rician noise. Comparison of the different overlap rates for CSF, GM and WM.

Methods	CSF	GM	WM
Classic FCM [10]	90.4635	84.3567	85.4812
RFCM without weights [9]	92.0914	91.1193	92.9095
RFCM with weights	92.7614	91.0874	92.4898
NL-FCM without weights	92.2247	92.2154	94.1175
NL-FCM with weights	93.6307	93.3486	94.7661

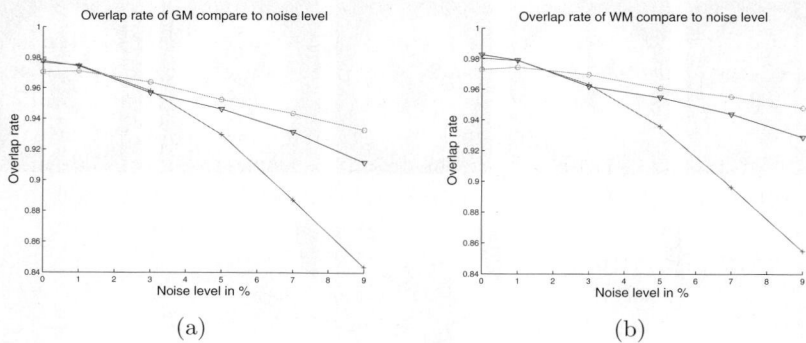

Fig. 4. Application of different techniques on the same Brainweb T1-weighted image with different noise level. (a) Overlap rate of GM, (b) overlap rate of WM. Legend: NL-FCM (∘), RFCM (▽) [9], FCM (+) [10].

3. RFCM with adaptive weights;
4. NL-FCM with fixed weights; and
5. NL-FCM with adaptive weights.

The results are reported in Table 1. The NL Regularisation approach improves the segmentation results with respect to classic FCM and RFCM. The comparison between RFCM and NL-FCM without weights shows that using a larger neighbourhood leads to significant improvements especially for GM and WM (approx. 1 %). Moreover, considering extended neighbourhood, introducing NL approach results in a better overlap rate.

Fig. 2 provides a visual insight of these improvements, especially on GM and CSF. This may be due to the low contrast between CSF and GM on a noisy image which can however be correctly handled by the NL regularisation framework. In addition, we observe that NL-FCM results resolve fine structure more clearly such as the borders between ventricles and GM, and around cortical sulci as shown by the zooms done on RFCM segmentation in Fig. 2(g) and on NL-FCM segmentation in Fig. 2(h) compared to the ground truth in Fig. 2(f).

We carried out complementary experiments to determine the robustness to noise for classic FCM [10], RFCM [9] and NL-FCM with Brainweb T1-weighted images with varying noise levels (see Fig. 4). It can be seen that NL-FCM begins to emerge as a strong approach at noise levels of 3 % and above, and becomes more accurate compared to RFCM approach at a 5 % noise.

5 Conclusion

In this article, an extension of the Robust Fuzzy C-Means Algorithm method [9] has been proposed, by introducing a non-local approach in the regularisation term, and by considering adaptive (*i.e.* possibly large) neighbourhoods for computing this term. The method depends on parameters which do not require a very fine setting. Experiments performed on several noisy brain MR images (up to 9 % Rician noise) from the Brainweb database emphasise the usefulness of this extension. Additional experiments are also needed to evaluate how the regularisation strength is related to the underlying spatial resolution of the reconstructed imaging data. Overall, this new approach may be particularly useful in more challenging imaging applications such as those limited by the imaging time, for example in imaging the moving human foetus.

Acknowledgement

The research leading to these results has received funding from the European Research Council under the European Community's Seventh Framework Programme (FP7/2007-2013 Grant Agreement no. 207667). This work is also funded by NIH Grant R01 NS055064 and a CNRS grant for collaboration between LSIIT and BICG.

References

1. Awate, S.P., Zhang, H., Gee, J.C.: A fuzzy, non parametric segmentation framework for DTI and MRI analysis: With applications to DTI-tract extraction. IEEE Trans. Med. Imaging 26(11), 1525–1536 (2007)
2. Bazin, P.-L., Pham, D.L.: Topology-preserving tissue classification of magnetic resonance brain images. IEEE Trans. Med. Imaging 26(4), 487–496 (2007)
3. Bougleux, S., Peyré, G., Cohen, L.: Non-local regularization of inverse problems. In: Forsyth, D., Torr, P., Zisserman, A. (eds.) ECCV 2008, Part III. LNCS, vol. 5304, pp. 57–68. Springer, Heidelberg (2008)
4. Buades, A., Coll, B., Morel, J.M.: A review of image denoising algorithms, with a new one. Multiscale Modeling & Simulation 4(2), 490–530 (2005)
5. Cocosco, C.A., Kollokian, V., Kwan, R.K.-S., Evans, A.C.: BrainWeb: Online interface to a 3D MRI simulated brain database. In: HBM 1997, Proceedings. NeuroImage, vol. 5(4 Pt 2), p. S425 (1997)
6. Coupé, P., Yger, P., Prima, S., Hellier, P., Kervrann, C., Barillot, C.: An optimized blockwise nonlocal means denoising filter for 3-D magnetic resonance images. IEEE Trans. Med. Imaging 27(4), 425–441 (2008)
7. Kwan, R.K.-S., Evans, A.C., Pike, G.B.: MRI simulation-based evaluation of image-processing and classification methods. IEEE Trans. Med. Imaging 18(11), 1085–1097 (1999)
8. Mignotte, M.: A non-local regularization strategy for image deconvolution. Pattern Recognition Letters 29(16), 2206–2212 (2008)
9. Pham, D.L.: Spatial models for fuzzy clustering. Computer Vision and Image Understanding 84(2), 285–297 (2001)
10. Pham, D.L., Prince, J.L., Dagher, A.P., Xu, C.: An automated technique for statistical characterization of brain tissues in magnetic resonance imaging. International Journal of Pattern Recognition and Artificial Intelligence 11(8), 1189–1211 (1996)
11. Rousseau, F.: Brain hallucination. In: Forsyth, D., Torr, P., Zisserman, A. (eds.) ECCV 2008, Part I. LNCS, vol. 5302, pp. 497–508. Springer, Heidelberg (2008)
12. Tikhonov, A.N.: Regularization of incorrectly posed problems. Soviet Mathematics. Doklady 4(6), 1624–1627 (1963)

Development of a High Resolution 3D Infant Stomach Model for Surgical Planning

Qaiser Chaudry[1], S. Hussain Raza[1], Jeonggyu Lee[2], Yan Xu[3], Mark Wulkan[4], and May D. Wang[1,5,*]

[1] Department of Electrical and Computer Engineering, Georgia institute of Technology, Atlanta, GA, USA
[2] Department of Computer Science, Georgia institute of Technology, Atlanta, GA, USA
[3] Microsoft Research, Redmond, WA, USA
[4] Division of Pediatric Surgery, Department of Surgery, Emory University School of Medicine
[5] The Wallace H. Coulter Department of Biomedical Engineering, Georgia Institute of Technology and Emory University, Atlanta, GA, USA
maywang@bme.gatech.edu

Abstract. Medical surgical procedures have not changed much during the past century due to the lack of accurate low-cost workbench for testing any new improvement. The increasingly cheaper and powerful computer technologies have made computer-based surgery planning and training feasible. In our work, we have developed an accurate 3D stomach model, which aims to improve the surgical procedure that treats the infant pediatric and neonatal gastro-esophageal reflux disease (GERD). We generate the 3-D infant stomach model based on in vivo computer tomography (CT) scans of an infant. CT is a widely used clinical imaging modality that is cheap, but with low spatial resolution. To improve the model accuracy, we use the high resolution Visible Human Project (VHP) in model building. Next, we add soft muscle material properties to make the 3D model deformable. Then we use virtual reality techniques such as haptic devices to make the 3D stomach model deform upon touching force. This accurate 3D stomach model provides a workbench for testing new GERD treatment surgical procedures. It has the potential to reduce or eliminate the extensive cost associated with animal testing when improving any surgical procedure, and ultimately, to reduce the risk associated with infant GERD surgery.

Keywords: Computer-Based Surgery Planning and Training, Image Guided Surgery, Medical Image Processing, Image Registration, Computer Tomography, Morphing, and Visualization.

1 Introduction

Gastroesophageal reflux disease (GERD) is caused by gastric acid flowing from the stomach into the esophagus. Under healthy conditions, a physiologic barrier called the lower esophageal sphincter (LES) prevents pathologic reflux of stomach contents into the esophagus. GERD is an extremely common disease, affecting between 60 and 70

---

* Corresponding author.

Development of a High Resolution 3D Infant Stomach Model for Surgical Planning   615

million people in the United States [2]. Chronic and extreme cases of GERD in infants can cause failure to thrive and damage the esophagus. When medical management fails, a surgical procedure called a fundoplication is performed. The most common fundoplication is the Nissen fundoplication, in which the fundus of the stomach is wrapped around the lower esophagus 360 degrees.. This procedure is often performed using laparoscopic (minimally-invasive) surgical techniques. The procedure is not perfect and there is a recurrence rate of 10-15%, especially in infants. The surgeons at Emory Children's Hospital would like to improve the long-term results [2] of fundoplication using low risk system for testing first. We have formed an interdisciplinary team to design such system to meet the medical needs.

We first build two 3D stomach models from in vivo CT scan of an infant, and Visible Human Project (VHP) data provided by the National Library of Medicine [1] (see Figures 1 and 2). Because CT has low spatial resolution, the infant stomach model is coarse. Thus, we improve CT model using the surface normal based morphing [5] and interpolation techniques with high resolution VHP model. Next, we use spring-mass system [4] to model stomach muscle deformation, and use virtual reality haptic device to control deformation upon touching force. To study infant stomach 3D deformation properties, we use videos of infant fundoplications.

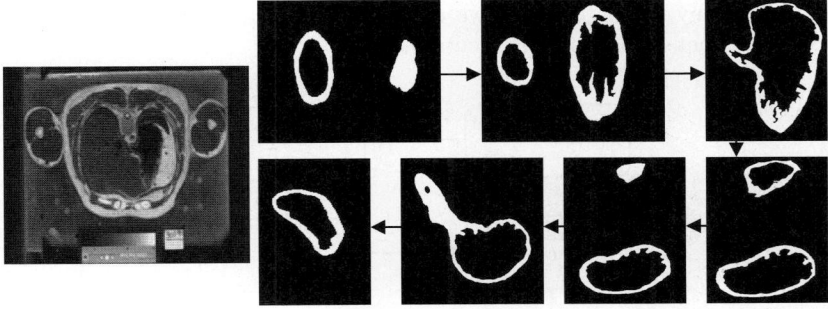

Fig. 1. Few images representing original axial image and segmented stomach walls from VHP. The Figure shows sequence of axial slices from top to bottom (actual data set contains more than 300 slices in stomach region).

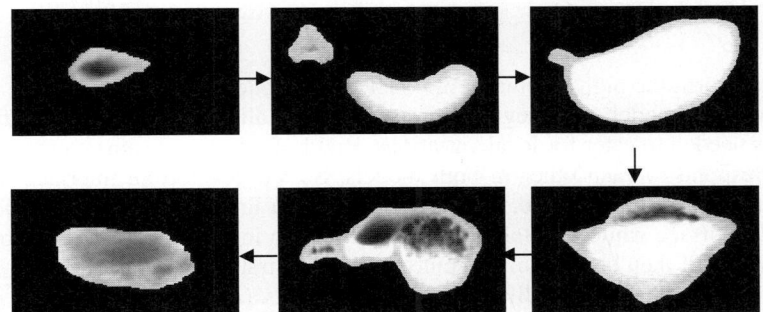

Fig. 2. Segmented stomach from infant CT scan data. The Figure shows sequence of axial slices from top to bottom (actual data set contains 21 slices in stomach region).

2 Materials and Methods

We use two datasets for our work. The first dataset contains 300+ axial thorax anatomical images from VHP [1] male, with a spatial resolution of 2048x1216 pixels in the cryo-section. We use standard image processing techniques to semi-automatically segment these images into binary representation of stomach walls as shown in Figure 1. The second dataset contains approximately 20 CT axial scans of an infant stomach. The normal configuration of CT generates images in DICOM format with a spatial resolution of 512x512 pixels. We also segment these images as shown in Figure 2.

Next, we use VTK [3] to develop 3D geometric models of stomach. That is, we use segmented images to create VTK volume and render the volume with smooth surface shading. Figure 3 shows the workflow of the whole process. Figure 4 shows the difference in model accuracy. The VHP model is created from 300+ images and is highly detailed, while the infant stomach model is created from ~20 CT scan images and suffers from the loss of details. Thus, the infant model cannot be directly used as a workbench to test new GERD surgical procedures, and has to be improved for more accuracy. In infant stomach CT scans, large variations exist among all successive slices. With many details missing, we decide to use high resolution VHP model to improve the infant stomach model accuracy.

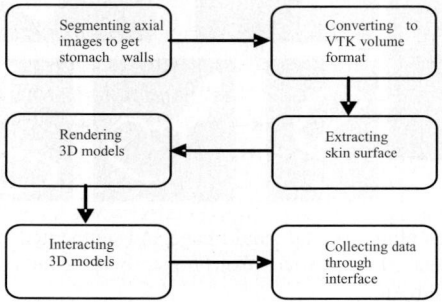

Fig. 3. VTK workflow for creation and interactivity with stomach model

To transform the high resolution 3D VHP male model to best approximate the 3D infant stomach model, rigid registration is needed to align both, and non-rigid registration is needed to cater for local variations. Rigid registration requires a set of points that correspond to each other in both models. So we develop an interface in VTK based on shooting ray method. Shooting ray draws a line from the mouse pointer to the model surface while keeping the camera direction in consideration. The points on the surface will then be used to compute rigid affine transformation. Because the two models originate from two different human subjects (a male adult and an infant), significant morphological differences exist and result in large variation in post-rigid transformation slices.

Development of a High Resolution 3D Infant Stomach Model for Surgical Planning 617

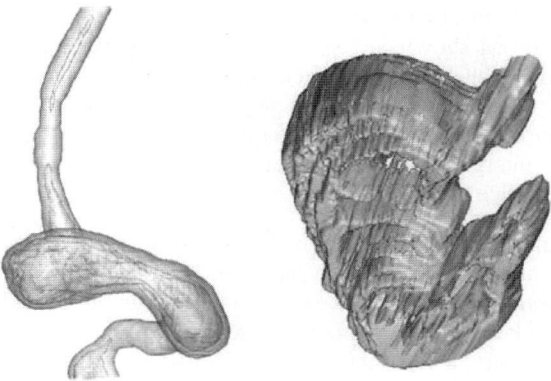

Fig. 4. (a) VHP stomach model; (b) Infant stomach model based on CT scans. VHP data-based model is highly detailed and smooth, and the CT-based model is coarse and distorted due to the lack of data and resolution.

As shown in Figure 5(a) and 5(b), without deforming the model, when we align one part of the model based on the contour centroid in Figure 5(b), the other parts in the model may be totally misaligned as in Figure 5(a). In addition, certain regions (e.g., stomach outer wall) represented by a single contour in one model, may appear to be two in another model as shown in Figure 5(c). Thus, the two models cannot be registered using only rigid and nonrigid registration techniques.

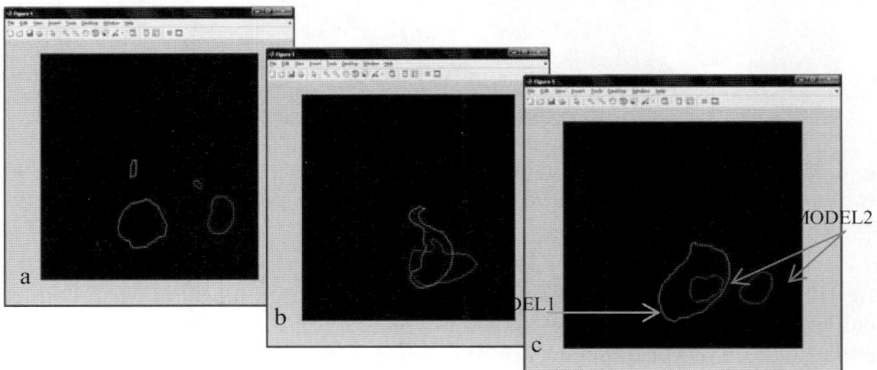

Fig. 5. (a) and (b) Sections of two models after rigid registration (c) Dissimilarity in two models resulting in single versus dual contour representation in certain regions

To address these problems, we have decided to use surface normal based morphing [5] plus information obtained from VHP data, to interpolate images between successive slices of the infant's stomach model. Because an object in one slice may correspond to two objects in the adjacent slice, based on the number of objects in consecutive slices, we classify the morphing into four cases as the following:

Case I: This is the simplest case where consecutive two slices have only one contour with shape variation as shown in Figure 6(a)). We first trace the boundary of both contours and compute the normal at each point on the contour as in Figure 7 in an between the two contours. In the next step, between each pair of corresponding points, we compute a series of points based on constant velocity linear interpolation to generate multiple interpolating images between the slices. This leads to a smooth transition from one slice to the other.

Case II: Considering the two consecutive CT images in Figure 6 (b), the first slice has only one contour and the second slice has two. From the accurate VHP model, we know that one contour slice contains esophagus only, while the other slice also contains main stomach body that just appears in addition to the esophagus. Therefore, we morph the first contour in the first slice to the first contour of the second slice to construct esophagus. For the second contour that does not exist in the first slice, we introduce the centroid of the second contour as the starting point in the first slice. Morphing from this centroid in the first slice to the boundary of the second contour in the second slice generates smooth transition similar to the VHP model.

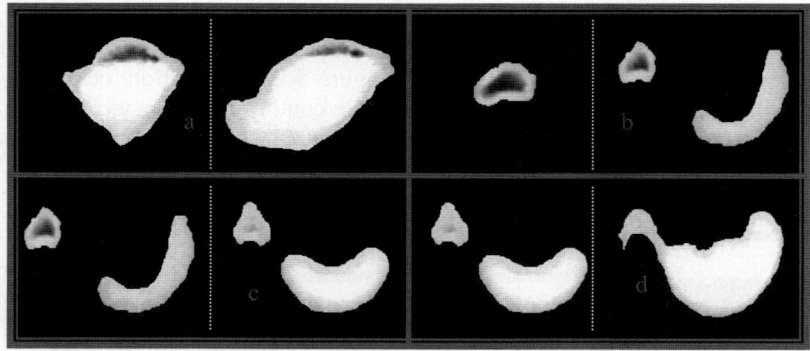

Fig. 6. Different cases for morphing based on number of objects in consecutive slices. Each case is treated separately as explained in relevant section a) Case I, b) Case II c) Case III d) Case IV.

Case III: The third case in developing the infant's stomach model is when both consecutive slices have two contours shown in Figure 6 (c). In this case, both contours of the first slice are morphed to the corresponding contours on the second slice.

Case IV: In the last case shown in Figure 6 (d), the first slice has two contours while the second contains only one contour. The knowledge from the VHP stomach model suggests that the two contours in first slice are going to merge into one contour. Therefore, before morphing, we merge the two contours into one by creating a line between the closest points on both contours as shown in Figure 7. Then we treat them as single contour to compute interpolating images as explained in Case I.

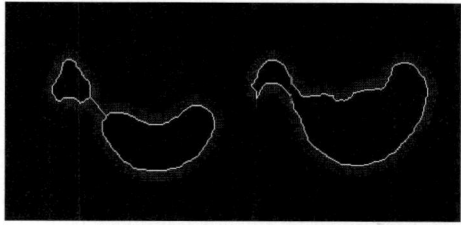

Fig. 7. For Case IV where two contours are mapped to one. The contours are merged first (L), treated as single contour and morphed to single contour (R) Surface normals are shown in blue.

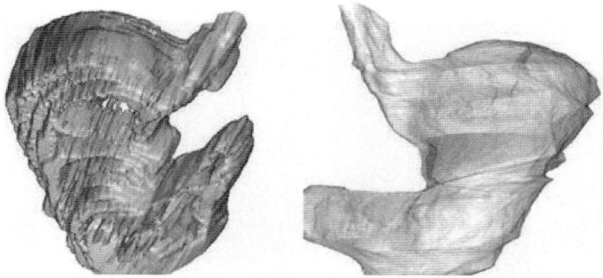

Fig. 8. (a) Original infant's stomach model, (b) Improved infant's stomach model by introducing intermediate slices based on morphing and interpolation

After computing all the interpolating images, we use VTK to visualize the high resolution version of the infant stomach model. Figure 8 shows the new model developed by morphed interpolation and its comparison with the initial infant stomach model.

Having developed a reasonably accurate model of the infant stomach, our next step is to create a 3D interactive model that reflects stomach deformation upon touching force. More specifically, we want to deform the model surface using 3D interactive devices like haptic device. We first step is to construct 3D stomach model by extracting isosurface. In order to achieve real-time interaction while maintaining smooth movement, we reduce the number of vertices in the model to 1,000. We establish and maintain the connectivity information in a vertex table, where each triplet in this table represents a triangle in the model. Using this information, we then construct a simplified mass-spring system for entire model as shown in Figure 9.

Every vertex is connected to its neighboring vertices with spring that has a rest-length measured by Euclidian distance between vertices. The force exerted on one vertex is a sum of forces from all its neighboring vertices. The total force F is determined by: $F = \sum_{i=0}^{n} f_i \quad f_i = k_s(L_c - L_r) + k_d v_s$

where k_s is the spring constant, L_c, L_r is the current and rest length of the spring respectively, k_d is the damping constant and v_s is the velocity of the spring. In our

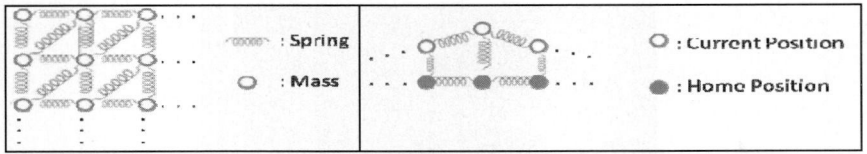

Fig. 9. (a) Mass-Spring model. Every vertex is connected to its neighboring vertices with springs that have rest-length measured by Euclidian distance between vertices. (b) Illustration of the return spring. The vertices moved to "Current Position" at application of force and will return back to "Home Position" when force is removed and return springs length become zero (rest length).

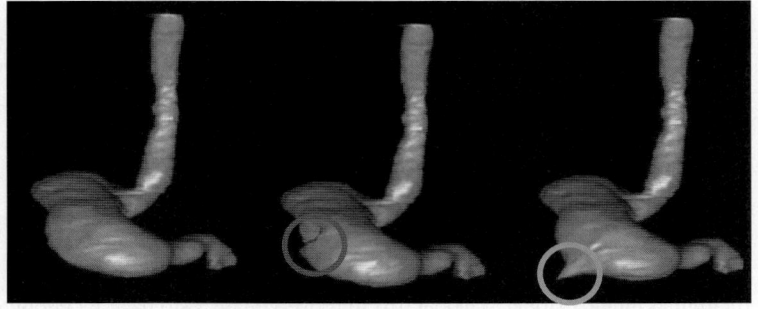

Fig. 10. (a) Stomach model with 1,000 vertices (b) Pushing force is applied and model deformation seen in the region marked red. (c) Pulling force is applied and model deformation seen in the region marked green.

model L_r and L_c are assigned as Euclidian distance between two vertices at the initial frame and the current frame of the simulation respectively. Figure 9 shows the deformable modeling using spring-damper mass system.

During the simulation, to prevent vertices from spreading out, we add "return spring" proposed by [4]. Return springs have "zero" rest length. They make every vertex return to their home position if there is any discrepancy between the home position and the current position. Figure 10 shows the stomach deformation under two different type of touching forces by haptic devices: pushing and pulling.

We execute the simulation on Intel Pentium Core 2 Duo 1.5Ghz CPU, with 2GB of RAM, Windows XP as the OS and Microsoft Visual C++ 8.0 as the development language. The haptic device we use is a Phantom Omni. We have achieved overall 60 fps with a stomach model consisting of 1,000 vertices.

3 Conclusion

In this paper, we have successfully developed a high quality infant stomach model from a low resolution CT scan data by incorporating anatomy information from high resolution VHP stomach data. We have also successfully simulated the stomach deformation upon touch force using a haptic device. By introducing more realistic surface properties to our models, we expect the future models to contain more detailed properties for use in surgical planning.

Acknowledgment

We appreciate Dr. Mitchell Parry, Todd Stokes, and Richard Moffitt from BioMIBLab for extensive discussions and preparation for this manuscript. This research has been supported by grants from National Institutes of Health (Bioengineering Research Partnership R01CA108468, P20GM072069, Center for Cancer Nanotechnology Excellence U54CA119338), Georgia Cancer Coalition (Distinguished Cancer Scholar Award to MDW), Hewlett Packard, and Microsoft Research.

References

1. Ackerman, M.J.: The visible human project. Proceedings of the IEEE 86, 504–511 (1998)
2. Tan, S.W., Wulkan, M.L.: Minimally Invasive Surgical Techniques in Re-operative Surgery for Gastroesophageal Reflux disease in Infants and Children Emory University School of Medicine, pp. 1–13 (2004)
3. http://www.vtk.org/
4. Zhang, S., Gu, L., Huang, P., Xu, J.: Real-time simulation of deformable soft tissue based on mass-spring and medial representation. In: Liu, Y., Jiang, T.-Z., Zhang, C. (eds.) CVBIA 2005. LNCS, vol. 3765, pp. 419–426. Springer, Heidelberg (2005)
5. Kaul, A., Rossignac, J.: Solid-interpolating deformations: construction and animation of PIPs, Techn. Rep. RC 16387 (#72685), IBM Research Division T.J. Watson Research Center Yorktown Heights, New York (December 1990)

Improved Arterial Inner Wall Detection Using Generalized Median Computation

Da-Chuan Cheng[1], Arno Schmidt-Trucksäss[2], Shing-Hong Liu[3], and Xiaoyi Jiang[4]

[1] Department of Radiological Technology, China Medical University, Taiwan
[2] Department of Prevention, Rehabilitation and Sports Medicine, TU München, University Hospital, Germany
[3] Department and Graduate Institute of Computer Science and Information Engineering, Chaoyang University of Technology, Taiwan
[4] Department of Mathematics and Computer Science, University of Münster, Germany

Abstract. In this paper, we propose a novel method for automatic detection of the lumen diameter and intima-media thickness from dynamic B-mode sonographic image sequences with and without plaques. There are two phases in this algorithm. In the first phase a dual dynamic programming (DDP) is applied to detect the far wall IMT and near wall IMT. The general median curves are then calculated. In the second phase, the DDP is applied again using the median curves as the knowledge to obtain a more informed search and to potentially correct errors from the first phase. All results are visually controlled by professional physicians. Based on our experiments, this system can replace the experts' manual work, which is time-consuming and not repeatable.

1 Introduction

Arterial IMT analysis. Common carotid artery intima-media thickness (**CCA-IMT**) measurements have been confirmed to be an early marker of atherosclerosis [1] and have been associated with a higher risk of stroke [2] and myocardial infarction [3]. The non-invasive sonographic examination has demonstrated its potential in early predicting cardiovascular diseases. The IMT is an important index in modern medicine and can be measured either manually [4] or automatically [5,6,7,8,9,10,11].

Arterial elasticity analysis. Moreover, the carotid artery stiffness (**CS**) (or elasticity) is one of the important factors for predicting cardio-vascular (CV) events [12]. The factor CS can be measured via measuring the systolic and diastolic carotid diameter on the distal wall of the CCA, 1 to 2 cm beneath the bifurcation with high-precision sonographic modality. Firstly, carotid distensibility (CDist) is estimated through the variations in arterial cross-sectional area and blood pressure (BP) during systole based on the assumption of a circular lumen. CDist is calculated as CDist=$\Delta A/A \cdot \Delta P$, where A is diastolic lumen area and ΔA is the maximum area change during a systolic-diastolic cycle, and ΔP is the local blood pressure change measured by an applanation tonometer. This pressure change can be approximated by the pressure change measured on the arm in case that the applanation tonometer is deficient. This can be easily done in the routine examination. The CDist can be converted to CS by giving CS=(CDist)$^{-1/2}$ [12].

The automatic methods have the potential in reproducing results and eliminating the strong variations made by manual tracings of different observers. Moreover, the processing time can be considerably reduced. The motivation of this study is to develop a confidential system which is able to detect the intima and adventitia of both near and far artery walls, with or without plaques, automatically even under strong speckle noises using dynamic B-mode sonographic image sequences. This system can identify not only the IMT but also the lumen diameter (**LD**) during systolic and diastolic cycles, from which the artery elasticity can be potentially calculated. Via this system, the dynamic process of carotid artery (**CA**) can be represented by some parameters such as IMT variation, lumen diameter variation, and IMT compression ratio.

This study provides a new technique for detecting the IMT and the LD changes along a section of CCA, which is in general different from previously published works. The proposed system contains two phases. In the first phase, a novel dual dynamic programming (**DDP**) combined with some anatomic knowledge makes the detection more robust against the speckle noises. In the second phase, the generalized median filter is applied and the median curves are calculated, which are fed backwards to the system and the DDP is applied again having the median curves as knowledge to correct its results fully automatically. The proposed scheme has the following steps:

Phase 1

1. Input image I_k; $1 \leq k \leq K$.
2. If $k=1$, manually select a rectangle r_1; else, track r_k using knowledge r_1. (Sec.2.3)
3. Extract feature image g_k from I_k having a rectangle r_k. (Sec.2.4)
4. Apply DDP on g_k to detect the dual curves (intima and adventitia). Output c_k^I and c_k^A. (Sec.2.5)
5. Goto Step 1 until $k > K$.

Phase 2

1. Input c_k^I and c_k^A, $1 \leq k \leq K$, calculate dual median curves M_I and M_A and their corresponding translation t_k^I and t_k^A. (Sec.2.6)
2. Apply DDP onto g_k with guides (M_I, M_A, t_k^I, t_k^A) and output the final intima and adventitia curves for each image I_k; $1 \leq k \leq K$. (Sec.2.7)

The rest part of this paper is organized as follows. In Section 2.1 we address how the image sequences are acquired. The problems of this study are illustrated in Sec. 2.2. The methods are described in Sec. 2.3 to 2.7. Then, results are demonstrated (Sec. 3), discussion and conclusion are given in Sec. 4 and Sec. 5, respectively.

2 Materials and Methods

2.1 Image Acquisition

After at least 15 minutes of rest in the supine position, the ultrasonic examinations of the right and left CCA were performed. An Esaote Mylab 25 ultrasound scanner with a high-resolution and digital beam former was used with a linear variable band 10-13 MHz transducer. The necks of the study subjects were turned slightly to the left or right side. The transducer was positioned at the anterior-lateral side of the neck. The lumen was maximized in the longitudinal plane with an optimal image of the near and the far vessel wall of the CCA or carotid bifurcation. Thus, typical double lines could be seen as the intima-media layer of the artery. Plaques were scanned in a longitudinal and cross-sectional plane showing the highest diameter. At least two heart cycles of every subject were acquired for measurement of the IMT or plaque, respectively. All sequences were stored digitally in the ultrasound device and transferred to a commercially available computer for further image analysis.

2.2 Problem Statement

Figure 1 shows a typical image made by our sonographic modality. The **first** problem is the impact of speckle noises. It is very common that there are speckle noises in the artery lumen. In general, the noises on the near-wall side are stronger than the noises on the far-wall side. It makes the detection of near wall intima much more difficult, comparing to the detection of far wall intima. **Second**, some diseases such as atherosclerotic plaques change the structure of IM complex or the intimal layer. Some changes result in strong echoes such as calcification. When the intima is damaged, there are nearly no echoes on the damaged part. In this case, the adventitia might be exposed on the artery lumen. **Third**, we found the plaque might cause in different echoes in the same places in the dynamic B-mode sonography, which might cause ambiguity in adventitia recognition. This problem is indicated in Fig.1(b), (c), and Fig.2(h). There is an echo near the adventitia, which is absent on the most images. The DDP detects it because it does exist some echoes. However, according to the human beings judgment, since the majority has no such echo so the majority wins. This detection as shown in Fig.1(b) and (c) would be judged to be false. **Finally**, the target we are processing is moving during the image sequence since it is made by dynamic B-mode sonography. Therefore we have to deal with the tracking problem. Fortunately, the target we are tracking does not change its shape in a large scale. We assume that the artery movement is only in longitudinal direction, although there is less movement in horizontal direction. This movement is due to the systolic and diastolic cycle. Moreover, there is no overlapping or shadow problems, which might happen very often in the camera videos. In order to conquer problems listed above, we propose the novel scheme as follows.

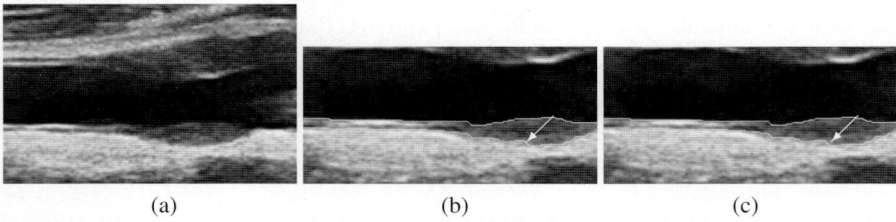

Fig. 1. A typical B-mode sonographic image. (a) A sub-image from a dynamic image sequence. (b) The DDP result superimposed on the sub-image of Frame No. 43. (c) The DDP result on Frame No. 45.

2.3 Artery Movement Tracking

Since the artery is moving during heart cycles in the dynamic B-mode sonography, the artery tracking is an important issue. It is assumed that the artery has 1D movement, i.e., in the vertical direction. This movement is actually an extension of the artery lumen in the heart systolic cycle. In addition, the whole artery might have a shift in the vertical direction in a whole heart cycle. This system needs only one single manual input in the beginning. The user has to select a rectangle area to cover the section area to be measured. However, the artery might move out of this given rectangle in the subsequent images. Therefore, a simple automated tracking algorithm is combined to help the system finding the correct artery position in each image. Due to page limitation, details are omitted here and the readers are referred to [13].

2.4 Image Features

The objective of this study is to detect the intima, adventitia of CA, and plaque outline if any. There are many kinds of methods achieving the same goal. Some used gray level or the gradient of gray level as features [7,11,6]. In this study, we use a simple feature extraction method which is able to detect the intima and adventitia of both near and far wall as well as plaques of CCA in dynamic B-mode sonography [11,14,13]. Here we do not repeat the feature extraction process.

2.5 Dual Dynamic Programming (DDP)

In our previous study [11] we have developed dual dynamic programming (DDP) for IMT detection. Some following works are based on this method which are able to detect the IMT and the lumen diameters [14,13]. Let $g_k(x,y)$ denote the k-th feature image grid of size $M \times N$, where M and N are the number of rows and columns, respectively. Assume the DDP running from left to right in order to find dual curves intima and adventitia of the far wall. The DDP intends to find the global maximum, which is the summation of the feature values on the grid where the dual curves go through. The dual curves can be denoted as a point set $\{(i,y1_i),(i,y2_i)\}, 1 \leq i \leq N$, and its corresponding feature values are $\{g_k(i,y1_i), g_k(i,y2_i)\}$. The cost function finding the k-th dual curves can finally be defined as:

$$J_i(y_1,y_2) = \min_{\substack{j_1,j_2 \in \\ \{-d_r,\cdots,d_r\}}} \{J_{i-1}(y_1+j_1, y_2+j_2) + g_k(i,y_1) \\ + g_k(i,y_2) + \lambda_1|w_i - w_{i-1}| + \lambda_2(|j_1|+|j_2|))\} \quad (1)$$

subject to $d_{min} \leq y_1 + j_1 - y_2 - j_2 \leq d_{max}$,
$d_r \geq |y_{1i} - y_{1\ i-1}|$, $d_r \geq |y_{2i} - y_{2\ i-1}|$ and $2 \leq i \leq N$.

where λ_1 and λ_2 are weighting factors of the curve smoothness. The parameters y_1, y_2 are the short form of y_{1i}, y_{2i}, respectively. All tuples (y_1,y_2) are tested if they fit the constraint given in Eq.(1). The following steps including the initialisation and backwards tracing the dual paths can be found in [13]. The output (y_1^*, y_2^*) which satisfies the global maximization is then redefined to (y_k^A, y_k^I) for following steps.

2.6 Generalized Median Filter

In this section, we address a clinical application of generalized median filter [15] to produce a median curve representing either detected intimal or adventitial curve. Suppose we have a sequence having K images, from each image the intimal layer is detected and represented by c_k^I, where $k = 1, 2, \cdots, K$ denotes the k-th image. Since the curve goes from left to right, the x-coordinate is in an ascending order. The important information is the y-coordinate which is represented by y_k^I of the corresponding curve c_k^I. The goal is to find a generalized median curve M_I which has the minimum error to all these detected curves c_k^I. However, since the artery has movement because the heart cycle, there is another parameter t_k^I representing the translation of the corresponding intimal curve on the k-th image. Therefore, the cost function to find out the generalized median curve of intima is defined as follows:

$$f(M_I, T_I) = \sum_{k=1}^{K} \sum_{i=1}^{N} (y_{ki}^I - y_i^I - t_k^I)^2 \qquad (2)$$

where

$M_I = \{y_i^I | 1 \leq i \leq N\}$ is the median curve, N is the curve length;
y_k^I is the y-coordinate of the k-th curve c_k^I which has n points;
$T_I = \{t_k^I | 1 \leq k \leq K\}$ is the composition of all translations for each curve c_k^I.
Through some derivations, one can easily obtain:

$$t_k^I = \frac{1}{N} \sum_{i=1}^{N} (y_{ki}^I - y_i^I) \qquad (3)$$

$$y_i^I = \frac{1}{K} \sum_{k=1}^{K} (y_{ki}^I - t_k^I) \qquad (4)$$

which can be solved iteratively by an EM algorithm. The calculations of $M_A = \{y_i^A | 1 \leq i \leq N\}$ and T_A for adventitia are similar.

2.7 Dual Dynamic Programming with Guides

Here we briefly describe how DDP uses median curves as guides. Let $g_k(x, y)$ denote the k-th feature image grid as defined in Sec.2.5.

The anatomic knowledge ($y_1 > y_2$, d_{min}, and d_{max}), the guides by the median curves (y^I for intima and y^A for adventitia) and the translations t_k^I and t_k^A are then embedded into the DDP structure. The cost function finding the k-th dual curves can finally be defined as:

$$J_i(y_1, y_2 | y^I, y^A, t_k^I, t_k^A) = \min_{\substack{j_1, j_2 \in \\ \{-d_r, \cdots, d_r\}}} \{J_{i-1}(y_1 + j_1, y_2 + j_2 | y^I, y^A, t_k^I, t_k^A) + g_k(i, y_1)$$
$$+ g_k(i, y_2) + \lambda_1 |w_i - w_{i-1}| + \lambda_2(|j_1| + |j_2|) + \lambda_3(|y_1 + j_1 - (y^A - t_k^A)|$$
$$+ |y_2 + j_2 - (y^I - t_k^I)|)\}$$

subject to $d_{min} \leq y_1 + j_1 - y_2 - j_2 \leq d_{max}$,
$d_r \geq |y_{1i} - y_{1\ i-1}|$, $d_r \geq |y_{2i} - y_{2\ i-1}|$, and $2 \leq i \leq N$. (5)

where λ_1 and λ_2 are weighting factors of the curve smoothness; λ_3 is the weighting for median curves. The parameters y_1, y_2, y^A, and y^I are the short form of y_{1i}, y_{2i}, y_i^A, and y_i^I, respectively. All tuples (y_1, y_2) are tested if they fit the constraint given in Eq.(2.7). The following steps including the initialisation and backwards tracing the dual paths can be found in [13]. The output (y_1^*, y_2^*) which satisfies the global maximization are the adventitia and intima of the corresponding k-th image.

3 Results

Figure 2 demonstrates the IMT and a plaque detection on the far wall. The problem we want to solve in this paper is indicated in Fig. 2(h). There is an echo existing near adventitia in the frames from frame number 40 to 46. We illustrate only frame number 41 and 44 as examples. With single DDP it can detect the adventitia as shown in

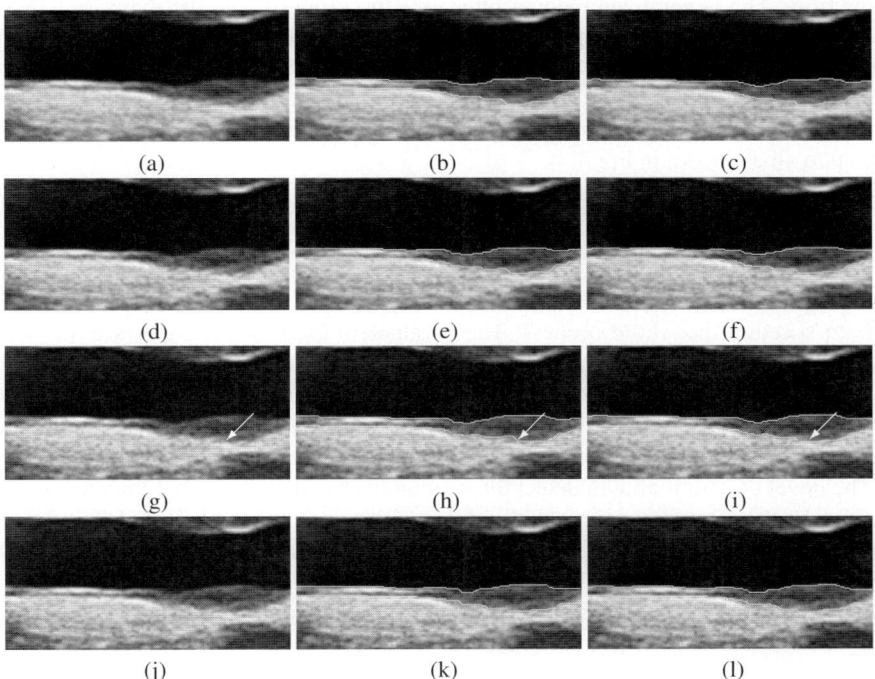

Fig. 2. The far wall IMT and plaque detection: The first column are the raw sub-images; they are frame No. 38, 41, 44, and 47, respectively. The second column are the results of DDP superimposed on the raw sub-images. The third column are the results of proposed scheme superimposed on the raw sub-images.

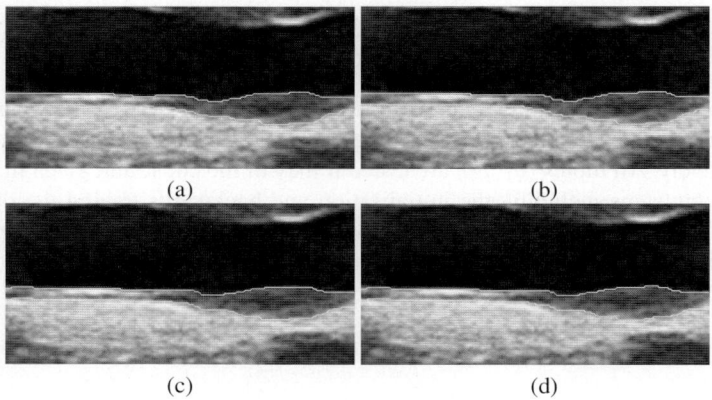

Fig. 3. Results of proposed method: From (a)-(d) are Frame No. 42, 43, 45, and 46

the second column. However, these results are judged to be wrong by an experienced physician. This is because in the rest frames there existed no such an echo, which are in majority. The generalized median filter uses the property that the majority wins so that it can simulate human beings judgment. It obtains the correct results as shown in the third column. Due to page limitation, all results cannot be displayed here. However, the results made by single DDP are similar to Fig.2(e) and (h) from frame No. 40 to 46. The results made by this proposed scheme are similar to Fig.2(f) and (i) from frame No. 40 to 46 as shown in Fig.3.

4 Discussion

The programs are setup on the Matlab platform. Some kernel functions are written in C to speedup the whole process. The parameters used in this paper are: $d_r = 1$, $d_{min} = 4$ in IMT detection, $d_{min} = 20$ in LD detection, $d_{max} = 40$ in IMT detection, $d_{max} = 0.9 \cdot M$ in LD detection, and $\lambda_1 = \lambda_2 = 0.1$, and $\lambda_3 = 0.05$. The computer has Intel Core(TM)2 T5600 CPU with 1.83GHz, 2GB RAM. The computation time for IMT detection is around 1.2 sec for the DDP with guides.

The novel system is able to detect the near and far wall IMT and the lumen diameter of CCA in the B-mode sonographic videos, with and without plaques. Having these results, we can provide physicians the CCA diameter changes during the heart cycle, the compression rate of IMT, the plaque thickness and shape morphology. Furthermore, in the future we are able to build the blood flow model to predict the shear stress on the artery wall, which is a critical index for the vascular diseases.

5 Conclusion

In this paper we propose an intelligent method to detect the near and far wall IMT as well as the LD of CCA in dynamic B-mode sonography, with and without plaques.

Based on the experiments, the detection results are correct and do not need any manual correction. This system is fully automated except it needs an initial rectangle area selected by the user. In the future work, we will explore the relationship between some diseases and the parameters extracted from the dynamic IMT and LD by our system.

Acknowledgment

This work is supported by the National Science Council (NSC), Taiwan, under Grant NSC 97-2218-E-039-001.

References

1. Bonithon Kopp, C., Scarabin, P., Taquet, A., Touboul, P., Malmejac, A., Guize, L.: Risk factors for early carotid atherosclerosis in middle-aged french women. Arterioscler Thromb. 11, 966–972 (1991)
2. O'Leary, D., Polak, J., Kronmal, R., Manolio, T., Burke, G., Wolfson, S.J.: Carotid-artery intima and media thickness as a risk factor for myocardial infarction and stroke in older adults. N Engl. J. Med. 340, 14–22 (1999)
3. Bots, M., Grobbee, D., Hofman, A., Witteman, J.: Common carotid intima-media thickness and risk of acute myocardial infarction. The role of lumen diameter. Stroke 36, 762–767 (2005)
4. Vemmos, K., Tsivgoulis, G., Spengos, K., Papamichael, C., Zakopoulos, N., Daffertshofer, M., Lekakis, J., Mavrikakis, M.: Common carotid artery intima-media thickness in patients with brain infarction and intracerebral haemorrhage. Cerebrovascular Diseases 17, 280–286 (2004)
5. Touboul, P., Elbaz, A., Koller, C., Lucas, C., Adrai, V., Chédru, F., Amarenco, P.: Common carotid artery intima-media thickness and brain infarction, the étude du profil génétique de l'infarctus cérébral (génic), case-control study. Circulation 102, 313–318 (2000)
6. Liang, Q., Wendelhag, I., Wikstrand, J., Gustavsson, T.: A multiscale dynamic programming procedure for boundary detection in ultrasonic artery images. IEEE Trans. on Medical Imaging 19, 127–142 (2000)
7. Cheng, D., Schmidt-Trucksäss, A., Cheng, K., Burkhardt, H.: Using snakes to detect the intimal and adventitial layers of the common carotid artery wall in sonographic images. Computer Methods and Programs in Biomedicine 67, 27–37 (2002)
8. Kanai, H., Hasegawa, H., Ichiki, M., Tezuka, F., Koiwa, Y.: Elasticity imaging of atheroma with transcutaneous ultrasound: Preliminary study. Circulation 107, 3018–3021 (2003)
9. Hasegawa, H., Kanai, H., Koiwa, Y.: Detection of lumen-intima interface of posterior wall for measurement of elasticity of the human carotid artery. IEEE Trans. on Ultrasonics Ferroelectrics & Frequency Control 51(1), 93–109 (2004)
10. Cardinal, M.H.R., Meunier, J., Soulez, G., Maurice, R., Therasse, É., Cloutier, G.: Intravascular ultrasound image segmentation: A three-dimensional fast-marching method based on gray level distributions. IEEE Trans. on Medical Imaging 25(5), 590–601 (2006)
11. Cheng, D.C., Jiang, X.: Detections of arterial wall in sonographic artery images using dual dynamic programming. IEEE Trans. on Information Technology in Biomedicine 12(6), 792–799 (2008)
12. Paini, A., Boutouyrie, P., Calvet, D., Tropeano, A.I., Laloux, B., Laurent, S.: Carotid and aortic stiffness: Determinants of discrepancies. Hypertension (2006), doi:10.1161/01.HYP.0000202052.25238.68

13. Cheng, D.C., Schmidt-Trucksäess, A., Pu, Q., Liu, S.H.: Motion analysis for artery lumen diameter and intima-media thickness of carotid artery on dynamic B-mode sonography. In: International Conference on Mass-Data Analysis of Images and Signals in Medicine, Biotechnology, Chemistry and Food Industry (July 2009) (accepted)
14. Cheng, D.C., Pu, Q., Schmidt-Trucksäess, A., Liu, C.H.: A novel method in detecting CCA lumen diameter and IMT in dynamic B-mode sonography. In: The 13th International Conference on Biomedical Engineering, pp. 734–737 (2008)
15. Wattuya, P., Jiang, X.: A class of generalized median contour problem with exact solution. In: Proc. of 11th Int. Workshop on Structural Pattern Recognition, Hong Kong (2006)

Parcellation of the Auditory Cortex into Landmark–Related Regions of Interest

Karin Engel[1], Klaus Tönnies[1], and André Brechmann[2]

[1] Otto von Guericke University Magdeburg, Germany
[2] Leibniz Institute for Neurobiology Magdeburg, Germany

Abstract. We propose a method for the automated delineation of cortical regions of interest as a basis for the anatomo–functional parcellation of the human auditory cortex using neuroimaging. Our algorithm uses the properties of the cortical surface, and employs a recent hierarchical part–based pattern recognition strategy for a semantically correct labelling of the temporal lobe. The anatomical landmarks are finally combined to obtain an accurate separation and parametrisation of two auditory cortical regions. Experimental results show the good performance of the approach that was automated using simplified atlas information.

1 Introduction

The variability of the sulco–gyral patterns of the human cortex remains a challenging issue in analysing the correspondence between brain anatomy and function, e.g. using anatomical and functional MRI. There is evidence that at least in some regions macro–anatomical landmarks are related to the individual underlying cytoarchitectonic and thus functional organisation of the brain [2,3,1]. Therefore, annotating brain regions of interest (ROI) based on anatomical landmarks is one promising approach to overcome the problem of inter–individual variation. However, the manual definition of ROI is tedious and time–consuming, and the reproducibility in highly variable brain regions, such as the auditory cortex (AC), is not satisfactory [2]. Hence there has recently been great interest within the brain imaging community in developing image analysis methods for identifying anatomical landmarks as a starting point for brain functional mapping. The popular warping methods map individual brains onto a labelled reference brain, or atlas, using image–based features [1,4,5,6], or manually labelled landmarks [7] to drive the registration. Topography–based parcellation methods use graph–based descriptions of an atlas brain and the individual cortices [10,11,12,13,8,9,14] for identifying regions of specific functionality. Even though some of the methods provide good results, the high inter–subject and inter–hemisphere variability in shape, topology and structural configurations of the cortical folds may prevent an automatic and semantically correct parcellation at the desired level of detail [2].

This paper aims at the detailed parcellation of the human auditory cortex. Parcellation of the AC is difficult, since it includes identification and discrimination of specific supratemporal structures, e.g. Heschl's gyrus, from similar shapes within each individual cortical hemisphere. As the cortical surfaces provide too little semantics in terms of curvature, an appropriate parcellation strategy must allow deformable object recognition, and must apply to surface–based representations of the data. In contrast to the brain warping approach and topography–based parcellation methods (e.g. [12]), our method directly employs a model of the variability in the AC folds and their configuration for mapping of cortical regions and their annotations. As shown in [16], anatomical labelling can for now only be partially automated using a contextual pattern recognition method. Hence our algorithm for parcellation relies on a two–stage strategy. First, we render the deformable prototype–based recognition method presented in [16] more application–specific by adding basic atlas information. In contrast to [13], this should not require extensive training, but constrain the search space to a certain portion of the cortical surface containing the temporal lobe, and automate the localisation of the desired anatomical landmarks (Sect. 2.1). Second, from the gyral and sulcal labels we parcellate the AC into the desired landmark–oriented ROI (Sect. 2.2). The local anatomical landmarks further provide an anatomically meaningful parametrisation of the surface–based ROI, which does, in contrast to e.g. [4], not require a warping of the individual surfaces.

2 Method

Our parcellation method uses the properties of the inner cortical surface (grey-white matter boundary) that is represented as a triangular mesh $\mathbf{V} = \{\varphi_k\}$, and exists in the folded and flattened states. The cortical surfaces were obtained from the T1–weighted anatomical MR data sets using segmentations of the white matter, corrected for topological errors and flattened using the software package BrainvoyagerQX (http://www.brainvoyager.com). By using the meshes in their flattened configurations, variability related to the 3D–embedding of the cortical landmarks is eliminated and labelling becomes less complicated (Fig. 1).

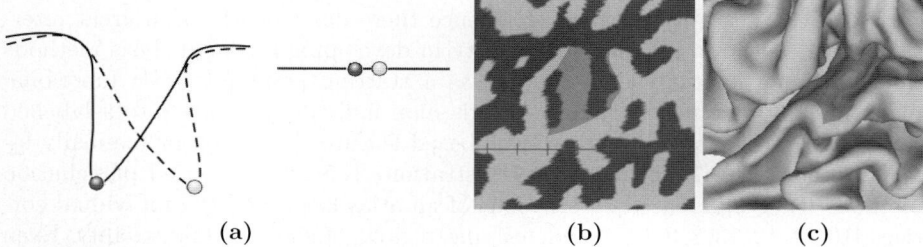

(a) (b) (c)

Fig. 1. Automatic labelling is easier in 2D, because the intrinsic anatomical variability can be estimated by comparing two folds on the planar projection space (a). Figure 1b shows a portion of a cortical flat map, where color indicates gyri (light) and sulci (dark) (cf. Eq. 1). Since each flat map vertex φ_k is associated with its position on the 3D folded cortical surface, assigned labels (b) can be projected back into 3D space (c).

2.1 Anatomical Labelling

An empirical system of landmark–related ROI serves to parcellate the AC into areas with known differential activation [15]. The ROI form adjacent territories in the individual temporal lobes in relation to anatomical landmarks. As depicted in Figures 2b and 2c, territory T1 follows the course of the most anterior transverse gyrus, or Heschl's gyrus (HG), on its anteromedial rim and extends on the lateral aspect of gyrus temporalis superior (lGTS). T2 is centered to Heschl's sulcus (HS) and borders the anterior planum temporale (PT).

Identification of the anatomical landmarks uses an abstract decomposition of the auditory cortical folding pattern into the different gyri and sulci in terms of a hierarchical shape model suitable for part–based object recognition [16]. As illustrated in Figure 2d, the lower level of the shape–structure hierarchy captures the morphology of the single folds, namely HG - which may or may not show a sulcus intermedius (SI) -, Sylvian Fissure (SF) and sulcus temporalis superior (STS), while their structural configurations are represented at the top level. The high inter–subject variability in shape and configuration of the folds of the temporal lobe [17] is accounted for first by combining non–specific morphological Finite Element Models (FEM) of the single folds to construct class–specific AC models that arrange HG (and HG plus SI, resp.) nearly orthogonal to the two surrounding parallel sulci. Second, the variable topology of SF and STS is adopted by using a single line–shaped FEM to represent a sulcus. This morphological shape model shall bridge over possible interruptions and must not follow the highly variable side branches of a specific sulcus.

Automatic labelling of the landmarks. In the individual flat maps is accomplished by combining the hierarchical model of the AC folding pattern with an evolutionary deformable shape search strategy, as described in [18]. At each step of the fit the instances of the morphological shape models are aligned to the top level model by top–down propagating the displacements of specific link nodes. Then by searching for the best values of the desired image features the FEM iteratively adapt to the local conditions in the data. The bottom–up flow

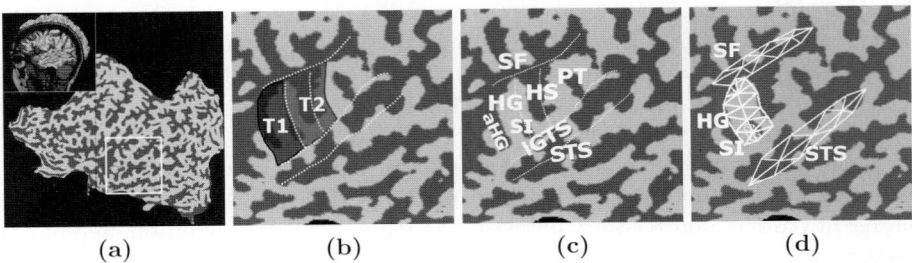

Fig. 2. The segmentation of the landmark–related auditory ROI T1 and T2 utilises the properties of 2D cortical flat maps. The hierarchical shape model (d) represents Heschl's gyrus (HG) as the central part of the auditory folding pattern located in the temporal region (detailed in b-d).

of information between the two levels of the model is implemented using the hierarchy of external forces derived from the underlying curvature maps. Optimisation uses an objective function that incorporates both internal (degree of deformation) and external energies (data mismatch) of the hierarchical model.

The curvature maps $\mathcal{K}(\varphi_k)$ are computed using an operator that separates convex regions (gyri) and concave regions (sulci) in the folded surfaces (Figs. 1b, 1c, Fig. 2). The boolean mean curvature operator relies on the weighted sum of vectors $\mathbf{r}_{ij} = \varphi_j - \varphi_i$ from vertex i to its neighbours $j \in N_i$, i.e. $\kappa(\varphi_i) = \sum_{j \in N_i}(\cot \alpha_j + \cot \beta_j)\mathbf{r}_{ij}$, where α_j and β_j are angles opposite to the edge through vertices i and j w.r.t. its adjacent faces [19], such that

$$\mathcal{K}(\varphi_i) = \begin{cases} 0, \kappa(\varphi_i) \cdot \mathbf{n}_i \leq 0 \text{ (sulcal region)}, \\ 1, \text{otherwise}, \end{cases} \quad (1)$$

where $\mathbf{n}_i$ is the surface normal at φ_i. Since discrete curvature estimates are sensitive to high–frequency details, the cortical regions are separated as desired only for smoothed curvature maps. We employ a discrete approximation of a heat diffusion kernel, whose weights g_σ are calculated based on geodesic inter–vertex distances $\mathbf{d}_{ij} = |\varphi_i - \varphi_j|^2$, approximated by the length of the shortest path between the vertices according to Dijkstra's algorithm. Let N_i^* be the set of φ_i and neighbours φ_j, for which $\mathbf{d}_{ij} < 3\sigma$. Assuming a sufficiently small kernel bandwidth σ (in our case $\sigma = 2mm$) and small inter–vertex distances,

$$g_\sigma(\varphi_i, \varphi_j) = \exp(-\mathbf{d}_{ij}(2\sigma)^{-2})\Big(\sum_{j \in N_i^*} \exp(-\mathbf{d}_{ij}(2\sigma)^{-2})\Big)^{-1}. \quad (2)$$

The discrete convolution $g_\sigma * \mathcal{K}(\varphi_i) = \sum_{j \in N_i^*} g_\sigma(\varphi_i, \varphi_j)\mathcal{K}(\varphi_j)$ is then repeated τ times, e.g. $\tau = 2$, to obtain a smoothed curvature map from which external model forces can be computed by linear filtering.

The Finite Element Method yields an algebraic function that relates the deformed positions of all finite element nodes to the forces acting on the deformable shape model [20],

$$\mathbf{f}_b(t) + \mathbf{f}(t) = \mathbf{K}(\mathbf{x}^t - \mathbf{x}^0). \quad (3)$$

$\mathbf{K}$ encapsulates the stiffness properties as well as the type of mesh and discretisation used, $\mathbf{x}^t$ denotes the nodal positions at time $t \geq 0$ of the simulation, and $\mathbf{f}_b$ represents the body forces. Using a map $\mathcal{D}(\mathcal{K}, \mathbf{x})$ of the distance of each point $\mathbf{x}$ to the closest relevant surface point (at which, e.g., the response to a linear filter is high), the dynamic load is $\mathbf{f}(t) \propto \lambda \mathcal{D}(\mathcal{K}, \mathbf{x}^t), \lambda > 0$ (Figure 3a).

Since we are interested in segmenting the ROI T1 and T2 (Fig. 2b), the labelling of HS, the anteromedial rim of HG (aHG) and lGTS completes our parcellation. HG and STS – which are segmented in the recognition step of our algorithm – define expectation maps for the segmentation of the adjacent HS, aHG and lGTS. These are likewise represented as morphological FEM to match the individual curvature pattern. Our algorithm initialises the FEM based on the parametrisation of the HG and STS model. More specifically, as HS represents the posterolateral border of HG, and STS defines the inferior border of

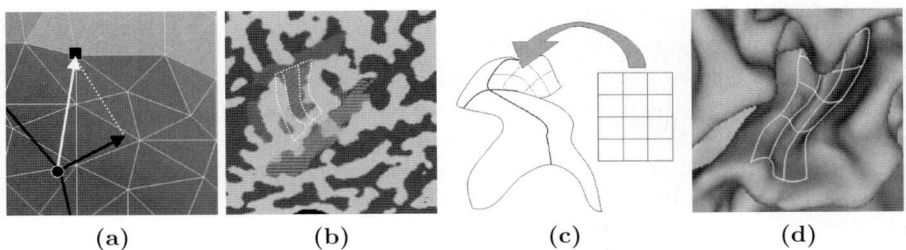

(a) (b) (c) (d)

Fig. 3. The local deformation of the morphological FEM depends on external forces computed from the curvature maps (a). For example, for each boundary node (black dot) we interpolate from $\mathcal{D}$ the vector to the nearest vertex (white arrow) with maximum gradient magnitude $|\hat{\nabla}\mathcal{K}|^2$, and use as nodal force a scaled version of its radial component (black arrow). Fig. 3b shows a flat map with a color–coded overlay of the segmented folds f, from which the landmark–related ROI T2 can be constructed. A local coordinate system is established by finite element mapping (c) of quasi landmarks that describe the ROI. Fig. 3d shows the deformed FEM $\mathcal{T}_{T2}$ (cf. Sect. 2.2).

GTS, we use the final position of their boundary nodes as displacement boundary conditions for the following shape fit. Using this segmentation procedure, geometrical and anatomical labels are combined, i.e. each flat map vertex which contributes to one of the folds $f \in \{\text{aHG, HG, SI, HS, SF, STS, lGTS}\}$, is assigned an additional label $\varphi_k = l_f$ (Figure 3b).

Atlas–based Restriction of the Search Space. Due to the difficulty of defining specific cortical folds from a purely geometric point of view, we cannot use the above strategy for a completely automated parcellation procedure (a correct labelling is provided with a probability of 60 − 70% [16]). As detailed in [18] model instances are initialised by an Euclidean transformation from the model coordinate frame to the flat map coordinate frame. Since it constrains the relative position of the folds, we only need to find a proper transformation of the top–level shape model. The search space (spanned by the affine parameters of the model instance in the image) is reduced using the 2D cartesian flat map coordinate system for storing prior statistics about the parcellation labels. We use $N = 10$ flat maps, which were constructed from MR data sets warped into Talairach reference space [21], and contain the parcellation labels for the sulcal landmarks SF and STS. The prior probability of parcellation label l_f occurring at atlas location $\psi(\mathbf{x})$ is

$$p(\mathcal{P}(\mathbf{x}) = l_f) = \text{card}(\{l_j | \psi(\mathbf{x})\}) N^{-1}, \qquad (4)$$

where card returns the set size, and ψ describes a Talairach transform.

2.2 Parcellation and Parametrisation of the ROI

A ROI–based coordinate system is established in two steps. First, we construct the landmark–oriented auditory ROI T1 and T2 based on the obtained segmentations. The deformed FEM identify the starting and ending points for each landmark as well as a number of border points where two or more landmarks meet. Based on these points, the surface patches labelled aHG, HG and HS are combined with portions of the patches labelled SF and lGTS. Finally, T1 and T2 can be separated by the medial axis nodes of the HG model (and supplementary SI, resp.), which extends from the medial tip of HG to its lateral intersection with lGTS (Figs. 2, 3b). Splitting curves between the border points are then parametrised with a predetermined number of points added between them. This results in a set of quasi landmarks $\mathbf{P}_i$, which can be used for assigning labels $\varphi_k = l_r, r = T1, T2$, to the vertices enclosed by the curves (Fig. 3d).

Flattening comprises unfolding the surface and mapping its vertices onto the plane. This introduces geometric distortions that affect the constrained parametrisation of the resulting surface patches based on the deformed shape models. To reduce this effect as much as possible we match a quadrilateral FEM $\mathcal{T}_r$ to the folded surface, using the point sets $\mathbf{P}_i$ to drive the registration. As a result the associated natural 2D–local coordinate system is related with the deformed shapes $\mathcal{T}_r$ embedded in 3D (Figs. 3c, 3d), allowing the interpolation of the associated field variables over the mesh [20].

3 Experimental Evaluation

The two auditory ROI in the left and right hemispheres were parcellated for 16 subjects. A gold standard was given in terms of manual segmentations of the ROI in the flat maps provided by two experts.

Typical results of our parcellation method are shown in Figure 4. In contrast to the results reported in [16] the anatomical landmarks were correctly identified with no false positives. From the small average and maximum boundary error of $1.3\pm0.2mm$ and $2.8\pm0.4mm$, respectively, in comparison with expert segmentations we can conclude that the individual parcellations were both accurate and reproducible. The boundary error was slightly higher in the right hemispheres, but at the order of the inter–rater difference (and local average inter–vertex distance, respectively). The simplified sulcal shape models allowed to bridge over interruptions and match the main branches as desired (Sect. 2.1). For example, STS is often splitted into two segments, while in these cases the

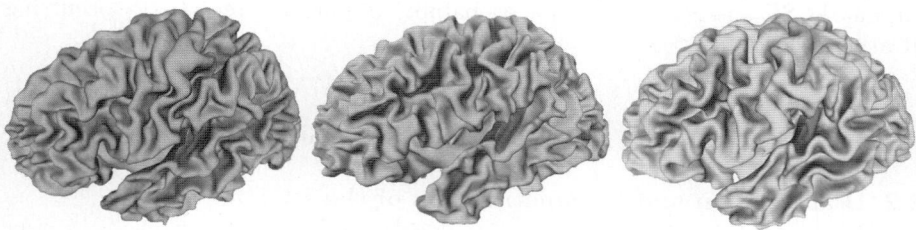

Fig. 4. Parcellation of the left auditory ROI T1 (blue) and T2 (red) for three different subjects

more prominent part, which defines the inferior border of GTS [2], was always segmented by our algorithm (Figure 3b). Our model–based approach further offers a characterisation of the fitted shapes and implies **classification**. We applied two different class–specific shape models to each data set and compared the quality of the resulting segmentations as described in [18]. This strategy allowed automatically deciding upon the existence of a sulcus intermedius in 94% of all cases. The almost automatic classification is of advantage since this specific landmark is considered an additional border between the primary (T1) and secondary auditory cortex (T2) [15] (see Figure 1c for an example).

Advantages of the atlas–based labelling. Our method produced similar results independent from the quality of the atlas, which was varied according to Sect. 2.1 using $N = 1, 10$ and 20. A Talairach transform into the atlas space not necessarily relates the semantics of coordinates across subjects. More specifically, in our case for $N = 10$ only a small portion of the atlas surface contained values $p(\mathcal{P}(\mathbf{x}) = l_f) > 0.5$ and no vertex was assigned the value 1. It was therefore sufficient to align the MR data sets w.r.t. the commissura anterior (i.e. the origin of the Talairach space [21]) instead of deforming the individual cortices to precisely match the atlas. The required user interaction was thereby reduced to identifying SF and STS in an example surface of each hemisphere and labelling the voxel representing the commissura anterior in all data sets. Nevertheless, this additional information rendered the deformable shape search more efficient compared to [16]. Our results indicated 60% savings in computation time.

4 Conclusion and Outlook

Individual landmark–related ROI can be segmented almost automatically and very precisely using a deformable model of the auditory cortical folding pattern to adapt to the low resolution curvature information provided by the cortical surfaces. By constraining the parametrisation of the shape models according to promising atlas regions of two sulcal landmarks, the shape search for the auditory temporal subregions was rendered both more efficient and successful over [16]. Since in our case prior information regarding spatial relationships between parcellation labels is incorporated into the shape model, we can use a very simple atlas. This reduces the required user interaction to a minimum.

Results shown in [22] indicate that a regional, anatomically constrained parametrisation supports functional localisation. Our method will now be validated for a larger set of anatomical data – and possible macro–anatomical variations – for assessing its utility in view of clarifying the correlation between the (individual) anatomical configuration and functional organisation of the brain.

References

1. Eickhoff, S., et al.: Testing anatomically specified hypotheses in functional imaging using cytoarchitectonic maps. Neuroimage 32(2), 570–582 (2006)
2. Kim, J., et al.: An MRI–based parcellation method for the temporal lobe. Neuroimage 11(4), 271–288 (2000)

3. Regis, J., et al.: Sulcal root generic model: a hypothesis to overcome the variability of the human cortex folding patterns. Neurol. Med. Chir. 45(1), 1–17 (2005)
4. Fischl, B., et al.: High–resolution inter–subject averaging and a coordinate system for the cortical surface. Human Brain Mapping 8, 272–284 (1999)
5. Jaume, S., Macq, B., Warfield, S.K.: Labeling the brain surface using a deformable multiresolution mesh. In: Dohi, T., Kikinis, R. (eds.) MICCAI 2002. LNCS, vol. 2488, pp. 451–458. Springer, Heidelberg (2002)
6. Joshi, A., Shattuck, D.W., Thompson, P., Leahy, R.M.: Brain image registration using cortically constrained harmonic mappings. In: Karssemeijer, N., Lelieveldt, B. (eds.) IPMI 2007. LNCS, vol. 4584, pp. 359–371. Springer, Heidelberg (2007)
7. Kang, X., et al.: Local landmark–based mapping of human auditory cortex. Neuroimage 22, 1657–1670 (2004)
8. Cachia, A., et al.: A generic framework for parcellation of the cortical surface into gyri using geodesic voronoi diagrams. Med. Imag. Anal. 7(4), 403–416 (2003)
9. Fischl, B., et al.: Automatically parcellating the human cerebral cortex. Cerebral Cortex 14, 11–22 (2004)
10. Le Goualher, G., et al.: Automated extraction and variability analysis of sulcal neuroanatomy. IEEE Trans. Med. Imag. 18(3), 206–217 (1999)
11. Lohmann, G., von Cramon, D.: Automatic labeling of the human cortical surface using sulcal basins. Med. Imag. Anal. 4(3), 179–188 (2000)
12. Rivière, D., et al.: Automatic recognition of cortical sulci of the human brain using a congregation of neural networks. Med. Imag. Anal. 6(2), 77–92 (2002)
13. Tao, X., Han, X., Rettmann, M.E., Prince, J.L., Davatzikos, C.: Statistical study on cortical sulci of human brains. In: Insana, M.F., Leahy, R.M. (eds.) IPMI 2001. LNCS, vol. 2082, pp. 475–487. Springer, Heidelberg (2001)
14. Vivodtzev, F., et al.: Brain mapping using topology graphs obtained by surface segmentation. In: Bonneau, G., Ertl, T., Nielson, G. (eds.) Scientific Visualization: the visual extraction of knowledge from data. Springer, Berlin (2005)
15. Brechmann, A., et al.: Sound–level–dependent representation of frequency modulations in human auditory cortex. J. Neurophys. 87, 423–433 (2002)
16. Engel, K., et al.: A two–level dynamic model for the representation and recognition of cortical folding patterns. In: Proc. IEEE ICIP, pp. 297–300 (2005)
17. Leonard, C., et al.: Normal variation in the frequency and location of human auditory cortex landmarks. Cerebral Cortex 8, 397–406 (1998)
18. Engel, K., Toennies, K.: Hierarchical vibrations for part-based recognition of complex objects. Technical Report OvGU Magdeburg FIN-03-2009 (2009)
19. Meyer, M., et al.: Discrete differential geometry operators for triangulated 2-manifolds. In: Hege, H., Polthier, K. (eds.) Visualization and Mathematics, pp. 35–58. Springer, Heidelberg (2003)
20. Pentland, A., Sclaroff, S.: Closed–form solutions to physically based shape modeling and recognition. IEEE Trans. Patt. Anal. Mach. Intell. 13(7), 715–729 (1991)
21. Talairach, J., Tournoux, P.: Co–planar stereotaxic atlas of the human brain. Thieme, Stuttgart (1988)
22. Engel, K., et al.: Model-based labelling of regional fMRI activations from multiple subjects. In: Proc. MICCAI Workshop Anal. Funct. Med. Imag., pp. 9–16 (2008)

Automatic Fontanel Extraction from Newborns' CT Images Using Variational Level Set

Kamran Kazemi[1,2], Sona Ghadimi[2,3], Alireza Lyaghat[1], Alla Tarighati[1], Narjes Golshaeyan[1], Hamid Abrishami-Moghaddam[2,3], Reinhard Grebe[2], Catherine Gondary-Jouet[4], and Fabrice Wallois[2,5]

[1] Shiraz University of Technology, Shiraz, Iran
[2] GRAMFC EA 4293, Faculté de Médecine, Université de Picardie Jules Verne, Amiens, 80036, France
[3] K.N. Toosi University of Technology, Tehran, Iran
[4] Department of Neuroradiology, Centre Hospitalier Universitaire d'Amiens, 80054, Amiens, France
[5] GRAMFC, EFSN Pediatrique, Centre Hospitalier Universitaire d'Amiens, 80054, Amiens, France
{kamran.kazemi,reinhard.grebe,fabrice wallois, hamid.abrishami}u-picardie.fr,
gondry-jouet.catherine@chu-amiens.fr, sona_ghadimi@ee.kntu.ac.ir

Abstract. A realistic head model is needed for source localization methods used for the study of epilepsy in neonates applying Electroencephalographic (EEG) measurements from the scalp. The earliest models consider the head as a series of concentric spheres, each layer corresponding to a different tissue whose conductivity is assumed to be homogeneous. The results of the source reconstruction depend highly on the electric conductivities of the tissues forming the head. The most used model is constituted of three layers (scalp, skull, and intracranial). Most of the major bones of the neonates' skull are ossified at birth but can slightly move relative to each other. This is due to the sutures, fibrous membranes that at this stage of development connect the already ossified flat bones of the neurocranium. These weak parts of the neurocranium are called fontanels. Thus it is important to enter the exact geometry of fontaneles and flat bone in a source reconstruction because they show pronounced in conductivity. Computer Tomography (CT) imaging provides an excellent tool for non-invasive investigation of the skull which expresses itself in high contrast to all other tissues while the fontanels only can be identified as absence of bone, gaps in the skull formed by flat bone. Therefore, the aim of this paper is to extract the fontanels from CT images applying a variational level set method. We applied the proposed method to CT-images of five different subjects. The automatically extracted fontanels show good agreement with the manually extracted ones.

Keywords: newborns, fontanel, source reconstruction, level set, segmentation.

1 Introduction

The electroencephalogram (EEG) measures ongoing electrical activity of the brain recorded from electrodes placed on the scalp. It is widely used in clinical setting for the diagnosis and management of epilepsy in adults and children. EEG provides the necessary data for non-invasive estimation of the location of electrical sources in the brain. For these estimates inverse methods are used where location, amplitude, and orientation of a dipole source is adjusted to a model of the head to obtain the best fit between the measured EEG's and those produced by the source in the model [1-3]. In addition to other factors such as noise, errors of location measurement, etc., the accuracy of these estimates depends highly on the accuracy of the head model [4]. Magnetic resonance imaging (MRI) and computed tomography (CT) are the two major neuroimaging modalities that can be used for multi-component head model construction.

While MRI provides an excellent tool for non-invasive study of brain tissues it is not suitable for accurate skull extraction especially in newborns which have a very thin skull in comparison to adults. CT scan uses X-ray beams passing through the sample at a series of different angles. Based on the attenuation of these beams, an image can be constructed. Due to high its attenuation, the bone appears bright in CT images. Therefore, CT provides an excellent non-invasive tool for skull study and modeling.

Most of the major bones of the neonates' skull are ossified at birth but can slightly move relative to each other. This is due to the sutures, fibrous membranes that at this stage of development connect the already ossified flat bones of the neurocranium. These weak parts of the neurocranium are called fontanels [5]. Thus, fontanels are the narrow seams of fibrous connective tissue that separate the flat bones of the skull. They can be identified as gaps between two cranial bones in newborns' CT images. Newborns have six fontanels: the anterior and the posterior, two mastoid, and two sphenoid ones [5]. These fontanels have different electrical properties, conductivities, in comparison to the cortical bone. This difference may have not neglectable influence on the localization of the electrical sources in the brain [6]. Thus, accurate modeling of the fontanels will improve the results of source localization for newborns' brain.

The present study aimed to extract the cortical bone and fontanels from newborns' CT images. The extracted tissues then can be combined with the extracted brain tissues from MRI to construct a complete realistic 3D head model for newborns.

The rest of the paper is organized as follows. Section 2 presents our method of cortical bone and fontanel extraction from CT images. The extracted fontanels and evaluation results are presented in section 3. Finally concluding remarks and discussion are given in section 4.

2 Materials and Methods

Figure 1 shows the block diagram of our method for bone and fontanel extraction from newborns' CT images. As shown, the cortical bone is extracted from CT images applying an automatic thresholding method. Then, the level set method of minimal surface reconstruction as proposed by Zhao et al., [7] is used to determine the inner and outer surface of a closed skull model based on the before extracted cortical bone. Finally, the fontanels can be extracted by removing the bone from the constructed closed skull model.

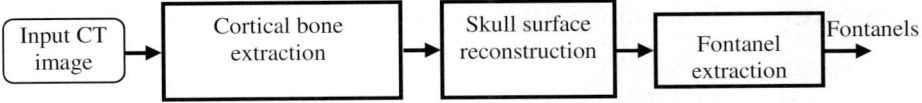

Fig. 1. Cortical bone and fontanel extraction method

2.1 Data Acquisition

The developed method has been applied to the 3D volumetric CT images of five newborns aged between 40 and 42 weeks at the date of examination (gestational age). The CT images have axial orientation with data matrix of 256×256 pixels and voxel size 0.75×0.75×1.25 mm^3.

2.2 Cortical Bone Extraction

Due to the high contrast between bone and adjacent tissues in CT images of the head, cortical bones can be extracted by simple automatic histogram thresholding [8]. The extracted bones are labeled V_{bone} and used to construct the inner and outer skull surface.

2.3 Skull Surface Reconstruction

In order to reconstruct the inner and outer skull (bone and fontanel) surfaces from extracted cortical bones, the variational level set method and tagging algorithm proposed by Zhao et al. [7] are applied. The method originally developed for surface reconstruction from unorganized data.

Minimum surface-like model. Let S denote the data set which includes the before determined cortical bone surface patches. So, the basic problem is determining an N dimensional surface Γ which minimizes the following energy functional:

$$E(\Gamma) = \left[\int d^p(x)ds \right]^{\frac{1}{p}}, \quad 1 \leq p \leq \infty \qquad (1)$$

where $d(\mathbf{x})$ is the distance function (dist($\mathbf{x}$, S)) from the point $\mathbf{x}$ to S, Γ is an arbitrary surface and ds is the surface area. Assuming $\Gamma(t)$ to be the zero level set of a function $\phi(x,t)$:

$$\Gamma(t) = \{x : \phi(x,t) = 0\} \qquad (2)$$

and satisfying the following conditions:

$$\begin{array}{ll} \phi(x,t) < 0 & \text{for } x \in \Omega \\ \phi(x,t) = 0 & \text{for } x \in \partial\Omega = \Gamma(t) \\ \phi(x,t) > 0 & \text{for } x \in \overline{\Omega} \end{array} \qquad (3)$$

where $\Omega(t)$ is the region enclosed by $\Gamma(t)$.

As derived in [7], the variational level set formulation for energy functional (1) is:

$$E(\Gamma) = \left[\int_\Gamma d^p(x)ds\right]^{1/p} = E(\phi) = \left[\int d^p(x)\delta(\phi(x))|\nabla\phi(x)|dx\right]^{1/p} \quad (4)$$

where δ(x) is the one-dimensional delta function and $\delta(\varphi(\mathbf{x}))|\nabla\varphi(\mathbf{x})|d\mathbf{x}$ is the surface area element at the zero level set of φ. With the corresponding gradient flow for φ and extending the motion to all level sets by replacing δ(φ) with $|\nabla\varphi|$, the level set formulation for gradient flow will obtain as follow:

$$\frac{\partial\phi}{\partial t} = \frac{1}{p}|\nabla\phi|\left[\int d^p(x)\delta(\phi)|\nabla\phi|dx\right]^{1/p-1}\nabla\cdot\left[d^p(x)\frac{\nabla\phi}{|\nabla\phi|}\right] \quad (5)$$

Though this level set can be used to smooth the implicit surface. For more and fast post smoothing, the level set proposed by Whitaker [9] is used using the variational level set formulation. Let φ_0 denote the initial level set function whose zero level set is the surface to be smoothed. The smoothed implicit surface is defined as the zero level set of φ that minimizes the following functional:

$$\frac{1}{2}\int(H(\phi)-H(\phi_0))^2 dx + \varepsilon\int\delta(\phi)|\nabla\phi|dx, \quad (6)$$

where $H(\mathbf{x})$ is the one dimensional Heaviside function. The first term in the above energy functional is a fidelity term that measures the symmetric volume difference between two closed surfaces. The second integral in the above functional is the surface area of the zero level set of ϕ, which is a regularization term that minimizes the surface area of the smoothed surface. The constant ε is a parameter that controls the balance between the fidelity and the regularization. We again find the minimizer by following the gradient flow of (6), whose level set formulation is:

$$\frac{\partial\phi}{\partial t} = |\nabla\phi|[\varepsilon\kappa - (H(\phi)-H(\phi_0))] \quad (7)$$

Initial surface. The surface reconstruction can be started with a simple initial surface such as cube or a sphere. However, if the initial surface is too far from the real shape, the PDE evolution takes a long time and computational cost. Therefore, a good initial surface helps to speed up convergence to the equilibrium surface. On a rectangular grid, an implicit surface can be used as an interface that separates the exterior grids from the interior grids.

The tagging algorithm proposed by Zhao [10] aims to identify as many correct exterior grids as possible and hence provides a good initial implicit surface. Since the size of the images in our application is rather big, applying this algorithm becomes very time consuming, therefore, we introduce here our modified tagging algorithm. We start from any initial exterior region that is a subset of the true exterior region. All grids that are not in the initial exterior region are labeled as interior grids. Those interior grids that have at least one exterior neighbor are labeled as temporary boundary grids. Now to march the temporary boundary, the Euclidean distance between all grids and the extracted newborns' cortical bones is computed. Then all temporary boundary voxels which are not on the data sets are checked one by one. If it has an interior neighbor with a large or equal distance, this temporary boundary point will be considered as a final boundary point. Otherwise, this temporary boundary point will be returned into an exterior point.

Reconstruction of inner and outer skull surface. The described level set method is applied to reconstruct the inner and outer skull surfaces (Γ_{in} and Γ_{out} respectively) based on the extracted cortical bones. The inner surface (Γ_{in}) is extracted by initializing the zero level set of function $\phi_{in}(x,t)$ inside the skull. According to equation (3), $\phi_{in}(x,t)$ is negative inside Γ_{in} and positive outside Γ_{in}. Thus a negative value for $\phi_{in}(x,t)$ indicates the intracranial volume which is named V_{IC}. Accordingly, the outer skull surface Γ_{out} is extracted by initializing the zero level set of function $\phi_{out}(x,t)$ outside the skull. Here a negative value of $\phi_{out}(x,t)$ indicates the intracranial (IC) and skull volume ($V_{IC\text{-}skull}$).

2.4 Fontanel Extraction

In order to extract the fontanels, the obtained V_{IC} and V_{bone} are combined and a morphological filling filter is applied to fill eventual holes ($V_{IC\text{-}bone}$). Then, by removing $V_{IC\text{-}bone}$ from $V_{IC\text{-}skull}$ the fontanels are extracted ($V_{fontanel}$).

$$V_{IC-bone} = fill(V_{IC} \mid V_{bone})$$
$$V_{fontanel} = XOR(V_{IC-bone}, V_{IC-skull}) \qquad (8)$$

2.5 Evaluation

Quantitative evaluation of the results has been performed by calculating the similarity between the automatically extracted fontanels (L_a) and the corresponding manually extracted ones (ground truth, L_m). The results were evaluated using the following similarity index (SI):

$$SI = \frac{2n(L_a \cap L_m)}{n(L_a) + n(L_m)} \qquad (9)$$

3 Results

The obtained results demonstrate that the developed method reliably extracts fontanels from newborns' CT images. Figure 2 shows the CT image of a selected subject, its extracted cortical bone, inner and outer skull surfaces and finally the extracted fontanel. As can be seen in Figures 2.c and 2.e, the inner and outer surfaces are closed surfaces. The fontanels are closing smoothly the gaps left by the before detected flat bones, both forming together the closed skull model. Therefore, by removing these bones from the closed skull, the fontanels can be obtained (Figure 2.h).

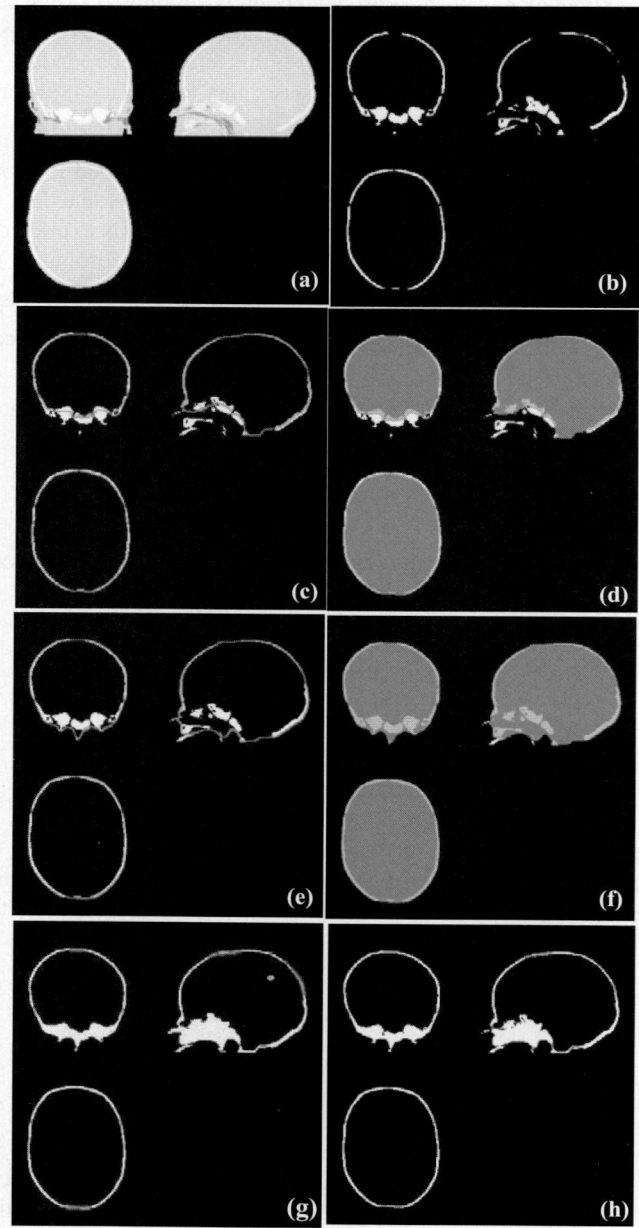

Fig. 2. Fontanel extraction. (a) Input CT image, (b) cortical bone after automatic thresholding, (c) final inner skull surface, (d) V_{IC}, (e) final outer skull surface, (f) $V_{IC\text{-}skull}$, (g) skull volume including extracted fontanels and bone, (h) manually extracted fontanels.

Table 1 shows the quantitative results for three subjects obtained according to the similarity index. The results show 76±4% similarity for the automatically extracted fontanels with the corresponding ground truth.

Table 1. Similarity index, providing a measure for the similarity between automatically and manually extracted fontanels from newborns' CT images applying variational level set

	Similarity Index
Subject 1	0.74
Subject 2	0.75
Subject 3	0.72
Subject 4	0.80
Subject 5	0.82

4 Discussion and Conclusion

In this paper we presented our approach for extracting fontanels from newborns' CT images based on variational level set method. Quantitative and qualitative results demonstrate the accuracy of the developed method for fontanel extraction from newborns.

The extracted fontanels and cortical bone in conjunction with the brain tissue models (after coregistration) may provide a realistic head model that can be used for applications such as source localization in newborns. This is important because of the different electrical conductivity of fontanels with respect to bone. In addition, the automatically extracted fontanels can be used for determining growth patterns for the newborns' skull.

Acknowledgment

This work was partially supported by the Center for International Research & Collaboration of Iran (ISMO) under the grant number 86/13 and EGIDE France under the grant number 18639PL (Jundi Shapour scientific collaboration program).

References

1. Mosher, J.C., Leahy, R.M.: Recursive MUSIC: A Framework for EEG and MEG Source Localization. IEEE Trans. on Biomedical Engineering 45(11) (November 1998)
2. Liu, H., Gao, X., Schimpf, P.H., Yang, F., Gao, S.: A Recursive Algorithm for the Three-Dimensional Imaging of Brain Electric Activity: Shrinking LORETA-FOCUSS. IEEE Trans. on Biomedical Engineering 51(10) (October 2004)
3. Baillet, S., Riere, J.J., Marin, G., Mangin, J.F., Aubert, J., Garnero, L.: Evaluation of inverse methods and head models foe EEG source localization using a human skull phantom. Physics in Medicine and Biology 46(1), 77–96 (2001)

4. Cuffin, B.N.: EEG Localization Accuracy Improvements Using Realistically Shaped Head Models. IEEE Trans. on Biomedical Imaging 43(3) (March 1996)
5. Kiesler, J., Richer, R.: The Abnormal Fontanel. American Family Physician 67(12) (June 2003)
6. Roche-Labarbe, N., Aarabi, A., Kongolo, G., Gondry-Jouet, C., Dümpelmann, M., Grebe, R., Wallois, F.: High-resolution EEG and source localization in neonates. Human Brain Mapping 29(2), 167–176 (2008)
7. Zhao, H.K., Osher, S., Merriman, B., Kang, M.: Implicit and non-parametric shape reconstruction from unorganized points using variational level set method. Computer Vision and Image Understanding 80(3), 295–319 (2000)
8. Otsu, N.: A threshold selection method from gray-level histograms. IEEE Trans. Sys. Man. Cyber. 9, 62–66 (1979)
9. Whitaker., R.: A level set approach to 3D reconstruction from range data. International Journal of Computer Vision (1997)
10. Zhao, H., Osher, S.: Contributed chapter. In: Osher, S., Paragios, N. (eds.) Geometric Level Set Methods in Imaging, Vision and Graphics. Springer, Heidelberg (2002)

Modeling and Measurement of 3D Deformation of Scoliotic Spine Using 2D X-ray Images*

Hao Li[1], Wee Kheng Leow[1], Chao-Hui Huang[1], and Tet Sen Howe[2]

[1] Dept. of Computer Science, National University of Singapore, Singapore
[2] Dept. of Orthopaedics, Singapore General Hospital, Singapore
{lihao,leowwk,huangch}@comp.nus.edu.sg, tshowe@sgh.com.sg

Abstract. Scoliosis causes deformations such as twisting and lateral bending of the spine. To correct scoliotic deformation, the extents of 3D spinal deformation need to be measured. This paper studies the modeling and measurement of scoliotic spine based on 3D curve model. Through modeling the spine as a 3D Cosserat rod, the 3D structure of a scoliotic spine can be recovered by obtaining the minimum potential energy registration of the rod to the scoliotic spine in the x-ray image. Test results show that it is possible to obtain accurate 3D reconstruction using only the landmarks in a single view, provided that appropriate boundary conditions and elastic properties are included as constraints.

Keywords: Scoliosis, spine, 3D reconstruction, modeling and measurement, deformation, Cosserat rod.

1 Introduction

Scoliosis is a disease that causes deformations such as twisting and lateral bending of the spine. To correct scoliotic deformations by surgery and spinal fixation, the extents of 3D spinal deformations need to be measured [1,2]. In principle, these measurements can be made on the 3D model reconstructed from the patient's CT volume image of the spine. However, the radiation dosage of the patient in such a CT scan is too high. Therefore, x-ray imaging is currently the imaging technique of choice for the diagnosis and treatment of scoliosis. Multiple views of the patient's spine can be taken at the same time using biplanar radiography [3] or at different time using conventional radiography, with the patient turning to the side. Biplanar radiographic machines are bulky and inflexible. As they have limited use in clinical practice, they have been replaced by CT scanners. Therefore, conventional radiography is more commonly used for capturing x-ray images of scoliotic spines. A comparison of radiographic and CT methods for measuring vertebral rotation is given in [4].

There is a wide spectrum of existing works on the measurement of 3D scoliotic deformations based on x-ray images. In one extreme, a 3D curve is regarded as a simplified model of the spine, and it is registered to the spine in the x-ray

---

* This research is supported by A*STAR SERC 0521010103 (NUS R-252-000-319-305).

image [5,6,7]. This approach is efficient but its accuracy is not guaranteed due to the simplicity of the model. In the other extreme, detailed patient-specific 3D models of vertebrae are reconstructed from biplanar radiography and then registered to the vertebrae in conventional radiographic images [8,9,10]. In theory, this approach should yield the most accurate measurements. However, it can be tedious and computationally very expensive.

From the anatomical point of view, the spine is a chain of rigid bodies (the vertebrae) that can rotate about their articulated joints within physical limits. Whereas a 3D curve may be too simplified a model for the spine, detailed patient-specific 3D models of vertebrae most likely contain too much redundant information for accurate measurement of scoliotic deformations. By determining the simplest model that produce accurate results, an efficient and less tedious approach for the measurement of scoliotic deformations can be identified. In this paper, we present an initial effort in determining the *optimal* model that balances accuracy and efficiency of measurement. Due to page limit constraint, this paper will focus on studying methods based on 3D curves. Methods based on detailed vertebra models will be studied in a follow-up paper.

2 Related Work

Existing work on the measurement of 3D deformations of scoliotic spine can be categorized according to the spine model used: (1) 3D curve and (2) detailed 3D model. For 3D curve-based approach, [5,6] provides a GUI for the user to manually fit a 3D Bezier curve with 18 control points to the centerline of the spine in biplanar x-ray images. The fitted 3D curve is then used to perform Lenke and King classification of the type of scoliotic deformities [11,12]. In [7], the user manually fits a 3D Bezier curve with 6 control points to the centerline of the spine in biplanar x-ray images. In addition, the user identifies landmarks on key vertebrae in the images, and the algorithm interpolates the landmarks of other vertebrae based on the fitted 3D curve. The landmarks of the vertebrae are then used to compute the 3D positions and orientations of the vertebrae. These methods are computationally efficient but they require a lot of user inputs and interactions. Their accuracy depends on the expertise of the user.

For detailed model-based approach, the method of [8,9] first extracts the centerline of the spine based on the segmentation algorithm of [13], which is used to compute the global positions of the vertebrae in the image. Then, it locally deforms a statistical vertebra model to register to the edges of individual vertebra in the x-ray image, subject to the constraint that the vertebra models form a smooth spine model. The statistical model is learned from a set of training scoliotic vertebrae [14]. The registered vertebra models can then be used to measure scoliotic deformations. Novosad et al. [10] fits patient-specific vertebra models reconstructed from biplanar radiography. The models are registered to the manually identified landmarks of the vertebrae in a single lateral bending x-ray image, subject to the constraint that the vertebra models form a smooth spine. In principle, this approach can be very accurate, with a reported mean accuracy of about 1mm on reconstructing a vertebra [3,15]. However, it is very tedious and computationally very expensive, especially for the reconstruction of patient-specific model of each vertebra of the spine.

3 Modeling Scoliotic Deformations

Our approach models a spine by a 3D curve with elastic properties called a *Cosserat rod* [16] (Sec. 3.1). The model is registered to the spine in x-ray images to recover the 3D structure of a scoliotic spine (Sec. 3.2). After registration, twisting and lateral bending of the model are computed and serve as the corresponding measurements of the patient's spine.

3.1 3D Spine Model Representation

The spine is represented by a sequence of points $\mathbf{r}_i$ on the spinal centerline. Each point $\mathbf{r}_i$ corresponds to the mid-point on the top end of a vertebra (Fig. 1(d)). When projected to 2D, this point remains in the middle of the top end of the vertebra, which is easy to identify (Fig. 1(e)).

Attached to each point $\mathbf{r}_i$ are the directors $\mathbf{d}_{ik}$, $k \in \{1,2,3\}$, which represent the 3D orientation of the vertebra (Fig. 1(d)). The director $\mathbf{d}_{i3}$ is the tangent direction of the centerline at $\mathbf{r}_i$. $\mathbf{d}_{i1}$ and $\mathbf{d}_{i2}$ are the frontal and lateral directions of the vertebra that are always orthogonal to $\mathbf{d}_{i3}$. The set of parameters $\{\mathbf{r}_i, \mathbf{d}_{ik}\}$ specifies the configuration of the model M.

According to the theory of Cosserat rod [16], the bending and twisting of the model can be defined based on the *linear strain vector* $\boldsymbol{v}_i$ and *angular strain vector* $\boldsymbol{u}_i$ such that

$$\boldsymbol{v}_i = \partial_s \mathbf{r}_i, \quad \partial_s \mathbf{d}_{ik} = \boldsymbol{u}_i \times \mathbf{d}_{ik}, \quad k \in \{1,2,3\}. \tag{1}$$

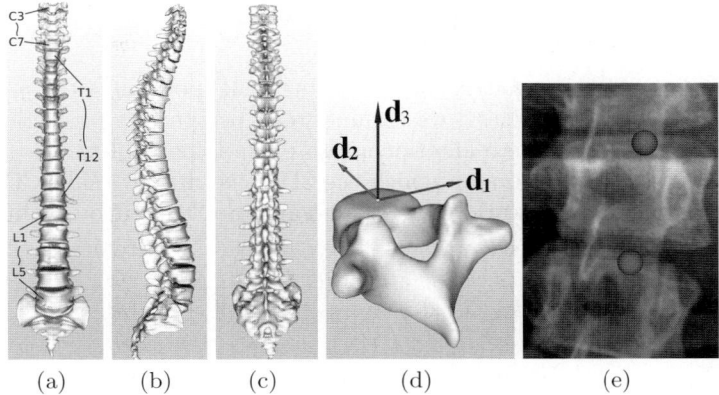

(a) (b) (c) (d) (e)

Fig. 1. Generic spine model. (a) Anterior view. (b) Lateral view. (c) Posterior view. (d) Vertebra with its directors. Triad indicates position and directors of vertebra. (e) Manually identified landmarks on the x-ray image. The landmarks are identified in the middle of the top end of each vertebra.

The strain vectors $\boldsymbol{u}_i$ and $\boldsymbol{v}_i$ can be resolved into three components by the directors $\mathbf{d}_{ik}$ to compute the *strain variables* u_{ik} and v_{ik}:

$$u_{ik} = \boldsymbol{u}_i \cdot \mathbf{d}_{ik}, \quad v_{ik} = \boldsymbol{v}_i \cdot \mathbf{d}_{ik}, \quad k \in \{1,2,3\}. \tag{2}$$

The components u_{i1} and u_{i2} are the curvatures along directors $\mathbf{d}_{i1}$ and $\mathbf{d}_{i2}$. They measure the postero-anterior bending and lateral bending of the vertebra. The component u_{i3} measures twisting, whereas v_{i1} and v_{i2} measure shear, and v_{i3} measures stretching.

3.2 3D-2D Non-rigid Registration

The most common radiographic imaging technique in clinical practice captures the postero-anterior and lateral views of a patient's spine at different time. It is humanly impossible for the patient to keep exactly the same posture during the two captures. To accommodate such clinical practice, landmarks $\mathbf{m}_i$ are manually placed only on the vertebrae in the postero-anterior view.

The directors $\mathbf{d}_{ik}$ of the first and last vertebrae derived from the postero-anterior and lateral views serve as the boundary conditions. In Section 4, we will show that these data from the two views are sufficient to yield an accurate reconstruction of the 3D model of the patient's spine.

Non-rigid registration is performed by determining the configuration $\{\mathbf{r}_i, \mathbf{d}_{ik}\}$ of the model M that minimizes the cost function:

$$E = E_f + E_I \tag{3}$$

subject to the boundary conditions. Quasi-Newton algorithm [17] is applied to optimize the cost function.

The term E_f is the 3D-2D registration error:

$$E_f = \sum_{i \in L} \|\mathbf{P}(\mathbf{r}_i) - \mathbf{m}_i\|_2^2, \tag{4}$$

where $\mathbf{P}$ is the projection matrix that projects the 3D point $\mathbf{r}_i = (x_i, y_i, z_i)$ onto the image plane. For simplicity, we assume weak perspective projection of the x-ray images and the postero-anterior image plane is parallel to the x-y plane of the world coordinate frame in which the 3D spine model resides. Therefore, the image plane and the world coordinate frame are related by scaling s and translation $\mathbf{T}$. The 3D-2D registration error is thus:

$$E_f = \sum_i \left\| s \begin{bmatrix} x_i \\ y_i \end{bmatrix} + \mathbf{T} - \mathbf{m}_i \right\|_2^2. \tag{5}$$

The potential energy E_I that constrains the bending and twisting of the spine is defined according to the Cosserat rod theory:

$$E_I = \sum_i \sum_{j=1}^{3} [\alpha_{ij}(u_{ij} - u_{ij}^0)^2 + \beta_{ij}(v_{ij} - v_{ij}^0)^2], \tag{6}$$

where u_{ij}^0 and v_{ij}^0 are the strain variables in the initial configuration of the spine. They represent the natural bending and twisting of a normal spine. The stiffness coefficients α_{ij} and β_{ij} of the corresponding strains are dependent on the elastic properties and the geometrical properties of the model. The elastic properties, including the Young's modulus Y and shear modulus G, are determined by applying Monte Carlo technique on a set of training data. In this way, the bending and twisting of the model will be consistent with those of actual spines.

4 Experiments and Discussion

Experiments were conducted to evaluate the proposed model, and to examine the necessary inputs for accurately reconstructing and measuring 3D scoliotic spine. Qualitative tests were performed on 30 pairs of real x-ray images (Sec. 4.1). For quantitative evaluation of the proposed model (Sec. 4.2), thirty sets of synthetic data were generated by manually adjusting the generic spine model to emulate different but realistic scoliotic spines in the real x-ray images.

Note that a spine has a total of 24 vertebrae. Medical assessment of scoliosis does not include the cervical vertebrae (Fig. 1, C3–C7). So, landmarks are not placed on them in the x-ray images. They are placed only on the thoracic vertebrae (T1–T12) and the lumbar vertebrae (L1–L5). In fact, some x-ray images do not contain cervical vertebrae.

4.1 Tests on Real x-Ray Images

In the qualitative tests, for each test set, a pair of postero-anterior and lateral x-ray images of scoliosis patient were provided. The landmarks on the vertebrae of the postero-anterior view were identified manually. The boundary conditions of the spine, i.e., the position $\mathbf{r}_i$ and directors $\mathbf{d}_{ik}$ of T1 and L5 were derived from the user specified points on both postero-anterior and lateral views. For each patient with a pair of x-ray image, the number of input points was 21 in total. The proposed model took on the average 15 seconds to fit the landmarks in the x-ray image, thus reconstructing the 3D patient-specific model of scoliotic spine. The average registration error (Eq. 4) is 0.228mm, which means the projection of the 3D model is very close to the landmarks in the postero-anterior view.

Three examples of the fitting results are illustrated in Figure 2, where the amount of lateral bending of the vertebrae are shaded in color. Note that although the lateral views provide only boundary conditions for the inputs, our reconstructed models can still fit the whole spines in the lateral views well. Therefore, our approach, with the trained elastic properties, is able to correctly model the bending and twisting of the actual spine. Since the lateral view was not scanned at the same time as the postero-anterior view, the posture of the patient may change between views. Thus, some of the reconstructed vertebra models do not align very well in the lateral views. On the other hand, since the landmarks in the postero-anterior views were given, the projections of the reconstructed spine models fit quite well with the postero-anterior x-ray images.

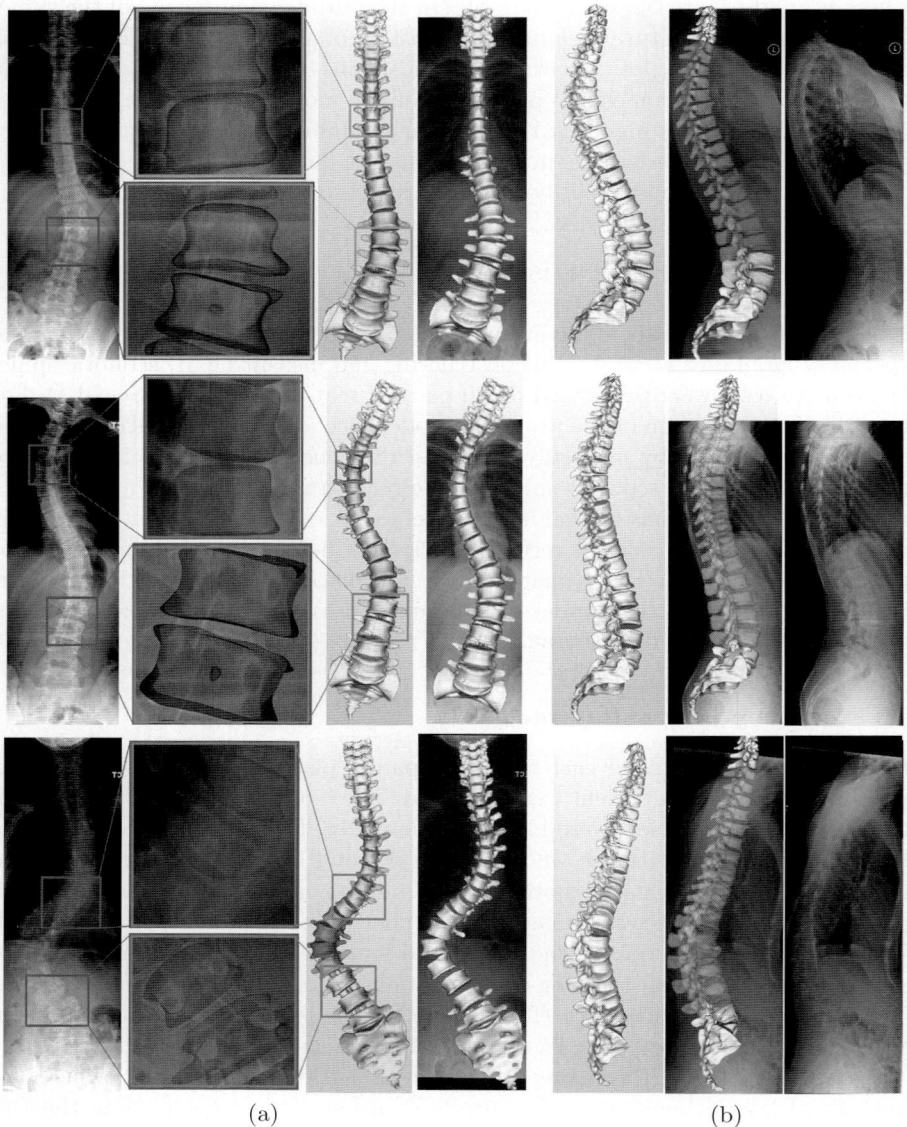

Fig. 2. 3D reconstruction results. Each row shows the results of a test set. (a) Posteroanterior view. (b) Lateral view. X-ray images and 3D models are overlapped and zoomed in to illustrate the fitting results. Zoomed-in 3D models in the overlapping images are set to transparent blue color. Vertebrae are shaded in red to illustrate the amount of lateral bending.

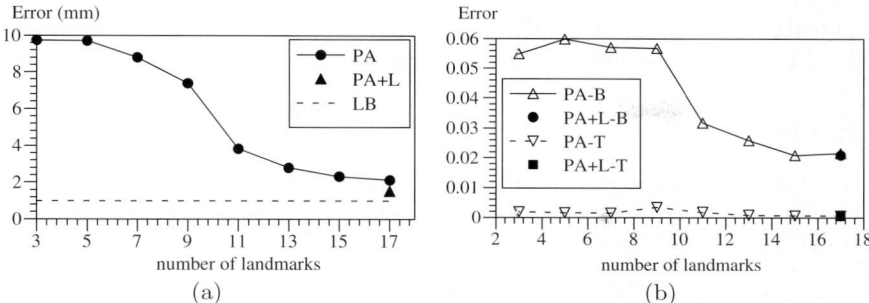

Fig. 3. Reconstruction errors for different number of input landmarks. (a) Mean fitting error of vertebra. PA: landmarks on postero-anterior view. PA+L: landmarks on PA and lateral view. LB: lower bound [3,15]. (b) Mean error of bending and twisting. -B: bending error. -T: twisting error.

4.2 Quantitative Tests with Synthetic Data

Thirty sets of synthetic data were generated for quantitative evaluation of the proposed model. For each test set, the ground truth spine model was obtained by manually adjusting the generic spine model to emulate a different but realistic scoliotic spine in the x-ray images. Eight test cases were performed with boundary conditions for T1 and L5 and varying number of postero-anterior landmarks (PA in Fig. 3(a)). For comparison, another test case was performed with 17 postero-anterior and 17 lateral landmarks (PA+L in Fig. 3(a)). For each test case, 3D reconstruction error, bending error, and twisting error averaged over 30 test sets were measured.

Figure 3(a) shows that the 3D reconstruction error decreases with increasing number of landmarks. For accurate reconstruction, e.g., error $\leq$ 2.3mm, 15–17 landmarks are required. With 17 postero-anterior landmarks, the reconstruction error is close to that obtained with 17 postero-anterior and 17 lateral landmarks. This indicates that the lateral landmarks are mostly redundant since the boundary conditions and elastic properties provide sufficient constraints to obtain accurate reconstruction. The reconstruction error of 2.114mm for the PA-17 case is also close to the theoretical lower bound of 1mm reported in [3,15].

Figure 3(b) shows that twisting error is not as strongly affected by the number of landmarks as the bending error. This is reasonable because the spine can bend more easily than it can twist. For accurate measurement of bending, sufficient number of landmarks, e.g., 15–17, are required.

5 Conclusion

This paper studies the modeling and measurement of scoliotic spine based on 3D curve model. The proposed method models a scoliotic spine as a 3D Cosserat rod. The reconstruction of the 3D scoliotic spine is achieved by obtaining the minimum potential energy registration of the rod to the scoliotic spine in the x-ray image. Experimental results show that it is possible to obtain accurate 3D reconstruction using only the landmarks in a simple view, provided that appropriate boundary conditions and elastic properties are included as constraints.

References

1. Bernhardt, M., Bridwell, K.H.: Segmental analysis of the sagittal plane alignment of the normal thoracic and lumbar spines and thoracolumbar junction. In: 23rd Annual Meeting of the Scoliosis Research Society (1988)
2. Panjabi, M., White, A.A.: A mathematical approach for three-dimensional analysis of the mechanics of the spine. Journal of Biomechanics 4, 203–211 (1971)
3. Dumas, R., Aissaoui, R., Mitton, D., de Guise, J.: Personalized body segment parameters from biplanar low-dose radiography. IEEE Trans. Biomed. Eng. 52(10), 1756–1763 (2005)
4. Lam, G.C., Hill, D.L., Le, L.H., Raso, J.V., Lou, E.H.: Vertebral rotation measurement: A summary and comparison of common radiographic and CT methods. Scoliosis 3, 16–25 (2008)
5. Lin, H., Sucato, D.: Identification of Lenke spine deformity classification by simplified 3D spine model. In: Proc. Int. Conf. IEEE/EMBS, pp. 3144–3146 (2004)
6. Lin, H.: Identification of spinal deformity classification with total curvature analysis and artificial neural network. IEEE Trans. Biomed. Eng. 55(1), 376–382 (2008)
7. Verdonck, B., Nijlunsing, R., Gerritsen, F.A., Cheung, J., Wever, D.J., Veldhuizen, A., Devillers, S., Makram-Ebeid, S.: Computer assisted quantitative analysis of deformities of the human spine. In: Wells, W.M., Colchester, A.C.F., Delp, S.L. (eds.) MICCAI 1998. LNCS, vol. 1496, pp. 822–831. Springer, Heidelberg (1998)
8. Benameur, S., Mignotte, M., Parent, S., Labelle, H., Skalli, W., de Cuise, J.: 3D/2D registration and segmentation of scoliotic vertebrae using statistical models. Computerized Medical Imaging and Graphics 27, 321–337 (2003)
9. Benameur, S., Mignotte, M., Labelle, H., Guise, J.A.D.: A hierarchical statistical modeling approach for the unsupervised 3-D biplanar reconstruction of the scoliotic spine. IEEE Trans. Biomed. Eng. 52(12), 2041–2057 (2005)
10. Novosad, J., Cheriet, F., Petit, Y., Labelle, H.: Three-dimensional (3-D) reconstruction of the spine from a single x-ray image and prior vertebra models. IEEE Trans. Biomed. Eng. 51(9), 1628–1639 (2004)
11. Lenke, L.G.: Adolescent idiopathic scoliosis. The Journal of Bone and Joint Surgery 83, 1169–1181 (2001)
12. King, H.A., Moe, J.H., Bradford, D.S.: The selection of fusion levels in thoracic idiopathic scoliosis. The Journal of Bone and Joint Surgery 65, 1302–1313 (1983)
13. Kauffmann, C., de Guise, J.A.: Digital radiography segmentation of scoliotic vertebral body using deformable models. In: Proc. SPIE (1997)
14. Benameur, S., Mignotte, M., Parent, S., Labelle, H., Skalli, W., de Guise, J.A.: 3D biplanar reconstruction of scoliotic vertebrae using statistical models. In: Proc. CVPR, pp. 577–582 (2001)
15. Mitulescu, A., Semaan, I., Guise, J.A.D., Leborgne, P., Adamsbaum, C., Skalli, W.: Validation of the non-stereo corresponding points stereoradiographic 3d reconstruction technique. Med. Biol. Eng. Comput. 39, 152–158 (2001)
16. Antman, S.S.: Nonlinear Problems of Elasticity. Springer, Heidelberg (1995)
17. Press, W.H., Teukolsky, S.A., Vetterling, W.T., Flannery, B.P.: Numerical Recipes in C++: The Art of Scientific Computing. Cambridge University Press, Cambridge (2002)

A Comparative Study on Feature Selection for Retinal Vessel Segmentation Using FABC

Carmen Alina Lupaşcu[1], Domenico Tegolo[1], and Emanuele Trucco[2]

[1] Dipartimento di Matematica e Applicazioni, Università degli Studi di Palermo, Palermo, Italy
[2] School of Computing, University of Dundee, Dundee, Scotland

Abstract. This paper presents a comparative study on five feature selection heuristics applied to a retinal image database called DRIVE. Features are chosen from a feature vector (encoding local information, but as well information from structures and shapes available in the image) constructed for each pixel in the field of view (FOV) of the image. After selecting the most discriminatory features, an AdaBoost classifier is applied for training. The results of classifications are used to compare the effectiveness of the five feature selection methods.

Keywords: Retinal images, vessel segmentation, AdaBoost classifier, feature selection.

1 Introduction

Automatic vessel segmentation in retinal images is very important, as the retinal vasculature may reveal vascular and nonvascular pathologies.

In the literature supervised methods have been used for vessel segmentation. Pixel classification based on supervised methods require hand-labeled ground truth images for training. Sinthanayothin et al. in [13] classify pixels using a multilayer perceptron neural net, for which the inputs were derived from a principal component analysis (PCA) of the image and edge detection of the first component of PCA. In [12] a simple feature vector is extracted for each pixel from the green plane and then a K-nearest neighbor (kNN) is used to evince the probability of being a vessel pixel. Another supervised method, called primitive-based method, was proposed in [16]. This algorithm is based on the extraction of image ridges (expected to coincide with vessel centerlines) used as primitives for describing linear segments, named line elements. Consequently, each pixel is assigned to the nearest line element to form image patches and then classified using a set of features from the corresponding line and image patch. The feature vectors are classified using a kNN-classifier. The method presented by Soares et al. in [14] produces also a segmentation after a supervised classification. Each image pixel is classified as vessel or non-vessel based on the pixel feature vector, which is composed of the pixel intensity and two-dimensional Gabor wavelet transform responses taken at multiple scales. A Gaussian mixture model (a Bayesian classifier in which each class-conditional probability density function is described as a linear combination of Gaussian functions) classifier is then applied to obtain a final segmentation. Feature selection is applied only by Staal et al. in [16]. The scheme used is the sequential forward selection method. The selection method starts with a null feature set and at each step, the best feature that satisfies the criterion function (the area under the receiver operating characteristic curve) is added to the feature set. The set with the best performance is chosen, after all features have been included.

Feature selection is used as a preprocessing step to machine learning, because is effective in reducing dimensionality, removing irrelevant data and increasing learning accuracy. Algorithms that perform feature selection can be divided into two categories: the filter model and the wrapper model. The filter model is computationally efficient especially when the number of features is very large, because it doesn't involve any learning algorithm when selecting features. It relies only on general characteristics of the training data, while the wrapper model needs one predetermined learning algorithm in feature selection and uses its performance in order to evaluate and determine which features are selected.

We have developed a new supervised method for retinal vessel segmentation called FABC. The method is based on computing feature vectors for every pixel in the image and training an AdaBoost classifier with manually labeled images. The feature vector is a collection of measurements at different scales taken from the output of filters (the Gaussian and its derivatives up to the 2 order, matched filters and two-dimensional Gabor wavelet transform responses), from the identification of edge and ridge pixels and from other local information which are extracted after computing the Hessian of the image for each pixel. The basic idea is to encode in the feature vector local information (pixel's intensity, Hessian-based measures), spatial properties (the gray-level profile of the cross-section of a vessel can be approximated by a Gaussian curve) and structural information (vessels are geometrical structures which can be seen as tubular). We used an AdaBoost classifier to divide pixels into two classes, i.e., vessels and non-vessel pixels.

In order to analyze the improvement of the computational time, as well as of the accuracy and the performance of the classification, we decided to perform a comparative study on feature selection methods applied as a preprocessing step to the AdaBoost classification of the vessel and non-vessel pixels. In this paper we present five feature selection heuristics, which are evaluating the goodness of features through feature subsets. All five methods are subset search algorithms based on the filter model (as the filter model is computationally less expensive than the wrapper model).

2 Database and Features

2.1 Database

The database we use for testing and evaluating the methods is the public database named DRIVE (Digital Retinal Images for Vessel Extraction). The photographs for the DRIVE database were obtained from a diabetic retinopathy screening program in The Netherlands. Each image has been JPG compressed. The images were acquired using a Canon CR5 non-mydriatic 3CCD camera with a 45 degree field of view (FOV). Each image was captured using 8 bits per color plane at 768 by 584 pixels. The FOV of each image is circular with a diameter of approximately 540 pixels. For this database, the images have been cropped around the FOV. For each image, a mask image is provided that delineates the FOV and also a ground truth segmentation of the vessels (Figure 1). The

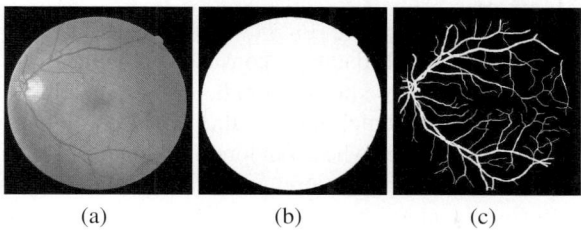

Fig. 1. Retinal image (a), mask image (b) and ground truth segmentation of the vessels (c)

data set includes 40 584 × 565 fundus images, divided into a training and test set, each containing 20 images. All images are available for download at the web site of the Image Sciences Institute of the University Medical Center Utrecht http://www.isi.uu.nl/Research/Databases/DRIVE/download.php.

2.2 Pixel Features

Features are extracted from the green plane of the retinal images, because in the green plane the contrast between vessel and background is higher than in the blue or red plane.

The feature vector consists of the output of filters (items 1 and 2 in the list below), vesselness and ridgeness measures based on eigen-decomposition of the Hessian computed at each image pixel (items 3, 4 and 5), and the output of a two-dimensional Gabor wavelet transform taken at multiple scales (item 6). Moreover the feature vector includes the principal curvatures, the mean curvature and the values of principal directions of the intensity surface computed at each pixel of the green plane image. The value of the root mean square gradient and the intensity within the green plane at each pixel are also included in the feature vector (items 7 and 8). The total number of features composing the feature vector at each image pixel is 41. Four scales are used to detect vessels of different width: $\sqrt{2}$, 2, $2\sqrt{2}$ and 4.

We give below the list of components of the feature vector.

1. Gaussian and its derivatives up to order 2. (features $1^{st} - 24^{th}$ in Table 3)
2. Green channel intensity of each pixel. (feature 25th in Table 3)
3. Multiscale matched filter for vessels using a Gaussian vessel profile. [15] (feature 26th in Table 3)
4. Frangi's vesselness measure. [3] (features $27^{th} - 28^{th}$ in Table 3)
5. Lindeberg's ridge strengths. [9] (features $29^{th} - 31^{st}$ in Table 3)
6. Staal's ridges. [16] (feature 32nd in Table 3)
7. Two-dimensional Gabor wavelet transform response taken at multiple scales. [14] (feature 33rd in Table 3)
8. Values of the principal curvatures (features $34^{th} - 35^{th}$ in Table 3), of the mean curvature (feature 36th in Table 3), of the principal directions (features $37^{th} - 40^{th}$ in Table 3) and of the root mean square gradient (feature 41st in Table 3) of the image.

3 Sample Selection

The training set within the DRIVE database consists of 20 images of size 584×565 pixels, hence for training we had to choose from $6,599,200$ pixels (but we considered only the pixels inside the FOV which are $4,449,836$). Due to the large number of pixels, only $789,914$ pixel samples where randomly chosen to train the classifier, i.e., $789,914/20$ pixels from each image, keeping the same proportion as in the ground-truth image between vessel and non-vessel pixels. The sample size was computed with a Z-test, considering a confidence level of 95% and a margin of error of 10%.

4 Feature Selection Heuristics

Feature selection is applied prior to classification, with the purpose to find a subset of features that optimizes the classification process, in terms of accuracy, performance and computational time.

As described in [6], Correlation-based Feature Selection (CFS) first calculates a matrix of feature-class and feature-feature correlations from training data. Based on these matrices a heuristic for evaluating the merit of a subset of features is computed. The heuristic takes into account the usefulness of individual features and in the same time the level of intercorrelation among them. The hypothesis on which the heuristic is based is: *Good feature subsets contain features highly correlated with the class, yet uncorrelated with each other.*

The following equation (as in [4]) formalizes the heuristic:

$$Merit_S = \frac{k\overline{r_{cf}}}{\sqrt{k + k(k-1)\overline{r_{ff}}}}$$

where $Merit_S$ is the heuristic merit of a feature subset S containing k features, $\overline{r_{cf}}$ is the average feature-class correlation, and $\overline{r_{ff}}$ is the average feature-feature intercorrelation. The numerator gives an indication of how predictive a group of features are; the denominator of how much redundancy there is among them.

4.1 Correlation-Based Feature Selection with Hill-Climbing Search Strategy (Heuristic H1)

In order to find the best subset that has the higher merit without trying all possible subsets, hill-climbing search strategy may be used. The algorithm starts with an empty set of features and generates all possible single feature expansions. The subset with the highest merit is chosen and expanded in the same way by adding single features until all features are added. The subset with the highest merit found during the search will be selected.

4.2 Correlation-Based Feature Selection with Best-First Search Strategy (Heuristic H2)

Usually CFS uses a best-first search strategy. Best-first search is similar to hill-climbing search, with the difference that if expanding a subset the merit doesn't maximize, the search drops back to the next best unexpanded subset and continues from there. CFS uses a stopping criterion of five consecutive fully expanded non-improving subsets. The subset with the highest merit, found in this way, is returned when the search terminates.

4.3 Consistency-Based Feature Selection with CFS (Heuristic H3)

An inconsistency is defined in [2] as two instances having the same feature values but different class labels. As the consistency measure is applied to each feature and not to a feature subset, in order to choose a minimum number of features that separate classes as consistently as the full set of features does, we sort the set of features by the number of inconsistencies in ascending order and apply a modified CFS to this sorted set. We apply the CFS in order to find the best correlated and consistent subset. The algorithm starts with the set containing only the feature with less inconsistencies. For this subset the merit is computed. The subset is expanded by adding one by one the features from the sorted set described above. The subset with the highest merit is returned as the best subset which is consistent and highly correlated.

4.4 Entropy-Based Feature Selection with CFS (Heuristic H4)

The entropy of each feature is defined as $-sum(plog_2(p))$, where p is the histogram counts of all the feature values. The smaller a feature's entropy, the more discriminatory the feature; hence we sort the set of features by the entropy values in ascending order and apply the modified CFS described above. The idea behind the modified CFS is not to decide a priori how many features to choose having low entropy, but to choose those highly correlated.

4.5 MIT Correlation with CFS (Heuristic H5)

The MIT correlation is based on the t-statistics and it is also known as signal-to-noise statistic. As described in [10] and [5], the method starts with a data set S containing m feature vectors: $X^i = (x_1^i, ..., x_n^i)$, where $1 <= i <= m$, m is the number of samples and n is the number of features. Each sample is labeled with $Y \in \{+1, -1\}$ (for classes, such as vessel pixels vs. non-vessel pixels). For each feature x_j, the mean μ_j^+ (resp. μ_j^-) and the standard deviation σ_j^+ (resp. σ_j^-) using only the samples labeled $+1$ (resp. -1) are calculated. A score $T(x_j)$ is then computed as

$$T(x_j) = \frac{|\mu_j^+ - \mu_j^-|}{\sqrt{\frac{(\sigma_j^+)^2}{n_+} - \frac{(\sigma_j^-)^2}{n_-}}},$$

where n_+ (resp. n_-) is the number of samples labeled as $+1$ (resp. -1). Features are then ordered by the score in descending order, as the features with the highest scores will be the most discriminatory features. In the same way as described above, we choose the best subset as the subset with the features highly correlated among, hence we compute the merit with the modified CFS.

5 Experimental Evaluation

The performances of the classification are measured using receiver operating characteristic (ROC) curves. ROC curves are represented by plotting true positive fractions versus false positive fractions as the discriminating threshold of the AdaBoost algorithm is varied. The true positive fraction (TPF) is determined by dividing the number of pixels correctly classified as vessel pixels (TP) by the total number of vessel pixels in the ground truth segmentation, while the false positive fraction (FPF) is the number of pixels incorrectly classified as vessel pixels (FP) divided by the total number of non-vessel pixels in the ground truth. The axes of the plot are rescaled so the true positives and false positives vary between 0 and 1. The area under the ROC curve (Az) measures discrimination, in our case, is the ability of the classifier to correctly distinguish between vessel and non-vessel pixels. An area of 1 indicates a perfect classification. We compute also the accuracy (ACC), which is degree of veracity of the classification, being the fraction of pixels correctly classified.

6 Experimental Results

As it can be seen in Table 2, after applying the heuristics described in Section 4, the number of features recommended by the first heuristic is 7, while the ones recommended by the second one are 4. The third heuristic selects 21 features, the fourth one 20, while for the fifth a set containing 16 features is doing the best separation between classes.

Table 1 shows the performance of different methods with or w/o feature selection that have been tested on DRIVE database. The performance is measured by the ROC index, defined as the area under the ROC curve and by the accuracy, defined as the fraction of pixels correctly classified. The performances shown here are those reported on www.isi.uu.nl/Research/Databases/DRIVE/.

Table 1. Overview of the performance of different methods. Az indicates the area under the ROC curve and ACC indicates the accuracy.

Segmentation method	Drive database		Segmentation method	Drive database	
	A_z	ACC		A_z	ACC
FABC	0.9560	0.9584	Niemeijer et al.	0.9294	0.9416
(w/o feature selection)			Zana et al.	0.8984	0.9377
Soares et al.	0.9614	0.9466	Jiang et al.	0.9114	0.9212
Human observer	-	0.9473	Martinez et al.	-	0.9181
Staal et al.	0.9520	0.9442	Chaudhuri et al.	0.7878	0.8773

Table 2. Performance of FABC with feature selection (H1, H2, H3, H4, H5) or w/o feature selection (All features). COLUMN 2: the area under the ROC curve A_z. COLUMN 3: the accuracy ACC. COLUMN 4: the computational time of constructing the model used by the AdaBoost classifier.

Heuristic	A_z	ACC	Computational time	Heuristic	A_z	ACC	Computational time
All features	0.9560	0.9584	$\approx 230\ min.$	H3 (21 features)	0.8970	0.9536	$\approx 24\ min.$
H1 (7 features)	0.9482	0.9554	$\approx 46\ min.$	H4 (20 features)	0.8953	0.9532	$\approx 21\ min.$
H2 (4 features)	0.9393	0.9551	$\approx 20\ min.$	H5 (16 features)	0.9544	0.9572	$\approx 85\ min.$

Our method was tested on an Intel(R) Core(TM)2 Duo CPU (3.16 GHz) with 3326 Mb memory. Feature generation for an image from the DRIVE database takes less than 2 minutes, while the classification of its pixels takes less than 5 seconds. The process of learning the AdaBoost model is computationally more expensive. It takes almost 4 hours when using the full set of features, while after feature selection it speeds up a lot as it can be seen in Table 2.

7 Discussion and Conclusion

When selecting features with the consistency-based feature selection, as well as with the entropy-based and MIT correlation feature selection, we don't use an arbitrary number to select top ranked features. In all cases we apply the CFS to strengthen the correlation among features and as well to determine automatically the number of discriminatory features.

We notice from Table 3, that the features that are playing an important discrimination role (the ones that were selected by all heuristics) are: feature 18 (the 2 order derivative of the Gaussian in the y direction at scale $2\sqrt{2}$), feature 26 (the maximum response of a multiscale matched filter using a Gaussian vessel profile) and feature 32 (the one containing information about Staal's ridges).

Even if after feature selection with the proposed heuristics the classification performance doesn't improve, the accuracies achieved outperforms the accuracies of the state-of-the-art approaches (even the one of the second human observer). As it can be seen in Table 2, of the five proposed selection heuristic the one using MIT correlation appeared to be the best one (it approached the performance obtained by the algorithm without feature selection). The performance of the fifth heuristic, as the one of FABC without feature selection, is still competitive to the best performance (the one achieved in [14] by Soares et al.).

Table 3. Features recommended by each heuristic

Feature	1	2	3	4	5	6	7	8	9	10	11	12	13	14	15	16	17	18	19	20	21
H1	X	X	X	X	✓	✓	X	X	X	X	X	X	X	X	X	X	X	✓	X	X	X
H2	X	X	X	X	X	X	X	X	X	X	X	X	X	X	X	X	X	✓	X	X	X
H3	X	X	X	✓	✓	✓	X	X	X	✓	✓	✓	X	X	X	✓	✓	✓	X	X	X
H4	X	X	X	✓	✓	✓	X	X	X	✓	✓	✓	X	X	X	✓	✓	✓	X	X	X
H5	X	X	X	X	✓	✓	X	X	X	X	X	X	X	X	X	X	✓	✓	X	X	X

Feature	22	23	24	25	26	27	28	29	30	31	32	33	34	35	36	37	38	39	40	41
H1	X	X	X	X	✓	X	X	X	✓	X	✓	X	✓	X	X	X	X	X	X	X
H2	X	X	X	X	✓	X	X	X	X	X	✓	X	✓	X	X	X	X	X	X	X
H3	✓	✓	✓	X	✓	X	✓	X	X	X	✓	X	✓	X	✓	✓	✓	✓	✓	X
H4	✓	✓	✓	X	✓	X	✓	X	X	X	✓	X	✓	X	✓	✓	✓	✓	✓	X
H5	X	X	✓	X	✓	X	✓	X	✓	✓	✓	X	✓	X	✓	X	X	X	X	✓

References

1. Chaudhuri, S., Chatterjee, S., Katz, N., Nelson, M., Goldbaum, M.: Detection of blood vessels in retinal images using two-dimensional matched filters. IEEE Transactions on Medical Imaging 8(3), 263–269 (1989)
2. Dash, M., Liu, H.: Consistency-based search in feature selection. Artificial Intelligence 151, 155–176 (2003)

3. Frangi, A.F., Niessen, W.J., Vincken, K.L., Viergever, M.A.: Multiscale vessel enhancement filtering. In: Wells, W.M., Colchester, A.C.F., Delp, S.L. (eds.) MICCAI 1998. LNCS, vol. 1496, pp. 130–137. Springer, Heidelberg (1998)
4. Ghiselli, E.E.: Theory of psychological measurement. McGraw-Hill, New York (1964)
5. Golub, T., Slonim, D., Tamayo, P., Huard, C., Gaasenbeek, M., Mesirov, J., Coller, H., Loh, M., Downing, J., Caligiuri, M., Bloomfield, C.: Molecular classification of cancer: class discovery and class prediction by gene expression monitoring. Science 286, 531–537 (1999)
6. Hall, M.A.: Correlation-based feature selection for discrete and numeric class machine learning. In: Proc. 17th International Conference on Machine Learning, pp. 359–366 (2000)
7. Hoover, A., Kouznetsova, V., Goldbaum, M.: Locating blood vessels in retinal images by piece-wise threshold probing of a matched filter response. IEEE Transactions on Medical Imaging 19(3), 203–210 (2000)
8. Jiang, X., Mojon, D.: Adaptive local thresholding by verification-based multithreshold probing with application to vessel detection in retinal images. IEEE Transactions on Pattern Analysis and Machine Intelligence 25(1), 131–137 (2003)
9. Lindeberg, T.: Edge detection and ridge detection with automatic scale seletion. Int. J. Comp. Vis. 30, 117–156 (1998)
10. Liu, H., Li, J., Wong, L.: A comparative study on feature selection and classification methods using gene expression profiles and proteomic patterns. Genome Informatics 13, 51–60 (2002)
11. Martínez-Pérez, M., Hughes, A., Stanton, A., Thom, S., Bharath, A., Parker, K.: Scale-space analysis for the characterisation of retinal blood vessels. In: Medical Image Computing and Computer-Assisted Intervention - MICCAI 1999, pp. 90–97 (1999)
12. Niemeijer, M., Staal, J., van Ginneken, B., Loog, M., Abràmoff, M.: Comparative study of retinal vessel segmentation methods on a new publicly available database. In: SPIE Medical Imaging, vol. 5370, pp. 648–656 (2004)
13. Sinthanayothin, C., Boyce, F., Cook, L., Williamson, H.: Automated localisation of the optic disc, fovea, and retinal blood vessels from digital colour fundus images. Br. J. Ophthalmol. 83, 902–910 (1999)
14. Soares, V.J., Leandro, J.J., Cesar, R.M.J., Jelinek, F.H., Cree, M.J.: Retinal vessel segmentation using the 2-d gabor wavelet and supervised classification. IEEE Transactions on Medical Imaging 25(9), 1214–1222 (2006)
15. Sofka, M., Stewart, C.V.: Retinal vessel centerline extraction using multiscale matched filters, confidence and edge measures. IEEE Transactions on Medical Imaging 25(12), 1531–1546 (2006)
16. Staal, J., Abràmoff, M.D., Niemeijer, M., Viergever, M.A., van Ginneken, B.: Ridge-based vessel segmentation in color images of the retina. IEEE Transactions on Medical Imaging 23(4), 501–509 (2004)
17. Zana, F., Klein, J.: Segmentation of vessel-like patterns using mathematical morphology and curvature evaluation. IEEE Transactions on Image Processing 10(7), 1010–1019 (2001)

Directional Multi-scale Modeling of High-Resolution Computed Tomography (HRCT) Lung Images for Diffuse Lung Disease Classification

Kiet T. Vo and Arcot Sowmya

School of Computer Science and Engineering,
University of New South Wales,
Sydney, NSW, 2052, Australia
{ktv,sowmya}@cse.unsw.edu.au

Abstract. A directional multi-scale modeling scheme based on wavelet and contourlet transforms is employed to describe HRCT lung image textures for classifying four diffuse lung disease patterns: normal, emphysema, ground glass opacity (GGO) and honey-combing. Generalized Gaussian density parameters are used to represent the detail sub-band features obtained by wavelet and contourlet transforms. In addition, support vector machines (SVMs) with excellent performance in a variety of pattern classification problems are used as classifier. The method is tested on a collection of 89 slices from 38 patients, each slice of size 512x512, 16 bits/pixel in DICOM format. The dataset contains 70,000 ROIs of those slices marked by experienced radiologists. We employ this technique at different wavelet and contourlet transform scales for diffuse lung disease classification. The technique presented here has best overall sensitivity 93.40% and specificity 98.40%.

Keywords: HRCT diffuse lung disease, texture classification, wavelet, contourlet, generalized Gaussian density.

1 Introduction

High Resolution Computed Tomography (HRCT) is generally considered to be the best imaging modality for assessment of the lung parenchyma in patients likely to have diffuse lung diseases [1]. However, the diagnosis of diffuse lung disease from HRCT images is a difficult task for radiologists because of the complexity and variation in the visual disease patterns on the images. Therefore, the construction of a computer-aided diagnosis system for diffuse lung disease is important in providing the radiologist with a "second opinion".

Texture classification has been a significant research topic in image processing, particularly in medical image analysis and many features has been proposed to represent a texture [2]. The method chosen for feature extraction is clearly critical to the success of texture classification. Five major categories of features for texture identification have been proposed: statistical, geometrical, structural, model-based, and signal processing features. Among these methods, the signal processing approach has advantages in the characterization of the directional and scale features of textures.

The most popular signal-processing feature extraction method based on wavelet transform offers computational advantages over other methods for texture classification [3-5]. In the literature of HRCT lung disease classification, Shamsheyeva et al. [6] applied a quincunx wavelet transform (QWF) along with SVM to classify 5 diffuse lung disease patterns. They used QWF decomposition combined with gray-level histogram features. The next work applying wavelet transform to detect only honeycombing pattern is Shojaii's [7], who utilized discrete wavelet transform to decompose a lung image into its directional sub-images, which were combined with histogram thresholding to extract honeycombing regions. In this work, the authors claimed that honeycombing region is best differentiable in vertical sub-image. Tolouee et al. [8] applied discrete wavelet frames (DWF) and rotated wavelet frames (RWF) to describe features. They used an energy-based method (L-2 norm) to measure the output of DWF and RWF along with SVM. Their best accuracy was obtained when combining DWF and RWF. Finally, experiments were conducted by Depeursinge's group using QWF combined with gray-level features, another feature called air-pix (number of air pixels in each ROI) and a set of clinical features [9-11]. They used mean and variance to represent the wavelet coefficients. In their research [11], SVM also proved to be the best classifier in diffuse lung disease categorization.

However, these wavelet-based methods suffer from the lack of directional information – a unique and important feature of multi-dimensional signals. Therefore, contourlet transform proposed by Do [12] with intrinsic multi-dimensional information (a different number of directions at each scale) and even less computational complexity has recently received increasing attention.

As mentioned above, the early applications of wavelet transform to lung disease classification on HRCT images are to use measures L-1 norm and L-2 norm (energy) to represent the coefficients of wavelet transform at each scale. Nevertheless, it has been shown in the literature that the distribution of both wavelet and contourlet coefficients in a given sub-band is highly non-Gaussian, heavily tailed and centered around zero [13, 14]. Hence, they cannot be modeled by normal Gaussian density function, and the more complex statistical probability distribution function called "generalized Gaussian density" that can adapt the marginal distribution of wavelet and contourlet coefficients at each sub-band most naturally is used widely.

In this paper, our approach to classification of diffuse lung disease is to use generalized Gaussian density with two parameters to model the wavelet and contourlet coefficients at each scale. In the next section, the methodology will be discussed in detail. The experiment and results are in section 3. The paper is concluded in section 4.

2 Methodology

In our method, at the feature extraction stage the texture images are transformed into directional sub-images at multi-scales by discrete wavelet and contourlet transform. The generalized Gaussian density (GGD) with two parameters is used to model the marginal distribution of wavelet and contourlet coefficients and the feature vector is built from these parameters. Finally, at the classification stage, support vector machines (SVMs) are used for disease pattern discrimination.

2.1 Wavelet Transform

In practice, the 2-D discrete wavelet transform (DWT) is computed by applying a separable filter-bank to the image [3] as seen in Fig. 1.

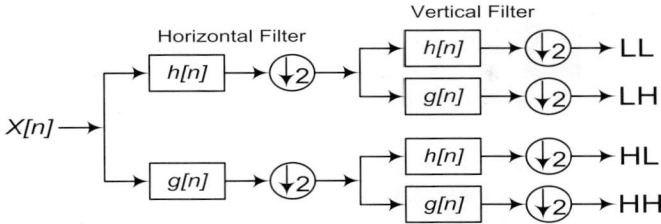

Fig. 1. Sub-band decomposition for one-level 2D DWT

In Fig. 1, $X[n]$ is the original image; $h[n]$ and $g[n]$ are a low-pass and high-pass filter, respectively. LL is obtained by low-pass filtering and is hence referred to as the low resolution image or approximation. LH, HL and HH contain directional detail information: horizontal, vertical and diagonal sub-images at each scale and only the coefficients from the directional information will be used to construct the feature vector. However, because of the down-sampling process, the performance of wavelet transform in texture description gets degraded.

Discrete wavelet frames (DWF) are used for texture retrieval and classification to overcome the limitation of standard DWT [5, 15]. Wavelet frames are simply the non-sampled version of the standard DWT; hence, they result in a texture description invariant with respect to translations of the input signal and yields a better estimation of texture characterization at region boundaries. In past research, wavelet frames have provided better results than the standard DWT. The performance of DWF will be tested in section 3 with different bases: Haar, Daubechies and Biorthogonal.

2.2 Contourlet Transform

Contourlet Transform is an efficient directional multi-scale image representation based on an efficient two-dimensional non-separable filter bank [12]. As seen in Fig.2, for contourlet transform, a Laplacian pyramid (LP) is first used to decompose the input image into multiple scale band-pass versions, and then followed by a non-separable directional filter bank (DFB) which decomposes each scale band-pass version into different numbers of directional sub-bands. In the contourlet transform in Fig. 3, a parent coefficient can have its children spread over two sub-bands. Therefore, contourlets possess not only the main features of wavelets, namely multi-resolution and time-frequency localization, but they also show a high degree of directionality and anisotropy.

In general, it can be easily seen that there are several ways to extract features based on contourlet transform. The difference between them is relied on the combination of different Laplacian pyramid filters and non-separable directional filter banks. The popular Laplacian pyramid filters are Haar and 9-7 filter; directional filter banks are

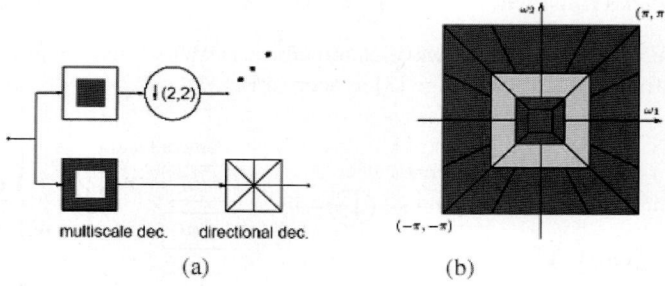

Fig. 2. Contourlet Transform

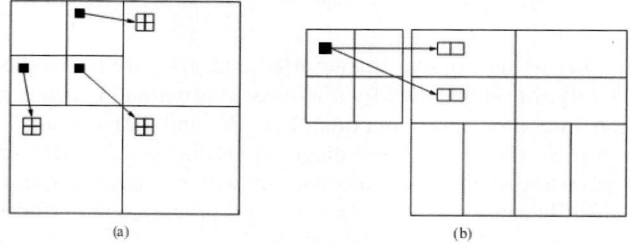

Fig. 3. Parent-children coefficient relationships for (a) Wavelets (b) Contourlets

CD filters and PKVA filters. In [14], 9-7 filters are claimed to be superior to Haar filters in terms of whitening the contourlet coefficients, while PKVA filters are more effective in localizing direction and should lead to better performance in applications. The performance of the combination of filters will be tested in the experiments.

2.3 Generalized Gaussian Density of Wavelet and Contourlet Coefficients

The generalized Gaussian density function is used model the output of both DWF and contourlet transform for each detail sub-image as follows:

$$p(x;\alpha,\beta) = \frac{\beta}{2\alpha\Gamma(1/\beta)} e^{-(|x|/\alpha)^{\beta}}$$

where $\Gamma(.)$ is the Gamma function, i.e. $\Gamma(z) = \int_0^{\infty} e^{-t} t^{z-1} dt$ and α, β are generalized Gaussian density model parameters indicating the width and the peak of the probability density function. Various estimators are used to estimate α and β, such as: moment-matching, entropy-matching and maximum likelihood. In this paper, an improved maximum likelihood estimator proposed by [16] is used to estimate α and β.

2.4 Classification Algorithm – Support Vector Machines (SVMs)

The support vector machine (SVM) has exploded in popularity within the machine learning literature and more recently has received increasing attention from the statistics community as well, and SVM has proved to be the best classifier in lung disease categorization [11]. The SVM paradigm was originally designed for the binary (two-class) classification problem [17]. To extend the original SVM to multi-class classification, popular methods include one-versus-all method using winner-takes-all strategy; one-versus-one method implemented by max-wins voting; and pair-wise coupling method (the extension of one-versus-one). The strategies are competitive to each other and there is no clear superiority of one method over others [18]. In our experiments, we use SVM with Gaussian radial basis function kernel, quadratic programming method for optimization and one-versus-all method for classification of different diffuse lung disease patterns because this configuration has been proven to be the most appropriate in experiments.

3 Experiments and Results

The dataset is constructed from a collection of 89 HRCT slices from 38 patients which have been investigated and labeled by experienced radiologists. From these labeled slices, 73,000 ROIs of size 32x32 are extracted. The size of ROIs is chosen to maintain the classification sensitivity with the smallest area and moreover, the size 32x32 is appropriate for 3-level wavelet decomposition (the size should not be less than 16x16 to maintain the accuracy of texture description and discrimination) and [2 3] contourlet transform ([4 8] directions). Distributions of the ROIs are detailed in Table 1.

Table 1. Distribution of ROIs over lung disease patterns

	Emphysema	GGO	Honeycombing	Normal
# of ROIs	20,000	18,000	15,000	20,000
# of patients	15	11	8	10

The performance of the experiment is evaluated through two measures: sensitivity and specificity where TP is true positives, FP false positives, TN true negatives and FN false negatives:

$$\text{Sensitivity}=\frac{TP}{TP+FN} \quad ; \text{Specificity}=\frac{TN}{TN+FP}$$

K-fold cross validation for the dataset is carried out to compute the classification sensitivity and specificity. The advantage of k-fold cross validation is that all the examples in the dataset are eventually used for both training and testing. For our size of dataset, 10-fold cross validation is appropriate to ensure that the bias is small and the computational time is acceptable. Moreover, in the experiments, use multiple run k-fold (10-run 10-fold) cross validation is also performed to ensure higher replicability and reliability of the results. Each run is executed with different random splits of the data set.

For the classifier, the initial configuration of SVMs is set as mentioned in section 2.4 with two parameters: cost C = 1000 and gamma $\gamma = 0.25$.

The results in Table 2 are generated from applying 3-scale discrete wavelet frames (18 features) with different bases: Haar, Daubechies and Biorthogonal. The results show that Haar is an appropriate filter for HRCT lung images in terms of overall sensitivity and specificity. It can be also seen that the generalized Gaussian density based (GGD) method always outperforms the energy-based method.

Table 2. Performance of Haar, Daubechies and Biorthogonal wavelet transform

	Haar		Daubechies 4-tap		Bior	
	Sensitivity (%)	Specificity (%)	Sensitivity (%)	Specificity (%)	Sensitivity (%)	Specificity (%)
Energy	85.28	93.05	86.35	93.10	85.13	92.95
GGD	**90.41**	**97.17**	89.93	96.95	85.58	93.95

Table 3 illustrates the performance of different combinations of filters: Haar, 9-7 and CD, PKVA for [2 3] contourlet transform ([4 8] directions at each scale) in diffuse lung disease classification. The experiments use generalized Gaussian density to model the coefficients of the output of the contourlet transform at each scale. Therefore, the feature vector consists of 24 parameters (two for each direction). The results are also computed in terms of overall sensitivity and specificity. The table shows that the best combination is using 9-7 and PKAV filters.

Table 3. Performance of contourlet transform

	Haar; CD	Haar; PKVA	9-7; CD	**9-7; PKAV**
Sensitivity (%)	88.78	85.69	90.23	**92.57**
Specificity (%)	96.89	94.55	96.58	**98.40**

Table 4. Comparison of DWF and Contourlet

	Wavelet		Contourlet		Combined	
	Sensitivity (%)	Specificity (%)	Sensitivity (%)	Specificity (%)	Sensitivity (%)	Specificity (%)
Emphysema	95.16	97.58	97.47	98.56	98.07	98.57
GGO	88.58	97.75	90.58	96.99	90.22	97.10
Honeycombing	83.98	98.15	90.11	98.85	89.93	98.58
Normal	93.92	95.20	92.11	99.20	95.38	99.34

The comparison of performance of discrete wavelet frames with 3-level and Haar basis and contourlet transform with [4 8] directions and [9-7 PKVA] filters for lung disease pattern classification is shown in Table 4. As seen, the contourlet-based method generally outperforms the wavelet-based method. The combination of wavelet and contourlet transform is tested as well. Although the performance of the combination is slightly better, the computational complexity increases dramatically.

Finally, some results are shown on HRCT lung images for easy visualization in Fig. 4.

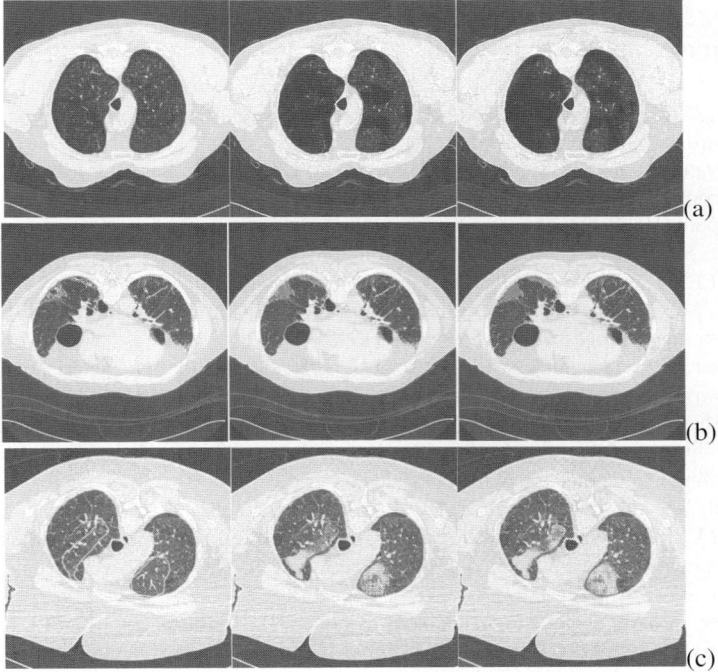

Fig. 4. (a) Emphysema (red) (b) Honeycombing (green) (c) GGO (yellow) (the first column is the labeled original image, the second and third show classification results by the proposed method for DWF and contourlet transform, respectively)

4 Conclusion

In this paper, wavelet-based and contourlet feature extraction strategy with the support of generalized Gaussian density model is employed to discriminate between four lung tissue patterns from the ROIs in an HRCT image database: normal, emphysema, ground glass opacity and honey-combing. These ROIs are decomposed to multi-directional multi-scales by wavelet and contourlet transforms. The strategy was tested with SVMs as classifier. The results prove that the GGD-based strategy outperforms the energy-based method using the same wavelet transform employed by Tolouee

[12]. The results also show that the contourlet-based method gives the better performance than the wavelet-based method. The best result is achieved with the combination of wavelet and contourlet. Although our method did not make comparisons with other traditional methods such as gray-level histograms, we believe the overall accuracy of our method is competitive and the computation cost is much lower. These comparisons will be given in detail in the future work.

Acknowledgements. The use of lung HRCT images from the LMIK database, School of Computer Science and Engineering, UNSW, is gratefully acknowledged.

References

1. Webb, W.R., Muller, N.L., Naidich, D.P.: High-Resolution CT of the Lung. Lippincott Williams & Wilkins, Philadelphia (2001)
2. Tuceryan, M., Jain, A.K.: Texture Analysis. In: Chen, C.H., Pau, L.F., Wang, P.P. (eds.) Handbook of Pattern Recognition and Computer Vision, pp. 235–276. World Scientific, Singapore (1993)
3. Mallat, S.G.: A theory for multiresolution signal decomposition: the wavelet representation. IEEE Transactions on Pattern Analysis and Machine Intelligence 11, 674–693 (1989)
4. Unser, M.: Texture classification and segmentation using wavelet frames. IEEE transactions on image processing 4, 1549–1560 (1995)
5. Wouwer, G.V., Scheunders, P., Dyck, D.V.: Statistical texture characterization from discrete wavelet representations. IEEE Trans. Image Processing 8, 592–598 (1999)
6. Shamsheyeva, A., Sowmya, A.: The anisotropic Gaussian kernel for SVM classification of HRCT images of the lung. In: Proc. Intelligent Sensors, Sensor Networks and Information Processing Conference, pp. 439–444 (2004)
7. Shojaii, R., Alirezaie, J., Babyn, P.: Automatic Segmentation of Abnormal Lung Parenchyma Utilizing Wavelet Transform. In: Alirezaie, J. (ed.) Proc. IEEE International Conference on Acoustics, Speech and Signal Processing ICASSP 2007, vol. 1, pp. 1217–1220 (2007)
8. Tolouee, A., Abrishami-Moghaddam, H., Garnavi, R., Forouzanfar, M., Giti, M.: Texture Analysis in Lung HRCT Images. In: Digital Image Computing: Techniques and Applications, 2008. DICTA 2008, pp. 305–311 (2008)
9. Depeursinge, A., Hidki, A., Platon, A., Poletti, P.-A., Unser, M., Muller, H.: Lung Tissue Classification Using Wavelet Frames. In: Proceedings of the 29th Annual International Conference of the IEEE EMBS Cité Internationale, Lyon, France, vol. 6259-6262 (2007)
10. Depeursinge, A., Iavindrasana, J., Cohen, G., Platon, A., Poletti, P.A., Muller, H.: Lung Tissue Classification in HRCT Data Integrating the Clinical Context. In: 21st IEEE International Symposium on Computer-Based Medical Systems. CBMS 2008, pp. 542–547 (2008)
11. Depeursinge, A., Iavindrasana, J., Hidki, A., Cohen, G., Geissbuhler, A., Platon, A., Poletti, P.-A., Muller, H.: A classification framework for lung tissue categorization. In: SPIE, vol. 6919 (2008)
12. Do, M.N., Vetterli, M.: The contourlet transform: an efficient directional multiresolution image representation. IEEE Transactions Image on Processing 14, 2091–2106 (2005)
13. Do, M.N., Vetterli, M.: Wavelet-based texture retrieval using generalized Gaussian density and Kullback-Leibler distance. IEEE Transactions on Image Processing 11, 146–158 (2002)

14. Po, D.D.Y., Do, M.N.: Directional multiscale modeling of images using the contourlet transform. IEEE Transactions on Image Processing 15, 1610–1620 (2006)
15. Do, M.N., Vetterli, M.: Wavelet-Based Texture Retrieval Using Generalized Gaussian Density and Kullback-Leibler Distance. EEE Trans. on Image Proc. 11 (2002)
16. Qu, H., Peng, Y., Sun, W.: Texture Image Retrieval Based on Contourlet Coefficient Modeling with Generalized Gaussian Distribution. Advances in Computation and Intelligence, 493–502 (2007)
17. Burges, C.J.C.: A Tutorial on Support Vector Machines for Pattern Recognition. Kluwer Academic Publishers, Boston (1998)
18. Duan, K., Keerthi, S.S.: Which Is the Best Multiclass SVM Method? An Empirical Study: Multiple Classifier Systems, 278–285 (2005)

Statistical Deformable Model-Based Reconstruction of a Patient-Specific Surface Model from Single Standard X-ray Radiograph

Guoyan Zheng

Institute for Surgical Technology and Biomechanics, University of Bern, CH-3014, Bern, Switzerland
guoyan.zheng@ieee.org

Abstract. In this paper, we present a hybrid 2D-3D deformable registration strategy combining a landmark-to-ray registration with a statistical shape model-based 2D-3D reconstruction scheme, and show its application to reconstruct a patient-specific 3D surface model of the pelvis from single standard X-ray radiograph. The landmark-to-ray registration is used to find an initial scale and an initial rigid transformation between the X-ray image and the statistical shape model. The estimated scale and rigid transformation are then used to initialize the statistical shape model-based 2D-3D reconstruction scheme, which combines statistical instantiation and regularized shape deformation with an iterative image-to-model correspondence establishing algorithm. Quantitative and qualitative results of a feasibility study on clinical and cadaveric datasets are given, which indicate the validity of our approach.

Keywords: point distribution model, statistical deformable 2D-3D registration, surface reconstruction, pelvis.

1 Introduction

Constructing a three-dimensional (3D) surface model from two-dimensional (2D) calibrated fluoroscopic image(s) is a challenging task. *A priori* information is often required to handle this otherwise ill-posed problem. In Fleute and Lavallée [1], a point distribution model (PDM) of distal femur was iteratively fitted to the bone contours segmented on the X-ray views by sequentially optimizing the rigid and non-rigid parameters. It utilizes the principle of the shortest distance between the projection ray of an image edge pixel and a line segment on the apparent contour to set up image-to-model correspondence. In Benameur et al. [2][3], a PDM of scoliotic vertebrae was fitted to two radiographic views by simultaneously optimizing both shape and pose parameters. The optimal estimation was obtained by iteratively minimizing a combined energy function, which is the sum of a likelihood energy term measured from an edge potential field on the images and a prior energy term measured from the statistical shape models. Previously, we proposed a 2D-3D reconstruction scheme combining statistical instantiation and regularized shape deformation with an iterative image-to-model correspondence establishing algorithm, and showed its application to reconstruct the surface model of the proximal femur [4]. Common to all these previous works are (a) at least two images are used as the input; and (b) all images are calibrated. However, in clinical practice, no radiograph-specific calibration is available for most cases. The only information that we can assume to know about the radiograph is the image scale (mm/pixel) and the distance from the focal point to the imaging plane or to the film. As long as the radiograph is acquired in a standardized way, which is performed in a clinical routine [5], both parameters can be directly retrieved from the DICOM image of the X-ray radiograph or can be estimated by performing one-time calibration [6].

This paper presents a hybrid 2D-3D deformable registration strategy combining a landmark-to-ray registration with a statistical shape model-based 2D-3D reconstruction scheme, and shows its application to reconstruct a patient-specific 3D surface model of the pelvis from single standard X-ray radiograph. The landmark-to-ray registration is used to find an initial scale and an initial rigid transformation between the X-ray image and the statistical shape model. The estimated scale and rigid transformation are then used to initialize the statistical shape model-based 2D-3D reconstruction scheme, which combines statistical instantiation and regularized shape deformation with an iterative image-to-model correspondence establishing algorithm.

This paper is organized as follows. Section 2 briefly presents the construction of the statistical shape model. Section 3 describes the statistically deformable 2D-3D registration approach. Section 4 presents the experimental results, followed by the conclusions in Section 5.

2 Construction of the Statistical Shape Model of the Pelvis

We chose the point distribution model (PDM) as the representation of our statistical shape model of the pelvis. The pelvic PDM used in this paper was constructed from a training database consisted of 14 segmented binary volumes (12 of them were segmented from CT scans of dry bones and the rest 2 were segmented from patient CT scans) where the sacrum was removed from each dataset. Demon's algorithm [7] as implemented in MedINRIA [8] was used to estimate the dense deformation fields between the reference binary volume and the other 13 binary volumes. Each estimated deformation field was then used to displace the positions of the vertices on the reference surface model, which was constructed from the reference binary volume, to the associated target volume. We thus obtained 14 surface models with established correspondences.

Following the alignment, the PDM is constructed as follows. Let $x_i, i = 0, 1, ..., m-1$, be m (here $m=14$) members of the aligned training surfaces. Each member is described by a vectors x_i with N (here $N=24994$) vertices:

$$x_i = \{x_0, y_0, z_0, x_1, y_1, z_1, ..., x_{N-1}, y_{N-1}, z_{N-1}\} \quad (1)$$

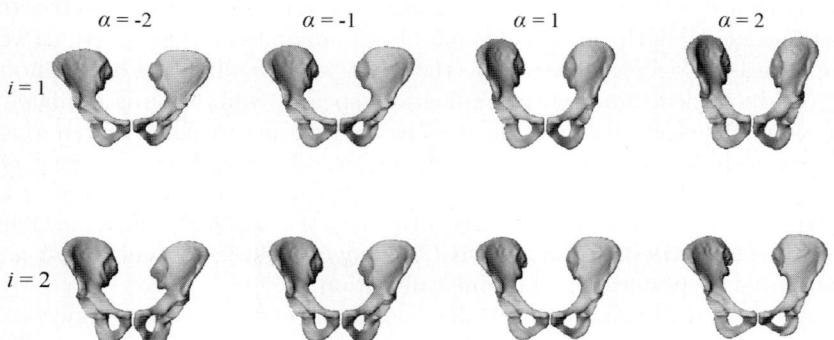

Fig. 1. The first two eigen modes of variation of our PDM of the pelvis. The shape instances were generated by evaluating $\bar{x} + \alpha\sigma_i \mathbf{p}_i$ with $\alpha \in \{-2, -1, 1, 2\}$.

The PDM is obtained by applying principal component analysis.

$$\begin{array}{c} D = ((m-1)^{-1}) \cdot \sum_{i=0}^{m-1}(\mathrm{x}_i - \bar{\mathrm{x}})(\mathrm{x}_i - \bar{\mathrm{x}})^T \\ P = (\mathbf{p}_0, \mathbf{p}_1, ...); \; D \cdot \mathbf{p}_i = \sigma_i^2 \cdot \mathbf{p}_i \end{array} \quad (2)$$

where $\bar{\mathrm{x}}$ and D are the mean vector and the covariance matrix, respectively.

Fig 1 shows the variability captured by the first two modes of variation of our PDM.

3 Statistically Deformable 2D-3D Registration

Our single image based surface model reconstruction technique is based on the algorithm that we introduced in [4], which combines statistical instantiation and regularized shape deformation with an iterative image-to-model correspondence establishing algorithm. The image-to-model correspondence is established using a non-rigid 2D point matching process, which iteratively uses a symmetric injective nearest-neighbor mapping operator and 2D thin-plate splines based deformation to find a fraction of best matched 2D point pairs between features extracted from the X-ray images and the projections of the apparent contours extracted from the 3D model. The obtained 2D point pairs are then used to set up a set of 3D point pairs such that we turn a 2D-3D reconstruction problem to a 3D-3D one. The 3D/3D reconstruction problem is then solved optimally in three sequential stages including iterative scaled rigid registration, statistical instantiation, and regularized shape deformation. For details, we refer to our previous works [4].

In our previous work, we asked for 2 or more X-ray images as the input and that all images should be calibrated. However, these requirements are the conditions for the application in our previous work rather than the constraints to our algorithm. Actually, the algorithm that we introduced in [4] can be directly applied to single image, as long as at least four non-colinear point pairs are found.

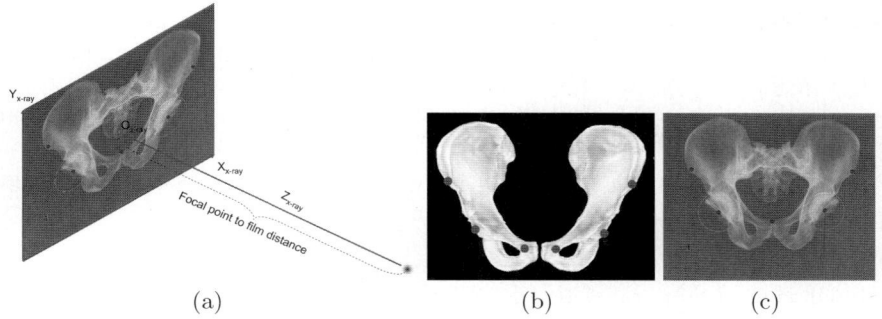

Fig. 2. (a) the radiograph coordinate system and the cone-beam projection model; (b) Landmarks extracted from the mean model of the PDM; and (c) landmarks extracted from radiograph

Similar to the situation when multiple images are used, the convergence of the single image based 2D-3D reconstruction also depends on the initialization. Thus, in the following we focus on the establishment of the projection geometry of the input radiograph, and on a landmark-based scaled registration for initializing the single image based 2D-3D reconstruction.

3.1 Establishment of Projection Geometry

The local coordinate reference and the cone-beam projection model of the radiograph is established as follows (see Fig. 2(a) for details). The image center is taken as the coordinate origin. The X-axis and the Y-axis of the image are taken as the X-axis and the Y-axis of the local coordinate reference of the radiograph. The central projection line is perpendicular to the radiograph plane and its opposite direction is regarded as Z-axis.

3.2 Landmark-Based Scaled Rigid Registration for Initialization

Initialization here means to estimate the initial scale and the rigid transformation between the mean model of the PDM and the input radiograph. For this purpose, we have adopted an iterative landmark-to-ray scaled rigid registration. The five anatomical landmarks that we used here are left and right ASIS, left and right acetabular centers, and pubic symphysis. Their positions on the mean model of the PDM are obtained through point picking (for left and right ASIS, and pubic symphysis) or sphere fitting (for left and right acetabular centers), while their positions on the radiograph are defined through interactive picking (for the projections of left and right ASIS, and pubic symphysis) or circle fitting (for the projections of left and right acetabular centers) (see Fig. 2(b) and 2(c) for details).

Let us denote those landmarks defined on the mean model of the PDM, i.e., the left and the right acetabular centers, the pubic symphysis, and the middle points of the left and the right ASIS, as $v^1_{Mean}, v^2_{Mean}, v^3_{Mean}$, and v^4_{Mean}, respectively; and their corresponding landmarks interactively picked from the radiograph as $v_{X-ray}^1, v_{X-ray}^2, v_{X-ray}^3$, and v_{X-ray}^4 (v_{X-ray}^4 is the middle point of the projections of the left and the right ASIS), respectively. And for each X-ray landmark, we can calculate a projection ray emitting from the focal point to the landmark. We then calculate the length between v_{Mean}^1 and v_{Mean}^2 and denote it as $l_{Mean}^{1,2}$. Using the known image scale, we also calculate the length $l_{X-ray}^{1,2}$ between v_{X-ray}^1 and v_{X-ray}^2. Then, we do:

Data Preparation. In this step, we assume that the line connecting the acetabular centers is parallel to the AP pelvic radiograph plane and is certain distance away from the imaging plane (in all the experiments reported in this paper, we used a fixed distance of 150 mm). Using this assumption and the correspondences between the landmarks defined in the CT volume and those from the radiograph, we can compute two points $\bar{v}_{X-ray}^1$ and $\bar{v}_{X-ray}^2$ on the projection rays of v_{X-ray}^1 and v_{X-ray}^2, respectively (see Fig. 3(a)), which satisfy:

$$\bar{v}_{X-ray}^1 \bar{v}_{X-ray}^2 // v_{X-ray}^1 v_{X-ray}^2; \text{ and } |\bar{v}_{X-ray}^1 - \bar{v}_{X-ray}^2| = l_{X-ray}^{1,2} \times \frac{F-d}{F} \quad (3)$$

where F is the known distance from the focal point to the imaging plane and d is the assuming distance from the acetabular centers to the imaging plane.

The current scale s between the mean model and the input image is then estimated as,

$$s = |\bar{v}_{X-ray}^1 - \bar{v}_{X-ray}^2|/l_{Mean}^{1,2} \quad (4)$$

Using s, we scale all landmark positions on the mean model and denote them as $\{\bar{v}_{Mean}^i; i=1,2,3,4\}$. We then calculate the distances from $\bar{v}_{Mean}^3$ and $\bar{v}_{Mean}^4$ to line $\bar{v}_{Mean}^1 \bar{v}_{Mean}^2$ and denote it as $\bar{l}_{Mean}^{3,1-2}$ and $\bar{l}_{Mean}^{4,1-2}$, respectively.

Next we find two points, point $\bar{v}_{X-ray}^3$ on the projection ray of v_{X-ray}^3 whose distance to the line $\bar{v}_{X-ray}^1 \bar{v}_{X-ray}^2$ is equal to $\bar{l}_{Mean}^{3,1-2}$, and point $\bar{v}_{X-ray}^4$ on the projection ray of v_{X-ray}^4 whose distance to the line $\bar{v}_{X-ray}^1 \bar{v}_{X-ray}^2$ is equal to $\bar{l}_{Mean}^{4,1-2}$. A paired-point matching [9] based on $\{\bar{v}_{Mean}^i; i=1,2,3,4\}$ and $\{\bar{v}_{X-ray}^i; i=1,2,3,4\}$ is used to calculate a updated scale s_0 and a rigid transformation $\bar{T}_{Mean}^{X-ray}$ (see Fig. 3(a) for details). From now on, we assume that all information defined in the mean model coordinate frame has been transformed into the radiograph coordinate frame using s_0 and $\bar{T}_{Mean}^{X-ray}$. We denote the transformed mean model landmarks as $\{\tilde{v}_{Mean}^i\}$.

Iteration. The following steps are iteratively executed until convergence:

1. For a point $\tilde{v}_{Mean}^i$, we find a point on the corresponding projection ray of v_{X-ray}^i which has the shortest distance to the point $\tilde{v}_{Mean}^i$ and denote it as $\tilde{v}_{X-ray}^i$ (see Fig. 3(b)). We then perform a paired-point matching [9] using the extracted point pairs to compute a scale $\tilde{s}$ and a rigid transformation $\Delta \tilde{T}_{Mean}^{X-ray}$.
2. We update the mean model coordinate frame using $\tilde{s}$ and $\Delta \tilde{T}_{Mean}^{X-ray}$.

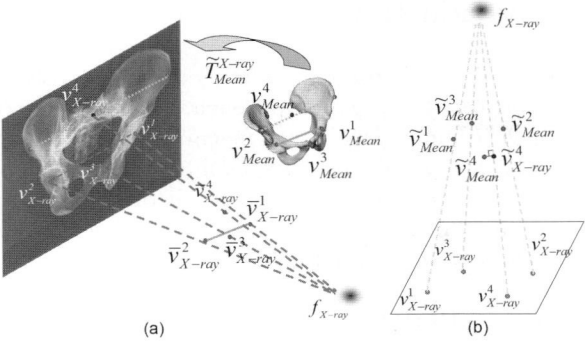

Fig. 3. Iterative landmark-to-ray registration. (a) schematic view of data preparation; and (b) schematic view of finding 3D point pairs.

3.3 2D-3D Reconstruction

The estimated scale and the rigid transformation between the mean model and the input image are then treated as the starting values for the algorithm that we introduced in [4]. As a feature-based 2D-3D reconstruction approach, our algorithm requires a pre-requisite image feature extraction. In this paper, observing the superimposition of the projections of different bone structures around the pelvis and the post-operative characteristic of the X-ray radiograph, we opt for an interactive way to identify contours of the pelvis. We thus developed a program allowing the user to define up to eight contours by interactively picking points from the radiograph. Each contour is then interpolated by a cubic-spline to have the same resolution as the image resolution. The extracted contours are then used together with the initial estimation of the scale and the rigid transformation as the input to our PDM based 2D-3D reconstruction scheme for an accurate reconstruction of a surface model of the pelvis. Fig. 4 shows different stages of reconstruction of a patient-specific surface model of the pelvis from single standard X-ray radiograph of a cadaver. The reconstructed surface model of the pelvis can then be used to determine the post-operative cup orientation.

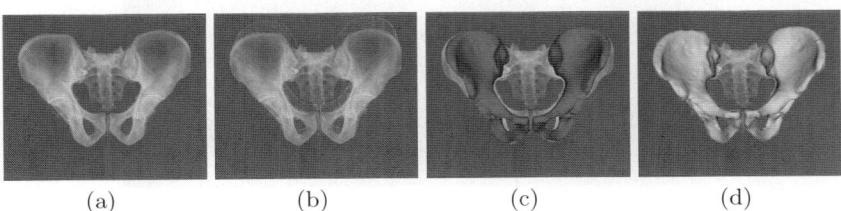

Fig. 4. (a) the image contours (white line); (b) establishment of the initial image-to-model correspondences (yellow points: projections of the apparent contours extracted from the mean model; green lines: visualization of the correspondences); (c) the result of the iterative scaled registration; and (d) the final reconstructed surface model

4 Experiments and Results

We designed and conducted experiments on 3 cadaveric pelvis datasets and 1 patient dataset to validate the present approach. Each dataset contains an X-ray radiograph and a CT volume. Two X-ray machines were used to acquire the X-ray radiographs. The X-ray radiographs for all 3 cadaveric pelvis were acquired by one X-ray machine with a focal point to film distance of 1200 mm and a pixel size of 0.143 mm while the X-ray radiograph for the patient was acquired by the other X-ray machine with a focal point to film distance of 1016 mm and a pixel size of 0.17 mm.

To evaluate the reconstruction accuracy, we established the ground truth for each dataset from the associated CT volume. A commercially available software package, AMIRA 5.0 (TGS Europe, Paris, France) was used for semiautomatic segmentation of the surface model of the pelvis from each CT volume. The derived ground truths were transformed to the associated reference coordinate systems of the reconstructed surface models by performing a surface-based scaled rigid registation [10]. After that, we used the open source tool *MESH* to compute the distances between the surface model reconstructed from each X-ray radiograph and its associated ground truth. The median and the mean reconstruction errors are presented in Table 1. An average mean reconstruction error of 1.6 mm was found. Such accuracy was regarded as accurate for surgical navigation applications according to Livyatan et al. [11]. Fig. 5 shows the error distributions of the reconstructed surface model of a pelvis.

Table 1. Difference between the estimated results estimated and the ground truths

Bone Index	cadaver_01	cadaver_02	cadaver_03	Patient_01	Mean
Median (mm)	1.3	1.2	1.4	1.3	1.3
Mean (mm)	1.6	1.4	1.7	1.6	1.6

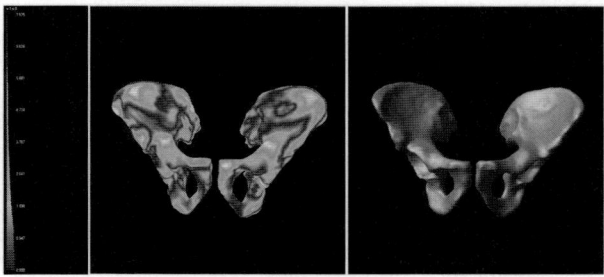

Fig. 5. Color-coded error distribution when the reconstructed surface model of a pelvis was compared to its ground truth. left: color-coded errors; middle: the reconstructed surface model with color-coded error distributions; right: the ground truth.

5 Conclusions

In this paper, we presented a statistically deformable 2D-3D registration approach to instantiate a patient-specific pelvis surface model from single standard X-ray radiograph. We designed and conducted feasibility experiments on three cadaver datasets and on one patient dataset to validate the accuracy of the present approach. Our experimental results demonstrate that it is feasible to reconstruct a patient-specific model from single standard X-ray radiograph for surgical navigation applications.

References

1. Fleute, M., Lavallée, S.: Nonrigid 3-D/2-D registration of images using statistical models. In: Taylor, C., Colchester, A. (eds.) MICCAI 1999. LNCS, vol. 1679, pp. 138–147. Springer, Heidelberg (1999)
2. Benameur, S., Mignotte, M., Parent, S., et al.: 3D/2D registration and segmentation of scoliotic vertebra using statistical models. Comput. Med. Imag. Grap. 27, 321–337 (2003)
3. Benameur, S., Mignotte, M., Parent, S., et al.: A hierarchical statistical modeling approach for the unsupervised 3D biplanar reconstruction of the scoliotic spine. IEEE Trans. Biomed. Eng. 52, 2041–2057 (2005)
4. Zheng, G., Ballester, M.Á.G., Styner, M.A., Nolte, L.-P.: Reconstruction of patient-specific 3D bone surface from 2D calibrated fluoroscopic images and point distribution model. In: Larsen, R., Nielsen, M., Sporring, J. (eds.) MICCAI 2006. LNCS, vol. 4190, pp. 25–32. Springer, Heidelberg (2006)
5. Della Valle, C.J., et al.: Primary total hip arthroplasty with a flanged, cemented all-polyethylene acetabular component: evaluation at a minimum of 20 years. J. Arthroplasty 19, 23–26 (2004)
6. The, B.: Digital radiographic peroperative planning and postoperative monitoring of total hip replacements - techniques, validation and implementation. Doctoral dissertations, University Medical Center Groningen, The Netherlands (2006)
7. Thirion, J.-P.: Image matching as a diffusion process: an analogy with Maxwell's demons. Med. Image Anal. 2, 243–260 (1998)
8. Toussaint, N., Souplet, J.-C., Fillard, P.: MedINRIA: DT-MRI processing and visualization software. In: MICCAI 2007 Workshop on Interaction in Medical Image and Visualization (2007)
9. Veldpaus, F.E., et al.: A least-square algorithm for the equiform transformation from spatial marker coordinates. J. Biomech. 21, 45–54 (1988)
10. Besl, P., McKay, N.D.: A method for registration of 3D shapes. IEEE Trans. Pattern Anal. Mach. Intell. 14, 239–256 (1992)
11. Livyatan, H., Yaniv, Z., Joskowicz, L.: Gradient-based 2-D/3-D rigid registration of fluoroscopic X-ray to CT. IEEE Trans. Med. Imaging 22, 1395–1406 (2003)

Plant Species Identification Using Multi-scale Fractal Dimension Applied to Images of Adaxial Surface Epidermis

André R. Backes[1], Jarbas J. de M. Sá Junior[1], Rosana M. Kolb[3], and Odemir M. Bruno[2]

[1] Instituto de Ciências Matemáticas e de Computação (ICMC)
Universidade de São Paulo (USP)
backes@icmc.usp.br
jarbas_joaci@yahoo.com.br
[2] Departamento de Ciências Biológicas
Universidade Estadual Paulista Júlio de Mesquita Filho
rosanakolb@hotmail.com
[3] Instituto de Física de São Carlos (IFSC)
Universidade de São Paulo (USP)
bruno@ifsc.usp.br

Abstract. This paper presents the study of computational methods applied to histological texture analysis in order to identify plant species, a very difficult task due to the great similarity among some species and presence of irregularities in a given species. Experiments were performed considering 300 × 300 texture windows extracted from adaxial surface epidermis from eight species. Different texture methods were evaluated using Linear Discriminant Analysis (LDA). Results showed that methods based on complexity analysis perform a better texture discrimination, so conducting to a more accurate identification of plant species.

Keywords: plant identification, complexity, multi-scale fractal dimension, texture analysis.

1 Introduction

Traditional methods used in taxonomy, which use arborized plants and have in the external morphology their main tool for the taxa identification, not always are adequate to solve taxonomic problems [1]. Although not so accessible as those used in external morphology, anatomical methods have been used increasingly, with the purpose of searching characteristics that may assist in solving taxonomic problems [2]. Another difficulty comes from the fact that taxa identification is widely based on morphologic characteristics of reproductive organs, not always present in the sample. Otherwise, anatomic characters have shown to be important in the identification of some taxa, even when the samples present a vegetative state [2]. Different types of stomata and trichromes [1], cell shape, cuticle presence/thickness of cuticle, proportion between palisade and spongy parenchyma, presence of tissues (such as hypodermis), secretion structures, crystals, etc [3], have been used in the characterization and taxonomic understanding of different groups. However, other relevant characteristics, such as color, texture, complexity of the anatomic cuts has not been considered.

Although there is no formal definition about the concept of texture, it is easily identified by humans, and it is rich in visual information. In general, textures are complex visual patterns composed by entities, or sub-patterns, with bright, color, orientation and size characteristics [4]. So, textures supply very useful informations for automatic recognition and interpretation of an image by a computer [5].

This work aims to assess computational methods of texture analysis in histological images of adaxial surface epidermis in order to identify plant species by its leaf tissue. For this purpose, the following eight species were considered: *Byrsonima intermedia* A. Juss., *Miconia albicans* (Sw.) Triana, *Tibouchina stenocarpa* (DC.) Cogn., *Vochysia tucanorum* Mart., *Xylopia aromática* (Lam.) Mart., *Gochnatia polymorpha* (Less.) Cabrera, *Miconia chamissois* Naudin and *Jacaranda caroba* (Vell.) A. DC., typical species of the neotropical savanna of Brazil, locally known as "cerrado". The approach proposed in this work is unedited uses of computational methods applied to histological images and uses of analysis of texture of foliar epidermis as a novel descriptor to be used in taxonomy.

2 Materials and Methods

All leaf samples were collected at Estação Ecológica de Assis, São Paulo State, Brazil, situated between 22°33'65" - 22°36'68"S and 50°22'29" - 50°23'00"W.

2.1 Sample Preparation and Image Acquisition

Each sample consisted of a middle fragment of completely expanded leaves, between the main vein and the leaf margin, collected from five randomly chosen individuals of each species. All samples were fixed in FAA70, dehydrated in a graded ethanol series, infiltrated and embedded in paraffin and cut into $8\mu m$ sections. The cross-sections obtained were stained with astra blue-basic fucsine and permanently mounted in entellan.

Six images of different regions of each sample were obtained using a trinocular microscope Leica, model DM-1000, coupled with a video camera Leica, DFC-280. These images were amplified by factor 200× (Figure 1) and windows 150 × 300 pixels were cropped out from them (Figure 2). Altogether, 30 texture windows were acquired per each species. When hypodermis was present, like in *T. stenocarpa*, this was also considered in the process. We did not discriminate it from the epidermis, and we named all these cell layers as adaxial surface epidermis.

To obtain the final set of images used to assess the success rates of the evaluated methods, the adaxial surface epidermis was automatically selected from the images. For this purpose, the Mumford-Shah algorithm [6] was used to segment

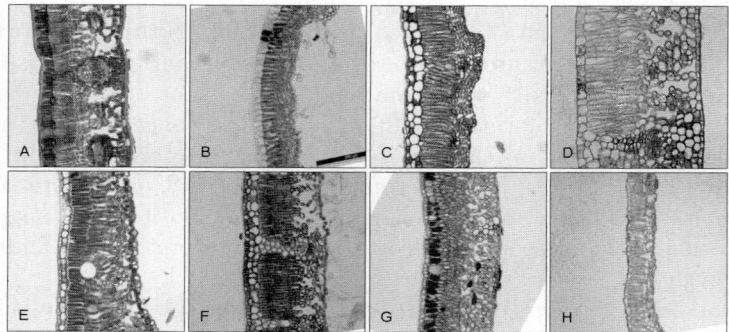

Fig. 1. Images of cross-sections of leaves of the species: A - *Byrsonima intermedia*, B - *Miconia albicans*, C - *Tibouchina stenocarpa*, D - *Vochysia tucanorum*, E - *Xylopia aromatica*, F - *Gochnatia polymorpha*, G - *Miconia chamissois* e H - *Jacaranda caroba*

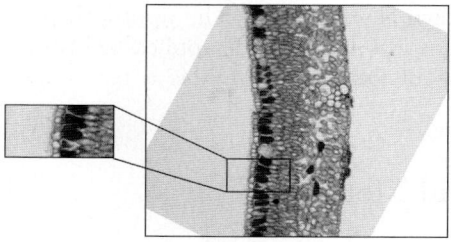

Fig. 2. Image of cross-section of *Miconia chamissois* leaf and respective window 150 × 300 pixels containing adaxial surface epidermis

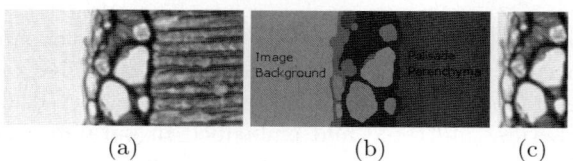

Fig. 3. Images of adaxial surface epidermis of *Tibouchina stenocarpa* species. a - original window (150 × 300 pixels); b - segmented window (150 × 300 pixels); c - segmented epidermis (150 × measure of adaxial surface epidermis thickness).

the palisade parenchyma and the image background in order to find the borders of the adaxial surface epidermis. Thus, new images containing only this region of the leaf were created, as shown by Figure 3.

However, these images presented different widths, which are determined by the adaxial surface epidermis thickness, what makes impossible the extraction of texture features. To solve this, it was adopted a standard window of 300 × 300

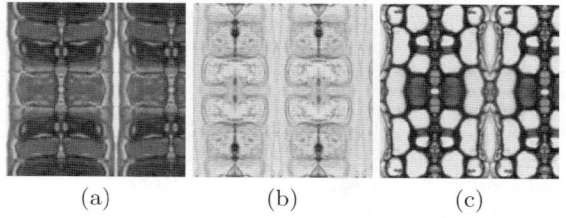

Fig. 4. Mosaic of 300 × 300 pixels of adaxial surface epidermis of the species: a - *Byrsonima intermedia*, b - *Miconia albicans*, c - *Tibouchina stenocarpa*

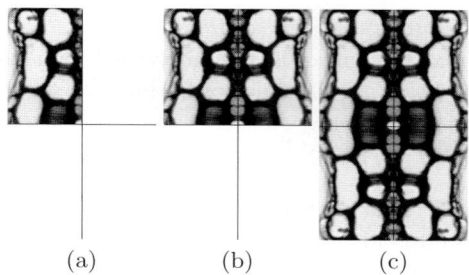

Fig. 5. Process of building a texture mosaic by copy and reflection: (a) Original texture; (b) Copy and reflection; (c) Copy and reflection of the previous step. This process continues until the texture of 300 × 300 is achieved.

pixels that is a mosaic composed by reflected images of the adaxial surface epidermis, as shown by Figure 4. So, the original image is copied and reflected over y axis. The resulting image is placed beside the original. A new copy and reflection are performed over the previous image. This time, the reflection is over x axis and the resulting image is placed under the previous. This process continues, alternating y and x axis, until an image with 300 × 300 of size be composed - see Figure 5.

3 Fractal Dimension

The fractal dimension is a property from fractal objects related to its complexity. It is widely used in literature in the characterization of objects and images in terms of space occupation and self-similarity [7,8,9]. Among the methods found in literature, the Bouligand-Minkowski Fractal Dimension is considered the most precise [8,9]. Using a disc of radius r, this method carries out object dilation in order to compute its influence region, which is very sensitive to structural changes.

For texture analysis, consider the texture as a set of points $S \in R^3$. Each element $s \in S$ is defined by the triple (y, x, z), where y and x are the pixel

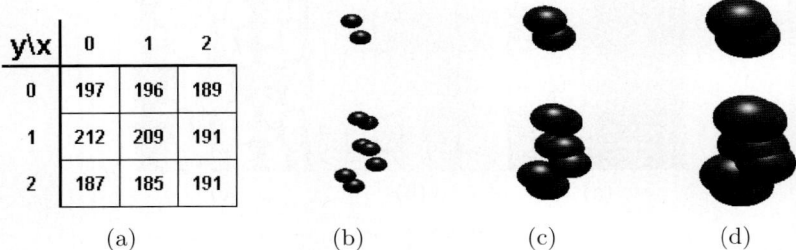

Fig. 6. Example of texture dilation: (a) Set of image pixels; (b) Pixels modeled as points in R^3; (c) Dilation using $r = 2$; (d) Dilation using $r = 3$

coordinates in the texture and z is the gray-level value at pixel (y, x). The influence volume, $V(r)$, for a given dilation radius r is defined as

$$V(r) = \left|\left\{s' \in R^3 | \exists s \in S : |s - s'| \leq r\right\}\right|,$$

where s' is a point in R^3 that dists r or less from s. As the r value increases, the spheres produced by different pixels start to interact with each other. This interaction disturbs the way the influence volume $V(r)$ increases, what makes it very sensitive to detect even small changes in the texture (Figure 6). From the influence volume $V(r)$, the Bouligand-Minkowski fractal dimension D is estimated as

$$D = 3 - \lim_{r \to 0} \frac{\log V(r)}{\log(r)},$$

where D is a number within $[0; 3]$.

4 Multi-scale Fractal Dimension

Besides the great importance of the fractal dimension in literature, natural objects are not real fractals, and so, its complexity goes to zero as the scale increases. For these objects, the fractal dimension is estimated by carrying out linear interpolation over the logarithm curve of the volume $(V(r))$ in terms of dilating radius. This yields a line with angular coefficient α, where $D = 3 - \alpha$ is the estimated fractal dimension. However, the log-log curve computed by Bouligand-Minkowski method presents a very rich degree of details along the scales that cannot be expressed by a single numeric value.

The Multi-Scale Fractal Dimension [10,11] has been proposed as a solution for this deficiency in the characterization of objects using complexity. Instead of using linear interpolation to estimate the angular coefficient of the log-log curve, this approach exploits the infinitesimal limit of the linear interpolation by using the derivative. As a result, a curve that expresses complexity value along the scale is yielded (Figure 7), what provides a more efficient characterization of the

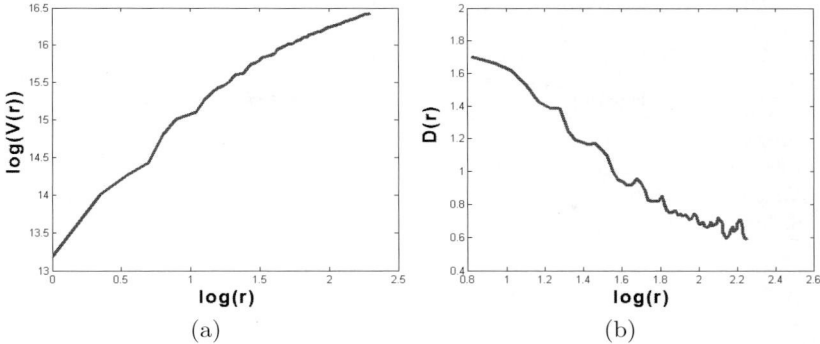

Fig. 7. (a) Log-log curve; (b) Multi-scale Fractal Dimension

object. From the logarithm of the influence volume, $\log V(r)$, the Multi-Scale Fractal Dimension $D(r)$ is computed as

$$D(r) = 3 - \frac{d \log V(r)}{d \log r},$$

where $D(r)$ represents the complexity of the object at scale r.

5 Experiment and Results

We carried out the complexity analysis by applying our Bouligand-Minkowski based approach over each texture sample. Finite Difference method [12] was applied over log-log curves, so resulting in a Multi-scale fractal dimension curve characteristic for that texture. Experiments showed that the best results are yielded when a dilation radius $r = 10$ is considered, what results in a Multi-scale curve containing 50 descriptors.

The data evaluation of Multi-scale curves was performed using Linear Discriminant Analysis (LDA) [13,14], a supervised statistical classification method. LDA objectives to find a linear combination of descriptors (independent variables) that minimizes the intra-classes variance of the samples while maximizes the inter-classes variance. The method also used the *leave-one-out cross-validation* strategy to define training and test sets during its execution.

Multi-scales curves were also compared with traditional texture analysis methods, so that, a better performance evaluation is accomplished. The following methods were used: Fourier descriptors (energy of the 63 most meaningful coefficients) [15], Wavelet descriptors (energy and entropy from details of 3 decompositions) [16,17], Co-occurrence matrices (energy and entropy from matrices computed using distances of 1 and 2 pixels with angles of $-45°$, $0°$, $45°$, $90°$) [4] and Gabor filters (energy from 4 rotations and 4 scales with frequencies 0.01, 0.0311, 0.0965, 0.3) [18,19]. Despite the many different versions existent in the literature, this paper considers the conventional implementation of the methods.

Table 1. Classification performance of different texture descriptors

Descriptor	No of Descriptors	Samples correctly classified	Success rate (%)
Gabor filters	16	205	85.42
Fourier descriptors	63	184	76.67
Co-occurrence matrices	16	221	92.08
Wavelet descriptors	36	210	87.50
M.S. Fractal Dimension	50	224	93.33

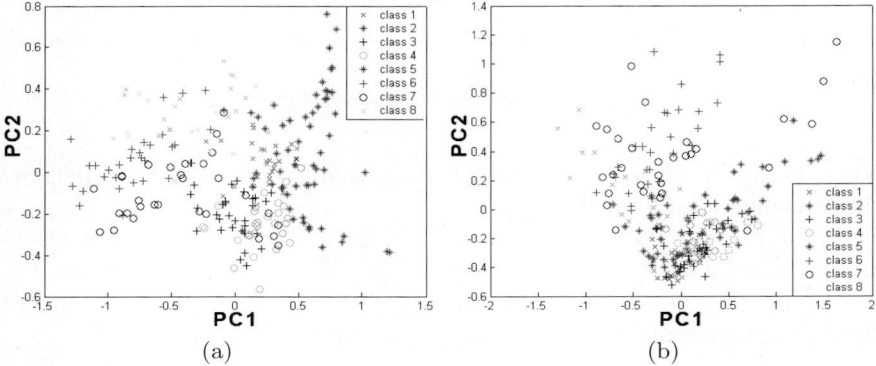

Fig. 8. Plot of the first and second components from PCA: (a) Multi-scale curves; (b) Co-occurrence matrices

Results (Table 1) demonstrates that Multi-scale curves are more robust in the classification of the histological texture patterns evaluated. This is due to the great sensitiveness and accuracy of the Bouligand-Minkowski method to detect small changes in the texture. This, combined with the Multi-scale Fractal dimension, allows a texture analysis at different scales, i.e., micro and macro texture are considered to provide a better texture analysis. An analysis using Principal Component Analysis (PCA) [13,14] was also performed over the methods that presented the best results. Figure 8 shows the plots of first and second components computed from Multi-scale curves and Co-occurrence matrices. We note higher dispersion inter classes in Multi-scale curves, i.e., there is a smaller intersection among classes. Otherwise, Co-occurrence matrices tend to converge in a specific point, so different classes are superimposed, what diminishes the quality of the descriptor.

6 Conclusion

This paper presented a comparison among different computational methods of texture analysis in a plant species identification task. Experiments were conducted using histological images of adaxial surface epidermis from different plant species. Statistical analysis of the methods was carried out using Linear discriminant Analysis (LDA).

Among the compared methods, the Bouligand-Minkowski method and Multiscale fractal dimension offered the best results. The first method allows the study of texture complexity in terms of the radius of the influence volume. The second one is applied over Bouligand-Minkowski method, and it allows to exploit characteristics of complexity present in the texture at different scales, so providing a more efficient plant texture characterization. Plant identification is a difficult task due to the great similarity among some species and irregularities present in a given species. However, results showed that complexity analysis methods play an important role in the analysis of histological images and that the adaxial surface epidermis is a feasible source of plant characteristics.

Acknowledgments

A.R.B. acknowledges support from FAPESP (2006/54367-9), J. J. M. S. Jr. acknowledges support from CNPq (135251/2006) and O.M.B. acknowledges support from CNPq (306628/2007-4).

References

1. Metcalfe, C.R., Chalk, L.: Anatomy of dicotyledons, 2nd edn. Oxford University Press, Oxford (1979)
2. Stace, C.A.: Plant taxonomy and biosystematics, 2nd edn. Cambridge University Press, Cambridge (1989)
3. Robinson, H.: A monograph on foliar anatomy of the genera connelia, cottendorfia and navia (bromeliaceae). Smithsonian Contributions of Botany 2, 1–41 (1969)
4. Haralick, R.M.: Statistical and structural approaches to texture. Proc. IEEE 67(5), 786–804 (1979)
5. Bala, J.W.: Combining structural and statistical features in a machine learning technique for texture classification. In: IEA/AIE, vol. 1, pp. 175–183 (1990)
6. Chambolle, A.: Image segmentation by variational methods: Mumford and Shah functional and the discrete approximations. SIAM J. Appl. Math. 55(3), 827–863 (1995)
7. Backes, A.R., Bruno, O.M.: A new approach to estimate fractal dimension of texture images. In: Elmoataz, A., Lezoray, O., Nouboud, F., Mammass, D. (eds.) ICISP 2008. LNCS, vol. 5099, pp. 136–143. Springer, Heidelberg (2008)
8. da, L., Costa, F., Cesar Jr., R.M.: Shape Analysis and Classification: Theory and Practice. CRC Press, Boca Raton (2000)
9. Tricot, C.: Curves and Fractal Dimension. Springer, Heidelberg (1995)
10. Gonzalez, R.C., Woods, R.E.: Digital Image Processing, 2nd edn. Prentice-Hall, New Jersey (2002)
11. de, O., Plotze, R., Falvo, M., Pádua, J.G., Bernacci, L.C., Vieira, M.L.C., Oliveira, G.C.X., Bruno, O.M.: Leaf shape analysis using the multiscale minkowski fractal dimension, a new morphometric method: a study with passiflora (passifloraceae). Canadian Journal of Botany 83(3), 287–301 (2005)

12. Smith, G.D.: Numerical Solution of Partial Differential Equations: Finite Difference Methods, 3rd edn. Oxford (1986)
13. Everitt, B.S., Dunn, G.: Applied Multivariate Analysis, 2nd edn. Arnold (2001)
14. Fukunaga, K.: Introduction to Statistical Pattern Recognition, 2nd edn. Academic Press, London (1990)
15. Azencott, R., Wang, J.P., Younes, L.: Texture classification using windowed fourier filters. IEEE Trans. Pattern Anal. Mach. Intell. 19(2), 148–153 (1997)
16. Sengür, A., Türkoglu, I., Ince, M.C.: Wavelet packet neural networks for texture classification. Expert Syst. Appl. 32(2), 527–533 (2007)
17. Huang, P.W., Dai, S.K., Lin, P.L.: Texture image retrieval and image segmentation using composite sub-band gradient vectors. J. Visual Communication and Image Representation 17(5), 947–957 (2006)
18. Jain, A.K., Farrokhnia, F.: Unsupervised texture segmentation using Gabor filters. Pattern Recognition 24(12), 1167–1186 (1991)
19. Idrissa, M., Acheroy, M.: Texture classification using gabor filters. Pattern Recognition Letters 23(9), 1095–1102 (2002)

Fast Invariant Contour-Based Classification of Hand Symbols for HCI

Thomas Bader[1], René Räpple[1], and Jürgen Beyerer[1,2]

[1] Universität Karlsruhe, Institut für Anthropomatik, Germany
[2] Fraunhofer IITB, Institut für Informations- und Datenverarbeitung, Germany
bader@ies.uni-karlsruhe.de, rene.raepple@gmail.com,
beyerer@iitb.fraunhofer.de

Abstract. Video-based recognition of hand symbols is a promising technology for designing new interaction techniques for multi-user environments of the future. However, most approaches still lack performance for direct application for human-computer interaction (HCI).

In this paper we propose a novel approach to contour-based recognition of hand symbols for HCI. We present adequate methods for normalization and representation of signatures extracted from boundary contours, which allow for efficient recognition of hand poses invariant to translation, rotation, scale and viewpoint variations, which are relevant for many applications in HCI. The developed classification system is evaluated on a dataset containing 13 hand symbols captured from four different persons.

1 Introduction

The paradigm of personal computers, which mainly consist of a single display with a mouse and a keyboard as input devices, dominated human-computer interaction for the past decades. In recent years, however, the variety of commercially available display types, sizes and form factors increased tremendously, which allows for designing completely new human-computer interfaces (HCIs). The main challenge in this field for the next years will be to design new input devices and interaction techniques, which allow for intuitive interaction in multi-user environments consisting of multiple heterogenous displays as shown in Fig.1(a).

Video-based recognition of user actions is a promising technology for creating input devices for such environments. It allows for the capturing of user input in a non-intrusive way, independent from display technologies. Especially recognition of hand gestures enables the design of many natural interaction techniques which do not require additional devices or tools.

In this paper we first formulate requirements regarding performance of video-based hand gesture recognition for HCI and review related literature in Sect. 2. In Sect. 3 we present a new approach to contour-based classification of hand symbols, invariant to influences relevant for many HCI applications. In Sect. 4 evaluation results for data from a scenario as shown in Fig.1 are presented and discussed.

 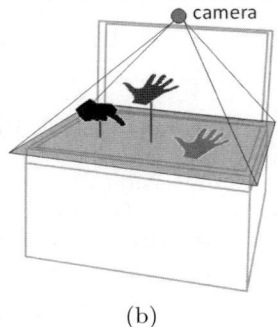

(a) (b)

Fig. 1. Multi-Display environment which was used for evaluation [1]. The gesture recognition system serves as video-based input device for interaction across multiple displays.

2 Video-Based Recognition of Hand Gestures for HCI

The task of video-based recognition of hand gestures can be divided into three subtasks, namely image segmentation, hand segmentation and classification. *Image segmentation* means the separation of regions in video images, which potentially contain a hand, from irrelevant background. *Hand segmentation* denotes the problem of separating the human hand from other parts in segmented image regions with similar reflectance properties as human skin (e.g. arm segments). *Classification* means the recognition of hand symbols, which represent certain states of the hand. This classification should be at least invariant to the basic geometrical transformations rotation, translation and scaling as well as to relevant changes of viewpoint and user dependent properties like hand size and individual realization of the different hand symbols. Another important requirement results from the application for HCI and concerns processing time for the whole process. The system reaction time to user inputs should be smaller than 50ms to not be recognized by the user.

This paper mainly focuses on the last mentioned task, namely hand symbol classification. Approaches to this problem roughly can be divided into two types: model- and appearance-based approaches. Model-based approaches use geometrical and/or kinematic models which are fit to the image data (e.g.[2,3]). However, due to the large number of degrees of freedom of the human hand, such approaches can hardly be implemented on current hardware satisfying the requirements regarding processing time. Appearance-based approaches only try to recognize certain characteristic views of a hand symbol, which are relevant for the application. The range of viewpoints considered in appearance-based approaches varies from full 3D [4,5,6] to one single viewpoint [7,8,9]. Recognition rate and processing time of full 3D approaches, however, are not adequate for HCI applications. A commonly used feature in appearance-based approaches are Hu's moments [5,8,9]. They deliver good results, at least for a small number of hand symbols and persons. In [7] an approach for contour-based classification of hand symbols captured from one viewpoint is proposed. A localized contour sequence is used for representing the boundary contour. However, calculation of the proposed distance measure is computationally very expensive. An extensive review of other approaches to hand pose estimation can be found in [10].

In this paper we present an new appearance-based approach to hand symbol classification meeting the requirements for use in HCI applications. As a sample application scenario we have chosen interaction in a multi-display environment as shown in Fig.1. We extend the approach of a localized contour sequence as it was used in [7] by additional signatures as features for classification. In contrast to [7], by appropriate normalization and representation of contour signatures, standard classifiers like support vector machines (SVM) [11] can be used instead of nearest-neighbor classification based on signature distance measures.

3 Robust Classification of Hand Symbols

3.1 Image Segmentation

In a first step regions potentially containing a hand are segmented in the image. Since we use images taken in near infrared (NIR), the highly dynamic content emitted by displays, in our case the tabletop display (see Fig.1), is filtered out. Therefore very good segmentation results can already be achieved by a simple background subtraction followed by a threshold filter. For further smoothing of the segmentation results we use morphological filtering.

3.2 Feature Extraction

For all consecutive steps, namely fingertip detection, hand segmentation and hand symbol classification, we use features which are derived from boundary contours of foreground regions. For each region its contour is extracted in clockwise direction (see Fig.2(a)) and represented by the parametric form

$$\boldsymbol{c}(i) = (x(i), y(i))^T \ , \tag{1}$$

where $x(i)$ and $y(i)$ are the coordinates of the i-th contour pixel in the image. From $\boldsymbol{c}(i)$ the one dimensional signatures

$$f_w(i) = \|\boldsymbol{s}_{lr}\| \tag{2}$$

$$f_h(i) = \langle \boldsymbol{s}_{l0}, \bar{\boldsymbol{s}}_{lr} \rangle \tag{3}$$

are calculated, where $\langle \cdot \rangle$ is the standard scalar product in $\mathbb{R}^2$. The arc width function f_w is equal to the length of the chord

$$\boldsymbol{s}_{lr}(i) = \boldsymbol{c}_r - \boldsymbol{c}_l \tag{4}$$

of a contour segment of length L lying between $\boldsymbol{c}_r := \boldsymbol{c}(i + L/2)$ and $\boldsymbol{c}_l := \boldsymbol{c}(i - L/2)$ (see Fig.2(a)). In (3) $\bar{\boldsymbol{s}}_{lr}$ is equal to $\boldsymbol{s}_{lr}$ rotated by 90 degrees counterclockwise and normalized to $\|\bar{\boldsymbol{s}}_{lr}\| = 1$. The other variable in (3) is $\boldsymbol{s}_{l0} = \boldsymbol{c}(i) - \boldsymbol{c}_l$.

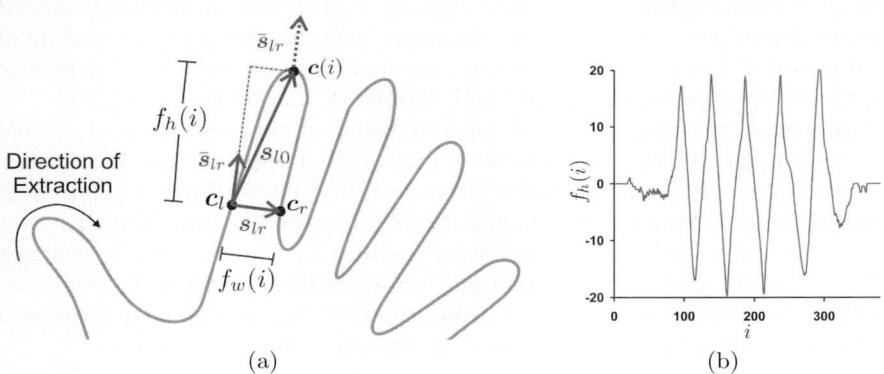

Fig. 2. Contour and feature extraction

Geometrically f_h is the height of the arc defined by $\boldsymbol{c}_r$ and $\boldsymbol{c}_l$, more precisely the orthogonal distance of the point $\boldsymbol{c}(i)$ to the line $\boldsymbol{c}_l + \lambda \boldsymbol{s}_{lr}$. Hence, it is an approximation of the curvature of a low-pass filtered version of the contour, where L determines the low-pass characteristic of the filter [7]. Additionally the two dimensional signature

$$\boldsymbol{f_n}(i) = \bar{\boldsymbol{s}}_{lr} \qquad (5)$$

is calculated, which is an approximation of the normal to the contour in $\boldsymbol{c}(i)$ at every contour pixel (see Fig.2).

All of the above signatures are invariant to *translation* of the whole contour. A *rotation* of the contour leads to a shift of the signatures along i. In later processing steps the signatures are therefore aligned to significant points on the contour of the human hand (see Sect.3.5). *Scaling* the contour has two different effects on f_w and f_h, while f_n is invariant to scaling. First, the distance between two contour points in 2-dimensional space increases linearly with the scale, likewise $\|\boldsymbol{s}_{lr}\|$ and $\|\boldsymbol{s}_{lo}\|$. Therefore the amplitudes of f_w and f_h are normalized by $\max f_w(i)$ and $\max |f_h(i)|$, respectively. The second effect is, that the arc length increases due to the increasing sampling rate (number of pixels per inch/cm). One option to normalize the signatures regarding this effect would be to normalize the total length of the contour before the signature calculation. This, however, would require approximately the same contour length for all hand symbols to be recognized. Otherwise the normalization could destroy properties of the signature, which are relevant for distinguishing the different hand symbols. Therefore we use the length of fingers for normalization of arc length, since it is independent from the hand symbol (see Sect.3.3). At the same time, by finger length normalization, interpersonal differences in finger length are normalized. After the feature extraction the lengths of the signatures are normalized to a constant length as already proposed in [7].

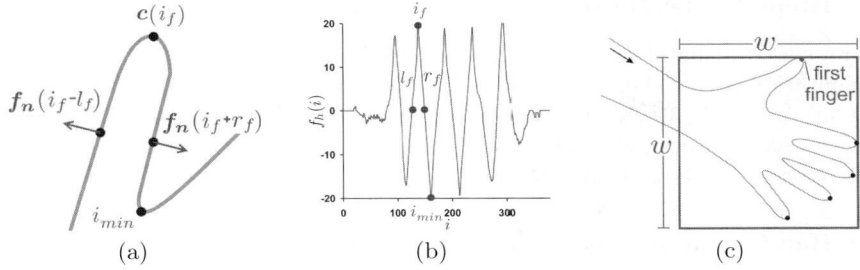

Fig. 3. Fingertip detection and hand segmentation

As a basis for the following processing steps, the local extrema in f_h are detected and will further on be denoted by

$$E = E^+ \cup E^- \ , \tag{6}$$

where $E^+ = \{i^0_{max}, ..., i^M_{max}\}$ and $E^- = \{i^0_{min}, ..., i^N_{min}\}$ are two sets that contain the indices of all the local maxima and minima.

3.3 Fingertip Detection

In this section we present an algorithm for scale invariant fingertip detection. The algorithm is based on unnormalized signatures and, by detecting fingers and calculating finger lengths, is the basis for the finger length normalization.

We exploit two characteristic properties of fingers which are represented in the f_h and f_n signatures. The first property is, that fingers always induce convex segments of the contour and fingertips correspond to points of maximal curvature within these segments and therefore to local maxima in f_h. The second property is, that edges along both sides of a finger are almost parallel. This leads to contour normals pointing into opposite directions for points lying on opposite edges of a finger (see Fig.3(a)).

The set of indices corresponding to fingertips is denoted as

$$E^f = \{i^0_f, ..., i^F_f\} \subset E^+ \ . \tag{7}$$

The length of a finger with index i_f is calculated as the distance to the closest local minimum on the contour, which represents the beginning of a finger (see Fig.3(a) and (b))

$$len(i_f) = \min_{i_{min} \in E^-} |i_f - i_{min}| \ . \tag{8}$$

The amplitude of the noise in f_h, induced by segmentation and quantization of the real hand silhouette, is approximately 0.5 pixel. As long as the amplitude of f_h at local extrema of the contour curvature is significantly higher, fingers can be detected robustly. In a range of scales relevant for practical applications this is the case.

3.4 Hand Segmentation

In our case the hand cannot be segmented from the arm robustly by image segmentation only. Therefore a bounding box of the hand to be segmented is calculated (see Fig.3(c)). The box is aligned to the position of fingertips detected in the previous step and scaled according to the finger length. Contour points lying outside this box are ignored in further processing steps.

3.5 Hand Symbol Classification

For the classification of hand symbols we evaluated different approaches to normalization and representation of the extracted signatures. Evaluation results on real data are presented in the next section.

In a first step the signatures are normalized with regard to the rotation of the whole contour. This is done by shifting the first finger on the contour which was found in clockwise direction to the beginning of the signatures (see Fig.3(c)). Normalization regarding contour length and scaling is done before the signature calculations by normalization of the finger length. The length of a finger can robustly be determined during fingertip detection as described in Sect.3.3 and is similar for a certain finger for all hand symbols.

Another question regarding classification is, how signatures are to be represented. We evaluated two different approaches. In the first one, f_h and f_w are represented as feature vectors of a constant length N, which contain equidistant samples of the two signatures. In the second approach, we use the first N components of the discrete cosine transform (DCT) for representation of the signatures. The two feature vectors representing f_h and f_w are combined to a vector of length $V_{len} = 2N$ for both of the two approaches, which will be referred to as 'samples' and 'DCT', respectively.

For classification we use a soft-margin SVM for multiple classes[11] with a radial basis function as kernel. As input for training and classification the feature vectors mentioned above are used.

4 Results

In this section we present results achieved with our approach on data of 13 different hand symbols as shown in Fig.4(a) from four different persons and compare them with the performance of features from literature [5,8,9,7]. For each hand symbol and person 300 unsegmented samples with constant scale were captured. The rotation of the hand symbols was varied over each set of samples. Since the distance of the hand to the optical axis of the camera was varied, the viewpoint varies across the sample sets (see Fig.4(b)). Additionally, for evaluating the robustness of the fingertip detection regarding scaling, we used a second dataset containing 200 samples of each of the 13 hand symbols from one person at different scales. The correct number of fingers was detected for 95.11% of the dataset. For the first dataset from four persons without scale variatons the correct number of fingers was detected for 96.10%.

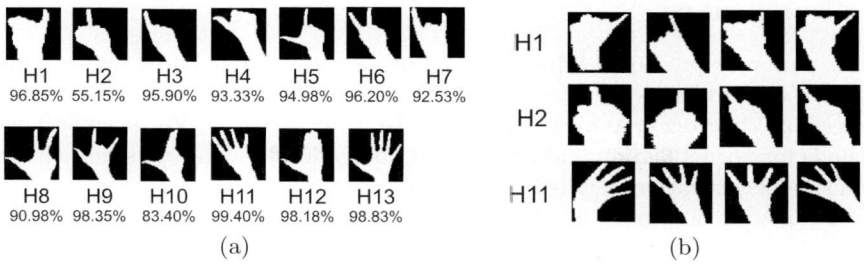

Fig. 4. Dataset for classification: (a) symbol set and recall for configuration 5, (b) variants of selected hand symbols

Table 1. Classification results

Config	Feature	Norm	Representation	V_{len}	recall	σ
1	f_h	SignLen	samples	20	86.90	5.37
2	f_h	FingerLen	samples	20	89.89	2.60
3	f_h, f_w	SignLen	samples	40	90.78	4.53
4	f_h, f_w	SignLen	DCT	40	79.16	18.74
5	f_h, f_w	FingerLen	samples	40	91.87	3.87
6	f_h, f_w	FingerLen	DCT	40	80.44	21.07
7	HuMom [5,8,9]	-	-	7	40.00	6.85

Table 1 shows the results of a 4-fold cross-validation for a dataset of 13 different hand symbols. For each of the for runs the data of a different person was used as testing dataset and was tested against the data of the remaining three persons as training dataset. The columns *recall* and σ show the means and the corresponding standard deviations of the correct classifications over the four runs and all hand symbols. For the 'samples'-representation the number of samples was set to $N = 20$ for each signature, according to the sampling theorem. For the 'DCT'-representation the first 20 components of the discrete cosine transform of each signature were used. Different configurations of features and representations where evaluated and compared to approaches from the literature (Config 1,7). However, in literature different classifiers were used. The combination of f_h and f_w represented by equidistant samples showed the best classification results. Finger length normalization of the f_h-only feature (Config 2) leads to a significant increase in classification performance compared to signature length normalization as proposed in [7]. Additionally, by using standard classifiers instead of non-linear distance measures as they are used in [7] the computational complexity could be reduced to fulfill real-time requirements. The good results using Hu's moments as presented in [5,8,9] could not be validated on our dataset. The overall process took approximately 12 milliseconds per image on a 1.6 GHz CPU. Image segmentation and contour extraction were implemented in C++, all other routines in MATLAB.

5 Conclusion

In this paper we presented a promising approach to contour-based hand symbol classification, which is invariant to translation, rotation, scale and viewpoint changes, as they occur typically in many HCI applications. We proposed new methods for normalization and representation of contour signatures that showed promising classification results on a large dataset.

Classification for most hand symbols is better than 92%, up to 99% (see Fig.4(a)). The average number of true positives over all hand symbols, however, is decreased by some hand symbols, which can only hardly be distinguished (e.g. H2 from H3, H10 from H3). In future work, we want to further improve classification performance by incorporating additional features which help to distinguish difficult hand symbols, e.g. texture. Additionally we plan to use more efficient classification structures like decision trees, which enable combination of multiple classifiers, each optimized to separate a certain subset of hand symbols.

References

1. Bader, T., Meissner, A., Tscherney, R.: Digital map table with fovea-tablett: Smart furniture for emergency operation centers. In: International Conference on Information Systems for Crisis Response and Management (2008)
2. Ueda, E., Matsumoto, Y., Imai, M., Ogasawara, T.: Hand pose estimation for vision-based human interface. IEEE Transactions on Industrial Electronics, 676–684 (2003)
3. Chen, W., Fujiki, R., Arita, D., ichiro Taniguchi, R.: Real-time 3d shape estimation based on image features analysis and inverse kinematics. In: 14th International Conference on Image Analysis and Processing (ICIAP 2007), pp. 247–252 (2007)
4. Athitsos, V., Sclaroff, S.: Estimating 3d hand pose from a cluttered image. In: IEEE Conference on Computer Vision and Pattern Recognition (CVPR), pp. 432–439 (2003)
5. Rosales, R., Athitsos, V., Sigal, L., Sclaroff, S.: 3d hand pose reconstruction using specialized mappings. In: IEEE Conference on Computer Vision (ICCV), pp. 378–385 (2001)
6. Guan, H., Chang, J.S., Chen, L., Feris, R.S., Turk, M.: Multi-view appearance-based 3d hand pose estimation. In: Conference on Computer Vision and Pattern Recognition Workshop (CVPRW), p. 154 (2006)
7. Gupta, L., Ma, S.: Gesture-based interaction and communication: automated classification of hand gesture contours. IEEE Transactions on Systems, Man and Cybernetics 13, 114–120 (2001)
8. Zobl, M., Nieschulz, R., Geiger, M.J., Lang, M., Rigoll, G.: Gesture components for natural interaction with in-car devices. In: Camurri, A., Volpe, G. (eds.) GW 2003. LNCS (LNAI), vol. 2915, pp. 448–459. Springer, Heidelberg (2004)
9. Akyol, S., Canzler, U., Bengler, K., Hahn, W.: Gesture control for use in automobiles. In: IAPR Workshop on Machine Vision Applications, pp. 349–352 (2000)
10. Erols, A., Bebis, G., Nicolescu, M., Boyle, R.D., Twombly, X.: Vision-based hand pose estimation: A review. Computer Vision and Image Understanding 108, 52–73 (2007)
11. Chang, C.C., Lin, C.J.: LIBSVM: a library for support vector machines (2001), http://www.csie.ntu.edu.tw/~cjlin/libsvm

Recognition of Simple 3D Geometrical Objects under Partial Occlusion

Alexandra Barchunova and Gerald Sommer

Bielefeld University, 33615 Bielefeld, Germany
{abarch}@cor-lab.uni-bielefeld.de

Abstract. In this paper we present a novel procedure for contour-based recognition of partially occluded three-dimensional objects. In our approach we use images of real and rendered objects whose contours have been deformed by a restricted change of the viewpoint. The preparatory part consists of contour extraction, preprocessing, local structure analysis and feature extraction. The main part deals with an extended construction and functionality of the classifier ensemble Adaptive Occlusion Classifier (AOC). It relies on a *hierarchical fragmenting* algorithm to perform a local structure analysis which is essential when dealing with occlusions. In the experimental part of this paper we present classification results for five classes of simple geometrical figures: prism, cylinder, half cylinder, a cube, and a bridge. We compare classification results for three classical feature extractors: Fourier descriptors, pseudo Zernike and Zernike moments.

1 Introduction

Contour-based recognition of partially occluded objects involves handling of several challenging issues. Contour acquisition and its quality improvement is the first task. In this work we have made use of some common techniques for noise cancelling like cautious Gaussian smoothing and B-Spline modelling [7]. Partial occlusion of shape poses a big challenge for algorithms with a global approach. An object is made invisible in a local environment. At the same time its shape in this environment is replaced by the shape of the occluding object. Recognition of occluded shapes by a human involves an analysis of the local structure, a search for a characteristic contour fragment allowing a clear assignment to the corresponding object class. Automatisation of the recognition process requires likewise a method that conducts such a local structural analysis of the object contours under partial occlusion. The algorithm that performs the local structure analysis in our work will be referred to as hierarchical fragmenting. For a given input contour this method generates several fragment levels, whereby the structural complexity of the fragments increases from level to level. The B-spline interpolation for noise cancelling and hierarchical fragmenting are both based on the segmentation that uses local maxima of the curvature function for extraction of the points of interest. The importance of such points in contour-based approaches has been investigated in [3].

In this paper we present classification results for two classical affine-invariant feature extractors based on Fourier descriptors and Zernike moments [10,9].

Within our experimental setup we allow a restricted perspective deformation of the contour. The scene is shot only from above while the camera is moving parallel to the surface. It has been shown in our work that it is possible to compensate a certain degree of perspective deformation by training. Experiments with rendered object images have shown that the larger the degree of perspective change of the contour the larger the classification error when using an affine-invariant feature extractor.

Multiple classifier systems have been employed in complex computer vision tasks starting in the eighties. Different aspects of development in this field have been discussed by T.K. Ho in [4]. In our approach each member of the ensemble specializes on its own degree of occlusion, defined by the corresponding level of the hierarchical fragmenting. The final hypothesis is generated by application of the class-related weighted average method. The determination of weights is formulated as a linear optimization problem being solved for an additional set of occluded shapes. The empirical results have shown that our approach delivers a considerable improvement of classification results compared to using a single classifier for recognition of partially occluded objects.

2 Contour Preprocessing and Feature Extraction

Factors like hardware noise, reflection or shadows result in an acquired contour that doesn't comply with the smooth shape of the original objects (see Figure 1(a)). Our goal is to improve the contour locally without loosing the information about the global structure. In our approach we describe the global structure by a set of *points of interest* (POI), which we later interpolate with B-splines. Here we use a common way to define a point of interest as a local maximum of the curvature function [5,2]. Let x and y be functions defining a discrete contour in a parametric representation. Then the curvature in the point $p := (x(t), y(t))$, $t \in \underline{T}$[1] is defined as follows:

$$k(p) := \frac{x'(t)y''(t) - x''(t)y'(t)}{\left(x'^2(t) + y'^2(t)\right)^{3/2}}, \quad (1)$$

where T denotes the number of discrete points in the contour. POI $p_0 = (x(t_0), y(t_0))$ is defined for $\epsilon \in \mathbb{N}$, $t_0 \in \underline{T}$ and an environment $U_0 = U(t_0, \epsilon)$ as follows:

$$|k(p_0)| = \max_{p \in P_u} |k(p)|, \quad (2)$$

where $P_u = \{(x(t), y(t)) | t \in U_0\}$. In order to obtain realistic values for $k(p)$, we first apply Gaussian smoothing with $\sigma = 5$ on the contour data. Then we calculate averaged derivatives that we use instead of regular discrete derivatives in evaluation of $k(p)$. Each derivative is built as an average over 5 neighbouring points, which reduces the influence of noise. In the second step we apply a procedure for POI extraction that can be schematically described as follows:

---

[1] Throughout this work $t \in \underline{T}$ denotes $t \in \{1, \ldots, T\}$ for a $T \in \mathbb{N}$.

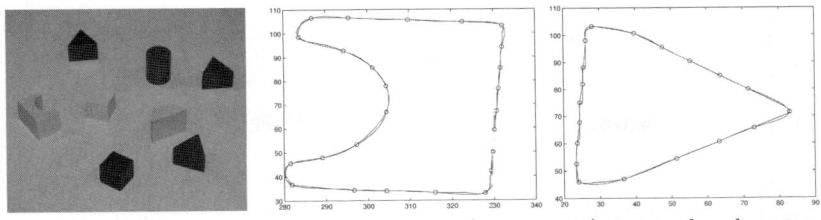

(a) Test objects in a typical scene.

(b) Extracted POI (blue points); interpolated contour (red) and original contour (blue).

Fig. 1. Typical test objects, preprocessed and segmented contours

1. Sort $\{(p, k(p))\}$ in descending order according to the values $k(p)$
2. Select the first available point from the sorted list to be the next POI; prohibit selection of further points belonging to the local environment of the selected point
3. Go to 2 if there are points available in the list, otherwise emit the chosen POI

The value of the local environment parameter in 2 depends on the structural complexity of the objects. In our experiments we have used the value 1/10 of the contour length. Finally, we conduct B-spline interpolation for the calculated POI. This results in an improvement of the local contour structure, while sustaining the global shape characteristics (see Figure 1(b)).

Contours of three-dimensional objects are rarely planar. Thus we can either try to reconstruct the three dimensional structure of the curve or we work with the two dimensional projection. Here we make use of the second option. In our tests (see Section 4) we have shown that it is possible to use affine-invariant feature extractors on such data and compensate the restricted perspective deformation by learning. For a given sampled contour we calculate a vector of normalised Fourier descriptors and (pseudo) Zernike moments. In our experiments we have used a constant number of points, $L = 64$, to represent any kind of contour data. The dependency between the dimensionality of the feature vector and the classification error will be described in Section 4.

3 Classifier Ensemble and Its Organisation

In the following sections we will describe the nature of Adaptive Occlusion Classifier (AOC) by looking at the following four main components: the data set, feature extractors, basis classifiers and the combination technique [6].

3.1 Data Set

The data pool consists of real camera images and rendered images[2] of the non-occluded test objects. In our experiments we investigate solely artificially

---
[2] In our experiments we have used POV-Ray [1] for rendering.

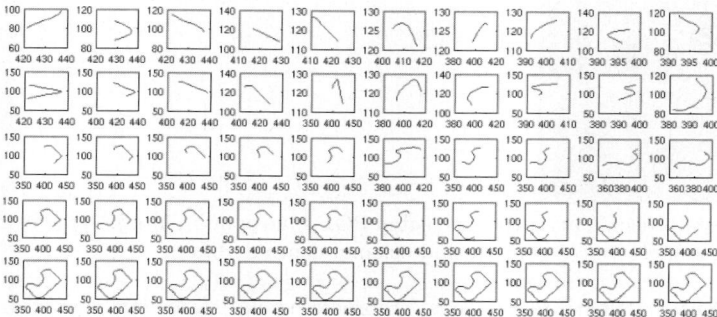

Fig. 2. Five data subsets of fragments generated by the hierarchical fragmenting of a bridge contour; rows correspond to the hierarchical levels

generated straight line boundary occlusion and make use of two methods for occlusion generation. The first method of *contour occlusion* simply deletes a given part of the contour data and connects the gap with a straight line. It is computationally efficient but can produce unrealistic contours when applied to non-convex shapes. The second method calculates the positions of the pixels within the contour that can be deleted to yield a linear occlusion of the object area. The later method of *area occlusion* is computationally more complex but delivers realistic linear occlusions independent of the shape convexity.

On the data level the members of the AOC are assigned to their personal data subsets. These are generated by the hierarchical fragmenting algorithm based on the set of POI (or control points) for the B-spline interpolation. Consider a contour fragment located between **three** neighbouring control points (see the example fragments in the first line of Figure 2). A set of all such fragments builds up the first hierarchical level or the first data subset. Note that the set of POI used in the algorithm contains a subset of structurally descriptive points, e.g. a corner of a prism. Analogously, the subset of the generated fragments contains a subset of local shape-descriptive fragments, e.g. a fragment, containing a corner of a prism. The hierarchical fragmenting algorithm on the i-th step connects two neighbouring segments of the (i-1)-th level to a new one. This generates levels of contour fragments of growing structural complexity, defining different levels of partial occlusion. In Figure 2 you can see some example fragments of five hierarchical levels for a bridge contour segmentation.

3.2 Feature Extractors and Basis Classifiers

In a perfect scenario classifier training results in a model that completely covers the feature space of the partially occluded objects. Because the set of all possible occlusions is vast, we have to choose training data that allows an approximate solution. In our work we use the data subsets generated by the hierarchical fragmenting which the feature extractor transforms into feature vector subsets.

As mentioned above, we use normalized Fourier descriptors, Zernike and pseudo Zernike moments. On the classifier level we use Local Credibility Criterion (LCC) classifiers [8]. LCC classifiers consists of multiple hypersphere shaped models. Their feasibility or credibility is determined by the ratio of correct responses to the number of overall responses: $\gamma = R_c/R_t$. Both the set of models and their number is dynamic.

3.3 Combination Technique and Weight Vector Estimation

Let $F \subset \mathbb{R}^n$ be a feature space and $C := \{1, \ldots, K\} \subset \mathbb{N}$ the set of class labels. Let $D := \{D_1, \ldots, D_L\}$ be the classifier ensemble, where a classifier D_i for $i \in \underline{L}$ can be described by the following map:

$$D_i : F \to [0,1]^K, \quad x \mapsto (d_{i1}(x) \ldots d_{iK}(x)). \tag{3}$$

In this work we have used the *class-related weighted average* method in order to combine the responses of the AOC members. For each $j \in \underline{K}$, $i \in \underline{L}$ the class-related weighted average for a sample x is defined by:

$$\mu_j(x) := \sum_{i=1}^{L} w_{ij} d_{ij}(x), \tag{4}$$

where the w_{ij}'s denote the class specific weights.

AOC was designed to allow contour-based classification of objects with different degrees of occlusion. Each member of the ensemble specializes on its own degree of occlusion during training as well as during testing. The main task of the weight vector is to integrate the individual classifier class responses to a final hypothesis according to their classification performance. For this purpose we use an additional set of contours with a random uniform area occlusion up to 80 percent, denoted by $\gamma_{max}^w = 0.8$. Further let $X := \{(x_n, c_n) | n \in \underline{N}\}$ be a labeled sample set. The response matrix $R_l \in \mathbb{R}^{N \times K}$ of the lth classifier to the sample set X is given by:

$$R_l := \begin{pmatrix} d_{l1}(x_1) \ldots d_{lK}(x_1) \\ \ldots \ldots \ldots \\ d_{l1}(x_N) \ldots d_{lK}(x_N) \end{pmatrix} \tag{5}$$

We define an auxiliary function that allows a correct building of a scalar product and adding up of columns of the matrix R_l, $l \in \underline{L}$ for correct solving of the minimization problem (see Eq. 8). For a $k \in \underline{K}$ we define:

$$f_k : \mathbb{R}^{N \times K} \to \mathbb{R}^{N \cdot K}, \quad R_l \mapsto (v_1, \ldots, v_p)^T, \tag{6}$$

where $p = N \cdot K$ and for $i \in \underline{p}$, $n \in \underline{N}$:

$$v_i := \begin{cases} d_{lk}(x_n), & \text{if } i = N \cdot (k-1) + n \\ 0, & \text{otherwise.} \end{cases} \tag{7}$$

Let $r = L \cdot K$ be the dimension of the weight vector. We combine the responses of the classifier ensemble in the following matrix $\hat{R} \in \mathbb{R}^{p \times r}$:

$$\hat{R} := (f_1(R_1), f_2(R_1), \ldots, f_j(R_i), \ldots, f_K(R_L)).$$

For a given sample set X the weight vector $w \in \mathbb{R}^r$

$$w := (w_{11}, w_{12}, \ldots, w_{lk}, \ldots, w_{LK})^T,$$

can be calculated by minimizing the distance between the optimal response matrix R_{opt} and the weighted (see Eq. 4) response matrix $\hat{R}$:

$$\min_{w \in \mathbb{R}^r} \|\hat{R}w - R_{opt}\|. \qquad (8)$$

4 Experimental Results

The data pool consists of 1000 camera images and 1600 rendered non-occluded images. For every type of geometrical figure (bridge, cylinder, half cylinder, cube and prism) the data pool provides the same number of images. The objects have been recorded with a perspective deformation through a change of the viewpoint or the POV-Ray configuration.

In the first experiment we analyse the dependency between the dimensionality of the feature vector and the classification error rate. For the training of the ensemble members we have randomly selected 370 contours of non-occluded objects out of the data pool. For the calculation of the weight vector we have used 250 contours with generated random area occlusion up to 30 percent, $\gamma_{max}^w = 0.3$. In all algorithms for random occlusion generation we use uniform distribution on the interval $(0, \gamma_{max}]$. On average random uniform area occlusion up to 30 percent is approximately equivalent to random uniform contour occlusion up to 50 percent in our experimental setup. We have tested 250 contours with $\gamma_{max}^t = 0.3$. In the Figure 3(a) you can see the dimension of the feature vector on the x-axis and the average classification error rate on the y-axis. Zernike moments (ZM), pseudo Zernike moments (PZM) and Fourier descriptors (FD) yield approximately the same results. The lowest average error is about 7 percent which can be explained by the ambiguity of the contours with partial occlusion.

In the next experiment (see Figure 3(b)) we have compared the performance of a single LCC classifier vs. AOC ensemble. For the calculation of the weight vector we have used contours with a random uniform area occlusion up to 80 percent, $\gamma_{max}^w = 0.8$. As a representation of the contour data we have chosen to use a 14-dimensional normalized vector of FD. On the x-axis we have plotted the constant area occlusion parameter γ_{const}^t. The value of this parameter indicates for all objects in the sample set a constant but randomly placed area occlusion specified by the value γ_{const}^t. On the y-axis you can see the recognition error. For all levels of occlusion AOC classifies considerably better than a single LCC. Data with a constant area occlusion of more than 80 percent cannot be classified better than by random guessing. This can be explained by a high degree of ambiguity of strongly occluded object shapes.

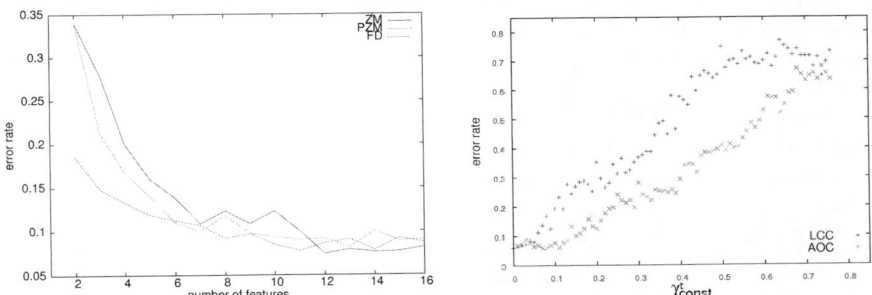

(a) Classification results for different feature extractors
(b) Classification results: AOC vs. single LCC

Fig. 3. Classification results for AOC ensemble and a single LCC classifier

Table 1. Comparison of error rates for different sets of synthetic data

	R-R-S	R-S-R	R-S-S	S-R-R	S-R-S	S-S-R	S-S-S
Error rate (Set 1)	0.28	0.20	0.22	0.21	0.18	0.24	0.15
Error rate (Set 2)	0.27	0.19	0.24	0.22	0.21	0.23	0.22

The third experiment was aimed at exploring the capacity of rendered data alone as well as its potential in combination with real camera images within our applications. We have trained AOC with 350 samples. For weight vector estimation we have used 250 contours with $\gamma_{max}^{w} = 0.5$. Table 1 shows the summary of the test results for different combinations of real (R) and synthetic (S) data in training, weight vector estimation and testing (R/S-R/S-R/S). Our test data consists of 100 samples with $\gamma_{max}^{t} = 0.5$. The in-plane translation of the objects with regard to the camera position within Set 2 is two times as large as within Set 1. The classification using real camera images (R-R-R) yielded an average error of about 17 percent. Consider the sixth column of the table corresponding to the S-S-R configuration. The test results are only about 5 percent worse compared to the case where only real data (R-R-R) has been used in training and weight estimation.

5 Conclusions

In this paper we have shown that the usage of the AOC ensemble for recognition of partially occluded shapes of three-dimensional objects considerably improves the results in comparison to a single LCC classifier. Classification for γ_{max}^{t} yielded an average error rate of about 17 percent. This can be explained by the ambiguity of the partially occluded data.

Our procedure of hierarchical fragmenting delivers levels or subsets of contour segments with growing structural complexity. By using this method in our application we have demonstrated that it is well suited for carrying out local structure analysis. In our tests we have compared the following classical affine-invariant feature extractors: Fourier descriptors, pseudo Zernike moments and Zernike moments. All three have yielded comparable results. It can be concluded that the usage of 12 to 14 dimensional feature vectors is sufficient for our application.

Tests with rendered images have revealed an automation potential of the ensemble training within our experimental setup. Note that in our tests the usage of rendered images in training resulted in an error rate increase of approximately 5 percent. More tests could be conducted in this area.

References

1. Persistence of vision raytracer, http://www.povray.org/
2. Ghosh, A., Petkov, N.: Robustness of shape descriptors to incomplete contour representations. IEEE Trans. Pattern Anal. Mach. Intell. 27(11), 1793–1804 (2005)
3. Ghosh, A., Petkov, N.: Effect of high curvature point deletion on the performance of two contour based shape recognition algorithms. International Journal of Pattern Recognition and Artificial Intelligence 20(6), 913–924 (2006)
4. Ho, T.K.: Hybrid methods in pattern recognition. Series in Machine Perception and Artificial Intelligence, ch. 5, vol. 47, pp. 367–382. Springer, Heidelberg (2002)
5. Kindratenko, V.: Development and Application of Image Analysis Techniques for Identification and Classification of Microscopic Particles. PhD thesis, University of Antwerpen (1997)
6. Kuncheva, L.I.: Combining Pattern Classifiers: Methods and Algorithms. Wiley-Interscience, Hoboken (2004)
7. Lim, K.B., Du, T., Zheng, H.: 2d partially occluded object recognition using curve moments. In: Proceedings of the Seventh IASTED International Conference Computer Graphics And Imaging, Kauai, Hawaii, USA (2004)
8. Prehn, H., Sommer, G.: Incremental classifier based on a local credibility criterion. In: Proceedings of the IASTED International Conference on Artificial Intelligence and Applications, AIA 2007, Innsbruck, Austria, pp. 372–377. ACTA Press (2007)
9. Teh, C.-H., Chin, R.T.: On image analysis by the methods of moments. IEEE Trans. Pattern Anal. Mach. Intell. 10(4), 496–513 (1988)
10. Zhang, D., Lu, G.: Review of shape representation and description techniques. Pattern Recognition 37, 1–19 (2004)

Shape Classification Using a Flexible Graph Kernel

François-Xavier Dupé* and Luc Brun

GREYC UMR CNRS 6072,
ENSICAEN-Université de Caen Basse-Normandie,
14050 Caen France,
{francois-xavier.dupe,luc.brun}@greyc.ensicaen.fr

Abstract. The medial axis being an homotopic transformation, the skeleton of a 2D shape corresponds to a planar graph having one face for each hole of the shape and one node for each junction or extremity of the branches. This graph is non simple since it can be composed of loops and multiple-edges. Within the shape comparison framework, such a graph is usually transformed into a simpler structure such as a tree or a simple graph hereby loosing major information about the shape. In this paper, we propose a graph kernel combining a kernel between bags of trails and a kernel between faces. The trails are defined within the original complex graph and the kernel between trails is enforced by an edition process. The kernel between bags of faces allows to put an emphasis on the holes of the shapes and hence on their genre. The resulting graph kernel is positive semi-definite on the graph domain.

Keywords: Shape, Skeleton, Kernel Machine, Graph Kernel.

1 Introduction

The medial axis being an homotopic transformation, the skeleton of a 2D shape is a 2D structure with as many holes as the shape. A natural way to encode such a structure by a graph consists in creating an edge for each branch of the skeleton and a node for each junction of branches or branch's extremity. The resulting graph is a non simple planar graph which may be enriched using information from the radius of the osculating circle along branches [1,2,3,4,5]. The shape comparison is thus transformed into a graph comparison problem. However, graph comparison methods robust against structural noise such as the maximal common sub-graph method or the related graph edit distance problem [6] have an exponential complexity on general graphs. Many authors use thus a simpler encoding of the skeleton leading to a comparison function with a reduced complexity.

Siddiqi [1] and Sebastian [7] transform the graph into a tree and apply a tree comparison scheme. Another method, introduced by Pelillo [8], transforms

---

* This work is performed in close collaboration with the laboratory Cyceron and is supported by the CNRS and the Région Basse-Normandie.

graphs into trees and then models the tree matching problem as a maximal clique problem within a specific association graph. A last method proposed by Bai and Latecki [4] matches end points (vertices with a degree one) and then compares paths between end-points. Contrary to the previous approaches, this last method can deal with closed structures and thus takes the holes of the shape into account.

Although these methods have been developed for indexation and classification tasks, they can not be readily used within the kernel machine framework. This limitation is related to the lack of mathematical tools inside the graph domain. Neuhaus and Bunke [9] proposed an elegant framework for the construction of graph kernels based on edit distances. Another solution consists in using graph kernels such as random walk or marginalized graph kernel [10] which are positive semi-definite on the graph domain. Though, these kernels are easier to use, they lack the flexibility and the noise robustness provided by the kernels based on graph edit distances.

This paper follows a first contribution [11] where we defined the notion of path rewriting within the graph kernel framework. However this method is defined on trees and thus does not encode properly the holes of the shapes. First, we recall some definitions and then extend our graph kernel framework to trails (Section 2). Second, we propose to extend the rewriting process, initially defined on trees, to graphs (Section 3). Then, we propose to combine our graph kernel with a closed paths kernel which compares graphs' faces (Section 4). Finally, an experiment with a multi-class classifier is proposed to highlight the relevance of holes inside holed shapes (Section 5).

2 Bag of Trails Kernel

Let us consider a *graph* $G = (V, E)$ where V denotes the set of vertices and $E \subseteq V \times V$ the set of edges. We define a *simple-graph* as a graph with no multiple edges between two vertices and no loops (an edge linking a vertex with itself). We define a *trail* as an alternating sequence of vertices and edges with distinct edges and a *path* as a trail with distinct vertices. A *closed path* is a path whose first vertex is equal to the last one. A *bag of trails* T associated to G is defined as a set of trails of G whose cardinality is denoted by $|T|$. We finally denote by K_{trail} a generic trail kernel.

2.1 Mean Kernels

By considering bags as sets, Suard [3] has proposed several kernels for bags of paths which are extensible to trails. Amongst these kernels, the mean kernel is proposed as a convolution kernel [12] between trails: let T_1 and T_2 denote two bags of trails, the *mean kernel* between these two bags is defined as:

$$K_{mean}(T_1, T_2) = \frac{1}{|T_1|} \frac{1}{|T_2|} \sum_{t \in T_1} \sum_{t' \in T_2} K_{trail}(t, t'). \tag{1}$$

This kernel is positive definite on the bag of trails domain if and only if K_{trail} is positive definite on the trail domain.

The major drawback of this kernel is the information averaging when bags are composed of many trails. Such a loss of information may be avoided using a weighted mean kernel [13]. The design of this kernel assumes that most of the relevant information of a bag is located near its mean trail. Let T_1 and T_2 denote two bags of trails, then the *weighed mean kernel* is defined as:

$$K_{weighted}(T_1, T_2) = \frac{1}{|T_1|} \frac{1}{|T_2|} \sum_{t \in T_1} \sum_{t' \in T_2} <K_{trail}(t,m), K_{trail}(t',m')>^d \quad (2)$$
$$\frac{\omega(t)}{W} \frac{\omega(t')}{W'} K_{trail}(t,t').$$

where $d \in \mathbb{R}^+$, m and m' denote the mean trails of T_1 and T_2, $\omega(t)$ (resp. $\omega(t')$) denotes the sum of the edge's weights of t (resp. t') and W (resp. W') the whole weight of the graph containing t (resp. t'). The trail kernel between a trail t and the mean trail m is defined as: $K_{trail}(t,m) = \frac{1}{|T|} \sum_{t_i \in T} K_{trail}(t,t_i)$. The weighted mean kernel is a convolution kernel based on a scalar product (the similarity with the mean trails) and the trail kernel K_{trail}. So it is positive definite if and only if K_{trail} is positive definite.

2.2 A First Trail Kernel

For its marginalized kernel, Kashima proposed a walk kernel based on a tensor product [14]. As trails are particular walks, the walk kernel remains available. Let t and t' denote two trails, the *trail kernel* denoted $K_{classic}$ is defined as 0 if $|t| \neq |t'|$ and as follows otherwise:

$$K_{classic}(t,t') = K_v(\varphi(v_1), \varphi(v'_1)) \prod_{i=2}^{|t|} K_e(\psi(e_{v_{i-1}v_i}), \psi(e_{v'_{i-1}v'_i})) K_v(\varphi(v_i), \varphi(v'_i)), \quad (3)$$

where $\varphi(v)$ and $\psi(e)$ denote respectively the vectors of features associated to the vertex v and the edge e. The terms K_v and K_e denote two kernels for respectively vertex's and edge's features. $K_{classic}$ is a tensor product kernel and so is positive definite if and only if K_e and K_v are two positive definite kernels. For the sake of flexibility and simplicity, we use Gaussian RBF kernels based on the distance between the attributes.

3 Edition Kernel on Trails

The main issue with skeleton based graphs is that two different graphs may encode similar shapes. Two different kind of structural noise may appear inside a skeleton: ligatures produced by noise on the boundary and elongations produced by a general deformation of the shape. Usually, this structural noise is tackled using edition operations on graphs. However, within a bag of trails framework we must consider edition operations on trails. The effect of the structural noise on a trail is twice: addition of edges and addition of vertices.

We suppose that the edges of our graph are associated to a weight which encodes their relevance. Torsello [15] has proposed such a relevancy measure: for each edge this measure approximates the length of the boundary associated to the skeleton's branch encoded by this edge. Using this weight, we compute the relevance of each vertex and edge inside a trail: the relevance of an edge corresponds to its weight and the relevance of a vertex corresponds to the weight of the sub-graph (i.e. the sum of the weight of all the sub-graph's edges) connected to the trail by this vertex. When graphs are trees [11], the sub-graphs correspond to sub-trees and the computation of the relevancy measure of vertices is unambiguous. Fig. 1a shows for example a path within a tree, where the sub-trees related to the two vertices of the path are clearly defined and so their weight.

However, with holed shapes, graphs are not trees and the definition of the relevancy of vertices is not straightforward. Indeed, sub-graphs may connect several vertices of the considered trail. We propose to solve this difficulty by using the random walker diffusion algorithm [16] where normalized edge's weights are considered as transition probabilities. For each vertex v_i of the trail, this diffusion algorithm associates to each vertex v_l of the graph the probability $p_{l,i}$ that a random walker starting at v_l first reach v_i. Each vertex of the graph is then associated with the vertex of the trail with the maximal probability. The sub graph induced by this set of vertices is called the influence zone of the trail's vertex. However, the random walker is designed for simple-graphs. We thus transform our non-simple graph into a simple one by defining the transition probabilities between vertices as follows: loops are removed and multi-edges between two vertices are transformed into a single edge whose weight is the sum of edges' weights. Single edges between vertices are kept unchanged. Note that this transformation is only used for the random walker algorithm. Our trails and the sub-graphs encoding the influence zones are both defined within the initial non simple graph.

The weight of the influence zone of a vertex v is defined as the sum of 1) the weight of the edges within the influence zone and 2) a ratio of the weight of the edges shared with another influence zone (i.e. edges whose vertices belong to two influence zones). For example, the dash-dotted edges within Fig. 1b are shared by the two influence zones. Let v_1 and v_2 be the two incident vertices of an edge of weight w, v_1 (resp. v_2) is associated to its influence zone by a probability p_1 (resp. p_2) then we define as $\frac{p_1}{p_1+p_2}w$ (resp. $\frac{p_2}{p_1+p_2}w$) the part of the weight associated to the influence zone related to v_1 (resp. v_2). Fig. 1b shows an example of influence zone of trail vertices (the trail is defined by the dashed line): remark the importance of the influence zone of vertex 1 compared to the one of vertex 2.

Given a relevancy measure of each vertex and edge of a trail we introduce two edition operations: vertex suppression followed by edge merging and edge contraction (or suppression for loops). The cost of an operation is defined as the relevancy measure of the removed edge or vertex. Finally, we defined an edition function κ which applies the cheapest edition. Then $\kappa^i(t)$ denotes the trail t after i editions. In addition, we denote by $\text{cost}_i(t)$ the cumulative cost of

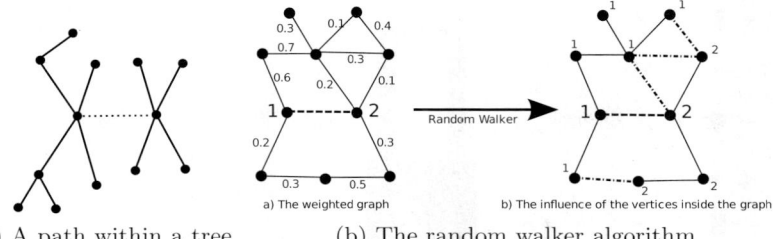

(a) A path within a tree   (b) The random walker algorithm

Fig. 1. Influence zones: (a) Example of tree with a selected path (dotted edge) (b) The influence of the vertices of the dashed trail between the vertices 1 and 2 using the random walker

operations leading to $\kappa^i(t)$. Finally, we construct the edition kernel as a weighted convolution kernel between the trails and their rewritings:

$$K_{edit}(t,t') = \frac{1}{D+1}\sum_{k=0}^{D}\sum_{l=0}^{D}\exp\left(-\frac{\text{cost}_k(t)+\text{cost}_l(t')}{2\sigma_{cost}^2}\right)K_{classic}(\kappa^k(t),\kappa^l(t')), \quad (4)$$

where D is the maximal number of editions and σ_{cost} the RBF parameter of the cost kernel which penalizes edition. This kernel is a convolution kernel [12] and is positive definite if and only if K_{trail} is positive definite. Experiments showing the insight of kernel K_{edit} for the robustness against structural noise are provided in [11].

4 Closed Paths Kernel

The faces of a skeletal graph encode the holes of a shape and represent as such important information about the shape. When using the previously defined kernels, faces are just encoded as trails. So when constructing a bag of trails, these particular trails may not appear in the bag or may be drowned with many other trails. Thus it is relevant to put an emphasis on faces when dealing with holed shapes.

Several kernels based on cycles have been proposed for graphs [17]. However these kernels are not designed for shape classification for two main reasons: they don't consider the orientation of faces and they are not restricted to cycles encoding faces.

An efficient comparison of faces within a shape recognition framework requires a kernel robust against structural noise. We propose to encode each hole by a unique closed path which describes the corresponding face. This path begins at the closest vertex to the gravity center of the shape and crosses the edges using a counter-clockwise orientation. For example, the hole of the padlock in Fig. 2b is described by the closed path "4 e3 3 e1 4". Finally, two closed paths encoding

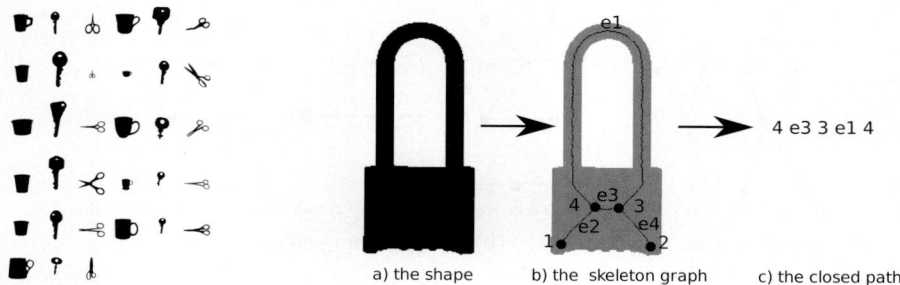

a) Holed shapes databases   b) Computation of the closed path of a padlock shape

Fig. 2. Holed shapes and closed paths computation

faces are simply compared using a trail kernel such as $K_{classic}$ (section 2.2) or K_{edit} (section 3).

However, while comparing two closed paths, we may have to face to alignment errors due to the selection of the initial vertex. In order to enforce the robustness of our kernel, shifted versions of the closed paths are also compared. For example, the face in Fig. 2b presents two vertices at an equal distance to the gravity center and the closed path "3 e1 4 e3 3" is thus an acceptable path which corresponds to a shifted version of the previous path. We define the function $\mu_i(t)$ which performs a circular shift of i edges of the path t clockwise if i is positive and counter-clockwise i is negative.

The *shift kernel* is then defined as the weighted convolution between paths and their shifted versions using a trail kernel denoted K_{trail}:

$$K_{shift}(t,t') = \frac{1}{(2p+1)^2} \sum_{i=-p}^{p} \sum_{j=-p}^{p} \exp\left(-\frac{|i|+|j|}{2\sigma_{closed}^2}\right) K_{trail}(\mu_i(t),\mu_j(t')), \quad (5)$$

where p is the maximal number of shifts. This kernel is positive definite if and only if K_{trail} is positive definite. Finally, the closed paths kernel is defined as the mean kernel between all the closed paths surrounding the faces of two planar graphs G_1 and G_2:

$$K_{closed\ paths}(G_1,G_2) = \frac{1}{|C(G_1)|} \frac{1}{|C(G_2)|} \sum_{t \in C(G_1)} \sum_{t' \in C(G_2)} K_{shift}(t,t'), \quad (6)$$

where $C(G_1)$ (resp. $C(G_2)$) denotes the set of closed paths encoding the faces of G_1 (resp. G_2) and $|C(G_1)|$ (resp. $|C(G_2)|$) denotes the size of the set $C(G_1)$ (resp. $C(G_2)$). This kernel is positive definite if and only if K_{shift} is positive definite.

Finally, a kernel denoted $K_{combined}$ is built using the two proposed kernels:

$$K_{combined}(G_1,G_2) = (1-\gamma)K_{weighted}(T_1,T_2) + \gamma K_{closed\ paths}(G_1,G_2), \quad (7)$$

where T_1 (resp. T_2) is the bag of trails associated to G_1 (resp. G_2), $\gamma \in [\,0,1\,]$ is a tuning variable, $K_{weighted}$ (2) denotes our bag of trails kernel and $K_{closed\ paths}$ our closed paths kernel (6). This kernel is positive definite on the union of the bag of trails and bag of faces domains as it is defined as the addition of two positive definite kernels multiplied by positive coefficients [18].

5 Experiments

We propose an experiment using a multi-class classifier [19]. The test database is built by adding shapes with holes (Fig. 2a) to the Kimia 99 shapes database [20]. Three kernels are used: the combination of the weighted mean kernel with the closed paths kernel denoted $K_{combined}$ (7) , the weighted mean kernel alone denoted $K_{weighted}$ (2) and the random walk kernel [10]. The trail kernel used within the weighted mean and the shift kernels (Section 4) is the edition kernel K_{edit} (Section 3).

For this experiment, the bags of trails (Section 2) were composed of 2 percent of the heaviest paths amongst all the trails with up to 9 edges. The maximal number of editions (Section 3) was set to 9 and the number of shifts (Section 4) to 5. The parameters of the involved RBF kernels have been manually tuned based on a first estimate provided by a cross validation algorithm. For efficiency reason, the random walk [10] is performed on an augmented version of the maximal spanning tree: while considering an edge which is implied in the formation of a cycle or a loop, we change one of its incident vertices into a new vertex (of degree 1) with the same characteristics in order to break the cycle or loop. Using this trick, the graph may be encoded by an adjacency matrix and efficient random walk kernels [10] based on such an encoding may be used. Note that alternative encoding using line graphs may also be considered.

The experiment consists in the classification of the whole database into 5 classes (2 classes from the Kimia databases and 3 classes of holed shapes). The training set was composed of 5 shapes of each class taken arbitrarily. The classifier algorithm [19] is based on kernel principal analysis and quadratic discriminant analysis and so considers both inter-classes and intra-classes properties. The computational times require to compute the Gram matrices associated to

Table 1. On the left, the confusion matrix on 5 classes of shapes: (1) Cups, (2) Keys, (3) Scissors, (4) Dudes and (5) Tools. On the right, the computational time of the Gram matrix for the three kernels.

Classes	$K_{combined}$					$K_{weighted}$					Random Walk				
	(1)	(2)	(3)	(4)	(5)	(1)	(2)	(3)	(4)	(5)	(1)	(2)	(3)	(4)	(5)
(1)	8	2	1			7	4				4	6			1
(2)		11					11				2	8	1		
(3)			11				2	9					10	1	
(4)				11					11					11	
(5)	1				10	1	3	2		5		2		1	8

Kernel	Times
$K_{combined}$	19s
$K_{weighted}$	8s
Random walk	9min

our three kernels are given in Tab. 1(right). Tab. 1(left) shows the confusions matrices of the three kernels. The $K_{combined}$ kernel shows very good results with some confusion on the cups. The $K_{weighted}$ kernel shows good results, but is very confused on tools. This confusion comes from the few trails contained inside the bag of trails which are not sufficient for a proper class separation. The random walk kernel shows good results too with confusion on tools and on cups. The confusion on the cups is due to the maximal spanning tree which conducts to a loss in the description of the faces of the graph.

6 Conclusion

We have defined in this paper a positive semi-definite kernel for shape classification which is robust to noise and takes holes into account. The experiments show taking into account such topological feature of the shape improve the classification performances. In the future, we plan to further improve the selection of the trails and the combination of the trail kernel results.

References

1. Siddiqi, K., Shokoufandeh, A., Dickinson, S.J., Zucker, S.W.: Shock graphs and shape matching. Int. J. Comput. Vision 35(1), 13–32 (1999)
2. Ruberto, C.D.: Recognition of shapes by attributed skeletal graphs. Pattern Recognition 37(1), 21–31 (2004)
3. Suard, F., Rakotomamonjy, A., Bensrhair, A.: Kernel on bag of paths for measuring similarity of shapes. In: European Symposium on Artificial Neural Networks, Bruges-Belgique (April 2007)
4. Bai, X., Latecki, J.: Path Similarity Skeleton Graph Matching. IEEE PAMI 30(7) (2008)
5. Goh, W.B.: Strategies for shape matching using skeletons. Computer Vision and Image Understanding 110, 326–345 (2008)
6. Bunke, H.: On a relation between graph edit distance and maximum common subgraph. Pattern Recognition Letters 18(8), 689–694 (1997)
7. Sebastian, T., Klein, P., Kimia, B.: Recognition of shapes by editing their shock graphs. IEEE Trans. on PAMI 26(5), 550–571 (2004)
8. Pelillo, M., Siddiqi, K., Zucker, S.: Matching hierarchical structures using association graphs. IEEE Trans. on PAMI 21(11), 1105–1120 (1999)
9. Neuhaus, M., Bunke, H.: Bridging the Gap between Graph Edit Distance and Kernel Machines. Machine Perception and Artificial Intelligence, vol. 68. World Scientific, Singapore (2007)
10. Vishwanathan, S., Borgwardt, K.M., Kondor, I.R., Schraudolph, N.N.: Graph kernels. Journal of Machine Learning Research 9, 1–37 (2008)
11. Dupé, F.X., Brun, L.: Edition within a graph kernel framework for shape recognition. In: GbRPR 2009, pp. 11–20 (2009)
12. Haussler, D.: Convolution kernels on discrete structures. Technical report, Department of Computer Science, University of California at Santa Cruz (1999)
13. Dupé, F.X., Brun, L.: Tree covering within a graph kernel framework for shape classification. In: ICIAP 2009 (accepted, 2009)

14. Kashima, H., Tsuda, K., Inokuchi, A.: Marginalized kernel between labeled graphs. In: Proc. of the Twentieth International conference on machine Learning (2003)
15. Torsello, A., Hancock, E.R.: A skeletal measure of 2d shape similarity. CVIU 95, 1–29 (2004)
16. Grady, L.: Random Walks for Image Segmentation. IEEE Transactions on Pattern Analysis and Machine Intelligence 28(11), 1768–1783 (2006)
17. Horváth, T.: Cyclic pattern kernels revisited. In: Ho, T.-B., Cheung, D., Liu, H. (eds.) PAKDD 2005. LNCS (LNAI), vol. 3518, pp. 791–801. Springer, Heidelberg (2005)
18. Berg, C., Christensen, J.P.R., Ressel, P.: Harmonic Analysis on Semigroups. Springer, Heidelberg (1984)
19. Wang, J., Plataniotis, K., Lu, J., Venetsanopoulos, A.: Kernel quadratic discriminant for small sample size problem. Pattern Recognition 41(5), 1528–1538 (2008)
20. LEMS: shapes databases, http://www.lems.brown.edu/vision/software/

Bio-inspired Approach for the Recognition of Goal-Directed Hand Actions

Falk Fleischer, Antonino Casile, and Martin A. Giese

Dept. of Cognitive Neurology, Hertie Institute for Clinical Brain Research, Tübingen, Germany

Abstract. The recognition of transitive, goal-directed actions requires a sensible balance between the representation of specific shape details of effector and goal object and robustness with respect to image transformations. We present a biologically-inspired architecture for the recognition of transitive actions from video sequences that integrates an appearance-based recognition approach with a simple neural mechanism for the representation of the effector-object relationship. A large degree of position invariance is obtained by nonlinear pooling in combination with an explicit representation of the relative positions of object and effector using neural population codes. The approach was tested on real videos, demonstrating successful invariant recognition of grip types on unsegmented video sequences. In addition, the algorithm reproduces and predicts the behavior of action-selective neurons in parietal and prefrontal cortex.

1 Introduction

The recognition of transitive actions requires additional computational mechanisms, compared to the recognition of human actions without goal objects. Recognition has to be invariant against changes in low-level image features, shifts in position, and shape transformations over time. At the same time, the distinction of different grip types (e.g. precision or power grip) requires a remarkable accuracy with respect to the detection of shape details (e.g. finger positions or their relationship to the grasped object).

This paper presents a physiologically-inspired model for the recognition of goal-directed hand actions. The model accomplishes the recognition of goal-directed hand actions from unsegmented gray-level videos. At the same time, it reproduces several biological findings in the mammal visual system, such as tuning properties of action-selective neurons in premotor cortex [1], view-dependence of recognition [2], and selectivity for the relationship between effector and object [3].

Related models have been discussed in robotics in the context of imitation learning. Many existing models in this domain are based on explicit three-dimensional shape models of effector and object (see [4] for an overview). Opposed to this work, we propose here an example-based approach that extends biologically-inspired models for the recognition of objects and actions [5,6,7,8,9].

Similar appearance-based approaches have been quite successful in object detection and recognition [10,11,12]. Opposed to previous work that has focused on the recognition of effector and body shapes from silhouettes (e.g. [13,14]), our system recognizes effector and object without previous segmentation. In contrast to other recent systems for action recognition, our system does not rely on combined space-time features (e.g. [15,14]) or motion features (e.g. [16,17]). Instead, spatio-temporal order is explicitly modeled by a dynamical interaction between shape representations using neural fields [18,5]. In contrast to many existing models for shape recognition that are characterized by complete position invariance, the proposed system exploits partially position-invariant detectors for the reconstruction of the spatial relationship between effector and goal object. This relationship is crucial for the detection of functional and dysfunctional grips.

In the following, we first present the architecture and its components (Section 2). We then show results of evaluating the different components of the system in Section 3. Finally, in Section 4 implications and further extensions of the approach are discussed.

2 Architecture for the Recognition of Transitive Actions

The architecture consists of three major components that correspond to cortical structures that seem to play a central role in visual action recognition: (1) a hierarchical neural system for the view-dependent recognition of object and effector shapes, (2) a circuit that is selective for temporal sequences of detector shapes, (3) a level that integrates the information about effector, object and their spatial relationship.

2.1 Neural Hierarchy for Shape Recognition

The first levels of the developed system are formed by a hierarchical neural architecture for shape recognition. Each layer of this hierarchy consists of a set of neural feature detectors that are inspired by the properties of real physiological neurons. Levels with neurons that are selective for individual features alternate with levels that increase invariance by pooling over detectors with different spatial and scale preference using a maximum operation [6,7]. The sequence of computations within each of the five layers of this hierarchy is given by: (i) feature detection through template matching, (i) maximum computation over detectors at neighboring spatial positions, (iii) application of a linear threshold function, and (iv) down-sampling by a factor of two. The parameters of the operations within each layer are summarized in Figure 1.

Layer V1/V2 - Local Orientation Detectors. Local orientations are extracted by simple cells that are modeled by a set of Gabor filters. To cover the structure of the hand, we use Gabor filters with 12 different preferred orientations θ and two different spatial frequencies ξ, as summarized in Figure 1.

Complex cells in the following layer integrate responses from simple cells with same orientation preference over position, scale and phase. Let $(x_1^{\text{even}_{\theta,\xi}}, \ldots,$

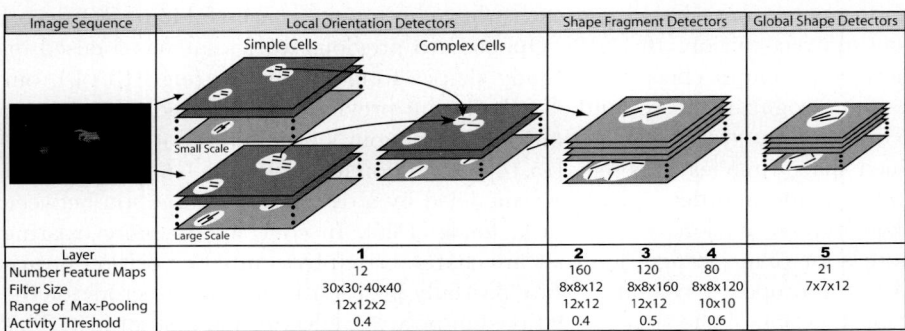

Fig. 1. Overview of the shape-recognition hierarchy

$x_m^{\text{even}_{\theta,\xi}}$) and $(x_1^{\text{odd}_{\theta,\xi}}, \ldots, x_m^{\text{odd}_{\theta,\xi}})$ denote the responses of the even and odd Gabor filters from the same local neighborhood S of size m and scale ξ. Then the response of a complex cell is given by $r^\theta = \max_{j \in S, \xi} \{(x_j^{\text{even}_{\theta,\xi}})^2 + (x_j^{\text{odd}_{\theta,\xi}})^2\}$. Above the second layer no distinction of different spatial frequency regimes was realized.

Layers V4/IT - Detectors for Shape Fragments. The neurons in the three intermediate layers represent detectors that extract features of increasing complexity. The feature detectors on the intermediate layer i were defined by Gaussian Radial Basis Functions (RBFs) with the form

$$r^i = exp\left(-\beta \left\| \frac{\tilde{\mathbf{r}}^{i-1}}{\|\tilde{\mathbf{r}}^{i-1}\|} - \frac{\tilde{\mathbf{p}}}{\|\tilde{\mathbf{p}}\|} \right\|^2 \right). \quad (1)$$

The centers $\mathbf{p}$ of the RBF functions were tuned to local combinations of input features from the previous layer $i-1$ that were specified by training patterns, and we chose $\beta = 0.5$.

During training, on each layer novel intermediate features $\mathbf{p}$ were extracted from the responses of the previous layer within a limited spatial region. Training images show individual hand configurations or objects. Over the training set, for dimensionality reduction, features were centered around the training mean $\mathbf{m}$ and their dimensionality was reduced by the mapping $\tilde{\mathbf{p}} = \mathbf{A}(\mathbf{p} - \mathbf{m})$, retaining only the PCA components that were necesary for explaining 99% of the variance. The transformed features $\tilde{\mathbf{p}}$ were then clustered based on their correlations, and the average feature of each cluster was retained. The number of remaining feature detectors on each intermediate layer is summarized in Figure 1.

Outputs again were thresholded, and responses within a local spatial neighborhood were pooled with a maximum operation, followed by a spatial downsampling with factor 2.

Layer IT/STS - Shape Templates for Hand and Object. The feature detectors on the highest level of the recognition hierarchy respond selectively

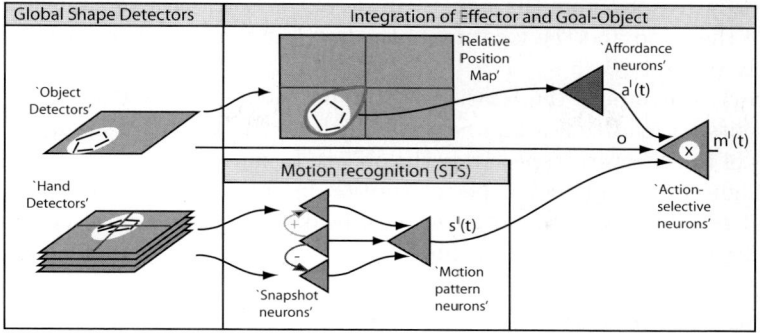

Fig. 2. Integration of hand and object information

to views of objects and hands, being sensitive to configuration, orientation and size. The response function is computed using a RBF as described before, while responses were not pooled and down-sampled. The responses of this level varied still partially with the object position, making it possible to read out the positions of object and effector by a simple population code.

2.2 Temporal Sequence Selectivity Exploiting Neural Fields

The outputs of the detectors for the effector shapes that correspond to a specific grip type l, signified by $z_k^l(t)$, provide input to *snapshot neurons* that are selective for the temporal order with which these shapes occur. This temporal order selectivity was implemented using a simple recurrent neural network, which can be interpreted as a direction-selective neural field [5,18]:

$$\tau_r \dot{r}_k^l(t) = -r_k^l(t) + \left(\sum_m w(k-m) \, [r_m^l(t)]_+ \right) + z_k^l(t) - h_r$$

where w is an asymmetric interaction kernel, h_r determines the resting level, and where τ_r is the time constant of the dynamics.

The responses of all snapshot neurons encoding the same action were integrated by *motion pattern neurons*, which smooth the activity over time. Their response depends on the maximum of the activities $r_k^l(t)$ of the corresponding snapshot neurons:

$$\tau_s \dot{s}^l(t) = -s^l(t) + \max_k \, [r_k^l(t)]_+ - h_s \tag{2}$$

The motion pattern neurons are active for individual grip sequences, independent of the presence of a goal object.

2.3 Integration of Object and Effector

The recognition of functional transitive action requires the detection of the correct match of object shape, effector configuration and relative position. For example, if a bottle is grasped from the side, the form of the bottle, the opening and orientation of the hand and the location of the hand at the side of the bottle need to be jointly recognized.

In order to compute the relative spatial positions of the effector and object, we computed a *relative position map (RPM)* from the activity maps $a_E(u,v)$ and $a_O(u,v)$ of the effector, respectively the object. In these maps, which corresponds to the highest layer of the shape recognition hierarchy described before, object and goal positions correspond to activity peaks. A simple neural network that can be described by the relationship

$$a_{RP}(u,v) = \int a_O(u',v')\, a_E(u'-u, v'-v)\, \mathrm{d}u'\, \mathrm{d}v'. \qquad (3)$$

realizes a coordinate transformation that results in an activity map, whose peak position corresponds to the position of the goal object in a coordinate system that is centered in the (retinal) position of the effector. This allows the definition of tuning functions $g_l(u,v)$ that are positive for all object positions relative to the effector (in this coordinate system) for which effector shape and position would result in an effective grip, and which are zero otherwise (cf. blue region indicated in Figure 2). The response of these detectors was given by:

$$a^l = \int a_{RP}(u,v)\, g_l(u,v)\, \mathrm{d}u\, \mathrm{d}v. \qquad (4)$$

Finally, the information about this spatial congruency between effector and object can be integrated with the information about the grip type that is indicated by the motion pattern neurons. The response of the neural detectors at the highest level of the hierarchy was simply given by the product of the responses on the previous layers:

$$m^l(t) = s^l(t) \cdot a^l(t) \qquad (5)$$

In consistency with action-selective cortical neurons (e.g. [1,3]), these top-level detectors show strong activity only if the grip type and effector position and orientation matches the grasped object.

3   Results

We tested the model on unsegmented video sequences (640x480 pixels, RGB, 30 frames/sec, 30 to 40 frames) showing a side view of a hand grasping a ball (8cm diameter, 30cm starting distance) either with a power or a precision grip. We evaluated the performance by leave-one-out cross-validation on 10 sequences per grip type. For training of the feature detectors, images (120x120 pixels) containing either the hand or the object were extracted from the training sequences. The video frames were converted to grayscale and preprocessed by removing background noise, performing local contrast normalization and image whitening.

3.1 Recognition of Grip Type

The performance of the hand-shape recognition was evaluated based on the output of the highest layer of the shape recognition hierarchy. Image frames were classified as representing either power or precision grip, according to the learned feature map with the maximum activity. Figure 3a shows the corresponding classification performance (percent correct) over time (blue bars). Averaged over the cross-validation test set, we achieve a perfect classification after approximately half of the sequence length. Figure 3b depicts the corresponding confusion matrix for power and precision grips. It is apparent that shapes usually are more likely to be confused with other shapes from the same grip type. Confusions are reduced by the sequence selection mechanism that results in an overall increase in recognition performance (see Figure 3a, red bars).

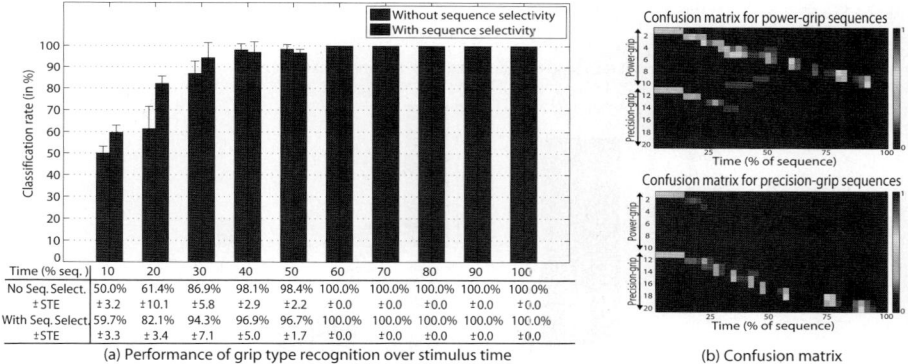

Fig. 3. (a) Recognition performance over time for the classification of power versus precision grips, with and without sequence selectivity; (b) corresp. confusion matrix

3.2 Position Estimation

Figure 4a shows that the object position can be reconstructed with high accuracy from the activity maps of the highest hierarchy layer (error: 8% in horizontal and 3% in vertical direction). This high accuracy form the basis of the reliable relative position estimation realized by Eq. (3). Figure 4b demonstrates the efficiency of this mechanism, showing the very small variation of responses if action stimuli are presented at different positions of the visual field (standard deviation ±0.61% of the response to the prefered and ±2.65% to the non-prefered stimulus).

3.3 Recognition of Functional vs. Dysfunctional Actions

The proposed system not only recognizes grip type, but also is suitable for distinguishing functional and dysfunctional grips, consistent with the properties of action-selective cortical neurons. Figure 5 depicts the average activity of the

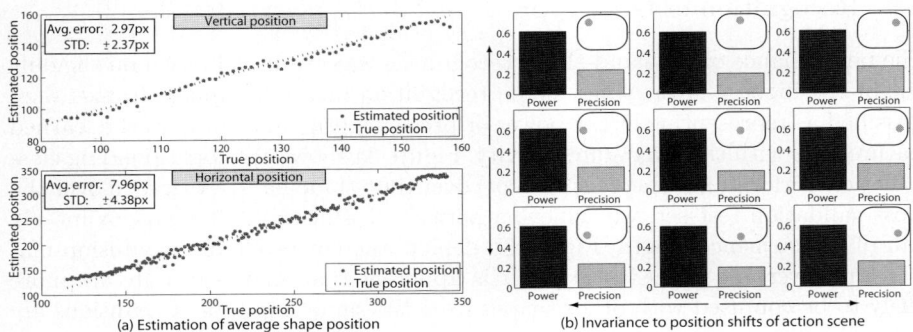

Fig. 4. (a) Accuracy of position estimation using neural population code; (b) test of position invariance presenting the same action in different positions of the visual field; responses of output neuron trained with power grip

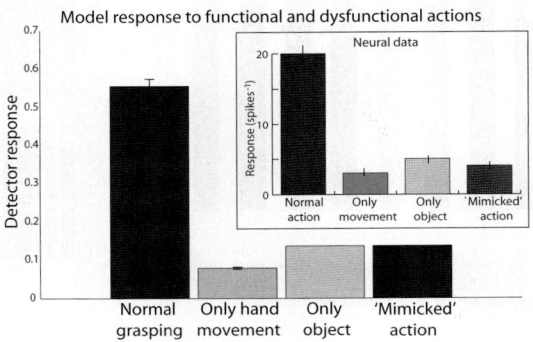

Fig. 5. Sensitivity of action-selective detectors (power grip) on the test set; inset shows neural data taken from [3]

power grip detectors on the top-level of the system (Eq. 5) over the set of cross-validation stimuli. Strong responses arise only for the correct grip type in presence of the object and if the object is placed correctly relative to the effector. 'Mimicked actions' where the object occurs next to the effector do not result in a significant response. This illustrates that the matching of grip affordances can be realized in an appearance-based framework without the assumption of three-dimensional representations. In addition, the behavior of the model closely resembles the ones of action-selective neurons in the superior temporal sulcus of monkeys ([3]see inset).

4 Conclusions

We have presented a biologically inspired architecture for the recognition of transitive actions. The system explicitly models the interaction between an effector and a goal-object without a detailed reconstruction of 3D structure. The system successfully classifies different grip types based on unsegmented video stimuli and provides estimates for the 2D positions of effector and object. Recognition was highly invariant against position changes, at the same time being quite selective against the small image changes that characterize the differences between precision and power grip. Ongoing work focuses on extending and testing the architecture on view-independent recognition tasks using an extended video data basis including a variety of object shapes.

References

1. Gallese, V., Fadiga, L., Fogassi, L., Rizzolatti, G.: Action recognition in the premotor cortex. Brain 119(Pt 2), 593–609 (1996)
2. Logothetis, N.K., Pauls, J., Poggio, T.: Shape representation in the inferior temporal cortex of monkeys. Curr. Biol. 5(5), 552–63 (1995)
3. Perrett, D.I., Harries, M.H., Bevan, R., Thomas, S., Benson, P.J., Mistlin, A.J., Chitty, A.J., Hietanen, J.K., Ortega, J.E.: Frameworks of analysis for the neural representation of animate objects and actions. J. Exp. Biol. 146, 87–113 (1989)
4. Oztop, E., Kawato, M., Arbib, M.: Mirror neurons and imitation: a computationally guided review. Neural Netw. 19(3), 254–271 (2006)
5. Giese, M.A., Poggio, T.: Neural mechanisms for the recognition of biological movements. Nat. Rev. Neurosci. 4(3), 179–192 (2003)
6. Riesenhuber, M., Poggio, T.: Hierarchical models of object recognition in cortex. Nat. Neurosci. 2(11), 1019–1025 (1999)
7. Serre, T., Wolf, L., Bileschi, S., Riesenhuber, M., Poggio, T.: Robust object recognition with cortex-like mechanisms. IEEE Trans. Pattern Anal. Mach. Intell. 29(3), 411–426 (2007)
8. Mutch, J., Lowe, D.G.: Multiclass object recognition with sparse, localized features. In: IEEE Conf. on Comp. Vision and Pattern Recognition (CVPR), vol. 1, pp. 11–18 (2006)
9. Prevete, R., Tessitore, G., Santoro, M., Catanzariti, E.: A connectionist architecture for view-independent grip-aperture computation. Brain Research 1225, 133–145 (2008)
10. Weber, M., Welling, M., Perona, P.: Unsupervised learning of models for recognition. In: Vernon, D. (ed.) ECCV 2000. LNCS, vol. 1842, pp. 18–32. Springer, Heidelberg (2000)
11. Leibe, B., Leonardis, A., Schiele, B.: Combined object categorization and segmentation with an implicit shape model. In: Proceedings of the Workshop on Statistical Learning in Computer Vision, Prague, Czech Republic (May 2004)
12. Fidler, S., Leonardis, A.: Towards scalable representations of object categories: Learning a hierarchy of parts. In: IEEE Conf. on Comp. Vision and Pattern Recognition (CVPR), pp. 1–8 (2007)
13. Bobick, A.F., Davis, J.W., Society, I.C., Society, I.C.: The recognition of human movement using temporal templates. IEEE Trans. on Pattern Anal. and Mach. Intell. 23, 257–267 (2001)
14. Blank, M., Gorelick, L., Shechtman, E., Irani, M., Basri, R.: Actions as space-time shapes. In: IEEE Int. Conference on Computer Vision (ICCV), pp. 1395–1402 (2005)

15. Jhuang, H., Serre, T., Wolf, L., Poggio, T.: A biologically inspired system for action recognition. In: Proc. IEEE Int. Conf. on Comp. Vision (ICCV), vol. 1, pp. 1–18 (2007)
16. Zelnik-Manor, L., Irani, M.: Event-based analysis of video. In: IEEE Conf. on Comp. Vision and Pattern Recognition (CVPR), vol. 2, p. 123 (2001)
17. Escobar, M.J., Masson, G.S., Vieville, T., Kornprobst, P.: Action recognition using a bio-inspired feedforward spiking network. Int. J. Comput. Vision 82(3), 284–301 (2009)
18. Zhang, K.: Representation of spatial orientation by the intrinsic dynamics of the head-direction cell ensemble: a theory. J. Neurosci. 16(6), 2112–2126 (1996)

Wide-Baseline Visible Features for Highly Dynamic Scene Recognition

Aram Kawewong, Sirinart Tangruamsub, and Osamu Hasegawa

Department of Computational Intelligence and Systems Science
Tokyo Institute of Technology
{kawewong.a.aa,tangruamsub.s.aa,hasegawa.o.aa}@m.titech.ac.jp

Abstract. This paper describes a new visual feature to especially address the problem of highly dynamic place recognition. The feature is obtained by identifying existing local features, such as SIFT or SURF, that have wide baseline visibility within the place. These identified local features are then compressed into a single representative feature, a wide-baseline visible feature, which is computed as an average of all the features associated with it. The proposed feature is especially robust against highly dynamical changes in scene; it can be correctly matched against a number of features collected from many dynamic images. This paper also describes an approach to using these features for scene recognition. The recognition proceeds by matching individual feature to a set of features from testing images, followed by majority voting to identify a place with the highest matched features. The proposed feature is trained and tested on 2000+ outdoor omnidirectional. Despite its simplicity, wide-baseline visible feature offers two times better rate of recognition (*ca.* 93%) than other features. The number of features can be further reduced to speed up the time without dropping in accuracy, which makes it more suitable to long-term scene recognition and localization.

1 Introduction

In recent years the problem of place recognition and localization has achieved much attention. It has been solved with various visual features [6], [7]. To obtain an ideal recognition system, one may need to overcome three main difficulties: dynamic places, changes of viewpoint, and scene categorization. For dynamic environment (i), places may look very different in time because of illumination changes and because of moved stuffs, i.e. parking-lots become empty on holidays. For the second difficulties (ii), different viewpoints often make the scene look different. An object's appearance taken from two different camera positions in exactly the same time could be various. The last sub-problem (iii), categorization, is a question of how the robot understand the scene so that it can categorize a new place to the one it knew. By the term "highly" dynamic changes, we define it as the situation where many objects have been changed because of some particular event (i.e. the university campus on Monday and Holiday, a city park with and without special event.

In this paper, we examined the first difficulty: the scene recognition in highly dynamic environment where major components of scenes can be changed over time because of moved objects. Among many solutions we have chosen the object-based approach [4], because we assume that, in highly dynamic place, finding the good landmarks is better than finding a good global representation. In object-based approach, a scene location is recognized by identifying a set of landmarks known to be included in the scene. Unfortunately, these approaches are prone to carrying over and amplifying low-level errors along the stream of processing. For instance, upstream identification of small objects (pixel-wise) is hindered by downstream noise inherent to camera sensors and by variable lighting conditions. This is problematic in spacious environment like the outdoors, where landmarks tend to be more spread out and possibly at farther distances from the agent.

The approach might become efficient if (i) the good landmarks could be found, and (ii) the error in capturing distant objects could be reduced. Therefore, we propose the feature which satisfies these two requirements. Wide-baseline visibility feature is developed upon the existing local features such as Scale-invariant Feature Transformation (SIFT) [6] or Speeded up Robust Features (SURF) [7]. Firstly, the local descriptors found in the sequential images are filtered to obtain only those descriptors which are slow-moving relatively to changes in camera positions. This satisfies the first requirement. Secondly, thanks to the descriptive power of the basic local descriptors, a good feature of even small distant objects can be captured precisely. This satisfies the second requirement. Despite its simplicity, wide-baseline visible feature works very well for highly dynamic outdoor scenes. We trained and tested its performance with 2000+ outdoor scenes collected from our campuses. For our dataset, training data is collected on *holidays*, while testing data is collected on *weekdays*.

2 Related Works

Place recognition has been addressed in the past by a variety of approaches. Many effective features have been proposed and used in various ways. Histogram of image properties, i.e. color [15], has been widely used in place recognition. However, after SIFT [6] is popularized in the vision community, it nearly dominates the feature choice in place recognition systems [2], [10], [11], [14]. SIFT features are invariant to scale and robust to rotation changes. The 128 dimensional SIFT descriptors have high discriminative power, while at the same time are robust to local variations [11]. It is shown that SIFT significantly outperforms edge points [3], pixel intensities [16], and steerable pyramids [12] in recognizing places.

Oliva and Torralba [13] suggested that recognition of scenes could be achieved by using "global configurations", without detailed object information. Thus statistical analysis of SIFT distribution becomes popular. Torralba *et al.* [8] use the global image features to generate the Gaussian Mixture Models for place recognition, using fixed variance. The method gives limited tolerance for appearance variation and is not invariant to translation or scale changes. Lazebnik *et al.* [3], used the k-means algorithm to cluster SIFT features, and the cluster centers were used as the codebook to solve the 15 classes scene recognition. Cummins and Newman [10] integrate the bag-of-visual-word (BoW) into the recursive Bayesian framework and achieved the performance beyond the localization; it can determine that a new image comes from a previously unseen place. Later, Angeli et al. [17], proposed the incremental BoW.

Starting from empty dictionary, the system can gradually collect new words while localizing the places. Lately, Wu and Rehg [9] proposed the spatial Principle Component Analysis on Census Transform (sPACT) as the feature for scene recognition and categorization. Its performance is proved to be better than the BoW method of [4]. The authors also report the highest accuracy over the KTH-IDOL dataset (http://cogvis.nada.kth.se/IDOL/) of indoor robots.

Unlike Gist or sPACT, wide-baseline visibility feature captures only some objects which have wide baseline visibility. Our own experiments show that global representation may yield a lower rate of accuracy when the scenes are highly dynamic. Simply averaging the slow-moving local features can be surprisingly effective for outdoors where distant objects are abundant. Furthermore, the features can be extracted very rapidly (faster than Gist and sPACT).

3 Wide-Baseline Visibility Feature Extraction

Wide-baseline visible feature is a single local feature that is robust to wide range of camera position along the path within the same place. The basic idea comes from observing that outdoor scenes generally include distant views or objects. These objects are useful to identify the place because their appearance is stable, irrespective of camera position changes. Precisely, a single wide-baseline visibility feature is computed as an average of the existing local descriptors which appear to belong to distant objects. The extraction requires an image sequence because it needs to identify all associated slow-moving local features from the sequence and compress them into a single wide-baseline visible feature. Many single wide-baseline visible features are gradually collected to form an individual dictionary of place (one place contains many sequential images). An individual dictionary is used as a signature of an individual place. In this paper, SIFT is used as the basic descriptor for wide-baseline visibility feature extraction. However, other local features, such as SURF [7], are also compatible with our method. To be simple in presentation, we use the term "*w*-feature" to designate the wide-baseline visible feature.

Given N as the current number of all visited places in an environment, n_i as the number of sequential images $\mathbf{I} = \{I_1, ..., I_{n_i}\}$ of the ith place, where $i \leq N$. Matching is performed sequentially for every pair of images; namely $(I_1 - I_2), ..., (I_{n_i-1} - I_{n_i})$. We use the same matching criteria as done in the work of [7]. The threshold value is set to 0.6. After every pair of images has been matched, the matching result is kept as the matching index vector, $\bar{m}_q^i = (m_{1,q}^i, ..., m_{k_q,q}^i)$, where $q < n_i$, k_q is the number of local features of I_q. For example, $\bar{m}_1^i = (10, 0)$ is interpreted as the first matching between I_1 and I_2. The first feature of I_1 matches to the 10th feature of I_2, while the second features of I_1 is not found in the image I_2.

Considering the $(n_i)^{th}$ image (the last image of the ith place), after $n_i - 1$ matching index vectors $\bar{m}$ are derived, a *w*-feature is then extracted. However, an object with a stable appearance irrespective of the changed position is hard to find because the path might be long or curved. Therefore, we instead extract the *w*-features of the *sub-place*.

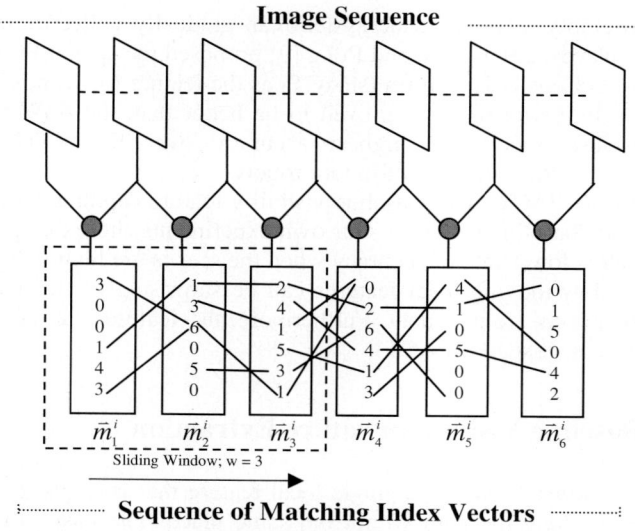

Fig. 1. The sample feature extraction of the ith place

Considering the sequence of vector $\bar{m}_q^i$ as the sequential input data, sliding windows feature extraction is performed to collect the w-features from many sub-places instead of the whole place. For example, if $w=3$, the first sub-place contains $\bar{m}_1^i, \bar{m}_2^i, \bar{m}_3^i$ corresponding to I_1, I_2, I_3, I_4 and the second sub-place contains $\bar{m}_2^i, \bar{m}_3^i, \bar{m}_4^i$ corresponding to I_2, I_3, I_4, I_5. The window size is w; the window is shifted by one, which means that, given D_j^i as the dictionary containing a set of w-features corresponding to the jth window (sub-place), there would be $n_i - w + 1$ dictionary for representing the place when the extraction is completed.

Fig. 1 illustrates the feature extraction of the ith place. Given the number of sequential images $n_i = 7$, and the size of sliding window $w = 3$. Every image pairs are compared by feature matching, resulting in six matching index vectors. An element of vector is the index of the corresponded feature in the next image. For example, for the first sub-place ($\bar{m}_1^i, \bar{m}_2^i, \bar{m}_3^i$), there is only three features which appear in all images; (1,3,6,1), (4,1,1,2), (6,3,6,1). Note that (1,3,6,1) is interpreted as the 1st, 3rd, 6th and 1th feature of the image I_1, I_2, I_3 and I_4 respectively. These four features (1,3,6,1) are interpolated to obtain a *single* wide-baseline feature. Therefore, there would be 3, 4, 4, 3 w-features for the 1st, 2nd, 3rd, and 4th sub-place respectively, resulting totally 14 wide-baseline visibility features for the whole i^{th} place. By repeat this extraction process for every place, N dictionaries $\Gamma = \{\mathbf{D}^1, ..., \mathbf{D}^N\}$ would be obtained, where each of them contains features used for representing an individual place $\mathbf{D}^i = (D_1^i; ...; D_{n_i-w}^i)$, where $D_j^i = (\bar{\psi}_{1,j}^i; ...; \bar{\psi}_{d_j,j}^i)$. $\bar{\psi}_{x,j}^i$ is the xth single wide-baseline visibility feature of the jth sub-place of the ith place, d_j is the number of wide-baseline feature in the dictionary of the jth subplace, $j \leq n_i - w + 1$, and

n_D^i is the total number of features in $\mathbf{D}^i$, $n_D^i = \sum_{p=1}^{n_i} d_p$. Γ can be used to represent all visited areas in the environment. The extraction is incremental because the new area can be added to the library. In addition, it is worth noting that extracting w-feature must match the images only $(\sum_{i=1}^{N} n_i) - 1$ times, while the spectral clustering (SC) requires $(\sum_{i=1}^{N} n_i) \times ((\sum_{i=1}^{N} n_i) - 1)/2$ times to form the affinity matrix.

4 Scene Recognition

Now that all N individual places, $\mathbf{P} = \{p_1, ..., p_N\}$, are well represented by a set of corresponding dictionaries $\Gamma = \{\mathbf{D}^1, ..., \mathbf{D}^N\}$, we describe how these dictionaries are used to recognize the places. Majority vote is selected as the recognition framework.

Majority voting (MV) is a very popular combination scheme because of its simplicity and its performance on real data. Its performance has been demonstrated experimentally in many studies such as handwriting recognition [1] and person authentication. The MV is our selection because of its main concept related to the independence of recognizers (classifiers). Based on the theoretical analyses, the MV seems to be effective if the recognizers are independent. Considering our problem, we assume that each place is independent. By applying the MV to our problem, each place vote for the matched descriptors found in the testing image. Additionally, MV is suitable to the task of incremental map-building in robotics as described in [2] because the similarity threshold for image comparison is not needed. The image is assigned to the place with the maximum number of votes.

Consider the problem in which the single omnidirectional image I is to be assigned to one of N possible existing places $(p_1, ..., p_N)$. First the image I is extracted and a set of descriptors, $\mathbf{Z} = (\vec{z}_1, ..., \vec{z}_n)$, is derived, where $\vec{z}$ is a single image descriptor, n is the number of descriptors. Of N places, each checks if the descriptors $\vec{z}_k, 1 \le k \le n$, is similar to any w-features in its dictionary $\mathbf{D}^i, 1 \le i \le N$. The vote is counted and the score is increased by one for every matching. That is, initializes $S_i \to 0$, for every i, $S_i = S_i + 1$ if $\min_{1 \le j \le n_D^i} |\vec{z}_k - \vec{\psi}_j^i| < \tau$, where τ is the similarity threshold for feature matching ($\tau = 0.6$ yields the best performance). The vote from places can be done in parallel, enabling rapid classification. After voting has been completed, the system classifies the image I as $I \to p_{\arg\max_i(S_i)}$ with confidence $c_{\arg\max_i(S_i)} = S_i / \sum_{j=1; j \ne i}^{N} (S_j)$.

5 Results and Experiment

The experiment is the scene recognition outdoor scenes. Two image databases were used in this study; *A-Campus* and *B-Campus*. We collected them by setting a tripod with height ca. 1.7 m. mounted with a camera (60D, DSLR; Nikon Corp.) with an omnidirectional lens. We walk along the road on campus while capturing omnidirectional images every few meters. For all images, the original solution are 3872×2592, but scaled down to 640×428 for using in experiments. For A-Campus, most of the

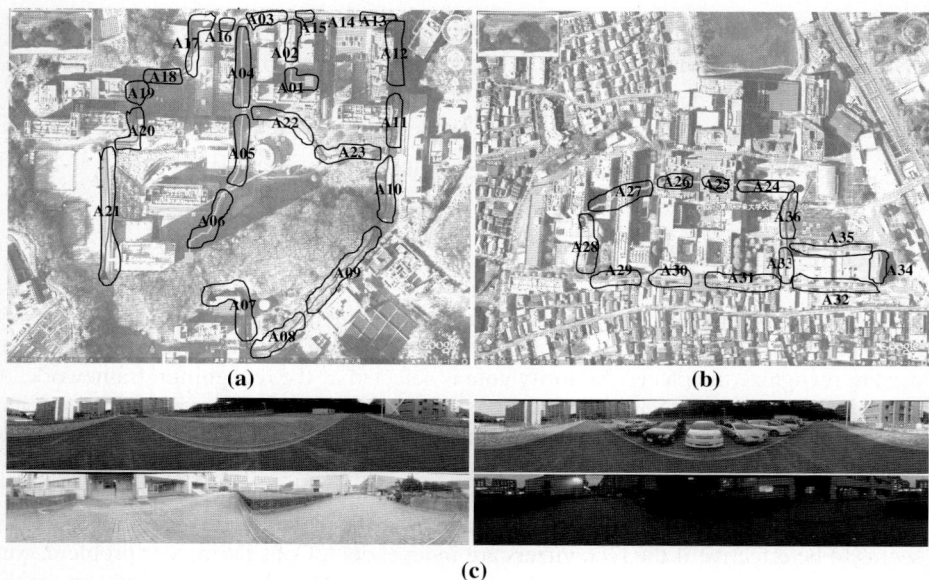

Fig. 2. Map of the outdoor experiment sites. (a) 23 places manually segmented by hand from campus A. (b) More 13 places of campus B are additionally added. (c) The sample of images from place A21 (top-row) and A01 (bottom-row) of A-campus. (Top-row) The training image was collected on holiday evening (left), while the testing image was obtained on weekday afternoon. (Bottom-row) The training image is collected in daytime, while the testing image is collected in nighttime. All images are unwrapped just for illustration.

training data are collected on *holidays* under clear weather, while the testing data are collected on *weekdays* under various weather conditions, resulting in 580 images for training, and 489 images for testing. All images are collected according to all three routes shown in Fig. 2(a). For B-Campus, we collect more images from places A24-A37 in respect to the path shown in Fig.2 (b). For this campus, people are crowded on both holidays and weekdays, so that all data are taken on weekdays under various time and weather condition. The data had been collected during 3 months, resulting in totally 450/493 images for training/testing. Fig.2 (c) shows the difference between training images and testing images.

Image data are manually segmented into 23 classes for A-Campus and 13 classes for B-Campus. Two baselines were used. The first baseline (i) is the 80-D gist vectors used in the work of Torralba *et al.* [8]. With 6 orientations of steerable pyramid and 4 scales applied to the monochrome image, 580 gist vectors are derived from 580 training images. However, we do not use the HMM as done in [8], because the transition matrix of labeled sequence data is not available. Therefore, we try to use the First Nearest Neighbour (1-NN) and Support Vector Machines (SVM) as the classifiers. For the second baseline (ii), the spatial Principal component Analysis on Cencus Transform (sPACT) proposed by [9] is our choice because of it's the reported highest result over KTH-IDOL database of. The classifiers used with sPACT are 1-NN and SVM, in the same way as done for the first baseline.

Results of recognition in outdoors are shown in Fig.3. *w*-feature obviously outperforms the others. The averaged accuracy of *w*-feature is about two times better than other features both for A-campus and B-campus. The rate for B-Campus is lower than A-Campus because the B-Campus is a very crowded environment with many buildings. These buildings often obscure the distant objects. On other words, *w*-feature will reach its highest performance in the environment where distant objects are abundant. That is to say, there are two main factors which affect the *w*-feature's efficiency. (i) Places with great number of objects blocking the distant view of camera inject bad *w*-features into the dictionary. (ii) A small number of image samples can fail the *w*-feature extraction in the sense that only a small number of wide-baseline features could be found. For example, A16 and A19 obtains low rate of accuracy (83.33% and 80.00% respectively) because their size are very small comparing to other places (see Fig.2 (a)), while A18 obtains low accuracy (85.00%) because of its high-slope blocks most of distant views. Time in *w*-feature extraction is fast. For every single image in Suzukakedai Campus, average time for creating CT histogram, gist and *w*-feature (including SIFT extraction) are 29.29s, 4.82s, and 3.23s respectively. In term of recognition time (per image), it is clear that *w*-feature is slower than Gist and sPACT since both of them encode an image into only one feature vectors. *w*-feature trades off the recognition time with better accuracy. Fig. 4 shows the accuracy for different number of *w*-features. Even 50% reduction of *w*-feature, the accuracy is still more than other baselines. Interestingly, parallel votes can reduce time to be less than a second per image (Blue Line in Fig. 4). Although the reduced time is still longer than that of Gist, it might be acceptable for robotic navigation in which the image capturing rate corresponds to the robot's motions.

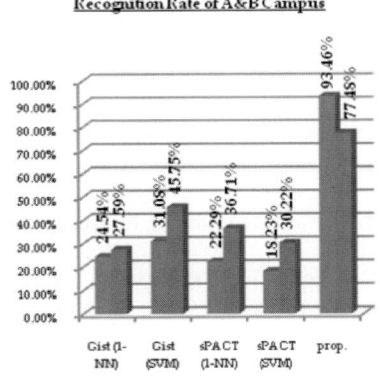
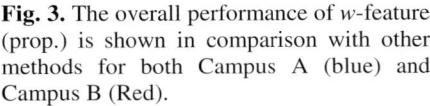
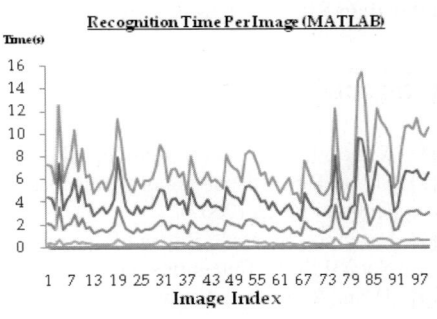

Fig. 3. The overall performance of *w*-feature (prop.) is shown in comparison with other methods for both Campus A (blue) and Campus B (Red).

Fig. 4. The recognition times per image by *w*-feature for first 100 testing images of A-Campus. From the upper line, the number of *w*-features is reduced by 75%, 50%, 25%, resulting in accuracy 92.6%, 89% and 76.7% respectively. The bottom blue line is the time when the recognition is done in parallel with 50% rate of reduction.

Generally, global feature which captures a whole scene, such as Gist, includes many sensitive objects. For the sPACT, the rate is also much lower than ours because of its basic nature of feature extraction. Like Gist, sPACT also includes many dynamic objects of the scenes. This explains why Gist and sPACT fail to recognize images taken from dynamically different conditions.

Another remarkable advantage of the w-feature is the reduction rate of memory. Since the w-features are sufficient to represent the place, the reference images are no longer needed. Most previous approaches work with a database of reference images [10], [14] which its size heavily depend on the size of the area. The memory required to store all 580+489 images (for both sites) is ~ 667MB, while the memory required for storing the w-features is 37.05~ MB (45647 features, no reduction). This means that using w-features reduces the size of memory by ~94%.

There are some limitations of w-feature and future research directions to improve it. First, w-feature is currently limited to only the recognition problem; its descriptive power is too great to be used in the problem of categorization or understanding. Second, in this paper, we use w-feature in a simple way to recognize scenes. Collecting a large number of w-feature from many places may finally face a problem of duplicated features. In addition to our w-feature reduction, vector quantization might be another good choice. Since w-feature is a highly distinctive feature in dynamic environment, bag-of- w-feature may become more stable for highly dynamic environments. Third, although time in w-feature extraction is faster than Gist and sPACT, its recognition time is slower. In this work, one place needs about 500-1000 w-features for representation. There is a room for improvement here; one may further compress these w-features to speed up the recognition time. Efficiency of w-features directly depends on the efficiency of the basic local descriptors it based on. If there are any descriptor better than SIFT, w-features will also become more efficient.

References

[1] Kittler, J., et al.: On Combining Classifiers. IEEE T PAMI 20(3), 226–239 (1998)
[2] Filliat, D.: A visual bag of words method for interactive qualitative localization and mapping. In: ICRA (2007)
[3] Lazebnik, S., et al.: Beyond Bags of Features: Spatial Pyramid Matching for Recognizing Natural Scene Categories. In: CVPR (2006)
[4] Thrun, S.: Finding Landmarks for Mobile Robot Navigation. In: ICRA (1998)
[5] Ullah, M.M., et al.: Towards Robust Place Recognition for Robot Localization. In: ICRA (2008)
[6] Lowe, D.G.: Distinctive Image Features from Scale-Invariant Keypoints. IJCV 60(2), 91–110 (2004)
[7] Bay, H., Tuytelaars, T., Van Gool, L.: SURF: Speeded up robust features. In: Leonardis, A., Bischof, H., Pinz, A. (eds.) ECCV 2006. LNCS, vol. 3951, pp. 404–417. Springer, Heidelberg (2006)
[8] Torralba, A., et al.: Context-Based Vision System for Place and Object Recognition. In: ICCV (2003)
[9] Wu, J., Rehg, J.M.: Where am I: Place instance and category recognition using spatial PACT. In: CVPR (2008)

[10] Cummins, M., Newman, P.: FAB-MAP: Probabilistic Localization and Mapping in the Space of Appearance. IJRR 27(6), 647–665 (2008)
[11] Mikolajczyk, K., Schmid, C.: A Performance Evaluation of Local Descriptors. IEEE T PAMI 27(10), 1615–1630 (2005)
[12] Kivinen, J.J., et al.: Learning Multiscale Representation of Natural Scenes Using Dirichlet Processes. In: ICCV (2007)
[13] Oliva, A., Torralba, A.: Modeling the Shape of Scene: A Holistic Representation of the Spatial Envelope. IJCV 42(3), 145–175 (2001)
[14] Goedeme, T., et al.: Omnidirectional Vision Based Topological Navigation. IJCV 74(3), 219–236 (2007)
[15] Ulrich, I., Nourbakhsh, I.R.: Appearance-based Place Recognition for Topological Localization. In: ICRA (2000)
[16] Fei-Fei, L., Perona, P.: A Bayesian Hierarchical Model for Learning Natural Scene Categories. In: CVPR (2005)
[17] Angeli, A., et al.: Fast and Incremental Method for Loop-Closure Detection Using Bags of Visual Words. IEEE Trans. Robotics 24(5), 1027–1037 (2008)

Jumping Emerging Substrings in Image Classification

Łukasz Kobyliński and Krzysztof Walczak

Institute of Computer Science, Warsaw University of Technology
ul. Nowowiejska 15/19, 00-665 Warszawa, Poland
{L.Kobylinski,K.Walczak}@ii.pw.edu.pl

Abstract. We propose a new image classification scheme based on the idea of mining jumping emerging substrings between classes of images represented by visual features. Jumping emerging substrings (JES) are string patterns, which occur frequently in one set of string data and are absent in another. By representing images in symbolic manner, according to their color and texture characteristics, we enable mining of JESs in sets of visual data and use mined patterns to create efficient and accurate classifiers. In this paper we describe our approach to image representation and provide experimental results of JES-based classification of well-known image datasets.

1 Introduction

Knowledge Discovery in Databases is a process concerning a broad range of types of data that needs to be processed every day. Originally, quantitative and textual data was in the center of interest for developing efficient and effective methods of finding interesting relationships. Today, analysis and understanding of enormous amounts of collected multimedia data seem to be the most pressing problem in the field of KDD.

As many methods for processing non-multimedia data have already been proposed, it is interesting to see how well they perform in the domain of visual data. Mining in such databases requires additional steps to represent visual information in symbolic form that is adequate for existing methods. In this paper we assess the performance of a data mining method, which has been developed focusing on textual data, in the task of image classification. For that purpose we propose an approach to image representation, a method of building a classifier and using it to perform classification of visual data.

Emerging substrings (ESs) [1] are patterns that can be used to differentiate classes of data consisting of sequences of symbols. The idea originates from emerging patterns (EPs) [2], a data mining method of extracting patterns that occur frequently in one class of data and seldom in another. Emerging patterns is an approach to KDD that proved to perform very well in the tasks of classification and prediction of large sets of data, many times much better than classical methods, such as rule- and tree-based classifiers. Emerging substrings allow additionally to reason about sequences of symbols or objects in data, which is an important feature of a visual data mining method. Specifically, we can reason about the spatial arrangement of objects on a particular image. On these grounds we expect an ES-based classifier to perform better in the task of image classification than previously proposed methods based on the idea of emerging patterns. In particular, we suggest using a subset of emerging substrings – jumping emerging substrings – to build classifiers capturing the most distinctive features of two data sets.

In what follows we first outline work conducted previously in the field of pattern-based image classification (Section 2), then give the necessary definitions of jumping emerging substrings (Section 3). Next, we describe image representation used in our experiments (Section 4), the proposed classification method (Section 5) and compare it with other known approaches (Section 6). Finally, we conclude with possibilities of further research (Section 7).

2 Previous Work

The idea of mining emerging substrings as means of capturing interesting relationships in textual data has been proposed in [1]. It was motivated by the earlier concept of emerging patterns, proposed in [2], which have been successfully used in classification of a variety of datasets. While the original algorithm for mining ESs was based on suffix trees, a generalized, linear-time solution has been proposed in [3]. This result, based on suffix arrays and longest common prefix (lcp) tables, has been later improved in [4].

To the best of our knowledge emerging substrings have not been previously studied in the context of image classification, while our own experiments concerning mining jumping emerging patterns in multimedia data have been presented in [5].

3 Jumping Emerging Substrings

Here we cite only the essential definitions of JESs, used in further parts of the paper. Please refer to [1] for complete formal definition.

A sequence is a non-empty string with finite length over an alphabet $\Sigma = \{a_1, a_2, \ldots, a_m\}$. The length of a sequence is the number of symbols contained in it. Having a string $s = s_1 s_2 \ldots s_k$ of length k and a sequence $T = t_1 t_2 \ldots t_l$ of length l, we say that s is a substring of T, denoted as $s \sqsubseteq T$ if $\exists i \in 1 \ldots (l-k+1)$ such that $s_1 s_2 \ldots s_k = t_i t_{i+1} \ldots t_{i+k-1}$. If $s \neq T$, s is a proper substring of T, denoted as $s \sqsubset T$.

A database D is a set of sequences T_i, each associated with a class label $c_{T_i} \in C = \{c_1, c_2, \ldots, c_n\}$, where C is the set of all labels. The support of a string s in a database D is the fraction of sequences in D that s is a substring of: $\text{supp}_D(s) = \frac{|\{T \in D: s \sqsubseteq T\}|}{|D|}$. Given two databases $D_1, D_2 \subseteq \mathcal{D}$ we say that a string s is a jumping emerging substring (JES) from D_1 to D_2 if $\text{supp}_{D_1}(s) = 0 \land \text{supp}_{D_2}(s) > 0$. The task of JES mining is to find all strings having a given minimum support θ in D_2, being a JES from D_1 to D_2. We will denote this set of strings as

Table 1. Example database and its jumping emerging substrings

class A	class B	JES	support class A	support class B	direction
acd	cde	b	0	1/2	A → B
ac	ab	e	0	1/2	A → B
		ab	0	1/2	A → B
		ac	1	0	B → A
		de	0	1/2	A → B
		acd	1/2	0	B → A
		cde	0	1/2	A → B

$JES(D_1, D_2, \theta)$. Furthermore, we can distinguish the set of only minimal JESs, that is sequences, for which no frequent substrings exist: $JES_m(D_1, D_2, \theta) = \{T \in JES(D_1, D_2, \theta) : \neg \exists s \in JES(D_1, D_2, \theta) \ s \sqsubset T\}$.

Table 1 shows a simple two-class database and its jumping emerging substrings. Based on the above definition, we look at all possible substrings of strings in class A and find these, which are not present in class B. Similarly, we check for JESs from class B to A. The string "ac" would be the only JES, if we were to find only jumping emerging substrings with minimum support of 1. Finally, we reduce the set of discovered patterns to only minimal JESs: $JES_m(D_A, D_B, 1/2) = \{b, e\}$, $JES_m(D_B, D_A, 1/2) = \{ac\}$.

4 Image Representation

We have compared two approaches to calculation of image features: using both a color descriptor and a texture descriptor based on Gabor filters (as in MPEG-7 standard), and a SIFT descriptor. In both cases we divide the images into a rectangular $x \times y$ grid and calculate features in each of the resulting tiles.

In the first approach color and texture features are calculated separately. Image colors are represented by a histogram calculated in the HSV color space, with the hue channel quantized to h discrete ranges, while saturation and value channels to s and v ranges respectively. In effect, the representation takes the form of a $h \times s \times v$ element vector of real values between 0 and 1. For the representation of texture we use a feature vector consisting of mean and standard deviation values calculated from the result of filtering an original image with a bank of Gabor functions. These filters are scaled and rotated versions of the base function, which is a product of a Gaussian and a sine function. By using m orientations and n different scales we get a feature vector consisting of mean (μ) and standard deviation (σ) values of each of the filtered images and thus having a size of $2 \times m \times n$ values. In our experiments a vector size of $2 \times 6 \times 4 = 48$ values has been used for texture and $18 \times 3 \times 3 = 162$ for color representation.

SIFT is a local feature descriptor, proposed in [6], which has been widely used for image representation in classification, recognition and retrieval tasks. Using

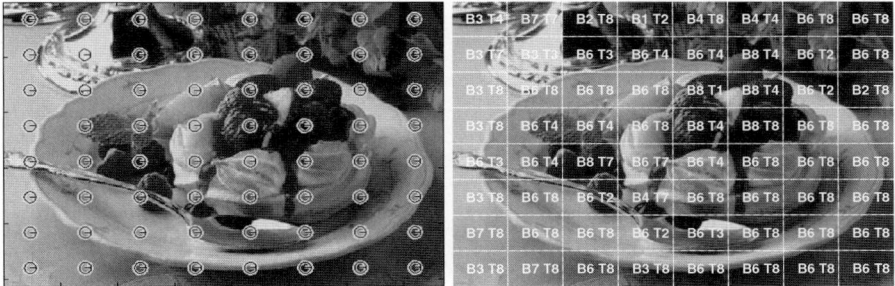

Fig. 1. An example of features calculation and symbolic image representation

the VLFeat open implementation [7], we have calculated SIFT features of the center point of each of the image tiles for H, S and V color channels, having a constant scale and orientation set for the descriptor. The feature vector size for every point is thus equal to 3×128 values.

Having calculated features of each of the images in both the training and testing set, we have created a visual dictionary of the most representative color and texture features. The dictionary is built by clustering corresponding feature values into a chosen number of groups. Resulting centroids become the elements of the dictionary and are labeled with unique symbols. These identifiers are then used to describe the images in the database by associating an appropriate label with every tile of each image. This is performed by finding the closest centroid to a feature vector calculated for a given image tile. The same dictionary is used during both the learning and classification phases.

Figure 1 illustrates the used method of image representation. A regular grid of points is used to calculate images features, which are then clustered to create the dictionary. In the case of MPEG-7 features, values representing color and texture are clustered separately and labeled $B_1, B_2, \ldots, B_n$ and $T_1, T_2, \ldots, T_n$ respectively. These labels are then used to describe each of the grid tiles.

5   JES-Based Classification

In our approach classification is a two-step process. The first phase consists of building a classifier on the basis of the learning dataset. We use image representation described in the previous section to associate sets of strings to each of the images in the dataset and then mine minimal jumping emerging substrings between respective classes in the database. The strings are formed by taking into account horizontal, vertical and diagonal sequences of symbols of representation of a particular image (see Fig. 2).

In the second phase, we use the created classifier to assign images from the testing set to respective categories. This is done by aggregating all minimal JESs that match the representation of a particular image and determining the majority class of the patterns. The winning category is then assigned to the example.

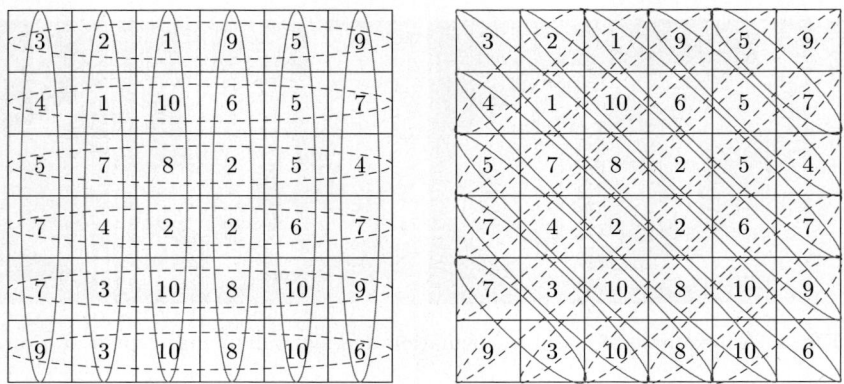

Fig. 2. Representation used to mine JESs between classes of images. Strings are formed by considering horizontal, vertical and diagonal sequences of symbols.

Formally, for a multi-class set of images, represented as a learning database of strings with associated class labels $\mathcal{D}_L = \bigcup_c C_c$, where C_c is a database containing images of class c and C'_c is its complementary database, and a test set $\mathcal{D}_T$, we can formally write the algorithm as follows:

1. For each $c \in C$:
   (a) Discover minimal $JES_m(C'_c, C_c, \theta)$
   (b) For each test image $T \in \mathcal{D}_T$: calculate $score(T, c) = \sum_X \text{supp}_{C_c}(X)$, where $X \in JES(C'_c, C_c)$ such that $X \sqsubseteq T$.
2. Assign image T to a class c, which has the maximum score.

6 Experimental Results

We have used two different datasets to assess the performance of the proposed JES-based image classification approach. Firstly, we have prepared a synthetic two-class set of images, which consists of photographs containing the same object, positioned randomly on a static background. On the images of class A the object is oriented vertically, while in class B – horizontally (see Fig. 3). Each of the classes contains ca. 60 images.

The database has been prepared to validate the idea behind using JESs for image classification and the chosen image representation method. While the object and background are exactly the same in each of the classes, in our approach we are able to capture more data about their relationship than using regular methods, which do not take spatial information into consideration. As presented in Table 2 JES-based classifier performs much better, regardless of used feature descriptor.

All experiments have been performed as a ten-fold cross validation, where the feature dictionary is recreated in every iteration. The images have been divided into 8×8 tiles and the dictionary size has been limited to 16 values. In the case

Fig. 3. A synthetic test database with two classes of images

Table 2. Classification accuracy of the synthetic dataset

method	minimum support	MPEG-7 features accuracy (%)	patterns no.	SIFT features accuracy (%)	patterns no.
	0.250	95.33	92	69.33	80
	0.200	96.67	156	79.67	140
	0.150	97.50	229	**87.00**	198
JES	0.100	98.33	352	86.00	574
	0.050	**99.17**	1175	85.67	1830
	0.025	99.17	3304	85.33	10934
	0.010	99.17	20797	85.33	10934
	0.005	99.17	20797	85.33	10934
C4.5	-	93.46	-	57.00	-
SVM	-	96.67	-	65.00	-

of this synthetic database, SIFT features have resulted in worse classification performance than the color and Gabor-based texture features, mostly because of the sparse grid used to calculate values in particular points. For comparison purposes, we have used the same locations for calculation of SIFT and MPEG-7 features. Classification with other methods than emerging substrings and emerging patterns has been carried out using the Weka package [8] and the LIBSVM library [9] with default parameter values.

Secondly, we have included results of classification of a dataset used in our earlier experiments in [5], namely the image database created by the authors of the SIMPLIcity CBIR system [10] (see Fig. 4). This set consists of 10 categories of photographs, 100 images in each class. As reported in Table 3, our current approach is in each case giving better results than any of the others. It may be noted that lowering the minimum support value when mining JESs improves the

Table 3. Classification accuracy of the SIMPLIcity dataset with MPEG-7 features

method	minimum support	accuracy (%)					
		flower/ food	flower/ elephant	flower/ mountain	food/ elephant	food/ mountain	elephant/ mountain
JES	0.250	92.26	93.68	96.37	30.50	83.50	58.00
	0.200	94.79	95.26	96.89	41.00	89.50	66.00
	0.150	96.37	97.89	96.89	63.50	93.00	74.50
	0.100	97.94	98.95	96.89	85.00	94.00	89.00
	0.050	**98.47**	98.95	96.89	**93.00**	95.50	92.00
	0.025	98.47	98.95	96.89	93.00	**96.00**	**93.50**
	0.005	98.47	98.95	96.89	93.00	95.50	93.50
occJEP [5]	-	97.92	**98.96**	**97.92**	88.00	91.00	88.50
JEP [5]	-	95.83	91.67	96.35	88.50	93.50	83.50
C4.5	-	93.23	89.58	85.94	87.50	92.50	82.00
SVM	-	90.63	91.15	93.75	87.50	84.50	84.50

Fig. 4. Example images from the SIMPLIcity test database

classification accuracy only to certain point, above which there is no additional gain of discovering greater number of patterns.

7 Conclusions and Future Work

In this paper we have proposed an approach to image classification that combines the methods used for sequence and text mining with image analysis and showed that such methodology may give promising results, surpassing the performance of other data mining methods. Using jumping emerging substrings to distinguish images of different classes in a database has a clear advantage over other pattern-based methods, thanks to its ability to capture spatial relationships between visual features. It is important to note that optimal (linear-time) algorithms exist to mine JESs between sets of sequential data. Furthermore, the proposed approach may be used in conjunction with different feature descriptors, as long as the images are expressed by a matrix of a finite number of symbols.

The following aspects of the described method could be enhanced in future work: invariance to scale by providing multiple layers of symbolic representation of an image, each calculated using a descriptor of a different scale; using a dense grid of points for SIFT and multiple orientations to achieve better results than the MPEG-7 approach.

References

1. Chan, S., Kao, B., Yip, C.L., Tang, M.: Mining emerging substrings. In: Proceedings of the Eighth International Conference on Database Systems for Advanced Applications, pp. 119–126 (2003)
2. Dong, G., Li, J.: Efficient mining of emerging patterns: Discovering trends and differences. In: KDD 1999: Proceedings of the fifth ACM SIGKDD international conference on Knowledge discovery and data mining, pp. 43–52. ACM, New York (1999)
3. Fischer, J., Heun, V., Kramer, S.: Optimal string mining under frequency constraints. In: Fürnkranz, J., Scheffer, T., Spiliopoulou, M. (eds.) PKDD 2006. LNCS (LNAI), vol. 4213, pp. 139–150. Springer, Heidelberg (2006)
4. Fischer, J., Mäkinen, V., Välimäki, N.: Space efficient string mining under frequency constraints. In: Proceedings of the 8th IEEE International Conference on Data Mining, pp. 193–202. IEEE Computer Society, Los Alamitos (2008)
5. Kobyliński, Ł., Walczak, K.: Efficient mining of jumping emerging patterns with occurrence counts for classification. In: Chan, C.-C., Grzymala-Busse, J.W., Ziarko, W.P. (eds.) RSCTC 2008. LNCS (LNAI), vol. 5306, pp. 419–428. Springer, Heidelberg (2008)
6. Lowe, D.G.: Distinctive image features from scale-invariant keypoints. International Journal of Computer Vision 60(2), 91–110 (2004)
7. Vedaldi, A., Fulkerson, B.: VLFeat: An open and portable library of computer vision algorithms (2008), http://www.vlfeat.org/
8. Witten, I.H., Frank, E.: Data Mining: Practical machine learning tools and techniques, 2nd edn. Morgan Kaufmann, San Francisco (2005)
9. Chang, C.C., Lin, C.J.: LIBSVM: a library for support vector machines (2001), http://www.csie.ntu.edu.tw/~cjlin/libsvm
10. Wang, J.Z., Li, J., Wiederhold, G.: SIMPLIcity: Semantics-sensitive integrated matching for picture libraries. IEEE Trans. on Patt. Anal. and Machine Intell. 23, 947–963 (2001)

Human Action Recognition Using LBP-TOP as Sparse Spatio-Temporal Feature Descriptor

Riccardo Mattivi and Ling Shao

Philips Research, Eindhoven, The Netherlands
{riccardo.mattivi,l.shao}@philips.com

Abstract. In this paper we apply the Local Binary Pattern on Three Orthogonal Planes (LBP-TOP) descriptor to the field of human action recognition. A video sequence is described as a collection of spatial-temporal words after the detection of space-time interest points and the description of the area around them. Our contribution has been in the description part, showing LBP-TOP to be a promising descriptor for human action classification purposes. We have also developed several extensions to the descriptor to enhance its performance in human action recognition, showing the method to be computationally efficient.

Keywords: Human action recognition, LBP-TOP, bag of words.

1 Introduction

Automatic categorization and localization of actions in video sequences has different applications, such as detecting activities in surveillance videos, indexing video sequences, organizing digital video library according to specified actions, etc. The challenge is how to obtain robust action recognition under variable illumination, background changes, camera motion and zooming, viewpoint changes and partial occlusions, geometric and photometric variations of objects and intra-class differences.

There are two main approaches: holistic and part-based representations. Holistic representations focus on the whole human body trying to search characteristics such as contours or pose. Usually holistic methods, which focus on the contours of a person, do not consider the human body as being composed of body parts but consider the whole form of human body in the analyzed frame. Efros et al. [1] use cross-correlation between optical flow descriptors and Shechtman et al. [2] use similarity between space-time volumes which allows finding similar dynamic behaviors and actions. Motion and trajectories are also commonly used features for recognizing human actions, e.g. Ali et al. [3] use trajectories of hands, feet and body. Holistic methods may depend on the recording conditions such as position of the pattern in the frame, spatial resolution, relative motion with respect to the camera and can be influenced by variations in the background and by occlusions. These problems can be solved in principle by external mechanisms (e.g. spatial segmentation, camera stabilization, tracking etc.), but such mechanisms might be unstable in complex situations and require more computational demand.

Part-based representations typically search for Space-Time Interest Points (STIPs) in the video, apply a robust description of the area around them and create a model based on independent features (Bag of Words) or a model that can also contain structural information. These methods do not require tracking and stabilization and are often more resistant to cluttering, as only few parts may be occluded. Different methods for detecting STIPs have been proposed, such as [11], [12]. The resulting features often reflect interesting patterns that can be used for a compact representation of video data as well as for interpretation of spatio-temporal events.

The paper is organized as follows. In section 2 we provide the methodology adopted for classification and in Section 3 we provide an introduction to the LBP and LBP-TOP descriptors on 3D data. Experimental results on human action recognition are shown and evaluated in Section 4. Finally, we conclude in Section 5.

2 Methodology

In the following sections we describe our algorithm in detail. In Section 2.1 we explain the classification scheme of our algorithm. In Section 2.2 we give a brief description about the detection of STIPs and the feature description method is introduced in Section 2.3. Section 2.4 explains the classifier used.

2.1 Bag of Words Classification

The methodology we adopt is a Bag of Words classification model [11]. As a first step, space-time interest points are detected using a separable linear filter and small video patches (named cuboids) are extracted from each interest point. They represent the local information used to learn and recognize the different human actions. Each cuboid is described using the LBP-TOP descriptor. The result is a sparse representation of the video sequence as cuboid descriptors. Having obtained all these data for the training set, a visual vocabulary is built by clustering using the k-means algorithm. The center of each cluster is defined as a spatial-temporal 'word' of which length depends on the length of the descriptor adopted. Each feature description is successively assigned to the closest (we use Euclidean distance) vocabulary word and a histogram of spatial-temporal word occurrence in the entire video is computed. Thus, each video is represented as a collection of spatial-temporal words from the codebook in the form of a histogram. For classification, we use non linear Support Vector Machines (SVM). As the algorithm has random components, such as the clustering phase, any experiment result reported is averaged over 20 runs. The entire methodology used is shown in Fig. 1.

2.2 Feature Detection

Several spatio-temporal feature detection methods have been developed recently and among them we chose Dollar's feature detector [11] because of its simplicity, fastness and because it generally produces a high number of responses. The detector is based on a set of separable linear filters which treats the spatial and temporal dimensions in different ways. A 2D Gaussian kernel is applied only along the spatial dimensions (parameter σ to be set), while a quadrature pair of 1D Gabor filters are applied only temporally (parameter τ to be set). This method responds to local regions which exhibit complex motion patterns, including space-time corners. For more implementation details, please refer to [11] as the feature detection part is beyond the scope of this paper.

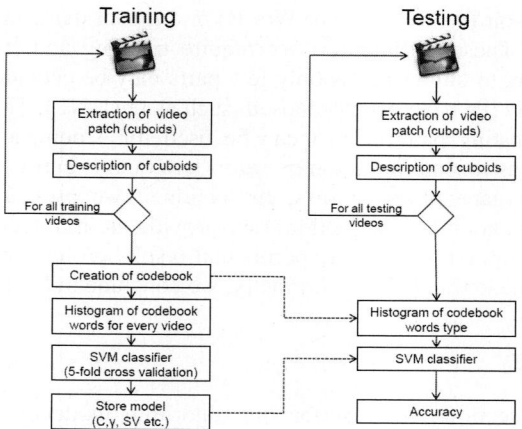

Fig. 1. Methodology adopted for action recognition

2.3 Feature Description

Once the cuboid is extracted, it is described using the LBP-TOP descriptor, which is an extension of LBP operator into the temporal domain. LBP has originally been proposed for texture analysis and classification [4]. Recently, it has been applied on face recognition [5] and facial expression recognition [6], [7]. While the original LBP was only designed for static images, LBP-TOP has been used for dynamic textures and facial expression recognition [8]. As a video sequence can not only be seen as the usual stack of XY planes in the temporal axis, but also as a stack of YT planes on X axis and as a stack of XT planes on Y axis, we prove that a cuboid can be successfully described with LBP-TOP for action recognition purposes.

2.4 Classification

Each video sequence is described as a histogram of space-time words occurrence which represents its signature. The dimension of the signature is equal to the size of the codebook and is given as input to the classifier (see Fig. 1). We chose to use non linear Support Vector Machines (SVM) with rbf kernel and the library libSVM [14] was adopted. The best parameters C and γ were chosen doing a 5-fold cross validation in a grid approach on the training data and one against one approach has been used for multi-class classification.

3 LBP-TOP and Its Extensions

The Local Binary Pattern (LBP) operator labels the pixels of an image by thresholding a circular neighborhood region [4]. The $LBP_{P,R}$ operator produces 2^P different output values, corresponding to the 2^P different binary patterns that can be formed by the P pixels in the neighbor set. The derived binary numbers encode local primitives such as curved edges, spots, flat areas etc. After the computation of the LBP for the whole image, an occurrence histogram of the labels is used as feature. It contains information about the distribution of local micro-patterns over the whole image and represents a statistical description of image characteristics. This descriptor has been proved to be successful in face recognition [5]. For more details about LBP operator, please refer to [4], [5], [6], [7]. Recently, LBP has been modified in order to be used in the context of dynamic texture description and recognition and for facial expression analysis [8]. LBP-TOP computes the LBP from Three Orthogonal Planes, denoted as XY-LBP, XT-LBP and YT-LBP. The operator is expressed as .

$$LBP-TOP_{P_{XY},P_{XT},P_{YT},R_X,R_Y,R_T} .\qquad(1)$$

where the notation (P_{XY}, P_{XT}, P_{YT}, R_X, R_Y, R_T) denotes a neighborhood of P points equally sampled on a circle of radius R on XY, XT and YT planes respectively. The statistics on the three different planes are computed and then concatenated into a single histogram. The resulting feature vector is of $3 \cdot 2^P$ length. Fig. 2 illustrates the construction of the LBP-TOP descriptor. In such a scheme, LBP encodes appearance and motion in three directions, incorporating spatial information in XY-LBP and spatial temporal co-occurrence statistics in XT-LBP and YT-LBP.

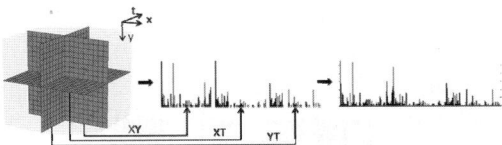

Fig. 2. LBP-TOP methodology

In our implementation, LBP-TOP is applied on each cuboid, as shown in Fig. 3, where XY, XT and YT planes are the central slices of it as can be seen in Fig. 4. Kellokumpu et al [8] have recently used LBP-TOP for human detection and activity description. However, their approach is based on background subtraction using LBP-TOP and a bounding volume has to be built around the area of motion. Their method can be categorize as holistic, since no space-time interest points have to be detected and differs from our part-based approach.

3.1 Modifications on LBP-TOP

As we described previously, the original LBP-TOP descriptor is the computation of LBP on the gray-level values of 3 orthogonal slices of each cuboid. We propose to extend the computation of LBP to 9 slices, 3 for each axis. Therefore, on the XY dimension we have the original XY plane (centered in the middle of the cuboid) plus other two XY planes located at 1/4 and 3/4 of the cuboid's length. The same is done for XT and YT dimensions. We named this method as Extended LBP-TOP. In this manner, more dynamic information in the cuboid can be extracted, as the 3 slices in one axis capture the motion at different times. We also exploit more information from the cuboid, dealing with 6 slices on each axis, located from 2/8 until 7/8 of the cuboid's length for each axis. In this case, a dimensionality reduction technique has to be applied since the final dimension of the descriptor vector would be too high.

Another modification we introduced is the computation of LBP operator on gradient images. The gradient image contains information about the rapidity of pixel intensity changes along a specific direction, has large magnitude values at edges and it can further increment LBP operator's performances, since LBP encodes local primitives such as curved edges, spots, flat areas etc. For each cuboid, the brightness gradient is calculated along x, y and t directions, and the resulting 3 cuboids containing specific gradient information are summed in absolute values. Before computing the image gradients, the cuboid is slightly smoothed with a Gaussian filter in order to reduce noise. LBP-TOP is then performed on the gradient cuboid and we name this method Gradient LBP-TOP. The Extended LBP-TOP can be applied on the gradient cuboid and we named this method as Extended Gradient LBP-TOP.

Fig. 3. Cuboid with XY, XT and YT planes

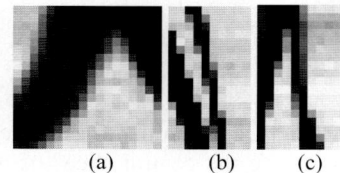

Fig. 4. Extracted XY *(a)*, XT *(b)* and YT *(c)* planes from the cuboid of Fig. 3

4 Experimental Results

For our action recognition experiments, we chose to use the KTH human action dataset [10]. This dataset contains six types of human actions: walking, jogging, running, boxing, hand waving and hand clapping. Each action class is performed several times by 25 subjects in different scenarios of outdoor and indoor environment. The camera is not static and the videos contain scale changes. In total, the dataset contains 600 sequences. We divide the dataset into two parts: 16 people for training and 9 people for testing, as it has been done in [10] and in [13]. We limit the length of all video sequences to 300 frames.

We extract the space-time interest points and describe the corresponding cuboids with the procedure described in Sections 2.2 and 2.3. The detector parameters are set to $\sigma=2.8$ and $\tau=1.6$, which gave better results in our evaluations, and 80 STIPs were detected for each sequence. The original LBP-TOP and the Extended LBP-TOP are computed on the original cuboid or on the gradient cuboid. The number of clusters used to build the codebook is chosen to maximize the classification accuracy on the testing data and best values have been achieved using 1000 visual-words.

The accuracy results for LBP-TOP with different parameters R and P are shown in Table 1. The notation of parameters is as illustrated in Equation (1). Better classification accuracy has been obtained with the parameter P greater than 6 and radius R equal to 2. The performance is generally slightly decreasing as the radius R is getting bigger, while it is increasing as the number of neighbors P is increased. This could be explained as more neighbors permit to take more information into account. However, the drawback is a higher computational cost and a higher dimensionality of the feature vector.

Table 1. Accuracy for different parameter values of LBP-TOP$_{P,P,P,R,R,R}$

		Neighbors (P)			
		4	6	8	10
Radius (R)	2	71.81%	85.65%	86.25%	86.32%
	3	84.54%	85.18%	85.12%	86.69%
	4	81.34%	85.12%	85.46%	83.82%

LBP-TOP$_{8,8,8,2,2,2}$ produces a 768 vector length, while LBP-TOP$_{10,10,10,2,2,2}$ has a descriptor dimension of 3072. The final descriptor of LBP-TOP$_{12,12,12,2,2,2}$ will be 12288 vector lengths. The use of uniform LBP operator decreases the performance results compared with the original operator, since less information is kept into account (see Table 2). Multiresolution LBP operator has also been tested, but the gain in performances is not considerable with the increase of the descriptor length and computational cost. The time calculated in the following tables is measured on a computer equipped with a 3 Ghz Pentium 4 CPU and 3 Gb RAM.

We choose to use LBP-TOP$_{8,8,8,2,2,2}$ for the following experiments as it is computationally more efficient and the accuracy is among the highest. As dimensionality reduction technique, we used Principal Component Analysis (PCA) and set the final dimension to 100.

In Table 2, the Extended LBP-TOP is evaluated and different number of slices is taken into account. As we can see, the Extended LBP-TOP descriptor performs better than the original one, since more information is taken into consideration at different times in XY planes and at different locations in the XT and YT planes. Although best result is obtained with 6 slices on each axis, the computational time is almost double than the Extended LBP-TOP version with 3 slices; because of this issue, in the following we are computing the Extended version on only 3 slices for each axis.

Table 3 is a summary of best results achieved for different enhancement of LBP-TOP. The usage of LBP-TOP applied to the gradient cuboid gives better results compared with the original one. The information extracted from the gradient calculated along x, y and t directions and combined into the gradient cuboid permits to have a better performance for LBP-TOP in the description of actions. Moreover, a slight increase in performances can be achieved by applying the Extended LBP-TOP on the gradient cuboids. The number of support vectors calculated by SVM is, in all methods tested, about 360. As a feature reduction method, we applied PCA and show that the classification accuracy is only decreased slightly in Extended LBP-TOP, while slightly increasing in the Extended Gradient LBP-TOP.

Table 2. Uniform and Extended LBP-TOP

Method	Accuracy	Descriptor length	Computational time (s)
LBP-TOP$_{8,8,8,2,2,2}$	86.25 %	768	0.0139
Uniform LBP-TOP$_{8,8,8,2,2,2}$	81.78 %	177	0.0243
Extended LBP-TOP$_{8,8,8,2,2,2}$ (3 slices on each axis)	88.19 %	2304	0.0314
Extended LBP-TOP$_{8,8,8,2,2,2}$ (3 slices on each axis) + PCA	87.87 %	100	0.0319
Extended LBP-TOP$_{8,8,8,2,2,2}$ (6 slices on each axis) + PCA	88.38 %	100	0.0630

Table 3. Accuracy for different LBP-TOP methods

Method	Accuracy	Descriptor length	Computational time (s)
Ext LBP-TOP$_{8,8,8,2,2,2}$	88.19 %	2304	0.0314
Ext LBP-TOP$_{8,8,8,2,2,2}$ + PCA	87.87 %	100	0.0319
Grad LBP-TOP$_{8,8,8,2,2,2}$	90.07 %	768	0.0788
Ext Grad LBP-TOP$_{8,8,8,2,2,2}$	90.72 %	2304	0.0992
Ext Grad LBP-TOP$_{8,8,8,2,2,2}$ + PCA	91.25 %	100	0.1004
HOG-HOF	89.88 %	162	0.2820
HOG-HOF + PCA	89.28 %	100	0.2894

As a comparison, we evaluate Laptev's method [13] with the same framework as illustrated in Section 2. We use Laptev's code publicly available on his website and recently being updated with the latest settings used in [13]. The combination of Laptev's extraction method and Laptev's HOG-HOF descriptor make us reach an accuracy of 89.88%. The time for Laptev's HOG-HOF descriptor in Table 3 is referred to both extraction and description parts, as the description part cannot be computed regardless of the extraction part in Laptev's provided executable. Therefore, we expect the description part to be about half the time shown in the table. The computational time for HOG-HOF is affected by the choice of the threshold and we have chosen a suitable threshold to have 80 detected STIPs for this comparison. There is also to mention that Laptev's executable code is compiled in C environment, while our LBP-TOP implementation is compiled in Matlab environment. Similar performance to Laptev's is achieved using the Extended LBP-TOP descriptor which is almost 3 times computationally faster than the Extended Gradient LBP-TOP descriptor.

5 Conclusion

In this paper, we have applied LBP-TOP as a descriptor of small video-patches used in a part-based approach for human action recognition and shown LBP-TOP to be suitable for the description of cuboids containing information about human actions. We have extended LBP-TOP considering the action at three different frames in XY plane and at different views in XT and YT planes. Furthermore, we applied LBP-TOP to gradient images. We have also shown that the performance of descriptor is quite stable when the PCA is applied. Regarding computational time, the Extended Gradient LBP-TOP descriptor, compared with HOG-HOF, is computationally more efficient and permits to reach better accuracy in our framework. The experimental results reveal that LBP-TOP and its modifications tend to be good candidates for human action description and recognition. The best accuracy has been obtained by using the Extended Gradient LBP-TOP$_{8,8,8,2,2,2}$ with PCA.

References

1. Efros, A.A., Berg, A.C., Mori, G., Malik, J.: Recognizing action at a distance. In: Proceedings of Ninth IEEE International Conference on Computer Vision, vol. 2, pp. 726–733 (2003)
2. Shechtman, E., Irani, M.: Space-time behavior based correlation. In: IEEE Computer Society Conference on CVPR, June 20-25, vol. 1, pp. 405–412 (2005)
3. Ali, S., Basharat, A., Shah, M.: Chaotic invariants for human action recognition. In: Proc. of IEEE International Conference on Computer Vision (ICCV), pp. 1–8 (2007)
4. Ojala, T., Pietikäinen, M., Mäenpää, T.: Multiresolution gray-scale and rotation invariant texture classification with local binary patterns. IEEE Transactions on Pattern Analysis and Machine Intelligence, 971–987 (2002)
5. Ahonen, T., Hadid, A., Pietikäinen, M.: Face recognition with local binary patterns. In: Pajdla, T., Matas, J.G. (eds.) ECCV 2004. LNCS, vol. 3021, pp. 469–481. Springer, Heidelberg (2004)
6. Ahonen, T., Hadid, A., Pietikäinen, M.: Face Description with Local Binary Patterns: Application to Face Recognition. IEEE Transactions on Pattern Analysis and Machine Intelligence 28(12), 2037–2041 (2006)
7. Shan, C., et al.: Facial Expression recognition based on Local Binary Patterns: A comprehensive study. Image and Vision Computing (2008)
8. Zhao, G., Pietikäinen, M.: Dynamic Texture Recognition Using Local Binary Patterns with an Application to Facial Expressions. IEEE Transactions on Pattern Analysis and Machine Intelligence 29(6), 915–928 (2007)
9. Kellokumpu, V., Zhao, G., Pietikäinen, M.: Human Activity Recognition Using a Dynamic Texture Based Method. In: British Machine Vision Conference (2008)
10. Schuldt, C., Laptev, I., Caputo, B.: Recognizing human actions: a local SVM approach. In: Proceedings of the 17th International Conference on Pattern Recognition, August 2004, vol. 3, pp. 32–36 (2004)
11. Dollar, P., Rabaud, V., Cottrell, G., Belongie, S.J.: Behavior recognition via sparse spatio-temporal features. In: Proc. of ICCV Int. work-shop on Visual Surveillance and Performance Evaluation of Tracking and Surveillance (VSPETS), pp. 65–72 (2005)
12. Laptev, I.: On space-time interest points. International Journal of Computer Vision (IJCV) 64(2-3), 107–123 (2005)
13. Laptev, I., Marszalek, M., Schmid, C., Rozenfeld, B.: Learning realistic human actions from movies. In: Proc. of IEEE International Conference on Computer Vision and Pattern Recognition (CVPR), pp. 1–8 (2008)
14. Chang, C.-C., Lin, C.-J.: LIBSVM: a library for support vector machines (2001), http://www.csie.ntu.edu.tw/~cjlin/libsvm

Contextual-Guided Bag-of-Visual-Words Model for Multi-class Object Categorization

Mehdi Mirza-Mohammadi[1], Sergio Escalera[1,2], and Petia Radeva[1,2]

[1] Dept. Matemàtica Aplicada i Anàlisi, Gran Via 585, 08007, Barcelona, Spain
[2] Computer Vision Center, Campus UAB, Edifici O, 08193, Bellaterra, Barcelona
me@memimo.net, {sergio,petia}@maia.ub.es

Abstract. Bag-of-words model (BOW) is inspired by the text classification problem, where a document is represented by an unsorted set of contained words. Analogously, in the object categorization problem, an image is represented by an unsorted set of discrete visual words (BOVW). In these models, relations among visual words are performed after dictionary construction. However, close object regions can have far descriptions in the feature space, being grouped as different visual words. In this paper, we present a method for considering geometrical information of visual words in the dictionary construction step. Object interest regions are obtained by means of the Harris-Affine detector and then described using the SIFT descriptor. Afterward, a contextual-space and a feature-space are defined, and a merging process is used to fuse feature words based on their proximity in the contextual-space. Moreover, we use the Error Correcting Output Codes framework to learn the new dictionary in order to perform multi-class classification. Results show significant classification improvements when spatial information is taken into account in the dictionary construction step.

1 Introduction

Multi-class object categorization is one of the most challenging problems in Computer Vision, which has been applied to a wide variety of applications. Usually, the problem of object categorization is split into two main stages: object description, where discriminative features are extracted from the object to represent, and object classification, where a set of extracted features are labeled as a particular object given the output of a trained classifier.

A general tendency in object recognition to deal with the object description stage is to define a bottom-up procedure where initial features are obtained by means of region detection techniques. These techniques are based on determining relevant image keypoints (i.e. using edge-based information [1]), and then to define a support region around the keypoint (i.e. looking for extrema over scale-space [1]). Several alternatives for region detection have been proposed in the literature [1]. Once a set of regions are defined, they should be described using some kind of descriptor (i.e. SIFT descriptor [2]), and the region-descriptions are related in some way to define a model of the object of interest.

Based on the previous tendency, a recent technique to model visual objects is by means of a bag-of-visual-words. The BOVW model is inspired by the text classification problem using a bag-of-words, where a document is represented by an unsorted set of contained words. Analogously, in object categorization problems, an image is represented by an unsorted set of discrete visual words, which are obtained by the object local descriptions.

Many promising results have been achieved with the BOVW systems in natural language processing, texture recognition, Hierarchical Bayesian models for documents, object classification [3], object retrieval problems [4,5], or natural scene categorization [6], just to mention a few. However, one of the main drawbacks of the BOVW model is that dictionary construction does not take into account the geometrical information among visual instances. Although this issue can be beneficial in natural language analysis, its adaptation to visual word description needs special attention. Note that based on the description strategy used to describe visual words, very close regions can have far descriptors in the feature space, being grouped as different visual words. This effect occurs for most of the state-of-the-art descriptors, even when coping with different invariance, and thus, a grouping based on spatial information of regions could be beneficial for the construction of the visual dictionary.

In this paper, we present a method for considering spatial information of visual words in the dictionary construction step, namely Contextual-Guided Bag-of-Visual-Words model (C-BOVW). Object's interest regions are obtained by means of the Harris-Affine detector and then described using the SIFT descriptor. Afterward, a contextual-space and a feature-space are defined. The first space codifies the contextual properties of regions meanwhile the second space contains the region descriptions. A merging process is then used to fuse feature words based on their proximity in the contextual-space. Moreover, the new dictionary is learned using the Error Correcting Output Codes (ECOC) framework [7] in order to perform multi-class object categorization. We compared our approach to the standard BOVW design and validated over public multi-class categorization data sets, considering different state-of-the-art classifiers in the ECOC multi-classification procedure. Results show significant classification improvements when spatial information is taken into account in the dictionary construction step.

The rest of the paper is organized as follows: section 2 describes the C-BOVW algorithm. Section 3 introduces the multi-class ECOC strategy used to learn the BOVW and C-BOVW dictionaries. Section 4 shows the experimental evaluation, and finally, section 5 concludes the paper.

2 Contextual-Guided Bag-of-Visual-Words

In this section we reformulate the BOVW model so that geometrical information can be taken into account in conjunction with the keypoint descriptions in the dictionary construction step.

The algorithm is split into four main stages: contextual and feature space definition, merging, representant computation, and sentence construction.

Space definition: Given a set of samples for a n-multi-class problem, a set of regions of interest are computed and described for each sample in the training set. Then, K-means is applied over the descriptions and the spatial locations of each region to obtain a K-cluster feature-space and a K-cluster contextual-space, respectively. In our case, we use the Harris-affine region detector and SIFT descriptor. The x and y coordinates of each region normalized by the height and width of the image in conjunction with the ellipse parameters that define the region are considered to design the contextual-space.

Merging: Let define a contextual-feature relational matrix M, where the position (i,j) of this matrix represents the percentage of points from the jth visual word of the feature-space that match with the points of the ith visual word of the contextual-space. Then, from each row of M, the two maximums are selected. These maximums correspond to the two words of the feature-space which share more percentage of elements for a same contextual word. In order to fuse relevant feature words, we select the contextual word which maximizes the minimums of all pairs of selected maximums. It prevents unbalanced feature words to be merged. Finally, the two feature words with maximum percentage in M for that contextual word (which have not been previously considered together) are labeled to be merged at the end of the procedure, and the process is iterated while an evaluation metric is satisfied or a maximum number of merging iterations is reached. Once the merging loop finishes, the pairs of feature words labeled during the previous strategy are merged and define the new visual words.

Representant computation: When the new C-BOVW dictionary is obtained, a set of representant for each final word is computed. In order to obtain an stratified number of representant related to the word densities, only one representant is assigned to the word with the minimum number of elements. Then, a proportional number of representant is computed for the rest of words by applying k-means and computing the mean vector for each of the word sub-clusters. With the final set of representant feature vectors, a normalized sentence of word occurrences is computed for each sample in the training set, defining its probability density function of C-BOVW visual words. The whole C-BOVW procedure is formally described in Algorithm 1. An example of a two-iteration C-BOVW definition for a motorbike sample is shown in Figure 1. At the top of the figure, the initial spaces are shown. In the second row, the shared elements from the two spaces which maximize the percentage of matches for a given contextual word are shown. The contextual-space just considers the x and y coordinates, and the 128 SIFT feature-space is projected into a two-dimensional feature-space using the two principal components. Note that the feature descriptions for the two considered words are very close in the feature space though they belong to different visual words before merging. On the right of the figure the new merged feature

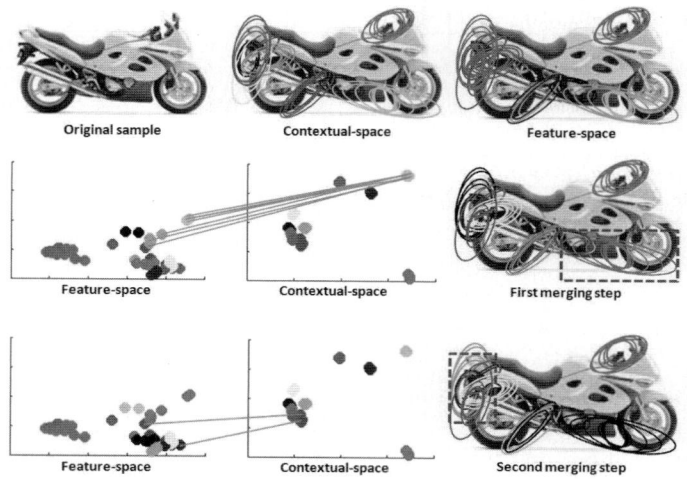

Fig. 1. Two iterations of C-BOVW algorithm over a motorbike sample

cluster is shown within a dashed rectangle. The same procedure is applied for the second iteration of the merging procedure in the bottom row of the figure.

Sentence construction: After the definition of the new dictionary, a new test sample can be simply described using the bag-of-visual-words without the need of including geometrical information since it is implicitly considered in the new visual words. The sentence for the new sample is then computed by matching its descriptors to the visual words with the nearest representant. Finally, the test sentence can be learned and classified using any kind of classification strategy.

3 Multi-class Extension

Error-Correcting Output Codes (ECOC) were defined as a framework to combine binary problems in order to deal with the multi-class case [7]. This framework is based on two main steps. At the first step, namely coding, a set of binary problems (dichotomizers) are defined based on the learning of different sub-partitions of classes by means of a base classifier. Then, each of the partitions is embedded as a column of a coding matrix M. The rows of M correspond to the codewords codifying each class. At the second step, namely decoding, a new data sample that arrives to the system is tested, and a codeword formed as a result of the output of the binary problems is obtained. This test codeword is compared with each class codeword based on a given decoding measure, and a classification prediction is obtained for the new object.

One of the most widely applied ECOC configurations is the one-versus-one design [8]. This strategy codifies the splitting of each possible pair of classes as a dichotomizer, which results in $N(N-1)/2$ binary problems for an N-class

Algorithm 1. Contextual-Guided Bag-of-Visual-Words algorithm

Require: $D = \{(\mathbf{x}_1, l_1), ..., (\mathbf{x}_m, l_m)\}$, where $\mathbf{x}_i$ is an object sample of label $l_i \in [1, .., n]$ for a n-class problem, K clusters, and I merging steps.
Ensure: Representant $R = \{(r_1, w_1), ..., (r_v, w_b)\}$, where r_v is a representant for word $w_i, i \in [1, ..b]$ for b words. Sentences $S = \{(s_1, l_1), ..., (s_m, l_m)\}$, where s_i is the sentence of sample $\mathbf{x}_i$.

1: **for** each sample $\mathbf{x}_i \in D$ **do**
2:   Detect regions of interest for sample $\mathbf{x}_i$:
     $X_i = \{(x_1, y_1, \rho_1^1, \rho_1^2, \rho_1^3), ..., (x_j, y_j, \rho_j^1, \rho_j^2, \rho_j^3)\}$, where x and y are spatial coordinates normalized by the height and width of the image, and ρ^1, ρ^2, and ρ^3 are ellipse parameters for affine region detectors.
3:   Compute region descriptors: $X_i^r = \{r_1, ..., r_j\}$, where r_j is the description of the jth detected region of sample $\mathbf{x}_i$.
4: **end for**
5: Define a contextual-space $C = \{(c_1, w_1^C), ..., (c_v, w_q^C)\}$ using K-means to define K contextual clusters, where w_i^C is the ith word of the contextual-space.
6: Define a feature-space $F = \{(f_1, w_1^F), ..., (f_v, w_q^F)\}$ using K-means to define K feature clusters, where w_i^F is the ith word of the feature-space.
7: Initialize a contextual-feature relational matrix M: $M(i, j) = 0, i, j \in [1, ..., K]$
8: Initialize $W = \emptyset$ the list of feature words to be merged
9: **for** I merging steps **do**
10: update M based on the contextual clusters and new feature clusters so that $M(i, j) = \frac{d(C, F, i, j)}{|w_j^F|}$, where $d(C, F, i, j)$ returns the number of points from contextual-space of word w_i^C that belong to the feature-space jth word w_j^F, and $|w_j^F|$ is the number of regions of the jth feature word.
11: Select the pair of positions with the maximum value for each row of M: $\max_{j,k} M(i, \_), j \neq k, \forall i$, where '$\_$' stands for all row positions.
12: $W = W \cup (w_j^F, w_k^F)$: Select the contextual word w_i^C and words w_j^F and w_k^F from the feature-space based on $\max_i (\min(M(i, j), M(i, k))), \forall j, k$
13: **end for**
14: **for** each pair (w_j^F, w_k^F) in W **do**
15: update F so that $w_j^F \leftarrow w_k^F$, and rename feature words so that $w_i^F, i \in [1, ..., p]$ becomes $w_i^F, i \in [1, ..., p-1]$
16: **end for**
17: Compute representant $R = \{(r_1, w_1), ..., (r_v, w_b)\}$ for the new F, where:
    $z_i = \text{round}\left(\frac{w_i}{\min |w_j|\forall j}\right)$
    is the number of representant for word w_i, computed using z_i-means, and $\{r_1, ..., r_{z_i}\}$ representant are computed as the mean value for each sub-cluster of w_i, obtaining an stratified number of representant respect the words densities.
18: Compute sentences $S = \{(s_1, l_1), ..., (s_m, l_m)\}$ for all training samples of all categories comparing with word representant of R.

problem. The one-versus-one ECOC technique is defined in the ternary ECOC framework $M^{N \times M} \in \{-1, 0, +1\}$, being M a coding matrix of N rows (as the number of classes), M the number of columns (dichotomizers to be trained, where $M = N(N-1)/2$ in the case of the one-versus-one design), $\{-1, +1\}$ symbols codify the class membership, and the zero symbol ignores a particular class for a given dichotomizer. Each column of the matrix M corresponds to the ith binary problem h_i, which splits a pair of classes using a given base classifier. Figure 2 codifies a one-versus-one coding matrix M for a 4-class problem. The black positions correspond to the symbol $+1$, the white positions to the symbol -1, and the gray positions to the zero symbol. Once the set of binary problems $h = \{h_1, ..., h_M\}$ is trained, a new test sample ρ that arrives to the system is tested applying the set h, and a test codeword $X^{1 \times M} \in \{-1, +1\}$ is obtained.

The decoding step was originally based on error-correcting principles under the assumption that the learning task can be modeled as a communication

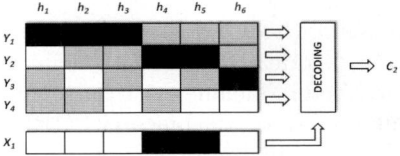

Fig. 2. One-versus-one ECOC coding design for a 4-class problem. A decoding function d for a new input test sample is performed, classifying by class C_2.

problem, in which class information is transmitted over a channel [7]. The first attempt for ECOC decoding is the Hamming Decoding (HD). The Euclidean Decoding (ED) is another of the most preferred decoding strategies in the literature. Still very few alternative decoding strategies have been proposed [8]. Then, after the codeword X for the test sample ρ is obtained, a decoding function $d(X, Y_j)$ applying any of the previous decoding strategies is used to compare the test codeword X with each codeword Y_j (jth row from M) codifying class C_j. Finally, the classification prediction corresponds to the class C_j which corresponding codeword Y_j minimizes d (C_2 in the case of the example of Figure 2).

4 Experimental Evaluation

Before the presentation of the results, first, we discuss the data, methods, and validation protocol of the experiments.

Data: The data used in the experiments consists of 15 categories from public Caltech 101 [9] and Caltech 256 [10] repository data sets. One sample for each category is shown in Figure 3. For each category, 50 samples were used, 10 samples to define the BOW and another 40 images to define new test sentences.

Methods: We compare the C-BOVW with the classical BOVW model. For both methods, the same initial set of regions is considered in order to compare both strategies at the same conditions. About 200±20 object regions are found by image using the Harris-Affine detector [1] and described using the SIFT descriptor [2]. The visual words are obtained using the public open source K-means software from [11]. After computing the final words and representant, multi-class classification is

Fig. 3. Considered categories from the Caltech 101 and Caltech 256 repositories

performed using an one-versus-one ECOC methodology with different base classifiers: Mean Nearest Neighbor (NMC), Fisher Discriminant Analysis with a previous 99% of PCA (FLDA), Gentle Adaboost with 50 iterations of decision stumps (G-ADA), Linear Support Vector Machines with the regularization parameter $C = 1$ (Linear SVM), and Support Vector Machines with RBF Kernel with C and γ parameters set to 1 (RBF SVM)[1]. Finally, we use the Linear Loss-weighted decoding to obtain the class label [8].

Validation protocol: We used the sentences obtained by the 50 samples of each category and performed stratified ten-fold cross-validation evaluation.

4.1 Caltech 101 and 256 Classification

In this experiment, we started classifying from three Caltech categories increasing by 2 up to 15. For each step, different number of visual words are computed: 30, 40, and 50. These numbers are obtained by performing ten iterations of the merging procedure (experimentally tested). In order to compare the BOVW and C-BOVW methods at the same conditions, the same detected regions and descriptions are used for all the experiments. The order in which the categories are considered is the following: (1-3) airplane, motor-bike, watch, (4-5) tripod, face, (6-7) ketch, diamond-ring, (8-9) teddy-bear, t-shirt, (10-11) desk-globe, backpack, (12-13) hourglass, teapot, (14-15) cowboy-hat, and umbrella. The obtained results applying ten-fold cross-validation are graphically shown in Figure 4 for the different ECOC base classifiers. Note that the classification error significantly varies depending on the ECOC classifier. In particular, Gentle Adaboost obtains the best results, with a classification error inferior to 0.2 in all the tests when using 30 C-BOVW words. Independently of the ECOC classifier, in most of the experiments the C-BOVW model obtains errors inferiors to those obtained by the classical BOVW. BOVW only obtains slightly better results in the case of Gentle Adaboost for eleven classes and 50 visual words.

An important remark of the C-BOVW model is about the selection of the number of merging iterations. This parameter has a decisive impact over the generalization capability of the new visual dictionary. First iterations of the merging procedure use to fuse very close feature-words which belong to different visual words whereas final merging iterations fuse more far regions of the feature-space. Thus, a large number of iterations could be detrimental since the new merged words could be too general for discriminating among sentences of different object categories. Thus, this parameter should be estimated for each particular problem domain (i.e. applying cross-validation over a training and a validation subset). In the previous experiment we checked that ten merging iterations obtains significant performance improvements, though we are aware that this parameter could be not optimal for all the data sets.

---

[1] We decided to keep the parameter fixed for the sake of simplicity and easiness of replication of the experiments, though we are aware that this parameter might not be optimal for all data sets.

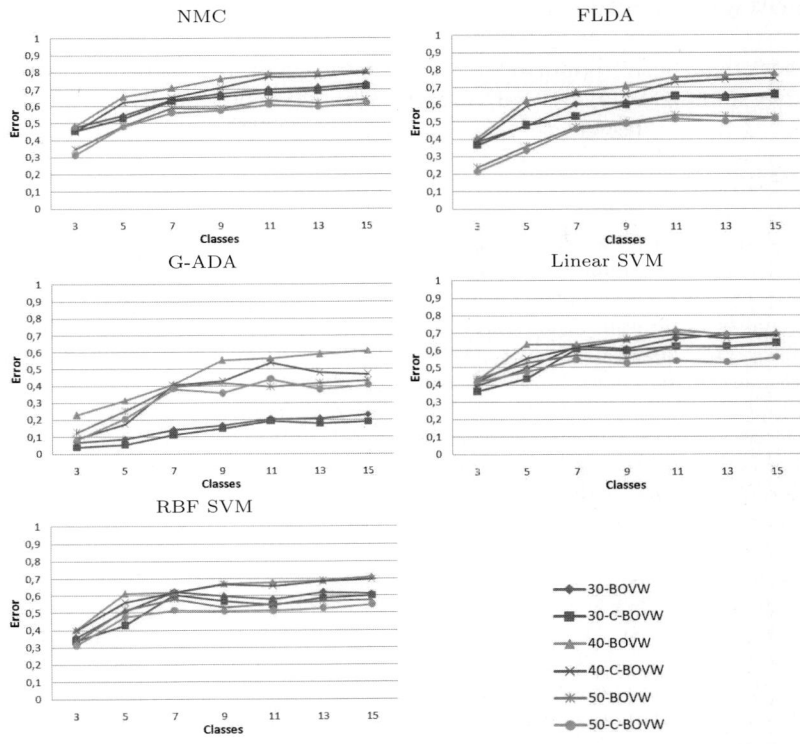

Fig. 4. Classification results for the Caltech categories using BOVW and C-BOVW dictionaries for different number of visual words and ECOC base classifiers

5 Conclusion

In this paper we re-formulated the bag-of-visual-words model so that geometrical information of significant object region descriptions are taken into account in the dictionary construction step. In this sense, regions which have slightly different descriptors because of small displacements in the region detection process can be merged together in a same visual word. The method is based on the definition of a contextual-space and a feature-space. The first space codifies the geometrical properties of regions meanwhile the second space contains the region descriptions. A merging process is then used to fuse feature words based on their proximity in the contextual-space. The new dictionary is learned in an Error-Correcting Output Codes design to perform multi-class object categorization. The results when spatial information is taken into account showed significant performance improvements compared to the classical approach for different number of object categories and visual words.

Acknowledgements

This work has been supported in part by projects TIN2006-15308-C02, FIS PI061290, and CONSOLIDER-INGENIO CSD 2007-00018.

References

1. Mikolajczyk, K., Tuytelaars, T., Schmid, C., Zisserman, A., Matas, J., Kadir, T., Van Gool, L.: A comparison of affine region detectors. IJCV 65(1-2), 43–72 (2005)
2. Lowe, D.: Distinctive image features from scale-invariant keypoints. International Journal of Computer Vision 60(2), 91–110 (2005)
3. Csurka, G., Dance, C., Fan, L., Willamowski, J., Bray, C.: Visual categorization with bags of keypoints. In: ECCV, pp. 1–22 (2004)
4. Philbin, J., Chum, O., Isard, M., Sivic, J., Zisserman, A.: Object retrieval with large vocabularies and fast spatial matching. In: CVPR, pp. 1–8 (2007)
5. Chum, O., Philbin, J., Sivic, J., Zisserman, A.: Automatic query expansion with a generative feature model for object retrieval. In: ICCV, pp. 1–8 (2007)
6. Carneiro, G., Jepson, A.: Flexible spatial models for grouping local image features. In: CVPR, vol. 2, pp. 747–754 (2004)
7. Dietterich, T., Bakiri, G.: Solving multiclass learning problems via error-correcting output codes 2, 263–282 (1995)
8. Escalera, S., Pujol, O., Radeva, P.: On the decoding process in ternary error-correcting output codes. Transactions in PAMI 99 (2009)
9. Caltech 101, http://www.vision.caltech.edu/image_datasets/caltech101/
10. Caltech 256, http://www.vision.caltech.edu/image_datasets/caltech256/
11. Clustering package, http://bonsai.ims.u-tokyo.ac.jp/~mdehoon/software/cluster/

Isometric Deformation Modelling for Object Recognition

Dirk Smeets*, Thomas Fabry**, Jeroen Hermans,
Dirk Vandermeulen, and Paul Suetens

K.U. Leuven, Faculty of Engineering, Department of Electrical Engineering,
Center for Processing Speech and Images,
Medical Imaging Research Center, Universitair Ziekenhuis Gasthuisberg,
Herestraat 49 bus 7003, B-3000 Leuven, Belgium

Abstract. We present two methods for isometrically deformable object recognition. The methods are built upon the use of geodesic distance matrices (GDM) as an object representation. The first method compares these matrices by using histogram comparisons. The second method is a modal approach. The largest singular values or eigenvalues appear to be an excellent shape descriptor, based on the comparison with other methods also using the isometric deformation model and a general baseline algorithm. The methods are validated using the TOSCA database of non-rigid objects and a rank 1 recognition rate of 100% is reported for the modal representation method using the 50 largest eigenvalues. This is clearly higher than other methods using an isometric deformation model.

1 Introduction

During the last decades, many developments in 3D modelling and 3D capturing techniques augmented the interest in the use of 3D objects for a number of applications. Examples of these are CAD/CAM, architecture, computer games, archaeology, medical applications and biometrics. Because of this growing use of 3D objects, we see the emergence of 3D databases, which leads to a new research question: 3D object retrieval. One witness of this are the yearly SHREC contests [1]. For the last 3 years already, the 3D SHape REtrieval Contest has the objective to evaluate the effectiveness of 3D-shape retrieval algorithms.

Our contribution considers 3D object recognition coping with non-rigid deformations in particular. A few examples of these kinds of deformations are the expression variations of a human face, the movement of different subparts of a fabrication robot or simply the movement of a walking human.

Based on the assumption that geodesic distances[1] remain approximately constant during natural non-rigid deformations, we propose a technique for non-rigid object recognition based on the geodesic distance matrix (GDM), a matrix summarizing all point-to-point geodesic distances on the object mesh.

---
* Corresponding author: dirk.smeets@uz.kuleuven.be
** Joint first author.
[1] The geodesic distance between two points is the length of the shortest path *on the object surface* between two points on the object.

2 Related Work

Some 3D object recognition methods dealing with non-rigid objects and making use of the geodesic distance matrix are already to be found in literature. The one that received the most attention is probably the method of Elad and Kimmel [2]. Here, the GDM is computed using the *fast marching on triangulated domains* (FMTD) method. Then, the GDM is processed using the multidimensional scaling (MDS) approach, converting the non-rigid objects into rigid *invariant signature surfaces*. These can be compared using simpler algorithms for rigid matching. We will use an implementation of this method for comparison.

Another 3D object recognition method that shows some similarity to one method we propose here is the *Geodesic Object Representation* of Hamza and Krim [3]. Here, the shape descriptor is a *global geodesic shape function*. This shape function is defined in each point on the surface and measures the normalized integral of squared geodesic distances to other points on the surface. These global geodesic shape functions are then used to construct *geodesic shape distributions*. These are kernel density estimates (KDE) made of the (discretized) global geodesic shape functions of a particular object. For the actual recognition, these KDEs are compared using the Jensen-Shannon divergence. The similarity of this method to our modal representation lies in the use of the geodesic distance matrix, which, in the method of Hamza and Krim, is used for the computation of the geodesic shape functions.

Finally, a similar method to our modal representation approach is the method shown in Jain and Zhang's work [4]. This method measures the inter-object distance by taking the χ^2-distance between the 20 largest eigenvalues of a weighted GDM. We will show that the weighting of the GDM has an adverse effect on the accuracy of the method.

3 Isometric Deformation Modelling

In mathematics, an isometry is a distance-preserving isomorphism between metric spaces. The basis of the isometric deformation model is therefore the invariance of distances measured along the surface, called geodesic distances. Therefore, an appropriate object representation to exploit the advantages of an isometric model is the geodesic distance matrix (GDM). We call G a GDM for a particular object if $G = [g_{ij}]$, with g_{ij} the geodesic distance between points i and j on the object surface. This matrix is a symmetric matrix and defined up to a random permutation of the points on the represented object surface. Figure 1 shows a 3D object and the associated GDM.

For the calculation of the GDM, a fast marching algorithm for triangulated meshes is used [5]. The algorithm computes the distance of the shortest (discrete) path between each pair of surface points. The complexity of this computation is $\mathcal{O}(n^2)$, with n the dimension of the GDM. Beside the geodesic distance matrix ($G_1 = [g_{ij}]$), also other affinity matrices, closely related to the GDM are examined. For example the squared GDM ($G_2 = [g_{ij}^2]$), the Gaussian weighted

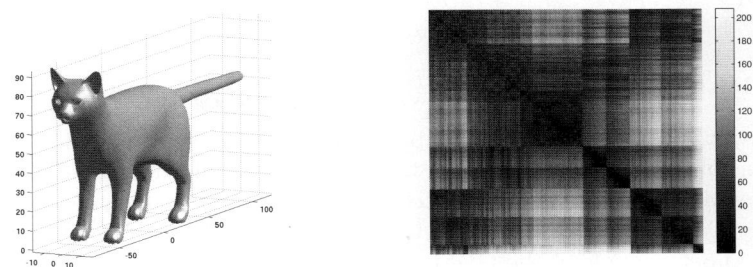

Fig. 1. 3D mesh of an object (a) and its geodesic distance matrix representation (b)

GDM ($G_3 = [\exp(-g_{ij}^2/(2\sigma^2)])$) and the *increasing weighting function* GDM ($G_4 = [1 + \frac{1}{\sigma} g_{ij}]^{-1}$) [6].

3.1 Multidimensional Scaling

Multidimensional scaling (MDS) is a technique that allows visualisation of the proximity between points with respect to some kind of dissimilarity (distance) measure matrix. For Euclidean distance matrix representations of a 3D object, three dimensional MDS provides the configuration of the original object. In [7], MDS is applied on the GDM in order to obtain a configuration of points where pointwise Euclidean distances approximately equal the original pointwise geodesic distances. Figure 2 shows the resulting 2D and 3D configurations, called canonical forms, calculated using classical MDS.

Because the geodesic distances remain constant under isometric transformations, the GDM of an object is invariant with respect to isometric transformations up to an arbitrary - simultaneous - permutation of rows and columns. However, the canonical forms have the same shape. Therefore, objects can be

Fig. 2. 2D (a) and 3D (b) canonical form of the same object as shown in Fig. 1

3.2 Histogram Comparison

We propose another way to compare deformable objects: by comparing histograms of the values contained in the geodesic distance matrices. The resulting representation is invariant for matrix permutations. Experiments were conducted with two kinds of histograms. The first are histograms calculated from all values in the upper triangle of the GDM. The second are histograms of mean geodesic distances per point. Examples of those histograms for the object in Fig. 1 are shown in Fig. 3. Other histogram variants are possible but are not considered here.

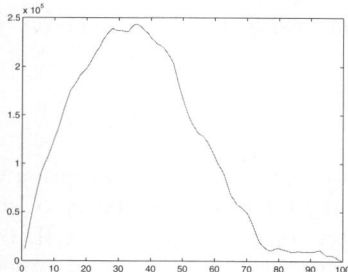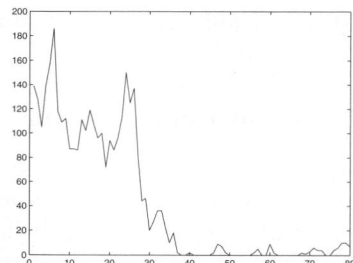

Fig. 3. Histograms of all (a) and point-wise averaged (PWA) (b) geodesic distances of the same object as shown in Fig.1

The histograms S^j ($j = 1, \ldots, n$), with n the number of objects, can be thought of as m-dimensional vectors, with m the number of bins. They can be compared with a plethora of dissimilarity measures. We have tested 8 different ones. Histograms can be compared using the Jensen-Shannon divergence [8]:

$$JSD(S^1, S^2, \ldots S^n) = H(\sum_{j=1}^{n} \pi_j S^j) - \sum_{j=1}^{n} \pi_j H(S^j), \qquad (1)$$

with π_j the weight for the histogram vector S^j and $H(S^j)$ the Shannon entropy, given by $H(S) = -\sum_{i=1}^{m} S_i \log_b S_i$. In this work only pair-wise comparisons are considered. Both histograms receive equal weighting ($\pi_1 = \pi_2 = 1/2$). The other dissimilarity measures need less explication and are listed in Tab. 1.

3.3 Modal Representation

A third approach for object comparison using the isometric model is based on a modal representation. Here, the information in the geodesic distance matrix is separated into a matrix that contains intrinsic shape information and a matrix with

Table 1. Dissimilarity measures

Dissimilarity measure	Formula
Jensen-Shannon Divergence	$D_1 = H(\frac{1}{2}S^k + \frac{1}{2}S^l) - (\frac{1}{2}H(S^k) + \frac{1}{2}H(S^l))$
Mean normalized Manhattan distance	$D_2 = \sum_{i=1}^{m} \frac{2|S_i^k - S_i^l|}{S_i^k + S_i^l}$
Mean normalized maximum norm	$D_3 = \max_i \frac{2|S_i^k - S_i^l|}{S_i^k + S_i^l}$
Mean normalized absolute difference of square root vectors	$D_4 = \sum_{i=1}^{m} \frac{2|\sqrt{S_i^k} - \sqrt{S_i^l}|}{\sqrt{S_i^k} + \sqrt{S_i^l}}$
Correlation	$D_5 = 1 - \frac{S^k \cdot S^l}{\|S^k\|\|S^l\|}$
Euclidean distance	$D_6 = \sqrt{\sum_{i=1}^{m}(S_i^k - S_i^l)^2}$
Normalized Euclidean distance	$D_7 = \sqrt{\sum_{i=1}^{m}(S_i^k - S_i^l)^2/\sigma_i^2}$
Mahalanobis distance	$D_8 = \sqrt{\sum_{i=1}^{m}(S^k - S^l)^T \text{cov}(S)^{-1}(S^k - S^l)}$

information about corresponding points. This is done with an eigenvalue decomposition (EVD) or a singular value decomposition (SVD) of the GDM. Both decompositions give similar results because the GDM is a symmetric matrix. The eigenvalues and singular values can be used as intrinsic shape descriptors, while the eigenvectors and singular vectors give information about correspondences. For numerical reasons, only the largest eigenvalues or singular values are computed.

Because we do not know anything about the order of the points, G and all possible simultaneous permutations of rows and columns of G determine the configuration of the object. Let P be a random permutation matrix, such that $G' = PGP^T$ is a GDM with rows and columns permuted, and $G = U\Sigma V^T$ a singular value decomposition, then

$$G' = PGP^T = PU\Sigma(PV)^T. \qquad (2)$$

Because PU and PV remain unitary matrices and Σ is still a diagonal matrix with non-negative real numbers on the diagonal, the right hand side of Eq. 2 is a valid singular value decomposition of G'. A common convention is to sort the singular values in non-increasing order. In this case, the diagonal matrix Σ is uniquely determined by G'. Therefore, $\Sigma = \Sigma'$, with Σ' the singular value matrix of G'.

From this, we can see that Σ contains the intrinsic information about geometry, while U and V contain the information about correspondences between points. This justifies our approach of object recognition using $S = \{\sigma_1, \sigma_2, \ldots, \sigma_k\}$, with $\sigma_1, \sigma_2, \ldots, \sigma_k$ the first k singular values of the GDM, as a shape descriptor. As such, the computational complexity singular value calculation is limited to $\mathcal{O}(k.n^2)$, with n the dimension of the GDM.

For comparing these singular value vectors, we can use the same dissimilarity measures as we described in Sect. 3.2 (see Tab. 1).

4 Experimental Validation

To examine the deviation to the isometric deformation assumption in a realistic situation, we looked at the change in geodesic distance between four finger tips in three situations with different configuration of a hand. This results in a mean coefficient of variation (CV) of 5.3% for the geodesic distances, while the CV for Eucledian distances is equal to 27.6%.

For the validation of the three proposed object recognition approaches, we use the TOSCA database [9]. This database consists of various 3D non-rigid shapes in a variety of poses and is intended for non-rigid shape similarity and correspondence experiments. We use 133 objects, i.e. 9 cats, 11 dogs, 3 wolves, 17 horses, 21 gorillas, 1 shark, 24 female figures, and two different male figures, containing 15 and 20 poses. Each object contains approximately 3000 vertices.

We compare the three GDM-based methods with a baseline algorithm: the standard iterative closest point (ICP) algorithm [10]. This is a well-known and extensively used rigid object registration method that minimizes the sum of squared Euclidean distances between closest points. After rigid registration the objects can be compared using the value of the employed registration objective function.

After roughly tuning the parameters, we used 100 bins for the histogram comparison with all values and 80 bins for the pointwise averaged (PWA) value histogram comparison. This number was determined experimentally.

The different approaches are validated using standard recognition experiments, i.e. the verification and the identification scenario. The performance of those scenarios is measured with the receiving operating characteristic (ROC) curve and the cumulative matching curve (CMC), respectively. The former is a curve plotting the false rejection rate (FRR) against the false acceptance rate (FAR), while the latter gives the recognition rate for several ranks. These curves can be found in Fig. 4. Here, we plotted the best combination of GDM weighting, dissimilarity measure and, for the modal representation approach, the optimal number of eigenvalues (see below).

The equal error (EER) and rank 1 recognition rate (R_1RR) are characteristic points on the ROC and CMC respectively. These are tabulated in Tab. 2.

Figure 5 plots the R_1RR against the number of eigenvalues (logarithmic scale) used in the shape descripor. A plateau of maximum recognition is observerd for shape descriptors using a number of eigenvalues between 35 and 430.

In Tab. 3, the different dissimilarity measures are compared, showing that he best results are obtained with the mean normalized absolute difference of square root vectors of the 50 largest eigenvalues.

Table 2. Results of different isometric deformation model methods on TOSCA database

experiment	R_1RR	EER
MDS	39.34%	29.49%
histogram of PWA values	63.11%	16.93%
histogram of all values	72.13%	14.90%
modal representation	100.0%	2.43%
ICP	35.29%	40.07%

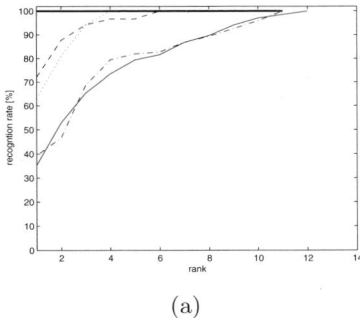 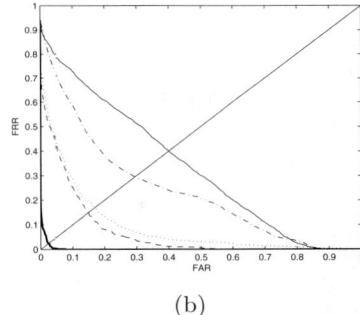

(a) (b)

Fig. 4. Results of standard validation experiments on the TOSCA-database with CMC (a) and ROC (b). Object recognition with a baseline algorithm (thin solid line) is compared to object recognition using MDS (dash-dot line), histogram comparison of PWA (dotted line) and all values (dashed line) and modal representation (thick solid line).

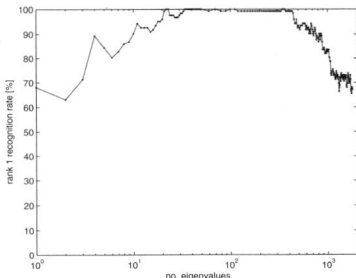

Fig. 5. The R_1RR is plotted against the number of eigenvalues (in log scale) used in the shape descriptor

Table 3. Comparison of different dissimilarity measures as defined in Tab. 1

Diss measure	D_1	D_2	D_3	D_4	D_5	D_6	D_7	D_8
PWA value Histogram comparison								
R_1RR	45.08%	54.92%	45.08%	54.92%	46.72%	56.56%	63.11%	20.49%
EER	18.68%	15.83%	25.31%	15.69%	34.68%	23.13%	16.93%	42.07%
All value Histogram comparison								
R_1RR	67.21%	69.67%	47.54%	69.67%	58.20%	72.13%	66.39%	20.49%
EER	14.95%	15.26%	21.01%	15.26%	19.63%	14.90%	16.94%	48.37%
Modal representation								
R_1RR	84.43%	100.0%	85.25%	100.0%	54.92%	76.23%	97.54%	33.79%
EER	10.11%	2.43%	10.09%	2.44%	20.33%	10.74%	7.74%	34.18%

To show the influence of different weightings of the GDM, we also tabulate the rank one recognition rate and the equal error rate for the different weighting functions as proposed in Sect. 3. This can be found in Tab. 4. The abbreviations used are the ones introduced in Sect. 3. We can clearly see that every weighting reduces the accuracy of both methods, sometimes quite drastically.

Table 4. Comparison of different weighting function of the GDM as defined in Sect. 3

	G_1	G_2	G_3	G_4
All Value Histogram comparison				
R_1RR	72.13%	70.49%	70.49%	69.67%
EER	14.90%	14.01%	14.14%	15.42%
Modal representation				
R_1RR	100.0%	97.54%	71.31%	90.98%
EER	2.43%	3.47%	17.79%	12.25%

All results clearly show that the modal representation of the geodesic distance matrices provides the highest performance. We also note that all methods using geodesic distance matrices perform better than the baseline algorithm.

5 Conclusions

In this article, different methods using geodesic distance matrices are compared. Amongst all the representations and methods, the modal approach outperforms the other methods, a geodesic histogram based representation, the MDS-approach and the baseline ICP algorithm. For the TOSCA database a rank 1 recognition rate of 100% is obtained.

As future work, we propose to further exploit the modal decomposition method in order to obtain correspondences between different objects. This can be done using the eigenvectors or singular vectors based on the method of Brady and Shapiro [11].

References

1. AIM@SHAPE: SHREC - 3D shape retrieval contest, http://www.aimatshape.net/event/SHREC
2. Elad, A., Kimmel, R.: On bending invariant signatures for surfaces. IEEE Transactions on Pattern Analysis and Machine Intelligence 25(10), 1285–1295 (2003)
3. Hamza, A.B., Krim, H.: Geodesic object representation and recognition. In: Nyström, I., Sanniti di Baja, G., Svensson, S. (eds.) DGCI 2003. LNCS, vol. 2886, pp. 378–387. Springer, Heidelberg (2003)
4. Jain, V., Zhang, H.: A spectral approach to shape-based retrieval of articulated 3D models. Computer-Aided Design 39(5), 398–407 (2007)
5. Peyré, G., Cohen, L.D.: Heuristically driven front propagation for fast geodesic extraction. Intl. Journal for Computational Vision and Biomechanics 1(1), 55–67
6. Carcassoni, M., Hancock, E.R.: Spectral correspondence for point pattern matching. Pattern Recognition 36, 193–204 (2003)
7. Bronstein, A.M., Bronstein, M.M., Kimmel, R.: Expression-invariant 3D face recognition. In: Kittler, J., Nixon, M.S. (eds.) AVBPA 2003. LNCS, vol. 2688, pp. 62–69. Springer, Heidelberg (2003)
8. Lin, J.: Divergence measures based on the shannon entropy. IEEE Transactions on Information Theory 37(1), 145–151 (1991)

9. Bronstein, A., Bronstein, M., Kimmel, R.: Numerical Geometry of Non-Rigid Shapes. Springer, Heidelberg (2008)
10. Besl, P.J., Mckay, H.D.: A method for registration of 3-d shapes. IEEE Transactions on Pattern Analysis and Machine Intelligence 14(2), 239–256 (1992)
11. Shapiro, L.S., Brady, J.M.: Feature-based correspondence: an eigenvector approach. Image Vision Comput. 10(5), 283–288 (1992)

Image Categorization Based on a Hierarchical Spatial Markov Model

Lihua Wang, Zhiwu Lu, and Horace H.S. Ip

Image Computing Group, Department of Computer Science
Center for Innovative Applications of Internet and Multimedia Technologies
(AIMtech Centre)
City University of Hong Kong, Hong Kong
{wlihua2@student.,lzhiwu2@student.,cship@}cityu.edu.hk

Abstract. In this paper, we propose a Hierarchical Spatial Markov Model (HSMM) for image categorization. We adopt the Bag-of-Words (BoW) model to represent image features with visual words, thus avoiding the heavy work of manual annotation in most Markov model based approaches. Our HSMM is designed to describe the spatial relations of these visual words by modeling the distribution of transitions between adjacent words over each image category. A novel idea of semantic hierarchy is exerted in the model to represent the composition relationship of visual words at semantic level. Experiments demonstrate that our approach outperforms Bayesian hierarchical model based categorization approach with 12.5% and it also performs better than the previous Markov model based approach with 11.8% on average.

Keywords: Image categorization, Hierarchical Spatial Markov Model, Visual words.

1 Introduction

In this paper, we focus on the problem of recognizing the semantic categories of unknown images via analyzing their visual features. In recent literatures of image categorization, researches have been progressing along two lines of researches [2]. The first line of researches considers the semantic of an image directly based on its low-level visual features [9][14] and the second line of researches employs intermediate representations that are generated from low-level features [2][7][13][5]. It has been shown by recent researches that the intermediate representation narrows the semantic gap between low-level visual features and high-level semantic and also improves the performance of image recognition. The BoW model has been developed and adopted for this purpose, especially in the fields of image classification and visual object recognition with good results [5][7]. However, the BoW model which fundamentally does not take into account the spatial relationships among the visual features, has given rise to certain limitations particularly when the spatial relationships among visual features provide important cues for many image classification problems. Developing effective approaches for representing and exploiting the spatial structure of an image within the BoW framework remains an open problem. Some previous works have yielded inspiring results [13][12][4]. Recently, the hierarchical image feature is increasingly used to represent the spatial layout of features in many schemes [7][1][11]. However, these approaches do not describe the relative spatial relations among visual features indeed. Instead, they merely bind the feature with its location. In this paper, we propose a new image categorization approach that combines the BoW model based image representation with a Hierarchical Spatial Markov Model, which represents the spatial relationship of features along different directions. Compared with relevant BoW model based approaches, the spatial layout within an image is adequately considered in our approach.

The Hierarchical Spatial Markov Model in this paper is a hierarchical generalization of a previously proposed 2-D hidden Markov model, the Spatial Hidden Markov Model (or SHMM) [14]. We inherit the fundamental assumption of SHMM in our model and extend the previous 2-D structure to be a hierarchy, while the hidden states in SHMM is abandoned. In recent years, there have been a few 3-D and hierarchical Markov models proposed for analyzing images [8][3][9]. Li described a multi-resolution hidden Markov model to speed up the procedure of image categorization in [8]. Bouman[3] developed a multi-scale random field for Bayesian image segmentation based on the Markov random fields. At the same time, some other hierarchical Markov models were proposed by extending the one-dimensional Markov models [6][10]. However, in most of the currently proposed hierarchical models, the connections of features in different layers are not adequately considered. And the requirement of manual annotation on image regions [9][14] is another drawback. In this paper, we present our HSMM, which differs from other Markov models with following two innovations. First, our model describes the semantic hierarchy within and across different scales of an image. The correlation between image features includes not only adjacency, like left-and-right, but also composition, such as part-and-integrity, which is built on the semantic level. For instance, the image of a car at coarse resolution can be partitioned into parts of window, wheel, and headlight at a finer resolution. Our HSMM is proposed to represent the composition relationship within an image by analyzing the correlation between features in adjacent scales of an image pyramid. Second, we avoid the heavy and error-proned tasks of manual annotation on individual image regions by taking the advantage of BoW model based intermediate image representation.

In this paper, we propose a Hierarchical Spatial Markov Model based approach to categorize images. Every image to be processed is constructed an image pyramid with quad-tree structure, with each block being mapped to a visual word. The HSMM is employed to describe the distribution of words over each image category, by means of capturing the neighboring relationship of visual words within the same layer (or resolution) as well as the containing relationship of visual words across different layers. To classify an unknown image, we develop its word-based representation and determine the model that best represents its word distribution among the constructed models. This approach avoids manual annotation, on the one hand, while making use of spatial correlation of features, and achieves better result than previous methods.

The remaining sections of this paper are organized as follows. In section 2, the definition of the proposed Hierarchical Spatial Markov Model and its application on image categorization is represented. We demonstrate the experimental results in Section 3.

2 Hierarchical Spatial Markov Model Based Image Categorization

The HSMM is a structured multi-level discrete stochastic process. Given several sets of images with each being represented by a set of visual words, the HSMM is built to describe the distribution of words over each image category.

The HSMM is constructed as a quad-tree structure with several uniformly portioned image layers. From top to bottom of the quad-tree structure, the image layers are partitioned into increasingly finer blocks, as each single block of current layer is divided into four child-blocks in the next finer layer (Figure 1(a)). Intuitively, the fine layer represents the high resolution of an image while the coarse layer represents what we see when overlooking it from somewhere far away. In each block, one visual word is used to represent its feature, which takes value from the codebook of corresponding layer, $O^n = \{o_1, o_2, \ldots, o_M\}$. With zigzag

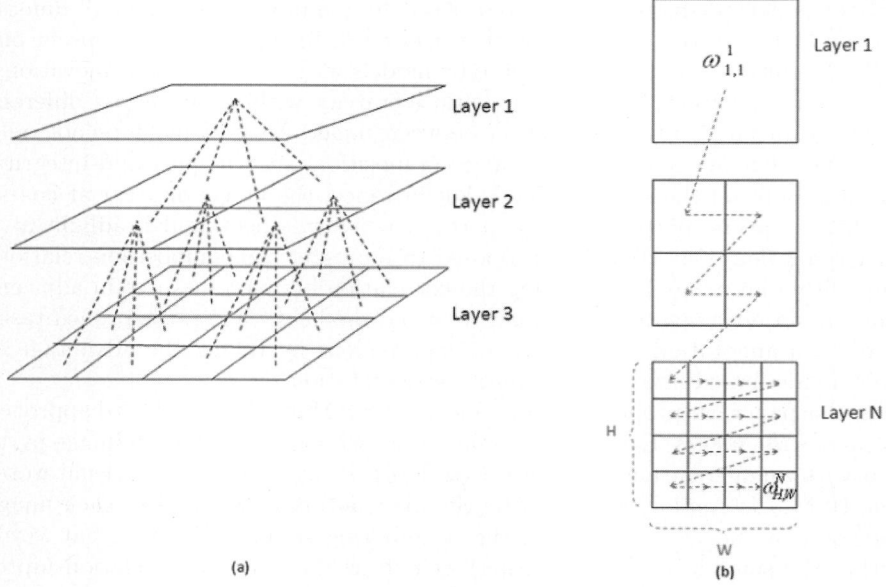

Fig. 1. (a) The quad-tree structure of Hierarchical Spatial Markov Model; (b) The visual word sequence $\Omega = (\omega_{1,1}^1, \omega_{1,2}^1, \ldots, \omega_{H,W-1}^N, \omega_{H,W}^N)$ is generated by zigzag traversing the image hierarchy

Image Categorization Based on a Hierarchical Spatial Markov Model

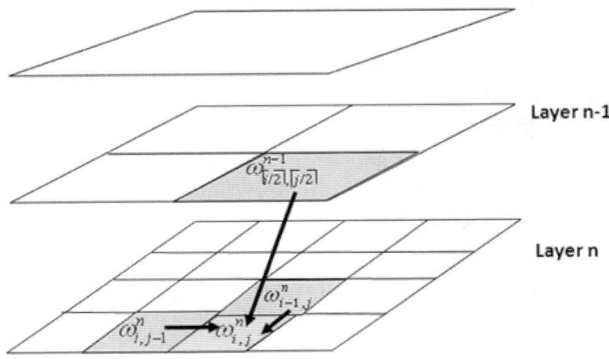

Fig. 2. Illustration of the fundamental assumption

traversing from the top-left block of the top layer to the bottom-right block of the bottom layer, the image hierarchy is represented by a block sequence $B = (b_{1,1}^1, b_{1,2}^1, \ldots, b_{H,W-1}^N, b_{H,W}^N)$, where $b_{i,j}^n$ denotes the block in the ith row and j^{th} column of the nth layer. And its corresponding sequence of visual words (or codewords) is denoted by $\Omega = (\omega_{1,1}^1, \omega_{1,2}^1, \ldots, \omega_{H,W-1}^N, \omega_{H,W}^N)$, as illustrated in Figure 1(b). Given a visual word sequence of an unknown image, Ω, the general framework to detect its category is to find out the image category whose model maximize the generation probability $P(\Omega \mid \lambda)$. The generation probability can be calculated as:

$$P(\Omega|\lambda) = P(\omega_{1,1}^1, \omega_{1,2}^1, \ldots, \omega_{H,W-1}^N, \omega_{H,W}^N|\lambda)$$
$$= P(\omega_{1,1}^1|\lambda) \cdot P(\omega_{1,2}^1|\omega_{1,1}^1, \lambda) \ldots P(\omega_{H,W-1}^N|\omega_{1,1}^1, \omega_{1,2}^1, \ldots, \omega_{H,W-2}^N, \lambda) \cdot$$
$$P(\omega_{H,W}^N|\omega_{1,1}^1, \omega_{1,2}^1, \ldots, \omega_{H,W-1}^N, \lambda) \tag{1}$$

where $\omega_{i,j}^n$ denotes the visual word of block (i,j) in the nth layer, and the model λ contains N layers, with the finest layer being divided into $H \times W$ blocks.

In our approach, we extend the memoryless property of Markov model to define the HSMM with following assumption:

• Given the visual words of blocks that precede current block $b_{i,j}^n$ in the sequence B, the visual word in block $b_{i,j}^n$ is only dependent of the visual words in its two neighboring blocks, $b_{i,j-1}^n$ and $b_{i-1,j}^n$, and in its parent block, $b_{\lceil i/2 \rceil, \lceil j/2 \rceil}^{n-1}$. As illustrated in Figure 2, the neighboring visual words, $\omega_{i,j-1}^n$, $\omega_{i-1,j}^n$ and $\omega_{\lceil i/2 \rceil, \lceil j/2 \rceil}^{n-1}$ are enough to determine the distribution of $\omega_{i,j}^n$. This fundamental assumption is formally indicated as:

$$P(\omega_{i,j}^n|\omega_{1,1}^1, \omega_{1,2}^1, \ldots, \omega_{i,j-1}^n) = P(\omega_{i,j}^n|\omega_{i-1,j}^n, \omega_{i,j-1}^n, \omega_{\lceil i/2 \rceil, \lceil j/2 \rceil}^{n-1}) \tag{2}$$

where $(\omega_{1,1}^1, \omega_{1,2}^1, \ldots, \omega_{i,j-1}^n)$ are the visual words of blocks $(b_{1,1}^1, b_{1,2}^1, \ldots, b_{i,j-1}^n)$ that are preceding blocks of block $b_{i,j}^n$ in the sequence B (Figure 1(b)).

To apply the model on solving real-life problem, it is necessary to estimate the generation probability $P(\Omega|\lambda)$ by deducing the equation 1. However, to estimate two and three dimensional Markov models with reasonable computational cost is still an open problem. In the proposed approach, we solve the computational problem by further assuming that, the right-hand part of equation 1 can be approximated by the multiplication of three conditional probabilities, which are the codeword transition probabilities in horizontal direction, vertical direction and inter-layer direction respectively.

$$P(\omega_{i,j}^n|\omega_{i-1,j}^n, \omega_{i,j-1}^n, \omega_{\lceil i/2 \rceil, \lceil j/2 \rceil}^{n-1})$$
$$\approx P(\omega_{i,j}^n|\omega_{i-1,j}^n) \cdot P(\omega_{i,j}^n|\omega_{i,j-1}^n) \cdot P(\omega_{i,j}^n|\omega_{\lceil i/2 \rceil, \lceil j/2 \rceil}^{n-1}) \quad (3)$$

The equation 1 therefore is transformed into:

$$P(\Omega|\lambda) = P(\omega_{1,1}^1|\lambda) \cdot \prod_{n=1}^{N} \prod_{i=1}^{H_n} \prod_{j=1}^{W_n} P(\omega_{i,j}^n|\omega_{i-1,j}^n, \omega_{i,j-1}^n, \omega_{\lceil i/2 \rceil, \lceil j/2 \rceil}^{n-1}, \lambda)$$
$$\approx P(\omega_{1,1}^1|\lambda) \cdot \prod_{n=1}^{N} \prod_{i=1}^{H_n} \prod_{j=1}^{W_n} [P(\omega_{i,j}^n|\omega_{i-1,j}^n, \lambda) \cdot$$
$$P(\omega_{i,j}^n|\omega_{i,j-1}^n, \lambda) \cdot P(\omega_{i,j}^n|\omega_{\lceil i/2 \rceil, \lceil j/2 \rceil}^{n-1}, \lambda)] \quad (4)$$

where the model λ contains N layers and the nth layer is divided into $H_n \times W_n$ blocks.

Here we introduce four new parameters.

• Initial codeword distribution $\alpha(x) = P(\omega_{1,1}^1 = o_x|\lambda)$, which represents the probability that we observe codeword o_x at block $(1,1)$ of the top layer, given the model λ. Therefore, variable $\alpha(\omega_{1,1}^1)$ denotes $P(\omega_{1,1}^1|\lambda)$.

• Horizontal word transition probability $h^n(x,y) = P(\omega_{i,j}^n = o_y|\omega_{i,j-1}^n = o_x, \lambda)$. It represents the probability that we observe codeword o_y in block (i,j) of the nth layer, given the codeword of its left block is o_x. Therefore, $h^n(\omega_{i,j-1}^n, \omega_{i,j}^n)$ denotes $P(\omega_{i,j}^n|\omega_{i,j-1}^n, \lambda)$, the probability that horizontal transition from codeword $\omega_{i,j-1}^n$ to codeword $\omega_{i,j}^n$ occurs in the nth layer.

• Vertical word transition probability $v^n(x,y) = P(\omega_{i,j}^n = o_y|\omega_{i-1,j}^n = o_x, \lambda)$. It represents the probability that we observe codeword o_y in block (i,j) of the nth layer, given the codeword of its above block is o_x. Consequently, variable $v^n(\omega_{i-1,j}^n, \omega_{i,j}^n)$ denotes $P(\omega_{i,j}^n|\omega_{i-1,j}^n, \lambda)$.

• Inter-layer word transition probability $t^n(x,y) = P(\omega_{i,j}^n = o_y|\omega_{\lceil i/2 \rceil, \lceil j/2 \rceil}^n = o_x, \lambda)$. It represents the probability that we observe codeword o_y in block (i,j) of the nth layer, given the codeword of its father block $b_{\lceil i/2 \rceil, \lceil j/2 \rceil}^n$ is o_x. And $t^n(\omega_{\lceil i/2 \rceil, \lceil j/2 \rceil}^n, \omega_{i,j}^n)$ is therefore used to represent $P(\omega_{i,j}^n|\omega_{\lceil i/2 \rceil, \lceil j/2 \rceil}^n, \lambda)$.

Substituting equation 4 with above parameters, we have:

$$P(\Omega|\lambda) \approx \alpha(\omega_{1,1}^1) \cdot \prod_{n=1}^{N} \prod_{i=1}^{H_n} \prod_{j=1}^{W_n} h^n(\omega_{i,j-1}^n, \omega_{i,j}^n) \cdot v^n(\omega_{i-1,j}^n, \omega_{i,j}^n) \cdot t^n(\omega_{\lceil i/2 \rceil, \lceil j/2 \rceil}^{n-1}, \omega_{i,j}^n) \quad (5)$$

The above two assumptions actually are derived from the conclusion of research [14]. The fundamental assumption is made based on the property that the texture of a region is strongly related with the texture in its directly adjacent regions and its relations with those unconnected regions are relatively weak.

To train a model for a specific image category, the probability distribution of initial word, horizontal word transition, vertical word transition, and interlayer word transition is estimated on training images with a simple maximum likelihood estimate (MLE) [15].

Given an unknown image σ with its codeword representation Ω_σ and K candidate image categories $\{\Psi_1, \Psi_2, \ldots, \Psi_K\}$, we estimate its most probable category by determining the model that is most likely to generate the sequence:

$$\sigma \in \Psi_X, with\ X = \arg\max_x [\alpha_{\lambda_x}(\omega_{1,1}^1) \cdot \prod_{n=1}^{N} \prod_{i=1}^{H_n} \prod_{j=1}^{W_n} h_{\lambda_x}^n(\omega_{i,j-1}^n, \omega_{i,j}^n) \cdot v_{\lambda_x}^n(\omega_{i-1,j}^n, \omega_{i,j}^n) \cdot t_{\lambda_x}^n(\omega_{\lceil i/2 \rceil, \lceil j/2 \rceil}^{n-1}, \omega_{i,j}^n)] \quad (6)$$

3 Experiments

The performance of the proposed HSMM-based image categorization approach is evaluated on a dataset of 13 categories of grayscale natural scenes [5] and the COREL dataset respectively. A Bayesian hierarchical model-based approach and a related Markov model based approach are compared with our method.

3.1 Comparison to Bayesian Hierarchical Model

The performance of proposed approach is evaluated on a dataset of 13 categories of natural scenes [5] and the result is compared with a Bayesian hierarchical model-based approach [5]. This approach is an extension of the LDA model. It also represents image patches with visual words and distinguishes image categories with the distribution of these words.

The dataset includes 3859 grayscale images from 13 categories of natural scenes. We randomly select N (N=5,10,20,40,60,80,100) images from each category to train the model and use the remaining for test. In the experiment, we employ the model of 3 layers and they are divided into blocks of 32×32, 16×16 and 8×8 pixels from top to bottom. According to experimental result, the capacity of the word dictionary of each layer is set to be 100, 150, and 200 accordingly. The descriptor of image features utilized in the experiment is the 128-dim SIFT vector that are employed in [5].

We conducted experiments with different ratios of training and testing data. As shown in Figure 3(a), the proposed approach achieves the classification rate of 77.7%, while we let the number of training data N take 100, which outperforms the result of 65.2% in [5] with 12.5%. As the number of training data decreases, the advantage goes more obvious. The result 62.8% that corresponds to 10 training data is almost double of the result of [5]. Figure 3(a) shows the classification rate by comparing with the result in Figure 10(a) of [5].

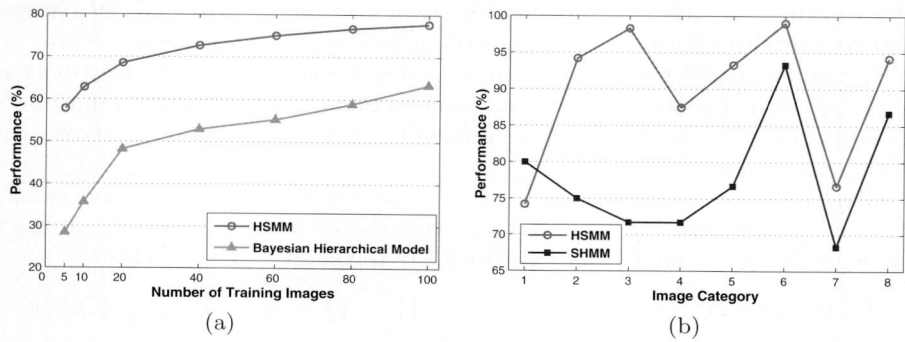

Fig. 3. (a) Classification rate of Hierarchical Spatial Hidden Markov Model and of Bayesian Hierarchical Model; (b) Classification rate of Hierarchical Spatial Markov Model and of Spatial Hidden Markov Model

3.2 Comparison to Previous Markov Model Based Approach

The Spatial Hidden Markov Model (SHMM) [14] extends the traditional Hidden Markov Model (HMM) to a 2-D model for categorizing and annotating images. It is also a block-based approach, which extracts image features in blocks and models the property of feature distribution in each image category. However, the SHMM is a supervised approach, which requires extensive annotation on image regions.

Experiments are conducted on 800 images from 8 classes of COREL image dataset. We randomly select N (N=40) images from each of the category to train the model and use the remaining images for test. As in [22], the 15-dimensional feature of average RGB values and 12-dimensional Gabor energy is used in the experiment. And the codebook is generated via K-means clustering. The 3 layers of the HSMM are under the same resolution and are partitioned in three different block sizes (32×32, 16×16, 8×8). And the dictionaries in size of 100, 150, and 250 for three layers are demonstrated to be appropriate.

As shown in Figure 3(b), the performance of the proposed approach achieves 89.7%. It outperforms the SHMM-based approach with 11.8% on average with even much less manual annotation for training models.

4 Conclusions

The proposed image categorization approach improves upon the traditional BoW model based approaches by utilizing the descriptive ability of the HSMM on spatial correlations of image features. This method also suggests a way for Markov model based approaches to adopt intermediate image representation with the advantage of avoiding the heavy workload of manual annotation on training data. Experimental results show that the spatial structure does have its advantages over other BoW based methods [5] and improves the performance of previous Markov model based approach [14] remarkably.

Acknowledgment

The work described in this paper was supported by a grant from the Research Council of Hong Kong SAR, China (Project No. CityU 114007) and a grant from City University of Hong Kong (Project No. 7002367).

References

1. Battiato, S., Farinella, G.M., Gallo, G., Ravi, D.: Scene categorization using bag of textons on spatial hierarchy. In: ICIP, pp. 2536–2539 (2008)
2. Bosch, A., Zisserman, A., Muñoz, X.: Scene classification via pLSA. In: Leonardis, A., Bischof, H., Pinz, A. (eds.) ECCV 2006. LNCS, vol. 3954, pp. 517–530. Springer, Heidelberg (2006)
3. Bouman, C.A., Shapiro, M.: A multiscale random field model for bayesian image segmentation. IEEE Transactions on Image Processing 3(2), 162–177 (1994)
4. Crandall, D.J., Huttenlocher, D.P.: Weakly supervised learning of part-based spatial models for visual object recognition. In: Leonardis, A., Bischof, H., Pinz, A. (eds.) ECCV 2006. LNCS, vol. 3951, pp. 16–29. Springer, Heidelberg (2006)
5. Fei-Fei, L., Perona, P.: A bayesian hierarchical model for learning natural scene categories. In: Proc. CVPR, vol. 2, pp. 524–531 (2005)
6. Fine, S., Singer, Y., Tishby, N.: The hierarchical hidden markov model: analysis and applications. Machine Learning 32, 41–62 (1998)
7. Lazebnik, S., Schmid, C., Ponce, J.: Beyond bags of features: Spatial pyramid matching for recognizing natural scene categories. In: CVPR, vol. 2, pp. 2169–2178 (2006)
8. Li, J., Gray, R.M., Olshen, R.A.: Multiresolution image classification by hierarchical modeling with two-dimensional hidden markov models. IEEE Transactions on Information Theory 46(5), 1826–1841 (2000)
9. Li, J., Joshi, D., Wang, J.: Stochastic modeling of volume images with a 3-d hidden markov model. In: ICIP, pp. 2359–2362 (2004)
10. Murphy, K.P., Paskin, M.A.: Linear time inference in hierarchical hmms. In: Proceedings of Neural Information Processing Systems, pp. 833–840 (2001)
11. Perina, A., Christani, M., Murino, V.: Nuatural scenes categorization by hierarchical extraction of typicality patterns. In: 14th International Conference on Image Analysis and Processing, pp. 801–806 (2007)
12. Sudderth, E.B., Torralba, A., Freeman, W.T., Willsky, A.S.: Describing visual scenes using transformed dirichlet processes. In: Advances in Neural Information Processing Systems, vol. 18, pp. 1299–1306 (2005)
13. Sudderth, E.B., Torralba, A., Freeman, W.T., Willsky, A.S.: Learning hierarchical models of scenes, objects, and parts. In: ICCV, vol. 2, pp. 1331–1338 (2005)
14. Yu, F., Ip, H.H.S.: Automatic semantic annotation of images using spatial hidden markov model. In: Proceedings of International Conference on Multimedia and Expo, pp. 305–308 (2006)
15. Yu, F., Ip, H.H.S.: Semantic content analysis and annotation of histological images. Computers in Biology and Medicine 38(6), 635–649 (2008)

Soft Measure of Visual Token Occurrences for Object Categorization

Yanjie Wang, Xiabi Liu*, and Yunde Jia

Beijing Laboratory of Intelligent Information Technology, School of Computer
Science, Beijing Institute of Technology
Tel.: +86-10-68913447; Fax: +86-10-86343158
{wangyanjie,liuxiabi,jiayunde}@bit.edu.cn

Abstract. The improvement of bag-of-features image representation by statistical modeling of visual tokens has recently gained attention in the field of object categorization. This paper proposes a soft bag-of-features image representation based on Gaussian Mixture Modeling (GMM) of visual tokens for object categorization. The distribution of local features from each visual token is assumed as the GMM and learned from the training data by the Expectation-Maximization algorithm with a model selection method based on the Minimum Description Length. Consequently, we can employ Bayesian formula to compute posterior probabilities of being visual tokens for local features. According to these probabilities, three schemes of image representation are defined and compared for object categorization under a new discriminative learning framework of Bayesian classifiers, the Max-Min posterior Pseudo-probabilities (MMP). We evaluate the effectiveness of the proposed object categorization approach on the Caltech-4 database and car side images from the University of Illinois. The experimental results with comparisons to those reported in other related work show that our approach is promising.

1 Introduction

In recent years, object categorization with bag-of-features image representation has become a hot topic in the field of compute vision and pattern recognition [1,2,3,4,5,6,7,8]. The bag-of-features method originated from the bag-of-words model for document analysis, which was firstly introduced to object categorization by Csurka et al. [1]. They cluster local features in images by k-means algorithm to generate a visual vocabulary. The image is then represented as a histogram over visual tokens in the vocabulary. After Csurka et al.'s work, statistical modeling of visual tokens has been advised to improve the effectiveness of bag-of-features representation [2,3,6]. The relations between local features and visual tokens can be described more accurately and reliably through statistical modeling of visual tokens. Furthermore, a local feature is allowed to be softly mapped to multiple visual tokens in this way, so the aliasing effects can be reduced. In existing

---

* The corresponding author.

methods of statistical modeling of visual tokens for object categorization, the Gaussian distribution is used to model each visual token and the set of visual tokens is considered as a Guassian Mixture Model (GMM) [2,3,6].

This paper proposes a new soft bag-of-features image representation based on the Gaussian Mixture Modeling (GMM) of visual tokens. The resultant object categorization approach includes four stages. Firstly, local features are extracted from an input image. Secondly, posterior probabilities of being visual tokens for local features are computed by using Bayes formula, where local features from each visual token are assumed to be of the distribution of GMM. The GMM is learned from the training data by the Expectation-Maximization (EM) algorithm with a model selection method based on the Minimum Description Length (MDL). Thirdly, the image is represented using one of three schemes: probabilities based hard histogram, classification based soft histogram, and completely soft histogram. Finally, the image is classified into one of object categories under a new discriminative learning framework of Bayesian classifiers, the Max-Min posterior Pseudo-probabilities (MMP) [9], where feature vectors of images from each object category is also assumed to be of the distribution of GMM. Following other related work, we evaluate the proposed object categorization approach on the Caltech-4 database and the car side images of the University of Illinois. Our approach experimentally outperforms some other related methods with the similar local features and achieves the comparable results to those reported by using more sophisticated local features.

2 GMM-MMP Classification Framework

In this section, we introduce posterior pseudo-probabilities based categorization approach with the MMP learning. The reader is referred to our paper for more details [9].

Let $\mathbf{X}$ be a feature vector, C be an object category, $p(\mathbf{X}|C)$ be the class-conditional probability density function, then the posterior pseudo-probability of being C for $\mathbf{X}$ is computed as

$$f(p(\mathbf{X}|C)) = 1 - \exp(-\lambda p^\theta(\mathbf{X}|C)), \tag{1}$$

where λ, θ are positive numbers. Consequently, $f(p(\mathbf{X}|C))$ is a smooth, monotonically increasing function of $p(\mathbf{X}|C)$, and $f(0) = 0$ and $f(+\infty)$. Given an input image, we use Eq. 1 to compute the posterior pseudo-probability for each object category. The category with maximum posterior pseudo-probability will be taken as the categorization result.

The MMP method is advised to learn unknown parameters in Eq. 1. Let $f(\mathbf{X}; \Lambda)$ be the posterior pseudo-probability measure function (Eq. 1) of an object category, where Λ denote the set of unknown parameters in it. Let $\hat{\mathbf{X}}_i$ be a feature vector of arbitrary positive sample of the category, $\bar{\mathbf{X}}_i$ be a feature vector of arbitrary negative sample of the category, m and n be the number of positive and negative samples, respectively. Then the objective function of MMP learning for estimating parameters is

$$F(\mathbf{\Lambda}) = \frac{1}{m}\sum_{i=1}^{m}[f(\hat{\mathbf{X}}_i;\mathbf{\Lambda})-1]^2 + \frac{1}{n}\sum_{i=1}^{n}[f(\bar{\mathbf{X}}_i;\mathbf{\Lambda})]^2. \quad (2)$$

$F(\mathbf{\Lambda}) = 0$ means the perfect classification performance on the training data. Consequently, we can obtain the optimum parameter set $\mathbf{\Lambda}^*$ of the posterior pseudo-probability measure function by using the gradient descent algorithm to minimize $F(\mathbf{\Lambda})$:

$$\mathbf{\Lambda}^* = \arg\min_{\mathbf{\Lambda}} F(\mathbf{\Lambda}). \quad (3)$$

The form of class-conditional probability density function $p(\mathbf{X}|C)$ in Eq. 1 should be provided for using MMP categorization framework, which is assumed to be the GMM in this paper. Let K be the component number of GMM, w_k, $\boldsymbol{\mu}_k$ and $\boldsymbol{\Sigma}_k$ be the weight, the mean, and the covariance matrix of the k-th Gaussian component, respectively. $\sum_{k=1}^{K} w_k = 1$. Then we have

$$p(\mathbf{X}|C) = \sum_{k=1}^{K} w_k N(\mathbf{X}|\boldsymbol{\mu}_k, \boldsymbol{\Sigma}_k), \quad (4)$$

where

$$N(\mathbf{X}|\boldsymbol{\mu}_k, \boldsymbol{\Sigma}_k) = (2\pi)^{-\frac{d}{2}} |\boldsymbol{\Sigma}_k|^{-\frac{1}{2}} \exp\left(-\frac{1}{2}(\mathbf{X}-\boldsymbol{\mu}_k)'\boldsymbol{\Sigma}_k^{-1}(\mathbf{X}-\boldsymbol{\mu}_k)\right). \quad (5)$$

So the set of unknown parameters in the posterior pseudo-probability measure function (Eq. 1) of each object category is

$$\mathbf{\Lambda} = \{\lambda, \theta, w_k, \boldsymbol{\mu}_k, \boldsymbol{\Sigma}_k\}, k = 1, \cdots, K. \quad (6)$$

3 Object Categorization by Soft Measure of Visual Token Occurrences

In bag-of-features image representation, a visual vocabulary consisting of visual tokens is generated to bridge local features and images. In this paper, we model the distribution of local features from each visual token as a GMM. The corresponding visual vocabulary can be seen as a set of visual token GMMs. According to visual token GMMs, we compute posterior probabilities of being visual tokens for local features. Then three corresponding image representation schemes are explored for object categorization under the GMM-MMP categorization framework.

3.1 Visual Token GMM with MDL-EM Training

We firstly cluster local features extracted from training images into designated number of groups. Each group of local features is corresponding with a visual token. This is the same as conventional bag-of-features methods. However, each group of local features is represented by a GMM, instead of its center, in this paper. The GMM is fitted to the group data by using the Expectation-Maximization (EM) algorithm [10] with a model selection method based on the Minimum Description Length (MDL) [11]. This strategy makes our method different from other statistical modeling based bag-of-features representations, where GMM is used to model the whole vocabulary and each visual token is corresponding with a Gaussian component.

Let t be the parameter number of each Gaussian component in the GMM, n be the number of training samples, $f(\mathbf{x}_1, \cdots, \mathbf{x}_n|\Theta)$ be the likelihood function over the training set. Then the training criterion with MDL-EM can be formalized to minimize [11]

$$-\log f(\mathbf{x}_1, \cdots, \mathbf{x}_n|\Theta) + \frac{t}{2}\log n, \qquad (7)$$

where the first and second terms stand for the objective of maximum likelihood and the simplest model, respectively.

After visual token GMMs are obtained from the training data, we employ Bayes formula with the assumption of the same prior probabilities for all the visual tokens to estimate the posterior probability of being visual token $\mathbf{v}_j$ for local feature $\mathbf{x}_i$:

$$P(\mathbf{v}_j|\mathbf{x}_i) = \frac{P(\mathbf{x}_i|\mathbf{v}_j)}{\sum_{k=1}^{N} P(\mathbf{x}_i|\mathbf{v}_k)}, \qquad (8)$$

where N is the number of all the visual tokens.

3.2 Image Representation

According to hard assignment of local features to visual tokens, it seems that we can only compute occurrence frequencies of visual tokens to obtain a hard histogram description of the image. Oppositely, $P(\mathbf{v}_j|\mathbf{x}_i)$ reflects the confidence of assigning $\mathbf{x}_i$ to $\mathbf{v}_j$. More reliable and accurate occurrence distribution can be defined based on this soft assignment. In this paper, we consider three corresponding representation schemes: Probability Based Hard Histogram (PBHH), Classification Based Soft Histogram (CBSH), and Completely Soft Histogram (CSH). In both PBHH and CBSH, local features are firstly classified into the visual token with maximum posterior probability. Then the image is represented as frequencies (PBHH) or mean probabilities (CBSH) of visual tokens. The CSH maps each local feature to all the visual tokens, and compute mean probabilities of visual tokens to represent the image. More formally, let I be an image, $\{\mathbf{x}_1, \mathbf{x}_2, \ldots, \mathbf{x}_M\}$ be local features extracted from I, $\{\mathbf{v}_1, \mathbf{v}_2, \ldots, \mathbf{v}_N\}$ be tokens in the visual vocabulary, M be the number of all the local features, $m_i|_{i=1}^{N}$ be the number of local features classified into visual token $\mathbf{v}_j$, then these three representation schemes are listed in Table 1.

After the image is represented by each of three schemes above, we assume that the feature vectors of the images from each object category are of the distribution of GMM. We then use GMM-MMP categorization framework descried in Section 2 to perform the image categorization.

Table 1. Three image representation schemes

Schema	Image Representation
PBSH	$\{m_i/M\}|_{i=1}^{N}$
CBSH	$\{\sum_{k=1}^{m_i} P(\mathbf{v}_i|\mathbf{x}_k)/m_i\}|_{i=1}^{N}$
CSH	$\{\sum_{k=1}^{M} P(\mathbf{v}_i|\mathbf{x}_k)/M\}|_{i=1}^{N}$

4 Experimental Results

4.1 Experimental Setup

Harris-affine detector [12] is adopted to extract local features from images. Then we use SIFT [13] as the feature descriptor, resulting a 128-dimentsional real vector ($4 \times 4 \times 8$) for each local feature. In order to fairly compare our approach to other related methods, the number of visual tokens is preset to 1000 as that used in [1].

In the MMP training for object categories, positive samples of each object category are images from this category, while negative ones are images from other categories. At first, we used the MDL-EM algorithm on positive samples to get the parameters in the GMM, and set λ and θ through experiments. Then the MMP training algorithm was used on all the samples including positive samples and negative samples to revise initial parameters obtained by the MDL-EM algorithm. For the MDL based model selection of the GMM, we evaluate the component numbers from 1 to 20 for visual tokens and object categories. The resultant numbers of components for visual tokens vary from 1 to 5, while those for object categorization are from 3 to 9.

4.2 Caltech-4 Database and Car Side Images

We conduct experiments of object categorization on the Caltech-4 database and car side images from the University of Illinois. The Caltech-4 database includes four object categories. The number of images from each category varies from 450 to 1074. Following the evaluation method used in other related work, we randomly select half of images from each object category for training, and the others for testing.

4.2.1 Comparative Evaluation

The proposed object categorization approach can be divided into three stages: visual token modeling, image representation, and discriminative learning for object categorization. Thus we design three groups of experiments to evaluate the influence of various factors in each stage.

In the first group, we tested the effectiveness of three visual token modeling methods including GMM, Gaussian Model, and traditional cluster center under hard histogram image representation setting. Gaussian Model is treated as 1-component GMM in our experiments. We also compared GMM and Gaussian Model under two types of soft histogram image representation setting, CBSH

Table 2. Comparing categorization accuracies for visual token modeling methods and image representation schemes

Methods	THH	PBHH		CBSH		CSH	
		GM	GMM	GM	GMM	GM	GMM
Airplanes	0.961	**0.968**	**0.968**	0.972	**0.980**	**0.985**	0.970
Cars(Rear)	0.960	0.964	**0.974**	0.945	**0.971**	0.945	**0.978**
Motorbikes	0.889	0.889	**0.910**	0.893	**0.910**	0.893	**0.932**
Faces	0.880	0.893	**0.907**	0.880	**0.889**	0.906	**0.933**
Cars(Side)	0.953	**0.942**	0.935	0.956	**0.960**	0.956	**0.964**
Mean	0.936	0.939	**0.947**	0.936	**0.952**	0.943	**0.960**

Table 3. Comparing categorization accuracies for MMP vs. MDL-EM training

Categories	MMP	EM
Airplanes	**0.980**	0.970
Cars(Rear)	**0.995**	0.978
Motorbikes	**0.960**	0.932
Faces	**0.947**	0.933
Cars(Side)	**1.000**	0.964
Mean	**0.977**	0.960

and CSH. In this group, only MDL-EM algorithm is used to learn the GMMs of object categories. The MMP algorithm is not triggered yet. Table 2 shows categorization results for the test data from 5 categories, where 'THH' denotes the Traditional Hard Histogram based on cluster centers, 'GM' denotes Gaussian Model. It demonstrates that the GMM behaved best and statistical modeling of visual tokens brings better performance than distance based vector quantization technique. In the second group, three proposed image representation schemes, PBHH, CBSH, and CSH, are compared under the GMM of visual tokens. The corresponding results are also reflected in Table 2, where CSH is shown to outperform other two schemes. We tested the effectiveness of the MMP discriminative learning algorithm in the third group. Under CSH image representation with the GMM of visual tokens, the training effects of the MMP and the MDL-EM for object categorization were compared and listed in Table 3. It shows that the mean categorization accuracy is improved from 96.0% (EM) to 97.7% (MMP).

4.2.2 Comparisons to Related Work

To confirm the effectiveness of our approach, we further compared our best categorization results achieved by using the GMM of visual tokens, CSH image representation, and MMP learning algorithm to those reported in other related work [1, 4, 14, 7, 15]. The comparisons of results are shown in Table 4. Among these work under comparison, Csurka et al. [1] and Sivic et al. [14] adopt the same local features as ours, namely, the Harris-affine detector with the SIFT descriptor; Fergus et al. [7] uses the Kadir-Brady local feature detector and the pixel descriptor; Kapoor et al. [15] employs the multiresolution local features;

Table 4. The comparisons between our approach and other related methods

Categories	Ours	[1]	[14]	[7]	[4]-1	[4]-2	[15]
Airplanes	0.980	0.963	0.953	0.902	0.889	0.975	0.980
Cars(Rear)	0.995	0.977	0.981	0.900	0.911	1.000	0.991
Motorbikes	0.960	0.927	0.936	0.925	0.922	0.943	0.970
Faces	0.947	0.940	0.940	0.964	0.935	1.000	0.995
Cars(Side)	1.000	0.996	–	–	0.830	1.000	–
Mean	0.977	0.961	–	–	0.897	0.984	–

Opelt et al. [4] tested two kinds of local features, including the affine invariant interest point detector with the moment invariant descriptor (denoted as [4]-1 in Table 4) and the similarity-measure-segmentation with the intensity distribution description (denoted as [4]-2 in Table 4). Our approach experimentally outperforms the methods using the similar local features [1,14], [4]-1, and achieved the comparable results to those reported by using more sophisticated local features in [15] and [4]-2.

5 Conclusions

In this paper, we explored the problem of soft histogram image representation based on Gaussian Mixture Modeling (GMM) of visual tokens for object categorization. The main contributions of this paper are summarized as follows: 1) The posterior probabilities of being visual tokens for local features are computed by assuming that local features from each visual token are of the distribution of GMM. Accordingly, three types of image descriptions are defined and compared for object categorization, including Probability Based Hard Histogram (PBHH), Classification Based Soft Histogram (CBSH), and Completely Soft Histogram (CSH). 2) A new discriminative learning framework of Bayesian classifiers, Max-Min posterior Pseudo-probabilities (MMP), is applied to object categorization.

We conducted three groups of comparative experiments on the Caltech-4 database and car side images from the University of Illinois. In the first group, GMM of visual tokens is compared to Gaussian modeling of visual tokens as well as traditional cluster center. The results show that the GMM outperforms other two strategies. In the second group, three types of histogram descriptions of the images are tested and CSH is shown to behave best. In the last group, we demonstrate that MMP is better than generative learning counterpart. To sum up, we achieved the best result by using the GMM of visual tokens, CSH image representation, and MMP learning for object categorization, which is better than those reported using similar local features and comparable to those obtained from more sophisticated local features.

The future developments of the proposed approach are described as follows. Firstly, visual token GMMs are currently learned by using the MDL-EM algorithm, since the number of visual token is 1000 and the MMP is not enough efficient to solve this huge classification problem. In the next work, we will improve the efficiency of MMP learning and apply it to the training of visual token GMMs for more accurate measure of visual token occurrence. Secondly, the experimental evaluation of our approach is planned to be performed on other widely used databases, including Caltech-101 and VOC 2008.

Acknowledgement

This research was partially supported by 973 Program of China (No. 2006CB303103), Beijing Key Discipline Program, and Excellent Young Scholars Research Fund of Beijing Institute of Technology (No. 2008YS1203).

References

1. Csurka, G., Dance, C., Fan, L., Willamowski, J., Bray, C.: Visual categorization with bags of keypoints. In: Workshop on Statistical Learning in Computer Vision in ECCV (2004)
2. Farquhar, J., Szedmak, S., Meng, H., Shawe-Taylor, J.: Improving "bag-of-keypoints" image categorisation: Generative models and pdf-kernels. Technical report, University of Southampton (2005)
3. Winn, J., Criminisi, A., Minka, T.: Object categorization by learned universal visual dictionary. In: IEEE International Conference on Computer Vision (2005)
4. Opelt, A., Fussengger, M., Pinz, A., Auer, P.: Generic object recognition with boosting. IEEE Transactions on Pattern Analysis and Machine Intelligence 28(3), 416–513 (2006)
5. Zhang, J., Marszalek, M., Lazebnik, S., Schmid, C.: Local features and kernels for classification of texture and object categories: a comprehensive study. International Journal of Computer Vision 73(2), 213–238 (2007)
6. Perronnin, F.: Univeral and adapted vocabularies for generic visual categorization. IEEE Transactions on Pattern Analysis and Machine Intelligence 30(7), 1243–1256 (2008)
7. Fergus, R., Perona, P., Zisserman, A.: Object class recognition by unsupervised scale-invariant learning. In: IEEE conference on Computer Vision and Pattern Recognition (2003)
8. Lazebnik, S., Schmid, C., Ponce, J.: Beyond bags of features: spatial pyramid matching for recognizing natural scene categories. In: IEEE conference on Computer Vision and Pattern Recognition (2006)
9. Liu, X., Jia, Y., Chen, X., Deng, Y., Fu, H.: Image classification using the maxmin posterior pseudo-probabilities method. Technical Report BIT-CS-20080001, Beijing Institute of Technology (2008),
   http://www.mcislab.org.cn/member/~xiabi/papers/2008_1.PDF
10. Dempster, A., Laird, N., Rubin, D.: Maximum likelihood from incomplete data via the em algorithm. Journal of the Royal Statistical Society 39(1), 1–38 (1977)
11. Hansen, M.H., Yu, B.: Model selection and the principle of minimum description length. Journal of American Statistical Association 96(454), 746–774 (1989)
12. Mikolajczyk, K., Shmid, C.: Scale and affine invariant point detectors. International Journal of Computer Vision 60(1), 63–86 (2004)

13. Lowe, D.G.: Object recognition from local scale-invariant features. In: IEEE International Conference on Computer Vision (1999)
14. Sivic, J., Russell, B.C., Efros, A.A., Zisserman, A., Freeman, W.T.: Discovering objects and their location in images. In: IEEE International Conference on Computer Vision (2005)
15. Kapoor, A., Grauman, K., Urtasun, R., Darrell, T.: Active learning with gaussian processes for object categorization. In: IEEE International Conference on Computer Vision (2007)

Indexing Large Visual Vocabulary by Randomized Dimensions Hashing for High Quantization Accuracy: Improving the Object Retrieval Quality

Heng Yang, Qing Wang, and Zhoucan He

School of Computer Science and Engineering, Northwestern Polytechnical University
Shaanxi Provincial Key Laboratory of Speech and Image Information Processing,
Xi'an 710072, P.R. China
qwang@nwpu.edu.cn

Abstract. The bag-of-visual-words approach, inspired by text retrieval methods, has proven successful in achieving high performance in object retrieval on large-scale databases. A key step of these methods is the quantization stage which maps the high-dimensional image feature vectors to discriminatory visual words. In this paper, we consider the quantization step as the nearest neighbor search in large visual vocabulary, and thus proposed a randomized dimensions hashing (RDH) algorithm to efficiently index and search the large visual vocabulary. The experimental results have demonstrated that the proposed algorithm can effectively increase the quantization accuracy compared to the vocabulary tree based methods which represent the state-of-the-art. Consequently, the object retrieval performance can be significantly improved by our method in the large-scale database.

Keywords: Object retrieval, Vocabulary tree, Randomized dimensions hashing.

1 Introduction

Object retrieval in large-scale databases has drawn a lot of attentions in research area of image understanding and computer vision. Given a query image which contains a particular object, our motivation is to return immediately from the large database a set of high related images in which that object appears. In general, the standard approach to solve this problem is firstly to represent the images by high-dimensional local features and then to match images by dealing with millions of feature vectors.

Several successful object based image and video retrieval systems have been recently reported [1-6], which mimicked the text-retrieval approaches using the analogy of visual words. Sivic and Zisserman [1] firstly introduced this bag-of-visual-words architecture for the video object recognition application. They apply flat k-means algorithm on feature vectors from the training frames to generate vocabulary of visual words. The feature vectors quantized to the same visual words are considered matched to each other. Then the standard TF-IDF (Term Frequency-Inverse Document Frequency) weighting scheme, which down-weights the contribution of the commonly occurred words, is used for scoring the relevance of an image to the query one. However, the vocabulary built by flat k-means is hard to be scalable to a large size. To address this issue, Philbin et al. [5] introduced an approximate k-means (AKM) algorithm to speed up the traditional flat k-means method by employing an approximate nearest neighbor (ANN) method based on randomized trees. They claimed that AKM can reduce computation complexity of the regular k-means greatly and thus scale the vocabulary to a large size. A more efficient way to build scalable large vocabulary is proposed by Nistér and Stewenius [2]. They designed a vocabulary tree structure based on hierarchical k-means (HKM) method [7]. The tree can efficiently define large visual vocabulary by recursively running k-means to partition features in current node to clusters with small k. Furthermore, it gave an efficient search procedure for quantization. Therefore, a much larger and more discriminatory vocabulary can be used efficiently which can improve the image searching quality dramatically. Schindler et al. [3] used the same data structure as the vocabulary tree for large-scale location recognition application. In particular, they presented a Greedy N-Best Paths (GNP) algorithm to improve the retrieval performance of the traditional vocabulary tree algorithm by considering more candidates instead of one at each level of the tree. Philbin et al. [6] introduced the soft-assignment technology which mapped a feature to a weighted combination of visual words rather than hard-assign to a single word. This method improved the performance of retrieval, since that it avoided the quantization lost of the hard-assigned method to some extent. Chum et al. [4] brought the query expansion technology, which is a standard method in text retrieval system, into the visual domain for improving the retrieval performance. They utilized the spatial constraint between query image and each returned image and used these verified images to learn a latent feature model which controlled the construction of expanded queries. In a word, there are mainly four key stages in current successful image retrieval systems based on the bag-of-visual-words model: 1) build large and discriminatory visual vocabulary; 2) quantize the feature vectors into large visual words efficiently; 3) use the TF-IDF scheme to score the similarity between images; 4) employ well-known technologies to further refine the retrieval results, such as spatial verification and query expansion.

In this paper, we focus on the quantization stage, since that it is the key factor to influence the retrieval quality based on the similar size vocabulary. In particular, we concentrate on solving the quantization issue of the vocabulary tree based methods. The tree structure based methods give us an efficient way to build large vocabularies, but low quantization accuracy (demonstrated in our experiment) which depresses the retrieval performance. The quantization process can be considered as the nearest neighbor searching in the visual word set and the quantization accuracy can be defined as same as the ANN searching accuracy. In order to address the quantization issue of tree based method, we employ a locality sensitive hashing (LSH) [9] based ANN algorithm for efficiently indexing and searching the large vocabulary to improve the quantization accuracy.

The remainder of this paper is organized as follows. Section 2 analyses the quantization limitations of the tree based methods and section 3 presents our approach in detail. The experimental results and related discussions are given in section 4. Finally, the conclusion is summarized in section 5.

2 Quantization in Vocabulary Tree Based Methods

The traditional vocabulary tree [2] is built by hierarchical k-means clustering on SIFT [8] feature vectors from the training data. The tree is a kind of k-way tree of depth L. Therefore, there are k^L leaf nodes (visual words) at the bottom of the tree. For quantization, a SIFT feature is down from the root node to a leaf node by at each level comparing the feature vector to the k candidate cluster centers and choosing the closest one. Then the path down the tree is encoded by an integer which represents a specific visual word. The quantization speed of the traditional vocabulary tree is very fast when k and L are both small, since it only needs comparing totally $k{\times}L$ candidates in the quantization process for one search. However, there exists an obvious drawback in the quantization process. At each level of the tree, the quantization error will be inevitable when the feature vectors locate on the boundaries that are defined by the cluster centers (see Figure 1 for illustration). Therefore, due to the accumulative errors at upper levels of the tree, the quantization accuracy at the leaf level will be depressed. This can be validated from our experimental results (see section 4.1).

The Greedy N-Best Paths (GNP) algorithm [3] improved the quantization method of [2]. It chooses the closest N nodes at each level by comparing the $k{\times}N$ candidates. The total comparisons of GNP is $k+k{\times}N{\times}(L-1)$ per search. GNP can increase the quantization accuracy, since it considers more nodes (when $N>1$) in traversing a tree. But the computation complexity is nearly linear with N, which will bring heavy computation when N is large.

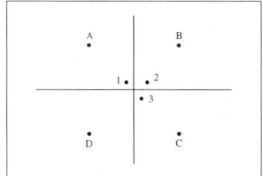

Fig. 1. Illustration of quantization error at an upper level of a vocabulary tree. Points A to D represent cluster center nodes and points 1, 2, 3 are feature vectors, respectively. The cross lines are the boundaries defined by cluster centers A to D together. Points 1, 2, 3 should be assigned to the same visual word at the bottom level for being close to each other, but they are assigned to different nodes at this level and thus are quantized to different visual words at last.

3 Our Approach

We follow the hierarchical k-means clustering scheme [2] to build large visual vocabulary, which is proved scalable and efficient. But we throw off the tree index structure and employ a LSH based ANN method to index the whole visual vocabulary in order to address the tree based quantization issue. Our quantization scheme, called randomized dimensions hashing (RDH), is given in detail in the section 3.1. At last, TF-IDF scheme [1] is used to get the relevance score between database image and the query one, which is accomplished using inverted files.

3.1 Quantization Using Randomized Dimensions Hashing

The target of quantization process is to map a feature vector to its nearest visual word in acceptable time. The exhaustive method is impracticable due to its low efficiency in the large vocabulary. Vocabulary tree based methods hold high speed, but low accuracy (see section 4.1). A natural consideration to this problem is to resort to the ANN searching method, considering both accuracy and speed.

Our algorithm belongs to a LSH based ANN searching method. LSH is widely used in high-dimensional feature searching for its independent to the feature dimensionality. LSH algorithm firstly project feature vectors to Hamming space which consists of super large binary bits. Then, a group hash functions are employed to randomly select m dimensions of vectors in Hamming space and thus compose the binary string as the hash value. Finally, p different hash tables are built. In searching phase, LSH only searches the features in the same buckets that have the same hash values to the query feature vector. In fact, two parameters m and p enable the designer to select an appropriate trade-off between accuracy and running time according to different request. However, LSH also has some drawbacks. First, the information of dimensions of the Hamming vector is very limited, which significantly decreases the local sensitivity of the algorithm. Secondly, projecting feature vectors into Hamming space consumes large memory cost. Both the two limitations decrease the performance of LSH algorithm.

Based on the LSH framework, we propose a randomized dimensions hashing (RDH) algorithm to efficiently index and search the large vocabulary. There are three major improvements in our algorithm. The first and most important one is that we project the feature sets into an equally-distributed space, not the Hamming space, which can significantly enhance the local sensitivity of search algorithm. The second one is that a specific designed hash function (see Algorithm 1) is employed to assign feature vectors to hash buckets discriminatively. The last one is that a constraint compare times is set to 200 to guarantee the search efficiency. The RDH algorithm is divided to three steps and described as follows:

Step 1: Calculate the data distributions for each dimension in the feature (visual word) set F. If the set containing the same dimension of all the features (D dimensions) is denoted as F_i ($i=1,2,...,D$), then a key value key_i is determined to equally divide F_i into two parts.

Algorithm 1. Hash Function

Input: m dimensions of a feature vector f_j

Output: hash value of f_j.

1 for $t:= 1$ to m
2   If $f_{jr(t)} > key_{r(t)}$, $bs[t] = 1$;
3   Elseif, $bs[t] = 0$.
4   end if
5 end for
6 Convert the binary string bs to an equivalent integer hash value hv_j.

Step 2: Establish p index structures of $\{I_1, I_2, ..., I_p\}$. For each index, we randomly choose m dimensions for each feature f_j, denoted as $f_{jr(t)}, t=1,2,...,m; 1 \leq r(t) \leq D$. Then for every feature f_j in F, we execute *Algorithm 1* to computer its hash value hv_j. Features with the same hash value are mapped into the same hash bucket.

Step 3: For a query feature q, *Algorithm 2* is executed to search the nearest neighbor as its quantization result.

Algorithm 2. Searching
Input: query feature q and the index structures $\{I_1, I_2, ..., I_p\}$
Output: nearest neighbor of q
1 *cmp_time*=0;
2 for l:=1 to p
3   Calculate the hash value of q using *Algorithm 1* and find out the corresponding hash bucket B_q.
4   Compare, but not repeatedly among different indexes, the features in the set B_q one-by-one to search the nearest neighbor, and accumulate the compare times to *cmp_time*.
5   If *cmp_time*>200
6   break;
7   end if
8 end for

3.2 Discussion on RDH

In the proposed algorithm, there are two important parameters which are the number of dimensions m for calculating hash value and the number of index structures p. Due to the fixed compare times in RDH, the computation complexity is fixed. However, different setting of m and p can bring the algorithm different performance on searching accuracy. For a specific vocabulary built in section 4.1 in our retrieval task, we find that $m=20$ and $p=30$ can reach the best accuracy performance. For different size vocabularies in different applications, the two parameters can be adjusted accordingly to reach the high performance.

In addition, employing RDH for quantization takes extra storage and computation at the index building step. But it is built for only one time and the index structure can be stored in memory for efficient online retrieval application. The benefit for doing this is the much higher quantization accuracy and further higher retrieval performance.

4 Experimental Results and Analysis

To evaluate the performance of the proposed quantization method and further its effect on object retrieval, we use the standard object recognition benchmark image database [2, 10]. It contains 10,200 images in groups of four that belong together. In each group, the same object is taken from different positions or under varying illumination conditions. We implemented SIFT algorithm to extract local features from images and create the invariant 128-D SIFT descriptor.

Our experiments can be divided into two parts. First, we compare the quantization accuracy and evaluate the quantization effect on the retrieval quality on the training data set. Second, we validate the retrieval superiority brought by our method on the whole large database. All the experiments are executed on a PC with Pentium IV dual-core 2.0 GHz processor and 2GB memory.

4.1 Quantization Accuracy Comparison on Training Data Set

The front 1,000 images from the object recognition database [2, 10] are used as training data set, from which 949,291 feature vectors are generated. We use the hierarchical k-means method [2] to train a vocabulary tree with 5 levels and a branching factor of 10, which will result in about 100K visual words. In our implementation, 96,994 visual words are created finally, since some null clusters are generated. Then the proposed RDH algorithm is employed to index the large visual words and perform ANN searching in quantization stage.

In order to compare the quantization accuracy, we run exhaustive algorithm on the image features from the group-to-group and record the exact nearest neighbors for each group as ground truth. Then the traditional vocabulary tree (denoted as VT for short), GNP, original LSH and RDH search algorithms are executed respectively on each group. The quantization accuracy of each algorithm can be obtained by comparing to the ground truth. The average compare times of the four algorithms are set to 50 (10×5), 210 (N=5, 10+10×5×4), 200, 200 respectively. It is worth noticing here, we also constrain the maximum compare times to 200 for LSH algorithm to make it hold similar computational complexity to GNP and our algorithm. Due to the space limitation, we only show the results of the front 10 groups as examples in Figure 2, from which we can see that traditional vocabulary tree algorithm has the lowest quantization accuracy which is averagely no more than 50%. GNP and LSH can achieve the similar performance, and RDH obtains the highest quantization accuracy which is averagely more than 90%.

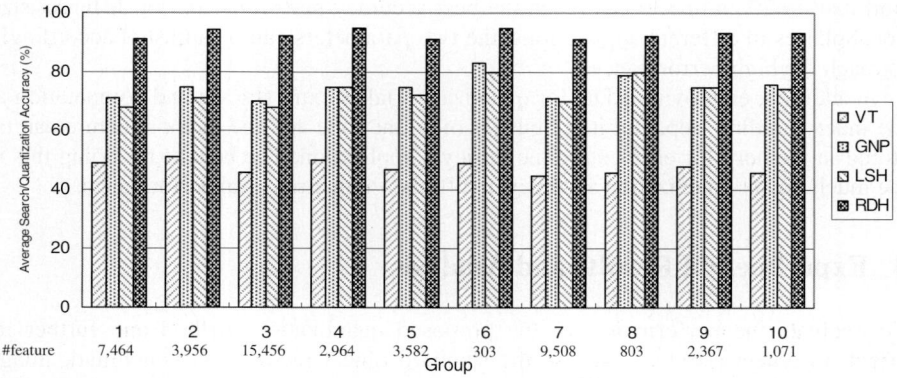

Fig. 2. The quantization performance of four algorithms searching on 96,994 visual words

Furthermore, Table 1 gives the average retrieval accuracy (ARA defined by (1)) of the VT, GNP, LSH and RDH algorithms on the training data set. It can be seen the higher quantization accuracy can further lead to higher retrieval accuracy and thus RDH reaches the highest retrieval quality.

Moreover, we use the AKM algorithm similar as [5] to recalculate the cluster centers. We take the visual words trained by HKM algorithm as the initial values, and employ the RDH algorithm to build index over the cluster centers at the beginning of each iteration to increase speed. The modified visual words are used for retrieval again, and the RDH is also employed for quantization. The recalculated result is ARA=0.917, which is nearly the same as that using the vocabulary built on HKM algorithm. This experimental result proves that both the AKM and HKM can give the similar results as long as they use the same quantization method and build vocabulary with the same size. In other words, with fixed size of visual vocabulary, the key factor to retrieval quality is the quantization accuracy.

Table 1. The average retrieval accuracy of the four algorithms on training data set

	VT	GNP	LSH	RDH
ARA	0.840	0.887	0.882	0.915

4.2 Object Retrieval Performance Comparison on the Large-Scale Database

We used the same visual vocabulary built by HKM method from the training data set, but assign image features of the whole database [2, 10] to visual words using different quantization methods – traditional vocabulary tree (VT), GNP and RDH respectively. The visual words are organized by inverted file structure for efficient retrieval, which keeps track of the number of times each visual word appears in each image in database. Finally, TF-IDF scheme is employed to give the relevance score between query image and database image. The retrieval quality in this database can be measured by Average Retrieval Accuracy (ARA) which is computed by (1), where cnt_i denotes how many of the first four most similar images in the same group as the i-th image (including the i-th image itself), and n is the number of the database images.

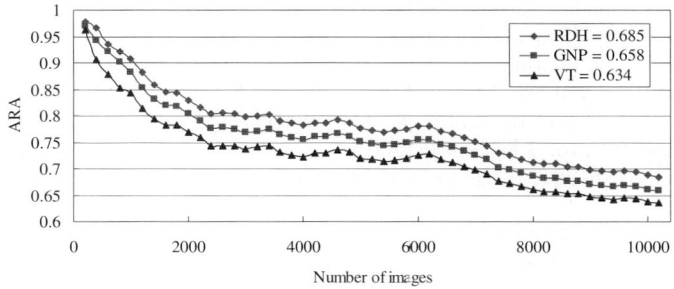

Fig. 3. Performance of our algorithm compared to the traditional vocabulary tree (VT) and GNP algorithms on the large-scale object recognition database

$$\text{ARA} = \frac{1}{n}\sum_{i=1}^{n}\frac{cnt_i}{4} \qquad (1)$$

The retrieval performances based on the three quantization algorithms along with the changing number of images of the whole large-scale database are shown in Figure 3. We can see that our algorithm obviously outperforms the other two tree-based methods due to the higher quantization accuracy. Even when the database size is up to 10K, our algorithm can reach 0.685 for average retrieval accuracy.

5 Conclusion

The main contribution of this paper is to propose an ANN method for efficient quantization in bag-of-visual-words model. Our quantization algorithm can achieve much higher quantization accuracy compared to the tree structure based methods. Furthermore, the quantization gain can lead to higher retrieval quality. Experimental results on the large-scale database have proved the effectiveness of our algorithm in the object retrieval application. In future work, we plan to apply our method to the other retrieval applications, such as image-based localization.

Acknowledgments

This work is supported by National Natural Science Fund (60873085) and National Hi-Tech Development Programs under grant No. 2007AA01Z314, P.R. China.

References

1. Sivic, J., Zisserman, A.: Video Google: A Text Retrieval Approach to Object Matching in Videos. In: ICCV, vol. 2, pp. 1470–1477 (2003)
2. Nistér, D., Stewénius, H.: Scalable Recognition with a Vocabulary Tree. In: CVPR, vol. 2, pp. 2161–2168 (2006)
3. Schindler, G., Brown, M., Szeliski, R.: City-Scale Location Recognition. In: CVPR (2007)
4. Chum, O., Philbin, J., Sivic, J., Isard, M., Zisserman, A.: Total recall: Automatic Query Expansion with a Generative Feature Model for Object Retrieval. In: ICCV (2007)
5. Philbin, J., Chum, O., Isard, M., Sivic, J., Zisserman, A.: Object Retrieval with Large Vocabularies and Fast Spatial Matching. In: CVPR, pp. 1–8 (2007)
6. Philbin, J., Chum, O., Isard, M., Sivic, J., Zisserman, A.: Lost in Quantization: Improving Particular Object Retrieval in Large Scale Image Databases. In: CVPR (2008)
7. Mikolajczyk, K., Leibe, B., Schiele, B.: Multiple Object Class Detection with a Generative Model. In: CVPR, vol. 1, pp. 26–36 (2006)
8. Lowe, D.G.: Distinctive Image Features from Scale-Invariant Keypoints. IJCV 60, 91–110 (2004)
9. Gionis, A., Indyky, P., Motwaniz, R.: Similarity Search in High Dimensions via Hashing. VLDB Journal, 518–529 (1999)
10. Object recognition database, http://www.vis.uky.edu/~stewe/ukbench/data/

Design of Clinical Support Systems Using Integrated Genetic Algorithm and Support Vector Machine

Yung-Fu Chen[1,*,**], Yung-Fa Huang[2,**], Xiaoyi Jiang[3,1,*], Yuan-Nian Hsu[4,5], and Hsuan-Hung Lin[6,***]

[1] Department of Health Services Administration, China Medical University, Taichung 40402
[2] Department of Information and Communication Engineering, Chaoyang University of Technology, Taichung 41349
[3] Department of Computer Science, University of Münster, D-48149 Münster, Germany
[4] Taichung Hospital, Department of Health, Executive Yuan, Taichung 40343
[5] Department of Health Care Administration and
[6] Department of MIS, Central Taiwan University of Science and Technology, Taichung 40601, Taiwan
yungfu@mail.cmu.edu.tw, yfahuang@mail.cyut.edu.tw,
xjiang@math.uni-muenster.de, taic@mail.taic.doh.gov.tw,
shlin@ctust.edu.tw

Abstract. Clinical decision support system (CDSS) provides knowledge and specific information for clinicians to enhance diagnostic efficiency and improving healthcare quality. An appropriate CDSS can highly elevate patient safety, improve healthcare quality, and increase cost-effectiveness. Support vector machine (SVM) is believed to be superior to traditional statistical and neural network classifiers. However, it is critical to determine suitable combination of SVM parameters regarding classification performance. Genetic algorithm (GA) can find optimal solution within an acceptable time, and is faster than greedy algorithm with exhaustive searching strategy. By taking the advantage of GA in quickly selecting the salient features and adjusting SVM parameters, a method using integrated GA and SVM (IGS), which is different from the traditional method with GA used for feature selection and SVM for classification, was used to design CDSSs for prediction of successful ventilation weaning, diagnosis of patients with severe obstructive sleep apnea, and discrimination of different cell types form Pap smear. The results show that IGS is better than methods using SVM alone or linear discriminator.

1 Introduction

Clinical decision support system (CDSS) provides knowledge and specific information for clinicians to enhance diagnostic efficiency. An appropriate CDSS can highly elevate patient safety, improve healthcare quality, and increase cost-effectiveness. In

---

[*] This work was supported by the NSC (NSC96-2912-I-039-001, NSC97-2911-I-039-001, and NSC98-2410-H-039-003-MY2) of Taiwan and DAAD (D/06/00327) of Germany.
[**] The first two authors contributed equally to this work.
[***] To whom corresponding should be addressed. No.666, Buzih Road, Beitun District, Taichung 40601, Taiwan; Tel.: 886-4-22391647 Ext. 7716.

order to achieve these objectives, American Medical Informatics Association recently identified and proposed three key points: best knowledge available when needed, high adoption and effective use, and continuous improvement of knowledge and CDS methods [1]. Several CDSSs have been developed for clinical applications in the past two decades.

In this study, integrated genetic algorithm (GA) and support vector machine (SVM), namely IGS, was used to design CDSSs for the prediction of successful ventilation weaning, diagnosis of patients with severe obstructive sleep apnea, and discrimination of different cell types form Pap smear, respectively. It is different from the traditional method that GA and SVM operate separately with the former first applied for feature selection and then the latter for classification [2]. The chromosome is consisted of features of clinical data and parameters of SVM.

Prediction of Successful Ventilation Weaning. Although modern mechanical ventilators are invaluable tools for stabilizing the condition of patients in respiratory failure, ventilator support should be withdrawn promptly when no longer necessary so as to reduce the likelihood of known nosocomial complications and costs [3,4].

Recently, closed-loop knowledge-based system and automated protocol-driven mechanical ventilator has been developed and used for more rapid extubation than the conventional protocol-driven ventilation [5]. The former is a real-time system which acquires and interprets the patient's clinical data and gradually adjusts the level of pressure to intubated or tracheotomized patients by keeping them at a comfortable state and trying to reduce the pressure to a minimal level. A trial of spontaneous breathing can be performed when the minimal pressure support is achieved. In contrast, the patients are automatically switched from mandatory to spontaneous ventilation mode if it detects two consecutive spontaneous breaths for the latter. If the event of continuously spontaneous breaths of a patient is not detected, it will switch back to the mandatory mode. However, before popularity and cost down of these devices it still needs a clinical decision support system (CDSS) to identify the earliest time when the patients can be weaned from the ventilators.

Diagnosis of Severe Obstructive Sleep Apnea. Obstructive sleep apnea (OSA) is a general sleep disorder and is commonly seen in 24% of men and 9% of women [6]. The severity of the respiratory events is measured by the frequency and duration of apneas and hypopneas per hour of sleep, namely apnea-hypopnea index (AHI), using polysomnography (PSG). Subjects with AHI smaller than 5 are considered as normal while AHI greater than 5 and smaller than 15 as mild. The patients with AHI between 15 and 30 and greater than 30 are diagnosed as moderate and severe, respectively.

Although PSG is treated as the gold standard for the diagnosis of OSA, it has several limitations, such as technical expertise is required and timely access is restricted [7]. Hence, home pulse oximetry was proposed as a valuable and effective tool for screening OSA patients, but its efficiency has been debated for several years [8]. Recently, a comprehensive evaluation of representative oxyhemoglobin indices for predicting severity of OSA was investigated by Lin et al [9] who concluded that oxygen desaturation index (ODI) alone had a better diagnostic performance than the time-domain and frequency-domain indices with a sensitivity/specificity achieving 84.0%/84.3% and 87.8%/96.6% using AHI=15/h and 30/h as thresholds, respectively [9]. Other variables might provide better predicting power than ODI alone.

Discrimination of Cell Types Using Pap Smears. Cytology evaluation is a safe, efficient, and well-established technique for the diagnoses of many diseases. The most famous success in cytology is its ability to reduce the mortality and morbidity of cervical cancer through mass screening. Classical cytologic diagnosis is based on microscopic observation of specialized cells and qualitative assessment by using descriptive criteria, which may be inconsistent because of subjective variability of different observers [10]. To lower the false negative rate in screening, many advanced technologies involving sampling, smear preparation, and screening quality control have recently been developed and introduced. However, most of the devices do not assist objective diagnosis by providing the calculable parameters that would eliminate interpretation errors and inter-observer discrepancy [11]. The technique of computerized image analysis used to assist diagnosis of cell abnormality or tumors in cytopathology or histopathology also can provide accurate and objective evaluation of nuclear morphology. Quantitative methods for estimating a cytological specimen can be traced back some 30 years ago and are still continuing developing [12,13]. Selection of salient morphometric parameters might be useful for classification and prediction of cervical cell types.

2 Materials and Methods

2.1 Subjects and Data Collection

Ventilation Weaning. Data collected from a total of 287 patients, who were recruited from two all-purpose respiratory care centers of a hospital in Taichung area and who had been on mechanical ventilation for longer than 21 days and were clinically stable that their primary physicians considered ready to undergo a weaning trial, were used for study. The first dataset is consisted of 189 data collected during the period from Nov. 2002 to Nov. 2003(D3). The second and the third datasets were collected from 99 patients and 21 patients from the periods within June-May in 2007 (D5) and Feb.-Nov. in 2008 (D8), respectively. Gender, age, APACHE II score at the time of permission, coma scale, and the biochemistry examination variables, including blood urea nitrogen (BUN), creatinine (Cr) albumin (Alb), and hemoglobin (Hb), were recorded and collected. Pulmonary diseases were classified based on the causes inducing respiration failure. These causes include pulmonary, cardiac, and brain vessel (not including trauma) diseases. Other causes related to internal medicine, multiple-organ failure, historical respiratory disease, trauma, ARDS, brain surgery, and other kinds of surgeries were also considered.

Obstructive Sleep Apnea. Retrospective data of 699 suspected OSA patients tested using PSG equipment for overnight attending recording at the Sleep Center of a University Hospital from Jan. 2005 to Dec. 2006 were collected. In which, data of 48 subjects with ages less than 20 or more than 85 years old and data acquired from 85 subjects with sleeping time less than 4 hours were excluded [14]. Hence only data obtained from 566 patients were used for further investigation. Alice 4 PSG recorder was used to monitor and record PSG during sleep with a number of physiologic variables measured and recorded, including (1) EEG for detecting brain electrical activity and sleep stages, (2) EOG and submental EMG for detecting eye and jaw muscle movement, (3) tibia EMG for monitoring leg muscle movement, (4) airflow for detecting breath interruption, (5) inductance plethysmorgraphy for estimating respiratory effort, (6) ECG for measuring heart rate, and (7) arterial oxygen saturation for inspecting blood oxygen. Demographic (age, gender, etc.), anthropometric (weight, height, BMI, waist and neck circumferences, etc.), and symptomatic (diabetes, hypertension, asthma, smoking, alcohol consumption, etc.) data were measured. Questionnaires, including Epworth scaling score (ESS) and the sleeping disorder, were also filled before PSG recording.

Pap-Smear Images. Cytological images were captured using a high-resolution digital camera (Olympus C-5060) mounted on a microscope (Olympus BX 51) and stored as digital format with a resolution of 1024x768 pixels and 32 bits color depth. Forty-two Papanicolaou-stained liquid-based cervical smears (Thin-Prep) were used for this investigation. Among them, 1814 cell images (1556×1076 pixels) were captured with different scales of magnification (100x, 200x, and 400x), in which 477 were classified as superficial, 499 as intermediate, 478 as parabasal, and 360 as abnormal cells. The abnormal cells were further divided into low grade (L) and high grade (H) squamous intraepithelial lesion (SIL). The digitized cellular images were reviewed by 3 certificated cytopathologists and 6 certificated cytotechnologists. By excluding the images with minimal magnification, extensive cellular overlapping, interference by other inflammatory cells or debris, and peer disagreement, only 503 images were selected for further investigation. Among them, 139 images were classified as superficial (S), 178 as intermediate (I), 128 as parabasal (P), and 58 as low-grade and high-grade squamous intraepithelial lesion (SIL) cells. Classification of cell types was based on peer agreement which is the gold standard for evaluating the efficiency of the designed CDSS.

Morphometric parameters including axle center, center of gravity, perimeter, maximum length, and maximum width, were calculated. With the help of internal calibration using the micrometer image (400x), various parameters were obtained for the evaluation of the nuclear size and shape irregularity of a cell including nuclear perimeter, area, maximum length, maximum width, ratio of nucleus and cytoplasm areas (N/C ratio), maximum length from axel center to perimeter (MAP), average length from axel center to perimeter (AAP), maximum length from center of gravity to perimeter (MGP), and average length from center of gravity to perimeter (AGP). Other parameters, including entropy of the co-occurrence matrix (ECM), contrast of co-occurrence matrix (CCM), coarseness and contrast of Tamura features were also applied to analyze the textural features of the nuclei [15].

2.2 Integrated Genetic Algorithm and Support Vector Machine

It is believed that SVM is superior to traditional statistical and neural network classifiers. However, it is critical to determine suitable combination of SVM parameters ($\log_2 C$ and $\log_2 \gamma$) regarding classification performance. Genetic algorithm can find optimal solution within an acceptable time, which is faster than greedy algorithm using exhaustive searching strategy. By taking the advantage of GA in quickly searching the optimal features and parameter, a nonlinear hyperplane with a maximum margin can be obtained by using SVM to classify two clusters. Classification of multiple clusters can be easily expanded. The freeware LIBSVM [16], a library for SVM, was adopted to design the SVM, while the genetic algorithm was modified to combine with LIBSVM to achieve best performance.

The IGS method is different from the traditional method which first applies GA for feature selection and then SVM for classification [2]. The values of SVM parameters, i.e. regularization parameter ($\log_2 C$) and kernel parameter ($\log_2 \gamma$), are critical in optimizing classification performance. Traditionally, regular grid search strategy was used to perform model selection, which is time-consuming with regards to computational complexity. In contrast, the IGS can converge to optimal solution in a reasonable time. As shown in Fig. 1(a), a chromosome is composed of clinical variables and SVM parameters. In Fig. 1(b), a model which integrates GA and SVM used to construct individual CDSSs is presented. The fitness value is defined as the accuracy of SVM classification. After several iterations, the best solution with optimal SVM parameters and selected features can be obtained. Detailed operation and theory of support vector machine can be found in [17].

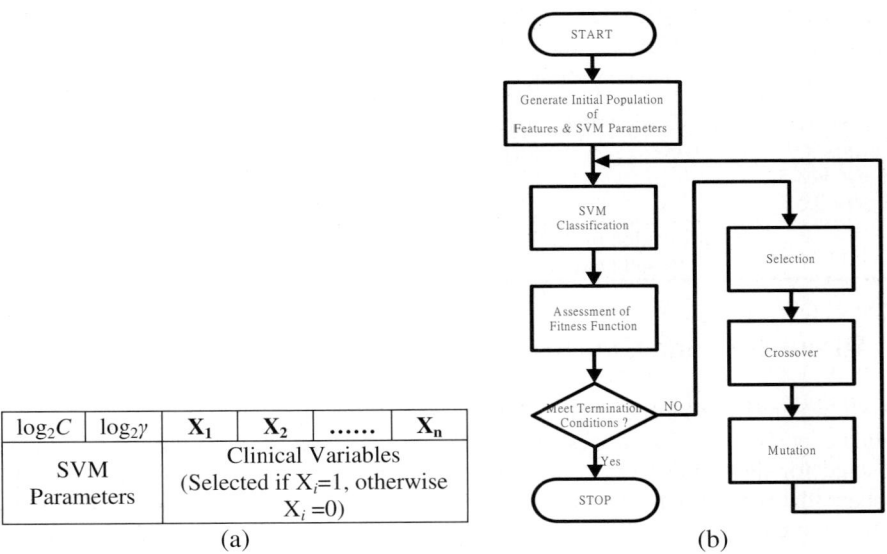

Fig. 1. (a) Chromosome and (b) flowchart of the integrated GA and SVM method

3 Experimental Results

As shown in Table 1, the classification results using IGS for constructing three CDSSs are demonstrated. The accuracies of cross-validation are within 84%-90% for ventilation weaning, 89%-92% for OSA diagnosis, and 96%-99% for cell classification, which are better than CDSSs designed using other methods. Table 2 compares the proposed method with the method using SVM alone or linear discriminator. As depicted in Table 2, the accuracy of ventilation weaning is 7.8% higher than a previous report [18], whereas it is 2.4% higher than a previous investigation [15] for cell classification. Both used statistical methods for feature selection and SVM for classification and validation [15, 18]. Regarding diagnosis of obstructive sleep apnea, the sensitivity for the CDSS designed using IGS is also 4%-7% higher than a previous study [9] using linear discriminator with respect to two different AHI thresholds (AHI=15/h and AHI=30/h) for discriminating severe from non-severe patients.

Table 1. Classification results using IGS for three CDSSs

CDSS	Data		Accuracy	Sensitivity	Specificity
Ventilation Weaning	D3 ($\log_2 C$=1.2, $\log_2\gamma$= 0.6)		83.6%	91.4%	73.8%
	D3+D5 ($\log_2 C$=-4.5, $\log_2\gamma$=2.1)		89.2%	93.7%	83.7%
	D3+D5+D8 ($\log_2 C$=13.3, $\log_2\gamma$=1.2)		89.7%	95.2%	83.2%
Obstructive Sleep Apnea ($\log_2 C$=20.9, $\log_2\gamma$=-1.9)	Variable	OSA Severity			
	ODI	AHI > 15/h	89.1%	86.6%	94.9%
		AHI > 30/h	91.3%	90.4%	92.2%
	ODI+ESS	AHI > 15/h	90.2%	90.9%	88.5%
	ODI+BMI	AHI > 30/h	91.7%	91.9%	91.5%
Pap-Smear Images ($\log_2 C$=18.5, $\log_2\gamma$=-3.5)	Classification of 4 cell types	Cell Type			
		Superficial cell	94.2%		
		Intermediate cell	98.9%		
		Parabasal cell	97.7%		
		Dysplastic cell	94.8%		
		All cells	96.8%		
	Classification of 2 cell types: Normal (S+I+P) and Dysplastic (SIL)		99.6%	100%	99.5%

4 Discussions and Conclusions

As shown in Table 3, the data collected from different periods were used to cross-validate performance of a designed CDSS for ventilation weaning. The variables selected for designing CDSSs are different using different aggregated datasets. We suspect that the dataset size might be the factor causing such variation. Additionally, different methods used for feature selection is another factor causing such a variation. For example, by using dataset D3 to construct a CDSS for ventilation weaning, only 7 variables were selected [18], while 14 variables were chosen in this study. By increasing the sample number to 288 (D3+D5) and 309 (D3+D5+D8), the selected features both decrease to 13 variables with 2 selected variables are different. A comparison of ROC curves for CDSSs constructed using different datasets is illustrated in Fig. 2. The CDSS constructed using more samples has larger area under curve (AUC), Fig. 2, and higher accuracy, sensitivity, and specificity, Table 1.

In conclusion, the integrated GA and SVM used to construct individual CDSSs demonstrates great efficiency for the prediction of successful ventilation weaning, diagnosis of patients with severe obstructive sleep apnea, and discrimination of different cell types from Pap smear images. Future works will focus on clinical testing of the designed CDSSs.

Table 2. Comparisons of different methods for individual CDSSs using the same dataset

CDSS	Method		Accuracy	Sensitivity	Specificity
Ventilation Weaning	SVM [18]	Dataset: D3+D5	81.4%	-	-
	IGS		89.2%	93.7%	83.7%
Obstructive Sleep Apnea	Linear Discriminant [9]	AHI > 15/h	-	84.0%	84.3%
		AHI > 30/h	-	87.8%	96.6%
	IGS	AHI > 15/h	90.2%	90.9%	88.5%
		AHI > 30/h	91.7%	91.9%	91.5%
Pap-Smear Images	SVM [15]	All Cells	94.4%	-	-
	IGS		96.8%	-	-

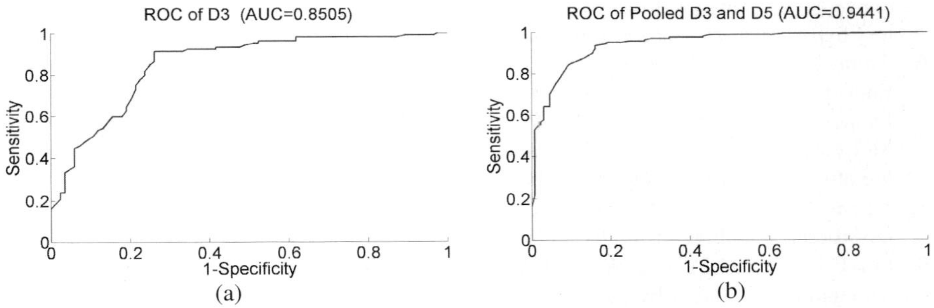

Fig. 2. ROC curves of CDSSs constructed for the prediction of ventilation weaning using (a) D3 (N=189) with AUC= 0.8505 and (b) D3+D5 (N=288) with AUC=0.9441

Table 3. Variables selected from different combinations of datasets for ventilation weaning

Variable	D3[a]	D3[b]	D3+D5[c]	D3+D5+D8[d]
Age	0	1	1	1
APACHE II score when hospitalized	1	1	1	1
Coma scale when hospitalized	1	1	1	1
Other causes related to internal medicine	0	1	1	1
Mutiple organ failure	0	1	1	1
Trauma	0	1	0	1
Brain surgery	0	1	1	1
Creatinine	0	1	1	0
Hemoglobin	0	1	0	0
Tracheotomy	0	1	1	1
Coma csale after weanning	1	1	1	1
RSBI after weanning	1	1	1	1
Length of ICU admission	0	1	0	0
Days using ventilator	1	1	1	1
Ventilator associated pneumonia	1	0	1	1
Urinary tract infection	1	0	1	1

[a]Using data collected before Nov. 2003 by Yang et al. [18]. Using data collected before [b]Nov. 2003, [c]May 2007, and [d] Dec. 2008, respectively, for this study.

References

1. Osheroff, J.A., Teich, J.M., Middleton, B., Steen, E.B., Wright, A., Detmer, D.E.: A roadmap for national action on clinical decision support. JAMIA 14, 141–145 (2007)
2. Osowski, S., Siroic, R., Markiewicz, T., Siwek, K.: Application of support vector machine and genetic algorithm for improved blood cell recognition. IEEE Trans. Instrument & Measurement (2008)
3. MacIntyre, N.R., Cook, D.J., Ely, E.W.J., Epstein, S.K., Fink, J.B., Heffner, J.E.: Evidence-based guidelines for weaning and discontinuing ventilatory support. Chest 120, 375–395 (2001)
4. Meade, M., Guyatt, G., Cook, D.J., Griffith, L., Sinuff, T., Kergl, C.: Predicting success in weaning from mechanical ventilation. Chest 120, 400–424 (2001)
5. Hendrix, H., Kaiser, M.E., Yusen, R.D., Merk, J.: A randomized trial of automated versus conventional protocol-driven weaning from mechanical ventilation following coronary artery bypass surgery. European Journal of Cardio-Thoracic Surgery 29(6), 957–963 (2006)
6. Young, T., Palta, M., Dempsey, J.: The occurrence of SDB among middle-aged adults. N Engl. J. Med. 328, 1230–1235 (1993)
7. Flemons, W., Littner, W.M.R.J., Rowley, A., Gay, P., Anderson, W.M., Hudgel, D.W., McEvoy, R.D., Loube, D.I.: Home diagnosis of sleep apnea: A systematic review of the literature. Chest 124, 1543–1579 (2003)
8. Netzer, N., Eliasson, A.H., Netzer, C., Krisco, D.A.: Overnight Pulse Oximetry for Sleep-Disordered Breathing in Adults-A Review. Chest 120, 625–633 (2001)
9. Lin, C.L., Yeh, C., Yen, C.W., Hsu, W.H., Hang, L.W.: Comparison of the indices of oxyhemoglobin saturation by pulse oximerty in obstructive sleep apnea hypopnea syndrome. Chest 135(1), 86–93 (2009)
10. DeMay, R.M.: Common problems in Papanicolaou smear interpretation. Arch. Pathol. Lab Med. 121(3), 229–238 (1997)
11. Doornewaard, H., van der Schouw, Y.T., van der Graaf, Y., Bos, A.B., van den Tweel, J.G.: Observer variation in cytologic grading for cervical dysplasia of Papanicolaou smears with the PAPNET testing system. Cancer 87(4), 178–183 (1999)
12. Tucker, J.H.: CERVISCAN: An image analysis system for experiments in automatic cervical smear prescreening. Comput. Biomed. Res. 9(2), 93–107 (1976)
13. Nunobiki, O., Sato, M., Taniguchi, E., Tang, W., Nakamura, M., Utsunomiya, H., Nakamura, Y., Mori, I., Kakudo, K.: Color image analysis of cervical neoplasia using RGB computer color specification. Anal. Quant. Cytol. Histol. 24(5), 289–294 (2002)
14. Lam, J.C.M., Lam, B., Lam, C.L., Fong, D., Wang, J.K.L., Tse, H.F., Lam, K.S.L., Ip, M.S.M.: Obstructive sleep apnea and the metabolic syndrome in community-based Chinese adults in Hong Kong. Respiratory Medicine 100, 980–987 (2006)
15. Huang, P.-C., Chan, Y.-K., Chan, P.-C., Chen, Y.-F., Chen, R.-C., Huang, Y.-R.: Quantitative assessment of pap smear cells by PC-based cytopathologic image analysis system and support vector machine. In: Zhang, D. (ed.) ICMB 2008. LNCS, vol. 4901, pp. 192–199. Springer, Heidelberg (2008)
16. Chang, C.C., Lin, C.J.: LIBSVM: a library for support vector machines (2001), http://www.csie.ntu.edu.tw/~cjlin/libsvm
17. Theodoridis, S., Koutroumbas, K.: Pattern Recognition, 2nd edn. Academic Press, San Dieago (2003)
18. Yang, H.Y., Hsu, J.C., Chen, Y.F., Jiang, X.Y., Chen, T.S.: Using Support Vector Machine to Construct a Predictive Model for Clinical Decision-Making of Ventilation Weaning. In: 2008 International Joint Conference on Neural Network, pp. 3980–3985. IEEE Press, New York (2008)

Decision Trees Using the Minimum Entropy-of-Error Principle

J.P. Marques de Sá[1], João Gama[2], Raquel Sebastião[3], and Luís A. Alexandre[4]

[1] INEB-Instituto de Engenharia Biomédica, Porto, Portugal
[2] LIAAD – INESC Porto, L.A. and Faculty of Economics, Porto, Portugal
[3] LIAAD – INESC Porto, L.A. and Faculty of Science, Porto, Portugal
[4] Informatics Dept., Univ. Beira Interior, Networks and Multim. Group, Covilhã, Portugal
jmsa@fe.up.pt, jgama@fep.up.pt, raquel@liadd.up.pt,
lfbaa@di.ubi.pt

Abstract. Binary decision trees based on univariate splits have traditionally employed so-called impurity functions as a means of searching for the best node splits. Such functions use estimates of the class distributions. In the present paper we introduce a new concept to binary tree design: instead of working with the class distributions of the data we work directly with the distribution of the errors originated by the node splits. Concretely, we search for the best splits using a minimum entropy-of-error (MEE) strategy. This strategy has recently been applied in other areas (e.g. regression, clustering, blind source separation, neural network training) with success. We show that MEE trees are capable of producing good results with often simpler trees, have interesting generalization properties and in the many experiments we have performed they could be used without pruning.

Keywords: decision trees, entropy-of-error, node split criteria.

1 Introduction

Decision trees are mathematical devices largely applied to data classification tasks, namely in data mining. The main advantageous features of decision trees are the semantic interpretation that is often possible to assign to decision rules at each tree node (a relevant aspect e.g. in medical applications) and to a certain extent their fast computation (rendering them attractive in data mining applications).

We only consider decision trees for classification tasks (although they may also be used for regression). Formally, in classification tasks one is given a dataset X as an $n \times f$ data (pattern feature) matrix, where n is the number of cases and f is the number of features (predictors) and a target (class) vector T coding in some convenient way the class membership of each case x_i, $\omega_j = \omega(x_i)$, $j = 1,...,c$, where c is the number of classes and ω is the class assignment function of X into $\Omega = \{\omega_j\}$. The tree decision rules also produce class labels, $y(x_i) \in \Omega$.

In automatic design of decision trees one usually attempts to devise a feature-based partition rule of any subset $L \subset X$, associated to a tree node, in order to produce m

subsets $L_i \subset L$ with "minimum disorder" relative to some m-partition of Ω, ideally with cases from a single class only. For that purpose, given a set L with distribution of the partitioned classes $P(\omega_i \mid L)$, $i = 1,\ldots,m$, it is convenient to define a so-called *impurity* (disorder) function, $\phi(L) \equiv \phi(P(\omega_1 \mid L),\ldots,P(\omega_m \mid L))$, with the following properties: a) ϕ achieves its maximum at $(1/m, 1/m,\ldots, 1/m)$; b) ϕ achieves its minimum at $(1,0,\ldots,0), (0,1,\ldots,0),\ldots,(0,0,\ldots,1)$; c) ϕ is symmetric.

We only consider univariate decision rules, $y_j(x_i)$ relative to two-class partitions ($m=2$), also known as Stoller splits (see [3] for a detailed analysis), which may be stated as step functions: $x_{ij} \leq \Delta$, $y_j(x_i) = \omega_k$; $\overline{\omega}_k$, otherwise (x_{ij} is one of the x_i features). The corresponding trees are binary trees. For this setting many impurity functions have been proposed with two of them being highly popularized in praised algorithms: the Gini Index (*GI*) applied in the well-known CART algorithm pioneered by Breiman and co-workers [2], and the Information Gain (*IG*) applied in the equally well-known algorithms ID3 and C4.5 developed by Quinlan [7, 8].

The *GI* function for two-class splits of a set L is defined in terms of

$$\phi(L) \doteq g(L) = 1 - \sum_{j=1}^{2} P^2(\omega_j \mid L) \in [0, 0.5];$$

namely, $GI_y(L) = g(L) - \sum_{i=1}^{2} P(L_i \mid L) g_y(L_i \mid L)$

In other words, *GI* depends on the average of the impurities $g_y(L_i)$ of the descending nodes L_i of L produced by rule y. Since $g(L)$ doesn't depend on y, the CART rule of choosing the feature which *maximizes* $GI_y(L)$ is equivalent to minimizing the average impurity.

The *IG* function is one of many information theoretic measures that can be applied as impurity functions. Concretely, it is defined in terms of the average of the Shannon entropies (informations) of the descending nodes of node set L:

$$IG_y(L) = info(L) - \sum_{i=1}^{2} P(L_i \mid L) info_y(L_i \mid L)$$

with $\phi(L) \doteq info(L) = -\sum_{k=1}^{2} P(\omega_k \mid L) \ln P(\omega_k \mid L) \in [0, \ln(2)]$

Again, maximizing *IG* is the same as minimizing the average Shannon entropy (the average disorder) of the descending nodes. In ID3 and C4.5 log_2 is used instead of ln but this is inessential. Also many other definitions of entropy were proposed as alternatives to the classical Shannon definition; their benefits remain unclear.

A fundamental aspect of these impurity measures is that they all are defined in terms of the probability mass functions of the class assignments $P(\omega_k \mid L)$ and node prevalences $P(L_i \mid L)$. The algorithms use the corresponding empirical estimates.

The present paper introduces a completely different "impurity" measure. One that does not directly depend on the class distribution of a node, $P(\omega_k \mid L)$, and the prevalences $P(L_i \mid L)$, but instead it solely depends on the errors produced by the decision rule:

$$e_i = \omega(x_i) - y(x_i),$$

with convenient numerical coding of $\omega(x_i)$ and $y(x_i)$.

We then apply as "impurity" measure to be minimized at each node the Shannon entropy of the errors e_i. This Minimum Entropy-of-Error (MEE) principle has in recent years been used with success in many different areas (regression, blind source separation, clustering, etc.); it has also been applied with success in neural network training for data classification (see e.g. [10]).

The present paper describes in section 2 how MEE decision trees can be implemented and how they perform in several real-world datasets in section 3. We also present a comparison of MEE and *IG* behaviors in section 4 and discuss the pruning issue in section 5. Finally we draw some conclusions and present future perspectives in section 6.

2 The MEE Approach

In accordance with [9] we consider $x_{ij} \in \Re$ (i.e. we do not consider categorical predictors), and at each node we assign a code $t \in \{-1, 1\}$ to the each candidate class ω_j. We thus have: $t = 1, \omega(x_i) = \omega_j$; $t = -1, \omega(x_i) = \overline{\omega}_j$ (t meaning $t(\omega(x_i))$). Likewise for $y(x_i)$.

The support of the error random variable E, associated to the errors $e_i = t(\omega(x_i)) - t(y(x_i))$ is therefore $\{-2, 0, 2\}$, with: 0 corresponding to a correct decision; 2 to a misclassification when x_i class is the candidate class and the splitting rule produces the complement; and -2 the other way around.

The splitting criterion is based on the Shannon entropy of E:

$$H_y(E \mid L) = -[P_{-1} \ln P_{-1} + P_0 \ln P_0 + P_1 \ln P_1] \in [0, \ln(3)],$$

where $P_{-1} = P_y(E = -2)$, $P_1 = P_y(E = 2)$ and $P_0 = P_y(E = 0) = 1 - P_{-1} - P_1$. Note that contrary to what happened with *GI*, *IG* (and other divulged impurity measures) there is here no room for left and right node impurities and subsequent average. One single function does it all.

Ideally, in the case of a perfect split, the error probability mass function is a Dirac function; i.e., the "errors" are concentrated at zero. Minimizing H_y corresponds to constraining the probability mass function of the errors to be as narrow as possible and usually around zero.

The main algorithmic operations for growing a MEE tree are simple enough and similar to what is done with other impurity measures:

1. At each tree node we are given an $n \times f$ feature matrix X and an $n \times m$ class matrix T (filled with -1, 1).
2. The error probabilities are estimated using:

$$P_2 = p\, n_{1,-1} / n; \quad P_{-2} = (1-p)\, n_{-1,1} / n; \quad P_0 = 1 - P_2 - P_{-2}$$

with $n_{t,t'}$ meaning the number of cases t classified as t' and $p = n_{\omega_j}/n$ the prevalence of the candidate class ω_j.

3. A univariate split y is searched for in the $f \times m$ space minimizing H_y.
4. If a stopping criterion is not satisfied the corresponding left and right node sets are generated and steps 1 through 3 are iterated.

Figure 1 illustrates two entropy-of–error curves relative to the Breast Tissue dataset presented later. In Figure 1a there is a clear class separation: the entropy curve is of the "convex" type and a global minimum corresponding to the interesting split is found. In Figure 1b the curve is of the "concave" type and the global entropy minimum is useless. As a matter of fact a reasonable split point for this last case would be located near the entropy maximum instead of the minimum. When we say "reasonable" (and later on, optimal) we mean from the probability-of-error point of view.

This phenomenon of the optimal working point for a Stoller split being located near the maximum of the entropy-of-error when there is a large class overlap, had already been studied in detail in [9]. This work also derives for a few class distribution settings the "turning point" when an entropy minimum turns into a maximum as the classes glide and overlap into each other.

In our algorithm we stick to the entropy minimum. This means that we do not consider the possibility of a "reasonable" split when there is considerable class overlap reflected by a "concave"-type entropy curve. This has an impact on the pruning issue as discussed in section 5. For that purpose our algorithm classifies every entropy curve as being of the "concave" or "convex"-type using a very crude rule: set a 100-point grid on the whole feature range and divide it into five equally sized intervals; compare the average of the three central intervals, m_c, with the average of the end intervals, m_e; if $m_c > m_e$ than the curve is classified as "concave", otherwise is classified as "convex". We tried other modifications of this basic scheme but didn't find any clear improvement.

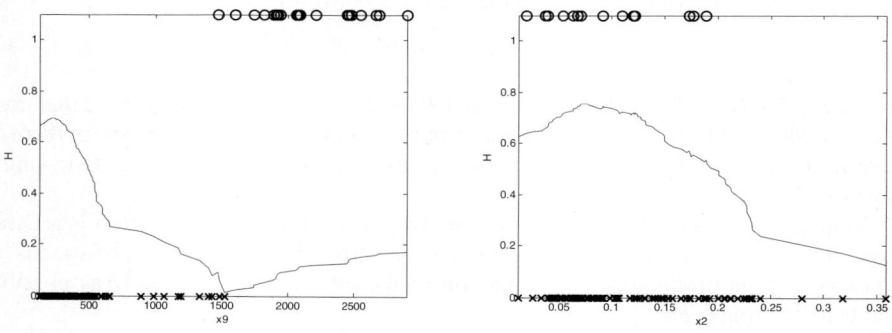

Fig. 1. Entropy-of-error curves for two splits of the Breast-Tissue dataset (splitting the balls from the crosses): a) feature x_9 with class 6; b) feature x_2 with class 2

When there is no valid split for any descendent node of a node L (all entropy curves are concave or the number of cases for any candidate class is very small), the node is considered a leaf.

In a large number of experiments performed with the MEE algorithm we found that one often found better splits (with lower H) when attempting to partition merged classes from the remaining ones. Such "multiclass" splits could even provide good solutions in cases where it was difficult or even impossible to obtain "convex" entropy curves. Figure 2 illustrates an example of a tree with multiclass splits. When evaluating the tree, cases falling into multiclass nodes are assigned to the class with the larger number of cases. The multiclass feature, considering combinations of classes up to $c/2$, is included in the MEE algorithm.

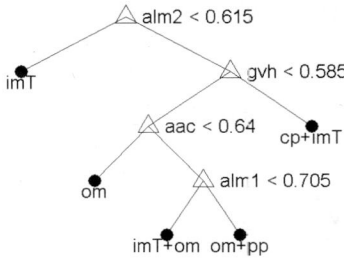

Fig. 2. Tree structure for the Ecoli4 dataset (see below) showing 2-class combinations

3 Application to Real-World Datasets

The MEE algorithm was applied to the datasets presented in Table 1 and its results confronted with those obtained with the CART algorithm implemented by Statistica (StatSoft, Inc.) and the C4.5 implemented by Weka (open source software).

Table 1. Datasets (main characteristics)

	Breast (a)	Breast4 (a)	Olive (b)	Ecoli (c)	Ecoli4 (c)	ImSeg (c)	Glass (c)
No. cases	106	106	572	327	327	2310	214
No. features	9	9	8	5	5	18	9
No. classes	6	4	9	5	4	7	6

(a) "Breast Tissue" dataset described in [6]. Breast4 is Breast reduced to 4 classes: merging {fad,mas,gla}.
(b) "Olive Oils" dataset described in [4].
(c) "E-coli", "Image Segmentation" and "Glass" datasets described in [1]. We removed classes omL, imL and imS from E-coli because they have a low number of cases (resp., 5, 2, 2). Ecoli4 is Ecoli reduced to 4 classes: merging {im, imU}.

All algorithms used unit misclassification costs (i.e., tree costs are misclassification rates). CART and C4.5 used, as is common practice, the so-called midpoint splits: candidate split points lie midway of feature points. In our algorithm we kept the original feature values as split candidate points.

CART was applied with the Gini criterion an cost-complexity pruning [2]. Weka C4.5 applied a postpruning scheme. The MEE algorithm was applied without pruning (justification below).

We applied cross-validation procedures to all datasets, namely leave-one-out with C4.5 and MEE and 25-fold cross-validation to CART (the leave-one-out method wasn't available for CART). Confusion matrices and estimates of the probability of error were computed as well as statistics regarding the tree size (number of nodes).

Table 2 shows the mean error rate and standard deviation (between brackets) for the cross-validation experiments. For the Breast and Ecoli datasets the errors for some classes were always quite high (also found with other classification methods). This led us to merge the poorly classified classes setting up the Breast4 and Ecoli4 datasets (see Table 1).

Table 2. Comparative table of results with mean (std) in cross-validation experiments

	Breast	Breast4	Olive	Ecoli	Ecoli4	Imseg	Glass
CART	<u>0.3679</u> (0.047)	<u>0.1698</u> (0.036)	**0.0962** (0.012)	<u>0.2049</u> (0.022)	**0.1040** (0.017)	0.0675 (0.005)	<u>0.3738</u> (0.033)
C4.5	**0.3396** (0.046)	0.1226 (0.032)	0.0979 (0.012)	**0.1743** (0.021)	<u>0.1498</u> (0.020)	**0.0290** (0.003)	0.3224 (0.032)
MEE	<u>0.3679</u> (0.047)	**0.0943** (0.028)	<u>0.1031</u> (0.013)	0.2110 (0.023)	0.1070 (0.017)	<u>0.1182</u> (0.012)	**0.2664** (0.030)

The three methods were compared using multiple comparison tests based either on the Oneway Anova or the Kruskal-Wallis test according to the p-value of a variance homogeneity test ($p < 0.05$ selects Kruskal-Wallis, otherwise selects Oneway Anova). Multiple comparison was performed at 5% significance level. In Table 2 the significantly best results are printed bold and the significantly worst results are underlined.

4 MEE versus Information Gain

In order to compare both entropy-based criteria, MEE and IG, we generated two-class datasets with an equal number of points, n, represented by 2 features (x_1, x_2) with

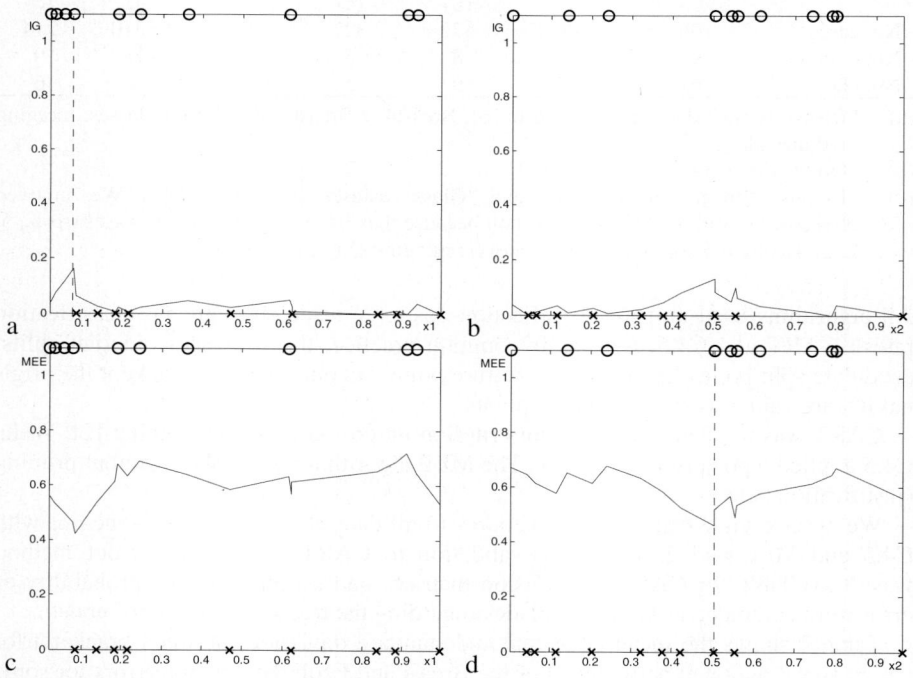

Fig. 3. Comparing IG (top figures) and MEE (bottom figures) in the separation of balls from crosses. IG prefers feature x_1 with IG_{max}=0.1639, whereas for x_2 IG_{max}=0.1325. MEE prefers feature x_2 with MEE=0.4609, whereas no valid minimum is found for x_1.

randomly and uniformly distributed values in [0,1[. One of the features was then selected according to MEE and to IG decision criteria.

For $n = 10$ and several batches of 1000 repetitions of the experiment we found that on average only 1% of the experiments where MEE found a solution that was different from the IG solution. Moreover, we found that all differences between MEE and IG were of the type illustrated in Figure 3. The error probability mass functions for Figure 3a (IG selects x_1) and Figure 3d (MEE selects x_2) are shown in Figure 4. From these figures one concludes that whereas MEE preferred a more "balanced" solution, resembling a Dirac function at zero, IG emphasized the good classification of only one of the two classes, even at the cost of increased errors of the other class.

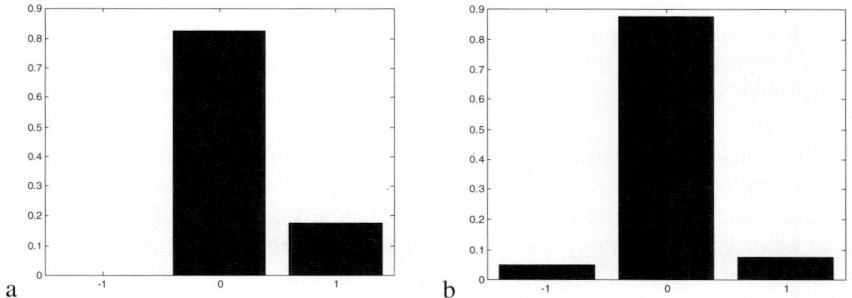

Fig. 4. Probability mass functions of the errors corresponding to: a) Figure 3a (IG selects x_1); b) Figure 3d (MEE selects x_2)

5 The Pruning Issue

Tree pruning is a means of obtaining simpler trees, i.e., simpler models, therefore with better generalization capabilities. CART, C4.5 and other tree design methods employ pruning techniques whenever some evidence of overfitting is found. The MEE method has an important characteristic: it doesn't attempt to find a split whenever the class distributions show a considerable degree of overlap. The quantification on theoretical grounds of what "considerable" means isn't easy. Taking into account the results in [9] one may guess that whenever the distance of the class means is below one pooled standard deviation the entropy-of-error curve will be "concave" and no valid split under the MEE philosophy is found. We believe that this characteristic is one of the reasons why the MEE algorithm always produced smaller trees, on average, than those produced by C4.5 (no tree size statistics were available for CART).

In our experiments MEE trees also showed a tendency to generalize better than those produced by other methods, as measured by the difference between resubstitution estimates of the error rate and the cross-validation estimates with significantly lower $R = |m_R - m_{CV}|/\bar{s}$, where m_R is the mean resubstitution error and m_{CV} the mean cross-validation error.

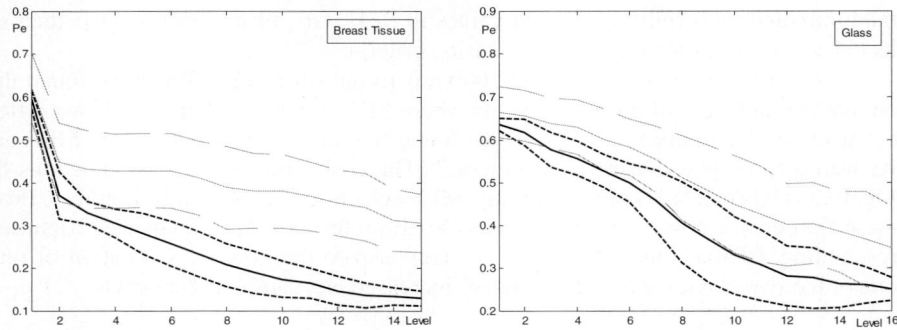

Fig. 5. Mean (solid) and mean±std (dashed) of the training set error (black) and test set error (grey) in 50 experiments on trees designed with 80% of the cases (randomly drawn) and tested in the remaining cases

We have also performed a large number of experiments with the MEE algorithm designing the tree with 80% of randomly chosen cases and testing in the remaining 20% cases, and plotted the mean and mean±standard deviation of the training and test set error estimates along the tree level for 50 repetitions of each tree design. The results of Figure 5 clearly indicate the absence of overfitting. The same conclusion could be drawn in all experiments (over 20 for each dataset) we have carried out.

6 Conclusions

The basic rationale of the MEE approach is that it searches for splits concentrating the error distribution at zero. For the classic approaches what the split is doing in terms of the error distribution is unclear.

From the large number of experiments we carried out we conclude that possible benefits of the MEE trees are the no need of applying a pruning operation and the obtaining of more interesting splits corresponding to errors distributed in a more balanced way as exemplified in section 4. This last aspect could be of interest for some datasets. The results obtained with MEE trees applied to real-world datasets, described in section 3, look quite encouraging especially taking into account that they were obtained with the first version of the algorithm and that there is still much space for improvements.

Besides of introducing obvious improvements in the algorithm (e.g. using midpoint splits) we also intend to study in more detail the following issues: the turning point from "convex" to "concave" behavior of the empirical error distribution; the stopping conditions of the algorithm; Generalization issues such as the evolution of training and test errors with the number of cases. We also intend to study in a comparative way the performance of MEE trees in a larger number of real-world datasets.

References

1. Asuncion, A., Newman, D.J.: UCI Machine Learning Repository. Univ. of California, SICS, Irvine, CA (2007),
   `http://www.ics.uci.edu/~mlearn/MLRepository.html`
2. Breiman, L., Friedman, J.H., Olshen, R.A., Stone, C.J.: Classification and Regression Trees. Chapman & Hall/CRC, Boca Raton (1993)
3. Devroye, L., Giörfi, L., Lugosi, G.: A Probabilistic Theory of Pattern Recognition. Springer, Heidelberg (1996)
4. Forina, M., Armanino, C.: Eigenvector Projection and Simplified Nonlinear Mapping of Fatty Acid Content of Italian Olive Oils. Ann. Chim. 72, 127–155 (1981)
5. Loh, W.-Y., Shih, Y.-S.: Split Selection Methods for Classification Trees. Statistica Sinica 7, 815–840 (1997)
6. Marques de Sá, J.P.: Applied Statistics Using SPSS, STATISTICA, MATLAB and R, 2nd edn. Springer, Heidelberg (2007)
7. Quinlan, J.R.: Induction of Decision Trees. Machine Learning 1, 81–106 (1986)
8. Quinlan, J.R.: C4.5 Programs for Machine Learning. Morgan Kaufmann, San Francisco (1993)
9. Silva, L.M., Felgueiras, C.S., Alexandre, L.A., Marques de Sá, J.: Error Entropy in Classification Problems: A Univariate Data Analysis. Neural Comp. 18, 2036–2061 (2006)
10. Silva, L.M., Embrechts, M.J., Santos, J.M., de Sá, J.M.: The influence of the risk functional in data classification with mLPs. In: Kůrková, V., Neruda, R., Koutník, J. (eds.) ICANN 2008, Part I. LNCS, vol. 5163, pp. 185–194. Springer, Heidelberg (2008)

k/K-Nearest Neighborhood Criterion for Improvement of Locally Linear Embedding

Armin Eftekhari[1], Hamid Abrishami-Moghaddam[1], and Massoud Babaie-Zadeh[2]

[1] K.N. Toosi University of Technology, Tehran, Iran
[2] Sharif University of Technology, Tehran, Iran
a.eftekhari@ee.kntu.ac.ir, moghaddam@eetd.kntu.ac.ir,
mbzadeh@ ee.sharif.edu

Abstract. Spectral manifold learning techniques have recently found extensive applications in machine vision. The common strategy of spectral algorithms for manifold learning is exploiting the local relationships in a symmetric adjacency graph, which is typically constructed using k-nearest neighborhood (k-NN) criterion. In this paper, with our focus on locally linear embedding as a powerful and well-known spectral technique, shortcomings of k-NN for construction of the adjacency graph are first illustrated, and then a new criterion, namely k/K-nearest neighborhood (k/K-NN) is introduced to overcome these drawbacks. The proposed criterion involves finding the sparsest representation of each sample in the dataset, and is realized by modifying Robust-SL0, a recently proposed algorithm for sparse approximate representation. k/K-NN criterion gives rise to a modified spectral manifold learning technique, namely Sparse-LLE, which demonstrates remarkable improvement over conventional LLE through our experiments.

Keywords: Local linear embedding, sparse representation, Robust-SL0.

1 Introduction

In the recent years, several algorithms have been developed to perform dimensionality reduction of low-dimensional nonlinear manifolds embedded in a high-dimensional space. In particular, due to technical advantages, local linear embedding (LLE) has found widespread applications in real-world problems [1, 2]. LLE is based on eigen decomposition of a special Gram matrix, which is designed to preserve the local structure of data. This local structure is typically defined using nearest neighborhood criterion in the Euclidean space by constructing a symmetric adjacency graph, in which the nodes represent the training samples and any pair of nodes are connected iff the corresponding data points are adjacent. Indeed, successful recovery of the low-dimensional structure of data highly depends on the construction of an accurate adjacency graph that gives a faithful representation of the local geometry of data [3]. In this regard, though widely used, k-NN criterion suffers from major drawbacks. In fact, since each sample is connected to its k direct nearest neighbors, k-NN rule is generally unable to exclude noisy samples or outliers in the neighborhood. In addition, k-NN criterion considers a fixed neighborhood size about each sample on the manifold. In this paper, with our focus on LLE, k-NN criterion is first represented as an optimization problem, which is then modified to yield k/K-nearest neighborhood (k/K-NN) criterion. As was the case in k-NN, new criterion searches for a small subset of samples in the neighborhood of each data point. However, unlike k-NN, this subset is not limited to k-nearest neighbors of each sample, but instead belongs to a larger neighborhood within the roughly linear patch on the manifold centered at that sample. Furthermore, size of this subset is chosen adaptively to include the minimum required samples among K ($> k$) nearest neighbors of each data point, which is often believed to give a more reliable representation of the manifold [4]. The proposed criterion involves finding the sparsest approximate representation of a sample in the dataset, and is realized by modifying the recently proposed Robust-SL0 algorithm for sparse approximate representation [5]. The modified spectral method, namely Sparse-LLE is then experimentally validated on several datasets, demonstrating remarkable improvement over the conventional LLE. The rest of this paper is organized as follows. Section 2 is devoted to a review of the LLE. In Section 3, shortcomings of k-NN are studied and k/K-NN criterion is introduced and justified. Implementation details are then discussed in Section 4 and, finally, experimental results are presented in Section 5.

2 Locally Linear Embedding

LLE, the local properties of the manifold are expressed by writing each sample as a linear combination of its nearest neighbors. LLE then attempts to preserve these local relationships in the low-dimensional space [6]. To be more specific, LLE first constructs the adjacency graph $\mathcal{G}(V, E)$, whose nodes V and edges E represent the data samples and neighborhood relations among samples, respectively. Denoting each sample by $x_i \in \mathbb{R}^N$, we will use $x_i \sim x_j$ to indicate that samples x_i and x_j are adjacent by some criterion, i.e. $x_i x_j \in E$. Similarly, $x_i \not\sim x_j$ will indicate $x_i x_j \notin E$. Furthermore, for each sample x_i, the subset of samples x_j satisfying $x_i \sim x_j$ will be denoted by $\{x_{i_j}\}$. In particular, for k-NN criterion, we have $\{x_{i_j}\} = \mathcal{N}_k(x_i)$, where $\mathcal{N}_k(x_i)$ denotes the subset of k-nearest neighbors of x_i. Additionally, $\mathcal{N}_k(i)$ would denote the corresponding subset of indices of $\mathcal{N}_k(x_i)$.

Once the adjacency graph is constructed, each sample x_i is written as a linear combination of its k nearest neighbors. This is achieved by solving:

$$\mathcal{L}: \min_{w_i} \|x_i - Xw_i\|_2^2 \text{ s.t. } \text{Supp}(w_i) \subset \mathcal{N}_k(i), w_i^T \mathbf{1} = 1 \qquad (1)$$

where $w_i \in \mathbb{R}^N$ contains the reconstruction weights and $\text{Supp}(\eta)$ denotes the support of η, i.e. subset of all indices j, for which η_j is nonzero. In addition, $\mathbf{1} = [1, \ldots, 1]^T \in \mathbb{R}^N$. The weight matrix $W = [w_1, \ldots, w_N] \in \mathbb{R}^{N \times N}$ is then constructed and the embeddings are found by computing the eigenvectors associated with the bottom nonzero eigenvalues of $M = (I - W)(I - W^T)$, where I is the identity matrix. To be more specific, denoting the resulting modal matrix by $V = [v_1, \ldots, v_N] \in \mathbb{R}^{N \times N}$, rows of V contain the embeddings $\{y_i\}_{i=1}^N$. In fact, up to a scaling factor that depends on the algorithm, the embedding of x_i, namely y_i, is a vector with $y_{i,j} = v_{j,i}, j \leq m$.

3 k/K-Nearest Neighborhood Criterion

k-NN criterion implies that $x_i \sim x_j$ iff $x_j \in \mathcal{N}_k(x_i)$, and is justified based on the notion that local geometry of the manifold at x_i is best represented by $\mathcal{N}_k(x_i)$ rather than by any other subset $V \subset \{x_i\}_{i=1}^N$ with $\#V = k$. In this section, with our focus on LLE, shortcomings of this notion are discussed. k/K-NN criterion is then introduced, which, to some extent, overcomes the shortcomings of k-NN.

As shown in the [11], (1) is asymptotically equivalent to:

$$\mathcal{J}: \min_{w_i} \lim_{c \to \infty} \|x_i - Xw_i\|_2 + c \sum_{j \neq i} u(w_{i,j}) \|x_i - x_j\|_2 \tag{2}$$

$$\text{s.t. } \|w_i\|_0 = k, w_i^T \mathbf{1} = 1, w_{i,i} = 0$$

where $u(\cdot)$ is the step function and $\|\eta\|_0$ is the ℓ^0-norm of the vector η, i.e. number of nonzero elements of η. It is observed that solving $\mathcal{J}$ primarily minimizes the second term of the functional by choosing $\{x_{i_j}\} = \mathcal{N}_k(x_i)$. Then, keeping $\{x_{i_j}\}$ fixed, $\mathcal{J}$ minimizes the reconstruction error $\|x_i - Xw_i\|_2$ by solving the linear system $x_i^\perp = Xw_i$ subject to $\text{Supp}(w_i) = \mathcal{N}_k(i)$, where $x_i^\perp$ is the projection of x_i onto $\text{Span}(\{x_{i_j}\})$. It is observed that, despite its importance, minimizing the reconstruction error does not contribute to the choice of $\{x_{i_j}\}$ in $\mathcal{J}$. Furthermore, $\#\{x_j\}$ is fixed to k in $\mathcal{J}$, while it is generally better to let the algorithm automatically decide on $\#\{x_j\}$ by selecting only necessary samples for representation of x_i [4]. To overcome these drawbacks, the following optimization problem is introduced:

$$\min_{w_i} \|x_i - Xw_i\|_2 + c_1 \sum_{j \neq i} u(w_{i,j}) \|x_i - x_j\|_2 + c_2 \|w_i\|_0 \tag{3}$$

$$\text{s.t. } w_i^T \mathbf{1} = 1, w_{i,i} = 0$$

where c_1, c_2 are finite positive scalars. By choosing $c_1 < \infty$, we ensure that, in contrast to $\mathcal{J}$, minimizing the reconstruction error $\|x_i - Xw_i\|_2$ contributes to our choice of $\{x_{i_j}\}$. Moreover, (3) uses the minimum required number of samples to best represent x_i, and hence adaptively selects $\#\{x_{i_j}\}$ on the manifold. Note that for every pair c_1 and c_2, there always exist a pair ϵ_1 and ϵ_2, for which (3) is equivalent to:

$$\min_{w_i} \|w_i\|_0 \tag{4}$$

$$\text{s.t. } \|x_i - Xw_i\|_2 \leq \epsilon_1, \sum_{j \neq i} u(w_{i,j}) \|x_i - x_j\|_2 \leq \epsilon_2, w_i^T \mathbf{1} = 0, w_{i,i} = 0$$

Furthermore, we notice that $\sum_{j \neq i} u(w_{i,j}) \|x_i - x_j\|_2 \leq \epsilon_2$ sets an upper limit on $\|x_i - x_j\|_2$ for $x_j \in \{x_{i_j}\}$ and hence there exists some $\epsilon_3 > 0$, for which the second constraint in (4) can be safely replaced by $\|x_i - x_j\|_2 \leq \epsilon_3, \forall x_j \in \{x_{i_j}\}$ [7]. A closer look reveals that this in turn could be safely replaced by $\text{Supp}(w_i) \subset \mathcal{N}_K(i)$, for some integer K. Therefore, we can rewrite (4) as follows:

$$\mathcal{JJ}: \min_{w_i} \|w_i\|_0 \tag{5}$$
$$\text{s.t. } \|x_i - Xw_i\|_2 \leq \epsilon_1, \text{Supp}(w_i) \subset \mathcal{N}_K(i), w_i^T \mathbf{1} = 1$$

Let w_i^* and $\{x_{i_j}^*\}$ denote the solution of $\mathcal{JJ}$ and the subset of samples corresponding to the nonzero elements of w_i^*, respectively. Note that, as a result of the second constraint in $\mathcal{JJ}$, $\{x_{i_j}^*\} \subset \mathcal{N}_K(x_i)$. In order to preserve the computational advantages of working with highly sparse matrices, we further limit #$\{x_{i_j}^*\}$ to k, for some integer $k < K$. This is achieved by keeping at most k top nonzero elements of w_i^* and setting others (if any) to zero. $\{x_{i_j}^*\}$ is also modified by discarding the corresponding samples. The new criterion will be referred to as k/K-NN rule and is summarized in Fig. 1. Notice that, in k-NN, $\{x_{i_j}^*\}$ is the subset of first k nearest neighbors of x_i, whereas in $\mathcal{JJ}$, $\{x_{i_j}^*\}$ is the best subset $V \subset \mathcal{N}_K(x_i)$ with #$V \leq k$, that contains the minimum required samples to achieve a reconstruction error less than the error tolerance ϵ_1. When compared to k-NN, k/K-NN criterion is able to exclude noisy neighbors and outliers, which is achieved by the constraint on the reconstruction error in $\mathcal{JJ}$. On the other hand, when compared to K-NN, k/K-NN criterion adaptively selects #$\{x_{i_j}^*\}$ ($\leq k$) to best represent x_i with the minimum required number of samples. Now, using k/K-NN criterion to construct the adjacency graph, LLE is modified to obtain an improved spectral algorithm, dubbed Sparse-LLE. Note that the only difference between LLE and Sparse-LLE lies in the construction of the adjacency graph.

Given integers k and K, with $k < K$, solve $\mathcal{JJ}$ for each sample $x_i \in \{x_i\}_{i=1}^N$, and denote the answer by w_i^*. Then, nodes i and j in the adjacency graph $\mathcal{G}$ are connected iff $w_{i,j}^*$ is among the top k nonzero elements of w_i^*.

Fig. 1. k/K-NN criterion for construction of the adjacency graph

4 Implementation

In Section 3, k/K-NN criterion for construction of the adjacency graph was introduced and justified. In order to apply this criterion, we shall study the following optimization problem: $\mathcal{P}_{0,\epsilon,S}: \min \|s\|_0$ s.t. $\|b - As\|_2 \leq \epsilon$ and $\text{Supp}(s) \in S$, where $b \in \mathbb{R}^n$ and S is a given subset of indices of $s = [s_1, \ldots, s_m]^T \in \mathbb{R}^m$. Our implementation assumes $m > n$, which fairly happens almost always in real-world situations. As the starting point, we first consider the well-known sparse approximate representation problem $\mathcal{P}_{0,\epsilon}: \min \|s\|_0$ s.t. $\|b - As\|_2 \leq \epsilon$. Among available approaches, we opt for the recently proposed Robust-SL0 as a fast and accurate algorithm [5]. Briefly speaking, Robust-SL0 solves a sequence of problems of the form $\mathcal{Q}_{\epsilon,\sigma}: \max_s \sum_{i=1}^m e^{-s_i^2/2\sigma^2}$ s.t. $\|b - As\|_2 \leq \epsilon$, decreasing σ at each step, and initializing the next step at the maximizer of the previous (larger) value of σ. Each

$\mathcal{Q}_{\epsilon,\sigma}$ is solved approximately by few iterations of gradient ascent. Convergence analysis of Robust-SL0 has been thoroughly considered in [5] and it was shown that, under some mild conditions, the sequence of maximizers of $\mathcal{Q}_{\epsilon,\sigma}$ indeed converges to the unique minimizer of $\mathcal{P}_{0,\epsilon}$, whenever such answer exists. Moreover, Robust-SL0 runs significantly faster than the competing algorithms, while producing answers with the same or better accuracy [5]. The idea is now to modify $\mathcal{P}_{0,\epsilon,S}$ in a way that enables using Robust-SL0 algorithm to solve $\mathcal{P}_{0,\epsilon,S}$. This necessitates proper modification of the second constraint in $\mathcal{P}_{0,\epsilon,S}$, i.e. Supp(s) $\in S$. While this may be achieved by, for instance, setting $s_i = 0$ for $i \notin S$ at each iteration, we prefer to preserve the studied convergence properties of Robust-SL0 by replacing Supp(s) $\in S$ with a term in functional that smoothly favors small values for s_i when $i \notin S$. Therefore, $\mathcal{P}_{0,\epsilon,S}$ is modified to:

$$\lim_{\sigma \to 0} \max_s \sum_{i \in S}(1-\alpha)e^{-s_i^2/2\sigma^2} + \sum_{i \notin S}(1+\beta)e^{-s_i^2/2\sigma^2} \qquad (6)$$

$$\text{s.t. } \|b - As\|_2 \le \epsilon$$

where we take $0 \le \alpha, \beta < 1$. Convergence properties of (6) are obtained by minimal modifications in the proof presented in [5]. Note that Robust-SL0 algorithm is now applicable to (6) by merely using the gradient of the functional of (6) in the algorithm. The interested reader is referred to [5] for details.

5 Experiments

The objective of this section is to experimentally assess the merits of the proposed k/K-NN criterion for construction of the adjacency graph. To this end, the performance of LLE and Sparse-LLE are compared on several datasets. In each experiment, k (and if available K) are experimentally tuned for the best results. Other parameters of Sparse-LLE are fixed to: $\epsilon = 0.05, \alpha = \beta = 0.9$. As our first experiment, we compare the performance of LLE and Sparse-LLE for visualizing the Frey face dataset, which consists of 1965 gray-level images of a single individual acquired under different expression and pose conditions [2]. Few images in this dataset are depicted in Fig. 2(a). Fig. 2(b) depicts the first two components of these images discovered by LLE. Depicted in Fig. 2(c) are the visualization results obtained by Sparse-LLE, which may be interpreted as follows. We can recognize four pair of opposite branches in the embedded space, labeled from 1 to 4. It is observed that the main trend in branches 1 and 2 includes left pose or slightly left pose images, whereas images in branches 3 and 4 are mainly either right pose or slightly right pose. In particular, while containing opposite poses, both branches 1 and 3 are similar in that one of their ends includes happy faces and the other end includes either sad faces or faces with visible tongue. The main trend of images in each branch is represented in Fig. 3. It is observed that the main trend in branches 1 and 2 includes left pose or slightly left pose images, whereas images in branches 3 and 4 are mainly either right pose or slightly right pose. In particular, while containing opposite poses, both branches 1 and 3 are similar in that one of their ends includes happy faces and the other end includes either sad faces or faces with visible tongue.

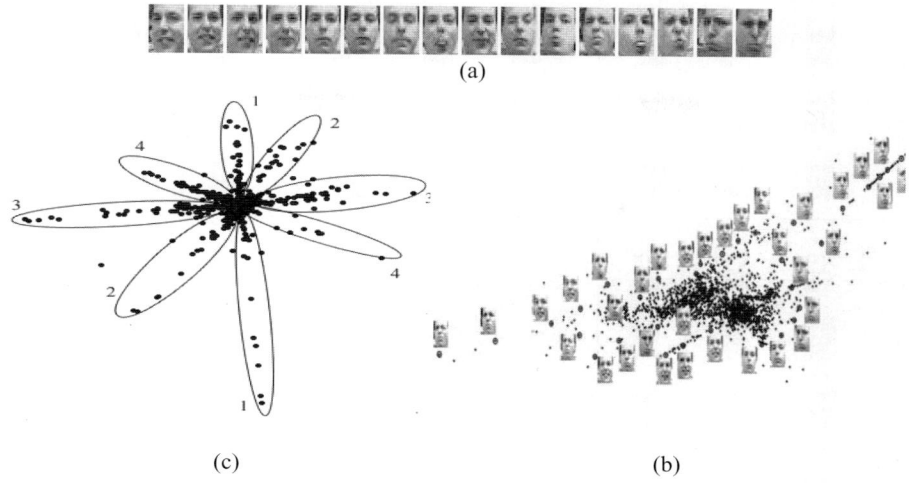

Fig. 2. A few samples of Frey dataset used in the first experiment (a). Images of faces mapped into the embedding space described by the first two coordinates of LLE with $k = 12$ (b), and sparse-LLE with $k = 5$ and $K = 12$ (c).

As our second experiment, the performance of LLE and Sparse-LLE is compared in face recognition task on the extended Yale face database. The dataset includes 2432 cropped frontal images of 38 individuals under expression and illumination variation [8], where the first 16 images of each individual are considered in this experiment. After vectoriation, using LLE and Sparse- LLE, dimension of image data is reduced to 10. Subsequently, motivated by the well-designed experimental setup in [6], quality of the resulting low-dimensional representations is evaluated by measuring the classification errors of 1 nearest neighbor classifiers trained on the low-dimensional representations using leave-one-out cross-validation. In other words, class of each sample is predicted by its nearest neighbor in the embedded space and the overall classification error is reported in Table 1.

Retinal biometrics refers to identity verification of individuals based on their retinal vessel tree pattern. Our third experiment is conducted on VARIA database containing 153 (multiple) retinal images of 59 individuals [9]. To compensate for the variations in the location of optic disc (OD) in retinal images, a ring-shaped region of interest (ROI) in the vicinity of OD is used to construct the feature matrix. To extract the ROI, using the technique presented in [10], OD and vessel tree are extracted. Then, a ring-shaped mask with proper radii centered at OD is used to form the feature vectors $X \in \mathbb{R}^{6 \times 8}$ by collecting the pixels along 8 beams of length 6 originating from OD. A special case is depicted in Fig. 4. After vectorization of feature matrices, dimension is reduced to 10 using LLE and Sparse-LLE. The performance of the resulting low- dimensional representations is then evaluated similar to the second experiment (Table 1).

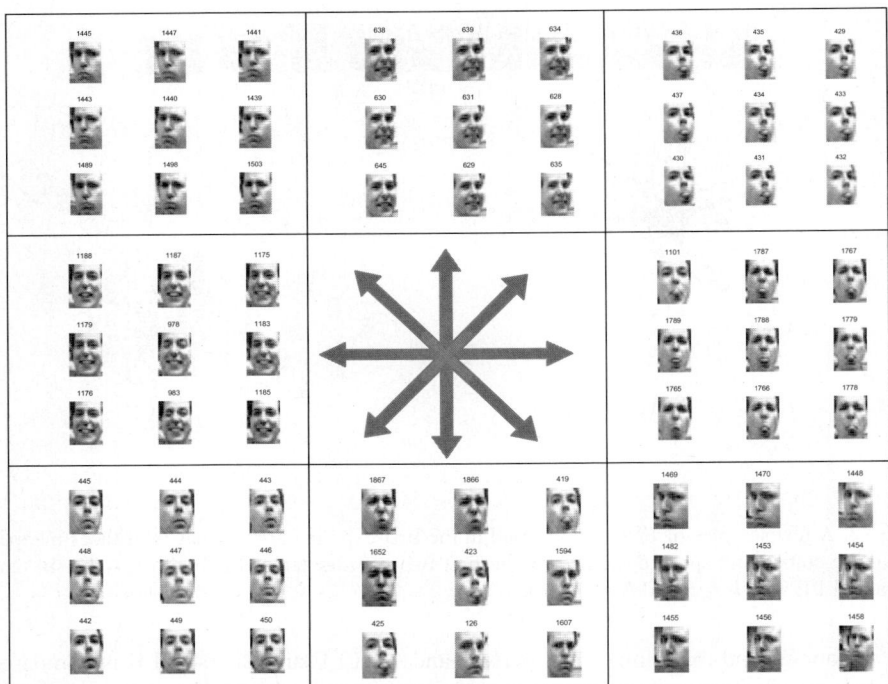

Fig. 3. Further study of the embedded space obtained by sparse-LLE in Fig. 2(c). Outer boxes are positioned similar to the distribution of branches in Fig. 2(c), where label of corresponding branches are indicated by the arrows. Each outer box contains few samples of the corresponding branch, which are selected to represent the main trend of the images inside the branch.

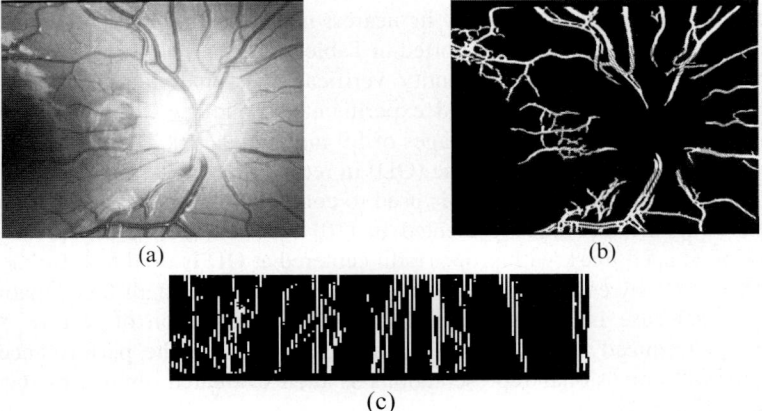

Fig. 4. (a) Retinal image; bright area is OD. (b) Vessel tree (in white) and mask (in blue). (c) Feature matrix obtained from 300 beams of length 100 pixels (images (a) and (b) are cropped).

Table 1. Generalization errors of 1-NN classifiers for different dimension reduction algorithms

Algorithm	Yale face database		VARIA database	
	Parameters	Generalization error of 1-NN	Parameters	Generalization error of 1-NN
PCA	-	35.5263	-	59.4771
LLE	$k=12$	29.9342	$k=4$	61.4379
Sparse-LLE	$k=5$ $K=12$	23.5197	$k=5$ $K=7$	56.2092

6 Conclusions

LLE is a well-known and powerful spectral dimension reduction algorithm. For successful recovery of the low-dimensional structure of data, however, LLE requires an adjacency graph, which is typically constructed using k-NN criterion. In this paper, deficiencies of k-NN for construction of the adjacency graph were first studied and k/K-NN criterion was then introduced to overcome the drawbacks. Implementation of k/K-NN involved a variant of Robust-SL0 algorithm for sparse approximate representation. The modified spectral method, namely Sparse-LLE, is experimentally validated on several datasets, demonstrating remarkable improvement over the conventional LLE.

References

1. Saul, L.K., Weinberger, K.Q., Ham, J.H., Sha, F., Lee, D.D.: Spectral methods for dimensionality reduction. In: Chapelle, O., Schölkopf, B., Zien, A. (eds.) Semisupervised Learning. MIT Press, Cambridge (2006)
2. Roweis, S., Saul, L.: Nonlinear dimensionality reduction by locally linear embedding. Science 290, 2323–2326 (2000)
3. Lebanon, G.: Riemannian geometry and statistical machine learning. Doctoral Thesis, School of Computer Science, Carnegie Mellon University (2005)
4. Lin, T., Zha, H., Lee, S.U.: Riemannian manifold learning for nonlinear dimensionality reduction. In: Leonardis, A., Bischof, H., Pinz, A. (eds.) ECCV 2006. LNCS, vol. 3951, pp. 44–55. Springer, Heidelberg (2006)
5. Eftekhari, A., Babaie-Zadeh, M., Jutten, C., Abrishami Moghaddam, H.: Robust-SL0 for stable sparse representation in noisy settings. In: Int. Conf. Acoustics, Speech, and Signal Proc. (accepted, 2009)
6. van der Maaten, L.J.P., Postma, E.O., van den Herik, H.J.: Dimensionality reduction: a comparative review. Submitted to Neurocomputing (2009)
7. Bernstein, M., de Silva, V., Langford, J.C., Tenenbaum, J.B.: Graph approximations to geodesics on embedded manifolds. Technical Report, Stanford University (2000)
8. Georghiades, A.S., Belhumeur, P.N., Kriegman, D.J.: From few to many: illumination cone models for face recognition under variable lighting and pose. IEEE Trans. Pattern Anal. Mach. Intell. 23, 643–660 (2001)
9. VARIA database, http://www.varpa.es/varia.html
10. Farzin, H., Abrishami, H.: A novel retinal identification system. EURASIP Jr. on Advances in Signal Proc. (2008)
11. Eftekhari, A., Abrishami Mohgaddam, H.: k/K-nearest neighborhood criterion for improvement of locally linear embedding, Technical report,
    http://nasim.kntu.ac.ir/MS/a_eftekhari

A Parameter Free Approach for Clustering Analysis

Haiqiao Huang, Pik-yin Mok*, Yi-lin Kwok, and Sau-Chuen Au

Institute of Textiles and Clothing
The Hong Kong Polytechnic University, Hunghom, Hong Kong

Abstract. In the paper, we propose a novel parameter free approach for clustering analysis. The approach needs not to make assumptions or define parameters on the cluster number or the results, while the clustered results are visually verified and approved by experimental work. For simplicity, this paper demonstrates the idea using Fuzzy C-Means (FCMs) clustering method, but the proposed open framework allows easy integration with other clustering methods. The method-independent framework generates optimal clustering results and avoids intrinsic biases from individual clustering methods.

1 Introduction

Clustering analysis is aimed at partitioning a large number of data or objects into different clusters (subsets, groups, or classes), while clustering must fulfill the requirements of homogeneity and heterogeneity. Homogeneity means that data in the same cluster should be as similar as possible and heterogeneity means that data in different clusters should be as different as possible [5].

The applications of clustering analysis are diverse, for instances in pattern recognition [1], face recognition [4], image understanding [2], and so forth. Many research work were focused on improving cluster algorithms, such as CURE [3], shifting grid [9], training neural networks [13], global FCM [11], and evidential C-Means (ECMs) [8]. However, these clustering algorithms are sensitive to the initial parameter settings, namely, the initialization. This means assumptions must be made about the data before the analysis is conducted, for example, the cluster number for FCMs and a static or dynamic threshold distance for KNN searching. The initial settings imply some assumptions are made towards the data shape or the cluster number. These assumptions would lead to misunderstanding and inaccuracy about the data. Therefore, a parameter free clustering approach is crucial to avoid misunderstanding and inaccuracy.

2 Research Motivation

In this paper, a flexible parameter free approach for clustering analysis is proposed. This approach has two advantages: (1) it makes no assumption like data

---

* Corresponding author. Email: tracy.mok@inet.polyu.edu.hk, Telephone: (852) 2766 4442 Fax: (852) 2773 1432.

shape or the cluster number; (2) it avoids prejudice on the data caused by adopting some particular clustering methods. The inspiration comes from human decision making process, which human usually make decision after integrating diverse information and comparing different alternatives. Similarly, in clustering analysis, it is proposed to make a clustering conclusion after synthetically analysing many opinions on a data set.

An opinion is defined as a clustered result from a clustering algorithm with certain parameters. This opinion is represented by a matrix, called *observation matrix*. Another matrix, called *judgement matrix* accumulates many observation matrices. Therefore, the judgment matrix contains comprehensive information about the data than any individual clustered result. By analysing the judgment matrix, more reliable results can be obtained. In this paper, for the purpose of simplicity and clarity, Fuzzy C-Means (FCMs) clustering algorithm is used to generate these observation matrices. Nevertheless, it is important to note that the observation matrices can be derived either from a specific clustering algorithm, like FCMs in this case, or from different algorithms.

3 Clustering Approach

3.1 Observation and Judgment Matrices

The objective of the paper is to make an optimal clustering judgment about any given data set according to different clustered results. These results are represented by observation matrices, judgment matrices and its graphs which are defined as follows:

Definition 1. For a data set X of m data points, the *observation matrix* J_c is a $m \times m$ matrix which is calculated from a cluster method, where c is the cluster number. Each element of the observation matrix, j_{uv}, represents the relationship of two data points, X_u and X_v.

Definition 2. The *judgment matrix* T is the sum of different observation matrices J_c with c ranging from 2 to k, and k is a number less than m.

Definition 3. The judgment graph G_T is defined as the graph of its adjacency matrix T.

3.2 Judgment Matrix Calculation

A data set X is given by m samples

$$X = \begin{bmatrix} X_1 & X_2 & \cdots & X_m \end{bmatrix}$$

where each sample X_j is a p-dimensional data point, i.e.

$$X_j = \begin{bmatrix} x_1 \\ x_2 \\ \vdots \\ x_p \end{bmatrix}$$

By clustering the data set X by FCMs, the following $c \times m$ membership matrix, U, can be obtained,

$$U = \begin{bmatrix} u_{11} & u_{12} & \cdots & u_{1m} \\ u_{21} & u_{22} & \cdots & u_{2m} \\ \vdots & \vdots & \ddots & \vdots \\ u_{c1} & u_{c2} & \cdots & u_{cm} \end{bmatrix}$$

where c is the cluster number; u_{ij} is the membership value of sample X_j belongs to the ith cluster. The sum of each column membership value equals to 1, i.e. $\sum_{i=1}^{c} u_{ij} = 1$.

Different c values result in different matrices of U. For each membership matrix U, a row vector, called decision vector, can be defined as follows,

$$D = \begin{bmatrix} l_1 & l_2 & \cdots & l_m \end{bmatrix}$$

where l_j is the row index of the largest membership value u_{ij} in each column of U. Therefore, l_j is also a label indicating which cluster the data point X_j is belong to.

In order to have a thorough understanding on the clustering results, a $m \times m$ observation matrix J is defined based on the decision vector D as follows,

$$J = \begin{bmatrix} j_{11} & j_{12} & \cdots & j_{1m} \\ j_{21} & j_{22} & \cdots & j_{2m} \\ \vdots & \vdots & \ddots & \vdots \\ j_{m1} & j_{m2} & \cdots & j_{mm} \end{bmatrix}$$

where $j_{ij}=1$ if X_i and X_j have the same label l in the decision vector D; otherwise, $j_{ij}=0$. J is thus a diagonal symmetric matrix, representing an observation result of clustering X into c clusters. Besides, J is also an adjacency matrix of a undirected graph, whose adjacent edges weights are all 1.

In order to make an optimal clustering decision, we need to accumulate many different observation results. A range of cluster numbers are used to compute different judgment matrices. The cluster number c can be any number in $[2, k]$, where k is an integer smaller than m. The decision on the cluster number upper bound k will be discussed in Section 4.1. The lower bound of c is 2 because a data set is usually clustered into more than one group. Accordingly, a judgment matrix T is defined as:

$$T = \sum_{c=2}^{k} J_c \tag{1}$$

where J_c is the observation matrix of clustering the data set into c clusters. The judgment matrix is also an adjacency matrix of a weighted graph. The adjacency matrix depicts the inter-relationships of data points within the data set.

3.3 Cluster Result Identification

After the judgment matrix T and adjacency graph are computed, a graph partitioning process is conducted, from which an optimal cluster number can be identified.

Graph Partitioning. The graph partitioning problem divides a graph into subgraphs according to the weight, length or distance between graph nodes. Kernighan-Lin(KL) algorithm is a well-known method for partitioning graphs [7]; a variation of the KL algorithm can be found in [6]. By defining the judgment matrix T and its graph G_T, our method of graph partitioning consists of a deducting step and a visualising step.

1. **Deducting step:** compute a new adjacency matrix T_{new} by deducting every element, t_{ij}, of the previous matrix $T_{previous}$ by one, i.e.

$$t_{ij}^{new} = \begin{cases} t_{ij}^{previous} - 1 & \text{if } t_{ij}^{previous} > 0, \\ 0 & \text{otherwise}. \end{cases} \qquad (2)$$

2. **Visualising step:** identify the relationship of data points as a new graph G_{new} for T_{new} by a breadth first searching (BFS)

$$G_{new} = BSF(T_{new}) \qquad (3)$$

The first 'previous' matrix, $T_{previous}$, is the judgment matrix T. The process of deducting and visualising repeats until all elements of T_{new} become zero. Accordingly, a collection of adjacency matrices and corresponding subgraphs are generated (as shown in the complete algorithm below). Subgraphs visualise the relationships between data points such that help identify cluster numbers. Fig. 1(a) is the graph visualising the judgment matrix T before deducting steps, and Fig. 1(b) shows obtained subgraphs after a few deducting steps. By visualising $G_{new}s$, cluster number is defined as the number of subgraphs of connecting nodes.

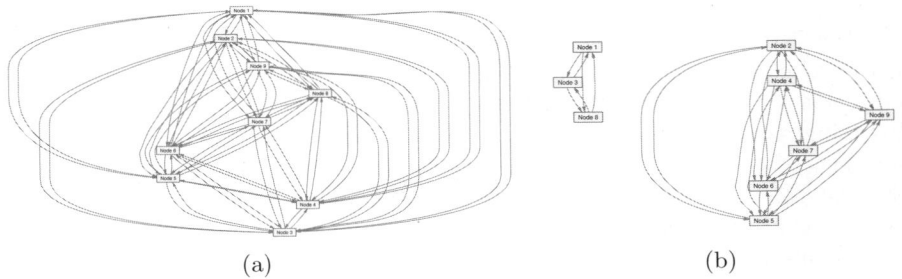

Fig. 1. Graph partitioning and visualization: (a) Graph of initial judgment matrix T; (b) Two connected subgraphs

Cluster result identification (group partitioning) algorithm

Input: The judgement matrix T
$n=0$;
$T_{new}(0)=T.\text{deducting}()$;
do until (all elements in $T_{new} == 0$)
  $G_{new}(n)=T_{new}(n).\text{BSFvisualising}()$;
  $\text{clusterNumber}(n)=G_{new}(n).\text{getConnectedGroupNumber}()$;
  $n = n+1$;
end

Output: A collection of matrices $T_{new}s$ and subgraphs $G_{new}s$, and corresponding cluster numbers

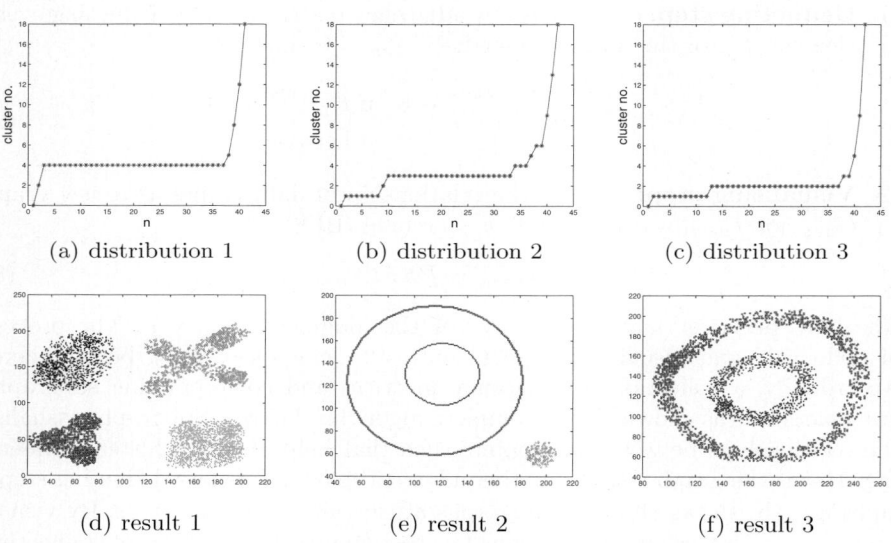

Fig. 2. FCMs clustering results ($k=50$)

Cluster Number Distribution. To carry out clustering analysis, most algorithms require the cluster number to be specified beforehand. Different initializations, namely, selecting different cluster numbers, can result in different clustering partitions. Some previous work, like [12] and [10], use validity indices to measure "goodness" of choices of cluster number In this paper, the *stability* of the clustering number distribution is examined so as to decide an optimal cluster number. During the deducting and visualising process, a collection of $G_{new}s$ and cluster numbers are generated. By examining the distribution of these cluster numbers, an optimal cluster result can be identified. The cluster number distribution is obtained by plotting the number of iterations, n, for group partitioning on the judgment matrix T against the cluster number (the resulted number of subgraphs from visualizing step) (see Fig. 2). It shows that the cluster result tends to be stable at certain levels and the cluster number will increase dramatically if further iterations of deducting and visualising continue. An optimal cluster number and clustered result can then be concluded according to the stability level of the distribution.

4 Results and Discussions

In this paper a parameter free approach for clustering analysis is proposed, which can integrate different opinions on a given data set in order to decide the optimal cluster result without defining initial parameters. The distribution of cluster numbers described above plays an important role in deciding the optimal cluster number. Fig. 2(a) and (c) indicate the distribution of cluster numbers are stable on 4 and 3. Therefore, the optimal cluster numbers are 4 and 3, respectively, for Fig. 2(b) and (d).

4.1 Clustering Stability and Cluster Number k

If data shape or boundaries of the given data set are complex, the cluster number distribution tends to be stable at several levels (see Fig. 3(b)) or even unstable. The decision on the cluster number is not trivial. In our approach, every clustering opinion about the data set is defined as a matrix J. The judgment matrix T is accumulated from these matrices J. Different upper bounds of cluster number,

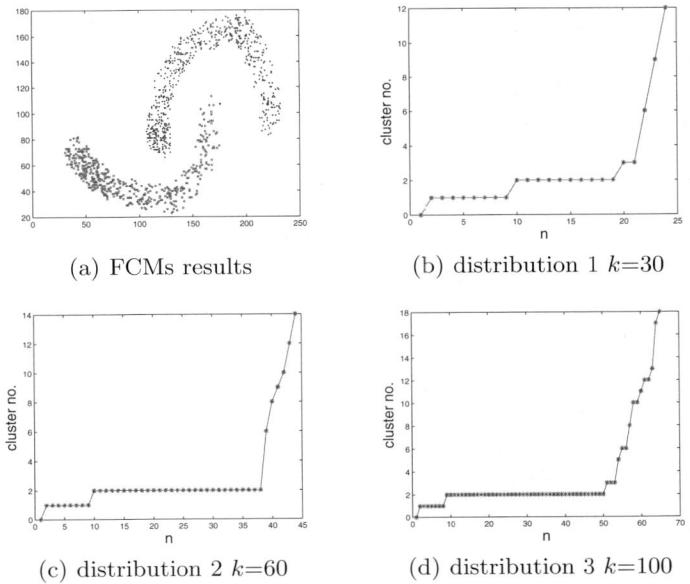

(a) FCMs results (b) distribution 1 $k=30$

(c) distribution 2 $k=60$ (d) distribution 3 $k=100$

Fig. 3. FCMs clustering results with different k values

k in equation (1), were tested. Increasing k implies more observation matrices J to generate, and the cluster number distribution was found to be more stable. Fig. 3 shows the distributions with different k values. Fig. 3(b) illustrates the distribution is more stable on 1 cluster than that on 2 clusters when $k=30$. However, when $k=60$ and $k=100$ the stability on 2 clusters become overwhelming. Therefore, increasing the number of observations k makes the cluster number distribution more stable, and thus decides the cluster number and result more easily. By experiments, it is found that $k=100$ can generate clear distribution of cluster numbers for most data sets and achieve promising results.

4.2 Integrating with Various Clustering Algorithms

Every approach for clustering or pattern recognition has advantages and disadvantages. Algorithms are normally proposed for solving particular problems. The clustering approach proposed in this paper is an open framework that can be easily integrated with different clustering algorithms. Fig. 4 shows the results of clustering the same data set in Fig. 3 by k-means algorithms. It is shown that k-means algorithm obtains similar cluster distribution and clustered results as that of FCMs. It is important to note that the values in decision vector D do not necessarily come from one clustering method, which can be a combination of FCMs and k-means. Consequently, the proposed method can also reduce bias caused by one individual clustering method.

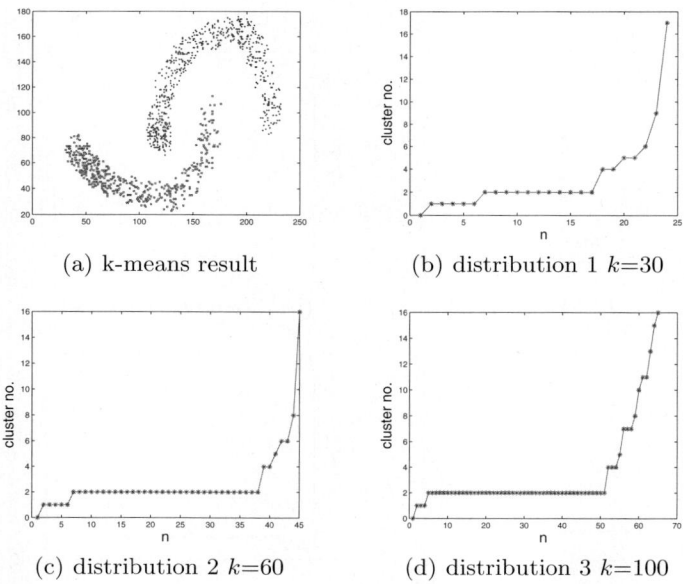

(a) k-means result

(b) distribution 1 $k=30$

(c) distribution 2 $k=60$

(d) distribution 3 $k=100$

Fig. 4. K-means clustering results with different k values

5 Conclusions

In this paper, a parameter free approach for clustering analysis has been proposed. For any given data set, an optimal cluster result is obtained without specifying the cluster number through analysing observation and judgment matrices. It has been demonstrated by experiments that the proposed method is effective in generating reliable and promising clustering results.

References

1. Bezdek, J.C.: Pattern Recognition with Fuzzy Objective Function Algorithms. Plenum Press, New York (1981)
2. Cai, W., Chen, S., Zhang, D.: Fast and robust fuzzy c-means clustering algorithms incorporating local information for image segmentation. Pattern Recognition 30(3), 825–838 (2007)
3. Guha, S., Rastogi, R., Shim, K.: Cure: an efficient clustering algorithm for large databases. Information Systems 26(1), 35–58 (2001)
4. Haddadnia, J., Faez, K., Ahmadi, M.: A fuzzy hybrid learning algorithm for radial basis function neural network with application in human face recognition. Pattern Recognition 36(5), 1187–1202 (2003)
5. Hoppner, F., Klawonn, F., Kruse, R., Runkler, T.: Fuzzy Cluster Analysis. Wiley Press, Chichester (1999)
6. Karypis, G., Kumar, V.: Multilevel-way Partitioning Scheme for Irregular Graphs. Journal of Parallel and Distributed Computing 48(1), 96–129 (1998)
7. Kernighan, B.W., Lin, S.: An efficient heuristic procedure for partitioning graphs. The Bell System Technical Journal, 291–307 (1970)
8. Masson, M., Denoeux, T.: ECM: An evidential version of the fuzzy c-means algorithm. Pattern Recognition 41(4), 1384–1397 (2008)
9. Ma, W.M.E., Chow, W.S.T.: A new shifting grid clustering algorithm. Pattern Recognition 37(3), 503–514 (2004)
10. Wang, W., Zhang, Y.: On fuzzy cluster validity indices. Fuzzy Sets and Systems 158(19), 2095–2117 (2007)
11. Wang, W., Zhang, Y., Li, Y., Zhang, X.: The Global Fuzzy C-Means Clustering Algorithm. Intelligent Control and Automation (2006)
12. Zahid, N., Limouri, M., Essaid, A.: A new cluster validity for fuzzy clustering. Pattern Recognition 32(7), 1089–1097 (1999)
13. Wen, J.H., Meng, K.W., Wu, H.Y., Wu, Z.F.: A novel clustering algorithm based upon a SOFM neural network family. In: 2nd International Symposium on Neural Networks, Chongqing, P.R. China (2005)

Fitting Product of HMM to Human Motions

M. Ángeles Mendoza, Nicolás Pérez de la Blanca,
and Manuel J. Marín-Jiménez*

Department of Computer Science and A.I.,
University of Granada, Spain
{nines,nicolas,mjmarin}@decsai.es

Abstract. The Product of Hidden Markov Models (PoHMM) is a mixed graphical model defining a probability distribution on a sequence space from the normalized product of several simple Hidden Markov Models (HMMs). Here, we use this model to approach the human action recognition task incorporating mixture-Gaussian output distributions. PoHMM allow us to consider context at different range and to model different dynamics corresponding to different body parts in an efficient way. For estimating the normalization constant Z we introduce the annealed importance sampling (AIS) method in the context of PoHMM in order to obtain no-relative estimates of Z. We compare our approach with one based on fitting a logistic regression model to each two PoHMMs.

Keywords: partition function, PoHMM, human action recognition.

1 Introduction

Probabilistic state-space models have been very successful in automatic human action recognition tasks [1], since states and state transitions can be associated to subject poses and transitions between poses. We approach this task using a PoHMM, which was introduced by Brown and Hinton [2] for modelling character strings. PoHMM approaches the full distribution as a product of experts, where each expert is represented by an individual HMM, which describes, through no-causal dependences, an observation accepted by all the HMMs. PoHMM provides a bigger representational capacity than traditional HMM where all the information is contained in a single k-state multinomial variable. In a PoHMM each individual HMM-factor tries to explain aspects of the data mainly associated to regularities present in different temporal scales. This allows to model the different dynamics present in an action from different HMMs. Moreover, many human activities share common poses, so that context to large or medium range needs to be considered for removing ambiguities. Thanks to its probabilistic dependence (see figure 1), in PoHMM given the observations the inference on each HMM is independent and therefore the full process is very efficient. In [2] simple Gaussian distributions are used, here we use mixture-Gaussian output distributions

---

* This work has been granted by the Consolider Ingenio MIPRCV project of the Spanish Minister of Science and Innovation, CSD2007-00018.

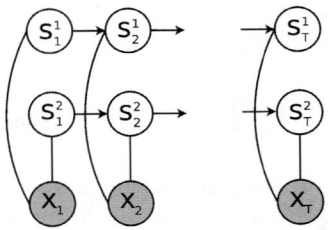

Fig. 1. PoHMM Graphical Model. Empty circles represent the hidden states (S) and shaded circles observed data (X).

since the human actions present too much variability for only one Gaussian. But the big issue when we model a multiclass classification task is the estimation of the normalization constant required to calculate probabilities on the global state-space. The computation of this constant Z is generally intractable. In [2] a logistic regression fitting was suggested to estimate the ratio Z_A/Z_B when only two models were considered. Here we estimate Z in the general case by adapting to the PoHMM the AIS method proposed by Neal in [3]. We compare the estimates from both approaches.

2 Products of Hidden Markov Models

PoHMM is defined by multiplying together the densities of the independent HMMs and normalizing by a constant Z called partition function. This is calculated summing over all the possible observation sequences $\mathcal{X}$.

$$P(X|\Theta) = \frac{\prod_{m=1}^{M} P^{(m)}(X|\theta^{(m)})}{Z(\Theta)} . \qquad (1)$$

$$Z(\Theta) = \sum_{X \in \mathcal{X}} \prod_{m=1}^{M} P^{(m)}(X|\theta^{(m)}) . \qquad (2)$$

M is the number of HMMs in the model. Θ is the set of all the model parameters, $\Theta = \{\theta\}_{m=1}^{M}$. $X = \{X_1, \ldots, X_T\} \in \mathcal{X}$ are the observation variables.

The unnormalized likelihood $P^{(m)}(X|\theta^{(m)})$ for each HMM can be approximated by only considering the most likely discrete state sequence $S_{1:T}^{(m)} = \{S_1^{(m)}, \ldots, S_T^{(m)}\}$, and factorizes as:

$$P^{(m)}(X|\theta^{(m)}) = max_{s^{(m)}} \left\{ P(S_1^{(m)}) \prod_{t=2}^{T} P(S_t^{(m)}|S_{t-1}^{(m)}) \prod_{t=1}^{T} P(X_t|S_t^{(m)}) \right\} \qquad (3)$$

$P(S_1)$ is the initial state probability usually denoted as K-component vector π, where K is the number of hidden states, $P(S_t|S_{t-1})$ the state transition probability denoted as $K \times K$ matrix A and $P(X_t|S_t)$ is the observation probability.

Here we introduced Gaussian mixture observation distributions. In common with most systems based on continuous HMMs:

$$P(X_t|S_t^{(m)}) = \sum_{g=1}^{G} C_{S_t^{(m)}}^g \mathcal{N}(X_t|\mu_{S_t^{(m)}}^g, \Sigma_{S_t^{(m)}}^g) \quad (4)$$

where $\mathcal{N}(.|\mu, \Sigma)$ denotes a Gaussian distribution, d is the dimension of feature vector, G the number of Gaussians, C^g is the K-component weight vector of the gth Gaussian, μ^g ($d \times K$) and Σ^g ($d \times d \times K$) represent the means and covariances characterizing the observed sequence. Therefore, the full set of unknown is: $\Theta = \{\pi_m, A_m, \mu_m^1, \ldots, \mu_m^G, \Sigma_m^1, \ldots, \Sigma_m^G, C_m^1, \ldots, C_m^G\}_{m=1}^M$.

Training and Recognition

In order to estimate the model parameters Θ, we get an initial solution estimating the parameters for each HMM independently using the Forward-Backward (F-B) algorithm [4]. Then, these estimates are updated minimizing the contrastive divergence given by Hinton [5] w.r.t the model parameters:

$$\Delta\theta^{(m)} \propto \frac{\partial}{\partial\theta^{(m)}} \log P(X^0|\theta^{(m)}) - \frac{\partial}{\partial\theta^{(m)}} \log P(X^1|\theta^{(m)}) \quad (5)$$

X_0 is an observation sequence taken from the experimental data and X_1 is a rebuilt sequence from the most likely state sequences (Viterbi algorithm [4]) for each one of the HMMs. $\frac{\log P}{\partial\theta^{(m)}}$ is calculated efficiently by F-B algorithm [2]. The probability that a sequence belongs to a class is given by equation (1).

3 Partition Function for PoHMMs

The value of the function Z, guaranteeing that equation (1) defines a probability distribution, is computationally intractable since we must sum on all the possible observation sequences $\mathcal{X}$. So, Z cannot be directly known in most cases, but it is possible to estimate ratios of Z values for two different models.

3.1 Estimating Ratio of Z

In [2], in order to compare two different PoHMMs the difference of their respective $\log Z$ is estimated discriminatively. Let P_A^* and P_B^* be the unnormalized probability distributions of two PoHMMs respectively, $(P = P^*/Z)$, then

$$\Delta \log P(X|\Theta) = \Delta \log P^*(X|\Theta) - \Delta \log Z(\Theta) \ . \quad (6)$$

In (6) $\Delta \log Z$ is considered a bias of the difference between the log-likelihood of both models and can be calculated by a simple logistic regression on the training data [5].

The estimation of Z in such way only allows us to calculate differences between log-likelihoods. But in classification task we usually want to compare more than two PoHMMs (classes) at the same time. For M PoHMMs we could calculate the $M(M-1)/2$ ratios of Z, but we cannot calculate the model giving the maximum probability.

3.2 Estimating Z by AIS

In [3], Neal proposes an Annealed Importance Sampling (AIS) technique to estimate partition function ratios Z_A/Z_B. In [6] good results were found estimating complex partition function, Z_B, starting from a simple model where Z_A can be calculated analytically. We adapt this idea in order to estimate Z in PoHMMs.

Let P_A^* and P_B^* be the unnormalized probabilities of two PoHMMs, and Z_A and Z_B their corresponding partition functions. Following [3] we define a sequence of intermediate probability distributions each one differing slightly to the next, $\{p_0^*, \ldots, p_K^*\}$, satisfying the condition $p_k^* \neq 0$ whenever $p_{k+1}^* \neq 0$:

$$p_k^*(X) \propto P_A^*(X)^{(1-\beta_k)} P_B^*(X)^{\beta_k} = \left(\prod_{m_a=1}^{M_A} P_A^{*m_a}\right)^{(1-\beta_k)} \left(\prod_{m_b=1}^{M_B} P_B^{*m_b}\right)^{\beta_k} \quad (7)$$

with $p_0^* = P_A^*$ and $p_K^* = P_B^*$, and $0 = \beta_0 < \beta_1 < \ldots < \beta_{K-1} < \beta_K = 1$ empirically fixed.

For each k-value, we generate a sample $X_k = \{X_{1_k}, \ldots, X_{T_k}\}$ from p_{k-1}^* and X_{k-1}. Using p_{k-1}^* we calculate the most likely hidden state sequence for the X_{k-1} sample using the Viterbi algorithm: $\{s_t^{(1)}, \ldots, s_t^{(M_A)}, s_t^{(1)}, \ldots, s_t^{(M_B)}\}$. Then we generate a sample X_k from the mixture of Gaussian conditioned on the hidden state configuration given at each time step of the HMMs.

$$\ldots \xrightarrow{p_{k-1}^*} \{s_t^{(1)}, \ldots, s_t^{(M_A)}, s_t^{(1)}, \ldots, s_t^{(M_B)}\} \xrightarrow{\mu, \Sigma_{\{s_t^{(1)}, \ldots, s_t^{(M_A)}, s_t^{(1)}, \ldots, s_t^{(M_B)}\}}} X_{t_k} \ldots$$

where the parameter estimation for each AIS-step is given by:

$$\Sigma_{\{s_t^{(1)}, \ldots, s_t^{(M_A)}, s_t^{(1)}, \ldots, s_t^{(M_B)}\}} = \left((1-\beta_k)\sum_{m_a=1}^{M_A}\left(\Sigma_{s_t^{(m_a)}}^{(m_a)}\right)^{-1} + \beta_k \sum_{m_b=1}^{M_B}\left(\Sigma_{s_t^{(m_b)}}^{(m_b)}\right)^{-1}\right)^{-1}. \quad (8)$$

$$\mu_{\{s_t^{(1)}, \ldots, s_t^{(M_A)}, s_t^{(1)}, \ldots, s_t^{(M_B)}\}} = \Sigma_{\{s_t^{(1)}, \ldots, s_t^{(M_A)}, s_t^{(1)}, \ldots, s_t^{(M_B)}\}} \times$$
$$\left((1-\beta_k)\sum_{m_a=1}^{M_A}\left(\Sigma_{s_t^{(m_a)}}^{(m_a)}\right)^{-1} \mu_{s_t^{(m_a)}}^{(m_a)} + \beta_k \sum_{m_b=1}^{M_B}\left(\Sigma_{s_t^{(m_b)}}^{(m_b)}\right)^{-1} \mu_{s_t^{(m_b)}}^{(m_b)}\right). \quad (9)$$

A run of AIS obtains the sequence $\{X_1^{(i)}, \ldots, X_K^{(i)}\}$, together with the corresponding i^{th} importance weight:

$$w^{(i)} = \frac{p_1^*(X_1^{(i)})}{p_0^*(X_1^{(i)})} \frac{p_2^*(X_2^{(i)})}{p_1^*(X_2^{(i)})} \cdots \frac{p_{K-1}^*(X_{K-1}^{(i)})}{p_{K-2}^*(X_{K-1}^{(i)})} \frac{p_K^*(X_K^{(i)})}{p_{K-1}^*(X_K^{(i)})}. \quad (10)$$

After performing N runs of AIS, the ratio of Z_B and Z_A can be estimated by averaging these weights [3]. To avoid overflow problems, we use $\log(w^{(i)})$.

$$\frac{Z_B}{Z_A} \approx \frac{1}{N} \sum_{i=1}^{N} w^{(i)} \quad (11)$$

Let Z_B be the partition function we want to evaluate, if we could assess Z_A, we can estimate directly Z_B from equation (11). We choose it as the probability distribution associated to a simple HMM, P_A is equal to equation (3) and $Z_A = 1$, then Z_B is given by $Z_B \approx \frac{1}{N} \sum_{i=1}^{N} w^{(i)}$.

4 Experimental Setup

4.1 Data

We ran our experiments on the well-known KTH database of human actions [7]. 25 different people of both genders carry out 10 different actions: *walking, jogging, running* and *boxing*, parallel to the camera in both directions; and *clapping* and *waving*, facing to the camera. The actions were performed in 4 different scenarios, outdoors and indoors, with different lighting conditions, appearing in the images shadows and compression artifices. In the scenario 2, during the recording of the *boxing, clapping* and *waving* sequences there is a continuous zoom, while in the rest of actions there are strong viewpoint changes.

4.2 Feature Vector

The optical flow is shown to be very discriminative when it is used in action recognition [8]. We estimate the optical flow by the Farnebäck´s method, the constrain on each pixel provides us a dense and more stable optical flow estimation with very low computation burden [9]. For improving the stability of the estimate a moving average of size 5 has been considered along the sequence. In order to obtain the bounding box enclosing the person carrying out the action, we learn the background distribution from the first frames on each scenario.

We split the bounding box into uniform non-overlapped meaningful tiles, which would increases the discriminant power of feature (e.g. walking and waving) and allow us to distinguish between oriented motions (left and right). We use 8 tiles: 2 horizontal sections, exploiting human symmetry; and 4 vertical sections corresponding to the principal segments of the body. On each tile we calculate one $2D$ histogram using orientation and magnitude. We used 4 bins for magnitude and 8 bins for orientation (from $0°$ to $360°$ in steps of $45°$). For each two adjacent frames, we obtain one $256D$ feature vector h (4 *magnitude bins* × 8 *angle bins* × 8 tiles). To reduce the h dimension we compute principal components by each tile on the training features. We select only the first 4 axes representing about 90% of information. Training and testing samples are projected on the new axes, given $32D$ feature vectors.

4.3 Model

Number of HMMs in a PoHMM. To fix the number of HMMs defining the PoHMM, we studied the spectral power of the components of the feature vectors. Figure 2 shows that *walking, jogging* and *running* present 2 well-differentiated peaks while the remaining actions present a single peak. This agrees with the

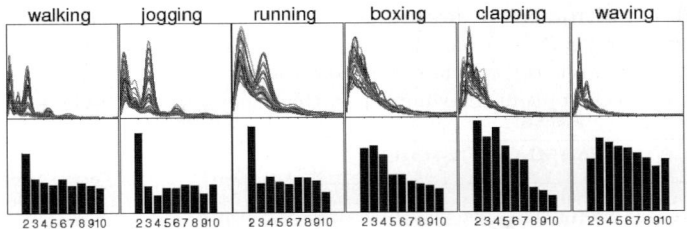

Fig. 2. Spectral power and features clustering. First row, spectral power of the components of the feature vectors for each action. Second row, separation values for a number of clusters ranging from 2 to 10 in the feature space.

studied actions, they are simple actions where the movement is principally addressed by the arms and/or legs. So that, we use PoHMM compound by 2 HMMs to model the 2 fundamental frequencies governing the action. We fixed the maximum number of states to 20, the number of frames of the shortest sequence. In order to analyze different temporal resolution with each HMM, we range the number of states from 2 to 10 in HMM-1 and from 11 to 20 in HMM-2.

Number of Gaussians
We construct clusters fitting Gaussian mixture with a specified number of components to the data, and measure the separation value between clusters, second row figure 2, the best separation values are obtained for a low number of components, so that 2 Gaussians are enough to model our features.

Number of $\beta's$. Other important issue is the discretization of the $\beta's$ values defining the intermediate probabilities. We try three different distributions: First, 1000 distributions uniformly spaced, $\beta = \{0 : 0.001 : 1\}$. Second, $\beta = \{0 : 0.01 : 0.5 \quad 0.5 : 0.001 : 0.9 \quad 0.9 : 0.0001 : 1.\}$, 1453 distributions with uniform spacing for $\log \beta_j$, this is an optimal schema according to [3]. Finally, we keep this schema but increasing the number of distributions, $\beta = \{0 : 0.001 : 0.5 \quad 0.5 : 0.0001 : 0.9 \quad 0.9 : 0.0001 : 1\}$, in total 5503 distributions.

5 Results

Results were validated by 3-fold cross validation on subject-independent tests, two third for training and the other third for testing, about 1590 and 800 samples respectively. One 2-Gaussians PoHMM with 2 chains (from 2 to 10 hidden states and from 11 to 20) was trained for each action. We obtained the best recognition score with a PoHMM 7 × 12-states. That is, 14854 $(7(\pi) + 7 \times 7(A) + 32 \times 7 \times 2(\mu) + 32 \times 32 \times 7 \times 2(\Sigma) + 7 \times 2(C))$ parameters for one HMM and 25524 $(12 + 12 \times 12 + 32 \times 12 \times 2 + 32 \times 32 \times 12 \times 2 + 12 \times 2)$ for the other.

Figure 3 shows the estimated $\log Z's$ and their variances for the three discretizations of β, which are similar for all the cases. We can observe that the variance in the activities of legs is much less in the test 1. Table 1 shows the results for each test of

Table 1. Recognition accuracy on KTH. First and second rows, log-likelihood using differences of $\log Z$ (section 3.1), third one the differences are calculated by AIS (section 3.2), fourth one directly normalizing by $\log Z$ (AIS). First row, we use simple HMMs with Gaussian output distribution, in the remaining ones 2-Gaussians mixture.

PoHMM 7 × 12-states	Acc (%)			
	Test1	Test2	Test3	Average
Gaussian output ($\Delta \log Z$ by regression)	91.6	84.6	89.3	88.5
2-Gaussian mixture output ($\Delta \log Z$ by regression)	91.8	90.1	90.6	90.8
2-Gaussian mixture output ($\Delta \log Z$ by AIS)	92.5	87.5	87.0	89.0
2-Gaussian mixture output ($\log Z$ by AIS)	92.5	88.2	86.9	89.2

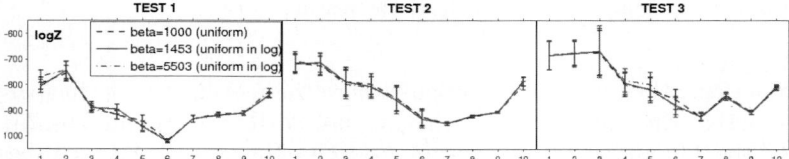

Fig. 3. Estimated $\log Z$ for each action. In each test of the 3-fold cross validation for the 3 described discretizations of β. 1:(l)walk, 2:(r)walk, 3:(l)jog, 4:(r)jog, 5:(l)run, 6:(r)run, 7:(l)box, 8:(r)box, 9:clap, 10:wave. (l) action to the left, (r) to the right.

Table 2. Confusion matrices of 2-Gaussian PoHMM 7 × 12-states. Left, matrix is calculated by the logistic regression method. Right, matrix is calculated by AIS method. Rows are the test classes and columns the recognized ones. Last column is correct % by action.

	(l)walk	(r)walk	(l)jog	(r)jog	(l)run	(r)run	(l)box	(r)box	clap	wave	%	(l)walk	(r)walk	(l)jog	(r)jog	(l)run	(r)run	(l)box	(r)box	clap	wave	%
(l)walk	198	0	1	0	1	0	0	0	0	0	99.0	195	0	3	0	2	0	0	0	0	0	97.5
(r)walk	1	195	0	1	0	2	1	0	0	0	97.5	1	191	0	4	0	3	1	0	0	0	95.5
(l)jog	8	0	178	0	14	0	0	0	0	0	89.0	2	0	151	0	46	1	0	0	0	0	75.5
(r)jog	0	1	0	189	0	10	0	0	0	0	94.5	0	0	0	136	0	64	0	0	0	0	68.0
(l)run	0	0	56	0	143	1	0	0	0	0	71.5	0	0	27	0	171	2	0	0	0	0	85.5
(r)run	0	1	0	43	0	156	0	0	0	0	78.0	0	0	0	17	0	183	0	0	0	0	91.5
(l)box	0	0	0	0	2	0	302	1	8	0	96.5	0	0	0	0	2	0	301	0	10	0	96.2
(r)box	2	0	1	0	2	0	17	44	18	0	52.4	1	0	0	0	3	0	18	37	25	0	44.1
clap	0	0	0	0	0	0	6	2	382	6	96.5	0	0	0	0	0	0	6	1	385	4	97.2
wave	0	0	0	0	0	0	4	0	10	384	96.5	0	0	0	0	0	0	0	1	0	16	381	95.7

the 3-fold cross validation. First and second row, we assess recognition rate using $\Delta \log Z$ fit by logistic regression ([2]), for Gaussian and 2-Gaussian mixture output respectively. We can observe that introducing HMMs with mixture-Gaussian observation distribution as factors in the PoHMM improves the accuracy in the action recognition task, so that this model was used for the rest of experiments. Third and fourth row, we use $\Delta \log Z$ and $\log Z$ respectively, both of them calculated by AIS. For calculating the difference between the $\log Z$ of two different PoHMMs we take as initial distribution one of the PoHMM and as final distribution the other PoHMM (equation 11), while that we calculated $\log Z$ taking as initial distribution a simple 2-state HMM ($Z = 1$) and as final distribution the PoHMM whose $\log Z$ we want to assess. We observe that for both cases we obtain similar results, but the computation in the second case is much less ($40\,\Delta \log Z$ vs. $10 \log Z$). The AIS performance increases in the test 1, but decays for the rest, the test 1 is the test with less variance in Z (see figure 3), the accuracy depends on the variability of the importance weights [3]. In the confusion matrices (table 2), we observe that the fall is principally due to *jogging* which is erroneously confused with *running*, in the graphic corresponding to the test 3 of the figure 3 *jogging* presents the largest variance. Instead, it increases the number of *running* actions correctly recognized with the AIS method. Both actions are similar and in this database are also confused by a human observer. Finally, table 2 shows that the total accuracy decreases by the poor performance of *boxing to the right*, which is the action with less training sequences, possibly the number of samples are not enough for these models.

6 Conclusions

We have shown that PoHMMs using mixture-Gaussian observations can be applied successfully to the human action recognition task. We have also presented a schema based on AIS which provides an estimation of the partition function Z for these models. The recognition scores obtained using this Z estimate are good at most of actions, however the experiment shows that a large enough number of samples is needed in order to get a stable value for Z. The classification score increases when the variance in Z decreases. The estimation of Z allows to evaluate different PoHMMs in a single classification framework. This fact combined with the linear computational complexity in the number of HMMs [2] would allow to handle large databases of complex actions.

References

1. Mendoza, M.A, Pérez de la Blanca, N.: Applying space state models in human action recognition: A comparative study. In: Perales, F.J., Fisher, R.B. (eds.) AMDO 2008. LNCS, vol. 5098, pp. 53–62. Springer, Heidelberg (2008)
2. Brown, A., Hinton, G.E.: Products of hidden markov models. Artificial Intelligence and Statistics, 3–11 (2001)
3. Neal, R.M.: Annealed importance sampling. Statistics and Computing 11(2), 125–139 (1998)
4. Rabiner, L.: A tutorial on hidden markov models and selected applications in speech recognition. Proc. of the IEEE 77(2), 257–286 (1989)
5. Hinton, G.E.: Training products of experts by minimizing contrastive divergence. Neural Computation 14(8), 1771–1800 (2002)
6. Salakhutdinov, R., Murray, I.: On the quantitative analysis of deep belief networks. In: Proceedings of the Int. conf. on Machine Learning, vol. 25, pp. 872–879 (2008)
7. Schuldt, C., Laptev, I., Caputo, B.: Recognizing human actions: a local svm approach. Pattern Recognition 3(1), 32–36 (2004)
8. Efros, A.A., Berg, A.C., Mori, G., Malik, J.: Recognizing action at a distance. Computer Vision 2, 726–733 (2003)
9. Farnebäck, G.: Two-frame motion estimation based on polynomial expansion. In: Bigun, J., Gustavsson, T. (eds.) SCIA 2003. LNCS, vol. 2749, pp. 363–370. Springer, Heidelberg (2003)

Reworking Bridging for Use within the Image Domain

Henry Petersen and Josiah Poon

University of Sydney, School of Information Technologies
hpet9515@it.usyd.edu.au, Josiah@it.usyd.edu.au

Abstract. The task of automated classification is a highly active research field with great practical benefit over a number of problem domains. However, due to the factors such as lack of available training examples, large degrees of imbalance in the training set, or overlapping classes, the task of automated classification is rarely straightforward in practice. Methods that adequately compensate for such difficulties are required. The recently developed bridging algorithm does just this for problems in the field of short string text classification. The algorithm integrates a collection of background knowledge into the classification process. In this paper, we have shown how the bridging algorithm was redesigned so it can be applied to image data. We also demonstrated it is effective to overcome a range of difficulties in the classification process.

1 Introduction

The task of automated classification is a highly researched field of great practical benefit over a number of domains. In it an algorithm is given an object represented by a set of descriptive features, and attempts to assign to it a label from a pre-determined finite set of classes. Typically these features describe data wholly contained within the object itself and thus the resultant classifier is required to effectively differentiate between classes using only the local properties of the objects. However in many cases the information contained in the objects may be insufficient to satisfactorily deal with the given classification task. It is possible that additional a-priori knowledge of the problem domain may be required in order to achieve the required standards of performance.

Within the field of short string text classification Zelikovitz et al [16] proposed the Bridging algorithm to include just this type of a-priori information (henceforth called background knowledge) into the classification process. When training classifiers on problems such as separating physics papers into sub-discipline (ie. Astrophysics, optics etc.) using the titles, and classifying company names by business function, improvements in both classifier stability and accuracy have been demonstrated when dealing with training data under the conditions of both high rates of imbalance and a lack of available labelled training examples [16,14].

The aim of this paper is to demonstrate the potential of the Bridging algorithm when used to compensate for a number of classification difficulties in domains other than short string text classification, namely the image domain and more specifically the problem of homogenous texture classification. Homogenous texture classification can be defined as classification where the training and test examples are comprised of homogenous textures. We will describe how we redesigned the Bridging algorithm for use in the image domain and present the results of an evaluation of its performance when dealing with classification problems including a lack of available training examples and high rates of imbalance in the training dataset.

This paper is organised as follows; in section 2 we provide an overview of the Bridging algorithm and the idea of background knowledge for classification. In Section 3 we introduce the idea of global context for homogenous textures and explain how it can be harnessed to implement Bridging for the problem of homogenous texture classification. In section 4 we present an overview of current methods for compensating classification problems such as imbalance and lack of training data. In section 5 we present an evaluation of our method and in section 6 we conclude and present possible directions for future research.

2 Definition of Bridging

The Bridging algorithm provides an extension to the standard classification paradigm by incorporating into the classifier a collection of unlabelled background knowledge related to the specific classification task at hand. In the standard classification model a test instance is classified by comparing it to a set of pre-labeled feature / class label pairs. The Bridging algorithm replaces the descriptive features with the results of comparisons between the training instance and each item in the unlabelled background knowledge using a similarity metric. The new feature vectors are then learnt by a classifier to build a model that approximates this derived space of instances. The same process is applied to the test instances, and the new feature vector is then evaluated by the model to derive the classification result.

This reasoning behind the inclusion of the background knowledge is that it is often able to provide information about the classification task that is not contained in the training examples themselves. This information is used by the algorithm to aid in separating instances and discerning the class to which they belong. Bridging could be said to provide a dimensionality increase for the instances in the problem domain. The additional information enables the classifier to better discern between classes and greatly clarifies the decision boundaries.

To understand exactly what comprises the background knowledge used with the Bridging algorithm two points must be noted:

- The background knowledge should be selected to contain additional information relevant to the problem domain that can be used to aid in providing better resolution between the test instances. This means that the background knowledge needs to be in some way similar to the test instances so results of a comparison between the two can be obtained. To this end it needs to be selected with the task at hand in mind (it is problem specific).

- The background knowledge does not need to be of the same form or size as the test instances; there only needs to exist some comparison method for the background items with the test instances (the size and form of the background knowledge do not need to be the same as for the test instances).

As such the background knowledge can be almost anything as long as it is related to the problem at hand, provides information pertinent to the classification problem that is not contained in the instances themselves, and is comparable to the test instances. Examples of problems / background knowledge pairings for Bridging from the literature include:

- Classifying physics papers by title into sub-disciplines of physics using paper abstracts as the background knowledge
- Classifying company names by business type (eg. Google would be IT, Westpac would be a bank etc.) employing the text of business related web sites as the background knowledge.

3   Current Strategies for Motivating Problems

Building classifiers with only small amounts of training data available is an actively researched topic within the classification community [4,2]. The majority of research uses the idea that a lack of available labelled instances for training does not necessarily imply a lack of unlabeled instances. This however is not always the case. In comparison Bridging requires only the existence of data containing some meaningful a-priori domain knowledge of benefit to the classifier. The relatively relaxed constraints on the form of the background knowledge mean that in many cases it may be much simpler to obtain background knowledge than unlabelled training examples.

Current data imbalance compensation strategies can be divided into several categories, the most well known being over-sampling, under-sampling, and weighting examples. Over-sampling involves duplicating examples to increase the number of minority class examples, under-sampling involves randomly removing examples from the majority class in order to rebalance the training set, and weighting methods involve the classifier assigning a positive bias to examples from the minority class. Comparisons have been made in several places throughout the literature [3,5,9,15] and current research indicates that all methods can be effective given the right circumstances.

4   Contextual Knowledge for Homogenous Textures

In Section 2 it was pointed out that for many classification tasks additional domain knowledge may be required to satisfactorily identify the class of a given test instance. Consider the problem of homogenous texture classification. Figure 1 shows a homogenous texture and the surrounding context in which it was found. When viewed alone the texture is difficult to identify due to the relative lack of information contained in the image. However when the texture is taken in conjunction

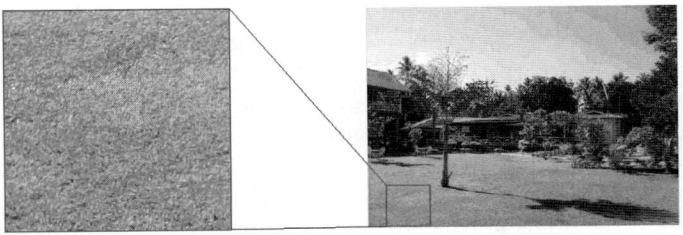

Fig. 1. Homogenous texture (grass) and surrounding context

with its surrounding context as shown in , the correct classification is significantly clearer.

The notion of contextual knowledge is the basis of our design for Bridging with homogenous textures. By their very definition homogenous textures contain no contextual data. Because of this we have chosen to use a collection of wider scene photographs as our background knowledge. A wider scene photograph can be interpreted as a collection of homogenous textures. By comparing test images to the scene photographs we hope to create some kind of surrounding context for the textures. This in turn can then be used to increase the performance of the resulting classifier.

As stated in Section 2 two tasks need to be accomplished in order to implement Bridging for homogeneous textures; choosing appropriate background knowledge and designing a suitable comparison metric. As we have elected to use scene photographs for the background knowledge, the remaining task is to determine how they are to be compared to the homogeneous textures. Given we can interpret scene photographs as collections of textures, we determined to calculate for each test and background image two types of information; how similar is the test image to each of the textures in the background image (represented as a value between 0 and 1), and where in the background image is the similarity strongest (represented by the X and Y values of a point on the background image with (0,0) being the centre of the image). The comparison metric used in

Algorithm 1. Comparison Metric for Bridging with Homogeneous Textures

1. The background image is divided into a uniform grid of equally sized sub-images.
2. Standard descriptive features are calculated for the test image and each sub-image tile.
3. The feature vectors for each sub-image tile are compared to the test image feature vectors using a simple metric (such as the dot product) to produce a value between 0 and 1. The metric is determined experimentally as a parameter of the algorithm.
4. Each tile in the background image grid is weighted with its corresponding comparison value and the centre of mass for the background image is calculated.
5. The result of the comparison between then test and background image is represented as the average of the comparison values from step 3, and the X and Y values of the centre of mass as calculated in step 4.

our experiments is shown in Algorithm 1. We use this algorithm to create new features for a given test image by feeding the classifier the values computed in running Algorithm 1 with the image and every item of background knowledge.

The division of the background image into a uniform grid in step 1 (as opposed to using a more complex segmentation algorithm) is done to simplify the process and also because a segmentation algorithm would not necessarily generate rectangular regions making it much harder to choose features to represent the textures in the items of background knowledge. The size of the uniform grid used in our experiments was 6x6. It should also be noted that the choice of descriptive features from step 2 and the choice of simple metric from step 3 are linked; some metrics may perform better with some features than others. We discuss our choice of descriptive features and simple metric in Section 5.

5 Evaluation

In order to evaluate the performance of our implementation of Bridging we tested it on both the problem of training with an imbalanced dataset and training with a small number of available training examples. To test the performance of our method on these problems we created a dataset using images from the Vistex Texture Database [1], a texture benchmark that has been employed a number of times throughout the literature [5,6,11,10]. The full database comprises 167 homogeneous textures from 19 distinct classes, however to create a dataset suitable for our purposes we have selected images from the 8 majority classes and divided each into 4 equal sub-images resulting in a balanced dataset of 432 images. The specifics of how the data was prepared for each set of tests is detailed below.

To evaluate results we employed an SVM classifier (from the WEKA machine learning toolkit [15]) with 10-fold stratified cross validation and measured performance using the weighted average f-score averaged across multiple runs on different random splits of the database. The weighted average f-score is the average f-score for each class in the dataset, with each score being weighted by the size of the class it represents. For each motivating problem the results of our method were compared against a baseline; the performance of the classifier when using only standard descriptive features without any background knowledge.

For standard descriptive features we have chosen the colour layout, edge histogram and scalable colour descriptors from the MPEG-7 standard for representation of multimedia content metadata [7]. This choice was made both for convenience and due to the benefits of utilising standardised descriptors which can be seen in a range of literature [12,8]. In addition to these descriptors we will also use the combined feature vectors of the colour layout, edge histogram and scalable colour as demonstrated by Spyrou et al [12]. For a more detailed description of this descriptor we refer you to the cited publication. The metrics used in our experiments to compare the test image feature vectors to the background image sub-tiles (in step 3 of Algorithm 1) are the dot product and the histogram intersection [13].

The background knowledge used in these evaluations was sourced from a number of photographs collected by the authors. There were 79 images in total collected both indoors and outdoors over a range of lighting conditions. They were stored in 2848x2134 resolution. In addition there were also 11 multi-texture scene images taken from the Vistex Texture Database [1], which were stored in either 786x512 or 512x786 resolution. It should be noted that the dataset we have chosen, while designed for use with textures is not specifically tailored for use with images from the Vistex database.

5.1 The Imbalanced Dataset

Performance of the bridging algorithm with an imbalanced dataset was evaluated by taking the average f-score of bridging with a balanced dataset, then examining any degradation in performance as the rates of imbalance were increased. To this end we first designated one of the majority classes in the test set as the target class and joined together the remaining 7 to form an 'other' class. Selecting 2, 4, 6, 8, 10, 15, 20, 30, 40, and 50 instances respectively from the target class, and then filling the remainder of the training set with random instances from the 'other' class we then created ten separate training sets of 100 instances, each representing varying degrees of imbalance. Any remaining images not used in any of the above training sets were then used as a test set.

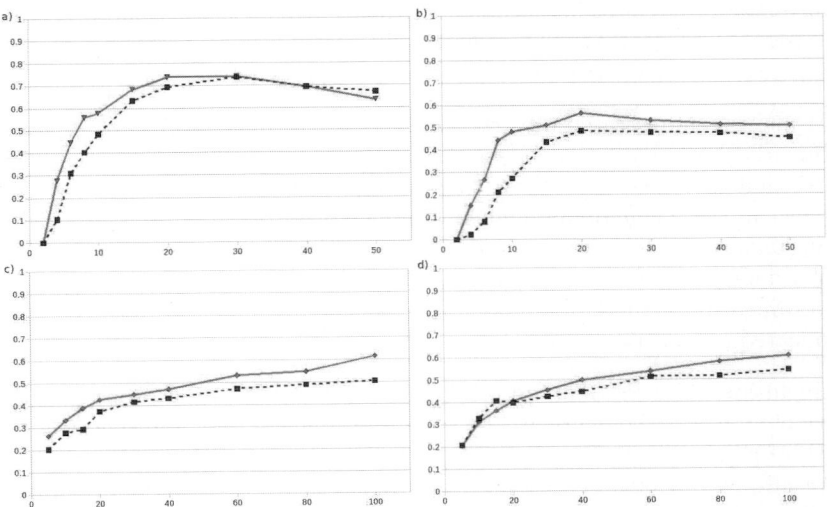

Fig. 2. Performance of descriptors with Bridging (solid line) verses without Bridging (dotted line). (a) - (b) Target F-Score vs. Number of target class examples for the Combined MPEG-7 Visual Descriptors and the Colour Layout descriptor using the Histogram-Intersection and Dot-Product metric for feature vector comparison respectively (c) - (d) Weighted Average F-Score vs. Percentage of database for Colour Layout and Edge Histogram using the Dot-Product for feature vector comparison.

Figure 2a and 2b are two examples of the results obtained when using bridging to compensate for the imbalanced dataset problem. They demonstrate that the application of bridging is able to provide a significant improvement in classifier performance, particularity at high rates of imbalance. The improvement in Figure 2b is more substantial and consistent than in Figure 2a, particularly when the datasets are more or less balanced.

5.2 Lack of Available Training Examples

In order to evaluate the performance of the Bridging algorithm with a lack of available training examples, we observed the changes in performance of the classifier on a progressively smaller dataset. A 50/16.6/33.3 train/dev/test split on the entire dataset was used, then the training split was used to create datasets comprising 100, 80, 60, 40, 30, 20, 15, 10 and 5% of the total database size respectively while maintaining the relative distribution of the classes. The results displayed in Figure 2c and Figure 2d demonstrate the ability of bridging to compensate for this problem. The performance of the bridging algorithm is equal to or better than the baseline.

6 Conclusions and Further Work

In this paper we have successfully redesigned and applied the Bridging algorithm to the problem of homogeneous texture image classification and we have demonstrated an improvement in classifier performance when faced with a range of difficulties. The contributions made by this paper are:

- We have demonstrated that the Bridging algorithm is applicable in domains outside short string text classification.
- We demonstrated that the Bridging algorithm can be applied to positive effect in the image domain when dealing with a range of problems including an imbalanced training set and a lack of available training examples.
- We demonstrated that within the image domain reasonable results can be obtained using the Bridging algorithm while employing a relatively uncomplicated comparison metric.

The inclusion of background knowledge into the classification process raises the question of from where are we to obtain this data. The Internet contains a wealth of knowledge on an almost limitless number of topics and is therefore a potentially endless source of background knowledge. Zelikovitz et al have previously used the Internet as a source of background knowledge in the domain of short string text classification [17]. One potential direction for further study would be automatically obtaining the corpus of background knowledge from the Internet.

References

1. Vistex texture database, http://www.white.media.mit.edu/vismod/imagery/VisionTexture/vistex.html
2. Blum, A., Mitchell, T.: Combining labeled and unlabeled data with co-training. In: COLT 1998: Proceedings of the eleventh annual conference on Computational learning theory, pp. 92–100. ACM, New York (1998)
3. Ertekin, S., Huang, J., Lee Giles, C.: Active learning for class imbalance problem. In: SIGIR 2007: Proceedings of the 30th annual international conference on Research and development in information retrieval, pp. 823–824 (2007)
4. Goldman, S.A., Zhou, Y.: Enhancing supervised learning with unlabeled data. In: ICML 2000: Proceedings of the Seventeenth International Conference on Machine Learning, pp. 327–334 (2000)
5. Japkowicz, N.: The class imbalance problem: Significance and strategies. In: Proceedings of the 2000 International Conference on Artificial Intelligence (ICAI), pp. 111–117 (2000)
6. Liapis, S., Tziritas, G.: Color and texture image retrieval using chromaticity histograms and wavelet frames. IEEE Transactions on Multimedia 6(5), 676–686 (2004)
7. Manjunath, B.S., Ohm, J.-R., Vasudevan, V.V., Yamada, A.: Color and texture descriptors. IEEE Transactions on Circuits and Systems for Video Technology 11(6), 703–715 (2001)
8. Park, D.K., Jeon, Y.S., Won, C.S.: Efficient use of local edge histogram descriptor. In: MULTIMEDIA 2000: Proceedings of the 2000 ACM workshops on Multimedia, pp. 51–54. ACM, New York (2000)
9. Seiffert, C., Khoshgoftaar, T.M., Van Hulse, J., Folleco, A.: An empirical study of the classification performance of learners on imbalanced and noisy software quality data. In: IRI, pp. 651–658 (2007)
10. Singh, M., Singh, S.: Spatial texture analysis: A comparative study. In: ICPR 2002: Proceedings of the 16th International Conference on Pattern Recognition (ICPR 2002), vol. 1, p. 10676 (2002)
11. Singh, S., Sharma, M.: Texture analysis experiments with meastex and vistex benchmarks. In: Singh, S., Murshed, N., Kropatsch, W.G. (eds.) ICAPR 2001. LNCS, vol. 2013, pp. 417–424. Springer, Heidelberg (2001)
12. Spyrou, E., Le Borgne, H., Mailis, T., Cooke, E., Avrithis, Y., O'Connor, N.E.: Fusing MPEG-7 visual descriptors for image classification. In: Duch, W., Kacprzyk, J., Oja, E., Zadrożny, S. (eds.) ICANN 2005, Part II. LNCS, vol. 3697, pp. 847–852. Springer, Heidelberg (2005)
13. Swain, M.J., Ballard, D.H.: Indexing via color histograms. In: Proceedings, Third International Conference on Computer Vision, pp. 390–393 (1990)
14. Weng, C.G., Poon, J.: A data complexity analysis on imbalanced datasets and an alternative imbalance recovering strategy. In: WI 2006: Proceedings of the 2006 IEEE/WIC/ACM International Conference on Web Intelligence, Washington, DC, USA, pp. 270–276. IEEE Computer Society, Los Alamitos (2006)
15. Witten, I.H., Frank, E.: Data mining: practical machine learning tools and techniques with Java implementations. ACM SIGMOD Record 31(1), 76–77 (2002)
16. Zelikovitz, S.: Using background knowledge to improve text classification. PhD thesis, New Brunswick, NJ, USA. Director-Hirsh, Haym (2002)
17. Zelikovitz, S., Hafner, R.: Automatic generation of background text to aid classification. In: FLAIRS Conference. AAAI Press, Menlo Park (2004)

Detection of Ambiguous Patterns Using SVMs: Application to Handwritten Numeral Recognition

Leticia Seijas and Enrique Segura

Departamento de Computación, Facultad de Ciencias Exactas y Naturales,
Universidad de Buenos Aires
Pabellón I, Ciudad Universitaria (C1428EGA) Buenos Aires, Argentina
{lseijas,esegura}@dc.uba.ar

Abstract. This work presents a pattern recognition system that is able to detect ambiguous patterns and explain its answers. The system consists of a set of parallel Support Vector Machine (SVM) classifiers, each one dedicated to a representative feature extracted from the input, followed by an analysing module based on a bayesian strategy in charge of defining the system answer. We apply the system to the recognition of handwritten numerals. Experiments were carried out on the MNIST database, which is generally accepted as one of the standards in most of the literature in the field.

Keywords: pattern recognition, support vector machine, ambiguous pattern, answer explanation, bayesian statistics.

1 Introduction

Optical character recognition (OCR) is one of the most traditional topics in the context of Pattern Recognition that includes as a key issue the automatic recognition of handwritten characters. The subject has many interesting applications, such as automatic recognition of postal codes, recognition of amounts in banking checks and automatic processing of application forms. Handwritten numeral classification is a difficult task because of the wide variety of styles, strokes and orientations of digit samples. One of the main difficulties lies in the fact that the intra-class variance is high, due to the different forms associated with the same pattern, because of the particular writing style of each individual. Many models have been proposed to deal with this problem, but none of them has succeeded in obtaining levels of response comparable to human ones.

The use of Support Vector Machines (SVMs) has provided good results in handwritten digit recognition due to its good generalization performance even in high dimensional space and under small training set conditions[1][2].

This work proposes a SVM based pattern recognition system with a probabilistic strategy in order to classify, detect ambiguous patterns and explain answers. This proposal is based on a model previously introduced by our group, consisting basically of a hybrid unsupervised, self-organising model, followed by a supervised stage [3]. We apply the system to the recognition of handwritten digits. Our experiments were carried out on the MNIST handwritten digit database, which is generally accepted as a standard in most of the literature in the field.

This work is organized as follows: in section 2 the recogniser structure is explained. In section 3 we present implementation details and experimental results. Concluding remarks are presented in section 4.

2 Recognition System

The first stage of the process consists of a pre-processing of input data in which relevant features are extracted from the patterns. This provides a more general and simple structure for the system, specifically oriented to classification rather than feature selection that is therefore independent from system architecture. The recogniser is composed of two levels. The first one is formed by a collection of parallel and independent classification elements implemented with SVM systems, each one specialised in a different feature. The second level consists of an analysing module in charge of defining and explaining the output of the system. This module is integrated by the following elements: the table of reliability and two parameters adjustable while running the system. Each classification element in the first level produces a response to an input pattern, as a judge who, only based on the analysis of the corresponding feature, decides which class the pattern belongs to. The connection between the first and second levels of the system is performed through this new representation of the pattern, formed with the answers of the "judges". The purpose of the table of reliability is to represent how trusty the "vote" of each classification element is. Using this data, the module of analysis of the second level has to produce the final answer. The system

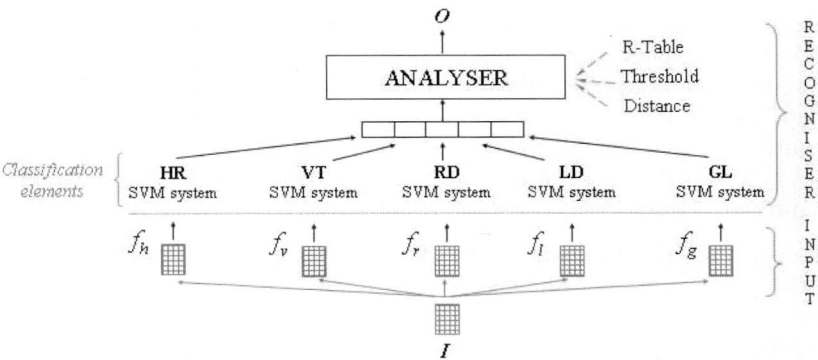

Fig. 1. Architecture of the proposed recogniser. Each SVM system is associated with a feature. In the analyser: table of reliability (R-Table), threshold of reliability and minimal distance parameter.

is able to explain its responses, indicating which class is most similar to the input pattern respecting each particular feature, on the base of the vote of each classification element/judge and the weight assigned to each one. As a part of the explanation, if a pattern is ambiguous for the system, we can know which other digits it could be identified with (i.e. which classes it has more features in common). The general structure of the system is depicted in Figure 1.

2.1 SVMs Level

The performance of a character recognition system strongly depends on how the features that represent each pattern are defined. In the first level of the recogniser, each classification element is trained on the base of a certain feature previously defined in the pre-processing stage according to the problem to deal with. In the context of handwritten numeral recognition, local detection of line segments and global detection of line structures seem to be an adequate feature extraction method. Kirsch masks [4] have been used as directional feature extractors by several authors [5][6], as they allow local detection of line segments. We used these masks to extract four directional features from the set of patterns: horizontal, vertical, right diagonal, left diagonal. In addition, we also considered the complete (original) pattern, which we call global feature. Then the dimension of each feature associated with a given pattern are considerably reduced applying the Cohen-Daubechies-Feauveau (CDF) 9/7 wavelet transform [7].

Hence we defined five SVM systems for the first layer of the recogniser, each one dedicated to a particular feature. The method of Support Vector Machine was proposed by Vapnik as a machine learning system based on statistical learning theory [8]. It has been sucessfully applied to the handwritten digit recognition problem by several authors [1] [2]. Instead of fitting nonlinear curves to the data, SVM uses a kernel function to map the data into a different space where a hyperplane can separate 2 classes. The main idea is to build the hyperplanes as decision boundaries by using the fitting kernel such as Gaussian, polynomial or linear classifiers. Then, the hyperplanes try to split the positive examples from the negative examples and maximize the distance of the marginal separation between classes. To implement our experiments we used TORCH [9], a machine-learning library specifically tailored for large-scale problems. We applied the Gaussian kernel function because of its high performance reported by several authors [2][6] for this kind of classification problem.

2.2 Analyser Level

Once the classification elements of the first level have been trained, the second level of the recogniser has to be constructed, being its parts: a reliability table and the parameters "confidence threshold" and "minimal distance". For constructing the table that expresses how reliable the answer of each SVM system is, a Bayesian probabilistic approach was chosen because of its better results. Using the definition of conditional probability and the multiplication rule, we define the probability of pattern p belonging to class C given that feature classification element f has responded C for input p as:

Table 1. Reliability Table: probabilities for each classifier and each class

Feat	0	1	2	3	4	5	6	7	8	9
HR	0.925	0.959	0.957	0.905	0.970	0.938	0.965	0.958	0.866	0.927
VT	0.978	0.972	0.965	0.952	0.960	0.966	0.985	0.932	0.945	0.923
RD	0.967	0.974	0.971	0.960	0.960	0.937	0.960	0.972	0.920	0.958
LD	0.975	0.983	0.961	0.944	0.955	0.979	0.967	0.957	0.939	0.928
GL	0.992	0.983	0.972	0.967	0.958	0.978	0.980	0.977	0.973	0.950

$$P(\{p,C\}/\{f,p,C\}) = \frac{P(\{f,p,C\}/\{p,C\})P(\{p,C\})}{P(\{f,p,C\})} \quad (1)$$

$P(\{p,C\})$: probability of input pattern p belonging to class C, estimated from the labeled training set. $P(\{f,p,C\})$: this probability is estimated from trained SVM system f. We assume that the response C of a SVM system f given an input pattern p is independent from responses of other classification elements. $P(\{f,p,C\}/\{p,C\})$: this probability is estimated from correct outputs of SVM system f given input patterns of class C.

In the classification stage, once the input pattern has been represented by the five votes of the SVM systems, a score is computed for each voted class, on the base of the reliability of each classification element, according to the values showed in Table 1. For this sake, such values are added for each class, so that a class with a greater score implies more reliable answers and more votes for the same class. The score s for each class was computed as:

$$s_C = \sum_{f \in F_C} r_{C,f} \quad (2)$$

where C indicates the selected class, f indicates SVM system associated with a feature, F_C the SVM systems that voted class C, $r_{C,f}$ reliability value taken from table for class C and feature f.

One of the main difficulties for classification is dealing with outliers, "ambiguous patterns", since the distortions they exhibit make difficult their correct classification (being far away from the mean value of its class, they could be incorrectly associated with another class closer in average). We have considered the distance from the pattern represented by its feature vector to the mean value of the class assigned by the SVM system: if the pattern is close to that mean value, we assume that it is well defined and belongs to that class. A pattern far away from the centroid might be consider as an outlier (according to the variance of the class) and hence a candidate to "ambiguous pattern". This information is used as a reinforcement factor for the score given by the table. Then, the score s assigned to each voted class C is calculated as

$$s_C = \sum_{f \in F_C} r_{C,f} \frac{1}{d(p_f, \mu_{f,C})} \quad (3)$$

where C indicates the selected class, f indicates SVM system associated with a feature, F_C the SVM systems that voted class C, $r_{C,f}$ reliability value taken from table for class C and feature f, p_f pattern represented by its feature vector associated with SVM system f, $d(p_f, \mu_{f,C})$ normalized distance between the feature vector of the pattern p and the mean value of class C for feature f.

As an example, the following vector V shows votes for each feature for a given test input: V = (5 2 5 0 0). Score associated with each voted class is computed using (3). In this case, scores are 2.84, 1.98 and 3.27 for classes "0", "2" and "5" respectively. Class "5" obtains the higher score.

In order to define the system output the class with the higher score is identified; this score is compared with the reliability threshold that determines which patterns are considered as ambiguous and which are not. If the total score for the winning class surpasses the threshold, then the system considers that pattern as well defined and the answer is that class. On the other hand, if the cumulative score is lower than the threshold, the system decides that the pattern is ambiguous. In this case it is necessary to determine the class it might be confused with. From all voted classes, the one closest to the winner is selected if distance between it and the winning class is lower than minimal distance parameter.

Values for the threshold of reliability and for the minimal distance are chosen empirically, on the base of information provided by the training set in the stage of adjustment of the classifier. Variation of these parameters permits to adjust the output of the system without need of a new training of the SVMs.

3 Experiments

3.1 The Data Set

The MNIST database of handwritten numerals is widely accepted as a standard benchmark to test and compare performances of pattern recognition and classification methods and it was used to perform our experiments. It contains 70,000 unconstrained handwritten numerals, including many different writing styles collected from a larger set available from NIST (National Institute of Standards and Technology) of the U.S. Department of Commerce. It has a training set of 60,000 samples and a test set of 10,000 samples. Each digit in the database is centered in a 28 x 28 graylevel image.

As a pre-processing stage Kirsch masks were applied on each image, as mentioned in Section 2.1. Then the CDF 9/7 wavelet transform [10][7] was applied in order to obtain a smaller descriptor associated with each feature for a given input image. The Discrete Wavelet Transform has been mainly used for image compression. We applied one step of the CDF 9/7 obtaining three detail subbands and one approximation subband LL from the vertical and horizontal convolutions with low and high pass filters. The LL subband constitutes a descriptor that preserves the structure and shape of the image in size a fourth of the original one. This means that each 28 x 28 image descriptor was reduced in size to 14 x 14. As the final step in the preprocessing stage, the coefficients of the LL image were thresholded, via rescaling them to the interval [0,1] and rounding the result. Hence we obtained a shape-preserving binarized smaller version of each digit in order to reduce the computational cost and to make the clasification process easier for large amounts of data.

3.2 Recognition Results

We have implemented the pattern recognition system described in previous sections. Results for different values of reliability threshold and minimal distance are shown in Table 2.

Table 2. Recognition results (%) - RT: reliability threshold - MD: minimal distance

RT	MD	Correct (includes ambiguous)	Correct (unique response)	Error
4.0	3.0	98.65	94.94	1.35
6.0	3.0	98.97	93.68	1.03
6.0	4.0	99.08	90.14	0.92
6.0	5.0	99.11	89.55	0.89

As threshold increases, patterns associated with greater values of scores of winning classes will result well defined, and the rest will be considered as patterns with a certain degree of similarity with elements of other classes. Using the minimal distance enables to introduce a second class of output for these patterns, as long as such class has a score near enough to that which obtained the maximum. If that is not the case, the output for this not clearly defined pattern is unique. Table 3 shows some data forming the output of the system for some digits in Figure 2, grouped by class. Digits in the first column are well defined, i.e. they are not ambiguous for the system. The first row of each group in Table 3 shows that all classifiers voted for the same class; however, in the definition of the output not only the number of votes takes part. The second and third columns in Figure 2 show patterns that the system considers as ambiguous. It can be observed in Table 3 that the output indicates two possible classes for the pattern, and one of them is the right one. The votes are distributed between different classes, hence the score of winning classes is lower than scores of well

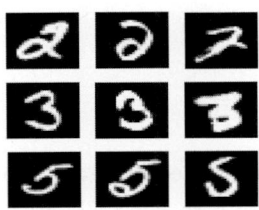

Fig. 2. Test patterns correctly classified. Each row: classes "2", "3" and "5" respectively.

Table 3. Some results over testing set: reliability threshold 6.0 - minimal distance 5.0

Class	Sys.Out.	Ambig	HR vote	VT vote	RD vote	LD vote	GL vote
2	2	No	2	2	2	2	2
2	2 or 8	Yes	3	2	8	2	2
2	7 or 2	Yes	2	7	7	9	2
5	5	No	5	5	5	5	5
5	5 or 0	Yes	5	2	5	0	0
5	5 or 8	Yes	3	8	5	5	8

Table 4. Recognition rates of different methods on MNIST database

Method	Recognition Rate %
PCA+Polynomial[11]	96.7
LeNet4[11]	98.3
Bhattacharyya Distance[12]	98.2
Proposed method	99.1
SVM Affine Distortion[1]	99.4

defined patterns. Visual analysis shows that the third "2" is in fact similar to a "7"; for the rest of the ambiguous patterns, forms can be observed that are not associated with a unique class. For example, the third pattern labelled as a "5" might well be an incomplete "8".

On the other hand, our strategy of combining the response of each individual classifier improves the error rate associated with each SVM system (error %:6.6, 4.4, 4.7, 4.1, 2.6 for features HR, VT, RD, LD and GL respectively) besides allowing an analysis of the system response. Table 4 compares published results of different methods on the MNIST database with the proposed method. Our classifier outperforms other approaches, as LeNet4 [11] based on a complex architecture, or the one based on the Bhattacharyya distance combined with a kernel approach [12]. In [1] a trainable feature extractor based on the LeNet5 convolutional neural network architecture with SVMs performing the classification task is presented. Additionally, new samples generated by affine transformations and elastic distortions are added to the training set. Our system is simpler than theirs and, although the error rate is higher, our strategy is quite effective and allows managing outliers and ambiguities, a remarkable property of the proposed recogniser.

4 Conclusions

We have presented a fully supervised system for pattern recognition that combines the use of SVM independent classifiers with a probabilistic bayesian approach. Besides of its simplicity, one of the notable features of our technique is the ability to manage outliers and ambiguities. The "reliability table" -estimating how reliable is the answer of each SVM system for an input pattern- and a few parameters are used to decide when a pattern is considered as ambiguous, and which class or classes it might be confused with.

The classifier was applied to the recognition of handwritten digits. Experiments were carried out on the MNIST handwritten digit database. The patterns were preprocessed extracting directional features from each digit, and reducing the input dimension to a fourth of the original one.

The recognition rate obtained with our method is high (99.11%). Our strategy of combining the response of each individual classifier improves the error rate associated with each SVM system, thus justifying the good overall performance obtained and allowing an analysis of the system response.

References

1. Lauer, F., Suen, C., Bloch, G.: A trainable feature extractor for handwritten digit recognition. Pattern Recognition 40, 1816–1824 (2007)
2. Oliveira, L., Sabourin, R.: Support vector machines for handwritten numerical string recognition. In: 9th IEEE International Workshop on Frontiers in Handwritten Recognition, pp. 39–44. IEEE Computer Society, Washington (2004)
3. Seijas, L., Segura, E.: Detection of ambiguous patterns in a SOM based recognition system: application to handwritten numeral classification. In: 6th International Workshop on Self-Organizing Maps. Bielefeld University, Germany (2007)
4. Pratt, W.: Digital Image Processing. Wiley, New York (1978)
5. Gorgevik, D., Cakmakov, D.: An efficient three-stage classifier for handwritten digit recognition. In: 17th Int. Conf. on Pattern Recognition, vol. 4, pp. 507–510 (2004)
6. Liu, C., Nakashima, K., Sako, H., Fujisawa, H.: Handwritten digit recognition: benchmarking of state-of-the-art techniques. Pattern Recognition 36, 2271–2285 (2003)
7. Skodras, A., Christopoulos, C., Ebrahimi, T.: JPEG 2000: The upcoming still image compression standard. Pattern Recognition Letters 22, 1337–1345 (2001)
8. Vapnik, V.: The Nature of Statistical Learning Theory. Springer, New York (1995)
9. Collobert, R., Bengio, S.: SVMTorch: Support vector machines for large-scale regression problems. Journal of Machine Learning Research 1, 143–160 (2001)
10. Daubechies, I.: Ten lectures on wavelets. Soc. Indus. Appl. Math. (1992)
11. LeCun, Y., Jackel, L., et al.: Comparison of learning algotirhms for handwritten digit recognition. In: Int. Conf. on Artificial Neural Networks, Paris, pp. 53–60 (1995)
12. Wen, Y., Shi, P.: A novel classifier for handwritten numeral recognition. In: IEEE Int. Conf. on Acoustics, Speech, and Signal Processing, pp. 1321–1324. IEEE Signal Processing Society, Las Vegas (2008)

Accurate 3D Modelling by Fusion of Potentially Reliable Active Range and Passive Stereo Data

Yuk Hin Chan[1], Patrice Delmas[1], Georgy Gimel'farb[1], and Robert Valkenburg[2]

[1] The University of Auckland, Auckland, New Zealand
ycha171@aucklanduni.ac.nz, {p.delmas,g.gimelfarb}@auckland.ac.nz
[2] Industrial Research Ltd. (IRL), Auckland, New Zealand
r.valkenburg@irl.cri.nz

Abstract. Possibilities of more accurate digital modelling of 3D scenes by fusing 3D range data from an active hand-held laser scene scanner developed in IRL and passive stereo data from stereo pairs of images of the scene collected during the scanning process are discussed. Complementary properties of two data sources allow for improving a 3D model by checking reliability of active range data and using it to adaptively guide passive stereo reconstruction. Experiments show that this avenue of the data fusion offers good prospects of error detection and correction.

1 Introduction

Geometrically and visually accurate 3D scene models are required in applications such as mobile robotics, surveying and mapping, computer-assisted training, video games, and crime scene examination. Active range scanners and passive stereo sensors that provide 3D measurements for scene modelling have different and often complementary strengths and weaknesses. Range scanner collects data sequentially, often by traversing a scene embedding all objects of interest. Stereo reconstruction uses either a single image pair acquired instantly or a sequence of images. Range data can be dense or sparse for different parts of a scene and is mostly accurate although prone to errors in some particular cases (e.g. for sharp edges, thin objects, and surfaces with poor reflection properties). Stereo data is dense but typically unreliable on textureless (uniformly coloured) surfaces and on repetitive textures. Both active and passive techniques may fail on large surface discontinuities caused by partial occlusions.

Fusion of active range and passive stereo data for the same scene is actively explored in various domains (mostly for autonomous navigation of robotic vehicles [1, 2, 3] and 3D scene modelling [4, 5, 6]) as a promising way to obtain more accurate and reliable 3D models. Initial approaches have combined image analysis and range data to separate objects of interest (e.g. road obstacles) and evaluate their spatial positions [7, 4, 8]. Better results were obtained by using range data to limit search zones of correlation-based passive stereo and combine both data sets into a unique depth map where points with inconsistent range and stereo measurements were labelled as outliers [9, 6]. These approaches use

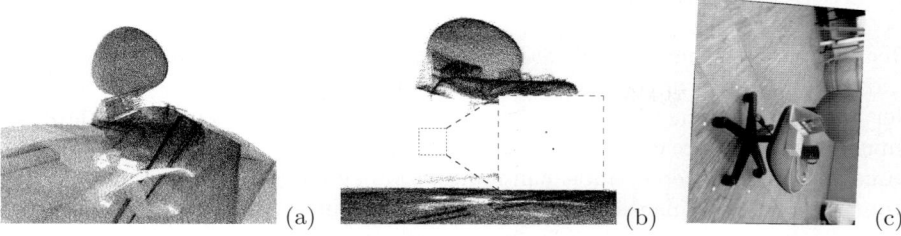

Fig. 1. Data from the IRL scene scanner: (a) the "cloud" of 3D points; (b) an example of the data missed: the marked area corresponds to a lever on the chair shown on the rectified left image of a stereo pair in (c). The lever is small and dark, so it returned only 2 range data points shown enhanced in the inset.

simple local stereo matching and eliminate range data contradicting stereo data by sequential heuristic criteria.

Considerably more advanced fusion strategy was outlined in [10] for a sensor system with a time-of-flight range scanner (200K points per field-of-view) and a pair of cameras. Complex photometric and geometric calibration of the system was verified for a test scene using ground truth generated by a separate structured lighting device. A depth map produced by a belief propagation based stereo matching was combined with the range data using a Markov random field model. The overall depth error of range data was reduced from 1.8% down to 0.6% comparing to the ground truth.

In principle, such a model-based fusion permits to not only use reliable stereo data and image analysis to fill in gaps and detect inaccuracies in range data but also guide stereo matching with reliable range data for improved accuracy and completeness. An assumption of only a pair of range and stereo measurements per field-of-view point in [10] does not hold for most of outdoor and large indoor scenes. Practical requirements to 3D scene models and physical limitations of range and stereo sensors dictate sequential scanning of a scene from different positions and in multiple directions. Thus the numbers of available range, stereo, and image data for each scene voxel may vary considerably.

This paper investigates possibilities of range and stereo data fusion for a hand-held 3D laser time-of-flight scene scanner developed in IRL [11]. Here, we focus on more accurate stereo reconstruction of visible 3D surfaces guided by reliable range data. Real-time calibration and registration of the scanner to a world coordinate system uses pre-placed beacons. The operator sweeps over a static scene covering more densely surfaces of interest and glossing over other scene features. The scanner produces a "cloud" of 3D points indicating a surface of interest (Fig. 1(a)). During scan sweeps, a colour camera mounted on the scanner outputs a video stream that captures scene texture. Both the scanner and camera are continuously calibrated. This allows us to build different rectified stereo pairs of the scene (Fig. 1(c)).

2 Reliability and Fusion of Range and Stereo Data

Real-world scenes have complex arrangements of objects with multiple occlusions and frequent uniform (non-textured) or poorly reflecting surfaces that may hinder range measurements or/and 3D stereo reconstruction. The range data and images of a scene are collected from many different positions and directions. The same surfaces of interest appear in multiple range scans and stereo pairs formed from the rectified images. The range measurements typically do not cover the whole scene and are dense only on objects of interest, whereas a large part of the scene is represented densely at different scales only by stereo data.

Formal statements and solutions of the data fusion problems depend on data collection scenarios. For definiteness, we restrict our consideration to relatively small 3D scenes formed by discontinuous surfaces of complex shape, covered by many different individual range scans, and depicted in many accompanying calibrated images while only a single stereo pair is used for 3D reconstruction in addition to the range data. Generally, range scanning is more accurate than stereo matching; it therefore seems natural to use the range data to guide stereo reconstruction in order to amplify its accuracy. Still, on some surfaces the situation is quite the opposite. Figure 1(b) shows inaccurate and missing (due to poor reflection and limited resolution) range data points near the hand lever of a chair. When a foreground object is not detected but the background is visible for the scans from other directions, the erroneous guidance affects considerably the accuracy of stereo reconstruction. As shown in Fig. 2(A), the range data from the background floor misguides the stereo algorithm for both the lever and the central column of the chair which have no range data available. Repeat scans of each surface from other directions may resolve some ambiguities at the expense of considerably larger time and complexity of data collection.

Fusing reliable and consistent range and stereo data holds more promise in accurate modelling of 3D scenes. Most of the known data fusion frameworks consider a range measurement reliable if it fits the results of stereo matching. Large difference between these data values may indicate not only a possible error from the range scanner, but also an erroneous stereo reconstruction. Moreover, similar values do not necessarily imply their reliability if colour inconsistencies exist between the corresponding points in stereo images.

When images taken from different directions are used to check the colour consistency of a particular 3D point, a mixture of colours will generally appear due to possible occlusions. The ambiguities can be resolved by comparing the mixture modes to the colours fetched for that point from the stereo pair assuming that the point is visible for more than a prescribed fraction of the images. In principle, such a comparison can account also for possible local contrast and offset deviations in the images. Differences between potentially equal colours allow us to build colour consistency indicators that evaluate the range data reliability. Then, reliable range measurements is used to improve the accuracy of a dense 3D model obtained by stereo matching. Comparing to the known fusion processes (e.g. in [9, 6, 10]), our approach not only accounts for multiple variants of the same 3D surfaces after the range and stereo measurements, but

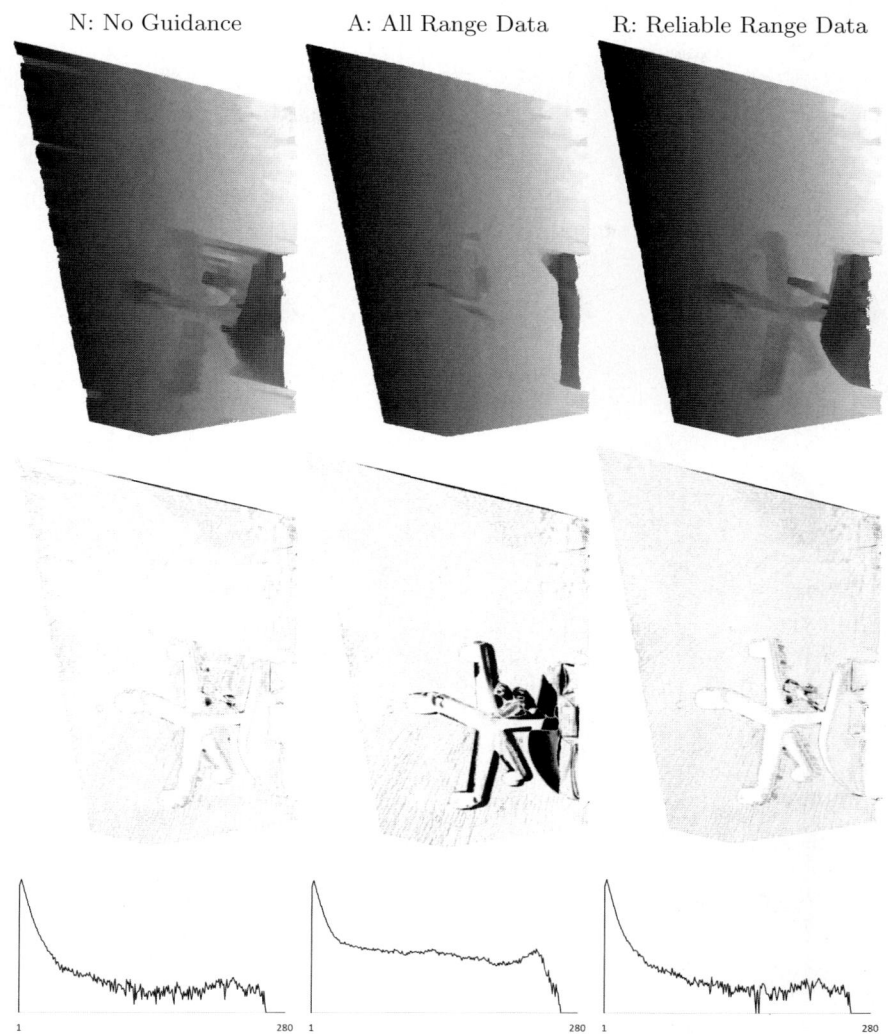

Fig. 2. Depth maps (top) obtained by stereo algorithm without and with guidance from all or only potentially reliable range data points and enhanced colour differences between corresponding pixels for the maps (middle) and corresponding histograms with logarithmic frequency scale (bottom)

also suggests the concurrent use of their reliability indicators to refine the final dense 3D model.

Our two-stage 3D scene modelling framework first conducts multiple independent range scans of a scene-of-interest, and a rectified stereo pair that depicts the scene is selected to conduct an independent stereo reconstruction. Next, the colour consistency indicators are evaluated for projections of all 3D surface

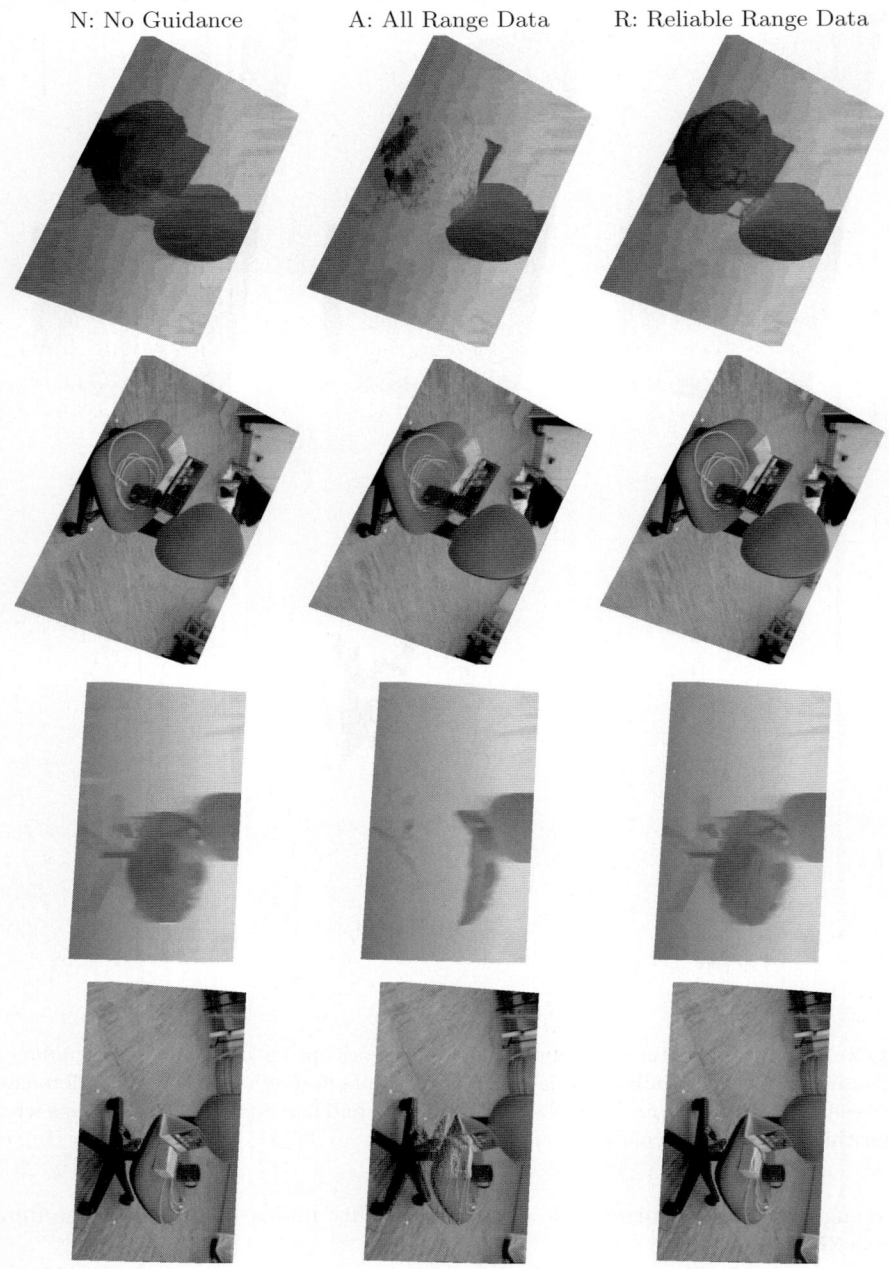

Fig. 3. Depth maps and cyclopean views obtained by stereo algorithm without and with guidance from all or only potentially reliable range data points

points produced by the range scanner, and a dense scene model is built by an adaptive stereo matching process constrained by the range data with due account of reliability.

The 3D data sets obtained initially from the stereo pair and the laser scanner are compared in order to update the stereo matching process using only the potentially reliable range data. The latter is deemed unreliable when the absolute difference between the colours that correspond in a stereo pair to that 3D point does not meet a matching criterion. The stereo matching is now guided by the range data with due account of its current reliability. The process can be iterated using the resulting absolute colour differences to update the reliability indicators. The low reliability indicates reconstruction errors as well as outliers.

To obtain the guidance map we select those 3D points which correspond to depth data produced by both range measurement and stereo reconstruction and project them onto the stereo pair. First, the histogram of the absolute colour differences between the pixels that correspond to all these points is computed. Next, the stereo matching process uses all the guidance points (i.e. points with both stereo and range data) to restrict the search for acceptable stereo matches. Then, the changes of the distributions of the colour differences are used to separate the unreliable range data and create a confidence mask of the guiding points. Finally, the stereo matching conducted with a reliable guidance map. Further assessment of the range data with respect to colour consistency, colour edges, uniformly coloured image segments, and textured segments are currently under investigation but not demonstrated here.

3 Experimental Results and Conclusions

We provide results for unguided and guided stereo-matching following the above reliability based framework on our test image (see Figs. 1(c), 2, and 3). The test scene contains a chair with various small objects that exhibit different textures and colour characteristics as well as a background floor. The stereo algorithm used is a version of Symmetric Dynamic Programming Stereo (SDPS) [12] modified to accept colour images and a (possibly sparse) depth map for per-pixel guidance. The initially obtained range and unguided stereo data are shown on Fig. 1(a), 2(N), and 3(N) respectively. The data sets have both the reliable and unreliable components as is shown by the unguided stereo matching (N) and the guided one using all the range data (A). The reliability of the range data was estimated from the colour difference histograms displayed in Fig. 2 (bottom) and the object colour. The much larger "tail" in Fig. 2(A)(bottom) indicates that using all range points for the guidance produces many more points with large colour differences, indicating the unreliable range data. Pure black regions with zero colour components were also considered as unreliable. The unreliability threshold for colour differences was set manually to 8 after analysing the histograms – in the future we are going to use a mixture identification technique such as in [13] for this setting.

The potentially reliable range data found allowed us to considerably improve the resulting 3D models shown in Figs. 2(R) and 3(R) . One more iteration of the algorithm with an updated guidance map did not change noticeably the obtained reliability indicators and the final model.

Data comparisons and enhancement as well as performance evaluations in the previous works often rely on the ground truth data acquired either with a separate optical measuring device [10] or by an user assessment [14,9]. Unfortunately, this ground truth is difficult to measure for experiments involving large natural scenes such as our test scene with objects lying on a chair. To obtain numerical comparisons, the ground truth data on points with both range and stereo data is obtained by using a graphics editing program to manually compare the disparity between the stereo pair. This process is a laborious but reliable way to produce ground truth data on complex real-world scenes where accurate measurement using other techniques is not available. The effort required to produce accurate ground truth data also limits the amount of numerical results in our experiments.

In terms of the 3 stereo pairs, the test scene was depicted on about 400,000 pixels, with only 50% of them having both the range and stereo matching data. The comparisons have been conducted on 153,414 points having both the range and stereo data as well as the manually found ground truth data. A criterion for "good pixels" was the difference from the ground truth within ± 1 disparity level. The percentage of such pixels in the obtained model increased from 87.3% originally to 93.8% after the proposed iterative refinement (the number of the "good" pixels increased by 9.1% among the guided ones and by 5.4% among the points excluded from the guidance). Qualitatively, the reliable guidance improved the reconstruction of the homogeneous areas such as the floor and the back of the chair as well as on small objects where range data is unreliable such as the hand lever and objects lying on the chair.

This clearly demonstrates that stereo matching guided by potentially reliable range data largely resolves the problem of "large differences" between the active range and passive stereo measurements first encountered. Checking the range data on consistency with the stereo data in terms of image correspondences and using estimates of data reliability to guide the stereo matching process does suppress the large stereo matching errors due to features exhibiting poor or inconsistent information.

References

1. Knoll, A., Schröder, R., Wolfram, A.: Generation of dense range maps by data fusion from active and passive colour stereo vision. In: Proc. 1996 IEEE/SICE/RSJ Int. Conf. Multisensor Fusion and Integration for Intelligent Systems (MFI 1996), Washington, USA, December 1996, pp. 410–415 (1996)
2. Scherba, D., Bajscy, P.: Depth Estimation by Fusing Stereo and Wireless Sensor Locations. Technical Report alg04-005, Automated Learning Group, National Center for Supercomputing Applications, Champaign, IL 61820 (December 12, 2004)

3. Kim, M.Y., Lee, H., Cho, H.: Dense range map reconstruction from a versatile robotic sensor system with an active trinocular vision and a passive binocular vision. Applied Optics 47, 1927–1939 (2008)
4. Biber, P., Fleck, S., Duckett, T.: 3d modeling of indoor environments for a robotic security guard. In: Proc. IEEE Computer Society Conf. Computer Vision and Pattern Recognition - Workshops, Advanced 3D Imaging for Safety and Security (SafeSecur 2005), San Diego, CA, USA, June 20-25, vol. III, pp. 124–124 (2005)
5. Curtis, P., Yang, C.S., Payeur, P.: An integrated robotic multi-modal range sensing system. In: Proc. Instrumentation and Measurement Technology Conf. (IMTC 2005), Ottawa, Canada, May 2005, pp. 1991–1996 (2005)
6. Kuhnert, K., Stommel, M.: Fusion of stereo-camera and pmd-camera data for real-time suited precise 3d environment reconstruction. In: IEEE/RSJ International Conference on Intelligent Robots and Systems, Beijing, China, October 2006, pp. 4780–4785 (2006)
7. Morgenthaler, D., Hennessy, S., DeMenthon, D.: Range-video fusion and comparison of inverse perspective algorithms in static images. IEEE Transactions on Systems, Man, and Cybernetics SMC-20(6), 1301–1312 (1990)
8. Frintrop, S.: VOCUS: A Visual Attention System for Object Detection and Goal-Directed Search. LNCS (LNAI), vol. 3899. Springer, Heidelberg (2006)
9. Perrollaz, M., Labayrade, R., Royere, C., Hautiere, N., Aubert, D.: Long range obstacle detection using laser scanner and stereovision. In: Proceedings of the IEEE Intelligent Vehicles Symposium, Tokyo, Japan, June 2006, pp. 182–187 (2006)
10. Zhu, J., Wang, L., Yang, R., Davis, J.: Fusion of time-of-flight depth and stereo for high accuracy depth maps. In: Proc. 2008 IEEE Computer Society Conference on Computer Vision and Pattern Recognition (CVPR 2008), Anchorage, Alaska, USA, June 24-26, pp. 1–8. IEEE Computer Society, Los Alamitos (2008)
11. Valkenburg, R.J., Penman, D.W., Schoonees, J.A., Alwesh, N.S., Palmer, G.T.: Interactive hand-held 3d scanning. In: Delmas, P., James, J., Morris, J. (eds.) Proc. Int. Conf. Image and Vision Computing New Zealand 2006, Great Barrier Island, Auckland, November 2006, pp. 245–250 (2006)
12. Gimel'farb, G.: Probabilistic regularisation and symmetry in binocular dynamic programming stereo. Pattern Recognition Letters 23, 431–442 (2002)
13. Farag, A.A., El-Baz, A., Gimel'farb, G.: Precise segmentation of multimodal images. IEEE Transactions on Image Processing 15(4), 952–968 (2006)
14. Lu, Z., Hu, Z., Uchimura, K., Kutoba, H., Ono, M.: Sensor fusion with occupancy fusion map for pedestrian detection in outdoor environment. In: Proc. IEEE International Conference on Vehicular Electronics and Safety (ICVES 2008), Columbus, Ohio, USA, September 22-24, pp. 199–204 (2008)

Rapid Classification of Surface Reflectance from Image Velocities

Katja Doerschner[1,*], Dan Kersten[2], and Paul Schrater[2,3]

[1] National Research Center for Magnetic Resonance (UMRAM) and Dept. of Psychology, Bilkent University
[2] Dept. of Psychology, University of Minnesota
[3] Dept. of Comp. Science & Eng., University of Minnesota
katja@bilkent.edu.tr, {kersten,schrater}@umn.edu

Abstract. We propose a method for rapidly classifying surface reflectance directly from the output of spatio-temporal filters applied to an image sequence of rotating objects. Using image data from only a single frame, we compute histograms of image velocities and classify these as being generated by a specular or a diffusely reflecting object. Exploiting characteristics of material-specific image velocities we show that our classification approach can predict the reflectance of novel 3D objects, as well as human perception.

Keywords: specular flow, rapid surface reflectance classification, velocity histogram, material perception, spatio-temporal filtering.

1 Introduction

Identifying the surface reflectance of an object is a fundamental problem in vision. Reflectance provides important information about the object's material and identity, and given known reflectance, algorithms for shape reconstruction exist for both, diffuse and specular surfaces [1]. However, because of the strong differences in the image motion generated by specular and diffuse surfaces, unknown reflectance is a serious problem for these methods. Previous work on diffuse vs. specular reflectance classification has relied on specific assumptions and conditions, such as the tracking of surface features during known camera motion [2], known surface shape [3], the use of structured lights [4], color [5], or a specific reflectance model [6].

Evidence from human vision, however, suggests that monocular image motion across a few frames provides sufficient information to classify a surface as diffuse or specular, e.g. [7] showed that static objects with ambiguous apparent reflectance could be unambiguously classified as shiny or matte when in motion. Additionally, [8] demonstrated that it is also possible to generate reflectance illusions from motion: under certain conditions, rotating specular objects look matte (also see [9]). What aspects of specular motion explain both, the rapid material classification and the perceptual errors? Although specular motion patterns

---

[*] This work has been supported in part by the EC FP7 Marie Curie IRG-239494.

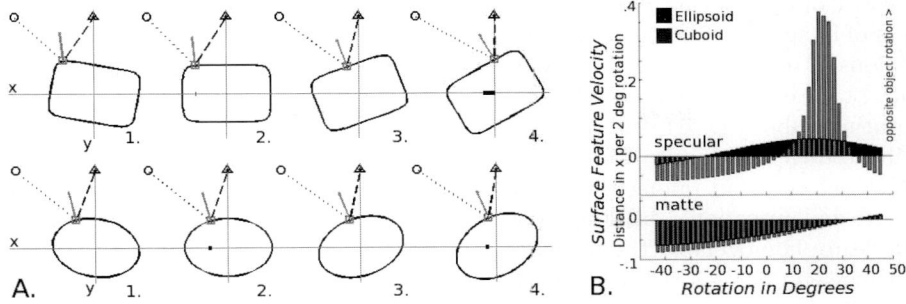

Fig. 1. Specular Velocity and Curvature Variability. A. Cross-sections through 3D scenes. The position of the 2D camera (triangle) and a point light source (circle) are fixed. We find the surface normal at the point on the object where the specular feature (square) will be visible to the camera. "Specular velocity" is measured as the distance traveled by the specular feature in x (indicated by fat black line) as the object rotates 10° counterclockwise around its origin. Consider the cuboidal cross-section: 1. The specular feature (sf) appears on a high curvature point and "sticks" to this region as the object rotates. 2. The sf moves some distance *in the direction* of object rotation. 3. The sf appears on a low curvature point. After a 10° rotation the distance that it has traveled, now in *opposite the direction* of object rotation, has nearly doubled. Compare this to the sf on the ellipsoid. **B.** Sf velocities for specular (upper plot) and surface feature velocities for diffusely reflecting (lower plot) objects per 2° rotation. See text for details.

can be quite complex, we will show that simple statistical measures on image velocities can be used to classify moving objects as specular or diffusely reflecting, without any additional assumptions or conditions. We will demonstrate that these classifiers can predict human perception, as well as the material of novel objects. Rapid methods for reflectance classification, such as the one proposed here, constitute an important step towards a fully automated vision system.

2 Specular Flow

The relative displacement of a specular feature or highlight due to camera or observer motion (or, conversely due to object motion relative to a stationary camera/observer), is negatively related to the magnitude of surface curvature [10,11], i.e. specular features "rush" across low curvature regions and "stick" to points of high curvature. In contrast, all points on a moving diffusely reflective surfaces stick. This suggests that the distribution of velocities across a moving object may contain important information about the object's material, because all specular surfaces with sufficient curvature variation undergoing a generic motion will have both low velocity "sticky" points and high velocity points, while diffusely reflective surfaces will have only "sticky" points. Moreover, except for rotations around the viewing axis, the flow generated by a rigid body motion will have a principle direction of motion.

For example, for an in-depth rotating specular object (Fig. 1A) the distribution of image velocities generated by the specular flow across the object will have regions of relatively high and low magnitude, whose specific range is directly related to the magnitude and range of surface curvatures. As an extreme case, a rotating cube, (0 curvature across sides and positive curvature at the corners) will produce two kinds of image velocities: high ones, opposite to the direction of object rotation (along the sides) and those congruent with object rotation speed and direction ("sticking" to corners). As an object increases in surface curvature homogeneity the resulting range of image velocities will decrease, the extreme end being a rotating specular sphere: it will produce image velocities of magnitude and range 0. This *velocity variability* can be exploited for reflectance classification: high image velocity variability, which can be easily identified from the image velocity histogram, appears to be crucial to induce the spatio-temporal characteristics associated with perceived shininess [8]. Conversely, specular objects with low curvature variability will, when rotated, generate low variability image velocity distributions which are, not surprisingly, not distinct from those generated by diffusely reflecting objects (Fig. 1B).

3 Implementation

General Strategy. To rapidly classify reflectance properties from image velocities our strategy was to 1) estimate velocities from rotating *specular* objects using spatio-temporal filters, 2) find the principal direction of motion, and 3) classify the velocity histogram in that principal direction using 3 different approaches: parametric, and non-parametric density estimation, as well as non-negative matrix factorization. We chose to classify movies on the basis of histogram velocities because we expected the velocity signature of specular or matte (appearing) reflectances to be largely object (identity) invariant (but see Section 2 for the special role of 3D curvature). Furthermore, by focusing on the principal direction of motion we achieve object motion invariance.

Spatio-temporal Filtering. We filtered image sequences by directionally selective filters G_2 (second derivative of a 3D Gaussian) and H_2 (and its Hilbert Transform) at orientations $(\alpha, \beta, \gamma)_i$ [12].

$$f^{\Omega}(x, y, z) = G(r)Q_N(x') \qquad (1)$$

are the even and odd filters formed by a nth order polynomial $Q_N(x')$[1] times a separable windowing function $G(r)$ (e.g. a Gaussian-like function), both of which are assumed to be rotationally symmetric. **R** is the transformation that these functions are rotated by such that their axis of symmetry points along the direction of cosines α, β and γ. We estimated velocities from the filter coefficients using the max-steering method of Simoncelli [13]. Subsequent analysis of these velocities was restricted to include velocity samples only from *within*

---

[1] $x' = \alpha x + \beta y + \gamma z$.

object boundaries in order to avoid contamination with boundary motion. Velocities were sampled from a grid indicated by the colored dots in Fig. 2C.

Parametric and Non-parametric Density Estimation. We performed principle components analysis on image velocities to estimate the dominant direction of motion for a given movie frame. Image velocities were projected onto this direction vector. To develop a statistical classifier for reflectivity we estimated the conditional probabilities of the projected velocities for both diffuse and specular objects. To verify our results did not depend on the details of a specific density estimation learning procedure, we used three different density learning approaches.

Histograms. Histogram densities were estimated with a generalized cross-entropy density estimator [14] that uses a gaussian kernel and data-driven bandwidth selection. To classify a given movie frame into shiny or matte we used histogram estimates of the conditional densities of velocity ξ given shiny S, $P(\xi|S)$, and matte M, $P(\xi|M)$, from image sequences judged shiny and matte in [8]. A sample velocity ξ' from a test image sequence was classified by comparing the likelihood ratio $P(\xi'|S)/P(\xi'|M)$ against a threshold k^2. Note, that we also used the value of the likelihood ratio as a graded material measure for the data set. Graded measures are particularly useful for comparisons to human perception, as discussed below.

Mixture of Gaussians. To confirm that the shape of a given histogram was indeed driven by "diagnostic" (high and low curvature) regions we fitted a Mixture of Gaussians with two components [15], and computed the posterior probability of each pixel given either Gaussian distribution. Pixel classifications are illustrated by mapping the samples back onto the frame they were taken from. From the two estimated Gaussian means (μ_1, μ_2) we compute the velocity contrast of the sample

$$C_b = \frac{|\mu_1 - \mu_2|}{\max(\sigma_1, \sigma_2)} \ . \qquad (2)$$

If $C_b > 1$ the sample is classified as specular, else as matte. The value of C_b also forms a graded material measure.

Mixture of Histograms Using Non-negative Matrix Factorization. To smooth the likelihoods and form a low-dimensional representation for the densities, we factorized the velocity histograms using convolutive non-negative matrix factorization (NNMF) [16]. We preserved 3 components based on an initial estimate that 3 components account for as much as 97% of the approximation error. Because the histogram of a test sequence can be represented as a weighted combination of the 3 components, these weights can be used to represent the velocity distributions of novel objects. To estimate the weights for a novel sequence, we maximized the likelihood of the total sample evaluated on the components with

---

[2] k was obtained by a bootstrapping procedure used to constrain the false alarm rate to 5%.

respect to the weights. The best fitting weight values were used to classify a sample as shiny or matte.

Movies. The test set consisted of 36 movies (6 shapes x 6 light probes) of rotating specular superellipoids (http://bilkent.edu.tr/~katja/g_run.html). Objects were constructed according to

$$1 = \left[\left| \frac{x}{r_x} \right|^{\frac{2}{n_2}} + \left| \frac{y}{r_y} \right|^{\frac{2}{n_2}} \right]^{\frac{n_2}{n_1}} + \left| \frac{z}{r_z} \right|^{\frac{2}{n_1}} . \qquad (3)$$

We set $r_x = 1$ and $r_y = r_z = 0.64$. Surface curvature was determined by setting n_1, n_2 to: $0.3, 0.5, 0.7, 0.8, 0.9$ or 1.0 (Fig. 2A). Each object rotated in depth. Its angular speed was adjusted $(0.1, 0.35, 0.61, 0.74, 0.87, 1.0°/frame)$ such that the resulting image velocities were in the range that our filters were sensitive to.

4 Experimental Results

Histograms. Figure 2B illustrates the characteristic changes that the velocity histogram undergoes as the object decreases in surface curvature variability (left to right). Table 1 shows normalized Log-Likelihood Ratios (LLR) for all histograms testing H_0 that a given histogram has been generated by a matte object.

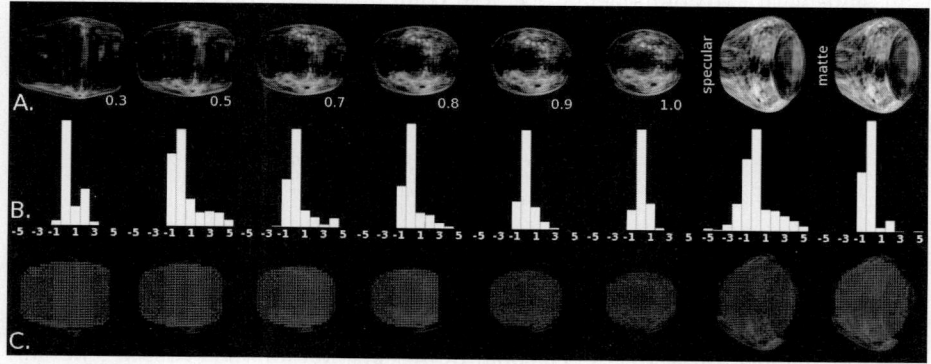

Fig. 2. Renderings, Histograms, and Pixel Classification. A. Sample frames for superellipsoids (SE) and for the specular and diffusely reflecting Utah Teapot. Numbers indicate values for n_1, n_2 in Eq.(3). SEs were rendered under 6 different light probes: 2 natural (L1 ("grace"), L3 ("uffizi") from http://gl.ict.usc.edu/Data/HighResProbes/), 2 partially- (L2, L4), 2 fully phase-scrambled (L3, L6) versions of L1 and L3, respectively. For each movie 40 512x512 images were rendered with *Radiance* [17]. **B.** Corresponding velocity histograms. **C.** Corresponding pixel classification results. See text for details.

Table 1. Normalized Log-Likelihood Ratios. Values larger than k ($k = 0.16$) (in bold) were classified as shiny with a predicted error rate of less than 5%. Training data are indicated by ^{T}.

Light Probe	Superellipsoid shape coefficient n_1, n_2					
	0.3	0.5	0.7	0.8	0.9	1.0
L1	**1.000**^{T}	**0.362**	0.145	0.153	0.114	0^T
L2	**0.961**	**0.362**	**0.184**	**0.215**	0.139	0.031
L3	**0.877**	**0.365**	**0.184**	**0.270**	0.103	0.011
L4	**0.749**	**0.267**	**0.178**	0.114	0.114	0.003
L5	**0.766**	**0.476**	**0.223**	**0.187**	0.142	0.014
L6	**0.805**	**0.368**	0.159	**0.187**	0.148	0.003
Average	**0.860**	**0.367**	**0.179**	**0.188**	0.127	0.010

Table 2. Average C_b. The average was computed across light probes for superellipsoids with shape coefficients $n_1 = n_2$ from 0.3 (cuboidal) to 1 (ellipsoidal). Values > 1 (in **bold**) indicate that the velocity histogram was classified as bimodal, which could be a rough predictor of material shininess. Compare the relative magnitudes of values to average observer ratings in Table 3.

Light Probe	Superellipsoid shape coefficient n_1, n_2					
	0.3	0.5	0.7	0.8	0.9	1.0
Average C_b	**1.658**	**1.4143**	0.6824	0.7247	0.4778	0.1341

Mixture of Gaussians Pixel Classification. Figure 2C shows that the simple velocity distribution measure was successful in roughly identifying image regions of high (blue pixels) and low (orange pixels) velocities. Purplish colors indicate that the sample could come from either Gaussian distribution. Note, that the distinctiveness of the high and low velocity regions decreases as the amount of the surface curvature variability decreases: in the corresponding two-Gaussian model fit, the two components approach a uni-modal mixture. The measure C_b exploits the bi-modality of specular velocity distributions to classify the material of test sequences (see Table 2).

Non-negative matrix factorization. The distribution of estimated weights across the stimulus set is shown in Fig. 3A. Ellipsoidal objects' velocity histograms (multiples of 6) tended to have high weights on component 2 (solid triangle) whereas most cube-like objects tended have high weights on components 1(circle) and/or 3(square). A very simple shininess criterion can be computed by taking the ratio of the weights of the 2 "specular components" and the weight of the "matte component" e.g. $C_w = 1/2(w_{f1} + w_{f3})/w_{f2}$, with values larger than 1 being classified as specular (see Fig. 3B).

Objective Classification of Material of Novel 3D Objects. To verify that the velocity distribution can be sufficient for objectively classifying material we tested an object with more complex shape variation. We generated 40 frames of a rotating version of the Utah "Teapot". This object was rendered with a diffuse

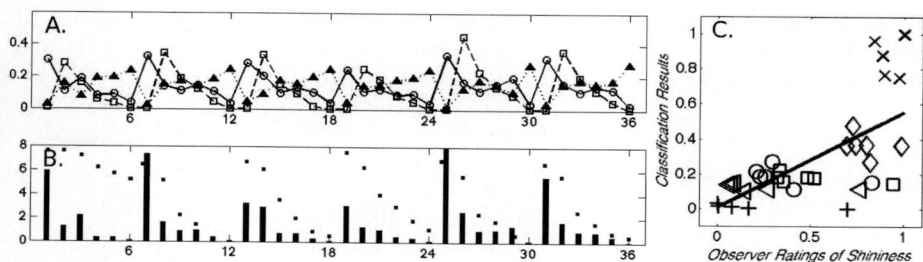

Fig. 3. NNMF of velocity histograms. A. Estimated weights for our test set. **B.** Average values of C_w : $5.4, 1.8, 1.0, 0.7, 0.5, 0.06$. The black square on top or next to each bar indicates average observer data for the same movie (note, observer values are plotted on a different scale). **C.** Regression of histogram classifications onto observer data. See text for details.

Table 3. Human Shininess Ratings. Shown are ratings for 2 light probes (those eliciting highest and lowest shininess ratings) as well the average data (across all light probes and observers). Differences in relative apparent shininess for different light probes is consistent with previous research [19]. In the experiment observers rated apparent shininess of all 36 light probe – shape combinations.

Light Probe	Perceived Shininess of Shape n_1, n_2					
	0.3	0.5	0.7	0.8	0.9	1.0
L1	0.9740	0.9635	0.9219	0.8125	0.7552	0.6927
L3	0.8229	0.6875	0.3385	0.2292	0.0938	0.0365
Average	0.8872	0.7830	0.4991	0.3837	0.2578	0.1962

[18] and with a specular reflectance (see Fig. 2A (right)). We evaluated the sequence using histograms, mixture of Gaussians, and NNMF approaches. Teapots were correctly classified as shiny and matte for all three methods. Histograms: LLR specular and diffusely reflecting teapot were 0.26 (classified as shiny) and 0.008 (classified as matte). Mixture of Gaussians: C_bs for specular and diffusely reflecting teapot were 1.16 (classified as shiny), and 0.87 (classified as matte). NNMF: The specular teapot classified as shiny $C_w = 33.2$, and the diffusely reflecting teapot was classified as matte $C_w = 0.7954$.

Predicting Human Perception. In the experiment 4 observers indicated via keyboard press on a scale from 1 (matte) - 7 (mirror reflection) how shiny a given superellipsoid appeared. A subset of results are reported in Table 3. Additional experimental details can be obtained from [8]. Regressing normalized LLRs (Table 1) onto normalized observer data (Fig. 3) yielded $R^2 = 0.45, p < 0.00001$. Repeating the analysis with only the most shiny and matte data points yielded $R^2 = 0.75, p = 0.0003$. Training data was excluded from the regression.

5 Discussion

We provide a first account of how to rapidly classify surface reflectance from a single frame of object motion, without any assumptions. We show that moving diffusely reflecting, and specular objects with sufficient curvature variability, generate distinct image velocity distributions whose respective characteristics can be captured by simple, invariant statistical measures. Our results account for the misperception of material in [8,9], demonstrating that diffusely reflecting and apparently matte objects, i.e. those that are specular but with insufficient surface curvature variability, share the same velocity histogram characteristics. Thus, we were able to correctly classify a diffusely reflecting object on the basis of a classifier that was trained on a matte-appearing (but physically specular) object. In future work we will extend our analysis to a velocity region-based approach.

References

1. Ihrke, I., Kutulakos, K., Lensch, H., Magnor, M., Heidrich, W.: State of the Art in Transparent and Specular Object Reconstruction (2008)
2. Oren, M., Nayar, S.: A Theory of Specular Surface Geometry. International Journal of Computer Vision 24(2), 105–124 (1997)
3. Roth, S., Black, M.: Specular Flow and the Recovery of Surface Structure. In: Proc. of the IEEE Conference on Computer Vision and Pattern Recognition (CVPR), vol. 2, pp. 1869–1876 (2006)
4. Healey, G., Binford, T.: Local shape for specularity. Jones and Bartlett Publishers, Inc., USA (1992)
5. Nayar, S., Fang, X., Boult, T.: Removal of specularities using color and polarization. In: 1993 IEEE Computer Society Conference on Computer Vision and Pattern Recognition. Proceedings CVPR 1993, pp. 583–590 (1993)
6. Chung, Y.-C., Chang, S.-L., Cherng, S., Chen, S.-W.: Dichromatic Reflection Separation from a Single Image. In: Yuille, A.L., Zhu, S.-C., Cremers, D., Wang, Y. (eds.) EMMCVPR 2007. LNCS, vol. 4679, pp. 225–241. Springer, Heidelberg (2007)
7. Hartung, B., Kersten, D.: Distinguishing shiny from matte. J. Vis. 2(7), 551–551 (2002)
8. Doerschner, K., Kersten, D.: Perceived rigidity of rotating specular superellipsoids under natural and not-so-natural illuminations. J. Vis. 7(9), 838–838 (2007)
9. Roth, S., Domini, F., Black, M.: Specular Flow and the Perception of Surface Reflectance. J. Vis. 3(9), 413–413 (2003)
10. Koenderink, J., Van Doorn, A.: Photometric invariants related to solid shape. Journal of Modern Optics 27(7), 981–996 (1980)
11. Blake, A.: Specular stereo. In: Proc. Int. J. Conf. on Artificial Intell., pp. 973–976 (1985)
12. Derpanis, K., Gryn, J.: Three-dimensional nth derivative of Gaussian separable steerable filters. In: IEEE International Conference on Image Processing (2005)
13. Simoncelli, E.: Distributed analysis and representation of visual motion. Ph.D. Thesis, Massachusetts Institute of Technology, Department of Electrical Engineering and Computer Science, Cambridge, MA (1993)

14. Botev, Z., Botev, Z.: A Novel Nonparametric Density Estimator. The University of Queensland (2006)
15. Nabney, I.: NETLAB: algorithms for pattern recognition. Springer, Heidelberg (2002)
16. O'Grady, P.D., Pearlmutter, B.A.: Convolutive non-negative matrix factorisation with a sparseness constraint. In: Proceedings of the IEEE International Workshop on Machine Learning for Signal Processing (MLSP 2006), Maynooth, Ireland, September 2006, pp. 427–432 (2006)
17. Larsen, G., Shakespeare, R.: Rendering with Radiance: The Art and Science of Lighting Visualisation (1998)
18. Fleming, R.: Rendering Sticky Reflections with Radiance. Personal Communication (2007)
19. Fleming, R., Dror, R., Adelson, E.: Real-world illumination and the perception of surface reflectance properties. Journal of Vision 3(5), 347–368 (2003)

Structure-Preserving Regularisation Constraints for Shape-from-Shading

Rui Huang and William A.P. Smith

Department of Computer Science, The University of York, UK
{rui,wsmith}@cs.york.ac.uk

Abstract. In this paper we present a new framework for shape-from-shading which relies on a novel regularisation term which preserves surface structure. The resulting algorithm is both robust and accurate. We show that it can recover stable surface estimates from both synthetic and real world images of complex objects, even under extreme illumination.

1 Introduction

Shape-from-shading is a classical problem in computer vision which has attracted over four decades of research [1,2]. The aim is to estimate surface shape, typically in the form of surface normals, given a single intensity image. The problem is underconstrained and proposed solutions have, in general, made strong assumptions in order to make the problem tractable. However, even when these assumptions are satisfied (for example in a synthetically produced image) existing shape-from-shading algorithms still fail to recover accurate surface shape from images of complex objects.

Minimization methods are a traditional and robust way to solve the shape from shading problem, first proposed by Horn [3]. These methods try to optimize the brightness error subject to additional regularisation constraints, such as surface smoothness or integrability. Worthington and Hancock [4] treated the image irradiance equation as a hard constraint. Their idea was to use robust regularisers to optimise the solution within the space of solutions which strictly minimise the brightness error. Prados and Faugeras [5] used viscosity solutions to solve the partial differential equation which arises from the shape-from-shading problem. Their method accounts for perspective projection effects but assumes frontal illumination.

More recently, several authors have posed shape-from-shading in terms of pairwise Markov Random Fields [6,7,8]. Haines and Wilson [7] describe surface normal direction probabilistically in terms of a Fisher-Bingham distribution for each pixel. Belief propagation is used to solve for the undetermined degree of freedom for each surface normal. Although their model provides an elegant formulation of the problem, in practice the results obtained are unconvincing. Potetz [8] also uses belief propagation but solves for the surface gradient at each pixel. The framework uses factor nodes to capture irradiance, smoothness and integrability constraints. Solving the resulting model is extremely difficult and results are only shown for a single synthetic image.

In this paper we present a new shape-from-shading algorithm which obtains a solution with zero brightness error while seeking to preserve surface structure. In practice, our algorithm provides improved results over existing methods on a wide range of imagery of complex objects.

2 Solving Shape-from-Shading

The aim of computational shape-from-shading is to make estimates of surface shape from the intensity measurements in a single image. Since the amount of light reflected by a point on a surface is related to the surface orientation at that point, in general the shape is estimated in the form of a field of surface normals (a needle-map). Assuming a normalised and linear camera response, the image intensity predicted by the simplest Lambertian reflectance model is given by

$$g(\mathbf{N}, \mathbf{L}, \rho_d) = \rho_d \mathbf{N} \cdot \mathbf{L}, \qquad (1)$$

where $\mathbf{N}$ is the local surface normal, $\mathbf{L}$ is a vector in the light source direction and ρ_d is the diffuse albedo which describes the intrinsic reflectivity of the surface.

For an image in which the viewer and light source directions are fixed, the radiance function reduces to a function of one variable: the surface normal. This image-specific function is known as the *reflectance map* in the shape-from-shading literature. The squared error between the observed image intensities, $I(x, y)$, and those predicted by the estimated surface normals, $\mathbf{N}(x, y)$, according to the chosen reflectance model is known as the *brightness error*:

$$\mathcal{E}_{\text{Bright}}(\mathbf{n}) = \sum_{x,y} (I(x, y) - g(\mathbf{N}(x, y), \mathbf{L}))^2. \qquad (2)$$

For typical reflectance models, this function does not have a unique minimum. In fact, there are likely to be an infinite set of normal directions all of which minimise the brightness error. In the case of a Lambertian surface there will be a set of normals lying on a cone, all of which have zero brightness error.

2.1 The Variational Approach

In order to make the shape-from-shading problem tractable, the most common approach has been to augment the brightness error with a regularization term, $\mathcal{E}_{\text{Reg}}(\mathbf{n})$, which penalises departures from a constraint based on the surface structure. A wide range of such constraints have been considered, such as surface smoothness and integrability. In some of the earliest work, Horn and Brooks [3] used a simple smoothness constraint in a regularization framework. They used variational calculus to solve the minimisation: $\mathbf{n}^* = \arg\min_{\mathbf{n}} \mathcal{E}_{\text{Bright}}(\mathbf{n}) + \lambda \mathcal{E}_{\text{Reg}}(\mathbf{n})$, where λ is a Lagrange multiplier which effectively weights the influence of the two terms. The resulting iterative solution is [4]:

$$\mathbf{n}_{t+1} = f_{\text{Reg}}(\mathbf{n}_t) + \frac{C(I - \mathbf{n}_t \cdot \mathbf{L})}{\lambda} \mathbf{L}, \qquad (3)$$

where $f_{\text{Reg}}(\mathbf{n}_t)$ is a function which enforces the regularising constraint (in this case a simple neighbourhood averaging which effectively smooths the field of surface normals). Note that the surface smoothness constraint is trivially minimised by a planar surface. The second term provides a step in the light source direction of a size proportional to the deviation of $\mathbf{n}_t$ from the image irradiance equation and seeks to reduce the brightness error.

The weakness of the Horn and Brooks approach is that for reasons of numerical stability, a large value of λ is typically required. The result is that the smoothing term dominates and image brightness constraints are only weakly satisfied. The recovered surface normals therefore lose much of the fine surface detail and do not accurately recreate the image.

2.2 The Geometric Approach

An approach which overcomes these deficiencies was proposed by Worthington and Hancock [4]. Their idea was to choose a solution which strictly satisfies the brightness constraint at every pixel but uses the regularisation constraint to help choose a solution from within this reduced solution space. If we make the assumption that the reflectance properties are homogenous across the surface (i.e. constant unit albedo), we obtain a simple relationship between observed intensity and the angle of incidence, $\theta_i = \angle \mathbf{NL}$, between the light source and surface normal:

$$I(x,y) = \mathbf{N}(x,y) \cdot \mathbf{L} = \cos\theta_i. \tag{4}$$

Geometrically, this means that the surface normal must lie on a right circular cone whose axis is the light source direction and whose half angle is $\theta_i = \arccos(I)$. By constraining the surface normal to lie on the cone, we satisfy the image irradiance equation and hence ensure the fullest possible use of the input image. In essence, the method applies the regularization constraint within the subspace of solutions which have a brightness error of zero: $\mathbf{n}^* = \arg\min_{\mathcal{E}_{\text{Bright}}(\mathbf{n})=0} \mathcal{E}_{\text{Reg}}(\mathbf{n})$. To solve this minimisation, Worthington and Hancock use a two step iterative procedure which decouples application of the regularization constraint and projection onto the closest solution with zero brightness error:

1. $\mathbf{n}'_t = f_{\text{Reg}}(\mathbf{n}_t)$
2. $\mathbf{n}_{t+1} = \arg\min_{\mathcal{E}_{\text{Bright}}(\mathbf{n})=0} d(\mathbf{n},\mathbf{n}'_t)$,

where $d(.,.)$ is the arc distance between two unit vectors and $f_{\text{Reg}}(\mathbf{n}_t)$ enforces a robust regularizing constraint. The second step of this process is implemented using $\mathbf{n}_{t+1} = \Theta\mathbf{n}'_t$, where Θ is a rotation matrix, determined by L, $\mathbf{n}'_t$, and I, which rotates a unit vector to the closest direction that satisfies $\theta_i = \arccos(I)$.

3 A Structure-Preserving Regularisation Constraint

Although the above framework is attractive in that it ensures strict satisfaction of the image irradiance equation, in practice its performance is critically determined by the choice of regularisation constraint. Worthington and Hancock experimented with several "robust" regularisers which sought to smooth surface normal estimates in such a way as to preserve surface structure, i.e. sharp changes in orientation. In other words, their approach was based on an assumption of piecewise smoothness. However, there are two key weaknesses in their approach:

1. Their robust kernels operate on the current surface normal estimates meaning that large changes in surface normal direction which are present at initialisation are exaggerated as the algorithm iterates.
2. The regularisation constraint is imposed as a one-shot, local process. Many normals simply alternate between two positions rather than converging towards a solution that satisfies both constraints.

We propose an alternative regularisation constraint which seeks to address both of these weaknesses. We detect rapid changes in surface orientation (i.e. finescale detail) in terms of change in the incident angle. Our assumption is that adjacent pixels with similar incident angles are likely to have similar surface normal directions. The influence of the regularisation constraint does not spread over discontinuities. We also pose regularisation as its own iterative process which is run to convergence. Although this is computationally expensive, it allows the influence of local surface features to diffuse over the surface.

For a pixel (x, y), we define the local neighbourhood as $\Omega(x, y) = \{(x + 1, y), (x - 1, y), (x, y + 1), (x, y - 1)\}$. We precompute the change in incident angle between all pairs of neighbouring pixels:

$$S((x_1, y_1), (x_2, y_2)) = \frac{|\arccos(I(x_1, y_1)) - \arccos(I(x_2, y_2))|}{\Delta S_{max}}, \quad (5)$$

where ΔS_{max}, which is used to normalize the term, is the largest change in incident angle over the image. We define a weight between adjacent pixels based on the magnitude of the change in incident angle: $W((x_1, y_1), (x_2, y_2)) = e^{KS((x_1,y_1),(x_2,y_2))}$, where the constant K determines the behaviour of the constraint (we use $K = 10$). For small values, the constraint reduces to local smoothness, for large values more structure is preserved at the cost of increased sensitivity to noise.

The total of the weights between a pixel and its neighbours is given by:

$$Z(x, y) = \sum_{(i,j) \in \Omega(x,y)} W((x, y), (i, j)). \quad (6)$$

We impose our structure-preserving regularisation constraint iteratively. The surface normal at pixel (x, y) at iteration $t + 1$ is given by the weighted average of its neighbouring normals at iteration t:

$$\mathbf{N}^{(t+1)}(x, y) = \frac{\boldsymbol{\mu}^{(t+1)}(x, y)}{\|\boldsymbol{\mu}^{(t+1)}(x, y)\|}, \quad (7)$$

where

$$\boldsymbol{\mu}^{(t+1)}(x, y) = \sum_{(i,j) \in \Omega(x,y)} \mathbf{N}^{(t)}(i, j) \frac{W((x, y), (i, j))}{Z(x, y)}. \quad (8)$$

This process is applied iteratively across the surface until convergence. In contrast to a simple surface smoothness update, this process retains surface structure since the weight function ensures smoothing does not occur across sharp changes in surface orientation.

3.1 Implementation

Our algorithm iteratively interleaves the process of imposing the hard constraint and enforcing the regularisation constraint. The update equation to impose the regularisation constraint typically requires 100 to 200 iterations to converge. The shape-from-shading algorithm typically requires around 5 iterations to converge. This means the regularisation update is applied on the order of 1000 times. The whole algorithm runs in around 2.5 minutes using unoptimised Matlab code.

Our implementation makes use of two further standard shape-from-shading constraints. We assume that surface normals at the boundary lie in the image plane and are orthogonal to the tangent to the object boundary. At critical points, where $I = I_{max}$, the surface nromal is constrained to a single direction and hence we can fix $\mathbf{N} = \mathbf{L}$. Boundary and critical point pixels are detected automatically and their normal directions remain fixed.

We initialise our algorithm using the negative gradient method of Worthington and Hancock [4]. This initialisation places each surface normal on its cone in the direction opposite to the local image gradient. This is consistent with an assumption of global convexity.

A summary of the algorithm is as follows:

1. Obtain $\mathbf{N}^{(0)}(x,y)$ using negative gradient initialisation [4]
2. Repeatedly apply (7) until convergence
3. Rotate normals back to cone: $\mathbf{N}(x,y) = \Theta \mathbf{N}^{(\text{final})}(x,y)$
4. Stop if converged, otherwise iterate to step 2

To obtain surface height estimates, we integrate the field of surface normals using the algorithm of Frankot and Chellappa [9].

4 Experiments

We now demonstrate the results of applying our algorithm to both synthetic images (drawn from the Stanford database) and real images (drawn from the

Fig. 1. Surfaces recovered from the input image shown in the top left panel of Fig. 2. From left to right: ground truth, proposed algorithm, [4], [7].

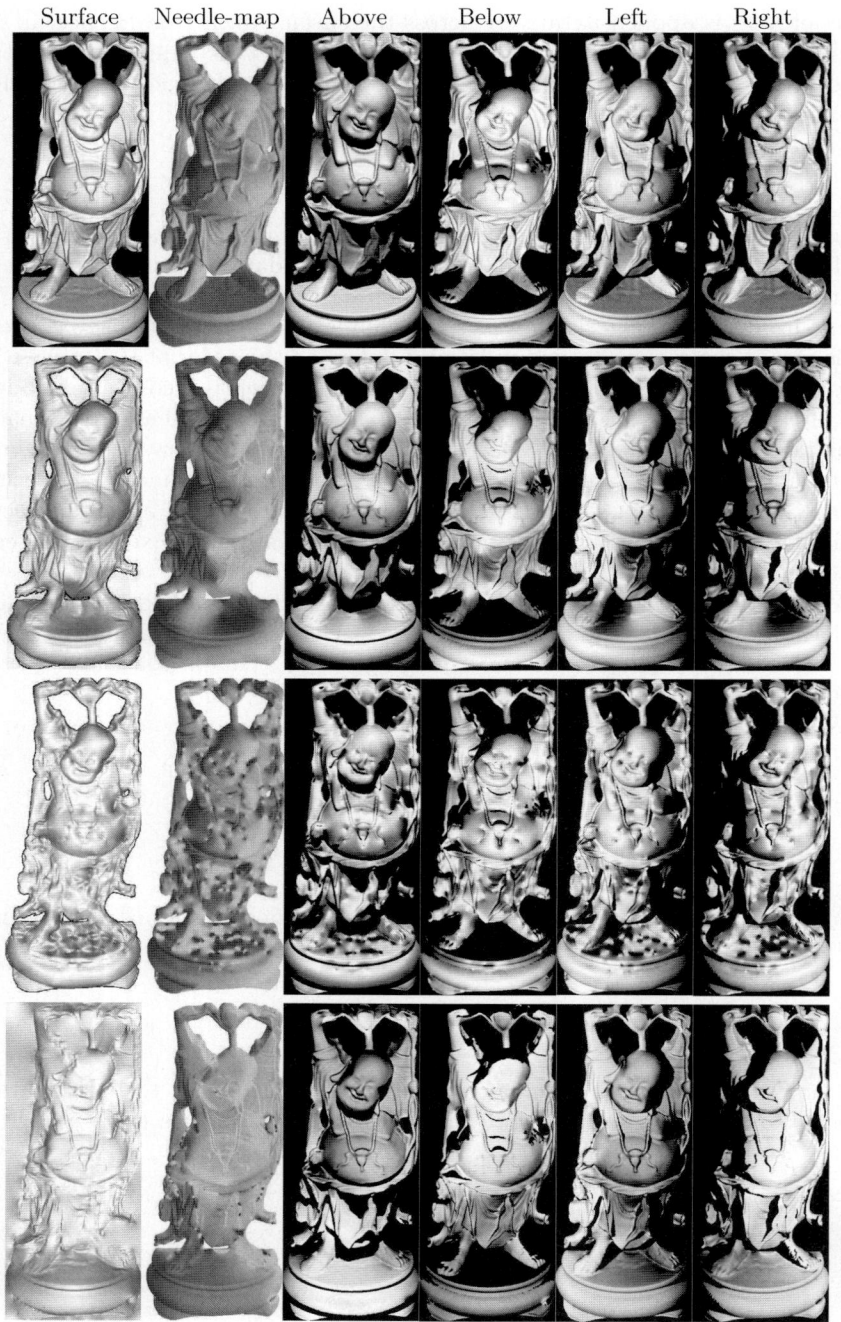

Fig. 2. Recovered surfaces, surface normals and reilluminations. Ground truth in first column, remainder show: proposed algorithm, [4] and [7] respectively.

Coil database). In Fig. 1, we show a novel view of the surfaces recovered using the proposed algorithm, Worthington and Hancock [4] and Haines and Wilson [7]. The corresponding view of the ground truth surface is shown in the first panel. The input image is shown in the top left panel of Fig. 2. Note that the surface recovered by the proposed algorithm has better global structure whilst still containing much of the finescale surface detail.

Fig. 2 shows the recovered surfaces and surface normals along with reilluminations of the surface normals under novel lighting. In column 1 we show the recovered surfaces rendered with frontal illumination. In column 2 we show the surface normals. The remaining columns show renderings under novel illumination. The top row shows ground truth images, the remaining rows show results from the proposed algorithm, Worthington and Hancock and Haines and Wilson respectively. We assume that the light source vector, L, is known. This is used in the initialisation and rotation back to the cone. Note that the quality of the estimated surface shape and reilluminations are considerably improved using the proposed algorithm.

In Fig. 3 we show the result of applying our method to real images from Coil database. The input is shown in the first column, the remaining columns show the surface from different viewpoints. Note that the surfaces in these images deviate from the Lambertian assumption and contain variations in albedo. Despite this, our algorithm recovers stable surface estimates which retain the fine surface detail (e.g. the wing of the duck).

Finally, we show results for non-frontal illumination. The ground truth surfaces in Fig. 4 are illuminated from an extreme angle resulting in much of the surface being in shadow. Note that our algorithm degrades gracefully, recovering unshadowed portions of the surface independently, whilst still retaining much of the global structure.

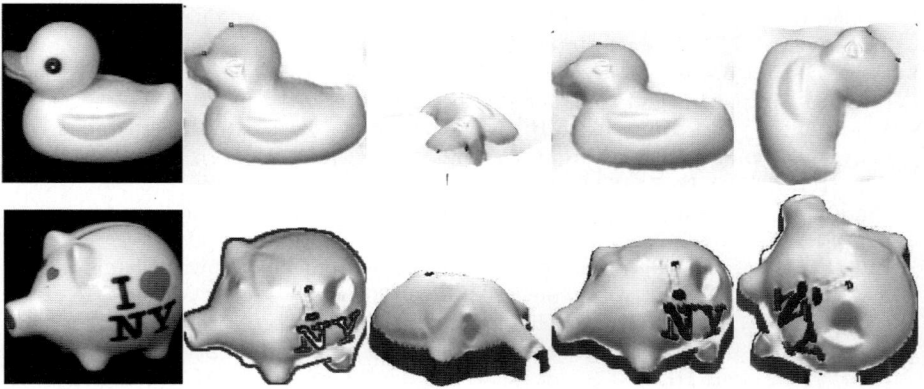

Fig. 3. Surfaces recovered from real images

Fig. 4. Surfaces recovered under extreme illumination

5 Conclusions

We have presented a practical and robust shape-from-shading algorithm which recovers stable surface estimates from a wide range of real and synthetic imagery. By using weights based on the change in incident angle, we are able to preserve structure such that the solution is not oversmoothed. This allows us to apply regularisation update iteratively until convergence, ensuring that the constraint is optimally satisfied. In future work we intend to investigate alternative initialisations and explore how to incorporate integrability constraints within our framework.

References

1. Durou, J.D., Falcone, M., Sagona, M.: Numerical methods for shape-from-shading: A new survey with benchmarks. Comput. Vis. Image Underst. 109(1), 22–43 (2008)
2. Zhang, R., Tsai, P.S., Cryer, J.E., Shah, M.: Shape–from–shading: a survey. IEEE Trans. Pattern Anal. Mach. Intell. 21(8), 690–706 (1999)
3. Horn, B.K.P., Brooks, M.J.: The variational approach to shape from shading. Comput. Vis. Graph. Image Process. 33(2), 174–208 (1986)
4. Worthington, P.L., Hancock, E.R.: New constraints on data–closeness and needle map consistency for shape–from–shading. IEEE Trans. Pattern Anal. Mach. Intell. 21(12), 1250–1267 (1999)
5. Prados, E., Faugeras, O.: Perspective shape from shading and viscosity solutions. In: Proc. ICCV, vol. 2, pp. 826–831 (2003)
6. Han, F., Zhu, S.C.: Cloth representation by shape from shading with shading primitives. In: Proc. CVPR, vol. 1, pp. 1203–1210 (2005)
7. Haines, T.S.F., Wilson, R.C.: Belief propagation with directional statistics for solving the shape-from-shading problem. In: Forsyth, D., Torr, P., Zisserman, A. (eds.) ECCV 2008, Part III. LNCS, vol. 5304, pp. 780–791. Springer, Heidelberg (2008)
8. Potetz, B.: Efficient belief propagation for vision using linear constraint nodes. In: Proc. CVPR, pp. 1–8 (2007)
9. Frankot, R.T., Chellappa, R.: A method for enforcing integrability in shape from shading algorithms. IEEE Trans. Pattern Anal. Mach. Intell. 10(4), 439–451 (1988)

3D Object Reconstruction Using Full Pixel Matching

Yuichi Yaguchi, Kenta Iseki, Nguyen Tien Viet, and Ryuichi Oka

The University of Aizu
Aizu-wakamatsu, Fukushima, 965-8580 Japan
{d8101109,m5121118,m5122105,oka}@u-aizu.ac.jp
http://iplpcx1.u-aizu.ac.jp

Abstract. This paper proposes an approach to reconstruct 3D object from a sequence of 2D images using 2D Continuous Dynamic Programming algorithm (2DCDP) as full pixel matching technique. To avoid using both calibrated images and fundamental matrix in reconstructing 3D objects, the study uses the same approach with Factorization but aims to demonstrate the effectiveness in pixel matching of 2DCDP compared with other conventional methods such as Scale-Invariant Feature Transform (SIFT) or Kanade-Lucas-Tomasi tracker (KLT). The experiments in this study use relatively few uncalibrated images but still obtain accurate 3D objects, suggesting that our method is promising and superior to conventional methods.

1 Introduction

Precise and compatible reconstruction of 3D real world objects from images or video sequences remains a great challenge in computer vision. There have been many image-based modeling methods developed, such as the stereo method [1], shape from shading [2], photometric stereo [3], baseline matching method using epipolar geometry [4,5], Factorization methods [6,7], and shape from silhouettes [8]. Most of these methods achieve their goals under specific conditions and require extra input information, such as internal and external camera parameters or light source position (see Table 1). All of them except Factorization need precise camera parameters, which are contained in a fundamental matrix. Normally, fundamental matrix can be extracted from several calibrated images [9]. An effective approach, Quasi-dense baseline matching, developed by J. Kannala and S.S. Brandt [10], uses the fundamental matrix and seeds provided by SIFT [11] or KLT tracker [12] to obtain more matching points to reconstruct 3D objects. However, materials such as movies and photos taken by ordinary people are difficult to calibrate. Another approach, Factorization, does not require the fundamental matrix and calibrated images to reconstruct 3D objects, so it is still a practical method from this situation.

In Factorization, SIFT and KLT are used as pixel matching techniques. They require small variation of angle in sequences of input images, and can obtain

Table 1. 3D reconstruction methods summary: 'x' indicates (a) Camera parameter, (b) Fixed camera position, (c) Fixed light position, (d) Camera distance, (e) Non-peculiarity scale matrix, (f) Corresponding points, (g) Minimum number of required images

Method	Characteristic	(a)	(b)	(c)	(d)	(e)	(f)	(g)	Note
Stereo method	Parallax + Triangular surveying	x			x		x	2	Principle of human eye
Shape from shading	Reflection coefficient map	x	x	x			x	1	Smooth object
Photometric stereo	Reflection coefficient difference	x	x				x	3	Lambertian surface model
Baseline matching	E/F Matrix + Camera motion	x					x	2	Weak matching noise
Factorization	Pixel correspondence + Motion separation					x	x	3	Affine camera model
Shape from silhouettes	Back projection + Voting	x	x		x			4+	Convex object only

only small number of matching pixel. For this reason, Factorization needs to use numerous input images to increase the amount of matching pixels.

The main objective of this paper is to propose an effective method to reconstruct 3D objects from few uncalibrated images by using the 2DCDP [13] algorithm and then Factorization to calculate points in 3D space. The 2DCDP algorithm in this paper is an advanced implementation of the 2DCDP developed by Yaguchi, Iseki, and Oka [13]. This algorithm preserves 2D pixel correlation and assures continuity and monotonicity in the input image, thereby giving a suitable full pixel matching ability. This fact is quite an advantage over Factorization, which requires many reliable matching points to calculate object shape and camera motion.

Section 2 comprises an overview of the 2DCDP algorithm. Section 3 describes 3D object reconstruction using the Factorization method. Section 4 shows the experimental results for nonparametric 3D object reconstruction. Finally, Section 5 is our conclusion.

2 Image Registration Algorithm

2.1 Definition of the 2DCDP Algorithm

2DCDP is an extension of CDP [14] to 2D correlation, and is an effective algorithm for full-pixel matching (Figure 1(a)). The pixel coordinates of input image S and reference image R are defined by:

$$S \triangleq \{(i,j) | 1 \leq i \leq I, 1 \leq j \leq J\}, R \triangleq \{(m,n) | 1 \leq m \leq M, 1 \leq n \leq N\}. \quad (1)$$

The pixel value at location (i,j) of input image S_p is $S_p(i,j) = \{r,g,b\}$, and the pixel value at location (m,n) of reference image R_p is $R_p(m,n) = \{r,g,b\}$, where r, g, and b are normalized red, green, and blue values respectively, and $0 \leq \{r,g,b\} \leq 1$. We define the mapping $R \to S$, $(m,n) \in R$ and $(\xi(m,n), \eta(m,n)) \in$

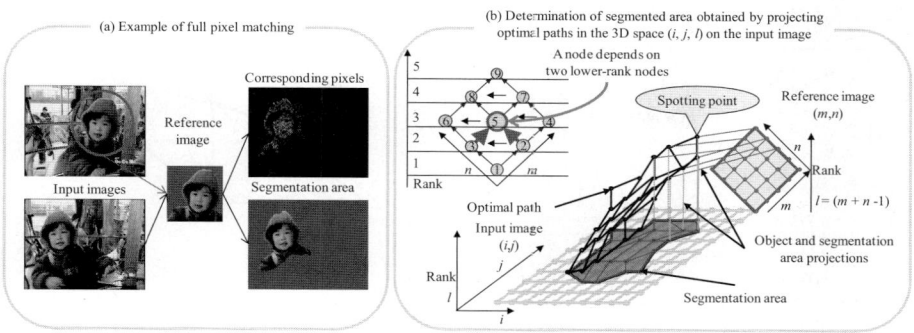

Fig. 1. Full pixel matching overview. (a) An example of full pixel matching; (b) Optimal paths are able to explain an 3D space (i, j, l) on the input image. l is rank, such that $l = m + n - 1$.

S by $(m, n) \Longrightarrow (\xi(m,n), \eta(m,n))$, setting the end location for pixel matching as $\hat{i} = \xi(M, N)$, $\hat{j} = \eta(M, N)$ and the point $(\hat{i}, \hat{j})$ as a nomination for the spotting point determined after $M + N - 1$ iterations of the proposed algorithm. Next, we set the local distance $d(i, j, m, n)$ as the difference between $S_p(i, j)$ and $R_p(m, n)$, and $w(i, j, m, n)$ as the weighted value of each local calculation. In this implementation, the local distance is determined by $d(i, j, m, n) = (S_p(i, j) - R_p(m, n))^2$, and weighted value sets as $w(i, j, m, n) = 1$ for all paths (Figure 2(c)). The accumulated local minimum $D(i, j, m, n)$ is used to evaluate the decision sequence, and is defined as:

$$D(\hat{i}, \hat{j}, m, n) = \tag{2}$$
$$\frac{1}{W} \min_{\xi, \eta} \{ \sum_{m=1}^{M} \sum_{n=1}^{N} w(\xi(m,n), \eta(m,n), m, n) d(\xi(m,n), \eta(m,n), m, n) \}.$$

Then $\xi^*(m, n)$ and $\eta^*(m, n)$ are used to represent the optimal solutions in $\xi(m, n)$ and $\eta(m, n)$ respectively, where W is the optimal accumulated weight $W = \sum_{m,n} w(\xi^*(m,n), \eta^*(m,n), m, n)$. To ensure continuity and monotonicity, $K(m, n) = \{\xi(m-1, n), \eta(m-1, n)\}$ and $L(m, n) = \{\xi(m, n-1), \eta(m, n-1)\}$ are used to define the sets of points that are movable in the i and j directions in the input image, taken from the movements in the m and n directions in the reference image. The following equation defines the relationship between two corresponding pixels $(m-1, n-1)$ and (m, n) (see Figure 1(b) and Figure 2):

$$(\xi(m-1, n-1), \eta(m-1, n-1)) \in$$
$$K(m, n) \otimes L(m-1, n) \cap L(m, n) \otimes K(m, n-1). \tag{3}$$

Here, the operator $\otimes$ represents the connection between a set of points on the left and a set of points on the right. To calculate accumulated local distance, each accumulated local $D(i, j, m, n)$ is derived from two previous accumulated local minimum $D(i', j', m-1, n)$ and $D(i'', j'', m, n-1)$. In this way, we define rank $l =$

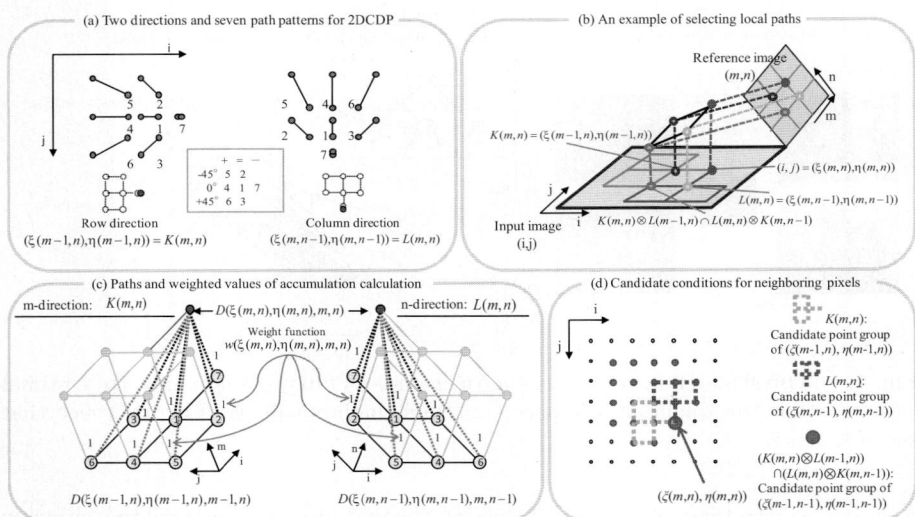

Fig. 2. Pixel corresponding overview. (a) Two directions and seven paths for selecting optimal path to accumulate value. (b) This figure shows only one case (linear matching) among the possible cases for optimal matching of local images, which include many different cases of nonlinear optimal matching of local areas. (c) Seven path directions and weighted values. Weighted value w is set to 1 in our implementation but it can change. (d) Each i and j direction can connect seven candidate pixels as (a) and (c). 2DCDP selects the node that has minimal accumulation value from among these paths, but a node depends on only two lower-rank nodes.

$m + n - 1$, as shown in Figure 1 (b), to smoothly calculate the accumulated local minima. A notice for the accumulation and back-tracking in the 2DCDP is that it selects two local paths that are needed to check the connection of the four points (m, n), $(m-1, n)$, $(m, n-1)$, and $(m-1, n-1)$ that form a quadrilateral (Figure 2(b)).

3 Method for 3D Reconstruction from Motion

Factorization [6,15] is used to factor a measurement matrix into a camera motion matrix and a 3D coordinates matrix without using camera parameters.

First, set the number of matching pixels between $I_f (f = 1, 2, \ldots F)$ input images and the reference image found by 2DCDP as P. The trajectory vector $\mathbf{T}_\alpha$ is defined as: $\mathbf{T}_\alpha = (x_{1\alpha} \ y_{1\alpha} \ldots x_{F\alpha} \ y_{F\alpha})^\top$, $\alpha = 1, 2, \ldots, P$. The centroid vector $\mathbf{T}_c = \frac{1}{P} \sum_{\alpha=1}^{P} \mathbf{T}_\alpha$ describes the origin for scene coordination. Using $\mathbf{T}_\alpha$ and $\mathbf{T}_c$, moment matrix C is calculated as:

$$C = \sum_{\alpha=1}^{P} (\mathbf{T}_\alpha - \mathbf{T}_c)(\mathbf{T}_\alpha - \mathbf{T}_c)^\top. \tag{4}$$

Eigenvalues $\lambda_1, \ldots, \lambda_{2F}$ and Eigenvectors $\mathbf{u}_1, \ldots, \mathbf{u}_{2F}$ are calculated by applying SVD to C. Then $\mathbf{T}_\alpha$ can be expanded as: $\mathbf{T}_\alpha = \mathbf{T}_c + c_1\mathbf{u}_1 + c_2\mathbf{u}_2 + c_3\mathbf{u}_3 + \ldots$. Eigenvalues $\lambda_1, \ldots, \lambda_{2F}$ are in non-increasing order. For this reason, $\mathbf{T}_\alpha$ can be approximated as a 3D Affine space by using the first three Eigenvectors $\mathbf{u}_1, \mathbf{u}_2$, and $\mathbf{u}_3$ if $\mathbf{T}_c$ is considered to be the centroid vector of that space. Alternatively, define the coordinates of the 3D point which corresponds to the $k - th$ scene of $\mathbf{T}_\alpha$ as $(X_\alpha, Y_\alpha, Z_\alpha)$, $\mathbf{t}_k$ as the origin, and $\{\mathbf{i}_k, \mathbf{j}_k, \mathbf{k}_k\}$ as the base coordinates of camera position. Then the position of $\mathbf{T}_\alpha$, in camera coordinates, is defined as $\mathbf{r}_{k\alpha} = \mathbf{t}_k + X_\alpha \mathbf{i}_k + Y_\alpha \mathbf{j}_k + Z_\alpha \mathbf{k}_k$. If a 3D affine camera model is used to approximate the perspective projection, the projection of $(X_\alpha, Y_\alpha, Z_\alpha)$ to a 2D image is $(x_\alpha \ y_\alpha)^\top = \mathbf{A}\mathbf{r}_\alpha + \mathbf{b} = \mathbf{m}_0 + X_\alpha \mathbf{m}_1 + Y_\alpha \mathbf{m}_2 + Z_\alpha \mathbf{m}_3$. In these equations, $\mathbf{A}$ is a 2×3 internal camera-parameter matrix that depends on the camera model, $\mathbf{b}$ is a 2D translation vector, and $\{\mathbf{m}_i | i = 0, 1, 2, 3\}$ are 2D vectors derived from internal camera parameters and the camera position in the scene. If $\mathbf{m}_0 = \mathbf{T}_c$, $\mathbf{T}_\alpha$ belongs to a 3D affine space, and this condition is called Affine space constraint [16]. This space is constructed from $\mathbf{m}_1$, $\mathbf{m}_2$, and $\mathbf{m}_3$, which can be expressed in terms of $\mathbf{u}_1, \mathbf{u}_2$, and $\mathbf{u}_3$: $\mathbf{m}_j = \sum_{i=1}^{3} A_{ij}\mathbf{u}_i$ $(j = 1, 2, 3)$. From this equation, camera motion and object shape can be calculated from the metric matrix $AA^\top$ using nonlinear least-square fitting.

4 Experiment

4.1 Experiment Specification

This section describes the results of using the proposed method to reconstruct 3D objects from just a few hard-deformed images. It also compares the results with those for other conventional methods. The computer used for the experiments was a Mac running OS X, with dual Xeon 3.0 GHz processors, 16 GB SDRAM, and a 300 GB HDD. The cameras comprised a Nikon D40 DSLR for Objects 1–3 and 5, and a Casio Exilim EX-Z1000 for Object 4. In these experiments, the background of the input images was removed manually, and the color of the background was set to (0,0,255) in RGB color space. In the contrast experiments, we used KLT tracker [12] and SIFT tracker [11] to extract correspondent pixels. Three images were used for Objects 1 and 3, five images for Object 2, and four images for Object 4, as indicated in Figure 3.

4.2 Results of 3D Reconstruction

Figure 3 shows reconstructed 3D objects using the proposed method with texture-mapped and mesh-structured objects using result of 2DCDP. Each of these objects is reconstructed either in dense mesh structure form alone or with texture mapping. Although some matching errors and occlusion occurred during the full pixel matching phase, the proposed method was still able to form the early shapes of objects precisely. Object 4 in Figure 3(d) was reconstructed from images captured by a person who did not mention either camera parameters or

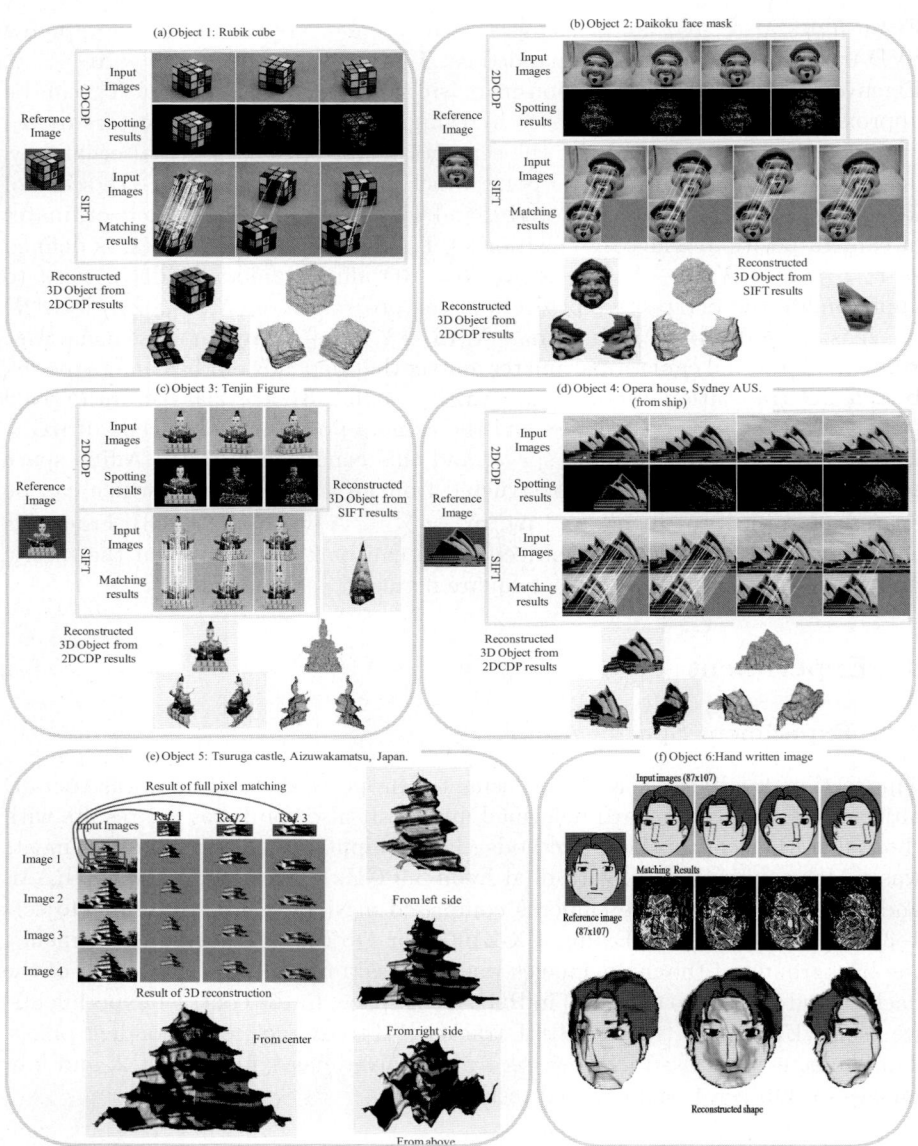

Fig. 3. Comparison Result of pixel matching and 3D reconstruction using 2DCDP and SIFT between an arbitrarily shaped reference image and another scene image. The reference image is drawn at (0,0,255) in RGB color space via user interaction for cut-out background because our method assumes template image is already known.

position, but the result was still acceptably close to the 3D object shape. Moreover, Table 2 indicates the relatively fast calculation time for 2DCDP which will guarantee the applicability of 2DCDP to real-world object reconstruction.

Table 2. 2DCDP calculation time, image size, number of images for reconstruction, accumulation of contributing rate from first to third Eigenvalues of equation (4) and the ratio of fourth and third Eigenvalues

	Object 1	Object 2	Object 3	Object 4
Input image size	200x132	200x132	150x150	200x129
Reference image size	96x104	89x107	150x150	125x102
Number of Images	3	5	3	4
Calculation time (second)	128.7685	198.706	227.3417	206.413
Accumulation of propotion rate	0.9985	0.9992	0.9993	0.9984
λ_4/λ_3	0.0622	0.1483	0.1950	0.2320

Table 3. Comparison between the proposed method and the KLT-based and SIFT-based methods. (0 means failure of point tracking or reconstruction.)

	Method	Object 1	Object 2	Object 3	Object 4
Matching points	2DCDP	7601	6799	7005	8115
	KLT	0	0	0	0
	SIFT	0	7	9	3
Number of polygons	2DCDP	7496	6731	6818	7981
	KLT	0	0	0	0
	SIFT	0	6	10	0

Table 3 shows the comparative numbers of matching points and polygons found by 2DCDP, SIFT, and KLT. Matching result of 2DCDP and SIFT are shown by Figure 3 (a)–(d). As the numbers indicate, SIFT and KLT could not find sufficient matching points in these large-variation images because they expected small-variation pixel movement in image sequences. For this reason, Factorization did not have enough corresponding pixels to form 3D objects. On the other hand, the proposed method demonstrated its ability to find corresponding pixels in these large-variation images. Since equation 4 suggests using principal component analysis [17], the bottom of Table 2 indicates the degree of separation between camera motion and object shape. The accumulated proportion rate using the first three Eigenvalues being almost equal to units means that the 3D object was well approximated by the 3D affine space. (λ_4/λ_3) represents the noise in the approximation calculation, and these noise values are sufficiently close to zero for the moment matrix rank to be considered as three. By comparing the noise values in the bottom of Table 2 and the reconstructed object shapes in Figure 3, the difference in quality between Objects 1 and 4 is recognizable because of the difference between the noise values for them. The Factorization method is able to reconstruct a 3D object shape not only from the images above but also from hand-drawn cartoons (Figure 3(f)) and large-scale object images (Figure 3(e)). These examples indicate that the proposed method can be considered effective for many purposes.

5 Conclusion

This paper has proposed a pixel matching-based 3D reconstruction method using 2DCDP and Factorization, which needs only a few images and no internal or external camera parameters. The proposed method can find optimal pixel correspondence between hard-deformed images using 2DCDP, and can reconstruct 3D objects using the Factorization method with corresponding pixels derived from 2DCDP. A problem of 2DCDP is that matching errors will occur if pixels are occluded or on texture-less curved surfaces. However, the Factorization method can overcome cover these matching errors because of the large number of corresponding pixels supplied by 2DCDP. To improve the proposed method, the matching errors should be reduced before applying Factorization, occluded points in matching results should be determined to estimate more suitable matching points between images, and, finally, a faster matching algorithm for optimal pixel matching should be developed.

References

1. Barnard, S., Fischler, M.: Computational Stereo. ACM CSUR 14(4), 553–572 (1982)
2. Horn, B., Brooks, M.: Shape from shading. MIT Press, Cambridge (1989)
3. Woodham, R.: Photometric method for determining surface orientation from multiple images. Optical Engineering 19(1), 139–144 (1980)
4. Mohr, R., Quan, L., Veillon, F.: Relative 3D Reconstruction Using Multiple Uncalibrated Images. The International Journal of Robotics Research 14(6), 619 (1995)
5. Zhang, Z.: Determining the Epipolar Geometry and its Uncertainty: A Review. IJCV 27(2), 161–195 (1998)
6. Tomasi, C., Kanade, T.: Shape and motion from image streams under orthography: a factorization method. IJCV 9(2), 137–154 (1992)
7. Mahamud, S., Hebert, M.: Iterative projective reconstruction from multiple views. In: Proc. of CVPR 2000, vol. 2 (2000)
8. Baker, H.: Three-dimensional modelling. In: IJCAI 1977, pp. 649–655 (1977)
9. Zhang, Z.: A flexible new technique for camera calibration. IEEE Trans. on PAMI 22(11), 1330–1334 (2000)
10. Kannala, J., Brandt, S.: Quasi-dense wide baseline matching using match propagation. In: Proc. of IEEE Conf. on CVPR, Minneapolis, MN, USA, pp. 1–8 (2007)
11. Lowe, D.G.: Distinctive image features from scale-invariant keypoints. IJCV 60(2), 91–110 (2004)
12. Tomasi, C., Kanade, T.: Detection and tracking of point features. Technical report, CMU-CS-91-132 (1991)
13. Yaguchi, Y., Kenta, I., Oka, R.: Optimal pixel matching between images. In: Wada, T., Huang, F., Lin, S. (eds.) PSIVT 2009. LNCS, vol. 5414, pp. 597–610. Springer, Heidelberg (2009)
14. Oka, R.: Spotting method for classification of real world data. The Computer Journal 41(8), 559–565 (1998)
15. Kanatani, K., Sugaya, Y.: Complete recipe for factorization. IEICE Technical Report. NC 103(391), 19–24 (2003)
16. Kurosawa, N., Kanatani, K.: Motion Segmentation by Affine Space Separation. IPSJ SIG Notes. CVIM-125-3 2001(4), 25–32 (2001)
17. Fujiki, J., Kurata, T.: An Mathematical Analysis of the Factorization Method for Generalized Affine Projection Model. Technical report of IEICE, PRMU 97(386), 101–108 (1997)

Rapid Inference of Object Rigidity and Reflectance Using Optic Flow*

Di Zang[1], Katja Doerschner[2], and Paul R. Schrater[1]

[1] Dept. of Computer Science & Engineering, University of Minnesota, USA
{zangx019,schrater}@umn.edu
[2] National Research Center for Magnetic Resonance (UMRAM) & Dept. of Psychology, Bilkent University, Turkey
katja@bilkent.edu.tr

Abstract. Rigidity and reflectance are key object properties, important in their own rights, and they are key properties that stratify motion reconstruction algorithms. However, the inference of rigidity and reflectance are both difficult without additional information about the object's shape, the environment, or lighting. For humans, relative motions of object and observer provides rich information about object shape, rigidity, and reflectivity. We show that it is possible to detect rigid object motion for both specular and diffuse reflective surfaces using only optic flow, and that flow can distinguish specular and diffuse motion for rigid objects. Unlike nonrigid objects, optic flow fields for rigid moving surfaces are constrained by a global transformation, which can be detected using an optic flow matching procedure across time. In addition, using a Procrustes analysis of structure from motion reconstructed 3D points, we show how to classify specular from diffuse surfaces.

Keywords: Optic flow, rigidity detection, specular motion, reflectance classification.

1 Introduction

For some computer vision applications like shape analysis from motion, it is typically required to know the material and rigidity of the objects. For instance, there would exist some difficulties to track highly reflective objects like cars without knowing if the object appearance remains constant across frames. Hence, most algorithms usually have strong assumptions about both the reflectivity and rigidity. For example, structure from motion algorithms assume rigidity and it is difficult to extract the point motion information needed without diffusely reflective and patterned objects [1]. Although there are methods to handle both nonrigid structure from motion and shape from specular flow, these methods are derived under the assumption that the rigidity and reflective properties of the object are known [2,3,4,5].

---

* This work has been supported in part by the European Commission Seventh Framework Programme Marie Curie International Reintegration Grant IRG-239494.

Detecting that an object is shiny and rigid would allow a tracking system to rely more on appropriate measurements and improve performance. Methods for rapidly classifying the reflectivity and rigidity of an object would provide the basis for automated recovery. Further, to be most useful, such methods should have minimal information demands. Ideally, we would like an assumption free, fast, image-based method for material and rigidity classification. In this paper we show how optic flow information from a single camera can be used to classify both rigidity of moving objects, and the reflectivity of rigid objects.

Previous methods for classifying material have largely relied on the ability to control the lighting in the scene, using multiple lights, structured lights, color, stereo, or combinations of these. For examples, see [6,7,8,9,10,11]. Oren and Nayar [12] develop a classification strategy to distinguish image points whose motions affected by specular reflectance from points behaving like diffuse reflectors based on caustic curves. To our knowledge, we are the first to suggest that rigidity can be classified for both diffuse and specular surfaces from optic flow information alone.

In this paper we develop an approach to classify the rigidity and reflectivity of a moving body using only optic flow information. Our approach consists of two parts. We show that rigidity produces characteristic transformations in optic flow that holds for objects with both diffuse and specular reflectance. We exploit this information to develop an optic flow matching algorithm for rigidity classification. We also show how an analysis of the consistency of structure from motion reconstruction can be used to identify diffuse rigid objects.

2 Rigidity from Optic Flow

To detect the rigidity of a specular or diffusely reflecting object from optic flow, we show a simple relationship exists between the optic flow fields at two time points for far-field environmental illumination and orthographic (or paraperspective) viewing. In particular, the flow fields generated by a rigid body motion that differ by a global transformation is derived below.

In order to derive a relationship between optic flow and rigid object motion, we assume that both the viewer and the environment are far from the object, approximated by orthographic viewing and illumination parameterized by direction on a sphere. These assumptions are not overly restrictive as [2] has shown that paraperspective is an exceedingly good approximation for most scenes. As shown in Fig. 1, the object surface $F(x,y) = (x,y,f(x,y))$ is represented as a function of image coordinates x, y, $\boldsymbol{n}(x,y) = S(\theta, \phi)$ indicates the surface normal at the surface point $F(x,y)$ with direction (θ, ϕ), S represents the mapping between spherical and cartesian coordinates, $\boldsymbol{u}(x,y)$ is the optic flow results from the rigid body transformation T. Because the viewing direction is $\boldsymbol{v} = (0,0,1)$, the mirror direction $\boldsymbol{r} = S(\theta, 2\phi)$ produces the image point at (x,y).

Rigid body transformation T can be applied to the surface F as $T[F(x,y)] = R[F(x,y)] + \boldsymbol{t}$, with R and $\boldsymbol{t}$ refer to the rotation matrix and the translation vector. This induces a motion field in spatial coordinates:

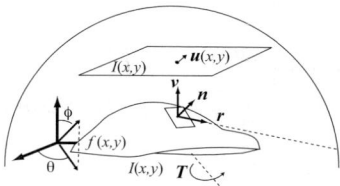

Fig. 1. Assumptions for our treatment of the rigidity from optic flow problem, adapted from [4]. A surface $f(x, y)$, reflecting a far-field illumination environment viewed orthographically to produce an image $I(x, y)$, undergoes a rigid body transformation T.

$$\begin{pmatrix} \frac{dx}{dt} \\ \frac{dy}{dt} \end{pmatrix} = I \left(-R \dot{R}^T F(x, y) + t \right), \qquad (1)$$

where $I = \begin{pmatrix} 1 & 0 & 0 \\ 0 & 1 & 0 \end{pmatrix}$ is the orthographic projection matrix, and $\dot{R}^T$ is the transpose of the cross product matrix $\dot{R}$ formed from the rotation axis ω, where $\dot{R}$ takes the following form:

$$\dot{R} = [\omega_\times] = \begin{pmatrix} 0 & -\omega_z & \omega_y \\ \omega_z & 0 & -\omega_x \\ -\omega_y & \omega_x & 0 \end{pmatrix}. \qquad (2)$$

For a fixed rotation axis, $\dot{R}^T F(x, y)$ is a constant flow. Thus, the optic flow pattern generated by a rigid-body transformation is an added translation and a global transformation that is the projection of the rotation onto the cartesian plane $-IR$: the flow is being rotated across time. This means that a global transformation of the motion field across time provides critical information about rigidity. For textured diffusely reflective objects, this motion field result translates directly into optic flow. After removing a global translation, we expect a rigid body motion to produce optic flow patterns that are projected rotations of an initial flow pattern.

We next show a similar result for specular surfaces, which reveals that the global transformations of optic flow patterns is a key piece of information about object rigidity. Because translations simply translate the flow under the viewing and illumination assumptions, we focus on rotations. For a specular surface, if the surface normals are rotated by a rotation R around an axis ω, then the transformation as a function of time is given by $R(t) [\omega_\times]$. In cartesian coordinates, $\frac{d\boldsymbol{n}}{dt} = R(t) [\omega_\times] \boldsymbol{n}$. This transformation of the normal field induces a specular flow field. Adapting the results in [5] to the case of object motion (rather than environment motion), an explicit relationship between the reflection direction and the first order derivatives of the surface can be used to relate differential changes in surface normals to optic flow, when the surface normals are expressed in spherical coordinates:

$$\begin{pmatrix} \frac{d\phi}{dt} \\ \frac{d\theta}{dt} \end{pmatrix} = \begin{pmatrix} \frac{1}{2|\nabla f|(1+|\nabla f|^2)} & 0 \\ 0 & \frac{1}{2|\nabla f|^2} \end{pmatrix} \begin{pmatrix} f_x & f_y \\ -f_y & f_x \end{pmatrix} \begin{pmatrix} f_{xx} & f_{xy} \\ f_{xy} & f_{yy} \end{pmatrix} \begin{pmatrix} \frac{dx}{dt} \\ \frac{dy}{dt} \end{pmatrix}. \quad (3)$$

To convert the normal flow between spherical coordinates and cartesian coordinates, we use the jacobian $\mathbf{J}$ of the cartesian to spherical coordinates mapping: $\left(\frac{d\phi}{dt}, \frac{d\theta}{dt}\right)^T = \mathbf{J}\frac{d\mathbf{n}}{dt}$. Chaining these relationships, the difference between a flow at an initial time $t = 0$ and a later time t is a rotation of the flow. This shows that specular flow patterns will differ by global transformations for rigid body motions.

Consequently, by matching optic flow patterns for motion sequences across time, classification can be made based on the measure of average angular error (AAE) [13]. The magnitude of AAE can be used to classify surface points as rigid, with small AAE indicating rigid and large AAE indicating nonrigid.

3 Distinguishing Specular and Diffuse Rigid Bodies

To distinguish rigid motions from diffusely reflective and specular objects, we use structure from motion [14] to reconstruct a candidate shape, and then assess the variation of the shape across time using Procrustes analysis [15]. For diffuse reflective and rigid objects, we would expect the variation in the reconstructed shape to be low and much higher for specular and nonrigid surfaces. Structure from motion is applied to a set of points that are tracked using normalized correlation [16]. To assess shape variation, we used a Procrustes analysis that removed the means of the set of tracked points within each time frame and aligned the points by finding a global rotation that minimized the least-squares difference between corresponding points. But unlike the normal Procrustes analysis, the scale is not removed. The average of the Euclidean distances between corresponding aligned points provides a measure of shape change that can be computed across time lags. Large values of this average shape change (ASC) measure indicate the surface is not both rigid and diffuse reflective. Combined with the optic flow matching measure, these optic flow based measures can distinguish rigid from nonrigid objects, and diffuse rigid from specular rigid motions.

4 Optic Flow Computation

We use a combined global local differential method (CLG) for optic flow computation based on Bruhn et al. [17]. CLG yields accurate, dense flow fields that are robust against noise. The method estimates the flow field by minimizing an energy function:

$$E(\boldsymbol{u}) = \int_\Omega (\psi_1(\boldsymbol{u}^T J_\rho(\nabla_3 f)\boldsymbol{u}) + \alpha\psi_2(1 - |\nabla \boldsymbol{u}|^2))dxdy, \quad (4)$$

where Ω denotes the image domain, α serves as regularization parameter, $\boldsymbol{u} = [u, v, 1]^T$ is the flow field, ∇ refers to the spatial gradient, and ∇_3 is the spatio-temporal gradient. The function J_ρ takes the form $J_\rho(\nabla_3 f) = K_\rho * (\nabla_3 f \nabla_3 f^T)$, where K_ρ means a Gaussian kernel with standard deviation ρ. Two nonquadratic penalisers $\psi_1(\cdot)$ and $\psi_2(\cdot)$ are computed as

$$\psi_i(z) = 2\beta_i^2 \sqrt{1 + \frac{z}{\beta_i^2}} \quad i \in \{1, 2\}, \tag{5}$$

with β_1 and β_2 as scaling parameters to handle outliers. For all the parameters, we take suggested values from [17].

5 Experimental Results

Test set. Our test set was comprised of novel 3D objects, generated by sinusoidally modulated spheres, which were organized into 4 categories according to

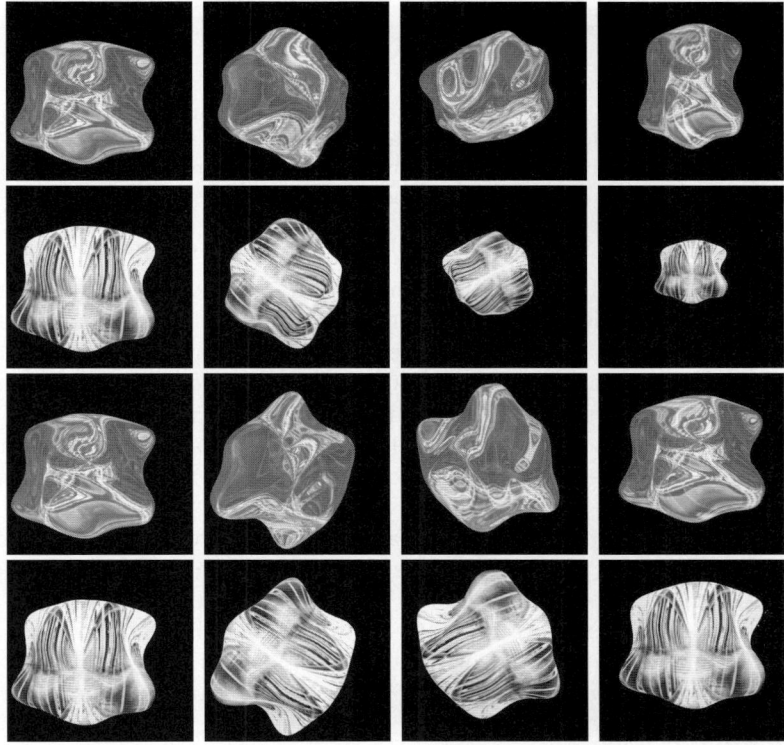

Fig. 2. Example frames (left to right: 1, 34, 67, 100) from our test for each of the 4 objects categories (top to bottom): specular nonrigid, diffuse nonrigid object, specular rigid and diffuse rigid. See text for details.

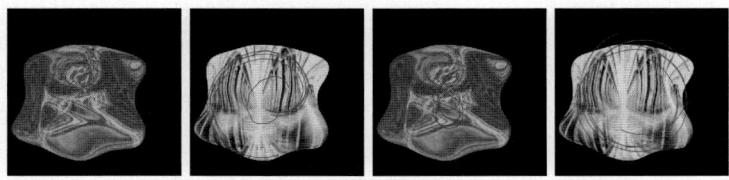

Fig. 3. Selected feature points tracked through 100 frames shown for all 4 object categories (left to right): specular nonrigid, diffuse nonrigid, specular rigid, and diffuse rigid

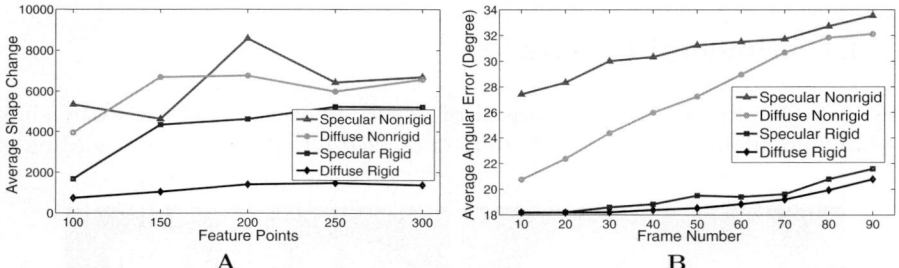

Fig. 4. A. ASC for all 4 object categories as a function of number of tracked feature points. **B.** Average angular errors between an initial flow field based on frames 1-2 and subsequent fields as a function of frame number. AAEs become larger with increasing lag, and are reliably small and stable for rigid objects of either reflectance.

their reflectivity (specular vs. diffuse) and rigidity (rigid vs. nonrigid). Nonrigid deformations were achieved by animating a phase shift of one sinusoidal modulator, in addition to scaling the object either in width (specular) or width and height (diffuse). For each measure (ASC, AAE) we generated 4 (1 per object category) 100-frame test sequences, some example frames are shown in Fig. 2. For ASC experiments, objects underwent a 90° rotation around the viewing direction and an xy-translation, whereas for AAE experiments, objects underwent a 90° rotation only.

Average shape change (ASC). We track object features across the duration of a sequence (see Fig. 3), and compute the ASC by comparing shape changes between the first and second 50-frame block. As shown in Fig. 4A, the ASC measure stabilizes when more than 100 feature points are tracked. Small ASC values reliably indicate the diffusely reflective, rigid object.

Average Angular Error (AAE). Fig. 5 shows sample optic flow fields for each object category. As expected, the flow fields generated by the specular rigid object are very similar between frames - up to a rotation (this is also true for the diffuse, rigid object - but not shown here). However, flow fields for nonrigid objects of either reflectance can vary in non-systematic ways. The AAE was computed by comparing the initial flow field (computed between frames 1 and

Table 1. Our method allows for a sequential classification approach: In step 1 diffuse rigid objects are successfully classified. In step 2, the AAE reliably distinguishes between rigid and non-rigid objects.

Step in Analysis	Object Class			
	specular		diffuse	
	rigid	nonrigid	rigid	nonrigid
1. ASC	large	large	small	large
2. AAE	small, stable	large, > diffuse	small, stable	large, <specular

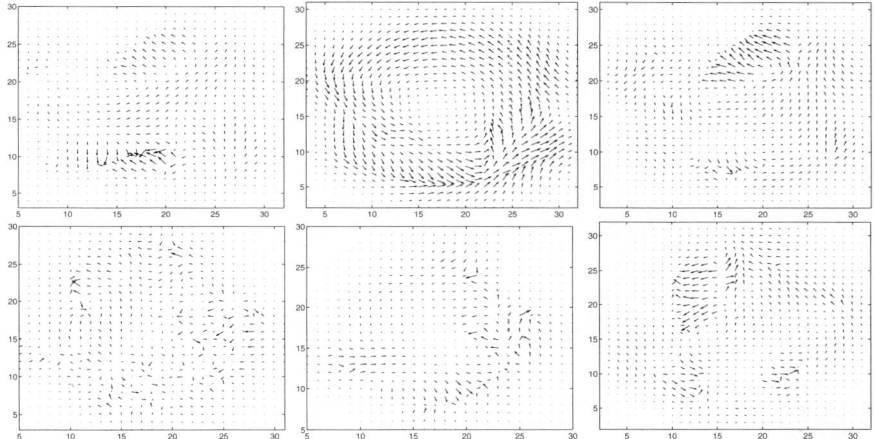

Fig. 5. The top row shows initial flow fields (see text) for specular nonrigid, diffuse nonrigid and specular rigid objects, respectively (see Fig 2 Column 1 for corresponding sequence frames). The bottom row shows optic flow fields between frames 51 and 52.

2) and the inverse rotated subsequent two-frame flow fields, incrementing frame counts by 10. As results in Fig. 4B illustrate, the AAEs for specular rigid and diffuse rigid objects are relatively stable and small compared to nonrigid objects of either reflectance. Thus it provides a reliable measure of the rigidity of an object.

Table 1 summarizes qualitatively results of each step (1.ASC, 2. AAE) in our approach.

6 Conclusions

We have shown that it is possible to distinguish the rigidity and reflectance of moving objects on the basis of the optic flow fields they generate. Rigidity for both specular and diffuse surfaces constrains the optic flow to follow a projected transformation across time. Using a structure from motion reconstruction criterion, it is possible to distinguish specular from diffuse reflectance of rigid motions. In future work it will be possible to formulate a statistical optic-flow based rigidity and reflectivity classifier and quantify the error rates.

References

1. Hartley, R.I., Zisserman, A.: Multiple View Geometry in Computer Vision, 2nd edn. Cambridge University Press, Cambridge (2004)
2. Zisserman, A., Giblin, P., Blake, A.: The information available to a moving observer from specularities. IVC 7(1), 38–42
3. Roth, S., Black, M.J.: Specular flow and the recovery of surface structure. In: CVPR 2006: Proceedings of the 2006 IEEE Computer Society Conference on Computer Vision and Pattern Recognition, pp. 1869–1876 (2006)
4. Adato, Y., Vasilyev, Y., Ben Shahar, O., Zickler, T.: Toward a theory of shape from specular flow. In: ICCV 2007, pp. 1–8 (2007)
5. Vasilyev, Y., Adato, Y., Zickler, T., Ben Shahar, O.: Dense specular shape from multiple specular flows. In: CVPR 2008, pp. 1–8 (2008)
6. Healey, G., Binford, T.: Local shape from specularity. CVGIP 42(1), 62–86 (1988)
7. Bhat, D., Nayar, S.: Binocular stereo in the presence of specular reflection. In: ARPA 1994, pp. 1305–1315 (1994)
8. Saito, M., Kashiwagi, H., Sato, Y., Ikeuchi, K.: Measurement of surface orientations of transparent objects using polarization in highlight. In: Proc. of IEEE Computer Society Conference on Computer Vision and Pattern Recognition, p. 1381 (1999)
9. Lin, S., Li, Y., Kang, S.B., Tong, X., Shum, H.-Y.: Diffuse-specular separation and depth recovery from image sequences. In: Heyden, A., Sparr, G., Nielsen, M., Johansen, P. (eds.) ECCV 2002. LNCS, vol. 2352, pp. 210–224. Springer, Heidelberg (2002)
10. Lellmann, J., Balzer, J., Rieder, A., Beyerer, J.: Shape from specular reflection and optical flow. International Journal of Computer Vision 80(2), 226–241 (2008)
11. Chow, S.K., Chan, K.L.: Removal of specular reflection component using multi-view images and 3d object model. In: Wada, T., Huang, F., Lin, S. (eds.) PSIVT 2009. LNCS, vol. 5414, pp. 999–1009. Springer, Heidelberg (2009)
12. Oren, M., Nayar, S.: A theory of specular surface geometry. International Journal of Computer Vision 24(2), 105–124 (1997)
13. Barron, J.L., Fleet, D.J., Beauchemin, S.S.: Performance of optical flow techniques. International Journal of Computer Vision 12(1), 43–77 (1994)
14. Tomasi, C., Kanade, T.: Shape and motion from image streams under orthography: a factorization method. International Journal of Computer Vision 9(2), 137–154 (1992)
15. Gower, J., Dijksterhuis, G.: Procrustes Problems. Oxford University Press, Oxford (2004)
16. Gonzalez, R.C., Woods, R.E.: Digital Image Processing. Addison-Wesley, Reading (1992)
17. Bruhn, A., Weickert, J., Schnörr, C.: Lucas/Kanade meets Horn/Schunck: Combining local and global optic flow methods. International Journal of Computer Vision 61(3), 211–231 (2005)

On the Recovery of Depth from a Single Defocused Image

Shaojie Zhuo and Terence Sim

School of Computing
National University of Singapore
Singapore,117417

Abstract. In this paper we address the challenging problem of recovering the depth of a scene from a single image using defocus cue. To achieve this, we first present a novel approach to estimate the amount of spatially varying defocus blur at edge locations. We re-blur the input image and show that the gradient magnitude ratio between the input and re-blurred images depends only on the amount of defocus blur. Thus, the blur amount can be obtained from the ratio. A layered depth map is then extracted by propagating the blur amount at edge locations to the entire image. Experimental results on synthetic and real images demonstrate the effectiveness of our method in providing a reliable estimate of the depth of a scene.

Keywords: Image processing, depth recovery, defocus blur, Gaussian gradient, markov random field.

1 Introduction

Depth recovery plays an important role in computer vision and computer graphics with applications such as robotics, 3D reconstruction or image refocusing. In principle depth can be recovered either from monocular cues (shading, shape, texture, motion *etc.*) or from binocular cues (stereo correspondences). Conventional methods for estimating the depth of a scene have relied on multiple images. Stereo vision [1,2] measures disparities between a pair of images of the same scene taken from two different viewpoints and uses the disparities to recover the depth. Structure from motion (SFM) [3,4] computes the correspondences between images to obtain the 2D motion field. The 2D motion field is used to recover the 3D motion and the depth. Depth from focus (DFF) [5,6] captures a set of images using multiple focus settings and measures the sharpness of image at each pixel locations. The sharpest pixel is selected to form a all-in-focus image and the depth of the pixel depends on which image the pixel is selected from. Depth from defocus (DFD) [7,8] requires a pair of images of the same scene with different focus setting. It estimates the degree of defocus blur and the depth of scene can be recovered providing the camera setting. These methods either suffer from the occlusion problem or can not be applied to dynamic scenes.

Fig. 1. The depth recovery result of the book image. (a) The input defocused image. (b) Recovered layered depth map. The larger intensity means larger blur amount and depth in all the depth maps presented in this paper.

Recently, approaches have been proposed to recover depth from a single image in very specific settings. Several methods [9,10] use active illumination to aid depth recovery by projecting structured patterns onto the scene. The depth is measured by the attenuation of the projected light or the deformation of the projected pattern. The coded aperture method [11] changes the shape of defocus blur kernel by inserting a customized mask into the camera lens, which makes the blur kernel more sensitive to depth variation. The depth is determined after a deconvolution process using a set of calibrated blur kernels. Saxena et al. [12] collect a training set of monocular images and their corresponding ground-truth depth maps and apply supervised learning to predict the value of the depth map as a function of the input image.

In this paper we focus on a more challenging problem of recovering the depth layers from a single defocused image captured by an uncalibrated conventional camera. As the most related work, the inverse diffusion method [13], which models the defocus blur as a diffusion process, uses the inhomogeneous reverse heat equation to obtain an estimate of the blur at edge locations and then proposed a graph-cut based method for inferring the depth in the scene. In contrast, we model the defocus blur as a 2D Gaussian blur. The input image is re-blurred using a known Gaussian function and the gradient magnitude ratio between input and re-blurred images is calculated. Then the blur amount at edge locations can be derived from the ratio. We also construct a MRF to propagate the blur estimate from the edge location to the entire image and finally obtain a layered depth map of the scene.

Our work has three main contributions. Firstly, we propose an efficient blur estimation method based on the gradient magnitude ratio, and we will show that our method is robust to noise, inaccurate edge location and interference from near edges. Secondly, without any modification to the camera or using additional illumination, our blur estimation method combined with MRF optimization can obtain the depth map of a scene by using only single defocused image captured by conventional camera. As shown in Fig. 1, our method can extract a layered depth map of the scene with fairly good extent of accuracy. Finally, we discuss two kinds of ambiguities in recovering depth from a single image using defocus cue, one of which is usually overlooked by previous methods.

2 Defocus Model

As the amount of defocus blur is estimated at edge locations, we must model the edge first. We adopt the ideal step edge model which is

$$f(x) = Au(x) + B, \tag{1}$$

where $u(x)$ is the step function. A and B are the amplitude and offset of the edge respectively. Note that the edge is located at $x = 0$.

When an object is placed at the focus distance d_f, all the rays from a point of the object will converge to a single sensor point and the image will appear sharp. Rays from a point of another object at distance d will reach multiple sensor points and result in a blurred image. The blurred pattern depends on the shape of aperture and is often called the circle of confusion (CoC) [14]. The diameter of CoC characterizes the amount of defocus and can be written as

$$c = \frac{|d - d_f|}{d} \frac{f_0^2}{N(d_f - f_0)}, \tag{2}$$

where f_0 and N are the focal length and the stop number of the camera respectively. Fig. 2 shows a thin lens model and how the diameter of circle of confusion changes with d and N, given fixed f_0 and d_f. As we can see, the diameter of the CoC c is a non-linear monotonically increasing function of the object distance d. The defocus blur can be modeled as the convolution of a sharp image with

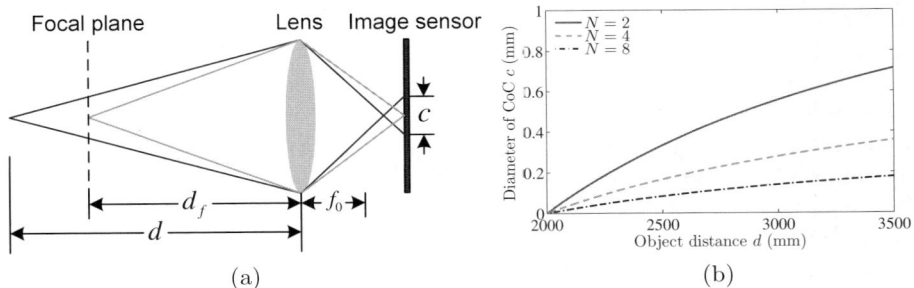

Fig. 2. (a) A thin lens model. (b) The diameter of CoC c as a function of the object distance d and f-stop number N given $d_f = 500mm$, $f_0 = 80mm$.

the point spread function (PSF). The PSF can be approximated by a Gaussian function $g(x, \sigma)$, where the standard deviation $\sigma = kc$ is proportional to the diameter of the CoC c. We use σ as a measure of the depth of the scene. A blurred edge $i(x)$ can be represented as follows,

$$i(x) = f(x) \otimes g(x, \sigma). \tag{3}$$

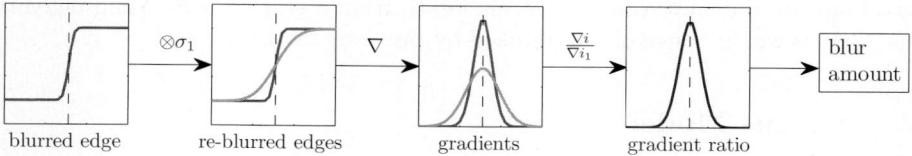

Fig. 3. Our blur estimation approach: here, $\otimes$ and ∇ are the convolution and gradient operators respectively. The black dash line denotes the edge location.

3 Blur Estimation

Fig. 3 shows the overview of our local blur estimation method. A step edge is re-blurred using a Gaussian function with know standard deviation. Then the ratio between the gradient magnitude of the step edge and its re-blurred version is calculated. The ratio is maximum at the edge location. Using the maximum value, we can compute the amount of the defocus blur of an edge.

For convenience, we describe our blur estimation algorithm for 1D case first and then extend it to 2D image. The gradient of the re-blurred edge is:

$$\begin{aligned}\nabla i_1(x) &= \nabla\big(i(x) \otimes g(x,\sigma_0)\big) \\ &= \nabla\big((Au(x) + B) \otimes g(x,\sigma) \otimes g(x,\sigma_0)\big) \\ &= \frac{A}{\sqrt{2\pi(\sigma^2 + \sigma_0^2)}} \exp(-\frac{x^2}{2(\sigma^2 + \sigma_0^2)}),\end{aligned} \qquad (4)$$

where σ_0 is the standard deviation of the re-blur Gaussian function. We call it the re-blur scale. The gradient magnitude ratio between the original and re-blurred edges is

$$\frac{|\nabla i(x)|}{|\nabla i_1(x)|} = \sqrt{\frac{\sigma^2 + \sigma_0^2}{\sigma^2}} \exp(\frac{x^2}{2\sigma^2} - \frac{x^2}{2(\sigma^2 + \sigma_0^2)}). \qquad (5)$$

It can be proved that the ratio is maximum at the edge location ($x = 0$). The maximum value is given by

$$R = \frac{|\nabla i(0)|}{|\nabla i_1(0)|} = \sqrt{\frac{\sigma^2 + \sigma_0^2}{\sigma^2}}. \qquad (6)$$

Giving the insight on (4) and (6), we notice that the edge gradient depends on both the edge amplitude A and blur amount σ, while the maximum of the gradient magnitude ratio R eliminates the effect of edge amplitude A and depends only on σ and σ_0. Thus, given the maximum value R, we can calculate the unknown blur amount σ using

$$\sigma = \frac{1}{\sqrt{R^2 - 1}}\sigma_0. \qquad (7)$$

For blur estimation in 2D images, we use 2D isotropic Gaussian function to perform re-blur. As any direction of a 2D isotropic Gaussian function is a 1D

Gaussian, the blur estimation is similar to that in 1D case. In 2D image, the gradient magnitude can be computed as follows:

$$\|\nabla i(x,y)\| = \sqrt{\nabla i_x^2 + \nabla i_y^2} \tag{8}$$

where ∇i_x and ∇i_y are the gradients along x and y directions respectively.

4 Layered Depth Map Extraction

After we obtain the depth estimates at edge locations, we need to propagate the depth estimates from edge locations to other regions that do not contain edges. We seek a regularized depth labeling $\hat{\sigma}$ which is smooth and close to the estimation in Eq. (7). We also prefer the depth discontinuities to be aligned with the image edges. Thus, We formulate this as a energy minimization over the discrete Markov Random Field (MRF) whose energy is given by

$$E(\hat{\sigma}) = \sum_i V_i(\hat{\sigma}_i) + \lambda \sum_i \sum_{j \in \mathcal{N}(i)} V_{ij}(\hat{\sigma}_i, \hat{\sigma}_j). \tag{9}$$

where each pixel in the image is a node of the MRF and λ balance the single node potential $V_i(\hat{\sigma}_i)$ and pairwise potential $V_{ij}(\hat{\sigma}_i, \hat{\sigma}_j)$ which are defined as

$$V_i(\hat{\sigma}_i) = M(i)(\sigma_i - \hat{\sigma}_i)^2, \tag{10}$$

$$V_{ij}(\hat{\sigma}_i, \hat{\sigma}_j) = \sum_{j \in \mathcal{N}(i)} w_{ij}(\hat{\sigma}_i - \hat{\sigma}_j)^2, \tag{11}$$

where $M(\cdot)$ is a binary mask with non-zeros only at edge locations. the weight $w_{ij} = exp\{-(I(i) - I(j))^2\}$ encodes the difference of neighboring colors $I(i)$ and $I(j)$. 8-neighborhood system $\mathcal{N}(i)$ is adopted in our definition.

We use FastPD [15] to minimized the MRF energy defined in Eq. (9). FastPD can guarantee a approximately optimal solution and is much faster than previous MRF optimization methods such as conventional graph cut techniques.

5 Experiments

There are two parameters in our method: the re-blur scale σ_0 and the λ. We set $\sigma_0 = 1$, $\lambda = 1$, which gives good results in all our examples. We use Canny edge detector [16] and tune its parameters to obtain desired edge detection output. The depth map are actually the estimated σ values at each pixel.

We first test the performance of our method on the synthetic bar image shown in Fig. 4(a). The blur amount of the edge increases linearly from 0 to 5. We first add noises to the bar image. Under noise condition, although the result of edges with larger blur amount is more affected by noise, our method can still achieve reliable estimation result (see Fig. 4(b)). We then create more bar images with different edge distances. Fig. 4(c) shows that interferences from neighboring edges increase estimation errors when the blur amount is large (> 3), but the

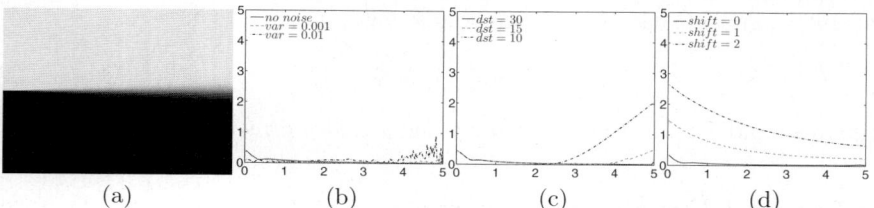

Fig. 4. Performance of our blur estimation method. (a) The synthetic image with blur edges. (b) Estimation errors under Gaussian noise condition. (c) Estimation errors with edge distances of 30, 15 and 10 pixels. (d) Estimation errors with edge shifts of 0, 1 and 2 pixels. The x and y axes are the blur amount and corresponding estimation error.

Fig. 5. The depth recovery results of flower and building images. (a) The input defocused images. (b) The sparse blur maps. (c) The final layered depth maps.

errors are controlled in a relative low level. Furthermore, we shift the detected edges to simulate inaccurate edge location and test our method. The result is shown in Fig. 4(d). When the edge is sharp, the shift of edge locations causes quite large estimation errors. However, in practice, the sharp edges usually can be located very accurately, which greatly reduces the estimation error.

Fig. 6. Comparison of our method and the inverse diffusion method. (a) The input image. (b) The result of inverse diffusion method. (c) Our result. The image is from [13].

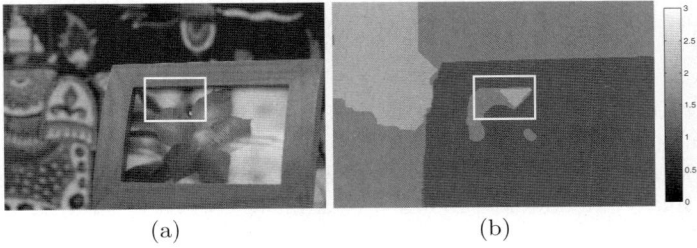

Fig. 7. The depth recovery result of the photo frame image. (a) The input defocused image. (b) Recovered layered depth map.

As show in Fig. 5, we test our method on some real images. In the flower image, the depth of the scene changes continuously from the bottom to the top of the image. The sparse blur map gives a reasonable measure of the blur amount at edge locations. The depth map reflects the continuous change of the depth. In the building image, there are mainly 3 depth layers in the scene: the wall in the nearest layer, the buildings in the middle layer, and the sky in the farthest layer. Our method extracts these three layers quite accurately and produces the depth map shown in Fig. 5(c). Both of the results are obtained using 10 labels of depth with the blur amount from 0 to 3. One more example is the book image shown in Fig. 1. The result is obtain using 6 depth labels with blur amount from 0 to 3. As we can see from the recovered depth map, our method is able to obtain a good estimate of the depth of the scene from a single image. In Fig. 6, we compare our method with the inverse diffusion method [13]. Both methods generate reasonable layered depth maps. However, our method has higher accuracy in local estimation and thus, our depth map captures more details of the depth. As shown in the figure, the difference in the depth of the left and right arms can be perceived in our result. In contrast, the inverse diffusion method does not recover this depth difference.

6 Ambiguities in Depth Recovery

There are two kinds of ambiguities in depth recovery from single image using defocus cue. The first one is the focal plane ambiguity. When an object appears blur in the image, it can be on either side of the focal plane. To remove this ambiguity, most of the depth from defocus methods including our method assume all objects of interest are located on one side of the focal plane. When taking images, we just put the focus point on the nearest/farthest point in the scene.

The second ambiguity is called the blur/sharp edge ambiguity. The defocus measure we obtained may be due to a sharp edge that is out of focus or a blur edge that is in focus. This ambiguity is often overlooked by previous work and may cause some artifacts in our result. One example is shown in Fig. 7. The region indicated by the white rectangle is actually blur texture of the photo in the frame, but our method treats it as sharp edges due to defocus blur, which results in error estimation of the depth in that region.

7 Conclusion

In this paper, we show that the depth of a scene can be recovered from a single defocused image. A new method is presented to estimate the blur amount at edge locations based on the gradient magnitude ratio. The layered depth map is then extracted using MRF optimization. We show that our method is robust to noise, inaccurate edge location and interferences of neighboring edges and can generate more accurate scene depth maps compared with existing methods. We also discuss ambiguities arising in recovering depth from single images using defocus cue. In the future, we would like to apply our blur estimation method to images with motion blur to estimate the blur kernels.

Acknowledgement. The author would like to thank the anonymous reviewers for their helpful suggestions. The work is supported by NUS Research Grant #R-252-000-383-112.

References

1. Barnard, S., Fischler, M.: Computational stereo. ACM Comput. Surv. 14(4), 553–572 (1982)
2. Dhond, U., Aggarwal, J.: Structure from stereo: A review. IEEE Trans. Syst. Man Cybern. 19(6), 1489–1510 (1989)
3. Dellaert, F., Seitz, S.M., Thorpe, C.E., Thrun, S.: Structure from motion without correspondence. In: Proc. CVPR, pp. 557–564 (2000)
4. Tomasi, C., Kanade, T.: Shape and motion from image streams under orthography: A factorization method. Int. J. Comput. Vision 9, 137–154 (1992)
5. Asada, N., Fujiwara, H., Matsuyama, T.: Edge and depth from focus. Int. J. Comput. Vision 26(2), 153–163 (1998)
6. Nayar, S., Nakagawa, Y.: Shape from focus. IEEE Trans. Pattern Anal. Mach. Intell. 16(8), 824–831 (1994)

7. Favaro, P., Favaro, P., Soatto, S.: A geometric approach to shape from defocus. IEEE Trans. Pattern Anal. Mach. Intell. 27(3), 406–417 (2005)
8. Pentland, A.P.: A new sense for depth of field. IEEE Trans. Pattern Anal. Mach. Intell. 9(4), 523–531 (1987)
9. Moreno-Noguer, F., Belhumeur, P.N., Nayar, S.K.: Active refocusing of images and videos. ACM Trans. Graphics, 67 (2007)
10. Nayar, S.K., Watanabe, M., Noguchi, M.: Real-time focus range sensor. IEEE Trans. Pattern Anal. Mach. Intell. 18(12), 1186–1198 (1996)
11. Levin, A., Fergus, R., Durand, F., Freeman, W.T.: Image and depth from a conventional camera with a coded aperture. ACM Trans. Graphics (2007)
12. Saxena, A., Sun, M., Ng, A.: Make3d: Learning 3d scene structure from a single still image. IEEE Trans. Pattern Anal. Mach. Intell., 1–1 (2008)
13. Namboodiri, V.P., Chaudhuri, S.: Recovery of relative depth from a single observation using an uncalibrated (real-aperture) camera. In: Proc. CVPR (2008)
14. Hecht, E.: Optics, 4th edn. Addison-Wesley, Reading (2001)
15. Komodakis, N., Tziritas, G., Paragios, N.: Performance vs computational efficiency for optimizing single and dynamic mrfs: Setting the state of the art with primal-dual strategies. Proc. CVIU 112(1), 14–29 (2008)
16. Canny, J.: A computational approach to edge detection. IEEE Trans. Pattern Anal. Mach. Intell. 8(6), 679–698 (1986)

Modelling Human Segmentation Trough Color and Space Analysis

Agnés Borràs and Josep Lladós

Computer Vision Center - Dept. Ciències de la Computació
UAB Bellaterra 08193, Spain
{agnesba,josep}@cvc.uab.es
http://www.cvc.uab.es

Abstract. This paper proposes an algorithm of color segmentation that models the human-based perception according to the Gestalt laws of similarity and proximity. We use the mean shift clustering to translate these laws into the analysis of the color layout of an image. Given a set of possible segmentations, the method uses a measure of stability to identify the most meaningful regions according to perceptual criteria. Quantitative results obtained on the Berkeley data set show that this approach outperforms state-of-the-art methods on human-based image segmentation.

1 Introduction

Image segmentation is one of the most common strategies to extract relevant information from images. Nevertheless, unsupervised segmentation turns into a very severe problem when it has to emulate human-based criteria. Processes involved in the human perception have been studied since decades from a phycological point of view. This way, the computer science community has attempted to translate the psychological framework into effective algorithms of image segmentation. The Gestalt school is a well-known organization of psychologists who modelled the perceptual process according to a set of rules called Laws of Organization. These rules explain the grouping processes of the image components into higher level patterns [1]. In this paper, we present a segmentation strategy inspired in two of the Gestalt laws: Similarity and Proximity. The Law of Similarity declares that the mind groups elements that share similar features. Moreover, the Law of Proximity asserts that humans perceive close elements as a collective (see Figure 1).

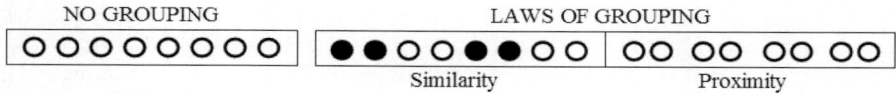

Fig. 1. Humans perceive as groups elements that are similar and elements that are placed close together

We present a segmentation method that translates the Similarity Law into the analysis of the color features and translates the Proximity Law into the analysis of their layout distribution. To perform a segmentation, we purpose to analyze a set of solutions that group the image pixels regarding to their color and their spatial position. We use a measure that evaluates the "stability" of these grouping possibilities and generates the final segmentation. We have called the proposed method Color Region Stability (CoReSt).

To evaluate the algorithm we have used the public benchmark provided by the Berkeley Segmentation Dataset [2]. This database is a collection of photographs which have been manually segmented by human subjects. We have used the Global Constancy Error to compare the algorithm against other state-of-the-art methods. Quantitative results show that our approach outperforms other strategies on human-based image segmentation.

The paper organization is as follows. In the section 2 we explain the main idea of our proposal and we detail its implementation. Then, in section 3, we present the experimental validation and, finally, in section 4 we expose the conclusions of the work.

2 The Color Region Stability Algorithm (CoReSt)

The goal of the CoReSt algorithm is to provide a segmentation solution that fits the human criteria. An image can be segmented in multiple ways according to the properties of color and position of the pixels. From this set of segmentations we use a "stability" measure to select those regions that are considered most relevant. This stability measure considers that a region is relevant if it remains most unchanged as possible along the collection of segmentation solutions.

In the literature we can find a huge variety of color segmentation algorithms [3]. Some proposals also study a set of segmentation solutions with regard to a stability function. The work of Heidemann [4] proposed a goodness function to optimize the parameters of color segmentation. The function evaluates the contrast on the boundary pixels of a set of segmented regions. This way, the segmentation that maximizes this goodness measure is taken as the optimal solution. Another different approach is the detector of regions of interest proposed by Matas [5]. The detector was designed to work in grayscale images so it analyzes the image segmentations in the intensity domain. The stability measure is computed by a function that evaluates the rate of area in which the regions vary across two consecutive segmentations. Then, a posterior work of Forssén [6] exported the concept to the color space using an agglomerative clustering process. In another direction, the work Wattuya combines multiple segmentations [7] and uses a random walker approach.

Notice that these proposals explore the color domain of the image but do not explore its layout configuration. This way, the obtained regions are always connected components but there is no grouping process that searches for relations between them. In this paper we introduce the spacial analysis of the image components in order to overcome occlusions and detect textured areas. In the next section we explain in detail the implementation of the proposal.

2.1 Implementation

Given an image, the measure of segmentation stability is computed in both color and spatial domains. This way, if we focus on the color analysis, we can obtain a set of segmentation instances increasing the threshold that controls the color similarity. Then, we observe that image pixels are merged into regions if their color distances are lower than this threshold value. The generated segmentations can be understood as the evolution of the image regions through the color domain. Then, if we analyze the evolution we see that the most contrasted regions remain most stable along a larger set of segmentation images. Otherwise, if we focus on the space analysis, we can generate a set of segmentations by increasing the threshold of the spatial similarity. The evolution of the segmentation shows that image regions are progressively merged according their spatial distance. From the stability viewpoint, we observe that the regions which are most isolated from other regions with similar color remain most unchanged along the segmentations. To obtain every segmentation we have used the mean shift clustering algorithm proposed by Comaniciu [8]. Thus, a pixel is understood like a point in a 5D space where its first three dimensions are related to the color values in the Luv space the other two represent the (x,y) coordinates in the image. The mean shift process defines for every pixel in the image a path that leads it to a local density maximum in the 5D space. A region is formed by all the pixels that belong to the same local density maximum despite they do not correspond to a single connected component in the image. We use an implementation of the mean shift clustering that depends on two thresholds, hc and hs, that control respectively the similarity constraints on the color and the space [9]. Then, as we show in Figure 2, we construct a bidimensional grid G filled up with the clustered images. Let us denote MSS the mean shift function and HC and HS the two sets of thresholds, $HC = \{hc_1, \ldots, hc_{NC}\}$ $HS = \{hs_1, \ldots, hs_{NS}\}$. Given a region R of a cell in the grid, we define as *analogous regions* the regions of the other grid cells that maximize the overlapping area with R. Analogous regions are therefore found along the color or space dimensions, varying the corresponding thresholds in HC and HS respectively. The intuitive idea of an analogous region of a region R is that it is the evolution, i.e. the closest region, to R in another segmentation scale. We denote $R_i^{(x,y)}$ the region i of the grid cell (x,y) and $AR_{(x,y)'}^{(x,y)_i}$ its analogous region of another cell $(x,y)'$.

Once we have the segmentation evolution along the color and the space dimensions, we need a function to evaluate the stability of the regions. The stability function models the shape variation of a region along the two dimensions of the grid. The features used in the computation are the first and the second central moments. These features visually correspond to the area of a region and the axis lengths of the minimum enclosing ellipse of the region. Then, given a region $R_i^{(x,y)}$ and another analogous one $AR_{(x,y)'}^{(x,y)_i}$ we calculate the stability S as a combination of the variation of the area rate and the axis length. Let us denote with A the function that computes the area of a region and with L and l the functions that compute the maximum and minimum lengths of the axis. For the sake of readability, we simplify the notation of $R_i^{(x,y)}$ to R_1 and $AR_{(x,y)'}^{(x,y)_i}$ to R_2. Thus, the stability measure between two regions R_1 and R_2 is defined as:

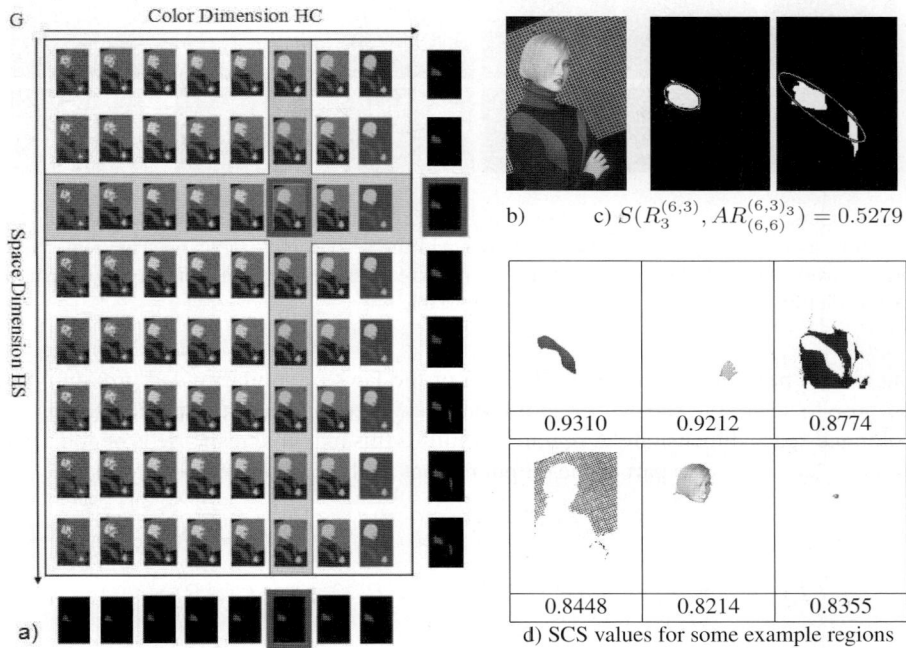

Fig. 2. a) Grid of segmented images using the MSS according to the parameters of color HC and space HS. We show an example of the region $R_3^{(6,2)}$ and its analogous ones. Observe how it grows trough the color, merging with similar pixels, and how it grows trough the space merging with similar regions. b) Original image c) Stability value S of $R_3^{(6,2)}$ according to the analogous region on the cell (6,6). d) Some selected regions and their stability value SCS.

$$S(R_1, R_2) = S_{area}(R_1, R_2) * 0.5 + S_{axis}(R_1, R_2) * 0.5$$

$$S_{area}(R_1, R_2) = \frac{min(A(R_1), A(R_2))}{max(A(R_1), A(R_2))}$$

$$S_{axis}(R_1, R_2) = min\left(\frac{min(L(R_1), L(R_2))}{max(L(R_1), L(R_2))}, \frac{min(l(R_1), l(R_2))}{max(l(R_1), l(R_2))}\right)$$

For each region of each cell we compute its stability along the two dimensions of the grid. The computation is done by the mean of the S values regarding the analogous regions. We name SC to the function that measures the stability along the color, and SS its equivalent in the space.

$$SC(R_i^{(x,y)}) = \sum_{X=1}^{\#HC} \frac{S(R_i^{(x,y)}, AR_{(X,y)}^{(x,y)_i})}{\#HC} \qquad SS(R_i^{(x,y)}) = \sum_{Y=1}^{\#HS} \frac{S(R_i^{(x,y)}, AR_{(x,Y)}^{(x,y)_i})}{\#HS}$$

At this point we have two measures of stability for every region of the segmented images of the grid. Nevertheless, we search for a representative subset of regions that fits

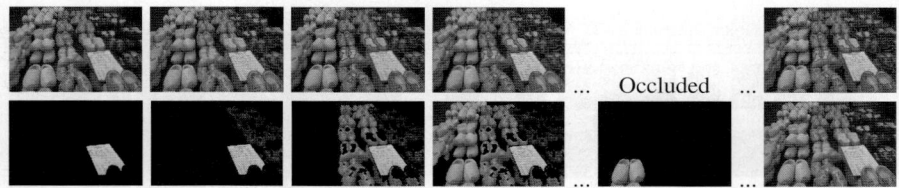

Fig. 3. Generation of the segmentation using the CoReSt regions ranked by the stability. The first row shows the progress of the boundaries and the second row shows how the regions are progressively incorporated. In the fifth step we observe an example of a region that is not included in the solution because it is occluded by a more stable one.

the human perception. Following the idea of the laws of similarity and proximity we propose to select the subset of regions that maximize the stability functions along the color and space dimensions. A region that fulfills this maximal response is denoted PR and is selected to form part of the output regions of the CoReSt method.

$$PR_i^{(x,y)} = R_i^{(x,y)} \mid PC(R_i^{(x,y)}) \text{ or } PS(R_i^{(x,y)})$$

$$PC(R_i^{(x,y)}) = SC(AR_{(x-1,y)}^{(x,y)_i}) \leq SC(R_i^{(x,y)}) > SC(AR_{(x+1,y)}^{(x,y)_i})$$

$$PS(R_i^{(x,y)}) = SS(AR_{(x,y-1)}^{(x,y)_i}) \leq SS(R_i^{(x,y)}) > SS(AR_{(x,y+1)}^{(x,y)_i})$$

Among the final output regions a global measure of relevance is also computed. This value combines the stability of color and space using the function SCS. Notice that all the computations we have presented work in the range $[0, 1]$, then the global stability is also in this range.

$$SCS(R) = (SC(R) + SS(R)) * 0.5$$

The result of SCS allow to rank the regions by its meaningfulness: the greater, the more meaningful. Taking into account this ranking we generate an image segmentation that combines the regions. The figure 3 provides an example of this segmentation construction.

The process consists in constructing a pile of regions ordered by stability. Then, in the deepest positions we find the less stable regions and in the most superficial positions we have the most stable ones. The segmentation generation can be understood as a z-buffering analysis of this pile of regions. The segmentation incorporates the boundaries of the regions following the priority order defined by the stability measure. Notice that a region PR_i will not be included in the final segmentation if its pixels are overlapped by another region PR_j that is more stable.

3 Experiments and Results

We have evaluated the performance of the proposed method with the public segmentation dataset of Berkeley [2]. The test set comprises 200 color photographs which have been manually segmented. For each of the images, at least 5 segmentations produced by

Table 1. Comparison of the CoReSt strategy against other state-of the-art methods: the Ridge based Distribution Analysis (RAD)[11], the Multiple Seed Segmentation (Seed)[12], the pairwise pixel affinity algorithm proposed by Fowlkes (Fow)[13] and the Normalized Cuts (nCuts)[14]. GCE results for the 200 Berkeley images (values taken from [11]).

	Human	CoReSt	RAD	Seed	Fow	MS	nCuts
GCE	0.0800	0.1946	0.2048	0.209	0.214	0.2598	0.336

Good segmentation results

Bad results achieving under-segmentation and over-segmentation

Fig. 4. Examples of the CoReSt segmentations

different people are available. We have generated a segmentation result on every original image of the dataset. The experimentation provides a numerical evaluation among the CoReSt solution and the manual benchmark.

In the literature we can find several measures to evaluate the segmentation results [10]. We have chosen to use the Global Consistency Error (GCE) since it is a standard framework in a number of a state-of-the-art methods. The GCE measure takes care of the refinement between two segmentations: being IS_1 the segmentation of the benchmark and IS_2 the segmentation we evaluate, it produces an error measure in the range $[0, 1]$ (the lower, the better). For each pixel p_i the GCE evaluates the difference between the regions of both segmentations $R \in (IS_1, p_i)$ and $R \in (IS_2, p_i)$ that contain this pixel. Let us denote n the number of pixels in a image, $\setminus$ difference operator, and $|\ |$ the cardinality one.

$$GCE = \frac{1}{n} min(\sum_i \frac{|R \in (IS_1, p_i) \setminus R \in (IS_2, p_i)|}{|R \in (IS_1)|}, \sum_i \frac{|R \in (IS_2, p_i) \setminus R \in (IS_1, p_i)|}{|R \in (IS_2, p_i)|})$$

The inter-variability among the human segmentations obtains a mean GCE value of 0.08. Fixing the method parameters for all the test set, we have obtained a score of 0.1946. The table 1 summarizes the GCE values obtained by other segmentation methods.

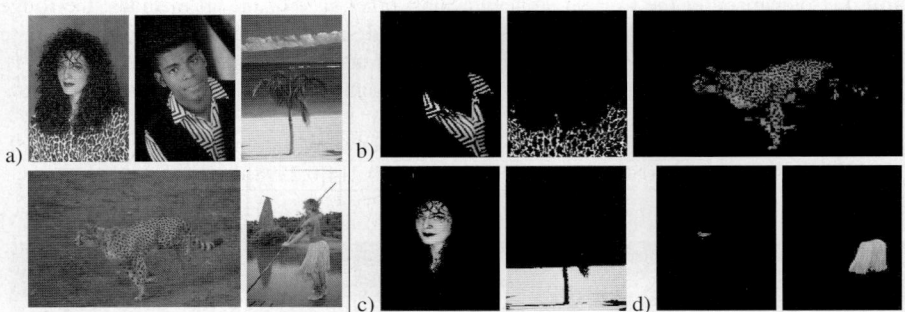

Fig. 5. a) Original Images. b) and c) present the grouping properties. b) Regions that conform a textured area. c) Regions that present occlusions. d) Regions detected as outstanding for having contrasted color and being isolated.

An interesting point in the comparative table is to see that the proposed method outperforms the results obtained by original Mean Shift algorithm (MS). This way, we can see that the stability measure helps on building a segmentation result that adapts better to the human based criteria. Some qualitative examples of the CoReSt segmentation are shown in the Figure 4. The images of the Figure 5 present some regions that illustrate the properties of the CoReSt method.

4 Conclusions

We have presented an algorithm to emulate the human-based image segmentation inspired in the Gestalt laws of Similarity and Proximity. The CoReSt method explores the color features of an image according to its spatial distribution. This twofold analysis allows the perceptual grouping of elements in a image an gives the segmentation process interesting properties. This way, we can detect entities that belong to be a textured element (a repetitive color pattern) or a plain surface that has been "broken" due to a partial occlusion. The method outstands big and homogeneous regions as well as isolated and contrasted ones. Due to this behavior, we can obtain a final segmentation that comprises a coarse description of the content but that also preserves the meaningful details. To prove the consistency of the perceptual segmentation we have test the method on the Berkeley benchmark of human-based segmentations. We have seen that the CoReSt method outperforms other state-of-the-art proposals according to the GCE measure. The proposal gives a very high degree of freedom on the shape of the output regions and can be applied in images of general purpose.

Future lines of research suggest potential applications in fields where the information needs to match the human representation. One example is the content based image retrieval field. Retrieval systems are meant to provide perceptually similar images according to a given query. The extreme application of the human perception can be found in those retrieval applications that have a drawing interface and allow the user to create its own query image. Then the regions drawn by a user could match the regions of the database extracted by CoReSt method.

Acknowledgments

This work has been partially supported by the grant 2002FI-00724 and the Spanish projects TIN2006-15694-C02-02 and CONSOLIDERINGENIO 2010 (CSD2007-00018).

References

1. Wertheimer, M.: Laws of Organization in Perceptual Forms, pp. 71–88. Routledge Kegan Paul Ltd. (1955)
2. Martin, D., Fowlkes, C., Tal, D., Malik, J.: A database of human segmented natural images and its application to evaluating segmentation algorithms and measuring ecological statistics. Technical report, EECS Department, University of California, Berkeley (2001)
3. Cheng, H.D., Jiang, X.H., Sun, Y., Wang, J.: Color image segmentation: advances and prospects. Pattern Recognition 34(12), 2259–2281 (2001)
4. Heidemann, G.: Region saliency as a measure for colour segmentation stability. Image Vision Comput. 26(2), 211–227 (2008)
5. Matas, J., Chum, O., Martin, U., Pajdla, T.: Robust wide baseline stereo from maximally stable extremal regions. In: Proceedings of the BMVC, vol. 1, pp. 384–393 (2002)
6. Forssén, P.E.: Maximally stable colour regions for recognition and matching. In: IEEE Conference on CVPR, Minneapolis, USA (June 2007)
7. Wattuya, P., Jiang, X., Rothaus, K.: Combination of multiple segmentations by a random walker approach. In: Rigoll, G. (ed.) DAGM 2008. LNCS, vol. 5096, pp. 214–223. Springer, Heidelberg (2008)
8. Comaniciu, D., Meer, P.: Mean Shift Analysis and Applications. In: Proceedings of the IEEE ICCV, Kerkyra, Greece, pp. 1197–1203 (1999)
9. Christoudias, C., Georgescu, B., Meer, P.: Synergism in low level vision. In: ICPR, pp. IV:150–IV:155 (2002)
10. Unnikrishnan, R., Pantofaru, C., Hebert, M.: Toward objective evaluation of image segmentation algorithms 29(6), 929–944 (2007)
11. Vazquez, E., van de Weijer, J., Baldrich, R.: Image segmentation in the presence of shadows and highlights. In: Forsyth, D., Torr, P., Zisserman, A. (eds.) ECCV 2008, Part IV. LNCS, vol. 5305, pp. 1–14. Springer, Heidelberg (2008)
12. Mičušík, B., Hanbury, A.: Automatic image segmentation by positioning a seed. In: Leonardis, A., Bischof, H., Pinz, A. (eds.) ECCV 2006. LNCS, vol. 3952, pp. 468–480. Springer, Heidelberg (2006)
13. Fowlkes, C., Martin, D., Malik, J.: Learning affinity functions for image segmentation: Combining patch-based and gradient-based approaches. In: IEEE Computer Society Conference on Computer Vision and Pattern Recognition, vol. 2, p. 54 (2003)
14. Shi, J., Malik, J.: Normalized cuts and image segmentation. IEEE Transactions on Pattern Analysis and Machine Intelligence 22(8), 888–905 (2000)

A Metric and Multiscale Color Segmentation Using the Color Monogenic Signal

Guillaume Demarcq*, Laurent Mascarilla, and Pierre Courtellemont

Laboratoire Mathématiques, Images, Applications
Université de La Rochelle, France
{gdemar01,lmascari,pcourtel}@univ-lr.fr

Abstract. In this paper, we use the formalism of Clifford algebras to extend the so-called Monogenic Signal to color images. This extension consists in a function with values in the Clifford algebra $\mathbb{R}_{5,0}$ that encodes color as well as geometric structure information. Using geometric calculus, such a mathematical object can be used to extend classical concepts of signal processing (filtering, Fourier Transform...) to color images in a consistent manner. Regarding this paper, a local color phase is introduced, which generalizes the one for grayscale image. As an example of application, we provide a new method for color segmentation. Based on our phase definition and the multiscale aspect of the Color Monogenic Signal, we provide a metric approach using differential geometry which reveals relevant on the Berkeley Image Dataset.

Keywords: Monogenic signal, Clifford algebras, color segmentation, color image processing, differential geometry.

1 Introduction

We propose in this paper a new framework for high dimensional signal processing based on Clifford algebras. The aim is to generalize in the context of color images the work of M. Felsberg [2] about the monogenic extension of the analytic signal to grayscale images. After some recalls on analytic and monogenic signals we first introduce the color monogenic signal of a color image as a scale-space signal using the Dirac operator and the Laplace equation. We show then how to define a color local phase that is parametrized by a vector of $\mathbb{R}_{5,0}$ containing color and geometric structures information. This color local phase can be used in many applications such as color optical flow or color object tracking (details will appear elsewhere). We focus here on defining a new color segmentation method based on a metric and multiscale approach. Segmentation in a chosen color can be done and experiments show accurate results on images from the Berkeley dataset [5].

---

* This work has been partially founded by Région Poitou-Charente and ONR Grant N00014-09-1-0493.

2 Dirac Operator and Cauchy-Riemann Equations

To a vector space E together with a quadratic form Q is associated a non-commutative algebra $Cl(E, Q)$ called the Clifford algebra of the couple (E, Q). In what follows we deal with the Clifford algebra of the euclidean vector space $\mathbb{R}^n$, usually denoted by $\mathbb{R}_{n,0}$. In this algebra, the product of two vectors a and b of $\mathbb{R}^n$, embedded in $\mathbb{R}_{n,0}$, is given by:

$$ab = a \cdot b + a \wedge b \tag{1}$$

where $a \cdot b$ is the inner product and $a \wedge b$, the wedge product of a and b, is a bivector. This product is usually called the geometric product of a and b. One could refer to [1] for further details.

2.1 The Clifford Algebra $\mathbb{R}_{3,0}$

As a vector space over $\mathbb{R}^3$, it is of dimension 8. A base of $\mathbb{R}_{3,0}$ is given by:
$\{1, \underbrace{e_1, e_2, e_3}_{\text{vectors}}, \underbrace{e_1 e_2, e_1 e_3, e_2 e_3}_{\text{bivectors}}, \underbrace{e_1 e_2 e_3}_{\text{trivector}}\}$ where $\{e_1, e_2, e_3\}$ is an orthonormal basis of $\mathbb{R}^3$. Given two vectors $u = ae_1 + be_2 + ce_3$ and $v = a'e_1 + b'e_2 + c'e_3$ of $\mathbb{R}_{3,0}$, the geometric product uv is:

$$uv = \underbrace{(aa' + bb' + cc')}_{\text{scalar part}}$$
$$+ \underbrace{(ab' - ba')e_1 e_2 + (ac' - ca')e_1 e_3 + (bc' - cb')e_2 e_3}_{\text{bivector part}}$$

One can recognize immediately the combination of the usual dot product and cross product of $\mathbb{R}^3$. In particular:
$\forall i, j \in \{1, 2, 3\}, e_i e_j + e_j e_i = 2\delta_{ij}$ where δ_{ij} is the delta function.

2.2 Generalized Cauchy-Riemann Equations

In $\mathbb{R}_{n,0}$, the Dirac operator is defined by $D = \sum_{k=1}^{n} e_k \frac{\partial}{\partial x_k}$, where $\forall i, j \in \{1, ..., n\}$, $e_i e_j + e_j e_i = 2\delta_{ij}$.

Let $f : \mathbb{R}^2 \rightarrow \mathbb{R}_{2,0}$ such that $f(x, y) = f_1(x, y)e_1 + f_2(x, y)e_2$. Applying the Dirac operator to this function gives:

$$Df(x, y) = D \cdot f(x, y) + D \wedge f(x, y)$$
$$= \frac{\partial f_1}{\partial x}(x, y) + \frac{\partial f_2}{\partial y}(x, y) + e_{12}\left(\frac{\partial f_2}{\partial x}(x, y) - \frac{\partial f_1}{\partial y}(x, y)\right).$$

Then, solving the Dirac equation $Df = 0$ in $\mathbb{R}_{2,0}$ is equivalent to find solution $f : \mathbb{R}^2 \rightarrow \mathbb{R}^2$ with $f(x, y) = (f_1(x, y), f_2(x, y))$ that satisfy the Cauchy-Riemann (CR) equations. Moreover, this can be extended to higher dimension and the CR equations are generalized by the Dirac equation.

3 Analytic Signal and Monogenic Signal

3.1 Analytic Signal

Let $s : \mathbb{R} \to \mathbb{R}$ be a real-valued signal and $f : \mathbb{R} \to \mathbb{R}_{2,0}$ be a vector-valued signal such that $f(x) = s(x)e_2$. The purpose is to construct a function fulfilling the Dirac equations (*i.e* an holomorphic function) whose real part is the real-valued signal. Thus it is equivalent to find the solution of a boundary value problem of the second kind (a Neumann problem):

$$\begin{cases} \Delta u = \dfrac{\partial^2 u}{\partial x^2} + \dfrac{\partial^2 u}{\partial y^2} = 0 \text{ if } y > 0 \\ e_2 \dfrac{\partial u}{\partial y} = f(x) \qquad \qquad \text{if } y = 0 \end{cases} \qquad (2)$$

with $D = e_1 \dfrac{\partial}{\partial x} + e_2 \dfrac{\partial}{\partial y}$ and $\Delta = D^2$.

The first equation is the 2D-Laplace equation restricted to the open domain $y > 0$. The second equation is called the boundary condition and the choice of the basis vector e_2 is coherent with the embedding of complex functions as vector fields (the real part is embedded as the e_2-component). Using the fundamental solution of the 2D-Laplace equation, the solution of the problem leads to:

$$f_A(x,y) = h_p * f(x,y) + h_p * h_H * f(x,y) \qquad (3)$$

where $h_p = \dfrac{y}{\pi(x^2 + y^2)}$ (1D-Poisson kernel) and $h_H = \dfrac{e_{12}}{\pi x}$ (Hilbert kernel).

The variable y is a scale parameter and setting it to zero[1], we obtain the classical analytic signal.

3.2 Monogenic Signal

Following the previous construction of the analytic signal, M. Felsberg has proposed an extension to 2D signals (such as grayscale images) [3]. Let $s : \mathbb{R}^2 \to \mathbb{R}$ be a real-valued signal and $f : \mathbb{R}^2 \to \mathbb{R}_{3,0}$ be a vector-valued signal such that $f(x,y) = s(x,y)e_3$. Generalizing 3.1 we are looking for a monogenic function (extension of holomorphic function) the e_3-component of which is the real-valued signal. The associated boundary value problem of the second kind is:

$$\begin{cases} \Delta u = \dfrac{\partial^2 u}{\partial x^2} + \dfrac{\partial^2 u}{\partial y^2} + \dfrac{\partial^2 u}{\partial z^2} = 0 \text{ if } z > 0 \\ e_3 \dfrac{\partial u}{\partial z} = f(x,y) \qquad \qquad \text{if } z = 0 \end{cases} \qquad (4)$$

where $D = e_1 \dfrac{\partial}{\partial x} + e_2 \dfrac{\partial}{\partial y} + e_3 \dfrac{\partial}{\partial z}$ and $\Delta = D^2$.

---

[1] In the Fourier domain, the Poisson kernel has an exponential form and equal to one for $y = 0$.

The first equation is the 3D-Laplace equation restricted to the open half-space $z > 0$ and the second equation is the boundary condition. Using the fundamental solution of the 3D-Laplace equation, the solution of this problem leads to:

$$f_M(x,y,z) = h_p * f(x,y) + h_p * h_R * f(x,y) \tag{5}$$

where $h_p = \dfrac{z}{2\pi(x^2+y^2+z^2)^{3/2}}$ is a 2D-Poisson kernel and $h_R = \dfrac{xe_1 + ye_2}{2\pi(x^2+y^2)^{3/2}}$ is the Riesz kernel, extension in 2D of the Hilbert kernel.

The variable z is a scale parameter and setting it to zero[2], we obtain the monogenic signal.

4 The Color Monogenic Signal

4.1 Construction

The aim of this paper is to construct a scale-space signal for color images seen as vectors in $\mathbb{R}_{5,0}$. Let $s : \mathbb{R}^2 \to \mathbb{R}^3$ be a real-valued signal and $f : \mathbb{R}^2 \to \mathbb{R}_{5,0}$ be a vector-valued signal such that $f(x_1, x_2) = f_3(x_1, x_2)e_3 + f_4(x_1, x_2)e_4 + f_5(x_1, x_2)e_5$. Thus a color image is decomposed in the $f_3 f_4 f_5$ space represented as the subspace spanned by $\{e_3, e_4, e_5\}$. Here, any orthonormal colorimetric system can be chosen for $f_3 f_4 f_5$ such as RGB, CIE XYZ or CIE L*a*b*. According to the previous construction, we need to find a function which is monogenic and the e_3, e_4 and e_5-component of which are the components f_3, f_4 and f_5 respectively.

$$\begin{cases} \Delta u = \dfrac{\partial^2 u}{\partial x_1^2} + \dfrac{\partial^2 u}{\partial x_2^2} + \dfrac{\partial^2 u}{\partial x_3^2} + \dfrac{\partial^2 u}{\partial x_4^2} + \dfrac{\partial^2 u}{\partial x_5^2} = 0 \\ e_3 \dfrac{\partial u}{\partial x_3} + e_4 \dfrac{\partial u}{\partial x_4} + e_5 \dfrac{\partial u}{\partial x_5} = f(x_1, x_2) \end{cases} \tag{6}$$

with $D = \sum_{i=1}^{5} e_i \dfrac{\partial}{\partial x_i}$ and $\Delta = D^2$.

The choice of $\mathbb{R}_{5,0}$ is related to the construction of the monogenic signal. Indeed, we want to define a scale-space signal which have independent scales in each component (x_3 for f_3, x_4 for f_4 and x_5 for f_5). We split the problem into three boundary value problems in $\mathbb{R}_{5,0}$ as follows:

$$(i = 3, 4, 5) \begin{cases} \dfrac{\partial^2 u}{\partial x_1^2} + \dfrac{\partial^2 u}{\partial x_2^2} + \dfrac{\partial^2 u}{\partial x_i^2} = 0 \text{ if } x_i > 0 \\ e_i \dfrac{\partial u}{\partial x_i} = f_i(x_1, x_2)e_i \quad \text{if } x_i = 0 \end{cases} \tag{7}$$

Solving each system is achieved by adapting the results of section 3.2, however we will not explain the construction in details. Each solution of the previous systems (7) leads to monogenic functions S_1, S_2, S_3: they satisfy the Dirac equation in

---

[2] Similar reason as footnote 1.

each subspace $E_i = span\{e_1, e_2, e_i\}$ ($i = 3, 4, 5$) and consequently the Dirac equation in $\mathbb{R}_{5,0}$ ($DS_i = 0$). Let $f_c = S_1 + S_2 + S_3$, then f_c is still monogenic in $\mathbb{R}_{5,0}$ (i.e. $Df_c = 0$) and satisfies the boundary conditions in (6). We call f_c the Scale-Space Color Monogenic Signal, it has the following form:

$$\begin{aligned}f_c =& h_p^3 * f_3 e_3 + h_p^4 * f_4 e_4 + h_p^5 * f_5 e_5 \\ &+ h_p^3 * h_R * f_3 + h_p^4 * h_R * f_4 + h_p^5 * h_R * f_5\end{aligned} \quad (8)$$

where $h_p^i = \dfrac{x_i}{2\pi(x_1^2 + x_2^2 + x_i^2)^{3/2}}$, ($i = 3, 4, 5$) is a 2D-Poisson kernel and $h_R = \dfrac{x_1 e_1 + x_2 e_2}{2\pi(x_1^2 + x_2^2)^{3/2}}$ is the Riesz kernel.

4.2 Local Color Phase

Let us first introduce some notations. We denote:

$$f_c = A_1 e_1 + A_2 e_2 + A_3 e_3 + A_4 e_4 + A_5 e_5 \quad (9)$$

where
$(i = 3, 4, 5)$
$\begin{cases} A_1 = h_p^3 * h_{R_{x_1}} * f_3 + h_p^4 * h_{R_{x_1}} * f_4 + h_p^5 * h_{R_{x_1}} * f_5, \\ A_2 = h_p^3 * h_{R_{x_2}} * f_3 + h_p^4 * h_{R_{x_2}} * f_4 + h_p^5 * h_{R_{x_2}} * f_5, \\ A_i = h_p^i * f_i \end{cases}$

The Color Monogenic Signal contains two kinds of information:

- A_1 and A_2 correspond to the smoothed vertical and horizontal structures with the meaning of Riesz transform.
- A_3, A_4 and A_5 correspond to the smoothed colors represented in the $f_3 f_4 f_5$ space spanned by $\{e_3, e_4, e_5\}$.

If $V = ue_1 + ve_2 + ae_3 + be_4 + ce_5 \in \mathbb{R}_{5,0}$ is a chosen vector containing structure information (u, v) and color information (a, b, c) then the geometric product $f_c V$ in $\mathbb{R}_{5,0}$ is given by:

$$f_c V = \langle f_c V \rangle_0 + \langle f_c V \rangle_2 \quad (10)$$

where the 0-graded part $\langle f_c V \rangle_0$ is the scalar part and the 2-graded part $\langle f_c V \rangle_2$ is the bivector part. We can explain this result in the context of Clifford algebra as follows. If B is any normalized bivector, the subspace spanned by $\{1, B\}$ is isomorphic to $\mathbb{C}$. Applying this to $f_c V = \langle f_c V \rangle_0 + \dfrac{\langle f_c V \rangle_2}{|\langle f_c V \rangle_2|} |\langle f_c V \rangle_2|$, allows to consider it as a complex number:

$$f_c V = \langle f_c V \rangle_0 + i|\langle f_c V \rangle_2| \quad (11)$$

This precisely means that $f_c V$ is a spinor which acts as a rotation in the plane spanned by the bivector $\langle f_c V \rangle_2$. Then the local color phase is the angle of the rotation and is given by:

$$\varphi = \arg(f_c V) = \arctan\left(\frac{|\langle f_c V \rangle_2|}{\langle f_c V \rangle_0}\right) \quad (12)$$

This phase describes the angular distance between f_c and a given vector V in $\mathbb{R}_{5,0}$, i.e it gives a correlation measure between a pixel fitted with color and structure information and a vector containing chosen color and structure.

5 A Metric and Multiscale Color Segmentation

5.1 Differential Geometry

A usual method of edge detection using metric information given by the first fundamental form [6] is to consider a multidimensional image, of components $(f_1, ..., f_n)$ defined on a domain D of $\mathbb{R}^2$ as a two-dimensional surface S parametrized by $\psi : (x, y) \rightarrow (x, y, f_1(x, y), ..., f_n(x, y))$ embedded into $\mathbb{R}^{n+2}$ fitted with the metric $g = \begin{pmatrix} 1 & 0 \\ 0 & 1 \end{pmatrix} \oplus \begin{pmatrix} \lambda_1 & & 0 \\ & \ddots & \\ 0 & & \lambda_n \end{pmatrix}$. This latter induces a metric on S called the first fundamental form of S, which takes the following form:

$$dS^2 = dx^2 + dy^2 + \lambda_1 df_1^2 + ... + \lambda_n df_n^2 \qquad (13)$$

Then variations on the image are assimilated to tangent vectors of S and a measure of these variations is given by dS^2. The rest of the method is devoted to select the strongest local variation, called edges. More precisely, let $I(q)$ be the matrix representation of the metric dS^2 at $q = \psi(p)$ ($p \in \mathbb{R}^2$) in the coordinates system given by $(d_p\psi(1, 0), d_p\psi(0, 1))$. Then $I(q)$ has the following form:

$$I(q) = \begin{pmatrix} E(q) & F(q) \\ F(q) & G(q) \end{pmatrix} \qquad (14)$$

Let $\lambda_+(q)$ and $\lambda_-(q)$, $\lambda_+ \geq \lambda_-$, be the two eigenvalues of $I(q)$ and $\theta_+(q), \theta_-(q)$ the corresponding eigenvectors. The edge measure is then given by:

$$w(q) = \sqrt{\lambda_+(q) - \lambda_-(q)} \qquad (15)$$

and we say that $p \in D$ is an edge point if the function w has a local maximum at $\psi(p)$ in the direction given by $\theta_+(\psi(p))$.

5.2 A Multiscale Approach

Due to the construction of the Color Monogenic Signal, this latter inherits a multiscale character in the Poisson Scale-Space. As shown in [4], this linear scale-space satisfies the axiomatic of Iijima and then is an alternative to the Gaussian Scale-Space. We will not extend the whole linear scale-space theory in this paper, the reader may refer to [4] for further details.

5.3 Segmentation Method

First of all, we need to choose a colorimetric system for our Color Monogenic Signal. In this application, we take the CIE L*a*b* space and considering each pixels of a color image as a vector in $\mathbb{R}_{5,0}$ is done by using the Color Monogenic Signal at a given scale s (the same for each component, i.e. $s = x_3 = x_4 = x_5$ in formula 8). Then, geometric calculus allows us to consider the geometric product between each vectors and a reference (chosen) vector $V = ue_1 + ve_2 + ae_3 + be_4 +$

ce_5. As in (4.2), a local color phase $\varphi(x,y)$ is obtained and associated with the scalar part $p(x,y) = \langle f_c V(x,y) \rangle_0$, we have an angle and magnitude information respectively.

Algorithmic scheme

1. Let $V = ue_1 + ve_2 + ae_3 + be_4 + ce_5$.
2. Compute $\varphi(x,y)$ and $p(x,y)$ as in (4.2).
3. Given S parametrized by $\psi : (x,y) \to (x,y,\varphi(x,y),p(x,y))$ in $\mathbb{R}^4$ with metric $M = \begin{pmatrix} 1 & 0 \\ 0 & 1 \end{pmatrix} \oplus \begin{pmatrix} \lambda_1 & 0 \\ 0 & \lambda_2 \end{pmatrix}$:
   - compute the matrix representation of the metric dS^2 as in (14);
   - calculate $w(q)$ as in (15).

Then variations in the local color phase and scalar part are assimilated to tangent vectors of S and we are able to measure edges in a chosen color given by V.

6 Experiments and Results

Some results obtained on well-known images and images from the Berkeley image dataset [5] are presented. We aim at showing the relevance of our approach on these highly regarded images. Firstly, we look at the Lab image (figure 1) and we study a yellow object segmentation.

Taking a vector V carrying yellow in the L*a*b* space without any structure, i.e. $V = ae_3 + be_4 + ce_5$, we apply the method described above with the euclidean metric ($\lambda_1 = 1$, $\lambda_2 = 1$) and obtain images in figure 1. As the reader can see, when the scale parameter is set to zero, we get not only yellow objects but all object with strong variations in this color. When the scale parameter increases, one can see in a first step that edges are smoothed and then that the blue block disappears. Finally we keep the desired yellow objects but also the green block, this is due to the strong variations in yellow between this object and the background.

Next, we choose two challenging images from the Berkeley image dataset: the Plane and Garden images (see figure 2). For the first image (first row), we would like to get the red edges. Using the multiscale aspect and taking the metric ($\lambda_1 = 2$,

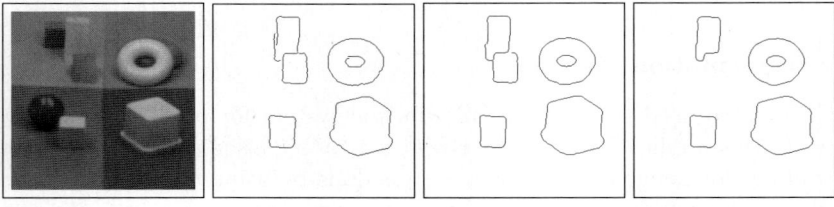

Fig. 1. From left to right: Lab image, segmentation at scale 0, 1.5 and 3.5 (hysteresis thresholds are the same for each scale.)

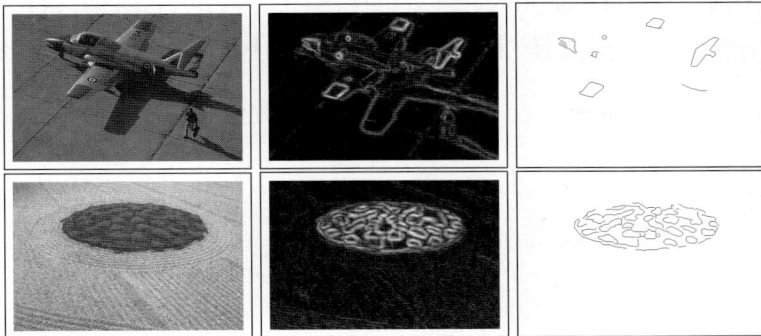

Fig. 2. From left to right. First row, the Plane image, the function w and the result at scale 3,5. Second row, the Garden image, the function w and the result at scale 3,5.

$\lambda_2 = 0.5$) in order to reinforce the phase relatively to the magnitude, we obtain results shown in figure 2. For the second image (second row), we would like to get green edges. Using the multiscale aspect and taking the euclidean metric, results in figure 2 show an interesting segmentation compared with segmentation in [5].

7 Conclusion

In this paper, we introduce a new theoretical framework to represent color images as scale-space signal. We use Clifford algebra formalism to encode structure and color information in a single vector-valued function. By geometric calculus, we define a local color phase that may be used in a wide range of applications. We have treated in this paper a method of color segmentation which reveals relevant for finding edges relative to a chosen color. Future work will be devoted to the analysis of the scale-space in the Color Monogenic Signal framework.

References

1. Sommer, G.: Geometric computing with Clifford Algebras. Theorical Foundations and Applications in Computer Vision and Robotics. SSVM. Springer, Heidelberg (2001)
2. Felsberg, M., Sommer, G.: The monogenic signal. IEEE Transaction on Signal Processing 49(12), 3136–3144 (2001)
3. Felsberg, M.: Low-level image processing with the structure multivector. PhD thesis, Christian Albrechts University Kiel (2002)
4. Felsberg, M., Sommer, G.: The monogenic scale-space: A unifying approach to phase-based image processing in scale-space. JMIV 21, 5–26 (2004)
5. Martin, D., Fowlkes, C., Tal, D., Malik, J.: A Database of Human Segmented Natural Images and its Application to Evaluating Segmentation Algorithms and Measuring Ecological Statistics. In: ICCV, vol. 2, pp. 416–423 (2001)
6. Di Zenzo, S.: A note on the gradient of a multi-image. Comput. Vis. Graph. Image Process. 33(1), 116–125 (1986)

An Interactive Level Set Approach to Semi-automatic Detection of Features in Food Micrographs

Gaetano Impoco[1] and Giuseppe Licitra[1,2]

[1] Co.R.Fi.La.C. - Consorzio Ricerca Filiera Lattiero-Casearia, Ragusa, Italy
[2] D.A.C.P.A., University of Catania, Italy
impoco@corfilac.it

Abstract. Microscopy is often employed in food research to inspect the microstructural features of food samples. Accurate detection of microscopic features is required for reliable quantitative analysis. We propose a user-assisted approach that can be easily integrated into a graphical interface. The proposed algorithm is based on a fast approximation of the common region-based level set equation, providing interactive computations. Experiments have been run on cheese micrographs acquired with electron and confocal microscopes.

Keywords: Level set, segmentation, feature detection, cheese microstructure.

1 Introduction

Rheological properties of complex food, such as taste, consistency, and texture, are closely related to food microstructure. Food research takes advantage from microscopy to study microstructural features in food samples, such as holes, protein, and fat. The potential of computer analysis of microscopic features has recently been recognised by several researchers [1], objective quantification and rapid data handling being the most prominent advantages over visual inspection.

Accurate detection of features in food micrographs is an important step for reliable quantitative analysis of food morphology [2]. The measurements computed on binarised images are deeply influenced by the segmentation accuracy, especially when shape descriptors of features are involved [3]. Although our approach is general enough to cover a broad range of applications, here we are interested in images from Scanning Electron Microscope (SEM) and Confocal Light Scanning Microscope (CLSM) (see Figure 1).

In the literature, segmentation of biological data is mostly concerned with detection of macroscopic structures. The extent of published work on segmentation of microscopy images is low compared to that devoted to the macroscopic world. Far from being caused by lack of interest, this phenomenon is the result of a number of causes (see [4]): the large diversity of imaging approaches, the lack of a priori knowledge about the image content, the lack of common evaluation procedures, the large variety of imaged objects for different disciplines. Moreover,

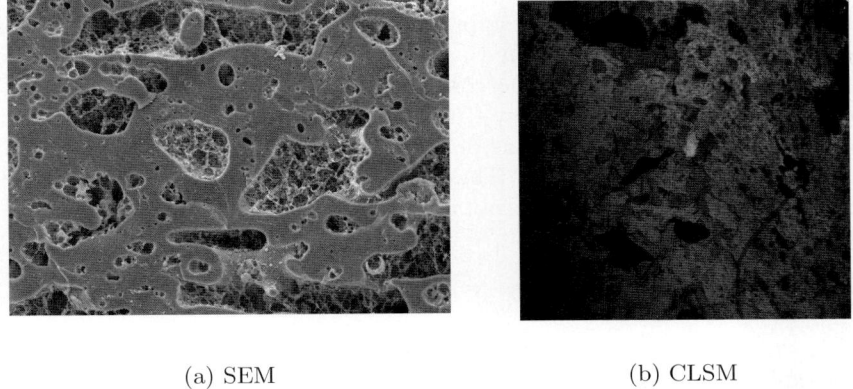

(a) SEM (b) CLSM

Fig. 1. Cheese micrographs acquired with different imaging devices: Scanning Electron Microscope (SEM), and Confocal Light Scanning Microscope (CLSM)

most of the literature in this field addresses biomedical data (e.g., analysis of cells) or the study of bacteria in biological images. Segmentation of structural features in food images is mostly carried out with time-consuming user interaction, using simple techniques such as thresholding and basic morphology [2].

In this paper, we propose a fast approximate level set algorithm for the detection of microstructural features in food micrographs. Our approximation method is in the spirit of [5]. Automatic segmentation methods offer little control to the user on the resulting segmentation. Hence, we chose a user-assisted approach. Approximating the exact evolution equation allows for interactive computations.

2 Algorithm

Level Sets [6] have been fruitfully employed for segmentation, thanks to their ability to give closed contours between segmented regions, avoiding the oversegmentation issues of region-based approaches. A binary segmentation is obtained evolving the boundary towards a rest position that minimises an energy functional. The boundary is implicitly represented by an embedding function, Φ, chosen as the signed distance from the contour. This formulation overcomes the implementation issues of active contours [7]. The *level set equation* is

$$\frac{\partial \Phi}{\partial t} = -\nabla \Phi \cdot \frac{d\boldsymbol{x}}{dt} = -\nabla \Phi \cdot \mathbf{F} \qquad (1)$$

where $\mathbf{F}$ is a function encompassing the partial derivatives of Φ evaluated at $\boldsymbol{x}$, and drives the evolution. Equation 1 is evaluated at the nodes of a finite grid. Resolution anisotropy is easily dealt with by using an appropriate grid spacing.

The level set framework can be easily employed for segmentation by defining a function $\mathbf{F}$ such that Equation 1 goes to zero close to the boundaries of interest objects in the image. Most of the level set segmentation algorithms in the literature are based on two different approaches: *edge-based,* and *region-based.* The edge-based method, introduced by Caselles et al. [8], employs an image gradient term to stop the evolution of the contour close to image intensity edges. The region-based method, proposed by Chan and Vese [9], employs global statistics about the inner and outer regions of the evolving contour to segment the image into two homogenous regions. These basic methods are intrinsically bimodal. Multi-region segmentation can be obtained by simultaneously evolving a number of different contours.

2.1 Evolution Equation

Our algorithm is intended as a smart tool to select homogeneous features in food micrographs. The user is asked to select a point inside the interest feature, that is used to initialise contour evolution. A circular contour is defined around the input point, and is grown towards the outside. We use the region-based framework. The original formulation [9] employed information about the inner and outer regions with respect to the contour. Namely, the deviation from the average intensity values in the inner and outer regions is penalised. In our application, the interest regions are reasonably homogenous, while the background can show high variability. Hence, we drop the homogeneity term for the outer region.

Our interest structures can present intensity gradients while moving from the centre to the border. However, they are locally homogenous. Hence, a term is added to favour homogeneity in local neighbourhoods, much in the same way as in [10]. This term is expressed as in the original region-based formulation, but over a local support determined by a binary function $\mathcal{B}(\overline{x}, x)$, where $\overline{x}$ is the interest pixel and x is defined over the image support.

Food images, both microscopic and macroscopic, often present globular structures due e.g., to fat, gas concentration, and so on. An elliptical template shape is used when looking for globular features. The evolution is guided by the equation:

$$\frac{\partial \Phi}{\partial t} = \delta_\varepsilon(\Phi) \left[-\nu - \lambda_G (\Phi - \mu_G^i)^2 - \lambda_L (\Phi - \mu_L^i)^2 \right.$$
$$\left. -\gamma(1 - |\mu_L^i - \mu_L^o|) + 2\alpha(\Phi - \Psi) \right] \qquad (2)$$

where Φ and Ψ are signed distance functions from the contour and from the prior shape, respectively, μ is an inflation force, and λ_G, λ_L, γ, and α are parameters. μ_G^i and μ_L^i are, respectively, the average pixel value inside the contour, and the local average inside the contour. μ_L^o is the local average of background pixels. These averages can be expressed as:

$$\mu_G^i = \frac{\int \mathcal{H}(\Phi(x)) I(x) dx}{\int \mathcal{H}(\Phi(x)) dx}$$

$$\mu_L^i(\overline{x}) = \frac{\int \mathcal{B}(\overline{x}, x) \mathcal{H}(\Phi(x)) I(x) dx}{\int \mathcal{B}(\overline{x}, x) \mathcal{H}(\Phi(x)) dx}$$

$$\mu_L^o(\overline{\boldsymbol{x}}) = \frac{\int \mathcal{B}(\overline{\boldsymbol{x}}, \boldsymbol{x})(1 - \mathcal{H}(\Phi(\boldsymbol{x})))I(\boldsymbol{x})d\boldsymbol{x}}{\int \mathcal{B}(\overline{\boldsymbol{x}}, \boldsymbol{x})(1 - \mathcal{H}(\Phi(\boldsymbol{x})))d\boldsymbol{x}}$$

Here, $\mathcal{B}(\overline{\boldsymbol{x}}, \boldsymbol{x})$ is defined as above, $\overline{\boldsymbol{x}}$ is the interest pixel, $I(.)$ is the image, and $\mathcal{H}(.)$ is the Heaviside function.

The homogeneity term $(\Phi - \mu_G^i)^2$ penalises the deviation of pixel values from the average values inside the contour. Similarly, $(\Phi - \mu_L^i)^2$ does the same in the neighbourhood defined by $\mathcal{B}(\overline{\boldsymbol{x}}, \boldsymbol{x})$. The similarity term $(1 - |\mu_L^i - \mu_L^o|)$ is maximum when the local average inside, μ_L^i, and outside, μ_L^o, the contour are identical, and goes to zero when they are extremely different. This term lets the contour go through homogeneous regions, stopping close to intensity edges. Finally, the term $(\Phi - \Psi)$ measures the difference of the contour from the prior shape Ψ. The function $\delta_\varepsilon(\Phi)$ is a smoothed Dirac delta such that $\delta_\varepsilon(\Phi) = 1$ at $\Phi(\boldsymbol{x}) = 0$, which prevents topology changes to occur far from the contour.

2.2 Approximate Equation

The computational burden of solving PDEs does not allow interactive segmentation using standard level set methods. On the other hand, a few applications require segmentation at sub-pixel precision. In order to speed up the evolution process, we are thus allowed to approximate the distance function such that the contour points have integer pixel coordinates. We do not even need to compute an exact distance function. Only its sign matters. In the same spirit as [5], we approximate Equation 2 as:

$$\frac{\partial \Phi}{\partial t} = \delta_{\text{box}}(\Phi)\left[-\nu - \lambda_G(\Phi - \mu_G^i)^2 - \lambda_L(\Phi - \mu_L^i)^2 \right.$$
$$\left. -\gamma(1 - |\mu_L^i - \mu_L^o|) - 2\alpha(\mathcal{H}(\Phi) - \mathcal{H}(\Psi))\right] \quad (3)$$

where $\mathcal{H}$ is the Heaviside function, and $\delta_{\text{box}}(\Phi) = 1$ for $|\Phi| \leq 1$ and zero elsewhere. The function δ_{box} is a sort digitised Dirac delta function that selects the pixels adjacent to the approximate contour. These points are the only candidates for a sign change. $\mathcal{H}(\Phi)$ and $\mathcal{H}(\Psi)$ are used in place of Φ and Ψ to avoid keeping track of the exact distance function, and store only its sign.

2.3 Implementation

Thanks to the approximations in Equation 3, a number of optimisations can be carried out to speed up the contour evolution. Since no signed distance function is defined, there is nothing to reinitialise. This saves most of the computational time of the exact level set method. Since the grid values can take only two values (i.e., the sign of the distance function) only pixels adjacent to the contour, selected by the function $\delta_{\text{box}}(\Phi)$, are candidate for a sign change. Hence, the narrow band only contains pixels whose distance from the contour is less than one pixel. The number of grid points involved for each iteration is thus considerably reduced.

The function $\delta_{\text{box}}(\Phi)$ can be easily implemented using the standard Moore boundary following algorithm. The computed pixel chain contains the candidate foreground boundary pixels. Notice that the boundary pixels found by this algorithm lie inside the contour. In order to get the background boundary pixels, we take care to store adjacent pixels during the execution of Moore's algorithm, at negligible additional cost. These two chains represent each of the two sides of the boundary, and are used to decide which inner (outer) pixels move outside (inside) the contour. The following expressions are used to implement Equation 3 for each side of the boundary (notice the sign of the homogeneity terms):

$$\underset{I \to O}{\Delta}(\boldsymbol{x}) = -\nu - \lambda_G(\Phi(\boldsymbol{x}) - \mu_G^i)^2 - \lambda_L(\Phi(\boldsymbol{x}) - \mu_L^i)^2$$
$$-\gamma(1 - |\mu_L^i - \mu_L^o|) - 2\alpha(\mathcal{H}(\Phi(\boldsymbol{x})) - \mathcal{H}(\Psi(\boldsymbol{x}))) \quad (4a)$$

$$\underset{O \to I}{\Delta}(\boldsymbol{x}) = -\nu + \lambda_G(\Phi(\boldsymbol{x}) - \mu_G^i)^2 + \lambda_L(\Phi(\boldsymbol{x}) - \mu_L^i)^2$$
$$-\gamma(1 - |\mu_L^i - \mu_L^o|) - 2\alpha(\mathcal{H}(\Phi(\boldsymbol{x})) - \mathcal{H}(\Psi(\boldsymbol{x}))) \quad (4b)$$

Equation 4a is used to shrink the contour, while Equation 4b is used for expansion.

2.4 Evolution of the Prior Shape

In our application, we are interested in blob-like objects that appear approximately elliptical in micrographs. Hence, an elliptical prior can be fit to the contour points using a fast ellipse fitting algorithm at each evolution step. We use a classical moment-based algorithm proposed in [11].

2.5 Colour

SEM micrographs are 8-bit greyscale images. Conversely, CLSM images are composite images that may contain more than three colour channels. Each channel maps the intensity of fluorescent light from stained structures to a 8-bit scale. A channel represents the density of e.g., fat, protein or other substances of interest. Microscopists take great care in order to choose colour stains that do not affect the response of the sensor for other stains. These micrographs can be viewed as a sort of multispectral images.

A mapping is needed from a colour micrograph with any number of colour channels to a unidimensional scale. Micrographs can contain more that three colour stains and colours may vary, depending on many factors (e.g., microscopist's choice, type of experiment, and so on). Most mapping schemes assume images with three colour channels, while mapping functions used for multispectral images are often bounded to the specific nature of the spectral components. In our experiments, we adopt the straightforward solution of summing-up all colour channels. Each new channel requires one more bit to represent pixel values (e.g., 9-bit values are used for images such as that shown in Figure 3).

3 Evaluation

Table 1. Parameter values for SEM and CLSM

	SEM	CLSM
λ_G	0.4	0.5
λ_L	1.0	2.0
γ	0.1	0.1
α	0.05	0.05
ν	0.0	0.0
W	5×5	3×3
d_{init}	20	15

We carried out a number of experiments using 20 cheese micrographs, 10 for each imaging device (SEM and CLSM). We used two sets of parameter values, one for each of the two imaging techniques.

Table 1 shows the values we used throughout all our experiments. The meaning of λ_G, λ_L, γ, α, and ν is as described in Section 2. W is the size of the active local neighbourhood in $\mathcal{B}(\overline{\boldsymbol{x}}, \boldsymbol{x})$ and d_{init} is the diameter of the initial curve, centred at the position selected by the user. Two different sets are used for SEM and

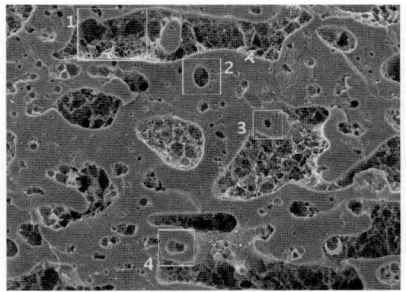

(a) Input SEM micrograph. Interest regions are marked and annotated.

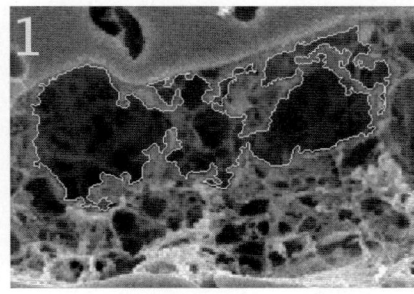

(b) A complex feature

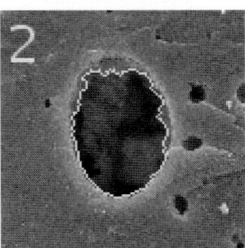

(c) A blob-like feature

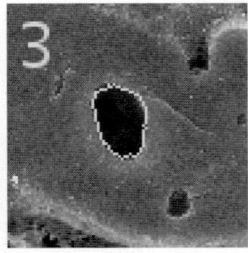

(d) A blob-like feature

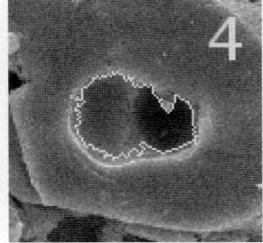

(e) A compound feature

Fig. 2. Examples of segmentation of SEM cheese micrographs. The segmentations are originated by user selections in the highlighted regions of the input images. Four features are shown: (b) a very complex feature corresponding to an agglomerate of fat, (c) a fat globule with complex background, (d) a globular structure, probably fat, with two small pores where whey can flow (e) two fat globules attached.

CLSM micrographs. This is mainly due to the different nature of the two imaging technologies. Parameter values are also influenced by the colour mapping scheme employed.

Figure 2 shows four cases of interest for food microscopists. In Figure 2(d) a fat globule was successfully segmented although in contains two small pores. Similarly, the fat globule in Figure 2(c) has a rough background, due to the cheese making process or to sample preparation. Most common segmentation algorithms fail in these cases. The case in Figure 2(e) shows a faintly distinguishable globule close to a clearly visible pore. This is another hard situation for most algorithms. A very complex feature, showing a highly variable background, is successfully detected in Figure 2(b).

Figure 3 shows four segmentation results on a CLSM cheese micrograph. Again, simple (Figure 3(c)) as well as complex (Figures 3(d) and 3(e)) features are successfully detected. Smooth transitions (Figure 3(b)) are also captured.

Our algorithm may converge to wrong segmentations in complex situations, such as that in Figure 2(b). This is due to the balance between global and local detection terms in Equation 3. Our method may not converge at all in very faint globules (e.g., the left globule in Figure 2(e)) or may leak through smooth transitions (e.g., Figure 3(b)). Clearly, this depends on the difficulty to mark inner and outer pixels when they present similar local statistics.

We cannot provide a more accurate quantitative analysis of segmentation results due to the lack of manually-labelled ground truth. However, we plan to get manual annotations for the near future. Improvements will concern the colour mapping scheme and a deeper adaptation to the input data.

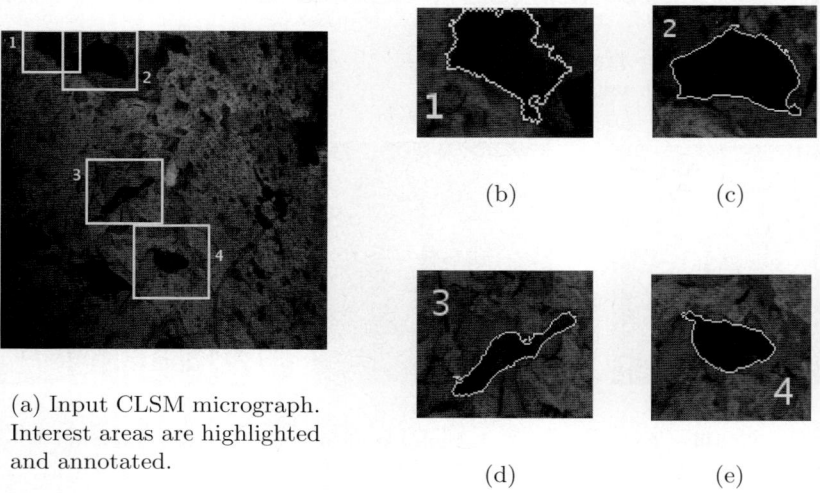

Fig. 3. Examples of segmentation of CLSM cheese micrographs. Four features are segmented: (b) a feature showing a smooth transition of the red component (fat) going out of focus, (c) a well distinguished feature, (d) and (e) features with complex internal background.

Acknowledgements

We wish to thank Laura Tuminello and Nicoletta Fucà for having provided test micrographs and for many helpful discussions. Financial support was provided by the Assessorato Agricoltura e Foreste della Regione Siciliana, Palermo, Italy.

References

1. Aguilera, J.M., Stanley, D.W.: Microstructural principles of food processing and engineering, 2nd edn. Aspen (1999)
2. Russ, J.C.: Image Analysis of Food Microstructure. CRC Press, Boca Raton (2004)
3. Impoco, G., Carrato, S., Caccamo, M., Tuminello, L.: Quantitative analysis of cheese microstructure using SEM imagery. In: SIMAI 2006, Minisymposium: Image Analysis Methods for Industrial Application (May 2006)
4. Nattkemper, T.W.: Automatic segmentation of digital micrographs: a survey. Studies in health technology and informatics 107(2), 847–851 (2004)
5. Fahmi, R., Farag, A.A.: A fast level set algorithm for shape-based segmentation with multiple selective priors. In: 15th IEEE International Conference on Image Processing (ICIP 2008), October 2008, pp. 1073–1076 (2008)
6. Osher, S., Fedkiw, R.P.: Level Set Methods and Dynamic Implicit Surfaces. Springer, Heidelberg (2002)
7. Kass, M., Witkin, A., Terzopoulos, D.: Snakes: Active contour models. International Journal of Computer Vision 1(4), 321–331 (1988)
8. Caselles, V., Kimmel, R., Sapiro, G.: Geodesic active contours. International Journal on Computer Vision 22, 61–79 (1997)
9. Chan, T., Vese, L.A.: An active contour model without edges. In: Nielsen, M., Johansen, P., Fogh Olsen, O., Weickert, J. (eds.) Scale-Space 1999. LNCS, vol. 1682, pp. 141–151. Springer, Heidelberg (1999)
10. Lankton, S., Tannenbaum, A.: Localizing region-based active contours. IEEE Transactions on Image Processing 17(11), 2029–2039 (2008)
11. Chaudhuri, B.B., Samanta, G.P.: Elliptic fit of objects in two and three dimensions by moment of inertia optimization. Pattern Recognition Letters 12(1), 1–7 (1991)

Shape Detection from Line Drawings by Hierarchical Matching

Rujie Liu[1], Yuehong Wang[1], Takayuki Baba[2], and Daiki Masumoto[2]

[1] Fujitsu Research and Development Center, Beijing, China
{rjliu,wangyh}@cn.fujitsu.com
[2] Fujitsu Laboratories, Kawasaki, Japan

Abstract. An object detection method from line drawing images is presented. In this method, the content of line drawing images are hierarchically represented, where a local neighborhood structure is formed for each primitive by grouping its nearest neighbors. The detection process is a hypothesis verification scheme. Firstly, the top k most similar local structures in the object drawing are obtained for each local structure of the model, and the corresponding transformation parameters are estimated. By treating each estimation result as a point in the parameter space, a dense region around the ground truth is then formed provided that there exists a model in the object drawing. At last, the mode detection method is used to find this dense region, and the significant modes are accepted as the occurrence of object instances.

1 Introduction

Object detection is usually required in content based engineering drawings management, where the purpose is to find the occurrences of a specific part from the database. Direct usage of object detection techniques often costs much time for each operation, therefore, they are not qualified under the real time requirement. Thanks to the specific characteristics of the drawings wherein lines and curves are the primary objects, a line/curve representation can be used instead for detection purpose [1][2][3].

Matching line/curve representations is not a trivial problem. A line segment itself is non-distinctive: a line in the model can be matched to any line of the image if affine invariance is allowed. The discriminative power of a line/curve representation lies in the structural information between line/curve segments. This naturally leads to graph matching approaches [4][5]. Although theoretically reasonable, the graph matching methods are difficult to be applied in real applications due to the high computational complexity.

Many attempts have been made to improve the graph matching speed. Huet and Hancock [6] proposed to integrate the geometric attributes of line/curve pairs into a histogram, and measure the similarity of two histograms by the Bhattacharyya distance. Similar to [6], histogram was also adopted in [3] to conglomerate the local structures of one image, and histogram intersection operations were employed for indexing, however, the local structure representation was constructed therein under the guidance of Gestalt psychology laws. To preserve the structural information in matching, a hypothesis-test scheme was used in [7] to realize the graphics recognition. The attributed graph of the model was firstly reduced into a spanning tree and a fixed traversal path was thus determined, next, this tree was used to direct the examination process to find all required components of this modal class.

Our method for object detection from line drawings commences by extracting lines and smooth curves from the row image data. These lines and curves, with the advantages of line/curve representation, are then adopted as primitives to construct local neighborhood structure by grouping several nearest neighbor primitives around one reference. Object detection is realized by the mode detection scheme. For each local structure of the model, we find the top k nearest structures in the drawing, and estimate the transformation parameters for each of the k candidates. If one candidate structure is actually in correspondence with the model structure, the estimation result will be close to the ground truth; otherwise, these values will be irregularly distributed. By treating each estimation result as a point in the parameter space, a dense area around the ground truth is then formed provided that there exists a model object in the drawing; otherwise, no dense area can be found. This dense area is finally detected using the mean shift technique [8].

2 Local Neighborhood Structure

The local neighborhood structure is constructed using the nearest neighbor criterion. Lines and curves are the primary components of line drawings, thus, it is a natural thing to adopt them as the primitives. Given a primitive as the reference, we find its neighbor primitives whose minimum distances to the reference are smaller than a threshold, and then group these neighbor primitives with the reference to form a local neighborhood structure. With this process, a local neighborhood structure is built for each primitive.

The local neighborhood structure is characterized by two kinds of attributes: (1) shape of the primitive, which is represented by its direction histogram. To realize rotation and scale invariance, the Fourier transform is applied to the histogram, and the magnitudes of the coefficients are used as the attribute; (2) the spatial relationship between neighbor primitives and the reference.

Four geometric cues are defined to describe the relationship between a neighbor primitive and the reference one [9]:

1. *Relative length*, which is defined as the length of the neighbor primitive divided by that of the reference;
2. *Relative distance*, which is defined as the length of the line segment connecting the centroids of these two primitives, divided by the length of the reference;
3. *Relative minimum distance*, similar to relative distance, we define the relative minimum distance as the minimum distance between the neighbor primitive and the reference, divided by the length of the reference;

4. *Relative angle*, which is defined as the acute angle between the neighbor primitive and the reference.

Measuring the angle of two lines is straightforward, but it is not an easy thing for two curves. To solve this problem, the earth mover's distance (EMD) of the direction histograms of two primitives is used to measure their relative angle. Intuitively, the angle between two curves can be interpreted as how much in average the angle is changed from one curve to another, which is fully consistent with the concept of EMD.

In summary, we can represent a local neighborhood structure P as $P = \{\mathbf{S}, (\mathbf{T}_i, RL_i, RD_i, RMD_i, RA_i), i = 1...u\}$, where, $\mathbf{S}$ and $\mathbf{T}_i$ are the shape descriptors of the reference and the i-th neighbor primitive, RL_i, RD_i, RMD_i, RA_i are the geometric cues between the reference and its i-th neighbor, u is the number of neighbors.

3 Matching Local Neighborhood Structures

Let $P^M = \{\mathbf{S}^M, (\mathbf{T}_i, RL_i, RD_i, RMD_i, RA_i)^M, i = 1...u\}$ and $P^G = \{\mathbf{S}^G, (\mathbf{T}_j, RL_j, RD_j, RMD_j, RA_j)^G, j = 1...v\}$ be local neighborhood structures of the model and the drawing image respectively. For the reason of simplicity, N_i^M and N_j^G are used to denote the neighbor primitive in P^M and P^G respectively. We will introduce in this section how to compute the distance of P^M and P^G, and estimate the transformation parameters from P^M to P^G.

Given one neighbor N_i^M in P^M and another neighbor N_j^G in P^G, their distance is defined as follows:

$$D_N(i,j) = \alpha \cdot d_T(i,j) + (1-\alpha) \cdot d_C(i,j) \quad (1)$$

where, $d_T(i,j)$ is the distance between shape descriptors T_i and T_j, $d_C(i,j)$ is the distance of the geometric cues, α is a weight coefficient between 0 and 1.

With this distance metric, we may calculate all the distances between the neighbors in P^M and those in P^G, which are represented as a matrix, $[\mathbf{D}_N]_{u \times v}$.

After that, a greedy searching process is used to obtain the distance of P^M and P^G. Firstly, find the minimum value $\mathbf{D}_N(x,y)$ in $\mathbf{D}_N$, and treat neighbors x and y as being processed; Next, search the minimum distance in the remained neighbors, and mark them as being processed. Repeat this process until no further operations can be made. In case that the number of the neighbors in P^M is larger than that in P^G, a punishment cost is assigned to each of the unprocessed neighbors in P^M.

Let D be the cumulative value of the distances obtained in the above searching process, the distance of P^M and P^G is then defined as:

$$D_{LNS}(P^M, P^G) = (D + d_T(\mathbf{S}^M, \mathbf{S}^G))/(2 * u + 1) \quad (2)$$

The transformation from P^M to P^G is restricted to be an affine one, i.e., only changes on scale, shift, and rotation are allowed. In this case, we may estimate the transformation parameters of the occurrence of the model in the drawing by two pairs of matched primitives.

Denote N_i^M and N_j^M be two primitives in the model, which are in correspondence with N_k^G and N_l^G in the drawing respectively. Let the vectors from the centers of N_i^M and N_j^M to the centroid of the model be $\vec{V_i}$ and $\vec{V_j}$, and $\mathbf{O}_k$ and $\mathbf{O}_l$ be the centers of N_k^G and N_l^G. Under the assumption of affine transformation, following relation holds:

$$\mathbf{O}_k + s \cdot \vec{V_i} \cdot (cos\theta + isin\theta) = \mathbf{O}_l + s \cdot \vec{V_j} \cdot (cos\theta + isin\theta) \quad (3)$$

Unfortunately, two main problems prevent the above method from being applied directly: (1) the assumption of affine transformation is too strict; (2) correspondences of the primitives are not known at all in the current stage, i.e., one neighbor in P^M may be matched to any neighbor in P^G.

To solve these problems, we rely on the most reliable matches of the primitives to estimate the parameters. Firstly, we select the reliable matches of the primitives from all possible matches in P^M and P^G. Next, the parameters are calculated with equation (3) for any two pairs of the reliable matches. Provided that P^G is actually in correspondence with P^M, most of the selected reliable matches will be correct, and thus a dense region of the calculated parameters will be formed.

Let $D_{\min}$ be the minimum value in the distance matrix $[\mathbf{D}_N]_{u \times v}$. A neighbor pair, N_x^M and N_y^G, are accepted as a reliable match if:

1. $D_N(x, y)$ is larger than $D_{\min}$ within a given threshold;
2. The ratio of $D_N(x, y)$ to $D_{\min}$ is smaller than a threshold.

With the parameters calculated from all the reliable match pairs, the Parzen Window method is used to search the dense clusters, and the 'true' transformation parameters are thus obtained.

4 Object Detection

4.1 Mean Shift Mode Detection

Obviously, many estimation results from the local neighborhood structures of the model and their k nearest candidates are not precise. At the first glance, it seems difficult to realize the detection task from these unreliable estimation results. However, it is not such a case. For a correct match pair of local neighborhood structures, where the reference primitives of the two structures are actually in correspondence with each other, the estimation result is usually close to the ground truth; otherwise, irregular and sparse values will arise. Thus, a dense area around the ground truth is formed by the correct match pairs if we treat each estimation result as a point in the parameter space. By searching this dense cluster, the occurrence of the object can be detected with its transformation parameters.

The variable bandwidth mean shift [8] is adopted to find the dense clusters from the estimation results. Let $\mathbf{x}_i \in R^d, i = 1...n$ be the estimated parameters by matching the local neighborhood structures, and each point $\mathbf{x}_i$ is associated with a bandwidth value h_i. The mean shift method is implemented by an iterative procedure, as follows:

$$\mathbf{Y}_{j+1} = \frac{\sum_{i=1}^{n} \frac{\mathbf{X}_i}{h_i^{d+2}} \mathcal{G}(\|\frac{\mathbf{Y}_j - \mathbf{X}_i}{h_i}\|^2)}{\sum_{i=1}^{n} \frac{1}{h_i^{d+2}} \mathcal{G}(\|\frac{\mathbf{Y}_j - \mathbf{X}_i}{h_i}\|^2)} \quad (4)$$

4.2 Verification

Let $\mathbf{C}$ be one mode obtained from the mean shift procedure, and $c_n = \{P_i^M, P_j^G, D_{LNS}(i,j), \boldsymbol{f}_{ij}\}_n, n = 1...N$ the points belonging to this mode, where P_i^M and P_j^G constitute a matched pair of local neighborhood structures, $D_{LNS}(i,j)$ is their distance, and $\boldsymbol{f}_{ij}$ is the estimated parameters from P_i^M and P_j^G. Denote the parameters corresponding to mode $\mathbf{C}$ be $\boldsymbol{f}$, i.e. the center of the dense region. Under the constraint that only one-to-one match is allowed for local neighborhood structures, we can evaluate each mode by solving the Hungarian assignment problem.

Firstly, a confidence value is assigned to each point c_n of mode $\mathbf{C}$:

$$Z_n = (1 - D_{LNS(i,j)})^2 \cdot exp(-(\boldsymbol{f}_{ij} - \boldsymbol{f})' \boldsymbol{W} (\boldsymbol{f}_{ij} - \boldsymbol{f})) \quad (5)$$

where, $\boldsymbol{W}$ is a diagonal matrix with the diagonal elements indicating the weights of the parameter components.

Two factors are considered in this definition: (1) the similarity of two local neighborhood structures; (2) the distance of the estimated parameters of these two structures to the mode center.

The value of the weight coefficients in $\boldsymbol{W}$ is calculated by solving a least squares problem with the manually labeled samples. For this purpose, some pairs of model and drawing images are firstly selected. A model and a drawing image compose a pair for manual labeling if the drawing contains one occurrence of the model and the mean shift mode detection result (the detected mode) is correct. Given these selected images, the correctly matched pairs of local neighborhood structures are manually labeled as positive samples while others are labeled as negative one.

With the training samples, the objective function used in the least squares problem is:

$$J = \sum_{n=1}^{N} \sum_{m=1}^{M} b_{nm} (e^{-(\boldsymbol{f}_{nm} - \boldsymbol{f}_n)' \boldsymbol{W} (\boldsymbol{f}_{nm} - \boldsymbol{f}_n)} - t_{nm})^2 \quad (6)$$

Here $n = 1...N$ denotes the selected pair of model and drawing images and $\boldsymbol{f}_n$ is the detected mode from these two images, $m = 1...M_n$ represents the matched pair of the local neighborhood structures in n and $\boldsymbol{f}_{nm}$ is the estimated parameters from these two structures, t_{nm} indicates the manual labeling result (1 for positive samples and 0 for negative ones), the weights b_{nm} are set to be the ratio of the number of positive samples to the negative ones if $t_{nm} = 1$, and 1 otherwise. These weights balance the relative contributions to the error function between positive and negative examples.

After the calculation of the confidence value z_n for the points of **C**, the optimal matches of local neighborhood structures can be selected therein by solving the Hungarian assignment problem, under the constraint that only one-to-one match is allowed. Thus, the confidence score of mode **C** can be easily obtained.

5 Experiments

5.1 Synthetic Data

The robustness of the proposed method is analyzed using randomly generated drawings, which are assumed to be composed only by straight lines. Firstly, one drawing image is created by randomly generating several straight lines; then, noise of various forms are added to the lines to get another drawing; at last, matching is run on the two drawings.

The ingredient lines of the first drawing are modified by two factors to generate another drawing, i.e., rotation, scale and translation.

- Rotation. Each ingredient line is rotated around its center with a random number between $[-R_{Max}, R_{Max}]$.
- Scale and translation. Given a random number S between $[-S_{Max}, S_{Max}]$, each end point of a line is moved to or away the line center by $S/2$ multiplied by the length of this line.

The evaluations are performed in two ways: the local neighborhood structure matching; and the whole drawing matching. In the first way, each local neighborhood structure of the first drawing is matched to its corresponding structure of another drawing, and the transformation parameters are estimated as described in section 3. In the second way, the two drawing are matched with the proposed method and the transformation parameters are then estimated.

In the experiments, the ingredients lines of the first drawing image are limited in the range [0, 0] to [100, 100], and a drawing image is composed by 10 lines. The distance from the estimated object center to the true object center is used as the measurement criterion. To analyze the robustness of the matching method, different levels of noise are added by changing the values of R_{Max} and S_{Max}. For each parameter value, 200 iterations are performed and the average value is computed. The experimental results are illustrated in Figure 1.

Compared with scale and translation variation, the proposed method is more robust to rotation variation. This phenomena is reasonable. In the matching algorithm, the centers of the constituent primitives (lines in this experiment) play a key role in parameter estimation, as shown in equation (3). Therefore, the estimated object centroid doesn't change at all in case of rotation variation,

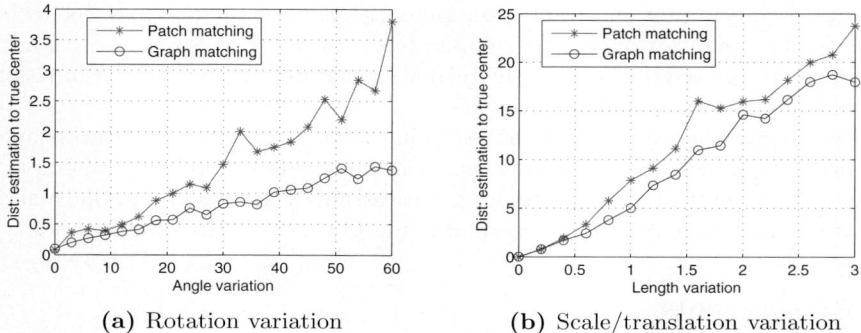

(a) Rotation variation **(b)** Scale/translation variation

Fig. 1. Estimation from k nearest candidates

provided that the two pairs of lines are in fact in correspondence. However, it is not the case for scale and translation change. With the scale and translation noise, the middle points of the primitives are changed, and as a result, the estimated object centroid is different from the true position.

5.2 Real-Life Engineering Drawings

We present in this section some detection results with real-life engineering drawings in manufacture area. As our algorithm is used for object detection under the framework of line/curve representation, raster images are firstly vectorized to extract the geometrical primitives such as segments and arcs. For the sake of conciseness, we do not thoroughly discuss the vectorization process. One well-know vectorization approach consists in computing the medial axis of the raster image and approximates it by the primitives. Although the basic raster-to-vector conversion problem is considered as solved, solutions are far from perfection in

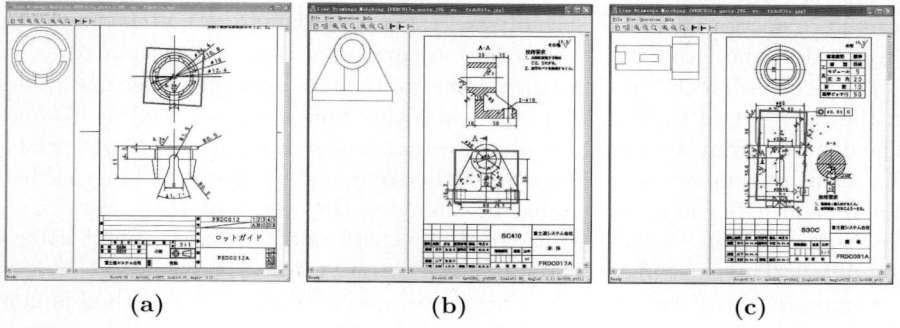

Fig. 2. Illustration of detection result. (a) Confidence=0.78; Angle=3.5; Scale=2.37; Time=12ms (b) Confidence=0.8; Angle=1.1; Scale=1.90; Time=10ms; (c) Confidence=0.71; Angle=272.1; Scale=3.99; Time=25ms.

terms of precision, robustness, and stability [7]. For example, the same object can be vectorized very differently when slight changes (e.g., resizing and/or rotation) are applied to the object.

Due to the lack of a benchmark engineering drawing database for performance evaluation, we only present in this paper some examples to illustrate the detection results, as shown in Figure 2(a)-(c). An off-line preprocessing stage is adopted, where the drawing images are vectorized and the local neighborhood structures are built. All the experiments are carried out with a computer of 3G CPU and 2G memory, and the time costs shown therein correspond to only the matching time. In each of the figures, a red rectangle is used to mark the detected object, while the red points are used to illustrate the estimated centroids from the local neighborhood structures of the model and their 6 nearest candidates.

6 Conclusion

In this paper, we present an object detection method to search shapes from line drawings. This method is in fact a hierarchical matching scheme, where the local neighborhood structure acts as the middle level representation. By grouping the elementary lines/curves into local neighborhood structure, both the appearance and the geometric structure of the line drawing are well described. For each local neighborhood structure of the model, its k-nearest neighbors in the drawing image are obtained and the transformation parameters are estimated. By a mean shift mode detection process, more precise parameters are got from these noisy estimations.

References

1. Ren, X.: Learning and matching line aspects for articulated objects. In: Proc. of CVPR, pp. 1–8. IEEE, Los Alamitos (2007)
2. Zuwala, D., Rendek, J.: Browsing graphics without prior knowledge. Pattern Recognition, 735–738 (2006)
3. Chi, Y., Leung, K.H.: Alsbir: A local-structure-based image retrieval. Pattern Recognition 40(1), 244–261 (2007)
4. Conte, D., Foggia, P., Sansone, C., Vento, M.: Thirty years of graph matching in pattern recognition. Int'l Journal of Pattern Recognition and Artificial Intelligence 18(3), 265–298 (2004)
5. Bunke, H.: Recent developments in graph matching. In: Proc. of ICPR, pp. 117–124. IEEE, Los Alamitos (2000)
6. Huet, B., Hancock, E.R.: Relational object recognition from large structural libraries. Pattern Recognition 35(9), 1895–1915 (2002)
7. Liu, W., Wan, Z., Luo, Y.: An interactive example driven approach to graphics recognition in engineering drawings. Int'l Journal on Document Analysis and Recognition 9(1), 13–29 (2007)
8. Comaniciu, D., Meer, P.: Mean shift: a robust approach toward feature space analysis. IEEE Trans. PAMI 24(5), 603–619 (2002)
9. Liu, R., Baba, T., Masumoto, D.: Attributed graph matching based engineering drawings retrieval. In: Marinai, S., Dengel, A.R. (eds.) DAS 2004. LNCS, vol. 3163, pp. 378–388. Springer, Heidelberg (2004)

A Fast Level Set-Like Algorithm with Topology Preserving Constraint

Martin Maška and Pavel Matula

Centre for Biomedical Image Analysis, Faculty of Informatics
Masaryk University, Brno, Czech Republic
xmaska@fi.muni.cz

Abstract. Implicit active contours are widely employed in image processing and related areas. Their implementation using the level set framework brings several advantages over parametric snakes. In particular, a parametrization independence, topological flexibility, and straightforward extension into higher dimensions have led to their popularity. However, in some applications the topological flexibility of the implicit contour is not desirable. Imposing topology-preserving constraints on evolving contours is often more convenient than including additional postprocessing steps. In this paper, we build on the work by Han et al. [1] introducing a topology-preserving extension of the narrow band algorithm involving simple point concept from digital geometry. In order to significantly increase computational speed, we integrate a fast level set-like algorithm by Nilsson and Heyden [2] with the simple point concept to obtain a fast topology-preserving algorithm for implicit active contours. The potential of the new algorithm is demonstrated on both synthetic and real image data.

1 Introduction

Implicit active contours [3,4] have been developed as an alternative to parametric snakes [5]. Their solution is usually carried out using the level set framework [6], in which the contour is represented implicitly as the zero level set (also called *interface*) of a scalar, higher-dimensional function ϕ. This representation has several advantages over the parametric one. In particular, it avoids parametrization problems, the topology of the contour is handled inherently, and the extension into higher dimensions is straightforward.

The contour evolution is governed by a partial differential equation (PDE):

$$\phi_t + F|\nabla\phi| = 0 , \qquad (1)$$

where F is an appropriately chosen speed function describing the motion of the interface in the normal direction. A basic PDE-based solution using an explicit finite difference scheme results in a significant computational burden limiting the use of this approach in near real-time applications.

Many approximations, aimed at speeding up the basic level set framework, have been proposed in last two decades. They can be divided into two groups.

First, methods based on the additive operator splittings scheme [7,8] have emerged to decrease the time step restriction. Therefore, a considerable lower number of iterations has to be performed to obtain the final contour in contrast to standard explicit scheme. However, these methods require for the implicit function to maintain in the form of signed distance function, which is computationally expensive. Second, since one is usually interested in the single isocontour – the interface – in the context of image segmentation, other methods have been suggested to minimize the number of updates of the implicit function in each iteration, or even to approximate the contour evolution in a different way. These include the narrow band [9], sparse-field [10], or fast marching method [11]. Other interesting approaches based on a pointwise scheduled propagation of the implicit contour can be found in [12,2].

The topological flexibility of the evolving implicit contour is a great benefit since it allows to detect several objects simultaneously without any a priori knowledge. However, in some applications this flexibility is not desirable. For instance, when the topology of the final contour has to coincide with the known topology of the desired object (e.g. brain segmentation), or when the final shape must be homeomorphic to the initial one (e.g. segmentation of two touching nuclei starting with two separated contours, each labeling exactly one nucleus). Therefore, imposing topology-preserving constraints on evolving contours is more convenient than including additional postprocessing steps.

The main motivation of our work is the need for a robust and fast segmentation method that would take advantages of the level set framework and be able to preserve the interface topology during the deformation. We build on the work by Han et al. [1] introducing a topology-preserving extension of the narrow band algorithm [9] involving simple point concept from digital geometry. Since the sparse-field method [10] can be considered as a special case of the narrow band algorithm (the band width is equal to 1) and based on our previous work [13] showing that a fast level set-like algorithm by Nilsson and Heyden [2] is about two orders of magnitude faster than the sparse-field method, we integrate the Nilsson and Heyden's algorithm with the simple point concept to obtain a fast topology-preserving algorithm for implicit active contours.

The organization of the paper is as follows. In Section 2, a review of implicit active contours with topology-preserving constraints as well as the simple point concept is presented. Section 3 describes the Nilsson and Heyden's algorithm and its topology-preserving extension. Section 4 is devoted to experimental results. We conclude the paper with a discussion and suggestions for future work in Section 5 and 6, respectively.

2 Topology-Preserving Constraint

This section briefly reviews the previous work devoted to imposing topology-preserving constraints on implicit active contours. Namely, the key ideas of existing approaches and a characterization of simple points are introduced.

2.1 Previous Work

Han et al. [1] proposed a first approach to preserve the topology of the implicit contour during the deformation. This preservation is achieved by maintaining the topology of the digital object enclosed by the implicit contour, for which the simple point concept is used. The topology of the digital object can change only if the implicit function changes the sign at a grid point (such a point is moved from inside the object to the outside or vice versa). However, not every point switching implies a topological change of the digital object. Its topology *will not* change if the considered point *is simple*. To summarize, Han et al. introduced an extension of the level set framework monitoring the sign changes of the implicit function and preventing the contour to be evolved at grid points that are not simple. The method guarantees that the final contour has exactly the same topology as the initial one *and* does not contain any self-interactions that may occur in the case of explicit contours.

For completeness, we refer the reader to other approaches [14,15], even though they are not related directly to the proposed method. Instead of modifying the level set framework as Han et al. do, the others integrate various penalization terms directly into the energy functional.

2.2 Simple Point Concept

One of the fundamental ideas of digital geometry is the simple point concept allowing topology-preserving deformations of digital images. A simple point is a point whose switching from the foreground to the background (or vice versa) does not change the image topology. In 2D, Klette and Rosenfeld [16] introduced the simple point characterization considering the number of n-connected foreground components and the number of $\bar{n}$-connected background components in the 8-neighborhood of the considered point, where $(n, \bar{n})$ is a pair of compatible connectivities avoiding a topological paradox. Since there are only 2^8 possible configurations, each of them can be evaluated in advance and a small look-up table can be used for the fast simple point detection.

An extension of this characterization into 3D is not straightforward. Bertrand and Malandain [17] proposed a characterization requiring the computation of two topologic numbers in the 26-neighborhood of the considered point. This approach can be implemented efficiently using the breadth-first search (BFS) algorithm [18]. Note that the configuration space contains 2^{26} possibilities in 3D. Therefore, it is not always convenient to use a precomputed look-up table, since its size may exceed the capacity of a workstation main memory. Instead, an evaluation using the BFS algorithm has to be performed repeatedly.

3 Proposed Algorithm

A principle of the proposed algorithm is explained in this section. We start with a brief description of the Nilsson and Heyden's algorithm, since the proposed one extends this approach. A mechanism involving the simple point concept and enforcing the implicit contour to preserve its topology is given in Sect. 3.2.

3.1 Nilsson and Heyden's Algorithm

Nilsson and Heyden proposed a fast approximation of the level set framework exploiting a pointwise scheduled propagation of the implicit contour in [2]. Instead of evolving the whole interface in a small constant time step, a point p of the interface with the minimal departure time is moved to the outside or inside of the interface depending on the sign of the speed function at this point. Simultaneously, its local neighborhood is updated accordingly. The departure time $T_d(p)$ of the interface point p (the time at which the propagation of p is expected to occur) is defined as

$$T_d(p) = T_a(p) + \frac{1}{\max\{|F(p)|, \varepsilon\}} \quad , \qquad (2)$$

where $T_a(p)$ is the arrival time (the time at which the interface arrived to p) and $F(p)$ is the speed function. The max-operation in the denominator avoids the division by zero (ε is a small number). Furthermore, considering the implicit function as a mapping of the set membership of each point, the need for its periodical reinitialization vanishes. This simplification also allows to roughly approximate the interface curvature in an incremental manner. These ideas in conjuction with a heap-sorted queue for the departure times of the interface points result in a near real-time algorithm for tracking implicit contours.

3.2 Topology-Preserving Extension

To enforce the Nilsson and Heyden's algorithm to preserve the interface topology during the deformation, we exploit the simple point concept. A local characterization of simple points allows a relatively straightforward and easy to implement modification of the pointwise scheduled interface propagation in each iteration.

Let p be a point of the interface with the minimal departure time. The behaviour of the new algorithm can be divided into two cases according to the sign of $F(p)$. First, assume that $F(p) < 0$ (Fig. 1a). The original algorithm removes p from the heap, transfers it to the exterior, and adds all its interior neighbors to the interface. Clearly, only p is switched from the foreground to the background, which could eventually change the interface topology. This could happen only if p is not simple. In order to preserve the interface topology, it is therefore sufficient to check whether p is simple or not. If p is simple, the new algorithm behaves in the same way as the original one. If p is not simple, its propagation is stopped. The implicit function remains unchanged, $T_a(p)$ is set to $T_d(p)$, and $T_d(p)$ is recomputed using (2), i.e. p is considered as a point which the interface is arriving to right now.

The second case, when $F(p) > 0$ (Fig. 1b), is more complicated than the first one. The original algorithm removes p from the heap, transfers it to the interior, and adds all its exterior neighbors (denote them by $E(p)$) to the interface. In this case, each point in $E(p)$ is switched from the background to the foreground and could eventually change the interface topology. Therefore, it is inevitable to check whether each point in $E(p)$ is simple or not (denote the set of simple ones

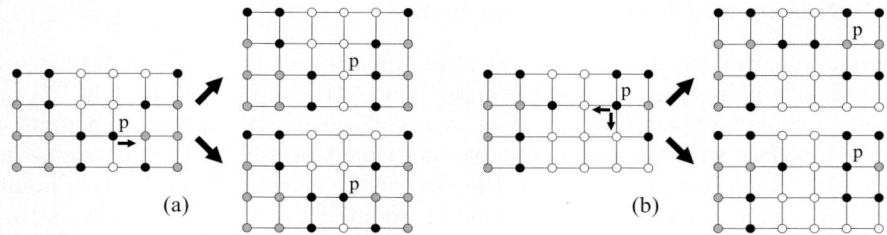

Fig. 1. Comparison of one iteration of the Nilsson and Heyden's algorithm (top right) and the proposed one (bottom right) in case of (a) $F(p) < 0$ and (b) $F(p) > 0$. The black points correspond to the interface, the white ones to the exterior, and the gray ones to the interior. The arrows from p correspond to the directions of possible propagations of the interface in this iteration.

by $E^+(p)$ and the set of non-simple ones by $E^-(p)$). For the points in $E^+(p)$, the new algorithm behaves in the same way as the original one. On the other hand, the propagation to points in $E^-(p)$ is stopped. Note that if $E^-(p)$ is not empty, p must be put back into the interface and consider it as a point which the interface is arriving to right now in order to preserve the interface connectedness.

For completeness, the case $F(p) = 0$ has already been solved in the original algorithm and does not require any special attention in the new one. In this case, $T_a(p)$ is set to $T_d(p)$, $T_d(p)$ is recomputed using (2), and the heap is updated according to the new value of $T_d(p)$.

4 Experimental Results

In this section, we present several results and comparisons on both synthetic and real image data to demonstrate the potential of the proposed algorithm. The experiments have been performed on a common workstation (Intel Core2 Duo 2.0 GHz, 2 GB RAM, Windows XP Professional).

We have solved a PDE related to the geodesic active contour model [4] and used two different inflation forces in the experiments. First, a constant inflation force over the whole domain is considered. We refer to this model as *standard geodesic model* (SGM). Second, the inflation force, defined as $(2 \cdot I(x) - 1)$ for a binary image $I(x)$, provides an expansion force inside the object and a contraction force outside. We refer to the second model as *binary geodesic model* (BGM). The original Nilsson and Heyden's algorithm is denoted as SGM or BGM algorithm depending on the geodesic model it implements. Similarly, the topology-preserving algorithm considering $(n, \bar{n})$ as a pair of compatible connectivities is denoted as TSGM$_{\bar{n}}^n$ or TBGM$_{\bar{n}}^n$ algorithm.

We start with a synthetic binary image containing two circles (Fig. 2a). We aim at finding both circles while the topology of the initial interface remains unchanged. In case of the BGM algorithm, the interface is splitted into two parts

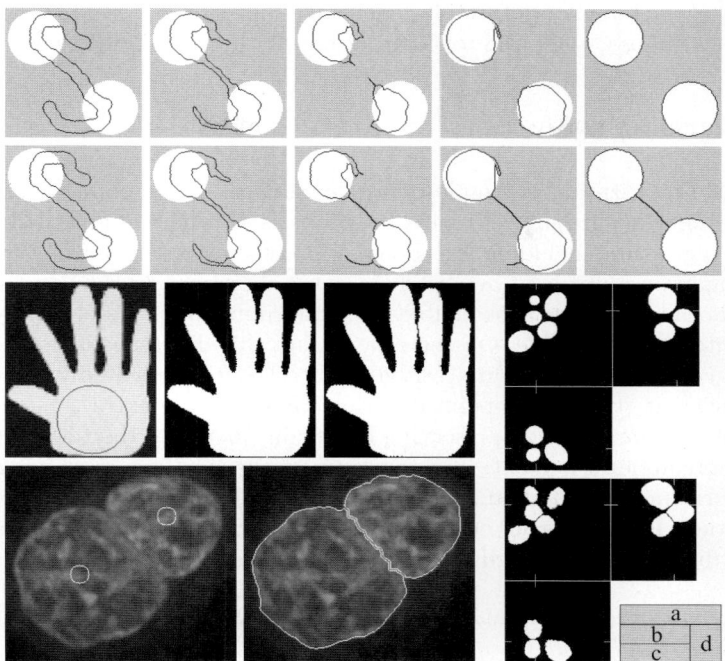

Fig. 2. (a) Segmentation of a synthetic image with two circles. Evolution of the BGM contour (upper row). Evolution of the $TBGM_8^4$ contour (lower row). (b) Segmentation of a hand-shaped digital phantom. Left: The original image including the initial interface. Centre: The result of the SGM algorithm. Right: The result of the $TSGM_4^8$ algorithm. (c) Segmentation of touching human lung cells. Left: The original image including the initial contour. Right: The final shape of the $TSGM_4^8$ contour. (d) Deformation of five ellipsoid-shaped digital phantoms. Top: Three mutually orthogonal cross sections of the original image. Bottom: The result of the proposed algorithm for the $(26, 6)$ compatible connectivity pair. Ticks show the positions of the other two cross sections.

and each circle is detected separately. On the other hand, the $TBGM_8^4$ algorithm preserves the contour topology and outputs only one 4-connected component.

The second experiment is aimed at finding the boundary of a hand-shaped digital phantom (Fig. 2b). Since two middle fingers are touching, the SGM contour changes its topology, which results in a hole inside the final object. On the other hand, the $TSGM_4^8$ algorithm keeps the boundary of each finger separated.

Figure 2c illustrates a segmentation of two touching cell nuclei. The image was acquired using a fluorescence microscope and the initial interface is made of two disjoint curves. The final contour is achieved by the $TSGM_4^8$ algorithm.

To conclude this section, we present an application of the proposed algorithm in 3D. We have developed a tool [19] that is able to generate 3D cell nuclei-like digital phantoms as well as to simulate the process of image formation in the fluorescence microscope. The task of the proposed algorithm has been to reduce a regularity of the boundary of simple geometric objects to conform better with the real nucleus shapes, while the topology of initial objects is preserved (Fig. 2d).

5 Discussion

A local characterization of simple points makes the simple point checking procedure straightforward and efficient. In comparison to the original Nilsson and Heyden's algorithm, including the topology-preserving constraint does not incur a significant computational overhead. In 2D, a difference in the computational speed of both algorithms is negligible since the simple point detection exploits a precomputed look-up table. On the other hand, this difference is not negligible in 3D, since the BFS algorithm [18] has to be executed repeatedly in each iteration to evaluate the simple point criterion. However, the extra time stemming from applying the topology-preserving constraint increases the total processing time by only about 4 percent. In comparison to the topology-preserving narrow band algorithm by Han et al. [1], the proposed algorithm is about two orders of magnitude faster (Fig. 3). On the other hand, it does not achieve subpixel accuracy due to the simplified representation of the implicit function.

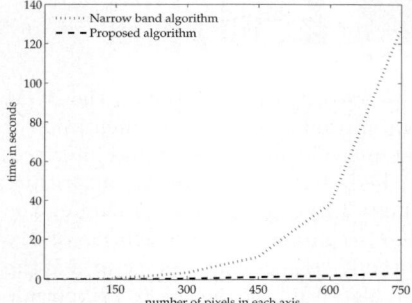
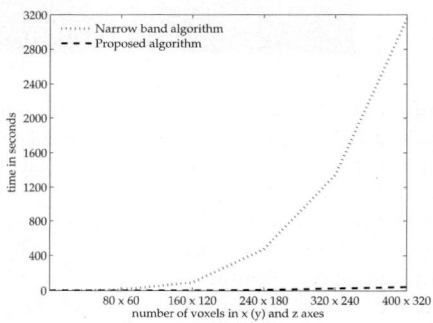

Fig. 3. Dependence of the computational time of the topology-preserving narrow band algorithm and the proposed one on the image size in 2D (left) and 3D (right)

The proposed algorithm preserves the topology of the contour as well as the topology of the background component. This behaviour results in a 4-connected path separating two touching cell nuclei in Fig. 2c. For segmentation of touching objects, it would be more convenient to omit such a path and let the contours evolve to touch themselves. This will be addressed in future work.

6 Conclusion

We have addressed the problem of imposing topology-preserving constraints on implicit contours. We have integrated a fast level set-like algorithm by Nilsson and Heyden with the simple point concept to obtain a fast topology-preserving algorithm for implicit active contours. A local characterization of simple points allows a relatively straightforward and easy to implement modification of the pointwise scheduled propagation of the implicit contour. The experiments verified topology-preserving properties of the proposed algorithm and indicated that it is fast enough to be used in near real-time applications.

Acknowledgments. This work has been supported by the Ministry of Education of the Czech Republic (Projects No. MSM-0021622419, No. LC535 and No. 2B06052).

References

1. Han, X., Xu, C., Prince, J.L.: A topology preserving level set method for geometric deformable models. IEEE Transactions on Pattern Analysis and Machine Inteligence 25(6), 755–768 (2003)
2. Nilsson, B., Heyden, A.: A fast algorithm for level set-like active contours. Pattern Recognition Letters 24(9-10), 1331–1337 (2003)
3. Caselles, V., Catté, F., Coll, T., Dibos, F.: A geometric model for active contours in image processing. Numerische Mathematik 66(1), 1–31 (1993)
4. Caselles, V., Kimmel, R., Sapiro, G.: Geodesic active contours. International Journal of Computer Vision 22(1), 61–79 (1997)
5. Kass, M., Witkin, A., Terzopoulos, D.: Snakes: Active contour models. International Journal of Computer Vision 1(4), 321–331 (1987)
6. Osher, S., Fedkiw, R.: Level Set Methods and Dynamic Implicit Surfaces. Springer, New York (2003)
7. Goldenberg, R., Kimmel, R., Rivlin, E., Rudzsky, M.: Fast geodesic active contours. IEEE Transactions on Image Processing 10(10), 1467–1475 (2001)
8. Kühne, G., Weickert, J., Beier, M., Effelsberg, W.: Fast implicit active contour models. In: Van Gool, L. (ed.) DAGM 2002. LNCS, vol. 2449, pp. 133–140. Springer, Heidelberg (2002)
9. Adalsteinsson, D., Sethian, J.A.: A fast level set method for propagating interfaces. Journal of Computational Physics 118(2), 269–277 (1995)
10. Whitaker, R.T.: A level-set approach to 3D reconstruction from range data. International Journal of Computer Vision 29(3), 203–231 (1998)
11. Sethian, J.A.: A fast marching level set method for monotonically advancing fronts. Proceedings of the National Academy of Sciences 93(4), 1591–1595 (1996)
12. Deng, J., Tsui, H.T.: A fast level set method for segmentation of low contrast noisy biomedical images. Pattern Recognition Letters 23(1-3), 161–169 (2002)
13. Maška, M., Hubený, J., Svoboda, D., Kozubek, M.: A comparison of fast level set-like algorithms for image segmentation in fluorescence microscopy. In: Bebis, G., Boyle, R., Parvin, B., Koracin, D., Paragios, N., Tanveer, S.-M., Ju, T., Liu, Z., Coquillart, S., Cruz-Neira, C., Müller, T., Malzbender, T. (eds.) ISVC 2007, Part II. LNCS, vol. 4842, pp. 571–581. Springer, Heidelberg (2007)
14. Alexandrov, O., Santosa, F.: A topology-preserving level set method for shape optimization. Journal of Computational Physics 204(1), 121–130 (2005)
15. Le Guyader, C., Vese, L.A.: Self-repelling snakes for topology-preserving segmentation models. IEEE Transactions on Image Processing 17(5), 767–779 (2008)

16. Klette, R., Rosenfeld, A.: Digital Geometry: Geometric Methods for Digital Picture Analysis. Morgan Kaufmann, San Francisco (2004)
17. Bertrand, G., Malandain, G.: A new characterization of three-dimensional simple points. Pattern Recognition Letters 15(2), 169–175 (1994)
18. Malandain, G., Bertrand, G.: Fast characterization of 3d simple points. In: Proceedings of 11th International Conference on Pattern Recognition, pp. 232–235 (1992)
19. Svoboda, D., Kozubek, M., Stejskal, S.: Digital cell phantom generation and simulation of image formation in 3d image cytometry. Cytometry Part A 75A(6), 494–509 (2009)

Significance Tests and Statistical Inequalities for Segmentation by Region Growing on Graph

Guillaume Née[1,2,*], Stéphanie Jehan-Besson[3], Luc Brun[1], and Marinette Revenu[1]

[1] GREYC - UMR CNRS 6072 - 14050 Caen Cedex, France
[2] General Electric Healthcare - 78140 Vélizy, France
[3] LIMOS - UMR CNRS 6158 - 63173 Aubière cedex, France
{gnee,jehan,brun,revenu}@greyc.ensicaen.fr

Abstract. Bottom-up segmentation methods merge similar neighboring regions according to a decision rule and a merging order. In this paper, we propose a contribution for each of these two points. Firstly, under statistical hypothesis of similarity, we provide an improved decision rule for region merging based on significance tests and the recent statistical inequality of McDiarmid. Secondly, we propose a dynamic merging order based on our merging predicate. This last heuristic is justified by considering an energy minimisation framework. Experimental results on both natural and medical images show the validity of our method.

1 Introduction

Segmentation of images into homogeneous regions is a fundamental low-level operation which is crucial for many applications such as video compression, image enhancement (different processing may be applied for different objects), object detection, etc. Spatial segmentation can be classified into two main categories, namely contour-based and region-based methods. In the first category, edges are computed and connected components can be extracted (see [1] for example). However, the connection of a set of disconnected edges in order to define an image partition remains a challenging problem. Moreover such a segmentation scheme can not take benefit of statistical properties of the considered image regions. The second category of methods, i.e. region-based, is then more often used. Such an approach may use features computed along the contours of the regions but uses the regions as basic elements within the segmentation scheme. We are interested here in a bottom-up segmentation approach. In such an approach, similar neighboring regions are merged according to a decision rule [2]. The initial regions can be defined from the grid of pixels or an oversegmentation of the image. The design of both the merging criterion and the merging order is crucial for segmentation purposes. When dealing with the merging predicate, the choice of the threshold is often difficult and can be crucial. Compared to other classical

---

[*] This work is funded by a grant co-financed by General Electric Healthcare and the Region Basse Normandie. MR Images are provided by Dr. M. Hamon (CHU Caen).

approaches, e.g. [3,4], the authors of [5,6] have proposed recently an adaptive threshold based on the use of statistical inequalities. Such a method provides good results with few parameters to tune. However, their merging predicate is based on the assumption that all the pixels of a given region have the same expectation of their intensities. This last assumption is only valid for piecewise constant images with a low level of noise. Let us also notice that the authors of [6] show that their algorithm provides an oversegmentation especially for small images. As far as the merging order is concerned, the authors propose two different distances between regions intensities computed once at the beginning of the merging process. These orders are not clearly related to the merging criterion.

In [7] we have revisited this statistical segmentation framework using a contrario principles. The a contrario approach is based on the perception theory and particularly the grouping law of the Wertheimer's theory. This grouping law states that "objects having a quality in common get perceptually grouped". The Helmholtz principle [8] which states that "an event is meaningful if its number of occurrences is very small in a random model" is a quantitative version of the previous law. More formally, let us consider an event E whose probability under an hypothesis $\mathbf{H_0}$ is bounded by a low threshold δ, the a contrario approach leads to reject the hypothesis $\mathbf{H_0}$ if such an event occurs. Using such a decision scheme, δ may be interpreted as an upper bound of the probability of a false alarm (rejection of $\mathbf{H_0}$ while $\mathbf{H_0}$ is actually true). The upper bound δ may be fixed a priori in which case the test $P(E|H_0) < \delta$ is called a significance test [9]. Desolneux [8] proposed to set δ according to the expected number of false alarms. Such a method provides an elegant way to fix the threshold but reduces the adaptability of the method to user requirements.

Following the work in [7], we propose to apply a contrario principles and significance tests for the design of both the merging order and the merging predicate. The general a contrario framework designed to compute merging predicates is given in section 2. The merging order and the whole algorithm are described in section 3. The influence of the merging order and the merging predicate are studied in section 4.

2 Statistical Merging Predicates

Due to the random part in image acquisition systems, an image I is classically considered as an observation of a perfect statistical image I^*. Using such an image model, an ideal region is defined as a vector $\mathbf{X} = (X_1, \ldots, X_n)$ of n random variables representing the pixel intensities. A "real" region is then considered as an observation of this random vector which takes its values in $\prod_{k=1}^{n} A_k$. In natural images, the set of admissible values A_k usually corresponds to $[0; M]$ where $M = 255$. However, in medical images (e.g. : MRI, Echography), the set A_k may be larger.

Using such a statistical model of regions, segmentation by region growing is realized through the definition of a merging predicate $P(X_i, X_j)$ and a merging order. The design of these two features determines the main properties of a segmentation algorithm.

2.1 Problem Statement Using a Contrario Approaches

Given two statistical regions $\mathbf{X}_1$ and $\mathbf{X}_2$ and a dissimilarity criterion $d(.,.)$, let us consider two observations R_1 and R_2 of respectively $\mathbf{X}_1$ and $\mathbf{X}_2$ and the event "E: the observed value $d(R_1, R_2)$ of $d(\mathbf{X}_1, \mathbf{X}_2)$ is greater than a threshold T". As mentioned in section 1, the a contrario approach is based on the estimation of the probability of this event under the similarity hypothesis $\mathbf{H_0}$. Let us consider an upper bound δ of this probability:

$$\mathbf{P}\{d(\mathbf{X}_1, \mathbf{X}_2) \geq T | \mathbf{H_0}\} \leq \delta \tag{1}$$

We can remark that the upper bound δ and the threshold T are dependent. Indeed if the threshold T is set to a high value, the event E corresponds to a non probable event under $\mathbf{H_0}$ and δ should then be small. On the contrary if the threshold T is set to a small value, E corresponds to a probable event under $\mathbf{H_0}$ and so δ must be large. More generally, one may usually assume that the threshold T is a decreasing function of δ which may be denoted as $T(\delta)$.

Using the a contrario approach, if we take δ as a low probability value, the event E is considered as not probable under the similarity hypothesis $\mathbf{H_0}$ and this hypothesis is then rejected. Given two observations R_1 and R_2 of statistical regions $\mathbf{X}_1$ and $\mathbf{X}_2$, our decision rule for region merging is thus defined as follows:

$$\text{if} \quad d(R_1, R_2) \geq T(\delta) \quad \text{then} \quad \mathbf{H_0} \text{ is rejected} \tag{2}$$

The rejection of $\mathbf{H_0}$ means that $\mathbf{X}_1$ and $\mathbf{X}_2$ are different and thus that the regions R_1 and R_2 must not be merged.

2.2 Computation of Thresholds Using Concentration Inequalities

The main difficulty of the above approach lies in the computation of the threshold $T(\delta)$. In this work, we propose to use the extension of the McDiarmid theorem [10] which allows to bound the probability of a large class of events. Let us remind this theorem:

Theorem 1. *Let $\mathbf{Y} = (Y_1, \ldots, Y_n)$ be a family of random variables with Y_k taking values in a set A_k, and let f be a bounded real-valued function defined on $\Omega = \prod_{k=1}^{n} A_k$. If μ denotes the expectation of $f(\mathbf{Y})$ we have for any $\alpha \geq \mu$:*

$$\mathbf{P}\{f(\mathbf{Y}) \geq \alpha\} \leq \exp\left(\frac{-2(\alpha - \mu)^2}{r^2}\right) + \mathbf{P}\{\mathbf{Y} \in C\} \tag{3}$$

Where C is a subset of Ω which corresponds to a set of outliers for $\mathbf{Y}$ and r^2 is the maximal sum of squared range [10] defined on $\overline{C} = \Omega \setminus C$.

Within our framework, we define $f(\mathbf{Y})$ as our dissimilarity measure $d(\mathbf{X}_1, \mathbf{X}_2)$ and $\mathbf{Y}$ as an appropriate combination of the two vectors $\mathbf{X}_1$ and $\mathbf{X}_2$.

Let us denote by $\Delta(\alpha)$, the bound provided by the McDiarmid's theorem:

$$\mathbf{P}\{f(\mathbf{Y}) \geq \alpha\} \leq \Delta(\alpha) \tag{4}$$

The threshold $T(\delta)$ introduced in (1) can then be computed by setting $\delta = \Delta(\alpha)$ and so $\alpha = \Delta^{-1}(\delta) = T(\delta)$.

2.3 Piecewise Constant Predicate

We measure the similarity between the two regions by the following dissimilarity measure:
$$f(\mathbf{X}) = d(\mathbf{X}_1, \mathbf{X}_2) = |U_1 - U_2| \tag{5}$$
where $\{U_j\}_{j=1;2}$ denote the random variables corresponding to the means of the statistical regions $\{\mathbf{X}_j\}_{j=1;2}$ of associated sizes $|\mathbf{X}_j|$.

Our goal is to compute a decision rule that indicates if two observations R_1 and R_2 of $\mathbf{X}_1$ and $\mathbf{X}_2$ are similar or not. We have thus to upper bound the probability that the function $f(\mathbf{X}) = d(\mathbf{X}_1, \mathbf{X}_2)$ is greater than a given threshold α using the McDiarmid's theorem (theorem 1). Such an upper bound is provided by the following proposition:

Proposition 1. *Using the previously defined notations, we have for any couple $(\mathbf{X}_1, \mathbf{X}_2)$ of statistical similar regions and any threshold $\alpha > 0$:*
$$\mathbf{P}\{d(\mathbf{X}_1, \mathbf{X}_2) \geq \alpha\} \leq \exp\left(-\frac{2|\mathbf{X}_1||\mathbf{X}_2|}{g^2(|\mathbf{X}_1| + |\mathbf{X}_2|)}(\alpha - \mu_{12})^2\right) + K \tag{6}$$

with $K = \mathbf{P}\{\mathbf{X} \in C\}$ where $C \subset \Omega$ is the set of outliers for $\mathbf{X}$ and $\mu_{12} = \mathbf{E}[d(\mathbf{X}_1, \mathbf{X}_2)]$. The parameter g comes from the computation of the maximum range r^2 and is equal to $N - N'$ when $\overline{C} = [N; N']$ ($N' > N$) defines the complementary of the set of outliers C in Ω.

See [11] for a similar proof of this proposition. Note that the McDiarmid's Theorem 1 doesn't require the independence of random variables, but we make this assumption. Rigorously, such an assumption is not valid, but allows to simplify the computation of r^2. In practice, we set $N = \min_{\mathbf{x} \in I}(I(\mathbf{x}))$ and $N' = \max_{\mathbf{x} \in I}(I(\mathbf{x}))$ which ensures a null value of the probability of outliers K. The parameter μ_{12} may be estimated using assumptions on the noise model. For example, in the case of a gaussian noise ($X_i \sim \mathcal{N}(m_i, \sigma)$ for $1 \leq i \leq N$), we obtain, for any couple $(\mathbf{X}_i, \mathbf{X}_j)$ of statistical regions, and after some calculus based on well-known properties for the combination of Gaussian models:
$$\mu_{ij} = \frac{2\sigma\left(\sqrt{|\mathbf{X}_i|} + \sqrt{|\mathbf{X}_j|}\right)}{\sqrt{2\pi|\mathbf{X}_i||\mathbf{X}_j|}} \tag{7}$$

Given two observations R_i, R_j of two different statistical regions $\mathbf{X}_i$ and $\mathbf{X}_j$, our merging criterion is defined by the following predicate:
$$P(R_i, R_j) = \begin{cases} \text{true} & \text{if } |\overline{R_i} - \overline{R_j}| < \alpha_{ij} \\ \text{false} & \text{otherwise} \end{cases} \tag{8}$$

with $\alpha_{ij} = g\sqrt{\dfrac{|R_i| + |R_j|}{2|R_i||R_j|}\ln\left(\dfrac{1}{\delta - K}\right)} + \mu_{ij}$.

According to Proposition 1, the probability that $P(R_i, R_j)$ is true under the hypothesis that $\mathbf{X}_i$ and $\mathbf{X}_j$ are parts of a same statistical region is bounded by δ. Note that for a fixed δ, the α_{ij} value ensures (1), but is not necessarily the largest value for which (1) holds.

This predicate can be understood as a generalization of the one proposed by [6,7]. The general version of McDiarmid's theorem provides an elegant way to reduce the range of the random variables via the parameter g and the probability K. It plays a similar role as the parameter Q introduced in [6]. Compared to this last approach, we do not make the assumption that in a same statistical region $\mathbf{E}\left[U_1 - U_2\right] = 0$. In fact, this is not the case for noisy images. Such an assumption is only valid under the law of large numbers and is therefore not verified for small regions. This last point is illustrated by our experimental results (Section 4).

3 Merging Algorithm

Given an image I, the regions adjacency graph (RAG) $\mathcal{G}$ is composed of a set of vertices $\mathcal{V}$ representing the observed regions and a set of edges $\mathcal{E}$ encoding the adjacency of regions in 4-connectivity. In our implementation, regions are initially reduced to a single pixel. A weighted edge is then a triplet composed of a couple of nodes (v_i, v_j) with their corresponding weight w_{ij}. In our work, this weight is defined as the ratio of the value of the criterion (left side of (8)) and the computed threshold (right side of (8)):

$$w_{ij} = \frac{|\overline{R_i} - \overline{R_j}|}{\alpha_{ij}} = \frac{|\overline{R_i} - \overline{R_j}|}{g\sqrt{\frac{|R_i|+|R_j|}{2|R_i||R_j|}\ln\left(\frac{1}{\delta - K}\right)} + \mu_{ij}} \quad (9)$$

Using the above formula, the predicate P between 2 regions R_i and R_j is true if and only if the weight of the edge e_{ij} between the associated vertices v_i and v_j is lower than 1. Our merging order on the edges e_{ij} corresponds to a decreasing order on the probability that $d(R_i, R_j) < \alpha_{ij}$ (6). Let us consider two distinct couple of regions (R_i, R_j) and $(R_{i'}, R_{j'})$ such that $w_{ij} \leq w_{i'j'} < 1$. Then $P(R_{i'}, R_{j'})$ is true implies that $P(R_i, R_j)$ is true. The probability of the event $P(R_i, R_j)$ is thus greater than the one of $P(R_{i'}, R_{j'})$. A merging order based on decreasing probability of our predicate is thus achieved by sorting our edge weights increasingly.

Our merging order differs significantly from the one proposed by [6,7]. Moreover it is updated at each merging step. We may also justify these choices using an energy minimisation scheme on the set of image partitions. Considering, for simplicity reasons, an image without noise (e.g. $\mu_{ij} = 0, \forall i, j$), we define the energy of a partition at instant t of the merging process as follows:

$$E(R_1, ..., R_N) = \sum_{k=1}^{N} \sum_{\mathbf{x} \in R_k} \frac{(I(\mathbf{x}) - \overline{R_k})^2}{g^2} + \lambda N \quad (10)$$

where the number of regions N is a regularization term balanced by a positive parameter λ. The merging of two selected regions R_i and R_j at step $t+1$ leads to an energy E_{t+1} and to the following energy difference:

$$\Delta E = E_{t+1} - E_t = \frac{|R_i||R_j|}{|R_i|+|R_j|} \frac{(\overline{R_i} - \overline{R_j})^2}{g^2} - \lambda \qquad (11)$$

This difference must be negative to ensure the minimization of the energy E. By setting $\lambda = 0.5 \ln \frac{1}{\delta - K}$ and so regarding δ as our regularisation parameter, ΔE becomes equal to $\frac{1}{2} \ln \frac{1}{\delta - K}(w_{ij}^2 - 1)$. Our merging predicate ($w_{ij} < 1$) thus ensures the negativity of ΔE. The merging order, if updated after each merging operation, ensures the selection of the couple of regions that provides the steepest energy descent for a fixed δ.

4 Experimental Results

In the first row of Fig. 1, the importance of the merging order is demonstrated. For each segmentation result, the merging predicate (8) is used and the value of the unique parameter δ is adjusted so as to obtain the same number of regions (i.e. 55). The merging order used for the second column of Fig. 1 is simply a scan-column, for the third column, we use a pre-computed order with $w_{ij} = (\mu_i - \mu_j)^2$ as in [7] and finally, segmentation using our dynamic update with the weight (9) is given in the fourth column. The parameter g is chosen as mentioned

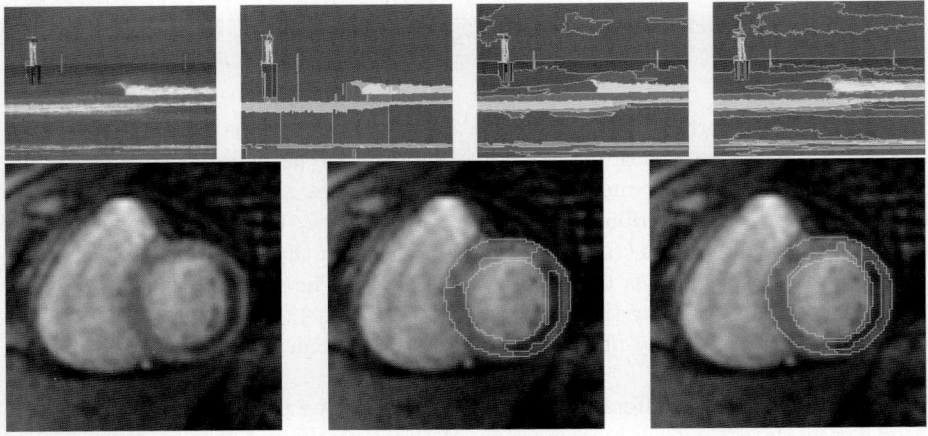

Fig. 1. Segmentation of the *Trouville's beach* image (first row). From left to right: original image - segmentation using a scan-column order ($\delta = 0.5$) - segmentation using a pre-computed order with $w_{ij} = (\mu_i - \mu_j)^2$ as in [7] ($\delta = 0.335$) - segmentation using a dynamic update and (9) ($\delta = 0.09$). The parameter δ is adjusted to obtain 55 regions in both reordering methods. Segmentation of an hypo-perfused region inside the myocardium in MRI perfusion imaging (second row) using respectively algorithm from [7] (second column) and our one (third column).

	EGBIS [3]	JSEG [4]	SRM [6]	TCVSEG [7]	Our algorithm
Mean	> 6	2.51	1.88	1.48	1
Minimum	> 3.5	1.85	1.51	1.29	1
Maximum	> 11	4.43	3.11	2.04	1

Fig. 2. Comparison with others algorithms on a random selection of 10 images from the Berkeley database [12]. For each image, the parameters of each algorithm have been chosen to obtain the same number of regions. The values represent the Mean Square Error (MSE) normalized by the minimum of MSE over all algorithms. Each line respectively presents the mean, the minimum, and the maximum of this normalized MSE on all the 10 images.

in section 2.3, which gives $g = \max_{\mathbf{x} \in I}(I(\mathbf{x})) - \min_{\mathbf{x} \in I}(I(\mathbf{x}))$. This parameter g is equal to 243 and the standard deviance of the noise has been estimated to $\sigma = 0.55$. The two last results are clearly better than the first. The small regions such as the lighthouse's antenna or the lighthouse's pillar are accurately recovered with our dynamical merging process. Such a dynamical merging order generally leads to a more accurate segmentation of small regions. This last point can be explained by the fact that we take into account the sizes of the regions in our merging order. The second row of Fig. 1 concerns the segmentation of hypo-perfused regions (darkest regions) inside the myocardium in perfusion cardiac MRI images. Accurate segmentation of pathological regions is a crucial task for practitioners to allow perfusion quantification inside these regions. The left segmentation result has been obtained by setting $\delta = 0.1$, it is composed of 5 regions. The right one has been obtained by setting $\delta = 10^{-4}$ and gives 4 regions, the standard deviance of the noise has been estimated to $\sigma = 35.36$ and the parameter $g = 7040$. We can see that the hypo-perfused region (low contrast on the right side of the myocardium) is accurately segmented by our method while the other method does not perfectly enclose the region. Generally, given an equal number of regions, our algorithm doesn't merge significantly different regions even if one of them is small. This last point is illustrated in fig. 2 which summarizes the results of our second experiment realized on 10 randomly chosen images from the Berkeley database [12]. In this second experiment, we have computed the MSE (Mean Square Error) normalized by the minimum of the MSE value obtained over the 5 algorithms. The MSE is computed between the original image and the image composed of segmented regions filled by their mean values. The table 2 presents the mean, the minimum and the maximum of this normalized MSE computed for each algorithm over all the 10 images. We can remark that our algorithm always gives the minimum value of MSE which is coherent with the fact that we ensure the steepest gradient descent of the criterion E (10).

5 Conclusion

We have proposed in this paper a new region merging algorithm. The merging predicate has been designed using an a contrario approach and a recent theorem about concentration inequalities. We have shown that merging criteria and merging orders are closely related and both contribute to the quality of the segmented image. Concerning the merging order, we have proposed an original sorting criterion based on the merging predicate and justified it within an energy minimisation scheme on the set of image partitions. Experimental results prove the applicability of our method, especially for segmentation of small regions in medical images.

References

1. Iannizzotto, G., Vita, L.: Fast and accurate edge-based segmentation with no contour smoothing in 2-D real images. IEEE TIP 9(7), 1232–1237 (2000)
2. Shi, J., Malik, J.: Normalized cuts and image segmentation. IEEE PAMI 22(8), 888–905 (2000)
3. Felzenszwalb, P.F., Huttenlocher, D.P.: Efficient graph-based image segmentation. IJCV 59(2), 167–181 (2004)
4. Deng, Y., Manjunath, B.: Unsupervised segmentation of colour-texture regions in images and video. IEEE PAMI 23(8), 800–810 (2001)
5. Fiorio, C., Nock, R.: Sorted region merging to maximize test reliability. In: International Conference on Image Processing, Vancouver, Canada, vol. 1, pp. 808–811. IEEE, Los Alamitos (2000)
6. Nock, R., Nielsen, F.: Statistical region merging. IEEE PAMI 26(11), 1452–1458 (2004)
7. El Hassani, M., Jehan-Besson, S., Brun, L., et al.: Time-consistent video segmentation algorithm designed for real-time implementation. VLSI Design (2008)
8. Desolneux, A., Moisan, L., Morel, J.M.: Computational Gestalts and perception thresholds. Journal of Physiology 97(2-3), 311–324 (2003)
9. Coupier, D., Desolneux, A., Ycart, B.: Image denoising by statistical area thresholding. Journal of Mathematical Imaging and Vision 22 (2-3), 183–197 (2005)
10. McDiarmid, C.: Concentration. In: Habib, M., McDiarmid, C., Ramirez-Alfonsin, J., Reed, B. (eds.) Probabilistic Methods for Algorithmic Discrete Mathematics. Springer, Heidelberg (1998)
11. Née, G., Jehan-Besson, S., Brun, L., Revenu, M.: Significance tests and statistical inequalities for region matching. In: da Vitoria Lobo, N., Kasparis, T., Roli, F., Kwok, J.T., Georgiopoulos, M., Anagnostopoulos, G.C., Loog, M. (eds.) S+SSPR 2008. LNCS, vol. 5342, pp. 350–360. Springer, Heidelberg (2008)
12. Martin, D., Fowlkes, C., Tal, D., Malik, J.: A database of human segmented natural images and its application to evaluating segmentation algorithms and measuring ecological statistics. In: Computer Vision, July 2001, vol. 2, pp. 416–423 (2001)

Scale Space Hierarchy of Segments

Haruhiko Nishiguchi[1], Atsushi Imiya[2], and Tomoya Sakai[2]

[1] School of Science and Technology, Chiba University, Japan
[2] Institute of Media and Information Technology, Chiba University, Japan
Yayoicho 1-33, Inage-ku, Chiba, 263-8522, Japan
{imiya,tsakai}@faculty.chiba-u.jp

Abstract. In this paper, we develop a segmentation algorithm using configurations of singular points in the linear scale space. We define segment edges as a zero-crossing set in the linear scale space using the singular points. An image in the linear scale space is the convolution of the image and the Gaussian kernel. The Gaussian kernel of an appropriate variance is a typical presmoothing operator for segmentation. The variance is heuristically selected using statistics of images such as the noise distribution in images. The variance of the kernel is determined using the singular point configuration in the linear scale space, since singular points in the linear scale space allow the extraction of the dominant parts of an image. This scale selection strategy derives the hierarchical structure of the segments.

1 Introduction

In this paper, we develop an algorithm of edge detection for segmentation using the deep structure in the linear scale space. The method of the Gaussian scale-space analysis [3,4,5,6,7,8] is an established multiresolution image analysis tool which provides the hierarchical expression of steel images and sequence of images [9]. An image in the linear scale space is the convolution of the image and the Gaussian kernel. The singular point configuration in the linear scale space is called the deep structure of scale space (the DSSS in abbreviation). The DSSS describes hidden topological nature of the original functions dealing with gray values of an n-variable function in the scale space as $(n+1)$-dimensional topographical maps [1,2,10,11,12,13]. The DSSS allows to extract dominant parts of an image and their topological relation.

For segmentation of an image, presmoothing for the image is usually operated. A typical presmoothing is the convolution of an image with a Gaussian kernel with an appropriate variance. Then, a class of differential operations are operated to the presmoothed image for the detection of steepest points as candidate of segment edges. In this process, the variance of the Gaussian kernel, which defines the bandwidth in the Fourier domain, is heuristically selected. We introduce a mathematical strategy for the selection of the variance of the Gaussian kernel using the DSSS. Since the stationary points on the stationary curves [1,2] define dominant parts and their topological relation, we use the topological properties of stationary curves for the selection of the variance of the presmoothing Gaussian kernel. This selection strategy derives the hierarchical structure of the segments in the linear scale space.

Kuijper et al [12,13] dealt with singular points whose second derivatives are zero as a DSSS feature. These singular point in the linear scale space is called the top points or critical points. A top point is a singular point in the scale space on which both first and second derivatives are zero. Iijima [3] defined the singular points whose first derivative is zero. These singular points in the scale space are called the stationary points [3,1,2]. These stationary points define the centroids of view fields which extract the dominate parts of an image for a fixed scale [3]. Zhao and Iijima proposed [1,2] a tree construction strategy in the linear scale space using the configuration of their stationary points to express the topological relations of dominant parts in the linear scale space. This paper is an application of Zhao and Iijima's treatment [1,2,3] of the DSSS for segmentation since we deal with the configurations of singular points whose first derivatives are zero for various scales.

2   Linear Scale Space and Structure Line

In the 2-dimensional Euclidean space $\mathbf{R}^2$, for an orthogonal coordinate system x-y defined in $\mathbf{R}^2$, a vector in $\mathbf{R}^2$ is expressed by $\boldsymbol{x} = (x,y)^\top$ where $\cdot^\top$ is the transpose of a vector. The solution of the linear diffusion equation

$$\frac{\partial}{\partial \tau} f(\boldsymbol{x}, \tau) = \Delta f(\boldsymbol{x}, \tau), \ \tau > 0, \ f(\boldsymbol{x}, 0) = f(\boldsymbol{x}) \tag{1}$$

defines the general image of the function $f(\boldsymbol{x})$ in the linear scale space. Setting $|\boldsymbol{x}|$ to be the length of $\boldsymbol{x}$, the solution of eq. (1) is obtained as

$$G *_2 f(\boldsymbol{x}) = f(\boldsymbol{x}, \tau) = \frac{1}{4\pi\tau} \int_{-\infty}^{\infty} \int_{-\infty}^{\infty} f(\boldsymbol{y}) \exp(-\frac{|\boldsymbol{x}-\boldsymbol{y}|^2}{4\tau}) d\boldsymbol{y}. \tag{2}$$

We define the following operators,

$$\nabla f(\boldsymbol{x},\tau) = \nabla_G f = \begin{pmatrix} G_x *_2 f \\ G_y *_2 f \end{pmatrix}, \ \boldsymbol{H}_G = \begin{pmatrix} G_{xx} *_2 f, G_{xy} *_2 f \\ G_{xy} *_2 f, G_{yy} *_2 f \end{pmatrix}. \tag{3}$$

Using these operators, zero point sets and zero-crossing sets are defined in the linear scale space.

Definition 1. *Stationary points [1,3] for the topographical maps in the scale space are the solutions of the equation $\nabla_G f = 0$.*

Denoting the signs of the eigenvalues of the Hessian matrix $\boldsymbol{H}_G$ of the function $f(\boldsymbol{x}, \tau)$ as $(-,-)$, $(+,-)$ and $(+,+)$ in the linear scale space, these labels of points correspond to the local maximum points, the saddle points, and the local minimum points, respectively. Using these labels, the stationary points are categorised into three types.

Definition 2. *The top points in the linear scale space are points which satisfy the conditions* $\nabla_G f = 0$ *and* $det \boldsymbol{H}_G = 0$.

Next, we introduce closed curves [14,15,17] for each τ in the linear scale space.

Definition 3. *The structure line in the linear scale space is*

$$E(\tau) = \{\boldsymbol{x} | \nabla_G f^\top \boldsymbol{H}_G \nabla_G f = 0\}. \tag{4}$$

For a fixed τ, $E(\tau)$ is the edge detected by the Canny edge-detection operator [15,16,17].[1] Canny [16] suggested the superposition of $E(\tau_i)$,

$$E_{Canny} = \cup_{i=1}^{n} E(\tau_i) \tag{5}$$

as the final edge for an appropriate set of parameters $\{\tau_i\}_{i=1}^{n}$, where $\tau_{i+1} > \tau_i$. In the next section, we introduce a mathematical method for the selection of τ_i using the linear scale space analysis.

3   Segmentation in Scale Space

3.1   Scale Space Tree

Definition 4. *The stationary curves in the linear scale space are the collections of stationary points.*

The trajectories of the stationary points which is expressed as $\boldsymbol{x}(\tau)$ is the solution of

$$\boldsymbol{H}_G \frac{d\boldsymbol{x}(\tau)}{d\tau} = -\nabla \Delta f(\boldsymbol{x}(\tau), \tau). \tag{6}$$

Since the Hessian matrix is always singular for singular points, this equation is valid for nonsingular points.

Definition 5. *For* $S(\boldsymbol{x}, \tau) = |\frac{d\boldsymbol{x}(\tau)}{d\tau}|$, *the stationary points on the stationary curves are the points which satisfy* $S(\boldsymbol{x}, \tau) = 0$ *or are isolated points under the conditions* $\frac{dS(\boldsymbol{x},\tau)}{d\tau} = 0$ *and* $\frac{d^2 S(\boldsymbol{x},\tau)}{d^2 \tau} = 0$.

Denoting a stationary point on the stationary curves as $(\boldsymbol{x}_i, \tau_i)$, the region

$$\boldsymbol{R}(\boldsymbol{x}_i, \tau_i) = \{\boldsymbol{x} | |\boldsymbol{x} - \boldsymbol{x}_i| \leq \sqrt{2\tau_i}\} \tag{7}$$

expresses a dominant part of $f(\boldsymbol{x}, \tau_i)$.

In refs. [1,2], an algorithm to define a unique tree whose nodes are the stationary points on the stationary curves is defined. This tree expresses a unique hierarchical relation of dominant parts in an image. The tree is constructed according to the order of the stationary points on the stationary curves. Since the

---

[1] In ref. [16], the zero-crossing $\frac{\partial^2}{\partial^2 \boldsymbol{m}}(G *_2 f) = 0$ for $\boldsymbol{m} = \frac{\nabla_G f}{|\nabla_G f|}$ is proposed as the segment edge. In ref. [15], the equality $\frac{\partial^2}{\partial^2 \boldsymbol{m}}(G *_2 f) = \frac{1}{|\nabla_G f|^2} \nabla_G f^\top \boldsymbol{H}_G \nabla_G f$ is proven.

stationary curves consist of many curves for $\tau > 0$, we call each curve a branch curve. The point $\boldsymbol{x}_\infty$ such that $\lim_{\tau \to \infty} \boldsymbol{x}(\tau) = \boldsymbol{x}_\infty$ is uniquely determined for any images. We call a curve on which the point $\boldsymbol{x}_\infty$ lies and a curve which is open to the direction of $-\tau$ the trunk and branch, respectively.

At the top of each branch, a top point exists. Therefore, for the construction of a unique hierarchical expression of stationary points, the following rule is proposed [1].

Tree Construction

1. The subroot of a branch is a top point.
2. A top point is merged with the closest maximal point on the scale which derives the top point.

This rule yields a monotonically branching curve from infinity to zero along the τ-axis in the linear scale space. Using this monotonically branching curve, we can define the order of scales [1,2].

Definition 6. *For the stationary points on the stationary curves which are merged using rule 1, the order of the stationary points is defined as $\boldsymbol{x}(\tau) \succ \boldsymbol{x}(\tau')$ if $\tau > \tau'$ on a branch.*

3.2 Segmentation Hierarchy

Using the geometrical properties of stationary points on stationary curves, we select a set $\{\tau_i\}_{i=1}^n$.

Definition 7. *We select scales $\{\tau_i\}_{i=1}^n$ for the Canny edge detection from scales at stationary points on stationary curves.*

Let $\sharp(\tau)$ be the number of extremals for the scale τ. Setting τ^* to be a scale which derives a top point, the difference between $\sharp(\tau^* + \varepsilon)$ and $\sharp(\tau^* - \varepsilon)$ is at least one for a small positive constant ε. $E(\tau)$ crosses at saddle points and a simple closed portion of $E(\tau)$ encircles at least one extremal [14]. These two geometric properties in the linear scale space and topology of $E(\tau)$ lead to the next assertion.

Assertion 1. *The difference between the number of simple closed curves for scales $\tau^* + \varepsilon$ and $\tau^* - \varepsilon$ is at least one.*

Figure 1 shows the topological and hierarchical relations of segments extracted from a simple image. Fig. 1(h) shows the merging of a pair of closed structure lines on a top point. From these geometrical properties we obtain the following proposition.

Proposition 1. *If a pair of branches of stationary curves is merged at a top point, a pair of simple closed curves in $E(\tau)$ which share a saddle point is merged into a simple closed curve.*

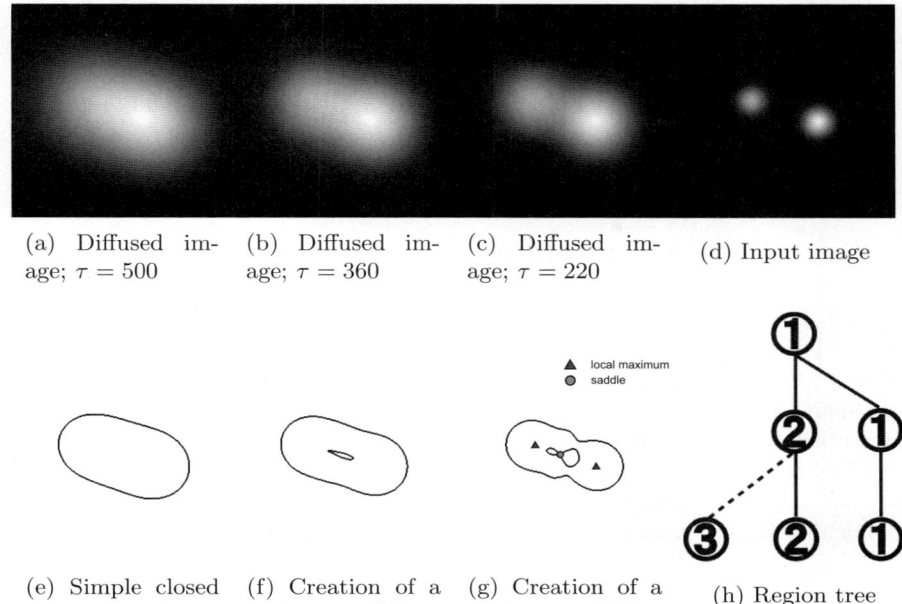

Fig. 1. Creation of regions with decreasing scale. Top row shows the diffused images of (d), and bottom row shows the edge lines. (e) A simple closed curve at a large scale, (f) creation of region inside existing region, and (g) creation of region which has a contact point. (h) is the extracted tree.

This property implies the next rule to select scales for the detection of $E(\tau)$ as the segment boundary.[2]

Parameter Selection for Edge Detection

1. Compute scales $\{\tau *_i\}_{i=0}^{k}$, which define top points, such that $\tau_i^* \leq \tau_{i+1}^*$ and $\tau_0 = 0$.
2. Select scales in the interval (τ_i^*, τ_{i+1}^*).

We define the hierarchical order of the segments extracted as the zero-crossing of $\nabla_G f^\top H_G \nabla_G$ for (τ_i^*, τ_{i+1}^*).

Region Tree

1. For $\tau = \infty$, set the region encircled with a closed loop to be the root of the tree.
2. While decreasing τ, operate the followings;

---

[2] For an appropriate function F, the set of points $F = \{x|F(x) = 0\}$ is expressed as a common set of the two sets $F = F_+ \cap F_-$ for $F_+ = \{x|F(x) \geq 0\}$ and $F_- = \{x|F(x) \leq 0\}$.

(a) If a new region encircled with a closed loop appears, set this region as the subroot of the node corresponding to the loop encircling the new loop and connect this subroot to the node using an inclusive relation edge.

(b) If a new region incident to a loop appears, set this region as the subroot of the node corresponding to the loop incident to this loop and connect this subroot using an incident relation edge.

4 Experimental Examples

Figure 2 shows edges and segments for the selected scales. The scales in Fig. 2 are selected based on the numbers of saddle points listed in Table 1. These numbers define τ^*. Figure 3 shows the tree extracted from the singular points in the linear scale space and the tree extracted from segments. The tree constructed from the segments may define a strategy for the unification of small segments to one large segments for the control of oversegmentation. Figure 4 shows the topological relations of regions. These results show that the zero point and zero-crossing in the linear scale space yield a hierarchical relation of image segments.

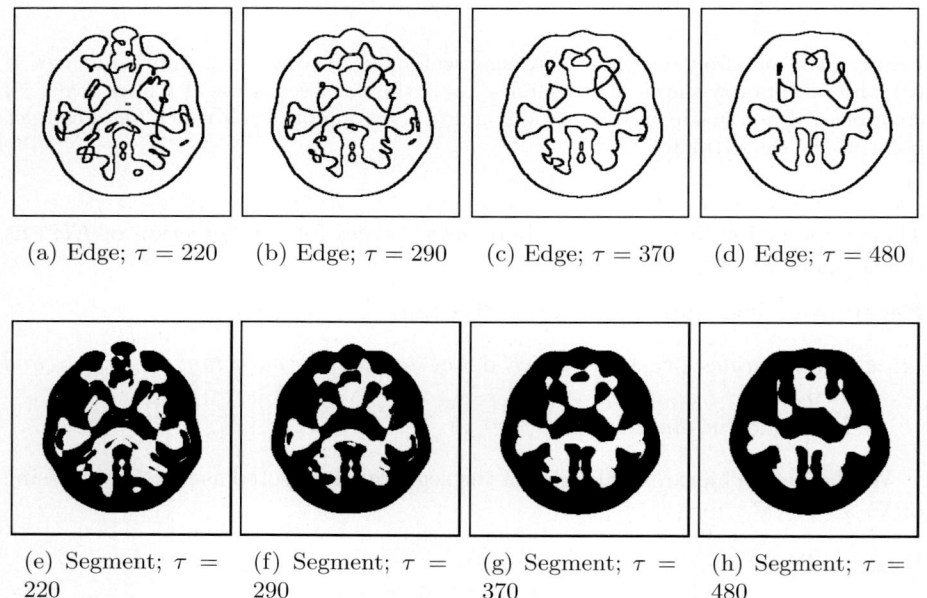

(a) Edge; $\tau = 220$ (b) Edge; $\tau = 290$ (c) Edge; $\tau = 370$ (d) Edge; $\tau = 480$

(e) Segment; $\tau = 220$ (f) Segment; $\tau = 290$ (g) Segment; $\tau = 370$ (h) Segment; $\tau = 480$

Fig. 2. Edges and segments in the linear scale space. From left to right, figures show edges and the segments for the selected scales.

Table 1. The number of saddle points for scales

scale	$730 > \tau \geq 400$	$> \tau \geq 380$	$> \tau \geq 360$	$> \tau \geq 230$	$> \tau \geq 220$
Number of points	4	5	6	7	8

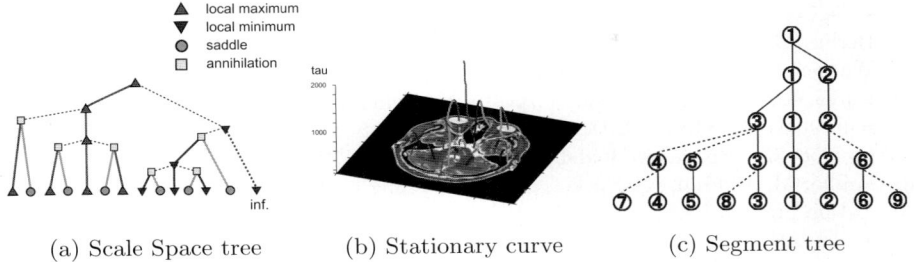

(a) Scale Space tree (b) Stationary curve (c) Segment tree

Fig. 3. Singular point tree and segment tree. The tree in (a) is extracted based on the curves in (b). The tree in (c) is extracted from the hierarchy of segments of Fig. 2.

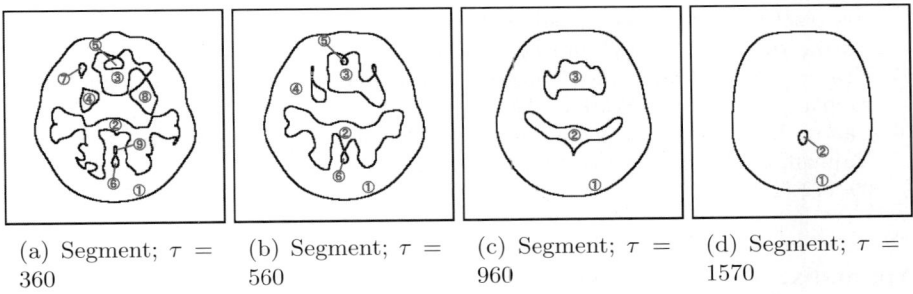

(a) Segment; $\tau = 360$ (b) Segment; $\tau = 560$ (c) Segment; $\tau = 960$ (d) Segment; $\tau = 1570$

Fig. 4. Edges and segments in the linear scale space. If a closed curve for a small scale encircles a collection of closed curves in a large scale, this relation defines a hierarchy of segments across the scale.

5 Conclusions

In this paper, we analysed the Canny operator using the Gaussian scale space framework and found a theoretical strategy on the determination of parameters involved in Canny operation. Furthermore, we extracted the hierarchical relation of segments using the configuration of the saddle points in the linear scale space.

References

1. Zhao, N.-Y., Iijima, T.: Theory on the method of determination of view-point and field of vision during observation and measurement of figure. IECE Japan, Trans. D J68-D, 508–514 (1985) (in Japanese)

2. Zhao, N.-Y., Iijima, T.: A theory of feature extraction by the tree of stable viewpoints. IECE Japan, Trans. D J68-D, 1125–1135 (1985) (in Japanese)
3. Iijima, T.: Pattern Recognition, Corona-sha, Tokyo (1974) (in Japanese)
4. Witkin, A.P.: Scale space filtering. In: Proc. of 8th IJCAI, pp. 1019–1022 (1993)
5. Lindeberg, T.: Scale-Space Theory in Computer Vision. Kluwer, Boston (1994)
6. Lindeberg, T.: Feature detection with automatic selection. IJCV 30, 79–116 (1998)
7. ter Haar Romeny, B.M.: Front-End Vision and Multi-Scale Image Analysis Multiscale Computer Vision Theory and Applications, written in Mathematica. Springer, Berlin (2003)
8. Weicker, J.: Anisotropic Diffusion in Image Processing. Teubner, Stuttgart (1998)
9. Imiya, A., Sugiura, T., Sakai, T., Kato, Y.: Temporal structure tree in digital linear scale space. In: Griffin, L.D., Lillholm, M. (eds.) Scale-Space 2003. LNCS, vol. 2695, pp. 356–371. Springer, Heidelberg (2003)
10. Pelillo, M., Siddiqi, K., Zucker, S.W.: Matching hierarchical structures using association graphs. PAMI 21, 1105–1120 (1999)
11. Yuille, A.L., Poggio, T.: Scale space theory for zero crossings. PAMI 8, 15–25 (1986)
12. Kuijper, A., Florack, L.M.J., Viergever, M.A.: Scale space hierarchy. Journal of Mathematical Imaging and Vision 18, 169–189 (2003)
13. Kuijper, A., Florack, L.M.J.: The hierarchical structure of images. IEEE Trans. Image Processing 12, 1067–1079 (2003)
14. Enomoto, H., Yonezaki, N., Watanabe, Y.: Application of structure lines to surface construction and 3-dimensional analysis. In: Fu, K.-S., Kunii, T.L. (eds.) Picture Engineering, pp. 106–137. Springer, Berlin (1982)
15. Krueger, W.M., Phillips, K.: The geometry of differential operator with application to image processing. PAMI 11, 1252–1264 (1989)
16. Canny, J.: A computational approach to edge detection. PAMI 8, 679–698 (1986)
17. Najman, L., Schmitt, M.: Watershed of a continuous function. Signal Processing 38, 99–112 (1994)

Appendix. The following algorithm [16] detects an approximation of E_{Canny}, since the gradient map of an original image approximates $E(\tau)$.

1. *Define the parameters τ^*, T_1 and T_2 such that $T_1 \geq T_2$.*
2. *Compute $h = G *_2 f$.*
3. *Mark $\theta(x,y) = \tan^{-1} \frac{h_x}{h_y} = \tan^{-1} \frac{G_x *_2 f}{G_y *_2 f}$ on points as the edge direction.*
4. *For $|\nabla h|$, select a point $|\nabla h| \geq T_1$ as the starting point of edge tracking.*
5. *Track peaks using $\theta(x,y)$ of $|\nabla h|$ as for as $|\nabla h| \geq T_2$.*

Point Cloud Segmentation Based on Radial Reflection

Mario Richtsfeld and Markus Vincze

Institute of Automation and Control
Vienna University of Technology
Gusshausstr, 27-29, Vienna, Austria
{rm,vm}@acin.tuwien.ac.at

Abstract. This paper introduces a novel 3D segmentation algorithm, which works directly on point clouds to address the problem of partitioning a 3D object into useful sub-parts. In the last few decades, many different algorithms have been proposed in this growing field, but most of them are only working on complete meshes. However, in robotics, computer graphics, or other fields it is not always possible to work directly on a mesh. Experimental evaluations of a number of complex objects demonstrate the robustness and the efficiency of the proposed algorithm and the results prove that it compares well with a number of state-of-the-art 3D object segmentation algorithms.

Keywords: point cloud segmentation, mesh segmentation, mesh decomposition.

1 Introduction

Cutting up an object into simpler sub parts has several benefits in modeling [10], robotics [13] or collision detection [16]. The presented work includes a new segmentation algorithm, based on radial reflection. The majority of the algorithms developed here can be applied with only trivial modifications to more complex shape matching problems.

1.1 Problem Statement and Contributions

Object segmentation and analysis, which can be interpreted as purely geometric sense are challenging problems in computer vision. An ideal shape descriptor should be able to find out the main features of an object and segment it into useful parts, which can be used for automatic processes such as matching, registration, feature extraction [12] or comparison of shapes. Different methods for mesh segmentation exist (e.g. Plumber [17], feature point and core extraction [14], Hierarchical Fitting Primitives (HFP) [3], spectral methods [22],...), but most of them are only able to work on a mesh and not a point cloud. This paper presents an algorithm which works directly on point clouds and is invariant under rotation, translation and scaling.

1.2 Algorithm Overview

Fig. 1[1] gives an overview of our segmentation algorithm. The proposed segmentation algorithm is based on radial reflection. At the beginning the algorithm calculates the internal center and the radius of the bounding sphere by computing the smallest enclosing sphere of points [11], see Fig. 1d. Then, all points are radial reflected inside in the direction to the center. Thus all points which are inside on the original point cloud are farthest out after this step. The algorithm uses the reflected point cloud to calculate the convex hull [18], Fig. 1e (yellow hull), whereby all adhering parts on the core part will be automatically cut off. To realize a hole free segmentation of the core part all vertices of the convex hull are transformed in the direction to the center depending on the distances of the neighboring points [2], see Fig. 1e (red hull). Based on these vertices an inner convex hull is calculated. These inner convex hull surrounds the rest parts of the object. Then our algorithm automatically segments the 3D point cloud into a set of sub-parts by recursive flood-filling [9] based on the segmented core part, see Fig. 1f. To realize a pose invariant object segmentation our algorithm generates a 3D mesh based on the power crust algorithm [1], see Fig. 1b, and uses multi-dimensional scaling (MDS) to get a pose-invariant model representation, see Fig. 1c. Thereby every vertex on the pose-invariant model corresponds to a vertex of the mesh and every point of the original point cloud corresponds to a vertex of the mesh.

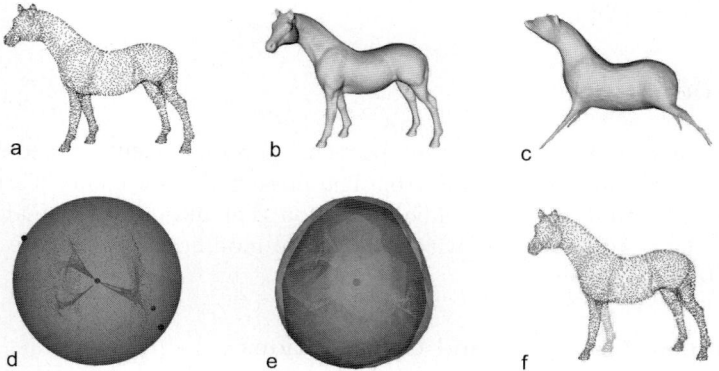

Fig. 1. Overview of our segmentation algorithm: **a** 3D point cloud (5360 points). **b** 3D mesh based on the power crust algorithm (58441 vertices). **c** Pose-invariant model representation based on multi-dimensional scaling (MDS) (58441 vertices). **d** Center and bounding sphere, the radial reflected point cloud (5360 points) is red colored, the original point cloud (5360 points) is green colored. The blue points (along the bounding sphere) correspond with the blue center of the radial reflected point cloud. **e** Outer convex hull (yellow), internal convex hull (red) to realize a hole free core part. **f** Segmented point cloud (2085 core points, 3275 rest points).

---

[1] All images are best viewed in color. The core part is in every case red colored.

1.3 Related Work

Different methods to automatic 3D object segmentation into meaningful parts have been published in the last few years.

3D Model Segmentation: algorithms can be categorized into two main classes. The first class is developed for applications like reverse engineering of CAD models [5]. The second class tries to segment natural objects into meaningful parts. Most work on mesh segmentation is based on iterative clustering. [20] segmented models into meaningful pieces using k-means clustering. Based on this idea [15] developed a fuzzy clustering and minimal boundary cuts method to achieve smoother boundaries between clusters. Unsupervised clustering techniques like mean shift can also applied to mesh segmentation [19]. [14] published a mesh segmentation algorithm based on pose-invariant models and extraction of core part and feature points. The method is able to produce consistent results. An computation intensive method is used to find feature points, to limit the complexity and number of parts of models.

Pose-Invariant Mesh Representation: To realize a pose-invariant mesh representation multi-dimensional scaling (MDS) is used. MDS is a generic name for a family of algorithms that construct a configuration of points in a target metric space from information about inter-point distances (dissimilarities), measured in some other metric space [8]. In our experiments, dissimilarities are defined as geodesic distances δ_{ij} between all vertices v_i on the mesh $\mathcal{M}$ in a symmetrical dissimilarities matrix $\Delta = \mathcal{N} \times \mathcal{N}$ between N points on a Riemannian manifold $\mathcal{S}$. We differentiate between metric and non-metric MDS (Shephard-Kruskal). Metric MDS preserves the intervals and the ratios between the dissimilarities and non-metric MDS only preserves the order of the dissimilarities. The goal is to minimize the embedding error, i.e. minimizing the sum of distances between the optimal scaled data $f(\delta_{ij})$ and the euclidean distances d_{ij}, where f is an optimal monotonic function (in order to obtain optimally scaled similarities). Thereby a stress function $\mathcal{F}_s$ will be used to measure the degree of correspondence of the distances between vertices. We use the scaled gradient-descent algorithm (SMACOF), as published by [8]. This algorithm is one of the most efficient at the moment and it allows real-time performance. Each vertex in MDS space corresponds to a vertex in euclidean space. In order to speed up the calculation time, the geodesic distances are calculated only on a reduced set of landmark points. Approximately the original points of the point cloud of the mesh vertices as landmark points has an optimal balance between accuracy

Fig. 2. Pose-invariance: each model was segmented separately

of representation and time. Fig. 2 illustrates our segmentation results based on pose-invariant model representation.

In our work the difference to the existing core extraction algorithm [14] is the radial reflection of the points in the direction to the center of the object and to calculate an internal convex hull to get a hole free core part, which is used to cut the 3D model. Additionally our algorithm works directly on point clouds, whereby no mesh generation is needed. The mesh generation with the power crust algorithm [1] is only needed to get a pose-invariant model representation.

2 Point Cloud Segmentation

This section describes each stage of the proposed segmentation algorithm for point clouds.

2.1 Core Extraction

The presented method is based on the principle of radial reflection. At the beginning the internal center $\mathcal{C}$ is calculated by computing smallest enclosing sphere of points [11]. The bounding sphere is defined by the maximum distance $\mathcal{R}$ between the center $\mathcal{C}$ and all points p_i:

$$\mathcal{R} = max\|p_i - \mathcal{C}\| \qquad (1)$$

Each point p_i of the point cloud with n points is radial reflected inwards in the direction to the calculated center $\mathcal{C}$, as illustrated in Fig. 1d and Fig. 1e.

$$p'_m = \mathcal{C} + (\mathcal{R} - \| p_i - \mathcal{C} \|)\frac{(p_i - \mathcal{C})}{\| p_i - \mathcal{C} \|} \qquad (2)$$

Thus all points which are farthest outside on the original point cloud are farthest in after this step, as illustrated in Fig. 1d. This way, the points of the core part reside on the outer convex hull $\mathcal{H}_{out}$ [18], whereby all adhering parts on the core part will be automatically cut off.

$$\mathcal{H}_{out} = ConvexHull\left(\bigcup_{i=0}^{n-1} p'_{m_i}\right) \qquad (3)$$

Every vertex v_m of the k vertices that reside on the outer convex hull $\mathcal{H}_{out}$ will be transformed in the direction to the center, depending on the distances of the neighboring points [2] with an offset o_{ff}. For that the algorithm calculates for each point of the original point cloud the distance to the nearest neighbor and then the minimum d_{min}, maximum d_{max} and average d_a of these distances. Then the algorithm finds out for every vertex v_m on the outer convex hull all neighboring points p'_m with the average distance d_a and calculates the offset o_{ff}, depending of the z point neighbors, see Equ. 5. This step is important to realize a hole free core part.

$$o_{ff} = \frac{\sum_{i=0}^{z-1} | p'_{m_i} - v_m |}{z} \quad (4)$$

The offset o_{ff} was calculated with all z neighboring points of the transformed point cloud of the vertex v_m on the convex hull $\mathcal{H}_{out}$. With the calculated offset o_{ff} the algorithm need no more connectivity analysis to realize a hole free core part. All vertices on the outer convex hull $\mathcal{H}_{out}$ will be transformed with an offset for every vertex:

$$v'_m = v_{m_i} - o_{ff} * \frac{(v_{m_i} - C)}{\| v_{m_i} - C \|} \quad (5)$$

This k transformed vertices v'_m are used to calculate an inner convex hull $\mathcal{H}_{in}$, as illustrated in Fig. 1e (red convex hull):

$$\mathcal{H}_{in} = ConvexHull \left(\bigcup_{i=0}^{k-1} v'_{m_i} \right) \quad (6)$$

The resulting inner convex hull $\mathcal{H}_{in}$ is used to cut the radial reflected point cloud into a core part and a rest part, as illustrated in Fig. 1f.

2.2 Cut Refinement

If the core part is found, all other segments of the point cloud are extracted by recursive flood-filling [9]. We define an object-part as a set of points, with distances between neighbors below a threshold d_{max}. We build a kd-tree [7] to find neighbors and use the recursive flood-filling function [9] to identify connected point sets. d_{max} is the maximum distance between the neighboring points, calculated by nearest neighbor search [2]. This step segments the point cloud into different components. An additional cut refinement was not arranged, because the main goal is to find out the core part. It is possible to improve the segmentation results with the help of a substantially curvature-based filter [21], mean shift, gaussian curvature or a feature point based approach [14]. It is also possible to improve the segmentation results with the calculation of the normal vector for every point, by fitting planes in a defined area d_a. Thus the angle α between the regarded point i and the considered point w can be used as weighting factor wg, as illustrated in Fig. 3.

$$\cos \alpha = \frac{\boldsymbol{n_i} \bullet \boldsymbol{n_w}}{\|\boldsymbol{n_i}\|\|\boldsymbol{n_w}\|} \quad (7)$$

$$wg = 1 - |\cos(\alpha)| \quad (8)$$

To belong to a fracture of the object the distance d between a fracture element w and the considered point i must be smaller than the average distance with the weighting factor.

$$d = \sqrt{(x_i - x_w)^2 + (y_i - y_w)^2 + (z_i - z_w)^2} \quad (9)$$

$$d < d_a \cdot wg \quad (10)$$

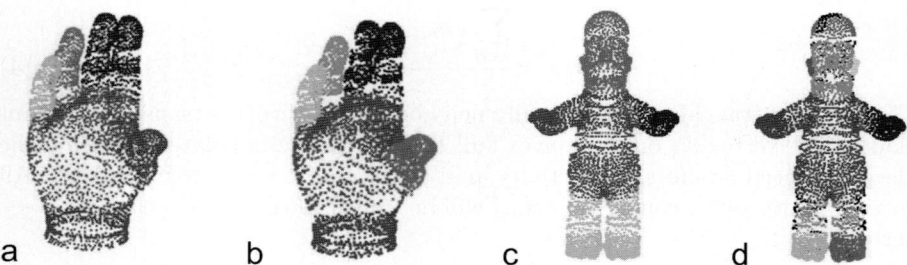

Fig. 3. Cut refinement: Improvement of the segmentation result by calculating an additional weighting factor. **a, c** Hand, Man: standard flood-filling. **b, d** Hand, Man: flood-filling with additional weighting function.

3 Results

We have created and collected at AIM@SHAPE repository[2] several challenging examples to test our segmentation algorithm, see Fig. 4. For similar segmentations of the same models in different poses, the segmentation based on pose-invariant models show almost best results. Our analysis shows that the position of the internal center of the models has a significant influence, as illustrated in Fig. 4g (dino) and h (elephant). It is important that the approximated center is inside the object.

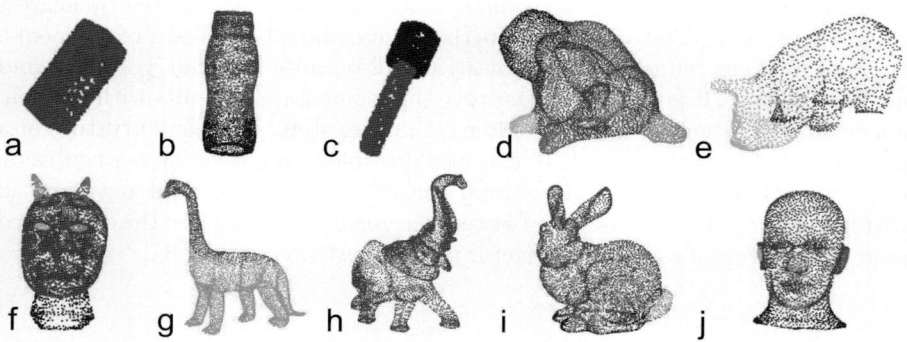

Fig. 4. Segmentation results: We analyzed different groups of models: **a** package, **b** coffee tin, **c** bolt, **d** frog, **e** pig, **f** oni, **g** dino, **h** elephant, **i** bunny, **j** mannequin

Fig. 4 and Fig. 5 show that the proposed algorithm is optimal to extract the core component and the surrounding parts. HFP performs good results, if the number of clusters is limited to a realistic number of parts of the analyzed object. Methods like Plumber perform good results with a a-priori knowledge about the

---

[2] http://shapes.aim-at-shape.net/index.php

Fig. 5. Comparison: **a** Proposed Algorithm, **b** HFP, and **c** Plumber

object. It performs good on features with elongation axis larger than section axis, whereby parts are correctly detected as a tubular feature. [5] described more exactly the properties of the Plumber and HFP algorithms and pointed out more exactly the segmentation results and differences based on their segmentation criteria. However, the presented results in this work confirm their work completely.

4 Conclusion

The proposed segmentation method represents a flexible and completely automatic way to segment a 3D object in a hierarchical manner, whereby the algorithm works directly on point clouds and shows high reliability. It is obvious from the results presented in this work that there exist no perfect segmentation algorithm. Each algorithm has his own benefits and drawbacks. Segmentation can neither be formalized nor measured mathematically [5]. If a pose-invariant model representation is needed, the algorithm generates a 3D mesh with the power crust algorithm [1] and use multi-dimensional scaling (MDS). We cut the object into sub-parts with an inner convex hull, which results from an outer convex calculated by radial reflection. This segmentation algorithm can be applied to a reasonable set of objects with different applications.

References

1. Amenta, N., Choi, S., Kolluri, R.: The power crust. In: Sixth ACM Symposium on Solid Modeling and Applications, pp. 249–260 (2001)
2. Arya, S., Mount, D.M., Netanyahu, N.S., Silverman, R., Wu, A.Y.: An Optimal Algorithm for Approximate Nearest Neighbor Searching in Fixed Dimensions. Journal of the ACM 45(6), 891–923 (1998)
3. Attene, M., Falcidieno, B., Spagnuolo, M.: Hierarchical Mesh Segmentation based on Fitting Primitives. The Visual Computer 22(3), 181–193 (2006)
4. Attene, M., Robbiano, F., Spagnuolo, M., Falcidieno, B.: Semantic annotation of 3D surface meshes based on feature characterization. In: Falcidieno, B., Spagnuolo, M., Avrithis, Y., Kompatsiaris, I., Buitelaar, P. (eds.) SAMT 2007. LNCS, vol. 4816, pp. 126–139. Springer, Heidelberg (2007)

5. Attene, M., Katz, S., Mortara, M., Patane, G., Spagnuolo, M., Tal, A.: Mesh Segmentation - A Comparative Study. In: IEEE International Conference on Shape Modeling and Applications, SMI, pp. 7–18 (2006)
6. Biasotti, S.: Computational Topology methods for Shape Modelling Applications. PhD thesis, University of Genoa, Italy (2004)
7. Bentley, J.L.: Multidimensional Binary Search Trees Used for Associative Searching. Communications of the ACM 18(19), 509–517 (1975)
8. Bronstein, M.M., Bronstein, A.M., Kimmel, R., Yavneh, I.: Multigrid multidimensional scaling. Numerical Linear Algebra with Applications (NLAA), Special issue on multigrid methods 13(2-3), 149–171 (2006)
9. Burger, W., Burge, M.: Digital Image Processing - An Algorithmic Introduction Using Java, 1st edn. Springer, UK (2007)
10. Funkhouser, T., Kazhdan, M., Shilane, P., Min, P., Kiefer, W., Tal, A., Rusinkiewicz, S., Dobkin, D.: Modeling by example. ACM Transactions on Graphics 23(3), 652–663 (2004)
11. Gärtner, B.: Fast and Robust Smallest Enclosing Balls. In: Nešetřil, J. (ed.) ESA 1999. LNCS, vol. 1643, pp. 325–338. Springer, Heidelberg (1999)
12. Gumhold, S., Wang, X., MacLeod, R.: Feature Extraction from Point Clouds. In: Proceedings of the 10th International Meshing Roundtable, pp. 293–305 (2001)
13. Huebner, K., Ruthotto, S., Kragic, D.: Minimum Volume Bounding Box Decomposition for Shape Approximation in Robot Grasping. In: IEEE International Conference on Robotics and Automation, ICRA, pp. 1628–1633 (2008)
14. Katz, S., Leifman, G., Tal, A.: Mesh segmentation using feature point and core extraction. The Visual Computer (Pacific Graphics) 21(8-10), 649–658 (2005)
15. Katz, S., Tal, A.: Hierarchical mesh decomposition using fuzzy clustering and cuts. ACM Transactions on Graphics 22(3), 954–961 (2003)
16. Li, X., Toon, T., Tan, T., Huang, Z.: Decomposing polygon meshes for interactive applications. In: Proceedings of the 2001 symposium on Interactive 3D graphics, pp. 35–42 (2001)
17. Mortara, M., Patanè, G., Spagnuolo, M., Falcidieno, B., Rossignac, J.: Plumber: a method for a multi-scale decomposition of 3D shapes into tubular primitives and bodies. In: Proceedings of the ninth ACM symposium on Solid modeling and applications, pp. 339–344 (2004)
18. O'Rourke, J.: Computational Geometry, 2nd edn. C. Univ. Press, Cambridge (1998)
19. Shamir, A., Shapira, L., Cohen-Or, D., Goldenthal, R.: Geodesic mean shift. In: Proceedings of the 5th Korea-Israel Conference on Geometric Modeling and Computer Graphics, pp. 51–56 (2004)
20. Shlafman, S., Tal, A., Katz, S.: Metamorphosis of Polyhedral Surfaces using Decomposition. Computer Graphics Forum 21(3), 219–229 (2002)
21. Trucco, E., Fisher, R.B.: Experiments in curvature-based segmentation of range data. IEEE Transactions on Pattern Analysis and Machine Intelligence 17(2), 177–182 (1995)
22. Zhang, H., Kaick, O., Dyer, R.: Spectral Methods for Mesh Processing and Analysis. In: Proceedings of Eurographics 2007, pp. 1–22 (2007)

Locally Adaptive Speed Functions for Level Sets in Image Segmentation

Karsten Rink and Klaus Tönnies

Department for Simulation and Graphics, University of Magdeburg, Germany
{karsten,klaus}@isg.cs.uni-magdeburg.de

Abstract. We propose a framework for locally adaptive level set functions. The impact of well-known speed terms for the evolution of the active contour is adjusted by parameterising them with functions based on pre-defined properties. This allows for the application of level set methods even if image features are subject to large variations or if certain properties of the model are only valid for parts of the segmentation process. We present a number of examples and applications for the proposed concept and also address advantages and drawbacks of combinations of locally adaptive speed terms.

Keywords: Image Segmentation, Level Set Methods.

1 Introduction

Level set methods have become very popular in recent years in many image processing domains. They allow for the segmentation of images even if the shape of the desired object is unknown or has large variances between data sets. Level sets can adapt to topological changes, usually only require a small number of parameters to be adjusted and their extension to higher dimensions is straightforward.

An implicit active contour $\phi_t + F|\nabla\phi| = 0$ is evolved either in a propagation process [11] or via energy minimisation [2]. Each point on the zero level set is moving along its surface normal with a speed F. This speed is calculated according to pre-defined properties of the desired object. In image processing these properties are usually image features like gradients [10], texture measures [12] or image intensities [2]. To compensate for image artefacts such as noise or imperfect boundaries a second class of speed terms is often needed for regularisation of the contour. Regularisation of the front ranges from smoothness terms based on mean curvature [11] to more sophisticated approaches such as incorporation of model-based knowledge based on shocks [1], geometric shapes like circles or ellipsoids [16] or the topology of the desired object [8]. Other approaches even incorporate explicit shape knowledge [3][9].

Unfortunately the use of sophisticated speed terms has a number of drawbacks in certain situations. The definition of level sets is very general, allowing for their application to data sets from many different domains. By incorporating highly specialised speed terms, the method becomes limited to specific applications.

Interesting properties of the method, such as the extension to any number of dimensions, are lost. Also, for a number of segmentation problems it is not possible to incorporate any of these advanced speed terms because the shape of the desired object is not known in advance. Examples in medical image analysis are the segmentation of the cerebral grey and white matter or segmentation of the vascular or bronchial tree. We propose a new framework for the design of level set speed functions that is also suitable for this kind of application.

2 Locally Adaptive Speed Terms

Many speed terms for level set functions can be incorporated for the segmentation of different kinds of images and data sets simply by adjusting a parameter that controls the impact of that particular speed term. For example, the often used speed function $F = F_g(F_\nu + F_\kappa)$ includes the well-known gradient-based speed term $F_g(\alpha) = (1 + \alpha|\nabla G_\sigma * I(\mathbf{x})|)^{-1}$ and the curvature-based term $F_\kappa(\epsilon) = \epsilon \nabla \frac{\nabla \phi}{|\nabla \phi|}$. In the first case, the value of α affects how large gradients need to be for the front to stop. In the second example, ϵ determines how smooth the contour is going to be and how much it will be affected by image noise.

Unfortunately, these parameter values are usually constant for the whole propagation process. If certain properties of the image differ substantially over image space, adjustment of the parameters is difficult. An example are field inhomogeneities in magnetic resonance imaging. Model-based properties may change in the same way. In the segmentation of the human brain, thickness of the cortical grey matter is different in various regions of the brain. Even the established coupled surface approach [7] does not account for that. More generally, smoothness of the contour may be desired only for certain parts of the objects but not for others. With other deformable models such as the active shape model [4] or mass spring models [15] it is possible to address this kind of variability by adjusting the modes of variation or the parametrisation of springs, respectively. To allow for the incorporation of such functionality in level sets and thus increase the application spectrum these methods, we propose the concept of locally adaptive speed terms for level set methods.

A speed term $F_i = F_i(\mathbf{f})$ is dependent on one or more features $\mathbf{f}$. Instead of controlling the influence of F_i on the active contour with a constant parameter, a function $\omega_i = \omega_i(\mathbf{g})$ is used for adjusting the impact of F_i based on a second set of features $\mathbf{g}$. The definition of a level set speed term $\hat{F}_i$ is then given by

$$\hat{F}_i = \omega_i F_i = \omega_i(\mathbf{g}) F_i(\mathbf{f}). \qquad (1)$$

Note that the properties $\mathbf{g}$ need not be based on image features. Below we will give examples of weighting functions ω based on distance-measures and position in image space. Using ω_i it is possible to adjust the influence of F_i, to apply it only for certain parts of the image or to set it to zero if it would otherwise impede the desired evolution of the active contour. If all ω_i and F_i are continuous functions the method is numerically stable and the combination of locally adaptive speed terms

$$F = \omega_1 F_1 + \omega_2 F_2 + \ldots + \omega_n F_n, \qquad (2)$$

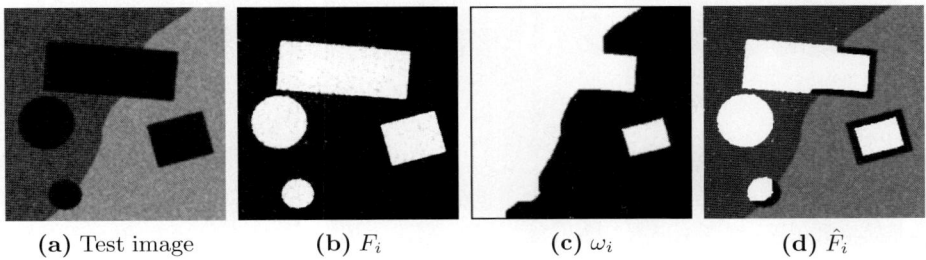

(a) Test image (b) F_i (c) ω_i (d) $\hat{F}_i$

Fig. 1. Example for a locally adaptive speed term based on distance. Bright intensities indicate a high velocity, dark intensities a low velocity. An image-based speed term F_i based on image intensities is parameterised by a function ω_i depending on distance and prevents the front from evolving if its distance to the light grey region on the right side is less than 15 pixels. The resulting speed term $\hat{F}_i$ is superimposed on figure (a) for better visualisation.

will be a continuous function as well. If for some speed terms no weighting function is necessary, ω_i can be chosen such that $\hat{F}_i = id(F_i)$. Figure 1 shows a simple example for a locally adaptive speed function based on distance. Similar to model-based speed terms, the influence of more sophisticated parameterising functions ω_j might also change over time.

The proposed concept allows for the use of comparatively basic speed terms even for challenging segmentation tasks. By using parameterising functions, well-known speed terms can be incorporated even if certain properties that define the object of interest do not hold for all parts of that object. Vice versa, they can also be employed if such properties are only needed for a correct segmentation of few or small parts of the object. A number of previously published examples that fit into this framework are summarised below. Applications are given in section 3.

2.1 Speed Terms Based on Distance

We demonstrated the possibilities of locally adaptive speed terms based on distance measures to introduce additional knowledge to the segmentation process in [13]. A number of segmentation results are depicted in figure 2. Given figure 2a, a segmentation process was startet with a single seed point within object A. Figure 2b shows a segmentation where the level set stops based on image-features but keeps a minimum distance of 15 pixels to region F (the corresponding speed function has been visualised in figure 1). In figure 2c a speed function that used a minimum *and* maximum distance criterion has been employed. The front keeps a minimum distance of $d_{min} = 15$ pixels to region F within objects A and D, where the properties of the desired object as defined in the speed function hold. The front does not stop at the boundary of A where the distance to F is too large. Instead it propagates into object E, but a maximum distance of $d_{max} = 30$ pixels is kept to region F. Again, the weighing functions are quite simple: the

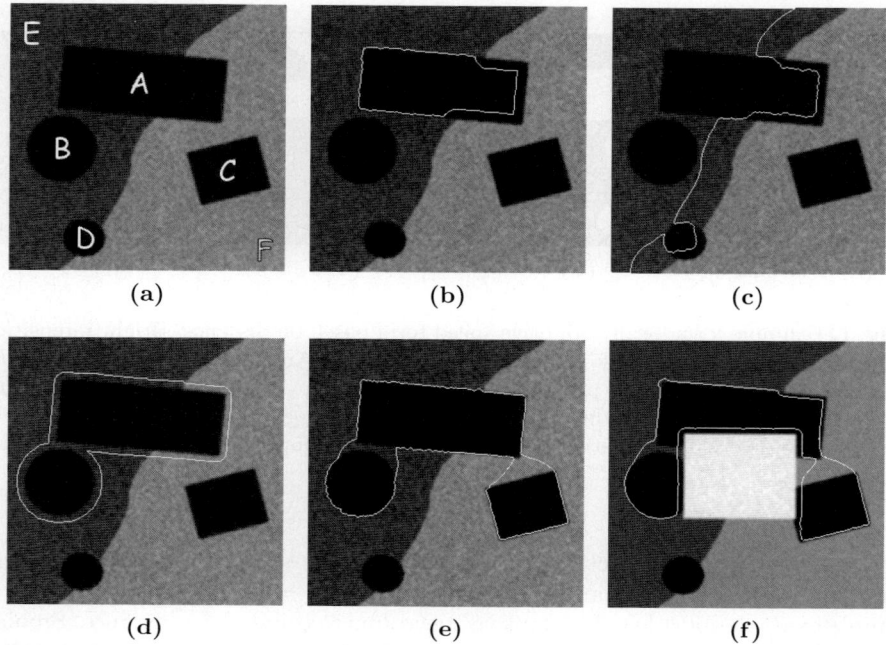

Fig. 2. Effects of distance-based speed terms for image segmentation

maximum distance is realised by a sigmoid function while the evolution of the front for $d_{min} < d < d_{max}$ is controlled by a regularised boxcar function (Such a function is used in section 3 for the segmentation of white matter in MR images of the brain). In figure 2d the front stops $d = 15$ pixels outside object A. Object B is segmented as well because its distance to A is smaller than d. Figure 2e visualises the segmentation result using an acceleration term that connects objects that have similar properties if their distance is smaller than a pre-defined distance d. That way, objects B and C have been connected to A. Note that the front propagated directly from A to the other objects and does not leak into the image background. Finally, figure 2f combines both approaches. Objects B and C are again connected to A but the front keeps a distance of 10 pixels to the white rectangle. For the definitions of the speed functions and further discussion the reader is referred to [13].

2.2 Speed Terms Based on Location

Speed terms can also be parameterised based on the absolute or relative position of the front in image space. In the first case the weighting function ω has the same size as the data set D that is being segmented. That is, for each pixel $\mathbf{x} \in D$ exists a parameter $\omega(\mathbf{x})$. This allows for the incorporation of information from an external source. We have applied this concept in [14] to ensure for the anatomical correct segmentation of cortical grey matter (see figure 4c). Also, Cremers et al. [5] incorporated a similar approach for allowing user interaction during the segmentation process.

If speed terms are dependent on local position, parameter values are adjusted based on pixels in the vicinity of front pixels. In [14] the image space has been subdivided into a grid of cubes of equal size to account for the effect of magnetic field inhomogeneities on the data. Within each subdivision D_i parametrisation was assumed to be constant. The transition between neighbouring subdivisions D_i and D_j have been smoothed based on the distance to the centers of D_i and D_j and the reliability of estimates for the intensity distributions. Again, the interested reader is referred to [14] for details. Obviously, it is not necessary to choose a static subdivision of image space for the calculation of parameters based on local position. A dynamically chosen set of pixel (for instance within a hyperball around $\mathbf{x}$) is also possible but computationally much more expensive.

2.3 Combinations of Speed Terms

Obviously weighting functions are not limited to the presented examples but can be based on any property that can calculated for each pixel on the front. The above examples have been chosen because they are available at nearly no additional computational cost. Other properties of the image or level set function that might also be incorporated in this way are image features or the speed of the active contour itself. If computational cost is not an issue, choices for weighting functions are only limited by properties of the desired object that can be formulated in mathematical terms.

Furthermore, locally adaptive speed terms can also be combined. In figure 2f a simple example has already been illustrated. In our experience a small set of weighting functions is useable for a surprisingly large number of applications. Regularised Heaviside- and Boxcar functions are easily incorporated examples. These functions also have the advantage that the set of parameters is limited to the slope of the function, which usually only needs to be slightly adjusted or not changed at all for different applications. Based on the above considerations this concept also allows for the creation of a construction kit for level set speed functions. By employing a set of weighting functions as well as a bank of well-known speed terms, the framework can thus be employed for a large number of segmentation problems.

Finally, a number of potential drawbacks of this framework should be addressed as well. Even though the number of parameters for each combination $\omega_i F_i$ is small, parameter space can get large when a number of locally adaptive speed terms are combined. In this case even slight adjustments to parameters might not be straightforward anymore. In the same way, the small computational offset introduced by each parameterising function ω_i might add up when a number of adaptive speed terms are combined. And finally, since it is possible to 'switch off' speed terms it might happen that for certain pixels in image space no speed is defined. Again, this is a problem that will usually only occur with the combination of more speed terms when parameter space becomes difficult to manage.

3 Applications

We will give a few examples of application of locally adaptive speed functions to 3D medical data sets to demonstrate their benefit to various segmentation tasks. Details on the algorithms and an evaluation of the results can be found in the respective publications.

In [13] we incorporated a distance-based acceleration term for the segmentation of dendrites in microscopic images. Due to partial volume effects, small spines attached to the dendrites do often appear unconnected in the data sets. Using a locally adaptive speed function, the propagating front is accelerated if spines are detected within a certain distance to the active contour. Segmentation results could thus be significantly improved in comparison to a conventional level set segmentation. Examples are given in figure 3.

We also employed an algorithm including locally adaptive speed functions based on local and global position in image space to guarantee for an anatomically correct segmentation of the cortical grey and white matter in MR data. Figure 4 shows a comparison between segmentation results for white matter using the commercial software BrainVoyager as well as our adaptive level set

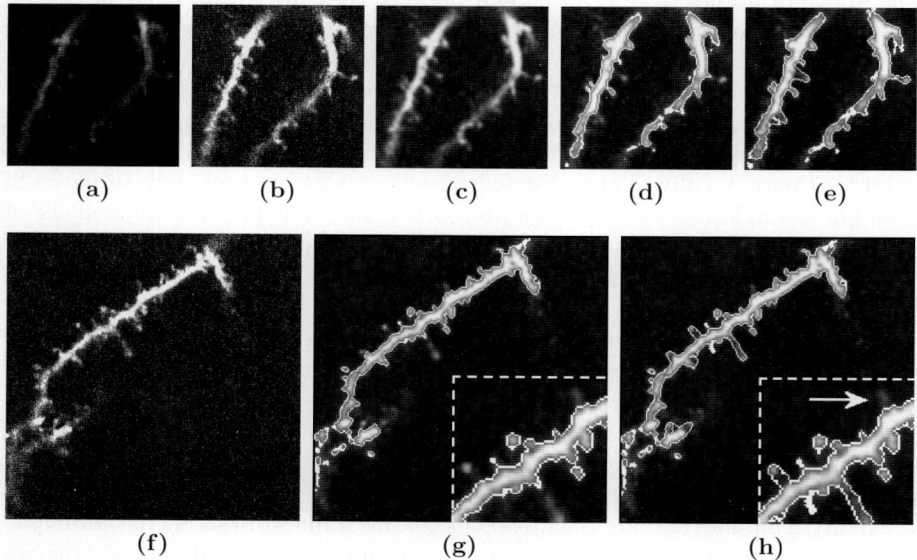

Fig. 3. Segmentation of dendritic spines using a distance based acceleration term. Figures 3a–3c shows pre-processing of data sets of the original data sets by contrast enhancement and low pass filtering. Figures 3d and 3e show segmentation results using a conventional level set speed function and the locally adaptive function, respectively. Figures 3g and 3h present results for a second example. Again most spines were found by the algorithm, although some have been missed since their image features are too similar to background noise (see enlarged region).

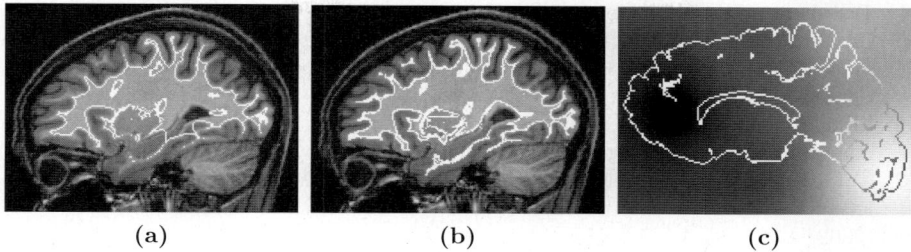

(a) (b) (c)

Fig. 4. Segmentation of the cortical white matter in the human brain brain. (a) Result using commercial software. (b) Result using our locally adaptive algorithm. (c) Cortical thickness map according to [6].

algorithm. A visual inspection by neurobiologists suggested that the boundary between grey and white matter found by our algorithm is usually more exact than the boundary found by the commercial software, which could often not provide a correct segmentation result in the presence of strong magnetic field inhomogeneities. Our algorithm uses a modified coupled surface approach [7]. The distance between inner and outer cortical surface varies between $1.5 - 2.5\,mm$ in the occipital lobe to about $4 - 5\,mm$ in the frontal lobe of the brain [6]. We employed a weighting function based on the absolute position within the brain to guarantee for a correct estimation of cortex thickness (see figure 4c). Furthermore, image-based speed terms are parameterised based on an analysis of local intensity distributions as briefly described in section 2.2.

4 Conclusions

We presented a novel framework for the design of level set speed functions using locally adaptive speed terms. By parameterising a level set speed term with a function ω_i its influence on the active contour can be adjusted depending on predefined properties. It is also possible to define speed terms for segmentation of parts of the desired object only and to switch them off when they are not needed. Examples for locally adaptive speed terms as well as possible applications have been presented. Future work includes the expansion of the concept to create a construction kit for level set speed functions.

References

1. Cao, Z., Dawant, B.M.: Edge completion from sparse data: A level set approach. In: Proc. of SPIE Medical Imaging, pp. 515–525 (2004)
2. Chan, T.F., Vese, L.A.: Active Contours Without Edges. IEEE Trans Image Process 10(2), 266–277 (2001)
3. Chan, T., Zhu, W.: Level set based shape prior segmentation. In: Proc. of IEEE Conf on Computer Vision and Pattern Recognition (CVPR), pp. 1164–1170 (2005)

4. Cootes, T.F., Cooper, D., Taylor, C.J., Graham, J.: Active Shape Models - Their Training and Application. Comput Vis Image Understand 61(1), 38–59 (1995)
5. Cremers, D., Fluck, O., Rousson, M., Aharon, S.: A Probabilistic Level Set Formulation for Interactive Organ Segmentation. In: Proc. of SPIE Medical Imaging, pp. 304–313 (2007)
6. Fischl, B., Dale, A.: Measuring the thickness of the human cerebral cortex from magnetic resonance images. Proc. Nat. Acad. Sci. 97(20), 11044–11049 (2000)
7. Goldenberg, R., Kimmel, R., Rivlin, E., Rudzsky, M.: Cortex Segmentation: A Fast Variational Geometric Approach. IEEE Trans. Med. Imag. 21(2), 1544–1551 (2002)
8. Han, X., Xu, C., Prince, J.L.: A Topology Preserving Level Set Method for Geometric Deformable Models. IEEE Trans. Pattern Anal. Mach. Intell. 25(6), 755–768 (2003)
9. Leventon, M.E., Grimson, W.E.L., Faugeras, O.: Statistical Shape Influence in Geodesic Active Contours. In: Proc. of IEEE Conf. on Computer Vision and Pattern Recognition (CVPR), pp. 316–322 (2000)
10. Malladi, R., Sethian, J.A., Vemuri, B.: Shape Modelling with Front Propagation: A Level Set Approach. IEEE Trans. Pattern Anal. Mach. Intell. 17(2), 158–175 (1995)
11. Osher, S., Sethian, J.A.: Fronts Propagating with Curvature Dependent Speed: Algorithms Based on Hamilton-Jacobi Formulation. J. Comput. Phys. 79, 12–49 (1988)
12. Paragios, N., Deriche, R.: Geodesic Active Regions and Level Set Methods for Supervised Texture Segmentation. Int. J. Comput. Vis. 46(3), 223–247 (2002)
13. Rink, K., Tönnies, K.: Distance-based Speed Functions for Level Set Methods in Image Segmentation. In: Proc. of British Machine Vision Conference (BMVC), pp. 283–292 (2008)
14. Sokoll, S., Rink, K., Tönnies, K., Brechmann, A.: Dynamic Segmentation of the Cerebral Cortex in MR Data using Implicit Active Contours. In: Proc. of Medical Image Understanding and Analysis (MIUA), pp. 184–188 (2008)
15. Terzopoulos, D., Fleischer, K.: Deformable Models. Visual Comput. 4(6), 306–331 (1988)
16. Van Bemmel, C.M., Spreeuwers, L.J., Viergever, M.A., Niessen, W.J.: A Level-Set-Based Artery-Vein Separation in Blood-Pool Agent CR-MR Angiograms. IEEE Trans. Med. Imag. 22(10), 1224–1234 (2003)

Improving User Control with Minimum Involvement in User-Guided Segmentation by Image Foresting Transform

T.V. Spina, Javier A. Montoya-Zegarra, P.A.V. Miranda, and A.X. Falcão

Institute of Computing – University of Campinas (UNICAMP),
C.P. 6176, 13084-971, Campinas, SP, Brazil
afalcao@ic.unicamp.br

Abstract. The image foresting transform (IFT) can divide an image into object and background, each represented by one optimum-path forest rooted at internal and external markers selected by the user. We have considerably reduced the number of markers (user involvement) by separating object enhancement from its extraction. However, the user had no guidance about effective marker location during extraction, losing segmentation control. Now, we pre-segment the image automatically into a few regions. The regions inside the object are selected and merged from internal markers. Regions with object and background pixels are further divided by IFT. This provides more user control with minimum involvement, as validated on two public datasets.

1 Introduction

User interaction is necessary in several image segmentation tasks. In publicity, for example, the edition of photos and video very often requires user-guided segmentation. A challenge is to minimize user involvement and the time required for segmentation without compromising accuracy and precision. This usually requires solutions that also provide complete user control over the process, such that the user's actions do not destroy parts already accepted as correct, making segmentation more effective [1].

In view of that, we have divided segmentation into three main tasks: recognition, enhancement and extraction. Recognition is the only interactive task, which consists of drawing markers inside and outside the object. During enhancement, discriminative image properties on marker pixels (Figure 1a) are used to increase the dissimilarities between object and background (Figures 1b–c). Finally, during extraction, the spatial extent of the object in the image is defined. Furthermore, we have shown that recognition and enhancement should be separated from recognition and extraction for more effective segmentation [2]. This is certainly a contribution with respect to interactive methods that usually consider enhancement and extraction as a single delineation task [3, 4, 5]. Given that, user interaction usually affects the entire image during *enhancement* and user corrections should only affect the wrong segmentation parts during *extraction*, markers used for extraction should never be used for enhancement. However, the absence of guidance in the location of markers during extraction also makes it less effective. In fact, it was suggested in [2] that the markers should be selected around

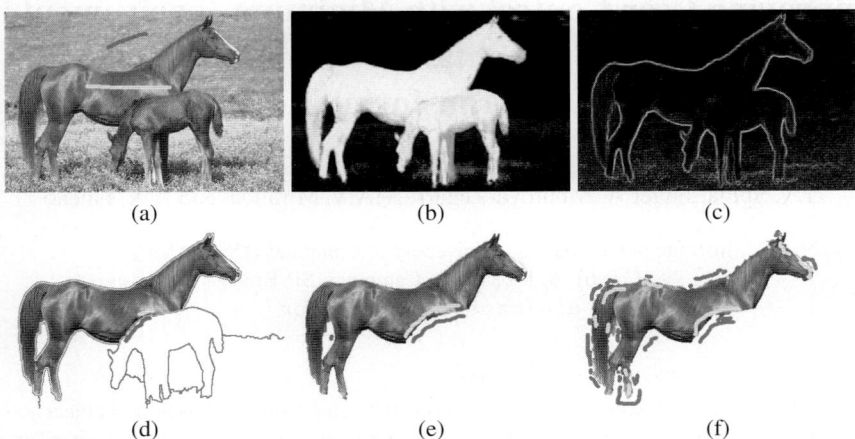

Fig. 1. (a) Markers are selected for enhancement. (b-c) Object membership map and the weight image resulting from enhancement. Segmentation results with (d) the proposed approach using two markers, where the color lines indicate the segmented regions, (e) the method in [2] using three markers, and (f) the watershed transform using several markers [7].

the weaker parts of the boundary, which could be visually identified from a weight image (Figure 1c). This is still too technical for a non-expert user and rather confusing. Furthermore, the weight image gives only a poor insight of how the extraction method will actually behave and it is difficult to predict from it what is the result of adding new markers.

To increase user control with minimum involvement, we pre-segment the image automatically into a few regions. The regions inside the object are selected and merged from internal markers, while regions with object and background pixels are divided from internal and external markers (Figure 1d). Region segmentation reduces the number of markers with respect to the previous approach [2] (Figure 1e) and the traditional approach [6, 7] (i.e. a watershed transform from markers), which does not use markers for enhancement (Figure 1f). Besides, the proposed solution incorporates both of these methods.

The *image foresting transform* (IFT) [8] is used to design image processing operators for object enhancement, region segmentation, and object extraction. In each case, a graph is derived from the image by taking some or all pixels as its nodes and defining some *adjacency relation* between them. A *connectivity function* assigns a value to any path in the graph, including *trivial* paths formed by a single node. Considering the minimum value among all possible paths with terminus at each node, the optimum path is trivial for some nodes, called *roots*, and the remaining nodes will have an optimum path coming from their most strongly connected root, partitioning the graph into an optimum-path forest (disjoint sets of optimum-path trees). The three operators differ in the parameters of the connectivity function, adjacency relation, and roots of the forest. The optimal connectivity values are used for enhancement while root labels are used for region segmentation and object extraction. Enhancement aims higher arc weights on the object's boundary than elsewhere. Under this condition, region segmentation by IFT can divide the object into a few regions, making extraction a trivial region merging task. When it fails and a region contains object and background pixels, they can be separated by internal and external marker competition. The connectivity function is such that the object pixels in this region are more strongly connected to the internal markers than to the external ones, completing the object extraction process.

Section 2 presents the concepts about the IFT, which will be used for enhancement (Section 3), region segmentation (Section 4), and object extraction (Section 5). In Section 6, experimental results on standard public datasets are given and are used to show the high accuracy of our framework. Conclusions are drawn in Section 7.

2 Image Foresting Transform (IFT)

A natural image $\hat{I}$ is a pair $(D_{\hat{I}}, \boldsymbol{I})$, where $D_{\hat{I}} \subset Z^2$ is the image domain and $\boldsymbol{I}(t)$ is the color vector $(I_1(t), I_2(t), I_3(t))$ in the *Lab* space. Multiscale feature extraction essentially transforms an image $\hat{I} = (D_{\hat{I}}, \boldsymbol{I})$ into the pair $\hat{F} = (D_{\hat{I}}, \boldsymbol{F})$ where $\boldsymbol{F}(t) = (F_1(t), F_2(t), \ldots, F_m(t))$ is a feature vector assigned to t. In this work, we use a cosine-based low-pass filter $\hat{L}$ applied to each of the Lab components in three different scales [9], followed by a leveling operation [10] to avoid border shifting. Considering the original Lab values and the additional filtered values, each pixel ends up with 12 features (i.e., $m = 12$).

A graph $(\mathcal{N}, \mathcal{A})$ may be defined by taking a set $\mathcal{N} \subseteq D_{\hat{I}}$ of pixels as nodes and an *adjacency relation* $\mathcal{A}$ between nodes of $\mathcal{N}$ to form the arcs. We use $t \in \mathcal{A}(s)$ or $(s, t) \in \mathcal{A}$ to indicate that a node $t \in \mathcal{N}$ is adjacent to a node $s \in \mathcal{N}$. A *path* $\pi_t = \langle t_1, t_2, \ldots, t \rangle$ is a sequence of adjacent nodes with terminus at a node t, being $\pi_t = \langle t \rangle$ a *trivial path*. A connectivity function f assigns to any path π_t a value $f(\pi_t)$. In all cases, we are interested in function $f_{\max}$:

$$f_{\max}(\langle t \rangle) = H(t) \qquad (1)$$

$$f_{\max}(\langle t_1, t_2 \ldots, t_n \rangle) = \max_{i=1,2,\ldots,n-1} \{H(t_1), w(t_i, t_{i+1})\}, \qquad (2)$$

where $H(t)$ is a handicap value, which is finite only to root candidates (i.e., *seed pixels*), and $w(t_i, t_{i+1}) \geq 0$ is an arc weight, both computed from $\hat{F}$ in different ways, depending on the operator.

Considering all possible paths with terminus at each node t, the optimum connectivity value map is $V(t) = \min_{\forall \pi_t \text{ in } (\mathcal{N}, \mathcal{A})} \{f(\pi_t)\}$. The IFT solves this minimization problem by computing an *optimum-path forest* — a function P which contains no cycles and assigns to each node $t \in \mathcal{N}$ either its predecessor node $P(t) \in \mathcal{N}$ in the optimum path with terminus t or a distinctive marker $P(t) = nil \notin \mathcal{N}$, when $\langle t \rangle$ is optimum (i.e., t is said *root* of the forest). The IFT algorithm is presented below for function $f_{\max}$. The root $R(t)$ of each pixel t can be obtained by following its optimum path backwards in P. However, it is more efficient to propagate them on-the-fly, creating a root map R.

Algorithm 1 – IFT ALGORITHM FOR $f_{\max}$

INPUT: Graph $(\mathcal{N}, \mathcal{A})$
OUTPUT: Optimum-path forest P, its connectivity value map V and its root map R.
AUXILIARY: Priority queue Q and variable tmp.

1. **For each** $t \in \mathcal{N}$, **do**
2. $\quad$ $P(t) \leftarrow nil$, $R(t) \leftarrow t$ and $V(t) \leftarrow H(t)$.
3. $\quad$ **If** $V(t) \neq +\infty$, **then** insert t in Q.
4. **While** $Q \neq \emptyset$, **do**
5. $\quad$ Remove s from Q such that $V(s)$ is minimum.
6. $\quad$ **For each** $t \in \mathcal{A}(s)$, such that $V(t) > V(s)$, **do**
7. $\quad\quad$ Compute $tmp \leftarrow \max\{V(s), w(s,t)\}$.
8. $\quad\quad$ **If** $tmp < V(t)$, **then**
9. $\quad\quad\quad$ **If** $V(t) \neq +\infty$, **then** remove t from Q.
10. $\quad\quad\quad$ Set $P(t) \leftarrow s$, $R(t) \leftarrow R(s)$, $V(t) \leftarrow tmp$.
11. $\quad\quad\quad$ Insert t in Q.

Lines 1–3 initialize maps for trivial paths. The minima of the initial map V compete with each other and some of them become roots of the forest. They are pixels with optimum trivial-path values, which are inserted in queue Q. The main loop computes an optimum path from the roots to every node s in a non-decreasing order of value (Lines 4–11). At each iteration, a path of minimum value $V(s)$ is obtained in P when we remove its last pixel s from Q (Line 5). Ties are broken in Q using first-in-first-out policy. The remaining lines evaluate if the path that reaches an adjacent pixel t through s is cheaper than the current path with terminus t and update Q, $V(t)$, $R(t)$ and $P(t)$ accordingly. The next sections show how to use this framework for object enhancement, region segmentation and object extraction.

3 Object Enhancement

Enhancement consists of feature extraction, fuzzy classification and arc-weight assignment, aiming higher weights to arcs on the object's boundary than elsewhere. Under this condition, the object can be extracted using $f_{\max}$ from only two marker pixels, one inside and one outside it. However, perfect arc-weight assignment is usually not possible, asking for more user involvement (marker selection).

Let $\mathcal{M}$ be a set of markers for enhancement, selected on parts where object and background have distinct properties. Note that, markers selected in Figures 1d–e for extraction should never be used for enhancement. We first randomly divide the labeled markers in $\mathcal{M} = \mathcal{T} \cup \mathcal{E}$ into a training set $\mathcal{T}$ and an evaluation set $\mathcal{E}$, with the same proportion of object and background markers. Set $\mathcal{T} = \mathcal{T}_b \cup \mathcal{T}_o$ is further divided into object markers in $\mathcal{T}_o$ and background markers in $\mathcal{T}_b$.

Now, consider a complete graph $(\mathcal{T}, \mathcal{A})$, where $t \in \mathcal{A}(s)$ for all $t \neq s$, with arc weights $w(s,t) = \|\mathbf{F}(t) - \mathbf{F}(s)\|$. The arcs (s,t), $s \in \mathcal{T}_o$ and $t \in \mathcal{T}_b$, or vice-versa, in a minimum-spanning tree of $(\mathcal{T}, \mathcal{A})$ define the closest nodes between object and background in the feature space. They represent key elements to protect each class, object and background, as seeds in an optimum-path forest classifier [11]. We use here a variant, by inserting s in a set $\mathcal{S}_o \subset \mathcal{T}_o$, t in a set $\mathcal{S}_b \subset \mathcal{T}_b$, and computing one optimum-path forest P_o for $f_{\max}$ on a complete graph $(\mathcal{T}_o, \mathcal{A})$ and one optimum-path forest P_b for $f_{\max}$ on a complete graph $(\mathcal{T}_b, \mathcal{A})$. In the first case, the handicap $H(t) = 0$, if $t \in \mathcal{S}_o$, or $H(t) = \infty$, otherwise. The second case is similar, changing $\mathcal{S}_o$ by $\mathcal{S}_b$.

A local processing operation can compute the optimum connectivity values $V_o(t)$ and $V_b(t)$ for any remaining pixel $t \in D_{\hat{f}} \backslash \mathcal{T}$ incrementally, as though t were part of the original graphs.

$$V_o(t) = \min\{\max\{V_o(s), w(s,t)\}\}, \; \forall s \in \mathcal{T}_o, \tag{3}$$
$$V_b(t) = \min\{\max\{V_b(s), w(s,t)\}\}, \; \forall s \in \mathcal{T}_b. \tag{4}$$

This allows fast propagation of the optimum connectivity values from $\mathcal{S}_o$ and $\mathcal{S}_b$ to the remaining image pixels.

An *object membership value* $M_o(t)$ (Figure 1b) can finally be assigned to each pixel $t \in D_{\hat{f}}$ as:

$$M_o(t) = \frac{V_b(t)}{V_o(t) + V_b(t)} \tag{5}$$

A binary classification of pixels $t \in \mathcal{E}$ is also possible by assigning to them the label $l(t) \in \{0,1\}$ of background or object ($l(t) = 1$ if $V_o(t) < V_b(t)$, and $l(t) = 0$ otherwise). If $\lambda(t) \in \{0,1\}$ is the correct label of $t \in \mathcal{E}$, then an error occurs when $\lambda(t) \neq l(t)$. Aiming to minimize the number of misclassifications, we can select better training nodes by replacing misclassified nodes in $\mathcal{E}$ with randomly selected nodes in $\mathcal{T} \backslash \mathcal{S}_o \cup \mathcal{S}_b$ [11].

Then, consider the image graph $(D_{\hat{f}}, \mathcal{A})$ where $t \in \mathcal{A}(s)$ if $t \neq s$ is 8-neighbor of s. Enhancement is represented by a weight image $\hat{W} = (D_{\hat{f}}, W)$ (Figure 1c).

$$W(s) = \gamma W_o(s) + (1-\gamma) W_f(s), \tag{6}$$

where $W_o(s)$ is an *object-based weight*, $W_f(s)$ is a *feature-based weight*, and $0 \leq \gamma \leq 1$ represents the importance of the object membership map in this estimation.

Given that $\boldsymbol{F}$ stores filtered maps F_b, $b = 1, 2, \ldots, 12$, for each Lab component in different scales, we estimate $W_f(s)$ as the maximum gradient magnitude among the feature-based gradients

$$\boldsymbol{W}_f(s) = \max_{b=1,2,\ldots,12} \{\| \sum_{\forall t \in \mathcal{A}(s)} (F_b(t) - F_b(s)) \boldsymbol{st} \|\}, \tag{7}$$

where $\boldsymbol{st}$ is the unit vector from s to t. The weight $W_o(s) = \|\boldsymbol{G}_o(s)\|$ is estimated in a similar way as the magnitude of an object-based gradient

$$\boldsymbol{G}_o(s) = \sum_{\forall t \in \mathcal{A}(s)} (M_o(t) - M_o(s)) \boldsymbol{st}. \tag{8}$$

4 Region Segmentation

A classical watershed transform can oversegment the image into all catchment basins of $W(t)$. We can simplify segmentation by closing basins of $W(t)$ with volume below a threshold [12] (i.e., a morphological volume closing). This requires the following parameter setting to obtain a few regions inside the object.

For the image graph $(D_{\hat{I}}, \mathcal{A})$, where $t \in \mathcal{A}(s)$ if $t \neq s$ is 8-neighbor of s, we define $w(s,t) = W(t)$ in Eq. 1 and $H(t)$ by

$$H(t) = \begin{cases} W(t) & \text{if } t \in \mathcal{R}, \\ VC(t) + 1 & \text{otherwise}, \end{cases} \quad (9)$$

where $\mathcal{R}$ is a root of the forest (minimum of $VC(t) + 1$) and $VC(t)$ is the resulting function from morphological volume closing of $W(t)$. Given that $VC(t) \geq W(t)$, we use $VC(t) + 1$ to guarantee that the roots in $\mathcal{R}$ will conquer all pixels in their basins. However, this requires a variant in Algorithm 1. Initially, all pixels start with $H(t) = VC(t) + 1$. When a pixel s is removed from Q in Line 5 and $P(s) = nil$, this implies that $s \in \mathcal{R}$ is a minimum of $VC(t) + 1$. By reducing $H(s)$ to $W(s)$ for all minima s, we partition the image into an optimum-path forest with roots in $\mathcal{R}$. The borders of the optimum-path trees rooted at each minimum are shown in Figure 1d to guide object extraction, by indicating which regions need to be selected.

5 Object Extraction

Object extraction is done by region merging inside the object and by internal and external marker competition inside regions with object and background pixels. This competition requires the following setting.

For the image graph $(D_{\hat{I}}, \mathcal{A})$, where $t \in \mathcal{A}(s)$ if $t \neq s$ is 8-neighbor of s, we now define $w(s,t) = \frac{W(s)+W(t)}{2}$ in Eq. 1 and $H(t)$ by

$$H(t) = \begin{cases} 0 & \text{if } t \in \mathcal{M}, \\ \infty & \text{otherwise}, \end{cases} \quad (10)$$

where $\mathcal{M}$ is the set of internal and external markers selected for extraction. The object consists of pixels 1 after a local operation, which assigns the correct label $\lambda(R(t)) \in \{0,1\}$ of the root to each pixel $t \in D_{\hat{I}}$. Note that the method includes the one in [2], when the entire image is one region, and the watershed approach from markers, when we do not select markers for enhancement ($\gamma = 0$ in Eq. 6).

6 Experiments and Results

We have evaluated three methods using the proposed framework: (M1) The watershed approach [7] ($\gamma = 0$ in Eq. 6), (M2) the method in [2] ($\gamma \neq 0$ in Eq. 6), and (M3) the proposed method with $\gamma \neq 0$ and region merging. Figure 2 shows examples of these approaches, which indicate that M3 usually requires less markers for segmentation.

A dataset with 50 natural images with ground-truths was obtained from [3] for the experiments. Furthermore, *accuracy* was estimated as the average of the F-measures [13] computed between the segmentation results and the ground truths of the 50 images. The standard deviation of these measures indicates *precision*. Two individuals used

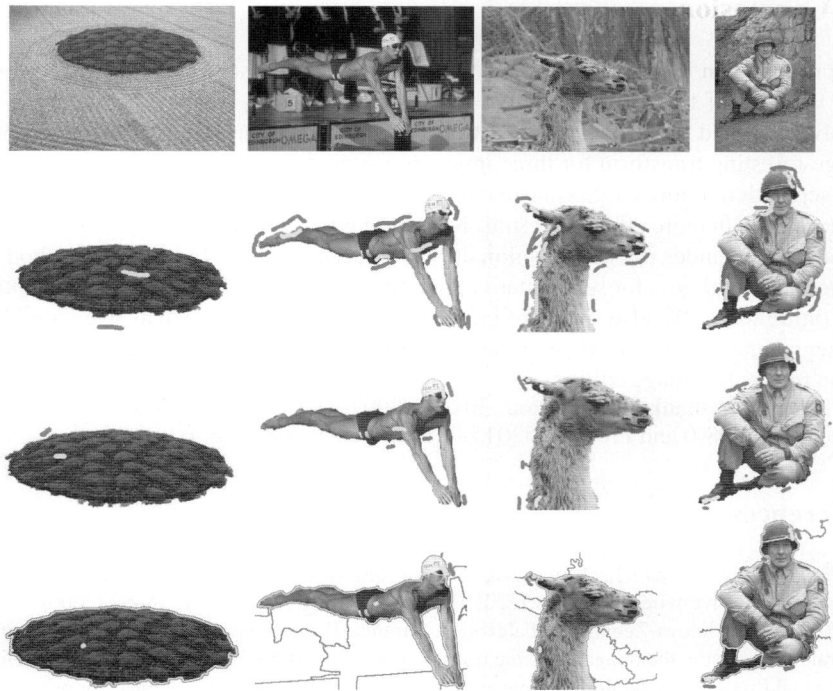

Fig. 2. The first row shows the original images and the subsequent ones present the segmentation results for methods M1, M2 and M3

Table 1. Average results of F-measure, number of internal and external markers with their respective standard deviations over the 50 images segmented by two users using M1, M2 and M3

	F-Measure			Markers		
	M1	M2	M3	M1	M2	M3
User 1	98.22 ± 0.88	98.48 ± 0.78	98.41 ± 0.82	10.78 ± 5.59	7.14 ± 5.26	5.32 ± 3.61
User 2	97.90 ± 1.00	98.04 ± 1.15	98.02 ± 0.95	9.52 ± 5.46	7.56 ± 3.90	6.80 ± 3.79

each method to segment each of the 50 images and the average and standard deviation of F-measure and number of internal and external markers (including markers for enhancement) were computed. All methods can obtain similar and high accuracy and precision, but user involvement (number of markers) is 40% less in M3 and 27.5% less in M2 with respect to M1. M3 also offers more user control than M2 and M1. In addition, our framework was also evaluated on 100 images from benchmark[1] which reports average results of F-measure 87.00 ± 0.01, while M3 achieved F-measure 96.04 ± 3.15 using 5.69 ± 3.72 markers.

---

[1] http://www.wisdom.weizmann.ac.il/~vision/GoodSegment.html

7 Conclusion

We presented an IFT-based approach for interactive segmentation which includes simpler versions of the method [2]. We have improved those versions by reducing user involvement and increasing user control over segmentation. The method exploits the image foresting transform for three image processing operations during segmentation. The separation of recognition and enhancement from recognition and extraction is a key aspect to obtain more effective results. More user control is provided by region segmentation, which guides marker selection during object extraction. Indeed, the method also reduces the total time for segmentation, but this aspect needs to be somehow quantified in a future work. We also intend to incorporate other methods, such as live wire [1], in the segmentation framework, and fuzzy classify border pixels to adequate the segmentation results for image edition.

The authors thank CNPq (Proc. 302617/2007-8), CNPq/PIBIC-PRP and FAPESP (Proc. 05/59808-0 and Proc. 07/52015-0) for the financial support.

References

1. Falcão, A., Udupa, J., Miyazawa, F.: An ultra-fast user-steered image segmentation paradigm: Live-wire-on-the-fly. IEEE Trans. on Medical Imaging 19, 55–62 (2000)
2. Spina, T., Montoya-Zegarra, J., Falcão, A., Miranda, P.: Fast interactive segmentation of natural images using the image foresting transform. In: DSP (International Conference on Digital Signal Processing), Santorini, Greece. IEEE, Los Alamitos (2009) (accepted for publication)
3. Rother, C., Kolmogorov, V., Blake, A.: "grabcut": interactive foreground extraction using iterated graph cuts. ACM Transactions on Graphics 23, 309–314 (2004)
4. Protiere, A., Sapiro, G.: Interactive image segmentation via adaptive weighted distances. IEEE Transactions on Image Processing 16, 1046–1057 (2007)
5. Vicente, S., Kolmogorov, V., Rother, C.: Graph cut based image segmentation with connectivity priors. In: IEEE Proc. of CVPR, Anchorage, Alaska, pp. 1–8 (2008)
6. Beucher, S., Meyer, F.: The morphological approach to segmentation: The watershed transformation. In: Mathematical Morphology in Image Processing, pp. 433–481. M. Dekker, New York (1993)
7. Lotufo, R., Falcão, A.: The ordered queue and the optimality of the watershed approaches. In: Mathematical Morphology and its Applications to Image and Signal Processing, vol. 18, pp. 341–350. Kluwer, Dordrecht (2000)
8. Falcão, A., Stolfi, J., Lotufo, R.: The image foresting transform: Theory, algorithms, and applications. IEEE Trans. on Pattern Analysis and Machine Intelligence 26, 19–29 (2004)
9. Portilla, J., Simoncelli, E.P.: A parametric texture model based on joint statistics of complex wavelet coefficients. Intl. Journal of Computer Vision 40, 49–70 (2000)
10. Meyer, F.: Levelings, image simplification filters for segmentation. Journal of Mathematical Imaging and Vision 20, 59–72 (2004)
11. Papa, J.P., Falcão, A.X., Suzuki, C.T.N., Mascarenhas, N.D.A.: A discrete approach for supervised pattern recognition. In: Brimkov, V.E., Barneva, R.P., Hauptman, H.A. (eds.) IWCIA 2008. LNCS, vol. 4958, pp. 136–147. Springer, Heidelberg (2008)
12. Salembier, P., Oliveras, A., Guarrido, L.: Antiextensive connected operators for image and sequence processing. IEEE Transactions on Image Processing 7, 555–570 (1998)
13. van Rijsbergen, C.: Information retrieval, 2nd edn. Wiley Interscience, London (1979)

3D Image Segmentation Using the Bounded Irregular Pyramid

Fuensanta Torres, Rebeca Marfil, and Antonio Bandera

Grupo ISIS, Dpto. Tecnología Electrónica,
Universidad de Málaga, Campus de Teatinos s/n 29071-Málaga, Spain
fuensantatorres@hotmail.com, rebeca@uma.es, ajbandera@uma.es
http://www.grupoisis.uma.es

Abstract. This paper presents a novel pyramid approach for fast segmentation of 3D images. A pyramid is a hierarchy of successively reduced graphs whose efficiency is strongly influenced by the data structure that codes the information within the pyramid and the decimation process used to build a graph from the graph below. Depending on these two features, pyramids have been classified as regular and irregular ones. The proposed approach extends the idea of the Bounded Irregular Pyramid (BIP) [5] to 3D images. Thus, the 3D-BIP is a mixture of both types of pyramids whose goal is to combine their advantages: the low computational cost of regular pyramids with the consistent and useful results provided by the irregular ones. Specifically, its data structure combines a regular decimation process with an union-find strategy to build the successive 3D levels of the structure. Experimental results show that this approach is able to provide a low–level segmentation of 3D images at a low computational cost.

1 Introduction

In many 3D image processing tasks, segmentation is an important step which can be defined as the process of decomposing a 3D image into regions which are homogeneous according to some criteria [5]. In these tasks, it constitutes a critical step towards content analysis and image understanding. Hence, 3D image segmentation plays an important role in different research fields such as reverse engineering or robotic vision. However, it must be noted that the most number of 3D image processing tasks are related to medical images: computerized tomographies (CT) or magnetic resonance images (MRI). Segmentation of medical images is still a difficult task because voxel intensities are not necessarily constant for each tissue class and the histograms of each tissue show great overlaps. Besides, noise artifacts in real data generate a fragmented segmented structure. Therefore, fully unsupervised segmentation approaches are far from satisfying in many real situations and it is usual that segmentation approaches follow a semi-supervised strategy. In these methods, input data are preprocessed as far as possible, and user interaction is allowed to control the final segmentation stage. A fast and effective unsupervised segmentation algorithm is then required to be employed as an initial preprocessing stage. For instance, in the MRI Brain Segmentation task, this initial stage is used to ease the separation of the brain from the scalp, the bone and other class of non-interesting tissue items.

This paper presents a novel, fast and effective volume segmentation approach. It extends the scheme of a two dimensional (2D) image segmentation approach, the Bounded Irregular Pyramid (BIP) [5,6], towards the processing of 3D images. Hence, our proposal, in the same manner as the original BIP, uses a hybrid pyramid structure to reduce the computational load associated to the segmentation process. The pyramid represents the input 3D image at different resolution levels. In this hierarchy, the bottom level contains the 3D image to be processed and each pyramid level is recursively obtained by processing its underlying level. The main advantage of the pyramid structure is that it can be used to reduce the time required to analyze an image. The more relevant features of the proposed 3D-BIP can be summarized as follows:

- It is a fast algorithm, which is able to segment a 3D image of 128 x 128 x 53 voxels in less than 16 seconds (in a Intel Core2 Duo CPU P8400 2.26GHz, 2268 Mhz PC). Although in medical applications the processing speed is less important than the final accuracy, it is interesting that preprocessing algorithms will be fast [7].
- The 3D Bounded Irregular Pyramid is an unsupervised segmentation method. It does not depend on seed points as in the Region Growing algorithms or on an initial spline curve as in Snake models.
- It is a general 3D segmentation algorithm. That is, it does not need training as in Atlas guided methods or as in Artificial Neural Network-based methods.

The rest of the paper is organized as follows: Section 2 describes the data structure and decimation process of the 3D-BIP. The experimental results revealing the efficacy of the method are described in Section 3. Finally, the paper concludes along with discussions and future work in Section 4.

2 Data Structure and Proposed Decimation Scheme

2.1 Definitions

The structure of a pyramid can be described as a graph hierarchy in which each level l is at least defined by a set of nodes N_l connected by a set of arcs E_l. These arcs define the horizontal relationships of the pyramid and they represent the neighborhood of each node at the same level (*intra-level arcs*). Another set of arcs define the vertical relationships by connecting nodes between adjacent pyramid levels (*inter-level arcs*). These inter-level arcs establish a dependence relationship between each node of level $l+1$ and a set of nodes at level l (*reduction window*). The nodes belonging to one reduction window are the *children* of the node which defines it. The value of each *parent* is computed from the one of its children using a *reduction function*. The ratio between the number of nodes at level l and the number of nodes at level $l+1$ is the *reduction factor*. Fig. 1 illustrates some of these terms. Using this framework, the general process to build $G_{l+1} = (N_{l+1}, E_{l+1})$ from level $G_l = (N_l, E_l)$ is the following:

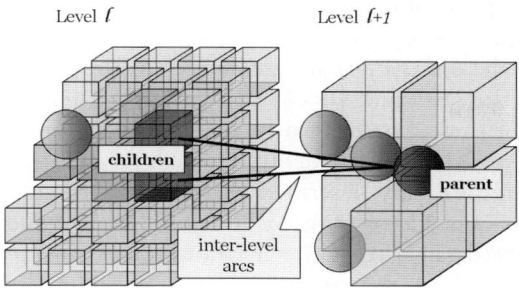

Fig. 1. Levels l and $l+1$ of a 3D-BIP hierarchy: spheres are related to irregular nodes and cubes to regular ones (see text)

1. Selection of the nodes of G_{l+1} among N_l: This selection step is a decimation procedure.
2. Definition of inter-level arcs: Each node of G_l is linked to its parent node in G_{l+1}. This step defines a partition of N_l.
3. Definition of intra-level arcs: The set of arcs E_{l+1} is obtained by defining the neighborhood relationships between the nodes N_{l+1}.

The parent-child relationship defined by the reduction window may be extended by transitivity down to the base level. The set of nodes in the base level linked to a pyramid node is named its *receptive field*. The receptive field defines the embedding of this node on the original image [2]. In the case of 3D images, the nodes of the bottom pyramid level (level 0) can be anything from an original 3D image voxel via some general numeric property to symbolic information, e.g. a node can represent an image voxel grey level or an 3D image edge. Corresponding to the generalization of the node contents, the intra-level and inter-level relations of the nodes are also generalized [4].

2.2 Data Structure of the Bounded Irregular Pyramid

The data structure of the 3D Bounded Irregular Pyramid (3D-BIP) is a combination of regular and irregular data structures: a 2x2x2/8 3D regular structure and a simple graph. A simple graph is a non-weighted and undirected graph containing no self-loops. It encodes the adjacency between two nodes by only one arc, although their receptive fields may share several boundary segments. The mixture of both regular and irregular structures generates an irregular configuration which is described as a graph hierarchy. In this hierarchy, there are two types of nodes: nodes belonging to the 2x2x2/8 structure, named *regular nodes* and *irregular nodes* or nodes belonging to the irregular structure.

Although regular pyramids can be explained as a graph hierarchy, it is more usual to represent them as a hierarchy of arrays due to their rigid structure. In the regular part of the 3D-BIP, each regular node is described by $\mathbf{x} = (i, j, k, l)$ where l represents the level and (i, j, k) are the x-, y- and z-coordinate within the level. In each of these arrays, two regular nodes are neighbors if they are placed in adjacent positions of the array in an 26-neighborhood. On the other hand, irregular nodes are only described by their level l and an index value i. In general, a node $\mathbf{x}$ is neighbor of other node $\mathbf{x}'$ if their reduction windows $w_\mathbf{x}$ and $w_{\mathbf{x}'}$ are connected. Two reduction windows are connected if there are at least two nodes at level l-1, $\mathbf{y} \in w_\mathbf{x}$ and $\mathbf{y}' \in w_{\mathbf{x}'}$, which are neighbors. Two nodes $\mathbf{x}_1$ and $\mathbf{x}_2$ which are neighbors at level l are connected by an intra-level arc $e = (\mathbf{x}_1, \mathbf{x}_2) \in E_l$.

2.3 Proposed Decimation Process

The proposed decimation algorithm runs two consecutive steps to obtain the set of nodes N_{l+1}. The first process generates the set of regular nodes of G_{l+1} from the regular nodes at G_l, meanwhile the second one determines the set of irregular nodes at level $l+1$. In this proposal, this second process conducts an union-find decimation algorithm which is simultaneously conducted over the set of regular and irregular nodes of G_l which do not present a parent in the upper level $l + 1$.

Let $G_l = (N_l, E_l)$ be a graph where N_l stands for the set of regular and irregular nodes and E_l for the set of intra-level arcs. Let $\varepsilon_l^{\mathbf{xy}}$ be equal to 1 if $(\mathbf{x}, \mathbf{y}) \in E_l$ and equal to 0 otherwise. Let $\xi_\mathbf{x}$ be the neighborhood of the node $\mathbf{x}$ defined as $\{\mathbf{y} \in N_l : \varepsilon_l^{\mathbf{xy}}\}$. It can be noted that a given node $\mathbf{x}$ is not a member of its neighborhood, which can be composed by regular and irregular nodes. Each node $\mathbf{x}$ has associated a $v_\mathbf{x}$ value. Besides, each regular node has associated a boolean value $h_\mathbf{x}$: the homogeneity [6]. At the base level of the hierarchy, G_0, all nodes are regular, and they have $h_\mathbf{x}$ equal to 1. Only regular nodes which have $h_\mathbf{x}$ equal to 1 are considered to be part of the regular structure. Regular nodes with an homogeneity value equal to 0 are not considered for further processing.

The proposed decimation process transforms the graph G_l in G_{l+1} such that the reduction factor is greater to 1. In our case, we focus on dividing the image into a set of homogeneous blobs. This aim is achieved using the pairwise comparison of neighboring nodes [3]. Then, a pairwise comparison function, $g(v_{\mathbf{x}_1}, v_{\mathbf{x}_2})$ is defined. This function is true if the $v_{\mathbf{x}_1}$ and $v_{\mathbf{x}_2}$ values associated to the $\mathbf{x}_1$ and $\mathbf{x}_2$ nodes are similar according to some criteria and false otherwise. The decimation process consists of the following steps:

1. Regular decimation process. The $h_\mathbf{x}$ value of a regular node $\mathbf{x}$ at level $l+1$ is set to 1 if the eight regular nodes immediately underneath $\{\mathbf{y}_i\}$ are similar according to some criteria and their $h_{\{\mathbf{y}_i\}}$ values are equal to 1. That is, $h_\mathbf{x}$ is set to 1 if

$$\{\bigcap_{\forall \mathbf{y}_j, \mathbf{y}_k \in \{\mathbf{y}_i\}} g(v_{\mathbf{y}_j}, v_{\mathbf{y}_k})\} \cap \{\bigcap_{\mathbf{y}_j \in \{\mathbf{y}_i\}} h_{\mathbf{y}_j}\} \qquad (1)$$

Besides, at this step, inter-level arcs among regular nodes at levels l and $l+1$ are established. If $\mathbf{x}$ is an homogeneous regular node at level $l+1$ ($h_\mathbf{x}==1$), then the set of four nodes immediately underneath $\{\mathbf{y_i}\}$ are linked to $\mathbf{x}$.

2. Irregular decimation process. Each irregular or regular node $\mathbf{x} \in N_l$ without parent at level $l+1$ chooses the closest neighbor $\mathbf{y}$ according to the $v_\mathbf{x}$ value. Besides, this node $\mathbf{y}$ must be similar to $\mathbf{x}$. That is, the node $\mathbf{y}$ must satisfy

$$\{||v_\mathbf{x} - v_\mathbf{y}|| = \min(||v_\mathbf{x} - v_\mathbf{z}|| : \mathbf{z} \in \xi_\mathbf{x})\} \cap \{g(v_\mathbf{x}, v_\mathbf{y})\} \tag{2}$$

If this condition is not satisfy by any node, then a new node $\mathbf{x}'$ is generated at level $l+1$. This node will be the parent node of $\mathbf{x}$. Besides, it will constitute a root node and its receptive field at base level will be an homogeneous set of pixels according to the specific criteria. On the other hand, if $\mathbf{y}$ exists and it has a parent $\mathbf{z}$ at level $l+1$, then $\mathbf{x}$ is also linked to $\mathbf{z}$. If $\mathbf{y}$ exists but it does not have a parent at level $l+1$, a new irregular node $\mathbf{z}'$ is generated at level $l+1$. In this case, the nodes $\mathbf{x}$ and $\mathbf{y}$ are linked to $\mathbf{z}'$.

This process is sequentially performed and, when it finishes, each node of G_l is linked to its parent node in G_{l+1}. That is, a partition of N_l is defined. It must be noted that this process constitutes an implementation of the union-find strategy. The union-find uses tree structures to represent sets [1]. A find operation looks for the parent of a node at level l. If two nodes at level l are similar, then a union operation will be performed by setting one of the two nodes to be the parent of both ones at level $l+1$.

3. Definition of intra-level arcs. The set of edges E_{l+1} is obtained by defining the neighborhood relationships between the nodes N_{l+1}. As it was described in Section 2.2, two nodes at level $l+1$ are neighbors if their reduction windows are connected at level l.

Fig. 2 shows an example of the described decimation process. Regular nodes are drawn as cubes meanwhile irregular nodes are drawn as spheres. The $v_\mathbf{x}$ values are represented by the color of the cells. Fig. 2a shows the regular part of the data structure after being built. The base level of the structure is composed by the 4x4x2 image voxels. The 8-to-1 regular decimation procedure generates a 2x2x1 level. Note that regular nodes with $h_\mathbf{x}$ equal to 0 are not depicted on the figure. Hence, the 2x2x1 level is reduced to a real 2x1x1 level. Only two regular nodes have been generated at level 1. Fig. 2a also shows the generation of a first irregular node at level 1 from the union of $\mathbf{x}_1^{(0)}$ and $\mathbf{x}_2^{(0)}$. Fig. 2b illustrates that the union-find process joins, to the same irregular parent node, those nodes linked to $\mathbf{x}_1^{(0)}$ and $\mathbf{x}_2^{(0)}$ which present the same color than them (7 nodes from level 0 are finally joined to $\mathbf{x}_2^{(1)}$). Two new irregular nodes are also generated. One of them is the parent node of a image region of three voxels at level 0. The other one presents the same color than the regular nodes generated at level 1. Inter-level arcs show that some voxels are linked to the regular nodes at level 1, but other voxels are connected to this new irregular node. The regular nodes at level 1 and this irregular node will be merged if level 2 is generated.

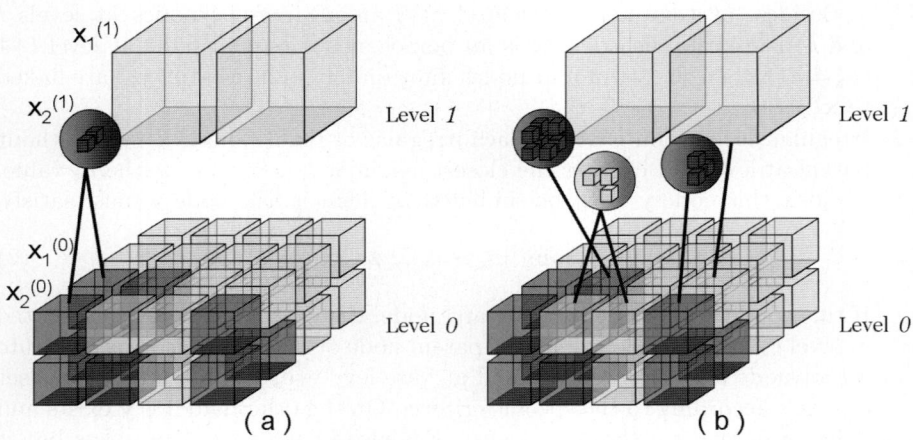

Fig. 2. Hierarchy generation: a) Regular nodes generated at level 1 and generation of a first irregular node; and b) nodes at level 1 when the decimation process between levels 0 and 1 has finished (see text for details)

3 Experimental Results

Fig. 3 shows the 3D image region obtained from the segmentation of the cranial CT image[1] at Fig. 3a. A cranial computed tomography (CT) scan is an imaging method that uses x-rays to create cross-sectional pictures of the head, including the cranium, brain, eye sockets, and sinuses. In this case, the size of the original CT image is 128 x 128 x 53 (868352 voxels), and it has been segmented in less than 16 seconds using a Intel Core2 Duo CPU P8400 2.26GHz, 2268 Mhz PC. Fig. 3b shows the segmentation region associated to the cranium, from different points of view. The $v_\mathbf{x}$ value of each node $\mathbf{x}$ at level 0 was the corresponding 3D image intensity value, and the pairwise comparison function, $g(v_{\mathbf{x}_1}, v_{\mathbf{x}_2})$ is defined as a simple thresholding of the difference of $v_\mathbf{x}$ values. These choices will allow us to qualitatively evaluate the 3D-BIP features. However, it must be noted that to apply this approach to segment medical images, more complex $v_\mathbf{x}$ values and pairwise comparison functions must be studied. As it is illustrated in Fig. 4a, the employed decimation process provides high reduction factors at low levels. This allows to reduce the storage requirements which could be associated to a 3D image representation using an irregular pyramid. The total number of nodes at each hierarchy level is represented at logarithmic scale in Fig. 4b. This figure also shows the presence of regular nodes at medium hierarchy levels. They are associated to large uniform image regions. On the contrary, irregular nodes are located at the region boundaries. With respect to the total number of nodes, it must be noted that the percentage of irregular nodes increases for higher

---

[1] http://www.vis.uni-stuttgart.de/ẽngel/pre-integrated/

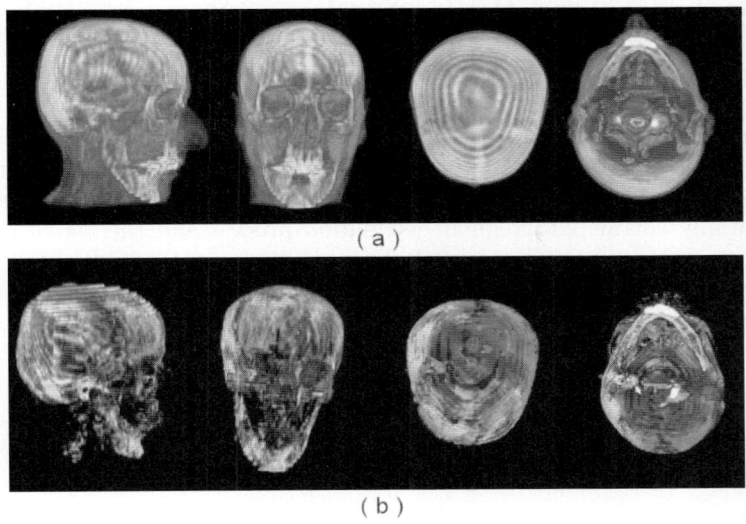

Fig. 3. 3D Segmentation example: a) Original CT image from different views; and b) segmentation region associated to the cranium (see text for details)

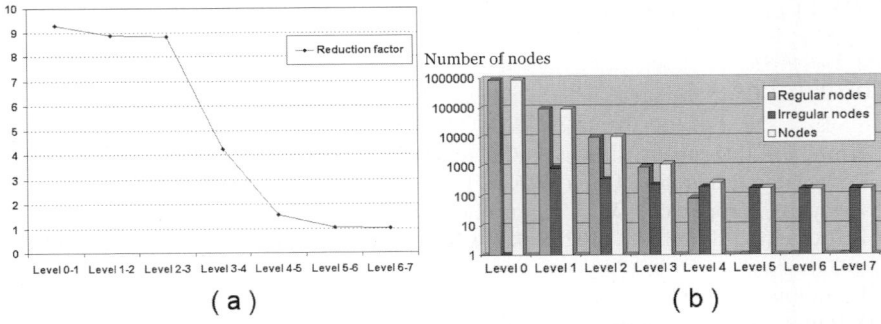

Fig. 4. Segmentation statistics associated to the segmentation example shown in Fig. 3: a) Reduction factors; and b) number of regular and irregular nodes with respect to the total number of nodes per level (see text for details)

pyramid levels. Finally, it must be also appreciated that the two last higher levels are only composed by irregular nodes. The union-find algorithm allows that irregular nodes have large reduction windows. In fact, the reduction factor is typically higher than the one imposed by the regular decimation (8-to-1) at low hierarchy levels (see Fig. 4a).

4 Conclusions and Future Work

This paper has proposed a fast approach for 3D image segmentation which extends the hybrid decimation process employed inside the BIP framework [5]. This strategy is specially suitable to deal with 3D images, as it is able to provide high reduction factors between pyramid levels if the original image present large uniform regions. The main drawbacks of the proposed approach are the use of a sequentially conducted irregular decimation process and the use of a simple graph to encode each pyramid level. Thus, in the employed data structure, a graph arc may encode a non-connected set of boundaries between the associated receptive fields. Moreover, the lack of self-loops in simple graphs does not allow to differentiate an adjacency relationship between two receptive fields from an inclusion relationship. Future work will be focused on employing a dual graph to encode each pyramid level and on studying the possibility to manage this hierarchy using a mixture of regular and irregular decimation processes which could run in a parallel way.

Acknowledgments

This work has been partially granted by the Spanish Junta de Andalucía under project P07-TIC-03106 and by the Spanish Ministerio de Ciencia y Tecnología (MCYT) and FEDER funds under project no. TIN2008-06196.

References

1. Brun, L.: Traitement d'images couleur et pyramides combinatoires. Habilitation à diriger des recherches, Université de Reims (2002)
2. Brun, L., Kropatsch, W.: Construction of Combinatorial Pyramids. In: Hancock, E.R., Vento, M. (eds.) GbRPR 2003. LNCS, vol. 2726, pp. 1–12. Springer, Heidelberg (2003)
3. Haxhimusa, Y., Glantz, R., Kropatsch, W.: Constructing stochastic pyramids by MIDES - maximal independent directed edge set. In: Hancock, E.R., Vento, M. (eds.) GbRPR 2003. LNCS, vol. 2726, pp. 35–46. Springer, Heidelberg (2003)
4. Kropatsch, W.G., Bischof, H., Englert, R.: Hierarchies. In: Kropatsch, W., Bischof, H. (eds.) Digital image analysis: selected techniques and applications, pp. 211–230. Springer, Heidelberg (2000)
5. Marfil, R., Molina-Tanco, L., Bandera, A., Rodríguez, J.A., Sandoval, F.: Pyramid segmentation algorithms revisited. Pattern Recognition 39(3), 1430–1451 (2006)
6. Marfil, R., Molina-Tanco, L., Bandera, A., Sandoval, F.: The construction of Bounded Irregular Pyramids with a union-find decimation process. In: Escolano, F., Vento, M. (eds.) GbRPR. LNCS, vol. 4538, pp. 307–318. Springer, Heidelberg (2007)
7. Nakib, A., Roman, S., Oulhadj, H., Siarry, P.: Fast brain MRI segmentation based on two-dimensional survival exponential entropy and particle swarm optimization. In: Proc. of the 29th Annual Int. Conf. IEEE EMBS, pp. 5563–5566 (2007)

The Gabor-Based Tensor Level Set Method for Multiregional Image Segmentation

Bin Wang[1], Xinbo Gao[1], Dacheng Tao[2], Xuelong Li[3], and Jie Li[1]

[1] School of Electronic Engineering, Xidian University, Xi'an, 710071, P.R. China
[2] School of Computer Engineering, Nanyang Technological University, 639798, Singapore
[3] State Key Laboratory of Transient Optics and Technology, Xi'an Institute of Optics and Precision Mechanics, Chinese Academy of Sciences, Xi'an 710119, Shaanxi, P.R. China

Abstract. This paper represents a new level set method for multiregional image segmentation. It employs the Gabor filter bank to extract local geometrical features and builds the pixel tensor representation whose dimensionality is reduced by using the offline tensor analysis. Then multiphase level set functions are evolved in the tensor field to detect the boundaries of the corresponding image. The proposed method has three main advantages as follows. Firstly, employing the Gabor filter bank, the model is more robust against the salt-and-pepper noise. Secondly, the pixel tensor representation comprehensively depicts the information of pixels, which results in a better performance on the non-homogenous image segmentation. Thirdly, the model provides a uniform equation for multiphase level set functions to make it more practical. We apply the proposed method to synthetic and medical images respectively, and the results indicate that the proposed method is superior to the typical region-based level set method.

Keywords: Gabor filter bank, tensor subspace analysis, image segmentation, geometric active contour, level set method.

1 Introduction

Geometric active contour model, implemented by the level set methods, becomes increasing popular in the field of image segmentation, and many methods have been developed [14]. These methods can be categorized into two groups, *i.e.*, edge-based ones [1][6][9][11][12] and region-based ones [3][4][18]. The former has to design an edge indicator to locate the edges in the image, while these edges are not always keeping closed and do not always correspond to the boundaries of objects. The latter seems to be a better choice and indeed it attracts more and more research interests in recent years.

Chan-Vese level set method [3] is a representative region-based level set method, and it gives the image a piecewise constant representation by introducing Mumford-Shah functional [8] into level set framework. This method constructs an energy functional by adding a regularization item and the fitting error between the piecewise constant representation and the image. By minimizing the energy functional, the object is separated from the image. The method is well extended in two ways generally as follows. Chan et al. [4] extended it to segment multi-channel images, and Wang and Vemuri [19] extended it to segment tensor diffusion MRI images. These extensions just take into account the pixel density depicted by a scalar, and that is not adequate to represent all image information. On the other side, Vese and Chan extended it to multiphase level set (MCV) method [18] which did not provide a practical formula to deal with the case more than two level set functions being used. Zhao et al. [20] extended it to multiple level set functions by adding a constraint into the energy functional to ensure one pixel just belongs to one level set function. Lie et al. [7] shared the same idea with [18] except for using binary level set functions which has to be regularized by using another constraint. These two extensions utilized an unnatural way to describe the partitioned regions, which results they had to append some regularization to energy functional, and that is not computational.

To overcome the above problems, we employ Gabor filter bank to extract the local geometrical features, e.g., gradient and orientation, and then build a high-order pixel tensor representation. This representation offers a more comprehensive description to pixels, by which the image corresponds to a tensor field. By applying multiphase level set functions to segment this tensor field, the original image can be fully segmented. The proposed method has four main advantages as follows. Firstly, the incorporation of the image smoothed by Gaussian filter makes the model is more robust against noise, especially the salt-and-pepper noise. Secondly, the pixel tensor representation is more accurate and comprehensive than the pixel density, which results in better segmentation performance on the image with the inhomogeneous background. Thirdly, the model extends MCV to the multiphase tensor level set method which is capable of evolving in tensor field, and does not need to add a constraint like [7][20]. Finally, the model utilizes the offline tensor analysis (OTA) [13], a general principle component analysis for high-order tensor, to reduce the dimensionality of the pixel tensor representation, and speeds up the process of execution.

2 MCV Level Set Method

The MCV method [18] makes use of two level set functions to evolve in and finally segment the image. Since two zero level curves partition the image domain into four sub-regions and one zero level curve only does two sub-regions, the MCV method gives a more accurate piecewise constant representation to the image than Chan-Vese method [3]. The following gives a brief description about this method.

Let $u_0 : \Omega \to R$ be a given image, and $\Omega \subset R^2$ be the image domain. $M = 2$ is the number of level set functions, and $N = 2^M$ is the number of the sub-regions partitioned by the zero level curves. The energy functional is defined as

$$E(c_1, c_2, c_3, c_4, \phi_1, \phi_2)$$
$$= \mu \sum_{i=1}^{2} \int_{\Omega} |\Delta \phi_i| dxdy + \int_{\Omega} |u_0(x,y) - c_1|^2 (1 - H(\phi_1))(1 - H(\phi_2)) dxdy$$
$$+ \int_{\Omega} |u_0(x,y) - c_2|^2 (1 - H(\phi_1)) H(\phi_2) dxdy + \int_{\Omega} |u_0(x,y) - c_3|^2 H(\phi_1)(1 - H(\phi_2)) dxdy \quad (1)$$
$$+ \int_{\Omega} |u_0(x,y) - c_4|^2 H(\phi_1) H(\phi_2) dxdy$$

where c_1, c_2, c_3, c_4 are the mean values of the four sub-regions respectively, ϕ_1 and ϕ_2 are the two level set functions, and $H(\cdot)$ is the Heaviside step function. This model just considers the pixel densities and that is not adequate for an accurate segmentation. For this purpose, in the next section we introduce Gabor filter bank to build a pixel tensor representation which is more comprehensive than gray value.

3 Multiphase Gabor-Based Tensor Level Set Method

This section firstly presents the construction of tensor representation for pixels, and then describes the evolution of the multiple level set functions for this tensor representation, including the energy functional and evolution equation.

3.1 Construction of the Tensor Representation

Marcelja [10] and Daugman [5] developed Gabor functions to model the response of the visual cortex, and it is usually used to give images a Gabor-based description [15][16][17]. Here, it is used to construct the pixel tensor representation. The tensor representation and tensor field, as shown in Fig.1, are constructed by following steps.

Step 1: The image smoothed by Gaussian filter is involved into the tensor representation as a matrix, and the process is formulated as

$$\left[t_{x,y}^{s,d,k=1} \right]_{S \times D} = \frac{1}{\sqrt{S \times D}} \left[u_{x,y}^0 * G_\sigma(x,y) \right]_{S \times D}. \tag{2}$$

Step 2: The original image is embedded into the tensor representation in the same way as above step, the matrix is written as

$$\left[t_{x,y}^{s,d,k=2} \right]_{S \times D} = \frac{1}{\sqrt{S \times D}} \left[u_{x,y}^0 \right]_{S \times D}. \tag{3}$$

Step 3: The local geometric features are incorporated into the tensor representation in following way

$$\left[t_{x,y}^{s,d,k=3} \right]_{S \times D} = \left[u_{x,y}^0 * GT_{s,d}(x,y) \right]_{S \times D}, \tag{4}$$

where $GT_{s,d}(\cdot)$ is the Gabor function. For more details, please refer to [15].

Though the pixel tensor representation provides a more comprehensive description to the image, it greatly increases the computational cost. To overcome this problem, we introduce the OTA [13], a kind of generalization of principle component analysis for tensors, to reduce the dimensionality of the tensor representation.

$$\left[t_{x,y}^{s,d,k=3} \right]_{S' \times D'} = \left[OTA \left(u_{x,y}^0 * GT_{s,d}(x,y) \right) \right]_{S' \times D'}. \tag{5}$$

where $OTA(\cdot)$ denotes the process of dimensionality reduction. After applying OTA, $S \times D$ is reduced to $S' \times D'$, where $S' < S$ and $D' < D$. The cost of computation is reduced from $O((S \times D \times K) \times N)$ to $O((S' \times D' \times K) \times N)$ per time step, where N is the numbers of the pixels in the image.

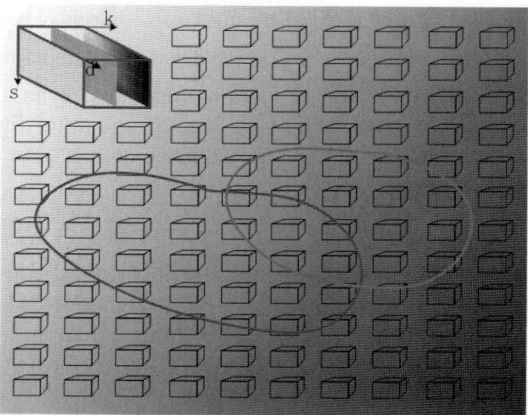

Fig. 1. The pixel tensor representation is zoomed out on the left-top corner, and the two zero level curves are denoted by red and green curve respectively. The two level set functions evolve in the tensor field composed of the elements in the form of the tensor representation.

Through steps 1-3, each pixel is depicted by a 3-order tensor in $R^{S \times D \times K}$, and the image will correspond to a 5-order tensor in $R^{M \times N \times S \times D \times K}$.

3.2 Multiphase Gabor-Based Tensor Level Set Method

Let us define a tensor T in $R^{M \times N \times S \times D \times K}$, then unfold T along the first two indices simultaneously. Thus, T develops a tensor field with elements in the form of tensor in $R^{S \times D \times K}$. We use M level set functions to segment the tensor field T, and the zero level curves of these level set functions are C_i in $\Omega \in R^{M \times N}$ which divide the tensor field into $N = 2^M$ sub-regions. The case of $M = 2$ is shown in Fig.1. The energy functional is composed of a regularization item and a fitting error item between tensor field T and the piecewise constant representation. The functional is defined as

$$\begin{aligned} E(\Phi, C) &= \sum_{i=1}^{M} \mu \operatorname{Length}(\phi_i) + \sum_{j=1}^{N} E_j(c_j, \chi_j) \\ &= \mu \sum_{i=1}^{M} \int_{\Omega} \delta(\phi_i(x,y)) |\nabla \phi_i(x,y)| dxdy + \sum_{j=1}^{N} \int_{\Omega} \operatorname{dist}_{x,y}^2 \left(t_{x,y}^{s,d,k}, c_j^{s,d,k} \right) \chi_j(\Phi) dxdy, \end{aligned} \quad (6)$$

where $\Phi = \{\phi_1, \ldots, \phi_M\}$, $C = \{c_1, \ldots, c_N\}$, $\delta(\cdot)$ is the Dirac delta function, $t_{x,y}^{s,d,k}$ is the element in T, and χ_j denotes the sub-region and formulated as

$$\chi_j(\Phi) = \prod_{k=1}^{M} \left((1 - b_k) - (-1)^{b_k} H(\phi_k) \right), \quad (7)$$

where $[b_k] = dec2binvec(j)$ and converts the denary number j into its binary notation. $c_j^{s,d,k}$ is the mean tensor inside the sub-region χ_j and defined as

$$c_j^{s,d,k} = \int_\Omega t_{x,y}^{s,d,k} \chi_j(\Phi)\,dxdy \Big/ \int_\Omega \chi_j(\Phi)\,dxdy. \tag{8}$$

The tensor distance function is defined as

$$\operatorname{dist}_{x,y}\left(t_{x,y}^{s,d,k}, c_j^{s,d,k}\right) = \sqrt{\sum_{s=1}^{S} \alpha_s \sum_{d=1}^{D} \beta_d \sum_{k=1}^{K} \gamma_k \left(t_{x,y}^{s,d,k} - c_j^{s,d,k}\right)^2}, \tag{9}$$

where $\alpha_s \geq 0, \sum_{s=1}^{S}\alpha_s = 1; \beta_d \geq 0, \sum_{d=1}^{D}\beta_d = 1; \gamma_k \geq 0, \sum_{k=1}^{K}\gamma_k = 1.$

The minimization of the energy functional is a calculus problem. By adding an artificial time variable t and computing the Euler-Lagrange equation for each unknown function ϕ_i, the gradient flow, *i.e.*, evolution equation, is formulated as

$$\begin{aligned}\frac{\partial \phi_i}{\partial t} &= \mu \delta_\varepsilon(\phi_i)\operatorname{div}\left(\frac{\nabla \phi_i}{|\nabla \phi_i|}\right) \\ &+ \sum_{j=1}^{N} \frac{\partial \chi_j(\Phi)}{\partial \phi_i} \sum_{s=1}^{S} \alpha_s \sum_{d=1}^{D} \beta_d \sum_{k=1}^{K} \gamma_k \left(t_{x,y}^{s,d,k} - c_j^{s,d,k}\right)^2,\end{aligned} \tag{10}$$

with the boundary condition $\partial \phi_i / \partial n = 0$ on $\partial \Omega$, where $\operatorname{div}(\nabla\phi_i / |\nabla\phi_i|)$ is the mean curvature of ϕ_i, and $\partial \chi_j(\Phi) / \partial \phi_i$ is the partial derivative with respect to the level set function ϕ_i.

4 Experimental Results and Analysis

We conduct three experiments on different images to illustrate the effectiveness of the proposed method compared with the MCV method [18]. All these experiments select two level set functions to segment the images, which are denoted by red and green curve respectively.

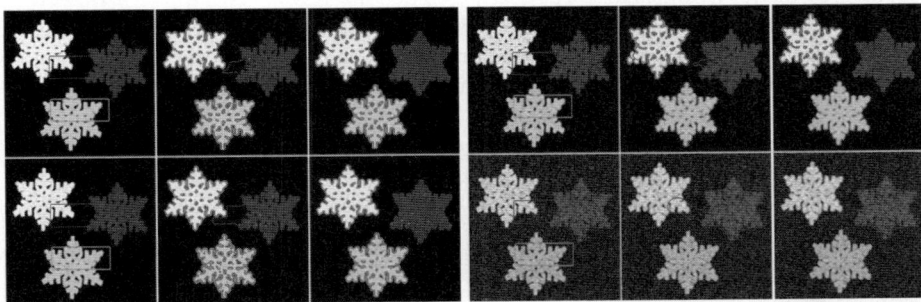

Fig. 2. The left two rows represent the evolutions applying MCV method on the images with the salt-and-pepper noise, and their noise density equals 0.01 and 0.05 respectively. The right two rows represent the evolution applying the proposed method on the same images, but the noise density equals 0.1 and 0.3 respectively.

Experiment 1 applies the MCV method and the proposed method on a synthetic image with salt-and-pepper noise. There are three paper-cut snowflakes with different gray value in the image. From the first row to the last row, the noise densities are set to 0.01, 0.05, 0.1 and 0.3 respectively. Fig. 2 shows that the MCV method wrongly classifies the positive impulse points as the object, and the proposed method correctly segments the snowflakes from the background even the noise density being 0.3.

Experiment 1 indicates the proposed method is more robust against the salt-and-pepper noise than the MCV method, since it incorporates the information of the Gaussian smoothed image.

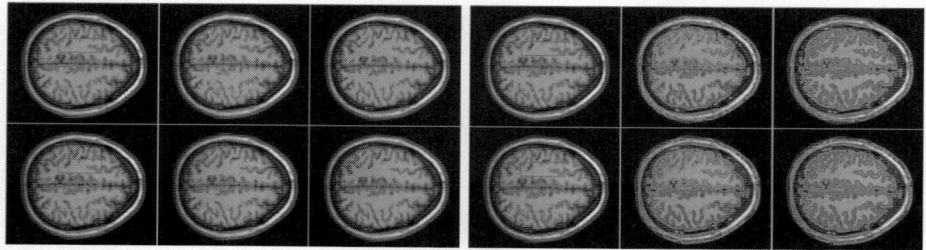

Fig. 3. The left two rows represent the evolutions applying the MCV method on the images with INU level being 20% and 40% respectively, and the right two rows represent the evolutions applying the proposed method on the image with the same INU level settings as MCV method

Experiment 2 applies the MCV method and the proposed method on the simulated magnetic resonance images with different density non-uniformity (INU) level. INU level is a kind of non-homogenous percentage and if INU is greater, the non-homogeneity is more serious. For more detail, please refer to [2]. Fig. 3 shows the proposed method detects more objects than the MCV method, e.g., the skin and skull. Meanwhile the proposed method obtains the same segmentation results whatever the INU level is 20% or 40%, but the MCV method treats the left-top part of white matter as the grey matter by mistake.

Experiment 2 indicates that the proposed method detects more objects than the MCV method, and is robust against the non-homogeneity, because the DC component of the image is removed by utilizing the Gabor filter bank in the proposed method.

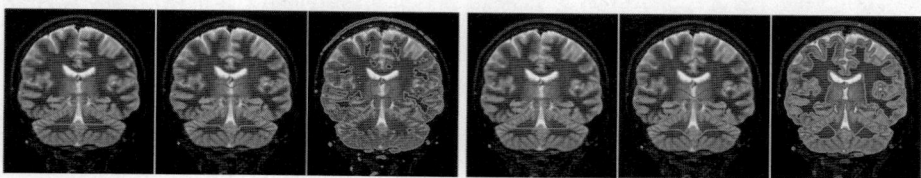

Fig. 4. The left row represents the evolution using MCV method, and the right row represents the evolution using the proposed method

In experiment 3, we apply the MCV method and the proposed method on a real magnetic resonance image, as shown in Fig. 4. The results indicate the MCV method cannot separate the cerebrospinal fluid from the grey matter, while the proposed one can make a correct segmentation.

Experiment 3 suggests that the proposed method employs the Gabor filter bank to essentially increase the weight of boundaries in energy functional, thus the more boundaries is detected and segmentation result is more natural.

5 Conclusions

In this paper, we utilize the Gabor filter bank to extract the local geometric features, *e.g.*, gradient and orientation, and combine these features with the original image and the image smoothed by Gaussian filter to build a comprehensive pixel tensor representation. By applying the multiphase tensor level set method on the tensor field corresponding to the given image, an accurate segmentation result is obtained. This method is more robust against noise and achieves better performance on the image with non-homogenous background. Additionally, the method provides a practical uniform formula to execute the multiphase level set method, and speeds up the process of execution by using OTA to reduce the dimensionality of the tensor representation. In the future, the proposed method will be applied on more images to verify its segmentation effectiveness.

Acknowledgments. We want to thank the helpful comments and suggestions from the anonymous reviewers. This research was supported by National Science Foundation of China (60771068, 60702061), and the Program for Changjiang Scholars and innovative Research Team in University of China (IRT0645).

References

1. Adalsteinsson, D., Sethian, J.A.: A Fast Level Set Method for Propagating Interfaces. J. Comput. Phys. 118(2), 269–277 (1995)
2. BrainWeb: Simulated Brain Database, http://www.bic.mni.mcgill.ca/brainweb/
3. Chan, T.F., Vese, L.A.: Active Contours without Edges. IEEE Trans. Image Process. 10(2), 266–277 (2001)
4. Chan, T.F., Sandberg, B.Y., Vese, L.A.: Active Contours without Edges for Vector-Valued Images. J. Vis. Commum. Image R. 11(2), 130–141 (2000)
5. Daugman, J.G.: Two-Dimensional Spectral Analysis of Cortical Receptive Field Profiles. Vision Res. 20(10), 847–856 (1980)
6. Li, C., Xu, C., Gui, C., Fox, M.D.: Level Set Evolution without Re-initialization: A New Variational Formulation. In: IEEE Conference on Computer Vision Pattern Recognition, pp. 430–436. IEEE Computer Society Press, Washington (2005)
7. Lie, J., Lysaker, M., Tai, X.: A Binary Level Set Model and Some Applications to Mumford-Shah Image Segmentation. IEEE Trans. Image Process. 15(5), 1171–1181 (2006)
8. Mumford, D., Shah, J.: Optimal Approximations by Piecewise Smooth Functions and Associated Variational Problems. Commun. Pur. Appl. Math. 42(5), 577–685 (1989)

9. Malladi, R., Sethian, J.A., Vemuri, B.C.: Shape Modeling with Front Propagation: A Level Set Approach. IEEE Trans. Pattern Anal. Machine Intell. 17(2), 158–175 (1995)
10. Marcelja, S.: Mathematical Description of the Responses of Simple Cortical Cells. J. Optical Soc. Am. 70(11), 1297–1300 (1980)
11. Osher, S., Sethian, J.A.: Fronts Propagating with Curvature-Dependent Speed: Algorithms Based on Hamilton-Jacobi Formulation. J. Comput. Phys. 79, 12–49 (1988)
12. Peng, D., Merriman, B., Osher, S., Zhao, H., Kang, M.: A PDE Based Fast Local Level Set Method. J. Comput. Phys. 155(2), 410–438 (1999)
13. Sun, J., Tao, D., Papadimitriou, S., Yu, P.S., Faloutsos, C.: Incremental Tensor Analysis: Theory and Applications. ACM Trans. Knowl. Disc. from Data 2(3), 11:1–11:37 (2008)
14. Suri, J.S., Liu, K., Singh, S., Laxminarayan, S.N., Zeng, X., Reden, L.: Shape Recovery Algorithms Using Level Sets In 2-D/3-D Medical Imagery: A State-Of-The-Art Review. IEEE Trans. Inf. Technol. Biomed. 6(1), 8–28 (2002)
15. Tao, D., Li, X., Wu, X., Maybank, S.J.: General Tensor Discriminant Analysis and Gabor Features for Gait Recognition. IEEE Trans. Pattern Anal. Machine Intell. 29(10), 1700–1715 (2007)
16. Tao, D., Song, M., Li, X., Shen, J., Sun, J., Wu, X., Faloutsos, C., Maybank, S.J.: Bayesian Tensor Approach for 3-D Face Modelling. IEEE Trans. Circuits and Systems for Video Technology 18(10), 1397–1410 (2008)
17. Tao, D., Li, X., Wu, X., Maybank, S.J.: Tensor Rank One Discriminant Analysis - A Convergent Method for Discriminative Multilinear Subspace Selection. Neurocomputing 71(10-12), 1866–1882 (2008)
18. Vese, L.A., Chan, T.F.: A Multiphase Level Set Framework for Image Segmentation Using the Mumford and Shah Model. Int. J. Comput. Vision 50, 271–293 (2002)
19. Wang, Z., Vemuri, B.C.: DTI Segmentation Using an Information Theoretic Tensor Dissimilarity Measure. IEEE Trans. Med. Imaging 24(10), 1267–1277 (2005)
20. Zhao, H., Chan, T.F., Merrian, B., Osher, S.: A Variational Level Set Approach to Multiphase Motion. J. Comput. Phys. 127, 179–195 (1996)

Embedded Geometric Active Contour with Shape Constraint for Mass Segmentation

Ying Wang[1], Xinbo Gao[1], Xuelong Li[2], Dacheng Tao[3], and Bin Wang[1]

[1] School of Electronic Engineering, Xidian University, Xi'an 710071, P.R. China
[2] State Key Laboratory of Transient Optics and Technology, Xi'an Institute of Optics and Precision Mechanics, Chinese Academy of Sciences, Xi'an 710119, P.R. China
[3] School of Computer Engineering, Nanyang Technological University, 639798, Singapore

Abstract. Mass boundary segmentation plays an important role in computer aided diagnosis (CAD) system. Since the shape and boundary are crucial discriminant features in CAD, the active contour methods are more competitive in mass segmentation. However, the general active contour methods are not so effective for some cases, because most masses possess very blurry margin that easily induce the contour leaking. To the end, this paper presents an improved geometric active contour for mass segmentation. It firstly introduces the morphological concentric layer model for automatically initializing. Then an embedded level set is used to extract the adaptive shape constraints. For refining the boundary, a new shape constraint function and stopping function are designed for the enhanced geometric active contour method. The proposed method is tested on real mammograms containing masses, and the results suggest that the proposed method could effectively restrain the contour leaking and get better segmented results than general active contour methods.

Keywords: Mass segmentation, active contour, shape constraint, embedded level set, mammogram.

1 Introduction

Breast cancer is the most common cancer among women [1], and early diagnosis is the only way for reducing the death rates. Mammography is currently the most effective tool to pronounce the abnormalities and detect early cancers. However, it is still a very difficult work for analyzing and diagnosing on mammography, because the serious impaction of the image quality and some subjective reasons. Computer-aided diagnosis (CAD) systems have been developed and proved to be a very useful tool for assisting radiologists by identifying the suspicious lesions. Mass, as one of the major indications of early breast cancer, its detection is still a challenging problem for many CAD systems. It is because mass always surrounded by density tissues, possess ill-defined margins, and also vary in their size and shapes. Accurate segmentation of the mass is very important for the CAD system, because it directly affects the detection performance. There has been several studies focus on the mass segmentation, such as pixel-based methods [2], edge-based methods [3-6], region-based methods [7]. Since the shape and boundary characteristics are more crucial in diagnosis, the edge-based methods attract more attention in the literature. Sahiner *et al.* [4] employed the active contour models as a final step for refining the segmentation. Timp and Karssemeijer [5] found the best contour of the mass based on dynamic programming. Ball and Bruce [6] studied the application of level set in mammograms.

However, there are still some problems that the active contour methods cannot easily handle. First, the initialization is a major drawback of general active contour methods. The second problem is the contour leaking of masses with blurry boundaries. Although the shape prior has been introduced to solve the similar problem [8,9], it does not work in mass segmentation because the mass possess various size and shapes that general shape prior could not adapt to any case. The stopping function designing is also a problem which should be set according to the particular image features.

To solve these problems, we proposed an improved geometric active contour (also known as level set) method for mass segmentation. The morphological concentric layer (MCL) method [10] is introduced first to locate the initial contour, according to which the level set function is automatically initialized. Then an embedded level set is applied to the smoothed image to adaptively extract the shape constraints corresponding to different mass regions. This shape constraint will make the segmented results more approached to the real size and shapes of various masses. Based on the shape constraints, we also design a new stopping function for further refining the boundary of masses and meanwhile preserve the gradient information within the masses, especially the real boundary of masses. Furthermore, the proposed method can also effectively avoid the contour leaking, because the final boundary will be restrained well by the shape constraints when the blurry boundary exists.

2   Active Contour Models and Their Limitations

There are two general types of active contour methods in literature [11]: parametric active contours and geometric active contours. In this section, we will give the overview of the two effective methods and their limitations on mass segmentation.

2.1   Parametric Active Contour

The parametric active contour, i.e., basic snake model [12], is a controlled continuous curve under the influence of internal and external forces. Representing the position of a snake parametrically by $\mathbf{v}(s) = (x(s), y(s))$, the energy function can be defined as

$$E_{snake} = \int_0^1 E_{int}(\mathbf{v}(s)) + E_{image}(\mathbf{v}(s)) + E_{con}(\mathbf{v}(s)) ds, \tag{1}$$

where E_{int} represents the internal energy, E_{image} gives rise to the image force, and E_{con} represents the external constraint force.

2.2   Geometric Active Contour

For capturing more complex topology difficult for parametric active contour methods, Osher and Sethian [13] introduced the concept of geometric active contour, and also provided an implicit formulation of the deformable contour in a level set framework.

In level set, the deformable contours can be denoted by C which are represented by $C(t)=\{(x,y) | \phi(\mathrm{x},t)=0\}$, the zero level set of a level set function $\phi(\mathrm{x},t)$. The evolution equation of ϕ can be written in a general form

$$\partial\phi/\partial t + F|\nabla\phi| = 0, \qquad (2)$$

which is called level set equation [13]. The function F is called the speed function. For the real image segmentation problems, Malladi et al. [14] multiplying the contour velocity by a "stopping" term $g(\mathrm{x})$ which is a monotonically decreasing function of the gradient magnitude of the image. In general case, $g(\mathrm{x})$ is given as

$$g(\mathrm{x}) = 1/\left(1+|\nabla G_\sigma(\mathrm{x})*I(\mathrm{x})|^2\right), \qquad (3)$$

where $|\nabla G_\sigma(\mathrm{x})*I(\mathrm{x})|$ is the absolute gradient of the convoluted image which is obtained by convolving the original image by the Gaussian function with a known standard deviation σ.

2.3 Challenging Issues

Although the parametric and geometric active contours models can work well on many segmentation problems, they still have certain limitations when applied on mammography. The parametric active models have two main limitations. Firstly, when the initial model and desired object boundary differ greatly in size and shape, the model must be re-parameterized dynamically to faithfully recover the object boundary, which means it is sensitive to the initialization. Secondly, it has difficulties to deal with the topological adaptation such as splitting or merging model parts, and also the sharp corners or pieces of the boundary intersect.

Then the geometric active contours which realized by level set methods is proposed to deal with these limitations. Even though numerous work have been proved that the level set is more effective on any of the cavities, concavities, splitting or merging, there are still some drawbacks when it is applied to our study. Firstly, the initial placement of the contour is a major problem, as it does not have either enough capture range or power to grab the topology of shapes. The second problem is the gaps in boundaries. Since the mass always possess so ambiguous margin that like a discontinuity in boundaries, it easily induces contour leaking. Therefore, more effective methods should be studied for handling the aforementioned problems.

3 An Improved Level Set for Mass Segmentation

For overcoming the above challenging issues, we work on an improved level set method. This new method incorporates the shape prior information as the shape constraint into the embedded level set process for restraining the contour leaking and refining the segmented results.

3.1 Initialization Based on Morphological Concentric Layer Model

The MCL model is worked on the smoothed image decomposed by morphological component analysis (MCA) first [15]. Since the mass region on smoothed image usually highlights in core and then gradually dimmer toward the margin, it can be viewed as a Gaussian area with some concentric density rings around the focal area. Then according to the adaptive density slicing, the focal area F within the mass region that satisfies the following conditions will be extracted as the initial contour

$$\{F|(F_{AR} \geq Th_{AR}) \& (F_{EC} \geq Th_{EC}) \& (F_{EX} \geq Th_{EX}), F \in R_{mass}\} \quad (4)$$

where F_{AR} and Th_{AR} are the *area* of F and its threshold respectively. Similarly, F_{EC} and F_{EX} are the *eccentricity* and the *extent*, while Th_{EC} and Th_{EX} are their thresholds. These limitations assure F can capture more information of various masses, and also avoid the unstable segmented results induced by manually marking.

3.2 Adaptive Shape Constraint

For adaptively extracting the meaningful shape constraints of masses with the above initial contour, an embedded level set is employed on the smoothed image (decomposed by MCA). Here, we employ a more effective evolution methods proposed by Li et al. [16] for eliminating the costly re-initialization procedure of classical level set. Since the smoothed component S_{MCA} already has smoothed variation on intensity, this component will substitute the convoluted image item in (3).

$$g(x) = 1/(1 + |\nabla S_{MCA}|^2). \quad (5)$$

After the curve evolving, we will extract the boundaries on the smoothed image according to various size, shapes and margin characteristics of the mass regions.

3.3 Improved Level Set with New Stopping Function

For introducing the shape constraint into the improved level set model, we convert the contour $\partial \Omega$ acquired by aforementioned method to a shape constraint function

$$H(x) = \begin{cases} 1 & x \subset \Omega \\ \max(\nabla S_{MCA}) & x \in \partial\Omega, \\ 0 & x \not\subset \Omega \end{cases} \quad (6)$$

where Ω represents the region within the boundary $\partial \Omega$.

According to this function, the gradient within the shape constraint of the original image will be preserved, and all the gradient variation out of the shape constraint will be set to zero for restraining the contour leaking. Furthermore, the shape constraint function will magnify the gradient on the margin of shape constraint to ensure the evolving curve could stop when the mass boundary is too ambiguous.

With this adaptive shape constraint item, we design a new stopping function for improving the performance of level set. It can be written as

$$g_s(x) = 1/(\varepsilon^2 + H(x) \cdot |\nabla G_\sigma(x) * I(x)|^2), \quad (7)$$

where the parameter ε is a real number (set to 3 on average) to adjust the whole contrast of the gradient based stopping condition. If the gradient varies too complex within the real mass regions, this parameter can automatically regulate the stopping function value to avoid the evolving contour stopping inside the regions.

Then the new stopping function will introduced into the total energy function [16]

$$\varepsilon(\phi) = \mu P(\phi) + \varepsilon_{g_s,\lambda,\nu}(\phi), \tag{8}$$

where the function $P(\phi)$ is a penalizing term to keep the level set function as a signed distance function, $\mu > 0$ and is the weight of the penalizing term. The external energy $\varepsilon_{g_s,\lambda,\nu}(\phi)$ will drive the zero level set toward the mass boundaries.

3.4 The Proposed Image Segmentation Algorithm

The proposed method includes initialization, shape constraints extraction, embedded level set segmentation, and so on. In this section, the detail procedures will be given by pseudo-code combined with equations as shown in Table 1.

4 Experimental Results and Analysis

The performance of the improved level set method is tested on a random selected sub-dataset from the Digital Database for Screening Mammography (DDSM) provided by the University of South Florida [17]. The mammograms were digitized with a LUMI-SYS laser scanner at a pixel size of 0.5mm and 12-bits per pixel. Since the malignant masses present more complex shape and margin characteristic, we first test the proposed method on 37 malignant masses. All of these mammograms containing abnormities are associated with ground truth information marked by experts.

Table 1. Algorithm flowchart of the improved level set segmentation method

---

Input: the original image to be segmented m^0
Output: the evolving curve C and the segmentation result R
Initialization: the initial position of evolving curve

Step1 Iterate $\{S_{Tex}, S_{MCA}\} = \arg\min_{\{S_{Tex}, S_{MCA}\}} \|T_{Tex} S_{Tex}\|_1 + \|T_{MCA} S_{MCA}\|_1 + \lambda \|S - S_{Tex} - S_{MCA}\|_2^2$

Step2 // extract the shape constraint by embedded level set
  Calculate $g(x) = 1/\left(1 + |\nabla S_{MCA}|^2\right)$, and iterate $\partial \phi / \partial t + g \cdot F |\nabla \phi| = 0$

Step3 // construct the shape constraint function and stopping function:

$$\text{Calculate } H(x) = \begin{cases} 1 & x \subset \Omega \\ \max(\nabla S_{MCA}) & x \in \partial\Omega \\ 0 & x \not\subset \Omega \end{cases}, \text{ and } g_s(x) = \frac{1}{\varepsilon^2 + H(x) \cdot |\nabla G_\sigma(x) * I(x)|^2}$$

Step4 Iterate $\partial \phi / \partial t = \mu[\Delta \phi - \text{div}(\nabla \phi / |\nabla \phi|)] + \lambda \delta(\phi) \text{div}(g_s \nabla \phi / |\nabla \phi|) + \nu g_s \delta(\phi)$
Step5 Calculate the evolving curve $C = \{(i,j) \in \Omega : \phi(i,j) = 0\}$
Step6 Calculate the segmentation result $R = m^0 H(\phi) + m^0(1 - H(\phi))$

All of the mammograms are processed firstly by MCA decomposition and MCL detection. Then the sub-images contains masses are cropped from the original and smooth image respectively. Fig. 1 gives the complete working procedure of the proposed scheme. As shown in Fig. 1, the malignant mass has ambiguous boundary in the top right corner, which easily brings the contour leaking problem. While, under the limitation of the shape constraint, the contour leaking problem can be restrained, and the segmented boundary is more approaching to the real mass margin.

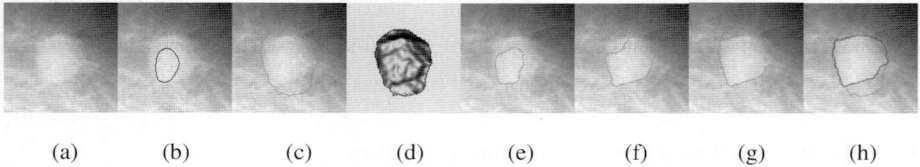

(a) (b) (c) (d) (e) (f) (g) (h)

Fig. 1. The segmentation result of the embedded level set with shape constraint. (a) The original image; (b) The initialized contour; (c) The shape constraint; (d) The new stopping function; (e) The contour of 20 iterations; (f) The contour of 180 iterations; (g) Segmentation result with 2000 iterations; (h) The segmentation result with shape constraint.

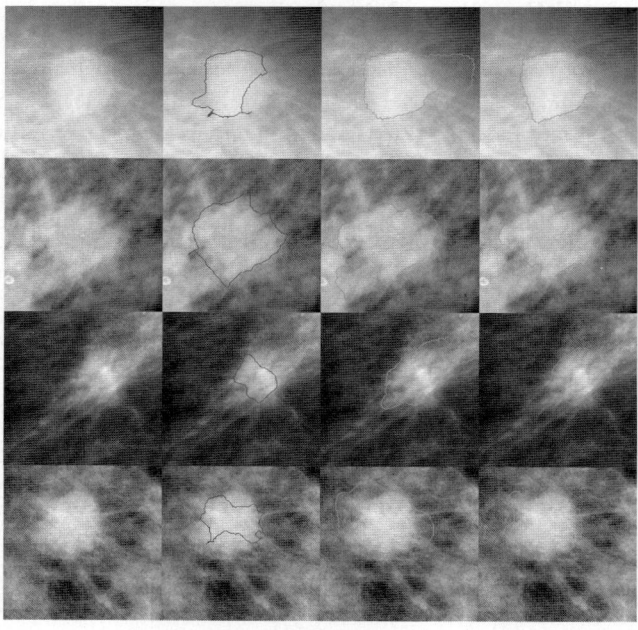

Fig. 2. The comparison of different active contour based segmentation methods. The first column gives the original images; the second column shows the segmental results using GVF snake; the third column is the level set segmental results without shape constraint; and the forth column presents the results of the proposed algorithm.

Then the comparison experiments are implemented for providing the competitive segmentation results between several active contour methods. As applied in the literature, we test those malignant masses using the GVF snake, level set without shape constraint, and the improved segmentation scheme respectively, and test results are given in Fig. 2. It indicates that the snake model can not effectively deal with the boundary concavities and complex intensity distribution within the mass regions. The level set method can find more meaningful boundary which is better than the parametric active contour. However, it still cannot handle the contour leaking problem when the boundaries of masses are ill-defined. The last column in Fig. 2 is the segmented results using the proposed segmentation method. With the adaptive shape constraint, the contours can be more approaching to the real mass margins.

5 Conclusion

In order to avoid the contour leaking and refine the segmentation results, this paper proposed an improved geometric active contour method for the mass segmentation. First, the evolving curve is initialized by the MCL method. Then, for adaptively extracting the shape constraints of various mass types, an embedded level set is operated on the smoothed image. According to the limitation of the shape constraints function, a new stopping function is constructed for the improved level set method to avoid the contour leaking and further refine the segmented results. The comparison experimental results show that the proposed method could more effectively on capturing complex topology. Furthermore, it also can avoid the contour leaking in weak margins and stopping earlier inside the regions.

For further validating the proposed methods, more malignant masses with more complex distribution and margins will be tested in the future. And more flexible shape constraints and stopping function should be considered for refining the segmentation.

Acknowledgments. We want to thank the helpful comments and suggestions from the anonymous reviewers. This research was supported by National Science Foundation of China (60771068, 60702061), and the Program for Changjiang Scholars and innovative Research Team in University of China (IRT0645).

References

1. American Cancer Society-Cancer facts and figures (2008), http://www.cancer.org/downloads/STT/008-CAFFfinalsecured.pdf
2. Li, H.D., Kallergi, M., Clarke, L.P., Jain, V.K., Clark, R.A.: Markov Random Field for Tumor Detection in Digital Mammography. IEEE Trans. Med. Imaging 14(3), 565–576 (1995)
3. Kobatake, H., Murakami, M., Takeo, H., Nawano, S.: Computerized Detection of Malignant Tumors on Digital Mammograms. IEEE Trans. Med. Imaging 18(5), 369–378 (1999)
4. Sahiner, B., Petrick, N., Chan, H.P., Hadjiiski, M., Paramagul, C., Helvie, M.A., Gurcan, M.N.: Computer-Aided Charaterization of Mammographic Masses: Accuracy of Mass Segmentation and Its Effects on Characterization. IEEE Trans. Med. Imaging 20(12), 1275–1284 (2001)

5. Timp, S., Karssemeijer, N.: A New 2D Segmentation Method Based on Dynamic Programming Applied to Computer Aided Detection in Mammography. IEEE Trans. Med. Imaging 31(5), 958–971 (2004)
6. Ball, J.E., Bruce, L.M.: Digital Mammogram Spiculated Mass Detection and Spicule Segmentation Using Level Sets. In: IEEE International Conference on Engineering in Medicine and Biology Society, pp. 4979–4984. IEEE Press, Lyon (2007)
7. Petrick, N., Chan, H.P., Sahiner, B., Helvie, M.A.: Combined Adaptive Enhancement and Region-Growing Segmentation of Breast Masses on Digitized Mammograms. Med. Phys. 26(8), 1642–1654 (1999)
8. Rousson, M., Paragios, N.: Shape Priors for Level Set Representations. In: Heyden, A., Sparr, G., Nielsen, M., Johansen, P. (eds.) ECCV 2002. LNCS, vol. 2351, pp. 78–92. Springer, Heidelberg (2002)
9. Ersoy, I., Bunyak, F., Palaniappan, K., Sun, M., Forgacs, G.: Cell Spreading Analysis with Directed Edge Profile-Guided Level Set Active Contours. In: Metaxas, D., Axel, L., Fichtinger, G., Székely, G. (eds.) MICCAI 2008, Part I. LNCS, vol. 5241, pp. 376–383. Springer, Heidelberg (2008)
10. Eltonsy, N.H., Tourassi, G.D., Elmaghraby, A.S.: A Concentric Morphology Model for the Detection of Masses in Mammography. IEEE Trans. Med. Imaging 26(6), 880–889 (2007)
11. Xu, C.Y., Prince, J.L.: Snakes, Shapes, and Gradient Vector Flow. IEEE Trans. Image Process. 7(3), 359–369 (1998)
12. Kass, M., Witkin, A., Terzopoulos, D.: Snakes: Active Contour Models. Int. J. Comput. Vis. 1, 321–331 (1987)
13. Osher, S., Sethian, J.: Fronts Propagating with Curvature-Dependent Speed: Algorithms Based on Hamiltons-Jacobi Formulations. J. Comput. Phys. 79(1), 12–49 (1988)
14. Malladi, R., Sethian, J.A., Vemuri, B.C.: Shape Modeling with Front Propagation. IEEE Trans. Pattern Anal. Machine Intell. 17(2), 158–175 (1995)
15. Starck, J.L., Elad, M., Donoho, D.: Redundant Multiscale Transforms and Their Application for Morphological Component Analysis. Adv. Imag. Elect. Phys. 132, 287–348 (2004)
16. Li, C.M., Xu, C.Y., Gui, C.F., Fox, M.D.: Level Set Evolution Without Re-initialization: A New Variational Formulation. In: IEEE Computer Society Conference on Computer Vision and Pattern Recognition, pp. 430–436. IEEE Press, San Diego (2005)
17. University of South Florida, Digital Database for Screening Mammography (DDSM), http://marathon.csee.usf.edu/Mammography/Database.html

An Efficient Parallel Algorithm for Graph-Based Image Segmentation

Jan Wassenberg[1], Wolfgang Middelmann[1], and Peter Sanders[2]

[1] FGAN-FOM, 76275 Ettlingen, Germany
jan.wassenberg@fom.fgan.de
[2] Institute for Theoretical Informatics, Universität Karlsruhe, Germany

Abstract. Automatically partitioning images into regions ('segmentation') is challenging in terms of quality and performance. We propose a Minimum Spanning Tree-based algorithm with a novel graph-cutting heuristic, the usefulness of which is demonstrated by promising results obtained on standard images. In contrast to data-parallel schemes that divide images into independently processed tiles, the algorithm is designed to allow parallelisation without truncating objects at tile boundaries. A fast parallel implementation for shared-memory machines is shown to significantly outperform existing algorithms. It utilises a new microarchitecture-aware single-pass sort algorithm that is likely to be of independent interest.

1 Introduction and Related Work

Segmentation (automatically partitioning an image into regions) is an important early stage of some image processing pipelines, e.g. object-based change detection. The final results of such applications are often strongly dependent on the quality of the initial segmentation. Since subsequent processing steps can use higher-level region information instead of having to examine all pixels, the segmentation may also be the limiting factor in terms of performance. Many algorithms have been proposed, but good quality results often come at the price of high computational cost.

One extreme example of this is a multi-scale watershed approach (MSHLK) [1]. Repeated applications of anisotropic diffusion smooth the image and reduce the oversegmentation caused by the watershed transform. The resulting subjective quality is very good, but its computational cost (1 second per kPixel) is unacceptable.

An alternative approach uses the Mean-Shift (MS) [2] procedure to locate clusters within a higher-dimensional representation of the image. This is guaranteed to converge on the densest regions in this space and yields good results in practice, but the processing rate (0.1 MPixel/s) is still inadequate.

Recent work has shown that Maximally Stable Extremal Regions (MSER) [3] within a gradient image are also suitable for image segmentation. While more efficient (2 MPixel/s), this scheme only detects high-contrast segments and does not provide full coverage of the image. It also seems ill-suited for parallelisation since the stability criterion depends on a global ordering of pixels.

Graph-based segmentation (GBS) [4] increases the amount of data to be handled (multiple edges per pixel) but has several attractive properties. Viewing pixels as nodes of a graph allows the reduction of segmentation to finding cuts in a Minimum Spanning Tree (MST). Defining edge weights as some function of the pixels' per-band intensity differences enables the use of colour information without having to compute gradients. Finally, an MST can be assembled from partial sub-trees, which provides the possibility of parallelisation.

The remainder of this article is structured as follows: Section 2 develops a new online graph-cutting heuristic for MST-based segmentation. Section 3 shows the promising results obtained on well-known images. Section 4 introduces 'PHMSF' (Parallel Heuristic for Minimum Spanning Forests), which we believe to be the first non-trivially-parallel segmentation algorithm. Perhaps most importantly, Section 5 shows it to significantly outperform existing segmentation techniques.

2 Segmentation Algorithm

Segmentation algorithms require (often application-dependent) definitions of 'image region'. We consider 'homogeneity' and high contrast to surrounding pixels to be reasonable criteria [5]. Homogeneity can be computed as distances between (vector-valued) pixels; we find the L2 norm to yield better results than L1 or pseudo-norms. Note that [4] advocates separate segmentation of the R/G/B component images and intersecting the results. Since object edges are not always visible in all multi-spectral bands [6], it is safer (and certainly faster) to segment once using all bands. Recalling the graph segmentation framework, the above homogeneity measure defines the weight of edges. It remains to be seen how an online graph-cutting heuristic should partition the MST depending on edge weight. A mere threshold is insufficient because it fails to account for noise or the overall homogeneity of a region. [4] suggests an adaptive threshold that is incremented by a linearly decreasing function of the region size[1]. The function's slope is a user-defined parameter that must be determined by experimentation since it has no physical explanation. This scheme also underestimates a region's homogeneity by defining it as the maximum weight in its MST, thus tending towards oversegmentation. We suggest the adoption of an idea from Canny's edge detection algorithm [7]. In the context of edge detection, pixels with large gradient magnitudes are likely to correspond to edges, but there is no single level at which this ceases to be the case. Applying a rather high limit finds likely candidates, which can be augmented by nearby pixels that lie above a second, lower threshold. Returning to segmentation terminology, regions connected by low-weight edges represent likely candidates that can subsequently be expanded by following adjoining edges with somewhat higher weights. To avoid potentially unbounded growth, we institute a 'credit' limit on the sum of edge weights that may be added to a candidate region. Since no shape can be more compact than a circle, the region's perimeter is bounded from below by the circumference

---

[1] This unduly penalizes the growth of large segments; we saw slightly better results when dividing by the logarithm of the region size.

$\sqrt{4\pi \cdot \text{regionSize}}$. Let us also assume additive white Gaussian noise with variance σ_n^2, for which several estimators have been proposed [8,9].s Defining contrast as the smallest edge weight along the border of any 'interesting' region minus $2 \cdot \sigma_n$ thus makes it likely that edges of total weight $\leq$ contrast $\cdot$ minPerimeter can be added to a region without inadvertently expanding beyond its bounds. This property is important because subsequent region merges are trivial, whereas splitting requires re-examination of the pixels or edges. However, the resulting regions are not necessarily too fine because pixels connected by low-weight edges are always merged. We have therefore averted global under- and oversegmentation of the image while using only local information. The algorithm first forms candidate regions by merging the endpoints of low-weight edges and then calls the following simple heuristic in increasing order of the remaining edges' weights:

Procedure EdgeHeuristic(*edge*)
  region$_1$, region$_2$:= Find(*edge*.endpoints);
  if region$_1 \neq$ region$_2$ **then**
    credit := min {region$_1$.credit, region$_2$.credit};
    if credit > *edge*.weight **then**
      survivor := Union(region$_1$, region$_2$);
      survivor.credit := credit − *edge*.weight;

---

3 Results

To demonstrate the usefulness of the new segmentation results, we compare them to the outputs of existing algorithms on standard images [10], the results of which are shown in Fig. 1:

MSHLK [1] is known for high-quality results and provides excellent smoothing of the walls (b) but merges the eaves into the sky segment. We also call

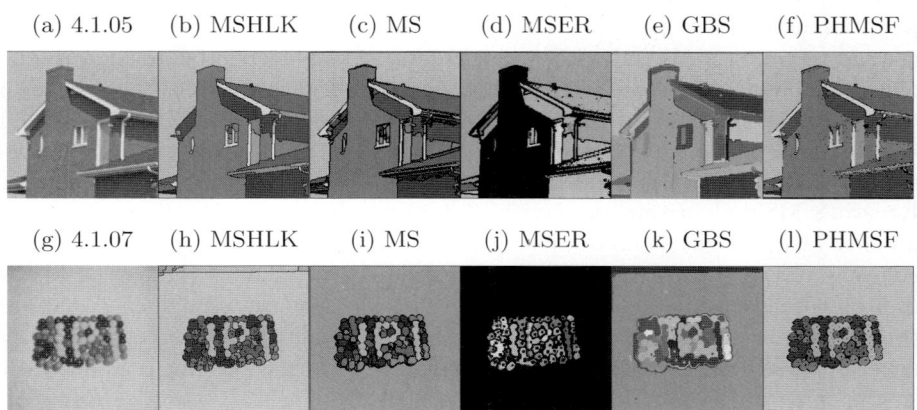

Fig. 1. Segmentation results of the new PHMSF algorithm and others on USC SIPI [10] images 4.1.05 ('House') and 4.1.07 ('Jelly beans')

attention to the oversegmentation of the second image and shock effects [11] in the background (h). MS [2] is more successful at merging the individual objects (i) but also splits some of them (e.g. below the P); spurious segments near edges (c) are its only visible flaws. As with MSHLK, segment borders are delineated by black pixels. MSER [3] produces mostly adequate label images, though the wall is not considered to be a stable region (d); the effects of the gradient filter are clearly visible (j). GBS [4] is satisfactory but results in undersegmentation near the roof lines and oversegmentation of the sky and wall (e). It also merges different-coloured objects (k) but fails to return a uniform background. Our new PHMSF algorithm provides results comparable to MSHLK and MS and requires only 1/4000 and 1/50 the computation time, respectively (cf. Sect. 5). The black pixels (f) indicate surface irregularities that resulted in regions smaller than the minimum size. The segmentation in (l) is quite accurate, correctly separating different-coloured objects without introducing spurious boundaries.

4 Parallel Algorithm

Despite the efficiency of the new segmentation algorithm, a highly-tuned sequential implementation is still far slower than the collection rates of commercial imaging satellites (e.g. IKONOS with up to 90 km^2/s [12]). Since a significant reduction of the algorithm's constant factors appears unlikely and sequential programs have seen less benefit from recent CPU advances [13], it appears our self-set performance goal of 10 MPixel/s can only be reached by means of parallelisation. Note that embarrassingly-parallel schemes that simply split the input into independent tiles are not acceptable because they do not correctly handle objects straddling a border. Nor are overlapping tiles sufficient because there is no upper bound on the size of objects of interest (e.g. rivers or roads). Our first attempt at parallelisation addressed the MST computation. The new Filter-Kruskal scheme [14] combines ideas from Quicksort and Kruskal's algorithm and discards non-MST edges without having to sort them. This 'filter' operation, partitioning and sorting can all be parallelised. However, the total speedup on a quad-core system is only 1.5 – chiefly due to the sequential portion of the algorithm, but also because our eight-connected grid graphs are too sparse to derive much benefit from discarding edges. Our second approach is designed to allow independent processing of image tiles, but still ensures consistent results irrespective of the number of processors P.[2] The key observation is that Kruskal's MST algorithm can run in a data-parallel fashion until encountering an edge that crosses a tile border. From then on, MST components using such edges and in turn their incident edges must be 'delayed' until the partial MSTs of both tiles are available. We accomplish this with per-tile edge queues that are processed in a subsequent sequential phase[3]. It remains to be seen how many edges are delayed – a long cross-border region of homogeneous pixels could affect

---

[2] We ignore the (negligible) effects of unstable edge sorting.
[3] This could be parallelised if edges indicate which border they cross, but our implementation cannot spare any space within the 32-bit representation.

a large proportion of a tile. However, high-weight edges at the boundary of such regions often serve as a 'firewall' because they can be discarded without affecting neighbouring regions. Only about 5 % of edges are delayed in practice, making Amdahl's argument less of a factor than real-world limits on memory bandwidth and P. The algorithm is described by the following pseudo-code:

Algorithm 2. Parallel Segmentation

parallel foreach tile **do**
  sort edges, merging those with weight < minWeight;
foreach *borderEdge* **do**   // connect and mark cross-border regions
  $\text{region}_1, \text{region}_2 := \texttt{Find}(borderEdge.\text{endpoints})$;
  survivor := $\texttt{Union}(\text{region}_1, \text{region}_2)$;
  Mark(survivor);
  tile.regions := tile.regions $\cup$ {survivor};
parallel foreach tile **do**
  foreach $r \in$ tile.regions **do** r.credit := $\texttt{ComputeCredit}(r.\text{size})$;
parallel foreach tile **do**
  foreach *edge* in ascending order of weight **do**
    $\text{region}_1, \text{region}_2 := \texttt{Find}(edge.\text{endpoints})$;
    if *edge* crosses border **then** Mark(region_1); Mark(region_2);
    else if $\texttt{IsMarked}(\text{region}_1, \text{region}_2)$ **then** tile.delayQ.push (*edge*);
      else EdgeHeuristic(*edge*);
foreach tile **do**
  foreach *edge* $\in$ tile.delayQ **do** EdgeHeuristic(*edge*);

To avoid scheduling and locality issues, the (manually partitioned) loops reside in a single OpenMP parallel region. A novel variant of counting sort uses paged virtual memory to simulate bins of unlimited size and thus dispenses with a separate counting phase. An explicit buffering technique further increases performance by enabling write-combining without cache pollution. Details are given in App. A within the expanded version of this work [15].

The algorithm outputs a Union-Find (UF) tree represented as an array of pointers to a parent pixel or region, as well as per-tile lists of regions, which each store size (number of pixels) and credit. Computing features for single-pixel regions would consume too much memory, so we only consider regions of size min..max. This requires relabeling the per-tile regions and replacing them with so-called 'accumulators' for the region features, which is accomplished by Alg. 3. Its separate and very efficient count phase seems preferable to updating the per-tile region count when cross-border merges are performed by the Kruskal algorithm. Since the desired output includes a label image, we 'collapse' the UF tree once all regions have been re-labeled. With all pieces in place, we can now compute the contribution of each pixel toward its region's features. The per-band intensities B_i and $\sum B_i^2$ are required for computing the band averages and standard deviations.

Algorithm 3. Parallel Relabeling

parallel foreach tile **do**　　　　　　　　　　　　　　　　// compress regions
　foreach $r \in$ tile.regions **do** r.isValid := r.size $\in [\min, \max]$;
parallel foreach tile **do**　　　　　　　　　　　　　　　　// count regions
　tile.numRegions := 0;
　foreach *pixel* **do**
　　if IsRepresentative(*pixel*) **and** Find(*pixel*).isValid **then**
　　　tile.numRegions := tile.numRegions + 1;
for $i := 0$ **to** $|\text{tiles}| - 1$ **do**
　tiles $[i]$.startIndex := $\sum_{0 \leq j < i}$ tiles $[j]$.numRegions;
parallel foreach tile **do**　　　　　　　　　　　　　　　　// re-label regions
　foreach *pixel* **do**
　　if IsRepresentative(*pixel*) **and** Find(*pixel*).isValid **then**
　　　parents $[pixel]$:= tile.startIndex;
　　　tile.startIndex := tile.startIndex + 1;

For pixel coordinates (Y, X), the six moments $\sum Y^p \cdot X^q$ ($p, q \in \mathbb{N}_0, p+q \leq 2$) are sufficient for estimating an ellipse [16]. Finally, counting the number of neighbour pixels belonging to different regions allows computing the region perimeter. Using 64-bit floating-point accumulators mitigates precision issues while still enabling vectorization via SSE2 instructions.

5 Performance Analysis

We first examine the complexity of the proposed algorithm. Counting sort is $O(N)$. Region merges via Union-Find are effectively $O(1)$ for all practical input sizes[4] [18]. All other operations are also constant-time and reside in loops with trip counts in $O(N)$, so the complexity is (quasi-)linear in the input size. Since this also applies to the MSER and GBS algorithms, we must compare their implementations. Table 1 lists the performance[5] of each algorithm for a representative 8.19 MPixel subset of a 16-bit, 4-component (RGB + NIR) Quickbird image of Karlsruhe.

Our PHMSF algorithm does more work (computing region features and processing the original four-component 16-bit pixels rather than an 8-bit RGB version), yet significantly outperforms the other algorithms. In this test it is 138 times as fast as MS [20], 28 times as fast as GBS [21] and 5 times as fast as our similarly optimised implementation of MSER. Note that (32-bit) MSHLK exhausted its address space after a single diffusion iteration. Our PHMSF

---

[4] We view the inverse Ackermann function as a constant ≤ 5 for $N < 10^{80}$. Note that an attempt at replacing Union-Find with a 'true linear algorithm' [17] introduces a constant factor of 8.

[5] Measured on a X5365 CPU (3.0 GHz, 32 GiB FB-DDR2 RAM) running Windows XP x64. Our implementation is compiled with ICC 11.0.066 /Ox /Og /Ob2 /Oi /Ot /fp:fast /GR- /Qopenmp /Qftz /QxSSSE3

Table 1. Performance comparison

Algorithm	MPixel/s
MSHLK	N/A
MS	0.09
GBS	0.45
MSER	2.53
PHMSF	12.80

Table 2. Performance on large images

Sensor	Preproc.[6]	Bits	MPixel	MPixel/s
IKONOS	PS	16×4	54	13.5
QuickBird	PS	16×4	219	14.3
JAS150s	BF	8×4	527	24.4

implementation requires much less memory: the working set is about 7.1 GB for a 1.97 GB image, which equates to 13.5 bytes/pixel. Table 2 shows measurements from processing large images of up to 527 MPixel. Performance improves with size due to increased parallelism – tile interiors grow faster than their borders. The parallel speedup varies between 2 and 3.2 when using 4 cores. In the latter case, sequential processing only accounts for 2 % of processing time; the limiting factor is memory bandwidth. RightMark Memory Analyzer [22] measures read and write throughputs of roughly 3500 MB/s and 2500 MB/s on this system. Having analysed the elapsed times and minimum amounts of data that must be transferred to/from memory during the credit computation, region compression/counting/relabeling and feature computation phases, we can conclude that each is at least 85 % efficient. Improving their performance or scalability is therefore contigent on increased bandwidth (e.g. via NUMA architecture or by adding further memory channels).

6 Conclusion

We have presented a new (quasi-)linear-time segmentation algorithm that provides useful results at previously unmatched speeds. Applications include automatic wide-area appraisal of the suitability of roofs for solar panels, object-based change detection, environmental monitoring and rapid updates of land-use maps. From an algorithm engineering standpoint, we believe this to be the first non-trivially-parallel segmentation algorithm. Its scalability is chiefly limited by the memory bandwidth of current SMP systems. Future work includes statistical estimation of the edge weight thresholds and efficiently computing a segment neighbourhood graph. We are also interested in applying this algorithm towards segment-based fusion of high-resolution electro-optical and hyperspectral imagery.

References

1. Vanhamel, I., et al.: Scale Space Segmentation of Color Images Using Watersheds and Fuzzy Region Merging. In: ICIP (1), pp. 734–737 (2001)
2. Comaniciu, D., Meer, P.: Mean Shift Analysis and Applications. In: ICCV, pp. 1197–1203 (1999)

---

[6] PS stands for pan-sharpening by an as-yet unpublished algorithm (56 MPixel/s), while BF denotes approximated per-channel Bilateral Filtering [19] with $\sigma_r = 15$ and $\sigma_s = 32$ (80 MPixel/s).

3. Wassenberg, J., Bulatov, D., Middelmann, W., Sanders, P.: Determination of Maximally Stable Extremal Regions in Large Images. In: Signal Processing, Pattern Recognition, and Applications (February 2008)
4. Felzenszwalb, P., Huttenlocher, D.: Efficient Graph-Based Image Segmentation. IJCV 59(2), 167–181 (2004)
5. Haralick, R., Shapiro, L.: Image Segmentation Techniques. CVGIP 29, 100–132 (1985)
6. Thomas, C., Ranchin, T., Wald, L., Chanussot, J.: Synthesis of Multispectral Images to High Spatial Resolution: A Critical Review of Fusion Methods Based on Remote Sensing Physics. IEEE Trans. Geoscience and Remote Sensing 46(5), 1301–1312 (2008)
7. Canny, J.: A Computational Approach to Edge Detection. In: RCV 1987, pp. 184–203 (1987)
8. Shin, D.H., Park, R.H., Yang, S., Jung, J.H.: Block-Based Noise Estimation Using Adaptive Gaussian Filtering. IEEE Trans. Consum. Electron. 51, 218–226 (2005)
9. Amer, A., Dubois, E.: Fast and Reliable Structure-Oriented Video Noise Estimation. IEEE Trans. Circuits Syst. Video Techn. 15(1), 113–118 (2005)
10. Weber, A.: The USC-SIPI Image Database, http://sipi.usc.edu/database/ (accessed 2008-10-06)
11. Buades, A., Coll, B., Morel, J.: The Staircasing Effect in Neighborhood Filters and its Solution. IEEE Trans. Image Processing 15(6), 1499–1505 (2006)
12. Dial, G., Bowen, H., Gerlach, F., Grodecki, J., Oleszczuk, R.: IKONOS satellite, imagery, and products. Remote Sensing of Environment 88(1-2), 23–36 (2003)
13. Sutter, H.: The Free Lunch is Over: A Fundamental Turn Toward Concurrency. Dr. Dobb's Journal (March 2005)
14. Osipov, V., Sanders, P., Singler, J.: The Filter-Kruskal Minimum Spanning Tree Algorithm. In: Finocchi, I., Hershberger, J. (eds.) ALENEX, pp. 52–61. SIAM, Philadelphia (2009)
15. Wassenberg, J., Middelmann, W., Sanders, P.: An Efficient Parallel Algorithm for Graph-Based Image Segmentation (June 2009), http://algo2.iti.uni-karlsruhe.de/wassenberg/wassenberg09parallelSegmentation.pdf
16. Zunic, J., Sladoje, N.: Efficiency of Characterizing Ellipses and Ellipsoids by Discrete Moments. IEEE Trans. Pattern Anal. Mach. Intell. 22(4), 407–414 (2000)
17. Nistér, D., Stewénius, H.: Linear Time Maximally Stable Extremal Regions. In: Forsyth, D., Torr, P., Zisserman, A. (eds.) ECCV 2008, Part II. LNCS, vol. 5303, pp. 183–196. Springer, Heidelberg (2008)
18. Harfst, G., Reingold, E.: A Potential-Based Amortized Analysis of the Union-Find Data Structure. SIGACT 31, 86–95 (2000)
19. Paris, S., Durand, F.: A fast approximation of the bilateral filter using a signal processing approach. In: Leonardis, A., Bischof, H., Pinz, A. (eds.) ECCV 2006. LNCS, vol. 3954, pp. 568–580. Springer, Heidelberg (2006)
20. Robust Image Understanding Lab: EDISON System, http://www.caip.rutgers.edu/riul/research/code/EDISON/doc/segm.html (accessed 2008-09-23)
21. Felzenszwalb, P.: Efficient Graph-Based Image Segmentation (March 2007), http://people.cs.uchicago.edu/~pff/segment/ (accessed 2008-01-11)
22. Besedin, D.: RightMark Memory Analyzer, http://cpu.rightmark.org (accessed 2009-01-09)

Coarse-to-Fine Tracking of Articulated Objects Using a Hierarchical Spring System*

Nicole Artner[1], Adrian Ion[2], and Walter Kropatsch[2]

[1] Austrian Research Centers GmbH - ARC, Smart Systems Division, Vienna, Austria
nicole.artner@arcs.ac.at
[2] PRIP, Vienna University of Technology, Austria
{ion,krw}@prip.tuwien.ac.at

Abstract. Tracking of articulated objects is a challenging task in Computer Vision. A highly target specific model can improve the robustness of the tracking by eliminating or reducing the ambiguities in the association task. This paper presents a flexible framework, which allows to build target specific, part-based models for arbitrary articulated objects. The rigid parts are described by hierarchical spring systems in form of attributed graph pyramids and connected via articulation points, which transfer position information between the adjacent parts.

1 Introduction

Tracking the parts of articulated objects in video sequences is still a challenging task with a lot of open problems. Promising approaches dealing with this task employ part-based models and match this models into the image with the help of statistics to maximize a probability function.

A possibility to build part-based models and describe the spatial relationships between the parts of the target object in a tolerant, deformable way are spring systems. Spring systems can be represented by graphs, where each part of a target object is a vertex and the edges encode their spatial relationships. Object recognition or tracking can be done by minimizing the energy in the spring system to find the most likely configuration of the object parts in an image. Spring systems have already been proposed in 1973 by Fischler et al. [1]. Felzenszwalb et al. employed this idea in [2] to do part-based object recognition for faces and articulated objects (humans). Their approach is a statistical framework minimizing the energy of the spring system learned from training examples using maximum likelihood estimation. Ramanan et al. apply in [3] the ideas from [2] in tracking people. In [4], Mauthner et al. present an approach using a two-level hierarchy of particle filters for tracking objects described by spatially related parts in a mass spring system.

In this paper we also employ spring systems to encode the relationships in a part-based model, but in comparison to the related work we try to stress solutions

---

* Partially supported by the Austrian Science Fund under grants P18716-N13 and S9103-N13.

that emerge from the underlying structure, instead of using structure to verify statistical hypothesis. The aim is to supply a flexible framework that allows to build part-based models for arbitrary objects with varying number of rigid parts and articulation points. Each rigid part is robustly tracked with the help of a hierarchical spring system encoding the spatial relationships of coarse and fine features. The articulation points in the model act as agents of the information transfer between the parts of the object. They transfer position information from reliable parts to ambiguous parts. The approach presented here refines and extends our previous work in [5]. Possible applications are action recognition, human computer interfaces, motion based diagnosis and identification, etc.

There is a vast amount of work in the field of tracking articulated objects and motion analysis [6,7,8]. It would go beyond the scope of this paper mentioning all of this work. In comparison to many related approaches our approach does not need any training and we do not employ motion models. The presented approach relies on the spatial relationships of object parts and their features, and hence resulting distance constraints.

The paper is organized as follows: Sec. 3 explains how the hierarchical spring systems are built. In Sec. 4 the task of articulation points is described. Sec. 5 sums up the presented concepts and describes their combination in tracking. Sec. 6 presents experiments to prove and qualitatively evaluate the concept of our approach and in Sec. 7 we draw conclusions.

2 The Building Blocks of the Spring Hierarchy

Articulated objects are made out of rigid parts connected through articulation points. On each rigid part, multiple features are tracked through a mixture of many independent trackers, one for each feature, and a spring system, one for each part. The final position of each feature is decided based on the offset vectors from the tracker and the spring system. This section recalls Mean shift, the method used for the independent trackers, and the spring system.

2.1 Mean Shift Algorithm

The Mean shift algorithm [9] is employed to associate the features of the object parts between consecutive frames. It does this by efficiently finding local maxima in a probability distribution, and generating an offset vector pointing to the corresponding position. The distribution encodes the probability that a given feature from the previous frame is in a certain position in the current frame. To compute the probability that a certain feature, *the target*, matches the feature at a certain position, the following similarity measure is used (see Eq. 3).

Region covariance was introduced by Porikli and Tuzel as a feature for detection, classification and tracking in [10,11]. It is invariant to scaling and rotation up to a certain degree (depends on the feature selection) and allows the combination of multiple features in an elegant way. Furthermore, compared to other region descriptors, region covariance is low-dimensional and can be efficiently calculated using integral images.

The covariance feature is extracted out of an one dimensional intensity or a three dimensional color image I. F is a $W \times H$ dimensional feature image extracted from I, encoding a feature vector of size d at each position $F(x, y)$:

$$F(x, y) = \phi(I, x, y), \qquad (1)$$

where the function ϕ can be any mapping including e.g. intensity, color, gradients and so on. A rectangular region of interest $R \subset F$ can be represented by the $d \times d$ covariance matrix

$$C_R = \frac{1}{n-1} \sum_{k=1}^{n} (z_k - \mu)(z_k - \mu)^T, \qquad (2)$$

where $\{z_k\}_{k=1..n}$ are the d-dimensional feature vectors of the points in R and μ is the mean over all points. The following distance measure is used to calculate the similarity between two covariance matrices [11]:

$$\rho(C_1, C_2) = \sqrt{\sum_{i=1}^{n} \ln^2 \lambda_i(C_1, C_2)}, \qquad (3)$$

where $\{\lambda_i(C_1, C_2)\}$ are the generalized eigenvalues of C_1 and C_2.

2.2 Spring System as a Graph

An attributed graph (AG) is a possible data structure for a spring system. The attributes of the vertices of the graph are the features and their corresponding positions. We use covariance matrices as the features (Sec. 2.1), but other features can also be used (e.g. 3D color histogram features [5]).

Given the features, the edges of the AG are obtained by a Delaunay triangulation. A fully connected graph (connected each vertex with each other vertex in the graph) could also be used but it would increase the complexity of the optimization process.

The elastic behavior (tolerance to variations in the structure) of a spring system can be modeled by graph relaxation. As the tracked object parts are rigid, the objective of the relaxation is to maintain the tracked structure as similar as possible to the initial structure. Thus the aim is to keep the edge lengths as similar as possible to the initial length. The total energy of the spring system is 0 in the initial state and increases with the deformation of the structure.

The variation of the edge lengths in the AG and their directions are used to determine a structural offset for each vertex. This offset vector is the direction where a given vertex should move such that its edges restore their initial length and the energy of the structure is minimized. This structural offset vector $\boldsymbol{O}$ is calculated for each vertex v as follows:

$$\boldsymbol{O}(v) = \sum_{e \in E(v)} k \cdot (|e'| - |e|)^2 \cdot (-\boldsymbol{d}(e, v)), \qquad (4)$$

where $E(v)$ are all edges e incident to vertex v, k is the elasticity constant of the edges in the structure, e is the edge length in the initial state and e' at a different point in time. $\mathbf{d}(e,v)$ is the unitary vector in the direction of edge e that points toward v. For more details see [5].

3 Building the Hierarchical Spring System

Each rigid part of a target object is described and tracked in a coarse-to-fine manner. Each part is described by a two level spring system represented by an attributed graph pyramid [12].

As shown in Fig. 1(a), the top level is described by one covariance feature C_t, extracted out of a region of interest (ROI) covering the whole object part. The bottom level consists of several features, which are from the same ROI (see Fig. 1(b)). A Harris corner detector is applied on the ROI to find promising positions for the region covariance features $\{C_b\}_{i=1..n}$ of the bottom level. Around each corner point a small ROI is built to calculate C_b (e.g. 9×9 pixels).

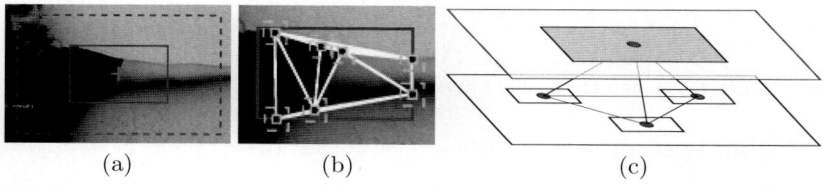

(a) (b) (c)

Fig. 1. Extracting region covariance features. (a) Feature of top level. (b) Features of bottom level. The white edges are the edges of the AG. (c) Attributed graph pyramid.

The **AG pyramid** is built as follows (see Fig. 1(c)). In the **top level** of the pyramid: one vertex to which the coarse feature C_t of the whole rigid part is assigned. In the **bottom level** all fine features C_b. The edges in the bottom level of the pyramid are inserted with a Delaunay triangulation. The vertex in the top level is connected with every vertex (child) in the bottom level. The spring system is initialized with the state in the first frame, meaning that the total energy of the spring system is considered 0 in this configuration.

4 Articulation Points: Agents of the Information Transfer

An *articulation point* connects several rigid parts. It allows them to move independent from each other, but forces them to always keep the same distance. From this follows that the movement of a rigid part in the image plane is constraint to the circle centered at the articulation point and spanned by the radius corresponding to its size (in 3D it is a sphere). Fig. 2 visualizes this concept. It would be possible to connect every point with the articulation point, but to

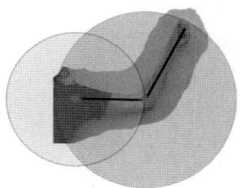

Fig. 2. Distance constraints imposed by articulation points

reduce the complexity we only built a spring system with two reliable vertices in the bottom level and with the top vertex.

Articulation points can be initialized manually or automatically by observing the articulated motion of the target object [5,13] parts.

Deriving the position of the articulation points. In the frame in which the position of the articulation point is initialized, for each adjacent part and each pair of features of the bottom level of the part, a local coordinate system is created. The coordinates of the articulation point in this coordinate systems is stored. Having the position of any two features is then enough to reconstruct the position of the articulation point, and thus at any time, each part can generate a hypothesis for the position of all adjacent articulation points. This hypothesis is produced with the local coordinate system of the two most reliable features (see Sec. 4.1) – further on named *reference vertices* – of each part.

The hypothesis of all connected parts of an articulation point are combined with a weighted sum, where the weight for each hypothesis depends on the reliability of the corresponding part (see Sec. 4.1). With this weighting, the influence of ambiguous parts (e.g. occluded parts) on the position of the articulation point is low and for reliable parts high.

After the position of the articulation point is computed, the articulation "transfers" position information from reliably to ambiguously tracked parts through its distance constraints (circles). This is done in spring systems, where the articulation point is connected to the reference vertices and the top vertices.

4.1 Computing the Reliability of Features and Parts

The reliability of a feature b depends on the number of incident edges in the spring system I_b, the energy of the incident edges in the spring system E_b and the similarity S_b (see Eq. 3) of the covariance feature C_b to the template covariance feature from the first frame:

$$R_b = I_b \cdot \alpha_I + E_b \cdot \alpha_E + S_b \cdot \alpha_S$$

$$I_b = \frac{E(v)}{E}, \quad E_b = \frac{\sum_{e \in E(v)} k \cdot (|e'| - |e|)^2}{T_{E_p}}, \quad S_b = \frac{\rho(C_b, C_t)}{\rho_{\max}}. \tag{5}$$

$E(v)$ are all edges incident to vertex v (feature), E is the number of edges in the spring system, T_{E_p} is the total energy of the spring system, and $\rho_{\max}$ is the highest similarity in the same part as feature b. In our experiments: $\alpha_I = 0.2$, $\alpha_E = 0.4$, $\alpha_S = 0.4$. The reliability of a part p is:

$$R_p = D_p \cdot \alpha_D + E_p \cdot \alpha_E + S_p \cdot \alpha_S$$

$$D_p = \frac{F_p}{F}, \quad E_p = \frac{T_{E_p}}{T_E}, \quad S_p = \frac{\sum_{b \in p} \rho(C_b, C_t)}{F_p} \tag{6}$$

is computed out of the size D_p of the part p, the energy of the part E_p, and the similarity S_p of the covariance features in comparison to their templates. F_p is the number of features of part p, F is the number of all features, and T_E is the energy of all spring systems. In our experiments: $\alpha_D = 0.2$, $\alpha_E = 0.4$, $\alpha_S = 0.4$. Intuitively the two measures model the mixture of "seeing" and "knowing".

5 Tracking as a Hierarchical Optimization Process

Tracking is done in a coarse-to-fine manner – from top to bottom level of each part (summarized in Algorithm 1).

Algorithm 1. Algorithm for tracking articulated objects

1: PROCESSFRAME
2:    associate top vertex of each part with Mean shift
3:    associate bottom vertices of each part with Mean shift and structural offsets
4:    select reference vertices for each part
5:    calculate current position of articulation points
6:    transfer position information over articulation points to top and bottom levels
7: **end**

The first step is to associate the top vertices of each part using the positions from the previous frame and applying Mean shift to a probability distribution built with the similarity measure in Eq. 3.

In the next step the bottom vertices of each part are associated by combining Mean shift offsets and structural offsets. The structural offsets are generated out of the spring systems of the bottom levels and the spring systems connecting each bottom vertices with the corresponding top vertices. For each feature (vertex) depending on its reliability value R_b a mixing gain $g = 0.5 - (R_b - 0.5)$ is computed and used to combine the offsets.

Then for each part the two vertices with the highest R_f are selected to generate the hypothesis for the positions of the articulation points. The hypothesis of the parts connected to a articulation point are mixed with a weighted sum depending on the reliability value R_p of each part.

In the last step, the position information between the parts is transfered over the articulation points to the top and reference vertices which forward the information to the vertices not directly connected. This transfer is again done in a combined iterative process with Mean shift and structure.

6 Experiments

In all experiments we use prior knowledge about the structure of the object to initialize the ROIs and the articulation points. The spring constant k is set for edges in the bottom level to 0.2 and for edges connecting to the articulated point or to the top vertex to 0.5.

In experiment 1, the lower and upper arm of a human are successfully tracked through articulated motion (see Fig. 3). Experiment 2 in Fig. 4 shows frames with scissors. One part of the structure is completely occluded, but the position of the articulation point (red star) is robust and the structure relaxes when the occlusion is gone. In experiment 3 one can see the tracking of 4 parts connected with 3 articulation points (see Fig. 5).

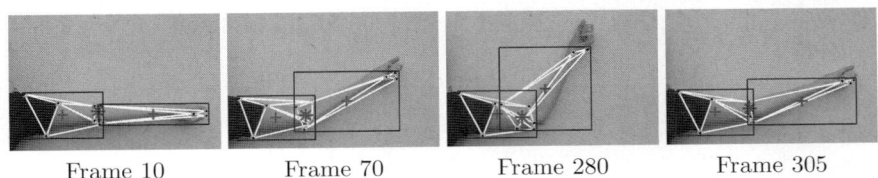

Fig. 3. Experiment 1: Tracking a human's upper and lower arm in articulated motion

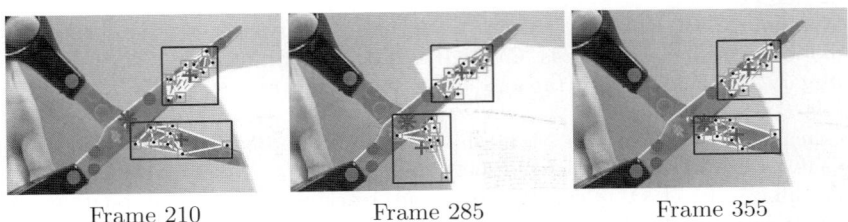

Fig. 4. Experiment 2: Tracking through occlusion

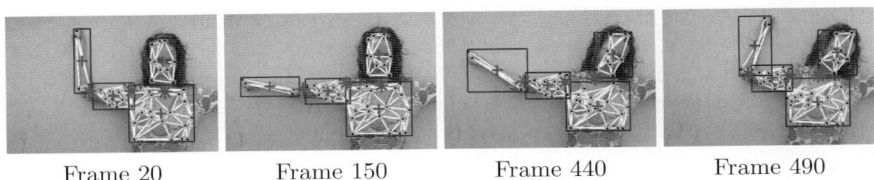

Fig. 5. Experiment 3: Tracking 4 parts of a human with 3 articulation points

7 Conclusion

This paper presented an approach for describing and tracking of articulated objects consisting of several rigid parts connected with articulation points. The object parts are described in a coarse-to-fine manner in an AG pyramid, where the features are region covariance matrices and the spatial relationships between the features are enforced during the tracking through a hierarchical spring system. Position information is transfered between the parts over the corresponding articulation points depending on the reliability of the parts and their features. Open issues are dealing with pose changes and the corresponding changes in the structure, optimizing the information transfer in big structures and automatically initializing the structure.

References

1. Fischler, M.A., Elschlager, R.A.: The representation and matching of pictorial structures. Transactions on Computers 22, 67–92 (1973)
2. Felzenszwalb, P.F.: Pictorial structures for object recognition. IJCV 61, 55–79 (2005)
3. Ramanan, D., Forsyth, D.: Finding and tracking people from the bottom up. In: CVPR, vol. 2, pp. 467–474. IEEE, Los Alamitos (2003)
4. Mauthner, T., Donoser, M., Bischof, H.: Robust tracking of spatial related components. In: ICPR, pp. 1–4. IEEE, Los Alamitos (2008)
5. Artner, N., Ion, A., Kropatsch, W.G.: Tracking objects beyond rigid motion. In: GbR. Springer, Heidelberg (2009)
6. Gavrila, D.M.: The visual analysis of human movement: A survey. CVIU 73(1), 82–980 (1999)
7. Moeslund, T.B., Hilton, A., Krger, V.: A survey of advances in vision-based human motion capture and analysis. CVIU 104(2-3), 90–126 (2006)
8. Aggarwal, J.K., Cai, Q.: Human motion analysis: A review. CVIU 73(3), 428–440 (1999)
9. Comaniciu, D., Meer, P.: Mean shift: A robust approach toward feature space analysis. PAMI 24(5), 603–619 (2002)
10. Porikli, F., Tuzel, O., Meer, P.: Covariance tracking using model update based on lie algebra. In: CVPR, June 2006, vol. 1, pp. 728–735 (2006)
11. Tuzel, O., Porikli, F., Meer, P.: Region covariance: A fast descriptor for detection and classification. In: Leonardis, A., Bischof, H., Pinz, A. (eds.) ECCV 2006. LNCS, vol. 3952, pp. 589–600. Springer, Heidelberg (2006)
12. Kropatsch, W.G., Haxhimusa, Y., Pizlo, Z., Langs, G.: Vision pyramids that do not grow too high. Pattern Recognition Letters 26(3), 319–337 (2005)
13. Mármol, S.B.L., Artner, N.M., Ion, A., Kropatsch, W.G., Beleznai, C.: Video object segmentation using graphs. In: Ruiz-Shulcloper, J., Kropatsch, W.G. (eds.) CIARP 2008. LNCS, vol. 5197, pp. 733–740. Springer, Heidelberg (2008)

Cooperative Stereo Matching with Color-Based Adaptive Local Support*

Roland Brockers

Jet Propulsion Laboratory - California Institute of Technology
4800 Oak Grove Drive, Pasadena, California
brockers@jpl.nasa.gov

Abstract. Color processing imposes a new constraint on stereo vision algorithms: The assumption of constant color on object surfaces used to align local correlation windows with object boundaries has improved the accuracy of recent window based stereo algorithms significantly. While several algorithms have been presented that work with adaptive correlation windows defined by color similarity, only a few approaches use color based grouping to optimize initially computed traditional matching scores. This paper introduces the concept of color-dependent adaptive support weights to the definition of local support areas in cooperative stereo methods to improve the accuracy of depth estimation at object borders.

1 Introduction

A closer look at recent publications in stereo vision shows that the old separation between slow algorithms, employing a large amount of computational power to achieve the most accurate results, and fast real-time algorithms concentrating on efficiency, becomes more and more blurred. Recent improvements in computer hardware seem to open space for an incorporation of additional computational power to real-time algorithms to increase accuracy by new improvement steps beyond the well known fixed window correlate-and-winner-takes-all scheme of former real-time approaches. Examples are algorithms using local improvement steps that follow the disparity estimation [1] or basic dynamic programming based optimization methods [2].

Another group of algorithms with the potential to significantly improve real-time approaches are algorithms using color grouping for an advanced adaptive correlation [3,2]. Surprisingly, grouping neighboring pixels that are assumed to be located on the same object surface by the similarity of their color is relatively new in stereo vision. Recent algorithms use this constraint to obtain local adaptive correlation windows which are better aligned to object borders, resulting in better correlation accuracy at disparity discontinuities [4,5,6,3]. Grouping is achieved either by color segmentation [4,5,6] or by calculating color-dependent correlation weights [3] to control the influence of pixels inside a correlation window on the matching score. Unfortunately, relatively large correlation windows are needed by these algorithms to eliminate ambiguities (in [3] the typical window size is 33x33 pixels, in [6] results are given for 51x51 windows), which results in a higher computational cost making these approaches unsuitable for

---
* This research was supported by DFG grant BR-3583/1-1.

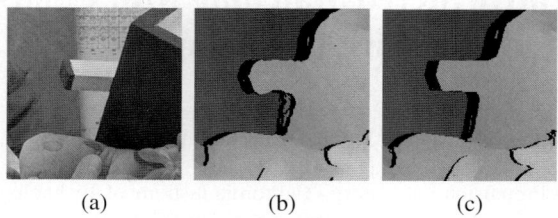

(a) (b) (c)

Fig. 1. Disparity maps for *Teddy* image detail (a): fixed local support leads to object blurring (b), while adaptive local support maintains accurate object borders (c)

real-time applications. To reduce the window size significantly and benefit from a faster implementation we transfer the idea of color grouping to a cooperative optimization algorithm which is highly suited to hardware implementation.

Cooperative approaches are a relatively old group of stereo methods, originally motivated by the biological model of the human visual cortex (cp. [7]). They optimize initially calculated matching scores in a three dimensional disparity label space $\phi(x, y, d)$ by iterative local cooperation of neighboring labels using mainly the stereoscopic continuity and uniqueness constraint coded in the variable update function. While the implementation of the uniqueness constraint differs among algorithms, all cooperative approaches implement the continuity constraint by coupling labels of neighboring pixels within local support areas. In traditional approaches these are fixed local support areas, which can be calculated very efficiently, but entail a common disadvantage: the coupling of labels across object borders, resulting in enlargement of foreground objects. When no adjustments are made, cooperative algorithms will always extend object surfaces from areas with high initial matches into regions with low matching scores, e.g. occluded areas or low textured regions of the input images (cp. Fig. 1). To cope with this general problem of fixed-window based algorithms, alignment mechanisms were proposed in the literature to get more accurate object borders during post-processing [1]. However, within an optimization algorithm, automated adjustment inside the optimization is preferable. Good results can be achieved e.g. with methods performing an initial color-based segmentation to group pixels belonging to the same object surface [4,5,6]. In this paper we show that a computationally expensive pre-segmentation is not necessary and that the grouping can be coded within pre-calculated weights of local support areas used in a cooperative optimization suitable for later hardware acceleration.

The proposed algorithm is described in section 2 and 3. In section 4 the new algorithm is applied to common test scenes with ground truth from the Middlebury stereo vision page [8] to demonstrate the quality of the approach and to compare the results with other related algorithms. Finally, section 5 summarizes the results.

2 Matching Costs

The structure of the proposed algorithm is similar to traditional stereo algorithms (cp. [9]). As an initial guess a correlation based similarity measure quantifies the similarities of potential corresponding pixel pairs. Matching scores are calculated using a sampling insensitive version of normalized cross correlation which correlates the grayscale intensities of the rectified input images to achieve tolerance to noise in local chromaticity (eq. 1-3). Because of the later optimization it is possible to choose a very small correlation window size (e.g. 3x3 pixels). This reduces calculation time and minimizes the influence of errors near disparity discontinuities caused by a non-constant disparity inside the correlation window.

$$s_0(x,d) = \frac{m_{x,d}}{s_{l,x} s_{r,x+d}} \quad (1)$$

$$m_{x,d} = \sum_{\tilde{x}} \varphi\left\{i_l(x+\tilde{x}), \bar{i}_l(x)\right\} \varphi\left\{i_r(x+d+\tilde{x}), \bar{i}_r(x+d)\right\} \quad (2)$$

$$s_{a,k} = \sqrt{\sum_{\tilde{x}} (\varphi\{i_a(k+\tilde{x}), \bar{i}_a(k)\})^2} \quad (3)$$

To achieve insensitivity to image sampling artifacts a difference operator φ is introduced to calculate the difference between the intensity of a particular pixel $i(x+\tilde{x})$ and the mean intensity $\bar{i}(x)$ inside the correlation window with an adapted version of the sampling insensitive dissimilarity measure of Birchfield and Tomasi [10]. As the intensity mean already includes an averaging among several pixel intensities, only $i(x+\tilde{x})$ is expected to be appreciably influenced by sampling effects. Therefore φ calculates the minimum difference between $i(x+\tilde{x})$, linearly interpolated in an interval of $\pm 1/2$ pixel, and the window mean (eq. 5) and uses this difference to calculate the deviation term for the cross correlation (eq. 4).

$$\varphi\{i(x_1), \bar{i}(x_2)\} = i(x_m) - \bar{i}(x_2) \quad (4)$$

$$\text{with } |i(x_m) - \bar{i}(x_2)| = \min_{x_1 - \frac{1}{2} \leq x \leq x_1 + \frac{1}{2}} |i(x) - \bar{i}(x_2)| \quad (5)$$

Note, that this definition of the difference operator minimizes the absolute differences while maintaining the sign of the difference as this is essential for the calculation of cross correlation.

3 Cooperative Optimization

In a second step, a cooperative optimization process adjusts the probability values of the similarity measure to calculate an optimal solution with respect to the implicitly coded stereoscopic constraints. For efficiency reasons, the cooperative optimization is formulated as an iterative cost minimization approach with a global cost function containing only squared cost terms (cp. [11]). To simplify equations, all labels in the disparity space are ordered in a single order parameter vector

$$\boldsymbol{\xi} = (\xi_{(1,d_{min})}, ..., \xi_{(n,d_{min})}, \xi_{(1,d_{min}+1)}, ..., \xi_{(n,d_{max})})^T \quad (6)$$

where the first index $i \in [1, ..., n]$ enumerates all pixels in the reference view.

The global cost function P defines two different cost terms for each variable $\xi_{(i,d)}$:

$$P(\boldsymbol{\xi}) = c_1 \sum_{d=d_{min}}^{d_{max}} \sum_{i=1}^{n} (\xi_{(i,d)} - \xi_{(i,d)_0})^2 + c_2 \sum_{d=d_{min}}^{d_{max}} \sum_{i=1}^{n} \sum_{j \in U_i} \gamma_{ij} (\xi_{(i,d)} - \xi_{(j,d)})^2 \quad (7)$$

Costs are generated when a variable $\xi_{(i,d)}$ differs from its initial value $\xi_{(i,d)_0}$ given by the similarity measure $s_0(x,d)$ (correlation confidence) or if the variable values within a local neighborhood U are diverging (continuity) (eq. 7).

The local support area U_i of a pixel i is defined by a local window surrounding i where the influences of neighboring pixels j are weighted by individual adaptive support weights γ_{ij} which are explained in the following section. c_1 and c_2 are positive constants to trim the final global cost function. Because of the squared cost terms, equation 7 has only one minimum, characterizing the global solution for the optimization problem (cp. [11]). It is calculated numerically with the gradient descent method, where the update function for the iteration is defined as:

$$\boldsymbol{\xi}_{k+1} = \boldsymbol{\xi}_k - \lambda \nabla P(\boldsymbol{\xi}_i), \quad \lambda > 0 \quad (8)$$

with

$$\frac{\delta P}{\delta \xi_{i,d}} = [2c_1 + 4c_2 \sum_{j \in U_i} \gamma_{ij}] \xi_{(i,d)} - 2c_1 \xi_{(i,d)_0} - 4c_2 \sum_{j \in U_i} \gamma_{ij} \xi_{(j,d)}. \quad (9)$$

After convergence, the valid disparity for each pixel in the reference view is selected by a winner-takes-all maximum search over all variables attached to the same pixel (uniqueness).

Computationally, this approach has some advantages compared to other cooperative approaches like [7] or [12]. The avoidance of local competition leads to a simple, linear first derivate of the cost function (eq. 9), making it possible to use a fast standard minimization method to compute the global cost minimum. All calculations are local, which implies a high potential for parallelization in a manner amenable to hardware implementations. Additionally, all calculations consist only of additions and multiplications which is particularly important for FPGA implementation. Finally, the independent definition of the local neighborhood U_i with its appropriate weights makes it possible to implement different kinds of local support. In the proposed algorithm, we define U_i as a circular 2D window in a constant disparity level with pre-calculated adaptive weights γ_{ij}.

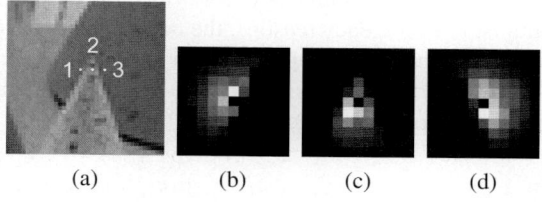

Fig. 2. Adaptive local support: (a) detail of *Cones* scene with marked window positions; (b-d) adaptive weights local support areas for position 1-3 ($D=11$, $\sigma_c=6$, $\sigma_r=2.04$)

3.1 Adaptive Local Support

The local support weights γ_{ij} define the amount of support a variable gets from its neighbors depending on color similarity c_{ij} and spatial distance r_{ij} of the underlying pixels (eq. 10). The definition of the individual weight terms essentially follows Tomasi and Manduchi's definition of bilateral filters [13]. To reflect perceptual metrics, the color difference is defined by the Euclidean distance of the two color vectors $\mathbf{c}_i$ and $\mathbf{c}_j$ of pixel i and j in CIELab color space, calculated from the reference view after applying a bilateral filter to remove local noise. The spatial distance is determined by the Euclidean distance between the image coordinate vectors of i and j. All distances are weighted by Gaussian weighting functions (eq. 11 and 12).

$$\gamma_{ij} = r_{ij} \cdot c_{ij} \tag{10}$$

$$r_{ij} = e^{-\frac{1}{2}\left(\frac{\delta_e(\mathbf{x}_i, \mathbf{x}_j)}{\sigma_r}\right)^2} = e^{-\frac{1}{2}\frac{(x_i-x_j)^2+(y_i-y_j)^2}{\sigma_r^2}} \tag{11}$$

$$c_{ij} = e^{-\frac{1}{2}\left(\frac{\delta_e(\mathbf{c}_i, \mathbf{c}_j)^4}{\sigma_c^2}\right)} = e^{-\frac{1}{2}\frac{((L_i-L_j)^2+(a_i-a_j)^2+(b_i-b_j)^2)^2}{\sigma_c^2}} \tag{12}$$

This differs from the weight definition in [3] where the authors choose a Laplacian kernel for the weighting function, which is less optimal because of lower compactness and smaller weights in the immediate proximity of the center pixel. Note that the color distance in (12) has an exponent of 4 for a sharper separation of color edges.

To define a local support area U_i for pixel i, adaptive support weights γ_{ij} are calculated for all neighboring pixels j inside a fixed local neighborhood of circular shape with diameter D to minimize the mean spatial distance between supporting pixels and the center. While all variables in U_i contribute to the development of the center variable i, the center variable itself is excluded from the support window to inhibit self amplification. Figure 2 shows an example of local support areas in an area of color transition in the *cones* scene of the Middlebury data set.

3.2 Occlusion Detection and Sub-pixel Precision

After the optimization process is complete, occlusions are explicitly detected by searching the disparity map for pixels that point to the same corresponding pixel in the non-reference view or are in succession in a cyclopean view, using the optimized correspondence probabilities (cp. [11]). To calculate sub-pixel precise disparities, the relaxation process is applied once again to the 2D pixel-precise disparity map similar to [11].

4 Experimental Results

In the following we evaluate our algorithm with images from the Middlebury test data set to demonstrate the capabilities of the new approach. Figure 3 illustrates the effect of adaptive local support on variable values before and after the optimization. After initialization with the similarity measure, variables linked to the correct object disparity

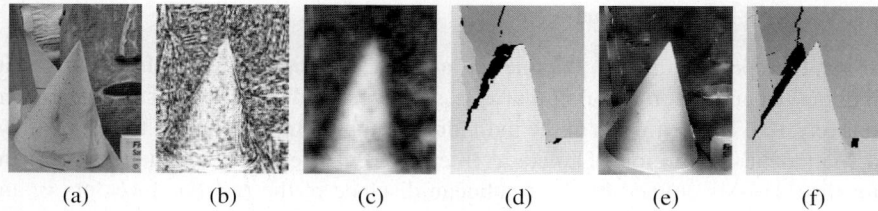

Fig. 3. Adaptive local support compared with fixed local support: (a) detail of reference view; (b) similarity measure in a constant disparity level of $d = 49$; (c) variable values after optimization with fixed local support (circ. Gaussian weighted, $D = 7$); (d) disparity map for fixed local support; (e) variable values after optimization with adaptive weights local support ($\sigma_c = 6$, $\sigma_r = 1.84$, $D = 7$); (f) resulting disparity map

already contain high values, but a large amount of noise can be observed as well due to the small correlation window size of 3x3 pixels (Fig. 3b). After the optimization most of the ambiguity is resolved. If using fixed local support, high variable values are propagated to neighboring variables disregarding of object boundaries (Fig. 3c), leading to misalignment of object borders and calculated disparity edges in the final disparity map (Fig. 3d). With adaptive local support, cross boundary coupling is significantly reduced, resulting in sharp object contours that are aligned with the real object boundaries in both the variable activity map (Fig. 3e) and the disparity map (Fig. 3f).

For quantitative comparison with related stereo algorithms, we applied our algorithm to all four Middlebury test scenes according to [9] ($\delta_d = 1.0$, constant parameter set and back filling of detected occlusions). The calculated disparity maps are shown in Figure 4 and the corresponding error percentages of false matches are illustrated in Table 1. Compared to the *CostRelax* approaches, which use basically a similar optimization method, adaptive local support clearly outperforms fixed local support. With the new similarity measure, accuracy can be further improved, most visibly in low resolution images like *Tsukuba*, which are more prone to sampling artifacts. Compared to other methods, which also use color-based adaptive grouping, the proposed algorithm achieves good results. In Figure 5 the error development during iteration is illustrated for the *Tsukuba* scene. Fig. 5b points out a major advantage of local optimization methods: The quick decrease of overall errors allows an early termination of the iteration

Table 1. Percentage of bad matching pixels for the Middlebury data set ($\delta_d = 1$) for non occluded pixels (non), all pixels (all) and near discontinuities (disc) (*cp. [8] visited in 03/2009)

Algorithm	Tsukuba			Venus			Teddy			Cones		
	non	all	disc	non	all	disc	non	all	disc	non	all	disc
SegmentSupport [6]*	1.25	1.62	6.68	0.25	0.64	2.59	8.43	14.2	18.2	3.77	9.87	9.77
AdaptiveWeight [3]*	1.38	1.85	6.90	0.71	1.19	6.13	7.88	13.3	18.6	3.97	9.79	8.26
Our Method (ncc+BT, 3x3)	**2.91**	**3.49**	**11.4**	**0.60**	**1.11**	**6.45**	**7.92**	**13.7**	**20.9**	**3.59**	**9.43**	**10.3**
Our Method (regular ncc, 3x3)	**4.60**	**5.10**	**13.0**	**0.69**	**1.62**	**6.11**	**8.06**	**16.0**	**21.0**	**3.60**	**12.3**	**10.4**
CostRelax [14] (3D fixed local supp.)*	4.76	6.08	20.3	1.41	2.48	18.5	8.18	15.9	23.8	3.91	10.2	11.8
CostRelax [11] (2D fixed local supp.)	6.33			1.44			9.60			5.24		

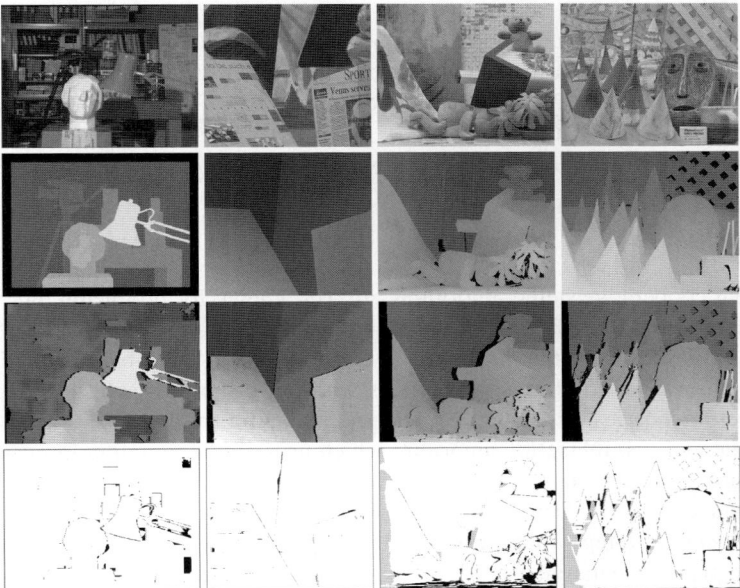

Fig. 4. Results for *Tsukuba*, *Venus*, *Teddy* and *Cones* scene; from top to bottom: left reference view; ground truth; disparity map calculated with proposed algorithm, with black labeled occlusions; error map with false matches (black) and unconsidered errors in true occluded areas (gray) ($c_1 = 0.5$, $c_2 = 10$, $c_3 = 0.5$, $c_4 = 0.2$, $\sigma_c = 6$, $\sigma_r = 8$, $D = 5$, 400 iterations)

in time critical applications, giving a calling process the ability to provide only the momentary available calculation time for stereo calculation while still benefiting from significantly improved disparity maps. When comparing fixed and adaptive local support, the biggest difference is visible at object boundaries (Fig. 5a). Where the old fixed support algorithm generates significant object blurring during iterations, adaptive local support decreases errors over time and maintains accurate object borders.

However, difficulties remain in un-textured areas. This is in part due to the relatively small window size of local support (e. g. $D = 5$ for results in Table 1) and the strict

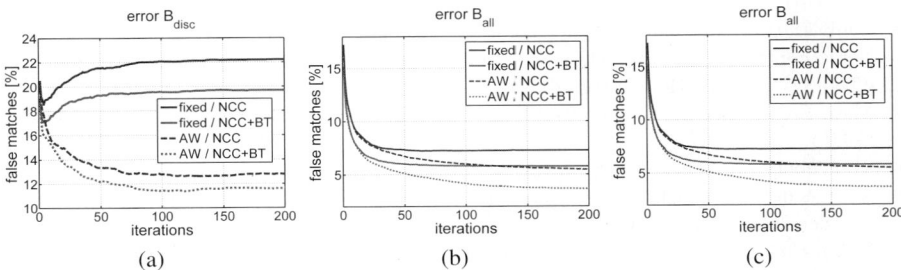

Fig. 5. Error during iteration for the *Tsukuba* scene: Adaptive weights local support (AW) compared with fixed local support (fixed); (a) error near discontinuities; (b) error for all pixels; (c) error in true non-occluded regions. For the definition of different regions see [8].

limitation on non-global optimization. In future work we plan to investigate variable window sizes that adapt to the local amount of texture to deal with these difficulties. In our tests, the algorithm was coded in standard non-optimized C++ code using full float precision and run on an 2.4GHz Intel Core2Duo T7700. The calculation time for the *Tsukuba* scene with 100 iterations was 20s on a single core and 11.5s on both cores. In a future FPGA based implementation we expect to run the algorithm in near real-time.

5 Conclusion

Traditional cooperative stereo algorithms are well known for producing blurry object borders due to a fixed coupling of neighboring pixels to implement the stereoscopic continuity constraint. In this paper we demonstrated that color-based adaptive local support can be used to align support areas with object borders and, thus keep accurate object boundaries throughout the cooperative optimization process. Implemented in a simple and fast relaxation algorithm which uses a new sampling insensitive normalized cross correlation for initial matching, our results show that optimization with local adaptive support generates results comparable with other algorithms that also use adaptive local grouping, while avoiding large correlation windows or computationally expensive pre-segmentation.

References

1. Hirschmüller, H., Innocent, P.R., Garibaldi, J.: Real-time correlation-based stereo vision with reduced border errors. Int. J. Comp. Vision 47, 229–246 (2002)
2. Wang, L., Liao, M., Gong, M., Yang, R., Nister, D.: High-quality real-time stereo using adaptive cost aggregation and dynamic programming. In: Proc. Int. Symp. on 3D Data Processing, Visualization, and Transmission (3DPVT), pp. 798–805 (2006)
3. Yoon, K.J., Kweon, I.S.: Adaptive support-weight approach for correspondence search. IEEE Trans. PAMI 28, 650–656 (2006)
4. Zhang, Y., Kambhamettu, C.: Stereo matching with segmentation-based cooperation. In: Heyden, A., Sparr, G., Nielsen, M., Johansen, P. (eds.) ECCV 2002. LNCS, vol. 2351, pp. 556–571. Springer, Heidelberg (2002)
5. Klaus, A., Sormann, M., Karner, K.: Segment-based stereo matching using belief propagation and a self-adapting dissimilarity measure. In: Proc. Int. Conf. Pat. Recogn. (ICPR), pp. III 15–18 III (2006)
6. Tombari, F., Mattoccia, S., Di Stefano, L.: Segmentation-based adaptive support for accurate stereo correspondence. In: Mery, D., Rueda, L. (eds.) PSIVT 2007. LNCS, vol. 4872, pp. 427–438. Springer, Heidelberg (2007)
7. Marr, D., Poggio, T.: Cooperative computation of stereo disparity. Science 194, 283–287 (1976)
8. Middlebury stereo vision page, http://vision.middlebury.edu/stereo
9. Scharstein, D., Szeliski, R.: A taxonomy and evaluation of dense two-frame stereo correspondence algorithms. Int. J. Comp. Vision 47, 7–42 (2002)
10. Birchfield, S., Tomasi, C.: A pixel dissimilarity measure that is insensitive to image sampling. IEEE Trans. PAMI 20, 401–406 (1998)
11. Brockers, R., Hund, M., Mertsching, B.: Stereo matching with occlusion detection using cost relaxation. In: Proceedings of the IEEE International Conference on Image Processing, ICIP, pp. III–389– III–392 (2005)

12. Zitnick, C.L., Kanade, T.: A cooperative algorithm for stereo matching and occlusion detection. IEEE Trans. PAMI 22, 675–684 (2000)
13. Tomasi, C., Manduchi, R.: Bilateral filtering for gray and color images. In: Proc. IEEE Int. Conf. on Comp. Vision (ICCV), pp. 839–846 (1998)
14. Brockers, R., Hund, M., Mertsching, B.: Stereo vision using cost-relaxation with 3d support regions. In: Proc. Image and Vision Comp. New Zealand (IVCNZ), pp. 96–101 (2005)

Iterative Camera Motion and Depth Estimation in a Video Sequence

Françoise Dibos[1], Claire Jonchery[2], and Georges Koepfler[2]

[1] LAGA, L2TI, Université Paris 13
99, avenue Jean-Baptiste Clément 93430 Villetaneuse, France
[2] MAP5, Université Paris Descartes,
45, rue des Saints-Pères 75270 Paris Cedex 06, France

Abstract. This paper addresses the problem of the joint determination of camera motion parameters and scene depth information. A video sequence obtained by a one hand moving camera is the input. It is well known that the movement of a pixel between two consecutive images depends on the motion parameters and also on the depth of the projected point. Based on a camera motion estimation which uses the registration group and an energy minimization based on the Belief Propagation algorithm, we propose an iterative method combining camera motion and depth estimation.

Keywords: camera motion, depth estimation, registration group, belief propagation.

1 Introduction

The estimation of scene depth based on two images is a difficult problem because of occlusions, discontinuities of the depth, texture and noise in the images. This problem, also called stereo matching, has been largely studied. Scharstein and Szeliski [9] propose a large comparison of methods. They conclude that good results are obtained when the depth map is modelled by Markov Random Fields and if for minimization Belief propagation and Graph Cut methods are employed. These minimization tools are compared, for efficiency and precision, by Tappen and Freeman [12].

In this paper the Belief Propagation algorithm will be used for estimating depth from two consecutive images of a video sequence. But rather than using rectified images, as do the methods in [5,9,11,12], an estimation of the camera motion parameters by a direct method is proposed. For a comparison of methods based on rectified/non rectified images see the work of Schreer, Brandburg and Kauff [10].

To estimate camera motion direct methods use the content of a couple of images. They are often based on the constraint of constant illumination, minimized by a least squares approach on the parameters of a motion model. Different assumptions are used to avoid estimating depths over all the points. Horn and Weldon in [4] and Bergen et al., in [1], assume that the depth map is locally constant. In [6], Negahdaripour and Horn consider that it is planar or quadratic. Following [2], we have precise constraints for applying the constant depth hypothesis in our algorithm.

In Section 2 we present our algorithm for camera motion estimation, Section 3 presents depth estimation based on Belief Propagation. Section 4 presents the iterative method for solving both problems and some experimental results.

2 Motion Estimation

In this section we will show that the deformation of an image due to camera displacement may be obtained thanks to a modelization with a six parameter group. These parameters will be used to find the displacement between two successive frames.

A camera projects a point in 3D space on a 2D image. This transformation can be described using the well-known pinhole camera model [3], see Figure 1, left. The camera is located in C, the optical center, and directed by k, the optical axis. The camera projects a point M of the 3D space on the plane $\mathcal{R} : \{Z = f_c\}$. The plane $\mathcal{R}$ is called the retinal plane and f_c the focal length. The projection m of M is then the intersection of the optical ray (CM) with $\mathcal{R}$.

Let c be the intersection of the optical axis with $\mathcal{R}$. If (X, Y, Z) are the coordinates of M in the camera coordinate system (C, i, j, k) and (x, y) the coordinates of m in the orthogonal basis (c, i, j), the relationship between (x, y) and (X, Y, Z) is the following

$$x = f_c \frac{X}{Z} \quad \text{and} \quad y = f_c \frac{Y}{Z}.$$

As f_c just acts as a scaling factor on the image and in order to simplify the presentation, we choose in this section without loss of generality, to set the focal length to one. Thus f_c will be the unit for the camera and image coordinate systems.

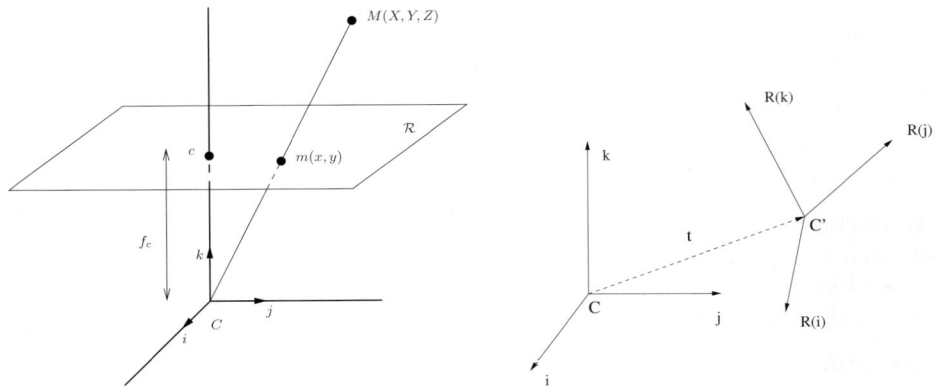

Fig. 1. Left: pinhole camera model. Right: displacement $D = (R, t)$.

Let D be a displacement of the camera or, in an equivalent way, a displacement of the plane $\mathcal{R}$. The movement D may be written in a unique way as $D = (R,t)$, where R is a rotation with axis containing C and t a translation. The set of displacements $D = (R,t)$ forms the Lie group of rigid transformations in $\mathbb{R}^3$ called $SE(3)$, which denotes the special Euclidean group.

The displacement $D = (R,t)$ transforms a point M belonging to $\mathbb{R}^3$ in $M' = RM + t$. Thus, the camera is identified before the displacement by (C, i, j, k) and after the displacement by $(C', R(i), R(j), R(k))$, with $CC' = t$, see Figure 1, right. In the following, denote

$$R = \begin{pmatrix} a_1 & b_1 & c_1 \\ a_2 & b_2 & c_2 \\ a_3 & b_3 & c_3 \end{pmatrix} \quad \text{and} \quad t = \begin{pmatrix} t_1 \\ t_2 \\ t_3 \end{pmatrix}.$$

Let f and g be two adjacent images in a sequence defined on rectangular domains K of $\mathcal{R}$ and K' of $\mathcal{R}'$ (with $f_c = 1$). Let M be a point in $\mathbb{R}^3$ such that its projections m and m' on $\mathcal{R}$ and $\mathcal{R}'$ belong to K and K'. Denote $m = (x,y)$ in (c, i, j) and $m' = (x', y')$ in $(c', R(i), R(j))$. Under the assumption of constant illumination, one has

$$g(x', y') = f(x, y) = f \circ \varphi(x', y')$$

thus

$$g(x', y') = f\left(\frac{a_1 x' + b_1 y' + c_1 + \widetilde{t_1}}{a_3 x' + b_3 y' + c_3 + \widetilde{t_3}}, \frac{a_2 x' + b_2 y' + c_2 + \widetilde{t_2}}{a_3 x' + b_3 y' + c_3 + \widetilde{t_3}} \right)$$

where $\widetilde{t} = t/Z(x', y')$, where $Z(x', y')$ is the depth of M in $(c', R(i), R(j))$.

In Theorem 1,[2], it has been proved that if the camera translation is small with respect to the mean scene depth, thus the above model can be simplified by assuming that $Z(x', y') = Z_0$=constant. The two images f and g are thus linked by $g = f \circ \varphi$ and $f = g \circ \psi$, where the applications φ and ψ are projective transformations associated to the following invertible matrices $\mathcal{M}_\varphi$ and $\mathcal{M}_\psi$

$$\mathcal{M}_\varphi = \begin{pmatrix} a_1 & b_1 & c_1 + \widetilde{t_1} \\ a_2 & b_2 & c_2 + \widetilde{t_2} \\ a_3 & b_3 & c_3 + \widetilde{t_3} \end{pmatrix} = R \begin{pmatrix} 1 & 0 & \langle \widetilde{t}, R(i) \rangle \\ 0 & 1 & \langle \widetilde{t}, R(j) \rangle \\ 0 & 0 & 1 + \langle \widetilde{t}, R(k) \rangle \end{pmatrix} = RH$$

and

$$\mathcal{M}_\psi = \begin{pmatrix} a_1 & a_2 & a_3 - \langle \widetilde{t}, R(i) \rangle \\ b_1 & b_2 & b_3 - \langle \widetilde{t}, R(j) \rangle \\ c_1 & c_2 & c_3 - \langle \widetilde{t}, R(k) \rangle \end{pmatrix} = R^{-1} \begin{pmatrix} 1 & 0 & -\widetilde{t_1} \\ 0 & 1 & -\widetilde{t_2} \\ 0 & 0 & 1 - \widetilde{t_3} \end{pmatrix} = R^{-1} \widetilde{H}.$$

Remark: In this formulation, the matrix $\mathcal{M}_\psi$ corresponds to the matrix associated to the inverse camera displacement. This leads us to introduce the registration group, in which the inverse deformation corresponds to the inverse camera displacement. This is not true in the projective group.

Definition of the registration group: Let $\mathcal{A}$ be the subset of projective applications

$$\mathcal{A} = \left\{ \varphi : \mathbb{R}^2 \to \mathbb{R}^2 \text{ so that } \forall (x,y) \in \mathbb{R}^2, \right.$$

$$\varphi(x,y) = \left(\frac{a_1 x + b_1 y + c_1 + A}{a_3 x + b_3 y + c_3 + C}, \frac{a_2 x + b_2 y + c_2 + B}{a_3 x + b_3 y + c_3 + C}\right), \quad (1)$$

$$\text{where } R = \begin{pmatrix} a_1 & b_1 & c_1 \\ a_2 & b_2 & c_2 \\ a_3 & b_3 & c_3 \end{pmatrix} \in SO(3) \text{ and } (A,B,C) \in \mathbb{R}^3 \Big\}.$$

The registration group is $(\mathcal{A}, \star)$, where the composition law $\star$ is deduced from the composition law $\circ$ of $SE(3)$ through the isomorphism $\mathcal{I} : \mathcal{A} \longrightarrow SE(3)$ defined for all $\phi \in \mathcal{A}$: $\mathcal{I}(\phi) = (R,t)$. Here R is the rotation defined above and $t = (A,B,C)$ is the translation.

Some properties of the registration group(see [2] for more details):

1. The registration group is a six parameter group, whereas the projective group is an eight parameter group.
2. A rotation may be decomposed into two rotations $R = R_{\theta,\alpha} R_\beta^k$. Where R_β^k is a rotation with axis k and angle β; $R_{\theta,\alpha}$ is a rotation with axis Δ and angle α, Δ belonging to the plane (C, i, j) and angle$(i, \Delta) = \theta$.
   This decomposition emphasizes the role of the purely projective deformation due to $R_{\theta,\alpha}$ which displaces the optical axis k.
3. The registration group will be used for the composition of elementary displacements. Let φ be given by (1) and $R = R_{\theta,\alpha} R_\beta^k$, then an elementary displacement is given by

$$\begin{cases} x' - x \approx -Cx + A - \beta y - \alpha x(y\cos\theta - x\sin\theta) + \alpha\sin\theta \\ y' - y \approx -Cy + B + \beta x - \alpha y(y\cos\theta - x\sin\theta) - \alpha\cos\theta \end{cases}$$

where (A,B,C) are the parameters of the elementary translation and (θ, α, β) are the parameters of the elementary rotation decomposed as above.

The movement parameters $(\theta, \alpha, \beta, A, B, C)$ are computed by adapting the parametric 2D registration algorithm "Motion2D" of Odobez and Bouthemy [7].

3 Depth Recovery

The preceding section showed how to obtain the camera motion parameters between two frames and thus the global registration between two frames.

The translation parameters depend on the mean depth of the scene which replaces the effective depths. Therefore, by comparing the second image to the deformation of the first image under the camera movement, we are able to determinate the relative depths of the scene.

Denote by $y = \{f, g, \theta, \alpha, \beta, t_1, t_2, t_3\}$ two consecutive images and the associated camera motion parameters. In the sequel, y is considered to be a realization of a random field Y. Indeed, we want to maximise the *a posteriori* probability

$$P(\mathbf{x}) = P(X = \mathbf{x} \mid Y = y)$$

where X is the depth field of the scene with respect to f. In the Bayesian context

$$P(\mathbf{x}) = P(X = \mathbf{x} \mid Y = y) \propto P(Y = y \mid X = \mathbf{x}) P(X = \mathbf{x}).$$

Where

$$P(Y = y \mid X = \mathbf{x}) = \prod_{s \in V} \psi(\mathbf{x}_s) \text{ and } P(X = \mathbf{x}) = \prod_{(s,t) \in E} \psi_{st}(\mathbf{x}_s, \mathbf{x}_t).$$

Here V is the set of pixels, $\mathbf{x}_s$ the relative depth at pixel s and $\psi(\mathbf{x}_s)$ the local likelihood of depth in pixel s. E is the set of edges and $\psi_{st}(\mathbf{x}_s, \mathbf{x}_t)$ enforces regularity of depth for neighbouring pixels s and t. The method used by Sun, Shum and Zheng [11] on rectified images has been adapted, taking

$$\psi_{st}(\mathbf{x}_s, \mathbf{x}_t) = (1 - e_1)e^{-\frac{|\mathbf{x}_s - \mathbf{x}_t|}{\sigma_1}} + e_1 \text{ and } \psi(\mathbf{x}_s) = (1 - e_2)e^{-\frac{|f(s) - g(s')|}{\sigma_2}} + e_2.$$

Here s' is the pixel obtained from s through the estimated camera movement $\{\theta, \alpha, \beta, t_1, t_2, t_3\}$ if the depth of the 3D point associated to s is equal to $\mathbf{x}_s$. If $s = (x, y)$ and $s' = (x', y')$, one obtains

$$g(x', y') = f\left(f_c \frac{a_1 x' + b_1 y' + f_c c_1 + \frac{t_1}{\mathbf{x}_s}}{a_3 x' + b_3 y' + f_c c_3 + \frac{t_3}{\mathbf{x}_s}}, f_c \frac{a_2 x' + b_2 y' + f_c c_2 + \frac{t_2}{\mathbf{x}_s}}{a_3 x' + b_3 y' + f_c c_3 + \frac{t_3}{\mathbf{x}_s}} \right).$$

Notice that here the focal length f_c is introduced again. With the preceding setup, it is easy to use the "max-product" version of the Belief Propagation algorithm. For more details on Belief Propagation see [8,13,14].

4 Algorithm

Based on the methods developed in the two preceding sections, an iterative algorithm for joint camera motion and depth estimation is now proposed. The estimated depths X are used to determine 2D movements of regions which are of similar depth, from this a new estimation of the camera movement is deduced. From this new displacement, depth can be estimated,...

More precisely, let Λ be the finite set of relative depths. In Section 2, we estimated the movement parameters by considering a mean scene depth Z_0. Let us now label Z_0 to depth 1. For $\mathbf{x}_s$ smaller than 1 the pixel is in front of the mean depth, whereas for $\mathbf{x}_s$ larger than 1 it is beyond.

The set Λ is partitioned into H subsets I_h, $\Lambda = I_1 \cup I_2 \cup \cdots \cup I_H$. This allows to partition the set of pixels according to depth. By $\hat{X}_h$ we denote the mean depth of I_h: the displacement for pixels belonging to the same class will be estimated with respect to the mean depth.

Algorithm:

1. Initialization
   - Estimate the displacement D_1 between images f and g (see Section 2).
   - Estimate the depth field $\mathbf{X}_1$ by using f, g and D_1 (see Section 3).
   - Denote by $\tilde{\varphi}$ the projective transformation associated to the displacement D_1 and depth $\mathbf{X}_1$.
2. For $i = 2, \ldots, N$
   - For $h = 1, \ldots, H$
     – Estimate the displacement for all pixels in the class $K \cap \{\mathbf{X}_{i-1} \in I_h\}$, the depth being set to $\hat{X}_h$;
     – Compute the weights $p_h = \#\{\mathbf{X}_{i-1} \in I_h\}/\#K$.
   - Compute D_i from the preceding H displacements and by using the weights p_h.
   - Estimate the depth field $\mathbf{X}_i$ by using f, g and D_i. Denote φ_i the projective transformation associated to the displacement D_i and depth Z_i.
   - If $\|f \circ \varphi_i - g\|_1 < \|f \circ \tilde{\varphi} - g\|_1$, then set $\tilde{\varphi} = \varphi_i$.

5 Experimental Results and Conclusion

In this section, we present numerical experiments in order to illustrate the improvements due to the iterations of depth and camera motion estimation.

Fig. 2. Top row: two images from a sequence; bottom left, depth map using only Belief Propagation (100 iterations); bottom right, the depth map obtained after 15 iterations of the iterative algorithm, (100 for each Belief Propagation)

Fig. 3. Top row: two images from a sequence; bottom left, depth map using only Belief Propagation (100 iterations); bottom right, the depth map obtained after 15 iterations of the iterative algorithm, (100 for each Belief Propagation)

For the experiments, the following parameter values have been used: In the Belief Propagation algorithm $e_1 = 0.01$, $\sigma_1 = 0.3$, $e_2 = 0.05$ and $\sigma_2 = 20$. We set $H = 3$, $I_1 = \{0.45, 0.5, 0.55, 0.6, 0.65, 0.7, 0.75\}$ $I_2 = \{0.8, 0.9, 1, 1.1, 1.2\}$ and $I_3 = \{1.4, 1.6, 1.8, 2\}$. Thus $\Lambda = I_1 \cup I_2 \cup I_3$, $\hat{X}_1 = 0.6$, $\hat{X}_2 = 1$ and $\hat{X}_3 = 1.7$. We consider only three "mean" depth planes, thus estimating the displacement in an independent way for each of these planes.

Figure 2 presents bottom left the result of applying 100 iterations of the Belief Propagation. As described above, the images are not rectified, instead the estimated camera movement is used. The bottom right image shows the result after 15 iterations of displacement and depth estimation. The bottle, the kettle and the boundaries of the table are much better identified.

In Figure 3 the initial result, bottom left, allows to identify the parasol in the first plane, the fronts of the buildings in the second plane and a region between the two buildings in the back plane. Nevertheless, for the left building some windows are wrongly detected to be in the first plane and for the right building, which is orthogonal to the optical axis, various depths are detected. These effects are attenuated after 15 iterative refinements of the depth estimation.

Conclusion. This paper presents an algorithm for depth estimation based only on the knowledge of 2D images from a sequence. Camera displacement and the depth of the 3D scene are iteratively estimated. This seems a natural way to address these closely linked problems. Each solution will be enhanced thanks to the refinement of the solution to the other problem.

Future work includes implementation of a block based Belief Propagation algorithm.

References

1. Bergen, J.R., Anandan, P., Hanna, K.J., Hingorani, R.: Hierarchical Model-Based Motion Estimation. In: Sandini, G. (ed.) ECCV 1992. LNCS, vol. 588, pp. 237–252. Springer, Heidelberg (1992)
2. Dibos, F., Jonchery, C., Koepfler, G.: Camera motion estimation through planar deformation determination. Journal of Mathematical Imaging and Vision 32(1), 73–87 (2008)
3. Faugeras, O., Luong, Q.T., Papadopoulo, T.: The Geometry of Multiple Images. MIT Press, Cambridge (2000)
4. Horn, B.K.P., Weldon, E.J.: Direct Methods for Recovering Motion. International Journal of Computer Vision 2, 51–76 (1988)
5. Larsen, E.S., Mordohai, P., Pollefeys, M., Fuchs, H.: Temporally Consistent Reconstruction from Multiple Video Streams Using Enhanced Belief Propagation. In: Proceedings of the 11th IEEE International Conference on Computer Vision, pp. 1–8 (2007)
6. Negahdaripour, S., Horn, B.K.P.: Direct passive navigation. IEEE Trans. on Pattern Analysis and Machine Intelligence 9(1), 168–176 (1987)
7. Odobez, J.M., Bouthemy, P.: Robust Multiresolution Estimation of Parametric Motion Models. Journal of Visual Communication and Image Representation 6(4), 348–365 (1995)
8. Pearl, J.: Probabilistic Reasoning in Intelligent Systems: Networks of Plausible Inference. Morgan Kaufmann Publishers Inc., San Francisco (1988)
9. Scharstein, D., Szeliski, R.: A taxonomy and evaluation of dense two-frame stereo correspondence algorithms. International Journal of Computer Vision 47(1-3), 7–42 (2002)
10. Schreer, O., Brandenburg, N., Kauff, P.: A comparative study on disparity analysis based on convergent and rectified views. In: Proceedings of the 11th British Machine Vision Conference, pp. 556–565 (2000)
11. Sun, J., Shum, H.-Y., Zheng, N.-N.: Stereo matching using belief propagation. In: Heyden, A., Sparr, G., Nielsen, M., Johansen, P. (eds.) ECCV 2002. LNCS, vol. 2351, pp. 510–524. Springer, Heidelberg (2002)
12. Tappen, M.F., Freeman, W.T.: Comparison of graph cuts with belief propagation for stereo, using identical MRF parameters. In: Proceedings of the Ninth IEEE International Conference on Computer Vision, vol. 2, pp. 900–906 (2003)
13. Weiss, Y., Freeman, W.: On the optimality of solutions of the max-product belief propagation algorithm in arbitrary graphs. IEEE Transactions on Information Theory 47(2), 723–735 (2001)
14. Yedidia, J., Freeman, W., Weiss, Y.: Generalized belief propagation. In: Advances in Neural Information Processing Systems, pp. 689–695 (2000)

Performance Prediction for Unsupervised Video Indexing

Ralph Ewerth and Bernd Freisleben

Department of Mathematics and Computer Science, University of Marburg
Hans-Meerwein-Str. 3, D-35032 Marburg, Germany
{ewerth,freisleb}@informatik.uni-marburg.de

Abstract. Recently, performance prediction has been successfully applied in the field of information retrieval for content analysis and retrieval tasks. This paper discusses how performance prediction can be realized for unsupervised learning approaches in the context of video content analysis and indexing. Performance prediction helps in identifying the number of detection errors and can thus support post-processing. This is demonstrated for the example of temporal video segmentation by presenting an approach for automatically predicting the precision and recall of a video cut detection result. It is shown for the unsupervised cut detection approach that the related clustering validity measure is highly correlated with the precision of a detection result. Three regression methods are investigated to exploit the observed correlation. Experimental results demonstrate the feasibility of the proposed performance prediction approach.

Keywords: Performance prediction, video indexing, video retrieval.

1 Introduction

Performance prediction is a powerful method to achieve optimal results for content analysis and retrieval tasks. Recently, proposals in the field of information retrieval addressing the issue of performance prediction have been made [1,2]. In this field, performance prediction is used to identify difficult queries that often lead to bad retrieval results [1,2]. He and Ounis [5] state that reliable prediction of query performance is a way of determining the best retrieval strategy for a given query.

In this paper, it is shown how performance prediction can be applied in the context of video content analysis. Here, performance prediction can help to identify the number of errors and can thus support post-processing. This is demonstrated exemplarily for the task of temporal video segmentation: an approach for automatically predicting the precision and recall of a video cut detection result is presented. The temporal segmentation of a video into particular shots is fundamental for video indexing and retrieval purposes. Shots in a video are separated by abrupt transitions ("cuts") and gradual transitions (such as dissolves or wipes). The prerequisite of our performance prediction is the application of unsupervised clustering to enable the evaluation of clustering quality using the silhouette coefficient [3]. It is shown that this clustering validity measure is highly correlated with the precision of a video cut detection result, and moderately correlated with recall. Three regression methods are investigated to exploit the observed correlation: linear regression, multilayer perceptron network, and support vector regression. Experimental results demonstrate the feasibility of the proposed performance prediction approach.

The paper is organized as follows. In Section 2, the underlying video indexing approach is presented, followed by the description of the proposed performance prediction approach. Experimental results for three regression methods are described in Section 3. Section 4 concludes the paper and outlines areas for future work.

2 Performance Prediction for Unsupervised Video Indexing

In this section, we demonstrate how performance prediction can be realized for unsupervised learning approaches in the context of video content analysis and indexing. This is shown exemplarily for the task of temporal video segmentation: an approach is presented for automatically predicting the precision and recall of a video cut detection result that is obtained via an unsupervised learning approach. In a previous paper [4], we have presented an unsupervised approach for video cut detection. For this approach, we show that there is an interrelationship between a clustering validity measure and precision or recall. Precision and recall are defined as follows:

$$precision = \frac{|C|}{|D|} \quad (1)$$

$$recall = \frac{|C|}{|X|} \quad (2)$$

X is the set of events of a particular class of interest (in our case cuts), D is the set of events for which a detection system decided that they belong to class X, and $C \subseteq D$ is the set of events for which this decision is correct.

The video cut detection approach is briefly summarized in Section 2.1, and our novel approach to predict precision and recall is presented in Section 2.2.

2.1 Unsupervised Video Cut Detection

To detect cuts, the dissimilarity of consecutive video frames is measured. Motion compensated pixel differences of DC-frames [10] are used in our approach. Feature vectors describe a frame position with respect to the possibility that a cut occurred at this position, the feature vector in our algorithm consists of two features. A feature vector is created only for a dissimilarity value, which is in the middle of a temporal neighborhood of size $2m+1$ and has the maximum value in this window. First, the dissimilarity measure itself is used as the first feature. The second feature is the ratio of the second highest and the highest dissimilarity value in the sliding window:

$$featureVec\ (t_i) = (max/d_{max},\ 1\text{-}sec/max) = (max',\ sec') \quad (3)$$

Here, $max>0$ and $d_{max}>0$. In practice, if $max=0$, then this frame position will not be considered as a cut candidate and no feature vector is created for this position. The range for both max' and sec' is [0,1]. The Euclidean distance is used as a distance metric, and the set of candidates is partitioned by the k-means algorithm in the subsequent steps for separating cuts from non-cuts. Since k-means is used to solve a classification problem, the number of clusters is known in advance (k=2).

However, detection results depend significantly on the sliding window size. To estimate the best sliding window size for a video automatically, we make use of a validity measure for clustering, the silhouette coefficient (SC) [3]. Normally, SC is used to measure the clustering validity for different numbers of clusters k to find the best suited k, in case it is not known in advance. The SC for a feature vector v in C can be computed by:

$$SC(v) = \frac{b(v) - a(v)}{\max\{b(v), a(v)\}} \qquad (4)$$

Here, $a(v)$ is the average distance of v to the members in the same cluster C, whereas $b(v)$ is the average distance of v to the members of the nearest other cluster, in our case it is the "non-cuts" cluster. The $SC(C)$ of the cuts cluster C is the average of the silhouette coefficients for all feature vectors in C. In our cut detection approach, the value $SC(C)$ is exploited to measure the clustering quality of the cuts cluster for a reasonable range of sliding window sizes, and the cut detection algorithm is modified as follows.

```
Input:     Min. and max. sliding window size (minSize,maxSize)
Output:    Estimated best sliding window size;
Algorithm
Find_Best_SlidingWindowSize()
  SC_max = 0; maxIndex = 1;
  for each window size m = minSize to maxSize
    Compute cuts and non-cuts cluster for window size m;
    Compute the quality SC(C) of the cuts cluster C;
    if SC(C) > SC_max then
      SC_max = SC(C);
      maxIndex=m;
  return window size maxIndex;
```

The higher the sliding window size parameter m, the fewer cut candidates are used in the clustering process. Furthermore, if (max'_m, sec'_m) and (max'_n, sec'_n) are feature vectors for the same frame position, but for different sliding window sizes m and n, with m<n, then the following holds (which is also illustrated by Figure 1):

$$max'_m = max'_n; \; sec'_m \geq sec'_n. \qquad (5)$$

2.2 Performance Prediction

In our cut detection experiments, we have observed that $SC(C_{cuts})$ is significantly correlated with the precision of a clustering result, and moderately with recall. For example, for the MPEG-7 [6] test set we have measured a correlation of 0.63 for precision and $SC(C_{Cuts})$, and a correlation of 0.26 for recall and $SC(C_{Cuts})$. Thus, we propose to estimate the interrelation between $SC(C_{cuts})$ and precision and recall, respectively. That is, a function $f(SC(C_{cuts}))$ is searched that approximates precision or recall, respectively, for a

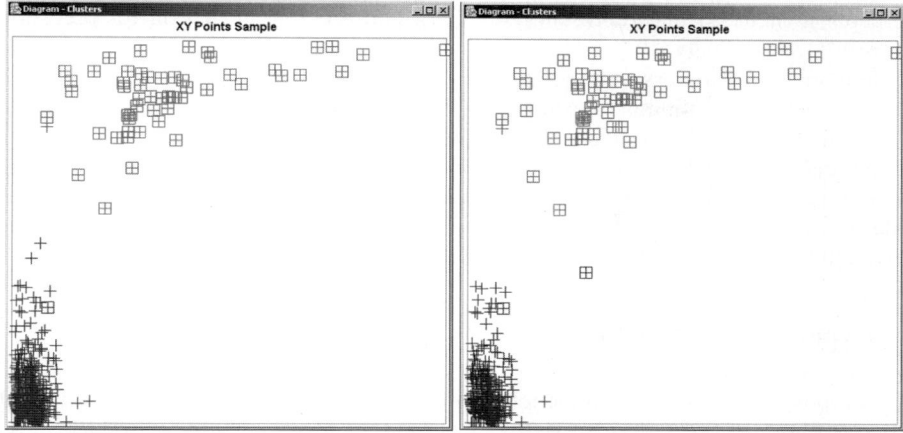

Fig. 1. Clustering result for the same video for different sliding window sizes $m=9$ (left) and $m=13$ (right). There are fewer feature vectors for $m=13$ (right) and some of the feature vectors "sink" down because feature sec' potentially decreases with increasing m. Dark (no cut) and bright (cut) crosses represent the cluster membership after k-means clustering, boxes indicate ground truth cuts.

given clustering result for a video. This is a simple numeric prediction problem that can be solved in several ways. In this paper, three methods are investigated to predict recall and precision automatically based on the silhouette coefficient $SC(C_{cuts})$:

- linear regression
- multilayer perceptron network
- support vector regression

The three methods are well known and the reader is referred to other publications [7,9] for details. For example, for linear regression, the following linear equations can be formulated to calculate precision and recall.

$$precision_{est} = p \cdot SC(C_{cuts}) + q \quad (6)$$

$$recall_{est} = r \cdot SC(C_{cuts}) + s \quad (7)$$

The parameters p and q, and r and s, respectively, are estimated using the method of least squares on a training set for which the outcomes of precision and recall are known (ground truth data).

A multilayer perceptron network arranges perceptrons in a number of layers [9]. These layers usually consist of an input layer (the attributes, in our case only SC), the hidden layer and the output layer that represents the outcome (in our case precision or recall). A perceptron itself is a simple classification approach that is able to solve linear classification problems. Learning in the multilayer perceptron is achieved by backpropagation. To predict precision or recall, there is one input perceptron (neuron) that represents $SC(C_{cuts})$, and one output neuron for precision (or recall, respectively).

Support vector regression is related to support vector machines [7]. The goal of support vector regression is to estimate a function that approximates the training data well. This function is expected to have at most a deviation of ε from the training data points, that is estimation errors up to a pre-defined parameter ε are discarded. At the same time, the regression function is expected to be as flat as possible. The parameter ε defines a tube around the regression function. The related optimization problem is solved in its dual formulation.

3 Experimental Results

The proposed performance prediction approach has been tested using two different test sets: the MPEG-7 [6] video test set and the shot boundary test set of TRECVID [8] 2005. The unsupervised video cut detection algorithm was run on the MPEG-7 test set and the best silhouette coefficient SC (for a number of sliding window sizes) for each video and the related precision or recall outcome were used to train each regression approach. Using this test set, the parameters p, q, r and s for linear regression were estimated as well as the corresponding support vector parameters and the perceptron network weights. The video cut detection algorithm was then run on the TRECVID shot boundary test videos and precision and recall were estimated for each video using the three different regression methods.

The experimental results are shown in Table 1 and 2 for the three competing regression approaches, where the estimated precision, real precision, estimated recall, real recall, and the corresponding estimation errors are presented. Assuming that the precision outcome for each video had been estimated with the mean precision obtained for the MPEG-7 test set (94.2%), the mean estimation error would have been about 5.7% (standard deviation 3.7); for recall the mean estimation error would have been 7.5% (standard deviation 6.2, based on the mean recall of 94.4 on the MPEG-7 test set). These errors are called reference errors in Table 1 and 2.

Using the proposed precision estimation via linear regression yields an average (absolute) estimation error of only 3.2% (standard deviation of 2.9). This estimate is nearly twice as precise as if the precision achieved for the MPEG-7 test set (94.2%) would be used to estimate precision. The multilayer perceptron achieves a slightly better result (estimation error of 3.0%, standard deviation of 2.9), whereas the result using support vector regression is slightly worse (estimation error of 3.9%). In particular, the perceptron network predicts the mean precision with respect to all test videos well with an estimation error of only 1.1%.

The recall estimate is not as precise as for precision, but this had to be expected due to the lower correlation. Using linear regression it is slightly better (average error: 6.5%) than a reference estimate that is based on the mean recall for the MPEG-7 test set (error: 7.5%, based on recall of 94.5% on the MPEG-7 test set), along with a lower standard deviation. Support vector regression and the multilayer perceptron network

Table 1. Estimated precision and the related error

Video ID	SC(C_{cuts})	Estimated precision (& error) [%]: Linear regression	Estimated precision (& error) [%]: Support vector regression	Estimated precision (& error) [%]: Multilayer perceptron	Reference error based on other test set (precision: 94.2)	Precision [%]
1	0.564	87.4 (7.0)	88.6 (8.2)	81.5 (1.1)	13.8	80.4
2	0.677	92.5 (1.6)	93.3 (2.4)	89.8 (0.9)	3.3	90.9
3	0.635	90.6 (3.4)	91.6 (4.4)	87.3 (0.1)	7.0	87.2
4	0.723	94.5 (6.0)	95.3 (6.8)	91.2 (2.7)	5.7	88.5
5	0.649	91.2 (0.8)	92.2 (0.2)	88.2 (3.8)	2.2	92.0
6	0.634	90.5 (1.9)	91.5 (2.9)	87.2 (1.4)	5.6	88.6
7	0.615	89.7 (0.7)	90.7 (1.7)	85.7 (3.3)	5.2	89.0
8	0.600	89.0 (0.4)	90.1 (0.7)	84.4 (5.0)	4.8	89.4
9	0.673	92.3 (9.2)	93.2 (10.1)	89.6 (6.5)	11.1	83.1
10	0.774	96.8 (0.2)	97.4 (0.8)	91.9 (4.7)	2.2	96.6
11	0.717	94.2 (4.6)	95.0 (5.2)	91.1 (1.3)	4.4	89.8
12	0.834	99.5 (2.8)	99.9 (3.2)	92.1 (4.6)	2.5	96.7
Average all (error)	-	92.4 (3.0)	93.2 (3.8)	88.3 (1.1)	5.7	89.4
Avg. error (std.dev.)	-	3.2 (2.9)	3.9 (3.1)	3.0 (2.0)	5.7 (3.7)	-

Table 2. Estimated recall and the related error

Video ID	SC(C_{cuts})	Estimated recall (& error) [%]: Linear regression	Estimated recall (& error) [%]: Support vector regression	Estimated recall (& error) [%]: Multilayer perceptron	Reference error based on other test set (recall: 94.4)	Recall [%]
1	0.564	91.6 (19.7)	93.8 (21.9)	94.9 (23.0)	22.5	71.9
2	0.677	93.7 (6.9)	95.8 (9.0)	95.3 (8.5)	7.6	86.8
3	0.635	92.9 (6.4)	95.1 (8.6)	95.1 (8.6)	7.9	86.5
4	0.723	94.5 (1.6)	96.7 (0.6)	95.8 (0.3)	1.7	96.1
5	0.649	93.1 (4.3)	95.3 (6.5)	95.1 (6.3)	5.6	88.8
6	0.634	92.9 (1.0)	95.1 (1.2)	95.1 (1.2)	0.5	93.9
7	0.615	92.5 (5.0)	94.7 (7.2)	95.0 (7.5)	6.9	87.5
8	0.600	92.2 (12.8)	94.5 (15.1)	95.0 (15.6)	15.0	79.4
9	0.673	93.6 (2.2)	95.8 (4.4)	95.2 (3.8)	3.0	91.4
10	0.774	95.5 (6.2)	97.6 (8.3)	97.4 (8.1)	5.1	89.3
11	0.717	94.4 (11.1)	96.5 (13.2)	95.7 (12.4)	11.1	83.3
12	0.834	96.6 (1.4)	98.6 (0.6)	100.0 (2.0)	3.6	98.0
Average all (error)	-	93.6 (5.9)	95.8 (8.1)	95.8 (8.1)	6.7	87.7
Avg. error (std. dev.)	-	6.5 (5.6)	8.1 (6.4)	8.1 (6.5)	7.5 (6.2)	-

cannot achieve a better prediction than the simple reference estimate based on the obtained recall using the MPEG-7 test set. The results for linear regression are also displayed in Figure 2 (Figure 3), where the silhouette coefficient, estimated precision (recall), and real precision (recall) are shown.

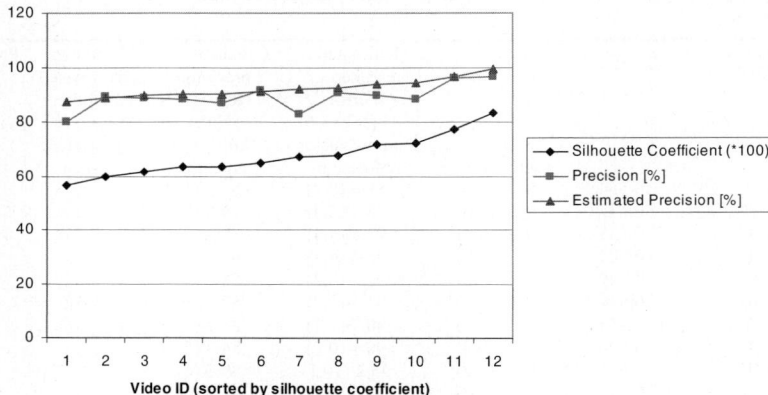

Fig. 2. This figure displays the precision, silhouette coefficients, and the estimated precision using linear regression for each video of the TRECVID 2005 shot boundary test set

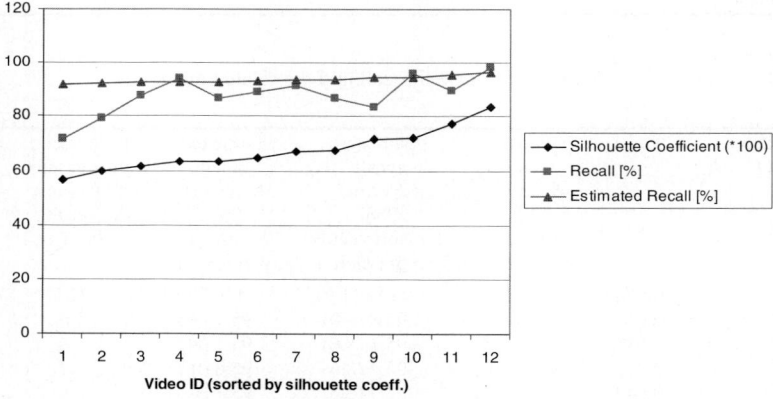

Fig. 3. This figure displays the recall, silhouette coefficient, and the estimated recall using linear regression for each video of the TRECVID 2005 shot boundary test set

4 Conclusions

In this paper, an approach to automatically predict the performance of an unsupervised video cut detection task in terms of recall of precision has been presented. It is based on the observation that cluster validity, in our case measured with the silhouette coefficient, is strongly correlated with precision and moderately with recall, though to a low degree. Exploiting this fact, it is suggested to learn the interrelationship between the silhouette coefficient and precision and recall, respectively. Three regression methods have been investigated to predict recall and precision: linear regression, multilayer perceptron network, and support vector regression. In the experiments, the comprehensive MPEG-7 video test set was used to learn the interrelationship and the

TRECVID 2005 shot boundary detection test set was used to test the prediction performance. Experimental results demonstrated that linear regression is sufficiently well suited to predict the performance and to outperform for both recall and precision the reference method that uses the results on another test set. The prediction yields an average error of only 3.0% for precision and 6.5% for recall. In both cases, the estimates are more precise than a prediction that would be based only on the overall performance on another test set. In case of precision, the prediction is nearly even twice as precise as the reference approximation. Both support vector regression and multilayer perceptron network also predict precision more precisely than the reference method, but they perform worse for recall.

Currently, we plan to consider three issues of future work. First, additional suitable features for performance prediction are sought for the investigated task of video cut detection. Second, the prediction approach should be transferred to other video content analysis approaches that use supervised learning. Finally, it will be investigated how performance prediction can be utilized to improve analysis results.

Acknowledgments. This work is financially supported by the Deutsche Forschungsgemeinschaft (German Research Foundation, SFB/FK615, Project MT).

References

1. Cronen-Townsend, S., Zhou, Y., Croft, W.B.: Predicting Query Performance. In: Proceedings of the 25th Annual International ACM SIGIR Conf. on Research and Development in Information Retrieval (SIGIR 2002), Tampere, Finland, pp. 299–306 (2002)
2. Cronen-Townsend, S., Zhou, Y., Croft, W.B.: Precision Prediction based on Ranked List Coherence. Information Retrieval 9(6), 723–755 (2006)
3. Ester, M., Sander, J.: Knowledge Discovery in Databases. Springer, Berlin (2000)
4. Ewerth, R., Freisleben, B.: Video Cut Detection Without Thresholds. In: Proceedings of the 11th International Workshop on Signals, Systems and Image Processing, Poznan, Poland, pp. 227–230 (2004)
5. He, B., Ounis, I.: Inferring Query Performance Using Pre-retrieval Predictors. In: Apostolico, A., Melucci, M. (eds.) SPIRE 2004. LNCS, vol. 3246, pp. 43–54. Springer, Heidelberg (2004)
6. MPEG-7: ISO/IEC 15938: Information Technology - Multimedia Content Description Interface Part 2: Description Definition Language. International Organization for Standardization (2002)
7. Smola, A., Schölkopf, B.: A Tutorial on Support Vector Regression. Statistics and Computing 14(3), 199–222 (2004)
8. TRECVID: TREC Video Retrieval Evaluation, http://www-nlpir.nist.gov/projects/t01v (March 27, 2009)
9. Witten, I.H., Frank, E.: Data Mining - Practical Machine Learning Tools and Techniques. Morgan Kaufmann Publishers, Elsevier Inc. (2005)
10. Yeo, B.L., Liu, B.: Rapid Scene Analysis on Compressed Video. IEEE Transactions on Circuits and Systems for Video Technology 5(6), 533–544 (1995)

New Lane Model and Distance Transform for Lane Detection and Tracking

Ruyi Jiang[1], Reinhard Klette[2], Tobi Vaudrey[2], and Shigang Wang[1]

[1] Shanghai Jiao Tong University, Shanghai, China
[2] The University of Auckland, Auckland, New Zealand

Abstract. Particle filtering of boundary points is a robust way to estimate lanes. This paper introduces a new lane model in correspondence to this particle filter-based approach, which is flexible to detect all kinds of lanes. A modified version of an Euclidean distance transform is applied to an edge map of a road image from a birds-eye view to provide information for boundary point detection. An efficient lane tracking method is also discussed. The use of this distance transform exploits useful information in lane detection situations, and greatly facilitates the initialization of the particle filter, as well as lane tracking. Finally, the paper validates the algorithm with experimental evidence for lane detection and tracking.

1 Introduction

Lane detection plays a significant role in driver assistance systems (DAS), as it can help estimate the geometry of the road ahead, as well as the lateral position of the ego-vehicle on the road. Lane detection and tracking have been widely studied for driving on a freeway or an urban road, for single or multiple lanes, with or without marks, based on region (texture or color) or edge features. Various models have been applied to describe the borders of a lane, such as *piecewise linear segments*, *clothoids*, *parabola*, *hyperbola*, *splines*, or *snakes*. For a complete review of lane detection algorithms, please refer to [4]. There are even some commercial lane detection systems available, working mainly on highways. Loose *et al* (Daimler AG) state in [3]: "Despite the availability of lane departure and lane keeping systems for highway assistance, unmarked and winding rural roads still pose challenges to lane recognition systems." Generally, it is a challenging task to robustly detect lanes in varying situations, especially in complex urban environment.

This paper introduces a new weak (i.e., with no assumption about the global shape of a lane) road model for lane detection and tracking. Instead of modeling global road geometry, this new model only constrains relations between points on the left and right lane boundaries. Tracking based on these points in the birds-eye image (using a particle filter) provides lane detection results. Furthermore, a modified version of a standard Euclidean Distance Transform (EDT) is applied on the edge map of the birds-eye image. Utilizing some beneficial properties of this distance transform for lane detection and tracking, this paper also specifies an innovative initialization method for the particle filter.

This paper is organized as follows: Section 2 describes a new lane model. Section 3 explains a modified version of a standard Euclidean distance transform, and its usefulness for lane detection. Lane detection and tracking methods are introduced in Section 4

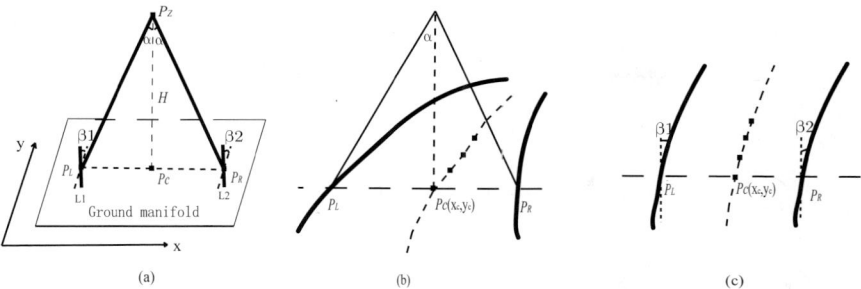

Fig. 1. Lane model as used in this paper. (a) 3D lane view; boundaries are drawn in bold. (b) Perspective 2D lane view in the input image. (c) Birds-eye image of the lane. Slope angles β_1 and β_2 are shown in the 3D view and the birds-eye images; the zenith angle α in the 3D and the projective view. See text for further explanations.

and Section 5. Experimental results are given in Section 6. Finally, conclusions are provided in Section 7.

2 A New Lane Model

The new lane model is illustrated in Figure 1. Five parameters x_c, y_c, α, β_1, and β_2 are used to model opposite points P_L and P_R, located on the left or right lane boundary, respectively. $P_C = (x_c, y_c)$ is the (virtual) *centerline point* of a lane in the ground plane. α is the *zenith angle* above P_C, defined by an upward straight line segment between P_C and the *zenith* P_Z of fixed length H, and a line incident with P_Z and either P_L or P_R. β_1 and β_2 are the *slope angles* between short line segments L_1 and L_2 and a vertical line in the ground plane; the two short line segments L_1 and L_2 are defined by a fixed length and local approximations to edges at lane boundaries (e.g., calculated during point tracking). Ideally, L_1 and L_2 should coincide with tangents on lane boundaries at points P_L and P_R; in such an ideal case, β_1 and β_2 would be the angles between tangential directions of lane boundaries at those points and a vertical line.

By applying this model, a lane is identified by two lane boundaries, and points are tracked along those boundaries in the birds-eye image. This model does not use any assumption about lane geometry, and applies to all kinds of lanes.

3 Distance Transform

The distance transform is applied to the binary edge map, which labels each pixel with the distance to the nearest edge pixel. The Euclidean distance transform (EDT) is, in general, the preferred option; using the Euclidean metric for measuring the distance between pixels. A modified EDT was proposed in [7], called *orientation distance transform* (ODT). This divides the Euclidean distance into a contributing component in the

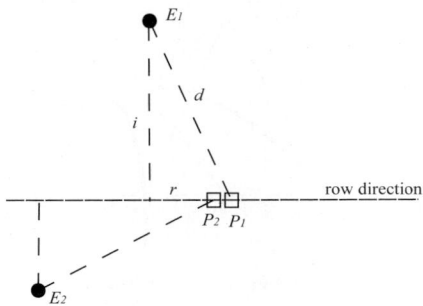

d: Euclidean distance of P_1 to the nearest edge point E_1
r: Real part of ODT of P_1 to the nearest edge point E_1
i: Imaginary part of ODT of P_1 to the nearest edge point E_1

Fig. 2. Euclidean distance and Orientation distance. P_1 and P_2 are two neighboring non-edge pixels. E_1 is the nearest edge point to P_1, and E_2 is the nearest edge point to P_2.

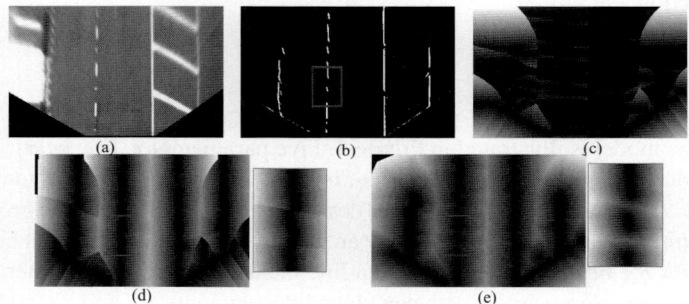

Fig. 3. EDT and ODT on a birds-eye road image. (a) Birds-eye road image. (b) Binary edge map. (c) Imaginary part of ODT. (d) Real part of ODT (absolute value). (e) EDT. (c)(d)(e) have been contrast adjusted for better visibility. The square root of distance value in (c) and (d) will be the Euclidean distance value as (e). The rectangle in (b) indicates a small window for comparisons of dashed lane mark with EDT and RODT(see below).

row and column direction. A complex number is assigned to each pixel by the ODT, with the distance component in the row direction as the real part, and the distance component in the column direction as the imaginary part. Note that distance component in row direction is signed, with a positive value indicating that the nearest edge point lies to the right, and a negative value if it is to the left. See Figure 2 for the relationship between ODT and EDT.

This paper uses only the Euclidean distance in the row direction, and we call this the *real orientation distance transform* (RODT). An example of EDT and ODT in our application is shown in Figure 3. The RODT of our edge map offers various benefits. First, the initialization of lane detection becomes much easier (to be discussed in

Section 4.1). Second, dashed lane marks will make no difference with continuous ones in the RODT, as illustrated in the comparison of the rectangle area in Figure 3. Third, more information about the centerline is provided by the distance transform compared to the edge map. Generally, a (non-edge) pixel on the centerline of a lane will have a local maximum in distance to the lane boundaries. Thus, combined with the lane model introduced in Section 2, a point with a high distance value is likely to be a centerline point P_C. The usefulness of these properties of the RODT will be discussed in the following sections.

4 Lane Detection Using a Particle Filter

For lane detection in a single image, a particle filter is used to track points along lane boundaries as in [5]. Furthermore, a novel initialization method is adopted based on a distance transform applied to the birds-eye edge map. The whole procedure of lane detection is illustrated in Figure 4 by an example.

The algorithm starts with mapping the perspective input image into a birds-eye view. An edge detection method, as introduced in [1], is then adopted to detect lane-mark-like edges. After binarization of the resulting edge map, a RODT is applied. The resulting distance map allows us to design a novel initialization method for finding the initial boundary points. These points are used to initialize the parameters of the particle filter, for tracking further boundary points through the whole image; a lane is finally detected.

Fig. 4. The overall work flow of lane detection. (a) Input image. (b) Birds-eye image. (c) Edge map. (d) Distance transform. (e) Lane detection results, shown in the birds-eye image (left and right lane boundaries, with lane center in the middle). (f) Lane detection results, shown in the input image.

4.1 Initialization

The aim of the initialization step is to find an initial value (e.g., the x-coordinate of a point P_L and point P_R in a selected image row) for the specified model. In [5], a clustered particle filter is used in order to find a start point on a lane boundary. In distinction to this, we fully utilize the distance map to find the first left and right boundary points.

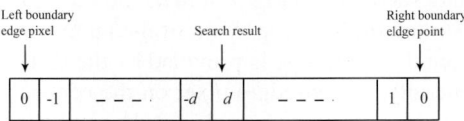

Fig. 5. Illustration of the search procedure in the start row of the distance map. Note that the distance values are signed, as described in Section 3.

In a pre-defined *start row* (near to the bottom) of the birds-eye image, a search is conducted, starting at the middle of the row, for a pixel which has a positive distance value but a negative distance value at its left neighbor (see Figure 5). When such a pixel is found, the left and right boundary points in the start row are instantly known using the distance value of the found pixel and of its left neighbor.

For the initial state $X_0(x_{c_0}, \alpha_0, \beta_{1_0}, \beta_{2_0})$ of the particle filter, x_{c_0} and α_0 are initialized by using the detected left and right start points, while β_{1_0} and β_{2_0} are simply set to be zero.

4.2 Particle Filter for Lane Detection

Particle filters are widely used for lane detection and tracking, such as in [5,6]. This section discusses particle filtering for our new lane model. The state vector $X = (x_c, \alpha, \beta_1, \beta_2)^T$ to be tracked is defined by the parameters of the lane model, without y_c, as y_c will be calculated incrementally by applying a fixed step Δ, starting at row y_{c_0} in the birds-eye image. For the application of a particle filter, two models are discussed in the following.

The dynamic model. The dynamic model A is used to define the motion of particles in the image. The prediction value $\hat{X}_n$ is generated from X_{n-1} by using $\hat{X}_n = A \cdot X_{n-1}$. We simply take A as being the identity matrix, because of the assumed smoothness of the lane boundary.

The observation model. The observation model determines each particle's importance factor for re-sampling. Based on the RODT information combining new model, it is reasonable to assume points (x_{c_n}, y_{c_n}) will have large distance values, and L_1 and L_2 coincide with short lines of pixels which all only have small distance values.

Tracking step n is identified by $y_{c_n} = (y_{c_0} + n \cdot \Delta)$. We calculate the lateral position of the left boundary point of the lane from the predicted state vectors, with $\hat{X}_n^i(\hat{x}_{c_n}^i, \hat{\alpha}_n^i, \hat{\beta}_{1_n}^i, \hat{\beta}_{2_n}^i)$ for the ith particle.

From now on, P_L and P_R only represent the lateral position of boundary points, for simplicity. The left position is calculated as follows:

$$P_L^i = \hat{x}_{c_n}^i - H \cdot \tan \hat{\alpha}_n^i$$

Next, the sum of the distance values along line segment L_1 is as follows:

$$S_{L_1}^i = \sum_{j=-L_1/2}^{L_1/2} \left| d\left(P_L^i + j \cdot \sin \hat{\beta}_{1_n}^i, \, y_{c_n} + j \cdot \cos \hat{\beta}_{2_n}^i\right) \right|$$

Here, $d(\cdot,\cdot)$ is the distance value of the RODT. Calculating $S_{L_2}^i$ in the analogous way, we obtain the ith importance factor

$$\omega_{dist}^i = \frac{1}{2\pi\sigma_1\sigma_2} \exp\left(-\frac{(S_{L_1}^i - \mu_1)^2}{2\sigma_1} - \frac{(S_{L_2}^i - \mu_2)^2}{2\sigma_2}\right)$$

For the centerline point $(x_{c_n}^i, y_{c_n})$, the importance factor is equal to

$$\omega_{center}^i = \frac{1}{\sigma_3\sqrt{2\pi}} \exp\left(-\frac{\left(\left|\frac{1}{d(x_{c_n}^i, y_{c_n})}\right| - \mu_3\right)^2}{2\sigma_3}\right)$$

where μ_k and σ_k are constants, for $k = 1, 2, 3$. The final observation model is given by the factors

$$\omega_i = \omega_{dist}^i \cdot \omega_{center}^i$$

5 Efficient Lane Tracking

Lane tracking uses information defined by previous results to facilitate the current detection. This section introduces an efficient lane tracking method. Note that when a lane is detected (as in Section 4), it is reasonably represented just by two sequences $\{P_{L_n} : n = 0, 1, \ldots, N\}$ and $\{P_{R_n} : n = 0, 1, \ldots, N\}$ of points on its left and right lane boundaries in the birds-eye image. Here, N is determined by the forward-looking distance. Tracking of a lane through an image sequence is then simplified as being tracking of these two point sequences.

Efficient lane tracking simply uses previously detected lane boundary points, adjusts them according to the ego-vehicle's motion model, and then offsets them according to values of the RODT on the current birds-eye edge map. The lane tracking scheme is summarized in Figure 6.

Sequences $\{P_{L_n}^{(t)}\}$ and $\{P_{R_n}^{(t)}\}$, detected in frame t, are already partially driven through by the ego-vehicle at time $t + 1$. The detection process of $\{P_{L_n}^{(t+1)}\}$ and $\{P_{R_n}^{(t+1)}\}$ at time $t + 1$ is composed of three steps: adjustment caused by the driven distance and the variation in yaw angle, new point detection, and offset specification according to the values of the RODT in the birds-eye edge map.

Because of the driven distance between frames t and $t + 1$, it holds (in principle) that

$$P_{L_n}^{(t+1)} = P_{L_{(n+k)}}^{(t)}, \qquad P_{R_n}^{(t+1)} = P_{R_{(n+k)}}^{(t)}, \qquad n = 0, 1, \ldots, N - k$$

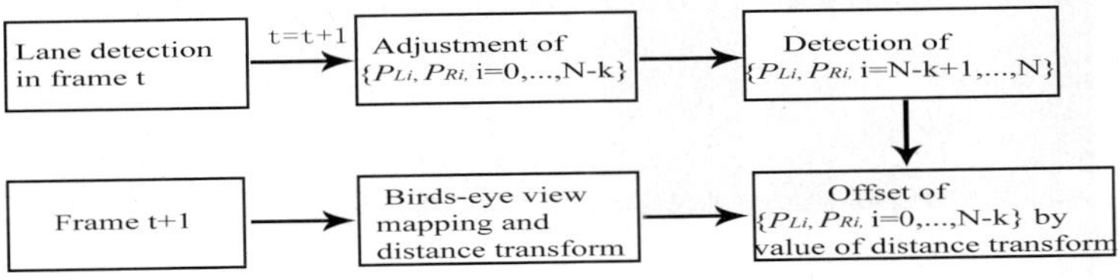

Fig. 6. Efficient lane tracking scheme

Here, k is determined by the driven distance between time t and $t+1$, and is usually a small number. Furthermore, points

$$\{P_{L_n}^{(t+1)} : n = 0, 1, \ldots, N - k\} \text{ and } \{P_{R_n}^{(t+1)} : n = 0, 1, \ldots, N - k\}$$

are obtained by adding some translation (according to n) caused by the variation in driving direction between t and $t+1$.

For the detection of $\{P_{L_n}^{(t+1)} : n = N - k + 1, \ldots, N\}$ and $\{P_{R_n}^{(t+1)} : n = N - k + 1, \ldots, N\}$, note that k is small and we also assume smoothness of lane boundaries. Thus, we simply start as follows:

$$P_{L_n}^{(t+1)} = P_{L_{n-1}}^{(t+1)}, \qquad P_{R_n}^{(t+1)} = P_{R_{n-1}}^{(t+1)}, \qquad n = N - k + 1, \ldots, N$$

For further refinement, those predictions $\{P_{L_n}^{(t+1)}\}$ and $\{P_{R_n}^{(t+1)}\}$ from the previous result at frame t are likely to be already located near the true points on the boundaries, as the variation of a lane is usually minor between two subsequent frames. – The adjustment

$$P_{L_n}^{(t+1)} = P_{L_n}^{(t+1)} + d(P_{L_n}^{(t+1)}, y_{c_n}), \qquad n = 0, 1, \ldots, N$$
$$P_{R_n}^{(t+1)} = P_{R_n}^{(t+1)} + d(P_{R_n}^{(t+1)}, y_{c_n}), \qquad n = 0, 1, \ldots, N$$

of all $N+1$ points is finally achieved by information available from values of the RODT of the current birds-eye edge map.

6 Experiments

Experiments were conducted on images and sequences recorded with the test vehicle "HAKA1" of the *.enpeda..* project [8].

Experimental results for lane detection are shown in Figure 7 and Figure 8. Different scenarios are considered. Note that detected lane boundaries are sometimes locally slightly curved. This is due to the fact that the distance transform of dashed lane marks is slightly unaligned in column direction in the birds-eye image (see Figure 3).

Fig. 7. Experimental results for lane detection. (a) Input images. (b) Lanes detected in the birds-eye image. Note that the centerline of a lane is also marked. (c) Lanes detected in input images.

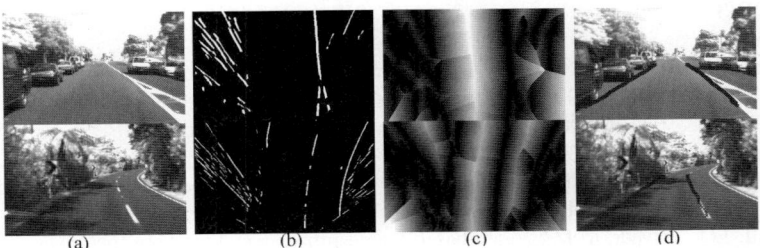

Fig. 8. Lane detection on roads with unmarked or blocked lane marks. (a) Input images. (b) Edge map. (c) RODT. (d) Lane detected. Note that RODT contains more information than edge map for lane detection as discussed in Sec. 3.

Fig. 9. Experimental results on two sequences using efficient lane tracking

Experimental results while using the efficient lane tracking method are illustrated in Figure 9. The results are really acceptable except for some outliers for lane boundaries. With an off-the-shelf computer, the computation time for the adjustment steps of efficient lane tracking is negligible. The only measurable time needed is that for birds-eye view mapping, edge detection and distance transform, and all three sub-processes can be computed highly efficiently.

7 Conclusions

This paper introduced a new weak model of a lane, and a possible lane detection scheme using a particle filter based on a monocular camera. Furthermore, an efficient lane tracking method was proposed and discussed.

A (simple and easy to calculate) distance transform was used in this paper for lane detection and tracking. It shows that the distance transform is a powerful method to exploit information in lane detection situations. The distance transform can deal with dashed lane marks, provides information for the detection of the centerline of a lane, finds initial values for the particle filter, and adjusts the tracking results conveniently.

Acknowledgments. This work is supported by the National Natural Science Foundation of China under Grant 50875169.

References

1. Bertozzi, M., Broggi, A.: GOLD - A parallel real-time stereo vision system for generic obstacle and lane detection. IEEE Trans. Image Processing 7, 62–81 (1998)
2. Kim, Z.: Robust lane detection and tracking in challenging scenarios. IEEE Trans. Intelligent Transportation System 9, 16–26 (2008)
3. Loose, H., Franke, U., Stiller, C.: Kalman particle filter for lane recognition on rural roads. In: Proc. IEEE Intelligent Vehicle Symp. (to appear, 2009)
4. McCall, J.C., Trivedi, M.M.: Video-based lane estimation and tracking for driver assistance: survey, system, and evaluation. IEEE Trans. Intelligent Transportation System 7, 20–37 (2006)
5. Sehestedt, S., Kodagoda, S., Alempijevic, A., Dissanayake, G.: Efficient lane detection and tracking in urban environments. In: Proc. European Conf. Mobile Robots, pp. 126–131 (2007)
6. Wang, Y., Bai, L., Fairhurst, M.: Robust road modeling and tracking using condensation. IEEE Trans. Intelligent Transportation Systems 9, 570–579 (2008)
7. Wu, T., Ding, X.Q., Wang, S.J., Wang, K.Q.: Video object tracking using improved chamfer matching and condensation particle filter. In: Proc. SPIE-IS & T Electronic Imaging, vol. 6813, pp. 04.1–04.10 (2008)
8. http://www.mi.auckland.ac.nz

Real-Time Volumetric Reconstruction and Tracking of Hands in a Desktop Environment

Christoph John[1,2], Ulrich Schwanecke[2], and Holger Regenbrecht[1]

[1] University of Otago, New Zealand
[2] University of Applied Sciences Wiesbaden, Germany

Abstract. A probabilistic framework for vision based volumetric reconstruction and marker free tracking of hand and face volumes is presented, which exclusively relies on off-the-shelf hardware components and can be applied in standard office environments. Here a 3D reconstruction of the interaction environment (userspace) is derived from multiple camera viewpoints which serve as input sources for mixture particle filtering to infer position estimates of hand and face volumes. The system implementation utilizes graphics hardware to comply with real-time constraints on a single desktop computer.

Keywords: Probabilistic Shape From Silhouette, Mixture Particle Filtering.

1 Introduction

Virtual and mixed reality environments rely on the implementation of (tele) presence: the perceived sense that a user's own body and body parts belong to the artificial world presented. Of paramount importance here is the efficient and accurate registration, tracking, reconstruction and display of the head and hands of a human operator.

In the following we present an approach for the reconstruction and tracking of hands and head (skin-colored objects) in a potential standard office environment which works with off-the-shelf hardware components. The supervised volume in our table-top environment (see figure 1) has a hand tracking volume size of $1.0\text{m} \times 1.0\text{m} \times 0.75\text{m}$. The lighting conditions have been constrained to controlled and reasonably well lit office room, following the recommendations of IEEE Std. 241 [1].

The system consists of a flock of six color cameras which are utilized to compute a volumetric reconstruction of the user-space. A variant of a *probabilistic Shape from Silhouette* (pSfS) algorithm, first introduced by Landabaso and Pardas [4] has been developed. Unlike *traditional SfS* (tSfS), which performs object segmentation in the image domain, pSfS utilizes a 3D probabilistic background model. This shifts object segmentation into the spatial domain and leads to improved segmentation results in presence of image noise or clutter. We have extended pSfS by imposing constraints on the 3D foreground process in terms of anticipated color and occupancy of hand and face volumes, thus limiting volumetric reconstruction to skin-colored foreground regions. This leads to more detailed reconstructions and an increased probabilistic distance to the background scene. In addition we allow dynamic per pixel on/off switching of cameras to allow the integration of occlusion masks and to stabilize reconstruction results in presence of occlusion.

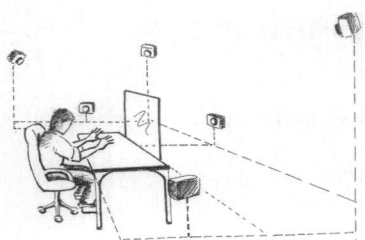 

Fig. 1. Proposed environment. Hands and face are tracked in front of the projection screen.

In [4] volumetric reconstructions have been projected into images to generate occlusion masks needed for background model update. We instead utilize the derived visual hulls as input source for a variant of mixture particle filtering [7] to estimate positions of hand and face volumes. Occlusion masks are then generated from tracked bounding boxes. This has the advantage that masks can be computed efficiently and that volumetric reconstruction errors do not degenerate the background model. Finally we present a GPU implementation of the presented system, which in contrast to [4] permits the whole system to run in real-time on a consumer graphics card and a single desktop computer.

2 Probabilistic Volume Reconstruction

The reconstruction algorithm presented below is based on probabilistic reasoning and can be subdivided into an image and volume based classification part. In the image based part a measure is assigned to each pixel which exhibits its probability of belonging to a skin-colored foreground silhouette. In the volume based part these silhouettes are utilized to derive volumetric reconstructions.

2.1 Image Based Likelihood Evaluation

The task of skin-colored foreground object segmentation can be formulated as a classification problem at pixel level. A pixel may belong to one of four groups which are given as the possible combinations of fore-/background and skin/non-skin color. Pixel likelihood evaluations are casted as *maximum a posteriori* (MAP) assignments in a discriminative model. I.e., the model expresses the per pixel probability of belonging to the foreground with the skin-colored class $P'(F, S|\mathbf{c})$ as a function of its observed color vector $\mathbf{c}$. The prime denotes augmentation with an outlier model which will be described in detail later. Here $\mathbf{c} = [r, g]^T$ is represented in the normalized-rg color space.

A combination of two classifiers constitutes our discriminative model (see figure 2). The first classifier $P'(F|\mathbf{c})$ is based on a model of the background process and estimates per pixel foreground probabilities by combining the MAP assignment of being foreground $P(F|\mathbf{c})$ with an outlier model. Equivalently the second classifier $P'(S|\mathbf{c})$

(a) $P'(F|\mathbf{c})$ (b) $P'(S|\mathbf{c})$ (c) $P'(F,S|\mathbf{c})$

Fig. 2. Left: foreground classification; Middle: skin classification; Right:combined classification

augments the MAP estimate of being skin color $P(S|\mathbf{c})$. The final per pixel classification scheme is thus given by:

$$P'(F,S|\mathbf{c}) = P'(F|\mathbf{c}) \cdot P'(S|\mathbf{c})$$

Our setup has been constrained to office environments with a fixed camera setup. This leads to a relatively static background scene which can be modeled with a *single Gaussian model* (SGM) [3]. A SGM is defined in the bivariate case with the mean normalized-rg color vector μ and covariance matrix Σ as:

$$P(\mathbf{c}|\mu, \Sigma) = \frac{1}{2\pi |\Sigma|^{\frac{1}{2}}} \cdot \exp\left(-\frac{1}{2}(\mathbf{c}-\mu)^T \cdot \Sigma^{-1} \cdot (\mathbf{c}-\mu)\right) \quad (1)$$

The background likelihood is obtained from equation (1) as $P(\mathbf{c}|\bar{F}) = P(\mathbf{c}|\mu_{\bar{F}}, \Sigma_{\bar{F}})$ and is used to derive the MAP foreground likelihood $P(F|\mathbf{c})$. Assuming equal likelihood of color appearance in the foreground i.e. $P(\mathbf{c}|F) = \frac{1}{256^2}$, we obtain:

$$P(F|\mathbf{c}) = \frac{P(\mathbf{c}|F) \cdot P(F)}{P(\mathbf{c})} = \frac{\frac{1}{256^2} \cdot P(F)}{\frac{1}{256^2} \cdot P(F) + P(\mathbf{c}|\bar{F}) \cdot P(\bar{F})}$$

Priors of fore-/background are derived from the expected volume occupancy of foreground objects and will be discussed in the next section.

For skin color classification we follow Caetano *et al.* [2] and model skin color with a static *Gaussian Mixture Model* (GMM) with $I = 2$ basis functions. This has been reported as a good tradeoff between accuracy and efficiency. The GMM is derived from equation (1) and associated weights w_i as $P(\mathbf{c}|S) = \sum_{i=1}^{I} w_i \cdot P(\mathbf{c}|\mu_i, \Sigma_i)$. The parameters have been trained with Expectation Maximization from a set of labeled skin-color images. Skin color classification can thus be cast in a Bayesian formulation by assuming equal likelihood of color appearance in non skin-colored regions $P(\mathbf{c}|\bar{S}) = \frac{1}{256^2}$ resulting in the MAP assignment:

$$P(S|\mathbf{c}) = \frac{P(\mathbf{c}|S) \cdot P(S)}{P(\mathbf{c})} = \frac{P(\mathbf{c}|S) \cdot P(S)}{P(\mathbf{c}|S) \cdot P(S) + \frac{1}{256^2} \cdot P(\bar{S})}$$

Notice that the models introduced so far do not permit for any type of classification error. Following [6], a more robust classification scheme is formulated by reverting to the prior in case of an outlier. Let $e_F, e_S \in [0,1]$ be the probabilities of being outlier in the foreground and skin color model respectively. Then the classifier augmentations are:

$$P'(F|\mathbf{c}) = e_F \cdot P(F) + (1-e_F) \cdot P(F|\mathbf{c}) \text{ and } P'(S|\mathbf{c}) = e_S \cdot P(S) + (1-e_S) \cdot P(S|\mathbf{c})$$

2.2 Volume Based Classification

pSfS has been adapted to combine the previously described image based classifiers. The difference between the presented algorithm and [4] is the definition of ϕ and β. In our setting ϕ describes a skin colored foreground and β a group of classes given as the remaining combinations of being fore-/background and skin/non-skin color. This leads to the introduction of multiple priors into pSfS.

Now let $\{\Gamma_1, \cdots, \Gamma_N\}$ be the set of super classes representing all $N = 2^S$ possible combinations of skin-colored foreground or background classifications of all S sensors.

$$\begin{aligned}
\Gamma_1 &= \{\phi, \quad \phi, \quad \phi, \quad \ldots, \quad \phi\} \\
\Gamma_2 &= \{\beta, \quad \phi, \quad \phi, \quad \ldots, \quad \phi\} \\
\Gamma_3 &= \{\phi, \quad \beta, \quad \phi, \quad \ldots, \quad \phi\} \\
&\vdots \\
\Gamma_{S+2} &= \{\beta, \quad \beta, \quad \phi, \quad \ldots, \quad \phi\} \\
&\vdots \\
\Gamma_n &= \{\Gamma_n[1], \Gamma_n[2], \Gamma_n[3], \ldots, \Gamma_n[S]\} \\
&\vdots \\
\Gamma_N &= \{\beta, \quad \beta, \quad \beta, \quad \ldots, \quad \beta\}
\end{aligned}$$

and let their group specific priors be given as $P(\Gamma_n) = \prod_{s=1}^{S} P(\Gamma_n[s])$ with projected priors:

$$P(\phi) = P(F) \cdot P(S) \qquad \text{and} \qquad P(\beta) = 1 - P(\phi)$$

In the absence of occlusion a voxel is assigned to be part of a visual hull $\mathcal{H}$ if all sensors classify the voxel as skin-colored foreground, with prior probability $P(\mathcal{H})$. That is:

$$\mathcal{H} = \Gamma_1 \qquad \text{and} \qquad P(\mathcal{H}) = P(\Gamma_1) \qquad (2)$$

$P(\mathcal{H})$ is defined as the occupancy ratio between the expected number of skin-colored foreground voxels and the total number of voxels. Projected skin priors can thus be derived from $P(\mathcal{H})$ as $P(S) = \frac{\sqrt[S]{P(\mathcal{H})}}{P(F)}$. Equivalently projected foreground priors can be derived from a visual hull of all foreground objects $\mathcal{H}_F$ with an expected volume occupancy ratio $P(\mathcal{H}_F)$ as $P(F) = \sqrt[S]{P(\mathcal{H}_F)}$. We have chosen $P(\mathcal{H}_F)$ and $P(\mathcal{H})$ statically from reference reconstructions which have been generated with a traditional SfS algorithm.

Cameras in SfS setups are usually mounted with wide stereo baselines leading to statistical independents between the camera views and Bayes theorem can be consulted to estimate class probabilities.

$$P(\Gamma_n|\mathbf{c}_1,\ldots,\mathbf{c}_S) = P(\Gamma_n) \cdot \prod_{s=1}^{S} \frac{P(\mathbf{c}_s|\Gamma_n)}{P(\mathbf{c}_s)} \qquad n = 1,\ldots,N \qquad (3)$$

Here $P(\mathbf{c}_s|\Gamma_n) = P(\mathbf{c}_s|\Gamma_n[s])$ is the conditional probability of the observation in sensor s, given a certain super class in its view. Conditional probabilities can be rewritten in means of posterior probabilities to plug in per pixel MAP assignments:

$$P(\Gamma_n|\mathbf{c}_1,\ldots,\mathbf{c}_S) = P(\Gamma_n) \cdot \prod_{s=1}^{S} \frac{P(\Gamma_n[s]|\mathbf{c}_s)}{P(\Gamma_n[s])} \qquad (4)$$

Here $P(\Gamma_n[s]|\mathbf{c}_s)$ conforms to the posterior probability of a certain superclass in sensor s given its observation $\mathbf{c}_s$. This is to say:

$$P(\Gamma_n[s]|\mathbf{c}_s) := \begin{cases} P'(F,S|\mathbf{c}_s) & \text{if } \Gamma_n[s] = \phi \\ 1 - P'(F,S|\mathbf{c}_s) & \text{if } \Gamma_n[s] = \beta \end{cases}$$

Finally the partitioning of voxels to super classes is obtained by following Bayes rule for minimum error. Therefore a voxel is assigned to the most probable super class Γ_m:

$$\Gamma_m = \arg\max_{\Gamma_n} P(\Gamma_n) \cdot \prod_{s=1}^{S} \frac{P(\Gamma_n[s]|\mathbf{c}_s)}{P(\Gamma_n[s])}$$

As computation of all class posteriors becomes computational intensive with growing number of sensors, it has been recommended in [5] to limit computation to the foreground class and set a threshold on its posterior instead. Our results suggest the same as we have obtained equivalent reconstruction results for both algorithmic variants.

The algorithm introduced so far does not account for systematic errors given through occlusion or segmentation errors. The assignment of multiple foreground classes is a common approach to resolve this issue in SfS type algorithms. In pSfS this can be done in two ways. First, by assigning multiple super classes to the visual hull. If for example, the appearance of a single systematic error was to be allowed, equation (2) would become:

$$\mathcal{H} = \bigcup_{s=1}^{S+1} \Gamma_s \qquad \text{and} \qquad P(\mathcal{H}) = \sum_{s=1}^{S+1} P(\Gamma_s) \qquad (5)$$

This approach has a serious disadvantage as all class posteriors now have to be computed. A more efficient procedure is given by assigning an active camera flag to each pixel and than limit the class computation to active projections. In the presence of occlusion masks these are the activity flags. If multiple foreground classes should be allowed, a given number of pixel projections with lowest foreground probability $P'(F,S|\mathbf{c})$ have to be disabled dynamically.

3 Mixture Particle Filtering

Detection of hand and face volumes within a reconstructed volume is usually a time consuming process which can be accelerated by incorporating temporal information through tracking. We assume independents of the movements of hands and head and therefore follow [7] and apply a 3D variant of mixture particle filtering for tracking. Here the joint distribution of object states is interpreted as a mixture in which each object is tracked with a dedicated particle filter. The prediction and update equations of the M-component mixture model are given with mixture weights $\sum_{m=1}^{M} \pi_{m,t} = 1$ as

predict: $\quad p(\mathbf{x}_t|\mathbf{Y}_{t-1}) = \sum_{m=1}^{M} \pi_{m,t-1} \cdot p_m(\mathbf{x}_t|\mathbf{Y}_{t-1})$

update: $\quad p(\mathbf{x}_t|\mathbf{Y}_t) = \sum_{m=1}^{M} \left(\frac{\pi_{m,t-1} \cdot p_m(\mathbf{y}_t|\mathbf{Y}_{t-1})}{\sum_{n=1}^{M} \pi_{n,t-1} \cdot p_n(\mathbf{y}_t|\mathbf{Y}_{t-1})} \right) \cdot \frac{p(\mathbf{y}_t|\mathbf{x}_t) p_m(\mathbf{x}_t|\mathbf{Y}_{t-1})}{p_m(\mathbf{y}_t|\mathbf{Y}_{t-1})}$

The first update term can be interpreted as the new mixture weight $\pi_{m,t}$ because the state $\mathbf{x}$ is not involved. Hence only the second term represents the component update. Component interaction is therefore limited to mixture weight computation which makes this particle filtering technique fast. Particle filters track 3D centroid positions of hand and face volumes. Hands and face are distinguished through their volume sizes. Particle states represent boxes in space with a fixed size, see right side of figure 1. We use the percentage of occupied skin-colored volume within these boxes as the source for weight evaluation. This average occupancy can be computed efficiently through utilization of a summed volume table for the reconstructed volume. The mixture particle formulation given above does not determine how a mixture is initialized or modified. In our setup initialization is done by spreading particles randomly until all expected objects are tracked. If a mode was found which is not already tracked, a new mixture is initialized on that mode. In cases in which object separation is impossible, a mixture update has to be enforced which provides merge and split operations. Here it is based on K-means analysis and similar to the one proposed in [7]. The difference is that we have to treat different particle types. Therefore we allow re-clustering only between mixtures of the same type, others are discarded.

4 Results

We have implemented pSfS as well as tSfS and compared both with respect to reconstruction quality and performance. Achieved reconstruction results favor pSfS over tSfS, see figure 3 for a comparison. Both pSfS variants achieved more detailed reconstructions than tSfS. Explicit computation of fore-/background classes and limited evaluation by thresholding the foreground class resulted in similar reconstructions. The similarity between outputs of both pSfS algorithms can be explained by detailing the impact of background class evaluation. Explicit evaluation of background classes leads to a less false positive rate for foreground class assignment in presence of highly ambiguous voxels. These false positives are known to have a low foreground probability, as they would not be assigned to a background class otherwise. This implies that they can be equivalently eliminated by enforcing a threshold on posterior probabilities.

The presented SfS variants were implemented on a GPU with NVIDIA CUDA to permit interactive frame rates. Here performance results of three different volume resolutions are presented. The runtime values were measured on an Intel Q6600 running at 2.4GHz with a NVIDIA GeForce 8800 GTX graphics card and are listed in table 1.

Table 1. Performance results of 3D reconstruction on a GPU

Reconstruction Type	Volume Resolution	Algo	Image eval.	GPU Readout	Total
tSfS	$64 \times 64 \times 48$ voxel	1.1ms	5.5ms	0.1ms	6.7ms
	$128 \times 128 \times 96$ voxel	5.7ms	5.5ms	0.8ms	12.0ms
	$256 \times 256 \times 192$ voxel	35.8ms	5.5ms	7.8ms	49.1ms
pSfS, foreground class	$64 \times 64 \times 48$ voxel	1.2ms	5.5ms	0.1ms	6.8ms
	$128 \times 128 \times 96$ voxel	5.9ms	5.5ms	0.8ms	12.2ms
	$256 \times 256 \times 192$ voxel	38.0ms	5.5ms	7.8ms	51.3ms
pSfS, all classes	$64 \times 64 \times 48$ voxel	4.3ms	5.5ms	0.1ms	9.9ms
	$128 \times 128 \times 96$ voxel	31.4ms	5.5ms	0.8ms	37.7ms
	$256 \times 256 \times 192$ voxel	241.8ms	5.5ms	7.8ms	255.1ms

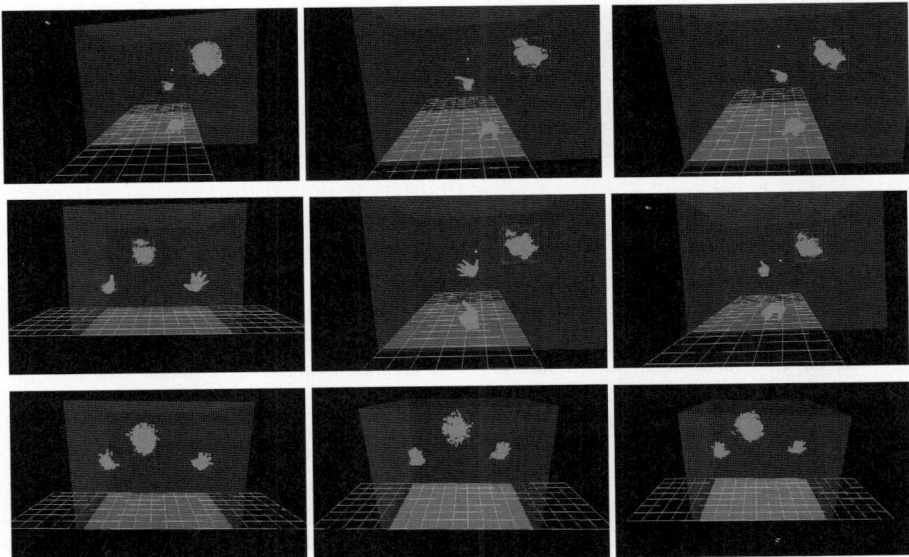

Fig. 3.

1. row: [left to right] tSfS, pSfS with fore-/background classes, pSfS with foreground class, all computations with 6 cameras in $128 \times 128 \times 96$ volume
2. row: pSfS, foreground thresholding with 6 cameras, $256 \times 256 \times 192$ voxels
3. row: [left to right] tSfS $P'(F, S|\mathbf{c})$ thresholded, pSfS multiple classes, pSfS active camera, all computations with 1 of 6 views occluded in $64 \times 64 \times 48$ volume

See also: www.mi.fh-wiesbaden.de/~cjohn/videos/psfsTracking.avi

The SfS performances vary between 1.1ms and 241.8ms, depending on the chosen algorithm and volume resolution. A performance comparison between tSfS and pSfS limited to foreground class evaluation resulted in similar runtimes. Both algorithms have a linear complexity $\mathcal{O}(S)$ where S is the number of sensors. In contrast explicit evaluation of fore-/background classes has an exponential complexity of $\mathcal{O}(2^S)$.

It is further essential to note how the presented algorithms behave in the presence of systematic errors like inter-object occlusion. tSfS and both pSfS variants cannot handle this and do not reconstruct partial occluded objects. The appearance of systematic errors therefore has to be explicitly modeled. As our particle filter is applied to low resolution volume reconstructions, we limit the following comparison to this type. Figure 3 depicts obtained reconstructions with 5 out of 6 cameras. Here. we compared tSfS and pSfS with explicit computation of multiple foreground classes and finally pSfS with the active camera concept. All algorithms achieve similar coarse reconstructions and can reconstruct the volume even in presence of occlusion.

5 Conclusion

We have presented a GPU based pSfS system which uses a cascade of classifiers for volumetric reconstruction of skin colored objects, i.e. to track hand and face volumes. Our GPU implementation makes the system suitable for the advanced HCI applications targeted, with a runtime of less than 15ms for coarse but reasonable volume resolutions.

Acknowledgments. We would like to thank Brendan McCane, Geoff Wyvill and Katrin Frank for their contributions. Part of this work has been funded by a University of Otago CALT research grant (JDLJ17400).

References

1. IEEE Recommended Practice for Electric Power Systems in Commercial Buildings, p. 388 (1990)
2. Caetano, T., Olabarriaga, S., Barone, D.: Performance evaluation of single and multiple-gaussian models for skin color modeling. In: XV Brazilian Symposium on Computer Graphics and Image Processing, Proceedings, pp. 275–282 (2002)
3. Koller, D., Weber, J., Huang, T., Malik, J., Ogasawara, G., Rao, B., Russell, S.: Towards robust automatic traffic scene analysis in real-time. In: Proceedings of the International Conference on Pattern Recognition (1994)
4. Landabaso, J., Pardas, M.: A unified framework for consistent 2-d/3-d foreground object detection. IEEE Transactions on Circuits and Systems for Video Technology 18(8), 1040–1051 (2008)
5. Landabaso, J.L., Pardàs, M.: Shape from Inconsistent Silhouette. Accepted for publication in Journal of Computer Vision and Image Understanding (2008)
6. Minka, T.: The 'summation hack' as an outlier model. Unpublished manuscript (2003), http://research.microsoft.com/~minka
7. Vermaak, J., Doucet, A., Perez, P.: Maintaining multi-modality through mixture tracking. In: ICCV 2003, vol. 2, p. 1110 (2003)

Stereo Localization Using Dual PTZ Cameras

Sanjeev Kumar, Christian Micheloni, and Claudio Piciarelli

Department of Mathematics and Computer Science
University of Udine, Via Della Scienze 206, Udine-33100, Italy
{sanjeev.kumar,christian.micheloni,claudio.piciarelli}@dimi.uniud.it

Abstract. In this paper, we present a cooperative stereo system based on two pant-tilt-zoom (PTZ) cameras that can localize a moving target in a complex environment. Given an approximate target position that can be estimated by a fixed camera with a wide field of view, two PTZ cameras with a large baseline are pointed toward the target in order to estimate precisely its position. The overall method is divided in three parts: offline construction of a look-up-table (LUT) of rectification matrices, use of the LUT in real time for computing the rectification transformations for arbitrary camera positions, and finally 3D target localization. A chain of homographic transformations are used for finding the matching between different pairs of wide baseline stereo images. The proposed stereo localization system has two advantages: improved localization on a partially occluded target and monitoring a large environment using only two PTZ cameras without missing significant information. Finally, through experimental results, we show that the proposed system is able to make required localization of targets with good accuracy.

1 Introduction

Modern video surveillance has been an active area of research. Nowadays, a number of research works are going to develop more intelligent and smart video monitoring systems according to the requirements and applicability [1], [2], [3]. The computation of reliable objects' trajectories by means of localization is really important for different contexts like traffic monitoring, behaviour analysis, suspicious event detection, sensor network configuration, etc. From the low-level to the high level techniques three main steps can be identified: a) detection and localization of interesting objects, b) frame-to-frame tracking of detected objects and c) behaviour recognition. To achieve all these goals visual surveillance systems usually exploit a network of cameras [4]. Existing non-stereo systems often localize objects in the environment by defining homographies between single cameras and a 2D map [4]. Such homographies are based on a ground plane constraint. When the detected object is occluded in such a way that its point of contact with the ground plane is not visible, such an approach introduces relevant localization errors. To overcome such a problem, stereo vision can be taken into account.

Stereo vision has the advantage that it is able to estimate an accurate and detailed 3-D representation of the position of an object with respect to a given co-ordinate system using its two or more perspective images [5]. Traditional stereo vision research usually uses static cameras for their low cost and relative simpleness in modelling. PTZ camera is a typical and the simplest active camera, whose pose can be fully controlled by pan, tilt and zoom parameters [3]. As PTZ cameras are able to obtain multi-view-angle and multi-resolution information (i.e. both global and local image information), they are used for many real applications specially in video surveillance. The PTZ camera based stereo system is able to cover large environments and, if overlapped fields of view are considered, to reduce the occlusions. However, PTZ cameras based stereo vision is much more challenging when compared to traditional static cameras based stereo vision as the intrinsic and external parameters of each camera can be changed in utility.

Recently, a novel stereo rectification method for dual-PTZ-camera system is presented to greatly increase the efficiency of stereo matching [6]. In this dual-PTZ-camera based stereo method, the problem related to inconsistency of intensities in two camera images is solved by addressing a two-step stereo matching strategy. An interesting approach to solve stereo vision problems by means of rotating cameras has been recently proposed with its analytic formulation [7]. An off-line initialization process is performed to initialize essential matrix using calibration parameters. During on-line operations the rotation angles of the cameras are retrieved and exploited to compute the essential matrix. When the zoom is considered, it would require the calibration for any zoom level of both cameras.

In this paper, we propose a stereo system based on two PTZ cameras from a network of cooperative sensors. The proposed solution is able to accurately localize a moving object in outdoor areas. Once a target is selected by the surveillance system, a pair of PTZ cameras are focused on the target with the required zoom to provide stereo localization of such a target. To solve stereo matching problem in case of dual PTZ camera, an uncalibrated approach that computes the rectification by interpolating the transformations contained in a LUT is proposed. Such a LUT is defined off-line by sampling the pan and tilt ranges of both PTZ cameras and using the same zoom level. The transformations contained in the LUT are computed on image pairs computed with a chain of homographies to solve the wide base-line problem. During on-line operations an interpolation based on neural network is proposed to estimate the rectification transformations for the given orientations of both cameras.

2 Pre-localization Steps

The localization is performed using various rectified pairs of stereo images. Therefore, few steps for real time rectification are needed before performing the task of localization. These steps involve wide baseline stereo matching, construction of LUT and learning of neural network using LUT data.

2.1 Construction of the Look-Up Table

A rectification transformation is a linear one-to-one transformation of the projective plane, which is represented by a 3×3 non-singular matrix. For a pair of stereo images $\mathbf{I}_l$ and $\mathbf{I}_r$, the rectification can be expressed in the following ways

$$\mathbf{J}_l = \mathbf{R}_l * \mathbf{I}_l \qquad \mathbf{J}_r = \mathbf{R}_r * \mathbf{I}_r$$

where $(\mathbf{J}_l, \mathbf{J}_r)$ are the rectified images and $(\mathbf{R}_l, \mathbf{R}_r)$ are the rectification matrices. These rectification transformations can be obtained by minimizing

$$\sum_i [(m_l^i)^T \mathbf{R}_r^T \mathbf{F}_\infty \mathbf{R}_l m_l^i] \tag{1}$$

where (m_l^i, m_r^i) are pairs of matching points between images $\mathbf{I}_l$ and $\mathbf{I}_r$ and $\mathbf{F}_\infty$ is the fundamental matrix for rectified pair of images. Generally, the minimization of (1) is time-consuming and therefore it is not possible to compute the rectifications in real time [8]. Here, an offline LUT containing rectification matrices corresponding to various image pairs captured at predefined pan and tilt angles is constructed. The rectification transformations can then be interpolated in real-time for any arbitrary orientation of both PTZ cameras by using this LUT data. The main steps to construct the LUT are:

1. Sample the different pan and tilt angles $(p_l^i, t_l^i)_{i=1:1:n_1}$ for the whole pan and tilt ranges of left PTZ camera into n_1 equal intervals. Similarly, sample the different pan and tilt angles $(p_r^i, t_r^i)_{i=1:1:n_1}$ for the right camera.
2. Capture $n_1 \times n_1$ different of images $(\mathbf{I}_l^{i,j})_{i=1:1:n_1}^{j=1:1:n_1}$ for left camera. Same time of instance capture their corresponding right stereo images $(\mathbf{I}_r^{i,j})_{i=1:1:n_1}^{j=1:1:n_1}$.
3. Compute the possible $k(> n_1 \times n_1)$ pairs of rectification transformations pairs $(\mathbf{R}_l^k, \mathbf{R}_r^k)$ for the different combination of these stereo images. Here, we use the constraint that the rectification transformations are computed for two images only if they share at least 30% of their field of view. This criterion is considered also during the sampling of pan and tilt angles for both cameras.
4. Store all these pairs of rectification transformations in a LUT in such a way that by choosing a combination of four independent variables (p_l, t_l, p_r, t_r), their corresponding rectification transformations $(\mathbf{R}_l, \mathbf{R}_r)$ can be easily computed. This is done through a neural network described in section 2.3.

The main problem to be addressed in the creation of the LUT is the automatic computation of the rectification transformations. Many works on rectification assume that the baseline (the distance between the two cameras) is small if compared to the distance of the object from the cameras, and thus the two images acquired by the cameras are similar. This allows the detection of the matching points using standard techniques such as SIFT matching [9]. However, in the proposed system this assumption is no longer valid for some combinations of pan-tilt values. The problem of finding matches in wide-baseline configurations is addressed in the next section.

2.2 Point Matching between Wide Baseline Stereo Images

SIFT matching [9] is a popular tool for extracting pairs of matching points between stereo images. However, this method fails to provide good results in case of wide baseline images. In this work, we have used a method based on a chain of homographic matrices for extracting pairs of matching points in these kinds of image pairs. In the case of wide baseline images, if the object is far enough along the optical axis then it is possible to extract pairs of matching points manually or using an approach proposed in [10]. Let $(\mathbf{I}_l^1, \mathbf{I}_r^1)$ be a pair of images of a 3D scene which is far from the cameras along their optical axis. An initial homography $\mathbf{H}^1$ is generated by using extracted pairs of matching points between $\mathbf{I}_l^1$ and $\mathbf{I}_r^1$ using standard approaches. Let $\mathbf{I}_l^n$ and $\mathbf{I}_r^n$ be a pair of images captures from left and right cameras of a scene/object near to cameras along their optical axis. The problem is to autonomously extract the pairs of matching points between the images $\mathbf{I}_l^n$ and $\mathbf{I}_r^n$. To solve such a problem, a set of n images is captured for each camera by moving the cameras from the initial position (the one at which $(\mathbf{I}_l^1, \mathbf{I}_r^1)$ are acquired) to the current position. Let these two sets of images be $(\mathbf{I}_l^1, \mathbf{I}_l^2 \ldots \mathbf{I}_l^n)$ and $(\mathbf{I}_r^1, \mathbf{I}_r^2 \ldots \mathbf{I}_r^n)$. Now we use the following steps to solve this matching problem for wide-baseline image pairs:

1. Perform the SIFT matching between image pairs $(\mathbf{I}_l^1, \mathbf{I}_l^2)$, $(\mathbf{I}_l^2, \mathbf{I}_l^3)$, ..., $(\mathbf{I}_l^{n-1}, \mathbf{I}_l^n)$ and use these sets of pairs of matching points for computing their respective homography matrices $\mathbf{H}_l^{1,2}$, $\mathbf{H}_l^{2,3}$, ..., $\mathbf{H}_l^{n-1,n}$.
2. Repeat the procedure given in above step on the sequences of images of right camera and compute $\mathbf{H}_r^{1,2}$, $\mathbf{H}_r^{2,3}$, ..., $\mathbf{H}_r^{n-1,n}$.
3. Compute the homography matrix $\mathbf{H}_l$ and $\mathbf{H}_r$

$$\mathbf{H}_l = \prod_{i=0}^{n-2} \mathbf{H}_l^{n-(i+1),n-i} \quad \text{and} \quad \mathbf{H}_r = \prod_{i=0}^{n-2} \mathbf{H}_r^{n-(i+1),n-i}$$

4. Compute the homography matrix $\mathbf{H}^n$ for the pairs of matching points between current images $\mathbf{I}_l^n$ and $\mathbf{I}_r^n$ as

$$\mathbf{H}^n = \mathbf{H}_r * \mathbf{H}^1 * (\mathbf{H}_l)^{-1} \qquad (2)$$

Figure 1 gives an intuitive interpretation of the procedure. The final homography matrix $\mathbf{H}^n$ can be computed for any value of n; however, the above procedure can accumulate errors in the final homography due to multiplication of several matrices. In order to minimize this error, we

1. keep the sampling step n as low as possible, with the constraint that we require at least a 30% image overlap for SIFT matching;
2. minimize errors due to bad matches by using a robust estimator for outlier detection and removal: we use the Iterative re-weighted least-square (IRLS) technique for computing the homography matrix from the pairs of matching points between any two images. IRLS provide a robust solution for homography computation when compared to other approaches like standard least square or Singular value decomposition.

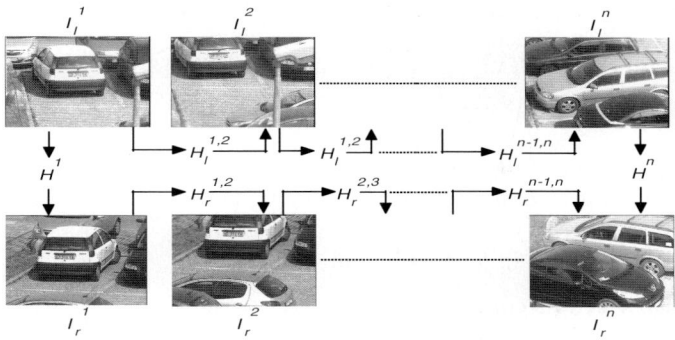

Fig. 1. Wide baseline stereo matching using a chain of homographic matrices

2.3 Neural Network Based Interpolation

Once the LUT is constructed, its content is used for the offline training of the neural networks using k combinations of pan and tilt angles $(p_l^i, t_l^i, p_r^i, t_r^i)_{i=1:1:k}$ as input and the elements of their respective rectification transformations $(\mathbf{R}_l^i, \mathbf{R}_r^i)_{i=1:1::k}$ as the network output. A multilayer feed-forward neural network containing one five-nodes hidden layer with backpropagation (BP)learning algorithm has been used in this work. The optimal input-to-hidden nodes weight matrix $\mathbf{W}$ and hidden-to-output nodes weight matrix $\mathbf{V}$ is stored and used for interpolating the rectification transformations for any arbitrary orientations of both cameras in real time.

Note that the LUT is built with constant and equal zoom levels for both cameras. Based on the requirements for the monitoring of the selected target, the zoom levels of the two cameras can actually be different, and the LUT data cannot be directly applied. In this case, compensation of unequal zoom settings is needed before stereo matching. This problem can be handled easily by using a focal-ratio-based methodology [11].

3 Localization

The 3D position of the target has to be computed in terms of its coordinates $[\mathbf{x}_w, \mathbf{y}_w, \mathbf{z}_w]$ in a world reference system. Once the pair of stereo images is rectified, the disparity between the matching pairs can be computed only for the pixels belonging to the target. Starting from the pixels in the left camera image, the search for its matching pixels is restricted only on the corresponding epipolar lines in the right camera image. In particular, for each pixel, starting from its x, y position, similarity scores are computed considering a normalized SSD measure that quantifies the difference between the intensity patterns as:

$$C(x,y,d) = \frac{\sum\limits_{(\xi,\eta)} [\mathbf{J}_l(x+\xi, y+\eta) - \mathbf{J}_r(x+d+\xi, y+\eta)]}{\sqrt{\sum\limits_{(\xi,\eta)} \mathbf{J}_l(x+\xi, y+\eta)^2 \sum\limits_{(\xi,\eta)} \mathbf{J}_r(x+\xi, y+\eta)^2}} \tag{3}$$

where $\xi \in [-n, n]$ and $\eta \in [-m, m]$ define a window centred in (x, y), while d is the disparity. The required disparity value is the one that minimizes the SSD error:

$$d_0(x, y) = \min_d C(x, y, |d|) \tag{4}$$

Once the disparity d is computed between the position of the target in the left and right images, the distance of the target $\mathbf{z}_w$ from the camera along optical axis is estimated by

$$\mathbf{z}_s = f_r \frac{B}{d} \tag{5}$$

where f_r is focal length for the rectified pair of images and B denotes the base line distance. Let $(\mathbf{x}_l, \mathbf{y}_l)$ be the position of the target in the left camera image, then its position in the plane orthogonal to the optical axis of camera is given by

$$\mathbf{x}_w = \frac{\mathbf{x}_l \mathbf{z}_w}{f_r} \qquad \mathbf{y}_w = \frac{\mathbf{y}_l \mathbf{z}_w}{f_r}$$

The location of target $(\mathbf{x}_m, \mathbf{Y}_m)$ in a ground plane map is given by

$$[\mathbf{x}_m, \; \mathbf{y}_m, \; 1]^T = \mathbf{H}_m^w [\mathbf{x}_w, \; \mathbf{y}_w, \; 1]^T$$

where $\mathbf{H}_m^w$ is the homography computed offline between the homogeneous coordinates of ground plane position $(\mathbf{x}_w, \mathbf{y}_w)$ of some selected points and their respective position in the map $(\mathbf{x}_m, \mathbf{y}_m)$. The iterative re-weighted least square (IRLS) algorithm allows to robustly estimate such a homography.

4 Experimental Results

The experimental results have been obtained from four different pairs of frames by considering different cases, i.e., partially occluded targets and using different zoom levels for both PTZ cameras. Six different pan and tilt angles have been selected in each direction by sampling with a step size of 3.0 degree along pan direction and 4.0 degree along tilt direction for both cameras to cover entire experimental outdoor environment. In this way, a total of 36 images have been captured by each camera. Out of these 36 × 36 combination of images, only $k = 120$ pairs of images have been selected for network training by considering the fact that at least 30% part of field of view should be common between both images. The rectification transformations have been computed using the matching pairs of feature points from these 120 pairs of images and stored in a LUT.

Experiments with real sequences have been carried out in order to test the performance of the proposed localization algorithm. Localization results are shown

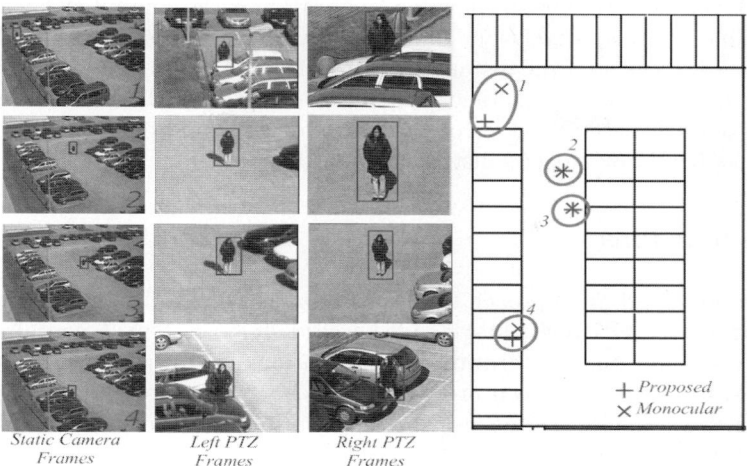

Fig. 2. Localization of a target in various stereo frames

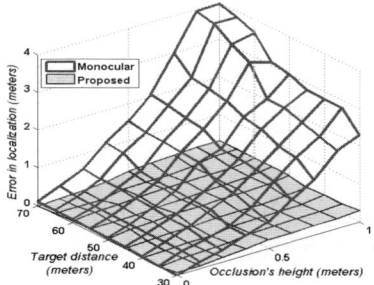

Fig. 3. Error in localization corresponding to the occlusion's height and object's distance

in Fig. 2 for four different pairs of frames captured at different camera settings (pan, tilt and zoom). The selection of these four frames have been performed to check the performance of proposed localization algorithm in different cases such as partially occluded target (frame pairs 1 and 4), images having unequal zoom (frame pairs 1 and 2) and non-occluded (frame pairs 2 and 3). Simultaneously, the localization results are computed based on a monocular camera based technique [4](here we have used a static camera having wide field of view) for making a comparison of the achieved results and for showing the superiority of proposed method on monocular camera based techniques in case of partially occluded targets (see localization for frame pairs 1 and 4). Localization has been made in a 2D ground-plane map (30 × 40) meters.

Fig. 3 represents a surface plot for localization error computed for a target with ground truth obtained from known marks. It can be seen from this plot that the error is increasing if the distance of object from the left camera or the height of occlusion is increasing in case of monocular camera based scheme. In the case of the proposed method, the error is almost constant and does not depend the occlusion's height or object's distance from camera.

5 Conclusions

We have presented an approach for the localization of an object in a given test map using dual PTZ camera based wide baseline stereo system. A neural network is used for finding the rectification transformations in real time using an offline LUT; and a method has been proposed for extracting pairs of matching points from wide base line stereo images. The required targets have been localized on a given test map using stereo based 3D position. Experimental results have proven that the proposed technique leads to better results than standard monocular camera based localization.

Acknowledgements

This work was partially supported by the Italian Ministry of University and Scientific Research and by the MUSAS Project A-380-RT-GC.

References

1. Abidi, B., Koschan, A., Kang, S., Mitckes, M., Abidi, M.: Automatic target Acquisition and Tracking with Cooperative Static and PTZ Video Cameras. Multisensors Surveillance Systems: The Fusion Perspective, 43–59 (2003)
2. Haritaoglu, S., Harwood, D., Davis, L.: W^4: Real-Time Surveillance of People and their Activities. IEEE Transactions on Pattern Analysis and Machine Intelligence 22(8), 809–830 (2000)
3. Jain, A., Kopell, D., Kakligian, K., Wang, Y.F.: Using Stationary-Dynamic Camera Assemblies for Wide-area Video Surveillance and Selective Attention. In: Proc. of IEEE Int. Conf. of Computer Vision and Pattern Recoginition, vol. 1, pp. 537–544 (2006)
4. Micheloni, C., Foresti, G.L., Snidaro, L.: A network of Co-operative Cameras for Visual Surveillance. IEE-proc. Vis. Image Signal Process. 152(2), 205–212 (2005)
5. Brown, M., Burschka, D., Hager, G.D.: Advances in Computational Stereo. IEEE Transactions on Pattern Analysis and Machine Intelligence 25(8), 993–1008 (2003)
6. Wan, D., Zhaou, J.: Stereo Vision Using Two PTZ Cameras. Computer Vision and Image Understanding 112(2), 184–194 (2008)
7. Hart, J., Scassellati, B., Zucker, S.W.: Epipolar Geometry for Humanoid Robotic Heads. In: Proc. of 4th International Cognitive Vision Workshop, pp. 24–36 (2008)
8. Isgro, F., Trucco, E.: On Robust Rectification of Uncalibrated Images. In: Proc. of 10th International Conference on Image Analysis and Processing, pp. 297–302 (1999)

9. Lowe, D.G.: Distinctive Image Features from Scale-Invariant Keypoints. International Journal of Computer Vision 2(60), 91–160 (2004)
10. Meltzer, J., Soatto, S.: Edge descriptors for robust wide-baseline correspondence. In: Proc. of IEEE Int. Conf. on Computer Vision and Pattern Recogination, pp. 1–8 (2008)
11. Kumar, S., Micheloni, C., Foresti, G.L.: Stereo Vision in Cooperative Camera Networks, Smart Cameras. Springer Science+Business Media, Inc., Heidelberg (in press, 2009)

Object Tracking in Video Sequences by Unsupervised Learning

R.M. Luque, J.M. Ortiz-de-Lazcano-Lobato, Ezequiel Lopez-Rubio, and E.J. Palomo

Department of Computer Science, University of Málaga, Málaga, Spain
{rmluque,jmortiz,ezeqlr,ejpalomo}@lcc.uma.es

Abstract. A Growing Competitive Neural Network system is presented as a precise method to track moving objects for video-surveillance. The number of neurons in this neural model can be automatically increased or decreased in order to get a one-to-one association between objects currently in the scene and neurons. This association is kept in each frame, what constitutes the foundations of this tracking system. Experiments show that our method is capable to accurately track objects in real-world video sequences.

1 Introduction

In video surveillance systems, accurate and real-time multiple objects tracking will greatly improve the performance of objects recognition, activity analysis and high level event understanding [1,2,3,4]. Segmentation and tracking of multiple objects are two important stages in visual surveillance. The most popular approach for visual tracking is the adaptive tracking of coloured regions, with techniques such as the particle filtering of coloured regions [5,6] and the Kalman/mean-shift [7], which uses the well known mean-shift algorithm [8] to determine the search region, and the Kalman filter to predict the position of the target object in the next frame.

In this paper, the use of growing competitive neural networks (GCNNs) to perform object tracking is proposed. These networks are derived from the usual competitive neural networks (CNNs) [9]. Their main particularity consists in that this kind of network is able to generate new process units (neurons) when needed, in order to get a better representation of the input space.

In general, CNNs are suitable for data clustering, since each neuron in a CNN is specifically designed to represent a single cluster. In the field of object tracking in video sequences, such clusters correspond to moving objects. Thus, it seems reasonable to use CNNs as trackers. However, due to the dynamic nature of a video sequence, objects are constantly appearing and disappearing from the scene, and the method used to track objects should take care of this situation. Consequently, the use of GCNNs become a good approach for tracking.

The rest of this paper is structured as follows: section 2 is devoted to the segmentation algorithm, in section 3 the tracking system is explained and finally, some experimental results and conclusions are presented in sections 4 and 5 respectively.

2 Object Segmentation

Detecting moving objects in video sequences is the first relevant stage of extracting information in most of the computer vision applications which are related to video analysis. Many works based on motion detection and, more concretely, on background subtraction using fixed cameras as the CCTV cameras installed in public transport can be found in the literature [4,2].

Our approach in this section is a modification of the algorithm proposed in [10] and can be considered as a pixel-based technique, since this kind of methods analyse each pixel separately. A classification problem has to be solved, in which a pixel (x, y) in an specific frame t only can belong to one of two classes: foreground or background. In order to perform this task, a neural network architecture based on Adaptive Resonance Theory (ART) is used. The inputs of this model are the colour components of the pixel according to the colour space in which the frames are obtained.

Each neuron is associated to one class (foreground or background) and each class can be composed of several neurons. Therefore, multimodal backgrounds (i.e. sea waves or waving trees) can be handled by our model, due to more than one neuron correspond to the background class. The neural network starts with an small number, M, of neurons or processing units. The network activates only one processing unit, called winning neuron, whose synaptic vector is closest to the input pattern x.

After that, the network checks if the input pattern is well represented by the synaptic vector of the winning neuron. In case of the test is failed, a new neuron k is created using the input pattern x as the synaptic vector of the neuron.

The neurons associated with background and foreground classes have to be determined. The B most activated neurons are used to model the background, whereas the rest of neurons correspond to foreground objects. This value B is computed as the amount of neurons whose number of activations $n_{a_1}, \ldots, n_{a_B}$ verify $\frac{n_{a_1}+\ldots+n_{a_B}}{N} > T$ for a prefixed threshold T, where N is the total number of activations of all neurons, as proposed in [4]. When the segmentation results have been obtained, it will be necessary to apply additional techniques to obtain clear foreground regions. Many shadow detection methods have been described in several reviews [11]. In our system, we develop the proposed technique cited in [12].

3 The Tracking Module

The tracking module is based on a growing competitive neural network (GCNNs), which follows an online training process based on a prediction-correction scheme. The number of neurons of the network is not fixed, and is changed depending on the amount of objects which must be tracked by the system in each time instant.

Every object appearing in the video frame is assigned to a neuron. This neuron becomes the responsible for identifying and representing the object, as well as predicting its features in future instants. New neurons are created when not previously detected objects appear in the image, whereas some neurons are killed when the objects associated to the neurons leave the scene.

In algorithm 1 the main steps of the algorithm are shown. Each step is described more thoroughly in the following sections.

3.1 Prediction-Correction Scheme

The proposed neural network is designed to follow a prediction-correction scheme during the training process. In the competition step every neuron predict the new state of the object that it is assigned to it. That is, where the object it is supposed to be in the current frame and how it is to be like, in terms of shape and colour. Each features vector corresponding to objects that appear in the current frame is compared with the estimated features vectors of the neurons. The neuron with the most accurate prediction is eligible as the winner. In the update step the winner neuron is the only neuron able to use the object features vector to correct the knowledge it has learnt, as it is explained later.

In order to calculate the object prediction the memory capacity of the neurons has been augmented. Each neuron, j, stores a log, H_j, which contains K entries with the known information about the object assigned to the neuron in some previous video frames. Every time an object is detected the current feature vector that represents it, H_j^w, and the frame in which the detection happens, H_j^f, are kept in the log of the neuron $H_j = \left(H_j^w, H_j^f\right)$.

The estimated pattern $\hat{x}_j(t)$ is obtained by summing the current object pattern, stored in the weight vector of the neuron, and an estimated change vector. This vector depends on the difference between the current frame t and the last frame in which the neuron was updated, and also on the averaged change observed in that pattern and computed for the last $P \leq K$ entries in the log,

$$\hat{x}_j(t) = w_j(t-1) + \left(t - H_j^w(K)\right) \sum_{i=K-P+1}^{K-1} \frac{H_j^w(i+1) - H_j^w(i)}{H_j^f(i+1) - H_j^f(i)} \quad (1)$$

with $H_j^w(i)$ the object features vector which was written down in the log of the j-th neuron in the frame $H_j^f(i)$. P is a user parameter which determines the number of log entries that are used for the object prediction and allows to adapt the prediction to the requirements of the specific application. Besides, low values of P allows to speed up the computation of the prediction.

Notice that the frame which appears in the last entry of the log coincides with the last frame in which the neuron was updated and, consequently, its weight vector was modified.

3.2 Competition Rule

Object tracking is a task which must be solved in real time. Therefore, the tracking system must be able to perform in an online way. In a time instant t the system is provided M training patterns (or input patterns) $x_i(t)$, $i \in \{1 \ldots M\}$. These features patterns correspond to M objects which were detected by the segmentation module in the video frame sampled in time instant t.

In the competition step, every time an input pattern $x_i(t)$ is provided to the network a competition process among the neurons starts. Each neuron predicts a features vector for the object it represents. This predicted feature vector represents the expected state of the object in the current frame t. The neuron whose predicted vector $\hat{x}_j(t)$ is the nearest in the input space to the input pattern is declared the winner.

$$c(t) = \arg \min_{1 \leq j \leq N} \{\|x_i(t) - \hat{x}_j(t)\|^2\} \qquad (2)$$

with $\hat{x}_j(t)$ computed by means of equation (1)

The proposed neural model also tries to detect which components of an object features vector are more reliable when we want to identify it. For that purpose the inverse of the standard deviation of the time difference of the data stored in the log is used

$$r_{jz}(t) = \frac{1}{\operatorname{var}\left(H_{jz}^w(t) - H_{jz}^w(t-1)\right)^{1/2}} \qquad (3)$$

where $H_{jz}^w(t)$ is the *z-th* component of the features vector $H_j^w(t)$. The variance is estimated as follows:

$$\begin{aligned}
\operatorname{var}\left(H_{jz}^w(t) - H_{jz}^w(t-1)\right) &= E\left[\left((H_{jz}^w(t) - H_{jz}^w(t-1)) - E[H_{jz}^w(t) - H_{jz}^w(t-1)]\right)^2\right] \\
&= \frac{1}{K} \sum_{h=1}^{K} ((H_{jz}^w(t-h) - H_{jz}^w(t-h-1)) - m_{jz}(t))^2
\end{aligned} \qquad (4)$$

where $m_{jz}(t)$ is the expectation of the difference

$$m_{jz}(t) = E\left[H_{jz}^w(t) - H_{jz}^w(t-1)\right] = \frac{1}{K} \sum_{h=1}^{K} (H_{jz}^w(t-h) - H_{jz}^w(t-h-1)) \qquad (5)$$

These measures for each component of an object features vector $x_j(t)$ are joined to form a vector of reliabilities $r_j(t)$.

Finally, a mask vector $m \in [0,1]^D$, with D dimension of the input space, has been added in order to let the user choose a weight for the object components based on their expert knowledge.

Then the resulting competition rule obtained when m and $r_j(t)$ are included is

$$c(t) = \arg \min_{1 \leq j \leq N} \{\|m \cdot r_j(t) \cdot (x(t) - \hat{x}_j(t))\|^2\} \qquad (6)$$

where $\cdot$ means the componentwise product.

3.3 Neuron Update

Once the winner neuron has been determined, the weight vector $w_{c(t)}$ must be updated in order to incorporate some knowledge from the pattern to the network. Only the winner neuron is updated in the instant time t.

$$w_i(t) = \begin{cases} w_i(t-1) + \alpha\,(x_i(t) - w_i(t-1)) & \text{if } i = c(t) \\ w_i(t-1) & \text{otherwise} \end{cases} \qquad (7)$$

where $\alpha \in [0\ldots 1]$ is named the *learning rate* and determines how important is the information extracted from the current input sample with respect to the background information that the neuron already known from previous training steps.

The proposed solution considers that each neuron represents an object in the frame and tracks this object through the input space frame by frame. The learning rate should be fixed to a large value, for example 0.9. Otherwise, the network cannot adequately detect changes in the object, and thus, the object may not be identified.

3.4 Neurons Birth and Death

The size of the neural network layer should not be fixed a priori because the number of objects which are present in the scene varies from one frame to another. Hence, the proposed network is formed by $n(t)$ neurons in a time instant t, and a mechanism to add new neurons to the network and to remove the useless neurons is needed.

When an unknown object appears in the scene, none of the existing neurons is able to represent it accurately and the error is expected to reach a high value, compared with the error obtained for correctly identified objects. Thus, a new neuron should be created in order to track that new object. A user-defined parameter $\delta \in [0, 1]$ has been utilised. It manages the neurons birth by means of the check

$$\forall j \in \{1 \ldots n(t)\} \quad \frac{\|\boldsymbol{x}(t) - \hat{\boldsymbol{x}}_j(t)\|}{\|\boldsymbol{x}(t)\|} > \delta \qquad (8)$$

with $\hat{\boldsymbol{x}}_j(t)$ computed by equation (1).

On the other hand, if an object leaves the scene then the neuron which represents it should be destroyed. For this purpose, each neuron has a counter C_{die} which means the *lifetime* of the neuron, measured in number of training steps. Each training step, the counter value is decreased by one and, if the value reaches zero then the corresponding neuron is removed. Every time a neuron wins a competition its counter value is reset. Therefore, only neurons associated to objects which are not longer in the scene are destroyed, since it is very unlikely for these neurons to win a competition.

4 Results

Several sequences have been used to prove the effectiveness of our tracking method, in which the objects are considered as rigid objects. The main objective is to demonstrate that this system is effective, reliable and robust in order to get valid trajectories to be analysed in the following behaviour stage. For this reason, two different kind of scenes are taken into account. The first one consists of typical traffic sequences provided by a video surveillance online repository generated by the Federal Highway Administration (FHWA) under the Next Generation Simulation (NGSIM) program.[1] In these sequences some common problems appear, such as occlusions, stopped car in the scene or errors happened in the segmentation phase, which must be satisfactorily solved by our tracking algorithm.

The second kind of scene corresponds to hand-generated sequences 1(c) and it is used to compare with other standard tracking techniques and check really that our method

---

[1] Datasets of NGSIM are available at http://ngsim.fhwa.dot.gov/

Algorithm 1. Main steps of the tracking algorithm

Input: Time instant t and the features of the segmented objects $x_i(t)$
Output: Labelling of the segmented objects
foreach *Segmented object $x_i(t)$* **do**
  Compute winner neuron by means of Eq. (6);
  if *Eq. (8) is satisfied* **then**
    | Create a new neuron. Initialise it;
  else
    | Update the network using equation Eq. (7);
  end
end
Refresh the counter values belonging to the neurons which win a competition;
Decrease all neurons counter values by one;
Check out neuron counters and destroy neurons whose counter value is zero;

(a)   (b)   (c)

Fig. 1. Different sequences are viewed in which the objects are identified and tracked. Subfigures 1(a) (frame 844) and 1(b) (frame 810) correspond to real traffic sequences while 1(c) (frame 90) is a hand-made sequence used to make the comparison.

can solve the aforementioned problems. The ground truth can be generated for these sequences in order to compare the performance of the tracking approaches, unlike the traffic sequences that do not provide this valuable information. For comparison purposes the kalman filter, which is one of the main reference algorithms for tracking objects, is chosen. This method uses the centroid to predict the position of the identified object in the next frame.

In figure 2(a), the errors in the x coordinate of the centroids obtained by several algorithms at each frame are shown. Two versions have been generated for each algorithm, in which an split object module is performed in order to divide the overlapped objects that appear in the scene. This module simply takes into account the mean and standard deviance of the object sizes appearing in the sequence and, if the size of a new object exceeds the normal distribution obtained using the mean and standard deviance, try to split into more than one. Comparisons between the ground truth of one object of the hand-made sequence 1(c) and the results obtained by using the proposed algorithms are shown in 2(b). The correct trajectories has been obtained using the neural networks models since kalman approaches confuses the trajectories with other objects.

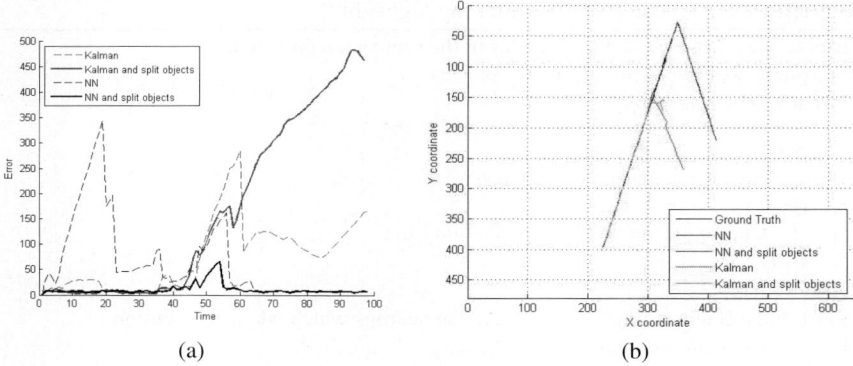

Fig. 2. 2(a) Comparative analysis of the centroid error at each frame using the ground truth. 2(b) Trajectory of the object with ID 2 in the scene showed in figure 1(c), is obtained using different algorithms.

Table 1. Comparative analysis of the success rate among the studied methods for the sequence observed in Fig. 1(c)

Method	Mean Error	Maximum Error	No. spurious objs	No. mixed trajs
Kalman	9.8133	45.3894	3	1
Kalman with split objects	21.4802	56.7362	2	3
NN	22.2467	97.2632	3	0
NN with split objects	**1.505**	**4.2302**	0	0

In the table 1 the mean and maximum errors of each trajectory is calculated for each algorithm. The last two columns represent the number of the spurious objects that appear in the scene and the number of mixed trajectories. It happens when two different objects swap their trajectories. This situation is not recommended due to the analysis of each trajectory will be done incorrectly. As we can observed, better results are obtained using both neural networks approaches.

5 Conclusions

A new algorithm for moving object detection and tracking in video sequences is presented. Tracking module is an important part of video surveillance systems, since it is necessary a good starting point to analyse object behaviour. With a reliable tracking algorithm, objects can be easily identified in the video sequence. Then, by using other analysis tools, the behaviour of these objects can be studied, and the system can determine whether they are suspicious/dangerous or not.

The algorithm proposed is based on the use of a type of the well-known competitive neural networks: the growing competitive neural network, which allows the creation and removing of neurons, which are assigned to the objects in the scene. Since the number of objects in a video sequence can change from frame to frame, it seems reasonable to permit a change in the number of process units of the network. Thus, a better representation of the foreground objects is obtained.

The new neural model is also able to predict the features of each object (location...), by using a log which stores all information known for every object during the last frames. This allows to deal with several problems produced at the segmentation phase, such as object occlusion or fusion. Experimental results show that our approach is a reliable and accurate method to detect objects in video sequences publicly available in Internet. In addition, segmentation derived problems can be robustly tackled by this method. Our future work covers aspects of behavioural analysis, as it is the next logical step in a surveillance system.

Acknowledgements

This work is partially supported by Junta de Andalucía (Spain) under contract TIC-01615, project name Intelligent Remote Sensing Systems.

References

1. Amer, A., Dubois, E., Mitiche, A.: Real-time system for high-level video representation: Application to video surveillance. In: Proceedings of the SPIE International Symposium on Electronic Imaging, pp. 530–541 (2003)
2. Haritaoglu, I., Harwood, D., Davis, L.: w^4: Real-time surveillance of people and their activities. IEEE Trans. Pattern Anal. Mach. Intell. 22(8), 809–830 (2000)
3. Lv, F., Kang, J., Nevatia, R., Cohen, I., Medioni, G.: Automatic tracking and labeling of human activities in a video sequence. In: Proceedings of the 6th IEEE International Workshop on Performance Evaluation of Tracking and Surveillance (2004)
4. Stauffer, C., Grimson, W.: Learning patterns of activity using real time tracking. IEEE Trans. Pattern Anal. Mach. Intell. 22(8), 747–767 (2000)
5. Grest, D., Koch, R.: Realtime multi-camera person tracking for immersive environments. In: IEEE 6th Workshop on Multimedia Signal Processing, pp. 387–390 (2004)
6. Nummiaro, K., Koller-Meier, E., Van Gool, L.: An adaptive color-based particle filter. Image Vision Comput. 21, 99–110 (2003)
7. Comaniciu, D., Ramesh, V.: Mean shift and optimal prediction for efficient object tracking. In: IEEE Int. Conf. Image Processing (ICIP 2000), pp. 70–73 (2000)
8. Comaniciu, D., Ramesh, V., Meer, P.: Real-time tracking of non-rigid objects using mean shift. In: IEEE Conference on Computer Vision and Pattern Recognition, pp. 142–149 (2000)
9. Ahalt, S.C., Krishnamurthy, A.K., Chen, P., Melton, D.E.: Competitive learning algorithms for vector quantization. Neural Networks 3, 277–290 (1990)
10. Luque, R.M., Valverde, F.L., Domínguez, E., Palomo, E.J., Muñoz, J.: Detecting critical situation in public transport. In: Proceedings of the 8th International Workshop on Pattern Recognition and Information Systems (PRIS), pp. 57–66 (2008)
11. Prati, A., Cucchiara, R., Mikic, I., Trivedi, M.: Analysis and detection of shadows in video streams: a comparative evaluation. In: Proceedings of the IEEE Computer Society Conference on Computer Vision and Pattern Recognition, CVPR, pp. 571–576 (2001)
12. Horprasert, T., Harwood, D., Davis, L.S.: A statistical approach for real-time robust background subtraction and shadow detection. In: Proceedings of International Conference on Computer Vision (1999)

A Third Eye for Performance Evaluation in Stereo Sequence Analysis

Sandino Morales and Reinhard Klette

The *.enpeda..* Project, The University of Auckland
Auckland, New Zealand

Abstract. Prediction errors are commonly used when analyzing the performance of a multi-camera stereo system using at least three cameras. This paper discusses this methodology for performance evaluation for the first time on long stereo sequences (in the context of vision-based driver assistance systems). Three cameras are calibrated in an ego-vehicle, and prediction error analysis is performed on recorded stereo sequences. They are evaluated using various common stereo matching algorithms, such as belief propagation, dynamic programming, semi-global matching, or graph cut. Performance is evaluated on both synthetic and real data.

1 Introduction

Assume a rectified stereo pair of a *left* and a *right* image, and a disparity map obtained by applying some stereo matching algorithm. In absence of ground truth data, one way to evaluate the performance of this matching algorithm is to calculate – from both the calculated disparity map and the given stereo image data – a *virtual* image as it would appear for a *virtual camera* at a defined pose, and compare this with a *third* image actually recorded at that pose. At pixels in the virtual image we assign either visible (in left and right image) surface textures, value 'black' for pixels occluded in the left image, and 'white' for pixels occluded in the right image. The comparison between virtual and third image takes those uncertainties into account.

This performance analysis is known as *prediction error evaluation* [15], and it is applied when at least three images of the same scene are available; see, for example, [1]. The third image is used as ground truth, and statistical analysis is performed to analyze the matching algorithms.

Fig. 1. Left to right: left and right images of frame 22 of the used synthetic sequence; third, left and right images of frame 95 of the used real-world sequence

We are recording video data with a three-camera system, and the described evaluation is not only done for a few triples of images but for trinocular ('long') image sequences, and (in this paper) for one 100-frame synthetic stereo image sequence where a third camera may be simulated based on available ground truth (see Figure 1). The use of long sequences allows us to observe the influence of varying *situations* (e.g., brightness differences between left and right image, or reflections) on the algorithmic performance. It allows to incorporate temporal filters into disparity calculation. To the best of our knowledge, the prediction error has not been used so far on long (especially real-world) stereo sequences.

The outline of this paper is as follows. We briefly recall a geometric approach that is commonly used to generate a virtual image from a disparity map and a pair of rectified images, and discuss poses of the third camera (Section 2). Section 3 informs about evaluated stereo algorithms, used quality metrics and experimental results. Conclusions are stated in Section 4.

2 Geometry of the Third View

Assume three cameras; the left and right camera are rectified [6] in such a way that their images satisfy the *standard stereo geometry* [9]. The left (right) camera is the *reference (matching)* camera and records the *reference (matching)* image. The third camera may be at an arbitrary pose and provides the *third image*. The predicted image is the *virtual image*.

2.1 Common Forward Equations

The coordinate system of the reference camera is identified with the world coordinate system. Image coordinates are defined by each camera individually. The reference, matching and third camera are positioned as sketched in Figure 2; the camera center of the reference camera is at $O = (0,0,0)$, that of the matching camera at $O_M = (b,0,0)$, and that of the third camera at $O_T = (b_1, b_2, b_3)$. Let $P = (X, Y, Z)$ be a scene point

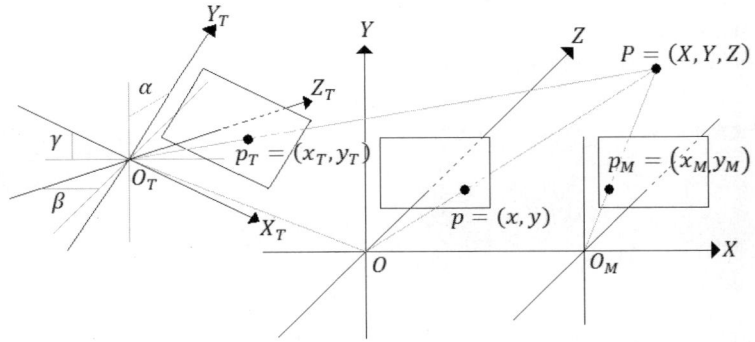

Fig. 2. Illustration of three-camera notations

visible for all the three cameras, and $p = (x, y)$, $p_M = (x_M, y_M)$, and $p_T = (x_T, y_T)$ are its projections on the reference, matching, or third image plane, respectively. In the virtual image we have $p_V = (x_V, y_V)$. Let b be the base-line distance, f the rectified focal length of reference and matching cameras, and d the disparity between p and p_M; disparity values have been calculated by some stereo matching algorithm. It follows [6] that

$$x_V = f_T \cdot \frac{m_{11}(bx - db_1) + m_{12}(by - db_2) + m_{13}(bf - db_3)}{m_{31}(bx - db_1) + m_{32}(by - db_2) + m_{33}(bf - db_3)} \quad (1)$$

$$y_V = f_T \cdot \frac{m_{21}(bx - db_1) + m_{22}(by - db_2) + m_{23}(bf - db_3)}{m_{31}(bx - db_1) + m_{32}(by - db_2) + m_{33}(bf - db_3)} \quad (2)$$

where m_{ij} are the elements in the rotation matrix (from third into reference camera system, defined by rotation angles α, β, and γ; see Figure 2), for $1 \leq i, j \leq 3$, and f_T is the focal length of the third camera. These two *forward equations* (e.g., see [8]) allow us to map any pixel (x, y) in the reference image into a point (x_V, y_V) in the image plane of the virtual image, to be shown at the nearest pixel position.

2.2 Poses of the Third Camera

Occluded points may cause a bias when evaluating the performance of a stereo algorithm. (This may be illustrated by examples generated for the used synthetic stereo sequence in Set 2 on [3]; occluded points vary depending on the pose of the third camera.)

By increasing differences between poses of the third and the other two cameras, more occluded areas occur in the virtual image. A point P may be visible in the reference and in the third view, but still possibly occluded in the matching image. Texture cannot be assigned in the virtual image if P is not visible in left or right image, as there is no depth information available in this case. Occlusions may be reduced (in general) by having the third view between left and right camera. Figure 3 shows three different occlusion cases. We may predict a virtual view at a new pose; we can also identify this 'new' pose with one of the two given poses of the left and right camera.

The Symmetric Pose. The symmetric pose of the third camera (i.e., focal point halfway on base line between reference and matching camera, with perpendicular bisector incident with optical axis) is expected in general to be the one which minimizes impacts of occlusions (i.e., the total number of either black or white pixels). In evaluations

Fig. 3. Stereo sequence in Set 2 of [3]. Left to right: calculated (from ground truth) image of a virtual camera positioned on the left of the reference camera; two calculated virtual images at poses of the left and right camera; ground truth for the left image. Used disparity code: light = close, dark = far, white = occlusion.

it would be ideal to separate the impact of occlusions from those of incorrect matching. Thus, the symmetric case seems to be a good choice. However, errors due to mismatches may not be as obvious when the pose of the third camera differs (much) from the symmetric case.

Collinear Poses. The focal point of the third camera is on (or close to) the base line of the left and right camera. For example (see left of Figure 3), if the third view is on the left of the reference camera, both kinds of occlusions (black and white) are present in the novel view.

For this paper we decided for the collinear case, having the third camera approximately 40 cm to the left of the reference camera. Rectified reference and matching camera are about 30 cm apart from each-other.

3 Evaluations Using the Third Sequence

For each rectified stereo image pair of a given sequence and its calculated depth map, we generate an image as it would be seen by a virtual camera in exactly the same collinear pose as our third camera (i.e., left of the reference camera). For this paper we use two sequences illustrated in Figure 1.

The gray-value synthetic sequence from Set 2 of [3] consists of 100 stereo pairs with available ground truth [16]. The usage of this sequence allows us to integrate results from a previous study [12] obtained for the same data set. We generate from available ground truth an image sequence (with occlusions) for a third camera being about 40 cm to the left of the reference camera.

We compare evaluation results for this synthetic example with those obtained for a trinocular real world sequence of 150×3 frames, taken with three calibrated cameras mounted in the research vehicle of the *.enpeda..* project. We selected the center and right camera to be the reference and matching camera, respectively. The focal point of the reference camera is considered to be the origin of the world coordinate system, and the other two cameras are calibrated with respect to this coordinate system.

Stereo Algorithms. We aimed at testing a representative collection of various stereo algorithms, and our selection is as follows:

Dynamic programming stereo. We compare a standard algorithm [13] (DP), against one with temporal (DPt), spatial (DPs), or temporal and spatial (DPts) propagation; see [11] for specifications of propagation.

Belief propagation stereo. We use a coarse-to-fine algorithm BP [4] with quadratic cost function, with parameter settings as reported in [5].

Semi-global matching. An SGM strategy [7] allows us to use different cost functions; we use mutual information (SGM MI) and Birchfield-Tomasi (SGM BT).

Graph Cut. For a detailed discussion of the GC method, see [2] and [10].

Quality Metrics. We use two common quality metrics. Let (x, y) be a pixel in the reference image I_R with intensity $I_R(x, y)$ and let (x_V, y_V) be the corresponding point in the virtual image I_V, with calculated intensity $I_V(x_V, y_V) = I_R(x, y)$. For each frame t of the given trinocular sequence, we compute the *root mean squared* (RMS) error between the third and the virtual image as follows:

$$R(t) = \frac{1}{|\Omega_t|} \left(\sum_{(x,y) \in \Omega_t} [I_T(x,y) - I_V(x,y)]^2 \right)^{\frac{1}{2}}$$

where $|\Omega_t|$ denotes the cardinality of the discrete domain Ω_t of non occluded pixels for frame t. The *normalized cross correlation* (NCC) is also used to compare third and virtual image, applying the following:

$$N(t) = \frac{1}{|\Omega_t|} \sum_{(x,y) \in \Omega_t} \frac{[I_T(x,y) - \mu_T][I_V(x,y) - \mu_V]}{\sigma_T \sigma_V}$$

μ_T and μ_V denote the means, and σ_T and σ_V the standard deviations of I_T and I_V, respectively.

Results for the synthetic sequence. For the RMS results (see Figure 4, left, and Table 1, left), the algorithm with the best overall performance was SGM BT, followed by BP and SGM MI; GC ranks fourth followed by the dynamic programming algorithms. Note that the order of rankings is about constant along the synthetic sequence.

The larger errors occurring in the first ten frames are a result of large occluded areas (Note that occluded points have influence on the calculation of the disparity map as well as on the generation of the virtual view.) caused by a close object - a car. The local error maximum around frame 45 is caused by a similar situation. However, we can conclude that the summarized ranking of the algorithms is not affected by those situations. The other local error maxima, around frames 15, 20 and 60, are due to errors in the calculated disparity maps, as there are no obvious changes in occluded areas in those frames.

The ranking of the algorithms resembles the one obtained in [12], where a different evaluation methodology was used (not a third camera but just a comparison with ground

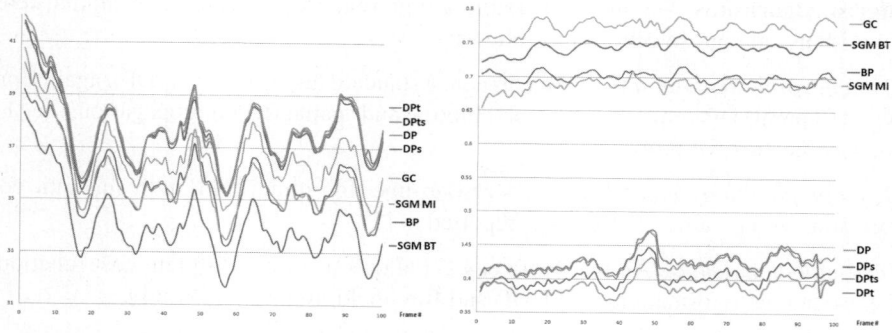

Fig. 4. Frame-by-frame results for the synthetic sequence. Left: RMS. Right: NCC; note that closer to 1.0 means "better".

Table 1. Summarizing results for the synthetic sequence. Left: RMS. Right: NCC.

Algorithm	Mean	Min	Max	Algorithm	Mean	Min	Max
SGM BT	34.05	13.67	30.68	GC	0.77	0.75	0.79
BP	35.69	14.72	31.59	SGM BT	0.74	0.72	0.76
SGM MI	35.72	14.24	29.85	BP	0.70	0.69	0.72
GC	36.67	17.30	34.19	SGM MI	0.69	0.65	0.71
DPs	37.55	13.75	32.43	DPt	0.43	0.38	0.45
DPt	37.68	12.99	32.43	DP	0.42	0.40	0.47
DP	37.70	12.95	32.53	DPs	0.40	0.38	0.45
DPts	37.70	13.03	28.98	DPts	0.39	0.37	0.43

Fig. 5. Examples of virtual images for the best four performing algorithms for both metrics. The black strip on the top is due to a minor tilt in the pose of the third view. Left to right: SGM BT, BP, SGM MI, and GC (RMS ranking).

truth). There, it was also stated that SGM BT performed best for this sequence. However, BP and SGM MI swapped their positions in the different evaluations in [12] and here. Another difference is that DPs ranked third among the four dynamic programming algorithms in that previous study, but shows now best performance out of those four.

The NCC measure ranking is different from the one derived from RMS. The GC algorithm performs best overall now, followed by SGM BT, BP and SGM MI. Figure 5 shows the virtual images of frame 22 for the top four performing algorithms for the both metrics. The four dynamic programming algorithms were the worst again; with DPt performing best for most of the frames, and DPts being the worst. For the top four algorithms it is evident that the performance on the first ten frames is again limited, impacted by occlusions; the four dynamic programming algorithms do not show this change in performance.

Results for the Real World Sequence. For RMS, all the eight algorithms behave pretty much the same! The difference in magnitude is not evident at all as the function graphs are highly overlapping; see Figure 6. Ranking at one particular frame may be totally different to a ranking at another frame! However, Table 2 shows that DPts appears to be the best algorithm with respect to RMS, followed closely by DPt. The worst algorithm by far is GC in this case. The local maxima correspond, also for this sequence, to frames where there are closer objects to the ego-vehicle, causing more occluded areas. – The NCC results show a totally different ranking. For this metric, BP performs the best, followed by DP and DPt; DPts was the worst algorithm for this metric, which tells us that it calculates inaccurate values at many pixels, but errors are fairly small. Note that

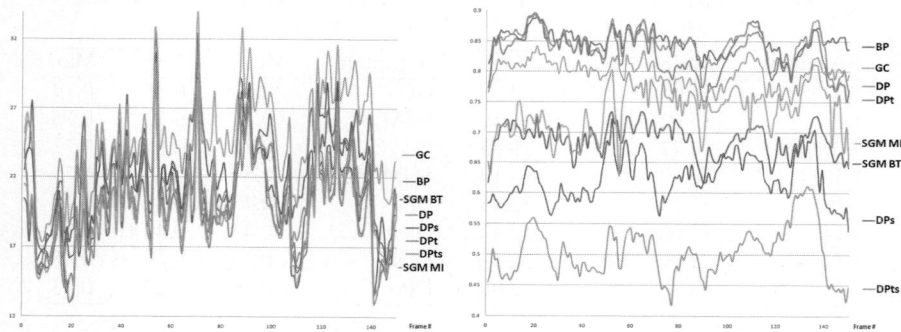

Fig. 6. Frame-by-frame results for the real world sequence. Left: RMS. Right: NCC.

Table 2. Results for the real world sequence. Left: RMS. Right: NCC.

Algorithm	Mean	Min	Max	Algorithm	Mean	Min	Max
DPts	19.74	13.03	28.98	BP	0.85	0.79	0.89
DPt	19.75	12.99	32.43	DP	0.84	0.76	0.89
SGM MI	20.09	14.24	29.85	DPt	0.83	0.75	0.72
SGM BT	20.09	13.67	30.68	GC	0.80	0.74	0.84
DP	21.19	12.95	32.53	SGM MI	0.73	0.63	0.81
BP	21.91	14.72	31.59	SGM BT	0.69	0.62	0.73
DPs	22.23	13.75	32.43	DPs	0.62	0.54	0.72
GC	24.79	17.30	34.19	DPts	0.50	0.42	0.61

Fig. 7. Examples of virtual images for the top four performing algorithms on the real-world sequence. Upper row (NCC), left to right: BP, DP, DPt, and GC. Lower row (RMS), left to right: DPts, DPt, SGM MI, and SGM BT.

SGM MI performs better than SMG BT, which confirms the ranking in [12], where SMG BT proved to be more sensitive to common real-world noise than SGM-MI. In general it seems that NCC ranking is more appropriate on real-world sequences than RMS-ranking; see Figure 7 for an example.

4 Conclusions

This paper evaluates the performance of several stereo algorithms, using the generation of a virtual image from the disparity map. We conclude that this prediction error analysis is a valuable tool to test the performance of stereo algorithms when no real-world ground truth is available. We notice a good correlation with RMS evaluations as previously obtained for the used synthetic sequence in [12], where the methodology was characterized by using the ground truth. Occlusions seem to have an influence on the magnitudes of the errors, but do not seem to affect the ranking of the algorithms very much. It is also evident that testing algorithms on real world sequences is necessary, as the rankings of algorithms may vary totally if used on either a synthetic or a real world sequence. The nearly constant ranking order on the synthetic sequence appears to be due to missing diversity of situations, as occurring in real-world sequences.

Future work may use different positions for the third camera and different metrics in order to widen the study about relationships between occlusions and accuracy for the presented approach.

Acknowledgement. The authors thank Stefan Gehrig for his implementations of SGM, Shushi Guan for his specification of the [4] implementation of BP, Joachim Penc for his GC implementation, and Tobi Vaudrey for his valuable comments.

References

1. Baker, S., Scharstein, S., Lewis, J.P., Roth, S., Black, M.J., Szelisky, R.: A database and evaluation methodology for optical flow. In: Proc. IEEE Int. Conf. Computer Vision, CD (2007)
2. Boykov, Y., Veksler, O., Zabih, R.: Fast approximate energy minimization via graph cuts. IEEE Trans. Pattern Analysis Machine Intelligence 23, 1222–1239 (2001)
3. .enpeda.. image sequence analysis test site (EISATS), http://www.mi.auckland.ac.nz/EISATS/
4. Felzenszwalb, P.F., Huttenlocher, D.P.: Efficient belief propagation for early vision. Int. J. Computer Vision 70, 261–268 (2006)
5. Guan, S., Klette, R., Woo, Y.W.: Belief propagation for stereo analysis of night-vision sequences. In: Wada, T., Huang, F., Lin, S. (eds.) PSIVT 2009. LNCS, vol. 5414, pp. 932–943. Springer, Heidelberg (2009)
6. Hartley, R., Zisserman, A.: Multiple View Geometry in Computer Vision, 2nd edn. Cambridge University Press, Cambridge (2004)
7. Hirschmüller, H.: Accurate and efficient stereo processing by semi-global matching and mutual information. In: Proc. Computer Vision Pattern Recognition, vol. 2, pp. 807–814 (2005)
8. Klette, R., Zamperoni, P.: Handbook of Image Processing Operators. Wiley, Chichester (1996)

9. Klette, R., Schlüns, K., Koschan, A.: Computer Vision. Three-Dimensional Data from Images. Springer, Singapore (1998)
10. Kolmogorov, V., Zabih, R.: What energy functions can be minimized via graph cuts? IEEE Trans. Pattern Analysis Machine Intelligence 26, 65–81 (2004)
11. Liu, Z., Klette, R.: Dynamic programming stereo on real-world sequences. In: Proc. ICONIP. LNCS. Springer, Heidelberg (to appear, 2009)
12. Morales, S., Vaudrey, T., Klette, R.: An in depth robustness evaluation of stereo algorithms on long stereo sequences. In: Proc. Intelligent Vehicles (to appear, 2009)
13. Ohta, Y., Kanade, T.: Stereo by two-level dynamic programming. In: Proc. IJCAI, pp. 1120–1126 (1985)
14. Scharstein, D., Szeliski, R.: A taxonomy and evaluation of dense two-frame stereo correspondence algorithms. Int. J. Computer Vision 47, 7–42 (2002)
15. Szeliski, R.: Prediction error as a quality metric for motion and stereo. In: Proc. Int. Conf. Computer Vision, vol. 2, pp. 781–788 (1999)
16. Vaudrey, T., Rabe, C., Klette, R., Milburn, J.: Differences between stereo and motion behavior on synthetic and real-world stereo sequences. In: Proc. Image Vision Computing New Zealand. IEEE, Los Alamitos (2008)

OIF - An Online Inferential Framework for Multi-object Tracking with Kalman Filter

Saira Saleem Pathan, Ayoub Al-Hamadi, and Bernd Michaelis

Institute for Electronics, Signal Processing and Communications (IESK)
Otto-von-Guericke-University Magdeburg, Germany
{Saira.Pathan,Ayoub.Al-Hamadi}@ovgu.de

Abstract. We propose an Online Inferential Framework (OIF) for tracking humans and objects under occlusions with Kalman tracker. The OIF is constructed on knowledge representation schemes, precisely semantic logic where each node represents the detected moving object and flow paths represent the association among the moving objects. A maximum likelihood is computed using our CWHI-based technique and Bhattacharyya coefficient. The proposed framework efficiently interprets multiple possibilities of tracking by manipulating the "propositional logic" on the basis of maximum likelihood at a time window. The logical propositions are built by formularizing facts, semantic rules and integrity constraints associated with tracking. The experimental results show that our novel OIF is able to track objects along with the interpretation of their physical states accurately and reliably under complete occlusion, illustrating its contribution and advantages over various other approaches.

1 Introduction

Ideally, it is expected in tracking to estimate the trajectory of the object to interpret its path. But practically it is not simple due to illumination variation, camera motion, and on field conditions. Besides, many other issues such as internal and external occlusion, abrupt change in orientation, and motion of the object make tracking a difficult problem. However, in this paper we are only addressing partial and full occlusions. Tracking has been extensively studied; a detailed review on visual tracking is given in [1],[2]. One of the prominent techniques is the data association approach such as Probabilistic Data Association (PDA) [3], Probabilistic Multi-hypothesis Tracker (PMHT) [4], and Multi hypothesis Tracker (MHT) [5]. Another approach is the online data association, where the decision of association is based on the computations of each consecutive frame which however is very sensitive to detected features and may result in false association. In [6], a solution is suggested by taking a time window T to reduce ambiguities which may arise due to occlusions. However, these methods have an edge over classical data association in which the association and estimation relies on the priors of previous frames.

Other authors use reasoning-based approaches to address the occlusion problem. For example, Elgammal et al.[7] presents occlusion-reasoning explicitly using segmentation when object interacts with each other. Each entered object is considered as isolated until any interaction is observed. As the objects interact, a color model is assigned which is then segmented in order to track object during occlusion. Wu and Nevatia [8] proposed a body part-based representation and detection of objects where the heuristics are used to overcome the inter-object occlusions. Recently, Rayoo and Aggarwal [9] proposes another approach based on Bayesian inference for long term occlusion where the object is entered in an explanation mode if the occlusion is observed. However, in the proposed technique with every successive frame a wide range of combination arises, thus computational complexity may raised. Therefore, despite of the many techniques that have been proposed to address problems and sub-problems related to tracking; yet, it still remains problematic and challenging.

In this paper, we present a generic framework for tracking objects using semantic logic-based framework and the maximum likelihood for data association. Our goal is to reliably track objects under severe occlusion with a motivation of tracking without any scene restriction and prior training. Our OIF extends the capabilities of Kalman tracker; so that it can handle occlusions explicitly. The paper is organized as: section 2 discusses tracking as a logical problem; The proposed OIF framework is introduced, conceptually in section 3 followed by the practical implementation of our tracking system in section 4. Section 5 represents the experimental results. Finally, the concluding remarks and the future directions of the work are sketched in section 6.

2 Tracking as Logical Problem

In our previous works, it is observed that online tracking of object during occlusion is a very generic problem but a generic solution is missing. This motivates us to interpret tracking as a logical problem and fuse it with typical predict-update model. The main contribution of this innovative idea is to prune the generation of wrong possibilities during data association through the "cognitive" state of the moving objects. For example if the object has just entered in the scene then it will be not considered when searching the "occluded" object. The data extracted from the pre-tracking stage (i.e. detection) at every time frame t is interpreted as a node which contains a property set. This property set represents two data structures which include: feature set and state set. The feature set is typically a set of physical features of detected object $F_i = \{h_i, a_i, \phi_i, c_i, b_i\}$ whereas the state set represents the cognitive state of the object $S_i = \{norm_i, occd_i, over_i, rapp_i, exit_i, new_i\}$. Each object has a fixed identity number which remains unchanged throughout tracking as presented in the illustration in Fig.(1).

During the propagation of inference models, these property sets are updated and stored in the knowledge base (i.e. hierarchical form) to generate next inferences. In this way, we represent our tracking problem on logical network (i.e. semantic inferences) to track objects. The propagation of flow path is based on maximum likelihood and integrity constraints. The search space criteria and integrity constraints are analogous to hypothesis pruning in this work. Thus, we

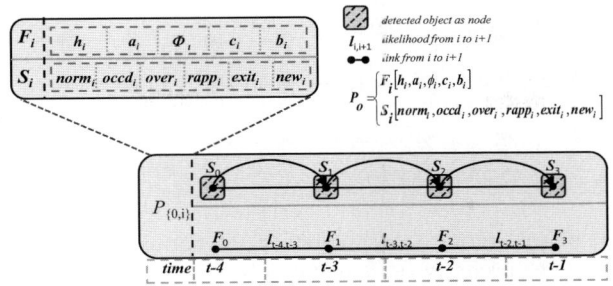

Fig. 1. The logical model of the object with its property set

are able to not only track objects efficiently by fusing logical framework in it, but also able to interpret the state of object.

3 OIF: A Conceptual Explanation

The main motivation behind OIF is to associate and incorporate the cognitive inference methodology (i.e. how a human-tracker would track multiple objects?) in tracking and manage the searching possibilities as mentioned in section2.

3.1 Logic Representation and Semantic Inference for Tracking

A semantic-logic based framework contains three components which builds our inferential framework. In this network, every detected moving object is a node, the relations among the nodes are built and inferred by manipulating logical premise, an elucidation of how human-tracker handles tracking?

Facts. Starting from the left: normal, occluded, overlaper, reappear, exit and new are the object's observation i at time t.

$$Facts = \{O_i^{norm}, O_i^{occd}, O_i^{over}, O_i^{rapp}, O_i^{exit}, O_i^{new}\} \tag{1}$$

Property set. The property set (see Fig.(1)) represents a unique property of each detected object in the scene. This data structure comprises of two sub structures as described in following:
- *Feature set:* These feature set are the result of post-detection response. The structure of the feature set is given below:

$$F_i = \{h_i, a_i, \phi_i, c_i, b_i\} \tag{2}$$

  where $h_i, a_i, \phi_i, c_i, b_i$ is the histogram, appearance, orientation, centroid and boundary of the detected object respectively.
- *State set:* The logical inference and tracking status of the object is built by keeping the cognitive interpretations under consideration. This state set can be expanded further but initially, the following states are associated to detect the objects:

$$S_i = \{norm_i, occd_i, over_i, rapp_i, exit_i, new_i\} \tag{3}$$

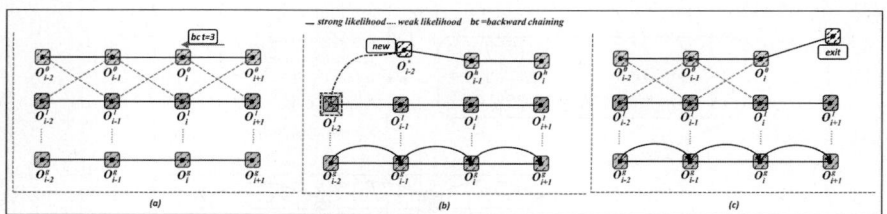

Fig. 2. A logical interpretation is presented. In (a) normal inference and tracking is shown in the scene, the property set is updated by OIF. In (b) the logical inference of "new" object is presented. In (c) The exit state inference is described.

3.2 Semantics and Semantic Events for Inference

Notations: These notations are used in following section: i is the observation instance, $1, ..., k$ represents the identities of the object, l is the likelihood.

For Normal $[O_i^{norm}]$. This status of moving object shows ideal behavior of detected object which is set during each detection when no appearance cluttering is observed as presented in Fig.(2a). This is the ideal situation of tracking when the property set of the objects are updated on the basis of maximum likelihood computation. The tracking system updates the location of the detected moving object by calculating the likelihood at time window T.
– Event: isNormal() Following is the semantic relations for "normal"

$$O_i^{\{1,...,k\},norm} \Rightarrow O_{i+1}^{\{1,...,k\},norm}$$

$$\textbf{iff } O_{i+1}^{\{1,...,k\},norm} \subseteq S_i^{\{1,...,k\},norm} \wedge \textbf{iff } max \left[\sum_{n=-3}^{0} l_{i+n,i+1} \right]$$

For New $[O_i^{new}]$. The decision of assigning "new" state is taken after considering two possibilities: the object search space must not fall in existing object space and the likelihood of the new detected object is minimum when compared to the existing objects. New detected object is assigned a new identity and a Kalman tracker is initialized to estimate its trajectory during the entire course of tracking as shown in Fig.(2b).
– Event: isNew() The logical premise is given below:

$$O_{i+1}^{\{h\},new?} \Rightarrow O_{i+1}^{\{h\},new} \quad \textbf{iff}\neg \left[O_{i+1}^{\{h\},new} \subseteq S_i^{\{1,...,k\},norm} \right]$$

For Exit $[O_i^{exit}]$. The object is set to "exit", if the object is outside the region S_I of the active scene. After this state, the tracker of the object is set to inactive and tracking is stopped. This object will not take part in any searching (i.e. dead node) as shown in Fig.(2c).
– Event: isExit() The interpretation is shown below:

$$O_i^{\{h\},norm} \Rightarrow O_i^{\{h\},exit} \quad \textbf{iff} \neg \left[O_{i+1}^{\{h\},norm} \subseteq S_I \right]$$

For Overlaper [O_i^{over}]. The object which retains its contextual information during occlusion is named as "overlaper" as presented in Fig.(3a). At first, both objects must fall in conflict search space, and then the maximum likelihood at time window T is computed and summed. The moving object with state "overlaper" contains its occluding object as child. As a result, two Kalman trackers are associated with one "conflicted" object. The feature set of the occluded object is updated using breadth-first search (BFS), an exhaustive searching technique whereas the state set reflects the conflicted-phase. The overlaper (i.e. parent) is our root node which expand and update its child nodes (i.e. occluded) until no child exits. The updation of overlaper and occluded object continues till split occurs.

– Event: isOverlap() The inference of overlaper state is given below:

$$O_i^{\{h\},norm} \Rightarrow O_i^{\{h\},over \wedge norm}$$

$$\textbf{iff } O_{i+1}^{\{1,\ldots,k\},over \wedge norm} \subseteq S_i^{\{1,\ldots,k\},norm} \wedge \textbf{iff } max\left[\sum_{n=-3}^{0} l_{i+n,i+1}\right]$$

For Occluded [O_i^{occd}]. During tracking, if an object lost its contextual information (i.e. features are affected by the influence of other moving object) due to occlusion, the object status is set to "occluded". As shown in Fig.(3a). After occlusion, the occluded object becomes a child of its overlaper and adapts the feature set of its parent, whereas the state set reflects its own behavior. Also it is not considered in any search till it reappears.

– Event: isOccluded() The state inference is based on the following:

$$O_i^{\{h\},norm} \Rightarrow O_i^{\{h\},occd}$$

$$\textbf{iff } O_{i+1}^{\{1,\ldots,k\},norm} \subseteq S_i^{\{1,\ldots,k\},norm} \wedge \textbf{iff } min\left[\sum_{n=-3}^{0} l_{i+n,i+1}\right]$$

For Reappear [O_i^{rapp}]. When the split occurs, we track the instance of occlusion in the past at time t using backward chaining and then computes the maximum likelihood of the object. The maximum likelihood sets the object state as "reappeared" as illustrated in Fig.(3b). When the split occurs, the child-parent relationship is killed and "occluded" state is updated to "normal".

– Event: isReAppear()

$$O_i^{\{1,\ldots,k\},occd} \Rightarrow O_i^{\{h\},rapp} \qquad \textbf{iff } max\left[\sum_{n=-3}^{0} l_{i+n-t,i+1}\right]$$

where t is the instance of time when occlusion was occurred.

3.3 Integrity Constraints

Following are the set of integrity constraints used in our logical framework:
- if $O_i^{new} \Rightarrow true$ then $\neg\left(O_i^{occd} \wedge O_i^{exit} \wedge O_i^{rapp}\right)$
- if $O_i^{exit} \Rightarrow true$ then $\neg\left(O_i^{occd} \wedge O_i^{over} \wedge O_i^{norm} \wedge O_i^{rapp}\right)$
- if $O_i^{occd} \Rightarrow true$ then $\neg\left(O_i^{norm} \wedge O_i^{rapp} \wedge O_i^{over}\right)$

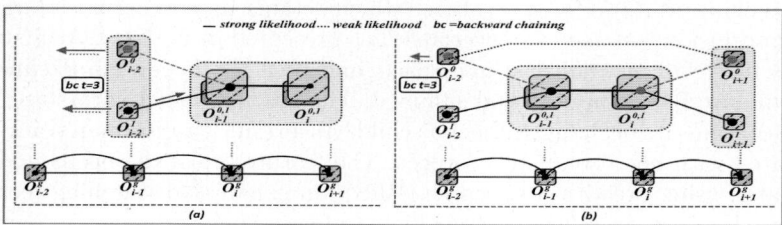

Fig. 3. In (a) state inferential model for occlusion is presented; In (b) the occluded object is reappeared afetr split

4 Implementation of Tracking System with OIF

In previous section 3, a conceptual and practical description is presented for our generic framework. Here in this section, we explain the likelihood computation for logical inferential model and tracking algorithm.

4.1 Maximum Likelihood Computation

During tracking, the states of objects are determined using maximum likelihood observation which is based on the online data association approach at time window T to avoid ambiguities of data association. Currently, we are using color features for maximum likelihood computation. The likelihood among objects is measured by integrating the CWHI-based technique [10] and Bhattacharyya coefficient, a general description is found in [11]. The formulations of these approaches are as follows:

CWHI-based likelihood. This approach is based on computing the normalized histogram correlation and use that correlation with conventional histogram intersection technique [12]. The relationship between color distance and the normalized correlation weights is formalized as defined below:

$$CWHI = \sum_{i=his_M} \sum_{j=his_T} min\left(h_M(i), h_T(j)\right) exp^{\left(-\frac{d}{2\rho_{fused}^2}\right)} \quad (4)$$

where d is the euclidean color distance, $h_M(i)$ represents histogram of the object at time t-1 and $h_T(i)$ represents histogram of objects at time t.

Bhattacharya coefficient. Bhattacharya coefficient approximates the normalized distance between the histograms of the objects $h_M(u)$ and $h_T^i(u)$ as defined in following:

$$BC\left(h_T^i(u), h_M(u)\right) = \sum_{u=0}^{m} \sqrt{h_T^i(u) h_M(u)} \quad (5)$$

where $h_T^i(u)$ and $h_M(u)$ are the histogram of object at time t and t-1, respectively, m is the dimension of the histogram, u is the index of each bin and i is

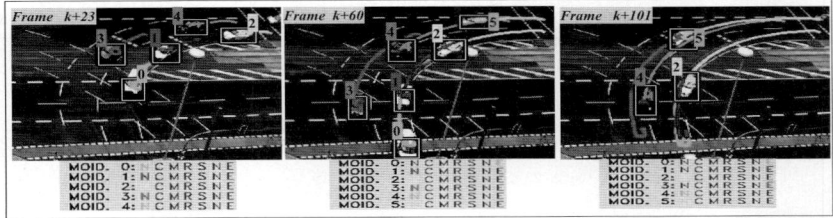

Fig. 4. The normal tracking of object with state window is presented. In the state window terms used are(from left to right): N = normal, C = overlaper, M= occluded, R = reappear; S = splitfrom, N = new, and E = exit. [Note: colored images are available online]

the object identity. The maximum $BC\left(h_T^i(u), h_M(u)\right)$ represents the maximum likelihood of the object i with the model object histogram.

Combined maximum likelihood. The likelihood is computed iteratively at time window T. Therefore, the combined likelihood $C_{bc,cwhi}$ from i to T is computed using following formulation:

$$l = C_{bc,cwhi} = BC + CWHI \qquad (6)$$

4.2 Tracking Using Kalman Filter

In whole tracking system, each Kalman tracker is associated with a specific object and responsible for the estimation of the object trajectories where the identities of object are managed by our OIF. Kalman tracker is defined in terms of its states and measurement equations. We consider the center of gravity of moving objects (i.e. the trajectories t_t^x and t_t^y) at time t as the states for Kalman tracker, hence the state vector and the measurement vector is as follows:

$$x_t = \begin{bmatrix} t_t^x & t_t^y \end{bmatrix}^T \qquad z_t = \begin{bmatrix} t_t^x & t_t^y \end{bmatrix}^T$$

A is the transition matrix and H is the measurement matrix of our tracking system along with the Gaussian process noise w_t and measurement noise v_t. These noise values are entirely dependent on the system that is being tracked and adjusted empirically. Finally, the equations of our tracking system are:

$$x_t = \text{A}x_{t-1} + w_t \quad and \quad z_t = \text{H}x_t + v_t$$

5 Experiments

We test our proposed approach on different cases which are observed in real scenes. In the Fig.(4), each detected object is assigned an identity called Moving Object Identity(MOID) and the trajectory is the estimated result of Kalman tracker. Besides, the state set window illustrates the mapping of cognitive state

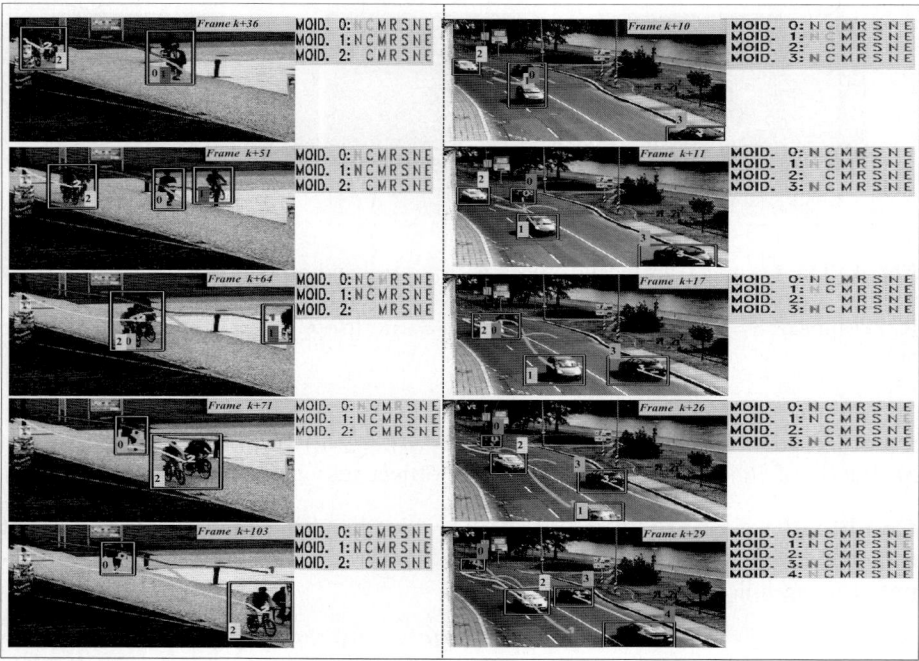

Fig. 5. In (a) (top to down), three objects are detected with normal state, the states are updated in frame k+36 when the occlusion occurs, similarly the tracking and state updation are observed in the entire scene. In the second sequence (b), cars are tracked under occlusions where the states of the objects are presented in state window.

of the object. When a new object is detected a symbol "N" is visible, similarly when object exits from the scene, the exit "E" status is active. It is noticeable that the objects size is very small and colors are very similar but our OIF is able to maintain their state accurately despite of the fuzzied likelihood measurements, thus tracking is reliable. In the second case which is shown in Fig.(5a), humans are tracked under full occlusions. It is observed that in frame k+36 two objects with MOID 0 and 1 occlude each other fully where the object with MOID 0 is the overlaper and the object with MOID 1 is the occluded object and is reappeared in frame k+51. In the state window, the respective states are active during occlusion and split. In the same video at frame k+64, another occlusion is observed due to object with MOID 2 which occludes the object 0. The occluded object is reappeared in frame k+71. The inference of object state is presented in state window, thus object under multiple occlusion can be tracked with OIF.

Fig.(5b) demonstrates tracking of object with crossing and parallel tracks. The example shows two objects occlude each other in frame k+10, where the occluded object's MOID is 0 which is reappeared in frame k+11. The same object is occluded by object with MOID 2 which is reappeared again in frame k+26 after split. The results present all the states of the object during tracking as explained in section 3. The experimental results shows efficient tracking using the proposed OIF which is based on online data association, therefore no prior threshold and empirical values are required.

6 Conclusion and Future Work

In this paper, the proposed novel framework handles occlusions, where property set interpretation is based on maximum likelihood computation. Later, each object links to its tracker. The tracking is very robust and efficient, however it is very pertenient to associate the correct Kalman with its object. In future, we will experiment our logical model with some reliable detection technique (i.e. AdaBoost) and will analyze more complex scenes.

Acknowledgement. This work is supported by Forschungspraemie (BMBF: FKZ 03FPB00213) and Transregional Collaborative Research Centre SFB/TRR 62 funded by the German Research Foundation (DFG).

References

1. Alper, Y., Omar, J., Mubarak, S.: Object tracking: A survey. ACM Computing Surveys 38(4), 13 (2006)
2. Blake, A.: Visual Tracking: A Short Research Roadmap. Springer, Heidelberg (2006)
3. Bar-Shalom, Y.: Tracking and data association. Academic Press Professional, Inc., London (1987)
4. Streit, R., Luginbuhl, T.: Probabilistic Multi-Hypothesis Tracking, Technical report (1995)
5. Reid, D.B.: An algorithm for tracking multiple targets. IEEE Trans. on Automatic Control 24, 843–854 (1979)
6. Khan, Z.: MCMC-based particle filtering for tracking a variable number of interacting targets. IEEE Trans. Pattern Anal. Mach. Intell. 27(11), 1805–1918 (2005)
7. Elgammal, A.M., Davis, L.S.: Probabilistic framework for segmenting people under occlusion. In: IEEE International Conference on Computer Vision, vol. 2, p. 145 (2001)
8. Wu, B., Nevatia, R.: Tracking of multiple, partially occluded humans based on static body part detection. In: IEEE Computer Society Conference on Computer Vision and Pattern Recognition, vol. 1, pp. 951–958 (2006)
9. Ryoo, M.S., Aggarwal, J.K.: Observe-and-explain: A new approach for multiple hypotheses tracking of humans and objects. In: IEEE Computer Society Conference on Computer Vision and Pattern Recognition, pp. 1–8 (2008)
10. Pathan, S., Al-Hamadi, A., Elmezain, M., Michaelis, B.: Feature-supported multi-hypothesis framework for multi-object tracking using kalman filter. In: Int. Conference on Computer Graphics, Visualization and Vision (accepted, 2009)
11. Kailath, T.: The divergence and bhattacharyya distance measures in signal selection. IEEE Trans. on Communications [legacy, pre - 1988] 15(1), 52–60 (1967)
12. Swain, M.J., Ballard, D.H.: Color indexing. Int. Journal of Computer Vision 7(1), 11–32 (1991)

Real-Time Stereo Vision: Making More Out of Dynamic Programming

Jan Salmen[1], Marc Schlipsing[1], Johann Edelbrunner[1], Stefan Hegemann[2], and Stefan Lüke[2]

[1] Institut für Neuroinformatik, Ruhr-Universität Bochum, 44780 Bochum, Germany
{Jan.Salmen,Marc.Schlipsing,
Hannes.Edelbrunner}@neuroinformatik.rub.de
[2] Continental AG, Division Chassis & Safety, Germany
{stefan.hegemann,stefan.lueke}@continental-corporation.com

Abstract. Dynamic Programming (DP) is a popular and efficient method for calculating disparity maps from stereo images. It allows for meeting real-time constraints even on low-cost hardware. Therefore, it is frequently used in real-world applications, although more accurate algorithms exist. We present a refined DP stereo processing algorithm which is based on a standard implementation. However it is more flexible and shows increased performance. In particular, we introduce the idea of multi-path backtracking to exploit the information gained from DP more effectively. We show how to automatically tune all parameters of our approach offline by an evolutionary algorithm. The performance was assessed on benchmark data. The number of incorrect disparities was reduced by 40 % compared to the DP reference implementation while the overall complexity increased only slightly.

1 Introduction

Stereo vision's task is to estimate depth by calculating a *disparity map* for two input images. It has actively been investigated for decades and still is a vivid topic in computer vision – mainly driven by a wide range of possible fields of application like robotics, automotive systems, surveillance and augmented reality just to name a few.

Depending on the application, very different requirements and objectives, i.e. runtime and performance, are relevant. If there are only weak constraints on these issues, a look at the Middlebury Stereo site[1] established by Scharstein and Szeliski [1,2], which compares results for a lot of different approaches, allows for finding a state-of-the-art algorithm.

However, if problem specifications become more restrictive, e.g. in terms of real-time performance on limited hardware resources, there are much fewer options as requirements may include low memory usage, mainly linear memory access, few complex operations (like floating point instructions), and parallelizability. DP approaches are perfectly suitable in this context, providing all the properties listed while their performance is sufficient for most real-world applications.

---
[1] http://vision.middlebury.edu/stereo/

In this work we will focus on how to tap the full potential of the basic, pixel-wise DP algorithm itself. We modify it to be better parameterizable and perform automatic offline optimization of those parameters. Additionally, we introduce a new idea to exhaustively benefit from the information gained by the DP – namely through a multi-path backtracking.

This article is organized as follows. In the forthcoming section, we present related work on DP stereo methods. The algorithm we chose as reference is presented in more detail in section 3. In section 4 modifications to this algorithm proposed by us are presented. Section 5 describes how the parameters of the new approach are optimized for a given task using the Middlebury benchmark images as an example. Finally we discuss our results in section 6.

2 Related Work

DP was first used for *edge-based* approaches, for example by Ohta and Kanade [3]. Geiger et al. [4] were among the first to propose a DP stereo method based on *pixel-wise* intensity differences. Bobick and Intille [5] presented the algorithm whose implementation serves as reference for this study. It is described in detail in section 3.

In [6] the usage of *MMX*, Assembler code, and some other techniques allow for building a real-time DP based stereo system. No results can be found in the Middlebury benchmark, but the results reported for two of the four current test images show a slightly better performance than the reference implementation does.

In [7] Gong et al. introduce a so called *reliability-based* DP algorithm. Tracing more than just the best path after the optimization step they provide a reliability measure for each disparity. A similar technique will be applied in this work.

Kim et al. [8] present a DP based algorithm which identifies possible disparities for each pixel by comparing orientation filter responses. The first (horizontal) DP optimization step is then performed only taking into account those candidates. Costs caused by that optimization step are incorporated in an energy function which is finally optimized in vertical direction. Computation takes several seconds on the Middlebury benchmark images. The idea of identifying candidates and choose among them can be found in our approach, too, but it is conducted in a different manner.

In [9] adaptive cost aggregation in the vertical direction (considering color information) is used to improve the DP results. Real-time capability is achieved with a GPU-based implementation.

Veksler introduces a DP algorithm that works on a tree structure instead of image rows [10]. Her implementation takes less than $1\ s$ and performs almost 20 % better on the Middlebury benchmark data than the basic method.

3 Reference Implementation

As reference and baseline for comparison, we consider the algorithm ranked 49th in the Middlebury evaluation under the name of 'DP'. The source code is freely available as part of the *StereoMatcher* framework. Because the goal in [1] was to evaluate the optimization technique itself, the DP approach presented in [11] was applied, but without *shiftable windows* and *ground control points*. In the course of this section the algorithm is described.

3.1 Cost Calculation

The matching costs d_{SAD} for two pixels P and Q are based on the *sum of absolute differences (SAD)* which is

$$d_{\text{SAD}}(P,Q) = d_R(P,Q) + d_G(P,Q) + d_B(P,Q) \tag{1}$$

for RGB images, where $d_R(P,Q) = |P_R - Q_R|$ with P_R as red component of pixel P etc. The cost for matching two pixels is interpolated in the range of a half pixel as proposed by Birchfield and Tomasi [5]. The result is the cost matrix C of size $X \times N$ for every image row, where X is the length of the row and N the number of allowed disparities.

3.2 Cost Aggregation

Values in the cost matrix are not aggregated. Thus, optimization is only based on the interpolated pixel-to-pixel intensity differences.

3.3 Optimization

The optimization is performed independently for each image row. A path with minimal total cost through the matrix C has to be found (cf. [11]). While the naive technique, i.e. calculating costs of all possible paths independently, would lead to a computational complexity of $O(N^X)$, DP solves this in $O(NX)$. The algorithm works in two phases: *forward step* and *backtracking*.

During the forward step, three possible states for each pixel and each disparity are managed simultaneously: M (matched), V (vertical occlusion) and D (diagonal occlusion). Pixels in matched state are assigned the corresponding cost from C, Pixels in states V or D are penalized by c_{occ}. Starting an occlusion is additionally penalized by $\rho_I(\Delta I)$ which depends on the actual intensity gradient ΔI and is realized as

$$\rho_I(\Delta I) = \begin{cases} c_{smooth} & \text{if } \Delta I < t_I \\ c_{smooth} \cdot p & \text{if } \Delta I \geq t_I. \end{cases} \tag{2}$$

The algorithm on the left side in figure 1 contains pseudo code for one part of the forward step and storage of the best transition leading to state M.

In the backtracking phase the minimal cost path is followed backwards along the stored transitions. At the same time, pixels in the result image are set to the corresponding disparity values or marked occluded.

3.4 Refinement

Pixels marked as occluded are filled from left to right. No sub-pixel refinement is performed.

4 Proposed Algorithm

Our implementation provides modifications of the reference algorithm. We add more flexibility and aim at utilizing the information gained from DP more efficiently.

4.1 Cost Calculation

We propose calculating a *weighted Euclidean distance* in the RGB-space

$$d_{\text{Euklid}}(P, Q) = \sqrt{w_R \cdot d_R^2(P, Q) + w_G \cdot d_G^2(P, Q) + w_B \cdot d_B^2(P, Q)}, \qquad (3)$$

where the three components are weighted individually by w_R, w_G, and w_B respectively.

The Birchfield cost measure is not applied because experiments in [1] did not show pay-off concerning efficiency.

4.2 Cost Aggregation

Slight aggregation in vertical direction, between neighboring scanlines, is performed. To realize that, we store the cost matrices for the current row and the two adjacent ones. A vertical Gaussian filter with weights $(1, 2, 1)$ is applied previous to optimization in order to calculate a smoothed match matrix.

4.3 Optimization

Extended parametrization. We introduce more parameters to the DP technique to allow for more flexibility: Different penalties for diagonal and vertical occlusions (c_D, c_V) and for starting these occlusions (p_D, p_V) are provided. Additionally, we introduce rewards r_D and r_V for transitions leading from occluded to matched state as counterpart to the penalties. As the penalty p_D and reward r_D e.g. would cancel each other out, this is only useful if all these costs are gradient-dependent. Therefore, for every parameter • named above, there is also a corresponding one used if $\Delta I \geq t_I$, marked as •̂.

Multi-path backtracking. It is noteworthy that a lot of information is generated during optimization, but most of it is disregarded in the classic approach: During the forward step only one transition per cell is chosen, not taking into account (almost) equally good solutions. For backtracking, just one possible path ending is considered.

We explicitly aim for incorporating more information for optimization. So, during the forward step of DP, *almost-optimal* transitions are stored. A transition is almost-optimal if its cost does not differ more than Δ_c from the best one. Figure 1 shows the resulting changes in pseudo code.

After the forward step, in the classic DP approach the only solution is found in the last column, corresponding to a path ending with minimal total costs m. We propose to additionally trace alternative path endings with costs $c \leq \tau \cdot m$ (with $\tau > 1$). Depending on the information from the forward step, more than one transition can be considered at every point during backtracking. All resulting possible paths through the DP array can be found by depth-first search, keeping track of already visited nodes in order to

```
1  cMM ← A(x − 1, d, M) + C(x, d);              1  cMM ← A(x − 1, d, M) + C(x, d);
2  cVM ← A(x − 1, d, V) + C(x, d);              2  cVM ← A(x − 1, d, V) + C(x, d) − r_D;
3  cDM ← A(x − 1, d, D) + C(x, d);              3  cDM ← A(x − 1, d, D) + C(x, d) − r_V;
4  cMin ← min(cMM, cVM, cDM);                   4  cMin ← min(cMM, cVM, cDM);
5  A(x, d, M) ← cMin;                           5  A(x, d, M) ← cMin;
6  if cMM = cMin then                           6  if cMM ≤ cMin + Δ_c then
7  |  optTrans(x, d) ← MM⃗;                      7  |  isPosTrans(x, d, MM⃗) ← true;
8  else if cVM = cMin then                      8  else
9  |  optTrans(x, d) ← VM⃗;                      9  |  isPosTrans(x, d, MM⃗) ← false;
10 else if cDM = cMin then                      10 if cVM ≤ cMin + Δ_c then
11 |  optTrans(x, d) ← DM⃗;                      11 |  isPosTrans(x, d, VM⃗) ← true;
                                                12 else
                                                13 |  isPosTrans(x, d, VM⃗) ← false;
                                                14 if cDM ≤ cMin + Δ_c then
                                                15 |  isPosTrans(x, d, DM⃗) ← true;
                                                16 else
                                                17 |  isPosTrans(x, d, DM⃗) ← false;
```

Fig. 1. Treatment of state M for pixel at x and disparity d during forward step of DP in the reference implementation (left) and in our approach (right)

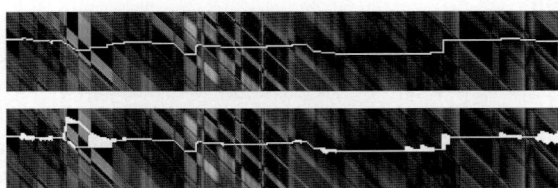

Fig. 2. Cost matrix for a single image row with path found by basic DP (top) and same matrix with results from multi-path backtracking (bottom)

avoid multiple processing. Figure 2 shows an example for optimization results in the reference implementation and in our approach.

If a pixel at (x, d) is traversed with state M, the disparity d is possible in this image row at position x and this information is stored, resulting in a sparse representation of candidate disparities for every pixel. As the assignment of disparities is not definite as in the reference implementation, an additional selection step is necessary.

Vertical optimization. To chose from the sparse number of candidate disparities, an additional optimization step is performed. The same idea as used for the horizontal optimization step is realized here: an assignment from the possible disparities in a column with minimal cost is chosen. If there was not found any possible disparity for a pixel at all during backtracking, every disparity is allowed. The cost function enforces smoothness by penalizing neighbor disparities that differ by 1 with λ and disparities that differ more with μ. Again, this optimization problem can be solved with DP, but here, it is significantly less complex due to the smaller amount of possible solutions. Resulting disparity values are accepted if the corresponding matching cost is less than $C_{\max}$. Otherwise, the pixel is marked as occluded.

4.4 Refinement

Occluded regions in the disparity image are filled in the same manner as in reference implementation. Furthermore, disparities for the left border of the reference image are not calculated, because not all matching costs can be computed there. This border region is filled from the right side.

5 Experiments

Along with our modifications and in favor of more flexibility the total number of parameters increase from four to 21, cf. table 1 for an overview. To tune the parameters offline, we use the *Covariance Matrix Adaption evolution strategy (CMA-ES)* [12], a variable-metric evolutionary algorithm which represents the "state-of-the-art in evolutionary optimization in real-valued optimization" [13]. As a baseline for comparison, we also optimize the four parameters of the reference approach the same way.

5.1 Experimental Setup

We optimized the parameters for the benchmark images from the Middlebury evaluation site as an example. As objective the *average number of bad pixels* as defined in [1] was minimized. Using the CMA-ES implementation from the *Shark* open-source machine learning library [14], five optimization trials were conducted for the basic approach and five for our modified approach.

5.2 Results

For the basic approach we obtained 14.5 % bad pixels as best solution with $c_{occ} = 28.8$, $c_{smooth} = 31.7$, $p = 1.5$ and $t_I = 5.1$. The perfomance is very similar to that of the reference implementation using the Birchfield-Tomasi measure.

For our modified approach, the best solution for the parameter settings given in table 1 results in 8.8 % bad pixels. Figure 3 shows results for two of the test images.

Table 1. Optimized parameter settings for our approach

Parameter	p_D	c_D	p_V	c_V	r_D	r_V	t_I	w_R	w_G	w_B
Best	30.7	27.4	-5.3	-12.9	-2.6	3.6	45.9	0.32	0.62	0.06

Parameter	$\hat{p}_D$	$\hat{c}_D$	$\hat{p}_V$	$\hat{c}_V$	$\hat{r}_D$	$\hat{r}_V$	Δ_c	τ	λ	μ	$C_{\max}$
Best	43.9	19.0	-16.7	-13.7	-1.9	4.0	1.95	1.17	22.6	57.5	76.2

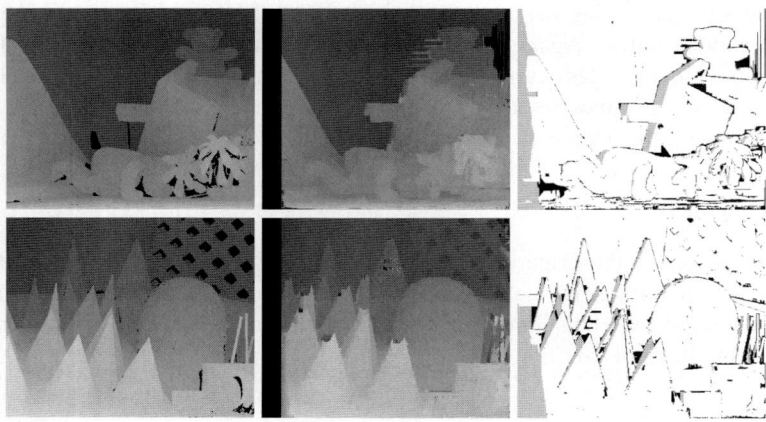

Fig. 3. Left to right: ground-truth disparity images, results obtained from our approach (border and occlusions shown black), and error images for Middlebury test images Teddy and Cones

Table 2. Results for Middlebury benchmark data sets

Rank	Algorithm	Tsukuba			Venus			Teddy			Cones			Avg. percent bad pixels
		nocc	all	disc	nocc	all	disc	nocc	all	disc	nocc	all	disc	
38	**Our approach**	2.0	3.8	9.8	3.3	4.7	13.0	6.5	13.9	16.6	5.2	13.7	13.4	8.83
⋮														
42	RealTimeGPU [9]	2.1	4.2	10.6	1.9	3.0	20.3	7.2	14.4	17.6	6.4	13.7	16.5	9.82
43	CostRelax [15]	4.8	6.1	20.3	1.4	2.5	18.5	8.2	15.9	23.8	3.9	10.2	11.8	10.6
44	ReliabilityDP [7]	1.4	3.4	7.3	2.4	3.5	12.2	9.8	16.9	19.5	12.9	19.9	19.7	10.7
45	TreeDP [10]	2.0	2.8	10.0	1.4	2.1	7.7	15.9	23.9	27.1	10.0	18.3	18.9	11.7
⋮														
50	DP [1]	4.1	5.0	12.0	10.1	11.0	21.0	14.0	21.6	20.6	10.5	19.1	21.1	14.2

See table 2 for detailed results and comparsions. Thus, our approach shows better performance in the Middlebury evaluation than most other algorithms using DP and it is the best one performing pixel-wise DP.

Processing test images *Tsukuba* takes $0.2\ s$, *Venus* $0.4\ s$, *Teddy* $0.8\ s$ and *Cones* $0.8\ s$ on a standard desktop PC with $1.8\ GHz$.

6 Conclusion

We considered a freely available implementation of a standard stereo algorithm based on DP. The technique is very popular due to its applicability to a large variety of real-world problems.

We showed how to modify and extend it in order to provide higher performance and more flexibility. In particular we focused on utilizing available information during DP more efficiently. At the same time the computational complexity did not increase significantly.

Offline optimization of all algorithm parameters for benchmark images as an example showed that an error reduction of 40 % compared to the reference implementation has been allowed. The algorithm proposed is widely applicable, escpecially for real-time applications.

References

1. Scharstein, D., Szeliski, R.: A taxonomy and evaluation of dense two-frame stereo correspondence algorithms. International Journal of Computer Vision 47, 7–42 (2002)
2. Scharstein, D., Szeliski, R.: High-accuracy stereo depth maps using structured light. In: Proceedings of the IEEE Conference on Computer Vision and Pattern Recognition, vol. 1, pp. 195–202 (2003)
3. Ohta, Y., Kanade, T.: Stereo by intra- and inter-scanline search using dynamic programming. IEEE Transactions on Pattern Analysis and Machine Intelligence 7, 139–154 (1985)
4. Geiger, D., Ladendorf, B., Yuille, A.L.: Occlusions and binocular stereo. In: Sandini, G. (ed.) ECCV 1992. LNCS, vol. 588, pp. 425–433. Springer, Heidelberg (1992)
5. Birchfield, S., Tomasi, C.: Depth discontinuities by pixel-to-pixel stereo. International Journal of Computer Vision 35, 1073–1080 (1999)
6. Forstmann, S., Kanou, Y., Ohya, J., Thuering, S., Schmitt, A.: Real-time stereo by using dynamic programming. In: Proceedings of the IEEE Conference on Computer Vision and Pattern Recognition Workshop, vol. 3, p. 29 (2004)
7. Gong, M., Yang, Y.H.: Near real-time reliable stereo matching using programmable graphics hardware. In: Proceedings of the IEEE Conference on Computer Vision and Pattern Recognition, pp. 924–931 (2005)
8. Kim, J.C., Lee, K.M., Choi, B.T., Lee, S.U.: A dense stereo matching using two-pass dynamic programming with generalized ground control points. In: Proceedings IEEE Conference on Computer Vision and Pattern Recognition, vol. 2, pp. 1075–1082 (2005)
9. Wang, L., Liao, M., Gong, M., Yang, R., Nister, D.: High-quality real-time stereo using adaptive cost aggregation and dynamic programming. In: Proceedings of the International Symposium on 3D Data Processing, Visualization and Transmission, pp. 798–805 (2006)
10. Veksler, O.: Stereo correspondence by dynamic programming on a tree. In: Proceedings of the IEEE Conference on Computer Vision and Pattern Recognition, vol. 2, pp. 384–390 (2005)
11. Bobick, A.F., Intille, S.S.: Large occlusion stereo. International Journal of Computer Vision 33(3), 181–200 (1999)
12. Hansen, N., Ostermeier, A.: Completely derandomized self-adaptation in evolution strategies. Evolutionary Computation 9(2), 159–195 (2001)
13. Beyer, H.G.: Evolution strategies. Scholarpedia 2(8), 1965 (2007)
14. Igel, C., Glasmachers, T., Heidrich-Meisner, V.: Shark. Journal of Machine Learning Research 9, 993–996 (2008)
15. Brockers, R., Hund, M., Mertsching, B.: Stereo vision using cost-relaxation with 3D support regions. In: Image and Vision Computing, New Zealand (2005)

Optic Flow Using Multi-scale Anchor Points

Pieter van Dorst, Bart Janssen, Luc Florack, and Bart M. ter Haar Romeny

Eindhoven University of Technology,
Den Dolech 2, Postbus 513, 5600 MB Eindhoven, The Netherlands
{P.A.G.v.Dorst,B.J.Janssen,L.M.J.Florack,B.M.terHaarRomeny}@tue.nl

Abstract. We introduce a new method to determine the flow field of an image sequence using multi-scale anchor points. These anchor points manifest themselves in the scale-space representation of an image. The novelty of our method lies largely in the fact that the relation between the scale-space anchor points and the flow field is formulated in terms of soft constraints in a variational method. This leads to an algorithm for the computation of the flow field that differs fundamentally from previously proposed ones based on hard constraints. We show a significant performance increase when our method is applied to the Yosemite image sequence, a standard and well-established benchmark sequence in optic flow research.

1 Introduction

Optic flow describes the apparent motion in an image sequence. A variety of approaches exists to estimate this motion. Survey papers include those by Barron et al. [1] and Mitchie et al. [2].

Differential methods are based on the most widespread approach, which uses spatiotemporal derivatives to describe the local image structure. The flow field is assumed to connect points in subsequent frames of the image sequence with similar structure. For example, in one of the earliest methods, proposed by Horn and Schunck [3], this "structure" is the image intensity, which leads to the well-known Optic Flow Constraint Equation. An overview of current developments in differential methods can be found in Bruhn et al. [4]. A problem that is encountered by these methods is that the structure does not always remain constant over time. For example, the global image intensity may vary over time. More complex terms to describe the structure can be used to overcome this problem [5][6]. A second problem is that many possible solutions exist, since points on level-sets have the same image intensity. This requires a so-called prior, which determines a unique solution based on prior knowledge. A prior usually is a regularization term, which can for example prefer an overall smooth solution with sparse discontinuities [7][8].

Another well performing approach is that of region matching, in which the image is split up into small blocks, each of which is translated to match the image neighborhood [9]. Because of their low computational cost, these methods are widely used in applications such as temporal up-scaling of video signals and video compression.

Our method can be placed in the category of feature-tracking methods. An overview of such methods can be found in [10]. However, in contrast to most feature-tracking algorithms, the features we use do not correspond to a specific point in the image sequence. Instead, we use anchor points that exist at different scales in scale-space, called toppoints (properly defined in section 2.1). Therefore, instead of corresponding to a point, the features we track actually represent entire regions in the image sequence. Using toppoints to extract the motion from an image sequence has been first proposed by Janssen et al. [11] and Florack et al. [12]. In these papers, the relation between the toppoint velocity and the flow field was implemented using a hard constraint, which means that this constraint has to be fulfilled exactly. The advantage is that their method is entirely parameter free, but the price of this is sensitivity to outliers. In the method presented in this paper, a 1-parameter soft constraint is used, giving more room for errors in the estimated toppoint velocity or deviations from the proposed relation between toppoint velocity and the flow field.

Toppoints are found throughout the scale-space of each frame of the image sequence as isolated entities. Therefore they are truly multi-scale, in contrast to other multi-scale features which are found by applying scale-selection to points that exist at every scale, such as corners. Another use of toppoints is to reconstruct an image from the values of derivatives taken at toppoint positions [13][14][15]. In these papers it is shown that features at the toppoint positions can be used to efficiently represent the information contained in an image. An important property is that the amount of toppoints found in a certain area of the image is proportional to the amount of information in that area.

2 Theory

In this chapter we will first explain how the scale-space representation of an image is defined and what toppoints are. Also important properties of toppoints are mentioned and we try to give toppoints a more intuitive meaning with some visualizations. Next we explain how to calculate toppoint velocities, and the method used to obtain the actual flow field from the toppoint velocities.

2.1 Scale-Space and Toppoints

The scale-space representation $f_s(x,y) = f(x,y;s) \in \mathbb{L}_2(\mathbb{R}^2 \times \mathbb{R}^+)$ of a static scalar image $f_0(x,y) \in \mathbb{L}_2(\mathbb{R}^2)$ is defined by the convolution of the image with a Gaussian kernel $\phi_s(x,y) = \phi(x,y;s) \in \mathbb{L}_2(\mathbb{R}^2)$, where $s \in \mathbb{R}^+$ denotes the scale (for tutorial books on scale-space see ter Haar Romenij [16] and Florack [17]):

$$f: \mathbb{R}^2 \times \mathbb{R}^+ \to \mathbb{R} : (x,y;s) \mapsto f(x,y;s) \stackrel{\text{def}}{=} (f_0 * \phi_s)(x,y),$$
$$\phi_s(x,y) = \phi(x,y;s) \stackrel{\text{def}}{=} \frac{1}{4\pi s} \exp\left(-\frac{x^2+y^2}{4s}\right). \qquad (1)$$

This results in a 3-dimensional function, where a slice of constant scale represents a blurred version of the original image.

The scale-space of an image fulfills the heat equation, since the Green's function of the Laplacian operator is a Gaussian kernel:

$$\partial_s f(x,y;s) = \triangle_{(x,y)} f(x,y;s) ,$$
$$\partial_s f(x,y;0) = f_0(x,y) . \qquad (2)$$

The Laplacian in the spatial directions x and y is denoted by $\triangle_{(x,y)}$.

A singular point in scale-space, also called a toppoint, occurs when the following conditions are fulfilled (see Gilmore et al. [18]):

$$\begin{bmatrix} \nabla_{(x,y)} f \\ \det \mathbf{H} \end{bmatrix} = \begin{bmatrix} f_x \\ f_y \\ f_{xx} f_{yy} - f_{xy}^2 \end{bmatrix} = \mathbf{0} , \qquad \mathbf{H} \stackrel{\text{def}}{=} \begin{bmatrix} f_{xx} & f_{xy} \\ f_{xy} & f_{yy} \end{bmatrix} . \qquad (3)$$

The gradient operator with respect to x and y is denoted by $\nabla_{(x,y)}$ and partial derivatives of f are indicated by self-explanatory subscripts. The condition states that the gradient is zero at toppoints, which in general occurs at extrema and saddle points in 2-dimensional images. These extrema and saddle points exist at every scale, and form so-called critical paths through scale-space. When two critical paths, corresponding to a saddle point and an extremum, collide as scale increases, an annihilation takes place. A pair of two critical paths can also be created when moving up in scale, which is called a creation. The points in scale-space where these events take place are called toppoints. As a consequence, toppoints are locations in scale-space where a topological change occurs. Figure 1 shows how two Gaussian blobs merge when scale increases, causing the maximum of the smallest blob to annihilate with the saddle point between the two blobs, creating a toppoint at the scale where this occurs.

A well-posed formulation of spatial derivatives of an image in scale-space is given by partially integrating the convolution product of a derivative of the image f_0 with a Gaussian filter ϕ_s, see eq. (1), using the property that ϕ_s is a Schwartz function:

$$\left(\partial_x^n \partial_y^m f_0 * \phi_s \right)(x,y) = \left(f_0 * \partial_x^n \partial_y^m \phi_s \right)(x,y) . \qquad (4)$$

In fact, because f_0 is often not $m+n$ times differentiable, we define the scale-space of an image derivative by the right hand side of eq. (4). This results in a lower-bound on the scale at which derivatives can be calculated numerically, which increases with derivative order. Derivatives with respect to scale can be calculated using only spatial derivatives by means of eq. (2).

2.2 Toppoint Velocity

If we consider a sequence of successive images, or a movie, in which objects move, the toppoints will move as well. The movement of toppoints in spatial and scale direction is defined as: $(\dot{x}, \dot{y}, \dot{s}) \in \mathbb{R}^3$. Note that e.g. $\dot{x}(t) = \partial_t x(t)$ represents the time derivative of the $x(t)$ position of the toppoint. An expression for this

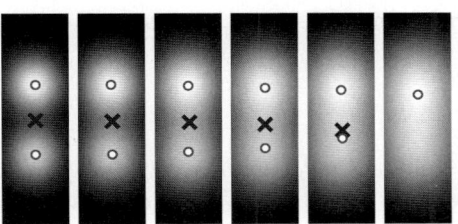 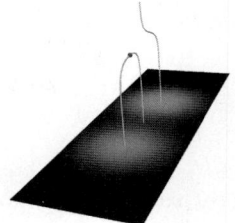

Fig. 1. (left) A series of scale-space slices of an image of two Gaussian blobs of different size, where scale increases to the right. Red circles denote maxima and blue crosses denote saddle points. A toppoint is located between the 5th and 6th slice, where a maximum and a saddle point annihilate. (right) The critical paths of the scale-space of the same image, where a toppoint is indicated by a red dot.

toppoint movement can be acquired by implicitly differentiating the definition of toppoints as stated in eq. (3) with respect to the time parameter t:

$$\frac{d}{dt}\begin{bmatrix}\nabla_{(x,y)}f \\ \det\mathbf{H}\end{bmatrix} = \begin{bmatrix} f_{xt} + \dot{x}f_{xx} + \dot{y}f_{xy} + \dot{s}f_{xs} \\ f_{yt} + \dot{x}f_{xy} + \dot{y}f_{yy} + \dot{s}f_{ys} \\ \partial_t\det\mathbf{H} + \dot{x}\,\partial_x\det\mathbf{H} + \dot{y}\,\partial_y\det\mathbf{H} + \dot{s}\,\partial_s\det\mathbf{H}\end{bmatrix} = \mathbf{0}$$

$$\Rightarrow \begin{bmatrix} f_{xx} & f_{xy} & f_{xs} \\ f_{xy} & f_{yy} & f_{ys} \\ \partial_x\det\mathbf{H} & \partial_y\det\mathbf{H} & \partial_s\det\mathbf{H}\end{bmatrix}\begin{bmatrix}\dot{x}\\ \dot{y}\\ \dot{s}\end{bmatrix} = -\begin{bmatrix} f_{xt}\\ f_{yt}\\ \partial_t\det\mathbf{H}\end{bmatrix}. \quad (5)$$

If the matrix is invertible, eq. (5) supplies us with a scheme to calculate the movement of toppoints in an image sequence. The notation for derivatives of $\det\mathbf{H}$ is abbreviated to avoid cumbersome notation. When we expand $\partial_s\det\mathbf{H}$ for example, we obtain (using eq. (2) to express scale derivatives in spatial derivatives):

$$\partial_s(f_{xx}f_{yy} - f_{xy}^2) = f_{xx}(f_{xxyy} + f_{yyyy}) + f_{yy}(f_{xxxx} + f_{xxyy}) - 2f_{xy}(f_{xxxy} + f_{xyyy}). \quad (6)$$

An estimation of the position of toppoints can be used to find a more accurate location. Florack and Kuijper [19][20] developed a method that iteratively refines the estimated position to the desired accuracy. Using the estimated toppoint velocity, we estimate the toppoint position in the next frame of the image sequence. Consequently, the position of the toppoint in the next frame is refined. This refined position is used to calculate a more accurate estimation of the movement of the toppoints from one frame the the next.

2.3 Optic Flow Using Toppoints

The velocity of the scale-space toppoints forms a sparse 3D flow field. In optic flow, the goal is to acquire a dense 2D flow field which describes the velocity in each pixel of the image sequence. In order to obtain the dense 2D flow field from the sparse 3D one the following assumption is made:

Assumption 1. *The velocity of toppoints in the scale-space of the image corresponds to the values at those points in the scale-spaces of $u(x,y)$ and $v(x,y)$:*

$$<u, \phi_i> = U_i ,$$
$$<v, \phi_i> = V_i , \quad (7)$$

where $U_i \in \mathbb{R}$ $(= \dot{x}_i)$ and $V_i \in \mathbb{R}$ $(= \dot{y}_i)$ are obtained by applying eq. (5) at the toppoint positions of the image sequence, and ϕ_i are Gaussian functions shifted to spatial position x_i, y_i and with scale s_i (recall eq. (1)). Here and henceforth, $<.,.>$ indicates a standard $\mathbb{L}_2$-inner product.

This assumption alone does not uniquely determine the flow field. Therefore, we use a flow driven isotropic prior, which allows for some discontinuities in the flow field. We combine this prior with the assumption regarding toppoint velocities in the following energy functional:

$$E(u,v) = \iint_{d\Omega} \gamma \sqrt{|\nabla u|^2 + |\nabla v|^2 + \epsilon^2} + \sum_{i=1}^{N}\left[(<u,\phi_i> - U_i)^2 + (<v,\phi_i> - V_i)^2\right] d\Omega, \quad (8)$$

where ϵ is a contrast parameter, γ determines the smoothness of the resulting flow field and N denotes the number of toppoints. Minimizing this energy functional leads to the dense flow field. Using variational calculus, we obtain the Euler-Lagrange equations corresponding to this energy functional. These are discretized using β-splines [21] and the resulting system of equations is solved using the BiCG-Stab algorithm [22].

Besides the toppoints of the regular image, we also calculate the toppoint positions and velocities in the gradient magnitude and Laplacian of the original image. This adds information on the movement of higher-order structures in the image, such as edges.

3 Numerical Evaluation

The error measure for flow fields used in literature is the Angular Error, as first proposed by Fleet and Jepson [23]. This measure describes the angle between the estimated 3D flow vector $\mathbf{v}_e = \{u_e, v_e, 1\}$ and the true flow vector $\mathbf{v}_t = \{u_t, v_t, 1\}$. In order to objectively compare different methods, the Average Angular Error, or AAE, is used.

Figure 2 shows the Yosemite image sequence, which is used in optic flow literature as a benchmark sequence. This sequence tests multiple aspects of the performance of optic flow methods: it contains spatial discontinuities, brightness change (the sky increases in brightness), rigid and non-rigid transformations.

In Table 1 different methods that introduced significant novelties can be found together with the improvement of the AAE of the Yosemite image sequence since Horn and Schunck introduced their method in 1981.

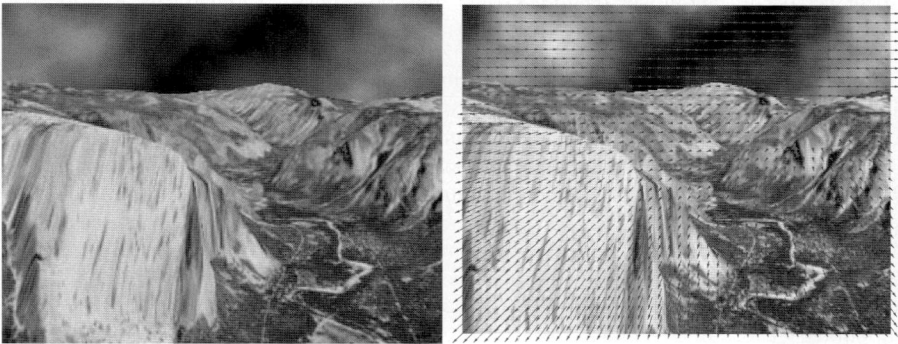

Fig. 2. (left) The first frame of the Yosemite image sequence. (right) The ground truth flow field of the Yosemite image sequence. The camera moves through the valley, and the clouds move to the right.

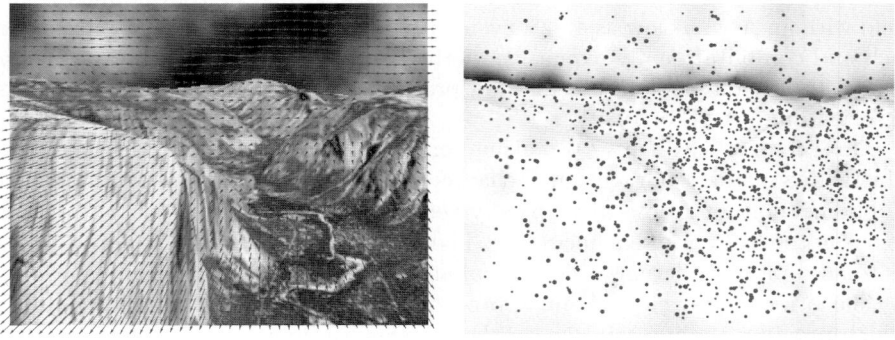

Fig. 3. (left) The flow field of the Yosemite image sequence calculated using our method. (right) The Angular Error of the flow field calculated using our method, displayed as shades of grey, with white = $0°$ and black = $102.7°$. Red dots indicate toppoints, the size of which is proportional to the scale.

The flow field that is obtained by our method can be found in Figure 3, together with the angular error and toppoint locations. We can see that, apart from the discontinuity at the border between the landscape and the sky, the flow field is fairly accurate, albeit not state-of-the-art. The AAE we obtained was $4.82°$. We can clearly see that the discontinuity between the landscape and the sky results in the largest error. This is partially caused by the low number of toppoints found in the low-texture sky, and partially by the suboptimal choice of prior. Using a smoothness term with better discontinuity-preserving properties may improve this result significantly.

Table 1. Yosemite sequence results of other methods. The first four results are obtained from Barron et al. [1].

Technique	AAE	Method description
Horn and Schunck [3]	32.43°	Original, only smoothness
Anandan	15.84°	Region matching
Singh	13.16°	Region matching and coarse-to-fine approach
Nagel [7]	11.71°	Discontinuity preserving smoothness
Alvarez et al. [8]	5.53°	Improvement of Nagel's method
Weickert and Schnörr[24]	4.85°	Spatio-temporal smoothness
Zang et al. [6]	2.67°	Monogenic curvature tensor constancy
Amiaz and Kiryati [25]	1.78°	Piecewise smoothness with level-sets
Papenberg et al. [5]	1.64°	High order data term, spatio-temporal smoothness
Brox et al. [26]	0.92°	Same as Papenberg, with level-sets

4 Conclusion and Future Work

We have shown that the information toppoint movement provides admits a fairly accurate estimation of the flow field of an image sequence. We obtained a flow field with an AAE as low as 4.82°, even without the use of a complex, parameter-rich and computationally expensive method to preserve discontinuities, such as level-sets. In comparison: the method proposed in [11], using hard constraints, resulted in an AAE of 19.19°.

This preliminary result is promising for several reasons: (i) Unlike superior sophisticated methods our method is characterized by only one global parameter. (ii) Toppoint representations are typically very sparse (1837 toppoints for the 79316 pixels in the Yosemite sequence). (iii) The method itself can be easily modified so as to account for different or additional anchor points and more effective priors.

The principle novelty of our approach is the term in the energy functional which provides the information on the flow field. Many improvements in differential methods have been made in the regularization term, which can also be incorporated into our method. Also other anchor points can be added, such as SIFT feature points [27].

Since the Yosemite image sequence is the benchmark sequence used in literature, for most methods only the AAE of this sequence is available. Our approach is very robust compared to differential methods, since it is inherently multi-scale and invariant under changing brightness, and rather generic, as it requires only a single global regularity parameter. Therefore it is expected to perform well on more challenging image sequences, such as those with opacity, reflections or a significant amount of noise. This is the subject of further research.

References

1. Barron, J.L., Fleet, D.J., Beauchemin, S.S.: Performance of Optical Flow Techniques. IJCV 12, 43–77 (1994)
2. Mitchie, A., Bouthemy, P.: Computation and Analysis of Image Motion: A Synopsis of Current Problems and Methods. IJCV 19, 29–55 (1996)

3. Horn, B.K.P., Schunck, B.G.: Determining Optical Flow. AI 17, 185–203 (1981)
4. Bruhn, A., Weickert, J., Kohlberger, T., Schnörr, C.: A Multigrid Platform for Real-Time Motion Computation with Discontinuity-Preserving Variational Methods. IJCV 70, 257–277 (2006)
5. Papenberg, N., Bruhn, A., Brox, T., Didas, S., Weickert, J.: Highly Accurate Optic Flow Computation with Theoretically Justified Warping. IJCV 67(2), 141–158 (2006)
6. Zang, D., Wietzke, L., Schmaltz, C., Sommer, G.: Dense Optical Flow Estimation from the Monogenic Curvature Tensor. In: Sgallari, F., Murli, A., Paragios, N. (eds.) SSVM 2007. LNCS, vol. 4485, pp. 239–250. Springer, Heidelberg (2007)
7. Nagel, H.H.: On the Estimation of Optical Flow: Relations Between Different Approaches and Some New Results. AI 33, 299–324 (1987)
8. Alvarez, L., Weickert, J., Sánchez, J.: Reliable Estimation of Dense Optical Flow Fields with Large Displacements. IJCV 39(1), 41–56 (2000)
9. de Haan, G., Biezen, P.W.A.C., Huijgen, H., Ojo, O.A.: True-Motion Estimation with 3-D Recursive Search Block Matching. IEEE TCSV 3(5), 368–379 (1993)
10. Shi, J., Tomasi, C.: Good Features to Track. In: IEEE CVPR, pp. 593–600 (1994)
11. Janssen, B.J., Florack, L.M.J., Duits, R., ter Haar Romeny, B.M.: Optic Flow from Multi-scale Dynamic Anchor Point Attributes. In: Campilho, A., Kamel, M.S. (eds.) ICIAR 2006. LNCS, vol. 4141, pp. 767–779. Springer, Heidelberg (2006)
12. Florack, L.M.J., Janssen, B.J., Kanters, F.M.W., Duits, R.: Towards a new paradigm for motion extraction. In: Campilho, A., Kamel, M.S. (eds.) ICIAR 2006. LNCS, vol. 4141, pp. 743–754. Springer, Heidelberg (2006)
13. Lillholm, M., Nielsen, M., Griffin, L.D.: Feature-Based Image Analysis. IJCV 52(2/3), 73–95 (2003)
14. Nielsen, M., Lillholm, M.: What do features tell about images? In: Kerckhove, M. (ed.) Scale-Space 2001. LNCS, vol. 2106, pp. 39–50. Springer, Heidelberg (2001)
15. Janssen, B.J., Kanters, F.M.W., Duits, R., Florack, L.M.J., ter Haar Romenij, B.M.: A Linear Image Reconstruction Framework Based on Sobolev Type Inner Products. IJCV 70, 231–240 (2006)
16. ter Haar Romenij, B.M.: Front-End Vision and Multiscale Image Analysis. Kluwer, Dordrecht (2003)
17. Florack, L.M.J.: Image Structure. Kluwer, Dordrecht (1997)
18. Gilmore, R.: Catastrophe Theory for Scientists and Engineers. Dover, New York (1993)
19. Florack, L.M.J., Kuijper, A.: The Topological Structure of Scale-Space Images. JMIV 12(1), 65–79 (2000)
20. Kuijper, A., Florack, L.M.J., Viergever, M.A.: Scale Space Hierarchy. JMIV 18(2), 169–189 (2003)
21. Janssen, B.J., Duits, R., Florack, L.M.J.: Coarse-to-fine Image Reconstruction based on Weighted Differential Features and Background Gauge Fields. To Appear in: Proc. of the SSVM. LNCS. Springer, Heidelberg (2009)
22. Barrett, R., Berry, M., Chan, T.F., Demmel, J., Donato, J., Dongarra, J., Eijkhout, V., Pozo, R., Romine, C., Van der Vorst, H.: Templates for the Solution of Linear Systems: Building Blocks for Iterative Methods, 2nd edn. SIAM, Philadelphia (1994)
23. Fleet, D.J., Jepson, A.D.: Computation of Component Image Velocity from Local Phase Information. IJCV 5(1), 77–104 (1990)
24. Weickert, J., Schnörr, C.: Variational Optic Flow Computation with a Spatiotemporal Smoothness Constraint. JMIV 14, 245–255 (2001)

25. Amiaz, T., Kiryati, N.: Piecewise-Smooth Dense Optical Flow via Level Sets. IJCV 68(2), 111–124 (2006)
26. Brox, T., Bruhn, A., Weickert, J.: Variational Motion Segmentation with Level Sets. In: Leonardis, A., Bischof, H., Pinz, A. (eds.) ECCV 2006. LNCS, vol. 3951, pp. 471–483. Springer, Heidelberg (2006)
27. Lowe, D.G.: Distinctive image features from scale-invariant keypoints. IJCV 60(2), 91–110 (2004)

A Methodology for Evaluating Illumination Artifact Removal for Corresponding Images

Tobi Vaudrey[1], Andreas Wedel[2], and Reinhard Klette[1]

[1] The *.enpeda..* Project, The University of Auckland, Auckland, New Zealand
[2] Daimler Research, Daimler AG, Stuttgart, Germany

Abstract. Robust stereo and optical flow disparity matching is essential for computer vision applications with varying illumination conditions. Most robust disparity matching algorithms rely on computationally expensive normalized variants of the brightness constancy assumption to compute the matching criterion. In this paper, we reinvestigate the removal of global and large area illumination artifacts, such as vignetting, camera gain, and shading reflections, by directly modifying the input images. We show that this significantly reduces violations of the brightness constancy assumption, while maintaining the information content in the images. In particular, we define metrics and perform a methodical evaluation to identify the loss of information in the images. Next we determine the reduction of brightness constancy violations. Finally, we experimentally validate that modifying the input images yields robustness against illumination artifacts for optical flow disparity matching.

1 Introduction

Previous studies have shown that when using correspondence algorithms (i.e., stereo and optical flow) to provide reliable information, the results on synthetically generated data (e.g., [10]) do not compare well with results on realistic images [16]. Further studies have shown that illumination artifacts (such as shadows, reflections, and vignetting) and differing exposures have the worst effect on the matching [11]. This effect is especially highlighted in driver assistance systems (DAS), where illumination can change drastically in a short amount of time (e.g., going through a tunnel, or the "dancing light" from sunlight through trees).

For dealing with illumination artifacts, there are three basic approaches: simultaneously estimate the disparity matching and model brightness change within the disparity estimation [5], try to map both images into a uniform illumination model, or map the intensity images into images which carry the illumination-independent information (e.g., using colour images [9,18]).

Using the first option, only reflection artifacts can be modelled without major computational expense. From experiments with various unifying mappings, the second option is near impossible. The third approach has more merit for research; we restrain our study to using the more common grey value images.

An example of mapping intensity images into illumination-independent images is the structure-texture image decomposition [1,12] (an example can be seen in Figure 1). More formally, this is the concept of *residuals* [7], which is the difference between

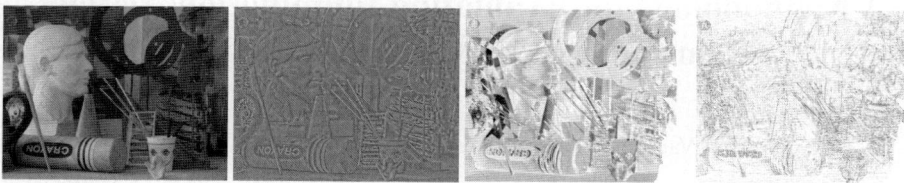

Fig. 1. Example for removing illumination artifacts due to different camera exposure in the *Art* image (left) by using its residual component (2nd from left). The brightness difference between the plain intensity images (3rd from left) shows laminar errors. The brightness difference of the residual images (right) contains spatially distributed noise but no large area illumination artifacts.

an intensity image and a smoothed version of itself. A subset of residual operators has been recently evaluated together with different matching costs in the context of stereo disparity matching in [6]. In this paper we systematically evaluate and compare residual operators as basic approach for preprocessing corresponding images, to reduce the effect of illumination variances.

The main contribution of this work is we provide a valid methodology for analysing information loss compared to illumination removal effects for an arbitrary filter. The methodology is based on first showing that information is not lost by applying the filter, using co-occurrence matrix [4] based measures. The second contribution is high-lighting these effects using correspondence images as validation. This is done by using ground truth correspondence data, comparing the differences in illumination, and summarising the information with an error metric. We go on to show that using residual images removes illumination artifacts, by using a mixture of synthetic and real-life images [3,10]. The illumination effects are highlighted more drastically when the illumination and exposure conditions of the corresponding images are not the same. The chosen filters are the TV-L^2 [12], median, mean, sigma [8], bilateral [14], and trilateral filter [2]. All are effectively "edge preserving" filters, except the mean filter.

2 Methodology

Here we define the methodology of our process. It is defined by two parts; firstly, identifying if the images loose information, and secondly, determining reduction of the effect of illumination artifacts.

Co-occurrence Matrix and Metrics. The co-occurrence matrix has been defined for analysing different metrics about the texture of an image [4]:

$$C(i,j) = \sum_{\mathbf{x} \in \Omega} \sum_{\mathbf{a} \in \mathcal{N} \setminus \{(0,0)\}} \begin{cases} 1, & \text{if } h(\mathbf{x}) = i \text{ and } h(\mathbf{x}+\mathbf{a}) = j \\ 0, & \text{otherwise} \end{cases} \quad (1)$$

where $\mathcal{N} + \mathbf{x}$ is the neighbourhood of pixel $\mathbf{x}$, $\mathbf{a} \neq (0,0)$ is one of the offsets in $\mathcal{N}$, and $0 \leq i, j \leq I_{\max}$, for maximum intensity $I_{\max}$. h represents any 2D image (e.g., f). All images are scaled $\min \leftrightarrow \max$ for utilizing the full $0 \leftrightarrow I_{\max}$ scale.

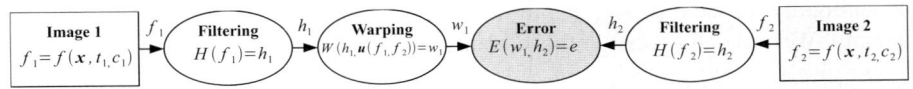

Fig. 2. Outline of the methodology used to obtain an error image. For our study $H = R$.

In our experiments we chose $\mathcal{N}$ to be the 4-neighbourhood, and we have $I_{max} = 255$. The loss in information is identified by the following metrics: *homogeneity* $T_{homo}(h) = \sum_{ij} \frac{C(i,j)}{1+|i-j|}$, *uniformity* $T_{uni}(h) = \sum_{ij} C(i,j)^2$, and *entropy* $T_{ent}(h) = -\sum_{ij} C(i,j) \ln C(i,j)$. An increase in homogeneity represents the image having more homogeneous areas, an increase in uniformity represents more uniform areas, and a decrease in entropy shows that there is less information contained in the image. To get a better representation of the effect of filters, we scale the result by the original image's metric, i.e., $T_*(h)/|T_*(f)|$, where h is the processed image (obviously, $h = f$ gives a value of 1). Previous studies [15] have shown that homogeneity and entropy can define the information loss in an image. In this study, we only use the homogeneity (see results section).

Testing Illumination Artifact Reduction. Correspondence algorithms usually rely on the brightness consistency assumption, i.e., that the appearance of an object (according to illumination) does not change between the corresponding images. However, this does not hold true when using real-world images, this is due to, for example, shadows, reflections, differing exposures and sensor noise. It is well known, that illumination artifacts propose the biggest problem for correspondence algorithms; a recent study has shown that illumination artifacts may, in fact, be the worst type of error [11]. Figure 2 shows our proposed approach for evaluating the effectiveness of a filter. In this paper, we chose the filtering operator H to be the residual image ($H = R$).

Image Warping. One way to highlight this (i.e., that the errors from residual images are lower than the errors obtained using the original images) is to warp one image to the perspective of the other (using ground truth) and compare the differences. The forward warping function W is defined by the following:

$$W\Big(h_1(\mathbf{x}), \mathbf{u}^*(\mathbf{x}, h_1, h_2)\Big) = w(\mathbf{x} + \mathbf{u}^*(\mathbf{x}, h_1, h_2)),$$

where $h(\mathbf{x})$ is the image value at $\mathbf{x} \in \Omega$, and $\mathbf{u}^*$ is the 2D ground truth warping (remapping) vector from h_1 to the perspective of h_2. In practice, the warping is performed using a lookup table with interpolation (e.g., bilinear or cubic). In the stereo case, $\mathbf{u}^*$ is the ground truth disparity map from left to right (all vertical translations would be zero). Another common example is optical flow, where $\mathbf{u}^*$ is the ground truth flow field from the previous to the current frame.

Image Scaling. For the purposes of this paper, h is discrete in the functional inputs ($\mathbf{x}$), but continuous for the value of h itself. For a typical grey-scale image, the information is discrete ($0 \leq h \leq 2^n - 1 \in \mathbb{N}^2$, where n is usually 8 or 16). However, we find it easier to represent image data continuously by $-1 \leq h \leq 1 \in \mathbb{Q}^2$, which takes away the ambiguity for the bits per pixel (as any n-bits per pixel image can be scaled to this domain). We scale all images to this domain using $h(\mathbf{x}) = h(\mathbf{x})/\max_{\mathbf{x} \in \Omega} |h(\mathbf{x})|$.

Error Images and Metrics. An error image e is the magnitude of difference between two images, $E(h, h^*) = e(\mathbf{x}) = \| \mathbf{h}(\mathbf{x}) - \mathbf{h}^*(\mathbf{x}) \|$, where, usually, $\mathbf{h}$ is the result of a process and $\mathbf{h}^*$ is the ground truth. For this paper, the error image is between $\mathbf{h}^* = h_2$ and the warped image $\mathbf{h} = W(h_1)$.

A common error metric is the *Root Mean Squared* (RMS) *Error*. The problem with this metric is that it gives an even weighting to all pixels, no matter the proximity to other errors. In practice, if errors are happening in the same proximity, this is much worse than if the errors are randomly placed over an image. Most algorithms can handle (by denoising or such approaches) small amounts of error, but if the error is all in the same area, this is seen as signal. We define the *Spatial Root Mean Squared Error* (Spatial-RMS) to take the spatial properties of the error into account:

$$RMS_S(e) = \sqrt{\frac{1}{M} \sum_{\mathbf{x} \in \Omega} \left(G(e(\mathbf{x}))^2 \right)} \qquad (2)$$

M is the number of pixels in the (discrete) non-occluded (when occlusion maps are available) image domain Ω, and G is a function that propagates the errors in a local neighbourhood $\mathcal{N}$. For our experiments, we chose a Gaussian error propagation using a standard deviation $\sigma = 1$.

Smoothing Operators and Residuals. Let f be any frame of a given image sequence (or stereo camera setup), defined on a rectangular open set Ω and sampled at regular grid points within Ω.

f can be defined to have an additive decomposition $f(\mathbf{x}) = s(\mathbf{x}) + r(\mathbf{x})$, for all pixel positions $\mathbf{x} = (x, y)$, where $s = S(f)$ denotes the *smooth component* (of an image) and $r = R(f) = f - S(f)$ the *residual* (Figure 1 shows an example of the decomposition). We use the straightforward iteration scheme:

$$s^{(0)} = f, \quad s^{(n+1)} = S(s^{(n)}), \quad r^{(n+1)} = f - s^{(n+1)}, \quad \text{for } n \geq 0.$$

The concept of residual images was already introduced in [7] by using a 3×3 mean for implementing S. We use the mean operator and also an $m \times m$ median operator in this study. The other operators for S are defined below.

TV-L^2 filter. [12] used the definition of $f = s + r$ (as above), where s is assumed to be in $L^1(\Omega)$ with bounded TV (in brief: $s \in$ BV), and r is in $L^2(\Omega)$. We use the residual image from this idea as implemented and exploited in [17].

Sigma filter. This operator [8] is effectively a trimmed mean filter; it uses an $m \times m$ window, but only calculates the mean for all pixels with values in $[a - \sigma_f, a + \sigma_f]$, where a is the central pixel value and σ_f is a threshold. We chose σ_f to be the standard deviation of f (to reduce parameters for the filter).

Bilateral filter. This edge-preserving Gaussian filter [14] is used in the spatial domain (using σ_2 as spatial σ), also considering changes in the colour domain (e.g., object boundaries). It therefore only takes into consideration values within a Gaussian kernel within the colour domain (σ_1 as colour σ).

Trilateral filter. This gradient-preserving smoothing operator [2] (i.e., it uses the local gradient plane to smooth the image) only requires the specification of one parameter σ_1, which is equivalent to the spatial kernel size. The rest of the parameters are self tuning.

All filters have been implemented in OpenCV, where possible the native function was used. For the TV-L^2, we use an implementation (with identical parameters) as in [17]. All other filters used are virtually parameterless (except a window size) and we use a window size of $m = 3$ ($\sigma_1 = 3$ for trilateral filter[1]). For the bilateral filter, we use color standard deviation $\sigma_1 = I_r/10$, where I_r is the range of the intensity values (i.e., $\sigma_1 = 0.2$ for the scaled images).

Datasets. We illustrate our arguments with the Middlebury dataset [10] and the EISATS [3] synthetic data (Set 2).

This highlights the major importance of removing illumination artifacts. For the Middlebury dataset we include both the 2005 and 2006 datasets (provided by [6,13]). This data has 3 different exposures and 3 different illuminations (for both the left and right images). This enables us to test the brightness consistency assumption under extreme conditions. Again, we only use images with ground truth available. For the 2005 set, that includes: *Art, Books, Dolls, Laundry, Moebius,* and *Reindeer*. For the 2006 set: *Aloe, Baby1-3, Bowling1-2, Cloth1-4, Flowerpots, Lampshade1-2, Midd1-2, Monopoly, Plastic, Rocks1-2,* and *Wood1-2*. We are not interested in "good quality" situations. Therefore, we only use images with differing exposure and illumination. To do this, for each image pair, we keep the left image with illumination $= 1$ and exposure $= 0$ (as defined by [10]). But for the right image, we make use of all all the differing illumination $(1, 2, 3)$ and exposure $(0, 1, 2)$ settings (excluding the exact same illumination $= 1$ and exposure $= 0$). This is a total of 8 different illumination/exposure combinations, for each image pair. That brings the total dataset to 216 (27×8).

3 Experimental Results

A previous study, has already pointed out that the results for slight illumination artifacts are improved using residual images [15]. We now show that these results get even better when illumination is a major issue (not just a minor one).

Co-occurrence Metrics. This subsection demonstrates that the important information for correspondence algorithms is contained in the residual image r. The residual image is, in fact, an approximation of the high frequencies of the image, and the smoothed image s is an approximation of a low-pass filter. Obviously, by iteratively running a smoothing filter, you will get a more and more smoothed image (i.e., you will be getting lower and lower frequencies, thus reducing the higher frequencies). In [15] the metrics were shown to represent this effect accurately.

---
[1] The authours thank Prasun Choudhury (Adobe Systems, Inc.) and Jack Tumblin (EECS, Northwestern University), for their implementation of the trilateral filter.

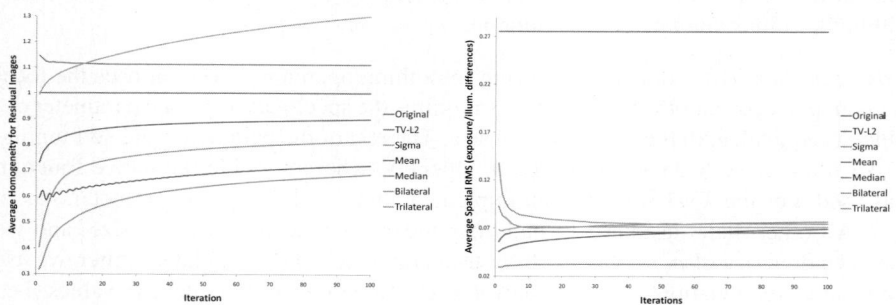

Fig. 3. Left: average scaled homogeneity of the residual images, averaged over dataset [10]. Right: Spatial-RMS graph for the same data. Notice the huge benefit from using residual images.

The residual of an image is an approximation of the high frequencies of the image, so the information should not be reduced. We average the co-occurrence metric results over dataset [10] to highlight this (Figure 3). This graph shows that the residual images do not lose information, in fact the homogeneity is slightly reduced (except for the trilateral and median filter). The increase in information could be seen as noise from the filter, or an increase in emphasis of the high-frequencies.

Illumination Differences. This subsection uses again the dataset [10]. A qualitative example of error images e can be seen in Figure 1. This specific error image is generated using the *Art* right image, with illumination and exposure both equal to 1 (left image is 1 and 0, respectively). The image is from [6], and has ground truth available (warping from left to right). The original error image (left) clearly shows how increasing the exposure (250 to 1000 ms) has very big consequences on the illumination differences between the left and right image. The error image using the TV-L^2 residual (right) reduces the error dramatically. Furthermore the magnitudes of the maximum errors are less; the original image is 1.83 and the TV-L^2 residual image is 1.25.

See again Figure 3. The trilateral filter was stopped at iteration 10. It is immediately obvious that the original images are far worse than residual images, around 3 times worse on average. This again highlights that with extremely different exposures and illuminations, the residual images provide the best information for matching.

Since most of the filters stabilize around iteration 40 (TV-L^2, sigma, bilateral, and median), we have presented statistical results of the RMS after 1 and 40 iterations. These results are shown in Table 1. You can see from these results that all the statistics for the original images are higher than any of the filters. The mean, trilateral, and median filter seem to be the most robust; showing the lowest standard deviation. The TV-L^2, mean, and median filters have the best average. The timing information provided in this table is the average time per iteration, on two sizes of images (470×370, 752×480 pixel resolution), this is to highlight the scalability of the filters. The tests were under Windows, the CPU was an Intel Core 2 Duo 3 GHz (multi-core processing not exploited), with 4GB memory.

Table 1. Average (Ave.) and Standard Deviation (S.D.) for the methodology performed on dataset [10]. Average running times per iteration are also included (right) for two image resolutions. The rank of the filters is also given for both evaluations.

# Its.	1		40		Time / iteration		Rank	
Filter	Ave.	S.D.	Ave.	S.D.	470 × 370	752 × 480	T_{homo}	RMS
Original	0.282	0.136	0.282	0.136	-	-	5	7
TV-L^2	0.068	0.028	0.080	0.030	30 ms	60 ms	2	3
Sigma	0.168	0.036	0.090	0.023	30 ms	100 ms	1	6
Mean	0.055	0.023	0.072	0.025	1 ms	2 ms	4	2
Median	0.039	0.020	0.041	0.022	7 ms	15 ms	6	1
Bilateral	0.113	0.021	0.086	0.026	160 ms	340 ms	3	5
Trilateral	0.085	0.017	0.082	0.016	5,000 ms	11,000 ms	7	4

Optical Flow on EISATS Dataset. For this subsection, we computed optical flow using TV-L^1 optical flow [19] (one of the top performing algorithms), on the EISATS dataset [3]; see [16] for Set 2. We altered the data to resemble illumination differences in time, as performed in [11]; the differences start high between frames, then go to zero at frame 50, then increase again. The flow field is computed using $U(h_1, h_2) = \mathbf{u}$. This is to show that a residual image r provides better data for matching, than for the original image f. Figure 5 shows an example of this effect, obviously the residual image vastly improves optical flow results. We calculated the end-point-error using the error image $e = E(\mathbf{u}, \mathbf{u}^*)$ and Spatial-RMS.

We computed the flow using $U(r_1^{(n)}, r_2^{(n)})$ with $n = 1, 2, 10$, and 40 to show how each filter behaves. The results are compared to optical flow on the original images $U(f_1, f_2)$. The results can be seen in Figure 4. It is immediately obvious that the original

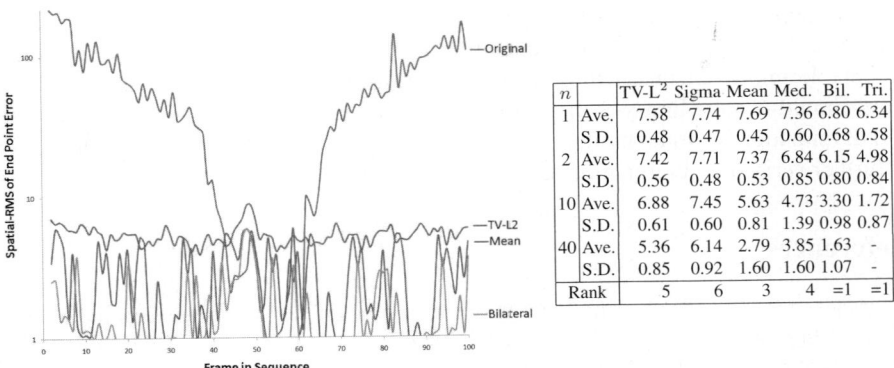

Fig. 4. Left: Flow end point error over entire EISATS sequence using number of filter iterations $r^{(40)}$, graph is logarithmically scaled ($\log_{10}$) (only TV-L^2, mean, and bilateral shown). Right: results using different number of filter iterations $r^{(n)}$, the original average (Ave.) is 61, and standard deviation (S.D.) is 53 (much higher than the rest).

Fig. 5. Top row: frame 1 (left) and 2 (middle) from EISATS scene. Ground truth flow with key (HSV circle for direction, saturation for vector length) is shown on the right. Bottom row: optical flow results using; original images (left) and residual images, TV-L^2 (middle) and trilateral (right), respectively.

image results are much worse quality than the residuum results. Residual images are more robust to illumination differences than standard images.

4 Conclusions and Future Research

We have identified a methodology for analysing the effect of illumination reducing filters using numerical comparisons. We went on to show that the results for this test do align with the optical flow performance, on a scene with drastic illumination variation. The tests showed that generating a simple mean residual image, produces acceptable improvements, while being the fastest (and easiest) to implement. Future work should test the limits of the proposed methodology. Other smoothing algorithms and illumination invariant models need to be tested. Finally, a larger dataset can be used to further verify the illumination artifact reducing effects of residual images.

References

1. Aujol, J.F., Gilboa, G., Chan, T., Osher, S.: Structure-texture image decomposition - modeling, algorithms, and parameter selection. Int. J. Computer Vision 67, 111–136 (2006)
2. Choudhury, P., Tumblin, J.: The trilateral filter for high contrast images and meshes. In: Proc. Eurographics Symp. Rendering, pp. 1–11 (2003)
3. .enpeda. dataset 2 (EISATS), http://www.mi.auckland.ac.nz/EISATS
4. Haralick, R.M., Bosley, R.: Texture features for image classification. In: Proc. ERTS Symposium. NASA, vol. SP-351, pp. 1219–1228 (1973)
5. Haussecker, H., Fleet, D.J.: Estimating optical flow with physical models of brightness variation. IEEE Trans. Pattern Analysis Machine Intelligence 23, 661–673 (2001)

6. Hirschmüller, H., Scharstein, D.: Evaluation of stereo matching costs on images with radiometric differences. IEEE Trans. Pattern Analysis Machine Intelligence (to appear)
7. Kuan, D.T., Sawchuk, A.A., Strand, T.C., Chavel, P.: Adaptive noise smoothing filter for images with signal-dependent noise. IEEE Trans. Pattern Analysis Machine Intelligence 7, 165–177 (1985)
8. Lee, J.-S.: Digital image smoothing and the sigma filter. Computer Vision, Graphics, and Image Processing 24, 255–269 (1983)
9. Mileva, Y., Bruhn, A., Weickert, J.: Illumination-robust variational optical flow with photometric invariants. In: Hamprecht, F.A., Schnörr, C., Jähne, B. (eds.) DAGM 2007. LNCS, vol. 4713, pp. 152–162. Springer, Heidelberg (2007)
10. Middlebury dataset, http://vision.middlebury.edu/stereo/data/
11. Morales, S., Woo, Y.W., Klette, R., Vaudrey, T.: A study on stereo and motion data accuracy for a moving platform. Technical report, MI-tech-32. University of Auckland (2009), http://www.mi.auckland.ac.nz/
12. Rudin, L., Osher, S., Fatemi, E.: Nonlinear total variation based noise removal algorithms. Physica D 60, 259–268 (1992)
13. Scharstein, D., Pal, C.: Learning conditional random fields for stereo. In: Proc. IEEE Conf. Computer Vision and Pattern Recognition (2007)
14. Tomasi, C., Manduchi, R.: Bilateral filtering for gray and color images. In: Proc. IEEE Int. Conf. Computer Vision, pp. 839–846 (1998)
15. Vaudrey, T., Klette, R.: Residual images remove illumination artifacts. In: Proc. Pattern Recognition - DAGM (to appear, 2009)
16. Vaudrey, T., Rabe, C., Klette, R., Milburn, J.: Differences between stereo and motion behaviour on synthetic and real-world stereo sequences. In: Proc. IEEE Image and Vision Conf., New Zealand (2008); Digital Object Identifier 10.1109/IVCNZ.2008.4762133
17. Wedel, A., Pock, T., Zach, C., Bischof, H., Cremers, D.: An improved algorithm for TV-L^1 optical flow. Post Proc. Dagstuhl Motion Workshop (to appear, 2009)
18. van de Weijer, J., Gevers, T.: Robust optical flow from photometric invariants. In: Proc. Int. Conf. on Image Processing, pp. 1835–1838 (2004)
19. Zach, C., Pock, T., Bischof, H.: A duality based approach for realtime TV-L^1 optical flow. In: Hamprecht, F.A., Schnörr, C., Jähne, B. (eds.) DAGM 2007. LNCS, vol. 4713, pp. 214–223. Springer, Heidelberg (2007)

Nonlinear Motion Detection*

Lennart Wietzke and Gerald Sommer

Cognitive Systems Group
Kiel University, Department of Computer Science
Christian-Albrechts-Platz 4, D-24118 Kiel, Germany
lw@ks.informatik.uni-kiel.de

Abstract. This work presents new ideas in multidimensional signal theory: an isotropic quadrature filter approach for extracting local features of arbitrary curved signals without the use of any steering techniques. We unify scale space, local amplitude, orientation, phase and curvature in one framework. The main idea is to lift up signals by a conformal mapping to the higher dimensional conformal space where the local signal features can be analyzed with more degrees of freedom compared to the flat space of the original signal domain. The philosophy is based on the idea to make use of the relation of the conformal signal to geometric entities such as hyper-planes and hyper-spheres. Furthermore, the conformal signal can not only be applied to 2D and 3D signals but also to signals of any dimension. The main advantages in practical applications are the rotational invariance, the low computational time complexity, the easy implementation into existing Computer Vision software packages, and the numerical robustness of calculating exact local curvature of signals without the need of any derivatives. Applications can be optical flow and object tracking not only limited to constant velocities but detecting also arbitrary accelerations which correspond to the local curvature.

1 Introduction

Low level image analysis is often the first step of many Computer Vision tasks. Therefore, local signal features determine the quality of subsequent higher level processing steps. It is important not to lose or to merge any of the original information within the local neighborhood of the point of interest. The constraints of local signal analysis are: to span an orthogonal feature space (split of identity) and to be robust against stochastic and deterministic deviations between the actual signal and the model. Image signals $f \in L^2(\Omega)$ with $\Omega \subset \mathbb{R}^n$ will be locally analyzed on a low level. The assumed local signal model is defined as a hyper-sphere

$$\mathcal{P}\{f\}(z;s) = (f*p)(z;s) = a \cos\left(\left\|z - \frac{1}{\kappa}\bar{o}\right\| + \phi\right), \quad (z,s) \in \Omega \times \mathbb{R}_+ \qquad (1)$$

with a as the local amplitude, $\phi \in [0, \pi)$ as the local phase, $\kappa > 0$ as the local curvature and $\bar{o}$ as the normal of the local orientation. For the special case of

---

* We acknowledge funding by the German Research Foundation (DFG) under the project *SO 320/4-2*.

Fig. 1. From left to right: a constant signal (i0D), an arbitrary rotated 1D signal (i1D) and an i2D checkerboard signal consisting of two simple superimposed i1D signals. A curved i2D signal and two superimposed curved i2D signals. Note that all signals displayed here preserve their intrinsic dimension globally.

$\kappa = 0$ the hyper-sphere degrades to a hyper-plane. One important local structural feature is the phase ϕ which can be calculated by means of the Hilbert transform [6]. Furthermore all signals will be analyzed in Poisson scale space $\mathcal{P}\{\cdot\}$ [3] since the Hilbert transform can only be interpreted for narrow banded signals. The Poisson kernel of the applied low pass filter reads

$$p_n(z;s) = \frac{2}{A_{n+1}} \frac{s}{(s^2 + \|z\|^2)^{(n+1)/2}}, \quad A_{n+1} = \frac{2\pi^{\frac{n+1}{2}}}{\Gamma\left(\frac{n+1}{2}\right)} \tag{2}$$

with A_{n+1} as the surface area of the unit sphere $\mathbb{S}^n$ [10] in Euclidian space $\mathbb{R}^{n+1}$.

2 Related Work

Local phase and amplitude of 1D signals can be analyzed by the analytic signal [6]. The generalization of the analytic signal to multidimensional signal domains has been done by the monogenic signal [2]. The nD monogenic signal is restricted to i1D signals for all dimensions. The monogenic signal replaces the classical one-dimensional Hilbert transform of the analytic signal by the generalized Hilbert transform [1] in Euclidean space

$$h_n(z) = \frac{2}{A_{n+1}} \frac{z}{\|z\|^{n+1}}, \quad z \in \mathbb{R}^n, \; n \in \mathbb{N} \setminus \{1\} \tag{3}$$

This paper shows, that signal analysis problems can be solved in higher dimensional conformal spaces [9,11]. Since the original signal will be analyzed in scale space it will be of advantage to summarize the Hilbert kernel and the Poisson to one unified kernel which will be called the conjugated kernel

$$q_n(z;s) = (h_n * p_n)(z;s) = \frac{2}{A_{n+1}} \frac{z}{(s^2 + \|z\|^2)^{(n+1)/2}} \tag{4}$$

To enable interpretation of the generalized Hilbert transform, its relation to the Radon transform is the key [12] of signal intelligence, see Figures (4) and (3). The generalized Hilbert transform can be expressed by a concatenation of the Radon transform, the inverse Radon transform and the well known classical 1D Hilbert transform kernel $h(t) = \frac{1}{\pi t}$. Note that the relation to the Radon transform is required solely for interpretation and theoretical results. Neither the

Fig. 2. Illustration of the conformal mapping of 2D signals to the 3D conformal space

Radon transform nor its inverse are ever applied to the signal in practice. Instead the generalized Hilbert transformed signal will be determined by convolution in spatial domain.

3 The Conformal Signal

In case of visual motion analysis a three dimensional isotropic quadrature filters are needed [7 4]. The conformal signal of a 3D signal $f \in L^2(\Omega)$ with $\Omega \subset \mathbb{R}^3$ delivers amplitude, 3D orientation, phase and curvature. For image sequences (3D signals) the concept of planes in 3D Radon space becomes the more abstract concept of hyper-planes in 4D Radon space. These 4D hyper-planes determine 3D spheres on the 4D hyper-sphere in 4D conformal space. Since 3D planes and 3D spheres of the three-dimensional signal domain are mapped to 3D spheres on the 4D hyper-sphere, the integration on these 3D spheres determines points in the 4D Radon space. The general inverse stereographic projection for any dimension $n \in \mathbb{N}$ which maps the Euclidian space $\mathbb{R}^n$ to the conformal space $\mathbb{R}^{n+1}$ reads

$$\mathcal{S}^{-1}(x_1, x_2, \ldots, x_n) = \frac{1}{1 + \sum_{\nu=1}^n x_\nu^2} \begin{bmatrix} x_1 \\ x_2 \\ \vdots \\ x_n \\ \sum_{\nu=1}^n x_\nu^2 \end{bmatrix} \qquad (5)$$

The inverse stereographic projection maps the Euclidian space $\mathbb{R}^n$ to the hyper-sphere in $\mathbb{R}^{n+1}$ with radius $\frac{1}{2}$ and the south pole of the hyper-sphere touching the origin $\mathbf{0} \in \mathbb{R}^n$ of the Euclidian space $\mathbb{R}^n$ and the north pole of the hyper-sphere with coordinates $(\mathbf{0}, 1) \in \mathbb{R}^{n+1}$. For the signal dimension $n = 3$ the inverse stereographic projection $\mathcal{S}^{-1}$ known from complex analysis [5] maps the 3D signal domain to the hyper-sphere. This projection is conformal and can be inverted by the general formula

$$\mathcal{S}(\xi_1, \xi_2, \ldots, \xi_n, \xi_{n+1}) = \frac{1}{1 - \xi_{n+1}} \begin{bmatrix} \xi_1 \\ \xi_2 \\ \vdots \\ \xi_n \end{bmatrix} \tag{6}$$

The back-projection $\mathcal{S}$ for all elements of the hyper-sphere reads with $\xi = (\xi_1, \xi_2, \xi_3, \xi_4)$. This mapping has the property that the origin of the 3D signal domain will be mapped to the south pole $\mathbf{0}$ of the hyper-sphere and both $-\infty, +\infty$ will be mapped to the north pole $(0, 0, 0, 1)$ of the hyper-sphere. 3D planes and spheres of the 3D signal domain will be mapped to spheres on the hyper-sphere and can be determined uniquely by hyper-planes in 4D Radon space. The integration on these hyper-planes corresponds to points $(t, \theta_1, \theta_2, \varphi)$ in the 4D Radon space. Since the signal domain $\Omega \subset \mathbb{R}^3$ is bounded, not the whole hyper-sphere is covered by the original signal. Anyway, all hyper-planes corresponding to spheres on the hyper-sphere remain unchanged. That is the reason why the conformal signal models 3D planes and all kinds of curved 3D planes which can be locally approximated by spheres. To provide the generalized Hilbert transform more degrees of freedom, the original three-dimensional signal will be embedded in an applicable subspace of the conformal space by the so called conformal signal $c \in \mathbb{R}^{(\mathbb{R}^4)}$ of the original 3D signal f

$$c(\xi) = \begin{cases} f(\mathcal{S}(\xi_1, \xi_2, \xi_3, \xi_4)^T) \;, & \sum_{\nu=1}^{3} \xi_\nu^2 + (\xi_4 - \frac{1}{2})^2 = (\frac{1}{2})^2 \\ 0 & , \text{ else} \end{cases} \tag{7}$$

by which the even signal part $c^e = (c * p_4)(\mathbf{0}; s)$ can by defined. Thus, the 4D generalized Hilbert transform can be applied to all points on the hyper-sphere. The center of convolution in spatial domain is the south pole where the origin of the 3D signal domain meets the hyper-sphere. At this point the generalized Hilbert transform will be evaluated in spatial domain by convolution for each test point

$$c^o = \begin{bmatrix} c_1^o \\ c_2^o \\ c_3^o \\ c_4^o \end{bmatrix} = (q_4 * c)(\mathbf{0}; s) = \frac{2}{A_5} \int_{z \in \mathbb{R}^4} \frac{z}{(s^2 + \|x\|^2)^{5/2}} c(z - \mathbf{0}) \, dz \tag{8}$$

The conformal signal for 3D signals is defined by the even part and the four odd parts of the 4D Hilbert transform. Note that the coordinates are relative to the local coordinate system for each test point of the original 3D signal and $\mathbf{0} = (0, 0, 0, 0)$ are the corresponding relative coordinates in conformal space, i.e. this is no restriction. The Hilbert transform of the 3D signal embedded in the conformal space can also be written in terms of the 4D Radon transform and its inverse

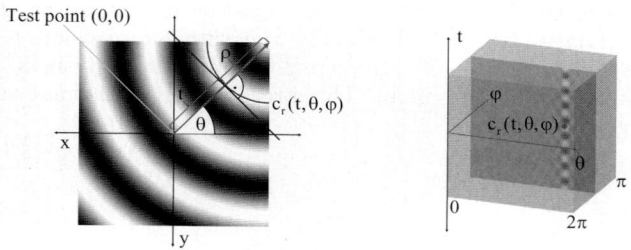

Fig. 3. Left figure: Curved i2D signal with orientation θ and curvature $\kappa = \frac{1}{\rho}$. Right figure: Corresponding 3D Radon space representation of the i2D signal spanned by the parameters t, θ and φ. Since the Radon transform on circles directly on the plane of the original 2D signal is not invertible, the Radon transform has to be done in higher dimensional 3D conformal space where circles correspond to planes.

$$c^o = \mathcal{R}^{-1} \left\{ \begin{bmatrix} \cos\varphi \sin\theta_1 \sin\theta_2 \\ \sin\varphi \sin\theta_1 \sin\theta_2 \\ \cos\theta_1 \sin\theta_2 \\ \cos\theta_2 \end{bmatrix} h(t) * \mathcal{R}\left\{(c * p_4)(z;s)\right\}(t;\theta_1,\theta_2,\varphi) \right\}(\mathbf{0};s) \quad (9)$$

This representation of the Hilbert transform is essential for the subsequent interpretation of the conformal signal. Remember that without loss of generality the signal will be analyzed at the origin of the local coordinate system of the test point of local interest. Compared to the monogenic signal the conformal signal is based on a Hilbert transformation in conformal space. Analogous to the interpretation of the monogenic signal in [12], the parameters of the hyperplane within the 4D Radon space determine the local features of the curved 3D signal. The conformal signal can be called the generalized monogenic signal for 3D signals, because the special case of planes in the original 3D signal can be considered as spheres with zero curvature. These planes are mapped to spheres passing through the north pole in conformal space. The 3D curvature corresponds to the parameter φ of the 4D Radon space,

$$\varphi = \arctan \frac{c_2^o}{c_1^o} \quad (10)$$

Besides, the curvature of the conformal signal naturally indicates the intrinsic dimension of the signal. The parameters (θ_1, θ_2) will be interpreted as the orientation of the signal in the original 3D space

$$\theta_1 = \arcsin \frac{\sqrt{[c_1^o]^2 + [c_2^o]^2}}{c_4^o} \quad \text{and} \quad \theta_2 = \arctan \frac{\sqrt{[c_1^o]^2 + [c_2^o]^2 + [c_3^o]^2}}{c_4^o} \quad (11)$$

The amplitude and phase are defined by

$$a = \sqrt{[c^e]^2 + \|c^o\|^2} \quad \text{and} \quad \phi = \arctan \frac{\|c^o\|}{c^e} \quad (12)$$

In all different intrinsic dimensions the phase indicates a measure of parity symmetry. Note that all proofs are analogous to those for the monogenic signal shown in [12].

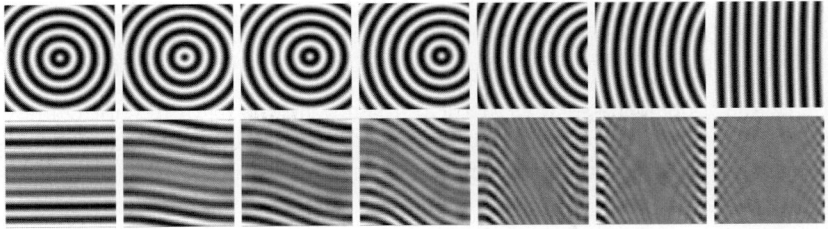

Fig. 4. From left to right: signal with varying curvature in spatial domain (top row) and in the corresponding 2D Radon space (bottom row). Obviously the 2D Radon space is too flat for analyzing and parameterizing the orientation and curvature of signals. Therefore the dimension of the 2D Radon space must be extended to 3D. This is the idea of the conformal signal.

3.1 Implementation

The computational time complexity is in $O(n^3)$ with n as the convolution mask size in one dimension.

```
//Input:   double Image3D(double x,double y,double z)
//Input:   double x,y,z (Local pixel test point for analysis)
//Input:   double Coarse > Fine > 0 (Bandpass filter parameters)
//Input:   double Size > 0 (Convolution mask size)
//Output:  Direction1, Direction2, Phase, Curvature, Amplitude

double Coarse=2,Fine=0.1; int Size=5;//e.g.
double rp=0,r1=0,r2=0,r3=0,r4=0;
for(double cx = -Size;cx <= Size;cx += 1)
for(double cy = -Size;cy <= Size;cy += 1)
for(double cz = -Size;cz <= Size;cz += 1)
{
    //Map points (cx,cy,cz) to conformal space (x1,x2,x3,x4)
    double d  = pow(cx,2)+pow(cy,2)+pow(cz,2)+1;
    double x1 = cx      / d;
    double x2 = cy      / d;
    double x3 = cz      / d;
    double x4 = (d-1)   / d;
    //Generalized Hilbert transform in conformal space
    double a  = pow(x1,2)+pow(x2,2)+pow(x3,2)+pow(x4,2);
    double pf = pow(pow(Fine   ,2) + a,-2.5);
    double pc = pow(pow(Coarse,2) + a,-2.5);
    double f  = Image3D(x + cx,y + cy,z + cz);
    double c  = f * (pf - pc);
    rp += f * (Fine*pf - Coarse*pc);
    r1 += x1 * c; r2 += x2 * c; r3 += x3 * c; r4 += x4 * c;
}
Curvature  = atan(r2/r1);
Direction1 = asin(sqrt(pow(r1,2)+pow(r2,2))/r4);
```

```
Direction2 = atan2(sqrt(pow(r1,2)+pow(r2,2)+pow(r3,2)),r4);
Phase      = atan2(sqrt(pow(r1,2)+pow(r2,2)+pow(r3,2)+pow(r4,2)),rp);
Amplitude  = sqrt(pow(rp,2)+pow(r1,2)+pow(r2,2)+pow(r3,2)+pow(r4,2));
```

In practical applications such as medical image analysis [4] or 3D optical flow (5) the convolution mask sizes must be DC-free. This can be achieved by removing their mean value after precalculating them.

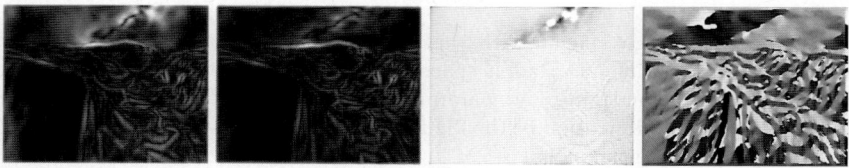

Fig. 5. The 3D conformal signal delivers four local features which can be used for image sequence analysis such as optical flow and motion analysis. From left to right: Curvature, phase and the two parts of the orientation information. 3D convolution mask size $5 \times 5 \times 5$ pixels.

4 Conclusion

In this paper a new fundamental idea for locally analyzing multidimensional signals has been presented. The n-dimensional domain of the original signal is always limited to its n-dimensional Radon space, which restricts the related feature space. To extend the dimension of the related feature space, this problem can be solved by embedding signals in higher dimensional conformal spaces in which the original signal can be analyzed by generalized Hilbert transforms with more degrees of freedom. Without steering and in a rotationally invariant way, local signal features such as amplitude, phase, orientation and curvature can be determined in spatial domain by convolution. The conformal signal can be computed efficiently and can be easily implemented into existing low level image processing steps of Computer Vision applications. Furthermore, exact curvature can be calculated with all the advantages of rotationally invariant local phase based approaches (robustness against brightness and contrast changes) and without the need of any partial derivatives. Hence, lots of numerical problems of partial derivatives on discrete grids can be avoided. All results can be proved mathematically as well as by experiments. Applications of the conformal signal such as object tracking [8] with arbitrary acceleration on three-dimensional data will be part of our future work.

References

1. Brackx, F., De Knock, B., De Schepper, H.: Generalized multidimensional Hilbert transforms in Clifford analysis. International Journal of Mathematics and Mathematical Sciences (2006)

2. Felsberg, M.: Low-level image processing with the structure multivector. Technical Report 2016, Kiel University, Department of Computer Science (2002)
3. Felsberg, M., Sommer, G.: The monogenic scale-space: A unifying approach to phase-based image processing in scale-space. Journal of Mathematical Imaging and Vision 21, 5–26 (2004)
4. Grau, V., Becher, H., Noble, J.A.: Phase-based registration of multi-view realtime three-dimensional echocardiographic sequences. In: Larsen, R., Nielsen, M., Sporring, J. (eds.) MICCAI 2006. LNCS, vol. 4190, pp. 612–619. Springer, Heidelberg (2006)
5. Gürlebeck, K., Habetha, K., Sprössig, W.: Funktionentheorie in der Ebene und im Raum. Grundstudium Mathematik. Birkhäuser, Basel (2006)
6. Hahn, S.L.: Hilbert Transforms in Signal Processing. Artech House Inc., Boston (1996)
7. Krause, M., Sommer, G.: A 3D isotropic quadrature filter for motion estimation problems. In: Proc. Visual Communications and Image Processing, Beijing, China, vol. 5960, pp. 1295–1306. The International Society for Optical Engineering, Bellingham (2005)
8. Lichtenauer, J., Hendriks, E.A., Reinders, M.J.T.: Isophote properties as features for object detection. In: CVPR (2), pp. 649–654 (2005)
9. Wietzke, L., Fleischmann, O., Sommer, G.: 2D image analysis by generalized hilbert transforms in conformal space. In: Forsyth, D., Torr, P., Zisserman, A. (eds.) ECCV 2008, Part II. LNCS, vol. 5303, pp. 638–649. Springer, Heidelberg (2008)
10. Wietzke, L., Fleischmann, O., Sommer, G.: Signal analysis by generalized hilbert transforms on the unit sphere. In: Simos, T.E. (ed.) International Conference on Numerical Analysis and Applied Mathematics, AIP Conference Proceedings, Melville, New York, vol. 1048, pp. 706–709 (2008)
11. Wietzke, L., Sommer, G.: The conformal monogenic signal. In: Rigoll, G. (ed.) DAGM 2008. LNCS, vol. 5096, pp. 527–536. Springer, Heidelberg (2008)
12. Wietzke, L., Sommer, G., Schmaltz, C., Weickert, J.: Differential geometry of monogenic signal representations. In: Sommer, G., Klette, R. (eds.) RobVis 2008. LNCS, vol. 4931, pp. 454–465. Springer, Heidelberg (2008)

Rotation Invariant Texture Classification Using Binary Filter Response Pattern (BFRP)

Zhenhua Guo, Lei Zhang, and David Zhang

Biometrics Research Centre, Department of Computing, the Hong Kong
Polytechnic University, Hong Kong
{cszguo,cslzhang,csdzhang}@comp.polyu.edu.hk

Abstract. Using statistical textons for texture classification has shown great success recently. The maximal response 8 (MR8) method, which extracts an 8-dimensional feature set from 38 filters, is one of state-of-the-art rotation invariant texture classification methods. However, this method has two limitations. First, it require a training stage to build a texton library, thus the accuracy depends on the training samples; second, during classification, each 8-dimensional feature is assigned to a texton by searching for the nearest texton in the library, which is time consuming especially when the library size is big. In this paper, we propose a novel texton feature, namely Binary Filter Response Pattern (BFRP). It can well address the above two issues by encoding the filter response directly into binary representation. The experimental results on the CUReT database show that the proposed BFRP method achieves better classification result than MR8, especially when the training dataset is limited and less comprehensive.

Keywords: Texture Classification, Texton, MR8, LBP.

1 Introduction

Texture analysis is an active research topic in the fields of computer vision and pattern recognition. Generally speaking, it involves four basic problems: classifying images based on texture content; segmenting an image into regions of homogeneous texture; synthesizing textures for graphics applications; and establishing shape information from texture cue [1].

In the early stage, researchers devoted themselves into extracting statistical feature to classify texture images, such as the co-occurrence matrix method [2] and the filtering based methods [3]. These methods could achieve good classification results if the training and testing samples have similar or identical orientation. In real applications, rotation invariance is a critical issue to be solved. Kashyap and Khotanzad [4] were among the first researchers to study rotation-invariant texture classification by using a circular autoregressive model. Later, many other models were explored, including the multiresolution autoregressive model [5], hidden Markov model [6], and Gaussian Markov random field [7]. Jafari-Khouzani and Soltanian-Zadeh proposed to use Radon transform to estimate the texture orientation and extract wavelet energy features for texture classification [8]. In [11], Ojala et al. proposed to use the Local Binary

Pattern (LBP) histogram for rotation invariant texture classification. LBP is a simple but efficient operator to describe local image patterns. Recently, Varma and Zisserman [9] proposed a statistical learning based algorithm, namely Maximal Response 8 (MR8), using a group of filter banks, where a rotation invariant texton library is first built from a training set and then an unknown texture image is classified according to its texton distribution. Under the same framework, Varma and Zisserman [10] proposed a new statistical learning based algorithm. Instead of using filter bank, pixel gray value is used directly to represent local region. Scale and affine invariant texture classification is anther active research topic, and some pioneer work have been done by using fractal analysis [13-14] and affine adaptation [12].

As a state-of-the-art rotation invariant texture classification method, MR8 has achieved good accuracy on public database [9]. However, this method requires a training step to learn a feature dictionary. The learned dictionary thus depends on training samples. For example, it may suffer from generality when the training sample set is limited. Furthermore, to obtain a statistical histogram for a given image, it requires an additional step to assign the local region with a texton by searching the closest one from the trained library. This step is time consuming when the library size is big.

To solve the above problems, in this paper we propose a new feature extraction operator, namely Binary Filter Response Pattern (BFRP), which could be viewed as a binary version of MR8. After filtering the input image, BFRP converts the filter response into binary strings directly, instead of preserving the real values of filter response. Hence, each local region is assigned with a predefined texton. Such a binary presentation is not only fast to compute, but also can retain more discriminant information for classification.

The rest of the paper is organized as follows. Section 2 introduces the proposed BFRP and the dissimilarity metric. Section 3 reports the experimental results on a representative texture database. Section 4 gives the conclusion and future work.

2 Binary Filter Response Pattern

The MR8 filter bank used in [9] is employed in the proposed binary filter response pattern (BFRP) method. The MR8 filter bank consists of 38 filters, which are shown in Fig.1. To achieve rotation invariance, the filters are implemented at multiple

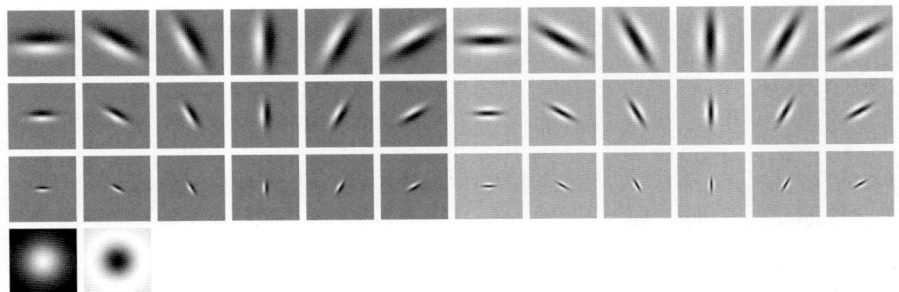

Fig. 1. The MR8 filter bank consists of a series of anisotropic filters (an edge and a bar filter at 6 orientations and 3 scales), and 2 rotationally symmetric ones (a Gaussian and a Laplacian of Gaussian) [9]

orientations and multiple scales. At each scale only the maximal response among the different orientations is kept. The final response at each position is an 8-dimension feature vector (3 scales for the edge and bar filters, plus 2 isotropic filters).

2.1 BFRP Feature Extraction

As shown in Fig.2, some local regions may have multiple dominant orientations. The magnitude of the filter response at each angle could be treated as a confidence measurement in the feature occurring at that orientation [9]. Thus we define the BFRP for multiple orientations as

$$BFRP(x,y) = \sum_{i=0}^{N-1} s(F_{\theta_i}(x,y))2^i, \ s(x) = \begin{cases} 1, x \geq 0 \\ 0, x < 0 \end{cases}, \ \theta_i = \frac{i}{N}\pi, \ i=0,1,\ldots,N-1 \quad (1)$$

$$F_{\theta_i}(x,y) = f_{\theta_i} * I(x,y) \quad (2)$$

where f_{θ_i} is the filter with orientation θ_i, * is the convolution operation. N is the number of filters at one scale (N=6 for bar and edge filters, N=1 for Gaussian and Laplacian filters). I is the input image.

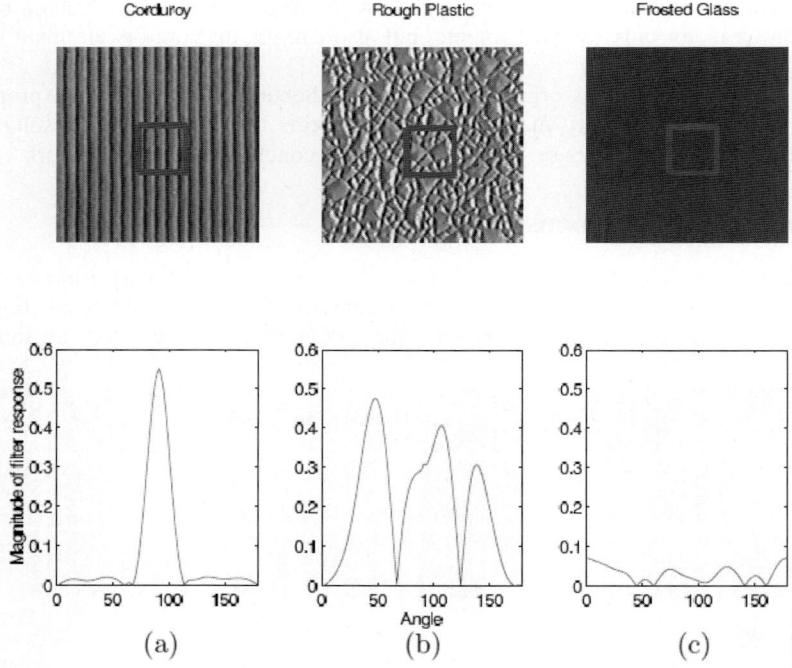

Fig. 2. The top row shows 3 texture images. The central image patch (highlighted by red rectangle) is matched with an edge filter at all orientations. The magnitude of the filter response versus the orientation is plotted in the bottom row [9].

To achieve rotation invariance, we shift the BFRP until the least value of the binary string is obtained. Similar to the definition of rotation invariant LBP code [11], we can define the rotation invariant BFRP and denote this binary code as $BFRP^{ri}$ (the superscript "ri" means the use of rotation invariant patterns). The rotation invariance of a binary code is achieved by shifting a number of binary bits, to find the least bit string value [11]. Thus, for the 6-bit binary code of bar and edge filters, there will be 14 kinds of rotation invariant strings (the 14 rotation invariant strings are 000000, 000001, 000011, 000101, 000111, 001001, 001011, 001101, 001111, 010101, 010111, 011011, 011111, and 111111). While for the 1-bit binary code of Gaussian and Laplacian filters, there will be 2 kinds of string only. With $BFRP^{ri}$, the filtering output at each position is a 8-dimensional vector, and there are 30,118,144 (14*14*14*14*14*14*2*2) kinds of patterns in total. Such a dimension is too large to build histogram and it will bring computation issue. To reduce the feature size, we empirically divide the 38 filters into 2 groups as show in Fig. 3. Thus for each image, only two 4-dimensional histograms need to be built and then the 2 histograms are concatenated. The final histogram size is reduced to 10,976 (14*14*14*2*2).

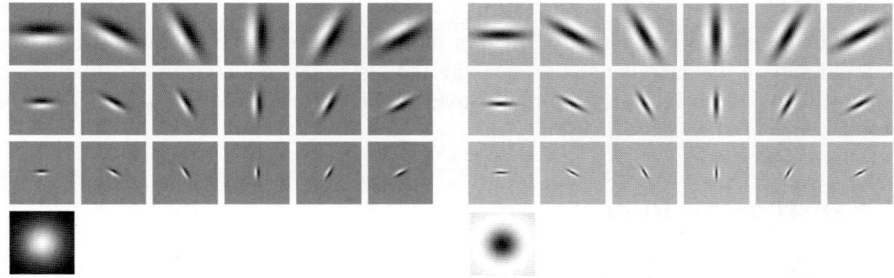

Fig. 3. Divide the MR8 filter banks into two equally groups

To further reduce the number of patterns at each scale, similar to the "uniform" LBP [11], the "uniform" BFRP is defined as:

$$BFRP^{riu2} = \begin{cases} \sum_{i=0}^{N-1} s(F_{\theta_i}) & if \ U(BFRP) \leq 2 \\ N+1 & otherwise \end{cases} \quad (3)$$

$$U(BFRP) = \left|s(F_{\theta_{N-1}}) - s(F_{\theta_0})\right| + \sum_{i=1}^{N-1} \left|s(F_{\theta_i}) - s(F_{\theta_{i-1}})\right| \quad (4)$$

The uniformity measure U is used to count the number of bit transition (bitwise 1/0 changes) in the bit string. By definition, exactly $N+1$ "uniform" binary patterns can occur in a circularly symmetric neighbor set of N binary bits, while the remaining (non-uniform) are grouped into a "miscellaneous" label (N+1).

The $BFRP^{riu2}$ (superscript "$riu2$" means the use of rotation invariant "uniform" patterns that have U value of at most 2), has 8 distinct output values for 6-bit binary strings and 2 distinct output values for 1-bit binary strings. Using $BFRP^{riu2}$, the feature size is 2,048 (8*8*8*2*2) which is comparable with that in MR8 [9].

2.2 Dissimilarity Metric

The dissimilarity of sample and model histograms is a test of goodness-of-fit, which could be measured with a nonparametric statistic test. There are many metrics to evaluate the goodness between two histograms, such as histogram intersection, log-likelihood ratio, and chi-square statistic [11]. In this study, a test sample T is assigned to the class of model L that minimizes the chi-square distance:

$$D(T,L) = \sum_{n=1}^{N} \frac{(T_n - L_n)^2}{T_n + L_n} \qquad (5)$$

where N is the number of bins, and T_n and L_n are the values of the sample and model image at the nth bin, respectively. In this paper, the nearest neighborhood classifier with chi-square distance is used to measure the dissimilarity between two histograms because it is equivalent to the optimal Bayesian classification [16] and get good performance for texture classification [17].

3 Experimental Results

In this section we will compare the proposed feature extraction scheme with MR8 and another non-training method (here, the "non-training" standing for the feature extraction is training free), LBP [11]. In MR8, 40 textons are clustered from each of the c texture classes using the training samples, and then a histogram based on the $c*40$ textons is computed for each model and sample image. To more comprehensively evaluate the proposed method, in the experiments we list both the classification rates by using the $BFRP^{ri}$ (rotation invariant BFRP) and $BFRP^{riu2}$ ("uniform" rotation invariant BFRP) schemes.

The CURet database [18] is one of the largest yet most difficult databases for rotation invariant texture recognition. It contains 61 textures and each texture has 205 images obtained under different viewpoints and illumination directions. There are 118 images whose viewing angles (θ_v) are less than 60^0. Out of the 118 images, the 92 images, from which a sufficiently large region could be cropped (200*200) across all texture classes, are selected [9]. Before feature extraction, all the cropped regions are converted into grey scale and are normalized to have a mean of 0 and a standard deviation of 1 [9, 11]. Here, instead of computing error bar (i.e. mean and standard deviations of results calculated over multiple splits), we performed experiments on four different settings to simulated four situations:

1. T46A: The training set for each class was selected by taking one from every two adjacent images. Hence, there are 2,806 (61*46) models and 2,806 testing samples. Because in CUReT, adjacent images have similar appearance than far images, this setting is used to simulate the situation of large and comprehensive training set.
2. T23A: The training set for each class was selected by taking one from every four adjacent images. Hence, there are 1,403 (61*23) models and 4,209 (61*69) testing samples. This setting is used to simulate the situation of small but comprehensive training set.
3. T46F: The training set for each class was selected as the first 46 images. Hence, there are 2,806 models and 2,806 testing samples. This setting is used to simulate the situation of large but less comprehensive training set.
4. T23F: The training set for each class was selected as the first 23 images. Hence, there are 1,403 models and 4,209 testing samples. This setting is used to simulate the situation of small and less comprehensive training set.

The performance is evaluated by using the classification rate with chi-square distance and the nearest neighborhood classifier. The classification results by different operators with its associated feature size are listed in Table 1. The column for "T46A", "T23A", "T46F", "T23F" represent the four different test setup as described above.

From Table 1, we could make the following findings. First, under similar feature size, the proposed BFRP has better recognition performance than MR8. For example, $BFRP^{riu2}$ (2,048 dimension) achieves 0.39%, 0.33%, 1.14%, and 2.66% higher recognition rates than MR8 with its best settings (c=61 with feature size being 2,440 dimensions) for T46A, T23A, T46F, and T23F, respectively.

Second, because MR8 requires a training stage, which depends on the training samples, when the training sample is few and not comprehensive, the accuracy will drops quickly compared with the proposed feature extraction. For example, the accuracy decreases 18.32% (from T23A to T23F) for MR8 (c=61) while decreasing only 15.99% for $BFRP^{riu2}$. So the proposed method is more suitable for real applications where training samples are limited and not comprehensive.

Table 1. Classification rate (%) for the four settings using different schemes

Method	Feature Size	T46A	T23A	T46F	T23F
$LBP_{8,1}^{riu2} + LBP_{16,3}^{riu2} + LBP_{24,5}^{riu2}$ [15]	54	95.47	93.09	85.64	78.50
MR8 (c=61)	2440	97.65	96.15	88.70	77.83
MR8 (c=20)	800	96.79	94.80	86.89	75.98
MR8 (c=10)	400	95.55	93.87	84.50	74.82
$BFRP^{ri}$	10976	98.33	96.58	90.02	80.97
$BFRP^{riu2}$	2048	98.04	96.48	89.84	80.49
Average		97.05	95.16	87.55	77.82

4 Conclusion

In the past decade, statistical texton has achieved great success for texture classification. However, it requires a training step and spends much cost on building the feature histogram. In this study, a training free rotation invariant feature extractor is proposed for texture classification. Using the same filter bank as in MR8, the proposed feature extractor requires much less computation cost on feature extraction but achieves better result than MR8. In the future we will investigate other statistical texton based operators, such as local fractal [14] and local patch [10], and try to extract their training free counterpart.

Acknowledgments

The work is partially supported by the CERG fund from the HKSAR Government, Hong Kong RGC General Research Fund (PolyU 5351/08E), the central fund from Hong Kong Polytechnic University, the Natural Science Foundation of China (NSFC) under Contract No. 60620160097, No. 60803090, and the National High-Tech Research and Development Plan of China (863) under Contract No. 2006AA01Z193.

References

1. Tuceryan, M., Jain, A.K.: Texture analysis. In: Chen, C.H., Pau, L.F., Wang, P.S.P. (eds.) Handbook of pattern recognition and computer vision, ch. 2, pp. 235–276 (1993)
2. Haralik, R.M., Shanmugam, K., Dinstein, I.: Texture features for image classification. IEEE Trans. on Systems, Man, and Cybertics 3(6), 610–621 (1973)
3. Randen, T., Husy, J.H.: Filtering for texture classification: a comparative study. IEEE Trans. PAMI 21(4), 291–310 (1999)
4. Kashyap, R.L., Khotanzed, A.: A model-based method for rotation invariant texture classification. IEEE Trans. on PAMI 8(4), 472–481 (1986)
5. Mao, J., Jain, A.K.: Texture classification and segmentation using multiresolution simultaneous autoregressive models. Pattern Recognition 25(2), 173–188 (1992)
6. Wu, W.R., Wei, S.C.: Rotation and gray-scale transform-invariant texture classification using spiral resampling, subband decomposition, and hidden Markov model. IEEE Trans. IP 5(10), 1423–1434 (1996)
7. Deng, H., Clausi, D.A.: Gaussian MRF rotation-invariant features for image classification. IEEE Trans. on PAMI 26(7), 951–955 (2004)
8. Jafari-Khouzani, K., Soltanian-Zadeh, H.: Radon transform orientation estimation for rotation invariant texture analysis. IEEE Trans.PAMI 27(6), 1004–1008 (2005)
9. Varma, M., Zisserman, A.: A statistical approach to texture classification from single images. International Journal of Computer Vision 62(1-2), 61–81 (2005)
10. Varma, M., Zisserman, A.: A statistical approach to material classification using image patch exemplars. IEEE Trans. PAMI (to appear)
11. Ojala, T., Pietikäinen, M., Mäenpää, T.T.: Multiresolution gray-scale and rotation invariant texture classification with Local Binary Pattern. IEEE Trans. PAMI 24(7), 971–987 (2002)
12. Lazebnik, S., Schmid, C., Ponce, J.: A sparse texture representation using local affine regions. IEEE Trans. PAMI 27(8), 1265–1278 (2005)

13. Xu, Y., Ji, H., Fermuller, C.: A projective invariant for texture. In: International Conference on Computer Vision and Pattern Recognition, pp. 1932–1939 (2005)
14. Varma, M., Garg, R.: Locally invariant fractal features for statistical texture classification. In: International Conference on Computer Vision (2007)
15. Pietikäinen, M., Nurmela, T., Mäenpää, T., Turtinen, M.: View-based recognition of real-world textures. Pattern Recognition 37(2), 313–323 (2004)
16. Varma, M., Zisserman, A.: Unifying statistical texture classification framework. Image and Vision Computing 22(14), 1175–1183 (2004)
17. Puzicha, J., Buhmann, J.M., Rubner, Y., Tomasi, C.: Empircal evaluation of dissimilarity measures for color and texture. In: International Conference on Computer Vision, pp. 1165–1172 (1999)
18. Dana, K.J., van Ginneken, B., Nayar, S.K., Koenderink, J.J.: Reflectance and texture of real world surfaces. ACM Trans. on Graphics 18(1), 1–34 (1999)

Near-Regular Texture Synthesis

Michal Haindl and Martin Hatka

Institute of Information Theory and Automation
of the ASCR, Prague, Czech Republic
{haindl,hatka}@utia.cz

Abstract. This paper describes a method for seamless enlargement or editing of difficult colour textures containing simultaneously both regular periodic and stochastic components. Such textures cannot be successfully modelled using neither simple tiling nor using purely stochastic models. However these textures are often required for realistic appearance visualisation of many man-made environments and for some natural scenes as well. The principle of our near-regular texture synthesis and editing method is to automatically recognise and separate periodic and random components of the corresponding texture. Each of these components is subsequently modelled using its optimal method. The regular texture part is modelled using our roller method, while the random part is synthesised from its estimated exceptionally efficient Markov random field based representation. Both independently enlarged texture components from the original measured texture are combined in the resulting synthetic near-regular texture. In the editing application both enlarged texture components can be from two different textures. The presented texture synthesis method allows large texture compression and it is simultaneously extremely fast due to complete separation of the analytical step of the algorithm from the texture synthesis part. The method is universal and easily viable in a graphical hardware for purpose of real-time rendering of any type of near-regular static textures.

1 Introduction

Physically correct virtual models require object surfaces covered with realistic nature-like colour textures to enhance realism in virtual scenes. Satisfactory models require not only complex 3D shapes accorded with the captured scene, but also realistic surface materials visualisation. This will significantly increase the realism of the synthetic generated scene. We define near-regular textures as textures that contain global, possibly imperfect, regular structures as well as irregular stochastic structures simultaneously. This is more ambitious definition than to view [1] a near-regular textures as a statistical distortion of a regular texture. Near regular textures are difficult to synthesise, however, these textures are ubiquitous in man-made environments such as buildings, wallpapers, floors, tiles, fabric but even some fully natural textures such as honeycomb, sand dunes or waves belong to this texture category. These textures can be modelled in simplified smooth or more precise rough (also referred as the bidirectional texture function - BTF [2]) representation. The rough textures do not obey the Lambert law and their reflectance is illumination and view angle dependent. Both types of such near-regular texture representations occur in virtual scenes models. The purpose of any synthetic texture approach is to reproduce and enlarge a given measured texture image so that ideally both natural and synthetic texture will be visually indiscernible. The related texture modelling approaches may be divided primarily into sampling and model-based-analysis and synthesis, but no ideal texture modelling method exists. Each of the existing approaches or texture models has its advantages and limitations simultaneously and it is applicable for a restricted subset of possible textures only. Model-based texture synthesis [3, 4, 5] requires non-standard multidimensional (3D for static colour textures or even 7D for static BTFs) models. Such models are non trivial and they suffer with several unsolved problems which have to be circumvented (e.g. optimal parameters estimation, efficient synthesis, stability). Model-based methods are also often too difficult to be implemented in contemporary graphical card processors. Sampling approaches [6, 7] rely on sophisticated sampling from real texture measurements. Sampling methods require to store original texture sample, thus they cannot come near the large compression ratio of the model-based methods.

Neither model-based or simple sampling algorithms alone can satisfactorily solve the difficult problem of near-regular texture modelling. Existing methods [1, 8, 9, 10, 11, 12, 13, 14, 15] usually try to overcome this problem by user assisted modelling of the regular structures and then rely on regular tiling. However Lin et al. [11] experimentally observed that several of these general purpose sampling algorithms fail to preserve the structural regularity on more than 40% of their tested regular textures. Tiling-based synthesis algorithms [9, 12] identify the underlying lattice of the input texture either automatically or by user selection of two translation vectors and use slightly modified image quilting method [7] for synthesis. Texture replacement method [10] can replace selected regular texture while preserving its lighting using a Markov random field model and slow iterative Markov chain Monte Carlo solution. Another interactive tiling method [1] requires user assistance to identify a coarse texture lattice structure. The method [15] separates the global regular structure from the irregular structure using fractional Fourier analysis similarly to our method. However the synthesis is performed by generating a fractional Fourier texture mask from the extracted global regular structure which is used to guide pixelwise and time consuming sample-based synthesis. All mentioned near-regular texture modelling methods suffer with drawbacks inherent to the tiling approach. They do not allow texture editing, near-regular BTF textures, unmeasured textures applications and have very limited compression ratio. Tiling approaches cannot eliminate visible repetitions even if they use several tiles which are randomly combined such as [2].

The presented fully automatic method proposes to combine advantages of both basic texture modelling approaches by factoring a texture into factors that benefit best from each of two basic different modelling concepts. The principle of the method is to separate texture regular and stochastic parts, to enlarge both parts separately and to combine these results (texture enlargement) or results from several different textures (texture editing) into the required resulting texture. The proposed solution is not only fully automatic, very fast due to strict separation of the analytical and very efficient synthesis steps, but it also allows significant data compression. Due to its stochastic modelling it completely eliminates visible repetitions (contrary to all mentioned tiling approaches) because there are never used two identical tiles in a scene. Finally the method can be easily used to near-regular texture editing by either combining texture parts from different measurement or by changing stochastic model parameters.

2 Periodic and Non-periodic Texture Separation

The prerequisite for the method is that near-regular input textures have distinct amplitude spectrum parts for both periodic and random components. Otherwise the method, schematised in Fig.1 and outlined in the following sections, would not be able to separate both texture parts. Periodic and non-periodic texture part are detected in the simplified monospectral texture space. The input colour texture is spectrally transformed using the principal component analysis (PCA). Let the digitised colour texture $\bar{Y}$ is indexed on a finite rectangular three-dimensional $M \times N \times d$ underlying lattice I, where $M \times N$ is the image size and d is the number of spectral bands. The original centered data space $\tilde{Y}$ is transformed into a new data space with PCA coordinate axes Y. This new basis vectors are the eigenvectors of the $d \times d$ second-order statistical moments matrix $\Phi = E\{\tilde{Y}_{r,\bullet} \tilde{Y}_{r,\bullet}^T\}$ where d is the number of spectral bands and the multiindex r has two components $r = [r_1, r_2]$ (the row and column index). The projection of random vector $\tilde{Y}_{r,\bullet}$ (the notation $\bullet$ has the meaning of all possible values of the corresponding index) onto the PCA coordinate system uses the transformation matrix $T = [u_1^T, \ldots, u_d^T]^T$ which has single rows u_j that are eigenvectors of the matrix Φ: $\bar{Y}_{r,\bullet} = T\tilde{Y}_{r,\bullet}$. The periodic texture part (Fig. 2) is detected on the most informative transformed monospectral factor, which corresponds to the largest Φ eigenvalue.

2.1 Textural Periodicity Direction

Near-regular measured textures can have arbitrary periodicity directions (Fig.1-top right), not necessarily simple axis aligned periodicity. The periodicity in two directions is detected from the spatial correlation field restricted with the help of Fourier amplitude spectrum (Fig.1-right). The method finds two largest Fourier amplitude spectrum coefficients provided that they do not represent parallel directions. Tolerance sectors (Fig.1- right), which accommodate for possible localisation imprecision of local amplitude spectra maxima, are specified and for all their indices the corresponding spatial correlations are evaluated. Local spatial correlation field maxima, larger than a threshold, are detected and the minimal periodicity maximum is selected. Detected periodicity $(\delta^{h^*}, \delta^{v^*})$ and its direction allows to rotate measured texture to have axis aligned periodicity which simplifies further analytical steps. Detected periodicity and directions specify a rhomboid which contains the largest periodic part from the input texture. The

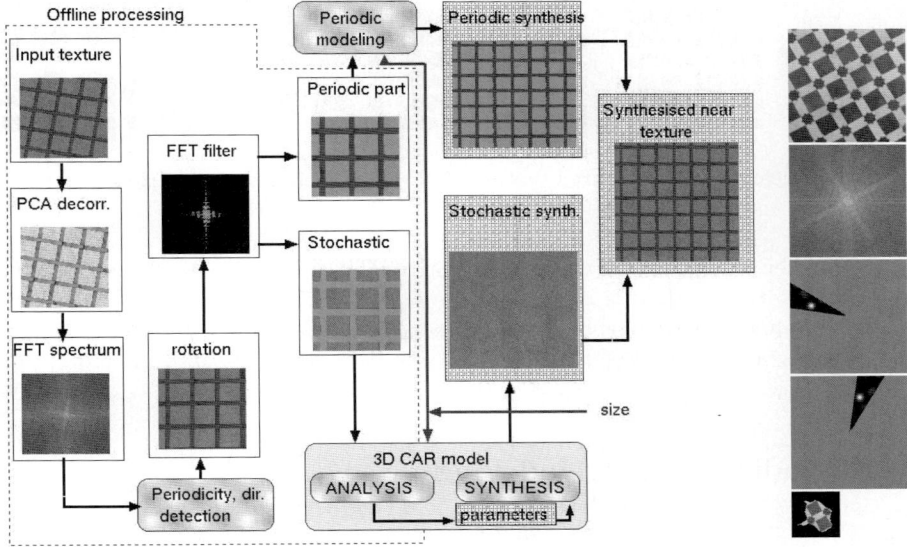

Fig. 1. Presented method overall schema, right from above - original measured texture, its amplitude spectrum, detected spatial correlation sectors, and the resulting toroidal tile (bottom)

rhomboid is further inscribed into the $\hat{M} \times \hat{N}$ rectangle which is cut out from the input texture. Although the double toroidal tile can be searched directly from the rhomboid the rectangular shape restriction simplifies this detection step.

2.2 Amplitude Spectrum Filter

The texture cutout is re-sampled to the lattice size of the power of two required by the fast Fourier transformation (FFT) based filter $\dot{M} \geq \hat{M}$, $\dot{M} = 2^i$, $\dot{N} \geq \hat{N}$, $\dot{N} = 2^j$, where i, j are minimum possible values. Let $A_{\max}$ is the Fourier amplitude spectrum maximum coefficient detected from the Fourier amplitude spectrum (Fig.1- right). The filter removes such coefficients, for which any of the following conditions holds: $A_r < k A_{\max}$, $A_r \notin \mathcal{M} \wedge r \notin I_m$, where $\mathcal{M}$ is a set of amplitude spectrum local maxima, $k \in \langle 0; 1 \rangle$ is a parameter and I_m is a contextual neighbourhood (we use the hierarchical neighbourhood of the first or the second order) of such a local maximum. Applying the inverse FFT and re-sampling the filtered tile back to the original $\hat{M} \times \hat{N}$ size we get the filtered cutout $\dot{Y}$ (Fig. 2- even images). FFT can be alternatively replaced by the rotated FFT from the section 2.1 but this option would introduce sampling errors into the filter. The filtered tile $\dot{Y}$ is binarized ($\hat{Y}$) using a threshold $t_{bin} \in \langle 0; 1 \rangle$. One label determines the periodic texture part and the other the stochastic part. To find the labels correspondence to both periodical and non-periodical parts of the original texture Fig.2 - odd img., the binary image $\hat{Y}$ is

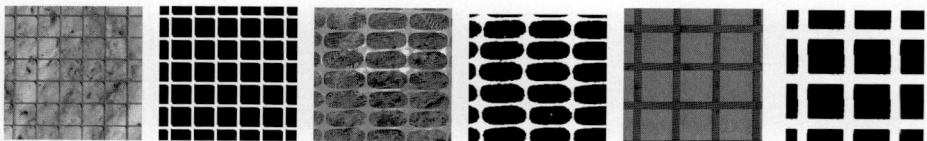

Fig. 2. Near-regular measured textures (odd) and their detected periodic parts

tested for periodicity δ^{h^*}, δ^{v^*}. The majority label complying to the periodicity test denotes the original texture periodic sites (Fig.2- even img.). When both periodic and stochastic parts are separated they can be independently modelled and enlarged to any required size as it is detailed in two following sections. The required near-regular texture is simple composite of both synthetic parts.

3 Periodic Texture Modelling

The regular part of the texture is enlarged using a simplification of our previously published [16] method. The roller method [2,16] is based on the overlapping tiling and subsequent minimum error boundary cut. One or several optimal double toroidal texture patches are seamlessly repeated during the synthesis step. This automatic method starts with the minimal tile size detection which is limited by the size of texture measurements, the number of toroidal tiles we are looking for and the sample spatial frequency content. The optimal horizontal and vertical edges cuts are searched using the dynamic programming method. These optimal vertical and horizontal cuts constitute a toroidal tile as is demonstrated on the Fig.1 - bottom right. Some textures with dominant irregular structures cannot be modelled by simple single tile repetition without clearly visible and visually disturbing regular artefact. Such textures exploit multiple toroidal tiles which share identical border but differ in their interior. Finally, the periodic texture enhancement is simple repetition of one or several randomly alternating double toroidal tiles in both directions until the required texture size is generated.

4 Random Texture Modelling

The random part of a texture is synthesised from the original input texture from where the detected periodic component was removed as described in section 2. If the stochastic texture patches are too small (few hundred pixels area) to reliably learn the random field model statistics, we replace occluded stochastic texture areas by using a modification of the image quilting algorithm [7]. The random part of the texture is synthesised using an adaptive probabilistic spatial model, a multiresolution 3D causal autoregressive model (CAR) [17], which is an exceptionally efficient type from the Markov random field (MRF) family of models. This model allows extreme compression (few tens of parameters to be stored only) and can be speedily evaluated directly in a procedural form to seamlessly fill an infinite texture space. The resulting near-regular texture is simple combination of both regular and stochastic synthesised factors.

5 Results

We have tested the presented method on near-regular textures from our extensive texture database, which currently contains over 1000 colour textures. Tested near-regular textures were either man-made such as two textures on Fig.4 or combinations of man-made structures with natural background (Fig.3) such as grass, wood, plants, snow, sand, etc. Both part of modelling were separately

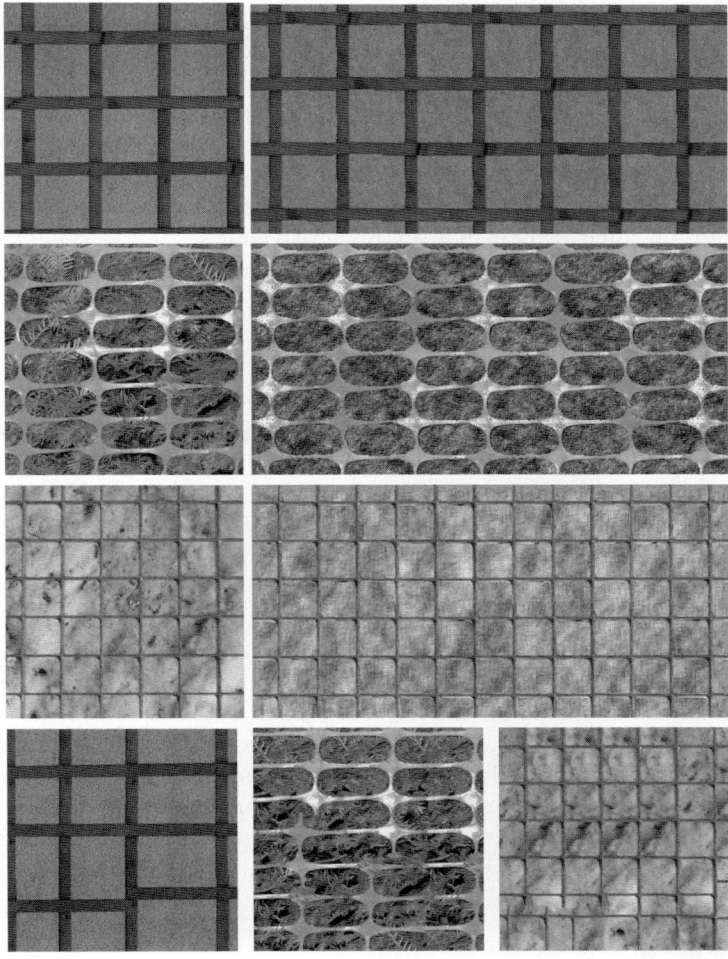

Fig. 3. Near-regular textures and their synthesis (right), image quilting [7] results (bottom row)

Fig. 4. Near-regular texture editing. Measured textures (three leftmost) and edited textures (three rightmost).

successfully tested on hundreds of colour or BTF textures with results reported elsewhere ([16]). Such unusually extensive testing was possible due to simplicity and efficiency of both crucial parts of the algorithm and it allowed us to get insight into the algorithm properties. The method is even capable to synthesise some near-regular textures combined from two distinctive types of regular structures provided they can be adequately separated in the Fourier domain. Resulting textures are mostly surprisingly good for such a fully automatic fast algorithm. Textures in Fig.3 were synthesised in real time ($\approx 1\,[s]$) while using the image quilting method [7] the synthesis took $90\,[s]$ on the same PC. Obviously there is no optimal texture modelling method and also the presented method fails on some near-regular textures with similar (and thus faultlessly unseparable) amplitude spectrum parts of both periodic and random components.

6 Conclusions

Our test results on available near-regular texture data are encouraging. The overall method is fully automatic and extremely fast due to strict separation of the analytical and very efficient synthesis steps. The regular part modelling is easily implementable even in the graphical processing unit. The method offers larger compression ratio than alternative tiling methods for transmission or storing texture information due to the periodic part modelling approach. The MRF based random part model can reach a huge compression ratio itself, hence its storage requirements are negligible, and simultaneously eliminates visible repetitions typical and unavoidable for tiling approaches. The overall method has negligible computation complexity for the periodic model and exceptionally efficient computational model for the random part as well. The method's extension for alternative texture types, such as BTF textures or some other spatial data such as the reflectance models parametric spaces is straightforward. Finally, the method can be easily used to near-regular texture editing by either combining texture parts from different measurement or by changing stochastic model parameters.

Acknowledgements

This research was supported by the grant GAČR 102/08/0593 and partially by the MŠMT grants 1M0572 DAR, 2C06019.

References

1. Liu, Y., Lin, W.C., Hays, J.: Near-regular texture analysis and manipulation. ACM Transactions on Graphics 23(3), 368–376 (2004)
2. Haindl, M., Hatka, M.: BTF Roller. In: Texture 2005. Proceedings of the 4th Int. Workshop on Texture Analysis, pp. 89–94. IEEE, Los Alamitos (2005)
3. Besag, J.: Spatial interaction and the statistical analysis of lattice systems. Journal of the Royal Statistical Society, Series B B-36(2), 192–236 (1974)
4. Haindl, M.: Texture synthesis. CWI Quarterly 4(4), 305–331 (1991)
5. Haindl, M., Havlíček, V.: A multiresolution causal colour texture model. In: Amin, A., Pudil, P., Ferri, F., Iñesta, J.M. (eds.) SPR 2000 and SSPR 2000. LNCS, vol. 1876, pp. 114–122. Springer, Heidelberg (2000)
6. Efros, A.A., Leung, T.K.: Texture synthesis by non-parametric sampling. In: Proc. Int. Conf. on Computer Vision (2), Corfu, Greece, pp. 1033–1038 (1999)
7. Efros, A.A., Freeman, W.T.: Image quilting for texture synthesis and transfer. In: Fiume, E. (ed.) ACM SIGGRAPH 2001, pp. 341–346. ACM Press, New York (2001)
8. Guo, C.E., Zhu, S.C., Wu, Y.N.: Modeling visual patterns by integrating descriptive and generative methods. Int. J. of Computer Vision 53(1), 5–29 (2003)
9. Liu, Y., Collins, R.T., Tsin, Y.: A computational model for periodic pattern perception based on frieze and wallpaper groups. IEEE Trans. Pattern Anal. Mach. Intell. 26(3), 354–371 (2004)
10. Tsin, Y., Liu, Y., Ramesh, V.: Texture replacement in real images. In: CVPR, pp. 539–544. IEEE Computer Society, Los Alamitos (2001)
11. Lin, W.C.C., Hays, J.H., Wu, C., Kwatra, V., Liu, Y.: A comparison study of four texture synthesis algorithms on regular and near-regular textures. Technical report, CMU Robotics Institute, Carnegie Mellon University (2004)
12. Liu, Y., Tsin, Y., Lin, W.C.: The promise and perils of near-regular texture. International Journal of Computer Vision 62(1-2), 145–159 (2005)
13. Karthikeyani, V., Duraiswamy, K., Kamalakkannan, P.: Texture analysis and synthesis for near-regular textures. In: ICISIP, pp. 134–139. IEEE Computer Society, Los Alamitos (2005)
14. Lin, W.C., Hays, J., Wu, C., Liu, Y., Kwatra, V.: Quantitative evaluation of near regular texture synthesis algorithms. In: CVPR, pp. 427–434. IEEE Comp. Soc., Los Alamitos (2006)
15. Nicoll, A., Meseth, J., Müller, G., Klein, R.: Fractional fourier texture masks: Guiding near-regular texture synthesis. Computer Graphics Forum 24(3), 569–579 (2005)
16. Haindl, M., Hatka, M.: A roller - fast sampling-based texture synthesis algorithm. In: Proceedings of the 13th International Conference in Central Europe on Computer Graphics, Visualization and Computer Vision, Plzen, February 2005, pp. 93–96. UNION Agency - Science Press (2005)
17. Haindl, M., Havlíček, V.: A multiscale colour texture model. In: Proceedings of the 16th International Conference on Pattern Recognition, pp. 255–258. IEEE Computer Society, Los Alamitos (2002)
18. Hays, J., Leordeanu, M., Efros, A.A., Liu, Y.: Discovering texture regularity as a higher-order correspondence problem. In: Leonardis, A., Bischof, H., Pinz, A. (eds.) ECCV 2006. LNCS, vol. 3952, pp. 522–535. Springer, Heidelberg (2006)
19. Lin, W.C., Liu, Y.: A lattice-based mrf model for dynamic near-regular texture tracking. IEEE Trans. on Pattern Analysis and Machine Intelligence 29(5), 777–792 (2007)

Texture Editing Using Frequency Swap Strategy

Michal Haindl and Vojtěch Havlíček

Institute of Information Theory and Automation
of the ASCR, Prague, Czech Republic
{haindl,havlicek}@utia.cz

Abstract. A fully automatic colour texture editing method is proposed, which allows to synthesise and enlarge an artificial texture sharing anticipated properties from its parent textures. The edited colour texture maintains its original colour spectrum while its frequency is modified according to one or more target template textures. Edited texture is synthesised using a fast recursive model-based algorithm. The algorithm starts with edited and target colour texture samples decomposed into a multi-resolution grid using the Gaussian-Laplacian pyramid. Each band pass colour factors are independently modelled by their dedicated 3D causal autoregressive random field models (CAR). We estimate an optimal contextual neighbourhood and parameters for each of the CAR submodel. The synthesised multi-resolution Laplacian pyramid of the edited colour texture is replaced by the synthesised template texture Laplacian pyramid. Finally the modified texture pyramid is collapsed into the required fine resolution colour texture. The primary benefit of these multigrid texture editing models is their ability to produce realistic novel textures with required visual properties capable of enhancing realism in various texture application areas.

1 Introduction

Image editing remains a complex user-directed task, often requiring proficiency in design, colour spaces, computer interaction and file management. Editing provides the scene designer with tools which enable to control virtual scene objects, geometric surfaces, illumination and objects faces appearance in the form of their corresponding textures. Image editing software is often characterised [1] by a seemingly endless array of toolbars, filters, transformations and layers. Although some recent attempts [2,3,4,5,6,7,8] have been made to automate this process, automatic integration of user preferences still remains an open problem in the context of texture editing [9,10].

The primary contribution of our method is a simple intuitive and fully automatic tool for the scene designer to modify objects surface appearance by controlled texture modifications. Contrary to some other texture editing approaches such as the procedural textures, the edited texture visual appearance predictably corresponds to the anticipated projection.

Authentic and photo realistic appearance of natural materials covering surfaces of virtual objects in virtual or augmented reality rendered scenes requires nature-like colour textures covering visualised scene objects. Such textures can be either digitised natural textures or textures synthesised from an appropriate mathematical model. The former simplistic option suffers among others from extreme memory requirements for storage of a large number of digitised cross-sectioned slices through different material samples. Synthetic textures are more flexible, extremely compressed (few parameters have to be stored only), they may be evaluated directly in procedural form and can be designed to meet certain constraints or to secure some desirable properties (e.g., smooth periodicity, no visible discontinuities, etc.). The underlying mathematical models have besides presented texture editing also many other applications (e.g., image restoration, image and video compression, classification, segmentation, etc.).

Several monospectral texture modelling approaches were published, e.g., [11,12], among them also few colour models, e.g., [13,14,15,16] and some survey articles are available [17,18] as well. [13] introduced a fast multiresolution Markov random field based method. Although this method avoids the time consuming Markov chain Monte Carlo simulation so typical for applications of Markov models it still requires several simplifying approximations. Several alternative Markovian colour texture models such as the simultaneous 2D causal autoregressive random fields (2D CAR) [16], 2D Gaussian Markov models (2D GMRF) [19], or 3D CAR [20] were introduced as well and later generalised also for Bidirectional Texture Function (BTF) [21,22,23,24] or dynamic textures [25]. These models are appropriate for colour texture synthesis not only because they do not suffer from some problems of alternative options (see [17,18] for details) but they are also easy to analyze as well as to synthesise and last but not least they are still flexible enough to imitate a large set of natural and artificial textures.

2   Markovian Texture Model

We assume to have two colour textures Y_α, Y_δ which can be represented using a Markovian random field model (MRF). The texture Y_α is the input texture which will be modified according to a target template texture Y_δ. The edited colour texture maintains most of its original colour spectrum but changes its frequency to resemble the template texture Y_δ. Single frequency factors are modelled using the exceptionally fast 3D wide-sense Markov causal autoregressive random field model (3D CAR). Let the digitised colour texture Y is indexed on a finite rectangular three-dimensional $N \times M \times d$ underlying lattice I, where $N \times M$ is the image size and d is the number of spectral bands (i.e., $d = 3$ for usual colour textures). Let us denote a simplified multiindices r, s to have two components $r = [r_1, r_2], s = [s_1, s_2]$. The first component is row and the second one is column index, respectively.

2.1   Frequency Factorisation

The analyzed colour texture image is decomposed into a multi-resolution grid using Laplacian pyramid and the intermediary Gaussian pyramid. The benefit of the multigrid approach is the replacement of a large neighbourhood CAR model with a set of several simpler CAR models which are easy to estimate and synthesise. Each resolution data are independently modelled by their dedicated CAR. Each one generates a single spatial frequency band of the texture. The Gaussian pyramid $\ddot{Y}_\nu^{(k)}$ is a sequence of images in which each one is a low-pass down-sampled version of its predecessor where the weighting function (FIR generating kernel) is chosen subject to the following constraints:

$$w_s = \hat{w}_{s_1}\hat{w}_{s_2}, \qquad \sum_i \hat{w}_i = 1 \quad, \qquad \hat{w}_i = \hat{w}_{-i} \quad, \qquad \hat{w}_0 = 2\hat{w}_1 \;\; (\zeta = 1)$$

and $\nu \in \{\alpha, \delta\}$. The solution of the above constraints for the reduction factor 3 $(2\zeta + 1)$ is $\hat{w}_0 = 0.5, \hat{w}_1 = 0.25$ and the FIR equation is now

$$\ddot{Y}_{r,\nu}^{(k)} = \sum_{i,j=-\zeta}^{\zeta} \hat{w}_i \hat{w}_j \ddot{Y}_{2r+(i,j),\nu}^{(k-1)} \; . \tag{1}$$

The Gaussian pyramid for a reduction factor n is

$$\ddot{Y}_{r,\nu}^{(k)} = \downarrow_r^n \left(\ddot{Y}_\nu^{(k-1)} \otimes w \right) \qquad k = 1, 2, \ldots \; , \tag{2}$$

where $\ddot{Y}_\nu^{(0)} = Y_\nu$, $\downarrow^n$ denotes down-sampling with reduction factor n and $\otimes$ is the convolution operation.

The Laplacian pyramid $\dot{Y}_{r,\nu}^{(k)}$ contains band-pass components and provides a good approximation to the Laplacian of the Gaussian kernel. It can be constructed by differencing single Gaussian pyramid layers:

$$\dot{Y}_{r,\nu}^{(k)} = \ddot{Y}_{r,\nu}^{(k)} - \uparrow_r^n \left(\ddot{Y}_\nu^{(k+1)} \right) \qquad k = 0, 1, \ldots \; , \tag{3}$$

where $\uparrow^n$ is the up-sampling with an expanding factor n. Single orthogonal multispectral components are thus decomposed into a multi-resolution grid and each resolution data are independently modelled by their dedicated independent Gaussian noise driven autoregressive random field model as follows.

2.2 3D CAR Texture Model

Single frequency factors are modelled using the causal autoregressive random field (3D CAR) model [20] which is a family of random variables with a joint probability density on the set of all possible realisations Y of the $M \times N \times d$ lattice I, subject to the following condition:

$$p(Y \mid \gamma, \Sigma^{-1}) = \frac{|\Sigma^{-1}|^{\frac{(MN-1)}{2}}}{(2\pi)^{\frac{d(MN-1)}{2}}} \exp\left\{ -\frac{1}{2} tr\{\Sigma^{-1} \begin{pmatrix} -I \\ \gamma^T \end{pmatrix}^T \tilde{V}_{MN-1} \begin{pmatrix} -I \\ \gamma^T \end{pmatrix} \} \right\} \; ,$$

where the following notation is used

$$\tilde{V}_{r-1} = \begin{pmatrix} \tilde{V}_{yy(r-1)} & \tilde{V}_{xy(r-1)}^T \\ \tilde{V}_{xy(r-1)} & \tilde{V}_{xx(r-1)} \end{pmatrix} \; , \qquad\qquad \tilde{V}_{yy(r-1)} = \sum_{k=1}^{r-1} Y_k Y_k^T \; ,$$

$$\tilde{V}_{xy(r-1)} = \sum_{k=1}^{r-1} X_k Y_k^T \; , \qquad\qquad \tilde{V}_{xx(r-1)} = \sum_{k=1}^{r-1} X_k X_k^T \; .$$

The 3D CAR model can be expressed as a stationary causal uncorrelated noise driven 3D autoregressive process:

$$Y_r = \gamma X_r + e_r , \tag{4}$$

where γ is the $d \times d\eta$ parameter matrix $\gamma = [A_1, \ldots, A_\eta]$, $\eta = card(I_r^c)$, I_r^c is a causal neighbourhood, e_r is a Gaussian white noise vector with zero mean and a constant but unknown covariance matrix Σ (estimated by (7)) and X_r is a corresponding vector of Y_{r-s} (design vector).

Parameter Estimation. The selection of an appropriate CAR model support is important to obtain good results in modelling of a given random field. If the contextual neighbourhood is too small it cannot capture all details of the random field. Inclusion of the unnecessary neighbours on the other hand add to the computational burden and can potentially degrade the performance of the model as an additional source of noise. The optimal Bayesian decision rule for minimising the average probability of decision error chooses the maximum posterior probability model, i.e., a model M_i corresponding to $\max_j \{p(M_j|Y^{(r-1)})\}$ where $Y^{(r-1)}$ denotes the known process history $Y^{(r-1)} = \{Y_{r-1}, Y_{r-2}, \ldots, Y_1\}$. The most probable CAR model given past data $Y^{(r-1)}$, the normal-Wishart parameter prior and the uniform model prior is the model M_i for which $i = \arg\max_j \{D_{j(r-1)}\}$

$$D_{j(r-1)} = \frac{d^2\eta}{2} \ln \pi \sum_{i=1}^{d} \left[\ln \Gamma(\frac{\beta(r) - d\eta + d + 2 - i}{2}) - \ln \Gamma(\frac{\beta(0) - d\eta + d + 2 - i}{2}) \right]$$
$$- \frac{d}{2} \ln |V_{xx(r-1)}| - \frac{\beta(r) - d\eta + d + 1}{2} \ln |\lambda_{(r-1)}|$$

where $\beta(r) = \beta(0) + r - 1$, $\beta(0) > 1$, and

$$\lambda_{(r)} = V_{yy(r)} - V_{xy(r)}^T V_{xx(r)}^{-1} V_{xy(r)} . \tag{5}$$

Parameter estimation of a CAR model using the maximum likelihood, the least square or Bayesian methods can be found analytically. The Bayesian parameter estimations of the causal AR model with the normal-Wishart parameter prior which maximise the posterior density are:

$$\hat{\gamma}_{r-1}^T = V_{xx(r-1)}^{-1} V_{xy(r-1)} \tag{6}$$

and

$$\hat{\Sigma}_{r-1} = \frac{\lambda_{(r-1)}}{\beta(r)} , \tag{7}$$

where $V_{uz(r-1)} = \tilde{V}_{uz(r-1)} + V_{uz(0)}$ and matrices $V_{uz(0)}$ are the corresponding matrices from the normal-Wishart parameter prior. The estimates (5), (6),(7) can be also evaluated recursively if necessary.

Model Synthesis. The CAR model synthesis is very simple and a 3D causal CAR random field can be directly generated from the model equation (4) using a multivariate Gaussian generator. Single CAR models synthesise spatial frequency bands of the texture.

2.3 Laplacian Pyramid Swap

The synthesised Laplacian pyramid layers from the target texture target template texture $\dot{Y}_\delta$ are used instead of the corresponding input texture Laplacian pyramid layers $(\dot{Y}_\alpha)$, i.e.

$$\dot{Y}_{r,\alpha}^{(k)} = \dot{Y}_{r,\delta}^{(k)} \qquad \forall k \ . \tag{8}$$

The input texture Y_α Laplacian pyramid layers $(\dot{Y}_\alpha)$ are not needed and their corresponding 3D CAR models are neither estimated nor synthesised. On the contrary, the input Gaussian pyramid $\ddot{Y}_{r,\alpha}^{(k)}$ at the most coarse level contains original texture colour spectrum and is needed (and thus estimated) for the edited texture synthesis. If the Laplacian pyramids of both textures have similar numerical values, then the edited texture colour spectrum is unchanged, otherwise its colour spectrum is a compromise between both textures colour spectra. The edited fine-resolution synthetic colour texture is obtained from the pyramid collapse procedure (inversion process to (2),(3) modified to (8)).

3 Experimental Results

Figs.1,2 show six examples of different natural or man made colour textures edited using the presented algorithm. All original natural colour textures (upper

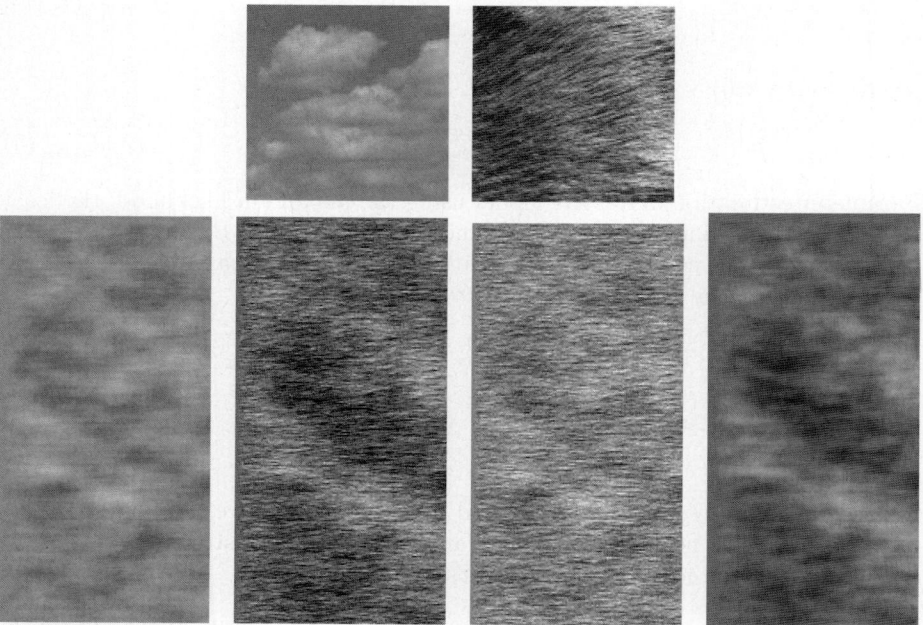

Fig. 1. Natural cloud and fur textures (upper row), their resynthesis using a set of 3D CAR models (bottom left) and their edited counterparts (bottom right)

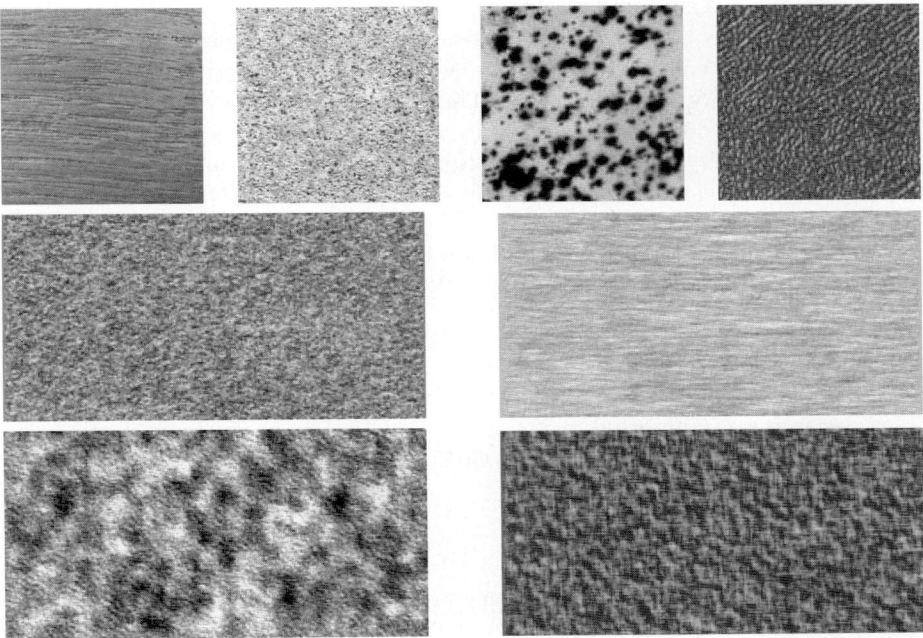

Fig. 2. Wood, tile, lichen, and leather natural textures and their resynthesised edited counterparts using the 3D CAR models (middle and bottom)

rows) are taken either from the VisTex [26] database or from our own extensive colour texture database. The images on Fig.1-bottom left show synthesised enlarged examples of the input textures while the Figs.1-bottom right,2-middle,bottom rows present results from the presented texture editing method with frequency modification using the alternate column texture as the template texture Y_δ with the reduction factor $n = 2$ and the number of pyramid layers $k \in \{2, 3\}$. The edited textures are generated fully automatically and they clearly demonstrate original texture frequency modified to resemble the template texture frequency. The method can be easily combined with some texture segmenter if we need to edit separately single textures appearing in the scene. The method allows very high compression ratio, because only tens parameters for every fractional 3D CAR model have to be stored regardless of the required texture enlargement. This extreme compression ration ($1 : 10^6$ for BTF modelling [21]) is the prerequisite for BTF editing applications where alternative texture editing methods cannot be used due to unsolvable memory requirements.

4 Conclusions

A simple fully automatic colour texture editing method is proposed. The method allows to synthesise and enlarge artificial textures which resemble both their parents textures. The edited texture inherits primarily spectral information from one parent and frequency information from the other one. This procedure can be repeated for more complex lineage trees which allows to inherit visual properties from more than two parent textures. The method allows very high compression ratio for transmission or storing texture information, while sometimes compromises visual quality of the resulting texture, similarly as any other adaptive texture model. The edited texture analysis as well as synthesis is extremely fast (due to complete analytical solution) and can be used in real-time applications. The method can be easily generalised also for other types of textures such as the Bidirectional Texture Function (BTF) or dynamic textures.

Acknowledgements

This research was supported by the grant GAČR 102/08/0593 and partially by the MŠMT grants 1M0572 DAR, 2C06019.

References

1. Brooks, S., Dodgson, N.A.: Integrating procedural textures with replicated image editing. In: Proceedings of the 3rd International Conference on Computer Graphics and Interactive Techniques in Australasia and Southeast Asia 2005, Dunedin, New Zealand, November 29 - December 2, pp. 277–280. ACM, New York (2005)
2. Ashikhmin, M.: Synthesizing natural textures. In: ACM Symposium on Interactive 3D Graphics, pp. 217–226 (2001)
3. Bar-Joseph, Z., El-Yaniv, R., Lischinski, D., Werman, M.: Texture mixing and texture movie synthesis using statistical learning. IEEE Transactions on Visualization and Computer Graphics 7, 120–135 (2001)
4. Liang, L., Liu, C., Xu, Y.Q., Guo, B., Shum, H.Y.: Real-time texture synthesis by patch-based sampling. ACM Transactions on Graphics (TOG) 20, 127–150 (2001)
5. Hertzmann, A., Jacobs, C.E., Oliver, N., Curless, B., Salesin, D.H.: Image analogies. ACM Trans. Graph., 327–340 (2001)
6. Wiens, A.L., Ross, J.: Gentropy: evolving 2d textures. Computers & Graphics 26, 75–88 (2002)
7. Wang, X., Wang, L., Liu, L., Hu, S., Guo, B.: Interactive modeling of tree bark. In: Proc. 11th Pacific Conf. on Comp. Graphics and Appl., pp. 83–90. IEEE, Los Alamitos (2003)
8. Brooks, S., Cardle, M., Dodgson, N.A.: Enhanced texture editing using self similarity. In: VVG, pp. 231–238 (2003)
9. Brooks, S., Dodgson, N.A.: Self-similarity based texture editing. ACM Trans. Graph 21, 653–656 (2002)
10. Khan, E.A., Reinhard, E., Fleming, R.W., Bülthoff, H.H.: Image-based material editing. ACM Trans. Graph 25, 654–663 (2006)
11. Besag, J.: Spatial interaction and the statistical analysis of lattice systems. Journal of the Royal Statistical Society, Series B B-36, 192–236 (1974)
12. Kashyap, R.: Analysis and synthesis of image patterns by spatial interaction models. In: Kanal, L., Rosenfeld, A. (eds.) Progress in Pattern Recognition, vol. 1. North-Holland, Elsevier (1981)

13. Haindl, M., Havlíček, V.: Multiresolution colour texture synthesis. In: Dobrovodský, K. (ed.) Proceedings of the 7th International Workshop on Robotics in Alpe-Adria-Danube Region, Bratislava, ASCO Art, pp. 297–302 (1998)
14. Bennett, J., Khotanzad, A.: Multispectral random field models for synthesis and analysis of color images. IEEE Trans. on Pattern Analysis and Machine Intelligence 20, 327–332 (1998)
15. Bennett, J., Khotanzad, A.: Maximum likelihood estimation methods for multispectral random field image models. IEEE Trans. on Pattern Analysis and Machine Intelligence 21, 537–543 (1999)
16. Haindl, M., Havlíček, V.: A multiresolution causal colour texture model. In: Amin, A., Pudil, P., Ferri, F., Iñesta, J.M. (eds.) SPR 2000 and SSPR 2000. LNCS, vol. 1876, pp. 114–122. Springer, Heidelberg (2000)
17. Haindl, M.: Texture synthesis. CWI Quarterly 4, 305–331 (1991)
18. Haindl, M.: Texture modelling. In: Proceedings of the World Multiconference on Systemics, Cybernetics and Informatics, Orlando, USA, vol. VII, pp. 634–639. International Institute of Informatics and Systemics (2000)
19. Haindl, M., Havlíček, V.: A simple multispectral multiresolution markov texture model. In: Texture 2002, The 2nd international workshop on texture analysis and synthesis, Copenhagen, pp. 63–66. Heriot-Watt University (2003)
20. Haindl, M., Havlíček, V.: A multiscale colour texture model. In: Proceedings of the 16th International Conference on Pattern Recognition, pp. 255–258. IEEE Computer Society, Los Alamitos (2002)
21. Haindl, M., Filip, J.: Fast BTF texture modelling. In: Texture 2003. Proceedings, Edinburgh, pp. 47–52. IEEE Press, Los Alamitos (2003)
22. Haindl, M., Filip, J., Arnold, M.: BTF image space utmost compression and modelling method. In: Proceedings of the 17th IAPR International Conference on Pattern Recognition, vol. III, pp. 194–197. IEEE, Los Alamitos (2004)
23. Haindl, M., Filip, J.: A fast probabilistic bidirectional texture function model. LNCS, pp. 298–305. Springer, Heidelberg (2004)
24. Haindl, M., Filip, J.: Extreme compression and modeling of bidirectional texture function. IEEE Transactions on Pattern Analysis and Machine Intelligence 29, 1859–1865 (2007)
25. Filip, J., Haindl, M., Chetverikov, D.: Fast synthesis of dynamic colour textures. In: Proceedings of the 18th International Conference on Pattern Recognition, ICPR 2006, vol. IV, pp. 25–28. IEEE Computer Society, Los Alamitos (2006)
26. Vision texture (vistex) database. Technical report, Vision and Modeling Group, http://www-white.media.mit.edu/vismod/

A Quantitative Evaluation of Texture Feature Robustness and Interpolation Behaviour[*]

Stefan Thumfart[1], Wolfgang Heidl[1], Josef Scharinger[2], and Christian Eitzinger[1]

[1] Profactor GmbH,
Im Stadtgut A2, 4407 Steyr-Gleink, Austria
stefan.thumfart@profactor.at
http://www.profactor.at

[2] Johannes Kepler University Linz, Department of Computational Perception,
Altenberger Str. 69, 4040 Linz, Austria
http://www.cp.jku.at

Abstract. Whenever an image database has to be organised according to higher level human perceptual properties, a transformation model is needed to bridge the semantic gap between features and the perceptual space. To guide the feature selection process for a transformation model, we investigate the behaviour of 5 texture feature categories.

Using a novel mixed synthesis algorithm we generate textures with a gradual transition between two existing ones, to investigate the feature interpolation behaviour. In addition the features' robustness to minor textural changes is evaluated in a kNN query-by-example experiment.

We compare robustness and interpolation behaviour, showing that Gabor energy map features are outperforming gray level co-occurrence matrix features in terms of linear interpolation quality.

1 Introduction

In addition to typical image processing applications such as fault detection or object background segmentation, texture became an integral part in recent content-based image retrieval (CBIR) systems [1]. Nowadays the field of CBIR must not be seen in the narrow context of e.g. example based image retrieval, but includes all technologies that facilitate the organisation of large digital image archives by their visual content [2].

The essential component of such systems is a distance measure, capable of representing high level image similarity concepts. Depending on the application scenario, it is very likely, that this distance measure cannot be computed directly from the low level feature set. The usage of specific higher level (e.g. perceptual) features partly solves this problem for a limited domain.

---

[*] This work was funded by the EC under grant no. 043157, project SynTex. It reflects only the authors' view.

A general solution can be achieved by transforming the low level feature space into another domain specific similarity space [3]. Within this space the image distance can be computed more easily. Furthermore machine learning enables us to build the transformation model automatically, e.g. based on psychological experiments or user relevance feedback. Regardless of the application domain the choice of the feature set and the related transformation model is essential.

We conducted two experiments to support this decision. The first experiment evaluates feature retrieval accuracy and robustness with respect to minor changes in the image, using a query-by-example image retrieval setup (section 3.2). The second experiment is dedicated to analyse feature changes in case of interpolation between given texture samples, using a novel texture mixing algorithm (section 3.3). Texture and colour features are discussed in section 2.

2 Texture Features

We use *statistical* (gray level co-occurrence matrix), *perceptual* (neighbourhood gray tone difference matrix, Tamura), *signal processing* (Fourier energy, Gabor energy map) and *Colour* features. See [4] for a complete list of all features including the parameter settings.

2.1 Statistical Features

Gray level co-occurrence. In 1973 Haralick et al. first proposed to use a gray level co-occurrence matrix, $GLCM$, to analyse the 2nd order statistical properties of textures [5].

As suggested in [5] we extract 4 GLCMs for a fixed displacement vector length $|d|$ and angle $\theta = \{0°, 45°, 90°, 135°\}$. For each of the resulting GLCMs we compute 11 statistical measures and comprise the final feature vector of length 22 by calculating the mean and range value for each measure. We are using the GLCM features computed for $|d| = \{1, 2, 4, 8\}$ for our experiments ($= 88$ features in total).

2.2 Perceptual Features

Tamura. Tamura et al. proposed 6 perceptual texture measures, namely: *coarseness, contrast, directionality, line-likeness, regularity* and *roughness* [6]. The computation of these features follows no general approach, but aims to achieve high correspondence to human judgements for the given texture properties.

Neighbourhood gray tone difference. In [7] Amadasun and King propose to extract texture features from a vector termed *neighbourhood gray tone difference matrix (NGTDM)*, describing the intensity difference between image pixels and their local neighbourhoods. Amadasun and King defined 5 measures which can be extracted from the NGTDM, all of them following from a visual deduction of perceptual texture properties. We compute these measures for neighbourhood sizes of $d = \{1, 2\}$.

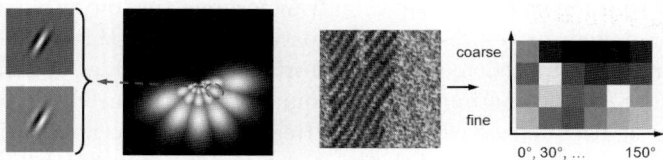

Fig. 1. F.l.t.r. the real and imaginary component of a single Gabor filter in spatial domain, a Gabor filter bank (4 scales, 6 orientations) in Fourier domain, a texture image, the Gabor energy map for the texture image using the given filterbank

2.3 Signal Processing Features

The use of signal processing based texture analysis methods is seconded by findings about the early human visual processing steps [8], [9].

Fourier energy features. The Fourier power spectrum is partitioned into circular rings or wedges to measure the energy present in these segments [1]. We use a partitioning into 12 log-scaled circular rings and 48 wedges. The ring energy distribution can be used to assess the coarseness of a texture, whereas the wedge energy distribution describes its directionality. Based on these energy features we compute several statistical measures such as the wedge energy standard deviation.

Gabor energy map features. Contrary to the Fourier transform the Gabor convolution is spatially localized. A two dimensional Gabor function is a sinusoidal plane wave modulated by a Gaussian envelope. An exhaustive comparison of Gabor energy based features for texture segregation and classification can be found in [10].

According to [11] we compute the Gabor energy map m, using a Gabor filter bank (4 scales, 6 orientations) as depicted in Fig. 1. Based on m Kim et al. use the *sum of Gabor orientation energy difference, SGOED* for object classification. Besides *SGOED*, we implemented several features such as *sum of Gabor scale energy difference, SGSED, maximum of Gabor orientation energy difference, MGOED*, which are modifications of the *SGOED*.

2.4 Colour Features

Colour has been used for image indexing and retrieval for almost two decades [12]. We use 3 measures (*f1, f2* and *f3*) as proposed in [13], to describe the images' *average intensity, colourfulness* and *average saturation*.

Furthermore we partition the hue component of the HSV colour space into 6 equally sized sectors (60° each), to count the relative frequency of image pixels within each sector. The frequencies are weighted by the average image value and saturation. Finally we end up with 6 measures describing e.g. the redness of an image.

3 Experimental Setup

We conduct two experiments to assess the robustness with respect to minor changes of the input image and the feature interpolation behaviour between given texture samples for the feature set outlined in section 2.

3.1 Texture Sample Selection

In total we selected 394 stationary [14] images from texture collections such as *Brodatz* [15], *Outex* [16], *Vistex* [17] and textures collected on our own, available at [4]. To unify the upcoming processing steps we cropped all 394 selected samples to 480 × 480 pixels (i.e. the minimum size of selected texture samples). Increasing the minimum size threshold for selecting texture samples, would have led to the exclusion of the above mentioned popular texture collections, lowering the value of the evaluation results.

3.2 Evaluation of Feature Robustness

We evaluate the robustness of the feature set with respect to minor texture changes, inspired by the *stationarity* criterion.

First we extract 16 patches $P_{i1}, ..., P_{i16}$ of size 128×128 pixels in scanline order for every selected texture T_i. We term a set of patches extracted from the same texture as *patch group* G_i. Given that T_i is stationary, the patches $P_{i1}, ..., P_{i16}$ are similar. If a feature returns instable results for many patch groups, it is likely that this feature is sensitive to minor texture changes. Prior to the actual robustness experiments we have to eliminate all images that are causing high variances for many different features because they are instationary. We use a voting approach to eliminate 81 instationary texture images.

kNN query-by-example results. Using the remaining 313 patch groups, we conduct a query-by-example *kNN classification experiment* to assess the retrieval performance of our feature set. For a random patch P_r the k nearest neigbhours among the other patches are retrieved, using euclidean distance. Based on the group memberships of the k selected candidates, the group for P_r is predicted. Table 1 shows the average retrieval accuracy for 1000 query patches per feature groups. Due to the low number of samples per class (16) the value $k = 3$ gives the best average retrieval accuracy. GLCM features are most robust, followed by Gabor energy map features.

3.3 Evaluation of Feature Changes

Pyramid based Texture Mixing. We use a novel texture mixing method, based on the idea of pixel based texture synthesis. We extend the method proposed in [14] to produce textures which are a weighted mixture of two input samples T_A, T_B. The algorithm of Wei and Levoy synthesizes a new texture of user defined size based on a single input texture. Starting with a random image (initialized with white noise) the resulting texture is synthesized pixel per

Table 1. kNN retrieval accuracy [%] for all feature groups, for 1000 query patches

k	1	3	5	7	10	15
avg. precision	78.0	78.2	75.1	74.1	73.0	70.3
GLCM $d=1$	91.4	90.8	88.5	88.4	86.5	83.5
GLCM $d=2$	91.5	91.9	86.4	89.2	84.2	84.9
GLCM $d=4$	84.1	86.7	82.5	81.4	82.8	77.7
GLCM $d=8$	79.6	78.1	77.1	77.0	75.9	75.5
NGTDM $d=1$	71.7	70.8	67.4	63.8	64.8	59.2
NGTDM $d=2$	71.4	71.3	64.7	65.9	61.5	59.8
Tamura	58.5	60.7	59.0	55.9	58.3	54.4
Fourier energy	73.1	75.9	73.5	72.7	73.2	69.9
Colour	78.3	76.8	75.3	71.9	70.7	66.5
Gabor energy map	80.2	79.0	76.9	74.5	71.8	71.8

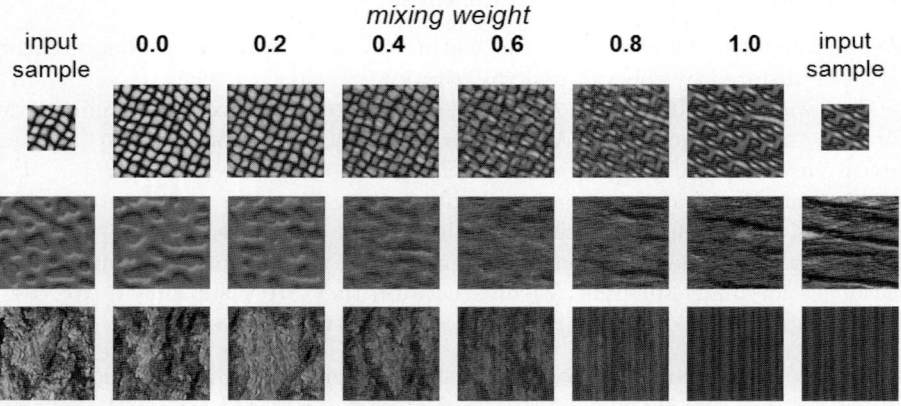

Fig. 2. The outer columns show the original mixing input samples. The 6 columns in the middle show the mixed texture samples for different mixing weights.

pixel in scanline order by comparing the local output pixel neighbourhood to all neighbourhoods of the input sample.

Instead of starting with a white noise image, we randomly select pixel values from T_A and T_B to build the starting image S_{rand}. Next, 2 Gaussian pyramids are built from S_{rand}. Separately, each pyramid is synthesized, using multi-resolution neighbourhood search [14]. By blending the highest resolution pyramid levels we get the mixing results as depicted in Fig. 2.

Texture sample mixing. We generate weighted (mixed) textures using *pyramid based texture mixing* to assess the interpolation behaviour of features between given texture samples. To select representative mixing inputs, we assigned the extracted patches (section 3.2) to 20 clusters, according to their normalized

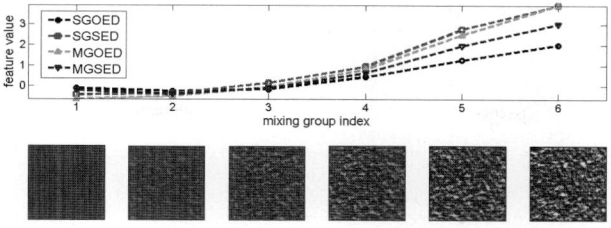

Fig. 3. Feature trajectories of 4 Gabor energy map features computed for a mixing group containing 6 textures

	mixing weights					
feature values	0.0	0.2	0.4	0.6	0.8	1.0
equidistant	-0.45	-0.08	0.30	0.67	1.04	1.415
computed	-0.45	-0.54	-0.37	0.08	0.58	1.415

(a) The feature trajectory values for mixing group 50.

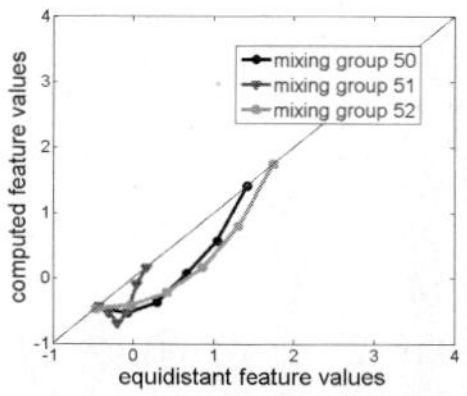

(b) Combined trajectory for mixing groups: 50, 51, 52.

Fig. 4. Building of a combined trajectory for the feature GLCM $d = 1$, mean(sum of squares)

average group variance. From each cluster, 1 patch is chosen randomly. We use the mixing weights $0.0, 0.2, 0.4, 0.6, 0.8$ and 1.0 to generate 6 textures (a *mixing group*) for each of the $\binom{20}{2} = 190$ possible combinations.

Experiment. We compute the features for all mixing groups to obtain 190 discrete trajectories of each feature for the transition from T_A to T_B. Fig. 3 shows the feature trajectory of 4 selected Gabor energy map features.

As the feature range is different for each mixing group, we shall first discuss how to compose a combined trajectory c_i for feature f_i.

For every mixing group j, we compute the equidistant feature values for each single trajectory $t_{i,j}$. Finally we get 190 matrices of size 2×6 per feature, with the first row storing the equidistant feature values and the second row containing the computed feature values (see Table 4(a)).

We superimpose all trajectories $t_{i,j}$, with $j \in \{1, ..., 190\}$ by plotting the equidistant values along the x-axis and the computed feature values along the y-axis. See Fig. 4(b) for a combined trajectory built from 3 mixing groups.

Results and Conclusions. The selection of a useful measure to assess the quality of the combined trajectories depends on the subsequently used model type (e.g. of a transformation model to bridge the semantic gap between low level features and human perception). For simplicity we assume a linear model

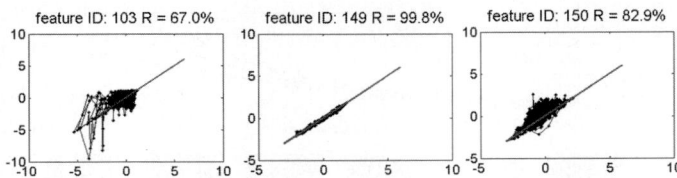

Fig. 5. Combined trajectories for *Tamura-regularity, ID: 103, Colour-average intensity, ID: 149* and *Colour-colourfulness, ID: 150*. The values of feature 149 are located close to the 1st median line. Therefore this feature is highly correlated to the equidistant feature steps. Feature 103 received the lowest correlation results. Feature 150 is the one which performed best in the retrieval experiment.

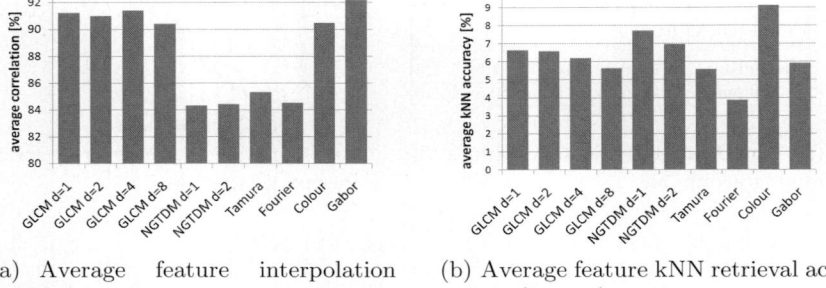

(a) Average feature interpolation quality.

(b) Average feature kNN retrieval accuracy ($k = 30$).

Fig. 6. Comparison of linear interpolation behaviour and *kNN* retrieval accuracy for all feature categories

(e.g. linear regression) and use Pearson's linear correlation coefficient to describe the linear interpolation quality of single features, based on combined trajectories as depicted in Fig. 5.

Fig. 6 shows the results of the *kNN* feature robustness and the interpolation behaviour experiment in terms of retrieval accuracy and linear correlation. Both diagrams contain the results averaged for all members of a separate feature category. Note that the results presented in Table 1 contain the retrieval accuracy obtained for an n dimensional feature space, with n representing the number of features per category.

As shown in Fig. 6(a), the perceptual features (*NGTDM, Tamura*) received low correlation results, indicating that they are not suited as input for a linear model. Also the Fourier features are outperformed by Gabor filters, both in terms of robustness and linear interpolation quality. We consider it interesting, that the robustness and the linear interpolation quality seem not to be correlated, as NGTDM features are among the robust ones, but received a low correlation score.

As expected, GLCM features gave a good retrieval accuracy. Surprisingly they were among the best in terms of linear interpolation quality, even though they mainly describe higher level statistical properties. Colour features performed well in both experiments, but could not outperform Gabor and GLCM features in the second experiment.

References

1. Sebe, N., Lew, M.S.: Texture features for content-based retrieval. Principles of Visual Information Retrieval, 51–85 (2001)
2. Datta, R., Joshi, D., Li, J., Wang, J.Z.: Image retrieval: Ideas, influences, and trends of the new age. ACM Comput. Surv. 40(2), 1–60 (2008)
3. Reyes, C., Durán, M.L., Alonso, T., Rodríguez, P.G., Caro, A.: Behaviour of texture features in a cbir system. In: Corchado, E., Abraham, A., Pedrycz, W. (eds.) HAIS 2008. LNCS (LNAI), vol. 5271, pp. 425–432. Springer, Heidelberg (2008)
4. Syntex: official web page, Section 'Public Material, CAIP 2009 (2009), http://www.syntex.or.at
5. Haralick, R.M., Shanmugam, K., Dinstein, I.: Textural features for image classification. IEEE Trans. Syst. Man Cybern. 3(6), 610–621 (1973)
6. Tamura, H., Mori, S., Yamawaki, T.: Textural features corresponding to visual perception. IEEE Trans. Syst. Man Cybern. 8, 460–473 (1978)
7. Amadasun, M., King, R.: Textural features corresponding to textural properties. IEEE Trans. Syst. Man Cybern. 19, 1246–1274 (1989)
8. Campbell, F.W., Robson, J.G.: Application of fourier analysis to the visibility of gratings. Journal of Physiology, 551–566 (1968)
9. Daugman, J.G.: Two-dimensional spectral analysis of cortical receptive field profiles. Vis. Res. 20, 847–856 (1980)
10. Grigorescu, S.E., Petkov, N., Kruizinga, P.: Comparison of texture features based on gabor filters. IEEE Trans. on Image Process. 11, 1160–1167 (2002)
11. Kim, M., Park, C., Koo, K.: Natural / man-made object classification based on gabor characteristics. In: Leow, W.-K., Lew, M., Chua, T.-S., Ma, W.-Y., Chaisorn, L., Bakker, E.M. (eds.) CIVR 2005. LNCS, vol. 3568, pp. 550–559. Springer, Heidelberg (2005)
12. Swain, M.J., Ballard, D.H.: Color indexing. Int. J. Comput. Vision 7(1), 11–32 (1991)
13. Datta, R., Joshi, D., Li, J., Wang, J.Z.: Studying aesthetics in photographic images using a computational approach. In: Leonardis, A., Bischof, H., Pinz, A. (eds.) ECCV 2006. LNCS, vol. 3953, pp. 288–301. Springer, Heidelberg (2006)
14. Wei, L.Y.: Texture Synthesis by Fixed Neighborhood Searching. PhD thesis, Stanford University (2001)
15. Brodatz, P.: A Photographic Album for Artists and Designers. Dover Publications, New York (1966)
16. Ojala, T., Mäenpää, T., Pietikäinen, M., Viertola, J., Kyllönen, J., Huovinen, S.: Outex - New framework for empirical evaluation of texture analysis algorithms. In: Proc. of ICPR 2002, vol. 1, pp. 701–706 (2002)
17. MIT Media Laboratory Cambridge: Vistex - Vision Texture Database (1995), http://vismod.media.mit.edu/pub/VisTex/

Nonlinear Dimension Reduction and Visualization of Labeled Data

Kerstin Bunte[1], Barbara Hammer[2], and Michael Biehl[1]

[1] University of Groningen, Mathematics and Computing Science,
9700 AK Groningen, The Netherlands
[2] Clausthal University of Technology, Institute of Computer Science,
D-38678 Clausthal-Zellerfeld, Germany

Abstract. The amount of electronic information as well as the size and dimensionality of data sets have increased tremendously. Consequently, dimension reduction and visualization techniques have become increasingly popular in recent years. Dimension reduction is typically connected with loss of information. In supervised classification problems, class labels can be used to minimize the loss of information concerning the specific task. The aim is to preserve and potentially enhance the discrimination of classes in lower dimensions. Here we propose a prototype-based local relevance learning scheme, that results in an efficient nonlinear discriminative dimension reduction of labeled data sets. The method is introduced and discussed in terms of artificial and real world data sets.

1 Intoduction

Dimension reduction techniques aim at finding a smaller set of features by reducing or eliminating redundancies. From a theoretical point of view the "curse of dimensionality" causes many difficulties in high-dimensional spaces, such that dimension reduction constitutes a valuable tool to deal with these problems [1].

In the last decades an enormous number of unsupervised dimension reduction methods has been proposed. In general, unsupervised dimension reduction is an ill-posed problem since a clear specification which properties of the data should be preserved, is missing. Standard criteria, for instance the distance measure employed for neighborhood assignment, may turn out unsuitable for a given data set, and relevant information often depends on the situation at hand.

If data labeling is available, the aim of dimension reduction can be defined clearly: the preservation of the classification accuracy in a reduced feature space. Supervised linear dimension reducers are for example the Generalized Matrix Learning Vector Quantization (GMLVQ) [2] and the Linear Discriminant Analysis (LDA) [3]. Often, however, the classes cannot be separated by a linear classifier while a nonlinear data projection better preserves the relevant information. Examples for nonlinear discriminative visualization techniques include, an extension of the Self Organizing Map (SOM) incorporating class labels [4]. Further supervised dimension reduction techniques are explained in [5,6].

In this contribution we propose a discriminative visualization scheme which is based on an extension of Learning Vector Quantization and relevance learning.

2 Supervised Nonlinear Dimension Reduction

For general data sets a global linear reduction to lower dimensions may not be powerful enough to preserve the information relevant for classification. In [1] it is argued that the combination of several local linear projections to a nonlinear mapping can yield promising results. We use this concept and learn local linear low-dimensional projections from labelled data. Alternatively to the direct usage of the local linear patches it is also possible to merge them into a global nonlinear embedding with a charting technique to obtain a smoother nonlinear projection. The following subsection gives a short overview over the algorithms.

Localized LiRaM LVQ. Learning vector quantization (LVQ) [7] constitutes a successful class of heuristic, prototype based classification algorithms. LVQ is intuitive, interpretable, fast, and easy to implement. It is distance based and a key issue is the selection of a suitable dissimilarity measure. However, the most frequent choice, i.e. standard Euclidean distance, is not necessarily suitable. Therefore, relevance learning schemes have been suggested which adapt more general metrics in the training process [8,9]. Recent extensions parameterize the distance measure in terms of a relevance matrix, the rank of which may be controlled explicitly. The algorithm suggested in [2] can be employed for linear dimension reduction and visualization of labeled data. The local linear version presented here provides the ability to learn local low-dimensional projections and combine them into a nonlinear global embedding. We consider training data $\boldsymbol{x}_i \in \mathbb{R}^N$, $i = 1 \ldots S$ with labels y_i corresponding to one of C classes respectively. A data point $\boldsymbol{x}_i$ is assigned to the class of the closest prototype $\boldsymbol{w}_j$ with $d(\boldsymbol{x}_i, \boldsymbol{w}_j)^{\Lambda_j} \leq d(\boldsymbol{x}_i, \boldsymbol{w}_k)^{\Lambda_k}$ for all $j \neq k$. During the training process LVQ adapts l prototypes $\boldsymbol{w}_j \in \mathbb{R}^N$ with class labels $c(\boldsymbol{w}_j) \in \{1, \ldots, C\}$ to represent the classification as accurately as possible. Generalized LVQ (GLVQ) [10] adapts prototypes by minimizing the cost function

$$E = \sum_{i=1}^{S} \Phi \left(\frac{d^{\Lambda_J}(\boldsymbol{w}_J, \boldsymbol{x}_i) - d^{\Lambda_K}(\boldsymbol{w}_K, \boldsymbol{x}_i)}{d^{\Lambda_J}(\boldsymbol{w}_J, \boldsymbol{x}_i) + d^{\Lambda_K}(\boldsymbol{w}_K, \boldsymbol{x}_i)} \right) , \qquad (1)$$

where $\boldsymbol{w}_J$ ($\boldsymbol{w}_K$) denotes the closest prototype with the same (a different) class label as $\boldsymbol{x}_i$ and Φ refers to a monotonic function, e. g. the logistic function or the identity, which is used in our experiments. Learning can take place by means of a stochastic gradient descent of the cost function E (Eq. (1) for details see [2]).

The localized generalized matrix LVQ (LGMLVQ) substitutes the squared Euclidean distance by a more complex dissimilarity measure which can take into account arbitrary pairwise correlation of features. This metric

$$d^{\Lambda_j}(\boldsymbol{w}_j, \boldsymbol{x}_i) = (\boldsymbol{x}_i - \boldsymbol{w}_j)^\top \Lambda_j (\boldsymbol{x}_i - \boldsymbol{w}_j) \qquad (2)$$

is defined through an adaptive symmetric and positive semi-definite matrix $\Lambda_j \in \mathbb{R}^{N \times N}$ locally attached to each prototype $\boldsymbol{w}_j$. By setting $\Lambda_j = \Omega_j^\top \Omega_j$ semi-definiteness and symmetry is guaranteed. $\Omega_j \in \mathbb{R}^{M \times N}$ with arbitrary $M \leq N$ transforms the data locally to an M-dimensional feature space. It can be shown that the adaptive distance $d^{\Lambda_j}(\boldsymbol{w}_j, \boldsymbol{x}_i)$ Eq. (2) equals the squared Euclidean distance in the transformed space $d^{\Lambda_j}(\boldsymbol{w}_j, \boldsymbol{x}_i) = [\Omega_j(\boldsymbol{x}_i - \boldsymbol{w}_j)]^2$. The target dimension M must be chosen in advance by intrinsic dimension estimation or suitable for the given task. For visualization purposes, usually a value of two or three is appropriate. We will refer to this algorithm as Limited Rank Matrix LVQ (LiRaM LVQ). After each training epoch (sweep through the training set) matrices are normalized to $\sum_i [\Lambda_j]_{ii} = 1$ in order to prevent degeneration. An additional regularization term in the cost function proportional to $-\ln(\det(\Omega_j \Omega_j^\top))$ can be used to enforce full rank M of the relevance matrices and prevent oversimplification effects, see [11]. At the end of the learning process the algorithm provides a set of prototypes $\boldsymbol{w}_j$, their labels $c(\boldsymbol{w}_j)$, and corresponding projections Ω_j. A low dimensional embedding of each data point $\boldsymbol{x}_i$ can then be defined by $P_j(\boldsymbol{x}_i) = \Omega_j \boldsymbol{x}_i$ using the projection Ω_j of its closest prototype $\boldsymbol{w}_j$, with $d^{\Lambda_j}(\boldsymbol{w}_j, \boldsymbol{x}_i) = \min_k d^{\Lambda_k}(\boldsymbol{w}_k, \boldsymbol{x}_i)$. For smoother visualizations the outcome of the classifier can also be mapped with a charting step.

Charting. The charting technique introduced in [12] provides a frame for unsupervised dimension reduction by decomposing the sample data into locally linear patches and combine them into a single low-dimensional coordinate system. For nonlinear dimension reduction we use the low-dimensional local linear projections $P_j(\boldsymbol{x}_i) \in \mathbb{R}^M$ for every data point $\boldsymbol{x}_i$ provided by localized LiRaM LVQ and apply only the second step of the charting method to combine them. The local projections $P_j(\boldsymbol{x}_i)$ are weighted by their responsibilities r_{ji} for data point $\boldsymbol{x}_i$. Here we choose the responsibilities

$$r_{ji} \propto \exp(-(\boldsymbol{x}_i - \boldsymbol{w}_j)^\top \Lambda_j (\boldsymbol{x}_i - \boldsymbol{w}_j)/\sigma_j) \ , \qquad (3)$$

with normalization $\sum_j r_{ji} = 1$ and an appropriate bandwith $\sigma_j > 0$. We set σ_j to a fraction of the Euclidean distance to the nearest projected prototype

$$\sigma_j = a \cdot \min_{k \neq j} [\Omega_j \boldsymbol{w}_j - \Omega_k \boldsymbol{w}_k]^2 \text{ with } 0 < a \leq 0.5 \ . \qquad (4)$$

The charting technique finds affine transformations $B_j : \mathbb{R}^M \to \mathbb{R}^M$ of the local coordinates P_j, such that the resulting points coincide on overlapping parts as much as possible in a least squares sense. An analytical solution can be found in terms of a generalized eigenvalue problem, which leads to a global embedding in $\mathbb{R}^M$. We refer to [12] for further details.

3 Unsupervised Nonlinear Dimension Reduction

We will compare this locally linear discriminative projection technique with some well-known unsupervised projection techniques which are based on different projection criteria.

Isomap. [13] is an extension of the metric Multi-Dimensional Scaling (MDS) and uses distance preservation as criterion for the dimension reduction. Whereas metric MDS frequently employs the Euclidean metric to compute this pairwise distances, Isomap incorporates the so called graph distances as an approximation of the geodesic distances. The weighted neighborhood graph is constructed by connecting points i and j if their distance is smaller than ϵ (ϵ-Isomap), or if i is one of the K nearest neighbors of j (K-Isomap). Isomap is guaranteed to find the global optimum of its error function in closed form. The approximation of the geodesic distances may be very rough and its quality depends on the number of data points, the noise and the parameters (ϵ or K). For details see [13]. For quantitative analysis we additionally compare the results of L-Isomap [14], which focuses on a small subset of the data, called the landmark points.

Locally Linear Embedding. (LLE) [15] uses the topology preservation criterion for dimension reduction. LLE aims at the preservation of local angles. The first step of the LLE algorithm is the determination of a number of neighbors for each data point, either by choosing the K nearest neighbors or all neighbors inside an ϵ-ball around the point. The idea is to reconstruct each point by a linear combination of its neighbors and to project data points such that this local representation of the data is preserved as much as possible. Advantages of this method are the elegant theoretical foundation which allows an analytical solution. From the computational points of view LLE requires the solution of an S-by-S eigenproblem with S being the number of data points. As reported in [16], the parameters must be tuned carefully, see [15] for further details.

Stochastic Neighbor Embedding. (SNE) [17] is closely related to Isotop [18]. It overcomes some limitations of the Self Organizing Maps (SOM) by separating the vector quantization and the dimensionality reduction in two steps. SNE is a variant, which follows a probabilistic approach to map high-dimensional data vectors into a low-dimensional space, while preserving the neighbor identities. Like Isotop it centers a Gaussian kernel on each data point to be embedded. The algorithm optimizes the approximation of a probability distribution over all potential neighbors if the same operation is performed on the low-dimensional representation of the data point. The minimization of the objective function is difficult and may stuck in local minima. Details can be found in [17]. In the quantitative analysis we additionally compare the results of the t-Distributed Stochastic Neighbor Embedding (t-SNE) [19], which uses a Studen-t distribution rather than a Gaussian and a different cost function.

4 Experiments

In this section we will compare the described dimension reduction techniques on two different data sets: an artificial data set and the segmentation data set from the UCI repository [20]. For visual comparison we reduce the dimension in both cases to two.

3 Tip Star. This dataset consists of 3000 samples in $\mathbb{R}^{10}$ with two classes (C1 and C2) arranged on three clusters respectively (see Fig. 1 top left). The first two dimensions contain the information whereas the remaining eight dimensions contribute high variance noise. Localized LiRaM LVQ was trained for $t = 500$

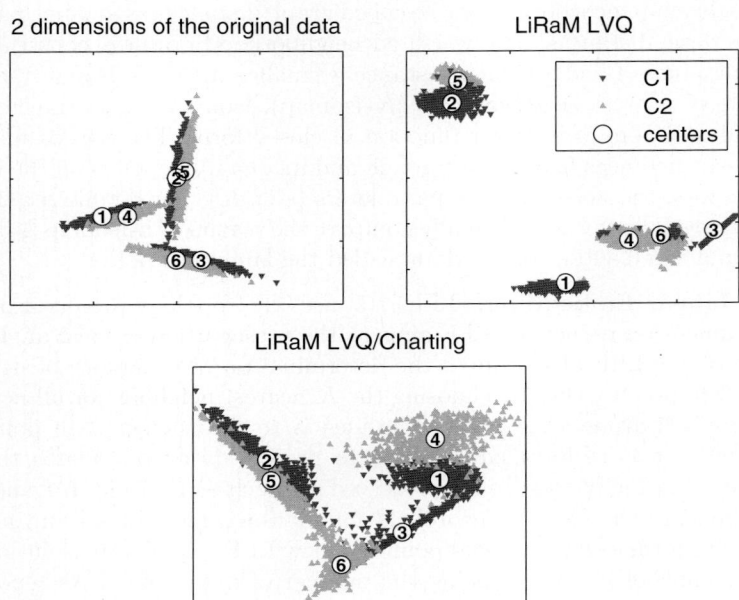

Fig. 1. Upper left: two informational dimensions of the original 3 Tip Star data set. Upper right: projection with LiRaM LVQ. Bottom: nonlinear projection based on the same LiRaM LVQ projections from the upper right figure combined with charting.

epochs, with three prototypes per class. Each of the prototype was initialized close to one of the cluster centers. The learning rate of the prototypes is set to $\alpha_1(t) = 0.01/(1 + (t-1) \cdot 0.001)$ and the metric learning starts at epoch $t = 50$ with a learning rate of $\alpha_2(t) = 0.001/(1 + (t-50) \cdot 0.0001)$. We run the localized LiRaM LVQ 10 times and one result of the local projected data is shown in Fig. 1 top right. Note that the aim of the LiRaM LVQ algorithm is not to preserve any topology or distances, but to find projections, which separate the classes as much as possible. So cluster four and six merge, because they carry the same class label. Nevertheless the different orientations and appearances of all six clusters are still visible. The bottom visualization in Fig. 1 shows the combination of the local projections shown in the top right after the charting step. Where the parameter a for σ_j (Eq. (4)) to fix the responsibilities for the local projection P_j is set to 0.4 (found by cross validation with values between [0.1 0.5]). The invariances inherited from the local linear projections of the LiRaM LVQ algorithm and the eigenvalue problem in the charting step lead to a flipped version of the original data, where cluster six and three are separated vertically but not horizontally. Fig. 2 shows the results of other dimension reduction methods on this data set. Principal Component Analysis (PCA) leads to very similar results like MDS in this problem. The classes are not well separated in two of the three modes. The other three figures show the results for SNE, and Isomap and LLE with

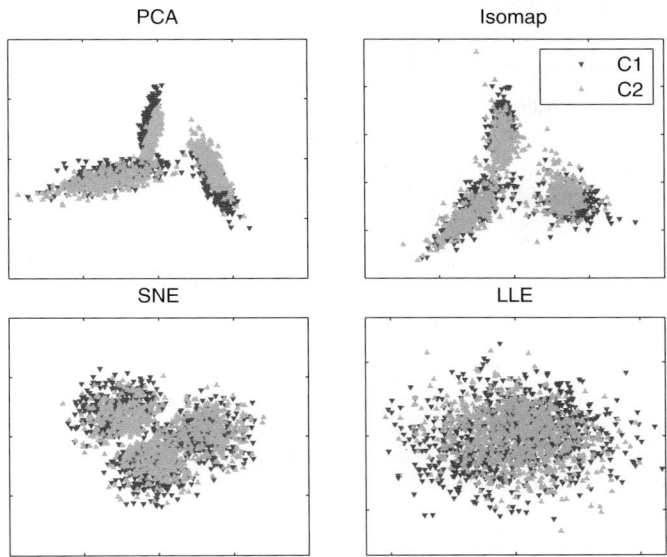

Fig. 2. Unsupervised projections of the 3 Tip Star data set from various methods

$K = 35$ neighbors each. Obviously, hardly any class structure is preserved in these projections. Note that, due to the presence of only two classes, standard linear discriminance analysis (LDA) yield a projection to one dimension only. Table 1 shows the Nearest Neighbor (NN) error on the projected data of the unsupervised methods and the mean NN error of the LVQ based projected data averaged over all 10 runs. The NN error of the LiRaM LVQ mapping in Fig. 1 is 0.06 and 0.09 with the charting step. We also tried kernel PCA with gaussian kernel and 9 different equidistant variances σ from the interval [1,5], and L-Isomap and t-SNE. The best results are shown in table 1.

Segmentation. The segmentation data set (available at the UCI repository [20]) consists of 19 features which have been constructed from regions of 3×3 pixels, randomly drawn from a set of 7 manually segmented outdoor images. Every sample is assigned to one of seven classes: brickface, sky, foliage, cement, window, path and grass (referred to as C1, ..., C7). The set consists of 210 training points with 30 instances per class and the test set comprises 300 instances per class, resulting in 2310 samples in total. We did not use the features 3,4 and 5, because they display zero variance over the data set. We use the same parameter

Table 1. Nearest neighbor errors on the mapped 3 Tip Star data set

LiRaM LVQ	0.12			kernel PCA (gauss kernel $\sigma = 4.5$)	0.41
PCA	0.29	Isomap	0.41	L-Isomap (20% landmarks, $K = 35$)	0.39
SNE	0.46	LLE	0.50	t-SNE (perplexity 30)	0.41

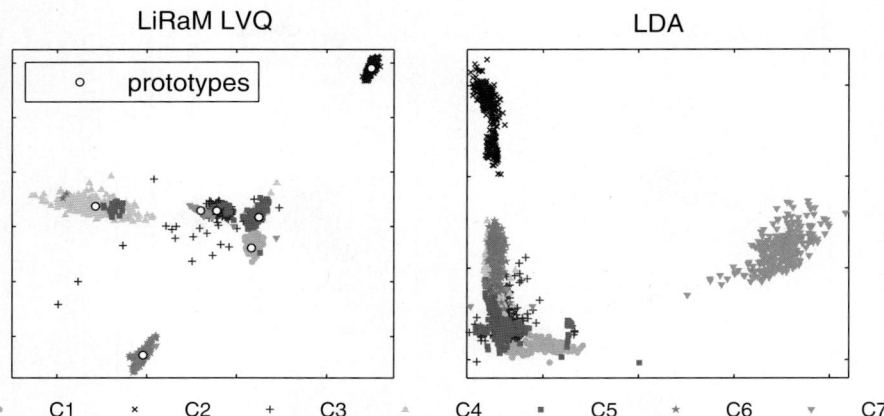

Fig. 3. Left: nonlinear supervised two-dimensional projection of the segmentation data set with LiRaM LVQ. Right: supervised two-dimensional projection of the same data with LDA.

settings as specified in the previous section. We set the parameter K of neighbors for Isomap and LLE to 108, according to the connectivity in the neighborhood graph. One example result for the localized LiRaM LVQ with a Nearest Neighbor error of 0.07 is shown in the top left panel in Fig. 3. For this seven class problem a supervised dimension reduction with LDA is also possible. The top right panel shows the result of dimension reduction with LDA. In particular, the classes C4 and C6 appear to be well separated in the LVQ based approach, whereas they fall together in LDA. We observed that Generalized Discriminant Analysis (GDA) [21] using gaussian kernels with 19 equidistant variances in the interval [1,10] or polynomial kernels with powers three to 10 and addition values between zero and 10 were a good deal worse than LDA (see table 2). Fig. 4 shows the results of the other dimension reduction techniques. Again PCA and MDS lead to nearly identical results only isolating one class: C2. For Isomap and LLE, even using the huge number of neighbors $K = 108$, unsatisfactory results are obtained. SNE yields to the best result compared to other unsupervised techniques, but some classes scatter in a circle around the zoomed area showed here.

For a quantitative analysis of the results obtained by the differend methods, we compute the leave-one-out estimate of the Nearest Neighbor (NN) classification error on the mapped segmentation data. The NN error of the localized LiRaM LVQ mapping is averaged over 10 random initializations of the algorithm. Additionally we evaluate the GDA, t-SNE and L-Isomap and list their best results together with the NN errors of all methods in table 2. t-SNE, GDA, PCA and LLE show the worst results with errors between 84% and 33%, followed by Isomap and L-Isomap with 27% and 25%. The supervised method LDA performs also not satisfactory with ca. 20%. SNE and localized LiRaM LVQ achieve the best mean error results with 11% and 9% respectively.

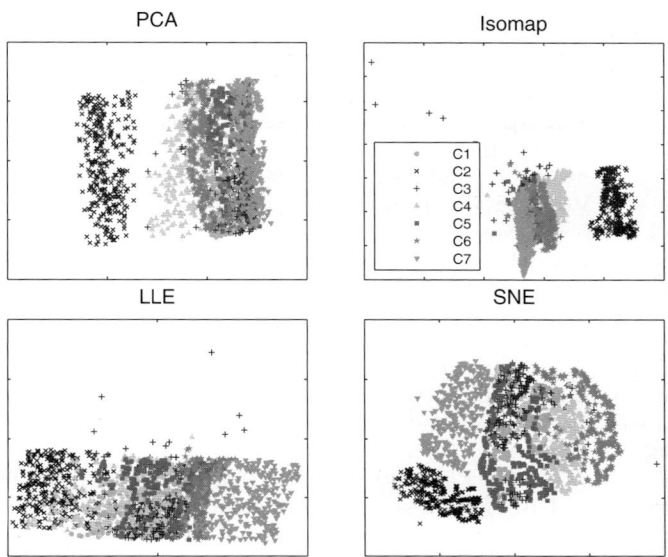

Fig. 4. Unsupervised projections of the segmentation data set from various methods. To see the structure of most samples the Isomap and SNE figures are zoomed, some samples spread widely.

Table 2. Nearest neighbor errors on the mapped segmentation data set

LiRaM LVQ	0.09	LDA	0.20	GDA (polynomial, pow. 3, offset 6)	0.70
Isomap	0.27	PCA	0.31	t-SNE (perplexity 40)	0.84
SNE	0.11	LLE	0.33	L-Isomap (20% landmarks, $K = 108$)	0.25

5 Conclusion

We proposed a supervised discriminative nonlinear dimension reduction technique based on a prototype-based classifier with adaptive distances and charting. Compared to other state-of-the-art methods it shows promising results in two examples. Unlike LDA this method provides a nonlinear embedding of the data. Its complexity is linear in the number of examples, which is an advantage especially in comparison with methods based on the construction of a neighborhood graph. The combination with a prototype based learning scheme additionally offers the possibility of data compression by embedding the prototypes. This is especially interesting for the processing of huge data sets. For the localized LiRaM LVQ combined with the charting we observe a small but non-negligible loss of classification accuracy which is due to the charting step. We will address the optimization of the latter in a forthcoming project.

Acknowledgment. This work was supported by the "Nederlandse organisatie voor Wetenschappelijke Onderzoek (NWO)" under project code 612.066.620.

References

1. Van der Maaten, L.J.P., Postma, E.O., Van den Herik, H.J.: Dimensionality Reduction: A Comparative Review (2007),
   http://ticc.uvt.nl/$\sim$lvdrmaaten/Laurens_van_der_Maaten/Matlab_Toolbox_for_Dimensionality_Reduction_files/Paper.pdf
2. Bunte, K., Schneider, P., Hammer, B., Schleif, F.-M., Villmann, T., Biehl, M.: Discriminative Visualization by Limited Rank Matrix Learning. Machine Learning Reports 2, 37–51 (2008),
   http://www.uni-leipzig.de/~compint/mlr/mlr_03_2008.pdf
3. Fukunaga, K.: Introduction to Statistical Pattern Recognition. Academic Press, New York (1990)
4. Villmann, T., Hammer, B., Schleif, F.M., Geweniger, T., Hermann, W.: Fuzzy classification by fuzzy labeled neural gas. Neural Networks 19(6-7), 772–779 (2006)
5. Kontkanen, P., Lahtinen, J., Myllymäki, P., Silander, T., Tirri, H.: Supervised model-based visualization of high-dimensional data. Intell. Data Anal. 4(3,4), 213–227 (2000)
6. Iwata, T., Saito, K., Ueda, N., Stromsten, S., Griffiths, T.L., Tenenbaum, J.B.: Parametric Embedding for Class Visualization. Neural Comp. 19(9), 2536–2556 (2007)
7. Kohonen, T.: Self-Organizing Maps, 2nd edn. Springer, Heidelberg (1997)
8. Hammer, B., Villmann, T.: Generalized relevance learning vector quantization. Neural Networks 15(8-9), 1059–1068 (2002)
9. Schneider, P., Biehl, M., Hammer, B.: Relevance Matrices in LVQ. In: Proc. of European Symposium on Artificial Neural Networks (ESANN), pp. 37–42 (2007)
10. Sato, A.S., Yamada, K.: Generalized learning vector quantization. In: NIPS, vol. 8, pp. 423–429 (1996)
11. Schneider, P., Bunte, K., Hammer, B., Villmann, T., Biehl, M.: Regularization in matrix relevance learning. Machine Learning Reports 2, 19–36 (2008), http://www.uni-leipzig.de/~compint/mlr/mlr_02_2008.pdf
12. Brand, M.: Charting a manifold. In: NIPS, vol. 15, pp. 961–968 (2003)
13. Tenenbaum, J.B., de Silva, V., Langford, J.C.: A global geometric framework for nonlinear dimensionality reduction. Science 290(5500), 2319–2323 (2000)
14. De Silva, V., Tenebaum, J.B.: Global versus local methods in nonlinear dimensionality reduction. In: Advances in Neural Information Processing System, pp. 705–712. MIT Press, Cambridge (2002)
15. Roweis, S.T., Saul, L.K.: Nonlinear Dimensionality Reduction by Locally Linear Embedding. Science 290(5500), 2323–2326 (2000)
16. Saul, L.K., Roweis, S.T.: Think globally, fit locally: unsupervised learning of nonlinear manifolds. Journal of Machine Learning Research 4, 119–155 (2003)
17. Hinton, G., Roweis, S.T.: Stochastic neighbor embedding. In: Advances in Neural Information Processing Systems, vol. 15, pp. 833–840 (2003)
18. Lee, J.A., Archambeau, C., Verleysen, M.: Locally linear embedding versus Isotop. In: 11th European Symposium on Artificial Neural Networks, pp. 527–534 (2003)
19. Van der Maaten, L.J.P., Hinton, G.E.: Visualizing High-Dimensional Data Using t-SNE. Journal of Machine Learning Research 9, 2579–2605 (2008)
20. Newman, D.J., Hettich, S., Blake, C.L., Merz, C.J.: UCI Repository of machine learning databases. University of California, Department of Information and Computer Science (1998), http://archive.ics.uci.edu/ml/ (last visit 19.04.2008)
21. Baudat, G., Anouar, F.: Generalized Discriminant Analysis Using a Kernel Approach. Neural Computation 12(10), 2385–2404 (2000)

Performance Evaluation of Airport Lighting Using Mobile Camera Techniques

Shyama Prosad Chowdhury, Karen McMenemy, and Jian-Xun Peng

School of Electronics, Electrical Engineering and Computer Science
Queen's University Belfast, UK
schowdhury01@qub.ac.uk, {k.mcmenemy,j.peng}@ee.qub.ac.uk

Abstract. This paper describes the use of mobile camera technology to assess the performance of Aerodrome Ground Lighting (AGL). Cameras are placed inside the cockpit of an aircraft and used to record images of the AGL during an approach to an airport. Subsequent image analysis, using the techniques proposed in this paper, will allow a performance metric to be determined for the lighting. This can be used to inform regulators if the AGL is performing to standards and it will also provide useful information towards the maintenance strategy for the airport. Since the cameras that are used to collect the images are mounted on a moving and vibrating platform (the plane), some image data may be effected by vibration. In the paper we illustrate techniques by which to quantify and remove the effects of vibration and illustrate how the image data can be used to derive a performance metric for the complete AGL.

Keywords: Photometrics, vibration, luminaire assessment.

1 Introduction

Aerodrome Ground Lighting (AGL) (Fig. 1(a)) is used to guide a pilot towards a runway for safe landing. It is important that the AGL is operating according to standards set by aviation governing bodies. One of those standards indicate that the AGL pattern should appear uniform to the pilot, that is, luminaires should exhibit similar performance. Other standards indicate luminaires should have a given color, projection and luminous intensity. However, because of the operating conditions for these lighting systems it is difficult to maintain the individual luminaires and check they are operating as intended. As yet no device exists which can monitor the performance of the complete AGL pattern [1]. Airports typically implement routine block change of luminaires to ensure they are operating as required, which is an expensive maintenance strategy.

In this paper, we propose camera based techniques to assess the overall performance of the AGL in terms of uniformity only. Future research will look at addressing the issues of color and luminous intensity assessment.

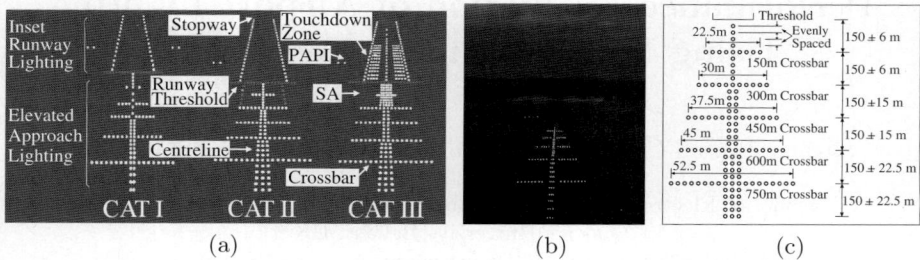

Fig. 1. (a) CAT I/II/III Lighting pattern layout (b) CATII Lighting system from aircraft (c) CATII template with physical details

2 Related Research

There are three categories of lighting patterns which are used in different visibilities with CAT I typically being used for reasonable visibility and CAT III for low visibility range. Each category of lighting consists of runway and approach luminaires. According to the International Civil Aviation Organisation standards [2], runway luminaires are inserted into the ground whilst the approach luminaries are placed at different elevation heights (0 to 15m) above the ground. To date, only ground based photometric assessment methods have been developed and as such, they are only suitable for the runway luminaires. A review of these techniques is available in [3] and [4].

Research at Queen's University Belfast has concentrated on developing assessment methods for the complete AGL pattern[5]. The developed system consists of a camera which is placed inside the cockpit of an aircraft and used to take images of the complete AGL pattern. Model based tracking [1] is used to identify luminaires in the collected images. One of the limitations of this work however is that no account is taken of image quality and more specifically the effect that vibration will have on the quality of the image data and subsequently the accuracy of the assessment. In addition, most of this work concentrated on extracting and tracking luminaires from the image data with limited work on determining a performance metric for the AGL.

3 Luminaire Performance Metric

Figure 1 (b) shows an image taken using the assessment system during an approach to an airport using a CAT II lighting pattern. This approach was undertaken during clear daylight conditions, with filters placed on the camera to prevent image saturation. One of the limitations of this system, is that the assessment should only be performed during clear daylight conditions. For a CAT II lighting pattern, the physical layout of the system is given in Fig. 1(c).

To determine a performance metric for the AGL, it is necessary to use the information presented in the image sequence [6]. Thus the pixel grey level per luminaire must be used to determine such a metric. In this research we propose using a comparison based approach in order to determine a performance metric for the complete pattern in terms of uniformity. It is known [7] that when using a camera to take a picture of a light source, the pixel grey level information recovered from the image is dependent on its luminous intensity and the distance between each. As such, for luminaires of similar luminous intensity, which are of similar or comparable distance away from the camera at the instant an image was taken, it is possible to use the pixel grey level information in order to compare their performance [8]. Using this theory, it is also possible to output a comparative performance metric per luminaire.

In order to do this accurately it is necessary for the image data to have a high quality. However, very often during an approach to the airport, the aircraft will suffer from vibration which will affect the quality of the acquired data [9]. It is necessary to determine how this vibration will distort the pixel grey level information recovered for each luminaire and therefore what impact it will have on the determination of a performance metric.

The pixel grey level information recovered for each luminaire will be used in order to estimate a performance metric for that luminaire [10]. Since any given luminaire will appear in a number of images (approximately 50-100 images for a camera acquiring images at 12 frames per second), it is necessary to uniquely identify a luminaire in the image it first appears and then track this luminaire through subsequent images. To achieve this we used a technique developed by Niblock et.al [5] which uses a template matching approach to identify and track the luminaires.

3.1 Three Layer Topography Model (TLTM)

Using this position of each luminaire, output by the tracking algorithm, it is possible to extract the pixel grey level information for this luminaire. For each luminaire it is necessary to decide which pixels belong to that luminaire and should be used for the performance estimate. Previous techniques to do this have used a single threshold where all pixels above a given value are assumed to belong to that luminaire. This single threshold technique can lead to considerable errors especially for those luminaires which only cover a small number of pixels within the image. Rather a TLTM was implemented as a more intuitive technique to determine the pixels which represent a given luminaire. In this TLTM, three thresholds are used, namely the $SeedPointThreshold$ (SPT), $MoveUpwardsThreshold$ (MUT) and $MinimumIntensityThreshold$ (MIT), where $SPT \geq MUT \geq MIT$. OTSU algorithm [11] finds out a suitable threshold to classify all pixel values into two classes. The recursive use of the OTSU algorithm allows us to generate the three required threshold values. Essentially TLTM is a controlled region growing algorithm where a new region starts for a point having intensity of at least SPT and then there is a recursive search of all neighborhood pixels. Once the grey level value of a pixel falls below the MUT, then the TLTM will prevent any successive pixels with a higher value being included in the region. It is more likely that these high value pixels belong to the next luminaire. No pixels having a value less than MIT will be included in any luminaire region. Figure 2(a) illustrates the output from the TLTM. The grey level value of any pixels, in a given column, which are assigned to a given luminaire, are summed.

In summary, the input to the TLTM algorithm is a two dimensional image function, F_q where q represents the image number within the video sequence. Thus, $F_q(i,j)$ represents the grey level on a pixel in i^{th} row and j^{th} column in image number q. The output from the TLTM algorithm is a function T which will identify all assigned pixels, where $T(i,j,p,q) = 1$ if the pixel at location (i,j), in the image F_q, belongs to a luminaire p, where p refers to the unique number used to represent each luminaire in a pattern. Otherwise the value will be zero.

3.2 Measurement of the Streak Vector of a Luminaire (SVL)

When vibration of the camera effects the acquired image data, the luminaire images have a number of prominent features. Rather than appear as a circular object in the image data, the luminaires develop a 'streak'. This is evident in figure 2(b). In addition, all streaks normally have the same direction and approximately the same length for a given amount of vibration. Thus it is necessary to determine both the direction and length of each streak, namely the SVL, in order to remove the effect vibration will have on the pixel grey level distribution profile of a luminaire.

For each luminaire p in each image F_q, we applied principal component analysis (PCA) on all the pixels (i,j) where $T(i,j,p,q) = 1$. In order to calculate a four element covariance matrix m, define m_{rc} as the rth row and c^{th} column element. Then $m_{11} = \eta_{20}$, $m_{12} = m_{21} = \eta_{11}$ and $m_{22} = \eta_{02}$, such that

$$\eta_{gh} = \sum_{\forall(i,j)} T(i,j,p,q)F_q(i,j)(i-\gamma_{10})^g(j-\gamma_{01})^h/\alpha, \quad g,h = 0,1,2 \ . \tag{1}$$

Here $\alpha = \sum_{\forall(i,j)} T(i,j,p,q)Fq(i,j)$ and product $T(i,j,p,q)Fq(i,j)$ gives the pixel value of the pixel (i,j) in image frame q if it belongs to the luminaire p. Let us denote γ_{01} and γ_{10} in its generic form γ_{de} where,

$$\gamma_{de} = \sum_{\forall(i,j)} T(i,j,p,q)F_q(i,j)i^d j^e/\alpha, \quad d,e = 0,1 \ . \tag{2}$$

Eigenvalues of the matrix m are calculated as e_1 and e_2. For the luminaire p on the image q, $EVec(p,q)$ denotes the eigenvector corresponding to the principal component or the largest eigenvalue me ($me = MAX(e_1, e_2)$) for a given luminaire in a given image. The direction of the streak is the same as directed by the vector $EVec(p,q)$ corresponding to me, whereas the length of streak, $EVal(p,q)$, is a function of me. $EVal(p,q)$ can be calculated after the SVL directional field is normalized, that is, when the streak is orientated to take a vertical direction. See Fig. 2(c) for an example of SVL-directional field.

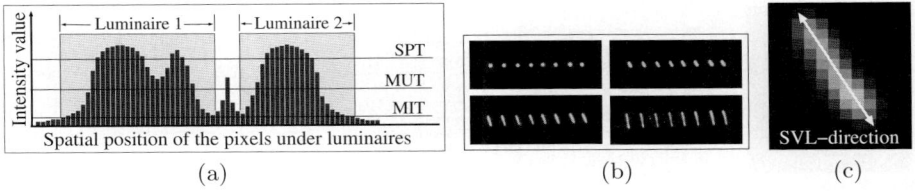

Fig. 2. (a) Three Layer Topography Model TLTM (b) Effect of different amount of vibration on luminaires in increasing order (c) SVL directional field

3.3 SVL-Directional Field Normalization and Recalculation of SVL-Intensity Field

Using the deviation angle between the vertical direction and the direction of vector $EVec(p,q)$, the luminaire image is rotated so that the corresponding $EVec(p,q)$ is parallel to the vertical axis (Fig. 3(a)). Let us denote the rotated p^{th} luminaire in q^{th} image frame by F_{pq}^R. For F_{pq}^R, let $\overline{F}$ denote the pixel grey level value and $\overline{T}$ denote the existence of the luminaire in a given pixel. $\overline{T}(i,j,p,q) = 1$ means pixel (i,j) is assigned to the luminaire p in image frame q of the rotated luminaire image F_{pq}^R. As a result of this rotation operation, the SVL-directional field will be normalized and the streak will occur in the vertical direction. PCA is again applied on this rotated image. The SVL-intensity field or length of the streak, $EVal(p,q)$ is then assigned by the principal component (Fig. 3(a)).

3.4 Peak of the Projection of the Pixel Grey Level (P3GL)

SVL-directional field normalization determines the vertical streak of the luminaire. After normalization, the pixel grey levels per luminaire per column can be summed to produce a normal-like distribution. The maximum among the column-wise sum of the grey level values is given as the P3GL. A visual P3GL is shown in figure 3(b). The peak P3GL for luminaire p, in the image frame q is given by $P3GL_{pq}$ and it is calculated as

$$P3GL_{pq} = MAX((\sum_i \overline{T}(i,j,p,q)\overline{F}_{pq}(i,j)), \forall j) \ . \tag{3}$$

3.5 Normalizing P3GL (nP3GL) Using SVL-Intensity Field

Fig. 3(c) illustrates the relation of P3GL and SVL-intensity field (length of the streak) per image for a given luminaire. For each of these images, the actual luminous intensity of the luminaire was constant. The diagram therefore indicates that the SVL intensity field is directly related to the vibration factor and is also directly proportional to the P3GL. For example, some images are not affected by vibration and the SVL-intensity field is small. Since we know the actual luminous intensity of the luminaire remains constant we would not expect such a

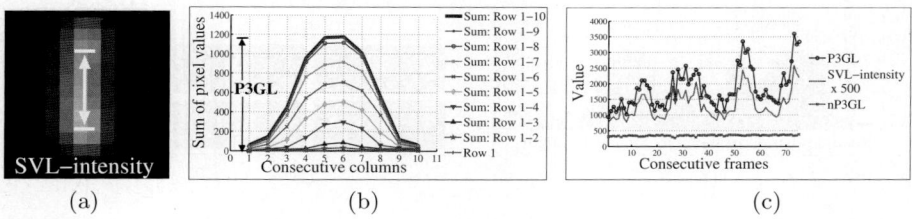

Fig. 3. (a) SVL intensity field on SVL directional field normalized rotated image (b)Consecutive vertical projection of the pixel grey values (c)Relation of P3GL, SVL-intensity field and nP3GL (Since the SVL-intensity field is small compared to the P3GL value, it has been multiplied by a factor of 500 for display purposes only)

large variation. Thus vibration, when present, clearly increases the value of the P3GL. Thus to remove the effect of vibration from the P3GL measurement, it is divided by the magnitude of the SVL-intensity field. This results in a normalized P3GL value or nP3GL.

4 Results from Luminaire Performance Comparison

In any given AGL system, it is realistic to expect that a number of the luminaires within the pattern will not operate as expected due to complete luminaire failure, or dust/dirt on the lens or a wrong orientation setting. By comparing the performance of luminaires with the pattern, it is possible to detect luminaires which are under-performing in relation to other luminaires.

Currently the nP3GL value recorded per luminaire per image is dependent of the distance variation between the camera and the luminaire under test. However this displacement is an important factor since the pixel information recorded for an image is dependent on this displacement [1]. Fig. 4(a) shows the variation of nP3GL recorded for each luminaire against the distance this measurement was taken. The distance is known because it is calculated as part of the tracking algorithm [5]. It can be seen that nP3GL decreases non-linearly with respect to the distance. A best fit least squares parabolic curve (LSP) was determined using the data set to show the average change of nP3GL with distance. A simplistic approach to determining a performance metric is to use this expected change.

For example, each luminaire will have a number of nP3GL measurements, the value of which will vary depending on the distance between the luminaire and the camera. Assuming that a luminaire p produces total of t_p nP3GL values during the complete approach (e.g. $nP3GL_1^p$ to $nP3GL_{t_p}^p$). The function $DT(p,v)$ is used to store the distance between camera and luminaire p for each of the vth instances ($1 \leq v \leq t_p$) of nP3GL. Let us denote the minimum and maximum value of nP3GL for a luminaire p as $MX(p)$ and $MN(p)$. Then the ratio of the fluctuation between both can be represented by $FLC(p)$ where, $FLC(p) = (MX(p) - MN(p)) / MN(p)$. For a luminaire p, the nP3GL values vary around the LSP. At any instance v, luminaire p produces a deviation DV calculated as

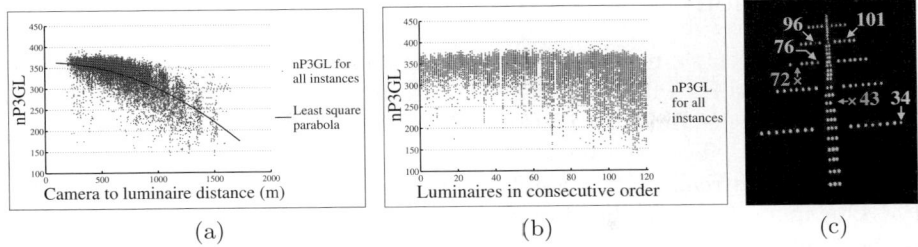

Fig. 4. (a)nP3GL versus distance measurement (b)Plotting of different nP3GL values consecutively for all the luminaires (c)Marked under performing luminaires in a non vibrating image

$$DV(p,v) = nP3GL_v^p - LSP(DT(p,v)) \qquad (4)$$

where, $LSP(d)$ is the value of LSP at distance d. Let us define GT_p and LS_p for a luminaire p which represents the squared sum of the deviations.

$$GT_p = \sum_{v=1}^{t_p} (DV(p,v))^2 \mid DV(p,v) > 0 \qquad (5)$$

$$LS_p = \sum_{v=1}^{t_p} (DV(p,v))^2 \mid DV(p,v) < 0 \ . \qquad (6)$$

Using GT_p and LS_p, a fluctuation normalized ratio FNR_p can be determined;

$$FNR_p = \frac{GT_p}{LS_p} (FLC(p))^\psi, \ \psi = \begin{cases} -2, & GT_p > LS_p \\ 0, & GT_p = LS_p \\ 2, & \text{otherwise} \ . \end{cases} \qquad (7)$$

Thus, FNR_p gives us a measure of how a luminaire's performance deviates with respect to the average performance of luminaires within the pattern.

For any luminaire p, FNR_p describes the comparative estimation of a luminaires performance. If one luminaire is under performing (less bright) in comparison to others then its FNR_p value will be less than 1. Fig. 4(c) shows the under-performing luminaires with a pointer and absent luminaires with a cross and pointer. Using this simplistic technique it is possible to identify luminaires which do not perform as well as others. This technique does not give a definitive measure of performance for a given luminaire in terms of luminous intensity. However, it would be possible for the airports to check the flagged luminaires. If the luminaires pass the airports test, then of course all remaining luminaires in the pattern will pass. Future work will look at the use of coupling this technique will definitive luminous intensity estimation using a pre-calibrated camera.

For this assessment, no information is available regarding how the pilot sensed missing or less bright luminaires. However, in future experiments this will be recorded and compared to the findings from the mobile assessment system.

5 Conclusion

This paper presents a new and robust methodology which compares the performance of luminaires within an AGL pattern using image data captured from a vibrating high speed moving platform. It has been shown that when vibration of the platform effects the image data, the luminaires in the image data will appear with a 'streak'. It is possible to remove the effect of vibration from the P3GL by quantifying the direction and length of the streak. Precise photometric calibration of the camera and accurate estimation of the surrounding visibility factor are quite challenging. Thus the measure of exact luminous intensity of a luminaire is not considered in this current research. Rather, we use comparative measurement to rank the performance of luminaires within an AGL pattern. Thus mobile camera technology can be used to provide a robust and fast technique by which to determine the performance of the AGL which airports can utilize to target their maintenance strategy.

Acknowledgements. The authors would like to thank the EPSRC (Grant: EP/D05902X/1) for financial backing. The contribution of Flight Precision and Belfast International Airport for providing flight time in order to collect airport lighting data is also gratefully acknowledged.

References

1. Niblock, J.H., Peng, J.X., McMenemy, K.R., Irwin, G.W.: Autonomous Model-based Object Identification and Camera Position Estimation with Application to Airport Lighting Quality Control. In: 3rd International Conference on Computer Vision Theory and Applications, VISAPP, Funchal (2008)
2. International Civil Aviation Organization: Aerodrome Design and Operations, vol.1, 4th edn. Annex 14 (2004)
3. TMS Photometrics, http://www.tmsphotometrics.com
4. FB Technology, http://www.fbtechnology.com
5. Niblock, J.H., McMenemy, K.R., Irwin, G.W.: Autonomous Tracking System for Airport Lighting Quality Control. In: 2nd International Conference on Computer Vision Theory and Applications, pp. 317–324. VISAPP, Barcelona (2007)
6. Morimoto, T., Kiriyama, O., Harada, Y., Adachi, H., Koide, T., Mattausch, H.J.: Object Tracking in Video Pictures Based on Image Segmentation and Pattern Matching. In: IEEE Int. Symp. on Circuits and Systems, pp. 3215–3218 (2005)
7. Inanici, M.N., Navvab, M.: The Virtual Lighting Laboratory: Per-pixel Luminaire Data Analysis. Leukos 3(2), 89–104 (2006)
8. Chen, S.Y., Zhang, J., Zhang, H., Wang, W., Li, Y.F.: Active Illumination for Robot Vision. In: IEEE International Conference on Robotics and Automation, Roma, pp. 411–416 (2007)
9. Matsushita, Y., Ofek, E., Ge, W., Tang, X., Shum, H.Y.: Full-frame Video Stabilization with Motion Inpainting. IEEE PAMI 28(7), 1150–1163 (2006)
10. Lewin, I., OFarrell, J.: Luminaire Photometry using Video Camera. Journal of the Illuminating Engineering Society 28(1), 57–63 (1999)
11. Otsu, N.: A Threshold Selection Method from Gray-level Histogram. IEEE Trans. SMC 9(1), 62–66 (1979)

Intelligent Video Surveillance for Detecting Snow and Ice Coverage on Electrical Insulators of Power Transmission Lines

Irene Y.H. Gu[1], Unai Sistiaga[1], Sonja M. Berlijn[2], and Anders Fahlström[3]

[1] Dept. of Signals and Systems, Chalmers Univ. of Technology, Sweden
[2] Statnett, Box 5192, Majorstuen, 0302, Oslo, Norway
[3] STRI AB, Box 707, 77180, Ludvika, Sweden
irenegu@chalmers.se, sonja.berlijn@statnett.no, anders.fahlstrom@stri.se

Abstract. One of the problems for electrical power delivery through power lines in northern countries is when snow or ice accumulates on electrical insulators. This could lead to snow or ice-induced outages and voltage collapse, causing huge economic loss. This paper proposes a novel real-time intelligent surveillance and image analysis system for detecting and estimating the snow and ice coverage on electric insulators using images captured from an outdoor 420 kV power transmission line. In addition, the swing angle of insulators is estimated, as large swing angles due to wind cause short circuits. Hybrid techniques by combining histogram, edges, boundaries and cross-correlations are employed for handling a broad range of scenarios caused by changing weather and lighting conditions. Experiments have been conducted on the captured images over several month periods. Results have shown that the proposed system has provided valuable estimation results. For image pixels related to snows on the insulator, the current system has yielded an average detection rate of 93% for good quality images, and 67.6% for images containing large amount of poor quality ones, and the corresponding average false alarm ranges from 9% to 18.1%. Further improvement may be achieved by using video-based analysis and improved camera settings.

Keywords: electric insulator surveillance, snow detection, ice detection, swing angle, insulator image analysis.

1 Introduction

Northern countries, e.g. Scandinavian, north Canada, Russia and China often encounter snow and ice during cold winter or in high areas. One of the problems for electrical power delivery through power lines is when snow or ice accumulates on electrical insulators. When the accumulated snow melts and freezes or in case of freezing rain, long ice bars hanging down along the edge of insulators could be formed. Also, the coverage of snow on insulators could be thick. When the ice or snow melts, a conducting layer is formed on the insulator or on the outside of the ice, and short circuit or flashover may occur. This may lead to ice-induced outages and voltage collapse, causing huge economic loss for the power company and the related users. For example, Norwegian power companies have observed ice-induced outages especially during 1987 and 1993. In Sweden, it has recently caused a number of large blackouts. In Canada it has led to large problems and sever blackout in the end of 1990's. Because snow and ice related outages happen during severe weather conditions, very little information and knowledge are available about the process of ice and snow accretion finally leading to flash over. Further, when upgrading power lines, ice performance is one of the important aspects of the insulation selection process. A better understanding of insulator's sensitivities to snow and ice would be useful to help improving future design. In Norwegian environment, classical freezing rain rarely occurs. Hence, it is unlikely that rain would lead to outages in their networks. However, it is assumed that accretions from wet snow (possibly in combination with heavy rime icing) that occur regularly could result in ice accretions with electrical properties similar to those of freezing rain.

So far, there is no benchmark method for solving this problem. Efforts so far for finding insulator's snow coverage have been relied heavily on weather predictions and human observations. Other efforts for insulators include, e.g. finding contamination of sea salts [?]. Our efforts in this investigation include arranging surveillance cameras along remotely located power lines, transferring captured videos through the Internet to the utilities or power companies followed by automatic analyzing the situations. Our aim is to automatic monitor and detect possible snow/ice accretions on electrical insulators. Once the snow or ice are detected, automatic estimation of the percentage of snow/ice coverage related to the distance between two neighboring shells of an insulator are then performed. The analysis results can be fed to network operators if the snow/ice coverage reaches a risk level, and necessary intervention can then be taken before short circuits occur. We proposes a full automatic image analysis system for detection and analysis of snow/ice coverage on electrical insulators of power lines using images captured by visual cameras in a remote outdoor laboratory test bed [1,2]. *To the best of our knowledge, this is the first successful insulators snow and ice surveillance system that is entirely based on automatic image analysis.* It is worth mentioning that as a by-product, such results may also provide power system experts with a better understanding of snow/ice bridging phenomena hence possible improvement in future insulator design [2].

2 Settings of Web-Based Surveillance

The measurements are performed for insulators on a 420 kV power transmission line that is set for remote outdoor tests. Basic components installed in the system include: (a) visual cameras and lamps; (b) a weather station; (c) a communication system between the remote test site and network operator; (d) a web-based database; (e) a real-time automatic image analysis system. Measurements (including videos) are done in the remote site (where 230 V supply is *not* available) in a fix time interval (10 minutes) and during severe conditions

Fig. 1. Insulators used for our tests. From the left to right: type (a), (b), (c).

such as low temperatures and darkness, where the power is sufficient for heating the lamps and the cameras. Weather and wind conditions are recorded. The communication system is established between the remote site and the power network operator, where measured data are transferred from the test site to the database and stored, which are accessible via a web interface immediately after the measurement. Meanwhile, real-time image analysis is performed for each newly captured image and the results are displayed and stored in the system. Other data, such as voltage, power, current, reports and results of analysis, can be added to the database. More details on system settings can be found in [1,2].

As shown in Fig. 1, three different types of insulators are monitored in our outdoor tests. type (a): composite insulator (diameter d=168 and 129 mm) consists of 11 large and 33 small shells. The insulator length (include corona rings placed above and below) is 2.12 meters. type (b): desert type insulator (d=420 and 280 mm) of 2 meters long, consisting of 6 large diameter shells and 7 small diameter shells. type (c): coastal type insulator (d=330 mm) of 2 meters long, consisting of 13 shells, with a corona ring placed below.

3   Complex Outdoor Image Scenarios

Despite rigid insulators and stationary cameras, automatic detection and analysis of snow coverage is non-trivial. The lack of electrical power supply (230v) in the remote area also puts a constraint on improving the insulators' visibility where natural daylight is usually short ($\approx$ 5 hours) in the winter. Since images were captured from natural scenes, not only lighting conditions and background may change abruptly (e.g. sunshine, cloudy, foggy, drizzle, raining and snow; moving clouds, unexpected moving objects e.g. airplanes or birds within camera views), but also cameras are often slightly moved due to strong wind (causing insulator positions in images drifting with time), see example scenarios in Fig. 2. Among them, some events that significantly impact the image analysis are:

- strong wind: may cause camera movement hence the insulator position in the image may drift;
- dark weather: may lead to low visibility or low contrast in images. This not only includes images at nights, but also during dark morning and afternoon times;

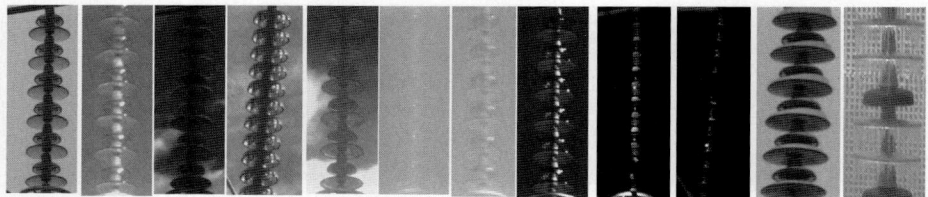

Fig. 2. Scenarios of outdoor captured insulators (only insulator areas are shown). Left to right: (a) sunny with clear sky; (b) insulator with reflections from sunlight; (c)-(e) clouds form non-uniform background and may change fast; (f)-(g) blurred images due to the fog; (h)-(j) dark and night images; (k) snow on shells; (l) ice on shells.

- cloudy weather: may lead to non-uniform fast changing background;
- foggy weather: may lead to low visibility / severely blurred images;
- dark night: images may vary significantly, depending on the snow, reflection of lighting and camera incident angle.
- strong sun: depending on the incident angle of camera, images may contain bright regions due to the reflection from the insulator.

4 Registration for Extracting ROI

To limit the computation in insulator image analysis, a small region containing the insulator (or, the region of interest - ROI) is extracted. Observing insulator's positions in images drift with time mostly due to minor camera movement but also from the swing of insulator, image registration is required for extracting ROI. Since the image size of insulator remains a constant, it is used as a priori information. Separate processing methods are applied to daytime and nighttime images due to significantly different nature of these images.

Nighttime images. Under the lighting condition of current system setting, only the central axis of insulator images is most visible (see Fig. 3(a)). Hence, a histogram-based accumulation method is proposed. Since histograms from night images contain a narrow sharp peak, a binary image B is generated by thresholding the histogram. To determine the ROI, vertical and horizontal accumulations are performed respectively by $a_v(i) = \sum_{j=1}^{N} B(i,j)$, $a_h(j) = \sum_{i=1}^{M} B(i,j)$. The vertical accumulated curve $a_v(i)$ usually shows one narrow peak (see Fig. 3(c)) related to the central axis of insulator. This peak position is assigned to the x-coordinate center of the ROI. The horizontal accumulated curve $a_h(j)$ usually shows two large peaks (see Fig. 3(d)), corresponding to the top and bottom frame where an insulator is fixed. The valley region between the two peaks in $a_h(j)$ is related to the insulator, the center position in this valley is hence assigned as the y-coordinate center of the ROI. The width and height for the ROI are then assigned according the pre-determined values (fixed constants for each type of insulator in respect to its camera setting). This results in the ROI (see Fig. 3(e)).

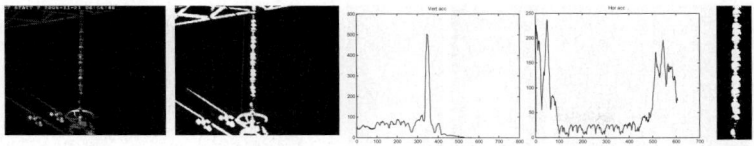

Fig. 3. Extract a ROI from a large image. Left to right: (a) night image; (b) binary image after histogram thresholding; (c) vertical accumulations a_v; (d) horizontal accumulations a_h; (e) extracted ROI.

Daytime images. Daytime images refer to all images captured during the daytime, including dark images captured during mornings and afternoons. Extracting the ROI from daytime images is more difficult due to a broad range of possible complex scenarios. Individual image analysis methods (e.g. segmentation [4], corner detection, histograms) often require tuning parameters and do not work well for a broad range of scenarios. We propose to use a priori information (a pre-stored template) and cross-correlations. The template E_T, containing broadened outer boundaries (width $w = 7$ in our tests) of insulator from an ideal image, is stored beforehand. For each new image frame, a binary edge image E is created from a simple edge detector. The following normalized cross-correlations are then computed,

$$\rho(u,v) = \frac{\sum_{x,y} E(x-u, y-v) E_T(x,y)}{\sqrt{\left(\sum_{x,y} E(x,y)^2 \sum_{x,y} E_T(x,y)^2\right)}} \quad (1)$$

where $u, v \in R_E$ are the lags for cross-correlation, the range of u, v is within the size of image E. The reason of using broadened boundaries in the template is to avoid sensitivity in the correlation when edges and boundaries from two images are slightly shifted. The best position is found by $(u^*, v^*) = \text{argmax}_{u,v} \rho(u,v)$. The extracted ROI is then further refined by applying horizontal and vertical accumulations (in the similar way as for the nighttime images).

Tests of these methods over 3 months of images have resulted in about 88.5% of success for extracting ROIs.

5 A Hybrid Method for Snow Detection

Once a ROI image is successfully extracted, detection of snow and subsequently analysis of snow (or, ice) coverage is performed. Since the emphasis is on the fully automatic analysis of captured images with a large dynamic range and different stochastic natures, many robust image analysis methods (e.g. mean shift [4], graph cuts [5]) requiring fine parameter tuning are not suitable. Noting that snow (or, ice) scenarios can vary significantly (see Fig. 4(a)-(e)), we exploit the following a priori information to achieve robustness:

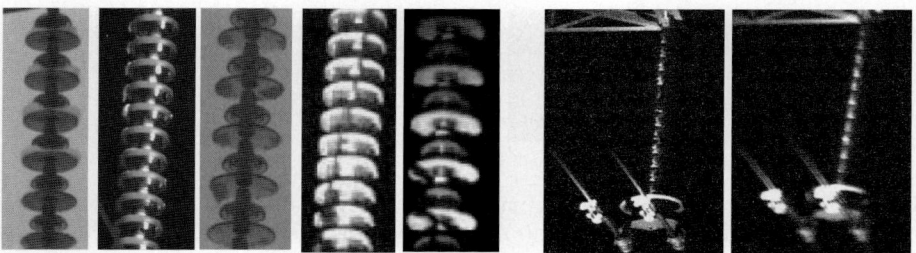

Fig. 4. Variety of scenarios of: snow on insulator and swing of insulator. From the left to right: (a)(b) snow; (c) melting snow; (d)(e) Rim frost; (f): night image of insulator with almost no wind. (g) night image of insulator with a relatively large swing (measured wind speed 10.4m/sec).

- An insulator is a rigid object, its size and shell edges and outer boundaries are fixed and known. These may be changed if an insulator is covered by something, e.g. snow or ice.
- There exists intensity difference between the snow, insulator or background.
- The intensity differences between snow, shells (or, background) cause *extra* edge curves on the top half of insulator shells.

Further, based on observations we assume that snow is only accumulated on the top or along the side of insulator's shells.

5.1 Detect and Analyze Snow Regions

To determine the snow (or ice) regions from images with a range scenarios, a joint boundary and insulator analysis scheme is employed as follows.

(a) Detect extra regions. Observing that snow may generate extra image edges and regions, an edge detector is first applied to the median filtered ROI image, followed by edge closing. The median filter is used for obtaining a ROI image with a smoother background, hence less edge noise. Each enclosed area surrounded by edge curves forms a region.

(b) Find extra regions above the shells. Snow on insulator shells, and other changes (e.g. local clouds, illuminations, reflections) could generate new extra regions. Using the prior information of standard shell positions and the 'ellipse' shell shape regions as the reference, these extra regions (including split regions) can be found and require further analysis. Since snow/ice is more likely to accumulate on the top and/or side part of insulator shells, only extra regions related to these locations are considered and analyzed.

(c) Tighten the width of ROI. To further limit the areas, a ROI is narrowed down by tightening the width determined by two *parallel lines* touching the outer sides of shells (using extremal left/right points of shells). These extremal points are detected either from the outer boundaries or from the silhouette of

insulator shells (see Section 6). To make the boundary or silhouette estimate more robust, cross-correlations with a pre-stored template (containing broadened outer boundaries or silhouette) from the ideal insulator can be applied.

(d) Compare the intensities. Region analysis is then performed by comparing the range of intensity values in each extra region with those of the shell and of the background. Decision on snow area is then made by combining the comparison results and the prior knowledge of snow intensity.

5.2 Compute the Snow Coverage

Once snow regions are determined, a narrow-width vertical bar parallel to the vertical center axis of insulator, is then placed and swept from the left to the right side of the insulator. For each area under the sweeping bar, the heights of detected snow regions are accumulated, and then compared with the total length of insulator shells under analysis, resulting in the percentage of snow coverage. Further, the maximum snow coverage between any two neighboring shells is computed.

6 Estimate Swing Angles

Due to the camera view angle, the angle of insulator in an image does not have to be $0°$ with respect to the vertical image axis (see example scenarios in Fig. 4(f)). The swing angle is hence defined as the relative angle, computed from the difference between the absolute insulator angle (with respect to the frame that hangs the insulator) in the given image and the reference insulator angle in an image captured when no wind is present. It is worth mentioning that camera movement does not cause absolute angle changes. Since computing a relative angle is straightforward, only the method for estimating the absolute angle is described. The proposed method is based on using the estimated outer boundaries or estimated silhouette of insulator shells. The basic idea is very similar to the cross-correlation used in Section 4, where the edges from the ROI is correlated with a template containing the broadened outer boundaries (or silhouette) of insulator. However, instead of translating the template, the template is now rotated in order to find the maximum correlation with the insulator in the ROI, using orientation-based cross-correlations $\rho(0, 0, \theta_k) = \frac{\sum_{i,j} E(i,j) E_T(i,j,\theta_k)}{\sqrt{(\sum_{i,j} E^2(i,j) \sum_{i,j} E_T^2(i,j,0))}}$, where $\theta_k = k\Delta\theta_k$ is the rotation angle of the template ($\Delta\theta_k = 0.25°$, $\theta_k \in [0°, 1.5°]$ used in our tests), and $\theta_k \in [0, \theta_1]$. The best angle is found from $\theta^* = \mathrm{argmax}_{\theta_k} \rho(0, 0, \theta_k)$. The original thin outer boundaries (or boundaries of silhouette) from the template at matched positions are then assigned as the outer boundaries for the given ROI image.

Once shell outer boundaries are found, two *parallel* vertical lines are determined by shifting lines to touching the outer most (extremal) points on shell boundaries or silhouette. The central axis is then determined from the middle of these two outer lines. From the central axis, the absolute angle is computed.

For dark/night images, the central axis is directly estimated through the vertical accumulations a_v of ROI image.

7 Experimental Results and Evaluation

Test Results. The proposed system is implemented with a graphical user interface (GUI), and tested using images measured from one winter period. Fig. 5 (left) shows 5 examples of several good and not so good results from snow analysis. It is observed that detection results are significantly affected by the variety of background scenarios, also, they are affected by the setting of camera view angle and image resolution for an insulator of concern. For example, the camera setting for the insulator (b) generates better images as compared with those for insulator (c) (not so good view angle) and for insulator (a) (too low resolution).

Ground Truth. To estimate the 'ground truth' of insulator snow regions, a semi-automatic assisted analysis is performed (see the right part of Fig. 5). In semi-automatic analysis, each shell is extracted and analyzed separately, allowing e.g. manually select thresholds for histogram, edge closing. This process is repeated over all shells.

Evaluation. The performance of automatic analysis results are evaluated for insulator type (b) by using the corresponding insulator snow 'ground truth' (generated from the above semi-automatic way) as the reference.

Our preliminary evaluation shows that: For good quality images, the average detection rate is about 93% with false alarm rate about 9%, (defined by the correctly and falsely detected pixels related to the snow, respectively). However, it is noticed that the average performance is significantly dragged down by poor

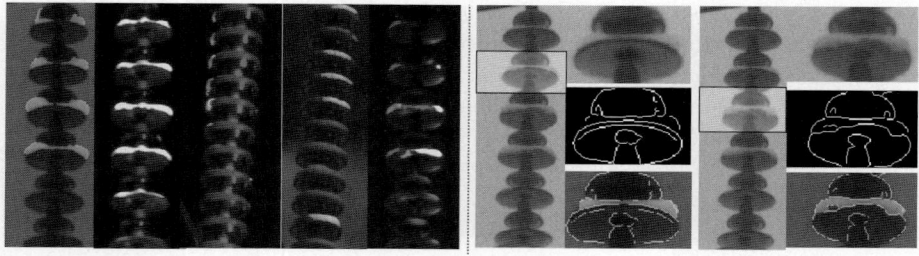

Fig. 5. *Left part*: Automatic analysis results. This part contains 5 results where the detected snow regions are visually enhanced: the first 3 insulators: good results from image with one clear and 2 dark backgrounds; the next 2 insulators: not so good results where only partial snow areas are detected from images with cloudy and dark background. *Right part*: Semi-automatic analysis to find snow ground truth for shell 2 and 3. From the left to right, top to bottom in each column: Selecting a shell (inside green box) from ROI, the selected shell, closed edge curves after modification, resulted snow region ground truth with visual enhancement.

quality images with low visibility, very weak edges and dark snow. The average detection rate is dropped to 61.03% for bright background images and 74.22% for dark background images. Meanwhile, the average false alarm also increased to 21.5% for bright background, and 18.13% for dark background. The highest snow coverage during that period was 14.73%.

8 Conclusion

The proposed insulator video surveillance system for automatically detecting and estimating insulator snow (or, ice) coverage and insulator swing angles, has been tested. Our test results on one year (winter season) measurements showed that the proposed hybrid method is relatively robust for a broad range of complex images, with an average detection rate ranging from 93% to 67.6% and averaging false alarms from 9% to 18.1% depending on image quality. Such results, though far from ideal, are encouraging since this implies a new approach entirely based on image analysis, can be a rather promising choice for insulator snow surveillance. Further improvement shall be made by exploiting temporal information in videos and by improving the settings of image capture system. The system has also increased the interest for a long term research, as our test results have demonstrated that automatically monitoring ice and snow phenomena, previously considered as not feasible, is now possible.

References

1. WAP project website, http://wap.stri.se
2. Berlijn, S.M., Gutman, I., Halsan, K.A., Gu, I.Y.H.: Laboratory Tests and a Web Based Surveillance to determine the Ice- and Snow Performance of Insulators. IEEE Trans. on Dielectrics and Electrical Insulation, Special Issue on Flashover of Ice- or Snow-Covered Insulators 14(6), 1373–1380 (2007)
3. Richards, C.N., Renowden, J.D.: Development of a Remote Insulator Contamination Monitoring System. IEEE Trans. Power Delivery 12, 389–397 (1997)
4. Sistiaga, U.: Automatic image analysis methods for estimating snow coverage and swing angle of insulators in power transmission lines, M.Sc. thesis, Dept. of Signals and Systems, Chalmers Univ. of Technology, Sweden (2007)
5. Gu, I.Y.H., Gui, V.: Joint space-time-range mean shift-based image and video segmentation. In: Zhang, Y.-J. (ed.) Advances in Image and Video Segmentation, pp. 113–139. Idea Group Inc. Pub. (2006)
6. Boykov, Y., Veksler, O., Zabih, R.: Fast approximate energy minimization via graph cuts. IEEE trans. on PAMI 23(11), 1222–1239 (2001)

Size from Specular Highlights for Analyzing Droplet Size Distributions

Andrei C. Jalba[1], Michel A. Westenberg[1], and Mart H.M. Grooten[2]

[1] Department of Mathematics and Computer Science,
[2] Department of Mechanical Engineering,
Eindhoven University of Technology, P.O. Box 513,
5600 MB Eindhoven, The Netherlands
{a.c.jalba, m.a.westenberg, m.h.m.grooten}@tue.nl

Abstract. In mechanical engineering, heat-transfer models by dropwise condensation are under development. The condensation process is captured by taking many pictures, which show the formation of droplets, of which the size distribution and area coverage are of interest for model improvement. The current analysis method relies on manual measurements, which is time consuming. In this paper, we propose an approach to automatically extract the positions and radii of the droplets from an image. Our method relies on specular highlights that are visible on the surfaces of the droplets. We show that these highlights can be reliably extracted, and that they provide sufficient information to infer the droplet size. The results obtained by our method compare favorably with those obtained by laborious and careful manual measurements. The processing time per image is reduced by two orders of magnitude.

1 Introduction

Since its introduction, the heat-transfer model by dropwise condensation [1] has drawn considerable interest in the field of mechanical engineering, as it allows higher heat transfer compared to filmwise condensation, which in turn offers opportunities for design improvements of compact industrial condensers.

Vapor condenses onto a surface if the temperature of the surface is below its saturation temperature. The condensate forms *droplets* on the surface, which grow and coalesce with adjacent droplets. Bigger droplets fall or roll from the surface (drainage) due to aerodynamic and/or gravity forces. Then, new droplets appear on the clear surface with the release of condensation enthalpy during nucleation inception of the droplets. Modeling heat transfer by dropwise condensation requires knowledge of the drop-size distribution and area coverage [2]. Although some research on nucleation and growth of water drops has been carried out [3], it is only recently that dropwise condensation from flowing steam/air mixtures at various process conditions has been observed in practical compact condensers [4]. Unfortunately for large data sets (routinely about 400 images are collected during an experiment), the task of determining droplet size distributions and area coverage becomes tedious and prohibitive (about 2 hours

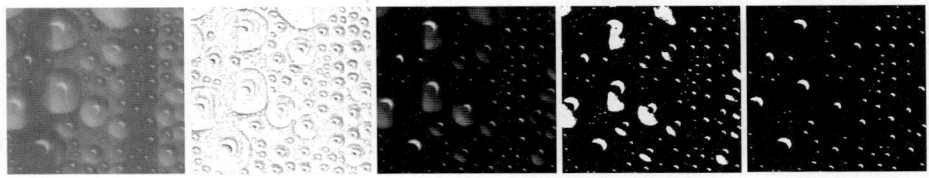

Fig. 1. *Left-to-right*: cropped input image, gradient magnitude, result after background subtraction (BS), BS + thresholding, BS + top-hat + thresholding, see Fig.2

per image), as one has to manually assess droplet sizes. Thus, automating the process is highly desirable, and the purpose of this paper is to propose an image processing approach to accomplish this.

The experimental rig used to acquire the input images is briefly described below, see [4] for full details. The condenser is optically accessible through a sapphire window of 66 mm in diameter. A horizontally-placed video camera measures condensate drops on one of the outer vertically-mounted condenser plates. Every image is 512×512 pixels, which corresponds to 6.5×6.5 mm^2, giving a spatial resolution of 13 μm. The time interval between two consecutive images is 0.2 s. All images are recorded with the air/steam mixture flowing from right to left over the plate with a mean velocity of 6.2 m/s.

A part of a typical image is shown in the first picture of Fig. 1. Note that classical image processing tools (gradient and automatic thresholding, even after background subtraction) fail to detect the boundaries of the droplets shown in the input image, see second and fourth pictures in Fig. 1. Moreover, in darker areas, as the contrast between the droplet boundary and the background plate is very low, such methods manage to only extract brighter *parts* of the droplet surfaces. Here we focus solely on using the information within the *specular highlights* visible on the surfaces of the droplets, as a means to infer their sizes. This is advantageous, as these specular highlights can be reliably extracted (see last picture of Fig. 1), and thus the whole process can be automated, see Section 2. That is, since droplets (should in theory) have hemispherical shapes [4], we propose to infer their sizes using a "size-from-specular-highlights" approach. Note that classical shape-from-shading methods (see [5, 6, 7] and references therein) not only assume that a single surface is present in the input image, but also postulate that the surface to be reconstructed exhibits Lambertian reflectance (i.e. there is no specular reflection). However, the droplets are translucent, so most of their surface is not visible, and they exhibit strong specular reflection. Moreover, some shape-from-shading methods assume that certain values are known (e.g. normals or height values at image boundaries), make smoothness assumptions, or postulate noise-free data.

2 Proposed Method

The computational flow diagram of our method is shown in Fig. 2. A background image was constructed by merging droplet-free regions from 3 images out of the

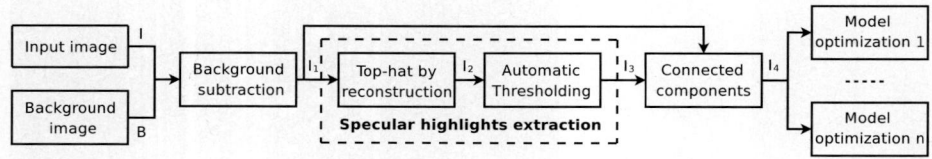

Fig. 2. Flow diagram of the proposed method

600 images collected during an experiment. Let I be the input image, and B denote the background image, with $I, B : D \subset \mathbb{Z} \to \{0, 1, \cdots, N-1\}$ and N the number of gray levels.

2.1 Extraction of Specular Highlights

The first step of our method consists in subtracting the background B from the input image I, which is based on the concept of *morphological grayscale reconstruction*, see [8]. Given threshold sets $T_k(I) = \{p \in D \mid I(p) \geq k\}$ at level k, the reconstruction $\rho_I(B)$ of I from B is given by

$$\forall p \in D, \ \rho_I(B) = \max\{k \in [0, N-1] \mid p \in \rho_{T_k(I)}(T_k(B))\}, \tag{1}$$

see [8] for further details and efficient algorithms. Using the above definition of grayscale reconstruction, image I_1 resulting after background subtraction is given by

$$I_1 = I - \rho_I(B). \tag{2}$$

The advantage of this method for background subtraction is that no grayscale levels (other than those already present in the input image) are introduced. Moreover, some intrinsic smoothing is also carried out during the process.

Extraction of the specular highlights is a two-step process. In our approach, it is essential to preserve their shape and graylevel (after background subtraction), as our method solely relies on this information to estimate the size of the droplets. As it is impossible to directly obtain the highlights by simple thresholding (see Fig. 1), one obvious idea would be to use a morphological *top-hat transform* [9] to extract the highlights, since they form (small) compact light spots on the droplets' surfaces. Unfortunately, using a classical top-hat transform (i.e. difference between original image and its *morphologically opened* result), the shape of those *connected components* which survive the opening is dramatically modified. However, if one uses morphological reconstruction, leading to *opening by reconstruction*, the original shape of the surviving regions is restored. More formally, I_2 is computed by *top-hat by reconstruction*, i.e.,

$$I_2 = I_1 - \rho_{I_1}(I_1 \circ S), \tag{3}$$

where $I_1 \circ S$ denotes morphological opening by a flat structuring element S (we fixed S to be the 7×7 square, the dimensions of the largest highlight).

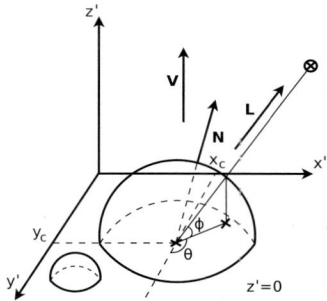

Fig. 3. Illustration of the global model. Two of the n droplets are shown in the image plane ($z' = 0$), one with center in $(x_c, y_c, 0)$; eye vector is $\hat{\boldsymbol{V}} = (0, 0, 1)$, and unknown light vector is $\hat{\boldsymbol{L}} = (\sin\theta\cos\phi, \cos\theta\cos\phi, \sin\phi)$, parametrized by angles ϕ and θ

Binary image I_3 denoting the extracted highlights is obtained by automatically thresholding I_2 [10]. Finally, after labeling the connected components of I_3, image I_1 is used to restore the grayscales at the locations of the specular highlights to yield I_4.

2.2 Size Distributions from Specular Highlights

Now that we have extracted one specular-highlight region per droplet, we propose to infer their sizes using Phong's specular reflectance model [11]. As the pictures are essentially taken through a microscope (with zoom factor of about 80×, see Section 1), one can assume that the viewer is infinitely far, and thus the perspective projection can be approximated by an orthographic one. Further, droplets can be assumed to have hemispherical shapes [4]. In *image coordinates* (x', y', z'), an illustration of the model assumed here is shown in Fig. 3. Note that, as the light location is not known from the experimental setup, the unit light vector $\boldsymbol{L}$ pointing in the direction of the light source is also assumed unknown. Further, since we aim at estimating both the radius r and the location $(x_c, y_c, 0)$ of each droplet, and as the position of the specular highlight changes with the droplet's position, we can assume that $\boldsymbol{L}$ is unknown but fixed in the *local coordinate system* $(x, y, z) : (x' - x_c, y' - y_c, z')$ placed in the center of a droplet at $(x_c, y_c, 0)$.

Within the local coordinate system, considering that droplets are translucent (see Fig. 1), the brightness of a droplet surface restricted to the highlight area is given exclusively by Phong's specular reflectance model [11], i.e.,

$$I = I_s(\hat{\boldsymbol{N}} \cdot \hat{\boldsymbol{H}})^n, \quad \hat{\boldsymbol{H}} = (\hat{\boldsymbol{L}} + \hat{\boldsymbol{V}})/\|\hat{\boldsymbol{L}} + \hat{\boldsymbol{V}}\|, \tag{4}$$

with I_s and n assumed fixed and known parameters ($I_s = 255$, $n = 10$ in our implementation) depending on the properties of the surface material. To reduce the number of unknowns, we parametrize the unit light vector by two angles, i.e. $\hat{\boldsymbol{L}} = (\sin\theta\cos\phi, \cos\theta\cos\phi, \sin\phi)$, see Fig. 3. Since the height of the droplet surface above the $z = 0$ plane is expressed as a function $S(x,y)$, i.e. $z = S(x,y)$, the unit (outer) surface normal is given by

$$\hat{N} = \frac{[-S_x, -S_y, 1]}{\sqrt{S_x^2 + S_y^2 + 1}}, \tag{5}$$

where S_x and S_y denote first-order partial derivatives of S with respect to x and y, respectively. Thus Eq. (4) becomes

$$I = I_s \left(\frac{-S_x \sin\theta\cos\phi - S_y \cos\theta\cos\phi + \sin\phi + 1}{\sqrt{(2 + 2\sin\phi)(S_x^2 + S_y^2 + 1)}} \right)^n, \tag{6}$$

with $S(x,y) = \sqrt{r^2 - x^2 - y^2}$. In the image space, the problem of determining for each droplet i a minimizer $\hat{k}^i = [r^i, x_c^i, y_c^i, \theta^i, \phi^i]$ can be regarded as a (non-linear) *least squares* minimization. Accordingly, one determines $\hat{k}^i$ such that the squared l_2 norm of the differences between the estimated values $I^i(x^i, y^i, k^i)$ and those of image I_4 (see Fig. 2) is minimized, i.e.,

$$\hat{k}^i = \arg\min_{k^i} \left\| I_4(x^i, y^i) - I^i(x^i, y^i, k^i) \right\|_2^2, \tag{7}$$

where $(x^i, y^i) \in D^i \subset D$ is the domain restricted to the highlight region of droplet i. Similar to [6] we use the Levenberg-Marquardt algorithm [12], as implemented in the *levmar* software package [13].

Thus, for each droplet i with $|D^i| \geq 5$ a separate minimization is conducted. As it is known that, especially for non-linear problems similar to ours, the minimizer $\hat{k}^i$ found by (7) can be just a local one, we repeat the whole minimization process a couple of times, if necessary. Between minimization steps, we slightly and randomly perturb the estimated paramaters such that the solution can be advanced towards a global minimizer. The convergence criterion is used to decide whether the minimization for a given droplet has to be ended.

For each droplet i, an initial estimate of the parameter vector k^i (see Section 2.2) is obtained as follows. Assuming that the area of the highlight $|D^i|$ is f times smaller than the area of the hemispherical surface of droplet i, it follows that an estimate for radius r^i is given by $r^i = \sqrt{f|D^i|/2\pi}$; in our implementation we set $f = 50$. Parameters (x_c^i, y_c^i) are set to the coordinates of the centroid of D^i, and initially the light vector $\hat{L}$ is (globally) parametrized using $\theta = 135°$ and $\phi = 80°$.

3 Results

We selected two images to compare the result of manual counting with the automatic method proposed above. Both images contain a sufficiently large variation in droplet sizes, and both have problematic regions where droplets are partially in a bright and dark region of the plate, see Fig. 1 for an example.

Manual counting was performed by two users on contrast-enhanced printouts of the images, sized at 18 by 18 centimeters. The users measured the extents of the droplets in both the x- and y-dimensions with a standard ruler having a millimeter scale. It took about two hours per image to perform the measurements and enter the data into a spreadsheet program. The radii were obtained from these measurements by taking the maximum of the x- and y-extents and multiplication with the appropriate scaling factor. Taking the maximum ensures that the droplet falls completely inside the circumscribing circle, and that no parts will be cut from it.

The automatic method performs the measurements in about two seconds on a machine equipped with a 2.4 GHz CPU; in this experiment about 400 non-linear least-squares minimizations were performed.

The results of the counting procedure are listed in Table 1. The radii were binned according to the size intervals proposed in [4], which are shown in the table together with their bin number. The symbol · indicates that the value of the lower limit of an interval is the same as the upper limit of the previous interval; this was done for space considerations. The first row, labeled 'auto', contains the droplet counts obtained by our method, and the remaining rows contains those obtained by the two users. The numbers are comparable, though there are disagreements between the manual counts. Some of the differences can be explained by the discretization at the one-millimeter scale inherent to the manual measurement. Some droplets were not measured, because they were on the border of the images. Others were simply missed because there are over 300 droplets visible in the images. The boundaries of the droplets in dark regions of the plate are very hard to see, thus causing yet another source of errors. Those droplets j with $|D^j| < 5$ are too small for the minimization procedure, however, they can be safely assigned to bin 1 directly.

We also compared the distributions of the radii in each bin for the results in Table 1. The comparison is best made by box-and-whisker plots, which are shown in Fig. 4. The top graph corresponds to image 1 and the bottom one to image 2. The left and right borders of a box correspond to the 25th percentile and 75th percentile, respectively. The median is shown as an encircled dot. The whiskers extend to 1.5 times the length of the box on either side, though they go never beyond the minimum or maximum data value. Outliers in the data are shown by the symbol '+'. The boxes are grouped according to bin number (shown at the

Table 1. Droplet counts by our method and manual counting. See text for details on the binning used.

	Image 1				Image 2			
Size interval (mm)	[0.02, 0.1)	[·, 0.15)	[·, 0.35)	[·, 0.85)	[0.02, 0.1)	[·, 0.15)	[·, 0.35)	[·, 0.85)
Bin number / Method	1	2	3	4	1	2	3	4
auto	174	70	95	8	128	113	138	8
user 1	173	73	94	7	133	109	118	8
user 2	169	70	95	7	86	92	145	6

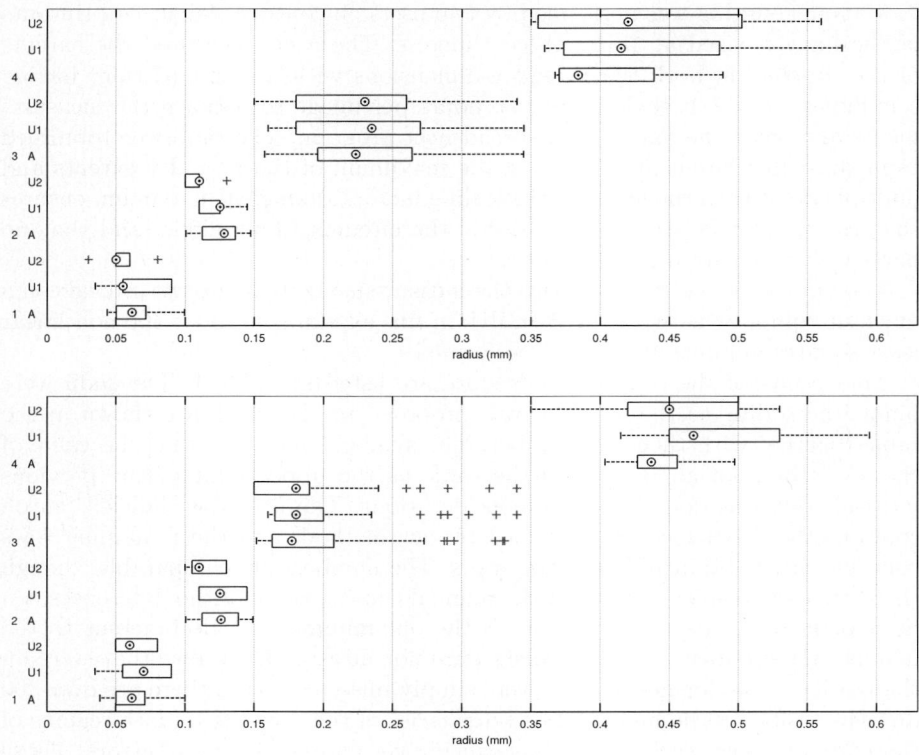

Fig. 4. Box-and-whisker plots of the distributions of radii per bin and per method. Top: results for image 1. Bottom: results for image 2. The boxes are grouped according to bin number (see left axis). Per bin, boxes for the automatic method ('A') and the user measurements ('U1' and 'U2') are shown. Median values are indicated by an encircled dot. For futher details, see text.

left axis). Per bin, our method (labeled 'A') and the user measurements (labeled 'U1' and 'U2') are put above each other. It can be seen that the medians in each bin are close to each other for all methods. Note that the counts in bin 4 (large radius) are very small, which makes a comparison hard. From these graphs, we can also conclude that the bin intervals can be chosen in a better way, as there appears to be a bias towards the lower limit of the bin interval. We could also replace the binning altogether by kernel density estimation [14]. This issue is beyond the scope of the current paper, however. Our method produces more symmetrical distributions of droplet radii, which is most likely due to the fact that it is not constrained by the discretization employed in the manual approach.

4 Conclusions

Considering that manually estimating droplet size distributions is a very tedious and time consuming task, extensive validation of the proposed automatic method is unfortunately difficult to achieve. However, we have shown that at least on a small sample of images (five images were considered, but due to the lack of space, we only reported results for two of them), the automatic method delivers results which are comparable to those obtained by laborious and careful manual counting. Further extensions and improvements of the method proposed here are the subject of ongoing research. Additionally, we also consider improving the experimental setup, such that the quality of the results by the automatic method can be assessed easier.

References

1. Schmidt, E., Schurig, W., Sellschop, W.: Versuche über die kondensation von wasserdampf in film - und tropfenform. Tech. Mech. Thermodyn (Forsch. Ing. Wes.) 1, 53–63 (1930)
2. Vemuri, S., Kim, K.: An experimental and theoretical study on the concept of dropwise condensation. International Journal of Heat and Mass Transfer 49, 649–657 (2006)
3. Leach, R., Stevens, F., Langford, S., Dickinson, J.: Dropwise condensation: experiments and simulations of nucleation and growth of water drops in a cooling system. Langmuir 22, 8864–8872 (2006)
4. Ganzevles, F.: Drainage and condensate heat resistance in dropwise condensation of multicomponent mixtures in a plastic plate heat exchanger. PhD thesis, Eindhoven University of Technology, The Netherlands (2002)
5. Zhang, R., sing Tsai, P., Cryer, J.E., Shah, M.: Shape from shading: A survey. IEEE Trans. Pattern Anal. Machine Intell. 21, 690–706 (1999)
6. Pong, T.C., Haralick, R.M., Shapiro, L.G.: Shape from shading using the facet model. Pattern Recognition 22(6), 683–695 (1989)
7. Prados, E., Faugeras, O.: Shape From Shading. In: Handbook of Mathematical Models in Computer Vision, pp. 375–388. Springer, Heidelberg (2006)
8. Vincent, L.: Morphological grayscale reconstruction in image analysis: Applications and efficient algorithms. IEEE Trans. Image Processing 2, 176–201 (1993)
9. Serra, J.: Image Analysis and Mathematical Morphology. Academic, London (1982)
10. Ridler, T.W., Calvard, S.: Picture thresholding using an iterative selection method. IEEE Trans. Syst. Man, Cybern. SMC-8, 629–632 (1978)
11. Phong, B.T.: Illumination for computer generated pictures. Commun. ACM 18(6), 311–317 (1975)
12. Gill, P.E., Murray, W.: Algorithms for the solution of the nonlinear least-squares problem. SIAM Journal on Numerical Analysis 15, 977–992 (1978)
13. Lourakis, M.I.A.: levmar: Levenberg-marquardt nonlinear least squares algorithms in C/C++ (July 2004),
    http://www.ics.forth.gr/~lourakis/levmar/
14. Silverman, B.W.: Density Estimation for Statistics and Data Analysis. Chapman and Hall, London (1986)

Capturing Physiology of Emotion along Facial Muscles: A Method of Distinguishing Feigned from Involuntary Expressions

Masood Mehmood Khan[1], Robert D. Ward[2], and Michael Ingleby[3]

[1] Faculty of Science & Engineering, Curtin University of Technology
GPO Box U1987, Perth, Western Australia 6845
Masood.Khan@curtin.edu.au
[2] Department of Behavioral Sciences
[3] Applied Criminology Centre, University of Huddersfield,
Queensgate, HD1 3DH, England

Abstract. The ability to distinguish feigned from involuntary expressions of emotions could help in the investigation and treatment of neuropsychiatric and affective disorders and in the detection of malingering. This work investigates differences in emotion-specific patterns of thermal variations along the major facial muscles. Using experimental data extracted from 156 images, we attempted to classify patterns of emotion-specific thermal variations into neutral, and voluntary and involuntary expressions of positive and negative emotive states. Initial results suggest (i) each facial muscle exhibits a unique thermal response to various emotive states; (ii) the pattern of thermal variances along the facial muscles may assist in classifying voluntary and involuntary facial expressions; and (iii) facial skin temperature measurements along the major facial muscles may be used in automated emotion assessment.

1 Introduction

Darwin, some 140 years ago, proposed the basic principles of expressing emotions [1]. As stated in [2], he suggested "muscles that are difficult to activate voluntarily might escape efforts to inhibit or mask expressions, revealing true feelings." About 108 years later Ekman and others [3] observed that less than 25% of the people they examined could produce deliberately what they termed *facial actions* on certain facial muscles: *orbicularis oris triangularis, depressor labii inferioris; frontalis pars medialis; frontalis pars lateralis; risorius* and *orbicularis oculi pars lateralis*. Ekman and other's observations [3], though initially based on limited psychophysiological evidence, were widely accepted in the scientific community.

The information-processing approach to facial expression classification is considered useful in investigating and treating neuropsychiatric and affective disorders such as schizophrenia and autism and in detecting malingering and dissimulation. Relying mostly on vision, earlier researchers discovered that sincerity of emotion was detectable [2]. The morphological differences in expressions: the differences in timing, duration and speed of onset of facial expressions; the analyses of the symmetry of facial features; and factors such as ballistic trajectory and apex overlap distinguish between evoked and pretended expressions of emotions to a limited extent [2]. Automation of these distinguishing schemas for real-world applications has seemed daunting, however, given the functional limitations and complexity of the patterns [4]. Similarly, several digital voice-analysis tools aimed at diagnosis of sincerity have encountered limitations to use in real-life scenarios [4].

The emotion-specific facial hæmodynamic and thermal variations of skin surface have been used to classify the expressions of emotive states [5,6]. Studies have demonstrated that pixel grey-levels in the thermal images would provide a reliable measure of skin surface radiance [7,8]. Investigators have consequently recognized stress levels and deceit and have classified positive and negative expressions of emotive states using the pixel grey-level information extracted from the thermal images [4-6, 9-11].

For the first time, this work investigates whether the thermal variations on the facial skin along those muscles that Ekman's subjects found difficult to activate during the voluntary expressions [3] differ under the influence of voluntary and involuntary expressions of emotive states. We attempt to use thermal intensity values (TIVs) measured on a set of significant facial thermal feature points (FTFPs) along the major facial muscles to classify voluntary and involuntary facial expressions of emotive states. Our goal is to determine whether psychophysiological information processing can assist in person-independent recognition of true emotive states. The reported results encourage the application of physiological information in psychology and psychiatric practice.

This paper presents in sequence our experimental design, feature measurement and feature extraction methods, data analysis, results, and conclusions.

2 Experimental Design

The infrared images were acquired under a normal, controlled and comfortable environment with the room temperature kept between 19-22 °C. Participants were given 20 minutes or more to acclimatize with the environment. Common pathological conditions such as fever and inflammation were ruled out. A low emissivity background (concrete wall with emissivity $\varepsilon = 0.54$) was used to ensure better separation of the background from the desired regions of the thermal infrared images (TIRIs).

A thermal infrared imaging camera capable of capturing an image array of 320 × 240 pixels, with an uncooled microbolometer (a special purpose temperature sensitive electrical resistor) mounted for measuring the incident radiation was used. The camera had a focal plane sensor array with a high thermal sensitivity in the wavelength range of 7.50-14.00 μm. The skin surface emissivity (ε) was set between 0.97 and 0.99. In order to capture the frontal view of a participant, the central vertical line on the camera viewfinder was aligned to the center of each participant's face.

A set of 156 images was used. The first subset included 96 images of 16 undergraduate students, 12 males and 4 females, with a mean self-reported age of 20 years 9 months who pretended facial expressions. A second subset of 60 visible and thermal infrared images with naturally evoked expressions of emotion from 7 male and 3 female participants (mean age of 21 years 2 months) was used. The practices suggested in [12] were followed for evoking expressions using carefully selected images and video clips, avoiding violent, disturbing and unethical images to evoke happiness and sadness.

3 A New Thermal Feature Measurement Space

Infrared cameras usually operate in a temperature range of -20 to 500 °C and cover this range using between 4096-16384 grey levels of pixels in a thermogram [13]. The built-in radiance measurement and image digitization mechanisms in a thermal camera cause addition of undesired noise in the TIRIs, and this affects severely the narrow band of temperature-differences involved in facial imaging. Many convolution methods are recommended for noise reduction and edge detection to minimize the influence of noise factors in the TIRIs [8,13]. We found that the median smoothing filter [14] was an effective noise reducer for our images as instead of averaging the pixels, it would pick one of the pixels being analyzed. The Sobel operator-based edge detection algorithm, generally considered a robust edge-detector, was invoked for extracting segment contours within our infrared images. For the selected 3x3 neighborhood the Sobel gradient operators were calculated according as [14]:

$$Gs = [f(i-1, j-1) + 2f(i-1, j) + f(i-1, j+1)] - [f(i+1, j-1) + 2f(i+1, j) + f(i+1, j+1)], \quad (1)$$

and

$$Gt = [f(i-1, j-1) + 2f(I, j-1) + f(i+1, j-1)] - [f(i-1, j+1) + 2f(I, j+1) + f(i+1, j+1)]. \quad (2)$$

The gradient magnitude was computed as

$$G[f(s,t)] = \sqrt{G_s^2 + G_t^2} \quad (3)$$

1	Symmetrically divide thermal image into N squares
2	Set Correlation$_{ST}$ = 0
3	Set Variance$_{ST}$ = 0
4	Set the list of FTFPs = Empty
5	**For** squares 1 to N,
6	Find the highest level of grey in the square
7	Measure the corresponding temperature of the discovered highest grey level
8	Add the discovered highest temperature point to the list of FTFPs
9	Calculate the correlation between the FTFPs
10	Set the FTFP Correlation = Cor$_{new}$
11	Calculate the Variance between the FTFPs
12	Set the FTFP Variance = Var$_{new}$
13	**If** {(Cor$_{new}$ > Correlation$_{ST}$) and (Var$_{new}$ > Variance$_{ST}$)}
14	Then ((Correlation$_{ST}$ = Cor$_{new}$)And (Var$_{new}$ = Variance$_{ST}$))
15	And Keep the newly discovered FTFP in the list of FTFPs
16	Else, Remove the newly discovered FTFP in the list of FTFPs
17	**End If**
18	**End For**

Fig. 1. Significant Facial Thermal Feature Point Selection Procedure

As suggested in [13], the time-sequential TIRIs were analyzed through comparing the temperature measurements at the points of registration within a series of images to discover the temporal changes on the faces with different facial expressions. The most significant facial thermal feature points were discovered by minimizing the correlation between corresponding regions of interest in a sequence, and maximizing variance. The optimization procedure is summarized in Figure 1.

Table 1. The Muscular Alignment of Facial Thermal Feature Point (FTFPs)

Facial Muscle	FTFPs
Frontalis, pars medialis	1, 3, 6, 8, 13, 15
Frontalis, inner center edges of pars medialis and pars lateralis	2, 7
Frontalis, pars lateralis	4, 5, 9, 10, 11, 12, 16, 17
Procerus/ Levator, labii superioris alaquae nasi	21
Depressor, supercilii	14
Orbicularis Oculi, pars orbital	18, 19, 20, 22, 23, 24, 25, 26, 27, 29, 30, 31
Orbicularis Oris	45, 51, 64, 65, 66
Levator, labii superioris alaquae nasi	28, 35, 36
Levator, labii superioris	33, 34, 37, 38, 44, 46
Masseter, superficial	40, 41, 49, 50
Levator, anguli oris	43, 47
Zygomaticus major	32, 39, 42, 48
Risorious/ Platysma	52, 53, 54, 59, 60, 61
Depressor anguli oris	55, 58
Buccinator	56, 57
Platysma	62, 63, 67, 68
Depressor Labii Inferioris	69, 70, 71, 72
Mentalis	73, 74, 75

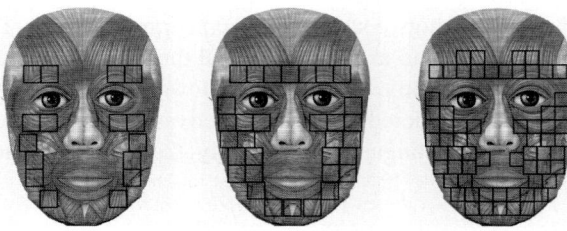

Fig. 2. Left to right: The highest thermal intensity values were measured within the shown 16, 32 and 64 square segments on the face

Figure 2 exhibits the progressively selected symmetrical regions of interest during this analysis.

The most significant thermal variations were discovered at 75 physical sites located all over the face along the major facial muscles within the TIRIs. The TIV data gathered from these 75 facial thermal feature points (FTFPs) were used for feature classification. Fig. 3 above represents the geometric profile of the facial thermal feature points showing a muscular map of a human face and the 75 FTFPs on an emotionally neutral human face. The 75 FTFPs and their muscular alignments are listed in Table I. The TIV data recorded at the 75 FTFP sites were used to represent each thermal face as a 75-dimensional thermal feature vector for the subsequent investigation and analysis.

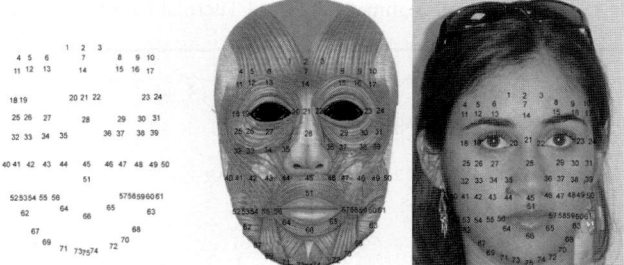

Fig. 3. Left to right: Geometric profile of FTFPs, FTFPs on a facial muscle map and FTFPs on a human face

4 Thermal Data Analysis

Figure 4 exhibits the observed mean of thermal variations along the facial muscles in °C. As evident in Figure 4, *orbicularis oris triangularis, depressor labii inferioris, frontalis pars medialis* and *orbicularis oculi pars lateralis* experienced more thermal variations during the involuntary expression of emotive states than voluntary expressions of the same affects. More thermal variations were observed, however, on *risorius* during the voluntary expression of positive affects than during the involuntary expression. We observed similar increased variation during voluntary expression of both negative and positive emotive states on *frontalis pars lateralis*. Looking at these initial results, it would be safe to assume that facial thermal features of voluntary and involuntary expressions of emotive states differ. However, the TIVs measured at the FTFPs on these muscles considered "difficult to activate during the voluntary expression of emotion" did not distinguish successfully the emotive states in a person-independent way.

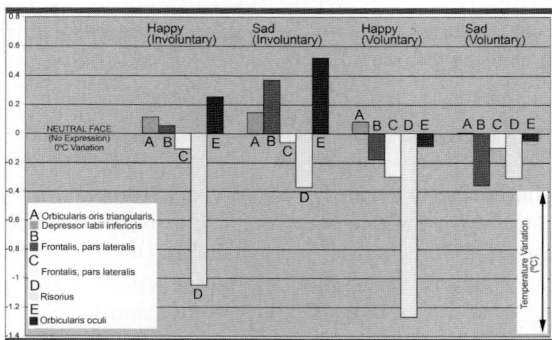

Fig. 4. Differences in thermal variations along certain facial muscles when emotive states were expressed voluntarily and involuntarily

In a follow-up analysis, using the TIVs measured on all the 75 FTFPs, fairly successful separation of emotional states was achieved (Figures 5 and 6). The separation was more successful for voluntary expressions (Figure5) than for involuntary expressions (Figure 6).

The combinations of FTFPs used to illustrate separation power were selected using principle component analysis (PCA) [15].

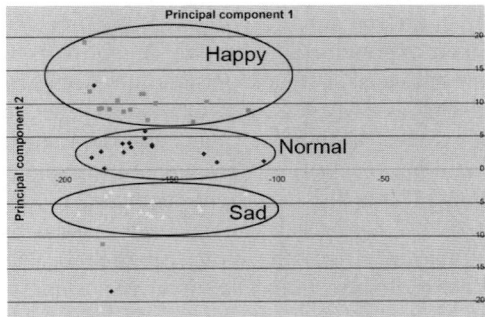

Fig. 5. Separation between the neutral, happy and sad voluntary facial expression in a 2-principal component eigenspace

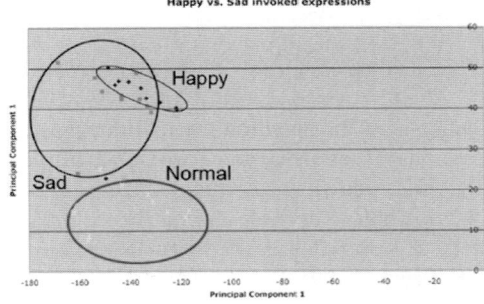

Fig. 6. Separation of the neutral faces and involuntary expression of happiness in a 2-principal component eigenspace

5 Results and Discussion

The investigation suggested significant and detectable thermal differences between the neutral, voluntary and involuntary facial expressions of emotive states. The observed between-group separation patterns might be understood in the light of the previous works on the correlation between emotions and musculo-physiological activities [16-19]. There seems to be some overlap between the neutral faces and faces with intentional expressions of sadness and happiness in a 2-component eigenspace obtained by PCA. However, the neutral faces do not seem to be overlapping the evoked expressions of happiness. Previous studies found that *Zygomaticus Major, Orbicularis Oris, Orbicularis Oculi, Mentalis* and *Platysma* contribute to the expression of positive evoked emotional experiences. The hæmodynamic and thermal characteristics of these muscles probably did not change when the expressions were being simulated, hence the intentionally happy faces overlapped with the neutral faces. When emotions were evoked, however, some significant musculo-physiological and hæmodynamic activities took place along these muscles, so evoked expressions of happiness did not overlap the neutral faces.

As reported in [20] the *Orbicularis Oculi, Mentalis,* and *Depressor Anguli Oris* contribute to the evoked expression of negative emotions. Simulating the negative emotion probably does not cause as much musculo-physiological and hæmodynamic activities in these muscles as evoked expressions do. So the simulated sadness was confused with the neutral face. In summary, there appears to be a traceable difference in the activities of certain muscles between emotional expressions that are evoked and those that are simulated. The apparent tendency is that activities along *Zygomaticus Major, Orbicularis Oris, Mentalis* and *Platysma* are evoked by positive emotional experiences, while the muscular activities along *Corrugator, Masseter, Triangularis, Orbicularis Oculi Palpabraeous, Platysma,* and *Bucccinator* are more strongly evoked by negative emotional experiences. Also, the cited studies in [20] report some musculo-physiological activities around the *Orbicularis* during the facial expression of both positive and negative emotions. The musculo-physical activities along this particular muscle may possibly have caused the overlap between the positive and negative facial expressions in our study.

6 Conclusion

The pixel grey-level analysis of the acquired thermal images suggested that voluntary and involuntary expressions of emotive states would result in dissimilar patterns of facial skin temperature variations along the major facial muscles. The facial muscles that were known to be "difficult to activate during the voluntary expressions of emotive states", experienced a lesser degree of thermal variation during the voluntary expressions than those observed during the involuntary expressions. Though further work is needed to consolidate these results, the work reported here suggests that facial skin temperature measurements taken along appropriate major facial muscles might help in distinguishing between the voluntary and involuntary facial expressions of positive and negative emotive states and help in automated detection of concealed emotions during the psychiatric investigations. If confirmed by further study, this finding is also significant to the emotion assessment, detection of dissimulation and deceit in forensic and security applications.

References

1. Darwin, C.: The expression of emotion in man and animals. Murray, London (1872)
2. Ekman, P., O'Sullivan, M.: From flawed self-assessment to blatant whoppers: The utility of voluntary and involuntary behavior in detecting deception. Behav. Sci. Law 24, 673–686 (2006)

3. Ekman, P., Roper, G., Hager, J.C.: Deliberate facial movement. Child Development 51, 886–891 (1980)
4. Khan, M.M.: Cluster-analytic classification of facial expressions using infrared measurements of facial thermal features. Ph.D. Thesis, School of Computing & Engineering, University of Huddersfield, UK (2008)
5. Pollina, D.A., Dollins, A.B., Senter, S.M., Brown, T.E., Pavlidis, I., Levine, J.A., Ryan, A.H.: Facial skin surface temperature changes during a concealed information test. Annals of Biomedical Engineering 34(7), 1182–1189 (2006)
6. Pavlidis, I., Levine, J.: Thermal image analysis for polygraph testing. IEEE Engineering in Medicine and Biology 21(6), 56–64 (2002)
7. Bales, M.: High-resolution infrared technology for soft –tissue injury detection. IEEE Engineering in Medicine and Biology 17, 56–59 (1998)
8. Otsuka, K., Okada, S., Hassan, M., Togawa, T.: Imaging of skin thermal properties with estimation of ambient radiation. IEEE Engineering in Medicine and Biology 21(6), 49–55 (2002)
9. Khan, M.M., Ingleby, M., Ward, R.D.: Automated facial expression classification and affect interpretation using infrared measurement of facial skin temperature variation. ACM Transactions on Autonomous and Adaptive Systems 1, 91–113 (2006)
10. Khan, M.M., Ward, R.D., Ingleby, M.: Classifying pretended and evoked facial expression of positive and negative affective states using infrared measurement of facial skin temperature. ACM Transactions on Applied Perception 6(1), 6:1–22 (2009)
11. Rimm-Kaufman, S.E., Kagan, J.: The physiological significance of changes in skin temperature. Motivation and Emotion 20(1), 63–78 (1996)
12. Wild, B., Erb, M., Bartels, M.: Are emotions contagious? Evoked emotions while viewing emotionally expressive faces: quality, quantity, time course and gender differences. Psychiatry Research 102, 109–124 (2001)
13. Jones, B.F., Plassmann, P.: Digital infrared thermal imaging of human skin. IEEE Engineering in Medicine and Biology 21(6), 41–48 (2002)
14. Gonzalez, R.C., Woods, R.E.: Digital Image Processing. Addison-Wesley, New York (2002)
15. Jolliffe, I.T.: Principal Component Analysis. Springer, New York (2000)
16. Dimberg, U.: Facial electromyography and emotional reactions. Psychophysiology 27(5), 481–494 (1990)
17. Ekman, P., Levenson, R.W., Friesen, W.V.: Autonomic nervous system activity distinguishes among emotions. Science 221, 1208–1210 (1983)
18. Bradley, M.M., Sabatinelli, D., Lang, P.J., Fitzsimmons, J.R., King, W., Desai, P.: Activation of the visual cortex in motivated attention. Behavioral Neuroscience 117, 369–380 (2003)
19. Wright, P., He, G., Shapira, N.A., Goodman, W.K., Liu, Y.: Disgust and the insula: fMRI responses to pictures of mutilation and contamination. NeuroReport 15, 2347–2351 (2004)
20. Ekman, P., Friesen, W.V.: Facial Action Coding System: A technique for the measurement of facial movement. Consulting Psychology Press, Pal Alto (1978)

Atmospheric Visibility Monitoring Using Digital Image Analysis Techniques

Jiun-Jian Liaw[1], Ssu-Bin Lian[1], Yung-Fa Huang[1], and Rung-Ching Chen[2]

[1] Department of Information and Communication Engineering,
ChaoYang University of Technology, Taichung, Taiwan
[2] Department of Information Management,
ChaoYang University of Technology, Taichung, Taiwan
{jjliaw,s9730602,yfahuang,crching}@cyut.edu.tw

Abstract. Atmospheric visibility is a standard of human visual perception of the environment. It is also directly associated with air quality, polluted species and climate. The influence of urban atmospheric visibility affects not only human health but also traffic safety and human life quality. Visibility is traditionally defined as the maximum distance at which a selected target can be recognized. To replace the traditional measurement for atmospheric visibility, digital image processing schemes provide good visibility data, established by numerical index. The performance of these techniques is defined by the correlation between the observed visual range and the obtained index. Since performance is affected by non-uniform illumination, this paper proposes a new procedure to estimate the visibility index with a sharpening method. The experimental results show that the proposed procedure obtains a better correlation coefficient than previous schemes.

Keywords: atmospheric visibility, digital image processing, homomorphic filtering.

1 Introduction

Many researchers have evaluated air quality by means of various indices and frequently use atmospheric visibility as an air quality indicator. The influence of urban atmospheric visibility not only effects human health, but also traffic safety and human life quality. Low visibility strongly implies air pollution by ambient pollutants, particulate matter, or gaseous species [1].

Atmospheric visibility is a standard of human visual perception of the environment. The traditional measurement defines visibility as the maximum distance at which the selected target can be recognized [2]. Since the variation in human-eye observation could be high due to different personal characteristics, researchers have applied some photograph processing methods for measuring visibility [3]. Moreover, several meters, such as the telephotometer [4], the nephelometer and the athelometer [5], have also been developed to monitor visibility.

Since the rapid development of information technologies, we have widely applied digital image processing schemes to meteorology, pattern recognition, biology, geographic information system, environmental monitoring and machine vision [6], etc. Digital camera can be used to measure the predetermined distance of target for computation of visibility [7]. Expensive digital panorama camera is also used to determinate the visibility with Sobel operator [8]. Digital image data can be translated to specific brightness value, the difference between the building and its background [9]. However, the pixel value difference and the digital image analysis techniques (both in spatial domain and frequency domain) have also been used to estimate atmospheric visibility [10, 11].

Luo et al. proposed the high pass filter concept to measure atmospheric visibility. They extracted high frequency data by the Fourier transform and the Sobel operation in frequency domain and spatial domain, respectively [12]. They used the ideal high pass filtered data to establish the visibility index and defined the performance of these techniques by the correlation between the observed visual range and the obtained index. The correlation is the coefficient of determination, R^2, which includes linear regression [13]. With the ideal filter, the R^2 result is less than 0.8. However, ideal high pass filters do not easily reveal the fine structures of a non-uniform illumination area, such as the shadow, of the image.

Digital image data can be characterized by illumination and reflectance components. Illumination is the amount of viewed source illumination incident on the scene and reflectance is the amount of illumination reflected by objects in the scene. Slow and fast spatial variations characterize illumination and reflectance, respectively. The homomorphic filter operates these components separately with two or more parameters. The homomorphic filter can be used to advantage the ideal filter [6]. The R^2 of measuring the visibility index via the homomorphic filter technology is close to 0.9 [14]. However, using the homomorphic filter is complex because it needs to decide the parameters.

Yang proposed an algorithm with the Harr function to improve non-uniform illumination sharpness [15]. Compared to the homomorphic filtering approach, this method helps reveal the fine structures of the non-uniform image. Since objects in the urban atmospheric image may be shaded and outlines in the dim area are not illuminated enough to be recognized, this scheme is suitable for measuring visibility.

This paper introduces a series of digital image analysis techniques with the ideal high pass filter and homomorphic filter. The current work also proposes a new procedure to estimate the visibility index with the Harr function. The experiments compare the proposed procedure with previous schemes. According to experiment results, when the proposed procedure obtains the indices, the correlation between the observed visual range and the obtained indices is better.

2 Methods and Experiments

This study uses the same image database in [12], to show practical application performance. This study shot the actual urban images on the top of the Linden Hotel, a 42-floor building in Kaohsiung, Taiwan. This study took a total 172 images with observed visibility in four months. At the time of shooting the image, a trained investigator also recorded observed visibility. The investigator stands on the top of the hotel and decides the farthest building that he can see. The visibility is measured as the pre-measured distance (using GPS).

We compute the visibility indices of 172 images with three methods (ideal high pass filter, homomorphic filter and sharpness with Harr function). Relationships between observed visibility values and the computed indices show the performance. The correlation, which is the coefficient of determination, R^2 (which includes linear regression [13]), defines the relationship. If the Relationship between observed visibility values and the computed indices is close to a line, the R^2 value will be close to 1. The following describes summaries of the methods and experiment results.

2.1 Fourier Transform and Ideal High-Pass Filter

Since impaired visibility can not appear detailed, the high-frequency components in the scene decrease. The high-frequency information can be used to estimate the visibility. We can use the Fourier transform and the ideal high-pass filter to separate the high frequency components [12].

The digital image can be described by a function, $f(x, y)$, where the value of the function presents the brightness level at location of (x, y). The Fourier transform of the image with $M \times N$ size can be written as [6]

$$F(u,v) = \frac{1}{MN} \sum_{x=0}^{M-1} \sum_{y=0}^{N-1} f(x,y) \exp[-j2\pi(\frac{ux}{M} + \frac{vy}{N})]. \quad (1)$$

The ideal 2-D high-pass filter passes high frequencies and cuts off low frequencies with the cutoff distance, D_0. The ideal high-pass filter function is defined as

$$H(u,v) = \begin{cases} 0 & \text{if } D(u,v) \leq D_0 \\ 1 & \text{if } D(u,v) > D_0 \end{cases}, \quad (2)$$

where D_0 is a positive integer and $D(u, v)$ is the distance between (u, v) and the center of the frequency domain. Finally, the filtered image can be obtained by the inverse Fourier transform. The visibility index with this method, I_F, is the average gray level of the filtered image. Figure 1 shows the relationships of the Fourier transform (with ideal high pass filter) where the line equation of the linear regression and R^2 are shown on the figure.

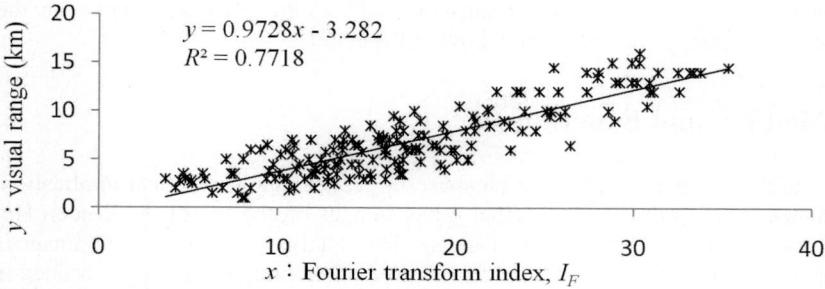

Fig. 1. The result of actual urban images by the Fourier transform

2.2 Homomorphic Filter

The ideal high pass filters do not easily reveal the fine structures of a non-uniform illumination area, such as the shadow, of the image. The homomorphic filter is useful for improving the non-uniform illumination area by brightness range compression and contrast enhancement. The homomorphic filter controls the filter function that affects low frequency and high frequency in different ways by choosing the parameters L and H. A homomorphic filter function can be defined as [14]

$$H(u,v) = (H-L)[1-\exp(-\frac{D^2(u,v)}{D_0^2})] + L. \qquad (3)$$

If we choose that $L<1$ and $H>1$, then the filter reduces the illumination (low frequencies) and increases the reflectance (high frequencies). In this paper, we set $L=0.5$ and $H=1.1$. After the homomorphic filter, the data is filtered again by a 2-D Gaussian high-pass filter which is defined as

$$H(u,v) = 1 - \exp(-\frac{D^2(u,v)}{2D_0^2}). \qquad (4)$$

Similarly, the filtered image can be obtained by the inverse Fourier transform and the visibility index, I_H, is the average gray level of the filtered image. Figure 2 shows the results of the Homomorphic filter.

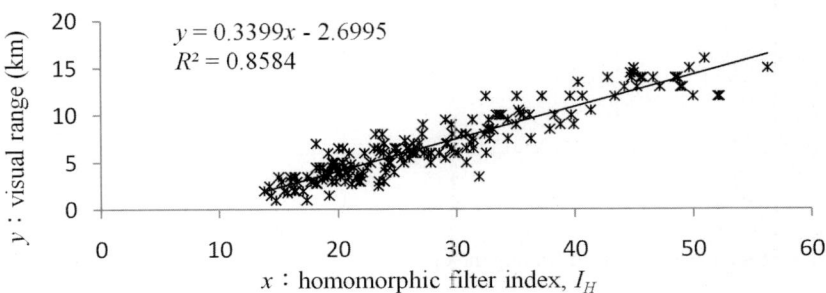

Fig. 2. The result of actual urban images by the Homomorphic filter

2.3 Sharpness with Harr Function

The Harr function is an orthogonal base of the wavelet transform [15]. In general, the orthogonal Harr filter coefficient is typically $1/\sqrt{2}$. In other words, when two data (p_1 and p_2) are inputted, they can be separated as ($p_1 - p_2$)/$\sqrt{2}$ and ($p_1 + p_2$)/$\sqrt{2}$, respectively.

We assume that the input image is described by a function, $f(x,y)$, where the ranges of x and y are $0 \le x \le M-1$ and $0 \le y \le N-1$, respectively. We compute the image in x direction and y direction. In the x direction, we set $p_1=f(2m, y)$ and

$p_2=f(2m+1, y)$ to be the input of the Harr function (where $0 \le m \le M/2$). We also extract one pixel shifted, $p_1=f(2m+1, y)$ and $p_2=f(2m+2, y)$, to operate the transformation. Similarly, the y direction proceeds two times with one pixel shifted. Two input of the y direction are $p_1=f(x, 2n)$, $p_2=f(x, 2n+1)$ and $p_1=f(x, 2n+2)$, $p_2=f(x, 2n+2)$, respectively (where $0 \le n \le N/2$). We can see that the original image is separated into four subsets. In the separated subset, we set the original data (p_1 and p_2) into the sharpened data (p_1' and p_2') by the value of ($p_1 - p_2$)/$\sqrt{2}$. If the value is positive, we set $p_1'=14\log(p_1 - p_2+1)$ and $p_2'=0$, otherwise, we set $p_1'=0$ and $p_2'=14\log(p_2 - p_1+1)$. Finally, the sharpen image can be obtained by the sum of the results of four modified sub-images.

To estimate the visibility index, we combine the sharpen image with the homomorphic filtered image (without Gaussian high-pass filter). The combination is (the sharpen image) + $c \times$(the homomorphic filtered image), where c is a combination weight. The visibility index with the proposed procedure, I_P, is the average gray level of the combined image. To find the optimum value of c, we compute indices of all 172 mages with variant c. The relation between R^2 and c is shown in Figure 3. The best value of R^2 is 0.9146 with $c=0.78$. Figure 4 shows the results of the proposed procedure.

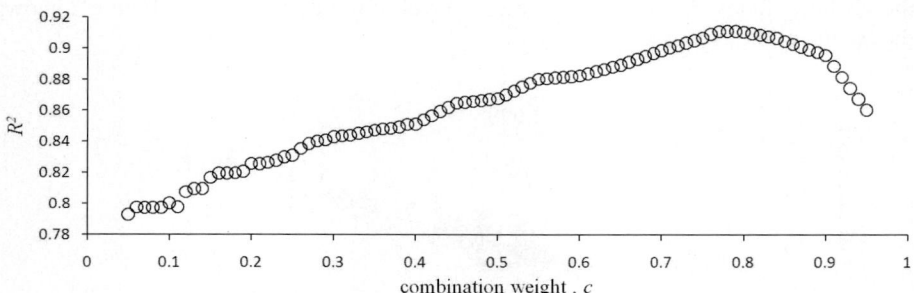

Fig. 3. The results of finding the optimum c value (the combination weight)

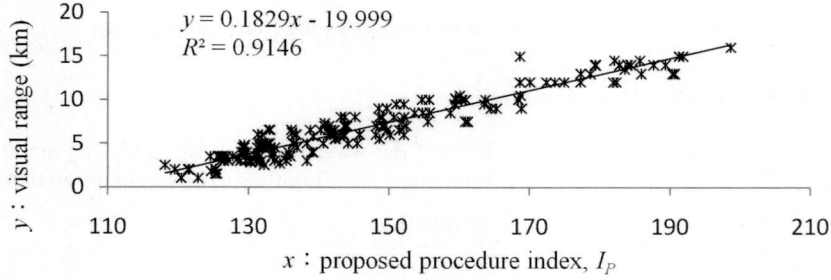

Fig. 4. The result of actual urban images by the proposed procedure

3 Comparisons

From the experiment results, the performances of the Fourier transform, the homomorphic filter and the proposed procedure are $R^2=0.77$, $R^2=0.86$, and $R^2=0.91$, respectively. The proposed procedure obtains higher relationship with human observed visibility than the other methods. We would like to show that the proposed procedure is better because it reduces the effect of shadow.

We compare of the three methods with shaded buildings compared to no shaded buildings in the same observed visibility. Since the shadow should not affect visibility index, with the same method, the indices of the two images should be the same. We use similarity to show performance comparisons. With two indices, i_1 and i_2, we define the similarity as

$$similarity = \frac{2\,i_1\,i_2}{i_1^{\,2}+i_2^{\,2}} \times 100\% . \tag{5}$$

Figure 5(a) shows an urban image obtained on a cloudy day. Figure 5(b) shows another image shot on a sunny day. The observed visibilities of Fig. 5(a) and (b) are the same (5km). Figure 5(a) shows less shadow than Fig. 5(b). Since human observed visibilities are the same, indices similarity with the same method should be close to 100%. Table 1 shows the results of computing visibility indices with similarity by the above methods. The similarity of the proposed procedure is the best.

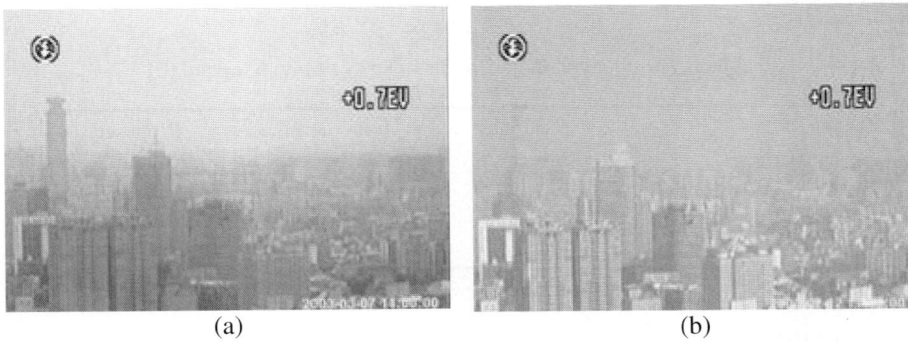

Fig. 5. Two actual urban images (with the same 5km observed visibility) which are shot on (a) cloudy day and (b) sunny day

Table 1. The obtained indices and similarity of Fig. 5 (a) and (b)

Image	Fourier transform	Homomorphic filter	Proposed procedure
Fig. 5(a)	19.7562	29.3877	146.2156
Fig. 5(b)	10.6512	24.2651	143.3134
similarity	83.54%	98.19%	99.98%

We also use two images both with 12km observed visibility to compare the similarity. Figure 6(a) and Figure 6(b) show that the non-shadow buildings and shadow buildings, respectively. Table 2 shows the results of computing visibility indices with similarity by the above methods. Similar to the above results, the similarity of the proposed procedure is the best.

Results of the above comparisons show that, the proposed scheme reduces the influence of non-uniform illumination. Obtaining the visibility indices with the proposed procedure can increase the similarity when the observed visibilities are the same.

Fig. 6. The same 12km observed visibility images with (a) non-shadow buildings and (b) shadow buildings

Table 2. The obtained indices and similarity of Fig. 6 (a) and (b)

Image	Fourier transform	Homomorphic filter	Proposed procedure
Fig. 6(a)	21.172	37.842	165.3027
Fig. 6(b)	32.847	49.673	179.5326
similarity	91.07%	96.41%	99.66%

4 Conclusion

This paper introduces a series of digital image analysis schemes, based on high pass filters. Since shadow may influence urban images, this paper also proposes a procedure based on sharpness of the non-illumination image. The experiments in this study use actual atmospheric images to show the comparison between the Fourier transform, the homomorphic filter and the proposed procedure. Results show that the proposed procedure obtains higher relationship with human observed visibility than the other methods. We also show that obtaining the visibility indices with the proposed procedure can increase the similarity when the observed visibilities are the same.

References

1. Sequeira, R., Lai, K.H.: The effect of meteorological parameters and aerosol constituents on visibility in urban Hong Kong. Atmospheric Environment 32, 2865–2877 (1998)
2. Horvath, H.: Atmospheric visibility. Atmospheric Environment 15, 1785–1796 (1981)
3. Larson, S.M., Cass, G.R., Hussey, K.J., Luce, F.: Verification of image processing based visibility models. Environmental Science and Technology 22, 629–637 (1988)
4. Agarwala, A., Dontcheva, M., Agrawala, M., Drucker, S., Colburn, A., Curless, B., Salesin, D., Cohen, M.: Interactive digital photomontage. ACM Transactions on Graphics 23, 294–302 (2004)
5. Jayaraman, A., Gadhavi, H., Ganguly, D., Misra, A., Ramachandran, S., Rajesh, R.A.: Spatial variations in aerosol characteristics and regional radiative forcing over India: measurements and modeling of 2004 road campaign experiment. Atmospheric Environment 40, 6504–6515 (2006)
6. Gonzalez, R.C., Woods, R.E.: Digital Image Processing. Pearson Prentice Hall, U.S.A (2008)
7. Caimi, F., Kocak, D., Justak, J.: Remote visibility measurement technique using object plane data from digital image sensors. In: IEEE International Geoscience and Remote Sensing Symposium, Alaska, USA, pp. 3288–3291 (2004)
8. Baumer, D., Versick, S., Vogel, B.: Determination of the visibility using a digital panorama camera. Atmospheric Environment 42, 2593–2602 (2008)
9. Luo, C.-H., Liu, S.-H., Yuan, C.-S.: Investigate atmospheric visibility by the digital telephotography. Aerosol and Air Quality Research 2, 23–29 (2002)
10. Kim, K., Kim, Y.: Perceived visibility measurement using the HSI color different method. Journal of the Korean Physical Society 46, 1243–1250 (2005)
11. Xie, L., Chiu, A., Newsam, S.: Estimating atmospheric visibility using general-purpose cameras. In: Bebis, G., Boyle, R., Parvin, B., Koracin, D., Remagnino, P., Porikli, F., Peters, J., Klosowski, J., Arns, L., Chun, Y.K., Rhyne, T.-M., Monroe, L. (eds.) ISVC 2008, Part II. LNCS, vol. 5359, pp. 356–367. Springer, Heidelberg (2008)
12. Luo, C.-H., Wen, C.-Y., Yuan, C.-S., Liaw, J.-J., Lo, C.-C., Chiu, S.-H.: Investigation of urban atmospheric visibility by high-frequency extraction: Model development and filed test. Atmospheric Environment 39, 2545–2552 (2005)
13. Kutner, M., Nachtsheim, C., Neter, J.: Applied Linear Regression Models. McGraw-Hill, USA (2004)
14. Sun, Y.-C., Liaw, J.-J., Luo, C.-H.: Measuring atmospheric visibility index by different high-pass operations. In: 2007 Computer Vision, Graphics and Image Processing Conference, Miao-Li, Taiwan, pp. 423–428 (2007)
15. Yang, C.C.: Improving the sharpness of an image with non-uniform illumination. Optics & Laser Technology 37, 235–238 (2005)

Minimized Database of Unit Selection in Visual Speech Synthesis without Loss of Naturalness

Kang Liu and Joern Ostermann

Institut für Informationsverarbeitung, Leibniz Universität Hannover
Appelstr. 9A, 30167 Hannover, Germany
kang@tnt.uni-hannover.de, ostermann@tnt.uni-hannover.de

Abstract. Image-based modeling is very successful in the creation of realistic facial animations. Applications with dialog systems, such as e-Learning and customer information service, can integrate facial animations with synthesized speech in websites to improve human-machine communication. However, downloading a database with 11,594 mouth images (about 120MB in JPEG format) used by talking head needs about 15 minutes at 150 kBps. This paper presents a prototype framework of two-step database minimization. First, the key mouth images are identified by clustering algorithms and similar mouth images are discarded. Second, the clustered key mouth images are further compressed by JPEG. MST (Minimum Spanning Tree), RSST (Recursive Shortest Spanning Tree) and LBG-based clustering algorithms are developed and evaluated. Our experiments demonstrate that the number of mouth images is lowered by the LBG-based clustering algorithm and further compressed to 8MB by JPEG, which generates facial animations in CIF format without loss of naturalness and fulfill the need of talking head for Internet applications.

1 Introduction

Visual speech synthesis (talking head) is studied by researchers in computer graphics, image processing, speech processing, artificial intelligence, communication, psychology, etc. Different competing talking head systems have been presented in the first visual speech synthesis challenge LIPS 2008 [1]. The image-based talking head system [2] achieved the most natural animations in terms of audio-visual consistency [3]. An image-based talking head may be combined with dialog systems, such as desktop agents on personal computers, e-Learning and human-car-entertainment services [4], which will open many opportunities in modern human-machine communications.

A typical architecture of a talking head for an Web-based customer information service is shown in Fig. 1(a). A Web server will forward any questions from client to a dialog system, which sends the answer to a TTS (Text-To-Speech) synthesizer. The TTS converts the text of the answer to the corresponding spoken audio track as well as the phonetic information and their duration which is required by the talking head plug-in embedded in the website. The talking

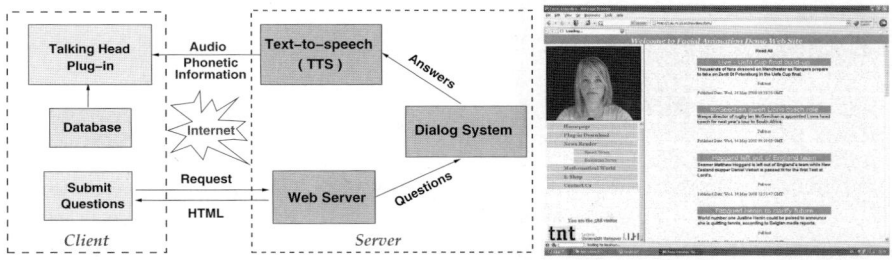

(a) Architecture for customer information service.   (b) Snapshot of a website

Fig. 1. Web-based applications with talking heads

head plug-in selects appropriate mouth images from the database to animate the talking head at the client. A snapshot of the Website is shown in Fig. 1(b).

For a talking head in PAL format, the database with about 12000 mouth images in JPEG format requires 120MB storage space and needs a long time to be downloaded from Internet. Therefore, the database should be minimized to realize the talking head for Internet applications. This paper proposes a prototype framework which can efficiently minimize the database and focuses on MST, RSST and LBG-based clustering algorithms.

The rest of the paper is structured as follows. Section 2 describes the creation of the database. Section 3 presents the proposed framework of database minimization. Experimental results are shown in Section 4, and concluding remarks are drawn in the final section.

2 Database Creation

Our image-based talking head system includes two parts: analysis and synthesis. The audio-visual analysis part creates a database, which is available for the synthesis part to generate animations. The details of the analysis and synthesis were presented in [2] and this section describes the database creation briefly.

A subject is recorded while reading a predefined corpus including about 300 sentences. The motion parameters of recorded subject are estimated by a gradient-based approach [5], which is used to compensate head motion such that mouth images can be cropped from the normalized face sequences. A snapshot of the database with a large number of mouth images is shown in Fig. 2(a).

Since the dimension of normalized mouth image is very high and computation using the original mouth image is inefficient, the dimensionality should be reduced before clustering and compression. PCA (Principal Component Analysis) [6] is very efficient to reduce the dimension of the mouth images. In the PCA space, each base is an eigen mouth and each mouth image is the sum of weighted eigen mouths. The weight is the coordinate of the mouth image in the PCA space. Therefore, the problem of mouth image classification is simplified to cluster the data set in the low dimensional PCA space of the database.

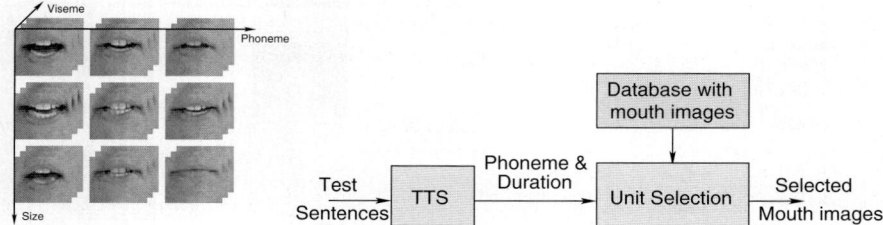

(a) Database with mouth images (b) Determining selection frequency of mouth images

Fig. 2. Frequency determination of using mouth images

3 Database Minimization

This section proposes a framework to minimize the database. First, probability of using a mouth image is determined and only used mouth images are retained in the database. Second, the key mouth images are identified in the PCA space of the retained mouth images by clustering algorithms. For each cluster, one image is selected as a representative image from the cluster. All the representative images build a final database. Last, the final database is further compressed by JPEG. In order to cluster the database, three clustering algorithms are developed and an objective performance measurement of the clustering algorithms is defined.

3.1 Probability Determination of Using Mouth Images

In order to evaluate the relevance of each image in the database, the test corpus including 1457 sentences from different sources is comprised of:

- 400 titles of top news in different categories from BBC website;
- 100 sentences from the story "The Tale of Two Cities";
- 657 sentences from the corpus used for speech synthesis to cover all diphones in English;
- 300 sentences from AT&T research lab.

Using this test corpus, the relative frequency p_i of using the mouth image x_i is determined. By doing so, only the mouth images with $p_i > 0$ are retained and the mouth images with $p_i = 0$ are discarded.

Fig. 2(b) shows the process of the frequency determination, which is part of the synthesis system. Depending on the phonetic information from TTS, the unit selection selects mouth images from the database and assembles them in an optimal way to produce the desired animation. The unit selection balances two competing goals: lip synchronization and smoothness of the transition between consecutive images. Lip synchronization considers the co-articulation effects by matching the distance between the phonetic context of the synthesized phoneme and the phonetic context of the mouth image in the database. The goal of smoothness is to reduce the visual distance at the transition of images in the final animation, favoring transitions between consecutively recorded images. The probability of using mouth images is derived from the selection frequency of mouth images.

3.2 Clustering Algorithms

The goal of the clustering algorithms is to identify key mouth images and discard similar mouth images from the database. For example, a lot of closed mouths are similar and only one is necessary for animations. Similarity depends on the distance d between two points in the PCA space. d is calculated as the weighted Euclidean distance between the two mouth images. Therefore, a threshold T is required to measure the similarity. $d(u,v) < T$ means that mouth image u and v are similar and should be classified in one cluster.

MST-Based Clustering

A spanning tree is an acyclic subgraph of a graph G, which contains all the vertices from G. The MST of a weighted graph is the minimum weight spanning tree of that graph. Prim's algorithm [7] is known to be a good algorithm to find an MST. The algorithm continuously increases the size of a tree starting with a single vertex until it spans all the vertices. The cost of constructing an MST is $O(mlogn)$, where m is the number of edges in the graph, n is the number of vertices. The classical MST based clustering algorithms begin with any point in the PCA space to construct an MST. From this tree, any edge with a weight $d(u,v) \geq T$ is removed from the tree. This leads to a set of disjoint subtrees $S_C = \{C_1, C_2, ...\}$. Each of the subtrees C_t is treated as a cluster, for which a representative point should be found. All representative points are collected to build a final database.

Each image in the database is a vertex of the graph. Prim's algorithm takes a long time and it becomes impossible to construct an MST for our database, since the database consists of a large number of mouth images, even though some non-used mouth images with $p_i = 0$ are discarded. In order to speed up MST construction for our database, we modify the algorithm by combining the two steps as follows. Once the weight of a new edge of the subtree is bigger than the threshold T, we stop constructing the subtree and the subtree is treated as a cluster. The remaining vertices of the database are treated the same to build clusters until all the vertices are part of the subtrees. The modified MST based clustering algorithm can build the same subtrees according to the threshold T as the classical MST based clustering does, but the computing time for our database is reduced to maximal 3 hours from several months or more.

To find the representative image, we assume that there are n_t points in the cluster C_t, the average node c_t and its standard deviation σ_t are computed considering the unit probability in the following way:

$$c_t = \sum_{j=1}^{n_t} p_j V_j \quad ; \quad \sigma_t = \sqrt{\sum_{j=1}^{n_t} p_j (V_j - c_t)^2} \quad ,$$

where V_j is a vector with the PCA weights of mouth image j in C_t and p_j is the probability of using the mouth image j. If the condition $\sigma_t < T/6$ is fulfilled, the nearest point to c_t is selected as the representative point in the cluster C_t. Otherwise, cluster C_t will be approximated by two Gaussian mixture distributions or more, till the condition is fulfilled.

RSST-Based Clustering

RSST-based clustering algorithm presents a powerful solution to the problem of incorporating global information into clustering algorithm [9]. RSST begins with the shortest link in the graph and merges the two vertices joined by this link. A new vertex and link weights are recalculated in the region. The region represents a vertex or many vertices in the same partition. The process will be repeated until the number of regions are enough for the clustering.

In the case of mouth image clustering, we define the region as a mouth image in the PCA space at the initialization stage or many mouth images clustered in a partition. The link weights are calculated by the weighted Euclidean distance of two mouth images. The RSST-based clustering algorithm begins with finding the least link in the graph and merging the two mouth images adjoined by this link into a region. The average PCA weight of the region is calculated to represent the new vertex of the region and the link weights of the region are updated. This process will be done recursively until the desired regions are obtained. We define a threshold T that controls the building of RSST. If the next least link weight is bigger than T, we stop the construction of RSST. The mouth image, which is the nearest to the average PCA weight, is treated as the representative mouth image of the region.

LBG-Based Clustering

We assume that there is a training data set consisting of M vectors: $\tau = \{x_1, x_2, ..., x_M\}$ and the vectors are K-dimensional: $x_m = (x_{m,1}, x_{m,2}, ..., x_{m,K})$, $m = 1, 2, ..., M$. In our case, the vector represents the PCA weight of a mouth image. The LBG VQ design algorithm [8] is an iterative algorithm to find the partition $S = \{s_1, s_2, ..., s_P\}$ and their representative vectors $r_p = (r_{p,1}, r_{p,2}, ..., r_{p,K})$, $p = 1, 2, ..., P$, which are subject to $Q(x_m) = r_p$, $x_m \in s_p$, such that the global average distortion is minimized in the following way, considering the probability p_m of using the mouth image x_m:

$$D_{ave} = \frac{\sum_{m=1}^{M} p_m \cdot \|x_m - Q(x_m)\|^2}{K \cdot \sum_{m=1}^{M} p_m}.$$

However, the classic LBG does not consider the maximum distortion of the partition, which results in jerky animations, when the mouth images are selected from the partitions with a large distortion. To overcome this problem, we define a threshold R that controls the size of the partitions. The partitions are further split, if the maximal distortion of the partitions is larger than the threshold. The clustering algorithm is repeated until all the partitions fulfill the threshold condition. Considering the probability of using a mouth image, the distortion $D(s_p)$ between any point and the representative point r_p in the partition s_p fulfills the following condition: $D(s_p) = \left\| x_m - r_p \right\|^2 < R, \ \forall \ x_m \in s_p$.

Objective Evaluation of Clustering Algorithms

To evaluate the clustering algorithms, PSNR is chosen as the objective measurement. The PSNR between the original database and the final database is defined as:

$$PSNR = 10 \cdot log_{10} \frac{255^2}{MSE} \quad ; \quad MSE = \frac{\sum_{m=1}^{M} P_{x_m} \cdot \left\| I_{x_m} - I_{Q(x_m)} \right\|^2}{w \cdot h \cdot \sum_{m=1}^{M} P_{x_m}}$$

where MSE is the mean square error, I_{x_m} is the pixel value vector of mouth image x_m, w and h are the width and height of the mouth image, $Q(x_m)$ is the representative image of x_m, P_{x_m} is the probability of using the mouth image x_m.

3.3 Compression of Final Database

JPEG and H.264 are the most efficient coding methods for pictures and sequences. Due to the discontinuity of the mouth images in the final database, inter coding of images by H.264 is not useful. The efficiency of the intra coding of H.264 is similar to the JPEG efficiency. In practice, JPEG is the most popular and efficient coder for pictures and very suitable for our case. The size of compressed database in JPEG is proportional to the number of mouth images in the final database.

4 Experimental Results

In order to determine the probability of using mouth images, we have built an image-based talking head system [2] with 11594 mouth images in the database. The resolution of the talking head image is 720×576 and the cropped mouth image size is 176×208. This means the original database is about $640MB$ in YUV format and $120MB$ in JPEG format.

Fig. 3(a) shows the number of mouth images that are not selected by the unit selection given the number of test sentences. 4230 mouth images are never used by the test corpus. Almost 36.6% of the mouth images are not selected by the unit selection. 93.7% of the non-used mouth images are labeled with silence. Given 1457 test sentences, the logarithm of the selection frequency of the mouth images is shown in Fig. 3(b), where the mouth images are sorted in the order of selection frequency.

The performance of the clustering algorithms is measured by PSNR between the final database and the original database as shown in Fig. 4. The black curve represents the PSNR between the final database clustered by LBG and the original database. The red one corresponds to RSST-based clustering and the blue

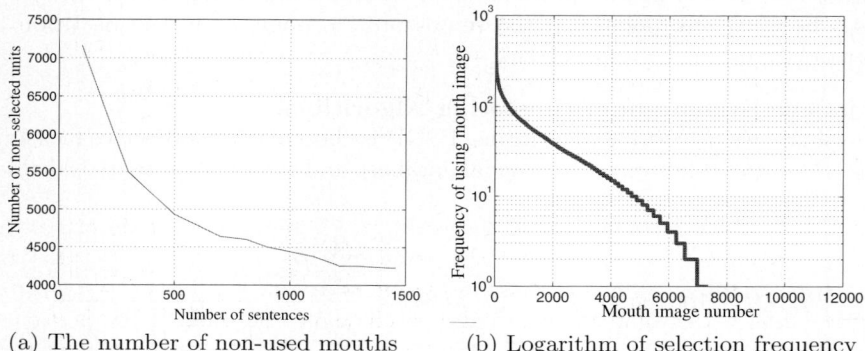

(a) The number of non-used mouths   (b) Logarithm of selection frequency

Fig. 3. Results of probability determination. (a) The number of non-used mouth images given the number of input sentences. (b) Logarithm of the selection frequency determined by unit selection from 1457 test sentences.

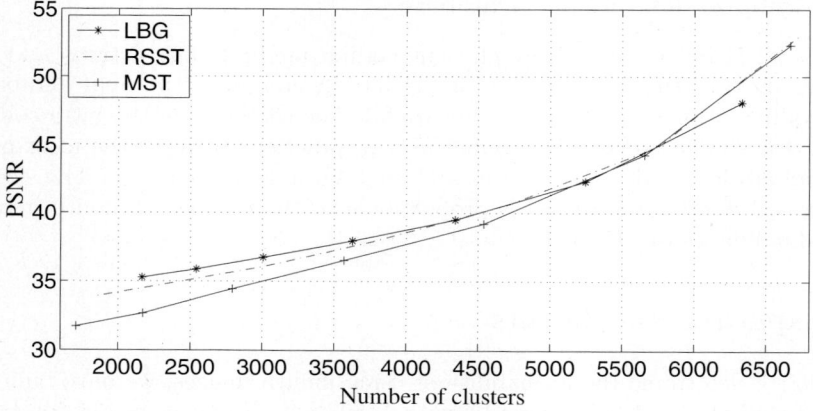

Fig. 4. PSNR Performance of clustering algorithms

one corresponds to MST-based clustering. LBG-based clustering performs better than the others, if the clustered database contains less than 5400 mouth images.

After clustering the database, the final database is compressed by JPEG. The size of the final database depends on the number of mouth images in the final database and the resolution of the talking head image. For example, 3000 mouth images need $29MB$ storage space in PAL format or $8MB$ in CIF format.

In order to evaluate the proposed clustering algorithms subjectively, we generated animations by using the original database and the final database clustered by MST, RSST and LBG with different size. These animations are shown to the viewers and they were asked to score the quality of the animations generated by using the final databases. The results of the informal subjective tests show that the database with not less than 3000 mouth images clustered by LBG, can synthesize animations without loss of naturalness.

To see the animations according to Fig. 4, the reader is encouraged to visit http://www.tnt.uni-hannover.de/project/facialanimation/demo/minidb/

5 Conclusion

In this paper we have presented a prototype framework for minimizing the database of unit selection so that the real-time talking head for Internet applications is possible. The database reduction is carried out in two steps: First, the database with useful mouth images is clustered; Second, JPEG is used to compress the final database. MST, RSST and LBG-based clustering algorithms are proposed and evaluated.

Experimental results show that the proposed methods can reduce the database efficiently. LBG-based clustering algorithm performs better than others given a small size of database. According to the subjective tests, the animations can be generated by a small database with at least 3000 mouth images without loss of naturalness.

Furthermore, because the non-used mouth images are discarded from the database, the number of candidates are reduced in the Viterbi search, so that the unit selection performs faster and more efficiently.

References

1. Theobald, B., Fagel, S., Bailly, G., Elisei, F.: LIPS2008: Visual Speech Synthesis Challenge. In: Proc. Interspeech 2008, Brisbane, Australia, September 2008, pp. 2310–2313 (2008)
2. Liu, K., Ostermann, J.: Realistic Facial Animation System for Interactive Services. In: Proc. Interspeech 2008, Brisbane, Australia, September 2008, pp. 2330–2333 (2008)
3. LIPS 2008: Visual Speech Synthesis Challenge (2008), http://www.lips2008.org/
4. Liu, K., Ostermann, J.: Realistic Talking Head for Human-Car-Entertainment Services. In: Proc. IMA 2008 Informationssysteme fuer mobile Anwendungen, Braunschweig, Germany, September 2008, pp. 108–118 (2008)
5. Weissenfeld, A., Urfalioglu, O., Liu, K., Ostermann, J.: Robust Rigid Head Motion Estimation based on Differential Evolution. In: IEEE Proc. ICME 2006, Toronto, Canada, July 2006, pp. 225–228 (2006)
6. Jolliffe, I.: Principal Component Analysis. Springer, New York (1989)
7. Prim, R.C.: Shortest connection networks and some generalizations. Bell System Technical Journal 36, 1389–1401 (1957)
8. Linde, Y., Buzo, A., Gray, R.M.: An algorithm for vector quantizer design. IEEE Trans. Commun. COM-28, 84–95 (1980)
9. Morris, O.J., Lee, M.J., Constantinides, A.G.: Graph theory for image analysis: an approach based on the shortest spanning tree. In: Proc. Inst. Electr. Eng., vol. 133, pp. 146–152 (1986)

Analysis of Speed Sign Classification Algorithms Using Shape Based Segmentation of Binary Images

Azam Sheikh Muhammad, Niklas Lavesson, Paul Davidsson, and Mikael Nilsson

Blekinge Institute of Technology, SE-372 25 Ronneby, Sweden
{azam.sheikh.muhammad,niklas.lavesson,paul.davidsson,
mikael.nilsson}@bth.se

Abstract. Traffic Sign Recognition is a widely studied problem and its dynamic nature calls for the application of a broad range of preprocessing, segmentation, and recognition techniques but few databases are available for evaluation. We have produced a database consisting of 1,300 images captured by a video camera. On this database we have conducted a systematic experimental study. We used four different preprocessing techniques and designed a generic speed sign segmentation algorithm. Then we selected a range of contemporary speed sign classification algorithms using shape based segmented binary images for training and evaluated their results using four metrics, including accuracy and processing speed. The results indicate that Naive Bayes and Random Forest seem particularly well suited for this recognition task. Moreover, we show that two specific preprocessing techniques appear to provide a better basis for concept learning than the others.

Keywords: Traffic Sign Recognition, Segmentation, Classification.

1 Introduction

Automatic Traffic Sign Recognition (TSR) systems attempt to detect and recognize traffic signs from live images captured by a video camera mounted on a vehicle. The development of such a visual pattern recognition system is not a trivial task [1]. It has attracted the research community since the eighties [2]. There are a number of issues associated with the automatic detection and classification of traffic sign patterns from real-world video or images. Visibility is the key issue in the performance of a TSR system because it determines the quality of the captured images and hence affects classification performance. A TSR system can only attract the transport community if it can outperform, or at least perform comparably to, humans in correctly locating and recognizing signs at low visibility. Visibility issues can arise due to many reasons. For instance, the various lighting conditions at different hours of the day and the different seasons have a strong impact on visibility. Additionally, because of trees or shadows of nearby buildings the signs may be partially hidden. These and other common issues have been discussed in related studies [1]. Researchers have applied a variety of preprocessing techniques based on color processing [3, 4], shape analysis [3-5], along with numerous classification and recognition techniques ranging from template matching algorithms [3, 6], Radial Basis Function (RBF) Network [7, 8], Multilayer

Perceptron and other Neural Networks [7, 8], Genetic Algorithms [9], Fuzzy Logic [10] to the applications of Support Vector Machines [11] and signal processing-based transformation-specific classifiers.

Accuracy and processing speed are two important performance metrics in the development of TSR systems. Some studies present detection and/or classification results around 90% [12, 13] or in some cases 95% [10] or more. However, few studies compare with other databases than their own and the used database is seldom made public. Moreover, the level of detail in describing the preprocessing and other steps is usually low, which makes it difficult to reproduce the experiment. Besides related work and also some commercial implementations of TSR systems [14] there are practically no significant traffic sign databases available. We have only managed to find a small database, consisting of 48 images of three different traffic signs, available online [15]. This study aims to remedy this situation by making our initial database publicly available. We also expect to enlarge the database in future work. For the initial database, we have collected 1,300 images from Swedish roads. Detection and recognition of speed signs in particular have been extensively studied [13, 16, 17]. In this work our main focus is to conduct an extensive study on various algorithms for the classification of this particular type of sign. For this purpose we extract the shape based binary information from the sign images to generate training and testing data sets for supervised concept learning. A number of classifiers are evaluated for speed limit recognition under various performance metrics. In order to make the preprocessing, segmentation and classification experiments reproducible, standard tools are used and each step is described in detail. Our database of traffic sign images and the implementation results are available online[1].

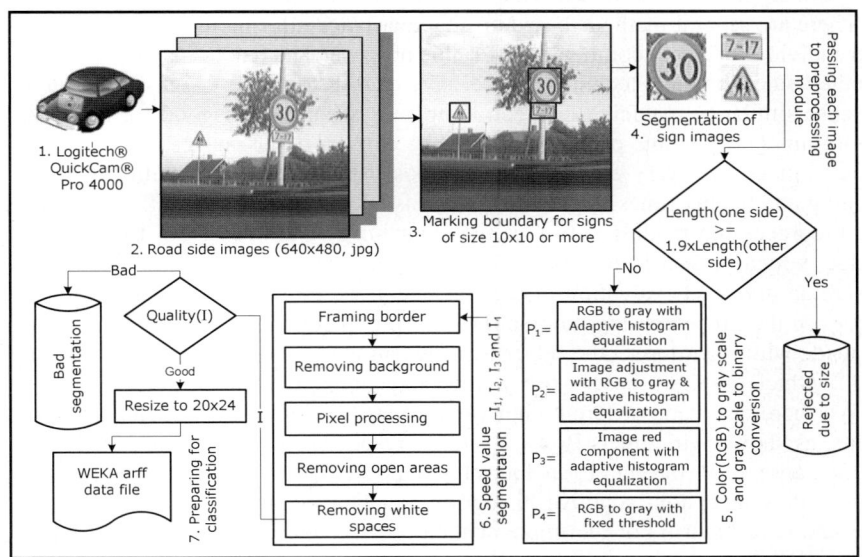

Fig. 1. Traffic signs collection & processing

---

[1] www.bth.se/tek/nla.nsf/sidor/resources

2 Data Collection, Size Analysis and Preprocessing

We collected 1,300 raw images of the traffic signs using a Logitech Quick Cam Pro 4,000 web camera with automatic settings and a resolution of *640×480* for each image. The collection of images along road sides was performed during different hours of the day, and thus under various lighting conditions to maximize visibility disorders. Since this study is concerned with speed signs only, each such sign (30, 50, 70, 90 and 110) is marked and associated with a corresponding class. We also considered multiple speed signs present in a single image and included a reasonable collection of non-speed signs as well. We labeled such signs as belonging to the Noise class. In Fig. 1, steps 1 to 4 describe the image capturing, boundary marking and the segmentation of sign images from the original image for further processing. During boundary marking (step 3 in Fig. 1) we have marked signs of size *10×10* pixels or more.

A traffic sign detection algorithm has to process images that include rectangular shapes, as opposed to the circular shape of the speed signs. Thus, we also marked rectangular signs as well as some non-sign rectangular shapes. To avoid unnecessary processing of rectangular shapes and to make the system robust against non-speed signs, a basic size analysis is conducted before passing the sign image to the preprocessing module. We have analyzed the height and width of the images for each class in our database, and can conclude that, for all of the properly segmented speed signs as well as other traffic sign images, the height and width are almost identical. For the other images the height is roughly *1.5* times the width. So, in our size analysis heuristic approach, we simply discard an image if the width of the image is *1.9* times the height or vice versa. We have kept this threshold slightly flexible just to ensure that no speed sign image is rejected. This check is shown after step 4 in Fig. 1.

There are several methods based on shape or color information that can be used either individually, or in combination, for this purpose. The reader may refer to [1] for a detailed study on such techniques. We have chosen to process signs based on shape because similarity matching between templates is considered good for classification algorithms [1]. For this purpose we need to convert color images to binary images. Binary images are very economical as opposed to color images with respect to the training and testing times required for the classification algorithms. Thus the scope of our preprocessing module is to take the color image and to convert it to a gray scale image, which is then further processed to get a binary (black and white) image. To ensure an unbiased evaluation of the classifiers, we implemented four different techniques in the preprocessing module. For our preprocessing, critical decisions need to be made with regard to 1) the selection of technique for color to grayscale conversion and 2) which gray threshold to use in order to generate the binary image.

We therefore made different choices for these decisions and developed four invariants, as shown with labels P_1, P_2, P_3 and P_4 respectively in step 5 of Fig. 1. Our implementation of P_1 through P_4 includes regular MATLAB functions for color image to gray scale conversion, adaptive histogram equalization and gray level threshold for conversion to binary. For the implementation of P_1 *rgb2gray* is used to convert from color image to gray image then *adapthisteq*, *graythresh*, and *im2bw* are applied in sequence to get the binary image. In P_2 the input image is first adjusted with regard to color intensities. For this purpose the function *imadjust* is used and the input parameter *[low_in; high_in]* is given as *[0.3 0.2 0.1; 0.8 0.6 0.8]*. We have analyzed the results using various sets of values for this parameter and heuristically we have concluded that the above given set of values produces sufficiently good results. The adjusted image is then processed similar to P_1. For P_3 we used the intensity of the red component from the color image as the gray level to get a gray image. The remaining steps are the same as for P_1. In P_4, the same method as the one used for P_1 is followed, but the gray threshold is further processed so that, if the threshold θ is greater than ½, it is recalculated as defined below:

$$\theta = \begin{cases} \frac{7}{10}\theta, & \theta > \frac{5}{8} \\ \frac{8}{10}\theta, & else \end{cases} \quad (1)$$

We have developed a technique consisting of six simple steps of sequential processing for possible segmentation of the speed limit value as shown in step 6 of Fig. 1. Our segmentation logic is independent of the preprocessing techniques discussed above. Output binary images I_1, I_2, I_3 and I_4 from the preprocessing P_1 through P_4 form the input to the segmentation module. The same segmentation steps are applied on each of I_1 through I_4, separately. The first step is the framing border, which ensures maximum reduction of the background noise. Black borders are added in the background with a size of *10%* of the length of each side. Then, in the next step the background noise is filtered by removing *4*-connected neighborhood pixels. It also removes the frame generated during the previous step. In the third step, the isolated black pixels and single pixel connected black lines are converted to white. This process removes small clusters of noise and separates single pixel connected black regions. After step three, some large clusters of noise in the form of groups of connected black pixels still exist. We call them *open areas* since they are not part of the shape of the speed limit value. We remove them in the fourth step. This time we search for *8*-connected neighbors and remove those open areas which have a pixel count below a certain threshold. The threshold pixel count (*Pc*) is taken as the percentage of black pixels in the image plus some constant integer value such as *7* or *8*. This step also ensures that the speed limit value is not affected. Next, we remove white spaces from all four sides and define a bounding box for the segmented speed limit value. After step five, the segmentation process is over. The last step is basically to determine the quality of the segmented image. We define segmentation quality as either *good* or *bad*. If the quality is *good*, the image is resized to a predefined width and height, so that each instance given to the classifier is of the same size. The dimension of the resized images is *20×24*, a dimension which was determined by taking the average size of the segmented images based on the output of the four preprocessing methods. The quality analysis is the most important (and the most technical) step in the segmentation algorithm. We will now discuss why quality analysis is important and describe our quality analysis approach.

Normally, on the road sides there are a lot of different signs and speed signs appear rarely. For example, in our collection only *20%* of the images represent speed signs. Since we are only concerned with the speed sign images, we classify each non-speed sign as Noise. For this purpose, a proper shape segmentation of the speed limit value is desired. At the same time, it is important to design robust criteria to deal with non-speed signs. It is not a good idea to simply use every non-speed sign image as a member of the Noise class. Non-speed signs do not follow any predefined pattern and thus it will be difficult for a classifier to properly learn the Noise class. During segmentation, any prior information about the class of the input image is not present and it is upon the segmentation method to reject as many of the noisy images as possible. Thus, during the last step of the segmentation process we analyze the images based on the quality criteria shown in Eqn. 2. A binary decision in terms of *good* or *bad* quality is made for each segmented image. Bad segmentations are discarded whereas good segmentations are normalized with regard to size and used for the further preparations for classification.

$$[r\ c] \leftarrow Size(I)$$
$$nbr \leftarrow numBlackRegions(I)$$
$$ws \leftarrow Sum(I, white)$$
$$Quality(I) = \begin{cases} Bad, & \begin{cases} r < 13, or\ c < 13,\ or \\ 3 > nbr \leq 1, or \\ ws \leq \frac{15}{100}(r \times c), or \\ ws \geq \frac{85}{100}(r \times c) \end{cases} \\ Good, & else \end{cases} \quad (2)$$

Our quality analysis is independent of the preprocessing and segmentation techniques. The *quality* check serves two main purposes. It allows the selection of all well segmented speed sign images by marking them as *good quality* and rejecting poor segmentation results by labeling them as *bad quality* segmentation to achieve a high hit ratio. Secondly, when we apply the check on the segmentation results of noisy images, most of them are simply rejected as being bad segmentations. Thus, the non-speed signs that are regarded as good quality after the segmentation process are the only non-speed signs that will be included in the Noise class. Therefore a classification as *good* is expected for each speed sign while *bad* is desired for every non-speed sign. The segmentation quality analysis is also robust against visibility disorders of the images and tries to reject most of them to avoid misclassifications at the later stage. The experimental results seem to indicate that all four preprocessing techniques perform well in rejecting non-speed signs, since 95% to 98% of these signs are put into the bad segmentation collection.

3 Analysis of Classification Algorithms

From the segmentation results we get the shape representation of the sign as binary images. To conduct the experiments we use the Weka [18] machine learning workbench. To examine the classification results separately for each preprocessing technique, we constructed four ARFF-based data sets. These data sets include six different classes: the speed limit values 30, 50, 70, 90, 110, and the Noise class. Segmentation results for each preprocessing technique are associated with the respective class labels and thus a training data set of binary strings for each preprocessing is generated. To convert the training set into Weka ARFF format we defined a relation consisting of 480 nominal attributes, each corresponding to the binary value of a pixel.

3.1 Algorithm Selection and Performance Analysis

The main objective of our experiments is to evaluate the performance of different classification algorithms for the speed sign recognition problem. The generated classifiers are evaluated based on the results from all four data sets. We have selected a diverse population of 15 algorithms from different learning categories (e.g., Perceptron and Kernel functions, Bayesian learners, Decision trees, Meta-learners, and Rule inducers), see Table 1 column one. We used Weka algorithm implementations and tried various parameter configurations but observed no significant difference in classifier performance as compared to the default configuration, except for Multilayer Perceptron. We observed that, with *8* neurons in the hidden layer and the learning rate set to *0.2* (the default is *0.3*) the same recognition performance was achieved, but a significant decrease in training time was achieved.

Table 1. Experiment results for accuracy and AUC

Algorithm	Accuracy (% correct) Dataset				Avg.	Area under ROC curve Dataset				Avg.
	P1	P2	P3	P4		P1	P2	P3	P4	
Multilayer Perceptron	94.75 ± 4.78	94.4 ± 5.68	94.56 ± 5.08	91.36 ± 7.24	93.8	1.00 ± 0.00	0.94 ± 0.19	1.00 ± 0.01	0.87 ± 0.33	0.95
RBF Network	93.93 ± 5.50	93.87 ± 6.39	95.96 ± 4.90	90.82 ± 8.22	93.7	0.99 ± 0.02	0.94 ± 0.16	0.99 ± 0.01	0.95 ± 0.11	0.97
Ridor	80.85 ± 9.95	79.03 ± 9.41	81.08 ± 8.31	78.64 ± 10.61	79.9	0.87 ± 0.17	0.85 ± 0.18	0.82 ± 0.18	0.67 ± 0.24	0.8
PART	82.63 ± 9.09	83.62 ± 9.63	83.9 ± 7.44	80.64 ± 10.31	82.7	0.88 ± 0.18	0.83 ± 0.19	0.79 ± 0.22	0.66 ± 0.30	0.79
NNge	78.44 ± 9.67	75.27 ± 11.89	78.57 ± 9.30	81.64 ± 11.77	78.5	0.75 ± 0.21	0.7 ± 0.20	0.73 ± 0.21	0.68 ± 0.24	0.72
JRip	80.13 ± 9.02	74.01 ± 10.39	78.14 ± 9.31	75.00 ± 11.73	76.8	0.89 ± 0.16	0.81 ± 0.21	0.87 ± 0.19	0.61 ± 0.33	0.80
SMO	97.6 ± 3.33	96.73 ± 4.15	98.86 ± 2.37	95.36 ± 5.85	97.1	0.97 ± 0.10	0.91 ± 0.21	1.00 ± 0.01	0.88 ± 0.32	0.94
J48	84.27 ± 7.98	85.11 ± 10.96	89.19 ± 6.98	79.55 ± 10.53	84.5	0.9 ± 0.16	0.82 ± 0.18	0.84 ± 0.21	0.72 ± 0.30	0.82
Random Forest	93.22 ± 6.33	92.82 ± 6.27	95.28 ± 4.72	88.91 ± 8.23	92.6	0.98 ± 0.08	0.95 ± 0.13	0.99 ± 0.04	0.91 ± 0.21	0.96
Bagging	89.77 ± 7.20	87.27 ± 7.67	88.76 ± 6.94	82.55 ± 8.92	87.1	0.99 ± 0.06	0.96 ± 0.10	0.99 ± 0.02	0.85 ± 0.27	0.95
Dagging	81.28 ± 6.20	80.98 ± 6.27	82.89 ± 4.81	70.27 ± 5.90	78.9	0.99 ± 0.03	0.92 ± 0.15	0.98 ± 0.05	0.79 ± 0.33	0.92
END	90.34 ± 7.58	90.3 ± 8.42	92.22 ± 7.05	87.27 ± 9.76	90.0	0.99 ± 0.06	0.95 ± 0.13	0.98 ± 0.07	0.86 ± 0.29	0.95
Random Committee	89.44 ± 6.81	89.10 ± 7.78	90.87 ± 6.49	85.09 ± 8.62	88.6	0.96 ± 0.08	0.91 ± 0.15	0.95 ± 0.11	0.89 ± 0.22	0.93
Logit Boost	91.27 ± 6.58	93.00 ± 6.77	94.12 ± 5.43	87.09 ± 8.78	91.4	1.00 ± 0.00	0.94 ± 0.18	1.00 ± 0.01	0.85 ± 0.32	0.95
Naïve Bayes	97.46 ± 3.94	94.54 ± 6.59	97.4 ± 3.44	93.55 ± 6.87	95.7	1.00 ± 0.00	0.97 ± 0.10	1.00 ± 0.00	0.89 ± 0.28	0.97
Average	83.25	82.00	84.10	79.85	82.3	0.89	0.84	0.87	0.75	0.84

Multilayer Perceptron is the most widely used [7, 9] algorithm in TSR applications, especially for speed sign recognition. Consequently, we used it as the base classifier in Weka (with optimized parameters) and all other classifiers (generated with default configurations) are evaluated against it. In addition to Multilayer Perceptron, RBF Network and Support Vector Machines are also used in a number of related studies, as described earlier. Thus, it was natural to include these algorithms in our experiments. We performed ten *10*-fold cross-validation tests and used the corrected paired t-test (confidence *0.05*, two-tailed) for all four data sets. We compared the performance based on four evaluation metrics; accuracy, the Area under the ROC curve (AUC), training time, and testing time. The experimental results are shown in Table 1. The average performance over all data sets is also presented in the Table. With respect to accuracy and AUC, we observe that the best performing algorithms are: Multilayer Perceptron, RBF Network, SMO (Weka implementation of SVM), Random Forest, and Naive Bayes. Now we consider the training and testing time. Besides accuracy, the elapsed time is also very important in the performance evaluation of classifiers, with respect to the application at hand. An analysis of training and testing times demonstrates that, among our best algorithms with respect to accuracy and AUC, Random Forest and Naive Bayes are by far the fastest algorithms with regard to both training and testing. From the results we can conclude that, aside from good performance by the commonly used classifiers for this problem, Naive Bayes and Random Forest have achieved quite promising results in terms of accuracy and significantly better training and testing times than the other algorithms. We have also analyzed the results of the individual preprocessing techniques. For almost all of the 15 algorithms and specially the above mentioned five algorithms, P_1 and P_3 have shown consistently a higher accuracy than P_2 and P_4. We can also observe that both P_1 and P_3 have better AUC values than the other two techniques.

4 Conclusion and Future Work

Our proposed size analysis criterion is able to properly differentiate the detected traffic signs from those of under or over segmented signs and other noisy segmentations. The criterion is database independent and hence it can be applied to any collection of traffic sign images. Due to the real-time nature of TSR applications, a short recognition time in conjunction with good accuracy is always desirable. This is indeed an important trade-off for most real-time recognition systems. The experimental results indicate that Naive Bayes and Random Forest are quite accurate and have better training and testing times than the other studied algorithms. We conclude that our approach is suitable, at least for the studied database. In addition, we have evaluated four preprocessing techniques. Experiments indicate that P_1 and P_3 are more suitable for preprocessing speed signs. Moreover, we have introduced the concept of segmentation quality analysis and proposed a general speed limit value segmentation technique. The two preprocessing techniques together with the segmentation algorithm provide a good basis for the development of a TSR system using the proposed classifiers. For future work, we intend to collect a larger database incorporating all types of traffic signs.

References

1. Nguwi, Y.Y., Kouzani, A.Z.: A study on automatic recognition of road signs. In: 2006 IEEE Conference on Cybernetics and Intelligent Systems, pp. 1–6 (2006)
2. Mace, D.J., Pollack, L.: Visual Complexity and Sign Brightness in Detection and Recognition of Traffic Signs. Transportation Research Record HS-036 167(904), 33–41 (1983)
3. Malik, R., Khurshid, J., Ahmad, S.N.: Road Sign Detection and Recognition using Colour Segmentation, Shape Analysis and Template Matching. In: 6th International Conference on Machine Learning Cybernetics, pp. 3556–3560 (2007)
4. Andrey, V., Hyun, K.: Automatic Detection and Recognition of Traffic Signs using Geometric Structure Analysis. In: SICE-ICASE International Joint Conference, pp. 1451–1456 (2006)
5. Cyganek, B.: Real-time detection of the triangular and rectangular shape road signs. In: Blanc-Talon, J., Philips, W., Popescu, D., Scheunders, P. (eds.) ACIVS 2007. LNCS, vol. 4678, pp. 744–755. Springer, Heidelberg (2007)
6. Cyganek, B.: Road Signs Recognition by the Scale-space Template Matching in the Log-polar Domain. In: Martí, J., Benedí, J.M., Mendonça, A.M., Serrat, J. (eds.) IbPRIA 2007. LNCS, vol. 4477, pp. 330–337. Springer, Heidelberg (2007)
7. Nguwi, Y.Y., Kouzani, A.Z.: Automatic Road Sign Recognition using Neural Networks. In: IEEE International Joint Conference on Neural Networks, pp. 3955–3962 (2006)
8. Zhang, H., Luo, D.: A new method for traffic signs classification using probabilistic neural networks. In: Wang, J., Yi, Z., Zurada, J.M., Lu, B.L., Yin, H. (eds.) ISNN 2006. LNCS, vol. 3973, pp. 33–39. Springer, Heidelberg (2006)
9. Aoyagi, Y., Asakura, T.: Detection and Recognition of Traffic Sign in Scene Image using Genetic Algorithms and Neural Networks. In: 35th SICE Annual Conference, pp. 1343–1348 (1996)
10. Fleyeh, H., Gilani, S.O., Dougherty, M.: Road Sign Detection and Recognition using Fuzzy ARTMAP: a case study Swedish speed-limit signs. In: 10th IASTED International Conference on Artificial Intelligence and Soft Computing, pp. 242–249 (2006)
11. Maldonado-Bascon, S., Lafuente-Arroyo, S., Gil-Jimenez, P., et al.: Road-Sign Detection and Recognition based on Support Vector Machines. IEEE Transactions on Intelligent Transportation Systems 8(2), 264–278 (2007)
12. Soetedjo, Y.K.: Traffic Sign Classification using Ring Partitioned Method. IEICE Trans. on Fundamentals E88(9), 2419–2426 (2005)
13. Torresen, J., Bakke, J.W., Sekanina, L.: Efficient Recognition of Speed Limit Signs. In: 7th International IEEE Conference on Intelligent Transportation Systems, pp. 652–656 (2004)
14. Siemens: Siemens VDO 2007 (2007), http://usa.vdo.com/products_solutions/cars/adas/traffic-sign-recognition/
15. Grigorescu, P.N.: Traffic Sign Image Database, http://www.cs.rug.nl/~imaging/databases/traffic_sign_database/traffic_sign_database_2.html
16. Cao Tam, P., Elton, D.: Difference of Gaussian for Speed Sign Detection in Low Light Conditions. In: International MultiConference of Engineers and Computer Scientists, pp. 1838–1843 (2007)
17. Moutarde, F., Bargeton, A., Herbin, A., et al.: Robust On-vehicle Real-time Visual Detection of American and European Speed Limit Signs with a Modular Traffic Signs Recognition System. In: IEEE Intelligent Vehicles Symposium, pp. 1122–1126 (2007)
18. Witten, H., Frank, E.: Data Mining: Practical Machine Learning Tools and Techniques, 2nd edn. Morgan Kaufmann, San Francisco (2005)

Using CCD Moiré Pattern Analysis to Implement Pressure-Sensitive Touch Surfaces

Tong Tu and Wooi Boon Goh

School of Computer Engineering,
Nanyang Technological University,
Singapore 639798
{tuto0001,aswbgoh}@ntu.edu.sg

Abstract. The Moiré fringe patterns obtained when a CCD camera views a repetitive line grating can be exploited to measure small changes to surface displacements. We describe how curved surfaces with line grating patterns can be reconstructed by analysing the instantaneous frequency of the extracted 1D Moiré waveform. Experimental results show that monotonically increasing displacements of a stretched canvas of less than 1mm can be clearly separated, suggesting the possibility of using the proposed Moiré-based vision technique to construct accurate pressure-sensitive touch surfaces.

Keywords: Moiré patterns, Image-based metrology, Surface deformation analysis, Human computer interface, Vision-based interface.

1 Introduction

Touch surfaces enable computer interaction using natural finger gestures. The trend in multi-touch interaction is fast gaining widespread acceptance with products like iPhone [1], Microsoft Surface [2], etc. The technologies behind these touch-sensitive surfaces include electronic-based [1,3] and vision-based [2,4,5,6,7]. Vision-based touch surfaces are relatively pervasive due to the affordability of imaging devices and the improved vision algorithms and its computational speed. Such system normally detect finger touches using reflection from non-visible infrared light sources [6,7] or shadows [5]. However, few systems are able to accurately detect the pressure being asserted by the fingers. Some like [6] rely on uncorrelated finger contact time or unreliable increase in the diameter of the detected finger contour [7] to infer increasing pressure. Others like [4] require special deformable surfaces with embedded color markers.

This paper explores the possibility of addressing pressure-sensitive touch detection by analyzing the changing Moiré patterns that result when a CCD camera is used to observe the small local deformations on a deformable interactive surface that is printed with fine line grating. Moiré patterns are the results of the interference fringes produced by superimposing two or more sets of repetitive gratings. These patterns have been used in metrology for tasks such as strain measurements, vibration analysis and the recovery of the surface geometry of 3-dimensional objects [8,9,10,11]. Traditionally, Moiré images are obtained by using a camera to capture the patterns generated by superimposing two alternating opaque-transparent gratings called Ronchi gratings [9] or the resulting interaction between two projected light patterns. A simpler approach is to use a camera's regular CCD cell arrangement as a reference grating. In [12], it was shown that the sinusoidal-like Moiré patterns obtained when a camera views a line grating can be easily extracted and modeled with the steps shown in Fig. 1c-1f. We extended the work in [12] to extract camera-to-surface distances of smooth curved surfaces. We were able to measure small local displacements in a deformable patterned surface, which suggest its potential for implementing a pressure-sensitive touch surface.

2 Moiré Patterns

Moiré patterns in our study are generated by imaging line gratings on a specimen surface using a standard CCD camera with regularly arranged CCD cells. The pitch S of the line gratings on the surface is reduced to pitch p_s' on the CCD's image through an approximate camera lens geometry shown in Fig. 1a, where the distance d, between line grating surface and the centre line of the camera's lens is much larger than the camera's focal length given by F (i.e. $d >> F$).

$$p_s' = \frac{FS}{d} \tag{1}$$

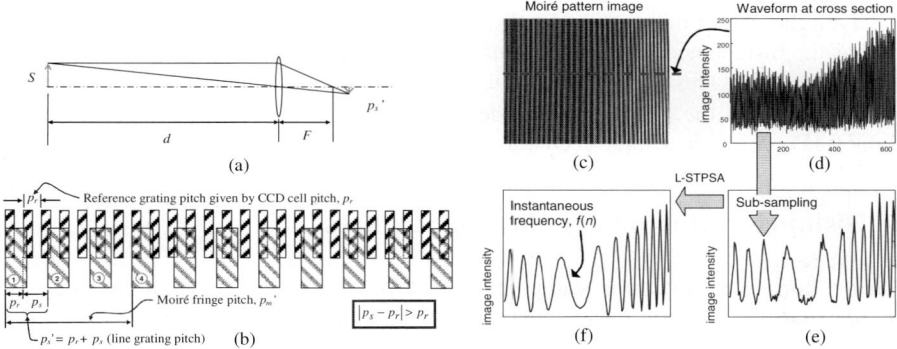

Fig. 1. (a) The camera geometry relating the line grating's pitch S to p_s' given in (1). (b) Moiré fringe formation when the pitch of the line grating is much larger than the pitch width of the CCD cell array. (c) CCD image of Moiré fringe pattern. (d) Moiré waveform extracted at selected 1-D cross-section that is orthogonal to line grating. (e) Waveform after removing line artifacts by sub-sampling every ($m=4$) pixel. (f) The AM-FM sinusoidal waveform with known instantaneous frequency $f(n)$, modeled using the L-STPSA algorithm [13].

Fig. 1b shows a common situation where the observed specimen line grating pitch, p_s' is some integer factor larger than the pitch width of the regular CCD cell array p_r. It has been shown in [12] that the generated Moiré pattern pitch p_m' satisfies

$$p_m' = \frac{p_r p_s'}{(1+m)p_r - p_s'}, \quad \text{where } m = \left\lfloor \frac{p_s'}{p_r} \right\rfloor \tag{2}$$

The factor m, which is the integer multiple of times the specimen line pitch is larger than the CCD cell array pitch, plays an important role in removing the line artifacts within the Moiré pattern images. As discussed in [12], the clever use of the CCD array as a reference grating permits the annoying high-frequency line grating artifacts to be removed by simply sub-sampling every other m pixels along the 1D profile (see Fig. 1d and 1e). By modeling the Moiré waveform in Fig. 1e as a AM-FM sinusoid using the L-STPSA algorithm [13,14], the instantaneous frequency $f(n)$ at any given point n, in the Moiré waveform along the 1D profile can be computed. Since $f(n)$ is the inverse of the instantaneous Moiré fringe pitch, the distance $d(n)$ (see Fig. 1a) on any point n along the 1D profile can be obtained by substituting (1) into (2).

$$d(n) = \frac{FS}{(1+m)}\left(\frac{1}{p_m{'}(n)} + \frac{1}{p_r}\right) = kf(n) + b \qquad \text{where} \qquad \frac{1}{p_m{'}(n)} = f(n) \qquad (3)$$

For a given system setup, the combination of camera focal length, line grating pitch width and CCD cell spacing can be represented by the constants k and b as shown in (3). Using a similar setup as shown in Fig. 3a, the system constants k and b were obtained through a simple calibration procedure. A flat line-patterned surface (printed using a 600 dpi laser printer) was imaged from distances d of 70.0cm to 75.0 cm at intervals of 0.5cm. A plot of the average Moiré pattern frequency f versus the distances d was produced. The linear relationship between these two measures as highlighted in (3) was observed and the straight line graph allowed the constants k and b to be extracted at values of -11.5 and 76.9 respectively [12].

Given the values of k and b, the distance d on any flat surface can be computed as long as the instantaneous frequency f of the Moiré waveform is known. This however, is not true of surfaces that exhibit local inclination such as a curved surface. The next section describes the additional consideration required to compute the distance $d'(n)$ along a 1D profile on a smooth curved line grating surface.

3 Distance Estimation of Curved Surfaces

When a horizontal surface is slanted at an angle θ, as shown in Fig. 2a, the pitch S of the line gratings viewed by the camera will be shortened to

$$S' = S \cos \theta \qquad (4)$$

Since a curved 1D profile can be locally approximated by an incline (see Fig. 2b), the same relationship in (4) is also applicable for smooth curved line grating surfaces. The distance $d'(n)$ of a point n on an incline surface to the camera is given by

$$d'(n) = \frac{FS'}{(1+m)}\left(\frac{1}{p_m{'}(n)} + \frac{1}{p_r}\right) = \frac{FS}{(1+m)}\left(\frac{1}{p_m{'}(n)} + \frac{1}{p_r}\right)\cos\theta = d(n)\cos\theta \qquad (5)$$

Even with known values of system constants k, b and the instantaneous frequency $f(n)$ on point n on the Moiré pattern waveform, it is not possible to compute the camera-to-surface distance on an inclined surface unless angle θ is known. The angle θ can be estimated by using geometry on the triangle OAB, where points A and B are immediate pixel neighbors of our point of interest C (see Fig. 2c). If the distances of points A and B to the camera are denoted as $d_A{'}$ and $d_B{'}$, then from (5) we get

$$\tan\theta = \frac{|OA|}{|OB|} = \frac{\Delta d'}{\Delta L} = \frac{|d'_A - d'_B|}{\Delta L} = \frac{|d_A - d_B|}{\Delta L}\cos\theta \tag{6}$$

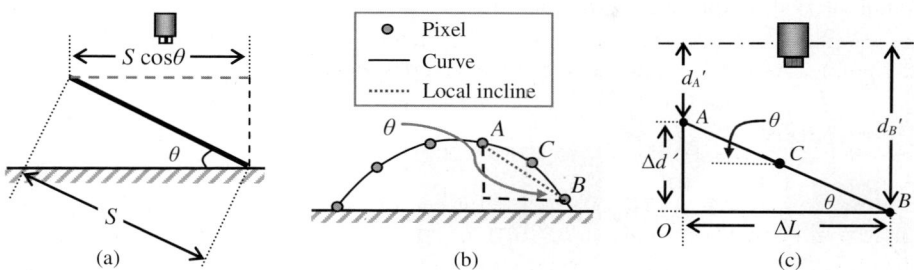

Fig. 2. (a) An incline shortens the line gratings pitch S by a factor $\cos\theta$. (b) The inclination on any point on a smooth curve can be approximated using the inclination produced by its two adjacent pixel points. (c) The parameters related to an incline measurements.

The camera-to-surface distances d_A and d_B (based on a flat surface) can be obtained using the system constants b, k and the instantaneous frequencies of the Moiré waveform at $f(A)$ and $f(B)$ respectively. This means we can then compute the local inclination angle θ at point C using

$$\theta = \sin^{-1}(\frac{-\Delta L + \sqrt{\Delta L^2 + 4|d_A - d_B|^2}}{2|d_A - d_B|}) \tag{7}$$

With angle θ, the true camera-to-surface distance at incline point C can be obtained by substituting θ into (5). The horizontal inter-pixel length ΔL is obtained with the aid of appropriately located ruler placed on the flat imaging platform next to the edge of the patterned surface, as shown in Fig. 3a.

4 Experiments

Two sets of experiments were carried out. The first was to verify the ability to use the proposed Moiré pattern analysis technique described in section 3 to reconstruct a smooth curved surface. The second investigated the sensitivity of this technique for measuring small displacements on a deformable surface. All images were acquired using a Dragon Fly Express monochrome CCD camera from PointGrey Research Inc. and a lens with a 25mm focal length. In all the experiments, the distance to the camera was adjusted such that a value of $m=4$ given in (2) was used.

4.1 Curved Surface Reconstruction

Fig. 3a shows the experimental setup used to test the reconstruction of a smooth curved line grating surface. An A4-sized paper printed with 33 lines per inch of equal-width black and white lines was used as the specimen patterned surface. It was allowed to buckle upwards into a smooth curve with the aid of two stoppers at either ends. With the aid of a ruler, surface height measurements were made at regular horizontal intervals at the edge of the curved paper.

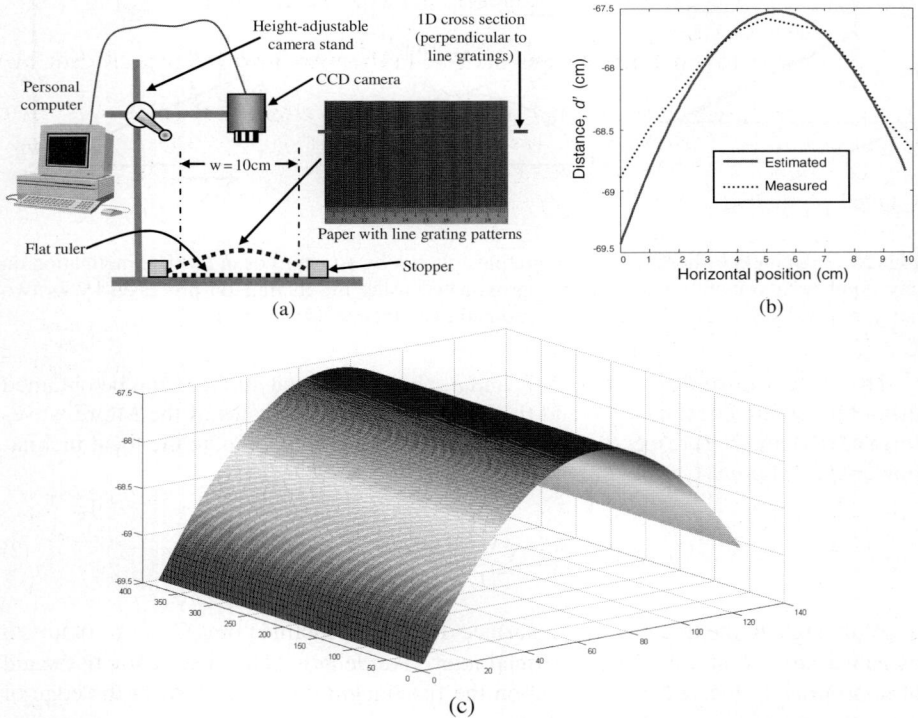

Fig. 3. (a) The experiment setup. (b) Estimated curve profile plotted against actual curve (interpolated from physically measured data). (c) The 3D curved surface plot obtained by combining consecutively reconstructed 1D profiles. Uniform bending of the paper resulted in a smooth 3D curve surface unlike the single-point surface depression example shown in Fig. 4a.

Based on the local incline approximation approach of smooth curved surface describe in section 3, the incline-compensated surface-to-camera distances $d'(n)$ given in (5) was computed from the estimated instantaneous frequency $f(n)$ of the Moiré waveform along a 1-D cross section (see Fig. 3a), and the earlier computed systems constants b and k given in (3). The result in Fig. 3b shows that the extracted smooth curve is reasonably close to that actual measured curve, given the limited accuracy of the crude ruler-based measurement technique employed. Fig. 3c shows the smooth 3D curved surface reconstructed by stitching together the consecutive 1-D cross sections taken orthogonal to the line grating.

In terms of absolute measurements, the highest point on the smooth curve was measured to be at 69.0cm from the camera and the estimated maximum height computed was 68.9cm, which was an error of 1mm or about 0.16% of the absolute height. The maximum depth of bend over the observed area was about 20mm. Based on this relative measure, the estimate error is higher at about 5% of maximum depth variation. With our proposed use of this technique to estimate pressure-induced small displacements on deformable surfaces, the absolute estimation of the camera-to-surface distance is not as critical as compared to the sensitivity of measuring small displacements on the surface. This issue is addressed in the next experiment.

4.2 Sensitivity to Small Displacements Due to Pressure on a Deformable Surface

The experimental setup shown in Fig. 4b was used to create very small mm-scaled displacements on a stretched canvas by applying horizontal pressure via a blunt-headed plunger. The deforming patterned surface of 18 lines per inch line grating was imaged from about 1.5 metres away (giving a m value of 4).

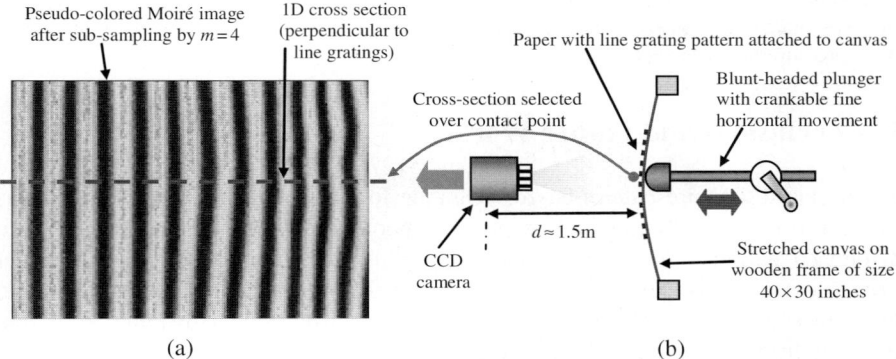

Fig. 4. (a) Moiré pattern image (pseudo-colored for clarity) with line artifacts removed by sub-sampling every other $m=4$ samples. The 1D cross-section analysed is shown in a red dashed line. (b) Setup used to investigate sensitivity to small displacements on a stretched canvas.

Fig. 5a shows plots of the extracted instantaneous frequencies, $f(n)$ along the 1D Moiré waveform at the cross-section line shown in Fig. 4a. The results suggest that small monotonic displacements of <1mm can be clearly separated by the proposed Moiré analysis technique using images captured from a distance of about >1.5m away. With a large imaging distance and small surface deformations, like in this case, the measure given by $d(n)$ in (3) would be a reasonable approximate of the camera-to-surface distance. If similar calibration procedures as describe in section 2 were carried out to determine the new system constants b and k in (3), the measured instantaneous frequency $f(n)$ in Fig. 5a can be used to determine the exact displacement of the canvas. However, if the task is merely to detect a measure proportional to the pressure applied on the canvas, then the frequency measure $f(n)$ would suffice. This is assuming the useful working range of the finger pressure to be detected by the proposed vision-based interface displaces the canvas in a proportional manner.

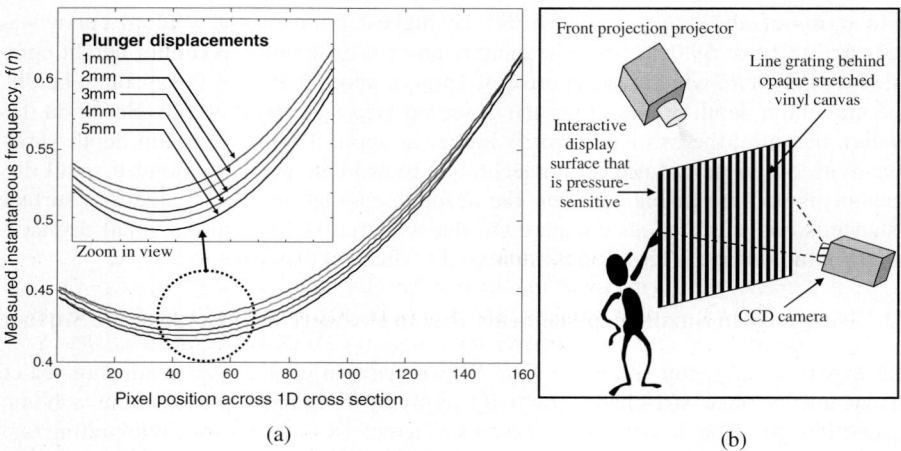

Fig. 5. (a) The extracted instantaneous frequencies along the 1D Moiré waveform for different plunger displacement positions. (b) A possible implementation of a front projection pressure-sensitive multi-touch display surface using the proposed Moiré pattern-based technique for touch pressure measurement.

5 Conclusions and Future Work

We proposed a Moiré pattern-based technique for surface-to-camera distance measurement. It uses a CCD camera to capture a smooth curved surface printed with regular-pitched parallel lines that can be easily generated with a 600dpi laser printer. Experimental results show that smooth 3D curved surfaces can be reconstructed with the proposed technique. More importantly, small displacement quantum of <1mm resulting from finger-like pressure applied to a stretched patterned canvas can be readily extracted and distinguished. This suggests the proposed technique would be a viable vision-based solution to implement pressure-sensitive touch surfaces.

We are currently in the process of developing a real-time pressure-sensitive touch display surface. The general idea for this interactive display is to have a stretched opaque vinyl canvas act as the interactive display on one side (with front projection). And on the other side that is printed with line grating, a vision-based system for pressure-sensitive multi-touch detection. Unfortunately, one of the side effects of employing frontal projection is that the projected imagery will annoyingly appear on the user's hands and this will also result in shadows on the display surface. A more complicated strategy would be to use an occlusion-free back-projection display that employs the imperceptible pattern projection technique of Cotting *et al.* [15] to time-slice the line grating pattern into pre-allocated synchronized temporal slots of the video frames. Unlike [15], the Moiré-based surface reconstruction technique proposed here is faster as there is no need to project a sequence of Gray-coded structured light pattern during surface depth reconstruction.

References

1. Apple Inc. (n.d.). Apple - iPhone - Feature - Multi-touch,
   http://www.apple.com/iphone/features/multitouch.html
   (retrieved April 5, 2009)
2. Microsoft Corporation. Microsoft Surface,
   http://www.microsoft.com/surface/Default.aspx
   (retrieved April 5, 2009)
3. Rekimoto, J.: SmarkSkin: An Infrastructure for Freehand Manipulation on Interactive Surfaces. In: 10th ACM Symposium on User Interface Software and Technology (2002)
4. Vlack, K., Mizota, T., Kawakami, N., Kamiyama, K., Kajimoto, H., Tachi, S.: GelForce: a Vision-based Traction Field Computer Interface. In: CHI 2005, ACM Press, New York (2005)
5. Echtler, F., Huber, M., Klinker, G.: Shadow tracking on multi-touch tables. In: Proc. of the working conference on Advanced visual interfaces, pp. 388–391. ACM, New York (2008)
6. Cassinelli, A.: Khronos Projector. In: SIGGRAPH 2005 - Emerging Technologies, Los Angeles (2005)
7. Smith, J.D., Graham, T.C., Holman, D., Borchers, J.: Low-cost Malleable Surfaces with Multi-touch Pressure Sensitivity. In: 2nd IEEE International Workshop on Horizontal Interactive Human-Computer System (2007)
8. Kafr, O., Glatt, I.: The Physics of Moiré Metrology. John Wiley & Sons, Chichester (1990)
9. Khan, A.S., Wang, X.: Strain Measurements and Stress Analysis. Prentice Hall, New Jersey (2001)
10. Walker, C.: Handbook of Moiré Measurement. Institute of Physics Publishing, Bristol (2004)
11. Creath, K., Schmit, J., Wyant, J.C.: Optical Metrology of Diffuse Surfaces. In: Malacara, D. (ed.) Optical Shop Testing, 3rd edn., pp. 756–807. John Wiley & Sons, Hoboken (2007)
12. Tu, T., Goh, W.B.: MOIRE Pattern from a CCD CAMERA – Are They Annoying Artifacts Or Can They Be Useful? In: 4th International Conference on Computer Vision Theory and Applications (2009)
13. Goh, W.B.: Noise Robust AM-FM Demodulation using Least-Squares Truncated Power Series Approximation. In: 6th International Conference on Information, Communications and Signal Processing (2007)
14. Goh, W.B., Chan, K.Y.: Amplitude Modulated Sinusoidal Modeling using Least-square Infinite Series Approximation with Applications to Timbre Analysis. In: IEEE International Conference of Acoustics, Speech and Signal Processing, pp. 3561–3564 (1998)
15. Cotting, D., Naef, M., Gross, M., Fuchs, H.: Embedding Imperceptible Patterns into Projected Images for Simultaneous Acquisition and Display. In: IEEE/ACM International Symposium on Mixed and Augmented Reality, pp. 100–109 (2004)

Enhanced Landmine Detection from Low Resolution IR Image Sequences

Tiesheng Wang[1,2], Irene Yu-Hua Gu[1], and Tardi Tjahjadi[3]

[1] Dept. of Signals and Systems, Chalmers Univ. of Technology, Sweden
[2] Institute of IPPR, Shanghai Jiao Tong University, Shanghai, China
[3] School of Engineering, University of Warwick, United Kingdom
{tiesheng,irenegu}@chalmers.se, T.Tjahjadi@warwick.ac.uk

Abstract. We deal with the problem of landmine field detection using low-resolution infrared (IR) image sequences measured from airborne or vehicle-borne passive IR cameras. The proposed scheme contains two parts: a) employ a multi-scale detector, i.e., a special type of isotropic bandpass filters, to detect landmine candidates in each frame; b) enhance landmine detection through seeking maximum consensus of corresponding landmine candidates over image frames. Experiments were conducted on several IR image sequences measured from airborne and vehicle-borne cameras, where some results are included. As shown in our experiments, the landmine signatures have been significantly enhanced using the proposed scheme, and automatic detection results are reasonably good. These methods can therefore be applied to assisting humanitarian demining work for landmine field detection.

Keywords: Landmine field detection, feature point, SIFT descriptor, bandpass filter, consensus of landmine candidates.

1 Introduction

Humanitarian demining is concerned with the detection and subsequent removal of mines. The process consists of identifying mine fields and reducing the suspected area by discriminating individual landmine-like object from clutter (e.g., bushes, rocks, petrified wood and animal burrows) in the suspected regions, and of the actual landmine clearance. Among the technologies for landmine detection (e.g. using metal detector, ground penetrating radar, chemical, and acoustic and optical sensors [1,2,3,4,5], the methods based on infrared (IR) sensors have drawn many interests [6,7,8,9].

The methods based on IR images utilize the property that the soil temperature on the ground above the buried objects (including landmines) are often different from that of unperturbed areas. For IR images of mine fields measured from airborne or vehicle-borne sensors, landmines are indicated by spatial difference from their surroundings due to digging, or thermal and material signatures. However, the background in images usually consists of various types of noise and clutter, e.g., thermal noise, sand, gravel road and vegetation, thus making the detection difficult. It was shown in [8] that IR landmine signatures in close-distance measured IR images over time can be modeled by a 3D Gaussian shaped function. Although landmines in IR images measured from close-distances to ground surfaces are significantly larger as compared with those measured from airborne, the principle of using thermal contrast differences to detect landmines is similar. [6] has shown that using a special type of multiscale isotropic bandpass filters in [7], or a multiscale matched filter whose profile is a spatially reversed replica of landmine patterns, is effective for detecting landmine candidates from airborne measured IR images. It has also shown that the profile from the filter (or, the landmine pattern) is consistent to the IR contrast model for landmines measured in close distance. Employing such a multiscale 2D isotropic bandpass filter, involving an automatic scale level selection, a post-processing of peak picking and inter-scale peak position tracing, has demonstrated its promising as a semi-automatic tool for mine field detection [6]. However, this method is based on detection using a single image frame. A common characteristic of landmines in low resolution images is that they all appear to be point-like features. False alarms and missed detections using image based detection are likely high. In order to remove outliers and improve the detection, jointly using multiple frames or temporal consistency is desirable for enhancing the landmine detection. This paper extends the previous method which uses spatial-frequency differences for detecting landmine candidates and locating landmine fields in low-resolution IR images, by adding an extra step based on temporal consistency of detected candidates over several frames. Hence, the proposed scheme is formulated as detecting landmine candidates from each frame followed by utilizing their temporal consistency constrained by a motion model to reducing false alarms and improve detections. The latter includes using landmine candidates that agree with the same motion model to reduce the outliers and use the frequency of coincident appearances of detected candidates for determining the inliers. The proposed scheme has been examined on IR landmine images measured from airborne and vehicle-borne sensors. Our experiments have shown that using the proposed scheme for landmine detection over image frames landmine signatures have been significantly enhanced, and reasonably good results are obtained from the automatic detection.

The remainder of the paper is organized as follows. In Section 2, a general description of the proposed method is presented. Section 3 provides the detection of landmine candidates in a single image. In Section 4, the enhanced landmine detection from image sequences is described. Section 5 demonstrates the performance of the proposed method with some experimental results included.

2 System Description

The block diagram of the proposed system is shown in Fig. 1. First, the landmine candidates are detected using a multi-scale 2D detector (or, a multiscale 2D matched filter) in each frame (block 1). Landmine candidates may not be reliable due to the changes of landmine and environmental temperature. A 2nd processing step is followed based on the temporal consistency of corresponding

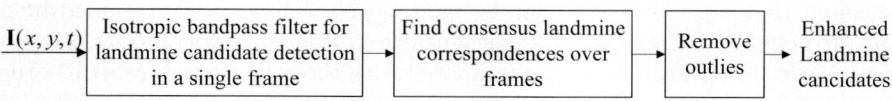

Fig. 1. Block diagram of the proposed system. From left to right: block 1 to 3.

landmine candidates over frames, agreeing with the same model of changes. This is realized by associating landmine candidates over frames using SIFT (Speeded Up Robust Features) descriptors (block 2) and then seeking the maximum consensus of corresponding landmine candidates over frames (block 3). In this way, an enhanced landmine detector is obtained that reduces false alarm and enhances the performance of the detector. The details of the proposed method are described in the subsequent sections.

3 Detect Landmine Candidates in a Single Image

Landmines are shown as point-like patterns in airborne or vehicle-borne IR images. To detect landmine candidates, a multi-scale anisotropic 2D bandpass filter in [6] is adopted. This feature point detector is similar to SIFT [11]. The multi-scale detector is formed by repetitively cascading (L times) of the same single-scale detector followed by thresholding and peak picking. Although many different types of bandpass filters satisfy the required specifications [7], a straightforward way to formulate a bandpass filter in a single-scale detector is to use the difference of two 2D isotropic lowpass filters, being the same type but with different bandwidths [10]:

$$g_k(x,y) = g_0^{(1)}(x,y) - g_0^{(2)}(x,y) \quad (1)$$

where $g_0^{(1)}(x,y)$ and $g_0^{(2)}(x,y)$ are 2D lowpass filter kernels with a wider and narrower frequency bandwidth, respectively, the suffix of g indicates the radial frequency f_k of the filter, and $g_0^{(i)}$ is a lowpass filter with an isotropic 2D Gaussian kernel, $g_0^{(i)}(x,y) = \frac{1}{2\pi\sigma_i^2} e^{\frac{-(x^2+y^2)}{2\sigma_i^2}}$, $i=1,2$, where $\sigma_2 = scale \cdot \sigma_1$, $scale > 1.0$. The 2D bandpass filter in (1) has an effective finite support despite its infinite support. To reduce computation, the kernel is truncated after the absolute value of normalized kernel envelope becomes insignificant (i.e. below a small value). It is worth mentioning that the parameter *scale* should be carefully selected so that the 2D filter magnitude spectrum attenuates all signal components at zero frequency (i.e., removes all d.c. elements). Failure to do so would lead to high false alarm in the detection. Empirically, this requirement can be monitored by observing the 1D cross-section of the 2D magnitude response.

In the multi-scale detector, detecting landmine candidates is done as follows: starting from the finest scale (level-1), if landmine candidates in the output image are not prominent against their surroundings, coarser scales are added by repetitively cascading one additional single-scale detector from the previous finer scale, until the 'best' scale level L_{max}, or a pre-determined maximum level $L_{optimum}$ is reached. Since significant landmine candidates become more pronounced in a coarse scale while most spurious noise is removed, it is easier to detect candidates from local peaks at a coarser level. Once local peaks in the coarsest level are detected, locations of mine candidates are determined by inter-scale peak position tracing where better localizations are found in finer scales.

4 Enhanced Detection by Seeking Temporal Consensus

The landmine candidates detected in Section 3 may contain spurious ones. The detection can be improved by considering the temporal consistency of the candidates over image frames under a given transformation model. This is done by first finding correspondences among detected candidates in consecutive frames using the point descriptors similar to SIFT and then applying RANSAC (RANdom SAmple Consensus) to find a subset of consensus candidates that best fit for a given model.

For each landmine candidate point obtained from Section 3, a feature descriptor (or, feature vector) is added. Although there are many different types of feature descriptors, we choose the SIFT keypoint descriptor in our application. Each SIFT feature vector (or keypoint descriptor) is of size 128. It is formed based on gradient magnitudes in orientation histograms for a 16x16 region centered at the corresponding keypoint [11]. The correspondences between the detected landmine candidates in two image frames are found by using the detected landmine candidates together with their features in the respective frames, where images may undergo a given transformation (e.g. affine model in this paper).

To find maximum consensus of landmine point correspondences between frames, RANSAC ([12]) is used to estimate parameters of a mathematical model from a set of observed data points containing outliers. Lately, RANSAC has been increasingly used ([13,14]). A basic assumption is that data consists of inliers and outliers. The distribution of inliers can be described by parameters, while outliers do not fit the model. The outliers are due to measurement noise or incorrect hypotheses. RANSAC also assumes that, given a small set of inliers, it may estimate the parameters of a model that optimally explains or fits the hypothetical inliers. For a given fitting problem and the model, RANSAC is performed as follows: 1) Randomly choose the minimum number of points needed to estimate the model. 2) Estimate the model parameters using these points. 3) Find the number of data points agreeing with the estimated model (within a pre-specified tolerance error). 4) Repeat Steps 1) to 3) for a predetermined number of times. 5) Choose the largest set of consensus points, and re-estimate the model parameters.

The joint scheme for removing outliers in detected landmine candidates can be described as follows. First, multi-scale landmine candidates are detected, and their feature descriptors are extracted from their surroundings in each image frame (Section 3). For each frame (e.g. frame t), a data set $\mathbf{S}_t = \{S_t, f_t\}$ is formed, where S_t is a subset containing landmine candidate positions, and f_t is a subset containing feature vectors. Using data sets from two frames, point correspondences of landmine candidates are established between these two frames under a given affine transformation. Next, RANSAC is employed to the subset S_t at frame t, and the subsets $S = \{S_{t-L}, \cdots, S_{t+L}\}$ from the nearby frames under a selected affine model with 4 unknown parameters,

$$\begin{bmatrix} x \\ y \end{bmatrix} = \beta \begin{bmatrix} \cos\theta & -\sin\theta \\ \sin\theta & \cos\theta \end{bmatrix} \begin{bmatrix} \tilde{x} \\ \tilde{y} \end{bmatrix} + \begin{bmatrix} dx \\ dy \end{bmatrix} \quad (2)$$

where (x, y) and $(\tilde{x}, \tilde{y})$ are point correspondences from 2 different frames, θ is the rotation angle, β is the scaling factor, and (dx, dy) are the translation between the 2 frames. The parameters can be estimated by the matrix inversion (for $k = 2$), or the least squares solution (for $k > 2$) of (3),

$$\begin{bmatrix} \tilde{x}_1 & -\tilde{y}_1 & 1 & 0 \\ \tilde{y}_1 & \tilde{x}_1 & 0 & 1 \\ \vdots & & & \vdots \\ \tilde{x}_k & -\tilde{y}_k & 1 & 0 \\ \tilde{y}_k & \tilde{x}_k & 0 & 1 \end{bmatrix} \begin{bmatrix} \beta\cos\theta \\ \beta\sin\theta \\ dx \\ dy \end{bmatrix} = \begin{bmatrix} x_1 \\ y_1 \\ \vdots \\ x_k \\ y_k \end{bmatrix} \quad (3)$$

where $\{(x_1, y_1)\}, \cdots, (x_k, y_k)\}$ and $\{(\tilde{x}_1, \tilde{y}_1), \cdots, (\tilde{x}_k, \tilde{y}_k)\}$ are subsets of k correspondence points selected from the 2 frames. Choosing the affine model in (2) is based on the assumption that images are measured straight downward. Under this model, we first use RANSAC to find the maximum consensus landmine candidates between the current frame t and one neighbouring frame t', $t' = t - L, \cdots, t - 1, t + 1, \cdots, t + L$. The total number for each landmine candidate being selected by RANSAC in $2L + 1$ frames is then accumulated. Those consensus candidate points, whose accumulated numbers exceed a given threshold, are then chosen as the refined landmine candidates. The remaining candidates are removed as being clutter points in order to reduce the false alarm.

5 Experiments and Results

Datasets: For test set (a), airborne IR images were captured by IR cameras on a helicopter at approximately 180m above the ground surface when scanning one of the landmine test site in Eksjo, Sweden. Each image contained more than a dozen surface-laid anti-tank mines between the two columns of man-made landmarks (i.e., the green panel, the astro turf and the red corn marks). For test set (b), vehicle-borne IR image sequences were measured by IR cameras mounted at 10m height above the ground surface in the test site of FOI laboratory during a 24-hour period. The desired target feature points correspond to surface-laid and shallowly-buried anti-personnel and anti-tank mines.

Results: For test set (a), the multi-scale detector was applied to airborne IR images, where the desired targets correspond to man-made landmarks. These landmarks were set for locating regions of interest. Fig. 2 shows some resulting

Enhanced Landmine Detection from Low Resolution IR Image Sequences

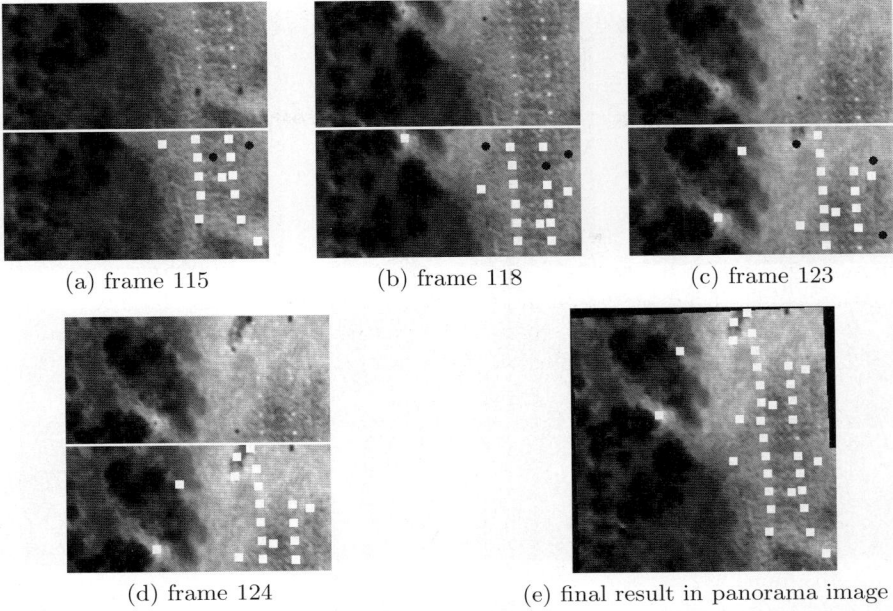

(a) frame 115 (b) frame 118 (c) frame 123

(d) frame 124 (e) final result in panorama image

Fig. 2. Row-1 in (a)-(d): 4 original images; Row-2 in (a)-(d): detected candidates of landmarks and landmines using the multi-scale detector marked by black circles plus white rectangles. Parameters used: $\sigma=0.9$, $scale = 2.0$, total scale levels $L = 4$, and $threshold = 0.24$. The enhanced scheme results in the selected maximum consensus candidates of landmarks and landmines marked by white rectangles. A candidate is selected as consensus point if the candidate appears in more than 45% of the scenes containing that candidate. (e): the final results based on detection over image frames, all detected landmine and landmark candidates are marked by white rectangles in a panorama image (through image registration of the 4 frames).

frames extracted from the airborne IR sequence. The results show that though the detector is effective in detecting man-made landmark candidates (marked by white rectangles and black circles) some clutter points are also detected. Further, it is noticed that almost none of the anti-tank mines are detected mainly due to too low spatial resolution of landmines using the given IR cameras at the measurement altitude. Further, as can be seen from the panorama image, after applying the enhanced detection over image frames, landmark and landmine candidates are chosen from the maximum consensus correspondences, which further improves the detection results (marked by white rectangles only).

For test set (b), the multi-scale detector was first applied to vehicle borne IR images. Fig. 3 shows some results from 10 image frames extracted from the vehicle-borne IR sequence. The contrast of landmines and background is quite small in these frames. The cross digging signs are weakly visible, and the signatures of landmines are very weak. from the results one observes that the multi-scale detector has successfully picked up some of the landmines, however,

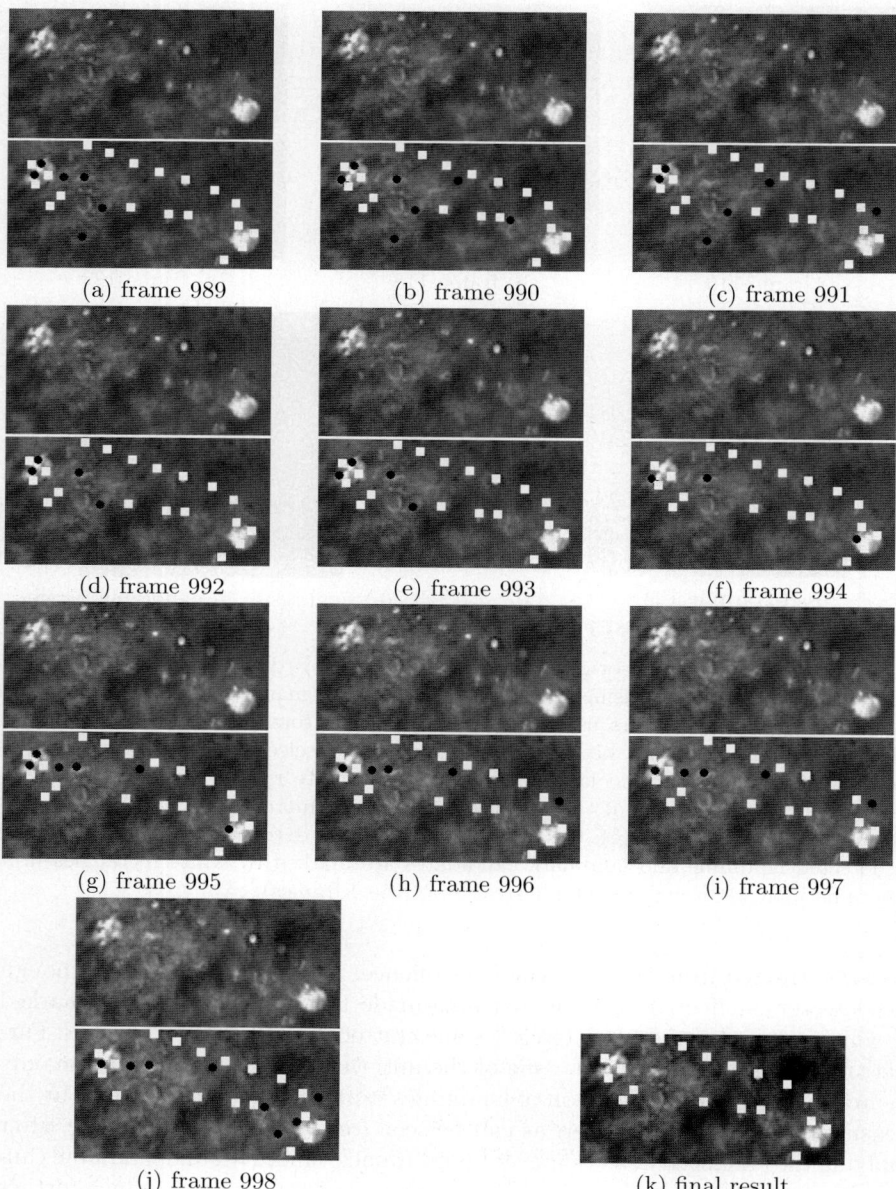

Fig. 3. Row-1 in (a)-(j): Ten original IR image frames. Row-2 in (a)-(j): detected landmine candidates using the multiscale detector marked by black circles plus white rectangles. Parameters used: $\sigma=1.4$, $scale = 2.0$, total scale levels $L = 4$, and $threshold = 0.15$. The enhanced scheme results in the selected maximum consensus candidates of landmarks and landmines marked by white rectangles. A candidate is selected as a consensus point if it appears in $\geq 80\%$ of the scenes under consideration. (k) The final candidates based on consensus information over the 10 frames.

some clutter points are also included as landmine candidates probably due to their similar pattern to landmines. Observing the results in Fig. 3 show that the multi-scale detector in this case has yielded reasonable good results in detecting anti-personnel mines on the top row (with intensity difference), and also located several buried anti-tank mines on the bottom row (with cross digging signs). However, some clutter points are also included. It is observed that after the enhanced detection using maximum consensus landmine candidates through frames, the final results have been further improved (indicated by white rectangles).

6 Conclusions

We propose an enhanced landmine candidate detection method for airborne and vehicle-borne measured IR images. The method utilizes the consistency of the landmine candidates detected through image frames to further reduce false alarms and improve detection rate. This is obtained by matching feature descriptor and seeking the maximum number of consensus correspondences over image frames. Experiments has been conducted on two IR landmine datasets. And experimental results show that the proposed method has improved the performance of image based landmine detection.

Acknowledgment

The landmine data were kindly provided by the Swedish Defence Research Agency (FOI), Sweden.

References

1. Earp, S., Hughes, E., Elkins, T., Vickers, R.: Ultra-wideband ground-penetrating radar for the detection of buried metallic mines. IEEE Aerospace and Electronics System Magazine 11(9), 14–17 (1996)
2. Filippis, A., Jain, L., Martin, N.: Using generic algorithms and neural networks for surface land mine detection. IEEE Trans. Signal Processing 27, 176–186 (1999)
3. Paik, J., Lee, C.P., Abidi, M.: Image processing-based mine detection techniques: a review. Subsurface Sensing Technologies & Applications 3(3), 153–202 (2002)
4. Faust, A.A., Chesney, R.H., Das, Y., McFee, J.E., Russell, K.L.: Canadian teleoperated landmine detection systems. Part I: The improved landmine detection project. Part II: Antipersonnel landmine detection. Int'l Journal of Systems Science 36(19), 511–543 (2005)
5. Sahli, H., Bruschini, C., Crabble, S.: Catalogue of advanced technologies and systems for humantarian demining. EUDEM 2 Technology Report, Dept. Electron. Inf., Vrij Univ. Brussel, Belgium, v.1.3 (2006)
6. Gu, I.Y.H., Tjahjadi, T.: Detecting and locating landmine fields from vehicle- and air-borne measured IR images. Pattern Recognition 35, 3001–3014 (2002)
7. Gu, I.Y.H., Tjahjadi, T.: Multiresolution feature detection using a family of isotropic bandpass filters. IEEE Trans. SMC (part B) 32(4), 443–454 (2002)

8. Lundberg, M., Gu, I.Y.H.: A 3D matched filter for detection of land mines based on spatio-temporal thermal model. In: Proc. SPIE Conf. Detection and Remediation Technologies for Mines and Minelike Targets V. SPIE, vol. 4038(I), pp. 179–188 (2000)
9. Lundberg, M., Gu, I.Y.H.: Infrared detetion of buried land mines based on texture modeling. In: Proc. of SPIE conf. Detection and Remediation Technologies for Mines and Minelike Targets VI. SPIE, vol. 4394, part one, pp. 275–283 (2001)
10. Lu, W.S., Antoniou, A.: Two-dimensional digital filters. Marcel Dekker, Inc., New York (1992)
11. Lowe, D.G.: Distinctive image features from scale-invariant keypoints. IJCV 20, 91–110 (2004)
12. Fischler, M.A., Bolles, R.C.: Random sample consensus: a paradigm for model fitting with applications to image analysis and automated cartography. Commun. of the ACM 24, 381–395 (1981)
13. Khan, Z.H., Gu, I.Y.H., Wang, T., Backhouse, A.: Joint anisotropic mean shift and consensus point feature correspondences for object tracking in video. In: IEEE Int'l conf. Multimedia & Expo. ICME 2009 (2009)
14. Wang, T., Gu, I.Y.H., Khan, Z.H., Shi, P.: Adaptive particle filters for visual object tracking using joint PCA appearance model and consensus point correspondences. In: IEEE Int'l conf. Multimedia & Expo., ICME (2009)

Author Index

Abrishami-Moghaddam, Hamid 639, 808
Akaho, Shotaro 149
Al-Hamadi, Ayoub 1087
Alamri, Huda 165
Aleksić, Tatjana 369
Alexander, S.K. 590
Alexandre, Luís A. 799
Artieres, Thierry 189
Artner, Nicole 1011
Astola, Laura 419
Au, Sau-Chuen 816
Azencott, R. 590

Baba, Takayuki 922
Babaie-Zadeh, Massoud 808
Backes, André R. 253, 680
Bader, Thomas 689
Bailly, Kevin 25
Bandera, Antonio 979
Barchunova, Alexandra 697
Bardají, I. 342
Batard, Thomas 394
Becciu, Alessandro 598
Beetz, Michael 99
Belaïd, Abdel 237
Bennamoun, Mohammed 410
Berlijn, Sonja M. 1179
Bermak, Amine 410
Berthier, Michel 394
Beyerer, Jürgen 689
Biehl, Michael 1162
Bodmann, B.G. 590
Boom, Bas 33
Borràs, Agnés 898
Bouamrani, A. 590
Brechmann, André 631
Breuel, Thomas M. 173
Brockers, Roland 1019
Brun, Luc 705, 939
Brune, Christoph 533
Bruno, Odemir M. 253, 680
Bukhari, Syed Saqib 173
Bunke, Horst 181, 189, 342, 377

Bunte, Kerstin 1162
Burger, Martin 533
Burnat, Mireia 427

Caldairou, Benoît 606
Casile, Antonino 714
Chan, Yuk Hin 848
Chaudry, Qaiser 614
Chen, Jie 41
Chen, Rung-Ching 1204
Chen, Yung-Fu 791
Cheng, Da-Chuan 622
Cheng, Ming-Shao 402
Chiappini, C. 590
Chowdhury, Shyama Prosad 1171
Chung, Kuo-Liang 402
Collet, Christophe 410
Courtellemont, Pierre 906

Davidsson, Paul 1220
Delgado Gomez, David 91
Delmas, Patrice 848
Demarcq, Guillaume 906
de Sá, J.P. Marques 799
Dibos, Françoise 1028
Do, Trinh-Minh-Tri 189
Doerschner, Katja 856, 881
Dord-Crouslé, Stéphanie 221
Dosil, Raquel 261
Dupé, François-Xavier 705

Edelbrunner, Johann 1096
Eftekhari, Armin 808
Eglin, Véronique 221
Eitzinger, Christian 1154
Emptoz, Hubert 221
Engel, Karin 631
Escalera, Sergio 748
Ewerth, Ralph 1036

Fabry, Thomas 757
Fahlström, Anders 1179
Falcão, A.X. 971
Faure, Claudie 213
Fdez-Vidal, Xosé R. 261

Feng, Zhilin 468
Ferrari, M. 590
Ferrer, M. 342
Fink, Gernot A. 83
Fischer, Andreas 181, 189
Fleischer, Falk 714
Flitti, Farid 410
Florack, Luc 419, 598, 1104
Florindo, João B. 253
Flusser, Jan 334
Francos, Joseph M. 549
Frangi, Alejandro F. 91
Freisleben, Bernd 1036
Frinken, Volkmar 189
Fujiki, Jun 149
Fuster, Andrea 419

Gama, João 799
Gao, Xinbo 987, 995
Gao, Yahui 277
Garcia, Miguel Angel 492
Garcia-Diaz, Antón 261
Ghadimi, Sona 639
Giese, Martin A. 714
Gil, Debora 427
Gilani, Zulqarnain 435
Gill, Gurman 269
Giménez, Adrià 197
Gimel'farb, Georgy 848
Goh, Wooi Boon 1228
Golshaeyan, Narjes 639
Gondary-Jouet, Catherine 639
Gong, Yuanhao 277
Grebe, Reinhard 639
Grooten, Mart H.M. 1188
Gros, Patrick 565
Große, Andree 285
Gu, Irene Yu-Hua 1179, 1236
Gunsel, Bilge 452
Guo, Dong 444
Guo, Yimo 41
Guo, Zhenhua 50, 1130
Gursoy, Ozan 452

Habas, Piotr 606
Haindl, Michal 1138, 1146
Hammer, Barbara 1162
Hancock, Edwin R. 369, 385
Hasegawa, Osamu 723
Hatka, Martin 1138

Havlíček, Vojtěch 1146
He, Chun Lei 165
He, Zhoucan 783
Hegemann, Stefan 1096
Heidl, Wolfgang 1154
Heinrich, Christian 606
Hermans, Jeroen 757
Hernàndez-Sabaté, Aura 427
Hino, Hideitsu 149
Hoang, Thai V. 205
Howe, Tet Sen 647
Hsu, Yuan-Nian 791
Huang, Chao-Hui 647
Huang, Fay 157
Huang, Haiqiao 816
Huang, Rui 865
Huang, Yong-Huai 402
Huang, Yung-Fa 791, 1204
Huynh, Du 410

Imiya, Atsushi 947
Impoco, Gaetano 914
Ingleby, Michael 1196
Ion, Adrian 1011
Ip, Horace H.S. 766
Iseki, Kenta 873

Jain, Brijnesh J. 351
Jalba, Andrei C. 1188
Jansen, Steven 427
Janssen, Bart 598, 1104
Jehan-Besson, Stéphanie 939
Jia, Yunde 774
Jiang, Ruyi 1044
Jiang, Xiaoyi 285, 622, 791
John, Christoph 1053
Jonchery, Claire 1028
Jordan, Tadeusz 293
Jouili, Salim 360
Juan, Alfons 197
Julià, Carme 492

Kawewong, Aram 723
Kazemi, Kamran 639
Kersten, Dan 856
Khan, Masood Mehmood 1196
Khoo, Wai L. 293
Khurshid, Khurram 213
Klette, Reinhard 157, 541, 1044,
    1078, 1113

Kobyliński, Łukasz 732
Koepfler, Georges 1028
Kolb, Rosana M. 680
Kropatsch, Walter 326, 1011
Kumar, Sanjeev 1061
Kwok, Yi-lin 816

Lavesson, Niklas 1220
Lee, Jeonggyu 614
Lenz, Reiner 573
Leow, Wee Kheng 647
Levine, Martin 269
Li, Cuihua 277
Li, Hao 647
Li, Jie 987
Li, Rongfeng 58
Li, Weiming 460
Li, Wenxin 58
Li, Xuelong 987, 995
Li, Youfu 460
Lian, Ssu-Bin 1204
Liaw, Jiun-Jian 1204
Licitra, Giuseppe 914
Lin, Hsuan-Hung 791
Liu, Jian 301
Liu, Jun 468
Liu, Kang 1212
Liu, Rujie 922
Liu, Shing-Hong 622
Liu, X. 590
Liu, Xiabi 774
Liu, Xiangyang 66
Liu, Xiaoming 468
Liu, Yuncai 301
Lladós, Josep 898
Long, Yangjing 74
Lopez-Rubio, Ezequiel 1070
Lovell, Brian C. 116
Lu, Hongtao 66
Lu, Zhiwu 766
Lüke, Stefan 1096
Luo, Heng 66
Lupaşcu, Carmen Alina 655
Luque, R.M. 1070
Lyaghat, Alireza 639

Mainberger, Markus 476
Malleron, Vincent 221
Marín-Jiménez, Manuel J. 824
Marfil, Rebeca 979

Martínez-Villalta, Jordi 427
Mascarilla, Laurent 906
Maška, Martin 930
Masumoto, Daiki 922
Matsakis, Pascal 309
Matsumoto, Takashi 229
Mattivi, Riccardo 740
Matula, Pavel 930
Mavridis, D. 484
Mayer, Christoph 99
McClean, Sally 557
McMenemy, Karen 1171
Mendoza, M. Ángeles 824
Michaelis, Bernd 1087
Micheloni, Christian 1061
Middelmann, Wolfgang 1003
Milgram, Maurice 25
Miranda, P.A.V. 971
Mirza-Mohammadi, Mehdi 748
Mok, Pik-yin 816
Molina-Abril, H. 326
Montoya-Zegarra, Javier A. 971
Morales, Sandino 1078
Moreno, Rodrigo 492
Morin, Annie 565
Morrow, Philip 557
Muhammad, Azam Sheikh 1220
Müller, Jahn 533
Muramatsu, Daigo 229
Murata, Noboru 149

Nasse, Fabian 83
Née, Guillaume 939
Ni, JingBo 309
Nilsson, Mikael 1220
Nishiguchi, Haruhiko 947

Obermayer, Klaus 351
Oka, Ryuichi 873
Okarma, Krzysztof 501
Ortiz-de-Lazcano-Lobato, J.M. 1070
Ostermann, Joern 1212
Ouwayed, Nazih 237

Palenichka, Roman M. 318
Palomo, E.J. 1070
Papamarkos, N. 484
Papari, Giuseppe 509
Pardo, Xosé M. 261
Passat, Nicolas 606
Pathan, Saira Saleem 1087

Pavani, Sri-Kaushik 91
Peng, Jian-Xun 1171
Pérez de la Blanca, Nicolás 824
Peter, Tim 189
Petersen, Henry 832
Petkov, Nicolai 509
Pham, Ngoc-Yen 205
Pham, Nguyen-Khang 565
Phothisane, Philippe 25
Piciarelli, Claudio 1061
Pietikäinen, Matti 41
Platero, Carlos 517
Poon, Josiah 832
Puig, Domenec 492

Radeva, Petia 748
Radig, Bernd 99
Ramella, Giuliana 525
Rao, Naveed Iqbal 435
Räpple, René 689
Raza, S. Hussain 614
Real, P. 326
Regenbrecht, Holger 1053
Régnier, Philippe 221
Ren, Peng 369
Revenu, Marinette 939
Riaz, Zahid 99
Richtsfeld, Mario 955
Riesen, Kaspar 377
Rink, Karsten 963
Robles, Oscar D. 581
Rodríguez, Ángel 581
Rodriguez, E. Patricia 557
Roode, Vivian 598
Rothaus, Kai 285
Rousseau, François 606

Sá Junior, Jarbas J. de M. 680
Saetzler, Kurt 557
Saito, Hideo 107
Sakai, Tomoya 947
Salmen, Jan 1096
Sanders, Peter 1003
Sanderson, Conrad 116
Sanguino, Javier 517
Sanniti di Baja, Gabriella 525
Sawatzky, Alex 533
Scharinger, Josef 1154
Scheibe, Karsten 157
Schlipsing, Marc 1096

Schmidt-Trucksäss, Arno 622
Schrater, Paul 856, 881
Schwanecke, Ulrich 1053
Sebastião, Raquel 799
Segura, Enrique 840
Seijas, Leticia 840
Sengor, Neslihan 452
Serratosa, F. 342
Shafait, Faisal 173
Shao, Ling 740
Shirato, Satoshi 229
Siddiqi, Imran 245
Sim, Terence 133, 444, 889
Sistiaga, Unai 1179
Smeaton, Alan F. 581
Smeets, Dirk 757
Smith, William A.P. 865
Smolic, Aljoscha 1
Solli, Martin 573
Sommer, Gerald 697, 1122
Sowmya, Arcot 663
Spina, T.V. 971
Spreeuwers, Luuk 33
Stork, David G. 9, 293
Studholme, Colin 606
Suen, Ching Y. 165
Suetens, Paul 757
Suk, Tomáš 334

Tabbone, Salvatore 205, 360
Tanaka, Hidenori 107
Tang, Darun 58
Tangruamsub, Sirinart 723
Tao, Dacheng 987, 995
Tarighati, Alla 639
Tasciotti, E. 590
Tegolo, Domenico 655
ter Haar Romeny, Bart M. 598, 1104
Thumfart, Stefan 1154
Thurau, Christian 83
Tjahjadi, Tardi 1236
Toharia, Pablo 581
Tönnies, Klaus 631, 963
Torres, Fuensanta 979
Trucco, Emanuele 655
Tu, Tong 1228

Usami, Yumi 149

Valkenburg, Robert 848
Valveny, E. 342

van Assen, Hans 598
Vandermeulen, Dirk 757
van Dorst, Pieter 1104
Vaudrey, Tobi 541, 1044, 1113
Velasco, Olga 517
Veldhuis, Raymond 33
Veltman, Melanie 309
Viet, Nguyen Tien 873
Vigdor, Boaz 549
Vincent, Nicole 213, 245
Vincze, Markus 955
Vo, Kiet T. 663

Walczak, Krzysztof 732
Wallois, Fabrice 639
Wang, Bin 987, 995
Wang, Jyun-Pin 402
Wang, Kuanquan 125
Wang, Lihua 766
Wang, May D. 614
Wang, Qicong 277
Wang, Qing 783
Wang, Shigang 1044
Wang, Tiesheng 1236
Wang, Yanjie 774
Wang, Ying 995
Wang, Yuehong 922
Ward, Robert D. 1196
Wassenberg, Jan 1003
Wedel, Andreas 1113

Weickert, Joachim 476
Westenberg, Michel A. 1188
Wietzke, Lennart 1122
Wilson, Richard C. 369
Wong, Yongkang 116
Wu, Xiangqian 125
Wu, Yihong 460
Wulkan, Mark 614

Xia, Shengping 385
Xu, Yan 614
Xu, Yong 125
Xu, Zhengguang 41

Yaguchi, Yuichi 873
Yang, Chenhui 277
Yang, Heng 783
Yasuda, Kumiko 229
Ye, Ning 133

Zang, Di 881
Zaremba, Marek B. 318
Zhang, David 50, 58, 125, 141, 1130
Zhang, Huaizhong 557
Zhang, Lei 50, 141, 1130
Zhang, Lin 141
Zhao, Guoying 41
Zheng, Guoyan 672
Zhu, Zhigang 293
Zhuo, Shaojie 889